DESIGN FOR MANUFACTURABILITY HANDBOOK

Other McGraw-Hill Books of Interest

Avallone and Baumeister • MARKS' STANDARD HANDBOOK FOR MECHANICAL ENGINEERS
Brady and Clauser • MATERIALS HANDBOOK
Considine • PROCESS/INDUSTRIAL INSTRUMENTS AND CONTROLS HANDBOOK
Fink and Beaty • STANDARD HANDBOOK FOR ELECTRICAL ENGINEERS
Fink and Christiansen • ELECTRONICS ENGINEERS' HANDBOOK
Freeman • STANDARD HANDBOOK OF HAZARDOUS WASTE TREATMENT AND DISPOSAL
Ganic and Hicks • THE MCGRAW-HILL HANDBOOK OF ESSENTIAL ENGINEERING INFORMATION AND DATA
Haines and Wilson • HVAC SYSTEMS DESIGN HANDBOOK
Harris • HANDBOOK OF ACOUSTIC MEASUREMENTS AND NOISE CONTROL
Harris and Crede • SHOCK AND VIBRATION HANDBOOK
Hicks • STANDARD HANDBOOK OF ENGINEERING CALCULATIONS
Higgins • MAINTENANCE ENGINEERING HANDBOOK
Hodson • MAYNARD'S INDUSTRIAL ENGINEERING HANDBOOK
Juran and Gryna • JURAN'S QUALITY CONTROL HANDBOOK
Karassik, et al. • PUMP HANDBOOK
Lingaiah • MACHINE DESIGN DATA HANDBOOK
Maidment • HANDBOOK OF HYDROLOGY
Merritt and Ricketts • BUILDING DESIGN AND CONSTRUCTION HANDBOOK
Perry and Green • PERRY'S CHEMICAL ENGINEERS' HANDBOOK
Rosaler • STANDARD HANDBOOK OF PLANT ENGINEERING
Shigley and Mischke • STANDARD HANDBOOK OF MACHINE DESIGN
Tuma • ENGINEERING MATHEMATICS HANDBOOK
Wadsworth • HANDBOOK OF STATISTICAL METHODS FOR ENGINEERS AND SCIENTISTS
Walsh • MCGRAW-HILL MACHINING AND METALWORKING HANDBOOK
Wang • HANDBOOK OF AIR CONDITIONING AND REFRIGERATION

DESIGN FOR MANUFACTURABILITY HANDBOOK

James G. Bralla Editor

Second Edition

MCGRAW-HILL
New York San Francisco Washington, D.C. Auckland Bogotá
Caracas Lisbon London Madrid Mexico City Milan
Montreal New Delhi San Juan Singapore
Sydney Tokyo Toronto

Library of Congress Cataloging-in-Publication Data

Design for manufacturability handbook / James G. Bralla, editor.—
2nd ed.
p. cm.
Rev. ed. of: Handbook of product design for manufacturing. 1986.
Includes index.
ISBN 0-07-007139-X (alk. paper)
1. Production engineering—Handbooks, manuals, etc.
2. Engineering design—Handbooks, manuals, etc. I. Bralla, James
G. II. Title: Handbook of product design for manufacturing.
TS176.H337 1998
670.42—dc21 98-23438
CIP

McGraw-Hill

A Division of The McGraw-Hill Companies

1 2 3 4 5 6 7 8 9 0 DOC/DOC 9 0 3 2 1 0 9 8

ISBN 0-07-007139-X

The sponsoring editors for this book were Robert Esposito and Linda Ludwig, the editing supervisor was Paul R. Sobel, and the production supervisor was Clare B. Stanley. It was set in Times Roman by Terry Leaden and Renee Lipton of McGraw-Hill's Professional Book Group composition unit.

Printed and bound by R. R. Donnelley & Sons Company.

This book is printed on acid-free paper.

CONTENTS

Section 2 Economical Use of Raw Materials

Chapter 3.2. Metal Stampings 3.13

Chapter 3.3. Fineblanked Parts 3.43

Chapter 3.4. Four-Slide Parts 3.57

Chapter 3.5. Springs and Wire Forms 3.75

Chapter 3.6. Spun-Metal Parts 3.91

Chapter 3.7. Cold-Headed Parts 3.103

Section 6 Nonmetallic Parts

Chapter 6.11. Ceramic and Glass Parts 6.175

Chapter 6.12. Plastic-Part Decorations 6.195

Section 7 Assemblies

Chapter 7.1. Design for Assembly (DFA) 7.3

Chapter 7.2. Arc Weldments and Other Weldments 7.35

PREFACE

Since its publication, the *Handbook of Product Design for Manufacturing* has enjoyed very good acceptance and sales in the United States and other countries. Editions have been printed overseas in the English and Spanish languages. Part of the reason for the book's widespread distribution has been the awakening of interest in the approach it advocated, which has become known as *design for manufacturability* (or *DFM*). (Prior to publication of the first edition, the word *manufacturability* was rejected for use in the title because it was felt to be too obscure. How times have changed!) The emergence of the DFM approach and the developments that have taken place since the first edition was published in 1986 make it advisable for a second edition of the handbook to be made available.

Because of the widespread acceptance of the terms *design for manufacturability* and *DFM*, it is appropriate that the new edition of the handbook be given a title consistent with these terms, hence *The Design for Manufacturability Handbook.*

Although the underlying principles of DFM are relatively unchanging, there are a number of significant additions and changes in this edition. Material on a number of subjects has been added to the handbook's contents. Additionally, new materials designations have been used, improved and newly developed processes have been added, and design recommendations have been modified as processes have become more versatile or more accurate. Most significant may be the enhanced material on design for Assembly (or DFA.) Added chapters cover the following subjects:

The history of DFM

DFM for limited production quantities (where emphasis and approaches are different even though the basic principles are the same)

DFX, the practice of designing a product for a series of desirable attributes such as safety, quality, reliability, user and environmental friendliness, and short time-to-market as well as manufacturability

DFM in electronics

The management of DFM projects

DFM terminology

Metal injection molded parts

I am indebted to many people who provided invaluable assistance that made the preparation of this book possible. My wife, Martha, deserves special thanks for tolerating my absence in libraries and our home office while working on this edition. Many others helped me to obtain the necessary data or referred me to specialists who either wrote the Handbook's chapters or gave me other assistance. Some of them are the following: Robert J. Johnson, Consultant, West End, NC, Stainless steel; Tom Prucher, Burgiss-Norton Manufacturing Co., Geneva, Il, Powder metals; Irving Laskin, Consultant, Sharon, MA, Plastic gear materials; David Towel, MEI, Lakewood, NJ, Magnesium; K. Savage, Magnesium Elecktron, Swinton, Manchester, England, Magnesium; Jim Hillis, Dow Chemical Company, Freeport Texas, Magnesium; Dale

Barsness, Consultant, Simsbury, CT, Adhesive bonding; Dr. Alan Lawley, Drexel University, Philadelphia, PA, Powder Metals; and Dan Mustico, Rubber Manufacturers Association, Washington, DC, Rubber Products.

Design for manufacturability is a powerful tool for reducing costs and improving products. All users of this handbook are extended best wishes for success with the technique.

Readers are invited to call to my attention any errors that may have crept into the data reproduced in this book.

James G. Bralla

CONTRIBUTORS

Donald A. Adams *The Singer Company, Elizabeth, New Jersey* (CHAP. 8.4)

Robert W. Bainbridge *Occidental Chemical Corp., Durez Resins & Molding Materials, North Tonawanda, New York* (CHAP. 6.1)

Ronald D. Beck *Fisher Body Division, General Motors Corp., Warren, Michigan* (CHAP. 6.12)

J. R. Casey Bralla *Takata Restraint Systems, Cheraw, South Carolina* (CHAP. 6.7)

Welsford, J. Bryan *INA Bearing Co., Inc., Cheraw, South Carolina* (CHAP. 4.20)

Boris Bryskin *Rhenium Alloys, Inc., Elyria, Ohio* (CHAP. 2.3, PART 4)

Roald Cann *Bryant Grinder Corporation, Springfield, Vermont* (CHAP. 4.12)

Jan Carlén *Rhenium Alloys, Inc., Elyria, Ohio* (CHAP. 2.3, PART 4)

E. N. Castellano *General Manager, Liqwacon Corporation, Thomaston, Connecticut* (CHAP. 3.14)

John Casteras *Alpha Metals, Inc., Jersey City, New Jersey* (CHAPS. 2.3, PART 4, 7.4)

John Cocco, P.E. *Loctite Corp., Rocky Hill, Connecticut* (CHAP. 7.5)

Glenn Coggins *Agie U.S.A. Ltd., Davidson, North Carolina* (CHAP. 4.18)

Calvin J. Cooley *Metallurgical Engineer, Committee of Stainless Steel Producers, American Iron and Steel Institute, Washington, D.C.* (CHAP. 2.2, PART 3)

Engineering Staff *Teledyne Landis Machine, Waynesboro, Pennsylvania* (CHAP. 4.8)

John L. Everhart, P.E. *Westfield, New Jersey* (CHAP. 3.8)

Joe K. Fischlin *American Feintool, Inc., White Plains, New York* (CHAP. 3.3)

Ivan A. Franson *Allegheny Ludlum Corp., Brackenridge, Pennsylvania* (CHAP. 2.2, PART 3)

Henry O. Fuchs *Stanford University, Stanford, California, and Consultant and Former President, Metal Improvement Co., Paramus, New Jersey* (CHAP. 8.7)

Laroux K. Gillespie *Allied Signal Aerospace, Kansas City, Missouri* (CHAP. 4.23)

Albert J. Gonas *Fisher Body Division, General Motors Corp., Warren, Michigan* (CHAP. 8.2)

Philip J. Guichelaar *Western Michigan University, Kalamazoo, Michigan* (CHAP. 3.13)

Stuart Haley *Agie U.S.A. Ltd., Davidson, North Carolina* (CHAP. 4.18)

Andrew W. Hanson *PCC-Advanced Forming Technology, Longmont, Colorado* (CHAP. 3.16)

Charles A. Harper *Technology Seminars, Inc., Lutherville, Maryland* (CHAP. 2.4)

Charles W. Hatmaker *Titanium and Alloys Corp., Warren, Michigan* (CHAP. 2.3, PART 4)

Paul M. Heilman *Sales Manager, Forgings, Bridgeport Brass Company, Norwalk, Connecticut* (CHAP. 3.13)

Nicholas S. Hodska *Stratford, Connecticut* (CHAPS. 6.5, 7.3)

Frederick W. Hornbruch, Jr. *Laguna Hills, California* (CHAP. 1.2)

International Magnesium Association *McLean, Virginia* (CHAP. 2.3, PART 3)

Fred H. Jay *Fisher Gauge Limited, Peterborough, Ontario* (CHAP. 5.4, MINIATURE DIE CASTINGS)

J. Franklin Jones *Springfield, Vermont* (CHAP. 4.22)

Theodore W. Judson *Professor of Process Engineering, GMI Engineering and Management Institute, Flint, Michigan* (CHAPS. 4.4, 4.6)

Konrad J. A. Kundig, PhD. *International Copper Associaton, New York, New York* (CHAP. 2.3, PART 2)

Russ W. Kirchner *High Performance Alloys, Tipton, Indiana* (CHAP. 2.3)

Kenneth Langlois *Torin Corporation, Torrington, Connecticut* (CHAP. 3.4)

Fred W. Lewis *Standard Locknut and Lockwasher, Inc., Carmel, Indiana* (CHAP. 4.3)

C. Richard Liu *School of Industrial Engineering, Purdue University, West Lafayette, Indiana* (CHAP. 4.17)

F. N. Longo, F.N. *Longo Associates, East Northport, New York* (CHAP. 8.3, PART 2)

John L. MacLaren *Vice President, Marketing, Western Die Casting Company, Emeryville, California* (CHAP. 5.4)

R. Bruce Macleod *Vice President, Taft-Peirce Supfina, Cumberland, Rhode Island* (CHAP. 4.15)

William B. McMullin *Process and Product Development Laboratory, Reynolds Metals Company, Richmond, Virginia* (CHAP. 2.3, PART 1)

R. W. Militzer, P.E. *Consulting Engineer, Fenton, Michigan* (CHAP. 4.16)

J. Gilbert Mohr *J. G. Mohr Co., Inc., Maumee, Ohio* (CHAPS. 6.4, 6.6, 6.11)

Wes Mowry *Senior Product Engineer, Vitrified Product Engineering, Abrasive Marketing Group, Norton Company, Worcester, Massachusetts* (CHAP. 4.13)

Alan J. Musbach *MCP Industries, Inc., Detroit, Michigan* (CHAP. 8.2)

B. W. Niebel *Professor Emeritus of Industrial Engineering, The Pennsylvania State University, University Park, Pennsylvania* (CHAP. 5.2)

Ralph A. Pannier *The Pannier Corporation, Pittsburgh, Pennsylvania* (CHAP. 8.6)

L. J. Piccinino *Head Instructor, School of Grinding, Norton Company, Worcester, Massachusetts* (CHAP. 4.14)

Robert Powell *Consultant, Bordentown, New Jersey* (CHAP. 3.12)

Robert Roseliep *General Broach and Engineering Corp., Chesterfield, Michigan* (CHAP. 4.9)

Harry Saperstein *Livingston, New Jersey* (CHAP. 3.6)

Gerald L. Schneberger *G. L. Schneberger and Associates, Grand Blanc, Michigan* (CHAPS. 7.5, 8.5)

Michael H. Skillingberg *Reynolds Metals Co., Richmond, Virginia* (CHAP. 2.3, PART 1)

Ted Slezak *Armstrong-Blum Mfg. Co., Chicago, Illinois* (CHAP. 4.2)

John G. Sommer *Research Division, GenCorp, Akron, Ohio* (CHAP. 6.10)

Paul Sopko *Airco Welding Products, Murray Hill, New Jersey* (CHAP. 4.11)

Robert J.Spinosa *Cooper Industries, Monroe, North Carolina* (CHAP. 5.3)

John Stein *The Singer Company, Elizabeth, New Jersey* (CHAP. 3.2)

Federico Strasser *Santiago, Chile* (CHAP. 3.2)

B. H. Swan *Industrial Powder Met (Consultants) Ltd., London, England* (CHAP. 3.12)

Dieter E. A. Tannenberg *Senior Vice President, AM International, Inc.; President, Multigraphics Division, Mount Prospect, Illinois* (CHAP. 3.15)

Technology Committee *Spring Manufacturers Institute, Wheeling, Illinois* (CHAP. 3.5)

James W. Throop *Professor of Process Engineering, GMI Engineering and Management Institute, Flint, Michigan* (CHAP. 4.19)

Daryl E. Tonini *Manager of Technical Services, American Hot Dip Galvanizers Association, Inc., Washington, D.C.* (CHAP. 8.3, PART 1)

William R. Tyrrell *Director, Plastics Processing, Branson Sonic Power Division, Branson Ultrasonics Corp., Danbury, Connecticut* (CHAP. 6.9)

Charles Wick *Consultant, Birmingham, Michigan* (CHAP. 3.7)

Jay C. Willcox *Toro Company, Minneapolis, Minnesota* (CHAP. 7.2)

Edward C. Zuppann *Meehanite Worldwide Division, Meehanite Metal Corporation, White Plains, New York* (CHAP. 5.1)

S • E • C • T • I • O • N • 1

INTRODUCTION

CHAPTER 1.1
PURPOSE, CONTENTS, AND USE OF THIS HANDBOOK

OBJECTIVE

This is a reference book for those practicing or otherwise having an interest in design for manufacturability (DFM). DFM principles and guidelines are many; no one person should be expected to remember them all nor the detailed information, such as suggested dimensional tolerances, process limits, expected surface finish values, or other details, of each manufacturing process. It is expected that those involved will keep this book handy for reference when needed.

Additionally, this handbook is intended to be an educational tool to assist those who wish to develop their skills in ensuring that products and their components are easily manufactured at minimum cost. Its purpose, further, is to enable designers to take advantage of all the inherent cost and other benefits available in the manufacturing process that will be used.

Like handbooks in other fields, it is a comprehensive summary of information which, piecemeal at least, is known by or available to specialists in the field. Although some material in this handbook has not appeared in print previously, the vast majority of it is a restatement, reorganization, and compilation of data from other published sources.

USERS OF THIS HANDBOOK

The subject matter of this book covers the area where product engineering and manufacturing engineering overlap. In addition to being directed to product designers and manufacturing engineers, this book is directed to the following specialists:

Operation sheet writers
Value engineers and analysts
Tool engineers
Process engineers
Production engineers
Cost-reduction engineers
Research and development engineers
Drafters

Industrial engineers

Manufacturing supervisors and managers

These specialists and any other individuals whose job responsibilities or interest involve low-cost manufactured products should find this handbook useful.

CONTENTS

This book contains summary information about the workings and capabilities of various significant manufacturing processes. The standard format for each chapter involves a clear summary of how each manufacturing process operates to produce its end result. In most cases, for added clarity, a schematic representation of the operation is included so that the reader can see conceptually exactly what actions are involved. In many cases, for further clarification, photographs or drawings of common equipment are presented. The purpose of this brief process explanation is to enable readers to understand the basic principles of the manufacturing process to determine whether it is applicable to production of the particular workpiece they have in mind.

To illustrate further the workings of each manufacturing process from the viewpoint of product engineers, descriptive information on typical parts produced by the process is provided. This book tells readers how large, small, thick, thin, hard, soft, simple, or intricate the typical part will be, what it looks like, and what material it is apt to be made from. Typical parts and applications are illustrated whenever possible so that readers can see by example what can be expected from the manufacturing process in question.

Since so many manufacturing processes are limited in economical application to only one portion of the production-quantity spectrum, this factor is reviewed for each process being covered. We want to help engineers to design a product for a manufacturing process that fits not only the part configuration but the expected manufacturing volume as well. We want to steer them away from a process that, even though it might provide the right size, shape, and accuracy, would not be practical from a cost standpoint.

To aid designers in specifying a material that is most usable in the process, information is provided on suitable materials in each chapter. Emphasis is on materials formulations that give satisfactory functional results and maximum ease of processibility. Where feasible, tables of suitable or commonly used materials are included. The tables usually provide information on other properties of each material variation and remarks on the common applications of each. Where available, processibility ratings are also included. All materials selection is a compromise. Functional considerations—strength, stiffness, corrosion resistance, electrical conductivity, appearance, and many other factors—as well as initial cost and processing cost, machinability, formability, and so on, must all be considered. When one factor is advantageous, the others may not be. Most of the materials recommendations included in this handbook are for run-of-the-mill noncritical applications for which processibility factors can be given greater weight. The purpose is to aid in avoiding overspecifying material when a lower-cost or more processible grade would serve as well. For many applications, of course, grades with greater functional properties must be used, and materials suppliers should be consulted.

The heart of this handbook (in each chapter) is the coverage of recommendations for more economical product design. Providing information to guide designers to configurations that simplify the production process is a prime objective of this handbook.

Design recommendations are of two kinds: general design considerations and detailed design recommendations. The former cover the major factors that designers should take into consideration to optimize the manufacturability of their designs. Such factors as shrinkage (castings and molded parts), machining allowances, the feasibility of undercuts, and the necessity for fillets and radii are discussed.

Detailed design recommendations include numerous specific tips to aid in developing the most producible designs with each process. Most of these are illustrated and are in the form of "do, don't," "this, not this," or "feasible, preferable" so that both the preferred and less desirable design alternatives are shown. The objective of these subsections is to cover each characteristic having a significant bearing on manufacturability.

Dimensional-tolerance recommendations for parts made with each process are another key element of each chapter. The purpose is to provide a guide for manufacturing engineers so that they know whether a process under consideration is suitable for the part to be produced. Equally important, these recommendations give product designers a basis for providing realistic specifications and for avoiding unnecessarily or unrealistically strict tolerances. The recommended tolerances, of course, are average values. The dimensional capabilities of any manufacturing process will vary depending on the peculiarities of the size, shape, and material of the part being produced and many other factors. The objective in this book has been to provide the best possible data for normal applications.

To give a fuller understanding of these tolerances and the reasons why they are necessary, most chapters include a discussion of the dimensional factors that affect final dimensions.

This handbook helps determine which process to use, but it does not tell how to operate each process, e.g., what feed, speed, tool angle, tool design, tool material, process temperature, pressure, etc., to use. These points are valid ones and are important, but of necessity, they are outside the scope of this book. To include them in addition to the prime data would make this handbook too long and unwieldy. This kind of material is also well covered in other publications. The emphasis in this book is on the *product* rather than the *process,* although a certain amount of process information is needed to ensure proper product design.

This book also does not contain very much functional design information. There is little material on strength of components, wear resistance, structural rigidity, thermal expansion, coefficient of friction, etc. It may be argued that these kinds of data are essential to proper design and that consideration of design only from a manufacturability standpoint is one-sided. It cannot be denied that functional design considerations are essential to product design. However, these factors are covered extensively and well in innumerable handbooks and other references, and it would be neither economically feasible nor practicable to include them in this book. This handbook is to be used in conjunction with such references. The subjects of functional design and design for manufacturability are complementary aspects of the same basic subject matter. In this respect, DFM is no different from industrial design, which deals with product appearance, or reliability design or anticorrosion design, to cite some examples of subsidiary design engineering disciplines that have been the subject of separate handbooks.

RESPONSIBILITIES OF DESIGN ENGINEERS

The responsibilities of design engineers encompass all aspects of design. Although functional design is of paramount importance, a design is not complete if it is functional but not easily manufactured, or if it is functional but not reliable, or if it has a good appearance but poor reliability. Design engineers have the broad responsibility to produce a design that meets all its objectives: function, durability, appearance, and cost. A design engineer cannot say, "I designed it. Now it's the manufacturing engineer's job to figure out how to make it at a reasonable cost." The functional design and the production design are too closely interrelated to be handled separately.

Product designers must consider the conditions under which manufacturing will take place, since these conditions affect production capability and costs. Such factors as production quantity, labor, and materials costs are vital.

Designers also should visualize how each part is made. If they do not or cannot, their designs may not be satisfactory or even feasible from the production standpoint. One purpose of this handbook is to give designers sufficient information about manufacturing processes so that they can design intelligently from a producibility standpoint.

RESPONSIBILITIES OF MANUFACTURING ENGINEERS

Manufacturing engineers have a dual responsibility. Primarily, they provide the tooling, equipment, operation sequence, and other technical wherewithal to enable a product to be manufactured. Secondarily, they have a responsibility to ensure that the design provided to the manufacturing organization is satisfactory from a manufacturability standpoint.

It is to the latter function that this handbook is most directly aimed. In the well-run product design and manufacturing organization, a team approach is used, and the product engineer and manufacturing engineer work together to ensure that the product design provides the best manufacturability.

Another function of manufacturing engineers, cost reduction, deserves separate comment. Manufacturing and industrial engineers and others involved in manufacturing under industrial conditions have, since the process began, made a practice of whittling away at the costs involved in manufacturing a product. Fortunes have been spent (and made) in such activities, and no aspect of manufacturing costs has been spared. No avenue for cost reduction has been ignored. In my experience, by far the most lucrative avenue is the one in which the product design is analyzed for lower-cost alternatives (value analysis). This approach has proved to provide a larger return (greater cost reduction) per unit of effort and per unit of investment than other approaches, including mechanization, automation, wage incentives, and the like.

HOW TO USE THIS HANDBOOK

This book can be used with any of three methods of reference: (1) by process, (2) by design characteristic, and (3) by material. Readers will use the first approach when they have a specific production process in mind and wish to obtain further information about the process, its capabilities, and how to develop a product design to take best advantage of it. Most of the handbook's chapters are concerned with a particular process, e.g., surface grinding, injection molding, forging, etc., and it is a simple matter to locate the applicable section from the Contents or Index.

The problem with the process-oriented book layout is that it is not adapted to designers (or manufacturing engineers) who are concerned with a particular product characteristic and do not really know the best way to produce it. For example, designers having the problem of making a nonround hole in a hardened-steel part may not be aware of the best process to use or even of all processes that should be considered. This is the kind of problem for which this handbook is intended to provide assistance. There are three avenues that readers can use to obtain assistance in answering their questions:

1. The handbook chapters, as much as possible, are aimed at a workpiece characteristic, e.g., "ground surfaces = flat," rather than a process, e.g., "surface grinding."
2. The Index has numerous cross-references under product characteristics such as "holes, nonround" or "surfaces, flat." It provides page listings for various methods of making such holes, e.g., electrical-discharge machining (EDM), electrochemical machining (ECM), broaching, etc.
3. This a chapter entitled, "Quick References" (Chap. 1.4), where readers can obtain comparative process-capability data for a variety of common workpiece characteristics such as round holes, nonround holes, flat surfaces, contoured surfaces, etc. A full listing of quick-reference subjects can be found in the Contents.

To aid readers interested in obtaining information about the manufacturability of particular materials, there is a section entitled, "Economical Use of Raw Materials" (Sec. 2), that summarizes applications of common metallic and nonmetallic materials and recommends certain material formulations or alloys for easy processibility with common manufacturing methods.

WHEN TO USE THIS HANDBOOK

This handbook can be used for reference at the following stages in the design and manufacture of a product:

1. When a new product is in the concept stage of product development, to point out, at the outset, potentially low-manufacturing-cost approaches. This is by far the best time to optimize manufacturability.
2. During the design stage, when prototypes are built and when final drawings are being prepared, particularly to ensure that dimensional tolerances are realistic.
3. During the manufacturability-review stage, to assist manufacturing engineers in ascertaining that the design is suitable for economical production.
4. At the production-planning stage, when manufacturing operations are being chosen and their sequence is being decided on.
5. For guidance of value-analysis activities after the product has gone into production and as production quantities and cost levels for materials and labor change, providing a potential for cost improvements.
6. When redesigning a product as part of any product improvement or upgrading.
7. When replacing existing tooling that has worn beyond the point of economical use. At this time it usually pays to reexamine the basic design of the product to take advantage of manufacturing economies and other improvements that may become evident.

METRIC CONVERSIONS

Most dimensional data in this handbook are expressed in both metric and U.S. customary units. Metric units are based on the SI system (International System of Units). In some cases, data have been rounded off to convenient values instead of following exact equivalents. This was done with design and tolerance recommendations when it was felt that easily remembered order-of-magnitude values were more important than precise conversions.

When dual dimensions are not given, Table 1.1.1 provides conversion factors that can be applied.

TABLE 1.1.1 Metric Dimensions Used in This Handbook

Measurement	Metric symbol	Metric unit	Conversion to U.S. customary unit
Linear dimensions	mm	millimeter	1 mm = 0.0394 in
	cm	centimeter	1 cm = 0.394 in
	m	meter	1 m = 39.4 in
Area	cm^2	square centimeter	1 cm^2 = 0.155 in^2
	m^2	square meter	1 m^2 = 10.8 ft^2
Surface finish	μm	micrometer	1 μm = 39.4 μin
Volume	cm^3	cubic centimeter	1 cm^3 = 0.061 in^3
	m^3	cubic meter	1 m^3 = 35.3 ft^3
Stress, pressure, strength	kPa	kilopascal	1 kPa = 0.145 lbf/in^2
	MPa	megapascal	1 MPa = 145 lbf/in^2
Temperature	°C	degree Celsius	$\text{degrees C} = \frac{\text{degrees F} - 32}{1.8}$

CHAPTER 1.2
ECONOMICS OF PROCESS SELECTION

Frederick W. Hornbruch, Jr.
Corporation Consultant
Laguna Hills, California

COST FACTORS

Design engineers, manufacturing engineers, and industrial engineers, in analyzing alternative methods for producing a part or a product or for performing an individual operation or an entire process, are faced with cost variables that relate to materials, direct labor, indirect labor, special tooling, perishable tools and supplies, utilities, and invested capital. The interrelationship of these variables can be considerable, and therefore, a comparison of alternatives must be detailed and complete to assess properly their full impact on total unit costs.

Materials

The unit cost of materials is an important factor when the methods being compared involve the use of different amounts or different forms of several materials. For example, the materials cost of a die-cast aluminum part probably will be greater than that of a sand-cast iron part for the same application. An engineering plastic for the part may carry a still higher cost. Powder-metal processes use a smaller quantity of higher-cost materials than casting and machining processes. In addition, yield and scrap losses may influence materials cost significantly.

Direct Labor

Direct labor unit costs essentially are determined by three factors: the manufacturing process itself, the design of the part or product, and the productivity of the employees operating the process or performing the work. In general, the more complex the design, the closer the dimensional tolerances, the higher the finish requirements, and the less tooling involved, the greater the direct labor content will be.

The number of manufacturing operations required to complete a part probably is the greatest single determinant of direct labor cost. Each operation involves a "pick up and locate" and a "remove and set aside" of the material or part, and usually additional inspection by the operators is necessary. In addition, as the number of operations increases, indirect costs tend to accelerate. The chances for cumulative dimensional error are increased owing to changing locating points and surfaces. More setups are required; scrap and rework increase; timekeeping, counting, and paperwork expand; and shop scheduling becomes more complex.

Typical of low-labor-content processes are metal stamping and drawing, die casting, injection molding, single-spindle and multispindle automatic machining, numerical- and computer-controlled drilling, and special-purpose machining, processing, and packaging in which secondary work can be limited to one or two operations. Semiautomatic and automatic machines of these types also offer opportunities for multiple-machine assignments to operators and for performing secondary operations internal to the power-machine time. Both can reduce unit direct labor costs significantly.

Processes such as conventional machining, investment casting, and mechanical assembly including adjustment and calibration tend to contain high direct labor content.

Indirect Labor

Setup, inspection, material handling, tool sharpening and repairing, and machine and equipment maintenance labor often are significant elements in evaluating the cost of alternative methods and production designs. The advantages of high-impact forgings may be offset partially by the extra indirect labor required to maintain the forging dies and presses in proper working condition. Setup becomes an important consideration at lower levels of production. For example, it may be more economical to use a method with less setup time even though the direct labor cost per unit is increased. Take a screw-machine type of part with an annual production quantity of 200 pieces. At this volume, the part would be more economically produced on a turret lathe than on an automatic screw machine. It's the total unit cost that is important.

Special Tooling

Special fixtures, jigs, dies, molds, patterns, gauges, and test equipment can be a major cost factor when new parts and new products or major changes in existing parts and products are put into production. The amortized unit tooling cost should be used in making comparisons. This is so because the unit tooling cost, limited by life expectancy or obsolescence, is very production-volume-dependent. With high production volume, a substantial investment in tools normally can be readily justified by the reduction in direct labor unit cost, since the total tooling cost amortized over many units of product results in a low tooling cost per unit. For low-volume-production applications, even moderate tooling costs can contribute relatively high unit tooling costs.

In general, it is conservative to amortize tooling over the first 3 years of production. Competition and progress demand improvements in product design and manufacturing methods within this time span. In the case of styled items, the period may need to be shortened to 1 or 2 years. Automobile grilles are a good example of items that traditionally have had a production life of 2 years, after which a restyled design is introduced.

Perishable Tools and Supplies

In most cost systems, the cost of perishable tools such as tool bits, milling cutters, grinding wheels, files, drills, taps, and reamers and supplies such as emery paper, solvents, lubricants, cleaning fluids, salts, powders, hand rags, masking tape, and buffing compounds are allocated as part of a cost-center manufacturing-overhead rate applied to direct labor. It may be, however, that there are significant differences in the use of such items in one process when compared with another. If so, the direct cost of the items on a unit basis should be included in the unit-cost comparison. Investment casting, painting, welding, and abrasive-belt machining are examples of processes with high costs for supplies. In the case of cutoff operations, it is more correct to consider the tool cost per cut as an element in a comparison. Cutting-tool costs for other types of machining operations also may constitute a major part of the total unit cost. The high cost and short tool life of carbide milling cutters for profile milling of "hard metals," such as are used in jet-engine components, contribute significantly to the cost per unit. The hard metals include Inconel, refractory-metal alloys, and superalloy steels.

Utilities

Here again, as with perishable tools and supplies, the cost of electric power, gas, steam, refrigeration, heat, water, and compressed air should be considered specifically when there are substantial differences in their use by the alternative methods and equipment being compared. For example, electric power consumption is a major element of cost in using electric-arc furnaces for producing steel castings. And some air-operated transfer devices may increase the use of compressed air to a point at which additional compressor capacity is needed. If so, this cost should be factored into the unit cost of the process.

Invested Capital

Obviously, it is easier and less risky for a company to embark on a program or a new product that utilizes an extension of existing facilities. In addition, the capital investment in a new product can be minimized if the product can be made by using available capacity of installed processes. Thus the availability of plant, machines, equipment, and support facilities should be taken into consideration as well as the capital investment required for other alternatives. In fact, if sufficient productive capacity is available, no investment may be required for capital items in undertaking the production of a new part or product with existing processes. Similarly, if reliable vendors are available, subcontracting may be an alternative. In this event, the capital outlays may be borne by the vendors and therefore need not be considered as separate items in the cost evaluation. Presumably, such costs would be included in the subcontract prices per unit.

On the other hand, there may be occasions when the production of a single component necessitates not only the purchase of additional production equipment but also added floor space, support facilities, and possibly land. This eventuality could occur if the present plant was for the most part operating near capacity with respect to equipment, space, and property or if existing facilities were not fully compatible with producing the component or product at a low unit cost.

When capital equipment costs are pertinent to the selection of a process, the unit-cost calculations should assign to each unit of product its share of the capital investment based on the expected life and production from the capital item. For example, a die-casting machine that sells for \$200,000, has an estimated production life of 10 years and an expected operating schedule of three shifts of 2000 h each per year, and is capable of producing at the rate of 100 shots per hour with a two-cavity mold, less a 20 percent allowance for downtime for machine and die maintenance and setups, would have a capital cost per unit as follows:

$$\text{Capital cost} = \frac{\$200{,}000}{10 \times 3 \times 2000 \times 100 \times 2 \times (100\% - 20\%)} = \$0.020 \text{ per piece}$$

This calculation assumes that the machine will be utilized fully by the proposed product or other production. Also, the computation does not include any interest costs. Interest charges for financing the purchase of the machine should be added to the purchase price. If interest costs of \$50,000 over the life of the machine are assumed, the capital cost per unit would be \$0.025 instead of \$0.020. This type of calculation is applicable solely to provide a basis for choosing between process alternatives and is simpler and different from the analysis involved in justifying the investment once the process selection has been made.

Other Factors

Occasionally, a special characteristic of one or several of the processes under consideration involves an item of cost that may warrant inclusion in the unit-cost comparison. Examples of this type might include costs related to packaging, shipping, service and unusual maintenance, and rework and scrap allowances. The important point is to recognize all the essential differences between the alternatives and to allow properly for these differences in the unit-cost comparison. Remember that the objective is to determine the most economical process for a given set of conditions, i.e., the process that can be expected to produce the part or product at the lowest total unit cost for the anticipated sales volume.

Also, in making a unit-cost comparison between several alternatives, it is necessary to include in the analysis only those costs which differ between alternatives. For example, if all choices involve the same kind and amount of material, the materials cost per unit need not be included in the comparison.

Further, when available capacity exists on production equipment used for similar components, the choice of process may be obvious. This is especially true when the production quantity for the new part or product is not high. The opportunity for utilizing available capacity makes an additional investment in an alternative process difficult to justify.

TYPICAL EXAMPLES

Exhibits 1.2.1 and 1.2.2 are examples showing a concise layout for comparing alternatives. Exhibit 1.2.1 compares sand-mold casting with die casting for one part. Exhibit 1.2.2 considers making a part on a turret lathe versus single-spindle and multispindle automatic screw machines. Neither of these examples attempts to justify the purchase of machines or equipment. These examples assume that the processes are installed and

EXHIBIT 1.2.1 Sand-Mold Casting versus Die Casting

Part: New model pump housing *Annual quantity: 10,000 pieces*
Expected product life: 5 years *Normal lot size: 2500 pieces*

Process	Gray-iron casting			Aluminum die casting		
Cost item	Cost of item	Frequency per piece	Unit cost	Cost of item	Frequency per piece	Unit cost
1. Tooling (jigs, fixtures, etc.)	$5000 (patterns)	1/50,000	$0.10	$35,000 (die)	1/50,000	$0.70
2. Material	$0.20/lb	6 lb	$1.20	$0.70/lb	2 lb	$1.40
3. Casting: setup	0.30 h at $8/h	1/2500	$0.00	4.0 h at $8/h	1/2500	$0.01
4. Casting: direct labor	0.08 h at $8/h	1	$0.64	0.04 h at $8/h	1	$0.32
5. Machining:setup	$50 (for 5 operations)	1/2500	$0.02	$25 (for 3 operations)	1/2500	$0.01
6. Machining:direct labor	0.05 h at $8/h (for 5 operations	1	$0.40	0.03 h at $8/h (for 3 operations)	1	$0.24
7. Total unit cost			$2.36			$2.68

EXHIBIT 1.2.2 Turret Lathe versus Single-Spindle and Multispindle Automatic Screw Machines (Excluding Secondary Operations)

Part: High-pressure hose fitting *Annual quantity: 500 pieces*
Expected product life: 2 years *Normal lot size: 500 pieces*

Cost item (machine)	Turret lathe		Automatic single-spindle		Automatic multispindle	
		Per unit		Per unit		Per unit
1. Tooling (chuck jaws, cams, form tools, other cutters	$350	$0.35	$680	$0.68	$1000	$1.00
2. Setup at $14/h	1 h ÷ 500 pieces	$0.03	2 h÷500 pieces	$0.06	3 h ÷ 500 pieces	$0.08
3. Direct labor and other overhead at $20/h	2 min (1 machine per operator)	$0.67	0.60 min (4 machines per operator)	$0.05	0.20 min (2 machines per operator)	$0.03
4. Total unit cost		$1.05		$0.79		$1.11

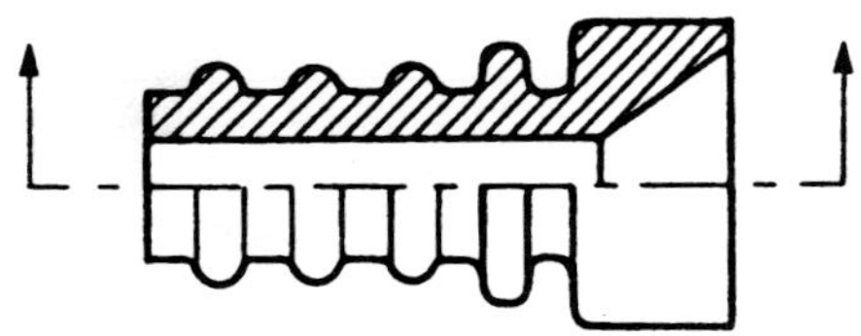

have available capacity for additional production. Note that the production quantity is an important factor in determining the most economical process. In both illustrations, as the production quantity increases, the unit-cost comparison begins to favor a different alternative.

CHAPTER 1.3
GENERAL DESIGN PRINCIPLES FOR MANUFACTURABILITY

BASIC PRINCIPLES OF DESIGNING FOR ECONOMICAL PRODUCTION

The following principles, applicable to virtually all manufacturing processes, will aid designers in specifying components and products that can be manufactured at minimum cost.

1. *Simplicity.* Other factors being equal, the product with the fewest parts, the least intricate shape, the fewest precision adjustments, and the shortest manufacturing sequence will be the least costly to produce. Additionally, it usually will be the most reliable and the easiest to service.

2. *Standard materials and components.* Use of widely available materials and off-the-shelf parts enables the benefits of mass production to be realized by even low-unit-quantity products. Use of such standard components also simplifies inventory management, eases purchasing, avoids tooling and equipment investments, and speeds the manufacturing cycle.

3. *Standardized design of the product itself.* When several similar products are to be produced, specify the same materials, parts, and subassemblies for each as much as possible. This procedure will provide economies of scale for component production, simplify process control and operator training, and reduce the investment required for tooling and equipment.

4. *Liberal tolerances.* Although the extra cost of producing too tight tolerances has been well documented, this fact is often not appreciated well enough by product designers. The higher costs of tight tolerances stem from factors such as (a) extra operations such as grinding, honing, or lapping after primary machining operations, (b) higher tooling costs from the greater precision needed initially when the tools are made and the more frequent and more careful maintenance needed as they wear, (c) longer operating cycles, (d) higher scrap and rework costs, (e) the need for more skilled and highly trained workers, (f) higher materials costs, and (g) more sizable investments for precision equipment.

Figure 1.3.1 graphically illustrates how manufacturing cost is multiplied when close tolerances are specified. Table 1.3.1 illustrates the extra cost of producing fine surface finishes. Figure 1.3.2 illustrates the range of surface finishes obtainable with a number of machining processes. It shows how substantially the process time for each method can increase if a particularly smooth surface finish must be provided.

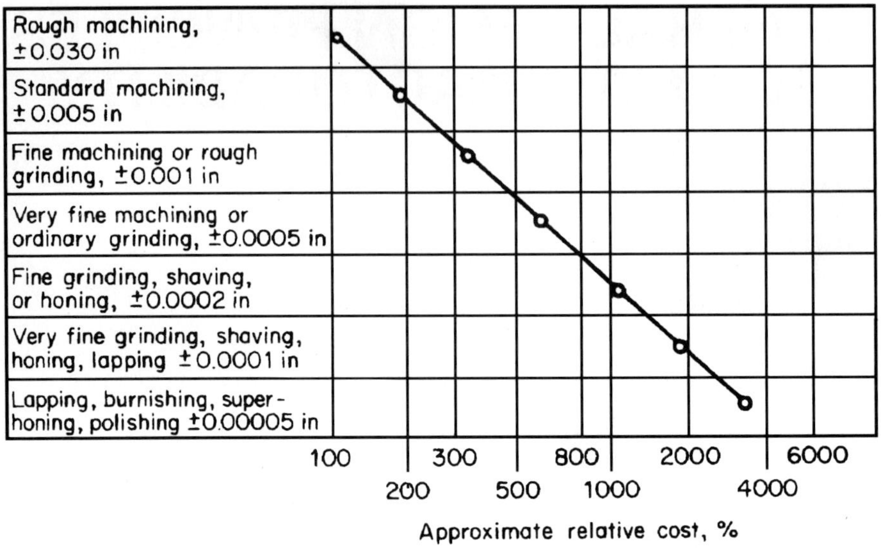

FIGURE 1.3.1 Approximate relative cost of progressively tighter dimensional tolerances. (*From N. E. Woldman,* Machinability and Machining of Metals, *McGraw-Hill, New York. Used with the permission of McGraw-Hill Book Company.*)

TABLE 1.3.1 Cost of Producing Surface Finishes

Surface symbol designation	Surface roughness, μin	Approximate relative cost, %
Case, rough-machined	250	100
Standard machining	125	200
Fine machining, rough-ground	63	440
Very fine machining, ordinary grinding	32	720
Fine grinding, shaving, and honing	16	1400
Very fine grinding, shaving, honing, and lapping	8	2400
Lapping, burnishing, superhoning, and polishing	2	4500

Source: N. E. Woldman, *Machinability and Machining of Metals,* McGraw-Hill, New York. Used with the permission of McGraw-Hill Book Company.

5. *Use of the most processible materials.* Use the most processible materials available as long as their functional characteristics and cost are suitable. There are often significant differences in processibility (cycle time, optimal cutting speed, flowability, etc.) between conventional material grades and those developed for easy processibility. However, in the long run, the most economical material is the one with the lowest combined cost of materials, processing, and warranty and service charges over the designed life of the product.

6. *Teamwork with manufacturing personnel.* The most producible designs are provided when the designer and manufacturing personnel, particularly manufacturing engineers, work closely together as a team or otherwise collaborate from the outset.

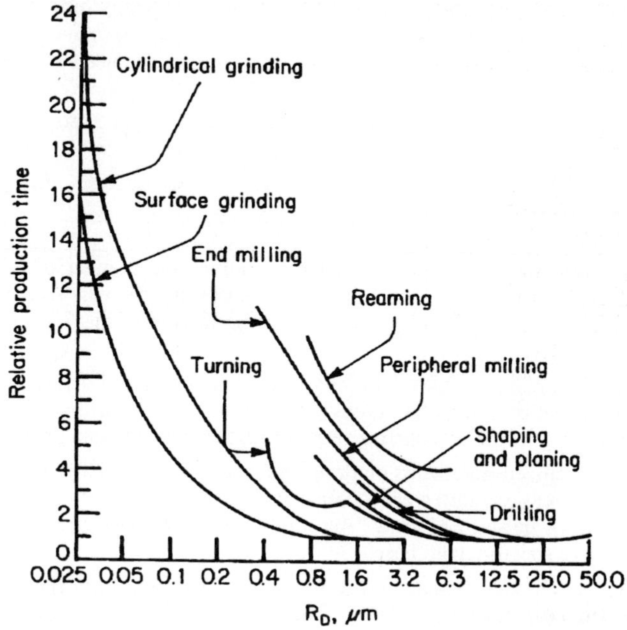

FIGURE 1.3.2 Typical relationships of productive time and surface roughness for various machining processes. (*From British Standard BS 1134.*)

7. *Avoidance of secondary operations.* Consider the cost of operations, and design in order to eliminate or simplify them whenever possible. Such operations as deburring, inspection, plating and painting, heat treating, material handling, and others may prove to be as expensive as the primary manufacturing operation and should be considered as the design is developed. For example, firm, nonambiguous gauging points should be provided; shapes that require special protective trays for handling should be avoided.

8. *Design appropriate to the expected level of production.* The design should be suitable for a production method that is economical for the quantity forecast. For example, a product should not be designed to utilize a thin-walled die casting if anticipated production quantities are so low that the cost of the die cannot be amortized. Conversely, it also may be incorrect to specify a sand-mold aluminum casting for a mass-produced part because this may fail to take advantage of the labor and materials savings possible with die castings.

9. *Utilizing special process characteristics.* Wise designers will learn the special capabilities of the manufacturing processes that are applicable to their products and take advantage of them. For example, they will know that injection-molded plastic parts can have color and surface texture incorporated in them as they come from the mold, that some plastics can provide "living hinges," that powder-metal parts normally have a porous nature that allows lubrication retention and obviates the need for separate bushing inserts, etc. Utilizing these special capabilities can eliminate many operations and the need for separate, costly components.

10. *Avoiding process restrictiveness.* On parts drawings, specify only the final characteristics needed; do not specify the process to be used. Allow manufacturing engineers as much latitude as possible in choosing a process that produces the needed dimensions, surface finish, or other characteristics required.

GENERAL DESIGN RULES

1. First in importance, simplify the design. Reduce the number of parts required. This can be done most often by combining parts, designing one part so that it performs several functions. There are other approaches summarized in Chap. 7.1. (Also see Figs. 6.2.2 and 5.4.2.)

2. Design for low-labor-cost operations whenever possible. For example, a punch-press pierced hole can be made more quickly than a hole can be drilled. Drilling, in turn, is quicker than boring. Tumble deburring requires less labor than hand deburring.

3. Avoid generalized statements on drawings that may be difficult for manufacturing personnel to interpret. Examples are "Polish this surface.... Corners must be square," "Tool marks are not permitted," and "Assemblies must exhibit good workmanship." Notes must be more specific than these.

4. Dimensions should be made not from points in space but from specific surfaces or points on the part itself if at all possible. This facilitates fixture and gauge making and helps avoid tooling, gauge, and measurement errors. (See Fig. 1.3.3.)

5. Dimensions should all be from one datum line rather than from a variety of points to simplify tooling and gauging and avoid overlap of tolerances. (See Fig. 1.3.3.)

6. Once functional requirements have been fulfilled, the lighter the part, the lower its cost is apt to be. Designers should strive for minimum weight consistent with strength and stiffness requirements. Along with a reduction in materials costs, there usually will be a reduction in labor and tooling costs when less material is used.

7. Whenever possible, design to use general-purpose tooling rather than special tooling (dies, form cutters, etc.). The well-equipped shop often has a large collection of standard tooling that is usable for a variety of parts. Except for the highest levels of production, where the labor and materials savings of special tooling enable their costs to be amortized, designers should become familiar with the general-purpose and standard tooling that is available and make use of it.

8. Avoid sharp corners; use generous fillets and radii. This is a universal rule applicable to castings and molded, formed, and machined parts. Generously rounded corners provide a number of advantages. There is less stress concentration on the part and on the tool; both will last longer. Material will flow better during manufacture. There may be fewer operational steps. Scrap rates will be reduced.

There are some exceptions to this "no sharp corner" rule, however. Two intersecting machined surfaces will leave a sharp external corner, and there is no cost advantage in trying to prevent it. The external corners of a powder-metal part, where surfaces formed by the punch face intersect surfaces formed by the die walls, will be sharp. For other corners, however, generous radii and fillets are greatly preferable.

9. Design a part so that as many manufacturing operations as possible can be performed without repositioning it. This reduces handling and the number of operations but,

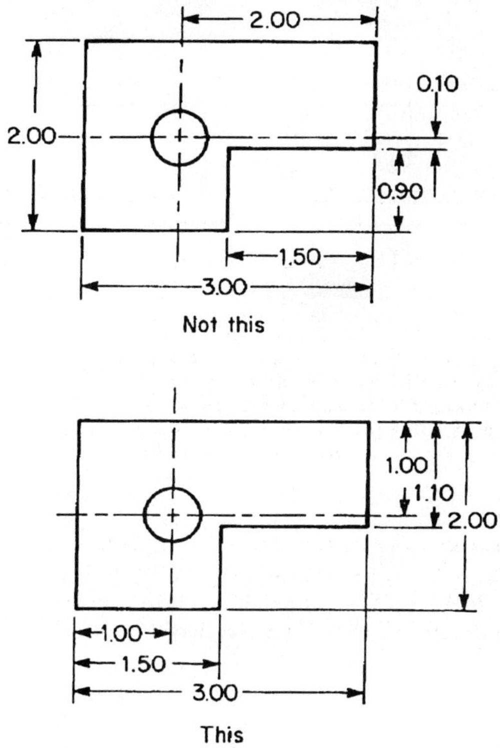

FIGURE 1.3.3 Dimensions should be made from points on the part itself rather than from points in space. It is also preferable to base as many dimensions as possible from the same datum line.

equally important, promotes accuracy, since the needed precision can be built into the tooling and equipment. This principle is illustrated by Fig. 4.3.3.

10. Whenever possible, cast, molded, or powder-metal parts should be designed so that stepped parting lines are avoided. These increase mold and pattern complexity and cost.

11. With all casting and molding processes, it is a good idea to design workpieces so that wall thicknesses are as uniform as possible. With high-shrinkage materials (e.g., plastics and aluminum), the need is greater. (See Figs. 6.1.5 and 5.1.21.)

12. Space holes in machined, cast, molded, or stamped parts so that they can be made in one operation without tooling weakness. Most processes have limitations on the closeness with which holes can be made simultaneously because of the lack of strength of thin die sections, material-flow problems in molds, or the difficulty in putting multiple machining spindles close together. (See Fig. 1.3.4.)

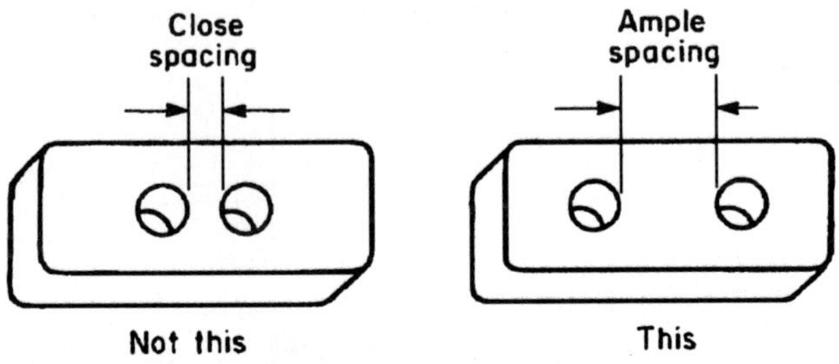

FIGURE 1.3.4 Most manufacturing processes for producing multiple holes have limitations of minimum hole spacing.

EFFECTS OF SPECIAL-PURPOSE, AUTOMATIC, NUMERICALLY CONTROLLED AND COMPUTER-CONTROLLED EQUIPMENT

For simplicity of approach, most design recommendations in this handbook refer to single operations performed on general-purpose equipment. However, conditions faced by design engineers may not always be this simple. Special-purpose, multiple-operation tooling and equipment are and should be the normal approach for many factories. Progressive designers must allow for and take advantage of the manufacturing economies such approaches provide whenever they are available or justifiable.

Types Available

Types of special-purpose and automatic equipment and tooling suitable for operations within the scope of this handbook include

1. Compound, progressive, and transfer dies for metal stamping and four-slide machines
2. Form-ground cutting tools
3. Automatic screw machines
4. Tracer-controlled turning, milling, and shaping machines
5. Multiple-spindle drilling, boring, reaming, and tapping machines
6. Various other multiple-headed machine tools
7. Index-table or transfer-line machine tools (which are also multiple-headed)
8. Automatic flame-, laser-, or other contour-cutting machines that are controlled by optical or template tracing or from a computer memory
9. Automatic casting equipment, automatic sand-mold-making machines, automatic ladling, part-ejection, and shakeout equipment, etc.
10. Automatic assembly and parts-feeding apparatus including both robotic equipment and that dedicated to a specific product
11. Program-controlled, numerically controlled (NC), and computer-controlled (CNC) machining and other equipment
12. Robotic painting and other automatic plating and/or other finishing equipment

Some high levels of automation are already inherent in methods covered by certain handbook chapters; for example, four-slide forming (Chap. 3.4), roll forming (Chap. 3.11), die casting (Chap. 5.4), injection molding (Chap. 6.2), impact extrusion (Chap. 3.8), cold heading (Chap. 3.7), powder metallurgy (Chap. 3.12), screw machining (Chap. 4.3), and broaching (Chap. 4.9) are all high-production processes.

Effects on Materials Selection

The choice of material is seldom affected by the degree to which the manufacturing process is made automatic. Those materials which are most machinable, most castable, most moldable, etc., are equally favorable whether the process is manual or automatic. There are two possible exceptions to this statement:

1. When production quantities are large, as is normally the case when automatic equipment is used, it may be economical to obtain special formulations and sizes of material that closely fit the requirements of the part to be produced and which would not be justifiable if only low quantities were involved.
2. When elaborate interconnected equipment is employed (e.g., transfer lines, index tables, multiple-spindle tapping machines), it may be advisable to specify free-machining or other highly processible materials, beyond what might be normally justifiable, to ensure that the equipment runs continuously. It may be economical to spend slightly more than normal for material if this can avoid downtime for tool sharpening or replacement in an expensive multiple-machine tool.

Effects on Economic Production Quantities

The types of special-purpose equipment listed above generally require significant investment. This, in turn, makes it necessary for production levels to be high enough so that the investment can be amortized. The equipment listed, then, is suited by and large only for mass-production applications. In return, however, it can yield considerable savings in unit costs.

Savings in labor cost are the major advantage of special-purpose and automatic equipment, but there are other advantages as well: reduced work-in-process inventory, reduced tendency of damage to parts during handling, reduced throughput time for production, reduced floor space, and fewer rejects.

Computer-controlled, numerically controlled, and program-controlled equipment noted in item 11 is an exception. The advantage of such equipment is that it permits automatic operation without being limited to any particular part or narrow family of parts and with little or no specialized tooling. Automation at low and medium levels of production is economically justifiable with numerical control and computer control. As long as the equipment is utilized, it is not necessary in achieving unit-cost savings to produce a substantial quantity of any particular part.

Effects on Design Recommendations

There are few or no differences in design recommendations for products made automatically as compared with those made with the same processes under manual control.

When there are limitations to automatic processes, these are generally pointed out in this handbook (e.g., design limitations of parts to be assembled automatically). In the preponderance of cases, however, the design recommendations included apply to both automatic and nonautomatic methods. In some cases, however, the cost effect of disregarding a design recommendation can be minimized if an automatic process is used. With automatic equipment, an added operation, not normally justifiable, may be feasible, with the added cost consisting mainly of that required to add some element to the equipment or tooling.

Effects on Dimensional Accuracy

Generally, special machines and tools produce with higher accuracy than general-purpose equipment. This is simply a result of the higher level of precision and consistency inherent in purely machine-controlled operations compared with those which are manually controlled.

Compound and progressive dies and four-slide tooling for sheet-metal parts, for example, provide greater accuracy than individual punch-press operations because the work is contained by the tooling for all operations, and manual positioning variations are avoided.

Form-ground lathe or screw-machine cutting tools, if properly made, provide a higher level of accuracy for diameters, axial dimensions, and contours than can be expected when such dimensions are produced by separate manually controlled cuts. Form-ground milling cutters, shaper and planer tools, and grinding wheels all have the same advantage.

Multiple-spindle and multiple-head machines can be built with high accuracy for spindle location, parallelism, squareness, etc. They have a definite accuracy advantage over single-operation machines, in that the workpiece is positioned only once for all operations. The location of one hole or surface in relation to another depends solely on the machine and not on the care exercised in positioning the workpiece in a number of separate fixtures. Somewhat tighter tolerances therefore can be expected than would be the case with a process employing single-operation equipment.

Automatic parts-feeding devices generally have little effect on the precision of components produced. They are normally more consistent than manual feeding except when parts have burrs, flashing, or some other minor defect that interferes with the automatic feeding action. No special dimensional allowances or changed tolerances should be applied if production equipment is fed automatically.

COMPUTER AND NUMERICAL CONTROL: OTHER FACTORS

Computer-controlled and numerically controlled equipment has other advantages for production design in addition to those noted above:

1. Lead time for producing new parts is greatly reduced. Designers can see the results of their work sooner, evaluate their designs, and incorporate necessary changes at an early stage.
2. Parts that are not economically produced by conventional methods sometimes are

quite straightforward with computer or numerical control. Contoured parts such as cams and turbine blades are examples.

3. Computer control can optimize process conditions such as cutting feeds and speeds as the operation progresses.
4. Computer-aided design (CAD) of the product can provide data directly for control of manufacturing processes, bypassing the cost and lead time required for engineering drawings and process programming. Similarly, the process-controlling computer can provide data for the production and managerial control system.
5. Setup and changeover times are greatly reduced. Processing times are also being reduced as high-velocity computer control is being developed.

To achieve these advantages, an investment in the necessary equipment is required, and this can be substantial. More vital and even more costly in many cases is the training of personnel capable of developing, debugging, and operating the necessary control programs.

CHAPTER 1.4

QUICK REFERENCES

TABLE 1.4.1 Guide to Surface Finishes from Various Processes

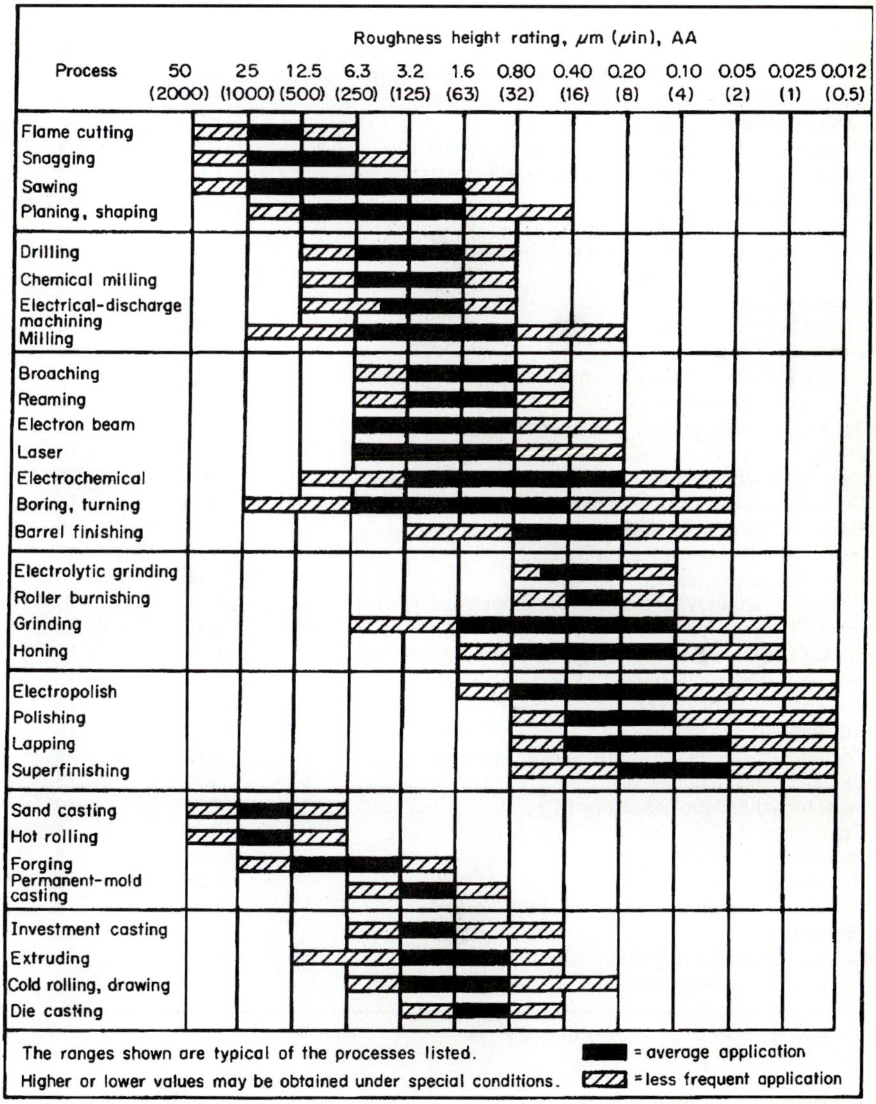

Source: From *General Motors Drafting Standards.*

TABLE 1.4.2 Normal Maximum Surface Roughness of Common Machined Parts

Surface or part	Maximum roughness	
	μm	μin
Chased threads	6.3	250
Clearance surfaces (machined)	6.3	250
Surfaces for soft gaskets	3.2	125
Housing fits (no gasket or seal)	3.2	125
Die- or tap-cut threads	3.2	125
Datum surfaces for tolerances over 0.025 mm (0.001 in)	3.2	125
Mating surfaces: brackets, pads, faces, bases, etc.	3.2	125
Teeth of ratchets and pawls	1.6	63
Gear teeth (ordinary service; diametral pitch over 10)	1.6	63
Datum surfaces for tolerances under 0.025 mm (0.001 in)	1.6	63
Milled threads	1.6	63
Pressed fits, general: keys and keyways	1.6	63
Rolling surfaces, general: cams and followers, etc.	1.6	63
Surfaces for copper gaskets	0.80	32
Slideways and gibs	0.80	32
Journal bearings, general	0.80	32
Push fits	0.80	32
Sliding surfaces of mating mechanisms or parts, general	0.80	32
Worm gears, general	0.80	32
Rotating surfaces, general: pivot pins and holes, etc.	0.80	32
Pistons	0.40	16
Cam lobs (automotive)	0.40	16
Cylinder bores: O-rings or leather packings	0.40	16
Ground screw threads and worms	0.40	16
Piston rods: O-rings or leather packings	0.40	16
Gear teeth (heavy loads)	0.40	16
Gear teeth (ordinary service; diametral pitch under 10)	0.40	16
Journal bearings (precision)	0.40	16
Sliding surfaces of mating mechanisms or parts (precision)	0.40	16
Worm gears (heavy loadings)	0.40	16
Valve stems (automotive)	0.40	16
Rotating surfaces (precision)	0.40	16
Friction surfaces: brake drums, clutch plates, etc.	0.40	16
Cylinder bores (automotive)	0.33	13
Seats for antifriction-bearing races	0.25	10
Crankpins	0.20	8
Valve seats	0.20	8
Rolled threads	0.20	8
Rolling surfaces (precision; heavy-duty)	0.20	8
Piston pins	0.13	5
Surfaces of fluid seals (sliding or rubbing)	0.13	5
Pressure-lubricated bearings	0.13	5

Source: Courtesy of *Machine Design.*

TABLE 1.4.3 Guide to Dimensional Tolerances from Various Machining Processes

Drilled holes			
Drill size, in	Tolerance, in	Drill size, in	Tolerance, in
0.0135–0.0420	+0.003 −0.002	0.2660–0.4219	+0.007 −0.002
0.0430–0.0930	+0.004 −0.002	0.4375–0.6094	+0.008 −0.002
0.0935–0.1560	+0.005 −0.002	0.6250–0.8437	+0.009 −0.003
0.1562–0.2656	+0.006 −0.002	0.8594–2.0000	+0.010 −0.003

Diameter or stock size, in	To 0.250	0.251 to 0.500	0.501 to 0.750	0.751 to 1.000	1.001 to 2.000	2.001 to 4.000
Reaming						
Hand	±0.0005	±0.0005	±0.0010	±0.0010	±0.0020	±0.0030
Machine	±0.0010	±0.0010	+0.0010 −0.0015	+0.0010 −0.0020	±0.0020	±0.0030
Turning		±0.0010	±0.0010	±0.0010	±0.0020	±0.0030
Boring		±0.0010	±0.0010	±0.0015	±0.0020	±0.0030
Automatic screw machine						
Internal			Same as in drilling, reaming, or boring			
External forming	±0.0015	± 0.0020	±0.0020	±0.0025	±0.0025	±0.0030
External shaving	±0.0010	±0.0010	±0.0010	±0.0010	±0.0015	±0.0020
Shoulder location (turning)	±0.0050	±0.0050	±0.0050	±0.0050	±0.0050	±0.0050
Shoulder location (forming)	±0.0015	±0.0015	±0.0015	±0.0015	±0.0015	±0.0015
Milling (single cut)						
Straddle	±0.0020	±0.0020	±0.0020	±0.0020	±0.0020	±0.0020
Slotting (width)	±0.0015	±0.0015	±0.0020	±0.0020	±0.0020	±0.0025
Face	±0.0020	±0.0020	±0.0020	±0.0020	±0.0020	±0.0020
End (slot widths)	±0.0020	±0.0025	±0.0025	±0.0025		
Hollow		±0.0060	±0.0080	±0.0100		
Broaching						
Internal	±0.0005	±0.0005	±0.0005	±0.0005	±0.0010	±0.0015
Surface (thickness)		±0.0010	±0.0010	±0.0010	±0.0015	±0.0015
Precision boring						
Diameter	+0.0005 −0.0000	+0.0005 −0.0000	+0.0005 −0.0000	+0.0005 −0.0000	+0.0005 −0.0000	+0.0010 −0.0000
Shoulder depth	+0.0010	+0.0010	+0.0010	+0.0010	+0.0010	+0.0010
Hobbing	±0.0005	±0.0010	±0.0010	±0.0010	±0.0015	±0.0020
Honing	+0.0005 −0.0000	+0.0005 −0.0000	+0.0005 −0.0000	+0.0005 −0.0000	+0.0008 −0.0000	+0.0010 −0.0000
Shaping (gear)	±0.0005	±0.0010	±0.0010	±0.0010	±0.0015	±0.0020
Burnishing	±0.0005	±0.0005	±0.0005	±0.0005	±0.0008	±0.0010
Grinding						
Cylindrical (external)	+0.0000 −0.0005	+0.0000 −0.0005	+0.0000 −0.0005	+0.0000 −0.0005	+0.0000 −0.0005	+0.0000 −0.0005
Cylindrical (internal)		+0.0005 −0.0000	+0.0005 −0.0000	+0.0005 −0000	+0.0005 −0.0000	+0.0005 −0.0000
Centerless	+0.0000 −0.0005	+0.0000 −0.0005	+0.0000 −0.0005	+0.0000 −0.0005	+0.0000 −0.0005	+0.0000 −0.0005
Surface (thickness)	+0.0000 −0.0020	+0.0000 −0.0020	+0.0000 +0.0030	+0.0000 −0.0030	+0.0000 −0.0040	+0.0000 −0.0050

Source: From Douglas C. Greenwood, *Engineering Data for Product Design,* McGraw-Hill, New York.

TABLE 1.4.4 Processes for Flat Surfaces

Process	Maximum size, flat area	Deviation from flatness	Surface finish	Other limitations	See Chap.
Sand-mold casting	3 × 3 m (10 × 10 ft)	4.2 mm/m (0.050 in/ft)	12–25 μm (500–1000 μin)	Grainy, irregular surface	5.1
Die casting	0.6 m^2 (1080 in^2)	1.1 mm/m (0.013 in/ft)	0.8–1.6 μm (32–63 μin)	Aluminum, zinc, brass	5.4
Injection molding	0.6 m^2 (1080 in^2)	1.7 mm/m (0.020 in/ft)	Per mold surface	Plastic materials; large flat surfaces normally not recommended	6.2
Planer and shaper machining	Planer to 1 × 4 m (3 × 12 ft)	0.4 mm/m (0.005 in/ft)	1.6–12.5 μm (63–500 μin) fore machining affecting	Stresses in workpiece be- flatness after machining	4.7
Milling-machine machining	Planer type to 1 × 4 m (3 × 12 ft)	0.4 mm/m (0.005 in/ft)	0.8–6.3 μm (32–250 μin)	Stresses in workpiece before machining affecting flatness after machining	4.6
Surface grinding	1.2 × 6 m (4 × 20 ft) but normally much smaller	0.08 mm/m (0.001 in/ft)	0.1–1.6 μm (4–63 μin)	Magnetically clamped, nonductile materials best	4.15
Sheet metal (cold-rolled)	6 × 10 ft and longer	1.3 mm/m (0.17 in/ft)	0.8–3.2 μm (32–125 μin)	Flatness tolerance applied to sheet stretched for flatness	2.2, 2.3
Sheet metal (hot-rolled)	8 × 12 ft	5.2 mm/m (0.65 in/ft)	12.5–25 μm (500–1000 μin)	—	2.2
Magnesium tooling plate	4 × 12 ft	0.8 mm/m (0.010 in/ft)	0.8 μm (32 μin)	Weight two-thirds that of aluminum; readily welded and machined	2.3

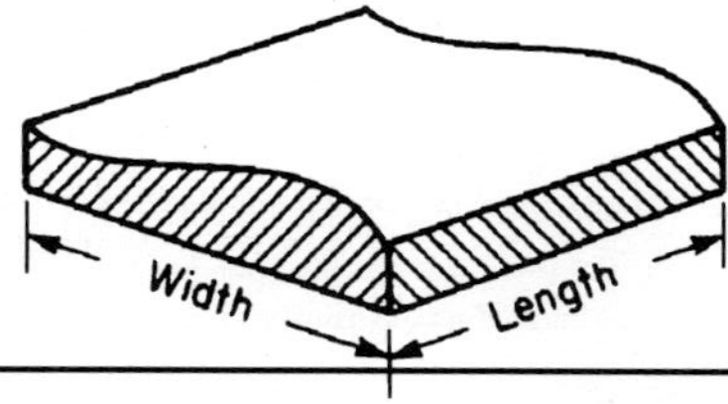

TABLE 1.4.5 Processes for Two-Dimensional Contoured Surfaces*

Process	Usual materials	Normal dimensional accuracy, mm (in)	Surface finish, μm (μin)	Normal maximum size	Remarks	See Chap.
Cold drawing and rolling	Steel and various nonferrous metals	±0.05 (0.002)	0.8–3.2 (32–125)	To 25 mm (1 in) wide × indefinite length	Intricate and nonsymmetrical shapes feasible	2.2
Powder metallurgy	Iron, bronze	±0.025 (0.001)	0.2–0.8 (8–32)	Cross section to 90 cm^2 (14 in^2)	Powder metallurgy most suitable if contour is the part's cross section as pressed	3.12
Broaching	Machinable materials	±0.4 (0.015)	0.8–3.2 (32–125)	30 cm (12 in) wide × 8 cm (3 in) deep	For high-production applications; broaches expensive	4.9
Extrusion	Aluminum, magnesium, plastics	±0.5 (0.020)	0.8–3.2 (32–250)	To 250 mm (10 in) wide × indefinite length	Undercuts feasible	3.1, 6.7
Shaper and planer machining	Machinable materials	±0.13 (0.005)	1.6–12.5 (63–500)	1 × 4 m (3 × 12 ft) (planer)	Tolerance shown for tracer feed under production conditions	4.7
Form-cutter milling machining	Machinable materials	±0.05 (0.002)	0.8–6.3 (32–250)	25 cm (10 in) × 3 m (10 ft) long	No undercuts; precise ground cutter required	4.6
Roll forming	Sheet metal	±0.13 (0.005)	0.8–1.6 (32–63)	To 0.5 m (20 in) wide × indefinite length	Undercuts feasible	3.11
Contour sawing	Machinable materials	±0.8 (0.030)	1.6–7.5 (63–300)	To 1.4 m (55 in) long	Sharpness of contours limited by blade width	4.10
Flame cutting	Usually steel	±1.5 (0.060)	12.5–25 (500–1000)	To 300 mm (12 in) long		4.11

*Processes listed in Table 1.4.6 also are applicable.

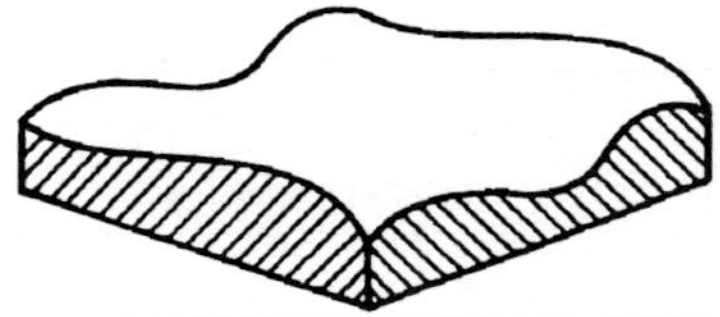

TABLE 1.4.6 Processes for the Three-Dimensional Contoured Surfaces

Process	Usual materials	Normal dimensional accuracy*	Surface finish, μm (μin)	Size range	Remarks	See Chap.
Sand casting	Iron, aluminum, brass, steel	±0.8 mm (0.030 in)	12–25 (500–1000)	To 4 m^2 (43 ft^2)	Undercuts possible with mold cores.	5.1
Die casting	Aluminum, zinc	±0.2 mm (0.008 in)	0.8–1.6 (32–63)	To 0.6 m^2 (1080 in^2)	Undercuts require considerably more costly dies.	5.4
Other casting	Various nonferrous metals	±0.25 mm (0.010 in)	1.6–3.2 (63–125)	10 cm^2 (1.5 in^2) to 2 m^2 (22 ft^2)	Plaster, permanent, ceramic molds.	5.2
Injection molding	Plastics	±0.2 mm (0.008 in)	Per mold surface	To 0.6 m^2 (1080 in^2)	For high production. A uniform fairly thin wall is best.	6.2
Injection molding	Metal powders	±0.5%	0.8–1.1 (32–45)	Smaller than a golf ball in volume	For high production; side-walls may be nonparallel.	3.16
Profile milling	Various machinable materials	±0.05 mm (0.002 in)	0.8–6.3 (32–250)	To approximately 3 m^2 (32 ft^2)	Only a roughing operation; requires hand finishing.	4.6
Electrical-discharge machining	Metals, especially when hardened	±0.05 mm (0.002 in)	1.6–6.3 (63–250)	To 2.2 m^2 (24 ft^2)	Normally used for dies and molds; slow metal removal; no undercuts.	4.18
Electrochemical machining	Metals, especially when hardened	±0.05 mm (0.002 in)	0.2–3.2 (8–125)	To 1.3 m^2 (14 ft^2)	Tools require development; no undercuts.	4.19
Press forming	Sheet metal	±0.3 mm (0.012 in)	0.8–3.2 (32–125)	To 3 m^2 (32 ft^2)	High production; no undercuts.	3.2
Forging	Steel, aluminum, brass, magnesium	±0.4 mm (0.015 in) or more	3.2–12.5 (125–500)	To 3 m^2 (32 ft^2)	For applications requiring high strength.	3.13

*Based on 50-mm (2-in) dimension.

TABLE 1.4.7 Processes for Embossed Surfaces

Process	Usual materials	Quality of definition of pattern	Tooling cost	Remarks	See Chap.
Press coining	Sheet metal	Excellent	High	High press tonnage required; for small parts	3.2
Press embossing	Sheet metal	Good	High	Less tonnage required; larger parts feasible	3.2
Sand casting	Iron, aluminum, brass	Poor	Low	For large designs	5.1
Die casting	Aluminum, zinc	Fairly good	High	For smaller to medium-size parts	5.4
Powder metallurgy	Iron, steel, bronze	Good	High	For small designs	3.12
Injection molding	Various thermoplastics or metal powders	Excellent	High	Capable of being vacuum metalized or plated	3.16, 6.2
Electroforming	Copper, nickel	Best	Low	Not suitable for high production	3.14
Roll embossing	Sheet metal	Good	High	For patterned sheet	
Plaster casting	Aluminum	Fairly good	Low	For larger designs	5.2

TABLE 1.4.8 Processes for Round Holes

Process	Usual materials	Normal dimensional accuracy of 25-mm (1-in) hole	Surface finish, μm (μin)	Maximum depth-to-diameter ratio	Remarks	See Chap.
Drilling	All machinable materials	+0.2, −0.05 mm (+0.008, −0.002 in)	1.6–6.3 (63–250)	8:1	Often more economical than casting, especially for small holes.	4.5
Reaming	All machinable materials	±0.025 mm (0.001 in)	0.8–3.2 (32–125)	8:1	Finishing operation for drilled holes.	4.5
Boring	All machinable materials	±0.04 mm (0.0015 in)	0.4–6.3 (16–250)	5:1	Solid carbide boring bars can be used in holes 8:1 in depth.	4.5
Gun drilling	All machinable materials	±0.05 mm (0.002 in)	0.8–2.5 (32–100)	300:1	Best with parts that can be rotated.	4.5
Sand casting	Gray iron, aluminum	±1.5 mm for holes to 75-mm diameter	6.3–25 (250–1000)	1.5:1	Minimum practical cast hole diameter 13 to 25 mm; draft required.	5.1
Die casting	Aluminum, zinc	±0.1 mm (0.004 in)	0.8–1.6 (32–63)	4:1	Draft required.	5.4
Injection molding	Thermoplastics	±0.1 mm (0.004 in)	0.2–3.2 (8–125)	2:1 (blind) 4:1 (through)	Draft preferable	6.2
Punching (piercing)	Sheet metal	±0.15 mm (0.006 in)	0.8–1.6 (32–63)	1:1	Breakaway area in lower portion of hole wall.	3.2
Powder metal	Iron, steel, bronze	±0.06 mm (0.0024 in)	0.2–1.6 (8–63)	From 1.35:1 (blind) to 4:1 (through)	Only holes in direction of pressing feasible.	3.12
Jig or inside-diameter grinding	Usually hardened steel	+0.0125, −0.0000 mm (+0.0005, −0.0000 in)	0.1–1.6 (4–63)	6:1	Finishing operation.	4.12
Flame cutting	Usually steel	±0.8 mm (0.032 in)	12.5–25 (500–1000)	3.2:1	For minimum size, see Table 4.11.2.	4.11
Investment casting	Various metals, most often ferrous	±0.38– ± 0.13 mm (0.015–0.005 in)	1.6–3.2 (63–125)	1:1 (blind) 4:1 (through)	Minimum diameter 1.5 mm (0.060 in).	5.3
Laser beam machining	Materials difficult or impossible to drill conventionally	±0.13 mm (0.005 in) (trepanned holes) ±0.025 mm (0.001 in) (very small holes)	3.2–6.4 (125–250)	50:1 (in 0.005-in diameter holes)	Holes smaller than 1.3 mm (0.050 in) and stock thinner than 5 mm (0.200 in) give best results.	4.21

TABLE 1.4.9 Processes for Nonround Holes

Process	Usual materials	Normal dimensional accuracy, mm (in)*	Normal surface finish, μm (μin)	Maximum depth-to-width ratio	Remarks	See Chap.
Sand casting	Iron, aluminum, brass, steel	±0.8 (0.030)	12–25 (500–1000)	1-½:1	Requires draft.	5.1
Die casting	Aluminum, zinc	±0.2 (0.008)	0.8–1.6 (32–64)	1:1 (blind) 4:1 (through)	Minimum width 1.5 mm, ferrous materials; 2.2 mm, nonferrous materials.	5.4
Other casting	Various nonferrous metals	±0.25 (0.010)	1.6–3.2 (63–125)	Approximately 4:1	Over 13 mm wide.	5.2
Contour sawing	Machinable materials	±0.8 (0.030)	1.6–6.3 (63–250)	Approximately 10:1	Requires starting hole, re-welding of blade, and large corner radii.	4.10
Punching	Sheet metal	±0.15 (0.006)	0.8–1.6 (32–64)	1:1	Through holes only.	3.2
Powder metallurgy	Iron, bronze	±0.025 (0.001)	0.2–0.8 (8–32)	†	In direction of pressing only.	3.12
Flame cutting	Steel plate	±1.2 (0.047)	12.5–25 (500–1000)	6:1	Through holes only.	4.11
Injection molding	Thermoplastics, powdered metals	±0.2 (0.008)	Per mold surface	2:1 (blind) 4:1 (through)	Depth-to-width ratio for blind holes 1.5 mm or less wide, 1:1.	6.2
Electrical-discharge machining	Hardened steel	±0.075 (0.003)	1.6–4.8 (63–190)	Over 20:1	Slight taper unavoidable.	4.18
Electrochemical machining	Hardened steel	±0.075 (0.003)	1.6–3.2 (63–125)	Over 20:1	Slight taper unavoidable; for quantity production.	4.19
Broaching	Machinable materials	±0.025 (0.001)	0.8–3.2 (32–125)	4:1‡	Requires starting hole.	4.9
Chemical machining	Various metals	±0.25 (0.010)	1.6–6.3 (63–250)	0.5:1	For large, shallow holes.	4.20

*Based on 50-mm (2-in) hole width.

†Powder-metal parts: through holes, 4:1; blind holes, top of part, 1.35:1; blind holes, bottom of part, 1.75:1.

‡Larger ratios for larger holes.

TABLE 1.4.10 Processes for Hollow Shapes

Process	Usual materials	Minimum wall thickness, mm (in)	Maximum size	Normal dimensional accuracy*	Remarks	See Chap.
Blow molding	Plastics, glass	0.4 (0.015)	0.22 m^3 (8 ft^3) and larger	±0.9 mm (0.035 in)	Mass production	6.5
Thermoforming	Sheet plastics	0.025 (0.001)	3 × 9 m (10 × 30 ft)	±6% of nominal dimension	Open one side; most suitable for shallow shapes	6.8
Rotational molding	Plastics	1.4 (0.055)	1.6 m^3 (58 ft^3)	±0.9 mm (0.035 in)	Granular surface; moderate production quantities	6.4
Deep drawing	Sheet metal	0.5 (0.20)	2 × 3 m (6.5 × 10 ft)	±0.4 mm (0.016 in)	Open one end; shallow shapes easier	3.2
Metal spinning	Sheet metal	0.1 (0.004)	4-m (12-ft) diameter	±0.4 mm (0.016 in)	Shapes of circular cross section only; open one end	3.6
Die casting Sand-mold casting Injection molding	}		See Table 1.4.6.			
Centrifugal casting	Cast iron, aluminum, brass, bronze	6.3 (0.25)	To 3-m (10-ft) diameter × 16 m (52 ft) long	±0.5 mm (0.020 in)	Pipe and similar shapes	5.2
Electroforming	Nickel, copper	0.025 (0.001)	0.5-m (20-in) diameter × 1.5 m (60 in) long	±0.025 mm (0.001 in)	Precise reproduction of Surface detail possible	3.14
Lay-up molding	Polyester plastic with fiberglass	1.5 (0.060)	12-m (40-ft) diameter × 12 m (40 ft) deep	±0.5 mm (0.020 in)	Low-quantity production feasible	6.6

*Based on 50-mm (2-in) dimension.

TABLE 1.4.11 Commonly Used Materials: Various Metal-Working Processes

Type of part \ Material	Iron	Carbon steel	Alloy steel	Stainless steel	Tool steel	Aluminum alloys	Copper alloys	Magnesium alloys	Nickel alloys	Zinc alloys	Tin alloys	Lead	Titanium	Precious metals
Extrusions	—	○	○	○	—	●	●	●	○	○	○	○	○	—
Metal stampings	—	●	●	○	—	●	●	○	○	○	—	—	—	●
Fine stampings	—	●	○	○	—	○	○	—	—	—	—	—	—	—
Four-slide parts	—	●	○	●	—	●	●	—	—	—	—	—	—	—
Spring and wire parts	—	●	●	●	—	—	●	—	○	—	—	—	—	—
Metal spinnings	—	●	○	●	—	●	●	○	●	○	○	○	—	—
Cold-headed parts	—	●	○	○	—	●	●	—	○	—	—	○	—	—
Impact extrusions	—	●	○	—	—	●	●	●	○	●	●	●	—	—
Swaged and bent tubing	—	●	●	●	—	●	●	○	●	○	○	—	○	—
Roll-formed sections	—	●	●	●	—	●	●	—	—	●	—	—	—	—
Powder-metal parts	●	○	○	○	○	○	●	—	○	—	—	—	○	—
Forgings	—	●	●	●	○	●	●	●	○	—	—	—	○	—
Electroformed parts	—	—	—	—	—	—	●	—	●	—	—	—	—	○
Electromagnetically formed parts	—	○	—	—	—	●	●	○	—	○	—	—	—	○

TABLE 1.4.11 Commonly Used Materials: Various Metal-Working Processes (*Continued*)

Screw-machine parts	○	•	○	○	—	•	•	○	○	○	—	—	○	—
Flame-cut parts	•	•	○	P	P	P	P	—	P	—	—	—	—	—
Electrical-discharge-machined parts	—	○	○	○	•	○	○	—	○	—	—	—	○	—
Electrochemically machined parts	—	○	•	○	•	—	○	—	•	—	—	—	•	—
Chemically machined parts	—	•	○	•	○	•	•	•	○	—	—	—	○	—
Castings made with sand molds	•	•	•	•	○	•	•	•	•	○	○	○	—	—
Permanent-mold castings	•	○	—	—	—	•	•	•	○	○	○	○	—	—
Ceramic-mold castings	•	•	•	•	•	○	•	○	•	○	—	—	—	—
Plaster-mold castings	—	—	—	—	—	•	•	○	—	•	○	○	—	—
Centrifugal castings	•	•	•	—	—	•	•	—	•	—	—	—	—	—
Investment castings	—	•	•	•	•	•	•	○	•	—	—	—	—	○
Die castings	—	—	○	○	○	•	○	○	—	•	○	○	—	—

Note: •, frequently processed with this method; ○, sometimes processed with this method; —, seldom or never processed with this method; P, cut with plasma.

Machining of carbide is also a frequent application of EDM.

TABLE 1.4.12 Formed Metal Parts: Summary of Processes

Type of component	Process description	Advantages	Limitations
Metal extrusions	Billet of metal is forced by a hydraulic ram through a die hole of the desired shape. The metal emerges from the die in solidified form and closely conforms in cross section to the shape and dimensions of the die opening.	Intricate cross-sectional shapes, including those with undercuts or those that are hollow, can be produced.Tooling costs are low.	Limited to ductile metals and also in the maximum size of cross section. Parts of nonuniform cross section require additional operations.
Metal stampings	Metal, in sheet form, is cut to the desired outline shape and is bent, formed, coined, or drawn to a different three-dimensional shape through a limited number of strokes of a power press to which a die (usually special for the part) is attached.	Rapid production of uniform parts which are sometimes quite intricate. Labor and materials costs are low.	Wall thickness must be essentially uniform. Edges may be rough. Tool costs may be high.
Fine stampings	This process is similar to conventional metal stamping except that special presses and dies are used. Material is tightly clamped during the operation. The punch-die clearance is very small, and the punch does not actually enter the die. The press motion is slower during shearing, which fully shears the stock without a breakaway.	Improved surface finish and squareness of sheared edges. Parts have higher dimensional accuracy and better flatness.	Tooling and equipment costs are higher than with conventional stamping. Production rates are ower.
Four-slide parts	This is a metal-stamping process which uses high-speed multiple-ram machines. Coiled strip stock is fed past one or more press units which perform conventional metal-stamping operations. The stock is then cut off and is formed by four opposed slides at 90° angles around a central forming area.	Suitable for complex parts with a high degree of forming. Reverse and overlapping bends are feasible.	Limited to smaller parts. Less depth of draw and less severe coining are feasible. Economic only for high production.
Spun-metal parts	The process uses a lathe-type machine which includes a chuck (forming block), rotating at the spindle, to which is clamped a rotating disk of sheet metal. A hand or power tool presses the metal against the chuck at a single point, gradually causing the sheet to take the shape of the chuck.	Very low tooling costs and short lead times make the process attractive for short runs, prototypes, etc. Surface finish is good.	Only parts symmetrical about an axis can be produced. Reentrant angles are difficult or not feasible. Labor content is higher than in stamping.

TABLE 1.4.12 Formed Metal Parts: Summary of Processes (*Continued*)

Type of component	Process description	Advantages	Limitations
Cold-headed parts	A blank of metal, cut from wire or rod, protrudes from a stationary die and is formed by a shaped punch, which moves axially against the blank, to a shape which is larger than the diameter of the blank.	Best suited to high-volume production of small parts. Parts have smooth surfaces and improved physical properties. There is no scrap loss.	Undercuts, thin walls. Reentrant shapes normally are not feasible. The size of a part is limited
Impact extrusions	This is the same process as cold extrusion. A blank of metal in a die is compressed by a punch, forming it either backward around the punch or forward through a die opening, or both. The shape of the punch and die determines the shape of the part.	Useful especially for long, hollow parts required in large quantities. Production is rapid. Surface finish is good.	Requires ductile material. There are shape and size limitations. (The process is generally limited to tubular shapes.)
Rotary-swaged parts	A workpiece of essentially cylindrical shape is held within a machine head which normally rotates. During its rotation, opposed dies repeatedly strike the workpiece, reducing its diameter and changing its shape.	Useful for changing the diameter or cross-sectional shape or for tapering tubular or rodlike parts. Provides a smooth surface and improved mechanical properties.	Limited size and shape of parts produced. Diameter reductions elsewhere than at the ends of parts are more difficult.
Bent tubing and sections	Tubing and members with other cross sections are clamped and bent over a bending form, often with hydraulic-powered bending machines.Mandrels or other means are used to support the part internally, preventing distortion and collapse in the bend area.	Tight bends are feasible with little loss of tubing diameter or strength. Large radius bends are the fastest and easiest.	Some wrinkling may occur at inside surfaces of bend, and tooling development time and cost may be required to control it. Outside wall becomes thinner.
Roll-formed sections	A strip of sheet metal is fed continuously through a series of contoured rolls in tandem. As the stock passes through the rolls, it is gradually formed into a shape with the desired uniform cross section.	The best applications are longer parts with complex cross-sectional shapes. Production is rapid. Surface finish and dimensional consistency are good.	Parts must have the same cross section for the whole length. Tooling and setup costs are high.

Powder-metal parts	Metal powder is placed in a die cavity and compacted by the application of high-tonnage pressure at room temperature. The compacted part is then heated to a temperature just below the melting point to fuse metal particles strongly together. A second pressing operation may be performed to provide greater dimensional precision.	Rapid production of parts with high dimensional accuracy, smooth surfaces, and excellent bearing properties. Parts can be somewhat intricate in shape. Scrap loss is low.	Size of parts is limited, and not all shapes can be produced. Tooling costs limit the process to high-production applications. Undercuts are not feasible. There are also strength limitations.
Metal injection molded parts	Fine metal powder with binders and flow agents is injected into a mold cavity using equipment very similar to that used for injection molding of plastics. The part is removed from the mold, heated or subjected to solvent vapors to drive off the binders, and then is sintered as in conventional powder metallurgy. Press sizing or other secondary operations may then be performed.	The process is suitable for small, complex metal parts. The more complex the shape, the more likely that metal injection molding will be advantageous. Sidewalls do not have to be parallel. Densities are normally high, between 95 and 99% of theoretical.	Parts are small—under 150 g normally. Shrinkage is considerable in sintering. Tolerances are broader than those needed with conventional powder metallurgy. Mold costs limit the process to high-production applications.
Forgings	A blank of material is heated to the softening point and then subjected to one or more blows by a shaped die. The resultant forging approximates shape of the finished part, but secondary machining is usually required.	Controlled grain structure provides enhanced mechanical strength to forged parts. Forgings are light in weight per unit of strength. Low loss of material; few internal flaws.	Machining is usually required to provide accurate finished dimensions; otherwise tooling and processing costs are high.
Electroformed parts	A mandrel or core piece is coated with a thick deposit of electroplated metal. The mandrel is then removed, leaving a part composed solely of the electrodeposited metal.	Intricate part with very fine detail and undercuts, reentrant angles, and reverse tapers are feasible.	The process is slow and rather expensive and is limited to nickel or copper materials. Differential wall thicknesses are difficult.

TABLE 1.4.13 Formed Metal Parts: Typical Characteristics

Type of part	Size and complexity	Dimensional characteristics	Typical uses	Cost factors	See Chap.
Extrusions	Constant cross sections of any length to about 7.5 m are feasible. Cross section can be large enough to occupy a circle of 250-mm diameter in aluminum or 150-mm diameter in steel and can be very complex.	Tolerances for cross-sectional dimensions range from ± 0.25 to ± 2.5 mm in aluminum and ± 0.5 to ± 1.6 mm in steel, depending on the nominal dimensions.	Building and automotive trim, window-frame members, tubing, aircraft structural parts, railings, furniture.	Tooling is low in cost, making the process advantageous for moderately low quantities or more. Economies are gained when machining is avoided by the use of extended shapes.	3.1
Stampings	Sizes range from small watch parts to large truck panels. Intricate shapes with holes, tabs, recesses, cavities, and raised sections are common. Wall thicknesses, essentially constant, range from 0.025 to 20 mm.	Tolerances of blanked dimensions range from 0.05 to 0.80 mm total, depending on the nominal dimension and stock thickness. Angles can be held to $\pm \frac{1}{2}$ to 2°; heights, to ± 0.2 to 0.4 mm.	All kinds of brackets and other mechanical parts, automobile body and internal parts, pans, cups, key blanks, hinges.	Tooling may be expensive, but production is very rapid, making the use of stampings advantageous for many mass-produced products.	3.2
Fineblanked parts	The usual size range is about 25 × 25 mm to 250 × 250 mm. Depth of draw is limited, but otherwise parts can be more intricate and irregular in shape than conventional stampings. Coined and thinned sections and semi-piercings are feasible.	There is greater precision and superior flatness than in conventional stampings. Sheared edges are square within less than 1°. Material thickness ranges from 0.8 to 8 mm. Tolerances range from ± 0.008 to ± 0.05 mm.	Mechanical parts such as levers, ratchets, gears, gear segments, brackets, and pawls.	Tooling and equipment are more expensive than for conventional stampings, necessitating high production levels. The cost advantage comes from eliminating machining operations.	3.3
Four-slide parts	Parts tend to be small and complex in shape. Most are made from stock narrower than 25 mm and thinner than 1.5 mm. The draw is shallow, and coining is limited, but numerous bends, including reverse bends, overlaps, and interlocks, are feasible. Welding, threading, reaming, etc., can be incorporated in the basic machine operation.	Tolerances are comparable with those of conventional stampings.	Picture hooks, hose clamps, electric switch, plug, and socket parts, contacts, bobby pins, cotter pins, and flat springs. Parts are normally formed and cut off from strip material or flattened wire.	Most suitable for high production because of the rapid production rate and need to amortize tooling, equipment, and setup costs	3.4

Spring and wire parts	Wire parts made on four-slide machines can be very complex with multiple multidirectional bends. Sizes range from miniature instrument springs to large coil and flat springs for construction and mining equipment.	Load-to-deflection ratios can vary as much as ±20% from nominal values. Coil-diameter tolerances range from ±0.01 to 1.5 mm.	Coil springs for both tension and compression, spring clips of various kinds, push rods, levers, hooks, etc., made from wire.	Small quantities of coil springs can be wound with lathe-type equipment, but labor costs are high. Most spring and wire parts are mass-produced on automatic four-slide or spring-winding equipment at low labor and materials costs.	3.5
Spun-metal parts	Circular parts range from 6 mm to 4 m in diameter, although the more common maximum sizes are considerably smaller. Relatively simple hemispherical, conical, or cylindrical shapes are most common, but reentrant shapes are possible.	Most common wall thicknesses are 0.6 to 1.3 mm. Diametral tolerances range from ±0.25 to 3 mm; depth tolerances, from ± 0.4 to 1.5 mm; and angles, from ±1 to 5°.	Lamp bases, reflectors, tankards, vases, pots, pans, bowls, nose cones, round trays, and aircraft parts.	Tooling costs are very low, making the process ideal for prototypes and short runs ,especially if lead times are short. For high production runs, conventional stamping is apt to be more economical.	3.6
Cold-headed parts	Parts are normally small, ranging from 0.5 × 0.5 mm to 230 × 20 mm diameter. However, 50-mm bars have been cold-headed. Shapes are relatively simple and essentially circular in cross section.	Tolerances for head diameters range from approximately ±0.2 to 0.9 mm; on head height, from approximately 0.1 to 0.5 mm. Tolerances on overall length are ±0.4 to 2.4 mm.	Fasteners of various types, valves, knobs, spacers, spark-plug bodies, hose fittings, gear blanks, spindles, shafts, stub axles.	Little loss of material and very low labor costs because of high-speed, automatic operation. Equipment and tooling costs are relatively high, making the process most economical for high production runs.	3.7
Impact- (cold-) extruded parts	Parts range in diameter from 12 to 160 mm. Lengths can reach 2 m but are normally much shorter. The process is most suitable when parts are essentially circular, hollow, and closed at one end.	Practical length-to-diameter ratios range from 2:1 (in steel) to 10:1 (in aluminum). Minimum wall thicknesses range from 0.25 to 10 mm. Diametral tolerances range from ±0.13 to 0.25 mm; bottom-thickness tolerances, from ±0.13 to 0.38 mm; length, ±0.8 mm.	Flashlight cases, aerosol cans, military projectiles, fire extinguishers, collapsible tubes.	Because of tooling, equipment, and setup costs, the process is best for mass-production levels. Material scrap loss is low.	3.8

TABLE 1.4.13 Formed Metal Parts: Typical Characteristics (*Continued*)

Type of part	Size and complexity	Dimensional characteristics	Typical uses	Cost factors	See Chap.
Rotary-swaged parts	The process is applicable to tubular or rodlike parts from 0.5 to 150 mm in diameter and of indefinite length. Complexity is limited, involving diameter reduction with or without shape change.	Diametral tolerances range from ±0.03 to 0.38 mm. Surface finish of parts after swaging ranges from 0.05 to 1.6 mm.	Furniture legs, golf clubs, fishing rods, pins, needles, punches, buttonhooks, bicycle spokes, screwdriver blades, automotive torque tubes, exhaust pipes, and steering posts, assembly of cable and hose fittings.	The process is fairly labor-intensive except when automated as in the production of sewing-machine needles. Best suited for moderate or higher production levels unless tooling is already on hand.	3.9
Bent tubing and sections	Draw bending is applicable to tubing from 12 to 250 mm in diameter. Wrinkle bending can used for diameters to 650 . mm. Multiple bends in one or more planes are feasible.	Bend-angle tolerances normally extend from ±2 to 4°, although ±¼° can be achieved with some methods. The minimum centerline bend radius ranges from 1 to 6 times diameter, depending on the bending method.	Automobile exhaust systems, aircraft hydraulic and fuel lines, outdoor and indoor furniture, railings, handles, heat-transfer and boiler tubing.	Use of tube bending avoids the cost of making and attaching separate fittings. Materials cost is low, but labor content, except for high-production press methods, is relatively high. Tooling cost is moderate.	3.10
Roll-formed parts	Parts are typically long, with a constant cross section that can be quite complex. Short pieces can be made by cutting longer forms. Widths of stock before forming normally range to 1 m and stock thickness to 4 mm.	100 mm is the normal maximum depth of form. Tolerances for cross-sectional dimensions extend from ±0.05 to ±0.8 mm; angles, from ±½ to ±1°.	Roof and siding panels for buildings, architectural trim, downspouts, window frames, stove and refrigerator panels and shelves, curtain rods, metal picture frames.	Production rates are rapid, so labor costs are low. Materials utilization is excellent. Tooling and setup costs are high, so best economies occur with mass production.	3.11
Powder-metal parts	Powder-metal parts are normally small (less than 75 mm in the largest dimension). Complex shapes are feasible, but sidewalls are parallel and undercuts and screw threads must be provided by secondary operations.	There is high dimensional accuracy. Tolerances range from ±0.006 mm in small bores after repressing to ±0.13 mm in larger dimensions on parts that are not repressed. Cross-sectional dimensions can be held to closer tolerances than those of dimensions in the direction of pressing.	Cams, slide blocks, levers, gears, bushings, ratchets, guides, spacers, splined parts, connecting rods, sprockets, pawls, and other mechanical parts for business machines, sewing machines, firearms, and automobiles.	Powder-metal parts are best suited for high-production operations in which tooling and equipment investments can be recovered. Labor costs are low, and materials are well utilized in the process. Other cost advantages come from eliminating much of the need for machining.	3.12

Metal injection molded parts	Parts are small—under 150 g normally and with a maximum dimension below 100 mm (4 in). However, they can be quite complex.	The high shrinkage during sintering (typically 19 to 25%) tends to limit the accuracy of the parts. Tolerances of 0.08 to 0.13 mm on small dimensions and 0.5% on larger dimensions are recommended, with larger values preferred for noncritical dimensions.	Rotors for stepper motors, gear housings for police pistols, hammers and links for printing machines, shotgun trigger guards, stainless steel orthodontic retainers, small hermetically sealable housings for electronic components, and various camera components.	Best suited for high-production applications because of tooling and other costs that must be amortized. Labor content, however, is low, materials are well utilized, and no or little machining is required on the parts produced.	3.16
Forgings	Closed-die forgings can be intricate, but secondary machining is normally required. The normal upper size limit is about 12 kg. Open-die forging can produce much larger parts (up to 5 tons), but shapes are more limited and greater secondary machining is required.	Typical tolerances across the parting line run from 0.8 mm for small forgings of easily forged materials like aluminum, brass, and mild steel to 9.5 mm on large forgings of refractory alloys. Tolerances of dimensions within one die half range from 0.4 to 1.2%.	Highly stressed mechanical parts used in aircraft engines and structures, land-based vehicles, portable equipment, connecting rods, crankshafts, valve bodies, gear blanks, tube and hose fittings, trailer hitches.	Tooling costs are moderate to high, depending on the complexity of the forging. Material loss is high because of flash and secondary machining. Labor costs are usually moderate. Forging is most economical for medium and high production levels.	3.13
Electroformed parts	The process is applicable to parts as small as 10 mm and as long as 1.8 m. Exact reproduction of surface detail is feasible, and parts can be quite complex. If a collapsible or dissolvable mandrel is used, undercuts, reverse tapers, and reentrant angles are feasible.	Wall thickness can range from 0.025 to 13 mm but may vary substantially within a complex part. Dimensions can be held to ±0.025 mm if the mandrel is accurate. Surface finish can also be held to values as low as 0.05 μm.	Molds, e.g., for phonograph records, paint masks, waveguides, nose cones, fountain-pen parts, surface-finish standards, reflectors, venturi nozzles, and rocket thrust chambers.	Production rates are very slow, and high skill levels are required. The choice of materials is limited. The optimum lot size is small to medium.	3.14
Stretch-formed parts	The process is suitable for large workpieces which have shallow or nearly flat shapes.	Tolerances of dimensions less than 500 mm with care can be held to ±0.25 mm. Larger or less controlled dimensions may require ±0.8 mm.	Body panels for vehicles, frame parts, aerospace parts, architectural forms.	Tooling and equipment investment costs are low. Labor costs are relatively high. The process is best for low-volume production.	3.15

TABLE 1.4.13 Formed Metal Parts: Typical Characteristics (*Continued*)

Type of part	Size and complexity	Dimensional characteristics	Typical uses	Cost factors	See Chap.
Explosively formed parts	The process is suitable for very large parts, up to 4 m in diameter. Complex shapes are possible, but concentric symmetrical shapes are best.	Tolerances for parts formed in dies can be held to ±0.25 mm, while free-formed parts (no die} may require ±6.3 mm.	Aircraft and jet-engine parts, housings, panels and ducts, tubing bulging.	Output rates are low, so unit labor costs are high. However, tooling costs are much lower than for conventional forming methods. The process is most economical for limited-quantity production.	3.15
Electromagnetically formed parts	The process is chiefly used to compress, expand, or form tubular shapes, usually in assembling a tubular part with another part. Occasionally flat material can also be formed with this method.	Normal tolerance for tubular or flat parts formed with a die is ±0.25 mm. Free-formed may require ±6.3 mm.	This process is used for universal-joint yokes, drive linkages, wheels, high-pressure-hose fittings, air-conditioner components, shock absorber dust covers oil-cooler heat exchangers, lighting fixtures aluminum baseball bats	The process is most advantageous for high production. Tooling costs are relatively high, but labor costs are low because of the rapid production rate.	3.15

TABLE 1.4.14 Machining Characteristics of Various Metals

Material		Machining characteristics				
Base metal	Representative alloy conventional designation and temper	UNS designation and temper	Cutting nature*	Expected finish†	Approximate ratio‡	Precautions§
Aluminum	Cast and wrought alloys					
	213.0-F (CII3-F)	A02120-F	F to G	E to G	400	2,3,4,7
	520.0-T4 (220-T4)	A05200-T4	F	E	400	2,4,7
	2017-T4	A92017-T4	F	E	300	2,4,7
	1100-HI2	A91100-H12	F to G	E to G	300	2,3,4,7
	1100-0	A91100-0	G to S	G	300	2,3,4,7
	520.0-F (220-F)	A05200-F	G	G	300	2,3,4,7
Beryllium	Unalloyed		A,B	F	50	6,7,8
Columbium (niobium)	Unalloyed, wrought		G	G		3
Copper	Wrought alloys					
	Brass, high-leaded (342)	C34200	F	D	220	4
	Brass, medium-leaded (340)	C34000	F	E	180	4
	Phosphor bronze, FC (544)	C54400	F	E	180	4
	Yellow brass (268)	C26800	G,S	G	120	3,4
	Naval brass (464)	C46400		G	120	4
	Aluminum bronze (614)	C61400		G	60	3,4
Ferrous	Carbon and alloy steels 12LI4	G12144	F	E	105–195	7
	C1212	G12120	F	E	100	7
	C1119	G11190	F	E	85–90	7
	C1114		F	E	75	7
	C1020	G10200	F	G	65–70	7
	C1040	G10400	F	G	50–55	7
	4140 resulfurized	G41400 resulfurized	F	G	60–70	7
	4140	G41400	F	G	55–65	7
	Stainless steels					
	416	S41600	F,R	G	95	1,2,3,4,5,7
	430 F	S42020	F,R	G	90	1,2,3,4,5,7
	203		F,R	G	85	1,2,3,4,5,7

TABLE 1.4.14 Machining Characteristics of Various Metals (*Continued*)

Material		Machining characteristics				
Base metal	Representative alloy conventional designation and temper	UNS designation and temper	Cutting nature*	Expected finish†	Approximate ratio‡	Precautions§
Ferrous	303	S30300	F,R	G	70	1,2,3,4,5,7
	410	S41000	S,R	G	60	1,2,3,4,5,7
	430	S43000	S,R	G	55	1,2,3,4,5,7
	431	S43100	S,R	G	50	1,2,3,4,5,7
Magnesium	ASTM-AZ-61	M11610	F	E	500	1,2,4,9
	ASTM-AZ-91	M11910	F	E	500	1,2,4,9
Molybdenum	Unalloyed, wrought		A	G	8	2,3,6,8
Nickel	Nickel 200 to 233	N02200 – N02233	H,S	G	55	5
	Monel alloy 400 to 404	N04400 –	H,S	G	55	5
	Monel alloy 501		H,S	G	40	5
	Monel alloy K500	N05500	H,S	G	35	5
	Inconel alloy 600	N06600	H,S	G	35	5
	Hastelloy B	N10001	H,S	G	35	5
	Inconel X-750	N07750	H,S	G	25	5
	Superalloys		H,R,S	G	6 to 10	3,5
Tantalum	Unalloyed, wrought		G	G		3
Titanium	Commercially pure		R	E	40	1,2,3,4,5,9
	TI-6AI-4V	R56401	R	E	20 to 30	1,2,3,4,5,9
	TI-13V-IICr-3AI	R59010	R	E	15	1,2,3,4,5,9
Tungsten	Unalloyed, wrought		R	G	8	3,6
	Unalloyed, sintered		A	F	8	
Zirconium	Commercially pure		G,R	G	35	1,2,3,4,5,9
Zinc	ASTM-AG40A (XXIII)	Z33520	F	E	200	

*F = free-cutting; G = gummy; H = strain-hardens excessively-; R = reactive to tools; S = stringy; A = abrasive; B = brittle.

†E = excellent; G = good; F = fair.

‡Based on B1112 steel as 100.

§Precautions: (1) Avoid dull tools to prevent distortion; (2) avoid overheating for best dimensional accuracy; (3) minimize galling, smearing, and tool buildup by using proper tool geometry, preparation, setup, and fluids; (4) minimize deflection by proper supports; (5) use positive feeds and avoid dwelling in the cut to minimize work hardening; (6) minimize spalling by proper setups and avoiding exit cuts; (7) minimize buildup of edge formation by proper speeds and fluids; (8) hood all machine tools and exhaust to dust collectors to avoid toxic dust; (9) chips are pyrophoric.

Source: Courtesy of *Machine Design.*

TABLE 1.4.15 Summary of Machining Processes

Process	Most suitable materials	Typical applications	Material-removal rate	Normal dimensional tolerances, mm (in)	Normal surface finish, μm (μin)	Remarks
Turning	Various machinable materials	Rollers, pistons, pins, shafts, rivets, valves, tubing and pipe fittings	With mild steel, up to about 21 cm^3 (1.3 in^3)/(hp · min)	±0.025 (0.001)	0.4–6.3 (16–250)	Both turning and facing performed with single point or form tools
Drilling	Various machinable materials	Holes for pins, shafts, fasteners, screw threads, clearance, and venting	Wild mild steel, up to 300 cm^3 (19 in^3)/min	±0.15, −0.025 (+0.006, −0.001)	1.6–6.3 (63–250)	Diameters from 0.025 to 150 mm (0.001 to 6 in)
Milling	Various machinable materials	Flat surfaces, slots, and contours in all kinds of mechanical devices	With mild steel up to 6000 cm^3 (365 in^3)/min with 300 hp	±0.05 (0.002)	0.8–6.3 (32–250)	Common and versatile at all levels of production
Planing and shaping	Various machinable materials	Primarily for flat surfaces such as machinery bases and slides but also for contoured surfaces	With mild steel, up to about 10 cm^3 (0.6 in^3)/(hp · min)	±0.13 (0.005)	1.6–12.5 (32–250)	Most suitable for low-quantity production
Broaching	Various machinable materials	Square, rectangular, or irregular holes, slots, and flat surfaces	Maximum of large-surface broaches about 1300 cm^3 (80 in^3)/min	±0.025 (0.001)	08–3 2 (32–125)	Most suitable for mass production
Center-type and center-less grinding	Virtually all materials including those that are hard and brittle but excluding soft, gummy materials	Shafts, pins, axles	To 164 cm^3 (10 in^3)/min in mild steel with 100 hp and high-speed grinding	+0, −0.013 (+0, −0.0005)	±0.1–1.6 (4–63)	Most common for finishing operations when close dimensional accuracy and smooth surface finish are required
Surface grinding		Dies, molds, gauge blocks, machine surfaces		+0, −0.1 (+0, −0.004)	±0.1–1.6 (4–63)	
Chemical machining	All common ferrous and nonferrous metals	Blank thin sheets; wide, shallow cuts	0.0025–0.13 mm (0.0001–0.005 in) depth of metal removed/min	±0.1 (0.004)	08.–1.6 (16–250)	Practical maximum depth of cut 6–13 mm (¼-½ in); burr and stress-free

TABLE 1.4.15 Summary of Machining Processes (*Continued*)

Process	Most suitable materials	Typical applications	Material-removal rate	Normal dimensional tolerances, mm (in)	Normal surface finish, μm (μin)	Remarks
Ultrasonic machining	Hard, brittle nonconductive materials	Irregular holes and cavities in thin sections	30–4000 cm^3 (1.8–240 in^3)/h	±0.025 (0.001)	1 (40)	Used for nonconductive materials not machinable by EDM
Abrasive-jet machining (AJM)	Hard, fragile, and heat-sensitive materials	Trimming, slotting, etching, drilling, etc.	1 cm^3 (0.06 in^3)/h	±0.13 (0.005)	1.3 (50)	For finishing operations
Abrasive-flow machining	All machinable materials, especially those of poor machinability	Primarily for deburring	Very low	±10% of stock to be removed	50% lower roughness than prior operation	Also used for removal of recast layer from EDM surfaces
Hydroynomic machining	Only soft nonmetallic materials	Contour sawing of sheet materials	Rapid for soft materials	±0.25 (0.010)	—	No dust or heat effect; noisy process
Abrasive water jet machining	Hard metals and nonmetals	Cutting reinforced plastics, honeycombed materials, metal sheets thicker than 13 mm (0.5 in)	Somewhat slower than with laser beam	±0.25 (0.010)	—	No heat-affected zone
Electron-beam machining	Any material	Fine cuts in thin workpieces	0.05–0.12 cm^3 (0.003–0.007 in^3)/h	±10% allowed on hole and slot dimensions	2.5 (100)	For very fine cuts (very small holes and slots); high investment
Laser-beam machining	Any material	Blanking parts from sheet material. Machining thin parts and small holes.	2.5 m/min (100 in/min) in mild steel with oxygen assist	±0.13 (0.005)	2.5 (100)	High investment
Electrochemical honing (ECH)	Hardened metals	Finishing internal cylindrical surfaces	3–5 times faster than conventional honing	±0.0060–0.0125 (0.00025–0.0005)	0.1–0.8 (4–32)	Comparable to ECG
Electrical discharge machining (EDM)	Hardened metals	Molds	49 cm^3 (3 in^3)/h	±0.05 (0.002)	1.6–3.2 (63–125)	Burr-free; heat-affected layer produced; intricate shapes possible

Wire EDM	Hardened metals	Blanking dies	130–140 cm^2/h ($20–22 in^2$/h) in 5-cm (2-in) thick material	±0.05 (0.002)		
Electrical discharge grinding (EDG)	Hard materials like carbide	Form tools	0.16–2.5 cm^3 (0.01–0.15 in^3)/h	±0.005 (0.0002)	0.4–0.8 (16–32)	
Electrochemical machining (ECM)	Difficult-to-machine metals	For making complex shapes and deep holes	1000 cm^3 (60 in^3)/h maximum	±0.05 (0.002)	1.6 (63)	Most rapid nontraditional machining process
Electrochemical-discharge machining (ECDM)	Hardened metals and carbides	Sharpening carbide cutting tools and machine honeycombs	With carbide 15 cm^3 (0.9 in^3)/h	±0.025 (0.001)	0.13–0.75 (5–30)	Suitable for intricate form grinding
Electrochemical grinding (ECG)	Hardened metals and carbides	Sharpening carbide cutting tools	100 cm^3 (6 in^3)/h	±0.025 (0.001)	0.4 (15)	

TABLE 1.4.16 Summary of Metal-Casting Processes

Type of casting	Process description	Advantages	Limitations
Sand-mold castings	A sand mixture (sand, clay binder, and other materials) is packed around a wood or metal pattern to form a mold. The pattern is removed, and the cavity left by it is filled with molten metal. Normally, the mold is in two halves which are fitted together before pouring.	Intricate shapes can be made over a wide size range. Many metals can be used, and tooling costs can be low.	There are accuracy and surface-finish limitations with this process which necessitate postmolding machining operations in most cases.
Lost-foam castings	This is similar to sand-mold casting except that the pattern is made of polystyrene foam and is vaporized when the mold is filled with molten metal. The pattern is coated with a water-soluble refractory material. The mold is one-piece and made of unbonded sand.	Casting shapes can be complex; no draft or flash; good surface finish. Cores may not be required.	Pattern costs may be high, and pattern is expended. The process requires different techniques from regular sand-mold casting.
Shell-mold castings	Sand mixed with thermosetting-plastic resin is flowed onto a heated metal pattern. The sand mixture adjacent to the pattern fuses together, forming a shell, which is removed from the pattern. Metal is poured into two shell halves fastened together. The shell is broken from the finished casting.	Best for high production; good surface finish and dimensional accuracy.	Limited casting size, equipment, and tooling require larger investment. Resin also adds to cost. Not all metals can be cast
Permanent-mold castings	Reusable metal molds are filled with molten metal by gravity or low air pressure. Metal or sand cores may be used.	Best for small and medium-size castings of nonferrous metals. Close tolerances, high densities, and smooth surface finish can be attained.	Castings are limited in complexity of shape. Molds are more costly than patterns for sand casting. Processes not suitable for high-melting-temperature metals.
Ceramic-mold castings	A thick slurry of ceramic material is poured over a reusable split and gated pattern The pattern is removed when the slurry gels. Two mold halves are assembled and preheated before the metal is poured.	Useful for ferrous and other high-melting-temperature alloys. Intricate shapes, high accuracies, and smooth surfaces can be achieved.	Process is more costly than sand-mold casting, and castings may have a coarse-grained structure with lower mechanical properties.

Plaster-mold castings	Similar to sand- and ceramic-mold casting except that mold and cores are made of plaster (or sand and plaster). Slurry is poured over the pattern, which is removed after the plaster sets. The mold is assembled and baked before pouring.	Castings have smooth surface and good dimensional accuracy; they can be intricate.	Usually limited to castings smaller than 10 kg; limited to nonferrous materials. The process is relativel costly.
Centrifugal castings	Molds, which may be permanent or sand-lined, are rotated during and after pouring. Centrifugal force presses molten metal against the mold walls or in mold cavities.	Useful for casting cylindrical parts, especially if they are long; applicable to almost all metals that are sand-cast.	Except for small, intricate parts, shapes that can be cast are limited. Spinning equipment is required.
Investment castings	A slurry of ceramic material is poured over a meltable or burnable pattern. After the slurry sets, the mold is baked at high temperature to remove the pattern. The mold is therefore one-piece. It is broken away after the casting solidifies.	Almost all metals can be investment-cast to high levels of dimensional accuracy and smooth finish. Shapes can be intricate.	Most suitable for smaller parts. The process is costly and is usually justifiable only when machining can be eliminated.
Die castings	Molten metal is injected under pressure into permanent metal molds which have water-cooling channels. When the metal solidifies, the die halves open and the casting is ejected. (The process is similar to plastics injection molding.)	High production rates are achievable with castings of smooth surface and excellent dimensional accuracy.	A large investment is required for equipment and tooling. The process normally is suited only to metals with lower melting temperature. The size of castings also is limited.

TABLE 1.4.17 Metal Castings: Typical Characteristics

Type of casting	Size and complexity	Dimensional characteristics	Typical uses	Cost factors	See Chap.
Sand-mold castings	Can be as small as 30 g and as large as 200 tons. Intricate shapes with undercuts, reentrant angles, and complex contours are practicable to cast.	Surfaces are irregular and grainy. Minimum wall thickness varies with the metal used but normally should not be less than 6 mm. Tolerances extend from ±0.6 to ±6 mm.	Engine blocks, machinery frames, compressor and pump housings, valves, pipe fittings, brake drums and disks, sewing-machine parts.	Depending on the degree of mechanization, the process can be economical. Costing and pricing are normally carried out on a weight basis.	5.1
Lost-foam castings	Originally used mainly for large one-of-a-kind castings, the prcess is now used for castings of the same size range as with conventional sand molds. More complex shapes, including undercuts and reentrant angles, are feasible.	Minimum wall thickness is about 2.5 mm. A draft allowance of ½° is preferred. Dimensional tolerances of 0.3 mm/10 cm are feasible.	Manifolds, cylinder heads, crankshafts, pump housings, machine bases, automobile-body dies, brake components.	Capital investment is lower than for conventional sand-mold castings. Scrap and core costs are reduced, but labor costs are higher than for sand-mold castings.	5.1
Shell-mold castings	Can be as heavy as 100 kg, but weights are normally less than 10 kg. Shapes can be complex, with bosses, undercuts, holes, and inserts	Section thicknesses of as little as 1.6 to 6 mm are feasible. Tolerances of 0.5 mm/10 cm are recommended.	Smaller mechanical parts like connecting rods, lever arms, etc.; gear housing, cylinder heads, suppoprt frames, etc. This process is used when greater precision is required.	Because tooling costs are higher and production is at a high rate, this process is used mostly for high-production situations. Sand cost is higher than for other processes owing to the use of resin.	5.1
Permanent-mold castings	Castings are normally small or medium-sized, rarely exceeding 100 kg (220 lb) in weight. Shapes are normally relatively simple, less intricate than those of die castings.	Minimum wall thickness is 3 mm for small permanent mold castings and 5 mm for larger castings. Machining allowances range from 0.8 to 2 mm.	Automotive pistons, gears, air-cooled cylinder heads, splines, wheels, gear housings, pipe fittings, hydraulic fittings.	The competitiveness of permanent-mold castings extends to about the 40,000-unit production level. Tooling cost is lower but labor cost higher than for die castings. Labor cost is lower than for sand-mold castings.	5.2

Centrifugal castings	Cylinders from 13-mm (½-in) to 3-m (10-ft) diameter with length to 16 m (52 ft) have been cast. Smaller parts, made with centrifuging-process variations, can be intricate.	Wall thicknesses range from 6 to 125 mm (½ to 5 in). Tolerances are comparable to those required for sand-mold or permanent-mold castings, depending on the centrifugal mold used.	Large rolls, gas and water pipe, engine-cylinder liners, wheels, nozzles, bushings, bearing rings, gears, etc.	Centrifugal equipment is not expensive, and with sand molds the process is competitive for small quantities. Metal or graphite molds require a moderate production quantity.	5.2
Plaster-mold castings	For small (up to 10-kg) nonferrous parts. 125 to 250 g, however, is a more typical size range. Complexity is average.	Surfaces have a satin texture and a smoothness of 0.8 to 1.3 μm (30 to 50 μin). Typical tolerances are ±0.10 mm/cm. The minimum wall thickness is from 1.0 to 2.4 mm.	Valves, fittings, gears, sprockets, ornaments, lock components, pistons, cylinder heads, pipe fittings, ornaments.	The process is relatively high in cost because of longer production cycles and higher pattern costs.	5.2
Ceramic-mold castings	Parts with complex geometry weighing as much as 700 kg (1500 lb) can be cast, but much smaller parts (down to 100 g) are feasible and more common.	Tolerances are ±0.1 mm for the first 25 mm and ±0.025 mm for each additional cm. Across parting line add ±0.3 mm. Surface finishes better than 3 μm can be held.	Stainless-steel and bronze pump impellers and impeller cores, molds for plastic parts, tool-steel milling cutters, cast-iron core boxes.	The process is expensive but is economically advantageous when it eliminates or reduces machining operations.	5.2
Investment castings	Normally small but can range from 1 g to 90 kg. Shapes can be very intricate, with contours, undercuts, bosses, recesses, etc.	Minimum wall thickness ranges from 0.75 to 1.8 mm, depending on the material. Recommended tolerances vary with the basic dimension from ±0.8 to ±1.5 mm.	Precision parts for sewing machines, typewriters, firearms, surgical and dental devices, wrench sockets, pawls, cams, gears, valve bodies, turbine blades, radar waveguides.	Savings come from elimina tion of machining operations. Otherwise, investment castings are more costly than other castings. The process is best for low and medium quantities.	5.3
Die castings	Maximum economic size for aluminum is 45 kg; it is somewhat smaller for other materials. Intricate shapes and details can be incorporated.	Minimum wall thickness extends from 0.38 mm (small zinc parts) to 2.8 mm (large aluminum or magnesium castings). Tolerances can be as low as ±0.08 mm, but ±0.25 is more suitable, with an extra allowance if the dimension crosses the die's parting line.	Intricately shaped parts for automobiles, appliances, outboard motors, hand tools, hardware, business machines, optical equipment, toys, etc.	Most advantageous at high production levels, at which the substantial cost of tooling and equipment can be amortized. Production rates are high. Major economies result from avoidance of machining operations.	5.4

TABLE 1.4.18 Ceramic and Glass Components: Summary of Processes

Process	Description	Advantages	Limitations
	Ceramics		
Wet pressing	Ceramic material in powder form is mixed with water and additives. The mixture, wet enough to flow under pressure, is placed in a mold and compressed under high pressure as the mold closes. The formed part is removed from the mold, trimmed as necessary, dried, and fired at high temperature to fuse powders into a homogeneous material. Machining operations may then be performed.	Intricate shapes can be produced Production rates are rapid.	Size of part is limited. Some materials cannot be used. Tooling may be expensive.
Dry pressing	The mixture placed in the mold has a minimum amount of moisture and behaves similarly to metal powders in the powder-metallurgy process; i.e., there is little lateral flow of material when it is compressed in the mold.	Close tolerances are possible. The process can be automatic for high production.	Parts with a high length-to-diameter ratio tend to have uneven density and dimensional distortion. Dies should be carbide for wear resistance and can be expensive. Presses should have multiple motions and may be expensive.
Casting	The ceramic mixture is more liquid and is poured into the mold, which it fills without pressure. The mold is made of plaster of paris, which absorbs moisture from the mixture, allowing it to solidify. Drying and firing follow the casting operation as with pressed parts. Machining may also take place.	Large parts are feasible. Equipment investment is modest.	Production rate is slower. Dimensional accuracy is limited.
Extrusions	The ceramic mixture is forced through a die orifice of the desired cross section. This cross section is imparted to the material, which is then dried and fired and machined as necessary.	Rapid production rates are possible. Tooling cost is low. Very small diameters are possible.	Limited to parts of constant cross section (unless secondary machining operations are performed). The cross section should be symmetrical, and there are wall-thickness limitations.
Jiggering	A rotating form, usually made of plaster, is used. A ceramic mixture is applied manually with a putty knife to the form. Separate pieces may be joined together before drying and firing.	Few or no tooling costs; short lead time.	Limited to round parts; dimensional accuracy limited. The process requires a high level of skill but has been mechanized.

Glass			
Blowing	Similar to blow molding of plastics. Air is blown into a hollow glob of heated glass, which expands against the interior walls of the mold, taking its shape.	Rapid, high-production method for the production of glass containers and other hollow objects.	Wall thickness varies in different sections of the part and is difficult to control. The process is best suited for high production.
Pressing	Similar to compression molding of plastics. A glob of heated glass is placed in an open mold. The mold closes, and the glass is forced into all portions of the mold. It cools and solidifies. The mold is opened, and the part is removed.	Rapid production of solid-glass parts, especially when rotating-table equipment is used. Parts can be moderately complex, with dimensions closely complex, with dimensions closely controlled.	Thin walls are not so feasible. Undercuts perpendicular to the direction of pressing are not feasible

TABLE 1.4.19 Plastics and Rubber Components: Summary of Processes

Process	Description	Prime type of material*	Advantages	Limitations
Compression molding	Molding material, usually preformed, is manually placed between the mold halves, which are heated and closed under pressure. The material then flows to fill the mold cavity, polymerizes from the heat, and hardens. The mold opens, and the part is ejected or removed.	TSP, R	Tooling relatively inexpensive; reduced flow lines and internal stresses. Materials used resist high temperatures, have few sink marks, and possess minimum creep.	Intricate shapes with side draws and undercuts are difficult. Phenolic, the common material, has limited molded-in color choice.
Transfer molding	Material is placed in a heating chamber from which, after softening, it is forced by a plunger into the mold cavity, where after further heating it polymerizes and solidifies.	TSP, R	More rapid production rates than with compression molding. Parts can be more intricate and delicate.	Tooling is more expensive than with compression molding. Material is lost in sprues and runners.
Injection molding	Material is fed into a heated cylinder, from which it is injected by plunger action, after it softens, into the mold cavities. With thermoplastics, the mold is cooled, causing the plasticized material to solidify. The process is normally automatic.	TP, TSP, R	Rapid production of intricate parts with molded-in color and little need for subsequent operations.	Tooling is costly. The process is less suitable for large parts or small quantities.
Structural-foam molding	There are several processes. With injection molding, pelletized thermoplastic material and a blowing agent are used. The blowing agent expands the material, causing the core of the part to be cellular while the surface (skin) is solid.	TP, TSP	Tooling is normally less expensive than for conventional injection molding, especially for large parts. Parts have a high stiffness-to-weight ratio and are free from sink marks.	The surface of the part tends to have a splay or swirl pattern which is costly to avoid or hide.
Rotational molding	A heated metal mold, rotating on two axes, is charged with plastic material, which coats its interior surface. The heat causes the material to fuse. As the mold is cooled, the fused material solidifies, forming a hollow part.	TP, TSP	Molds are inexpensive. Large parts are relatively inexpensive to produce.	Limited to hollow shapes, which cannot be too intricate. The process is slow and is not competitive for small parts.
Blow molding	Air pressure applied inside a small hollow plastic piece (called a parison) expands it against the walls of a mold cavity, whose shape it assumes. There it cools and hardens. The mold opens, and the part. is ejected	TP	An economical process for rapid mass production of containers and other hollow products.	Limited to hollow products; not suitable for small quantities. Tolerances are relatively broad, and wall thickness is difficult to control.

Extrusion	Material is fed into a heated cylinder, from which it is fed by a rotating screw through a die orifice of the desired cross section. The material is cooled and, after hardening, cut to length.	TP, R	Production is rapid, and tooling is inexpensive. Very complex cross sections can be produced.	Limited to parts of constant cross section. Tolerances are relatively broad.
Thermo-forming	The plastic sheet is heated and placed over a mold. A vacuum between the sheet and the mold causes the sheet to be drawn against the mold, taking its shape. Normally the mold is on only one side of the sheet, although air pressure and/or the second mold half can be used to assist the forming.	TP	Tooling and equipment are inexpensive. Production rates are good. Large parts can be produced.	Ribs, bosses, and other sections heavier than the basic sheet thickness are not feasible. Walls become thinner at deep draws.
Lay-up and spray-up	Resin and a reinforcement mixture (usually glass fibers) are placed by hand or spray gun against a single mold half. The material is pressed against mold by hand. Heat and time cause the plastic to set, forming a solid part.	TSP	Low tooling cost. Parts can be of any size. Nonplastic reinforcements can be incorporated.	Labor cost is high, and production rates are low. Choice of materials is limited. One surface of the part is rough.
Welding	Hot air or gas directed at the junction of two parts causes the material to melt and parts to fuse together. Material from a plastic rod may be added to joint. (The process is similar to the gas welding of metals.)	TP	Large, bulky parts can be fabricated. There is little or no tooling cost. Fabrication can take place on the site rather than in the factory.	Limited to low production because of high labor content; wide tolerances necessary.

*TP = thermoplastic; TSP = thermosetting plastic; R = rubber.

TABLE 1.4.20 Plastics and Rubber Components: Typical Characteristics

Component	Size and complexity	Dimensional characteristics	Typical uses	Cost factors	See Chap.
Injection moldings	Intricate parts can be produced advantageously from pinhead to bathtub size.	Tolerances vary with materials and dimensions from ±0.05 to ±0.5 mm (0.002 to 0.020 in).	Housings, containers, covers, knobs, tool handles, plumbing fittings, lenses.	Very favorable for high production because an injection-molded part can replace an entire assembly with no subsequent finishing or other operations.	6.2
Extrusions	Intricate cross sections, including hollow shapes of every length. Cross sections from thread thickness to 300 mm (12 in).	Cross-sectional tolerances vary with material and size from ±0.18 to ±3.8 mm (0.007 to 0.150 in).	Window frames and other architectural components, pipe and tubing, seals, wear strips.	Costs are very favorable because tooling costs are low. Extrusions have color and finish built in.	6.7
Thermo-formed parts	Shallow shaped articles with essentially uniformly thin walls. Size ranges from cups to parts 3 by 9 m (10 by 30 ft).	Part dimensions require a tolerance of ±0.35 to 1%. Wall thickness varies from ±10 to 30% of nominal dimension.	Refrigerator liners, vehicle interior liners, blister packages, meat trays, miscellaneous containers, appliance housings.	For low-cost packaging components, formed panels, and similar components with integral color and texture.	6.8
Blow moldings	For thin-walled and hollow objects with internal volumes from a few ounces to 55 gal.	Dimensions of blown portions require tolerances of from 0.5 to 2.3 mm (0.020 to 0.090 in).	Bottles and other containers; hollow toys and decorative objects.	Lowest-cost method of producing bottles and other closed containers at high production levels.	6.5
Rotational moldings	Hollow objects to 1.6 m^3 (58 ft^3) in capacity. Shape is normally fairly regular, and the size is fairly large.	Dimensional control is not as close as with other plastic processes and depends on the material used, the shape of the part, and various processing conditions. Tolerance for 100-mm (4-in) dimension is normally ±1.1 mm (0.044 in).	Containers, tanks, equipment housings, toys.	Lower investment than for injection molding, but the cycle is longer. The process is most cost-effective at medium-high production levels.	6.4

Compression moldings (thermoset plastics)	Parts can be intricate with undercuts, though these are more costly. Size ranges from that of miniature electronics components to large appliance housings.	Tolerances vary with material, nominal dimension, and processing conditions. Commercial tolerances for a 50-mm (2-in) dimension average about ±0.18 mm (0.007 in).	Electrical and electronic components, dishes, washing-machine agitators, utensil handles, container caps, knobs and buttons used for higher-temperature environments.	Molds are required which can withstand molding pressures and temperatures. Molding cycles are longer than with injection molding, but the cost of finished parts is low.	6.1
Fiberglass reinforced-plastics parts	Oceangoing vessels are the largest current reinforced-plastics products. The smallest parts are miniature gears injection-molded of glass reinforced thermoplastics.	Tolerances vary with the molding method, material, and size of the dimension. Hand-lay-up and sheet-molded parts should have ±0.010 in/ft.	Boats, vehicle bodies, tanks (filament-wound), architectural paneling, business-machine and appliance housings.	Cost depends on the molding method and quantity. Molds are required, but with lay-up methods they can be inexpensive and small-quantity production is justifiable.	6.6
Structural-foam-molded parts	Most common for large parts, which can be highly intricate if necessary. Density ranges from 40 to 100% of solid material. Parts have solid skin.	Tolerances similar to those required for regular injection-molded parts of equivalent size.	Business-machine housings, furniture components, pallets, storage-battery cases, musical-instrument parts, TV cabinets.	Particularly for lower quantities, costs can be lower than with conventional injection molding because, with some structural-foam processes, low pressure can be used and molds are simpler. However, cycle time is longer.	6.3
Compression molded rubber	Parts as small as miniature pads for instrument components and as large as tires for large mining trucks are produced. Parts can be irregular but normally are not highly intricate.	Tolerances vary with size, material, and processing conditions. Commercial tolerances average ±0.5 mm (0.020 in) for dimensions controlled by one mold half and ±0.8 mm (0.032 in) for dimensions across the parting line.	Cushioned and antiskid mounting pads, gaskets and seals, tires, diaphragms.	Molds are required, but equipment is not highly complex and the cost of parts is reasonable, especially for moderate quantities.	6.10

TABLE 1.4.21 Summary of Assembly Processes

Process	Typical applications	Advantages	Limitations	See Chap.
Mechanical assemblies	All mechanical devices, vehicles, and appliances.	Assembly can be disassembled for service; usually permits accurate adjustment of critical settings. Dissimilar materials can be joined.	Less permanent than welded assemblies. Can be simplified with DFA techniques.	7.1
Arc-welded assemblies	Structures and frames.	Great strength per unit weight; low tooling costs. Fabrication can take place at the final site.	Assembly distorts; not easily disassembled.	7.2
Resistance-welded assemblies	Housings, frames, cabinets, etc., produced in large quantities.	Assemblies are strong and light. Sheet-metal components can be more complex than is possible in one piece.	Normally suitable only for sheet metal, not for heavy structures; not easily disassembled.	7.3
Brazed assemblies	Intricate or hollow components.	Dissimilar materials can be joined. Assemblies can be light in weight. Need for costly forming or machining can be eliminated.	Size of assembly or joint is limited to that which can be heated at one time.	7.4
Soldered assemblies	Electrical assemblies or light assemblies of sheet-metal parts requiring a joint seal.	Faster and easier than brazing.	Joint strength is very limited. Some materials are not easily soldered.	7.4
Adhesively bonded assemblies	Furniture, packaging, aircraft honeycomb, automobile-brake shoes.	Metallic and nonmetallic materials can be joined. Light weight; no metallurgical effect on components. Galvanic insulation of components can be provided.	Careful joint precleaning is required for critical applications. Limitations on joint strength.	7.5

TABLE 1.4.22 Surface Coatings: Summary of Characteristics

Type of coating	Usual coating material	Usual substrate material	Normal coating thickness, mm (10^{-3} in)	Primary purpose*	Remarks	See Chap.
Electroplated	Various metals	Various metals, some plastics	0.005 (0.2)	PD	Hard chromium is also used for wear resistance.	8.2
Painted	Organic materials	Virtually all materials	0.013–0.1 (0.5–4)	PD	Easily applied; the most common surface coating.	8.5
Vacuum-applied	Aluminum	Plastics, glass	0.000025–0.025 (0.001–1)	DO	For mirrors and other optical uses; electronic and electrical applications.	8.3
Dip-applied	Zinc, tin, other metals	Low-carbon steel	0.086–0.173 (3.4–7)	P	Used for components exposed to the atmosphere or underground or under water.	8.3
Spray-metallized	Various metals	Steel	0.05–2.5 (2–100)	O	Most commonly used for wear resistance.	8.3
Anodized	Aluminum oxide	Aluminum	0.0025–0.005 (0.1–0.2)	PD	Coating results from chemical reaction with substrate material; can be dyed. Much harder than substrate material.	
Phosphate	Tertiary zinc phosphate	Steel or cast iron	0.0025–0.005 (0.1–0.2)	B	Coating results from chemical reaction with substrate material; used prior to painting.	
Fluidized-bed	Thermoplastics	Various metals	Over 0.2 (8)	PDO	Abrasive-resistant; more costly because of thick coat.	
Porcelain-enameled	Various glass and ceramic materials	Sheet steel	0.5 (20)	PDO	Resistant to corrosion, high heat, and abrasion.	

*B = base for coatings; D = decorative; O = other functions; P = protective.

TABLE 1.4.23 Major Principles of Design for Assembly

Minimize the number of parts in the assembly

Combine parts

Incorporate hinges, springs, bearings, guides, and other functions into the basic parts.

Eliminate fasteners by using snap fits to hold parts together.

Put electrical and electronic components in one location and consolidate components as much as possible.

Make an outright reduction For example, use fewer fasteners.

Make a full redesign of the assembly

Use a different technology For example, electronics instead of mechanical linkages.

The three-question approach to parts elimination:

Does the part move with respect to mating parts?

Must adjacent parts be made of different materials?

If parts were combined, would assembly of other parts or field service be more difficult or unfeasible?

If the answer to all three is "No," the part in question probably can be eliminated by putting its function into some other part of the assembly.

CHAPTER 1.5
THE HISTORY OF DFM*

The principles of design for manufacturability (DFM) and its application are not really new. Awareness of the importance of designing products for easy manufacture has existed in clever design and manufacturing engineers since product design and manufacturing activities originated. However, use of the term *design for manufacturability,* recognition of it as a worthwhile engineering approach, and development of an organized DFM methodology are more recent. The following are some highlights from the history of designing products for easier manufacture.

ELI WHITNEY

Eli Whitney is one person from an earlier period whose work is notable as an example of the use of some DFM approaches. Whitney was engaged in design for manufacturability over 180 years before use of the term became widespread. At the turn of the nineteenth century Whitney developed, for the U.S. government, a system for manufacturing muskets that incorporated the concept of interchangeable parts. Prior to his innovation, all U.S. muskets were handmade by individual craftsmen who each made the complete product. They used saws and files to shape each part and fit them together. No two muskets were ever exactly alike; parts from one musket could not fit into another. The craftsmen who fabricated the muskets shaped each part to fit only the mating parts of the same gun. Production rates were very low, depending on the availability of skilled craftsmen and their degree of skill.

Whitney's contribution was to "redesign" each part to a specific dimension with a limited tolerance. (He also developed manufacturing processes that depended on fixtures, machinery, and gauges, rather than the skill of a particular workman, to control the dimensions of each component.) Whitney made other design and process changes too. He changed one part from a sand-mold casting to a forging to gain the greater accuracy that production with a steel forging die provided. "The procedure enabled him to forge the guards of the locks quickly and so they will require less work in fitting up than could be cast in brass after a perfect pattern by an experienced founder.'"[1] (Note: Whitney's approach differed from what is now sometimes advocated, the loosening of dimensional tolerances where feasible. This is advocated in order to eliminate the need for some manufacturing operations necessary to meet a tight dimensional tol-

*This chapter is adapted from material in Chap. 1 of *Design for Excellence,* by James G. Bralla, McGraw-Hill, New York, 1996.

erance but not really necessary for the proper functioning of the product. Whitney went in the other direction. He tightened tolerances that were far too loose. This enabled him to organize and utilize a system of mass production that was far more economical—per unit of product produced—than the previous handcraft approach.*)

HENRY FORD

Henry Ford is undoubtedly most famous among engineers (as well as economists and social scientists) for his advanced and extensive use of the assembly line. This involved dividing manual assembly operations into short-cycle repetitive steps that could be carried out at high efficiency. However, the design concepts of his Model T, the car that revolutionized the automobile industry, probably were of equal or greater significance to the success it achieved.

From Burlingame[3] (p. 289):"...costs had been lowered by concentrating on the fewest possible standard parts and designing machines to turn them out—them and nothing else—automatically if could be." And (p. 292): "He had been studying parts, simplifying, estimating production methods, performance."

Ford's own book, *My Life and Work,*[4] states that his objectives in the Model T included simplicity in operation, absolute reliability, and high quality in materials used. He also had the objective of providing easy serviceability. "The important feature of the new model...was its simplicity. All [components] were easily accessible so that no skill would be required for their repair or replacement....[I]t ought to be possible to have parts so simple and so inexpensive that the menace of expensive hand repair work would be entirely eliminated....[I]t was up to me, the designer, to make the car so completely simple....[T]he less complex the article, the easier it is to make, the cheaper it may be sold....[T]he simplest designs that modern engineering can devise....Standardization, then, is the final stage of the process. We start with the consumer, work back through the design, and finally arrive at manufacturing."

John B. Rae writes that durability and simplicity were "achieved in 1907 with the Model T....Its 20-hp, four-cylinder engine was a marvel of mechanical simplicity, as was its planetary transmission."[5] It can be seen that much of what Ford accomplished is now referred to as DFM.

VALUE ANALYSIS

This approach first appeared in the late 1940s at General Electric[6] and got its largest impetus in the 1950s and 1960s. It involves a systematic review of the cost of producing a component or product and the evaluation of design alternatives that could produce the desired results—the desired "value"—at the lowest cost. Initially, it was

*While Whitney's advance, conceived in 1798, provided the basis for later mass-production industries in the United States, it should be noted that LeBlanc, a Frenchman, developed a similar system for the production of muskets in France 10 years earlier than Whitney. Whitney apparently did not know of LeBlanc's approach, which was actually broader than Whitney's, since it encompassed the barrel, mounting, and stock of the musket, while Whitney's involved only the lock. After LeBlanc's death, the system in France deteriorated and was abandoned.[2]

applied to existing products. However, Lawerence Miles and others recognized the advantages of performing the analysis during the design stage of a product rather than after it had been introduced or matured. The term *value engineering* was applied when the technique was used during the design phase of a product. However, even in the full-fledged application of value engineering, the emphasis has *not* been on such an early involvement of manufacturing-knowledgeable people in the design process as are now advocated in DFM/concurrent-design approaches. In addition to the fact that it is typically used later in a product's development cycle, value analysis has differed from DFM in that the use of an organized knowledge base has not been quite so well refined as it is with DFM. However, the philosophical approach of value analysis—questioning and comparing the value and cost of each feature and each element of a product's design—is compatible with the whole methodology of DFM. The brainstorming approach, frequently part of a value-analysis project, is another worthwhile technique.

FORMULATION AND DOCUMENTATION OF DFM

In 1941, the Metals Engineering Division of the American Society of Mechanical Engineers (ASME) conducted a survey that revealed a need for a reference for mechanical engineers and designers on the properties and characteristics of metals. A series of handbooks was published, and one of them, published in 1958, contained information on various manufacturing processes from the product designer's point of view. The book, *Metals Engineering Processes,*[7] provided a series of guidelines to aid the designer in enhancing the manufacturability of metal components made with a number of manufacturing processes such as casting, forging, extrusion, machining, joining, finishing, etc. The book was edited by Roger W. Bolz.

Bolz was actually one of the first persons to organize DFM methodology, though he did not use the term. In the 1940s, as associate editor of *Machine Design* magazine, he prepared a series of articles summarizing the capabilities of various manufacturing processes from a product designer's viewpoint. Included in these articles were design guidelines for components made with each process, dimensional tolerance recommendations, and data on characteristics of the parts produced. His articles were combined and supplemented to form a book, *Production Processes: The Producibility Handbook,* Industrial Press, NY, first published in 1947 and updated through at least five editions through 1981.

ORIGIN OF TERMS PRODUCIBILITY AND DFM

The terms *producibility* and *manufacturability* began to be used in the 1960s to indicate the ease with which a product or component could be manufactured. In 1960, General Electric developed for internal use in the company a *Manufacturing Producibility Handbook.* The term *producibility* is still used widely, particularly by the U.S. Department of Defense. The term *manufacturability* gradually became ascendant among those interested in the approach, and about 1985, *design for manufacturability* and its shortened form, *DFM,* came into wide use.

DESIGN FOR ASSEMBLY

Geoffrey Boothroyd and A. H. Redford studied automatic assembly, which led them to considerations of product and parts design to facilitate assembly. Their book, *Mechanized Assembly,*[8] published in 1968, contains in one chapter a series of design guidelines for facilitating assembly. It is noteworthy that even at that relatively early stage in the development of DFM, Boothroyd and Redford recognized that the element of design is far more important than the use of mechanization in reducing the cost of assembly. To quote from their book,

> Experience shows that it is difficult to make large savings in cost by the introduction of mechanized assembly in the manufacture of an existing product. In those cases where large savings are claimed, examination will show that often the savings are really due to changes in the design of the product necessitated by the introduction of the new process. It can probably be stated that in most of these instances even greater savings would be made if the new product were to be assembled manually. Undoubtedly, the greatest cost savings are to be made by careful consideration of the design of the product and its individual component parts.[8]

Over the years, but especially in recent years, various trade associations and vendors of parts have issued booklets that provide, for product designers, a series of design guidelines and tolerance and materials recommendations for parts made with processes of interest to the association or vendor. These kinds of publications have provided valuable and authoritative assistance to product designers.

OBJECTIVE MEANS OF EVALUATING DESIGNS

What can be very helpful to the designer/DFM practitioner is some objective means of evaluating design alternatives so that the best approach can be chosen easily with the assurance that it truly is the best one. The proof of whether one design is superior to another in manufacturability is its manufacturing cost compared with that of the alternative design. Computer systems that provide cost or time data on alternative designs can provide a highly useful measure of which design alternative is preferable, i.e., which one has the lowest combination of labor, materials, and overhead costs. Most convenient are approaches that permit the designer himself or herself to make the comparative evaluations. There are a number of such systems currently available to product designers.

More recently, others have taken steps toward developing systems—with the use of personal computers—that provide designers with both the cost implications of various design alternatives and redesign suggestions and the cost-reduction implications of those suggestions.

CURRENT INTEREST IN DFM

DFM has been accepted by industry as a valid element in the product design process. Seminars, conferences, and short-term training courses are frequently scheduled on the subject; the ASME has a Design for Manufacturability Committee, and many universities have added the subject to their engineering curriculum. A number of books are available that describe the process.

CONTINUING DEVELOPMENT

DFM is not a fixed system. This system is continually being developed, both in university research projects and by a number of consultants and within companies. The objectives of almost all developments are to make guidelines more accessible to designers and more easily applied. Additionally, and more important, evaluations are put on each guideline so that the designer can determine how much cost gain can be achieved if the guideline is incorporated. All these advances depend on the use of computers. Computerization is the developing movement in DFM.

REFERENCES

1. Jeanette Mirsky and Alan Nevans, *The World of Eli Whitney,* Macmillan, New York, 1952.
2. Constance Green, *Eli Whitney and the Birth of American Technology,* Little, Brown, Boston, 1956.
3 Roger Burlingame, *Backgrounds of Power: The Human Story of Mass Production,* Charles Scribner and Sons, New York, 1949.
4. Henry Ford, *My Life and Work,* 1922.
5. John B. Rae, *The American Automobile: A Brief History,* University of Chicago Press, Chicago, 1965, p. 59.
6. Arthur E. Mudge, *Value Engineering: A Systematic Approach,* McGraw-Hill, New York, 1971.
7. Roger W. Bolz (Ed.), *Metals Engineering Processes,* McGraw-Hill, New York, 1958.
8. G. Boothroyd and A. H. Redford, *Mechanized Assembly,* McGraw-Hill, London, 1968.

CHAPTER 1.6
MANAGING DFM*

It is entirely possible for a design engineer, aware of the advantages of designing for manufacturability to work independently in optimizing the product design for ease of manufacture. However, most current DFM specialists advocate a team approach in applying DFM. Essentially, this means that the design and manufacturing people work together to gain the benefits of manufacturing knowledge and experience that the designer may not have and to ensure that the product is both functional and manufacturable. When other objectives—as discussed in Chap. 9.2—are part of the project, a team approach becomes more important, even essential. This chapter deals primarily with the managerial techniques needed when a team approach is used.

MANAGEMENT'S ROLE

Strong management support is a minimum prerequisite for successful application of a team-based concurrent engineering approach to DFM. The company's product design staff and the manufacturing organization not only must cooperate in a successful DFM project but also must, to a substantial degree, overlap and combine functions.

Management's task with respect to a concurrent engineering team, once formed, is twofold: (1) to clearly define the team's task, objectives, and limits, and (2) to ensure that the participants function well as a team, that the reservations or anxieties of team members and their managers are overcome, and that true productive cooperation occurs.

Management also should anticipate that there may be resistance to change. When job duties and responsibilities are altered, employees' sense of security is affected.

PLANNED SEQUENCE[1]

The following is a recommended sequence for a company beginning a team approach to product design:

*Adapted from Chaps. 7, 8, 9, and 10 of *Design for Excellence,* by James G. Bralla, McGraw-Hill, New York, 1996.

1. An overview of the methodology, organization, benefits, and costs should be prepared and presented to the senior management group.
2. A pilot program should be selected.
3. A concurrent engineering coordinator should be appointed.
4. With the chief executive's concurrence, the objectives of the changed procedure should be defined and promulgated carefully. It is important that management make the team objectives very clear.
5. A team to carry out the pilot program should be appointed and organized.
6. The key people should be trained.
7. The pilot project should be carried out.
8. Management should follow up.
9. At the conclusion of the pilot project, the chief executive should announce the results to company employees with his or her comments.

REDIRECTION OF DESIGN EFFORTS

A major obstacle to the design engineer at present is the fact that he or she is now required to be an expert in many disciplines. In addition to the usual factors of concern—new product features and proper function and quality of the product—we now, with DFM, ask the designer to ensure manufacturability. DFX (Design for all desirable attributes) further expands the scope of needed skills dramatically. For many designers, this involves a complete redirection and refocus. It is no longer satisfactory to develop a product that simply functions well and has desirable features. The kind of teamwork inherent in concurrent engineering was not—and still in too many cases is not—a normal part of the product realization process for many companies. The following summarizes four possible levels of interaction:

1. Traditional approach: "over the wall." Designers and manufacturing engineers do not communicate about the design. Design documents are transmitted to manufacturing without any prerelease review by manufacturing engineers.
2. An improvement: The sign-off procedure. Manufacturing engineers approve and accept the design *after* it is completed but before it is released to production.
3. Further improvement: There is some collaboration between designers and manufacturing engineers during development of the design.
4. Current thinking: full concurrent engineering. Designers and manufacturing engineers work together on the design as a team.

THE TEAM

In addition to the important product design and manufacturing engineering members, the optimal design team should have representation from other functions that are important in the lifetime of the product. Ideal participants include purchasing and key vendors, safety engineering, reliability engineering, quality control, representatives of the manufacturing line organization, product service, environmental engineering, production planning and control, and product management—hence the sometimes used term *cross-functional design teams.*

Purchasing personnel should be in attendance regularly at team meetings as long as purchased components are a significant factor. Certain vendors should be called in whenever the component supplied by the vendor is critical to the design or the cost of the product.

The product manager is a key and frequent participant. He or she provides the input from the market and the company's sales force. As such, the product manager represents the customer's viewpoint, the most vital one to be considered if the product design is to be successful.

The very best—most talented, most motivated—people should be selected for the first try at concurrent DFM. The initial team members not only must be good performers but also should be respected by others in their groups and should have some ability to communicate the benefits of the team approach to their colleagues.

One key decision is how broad a charter to give to the concurrent engineering design team. Is the team's responsibility solely to provide the design for the proposed product such that it meets all prescribed objectives? Or does the team have a longer-lasting charter? Should it function as a team for the life of the product, monitoring and coordinating customers' reactions to the product, studying field-service problems and competitors' counterstrategy to the new or improved product, reviewing reliability and product safety statistics, etc.?

SOME COMMENTS ON TEAM BUILDING

Some questions that should be decided when a concurrent engineering approach is contemplated are the following:[2]

- What training is needed to prepare team members to work as a group rather than individually?
- To whom in the organization does the team report?
- What will happen to the team members when the project is completed? Are there other products for them to develop? If not, are there jobs waiting for them in their home departments?
- Will team members all move to a central team location and work together full time, or will they just meet periodically?
- How will the team properly weigh conflicting objectives and incorporate the best tradeoff in the final product design?
- How to ensure that it is productive and does not get bogged down with disagreements over conflicting objectives.

THE RISKS OF CONCURRENT ENGINEERING

Although it is the recommended approach, there are some risks to the concurrent engineering approach. Consider the following:

1. Teams are more difficult to manage than individuals. Team meetings can take excessive time if not managed well. Achieving teamwork may require a delicate and diplomatic approach.
2. Not all good designers and engineers are team players. Some of the best designs historically have come from brilliant individual achievements.
3. The cost of bringing team members together may not be insignificant, particularly if product design, production, and key staff functions are located at different facilities.
4. The cost of removing team members from their home departments to be part of the concurrent engineering team also must be considered.
5. There is the important factor of resistance to change that can arise when concurrent engineering is implemented and which must be allowed for and overcome.

RESISTANCE TO A CHANGE TO CONCURRENT ENGINEERING

Some of the factors that can cause resistance to a change to concurrent engineering on the part of managers and engineers are the following:[3]

- The basic conservatism and cautiousness of many managers, a preference for traditional rather than innovative approaches.
- The natural fear that people tend to have of the unknown.
- A lack of understanding or lack of clarification as to what job responsibilities each will have after the change is implemented.
- The need for team members to learn new job skills. They may not relish the prospect of training or may fear not being able to perform their changed duties well enough.
- The possibility of a loss of authority or status on the part of people affected by the changed system. For example, design engineers may fear loss of design authority to all the other team members, who have different objectives than they.
- Fear that one's job may be classified at a lower salary rate as a result of the changed system and, in the more extreme case, fear of eventual loss of one's job.
- The change to existing working relationships, which may be comfortable ones.
- Concern over loss of job satisfaction.
- Concerns that higher management may not manage the project properly. The underlying source of resistance has been stated by some to be a lack of information about the change.

OVERCOMING RESISTANCE TO A CHANGE TO CONCURRENT ENGINEERING

Four actions have been suggested[3] to overcome resistance to broad changes of this type. They are education/training, involvement/participation, information sharing, and role clarification.

Training can help overcome resistance stemming from job insecurities by making the participants more confident that they can perform their jobs under the new system. Lawrence[4] strongly emphasizes that resistance to changes can be overcome by getting the people involved in the change to participate in making it.

Role clarification is important. If management can define job responsibilities clearly, both for the team and for others involved in related work during the team project and after its completion, much of this apprehension will disappear.

MANAGING THE NEW SYSTEM

The Leader

To be successful, a concurrent engineering project should be guided by a person having authority, at least within the project, over all team participants. The role of the leader of a DFM or DFX design project is most critical to the project's success.

In addition to having the leadership and human relations skills needed to direct persons with varied skills, backgrounds, and attitudes, the leader must have sound technical judgment in matters of design, manufacturability, and the other key attributes wanted in the product. Much design engineering involves delicate compromises in which conflicting objectives have to be balanced. Knowledge and experience in the various functions involved in the realization and sale of the product are also important. The leader should be results-oriented and should be able to imbue the same attitude in the team. In addition, the leader must have persuasive powers to influence others both inside and outside the team.

System Tools to Aid in Managing the Team

To aid in managing the design project, consider the following.

CAD/CAM. One tool to aid the concurrent engineering project is a good CAD/CAM system, part of a system of computer-integrated manufacturing. A central design database, accessible by manufacturing engineers and other team members, can ensure that all team members are working with identical information. This implies that a workable computer network is also needed, and if the team participants are located some distance from one another, a wide-area network is required. Such facilities reduce the need for team members to carry out all team business face-to-face at the same location.

Standardization. Standardization is an important approach. As many elements as possible of the systems used to control engineering and production should be included.

Consider standardizing the following:

Engineering drawings

Design features

Parts (New parts should be as similar to existing parts as possible. Also, the total number of parts varieties used should be held to the smallest practicable number. Parts that have already been purchased or gone into production have proven functionality, quality, cost, and reliability.)

Fasteners, materials, and commercial parts

Modules

Design organization and procedures

Processes

Standardization has many advantages. Development costs are eliminated if existing components can be used. Training costs are reduced because when employees learn a procedure once, it is applicable to other situations. The need for additional training for other products, other locations, or other projects is eliminated. Mistakes are fewer because employees have developed the knowledge and skills that minimize mistakes. Startup costs are reduced because of the familiarity of workers, supervisors, and support personnel with the systems already in place. Quality is higher, lead time is shorter, and productivity is better because debugging has already taken place, at least partially. Tooling costs may be reduced because tools are already available from the manufacture of prior similar parts. Production quantities are higher, permitting many economies of scale, including the justifiability of investments in automated or otherwise more productive equipment and tooling. Just-in-time arrangements are easier because suppliers can concentrate more easily on scheduling and stocking the component involved as a result of its larger usage quantity and greater importance.

Brainstorming. Brainstorming is a technique for enhancing creativity. It is most useful in the early stages of a project when the design concept for the product is being developed. Its value lies in the fact that it can generate highly inventive solutions that may not be evident if design is performed in the usual manner by an individual designer.

The procedure involves the meeting of a group of persons, both those directly involved in the project and others. Having a group is an essential part of the procedure because it is based on the face-to-face interaction of ideas from a number of people. Participants are encouraged to put forth, in a group meeting, suggestions for the design of the component or product under review. All ideas, even if they seem silly, are to be mentioned. Participants are encouraged to respond to suggestions of others with further suggestions, perhaps built on one just voiced. A paramount rule in the procedure is to prohibit negative comments on any suggestion, no matter how impractical it may sound to some of the participants. (The meeting leader must police this strongly.) The idea is that even silly ideas may stimulate a similar but more practical idea from someone else. The objective is to create a free flow of ideas—to break away from existing patterns of thought that may limit innovation. The procedure originated in the advertising industry in the 1940s[5] and has been used for many different applications since then.

The composition of the brainstorming group is critical. Most critical is the leader or facilitator. He or she must act somewhat like a cheerleader, urging the team members on to further ideas. However, he or she also must exert control, particularly if inhibiting comments or actions come from some group members.

Only after the flow of ideas has subsided does the evaluation of the ideas expressed take place. This can be done by a subgroup from the team. They are assigned to sort through the ideas listed during the creative phase, selecting those which show promise for further investigation. The subgroup may request clarification on a proposal before rejecting it. Nevertheless, the subgroup does decide which suggestions merit further investigation or action.

The procedure, if successful, will provide design avenues that would not have been conceived of by designers working on their own or under a more conventional structure.

Benchmarking. Benchmarking is a procedure for making major improvements in some aspect of a company's operations—its product, system, process, method, organization, or procedure—by comparing it with the best known example elsewhere. Normally, the leader in the function being examined is selected, even if that leader is another company, either a competitor or a noncompetitor. The procedure is a sound one for product design as well as for developing operational improvements.

Aspects that could be compared by benchmarking include, but are by no means limited to, the following: the competitor's product or some component of it; some attribute of another product, its serviceability, reliability, environmental friendliness, etc.; some manufacturing process; some specific engineering procedure or system; a company's customer service operation; or a competitor's product line or business strategy.

The operation selected for comparison should be the "best in class." It can be in a competitor, an unrelated company, another division or plant of the company making the study, a nonprofit organization, a foreign company, etc. The only requirement is that the operation selected for study should offer a fresh perspective and have potential for providing improvement for the corresponding one in the organization doing the benchmarking. The amount of potential improvement should be sufficient to justify the cost and time required for the study.

Ideally, objective indicators of performance should be used as a basis for comparison of the operation studied. Indexes such as product development time, the number of customer complaints in the first year after product introduction, process yield, and fuel consumption per operating hour are examples.

There are many sources for information needed to conduct the benchmarking analysis. They include published data such as in newspaper and magazine articles, product brochures, testing of a product purchased from normal sources, and semipublic data such as credit reports, governmental filings, etc. The capability of estimating critical data that are not supplied directly may be a crucial factor in the success of a study.[6] Most useful are visits to the operation being studied and interviews with employees. This may not be feasible with direct competitors. For this reason, it is often most useful to visit a noncompetitive company. If the company has a superior operation in the function under study, the effect can be quite productive, even if the industry is a different one.

Desirable Sequence of DFM Activities

Effective DFM involves two aspects: (1) an analysis of the complete product in order to simplify its design (reduce the parts count, etc.), and (2) an analysis of each individual part to maximize its manufacturability.

It makes sense to perform the overall product analysis first—before individual components are analyzed—for several reasons:

1. Design for assembly (DFA) seems to provide a greater potential return—probably due to opportunities to simplify the whole product and eliminate parts—than does analyzing individual component parts.
2. The configuration of the overall assembly, decided on after a DFA analysis, will determine whether certain parts exist and what configuration they must have. It does not make sense to analyze the component parts until their existence and configuration are determined.

3. Final assembly is often the major operation in a factory, the one with the most cost for the company. The project, therefore, often will be more effective simply because the highest-cost area is attacked first.

Once the overall assembly is analyzed and the improved overall design is at least tentatively settled on, the next logical step is to analyze the proposed major component parts next. The design of each and every part in the assembly should be reviewed and simplified as much as possible, roughly in order of its cost or importance to the product.

TRAINING AND INDOCTRINATION

For a sound DFM team project, training/education of several types is needed:

1. *Attitudinal training.* It is necessary to redirect design efforts, to incorporate the objectives of DFM, to promote and facilitate a team approach to product design, and to help participants adapt to their new responsibilities. It is important that the group members have a team orientation and a team spirit.
2. *How-to training.* Development of the know-how needed to intelligently apply applicable DFM tools and guidelines is also essential.
3. *Information-sharing.* It is necessary to share information about the product and its place in the market.

Obviously, the design team members should be the primary recipients of indoctrination and training.

There is also a need for some "appreciation training" for others in the company. Most important, the managers of the functions that will be supplying team members must be included in the indoctrination sessions. Full information should be provided to them about team members' duties and responsibilities and team procedures.

On-the-Job Training

Unfortunately, many designers have little or no experience in manufacturing. Their exposure may be limited to a college course in manufacturing processes and a few plant tours. Actual shop floor troubleshooting, supervisory experience, or work as a manufacturing engineer in the types of operations likely to be involved in the company's products is very valuable, if not essential, in properly grounding the designer in the background understanding that will make DFM/DFX principles meaningful.

Sources of Instruction

Training instruction can come from the company's own personnel, vendors, consultants, or educational institutions. Each choice has its pluses and minuses.

Company personnel are preferred in most cases, especially when they have personal knowledge of the material being taught. This is the case for much of the subject matter mentioned earlier, such as product plans, design procedures, and company priorities, that is explained to the team. The best people to instruct and advise team members and others are the managers and specialists who have the most knowledge and familiarity with the material being presented.

Vendors can be useful for instruction if the subject matter concerns the vendor's product.

Consultants and local educational institutions may be good sources if what they offer happens to coincide with the needs of the company for the project.

Usually, some combination of teaching sources is advisable, with company-specific material best taught by company people and specialized techniques best taught by outside agents most knowledgeable about the techniques involved. In general, the more use that can be made of company instructors, the better.

Scheduling the Training

To ensure that all questions are answered at the start and that team members are fully aware of their roles and the tools available to them, the most important information sharing and training must be scheduled to take place at the very start of the project. It is essential that there exists in the team as high as possible a level of expertise in DFM/DFX methodology, i.e., that the team is able to shape a design to achieve DFM/DFX attributes. Developing such a level of expertise cannot take place in a short training session; it is really a prerequisite for sound DFM and DFX. If this level of specialization does not exist before the start of the project, some key team members should receive training in these principles prior to the start of the team's activities.

Some follow-up training and information sessions may well be required as the project progresses. These can be scheduled by the team leader, normally as part of regular team meeting.

Training Site

Almost all the information and training sessions probably can be conducted right in the meeting room used by the team. Special skills development for subgroups or individuals will best take place at some other location so as not to disrupt the team's work. Appreciation training for larger groups of nonteam personnel probably will require some larger meeting area.

Individual or Group Training

Almost all the information and training sessions for DFM and concurrent engineering will be on a group basis. The whole rationale of concurrent engineering, a team approach, precludes having much segmented or individual instruction. The exception may be some particular skill needed in some aspect of the project but not needed by all team members.

Sources of the Technical Expertise Needed

Traditionally, the knowledge base of manufacturability principles, the guidelines and rules of thumb that facilitate low-cost production, comes from the experience and expertise of the manufacturing engineer. Other sources of this information are the following:

1. The company's own design manuals often have useful information for families of parts common to the company's products.
2. Manufacturers' and trade associations' bulletins. These are often published by such groups to promote business for group members and to facilitate trouble-free, economical production.
3. More formal publications, such as this handbook.
4. Computer programs that evaluate the design of product assemblies are another source of guideline information. In these programs, the categorization of constituent parts by the designer using the program gives him or her implicit information as to what characteristics of parts are desirable and what characteristics are not.

REFERENCES

1. G. J. Burke and J. B. Carlson, "DFA at Ford Motor Company," *DFMA Insight* **1**(4): 1990.
2. J. R. Hackman and G. R. Oldham, *Work Redesign,* Addison Wesley, Reading, Mass., 1980.
3. A. Majchrzak, *The Human Side of Factory Automation,* Jossey-Bass, San Francisco, 1988.
4. P. R. Lawrence, "How to Deal with Resistance to Change," *Harvard Business Review* (May–June): 1954.
5. Society of Manufacturing Engineers, "Creative Problem Solving Techniques: Preliminary Design Issues," in *Tool and Manufacturing Engineers Handbook,* Dearborn, Mich., 1992.
6. M. J. Spendolini, *The Benchmarking Book,* Amacom Division, American Management Association, New York, 1992.

CHAPTER 1.7

EVALUATING DESIGN PROPOSALS*

Designers need some method for knowing if the new or redesigned product will meet its manufacturability and other objectives. The designer's general judgment may be very sound in weighing the design's conformance to planned design attributes, but an objective measurement almost always will be better. Every design variation has consequences in the properties of the product, including its manufacturability.

Evaluation is needed not only so that the design team or designer can know if the product's objectives have been met but also so that alternative designs can be compared and the most effective alternative selected.

Preferably, the design team should be able to carry out an evaluation early in the design process, ideally at the concept stage. Then the time-consuming and expensive development and detailed work does not take place unless it is verified that the proposed approach is really effective. The procedure should be one that can be applied easily and routinely by the product designers. It should employ some numerical rating, index, or cost so that as objective a comparison as possible can be made between alternative designs.

EVALUATING MANUFACTURABILITY

Manufacturing cost is the most complete measure of manufacturability. It can be expressed as a total cost for the product or component or can be approximated with some major cost element such as direct labor time.

Most progress of all has taken place with design for assembly (DFA). Providing an evaluation of assembly designs is somewhat simpler than evaluating the manufacturability of individual parts, where such factors as tooling cost, yield, and production quantity all weigh more heavily than they do with assemblies. Assembly evaluation systems can provide a rapid and easy comparison between several alternatives. Common measures, in addition to cost, are parts count, design efficiency, and assembly time.

Direct labor time is a straightforward indicator of manufacturing cost and is usable by itself in a large number of cases. (Exceptions are those in which materials costs,

*Adapted from Chap. 11 of *Design for Excellence,* by James G. Bralla, McGraw-Hill, New York, 1996.

labor rates, and overhead costs also vary significantly with different design variations.) Therefore, in many cases, manufacturability of a series of design choices can be evaluated by estimating and comparing the direct labor time required for production of each design. Eventually, however, a full cost estimate is the ultimate guide to the designer in knowing how well the product design has been engineered for manufacturability.

Conventional cost estimates are made by evaluating the materials content of a design and the labor content of the production operations involved. This is a valid and accurate way to estimate the manufacturing cost of a particular design, and hence its manufacturability, although it may be time-consuming. The time element can be reduced by using computer assistance.

ASSEMBLY EVALUATION SYSTEMS

What are most interesting and useful are cost-estimating computer programs developed specifically for DFM use. The longest-standing—and in many ways the most useful for DFM—are those applicable to assembly evaluation.

It should be noted that current programs evaluate the labor content of an assembly design, not the materials costs. Sometimes there are tradeoffs between materials and labor costs of design alternatives. For example, a complex part made by combining several simpler parts will reduce assembly costs, but the cost of the complex part could conceivably be higher than the cost of several simple parts. Fortunately, however, materials costs are easy and straightforward to estimate from per-pound or per-square-foot data. Materials cost differences can be combined with the labor cost differences of alternative designs to arrive at a more nearly total cost comparison. The programs may give a design efficiency rating, a ratio comparing the calculated assembly time with a theoretical ideal for the number of parts involved.

There is one other quite useful method to evaluate the manufacturability of assemblies. This is simply to count the number of parts that the design entails. Assemblies with fewer parts normally can be assembled in less time and have higher design efficiency ratings.

One powerful advantage of these computer programs and the parts-count approach is that they can be used by designers themselves. They provide an easy way for designers to gauge the effectiveness of their efforts. They help to eliminate "we versus they" feelings that can arise when manufacturing people are doing the evaluation and pressing the designer to simplify the designs. The programs also have guideline information implicit in their tables of time data in that the designers, in working to improve the rating of their designs, will apply guidelines that promote that objective. In this sense, they are also useful in training designers in principles applicable to better, more easily assembled products.

MANUFACTURABILITY EVALUATIONS OF INDIVIDUAL PARTS

One simple way to compare the manufacturability of alternative designs of a part is to count the number of process operations that each requires. Other factors being equal, the part with the fewest number of operations will be the simplest to manufacture and the lowest in cost. Of course, tooling complexity and materials cost often must be considered also. Nonetheless, this metric is often a useful one for comparing parts from a DFM standpoint.

Some companies have developed systems to facilitate the manufacturability evaluation of piece parts in a product. They are somewhat more complex than the assembly systems described earlier in that there are separate methods for each manufacturing process involved. For example, die castings, injection-molded plastic parts, machine parts, powder-metal parts, and metal stampings each are evaluated with separate systems, since design principles, rules of thumb, and manufacturing costs are different for each process. Current systems simplify and ease the task of making an estimate of the manufacturing cost of a part. They consider tooling cost and amortization, process labor, and materials costs. As in the case of assembly evaluation systems, comparisons can be made for different design concepts. Some systems are computerized and are programmed to request the input data needed to develop a cost estimate.

EVALUATING DFX ATTRIBUTES

All the systems described earlier deal with manufacturability and not the other DFX attributes that are discussed in Chap. 9.2. Evaluating designs for DFX requires different, more complex methods such as the following:

1. The attribute being evaluated can be expressed in terms of cost or some other monetary factor. This is difficult because of the intangible nature of many of the DFX attributes. The most intangible factor is the financial benefit that can accrue when a company incorporates desirable attributes into a product and gains additional sales, larger profit margins, or both. For instance, speed to market is emphasized in order to enhance product sales. How much additional profit margin can be generated if the product realization time is reduced by 3 months? Providing an accurate estimate for such a not easily predicted factor is difficult but remains a possible avenue for someone establishing an evaluation system for improved speed to market. Similarly, estimated cost or profit amounts could be related to different design concepts when evaluating a design for such attributes as safety, serviceability, reliability, etc.

2. Use a scoring system that rates or ranks design alternatives against some criterion. Ideally, such a system should provide a design efficiency or other numerical rating of the extent of the attribute. Normally, systems of this type must be somewhat generalized and rely quite heavily on the experience, knowledge, and judgment of the person making the evaluation.

3. Test the design. This involves making at least one prototype of the proposed product and subjecting it to whatever tests are appropriate for the objectives being evaluated. For example, if reliability is one of the design objectives, life tests are called for; if user friendliness is an objective, a number of user tests of the product with feedback information from the test users is needed.

Almost all DFM and DFX guidelines are still qualitative in nature and often conflicting. It would be ideal if the effect of any one design alternative, considered with respect to some DFX recommendation, could be evaluated.

There is one way that the DFX guidelines can be related to cost and thereby given quantitative evaluation. This is through use of the life-cycle cost concept—Taguchi's concept of overall product quality (see definition in Chap. 9.3). The lower the life-cycle cost for such factors as service, safety, the repair of quality defects, and so on, as well as the initial cost, the better is the DFX performance. As noted earlier, however, many of the life-cycle cost factors are highly intangible and not well suited to quantification. How, for example, can one predict the cost (or profit effect) of sales that are lost due to a poor reliability reputation? How can one predict the cost of product liability lawsuits resulting from safety defects? Broad overall projections of such costs may be possible, but relating them to specific design changes such as changing a sharp corner in a part to a radiused corner (sometimes a safety and sometimes a product reliability factor) is not really feasible. Even calculating the manufacturing cost effect of such a change may be somewhat lengthy and uncertain.

WHO SHOULD MAKE THE EVALUATION?

The best approach, when possible, is to have the designers themselves make the evaluation. This certainly has speed and accuracy advantages in that there is no need to transfer information about the design from one person to another. The time to prepare documentation and to explain the design concept to the specialist is avoided. One of the advantages of some of the current evaluation systems, particularly those involving assembly or other aspects of manufacturability, is that they make it relatively easy for the designers themselves to carry out the evaluation. In addition to the convenience and time advantages of this, there is also the learning factor that benefits the designer. Designers who conduct an evaluation with a prepared system tend to learn the design principles that underlie the system.

TESTING DESIGN PROPOSALS

Prototypes should be subjected to a variety of tests:

1. Use tests to verify that the product functions as it was designed to.
2. Life testing is important to determine the reliability and useful life of the product and its components.
3. Environmental testing, usually at various temperatures, humidities, and other conditions, confirms the product's performance under any extreme conditions that the product may face in use.
4. Field tests help to confirm the successful operation of the product under customer-use conditions. Another purpose of field tests is to verify that customers will understand and be able to use the product easily and as intended.
5. Shipping tests help to verify the effectiveness of packaging and the sturdiness of the product.

Good testing is a powerful and essential step in perfecting the design and in ensuring that it meets the varied objectives of the program.

S • E • C • T • I • O • N • 2

ECONOMICAL USE OF RAW MATERIALS

CHAPTER 2.1
INTRODUCTION

PROPER MATERIALS SELECTION

The choice of materials is a major determinant for the successful functioning and the feasible, low-cost manufacture of any product. The design engineer must tread a course (which is often a narrow one) between functional requirements on one hand and cost considerations on the other. In some cases, there will be a real conflict between function and cost in the selection of a material. In other cases, happily, the choice is easy: There is a particular material that does the job functionally and which minimizes the cost of production. It is the purpose of this section of the handbook to provide information about the properties, relative costs, and fabricability of materials so as to aid the designer in making such favorable choices.

There is a real need for economy in materials costs because, for most products, these costs account for a major portion of the total: 50 percent is a common order-of-magnitude percentage. However, the objective is not necessarily minimum materials cost but minimum overall cost, including the initial price of the material, the cost of processing and assembling it with other materials into a product, the cost of guaranteeing the durability of the product and servicing it, etc. Thus, in the long run, the lowest-cost material may not be the material with the lowest price tag.

MATERIALS AVAILABLE

The designer today is provided with a vast array of metallic and nonmetallic materials that have been engineered for various applications. The list of available materials is also growing as new alloys, plastics, and composite and clad materials are introduced.

Figure 2.1.1 is a master chart of engineering materials available to the product designer. Table 2.1.1 shows the forms in which a number of common materials are normally available.

FUNCTIONAL REQUIREMENTS

Tables 2.1.2 through 2.1.8 provide data on several functional properties of materials. Selected materials are listed in order of degree of possession of the property involved.

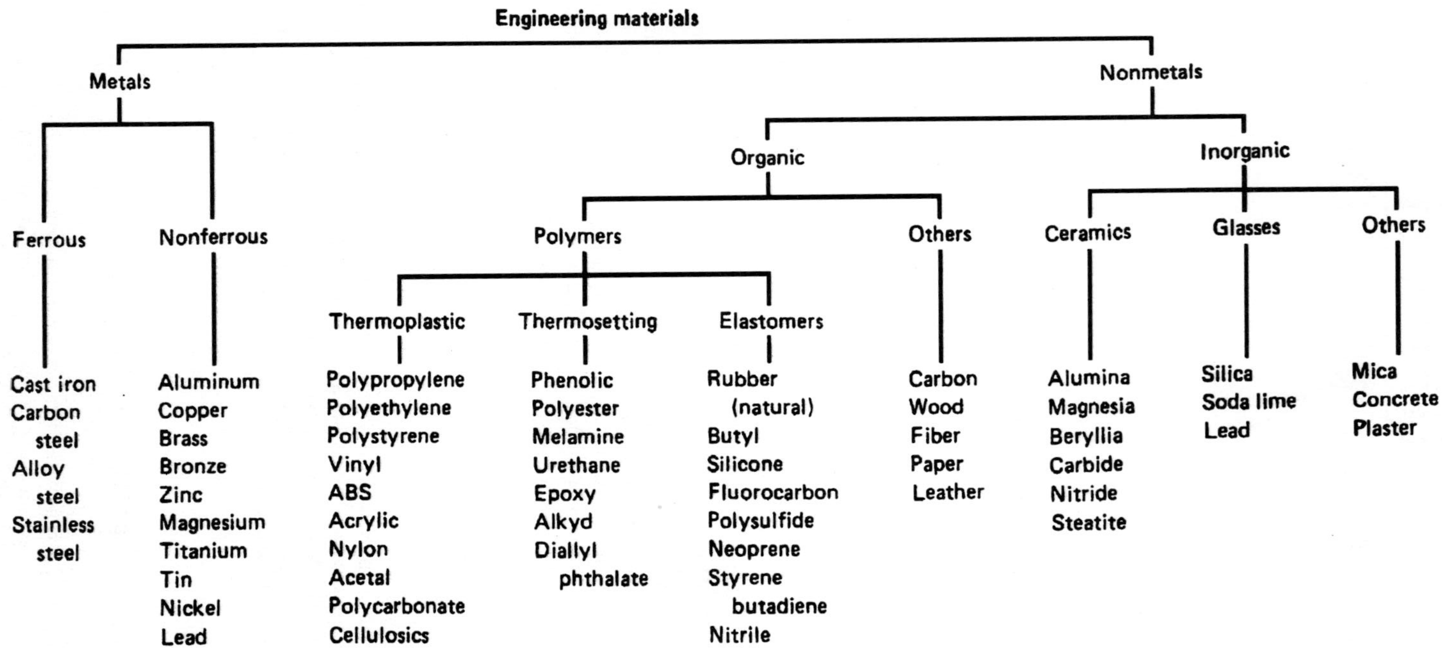

FIGURE 2.1.1 Classification of engineering materials.

TABLE 2.1.1 Common Commercial Forms of Selected Raw Materials[a]

Material	Ingots for casting	Molding compounds	Round bars	Rectangular bars	Tubing	Plate	Sheet	Foil or film	Wire	Structural shares	Powder for sintering
Cast iron	X[b]	—	—	—	X[c]	—	—	—	—	—	X
Carbon steel	X[b]	—	X	X	X	X	X	—	X	X	X
Alloy steel	X[b]	—	X	X	X	X	X	—	X	X	X
Stainless steel	X	—	X	X	X	X	X	—	X	X	X
Aluminum	X	—	X	X	X	X	X	X	X	X	X
Copper	X	—	X	X	X	X	X	L	X	L	X
Brass	X	—	X	X	X	X	X	L	X	L	X
Magnesium	X	—	X	X	X[c]	X	X	—	L	X	—
Zinc	X	—	X	L	—	X	X	X	X	—	—
Precious metals	X	—	X	L	L	—	X	X	X	—	—
Alumina ceramics	—	X	X	L	X	L	—	—	—	L	X
Paper	—	—	—	—	X	—	X	X	—	—	—
Fiber	—	—	X	—	X	X	X	X	—	—	—
Carbon graphite	—	—	X	X	X	X	—	—	X[d]	L	X
Glass	—	—	X	L	X	X	X	—	X[d]	L	—
Acetal	—	X	X	X	L	X	X	—	—	L	—
Nylon	L[e]	X	X	X	X	X	X	—	X[d]	—	—
Polyethylene	—	X	X	X	X	X	X	X	L	—	—
Polypropylene	—	X	X	L	X	X	X	X	L	—	—
Polystyrene	—	X	X	L	X	X	X	L	—	—	—
ABS	—	X	X	L	L	X	X	—	—	—	—
Polycarbonate	—	X	X	L	X	X	X	—	—	—	—
Phenolic	—	X	X	L	X	X	X	—	—	—	—
Wood	—	—	X	X	—	X[f]	X[f,g]	—	—	L	—
Natural rubber	—	X	L	L	X	X	X	—	—	—	—
Synthetic rubber	—	X	L	L	X	X	X	—	—	—	—

[a]X = usually available; L = of limited availability; — = rarely or never available.

[b]Ferrous metals are formulated in the melt rather than as prealloyed ingots.

[c]Pipe.

[d]Fine fibers.

[e]Liquid monomer.

[f]Plywood.

[g]Veneer.

TABLE 2.1.2 Ultimate Tensile Strength of Selected Materials

Material	Tensile strength kPa	Tensile strength 1000 lbf/in²
Steel: SAE 4340, drawn 205°C	2000	290
Steel: C.60–80, forged-rolled	860	125
Brass	275–825	40–120
Ductile iron: grade 90-65-02 as cast	655–725	95–105
Monel: 70 Ni, 30 Cu	690	100
Stainless steel: 18-8	585–655	85–95
Titanium: 99.0 Ti, annealed bar	655	95
Copper	220–470	32–68
Structural steel: ordinary grade	345–450	50–65
Aluminum: wrought and heat-treated	205–415	30–60
Cast iron	140–415	20–60
Magnesium alloys: wrought	220–380	32–55
Zinc: die-cast	215–355	31–52
Aluminum: die-cast	270–315	39–46
Polyimides: reinforced	35–185	5–27
Acetal copolymer: 25% glass-reinforced	125	18
Nylons	35–95	5–14
Lead alloys	12–90	1.7–13
Acrylics	35–80	5–12
Polystyrene	35–80	5–12
Acetal	55–70	8–10
Polycarbonate	55–60	8–9
Vinyl: rigid	35–60	5–9
ABS	27–55	4–8
Polyethylene: high-density	21–38	3.1–5.5
Polypropylene	20–38	2.9–5.5
Polyethylene: low-density	4–16	0.6–2.3

The properties covered are as follows: Table 2.1.2, tensile strength; Table 2.1.3, specific gravity; Table 2.1.4, melting point; Table 2.1.5, thermal conductivity; Table 2.1.6, coefficient of thermal expansion; Table 2.1.7, electrical resistivity; and Table 2.1.8, stiffness-weight ratio.

DESIGN RECOMMENDATIONS

Some general rules for minimizing materials-related costs are as follows:

1. Use commercially available mill forms so as to minimize in-factory operations. (See Fig. 4.1.3.)
2. Use standard stock shapes, gauges, and grades or formulations rather than specials whenever possible. Sometimes, larger or heavier selections of a standard material are less costly than smaller or thinner sections of a special material.

TABLE 2.1.3 Specific Gravity of Selected Materials

Material	Specific gravity	Material	Specific gravity
Platinum	21.3–21.5	Phenolic, molded	1.2–2.9
Gold	19.25–19.35	Aluminum	2.55–2.75
Tungsten	18.7–19.1	Glass, plate	2.4–2.7
Lead	11.28–11.35	Silicon	2.5
Silver	10.4–10.6	Porcelain	2.3–2.5
Molybdenum	10.2	Carbon, amorphous	1.9–2.2
Bismuth	9.8	Silicone, molded	0.99–1.50
Monel	8.8–9.0	Melamine	1.48–2.00
Copper	8.8–9.0	Epoxy	1.11–1.40
Cobalt	8.72–8.95	Magnesium	1.74
Nickel	8.6–8.9	Vinyl	1.16–1.45
Bronze: 8–14% Sn	7.4–8.9	Polycarbonate	1.2–1.5
Brass	8.4–8.7	Acetal	1.4
Steel	7.8–7.9	Rubber, neoprene	1.0–1.35
Iron, wrought	7.6–7.9	Ebony	1.2
Aluminum bronze	7.7	Acrylic	1.09–1.28
Tin	7.2–7.5	Nylon	1.04–1.17
Manganese	7.2–7.4	Polystyrene	1.04–1.10
Iron, cast	7.2	ABS	1.01–1.21
Zinc	6.9–7.2	Polyethylene, high-density	0.94–0.97
Chromium	6.9	Polyethylene, low-density	0.91–0.93
Antimony	6.6	Rubber, natural	0.93
Glass, high-lead	6.2	Polypropylene	0.89–0.91
Vanadium	5.5–5.7	Hardwood, North American	0.4–0.7
Alumina ceramics	3.5–3.9	Softwood, North American	0.4–0.5
Glass, flint	3.2–3.9		

TABLE 2.1.4 Melting Point of Selected Materials, °C

Material	Melting point	Material	Melting point
Carbon	3700	Manganese	1245
Tungsten	3400	Copper	1083
Tantalum	2990	Gold	1063
Magnesia	2800	Aluminum bronze	855–1060
Molybdenum	2620	Beryllium copper	870–980
Beryllia	2590	Silver	960
Alumina	2050	Salt (sodium chloride)	803
Vanadium	1900	Glass, soda-lime	695–720
Chromium	1840	Aluminum alloys	485–660
Platinum	1773	Magnesium	650
Titanium	1690	Antimony	630
Porcelain	1550	Glass, lead silicate	580–620
Glass (96% silica)	1540	Zinc alloys	385–419
Carbon steels	1480–1520	Lead	327
Alloy steels	1430–1510	Cadmium	320
Stainless steels, austenitic	1370–1450	Pewter	245–295
Nickel	1450	Bismuth	271
Wrought iron	1350–1450	Tin	231
Silicon	1420	India rubber	125
Cast iron, gray	1350–1400	Sulfur	112
Beryllium	1280	Paraffin	54

TABLE 2.1.5 Thermal Conductivity of Selected Materials, cal/(h · cm^2 · °C · cm)

Material	Value	Material	Value
Silver	3600	Phenolic	2.9–7.2
Copper	3400	Melamine	3.9–4.6
Gold	2500	Polyethylene, high-density	4.0–4.5
Aluminum	1900	Nylon	1.8–3.8
Magnesium	1400	Maple, hard	1.5–3.7
Graphite, cylindrical	1000–1300	Asbestos-cement board	3.3
Brass, 70–30	1000	ABS	1.6–2.9
Zinc	960	Polyethylene, low-density	2.9
Tin	560	Paper, kraft	2.3
Nickel	510	Acrylic	1.4–2.2
Steel (1% carbon)	390	PVC, flexible	1.9–2.1
Stainless steel	100–200	Acetal	2.0
Stone, typical	190	Polystyrene	0.9–1.9
Alumina ceramics	80–140	Polycarbonate	1.7
Carbon	45–89	Rubber, natural	1.1–1.4
Brick	74	PVC, filled rigid	1.4
Air at 38°C	23	Polypropylene	1.0–1.4
Steatite	22	Pine, white	0.89
Glass	5.0–10	Glass wool	0.32
Cinder block	9.2	Polyurethane foam	0.28

TABLE 2.1.6 Coefficient of Linear Thermal Expansion of Selected Materials, 10^{-6} cm/(cm · °C)

Material	Value	Material	Value
Silicones, flexible	80–300	Aluminum	24
Vinyl, flexible	70–250	Bronze (90 Cu, 10 Sn)	22
Polystyrene	34–210	Brass (66 Cu, 34 Zn)	19
Polyethylene, low-density	100–200	Silver	18
Vinyl, rigid	50–185	Copper	17
Nylon	80–150	Nickel alloys	12–17
Polyethylene, high-density	110–130	Phosphor bronze	17
ABS	60–130	Steel, carbon	10–15
Rubber, natural	125	Gold	14
Polypropylene	58–102	Concrete	14
Acrylic	50–90	Marble	4–14
Acetal	81–85	Titanium alloys	9–13
Polycarbonate	66	Cast iron	11
Epoxy	45–65	Glass	0.6–9.6
Ice	51	Brick	9.5
Phenolics, filled	8–45	Alumina ceramics	5.7–6.7
Zinc alloys	19–35	Oak wood	4.9–5.4
Lead	30	Tungsten	4.3
Magnesium alloys	25–29	Carbon	1.0–2.9
Tin	27		

TABLE 2.1.7 Electrical Resistivity of Selected Materials, $\mu\Omega \cdot cm$

Material	Resistivity	Material	Resistivity
Mica	10^{19}–10^{23}	Stainless steel, martensitic	40–72
Glass, lead silicate	10^{20}–10^{21}	Carbon steel	11–45
Ceramics, electrical	10^{19}–10^{21}	Monel	44
Polypropylene	10^{15}–10^{17}	Lead	22
Polystyrene	10^{13}–10^{17}	Magnesium alloys	5–16
Beryllia	10^{13}–10^{17}	Tin	12
Epoxy	10^{12}–10^{17}	Nickel	8.5
Polycarbonate	$2–5 \times 10^{16}$	Brass	6.2
Acetal	10^{14}–10^{16}	Zinc	6.0
Nylon	10^{11}–10^{14}	Beryllium copper	4.8–5.8
Phenolic	10^{10}–10^{13}	Aluminum	2.8
Carbon	3800–4100	Gold	2.4
Graphite	720–810	Copper	1.7
Cast iron	75–99	Silver	1.6
Stainless steel, austenitic	69–78		

TABLE 2.1.8 Stiffness-Weight Ratio for Various Materials

Material	Weight for equal stiffness
Steel	7.95
High-density polyethylene	6.40
Polypropylene	4.38
ABS	4.29
Polystyrene	3.90
Aluminum	3.66
Expanded ABS	3.06
Maple wood	1.62
Pine wood	1.00

3. Consider the use of prefinished material as a means of saving costs for surface-finishing operations on the completed component.
4. Select materials as much as possible (consistently with functional requirements) for processibility. For example, use free-machining grades for machined parts, easily formable grades for stampings, etc. It pays to take the time to determine which variety of the basic material is most suitable for the processing sequence to be used.
5. Design parts for maximum utilization of material. Make ends square or nestable with other pieces from the same stock. Avoid designs with inherently high scrap rates.

FINISHES AND TOLERANCES

Close tolerances and good finishes are found on metal shapes that are cold-rolled or cold-drawn. They are closer than those for hot-rolled or extruded metals. Cross-sectional dimensions of cold-finished bars or other shapes are usually held to within about ±0.06 mm (±0.0025 in). Surface finishes generally range from 0.8 μm (32 μin) to 3 μm (125 μin). Ground plates and bars, of course, are held to much closer tolerances, about ±0.025 mm (±0.001 in), but these are found primarily in materials for specialized applications. For example, tool steel may be finished in this way. Length tolerances for commercially stocked pieces of rod, tubing, bar, etc., are normally about ± 3 mm (±0.125 in) and sometimes greater.

For nonmetals, tolerances on dimensions of stock shapes are considerably broader than for metals because of flexibility, softness, or high shrinkage during manufacture. However, ceramics, carbon, and glass sections, if ground to size, have tolerances close to those of metals, e.g., ±0.05 mm (±0.002 in). Otherwise, sectional dimensions will vary by about ±0.25 mm (±0.010 in) or more on the average.

More detailed information on tolerances for stock shapes of specific materials are found in the chapters that follow.

COST DATA

Table 2.1.9 provides relative cost data for selected engineering materials.

TABLE 2.1.9 Relative Cost per Unit Weight and per Unit Volume for Various Materials (Compared with Hot-Rolled, Low-Carbon Steel)*

Material	Relative cost per unit weight	Relative cost per unit volume
Hot-rolled, low-carbon steel bar	1.0	1.0
Cold-rolled, low-carbon steel bar	1.3	1.3
Soft maple†	1.8	0.14
Poplar†	1.9	0.11
Red oak†	2.0	0.18
Cold-finished 4140 steel alloy	2.0	2.0
Low density polyethylene‡	2.2	0.26
Polypropylene‡	2.3	0.26
Polystyrene, general purpose‡	2.6	0.35
Lead sheet	3.2	4.6
Zinc alloy ingots for die casting	3.3	3.0
PVC‡	3.8	0.29
Phenolic‡	4.2	0.76
Aluminum ingot	4.6	2.0
Acrylic‡	5.1	0.78
ABS‡	5.4	0.82
Polycarbonate‡	5.6	1.6
400-series stainless steel bar	6.4	6.3
Copper wire bar	6.8	7.7
Acetal‡	8.2	1.5
Brass ingots	9.6	10.3
300-series stainless steel bar	9.9	10.2
Magnesium ingot	10.2	2.3
Tin ingots	15.1	14.0
Iron powder	18.5	18.5§
Nickel ingot	19.4	21.6
Polyphenylene sulfide (40% glass filled)‡	21.9	19.5
Polysulfone‡	24.0	3.8
Silver	429	572
Gold	30,600	75,500

*Based on published 1996 prices in various trade publications for large-quantity purchases.

†Top-quality, kiln-dried, rough-cut boards.

‡Figures for all plastic materials are based on prices for molding compound, not finished shapes or parts.

§At 100% density.

CHAPTER 2.2
FERROUS METALS

PART 1
HOT-ROLLED STEEL

THE PROCESS

Hot-rolled shapes are produced by passing a heated billet, bloom, or ingot of steel through sets of shaped rollers. Upon repeated passes, the rollers increase the length of the billet and change it to a cross section of specified size and shape. After rolling, the shape is sometimes pickled (by immersion in warm, dilute sulfuric acid) to remove scale and is then oiled.

TYPICAL CHARACTERISTICS OF HOT-ROLLED-STEEL SHAPES

Hot-rolled steel is produced in a variety of cross sections and sizes, of which the following are typical:

Round bars from 6 to 250 mm (1⁄4 to 10 in) in diameter

Square bars from 6 to 150 mm (1⁄4 to 6 in) per side

Round-cornered squares, 10 to 200 mm (3⁄8 to 8 in) per side

Flat bars from 5 mm (0.20 in) in thickness and up to 200 mm (8 in) in width but not over 80 cm2 (12 in2) in cross-sectional area

Angles, channels, tees, zees, and other sections that have a largest cross-sectional dimension of 75 mm (3 in) or less

Ovals, half-rounds, and other special cross sections

Sheets, 1.5 mm (0.060 in; 16 gauge) or thicker and plates

Figure 2.2.1 illustrates some common cross-sectional shapes in which hot-rolled steel is available from warehouse distributors. Figure 2.2.2 illustrates typical cross sections obtainable by special mill-run order.

Hot-rolled steel averages about 30 percent lower in price than cold-finished steel. This makes its use attractive whenever the application permits. However, hot-rolled steel has more dimensional variation, a rougher surface, mill scale (if not removed by pickling), less straightness, less strength in low-carbon grades, and somewhat poorer machinability. If these factors are not critical, the use of hot-rolled material may permit a substantial cost savings.

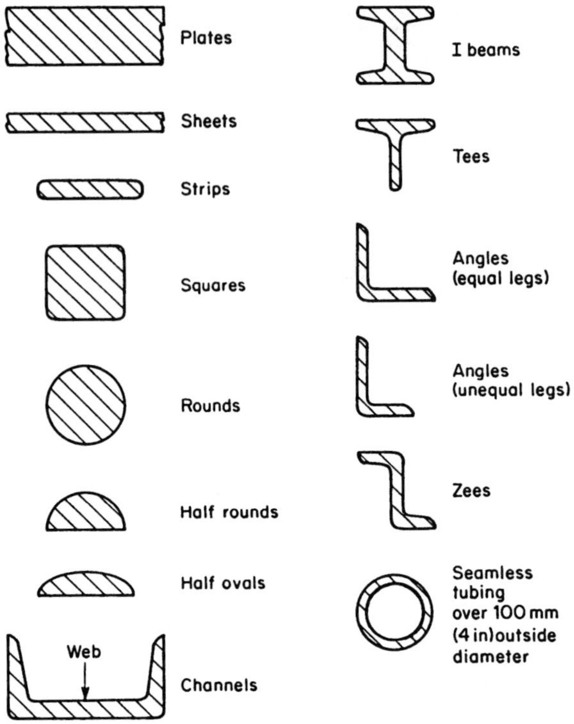

FIGURE 2.2.1 Hot-rolled sections commonly available from warehouse distributors.

Usually, hot-rolled steel is employed in applications for which only a small amount of machining is required and for which a smooth surface finish is not necessary. Some examples are tie rods, welded frames, lightly machined shafts, cover plates, riveted and bolted racks and frames, railroad cars, ships, storage tanks, bridges, buildings, and other applications for which heavy structural members are required.

The preponderance of steel sold in the hot-rolled form is of low-carbon content, 0.25 percent or less.

Hot-rolled steel can have good weldability, formability, and machinability. Although its machinability and formability may be slightly less than those of equivalent cold-finished alloys, the choice of cold-finished over hot-rolled material is usually a matter of surface finish, accuracy, or strength rather than ease of further processing. Hot-rolled material is generally used when surface finish is of secondary importance.

ECONOMIC QUANTITIES

As long as standard cross sections are involved, hot-rolled-steel shapes are suitable for all levels of production. Small quantities for maintenance purposes or for low-unit

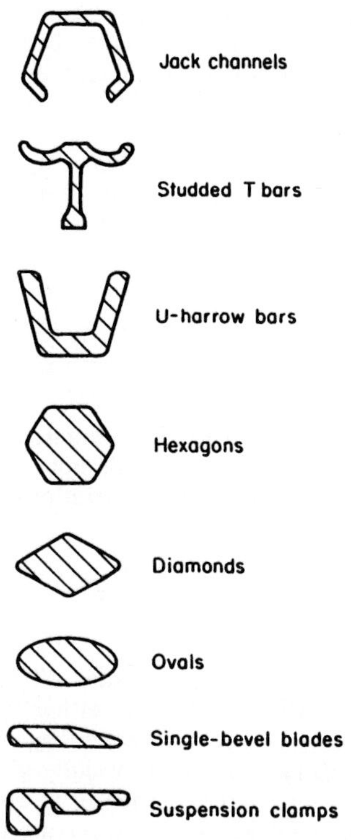

FIGURE 2.2.2 Typical hot-rolled sections that could be available on special mill order.

production can be purchased from steel-distribution warehouses. Large quantities for mass-production applications can be purchased directly from mills.

For special cross sections (such as those shown in Fig. 2.2.2), however, minimum mill quantities dictate use only for high production levels. Although some steel mills tend to specialize in shorter mill runs, the common minimum mill quantity for special shapes is on the order of 100 tons. A complex cross section is apt to have a larger minimum mill quantity than a simple one.

GRADES FOR FURTHER PROCESSING

Machining

Moderately low-carbon grades are best. Usually, alloy steels of the same carbon content and strength have poorer machinability than plain-carbon grades.

Hot-rolled steel with a carbon content of 0.15 percent or less tends to be gummy and to adhere to the cutting tool. If it is to be machined, it should first be hardened and tempered.

With carbon content in the range of 0.15 to 0.30 percent, machinability is good, especially in the upper part of the range, if the steel has not been hardened.

Machinability is good with the 0.30 to 0.50 percent carbon grades. Best results are achieved when there has been prior annealing such that the structure is partially spheroidized.

For higher-carbon grades (0.55 percent and more), annealing must provide a completely spheroidized structure; otherwise, machining will be difficult.

If heavy machining is required, free-machining grades containing sulfur or lead should be employed.

Forming

The low-carbon grades are best. The lower the yield strength and the more ductile the material, the more easily it can be formed. The scale on hot-rolled steel is usually not a significant deterrent to use of the material on press operations.

Welding

The best results are achieved with the low-carbon grades. Materials having 0.15 percent or less carbon are invariably easily welded. As either the carbon content, the thickness, or the alloy content increases, however, weldability decreases.

Hot-rolled steel with 0.15 to 0.30 percent carbon usually gives satisfactory results if the section is less than 1 in thick. For higher carbon or alloy content or heavy sections, preheating or postweld stress relieving may be necessary for best results.

DESIGN RECOMMENDATIONS

In deciding whether to use hot-rolled instead of cold-finished material and in choosing the grade, the product designer should consider the concept of designing for minimum cost per unit of strength. Often grades with higher carbon content or low alloy content will provide lower-cost parts that can be made from plain low-carbon grades. The reason is that lighter sections can be used.

When bending hot-finished-steel members, the bend line should be at right angles to the grain direction from the rolling operation. The bend radius also should be as generous as possible. Adhering to both these rules will help avoid fracturing the material at the bend. See Fig. 2.2.3 for a conservative set of rule-of-thumb values for minimum-bend radius.

When hot-rolled material is machined with the objective of providing a true surface, it is necessary to remove sufficient stock to get below the surface defects and irregularities. (These include seams, scale, deviations from straightness or flatness, and decarburization.) The American Iron and Steel Institute (AISI) recommended machining allowance per side is 1.5 mm (0.060 in) for finished diameters or thicknesses from 40 to 75 mm ($1\frac{1}{2}$ to 3 in) and 3 mm (0.125 in) per side for diameters or thicknesses over 75 mm (3 in). (See Fig. 2.2.4.) However, these values are considered

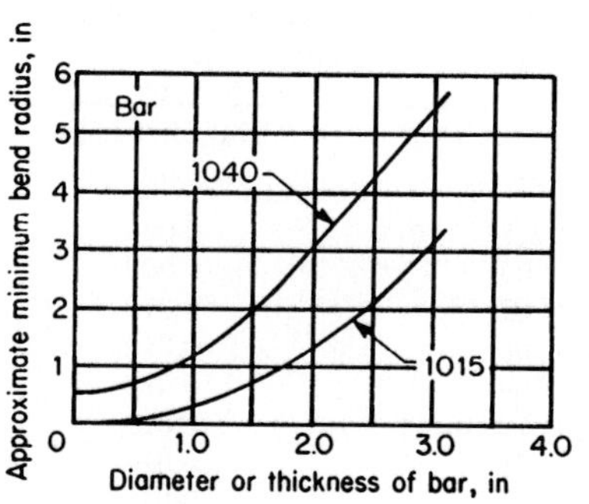

FIGURE 2.2.3 Recommended approximate minimum-bend radius for hot-rolled-steel members. (*Courtesy American Society for Metals,* Metals Handbook.)

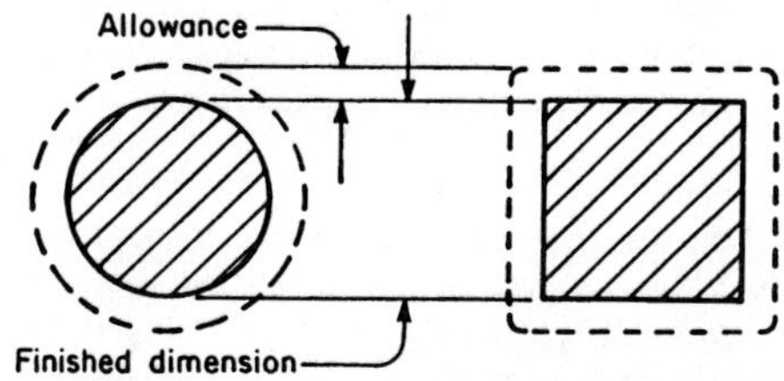

FIGURE 2.2.4 Recommended machining allowances per side to remove hot-rolling scale, surface inclusions, and irregularities.

Finished dimension	Allowance
Up to 38 mm (1.5 in)	0.8 mm (0.031 in)
38 to 75 mm (1.5 to 3 in)	1.6 mm (0.062 in)
Over 75 mm (3 in)	3.2 mm (0.125 in)

liberal. For moderate and high levels of production, it is worthwhile to test the actual condition of the steel being used.

DIMENSIONAL FACTORS

Since hot-rolled steel does not have the benefit of a secondary sizing operation, dimensional variations are considerably wider than with cold-finished material. The factors that lead to size, flatness, straightness, and twist deviations are heat, temperature, and cooling-rate variations, breaking off and movement of scale during rolling, sag of less supported bar areas when the material is red-hot, and variations in rolling equipment, tools, and reduction per pass.

Hot cutting of ends, usually by shearing, causes material at the ends to deviate dimensionally more than the limits stated in the tolerance tables. Ends can be trimmed by cold sawing if such deviations are objectionable.

TOLERANCES FOR HOT-ROLLED-CARBON-STEEL BARS

Tables 2.2.1 through 2.2.4 show standard tolerances as established by major steel mills. Mills customarily hold dimensions to closer limits than those shown, but there is no assurance that this will be the case for every lot rolled.

TABLE 2.2.1 Tolerances for Size and Out-of-Round or Out-of-Square Hot-Rolled-Carbon-Steel Bars: Round, Square, and Round-Cornered Square Bars

Specified sizes, in	Size tolerances, in		Out-of-round or out-of-square section, in
	Over	Under	
To $5/16$ inclusive	0.005	0.005	0.008
Over $5/16$ to $7/16$ inclusive	0.006	0.006	0.009
Over $7/16$ to $5/8$ inclusive	0.007	0.007	0.010
Over $5/8$ to $7/8$ inclusive	0.008	0.008	0.012
Over $7/8$ to 1 inclusive	0.009	0.009	0.013
Over 1 to $1\frac{1}{8}$ inclusive	0.010	0.010	0.015
Over $1\frac{1}{8}$ to $1\frac{1}{4}$ inclusive	0.011	0.011	0.016
Over $1\frac{1}{4}$ to $1\frac{3}{8}$ inclusive	0.012	0.012	0.018
Over $1\frac{3}{8}$ to $1\frac{1}{2}$ inclusive	0.014	0.014	0.021
Over $1\frac{1}{2}$ to 2 inclusive	$1/64$	$1/64$	0.023
Over 2 to $2\frac{1}{2}$ inclusive	$1/32$	0	0.023
Over $2\frac{1}{2}$ to $3\frac{1}{2}$ inclusive	$3/64$	0	0.035
Over $3\frac{1}{2}$ to $4\frac{1}{2}$ inclusive	$1/16$	0	0.046
Over $4\frac{1}{2}$ to $5\frac{1}{2}$ inclusive	$5/64$	0	0.058
Over $5\frac{1}{2}$ to $6\frac{1}{2}$ inclusive	$1/8$	0	0.070
Over $6\frac{1}{2}$ to $8\frac{1}{4}$ inclusive	$5/32$	0	0.085
Over $8\frac{1}{4}$ to $9\frac{1}{2}$ inclusive	$3/16$	0	0.100
Over $9\frac{1}{2}$ to 10 inclusive	$1/4$	0	0.120

Note: Out-of-round is the difference between the maximum and minimum diameters of the bar, measured at the same cross section. Out-of-square section is the difference in the two dimensions at the same cross section of a square bar between opposite faces.

Source: American Iron and Steel Institute.

TABLE 2.2.2 Tolerances for Thickness and Width of Hot-Rolled-Carbon-Steel Bars: Square-Edge and Round-Edge Bars

	Thickness tolerances, for thicknesses given, over and under, in							Width tolerances, in	
Specified widths, in	0.203 to 0.230 exclusive	0.230 to 1/4 exclusive	1/4 to 1/2 inclusive	Over 1/2 to 1 inclusive	Over 1 to 2 inclusive	Over 2 to 3 inclusive	Over 3	Over	Under
To 1 inclusive	0.007	0.007	0.008	0.010	—	—	—	1/64	1/64
Over 1 to 2 inclusive	0.007	0.007	0.012	0.015	1/32	—	—	1/32	1/32
Over 2 to 4 inclusive	0.008	0.008	0.015	0.020	1/32	3/64	3/64	1/16	1/32
Over 4 to 6 inclusive	0.009	0.009	0.015	0.020	1/32	3/64	3/64	3/32	1/16
Over 6 to 8 inclusive	*	0.015	0.016	0.025	1/32	3/64	—	1/8†	3/32†

*Flats over 6 in to 8 in inclusive in width are not available as hot-rolled-carbon-steel bars in thickness under 0.230 in.

†For flats over 6 in to 8 in inclusive in width and to 3 in inclusive in thickness.

Source: American Iron and Steel Institute.

TABLE 2.2.3 Angles for Hot-Rolled Steel: Tolerances for Thickness, Length of Leg, and Out-of-Square

Specified length of leg, in	Thickness tolerances, for thicknesses given, over and under, in			Tolerances for length of leg, over and under, in
	To 3/16 inclusive	Over 3/16 to 3/8 inclusive	Over 3/8	
To 1 inclusive	0.008	0.010		1/32
Over 1 to 2 inclusive	0.010	0.010	0.012	3/64
Over 2 to 3 exclusive	0.012	0.015	0.015	1/16

Note: The longer leg of an unequal angle determines the size for tolerance. The out-of-square tolerance in either direction is 1½°.

Source: American Iron and Steel Institute.

TABLE 2.2.4 Straightness Tolerances for Hot-Rolled-Steel Bars

Normal straightness tolerance	¼ in in any 5-ft length or ¼ × number of feet of length divided by 5
Special straightness tolerance	⅛ in in any 5-ft length or ⅛ × number of feet of length divided by 5

Note: Because of warpage, tolerances do not apply after any subsequent heating operation except stress relieving which has been performed after special straightening.

Source: American Iron and Steel Institute.

PART 2

COLD-FINISHED STEEL

DEFINITION

Cold-finished steel is a more physically refined product than hot-rolled steel. Its chemistry is identical except, in some cases, for the removal of surface decarburization. However, its surface finish, dimensional accuracy, and grain structure are superior, and it is free from scale. It also has improved tensile and yield strengths.

Cold-finished steel is available in a variety of grades under both "carbon steel" and "alloy steel" designations (for information about cold-finished stainless steel, see Part 3).

Cold-finished-steel products include bars (round, square, hexagonal, flat, or in special shapes), flat products (sheets, strips, or plates), tubular products, and wire of various cross sections.

TYPICAL APPLICATIONS

Cold-finished steel has applicability when

1. The greater accuracy and smoother surface finish of the material are necessary for the function of the product or for the elimination of additional machining or other operations which would otherwise be required. This advantage is particularly applicable when special cross sections are used or when a part is designed to conform to a standard shape.
2. The added mechanical properties resulting from cold working (higher yield and tensile strengths) are required. (Tensile strength may increase by 20 percent or more and yield strength by 60 percent or more when a bar is cold-drawn with a cross-sectional reduction of 12 percent or more.)
3. The improved machinability, improved formability, and freedom from surface scale facilitate manufacturing operations.

The following are typical applications of cold-finished steel: *sheet and strip:* automobile bodies, appliance housings, and sewing-machine parts; *bars:* cams, shafts, studs, screws, gears, pins, bushings, firearm parts, and machinery components; *tubular products:* boiler tubes, mechanical (structural) tubing, pipe for various applications, conduit, and pilings; *wire:* cold-headed parts, springs, cable, tire reinforcement, wire fences, and netting; *special cross sections:* pinions, ratchets, gun parts, lock parts, vanes, pawls, hinge parts, tool parts, valve parts, miscellaneous hardware, etc.

MILL PROCESSES

There are three basic processes that are used individually or in combination to produce cold-finished steel from hot-rolled material: cold drawing, cold rolling, and machining. They may involve some preliminary steps such as pickling or grit blasting, water washing, and dipping the bar in a solution of slaked lime.

Cold-finished steel also may be subjected to stress relieving, annealing, or normalizing; carbon restoration; and special inspection operations as the application dictates. Normally, sheet and strip stock intended for press bending, forming, or drawing is annealed.

GRADES FOR FURTHER PROCESSING

Machining

The improved machinability of cold-finished steel is due primarily to the increased hardness imparted by the drawing operation. Drawing causes the chips produced by the cutting tool to be more brittle so that they break away from the workpiece more easily. Power requirements and tool wear are both reduced. Cold-finished bars are commonly formulated specifically for screw-machining and other metal-cutting processes.

Sulfur, lead, and tellurium additives are frequently incorporated into cold-finished bars to optimize machinability. Table 2.2.5 lists some of the more common cold-finished-steel-bar formulations with particularly good machinability.

Stamping

Cold-rolled-steel sheet and strip generally are preferable to hot-rolled material for stampings because of the absence of scale, their greater uniformity of stock thickness, and their often better formability. The surface finish of cold-rolled material is superior, and its use is thus called for whenever dictated by the appearance specifications of the finished part. For thicknesses of 1.5 mm (0.060 in or 16 gauge) or less,

TABLE 2.2.5 Cold-Finished-Steel-Bar Formulations with Particularly Good Machinability

Common designation	UNS designation	Machinability rating, %
12L14	G12144	158
1215	G12150	137
1213	G12130	137
1212	G12120	100
1119	G11190	100
1211	G12110	94
1116	G11160	92
1117	G11170	89
1118	G11180	85
1144	G11440	85 (annealed)
1141	G11410	81 (annealed)
1108	G11080	80
1109	G11090	80

cold-rolled material is also apt to be more readily available and less costly than hot-rolled material.

When cold-rolled steel is specified for a stamping, the grade of steel required depends on the severity of the forming operation. Deep-drawn parts may require aluminum-killed (special-killed) or drawing-quality steel, the former having the greater formability. Lower carbon content, down to that of 1008 (maximum C, 0.10 percent) also usually provides better formability.

Slightly dulled surfaces are actually preferred over highly smooth surfaces if the part is produced by deep drawing or if the part is to be porcelain-enameled. The dull surface is produced by using etched rolls for the final rolling operation. The dull surface retains lubricant during drawing and gives better results.

Welding

Cold-finished steel is fully weldable, especially in the low-carbon, low-alloy grades. However, designers should bear in mind that distortion is inherent in arc-welded assemblies and that it may not be possible to take advantage of the superior surface finish and dimensional accuracy of cold-finished steel. Hence hot-rolled steel may be more appropriate and more economical.

For resistance weldments, cold-rolled material is more suitable, particularly since resistance-welded assemblies employ sheet metal that may require a smooth surface finish. Cold-rolled steel, by virtue of its freedom from scale, also resistance-welds more easily.

The preferred cold-finished steels for arc and resistance welding are the plain-carbon steels with a carbon content below 0.35 percent.

Brazing

Brazing is best accomplished with steels of lower carbon and alloy content. Ideal materials have carbon in the range of 0.13 to 0.20 percent and manganese in the range of 0.30 to 0.60 percent.

Heat Treating

Cold-finished steels are heat-treatable and are widely used in applications for which surface or through hardness is required. The following are recommended grades for various heat-treating processes:

Carburizing: 8620 (G86200), 4620 (G46200), 1020 (G10200), 1024, 9310

Nitriding: 4140 (G41400), 4340 (G43400), 8640 (G86400)

Flame hardening: Medium-carbon steels (0.35 to 0.70% carbon) are generally employed.

Cyaniding/carbonitiriding: 1020 (G10200), 1022 (G10220), 1010 (G10100)

Induction hardening: 1045 (G10450), 1038 (G10380), 1144 (G11440)

Other through-hardening: 4140 (G41400), 4130 (G41300)

Plating

Cold-finished bars of all formulations are suitable for plating. However, the normal cold-drawn finish, though smooth in appearance, is not usually plated without additional polishing. Such polishing is required because minor surface imperfections, though greatly smoothed and reduced by cold drawing, are highlighted by the plated finish. Cold-finished bars, which are turned or ground and polished as part of the cold-finishing operation, do not present this problem and provide a pleasing appearance after plating.

Painting

The painting of cold-finished steel is practical. Sometimes cold-finished material is specified for painted parts solely to avoid extensive cleaning prior to painting.

AVAILABLE SHAPES AND SIZES

Table 2.2.6 summarizes the common shapes, sizes, and lengths of cold-finished steel available from warehouse distributors in the United States. Special rolled and drawn

TABLE 2.2.6 Common Commercially Available Sizes of Cold-Finished Steel

	Minimum thickness or diameter	Maximum thickness or diameter	Size increments	Normal length
Round bars	0.125 in	12 in	32ds to 1 in, 16ths to 3 in, 8ths to 6 in	10–12 or 20–24 ft
Square bars	0.125 in	6 in	16ths to 1½ in, 8ths to 2¾ in	10–12 ft
Hexagonal bars	0.125 in	4 in	16ths to 2 in, 4ths to 4 in	10–12 ft
Flats	0.125 in thick × 0.25 in wide	3 in thick × 8 in wide	16ths to 11/16 in, 8ths to 1¾ in, 4ths to 3 in	10–12 ft
Sheet	0.015 in (28 gauge) × 13 in wide	0.179 in (7 gauge) × 48 in wide	Standard gauges	48, 60, 72, 96, 120 in or coil
Tubing, round seamless	0.125 in OD × 0.049 in wall	12 in OD × 0.5 in wall	32ds to 7/16-in diameter, 16ths to 2⅝-in diameter, 8ths to 5¾-in diameter, 4ths to 12-in diameter	17–24 ft, random lengths
Tubing, round welded	0.250 in OD × 0.035 in wall	6 in OD × 0.250 in wall	16ths to 1⅝-in diameter, 8ths to 3¾-in diameter, 4ths to 5-in diameter	17–24 ft, random lengths

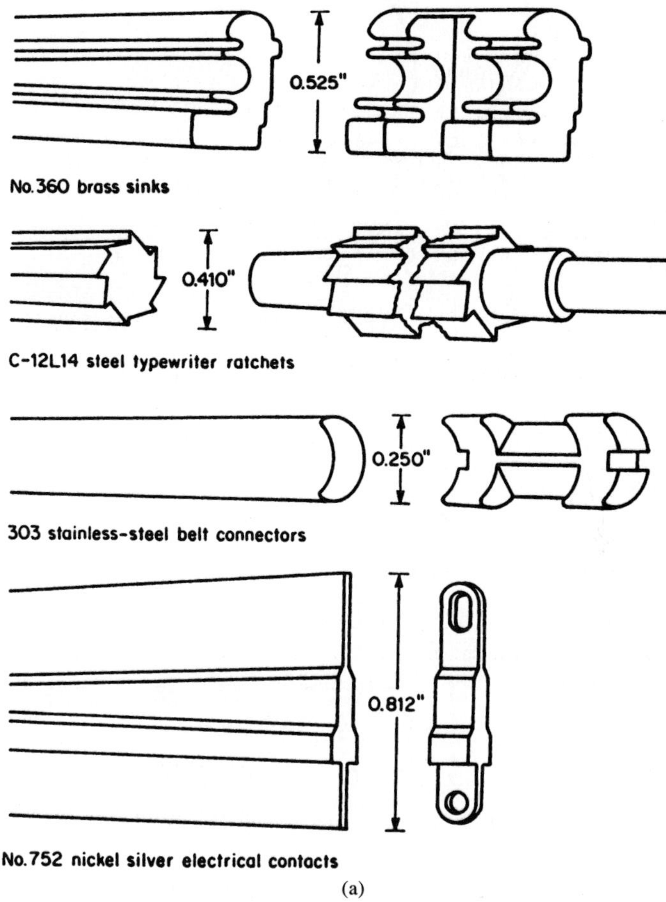

FIGURE 2.2.5 Typical special shapes which can be rolled or drawn from cold-finished wire or bar material. (*Courtesy Rathbone Corporation.*)

sizes and shapes are available from mills when necessary. Prices, of course, are higher for special orders. Lead time is also greater, and there is a minimum quantity or a minimum charge.

A number of firms specialize in shaping cold-finished wire or bars into special cross sections. Intricate and nonsymmetrical shapes can be produced as illustrated in Fig. 2.2.5. Some firms limit their output to sizes that can be produced from 13-mm ($\frac{1}{2}$-in) or smaller stock with width limits of 3 to 25 mm (0.125 to 1.00 in) and thickness limits of 0.5 to 7 mm (0.02 to 0.28 in). Other firms produce somewhat larger sections. Minimum order quantities from these fabricators are on the order of 225 kg (500 lb). Finished stock is supplied in coils or in standard lengths of 12 ft. Sections with holes (excluding tubing) are not produced.

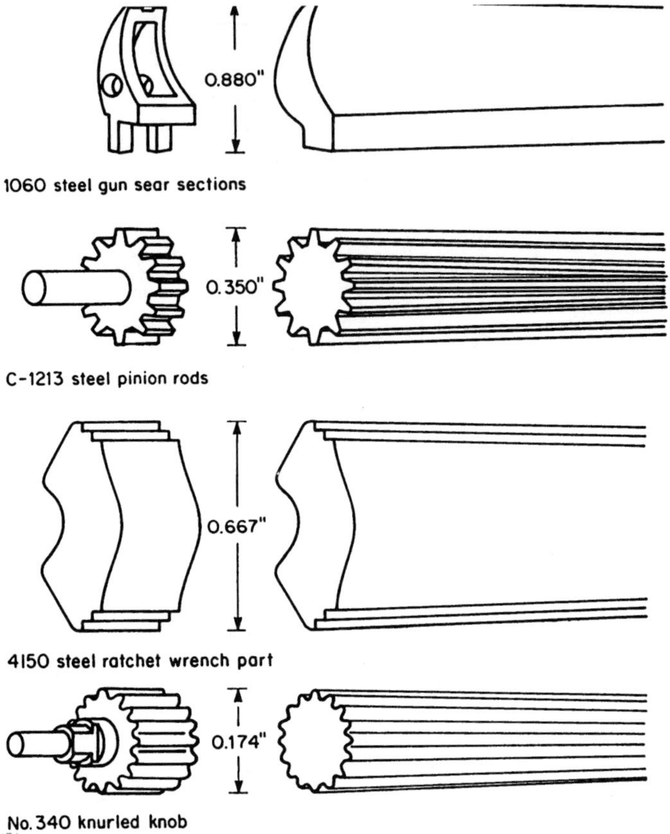

(b)

FIGURE 2.2.5 (*Continued*) Typical special shapes which can be rolled or drawn from cold-finished wire or bar material. (*Courtesy Rathbone Corporation.*)

DESIGN RECOMMENDATIONS

The basic approach of the design engineer in utilizing cold-finished steel most effectively is to specify a size and shape of material that minimize subsequent machining. This procedure involves the use of as-drawn or as-rolled surfaces and dimensions as much as possible and having the finished component conform as much as possible to the cross-sectional shape of the cold-finished-steel material.

Other primary rules are the following:

1. Use the simplest cross-sectional shape possible consistent with the function of the part. Avoid holes, grooves, etc. With special shapes, undercuts and reentrant angles can be produced, but they are more expensive than simpler shapes without these features. The latter, therefore, are preferable if functionally suitable. (See. Fig. 2.2.6.)

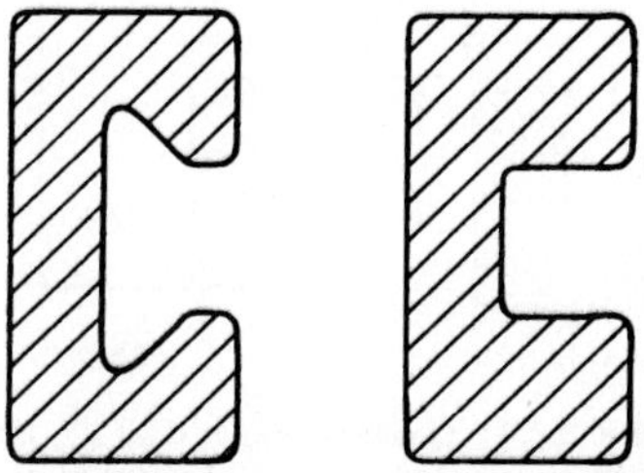

FIGURE 2.2.6 Avoid undercuts and reentrant angles in the cross section of special cold-finished-steel stock if possible, since these are more costly to produce.

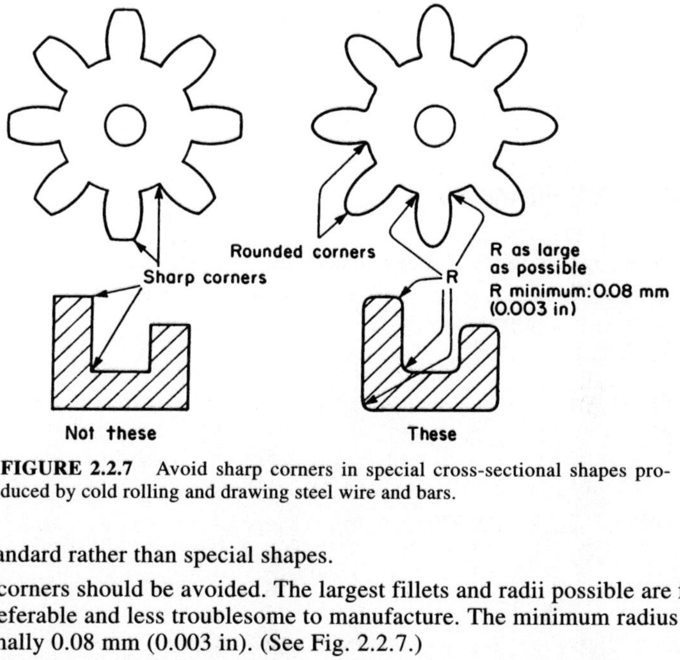

FIGURE 2.2.7 Avoid sharp corners in special cross-sectional shapes produced by cold rolling and drawing steel wire and bars.

2. Use standard rather than special shapes.
3. Sharp corners should be avoided. The largest fillets and radii possible are functionally preferable and less troublesome to manufacture. The minimum radius of fillets is normally 0.08 mm (0.003 in). (See Fig. 2.2.7.)
4. Grooves deeper than 1.5 times width are not feasible unless generous bottom radii are specified. (See Fig. 2.2.8.)
5. Abrupt changes in section thickness should be avoided since they introduce local stress concentrations.
6. Specify the most easily formed materials since these provide the lowest cost and the greatest precision.
7. With tubular sections, welded rather than seamless types are more economical, especially if drawing after welding is done without a mandrel. This type of tubing

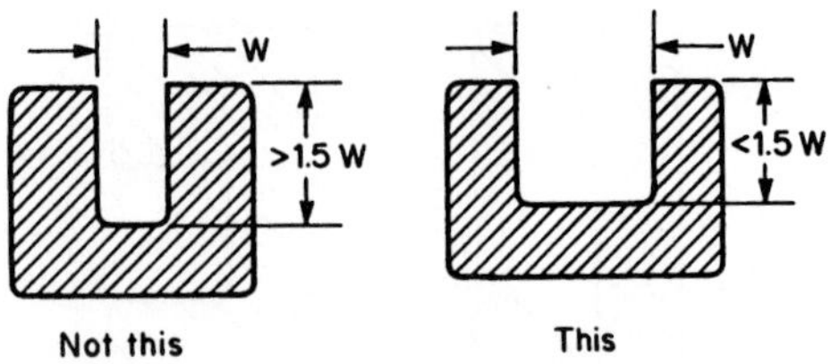

FIGURE 2.2.8 Grooves deeper than $1\frac{1}{2}$ times width are not feasible unless generous bottom radii are specified.

possesses internal weld fillets and internal shape variations, but for many applications it is fully usable.

The following recommendations apply to cold-drawn pinion-gear sections:

1. Top and bottom lands of the gear teeth should be rounded. (See Fig. 2.2.7.)
2. Avoid excessively thin and deep gear teeth, especially in difficult-to-draw materials.
3. Avoid undercuts in gear-tooth forms, if possible.

STANDARD TOLERANCES

Normal AISI tolerances for cold-finished-steel mill products are summarized in Tables 2.2.7 through 2.2.10.

TABLE 2.2.7 Normal Tolerances for Cold-Drawn Carbon Steel Bars: Rounds, Hexagons, Squares, and Flats

All tolerances are in inches and are minus.

Size, in	Maximum carbon 0.28% or less	Maximum carbon 0.28–0.55%	Maximum carbon to 55%, stress-relieved annealed	Maximum carbon over 0.55% and all grades quenched and tempered or normalized and tempered before cold finishing
		Rounds		
To 1.5 in inclusive	0.002	0.003	0.004	0.005
Over 1.5 to 2.5 in inclusive	0.003	0.004	0.005	0.006
Over 2.5 to 4 in inclusive	0.004	0.005	0.006	0.007
		Hexagons		
To 0.75 in inclusive	0.002	0.003	0.004	0.006
Over 0.75 to 1.5 in inclusive	0.003	0.004	0.005	0.007
Over 1.5 to 2.5 in inclusive	0.004	0.005	0.006	0.008
Over 2.5 to 3.125 in inclusive	0.005	0.006	0.007	0.009
		Squares		
To 0.75 in inclusive	0.002	0.004	0.005	0.007
Over 0.75 to 1.5 in inclusive	0.003	0.005	0.006	0.008
Over 1.5 to 2.5 in inclusive	0.004	0.006	0.007	0.009
Over 2.5 to 4.0 in inclusive	0.006	0.008	0.009	0.011

TABLE 2.2.7 Normal Tolerances for Cold-Drawn Carbon Steel Bars: Rounds, Hexagons, Squares, and Flats (*Continued*)

All tolerances are in inches and are minus.

Size, in	Maximum carbon 0.28% or less	Maximum carbon 0.28–0.55%	Maximum carbon to 55%, stress-relieved annealed	Maximum carbon over 0.55% and all grades quenched and tempered or normalized and tempered before cold finishing
		Flats		
Width:*				
To 0.75 in inclusive	0.003	0.004	0.006	0.008
Over 0.75 to 1.5 in inclusive	0.004	0.005	0.008	0.010
Over 1.5 to 3.0 in inclusive	0.005	0.006	0.010	0.012
Over 3.0 to 4.0 in inclusive	0.006	0.008	0.011	0.016
Over 4.0 to 6.0 in inclusive	0.008	0.010	0.012	0.020
Over 6.0 in inclusive	0.013	0.015		

*Width dimension governs tolerance for both width and thickness. The same tolerance applies to thickness as to width even though thickness is less than width.

Source: American Iron and Steel Institute. This table includes tolerances for bars that have been annealed, spheroidize-annealed, normalized, normalized and tempered or quenched, and tempered after cold finishing.

TABLE 2.2.8 Normal Tolerances for Cold-Rolled Sheets over 12 in in Width

Plus or minus tolerance for thickness, in

	Specified thickness, in							
Specified width, in	To 0.0112	0.0113 to 0.0194	0.0195 to 0.388	0.0389 to 0.0567	0.0568 to 0.0709	0.0710 to 0.0971	0.0972 to 0.1419	0.1420 and thicker
Over 12 to 24	0.0015	0.002	0.003	0.004	0.005	0.005	0.006	0.006
Over 24 to 40	0.0015	0.002	0.003	0.004	0.005	0.006	0.006	0.007
Over 40 to 60	—	0.002	0.003	0.004	0.005	0.006	0.007	0.008
Over 60 to 80	—	—	0.004	0.005	0.006	0.007	0.007	0.008
Over 80	—	—	—	0.006	0.006	0.007	0.008	0.010

Source: American Iron and Steel Institute.

TABLE 2.2.9 Normal Surface Finish for Cold-Finished Steel

Cold-rolled and cold-drawn surfaces	1.6 μm (63 μin)
Turned and polished bars	0.4 μm (15 μin)
Turned, ground, and polished bars	0.2 μm (8 μin)

Source: American Iron and Steel Institute.

TABLE 2.2.10 Normal Tolerances for Cold-Finished Steel: Special Bar and Wire Shapes

Cross-sectional dimensions	
Normal tolerance	±0.05 mm (0.002 in)
Tightest tolerance, normal for thickness	±0.013 mm (0.0005 in)
Angular tolerance	±0°30'
Surface finish	1.6 μm (63 μin)
Straightness	
Normal tolerance	0.7 mm maximum bow/m (0.010 in maximum bow/ft)
Tightest tolerance	0.35 mm maximum bow/m (0.005 in maximum bow/ft)
Twist	0°10'/m (0°30'/ft)

Source: American Iron and Steel Institute. The tightest tolerances shown entail a surcharge from suppliers.

PART 3

STAINLESS STEEL

Originally prepared by
Calvin J. Cooley

American Iron and Steel Institute
Washington, D.C.

Updated for this edition by
Ivan A. Franson

Allegheny Ludlum Corporation
Brackenridge, Pennsylvania

DEFINITION

Stainless steels are alloys of steel that possess unusual resistance to attack by corrosive media. They are produced to cover a wide range of mechanical and physical properties and corrosion resistance for particular applications at atmospheric, elevated, and cryogenic temperatures.

These steels may be subdivided into the following groups: (1) Iron-chromium alloys, which are hardenable by heat treatment (martensitic alloys such as type 410 [UNS number S41000]); they are ferromagnetic; (2) iron-chromium alloys, which cannot be hardened significantly by heat treatment or cold-working (ferritic alloys); all the stainless steels in this group, of which type 430 (S43000) is the general-purpose alloy, are ferromagnetic; (3) iron-chromium-nickel (300 series) and iron-chromium-nickel-manganese (200 series) alloys hardenable only by cold-working (austenitic group, of which type 304 [S30400] is the general-purpose alloy); these stainless steels are nonmagnetic in the annealed condition; (4) the precipitation-hardening stainless alloys consist of controlled compositions of iron, chromium, and nickel, together with other elements that enable hardening and strengthening by solution treating and aging; common precipitation-hardening stainless alloys are types UNS S13800, S15500, S17400, and S17700; and (5) the duplex alloys, which are two-phase austenitic and ferrite in roughly equal proportions; the most common duplex alloy is the 2205 material (UNS S31803). In addition, proprietary alloys belonging to each of the five alloy groups are available to the end user.

More recently, producers have developed a number of proprietary alloys for specialty and harsh-environment applications. They are included in ASTM specifications. Examples include the superferritic S44735 alloy and the superaustenitic 904L (NO8904) and AL-6XN (N08367) alloys.

APPLICATIONS

Common applications of stainless steels include aircraft, railway cars, trucks, trailers, food-processing equipment, sinks, stoves, cooking utensils, cutlery, flatware, architectural metalwork, laundry equipment, chemical-processing equipment, jet engine parts, surgical tools, furnace and boiler components, oil-burner parts, petroleum-processing equipment, dairy equipment, heat-treating equipment, and automotive exhaust systems, fuel lines, and trim, and power plant heat exchangers and flue gas scrubbers.

MILL PROCESSES

Stainless steels are normally produced by electric-arc melting followed by argon-oxygen decarburization (AOD) refining and continuous casting. A number of sheet finishes are available through pickling, bright annealing, and polishing or buffing processes.

GRADES FOR FURTHER PROCESSING

Stainless steels are generally selected, first, on the basis of corrosion resistance and, second, on the basis of strength or other mechanical properties. In some applications, end-use requirements may be so restrictive as to preclude a third-level consideration: fabricability. The production engineer, however, should keep in mind that many of the stainless steels, especially those in the 200 and 300 series, have fabrication characteristics different from those of carbon steels. For example, more force is required in the tools used to bend, draw, and cut, and slower cutting speeds and different tool geometries are required for machining.

Table 2.2.11 shows relative fabrication characteristics of 200 and 300 series types. Table 2.2.12 shows relative fabrication characteristics of 400 series types. The newer duplex superferritic and superaustenitic alloys generally require greater attention during fabrication. The producers of these materials should be contacted with regard to specific fabrication requirements.

Cold Forming

The bending characteristics of annealed 200 and 300 series stainless steels are considered excellent, in that many types will withstand a free bend of 180° with a radius equal to one-half the material thickness. In a controlled V block, the bend-angle limit is 135°. As the hardness of the stainless steel increases, bending becomes more restrictive. Bend characteristics of the 400 series types also are good. However, these types exhibit less ductility than 300 series types, and the minimum-bend radius is equal to the material thickness.

The 200 and 300 series stainless steels can be stretched more than carbon steel and, accordingly, have excellent stretch-forming characteristics. Types 301 or 201 are preferred for stretch-formed parts because they develop high strength during cold work-

TABLE 2.2.11 Relative Forming Characteristics of 200 and 300 Series (Not Hardenable by Heat Treatment)*

Forming method	201 S20100	301 S30100	302 S30200	303 303 Se S30300 S30323	304 S30400	304L S30403	305 S30500	309 S30900	310 S31000	310S S31008	316 S31600	316L S31603	321 S32100	347 S34700
Blanking	B	B	B	B	B	B	B	B	B	B	B	B	B	B
Brake forming	B	B	A	D	A	A	A	A	A	A	A	A	A	A
Coining	B,C	B,C	B	C,D	B	B	B	B	B	B	B	B	B	B
Deep drawing	A,B	A,B	A	D	A	A	A	B	B	B	B	B	B	B
Embossing	B,C	B,C	B	C	B	B	B	B	B	B	B	B	B	B
Forging, hot	B	B	B	B,C	B	B	B	B,C	B,C	B,C	B	B	B	B
Hardening by cold work, typical tensile strength annealed (1000 lbf/in^2)	115	112	92	90	88	83	85	90	95	90	88	85	90	88
25% reduction (1000 lbf/in^2)	170	175	142	145	138	137	125	130	126	125	130	125	136	140
50% reduction (1000 lbf/in^2)	224	230	180	195	178	176	155	169	165	166	169	160	167	175
Heading, cold	C,D	C,D	C	D,C	C	C	B	C	C	C	C	C	C	C
Roll forming	B	B	A	D	A	A	A	B	A	A	A	A	B	B
Sawing	C	C	C	B	C	C	C	C	C	C	C	C	C	C
Spinning	C,D	C,D	B,C	D	B	B	B	C	B	B	B	B	B,C	B,C

*A = excellent; B = good; C = fair; D = not generally recommended.

TABLE 2.2.12 Relative Forming Characteristics of 400 Series*

	Ferritic (not hardenable by heat treatment)						Martensitic (hardenable by heat treatment)							
	430	430F 430FSe	405	409	442	446	410	414	416 416Se	420	420F	440A	440B	440C
Forming method	S43000	S43020 S43023	S40500	S40900	S44200	S44600	S41000	S41400	S41600 S41623	S42000	S42020	S44002	S44003	S44004
Blanking	A	B	A	A	A	A	A	A	B	B	B	B,C		
Brake forming	A†	B,C†	A†	A†	A†	A†	A†	A†	C†	C†	C†	C†		
Coining	A	C,D	A	A	B	B	A	B	D	C,D	C,D	D	D	D
Deep drawing	A,B	D	A	A	B	B,C	A	B	D	C,D	C,D	C,D		
Embossing	A	C	A	A	B	B	A	C	C	C	C	C	D	D
Forging, hot	B	C	B	B	B,C	B,C	B	B	C	B	B	B	B	B
Hardening by cold work, typ. tensile str. (1000 lbf/in^2)														
Annealed	73	75	60	61		84	90		70					
25% reduction	96	95	79			110	120		90					
50% reduction	115	110	96			130	130		105					
Hardening by heat treatment	No	No	No	No	No	No	Yes	Yes	Yes	Yes	Yes	Yes	Yes	Yes
Heading, cold	A	D	A	A	B	C	A	D	D	C	C	C	C,D	C,D
Roll forming	A	D	A	A	A	B	A	C	D	C,D	C,D	C,D		
Sawing	B	A,B	B	B	B,C	B,C	B	C	B	C	C	C	C	C
Spinning	A	D	A	A	B,C	C	A	C	D	D	D	D	D	D

*A = excellent, B = good, C = fair, D = not generally recommended.

†Severe, sharp bends should be avoided

ing while retaining excellent ductility. Because of their work-hardening rate and relatively low ductility, the 400 series types cannot be stretched severely without thinning and fracturing. The low work-hardening austenitic grades such as type 305 (S30500) offer excellent deep drawability. The ferritic 400 series can be processed to provide deep drawability.

For cold-heading, types 304L (S30430), 305 (S30500), and 384 (S38400) are preferred within the 300 series because they work-harden less during cold working than other 300 series types. Straight chromium types such as 430 (S43000), 410 (S41000), and 409 (S40900) are readily cold-headed.

Hot Forming

Stainless steels are readily formed by hot operations such as rolling, extrusion, and forging.

Machining

The machining characteristics of stainless steels are substantially different from those of carbon or alloy steels and other metals. In varying degree, most stainless types are tough and rather gummy, and they tend to seize and gall.

The 400 series steels are the easiest to machine, although they produce a stringy chip that can slow productivity. The 200 and 300 series have the most difficult machining characteristics, primarily because of their gumminess and secondarily because of their propensity to work-harden at a very rapid rate.

There are three ways for production engineers to work around these conditions and achieve adequate machinability: (1) They can specify that the bar for machining be in a slightly hardened condition, which will result in a slight improvement in machining. (2) They can order a stainless steel type with an analysis that is "better suited for machining." In such cases they would receive the same type specified by the designer, in that the chemical composition, while altered slightly for better machining, would be within the specified composition range. (3) They can order a free-machining stainless steel.

In some of the stainless steels the sulfur, selenium, lead, copper, and/or phosphorus contents are varied to alter machining characteristics. These alloying elements reduce the friction between the workpiece and the tool, thereby minimizing the tendency of the chip to weld to the tool. Also, sulfur and selenium form inclusions that reduce the friction forces and transverse ductility of the chips, causing them to break off more readily. The improvement in machinability in the free-machining stainless steels, namely, types 303 (S30300), 303 Se (S30323), 430F (S43020), 430F Se (S43023), 416 (S41600), 416 Se (S41623), and 420F (S42020), is clearly evident in Fig. 2.2.9.

While the stainless steel types within each group in Fig. 2.2.9 have similar corrosion-resistance properties, they are not identical. The free-machining types have slightly less corrosion resistance than their counterparts. The selenium-bearing grades, types 303 Se (S30323), 430F Se (S43023), and 416 Se (S41623), are used when better surface finishes are required or when cold working, such as staking, swaging, spinning, or severe thread rolling, may be involved in addition to machining.

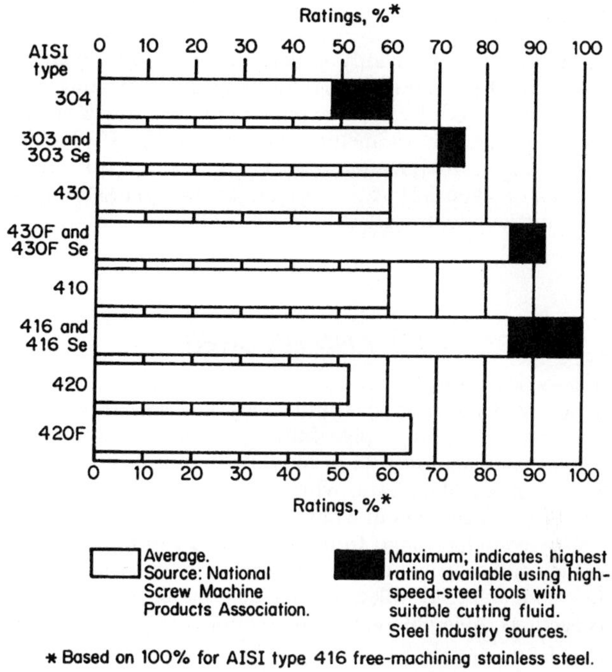

FIGURE 2.2.9 Comparative machinability of frequently used stainless steels and their free-machining counterparts. (*Courtesy American Iron and Steel Institute.*)

Welding

Weld-rod selection is important because the filler metal should have a composition equivalent to or more highly alloyed than the base metal; otherwise, the weld may not have the same corrosion-resistance properties. (The American Welding Society issues recommended practices.)

Chromium carbides can precipitate in the grain boundaries when stainless steels are heated and cooled (as in welding) through a temperature range of about 430 to 900°C (800 to 1650°F). This lessens corrosion resistance, and if the steel is used subsequently in corrosive environments, intergranular corrosion may occur. To prevent such an occurrence in welded fabrications, low-carbon or extra-low-carbon stainless steels are often used. These include types 304L (S30403), 316L (S31603), and 317L (S31703). Also, some stainless steels are stabilized with niobium or titanium to help prevent chromium carbide precipitation when service is in the 430 to 900°C (800 to 1650°F) range or when used in the as-welded condition. Among these are austenitic types 321 (S32100) and 347 (S34700) and most of the ferritic grades such as types 409 (S40900) and 439 (S43900).

Soldering and Brazing

Stainless steels can be soldered readily so long as proper techniques are employed. With some types, special precautions must be observed on heating and cooling. The 300 series alloys require special care to avoid chromium carbide precipitation during brazing, since the filler metal has a melting point above 430°C (800°F). If necessary, extra-low-carbon types or niobium-stabilized types can be used. Titanium-stabilized stainless types such as type 321 (S32100) or ferritic types 409 (S40900) or 439 (S43900) are more difficult to braze.

AVAILABLE SHAPES, SIZES, AND FINISHES

Stainless steels are available from steel producers and steel service centers as sheet, strip, plate, bar, wire, tubing, pipe, pipe fittings, fasteners, and other product forms. Material under $\frac{3}{8}$ in thick and over 24 in in width is classified as sheet, while that of the same thickness but less than 24 in wide is classified as strip. Plate stock is $\frac{3}{8}$ in thick and over and more than 10 in in width.

Surface finishes available range from as hot-rolled to highly reflective polished surfaces and include dull or bright cold-rolled and satin finishes.

Standard hot-rolled and cold-rolled bars include round, flat, square, hexagonal, and octagonal cross sections. Available wire cross sections are similar in shape.

DESIGN RECOMMENDATIONS

Stainless steels must be utilized efficiently if they are to be competitive with lower-cost metals. In some instances, it is possible to compensate for the higher cost per pound of stainless steel by utilizing its high strength and corrosion resistance to reduce the thickness of sections. For maximum economy, the following guidelines should be considered:

- Use the least expensive stainless steel and product form suitable for the application.
- Use rolled finishes.
- Use the thinnest gauge required.
- Use a thinner gauge with a textured pattern.
- Use a still thinner gauge continuously backed.
- Use standard roll-formed sections whenever possible.
- Use simple sections for economy of forming.
- Use concealed welds whenever possible to eliminate refinishing.
- Use stainless steel types that are especially suited to manufacturing processes, such as the free-machining grades.

DIMENSIONAL FACTORS AND TOLERANCES

Tolerances for the various mill forms of stainless steels are given in the *Steel Products Manual:* "Stainless and Heat-Resistant Steels" (Iron and Steel Society of AIME) and in American Society for Testing and Materials (ASTM) specifications as follows: ASTM A484 and A484M, *Bars, Billets, Forgings*; ASTM A480 and A480M, *Plates, Sheet, and Strip*; ASTM A555 and A555M, *Wire and Wire Rods*; ASTM A511, *Mechanical Tubing*; and ASTM A554, *Welded Mechanical Tubing.* Tolerances in these specifications are comparable with those applicable to carbon and alloy steels (see Parts 1 and 2).

BIBLIOGRAPHY

J. R. Davis (ed.), *Stainless Steel: ASM Specialty Handbook,* ASM International, Metals Park, Ohio, 1994.

D. Peckner and I. M. Bernstein (eds.), *Handbook of Stainless Steels,* McGraw-Hill, New York, 1977.

CHAPTER 2.3
NONFERROUS METALS*

PART 1

ALUMINUM

Originally prepared by
William B. McMullin

Reynolds Metals Company
Richmond, Virginia

Updated by
Michael H. Skillingberg

Reynolds Metals Company
Richmond, Virginia

DEFINITION

Aluminum is an abundant, slightly blue-white metal of high strength-to-weight ratio. Unalloyed, its melting point is 658.7°C (1218°F) and its specific gravity is 2.7, roughly 35 percent that of iron and 30 percent that of copper. It is nonmagnetic and has high thermal and electrical conductivity. Aluminum is resistant to corrosion and chemical attack in many environments at normal temperatures largely due to the very thin (0.025 μm) film of aluminum oxide that quickly forms on surfaces exposed to the atmosphere. Aluminum is highly malleable in the pure form as well as in many alloyed versions. Tensile strength ranges from 90 MPa (13,000 lbf/in^2) in the pure, annealed state to greater than 690 MPa (100,000 lbf/in^2) when highly alloyed and heat-treated.

*The data supplied in this chapter are believed to be reliable; however, no warrant is given, either expressed or implied, for the accuracy or completeness of this information. Neither the authors nor the publishers assume any responsibility or liability for the use of the information contained herein.

TYPICAL APPLICATIONS

Aluminum is used when some or all of the following properties are required: strength with light weight, corrosion resistance, nontoxicity, electrical conductivity, and easy machinability or formability. Aircraft structures and internal-combustion engine pistons are two notable applications for which light weight and high strength are important. Cooking utensils and beverage containers are examples where formability and nontoxicity are essential. Other examples, for common aluminum alloys, are shown in Table 2.3.1.

Corrosion resistance of higher-strength alloys can be improved by cladding with purer alloys or applying a chemical or electrochemical coating.

MILL PROCESSES

Mill forms of aluminum are made by rolling, extruding, forging, and cold drawing. Structural shapes such as I-beams, channels, angles, tees, and zees are generally made by the extrusion process (see Chap. 3.1). Sheet, plate, and foil are commonly made by a combination of hot and cold rolling large rectangular cast ingots to the desired thickness. Light-gauge plate, sheet, and foil also can be produced from continuously cast slab or coil stock depending on the exact product requirements. Rod and bar, other than extruded, are also made by rolling, usually hot. They also can be cold-finished to final dimensions to improve surface quality.

Wire is cold-drawn from redraw rod to its final form. Some tube products are also brought to their final dimensions by cold drawing from extruded tube stock. A number of hot- and cold-forging techniques are used to make a wide range of preforms and near-net-shape final products.

GRADES FOR FURTHER PROCESSING

Aluminum alloys suitable for processing with all major metal-working processes are available. For machining and forming operations, there are available grades of aluminum with processibility superior to that of steel or other materials. For other operations, techniques differ with aluminum but are not necessarily more difficult if carried out properly. Table 2.3.2 lists the preferred aluminum alloys, from a processing standpoint, for various manufacturing processes. Table 2.3.1 indicates the machinability, formability, and weldability of common aluminum alloys.

Machining

In general, machining speeds can be faster and require considerably lower forces than those for other materials of similar tensile strengths. Generally, the stronger, harder alloys are easier to machine than soft alloys that tend to become gummy and have longer, bigger chips.

TABLE 2.3.1 Comparative Characteristics and Applications of Aluminum Alloys

Alloy	UNS designation	Resistance to corrosion		Workability, cold	Machinability	Brazability	Weldability			Applications
		General¶	Stress corrosion cracking*				Gas	Spot and arc	Resistance seam	
1100	A91100	A	A	A,C	E,D	A	A	A	B,A	Sheet-metal work, spun hollowware, fin stock
1350	A91350	A	A	A,B	E,D	A	A	A	B	Electrical conductors
2008† 2010†	A92008 A92010	B,C	B	B	C	D	B	B	A,B	Auto body panel sheet
2011	A92011	D	D	C	A	D	D	D	D	Screw-machine products
2014	A92014	D	C	C,D	D	D	D	D	B	Truck frames, aircraft structures
2017	A92017	D	C	C	B	D	D	B	B	Screw-machine products, fittings
2018	A92018				B	D	D	C	B	Aircraft engine cylinders, heads, and pistons
2024 2124	A92024 A92124	D	C,D	C,D	B	D	D	C	B	Truck wheels, aircraft structures, screw-machine products
2036	A92036	B,C	B	B	C	D	D	B	B	Auto body panel sheet
2090†	A92090	D	C,D	D	B	D	B	B		Aircraft and aerospace structures
2095† 2195†	A92095 A92195	D	B,C	C	B	D	B	B		Aircraft and aerospace structures
2097† 2197†	A92097 A92197	D	C	C	B	D				Aircraft structures
2219	A92219	C	B	C	B	D	A	A	A	Aerospace structures, high-temperature structural uses
2618	A92618	D	C		B	D	D	C	B	Aircraft engines
3003	A93003	A	A	A,C	D,E	A	A	A	A	Cooking utensils, chemical equipment, pressure vessels, sheet-metal work, builders' hardware

TABLE 2.3.1 Comparative Characteristics and Applications of Aluminum Alloys (*Continued*)

Alloy	UNS designation	Resistance to corrosion		Workability, cold	Machinability	Brazability	Weldability			Applications
		General¶	Stress corrosion cracking*				Gas	Spot and arc	Resistance seam	
3004	A93004	A	A	A,C	D,C	B	B	A	A,B	Sheet-metal work, storage tanks
3105	A93105	A	A	B,C	D,E	A	A	A	A	Sheet-metal work, siding, rain-carrying goods
5005	A95005	A	A	A,C	D,E	B	A	A	A,B	Appliances, utensils, architectural work, electrical conductors
5050	A95050	A	A	A,C	C,E	B	A	A	A,B	Builders' hardware, refrigerator trim, coiled tubes
5052	A95052	A	A	A,C	C,D	C	A	A	A,B	Sheet-metal work, hydraulic tube, appliances
5056	A95056	A,B‡	B,D‡	A,D	D,B	D	C	A	A	Cable sheathing, rivets, screen wire, zippers
5083 5383†	A95083 A95083	A‡	B‡	B,C	C,D	D	C	A	A,B	Unfired welded pressure ves sels, marine service, automotive parts, aircraft
5086	A95086	A,B‡	B‡	B,C	C,D	D	C	A	A,B	Cryogenics, TV towers, drilling rigs, missile components, transportation equipment
5154	A95154	A‡	A‡	A,C	C,D	D	C	A	A,B	Welded structures, storage tanks, pressure vessels, saltwater service
5182†	A95182	A‡	B‡	A,B	C,D	D	A	A	A,B	Auto body panel sheet

5454 5754†	A95454 A95754	A	A	A,B	C,D	D	C	A	A,B	Welded structures, pressure vessels, marine service, automotive wheels and structural sheet
5456	A95456	A‡	B‡	B,C	C,D	D	C	A	A,B	High-strength welded structures, storage tanks, pressure vessels, marine applications
5657	A95657	A	A	A,C	D	B	A	A	A	Anodized automotive and appliance trim
6005 6105†	A96005 A96105	B		B,C	C,D	A	A	A	A	Structural extrusions
6022†	A96022	B	B	A		A	A	A	A	Automotive body panels
6053	A96053	A	A		C,E	A	A	A	A,B	Wire and rod for rivets
6061	A96061	B	A,B	A,C	C,D	A	A	A	A,B	Heavy-duty structures requiring good corrosion resistance, truck and marine, railroad cars, furniture, pipelines
6063	A96063	A	A	B,C	C,D	A	A	A	A	Pipe railing, furniture, architectural extrusions
6066	A96066	C	A,B	B,C	B,D	D	D	B	B	Forgings and extrusions for welded structures
6070	A6070	B	B	B,C	C	D	A	A	A	Heavy-duty welded structures, pipelines
6111†	A96111	B	B	B		A	A	A	A	Automotive body panels
6262	A96262	B	A	C,D	B	B	B	B	A	Screw-machine products
6351	A96351	A		C	C	C	B	B	A,B	Extruded shapes, structurals, pipe and tube

TABLE 2.3.1 Comparative Characteristics and Applications of Aluminum Alloys (*Continued*)

Alloy	UNS designation	Resistance to corrosion		Workability, cold	Machinability	Brazability	Weldability			Applications
		General¶	Stress corrosion cracking*				Gas	Spot and arc	Resistance seam	
6463	A96463	A	A	B,C	C,D	A	A	A	A	Extruded architectural and trim sections
7021† 7029	A97021 A97029	B	B	B,C	C,D			B	B	Automotive bumpers, bumper reinforcements, structural components
7049	A97049	C	B	D	B	D	D	D	B	Aircraft forgings
7050 7150	A97050 A97150	C	B	D	B	D	D	D	B	Aircraft and other structures
7055†	A97055	C		D	B	D	D	D	B	Aircraft and other structures
7075 7175	A97075 A97175	C§	B,C	D	B	D	D	C	B	Aircraft and other structures
8017	A98017	A	A	A	D	A	A	A	A	Electrical conductors
8176 8177	A98176 A98177	A	A	A	D E	A	A	A	A	Electrical conductors
8090†	A98090	D	C	C	B		B	B	B	Aircraft and aerospace structures

Note: All data and ratings are for comparison purposes only and should not be used by themselves for design purposes. No implied or expressed warranties are given by the Aluminum Association or member companies.

*Stress corrosion cracking ratings are based on service experience and on laboratory tests of specimens exposed to the 3.5% NaCl alternate immersion test.

†These alloys are not listed in *Aluminum Standards and Data, 1993,* and ratings are based on a limited amount of data and testing.

‡This rating may be different for material held at elevated temperatures for long periods.

§In relatively thick sections, the rating would be E.

¶Ratings A through E are relative ratings in decreasing order of merit, based on exposure to NaCl solution by intermittent spraying or immersion. Alloys with A and B ratings can be used in industrial and seacoast atmospheres without protection. Alloys with C, D, and E ratings generally should be protected at least on facing surfaces.

Source: Largely from *Aluminum Standards and Data,* Aluminum Association, Inc., Washington, DC, 1993; and *T8: Design for Aluminum: A Guide for Automotive Engineers,* Aluminum Association, Inc., Washington, DC.

TABLE 2.3.2 Aluminum Grades Recommended for Processing by Various Methods

Operation	Alloy recommended	UNS designation	Temper	Forms available	Remarks
Machining, grinding	2011	A92011	T3	Bar, rod	Machinability: 300–1500
Forming	1100	A91100	0	Sheet, plate	
Deep drawing	3003	A93003	0	Tubing	
Warm extrusion	1100	A91100		Billets	Extrudability: 150
	6063	A96063			Extrudability: 100
	6061	A96061			
Forging	2014	A92014		Forging bars, rods, billets	
Cold heading	2024	A92024	T4		Most common
	1350	A91350		Rod, bar, wire	
	1100	A91100	HI4		Most headable
	6061	A96061	T6		
Impact extrusion	1100	A91100	0	Rod and other forms	
	3003	A93003	0		
Die casting	360	A03600		Ingots	Most common and most castable
	380	A03800			
	413	A04130			
Other castings	B443	A24430		Ingots	Very good castability
Arc weldments	1060	A91060			
	1100	A91100		Various forms, castings	Wrought alloys
	1350	A91350			
	B443	A24430			
Resistance weldments	1100	A91100		Sheet, plate, etc.	
	5052	A95052			
Anodizing	5357	A95357		Sheet	
	5457	A95457			
	5557	A95557			
Brazing	1000	A91000		Various forms	Al-Si brazing alloys: 4047 (A94047) 4145 (A94145) 4343 (A94343)
	3003	A93003			

Forming

Aluminum sheet and plate can be formed by any of the conventional methods currently used. Softer tempers are normally preferred. Proper alloy selection is also important because some alloys will work-harden more rapidly than others and may require annealing between operations. Springback must be accounted for in all methods used. Since aluminum has a relatively low modulus of elasticity, it must be overformed more to accommodate springback.

When forming parts from aluminum sheet or plate, high elongation by itself is only a partial index of ductility. The strength and the total spread between yield and ultimate strength also must be considered. An alloy and temper that are soft and have a wide spread will take more severe forming than other alloys.

Joining

Fusion Welding. Gas metal arc (GMA) welding and gas tungsten arc (GTA) welding are two fusion welding techniques commonly used for joining structural assemblies. However, most of the structural alloys attain their strength by heat treating and/or strain hardening, and welding may cause local annealing, which produces a zone of lower strength along the weld seam. Some structural alloys are non-heat-treatable and generally do not lose as much strength when welded as others. High-energy techniques, such as laser-beam and electron-beam welding, can be used to reduce the extent of the heat-affected zone when welding relatively thin sections. In addition, new solid-state welding techniques, such as friction stir welding, have potential for almost complete elimination of the heat-affected zone and may allow for the welding of alloys previously classified as nonweldable. In all cases, proper joint preparation is required. It is important that the surfaces to be welded are clean and free from grease and oil. It is much better and easier to weld aluminum by a gas-shielded-arc method than to weld with an open-flame torch. This method prevents the formation of hard aluminum oxide on materials before they can be properly wetted by the liquid welding components.

Resistance Welding. Aluminum also can be joined using resistance spot or resistance seam welding using either ac or dc welding equipment. It is an ideal technique for high-strength, heat-treatable alloys, since they can be welded with little or no loss of strength. Because aluminum's electrical conductivity is much higher than that of steel, the current required is higher. While it is preferable that surfaces be clean and free from grease and oil, some forming lubricants may be compatible with subsequent spot-welding operations. Again, alloy selection is important, and the location and design of the joint are also important.

Brazing and Soldering. With proper technique, aluminum also can be soldered or brazed. The main difficulty is to break up the very tenacious aluminum oxide surface coating that is always present. This is readily accomplished if the proper equipment and procedures are used. Many special aluminum sheet alloys are available for brazing applications, and most have designations specific to the mill that produces them.

Mechanical Fastening and Adhesive Bonding. These methods can be used effectively with aluminum, including cases of joining dissimilar materials or gauges of material (see Chaps. 7.1 and 7.5).

AVAILABLE SHAPES AND SIZES

Forms of aluminum available from mills include ingots for casting; plate, 6.3 to 200 mm (0.250 to 8 in) thick; sheet, 0.15 to 6.3 mm (0.006 to 1/4 in) thick; foil, 0.0063 to 0.15 mm (0.00025 to 0.006 in) thick; rods, 9.5 mm (3/8 in) or more in diameter; bars, either square, rectangular, or hexagonal in cross section with one perpendicular dimension between opposite faces of more than 9.5 mm (3/8 in) and up to 100 mm (4 in); wire (by definition less than 9.5 mm or 3/8 in in diameter or distance between opposite faces) furnished with a round, square, rectangular, hexagonal, octagonal, or flattened cross section; structural shapes such as I's, H's, T's, channels, angles and Z's; tubing from 4.8 mm (0.187 in) inside diameter with 0.9-mm (0.035-in) wall to

270 mm (10.5 in) inside diameter with 19-mm (0.750-in) wall; and special forms like tread plate, tooling plate, mold plate, tapered plate, and patterned sheet.

Special grades of aluminum are available for operations that are enhanced by certain properties. Examples are anodizing sheet, porcelain enameling sheet, and lithographic sheet.

A typical mill length for rod and bar is 3.7 m (12 ft). Other lengths are available on special order. Aluminum distributors generally stock most of the preceding structural shapes, sheets, plates, bars, rods, and tubing.

ECONOMICAL PROCUREMENT QUANTITIES

Mill standard quantities in the United States are generally considered to be orders of 7000 lb or more for sheet and plate and 200 lb or more for wire, rod, and bar. For most mill items, the base price applies to orders in the range of 40,000 lb each. Order sizes between these amounts usually will be priced for quantity extras. Orders under the minimum size should be placed with distributors, whose minimum order size is smaller and is normally expressed in monetary terms rather than weight.

DESIGN RECOMMENDATIONS

When forming aluminum, use the largest bend radii possible in order to prevent tearing of the metal at the corners. The minimum allowable radius depends on the alloy, temper, stock thickness, and method of bending. Less ductile alloys, harder tempers, and thicker sections require larger radii. Table 2.3.3 provides summary rule-of-thumb minimum-bend radii for sheet and plate.

TABLE 2.3.3 Minimum Recommended Bend Radii for 90° Cold Forming of Aluminum Sheet and Plate

Radii in terms of stock thickness T.

	Stock thickness			
Temper and cold workability	To 0.8 mm (to 1/32 in)	Over 0.8 to 3 mm (over 1/32 to 1/8 in)	Over 3 to 6 mm (over 1/8 to 1/4 in)	Over 6 to 13 mm (over 1/4 to 1/2 in)
Soft tempers				
A–B*	0	0	1 T	1½ T
C–D*	0	1 T	1½–2½ T	2–4 T
Medium tempers				
A–B*	0–1 T	1–2 T	2–3 T	3–4 T
C–D*	3–4 T	4–5 T	5–6 T	6–8 T
Hard tempers				
A–B*	1–1½ T	2–3 T	3–4 T	4–6 T
C–D*	3–6 T	4–6 T	6–8 T	8–11 T

*Cold-workability rating from Table 2.3.1.

Source: *Aluminum Standards and Data, 1993,* Aluminum Association, Inc., New York, 1993.

If aluminum components are attached to parts made from other metals, the facing surfaces should be insulated so that galvanic corrosion is prevented. Zinc chromate or zinc phosphate may be used on the aluminum, and the other metal should be primed and painted with a good paint containing aluminum pigments. Precautions also should be taken when aluminum surfaces come in contact with wood, concrete, or masonry. A heavy coat of alkali-resistant bituminous paint should be applied. However, aluminum embedded in concrete does not need painting unless corrosive additives, such as some quick-setting compounds, have been added to the concrete.

Design recommendations in chapters covering the major manufacturing processes apply generally to aluminum to the same extent as to other materials.

STANDARD STOCK TOLERANCES

Standard tolerances for various mill forms of aluminum products are presented in Tables 2.3.4 through 2.3.7. Tighter tolerances than these may be obtainable, but most manufacturers will supply them only after special inquiry and mutual agreement between buyer and seller.

TABLE 2.3.4 Standard Thickness Tolerances for Aluminum Sheet and Plate

Note: Also applicable to alloys when supplied as Alclad.

Specified thickness, in		Specified width, in							
		Up thru 39.37	Over 39.37 thru 59.06	Over 59.06 thru 78.74	Over 78.74 thru 98.43	Over 98.43 thru 118.11	Over 118.11 thru 137.80	Over 137.80 thru 157.48	Over 157.48 thru 177.17
Over	Thru	Tolerances, in plus and minus							
0.0059	0.010	0.0010	0.0015						
0.010	0.016	0.0010	0.0015						
0.016	0.025	0.0015	0.0020	0.0030	0.0035				
0.025	0.032	0.0020	0.0025	0.0035	0.0040				
0.032	0.039	0.0020	0.0030	0.0035	0.0045	0.006			
0.039	0.047	0.0025	0.0035	0.0045	0.0055	0.007	0.008		
0.047	0.063	0.0030	0.0035	0.0050	0.006	0.007	0.009		
0.063	0.079	0.0035	0.0040	0.0055	0.007	0.008	0.010		
0.079	0.098	0.0035	0.0045	0.006	0.007	0.009	0.011		
0.098	0.126	0.0045	0.0055	0.007	0.009	0.011	0.013		
0.126	0.158	0.0055	0.007	0.009	0.011	0.013	0.015		
0.158	0.197	0.007	0.009	0.011	0.013	0.015	0.018		
0.197	0.248	0.009	0.011	0.013	0.015	0.018	0.022	0.027	
0.248	0.315	0.012	0.014	0.015	0.018	0.022	0.027	0.035	0.043
0.315	0.394	0.015	0.017	0.020	0.023	0.027	0.033	0.041	0.051
0.394	0.630	0.023	0.023	0.027	0.032	0.035	0.043	0.053	0.065
0.630	0.984	0.031	0.031	0.037	0.043	0.047	0.058	0.070	0.085
0.964	1.575	0.039	0.039	0.047	0.055	0.065	0.075	0.090	0.105
1.575	2.362	0.055	0.055	0.060	0.070	0.085	0.100	0.115	
2.362	3.150	0.075	0.075	0.085	0.100	0.105	0.125		
3.150	3.937	0.100	0.100	0.115	0.125	0.130	0.160		
3.937	6.299	0.130	0.130	0.145	0.165				

Source: Condensed from *Aluminum Standards and Data,* Aluminum Association, Washington, D.C., 1993.

TABLE 2.3.5 Standard Width Tolerances of Sheared Flat Aluminum Sheet and Plate

	Specified width, in					
	Up through 6	Over 6 through 24	Over 24 through 60	Over 60 through 96	Over 96 through 132	Over 132 through 168
Specified thickness, in	Tolerance, in					
0.006–0.124	±1/16	±3/32	±1/8	±1/8	±5/32	
0.125–0.249	±3/32	± 3/32	±1/8	±5/32	±3/16	
0.250–0.499	+1/4	+5/16	+3/8	+3/8	+7/16	+1/2

Source: Condensed from *Aluminum Standards and Data,* Aluminum Association, Inc., Washington D.C., 1993.

TABLE 2.3.6 Standard Diameter Tolerances of Round Aluminum Wire and Rod

	Tolerance, in; plus and minus except as noted; allowable deviation from specified diameter			
			Rolled rod	
Specified diameter, in	Drawn wire	Cold-finished rod	Plus	Minus
Up through 0.035	0.0005			
0.036–0.064	0.001			
0.065–0.374	0.0015			
0.375–0.500		0.0015		
0.501–1.000		0.002		
1.001–1.500		0.0025		
1.501–2.000		0.004	0.006	0.006
2.001–3.000		0.006	0.008	0.008
3.001–3.499		0.008	0.012	0.012
3.500–5.000		0.012	0.031	0.016
5.001–8.000			0.062	0.031

Source: Condensed from *Aluminum Standards and Data,* Aluminum Association, Inc., Washington D.C., 1993.

TABLE 2.3.7 Standard Tolerances for Distance across Flats: Square, Hexagonal, and Octagonal Aluminum Wire and Bar

Specified distance across flats, in	Tolerance, in, plus and minus, allowable deviation from specified distance across flats: Drawn wire	Cold-finished bar	Rolled bar
Up through 0.035	0.001		
0.036–0.064	0.0015		
0.065–0.374	0.002		
0.375–0.500		0.002	
0.501–1.000		0.0025	
1.001–1.500		0.003	
1.501–2.000		0.005	0.016
2.001–3.000		0.008	0.020
3.001–4.000			0.020

Source: Condensed from *Aluminum Standards and Data,* Aluminum Association, Inc., Washington D.C., 1993.

PART 2

COPPER AND BRASS

Konrad J. A. Kundig, Ph.D.

International Copper Association
New York, New York

DEFINITION

Wrought-copper mill products consist of sheets, strips, rods, bars, drawn shapes, wires, and tubes. Copper is either their major or virtually their sole ingredient. When the "copper" designation is used, these products consist of 99.3 percent or more of metallic copper. The balance may contain a number of elements, some intentional, some as impurities, including oxygen, phosphorus, silver, and other metals.

Brass is a copper alloy with zinc as the principal alloying element with or without small quantities of other elements. *Bronze* is a term that historically has referred to copper alloys that have tin as the principal or sole alloying element. In modern usage, the term *bronze* is seldom used alone. Phosphor bronze does utilize tin as the major alloying element, and aluminum bronze, silicon bronze, tin bronze, and nickel-tin bronze contain the respective named elements as the principal alloying addition. Contact bronze, bearing bronze, and commercial bronze are all really brass alloys (copper and zinc).

TYPICAL APPLICATIONS

Copper alloys, although considerably more expensive than ferrous materials on a purely materials-cost basis (except for some stainless alloys), nevertheless enjoy wide use. There are two major reasons for this:

1. Copper, brass, and bronze are superior for further processing: forming, machining, joining, and finishing. This property can reduce the total cost of the finished component.
2. These alloys are particularly useful when a combination of properties is required. Examples are electrical conductivity and spring action, corrosion resistance and weldability, heat conductivity and solderability, machinability and corrosion resistance, formability and nonmagnetic properties, and formability and wear resistance.

Typical applications of copper alloys in strip or sheet form are stampings, often of intricate form and shape, that require electrical conductivity, corrosion resistance, spring action, or joining to other components. Electrical connectors and spring contacts are common examples.

Bars, rods, and drawn shapes serve as machining stock, most frequently for use in automatic screw machines. Again, when electrical, anticorrosion, or antimagnetic properties or easy joinability is required in the finished part, copper alloys (usually brass) are more apt to be chosen. When machining is extensive, as in gear making, these alloys are also apt to be advantageous.

Wire (excluding wire for electrical transmission) is usually used when forming operations, including cold heading, impact extrusion, and simple bending, are used. Typical parts are springs, screws and rivets, and jewelry components.

Tubing is used for fluid-carrying applications and is often utilized when components formed from tubing by swaging, expanding, or bending are required. Water tube is by far the dominant product in this category, but the category also includes level-wound air-conditioning/refrigeration (ACR) tube, heat exchanger/condenser tube, and enhanced-surface tube.

MILL PROCESSES

Wrought products of brass and other copper alloys are made by hot and cold rolling, extrusion, and cold drawing.

GRADES FOR FURTHER PROCESSING

Copper alloys are purchased for applications that benefit from their combination of properties. Additionally, because of their high suitability to further processing, copper-alloy mill products are often selected for applications that require a considerable amount of further processing. Common operations include the following.

Cold Forming

Cold-forming operations performed on copper, brass, and bronze mill products involve strip, sheet, rod, wire, tube, and sometimes plate forms. The following operations may be employed: blanking, bending, and forming; drawing, including deep drawing; roll forming; swaging; cold heading; impact extruding; and spinning.

Table 2.3.8 lists some common copper alloys that have an excellent cold-formability rating. The table shows the Unified Numbering System (UNS) and the American Society of Testing and Materials (ASTM) specification numbers, the common name of the alloy, and its common applications. The alloys listed were chosen for relatively low materials cost as well as for formability. Note that the five-digit UNS numbers for copper metals are based on formerly used three-digit Copper Development Association (CDA) numbers.

Note: In selecting a material for punch-press blanking with little or no forming, the more machinable alloys are preferable to those listed in Table 2.3.8.

TABLE 2.3.8 Copper Alloys with Excellent Cold-Working Properties

UNS alloy designation	ASTM specification	Name	Common applications
C11000	B-152	Electrolytic tough-pitch copper	Electrical parts, switch gear
C10200	B-152	Oxygen-free copper	Electronic and radar equipment, connectors, glass-to-metal seals
C16200	—	Cadmium copper	Pole-line hardware
C22000	B-36	Commercial bronze	Electrical outdoor fixtures, costume jewelry
C23000	B-36	Red brass	Radiator cores, jewelry, fire extinguishers
C26000	B-36	Cartridge brass	Deep-drawn and spun parts, eyelets, small-arms cartridges
C26800	B-36	Yellow brass	Versatile: for general drawing, stamping, forming, spinning
C50500	B-103	Phosphor bronze	Spring contacts, diaphragms, instruments
C74500	B-206	Nickel silver	Silver-plated ware, costume jewelry, fishing reels, slide fasteners
C22600	B-36	Jewelry bronze	Jewelry findings, zippers
C24000	B-36	Low brass	Jewelry findings, parts with large upset heads, flexible hose
C66500	B-97	Silicon bronze A	Tanks, vessels for chemical and marine use

Source: Prepared from data provided by the Copper Development Association, Inc.

Hot Forming

Hot-forming operations include forging, extruding, and hot upsetting. Table 2.3.9 lists a number of alloys rated excellent or good in hot formability.

Machining

There are many highly machinable brasses. These materials are favored by screw-machine operators because they can be machined with high surface speeds, provide long tool life, hold dimensions well, and yield parts with excellent surface finish. However, not all copper alloys have good machinability. The alloys that are suitable for forming operations may be too soft and will cut less cleanly, requiring greater care in establishing machining parameters, tool geometries, etc. Cutting speed and tool life will be affected. Conversely, free-cutting alloys usually are not particularly well suited for use in cold-forming operations. Table 2.3.10 lists the specifications of the more machinable copper alloys.

TABLE 2.3.9 Copper Alloys with Excellent Hot-Working Properties

UNS alloy designation	ASTM specification	Name	Common applications
C10200	B-124, B-187	Oxygen-free copper	Electrical and electronic parts, heat sinks
C11000	B-152	Electrolytic tough-pitch copper	Switches, busbars, electronics parts, terminals, heat sinks
C28000	—	Muntz metal	Hot-forged parts
C38500	B-455	Architectural bronze	
C46400	B-21, B-124, B-283	Naval brass	Marine parts, bolts, nuts, hardware
C37700	B-124	Forging brass	Hot-forged and machined parts
C67500	B-124	Manganese bronze A	
C65100	B-98, B-99	Low-silicon bronze B	Electrical contacts, outdoor hardware, springs
C65500	B-99	Silicon bronze A	Springs, small tanks, pole-line hardware
C69400	B-371	Silicon red brass	
C27400	—	Yellow brass 63%	
C64200	B-150		Pump parts, valve stems, gears, worms, compression fittings

Source: Prepared from data supplied by the Copper Development Association, Inc.

TABLE 2.3.10 Copper Alloys with the Best Machinability

UNS alloy designation	ASTM specification	Name	Machinability rating	Common applications
C36000	B-16	Free-cutting brass	100	Versatile: meets 90% of applications where brass is indicated
C35600	—	Extra-high-leaded brass	100	Clock and watch gears, hardware, nuts
C38500	B-455	Architectural bronze	90	Outdoor uses, hinges, lock bodies
C14500	B-301	Tellurium	90	Soldering iron tips, current-carrying electrical parts, which require machining
C37700	B-124	Forging brass	80	Forgings of all kinds
C32000	B-140	Leaded red brass	90	Pump bodies, pipe fittings, valves
C14700	B-301	Sulfur copper	90	Blanked clock parts
C93700	B-22	Bushing and bearing bronze	80	High-load, high-speed applications, pumps, impellers, pressure-tight castings

Source: Prepared from data provided by the Copper Development Association, Inc.

Welding

The more weldable copper alloys are the following:

UNS alloy C12200 (ASTM B152) : Phosphorus deoxidized copper: This formulation is suitable for welding, brazing, and soldering. In tubing form, it is used for hydraulic lines for oil, gas, air, and other fluids.

UNS alloys C65500 and C65100 (ASTM B99 and BI05) : Silicon bronze: These alloys and related compositions are used as welding rod and wire for tanks and vessels and structures in the chemical and marine industries. Large quantities are consumed in automobile assembly and for the welding of galvanized steel.

Other alloys with good or excellent weldability using the GTAW or GMAW welding processes include oxygen-free copper, chromium copper, cadmium copper, red brass (85 percent), low brass (80 percent), phosphor bronze, copper nickels, and aluminum bronzes. Leaded alloys generally cannot be welded.

Casting

Copper alloys are commonly used when corrosion resistance, electrical conductivity, good bearing surface, or other properties of these materials are required in the function of a given product. Casting permits a freedom of shape not always possible with other processes. While many copper and brass alloys are castable, those with the highest castability in sand molds are high-lead tin bronze, leaded red brass, and leaded semired brass. Yellow brasses, silicon brasses, and other short-freezing-range alloys are particularly well suited to permanent mold-casting processes.

Soldering and Brazing

Almost all copper alloys are well suited for soldering and brazing. Exceptions are the aluminum bronzes and beryllium coppers, which may require plating or other special steps before soldering.

Springs

Copper-alloy springs are used when corrosion resistance, electrical conductivity, or some other special property is required in addition to the spring action. Copper alloys most suitable for springs are cartridge brass 70 percent, C26000, the most economical; phosphor bronze 5 percent A, C51000, the most widely used; phosphor bronze 8 percent C, C52100, used for more severe conditions; phosphor bronze 10 percent D, C52400, used for extreme conditions; nickel silver 55-18, C77000, used for corrosive environments; and nickel silver 65-12, C75700, which is less costly than C77000.

Demanding requirements of electronic, communications, automotive, and military equipment have prompted use of improved spring materials for electrical and sometimes mechanical uses. These include beryllium coppers C17700, C17400, and C17500 (automotive), nickel-copper C70250 (automotive, high duty), aluminum bronze C63800 and aluminum brass C68800 (properties similar to phosphor bronze but more formable), silicon bronze C65400 (peripheral connectors for computers), copper-iron-phosphorous-cobalt alloy C19500 (high conductivity), and high-strength spinodal copper-nickel-tin alloys such as C72500.

AVAILABLE SHAPES AND SIZES

Flat wire, strip, and *sheet* are terms used to designate brass and other copper alloys rolled or drawn to thicknesses of 0.188 in or less. The terms *bar* and *plate* refer to flat products of a greater thickness than 0.188 in. See Tables 2.3.11 and 2.3.12. Strip is available in thicknesses from 0.005 to 0.188 in and in widths to 20 in. Sheet is available in thicknesses from 0.010 to 0.188 in and in widths from 20 to 60 in.

The largest standard bar and plate thickness for most mills is 2 in. Round stock is referred to as wire if it is furnished wound or in coils and as rod if it is furnished in straight lengths. The standard lengths for rods are 6 to 14 ft. The minimum standard diameter for round rods is 0.25 in; the maximum standard diameter for most mills is 3 in. Wire (excluding electric wire) ranges, in standard sizes, from 0.010 to 0.75 in in diameter. The term *rod* is also applied to pieces with hexagonal and octagonal cross sections. See Tables 2.3.13 and 2.3.14.

TABLE 2.3.11 Thickness Tolerances of Flat Wire, Strip, and Sheet

	Width		
Nominal thickness	Flat wire to 31 mm (1¼ in)	Strip to 500 mm (20 in)	Sheets to 1.5 m (60 in)
Up to 0.33 mm (Up to 0.013 in)	±0.025 mm (±0.001 in)	±0.025 mm (±0.001 in)	±0.090 mm (±0.0035 in)
Over 0.33 to 0.52 mm (Over 0.013 to 0.021 in)	±0.033 mm (±0.0013 in)	±0.038 mm (±0.0015 in)	±0.088 mm (±0.0035 in)
Over 0.52 to 1.27 mm (Over 0.021 to 0.050 in)	±0.033 mm (±0.0013 in)	±0.05 mm (±0.002 in)	±0.13 mm (± 0.005 in)
Over 1.27 mm to 4.7 mm (Over 0.050 to 0.188 in)	±0.05 mm (±0.002 in)	±0.088 mm (±0.0035 in)	±0.2 mm (±0.008 in)

Note: Allow 50% greater tolerances to the following alloy families: beryllium copper, manganese bronze and inhibited admiralty bronze, copper nickel, nickel silver, and phosphor bronzes.

Source: Prepared by condensing data provided by the Copper Development Association, Inc.

TABLE 2.3.12 Thickness Tolerances of Plates and Bars (Rectangular)

	Width			
Nominal thickness	Bar to 300 mm (12 in) wide	Plate to 500 mm (20 in) wide	Plate to 900 mm (36 in) wide	Plate to 1.5 m (60 in) wide
Over 4.7 to 13 mm (Over 0.188 to 0.500 in)	±0.11 mm (±0.0045 in)	±0.13 mm (±0.005 in)	±0.25 mm (±0.010 in)	±0.35 mm (±0.014 in)
Over 13 to 25 mm (Over 0.500 to 1.00 in)	±0.18 mm (±0.007 in)	±0.23 mm (±0.009 in)	±0.45 mm (±0.018 in)	±0.60 mm (±0.024 in)
Over 25 to 50 mm (Over 1.00 to 2.00 in)	±0.60 mm (±0.024 in)	±0.60 mm (±0.024 in)	±0.70 mm (±0.028 in)	±0.90 mm (±0.036 in)

Note: Allow 30% greater tolerances for the following alloy families: beryllium copper, copper nickel, manganese bronze, inhibited admiralty bronze and nickel silver, phosphor bronze, silicon bronze, and Muntz metal.

Source: Prepared by condensing data provided by the Copper Development Association, Inc.

TABLE 2.3.13 Diameter Tolerances of Cold-Drawn Copper-Alloy Rod (Round, Hexagonal, and Octagonal)

Diameter or distance between parallel surfaces	Tolerance	
	Round	Hexagonal, octagonal
Up to 3.75 mm (Up to 0.150 in)	±0.033 mm (±0.0013 in)	±0.063 mm (±0.0025 in)
Over 3.75 to 12.5 mm (Over 0.150 to 0.500 in)	±0.038 mm (±0.0015 in)	±0.075 mm (±0.003 in)
Over 12.5 to 25 mm (Over 0.500 to 1.00 in)	±0.05 mm (±0.002 in)	±0.10 mm (±0.004 in)
Over 25 to 50 mm (Over 1.00 to 2.00 in)	±0.063 mm (±0.0025 in)	±0.13 mm (±0.005 in)
Over 50 mm (Over 2.00 in)	±0.15%	±0.30%

Note: For forging brass and the other materials listed in the note under Table 2.3.11, add 33% to these values.

Source: Prepared by condensing data provided by the Copper Development Association, Inc.

TABLE 2.3.14 Diameter Tolerances or Tolerances on the Distance between Parallel Surfaces of Extruded or Hot-Rolled Rods of Round or Other Shape

Diameter or distance between parallel surfaces	Tolerance, extruded round, square, rectangular, hexagonal, and octagonal sections
Up to 25 mm (Up to 1.00 in)	±0.25 mm (±0.010 in)
Over 25 to 50 mm (Over 1.00 to 2.00 in)	±0.38 mm (±0.015 in)
Over 50 to 75 mm (Over 2.00 to 3.00 in)	±0.63 mm (±0.025 in)
Over 75 to 88 mm (Over 3.00 to 3.50 in)	±0.88 mm (±0.035 in)
Over 88 mm to 100 mm (Over 3.50 to 4.00 in)	±1.5 mm (±0.060 in)

Note: For materials listed in the note under Table 2.3.11 or for hot-rolled rod, add 30% to these values.

Source: Prepared by condensing data provided by the Copper Development Association, Inc.

Copper-alloy tubing is available in a variety of diameters and wall thicknesses. Both thin- and heavy-wall varieties are available. Diameters range from 0.125 to 12 in and wall thicknesses from 0.010 to 0.625 in. Nonround tubing and round tubing with nonround holes are also available. Round copper tubing for refrigeration and automotive use comprises a major portion of the output of some mills. See Table 2.3.15.

Extruded shapes in brass and copper are available from mills and from custom extruders. Although brass is not the most readily extrudable metal, numerous sections

TABLE 2.3.15 Diameter and Wall-Thickness Tolerances for Round Seamless Tubing

Diameter	Diameter tolerance	Wall thickness and tolerance					
		To 0.45 mm (0.018 in)	0.45 to 0.88 mm (0.018–0.035 in)	0.88 to 2.08 mm (0.035–0.083 in)	2.08 to 4.13 mm (0.083–0.165 in)	4.13 to 9.5 mm (0.165–0.380 in)	Over 9.5 (0.380 in)
Up to 3 mm (Up to ⅛ in)	±0.05 mm (±0.002 in)	±0.05 mm (±0.002 in)	±0.075 mm (±0.003 in)	±0.075 mm) (±0.003 in)			
Over 3 to 16 mm (Over ⅛ to ⅝ in)	±0.05 mm (±0.002 in)	±0.025 mm (±0.001 in)	±0.063 mm ±0.0025 in)	±0.088 mm (±0.0035 in)	±0.11 mm (±0.0045 in)		
Over 16 to 25 mm (Over ⅝ to 1 in)	±0.063 mm (±0.0025 in)	±0.038 mm (±0.0015 in)	±0.063 mm (±0.0025 in)	±0.10 mm (±0.004 in)	± 0.14 mm (±0.0055 in)	±0.23 mm (±0.009 in)	
Over 25 to 50 mm (Over 1 to 2 in)	±0.075 mm (±0.003 in)	±0.05 mm (±0.002 in)	±0.075 mm (±0.003 in)	±0.10 mm (±0.004 in)	± 0.14 mm (±0.0055 in)	±0.25 mm (±0.010 in)	±5% ±5%
Over 50 to 100 mm (Over 2 to 4 in)	±0.11 mm (±0.0045 in)		±0.10 mm (±0.004 in)	±0.14 mm (±0.0055 in)	±0.19 mm (±0.0075 in)	±0.30 mm (±0.012 in)	±5% ±5%
Over 100 to 175 mm (Over 4 to 7 in)	±0.18 mm (±0.007 in)			±0.19 mm (±0.0075 in)	±0.24 mm (±0.0095 in)	±0.35 mm (±0.014 in)	±6% ±6%
Over 175 to 250 mm (Over 7 to 10 in)	±0.25 mm (±0.010 in)			±0.25 mm (±0.010 in)	±0.30 mm (±0.012 in)	±0.40 mm (±0.016 in)	±6% ±6%

Note: For the following alloys, 15% should be added to these values: inhibited admiralty and naval brass, aluminum brass, beryllium copper, copper nickel, high-silicon bronze, and phosphor bronze.

Source: Prepared by condensing data provided by the Copper Development Association, Inc.

can be produced, especially with commercially pure and modified copper, copper nickel 10 percent, C65100, and C65500. See Chap. 3.1 for further information on extruded copper-alloy sections.

DESIGN RECOMMENDATIONS

Because copper and brass alloys are easy to process, there are few design recommendations that would not be equally applicable to other materials. There are, however, some points that should be kept in mind by the designer:

1. Copper and brass are not inexpensive materials on an initial price basis compared, for example, with steel and aluminum. If the application does not demand the special properties of copper or brass (e.g., processibility, corrosion resistance, electrical conductivity), it may be more economical to use other materials, e.g., plated steel. Total cost of the part should be the determining factor.
2. Similarly, it is generally preferable, if possible, to use methods such as cold heading, extrusion, and press forming instead of machining to avoid the material loss (in the form of chips) that is inherent in machining, despite the fact that machining chips of copper alloys have a high salvage value.
3. Although further processing of copper and brass alloys is economical, it is advisable, if possible, to design parts so that they utilize stock sizes and shapes with a minimum of further operations.
4. For the lowest-cost further processing, it is important that the correct alloy be specified. Easily formable alloys are not usually easily machinable. Good alloys for cold forming may not be satisfactory for hot forming. Designers must consider the process involved as well as the other factors before specifying a particular copper alloy.

STANDARD STOCK TOLERANCES

Tables 2.3.11 through 2.3.15 provide tolerance data for commercial forms of wrought copper alloys.

PART 3
MAGNESIUM*

PROPERTIES

Magnesium is the lightest metal of those available in quantity and suitable for general structural and product-component use. Its specific gravity is 1.74 in the pure state, whereas alloys average 1.80. This is about 66 percent of the specific gravity of aluminum and only 24 percent of that of steel. Magnesium melts at 651°C (1204°F). Its tensile yield strength (when alloyed) ranges from about 75,000 kPa (11,000 lbf/in^2) to 260,000 kPa (38,000 lbf/in^2), with wrought forms having higher values than castings.

Magnesium has good damping properties, better thermal conductivity than steel, and lower electrical conductivity than copper and aluminum but higher than iron and steel. It is not very wear resistant, so bushings and wear pads should be of other materials unless the bearing pressure, velocity, and temperature are low.

Magnesium has good stability in normal atmospheric exposure. It is resistant to attack by hydrocarbons, alkalis, and some acids, but it is attacked by seawater and most salts and acids. Magnesium is flammable at high temperatures, but in practical use, a large part must be raised entirely to the melting point for it to burn freely. Care must be taken, however, to avoid ignition of fine particles such as powder, chips, or turnings. The cost of magnesium is higher than that of aluminum.

TYPICAL APPLICATIONS

The prime application for magnesium is lightweight structural components. Sound-resonator diaphragms, sound-damping and -shielding components, and plates for photoengraving are other applications, but these account for only a small portion of total use.

Since the yield strength of magnesium alloys diminishes above 230°C (450°F), its use is most common where atmospheric, indoor, or other moderate temperatures prevail. It is also most commonly found in applications in which low stresses are involved. Magnesium is used in machine parts with an oscillating motion in order to reduce vibration- or inertia-induced stresses.

Historically, the typical applications of magnesium have been aircraft-engine castings; aircraft and automotive wheels; high-speed impellers; ladders; loading ramps;

*Updated with the assistance of The International Magnesium Association.

sporting goods; housings for portable power tools such as chain saws, drills, and sanders; various aircraft structural parts; office equipment; hand trucks; grain shovels; gravity conveyors; foundry equipment; lawn-mower housings; electric-motor housings; and electronics mounting bases. More recently, automotive uses have expanded to such applications as clutch and brake pedal support brackets, steering column supports, seat frames, transmission housings, and instrument panel structures. Magnesium alloys are used increasingly for parts that operate at high speeds such as computer disk-drive armatures, where light weight minimizes inertial forces.

MILL PROCESSES

Magnesium is produced mainly by the electrolysis of magnesium chloride salts, which are obtained from seawater and other sources. It is then remelted and/or refined, alloyed, and cast into ingots, billets, or slabs. If wrought forms are involved, it can be extruded or rolled into many different shapes and sizes.

GRADES FOR FURTHER PROCESSING

Machining

All magnesium alloys are highly machinable. They can be processed at higher surface speeds and with greater depths of cut than other structural materials. Smooth surface finishes can be produced without the need for finish grinding and polishing. Power requirements per unit volume of metal removed are also the lowest among these metals. For fire protection, tools must be kept sharp and be constantly bathed in curring fluid.

The differences in machinability of magnesium alloys are not significant. Therefore, no specific alloy need be specified to ensure machinability. However, AZ31B (M11311) and AZ61A (M11610) are generally specified for screw-machine work. MIA (M15100) is not quite so freely cutting as other magnesium alloys. The surface finish of machined magnesium can be maintained at exceptionally fine values even with heavy cuts and high cutting speeds. Close tolerances can be held if care is exercised to prevent expansion of the workpiece from heat and if clamping distortion is avoided.

Forging

Magnesium alloys generally forge more satisfactorily by hydraulic press than by drop hammer. However, alloys MIA (M15100), AZ31B (M11311), and TA54A (M18540) are drop-hammer-forgeable, the last-named being developed expressly for this method of processing. Other forging alloys are ZK60A (M16600), AZ80A (M11800), HK31A (M13310), and HM21A (M13210); of these, ZK60A is the most easily forged. AZ80A provides strong forgings and is processible by hydraulic press, while HK31A and HM21A are used for higher-temperature applications.

Extrusion

AZ31B (M11311) is a good alloy for general-purpose extrusions; it has excellent extrudability. Other suitable alloys are AZ61A (M11610), AZ80A (M11800), and ZK60A (M16600). AZ61A has improved strength but slightly reduced formability, whereas ZK60A has further improved properties and is heat-treatable.

Press Forming

Magnesium sheet can be sheared, blanked, pierced, formed, spun, roll-formed, and deep-drawn. However, most forming operations are performed at elevated temperatures, usually up to about 290°C (550°F), with slow rates of deformation, particularly if the amount of working is severe. Annealed stock is preferred to unannealed or hard-rolled material if any substantial deformation is required. Casting alloys are not recommended for forming operations after casting. Alloys suitable for forming in order of decreasing formability are ZE1OA-0 (M16100-0), AZ31B-0 (M11311-0), HK31A-0 (M13310-0), ZE1OA-H24, AZ31B-H24, and HK31A-H24. Alloy AZ61A (M11610) provides cleanly sheared or blanked edges for parts requiring these operations without severe forming.

Sand Casting

The most common casting alloys are AZ63A (M11630) and AZ92A (M11920) because of their good castability and high strength. Other commonly used alloys are AM80A (M11800), AZ91C (M11914), and AM100A (M10100). Alloys containing rare earths (EK30A, EK33A, EK41A, ZE41A [M16410], ZK51A [M16510], and ZK61A [M16610]) have good properties and casting characteristics but are somewhat more costly. HK31A (M13310) is used for high-temperature applications.

Permanent-Mold and Plaster-Mold Casting

AZ91E (M11918) is the most commonly used alloy. Others frequently used are AZ63A (M11630), AMI00A (M10100), and AZ91C (M11914). Rare-earth alloys EK30A, EK41A, and EZ33A also can be used.

Investment Casting

AM100A (M10100) and AZ92A (M11920) are the recommended alloys.

Die Casting

AZ91D (M11916) is the most frequently used die-casting alloy. AZ91B (M11913) is lower in cost, whereas AZ91D has better saltwater-corrosion resistance. AM60B (M10602) and AM50B are also used commonly.

Arc Welding

Most magnesium alloys, both wrought and cast, are arc-weldable. The gas-tungsten-arc (GTAW) process is the best for magnesium. Some alloys, specifically the Mg-AI-Zn series and alloys that contain more than 1 percent aluminum, require postweld stress relieving to prevent cracking. MIA (M15100) and other alloys containing manganese, rare earths, zinc, or zirconium do not require stress relieving and hence are preferred for assemblies requiring extensive welding. (See also Table 7.1.3.)

Gas Welding

This process generally is not recommended for magnesium alloys because they are susceptible to corrosion from flux. Normally, gas welding of magnesium is limited to emergency repairs of thin-gauge stock.

Resistance Welding

AZ31B (M11311) and M1A (M15100) are the most common alloys to be resistance welded. Spot and seam welding can be employed. Spot welding is the usual process for sheet up to 0.5 mm (0.020 in) thick. AZ31B, AZ61A (M11610), and AZ80A (M11800) are suitable for flash welding. However, M1A is difficult.

Brazing

This process is not widely used with magnesium components but has been carried out with alloy M1A (M15100).

Soldering and Adhesive Bonding

These processes are feasible with all magnesium alloys. All commercial alloys are also paintable after normal precleaning.

AVAILABLE SHAPES AND SIZES

Wrought magnesium is available from commercial sources in the following forms: sheet, strip, plate, wire, rod, bar, tubing, and other shapes. Tooling plate—precision flat ground plate for tooling applications—and tread plate are two special forms that are widely stocked by distributors. Many special shapes are also readily procured from custom extruders.

Table 2.3.16 shows the normal size limits of commercially stocked magnesium alloys in wrought form.

TABLE 2.3.16 Size Range of Available Mill Forms of Magnesium Alloys

Form	Thickness range	Width	Length
Sheet	0.4 to 6.3 mm (0.016–0.250 in)	0.6 to 1.2 m (24–48 in)	2.4 to 5.5 m (96–216 in)
Plate	6.3 to 150 mm (0.250–6.0 in)	1.2 m (48 in)	2.4 to 3.6 m (96–144 in)
Bar and rod	3.2 to 90 mm (0.125–3.5 in)	25 to 150 mm (1–6 in)	3.6 m (12 ft)
Tube	1.6- to 6.3-mm wall (0.065- to 0.250-in) wall	12.7- to 100-mm (0.5- to 4-in) diameter	

SOURCE: From AMCD 706-100.

DESIGN RECOMMENDATIONS

The following recommendations arise from the special characteristics of magnesium.

Machined Parts

Sharp corners, notches, and other stress raisers should be avoided. This is particularly true with form tools and counterbores. Because magnesium has a relatively low modulus of elasticity (low stiffness) compared with other metals, particular care should be taken that the thickness of the part at the place where it is to be clamped for machining is sufficiently large. In other words, good, strong clamping points should be provided to avoid distortion. This is a universal rule for all materials, but it is particularly important for magnesium.

Forgings

Sidewalls should be fairly thick, ranging from 50 percent of height for walls of 6.3-mm (0.25-in) height to 13 percent of depth for walls of 50-mm (2-in) height. (See Fig. 2.3.1.)

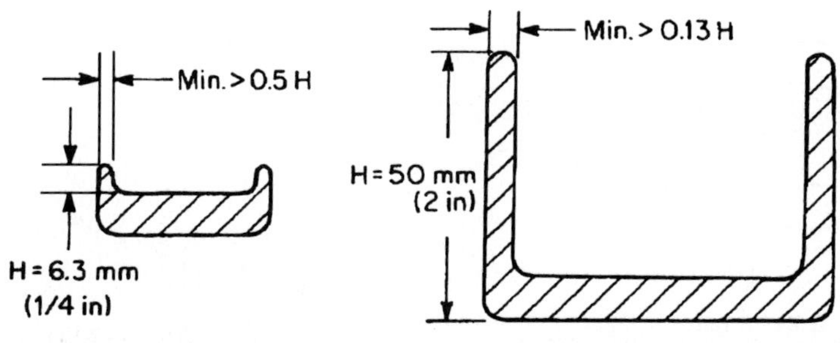

FIGURE 2.3.1 Sidewalls of magnesium forgings should be at least as heavy as indicated by these two examples.

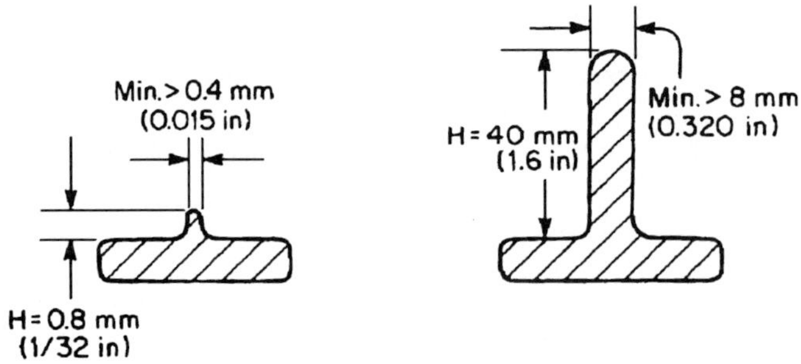

FIGURE 2.3.2 Ribs in magnesium forgings should be at least as heavy as the ribs in these two examples.

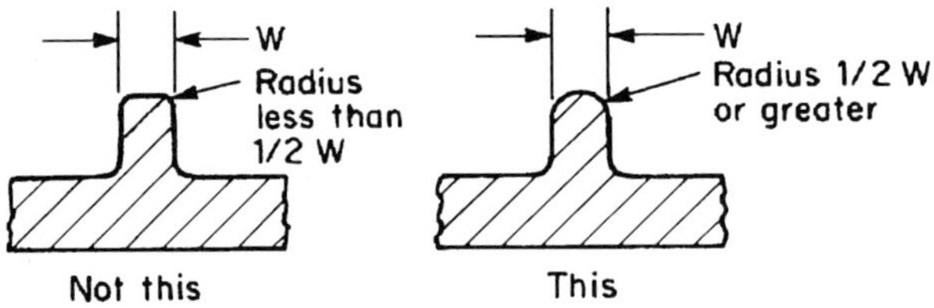

FIGURE 2.3.3 Minimum top radius of ribs in magnesium forgings.

Ribs also should be as thick as possible. Ribs 0.8 mm (0.030 in) in height can be as narrow as 0.4 mm (0.015 in). Ribs 40 mm (1.60 in) in height should be at least 8 mm (0.320 in) thick. (See Fig. 2.3.2.) Ribs should have a top radius equal to one-half the thickness. (See Fig. 2.3.3.) A generous radius should be allowed at all inside corners. See Table 3.13.3 for recommended minimum values. If surfaces are to be machined after forging, stock should be allowed for machining as in Table 3.13.4.

Extrusions

Very-thin-walled, large cross sections should be avoided. The length of a particular section should not be longer than 20 times its thickness. See Fig. 3.1.10 for an illustration. If the extrusion contains both a heavy section and a thin leg, the leg should not be too long in relation to the width. A rule of thumb is that the length should not exceed 20 times the thickness. (See Fig. 2.3.4.) Other items to avoid are illustrated in Fig. 3.1.9.

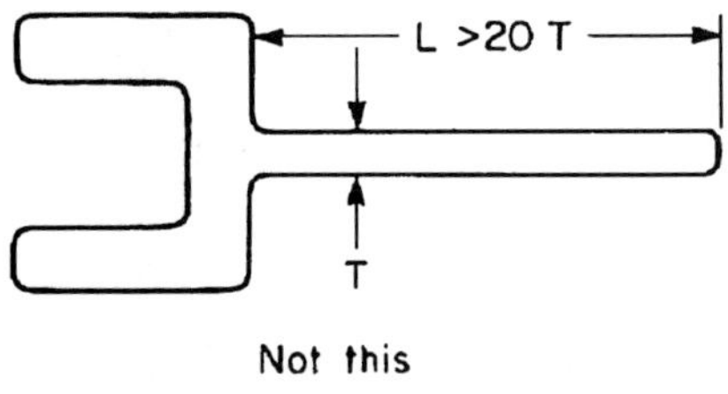

FIGURE 2.3.4 Limits of length-to-thickness ratio of thin legs in the cross sections of magnesium extrusions.

Press-Formed and Other Sheet Parts

Although most press forming of magnesium is performed at elevated stock temperatures, it is possible to form cold if the bend radii are generous and the deformation is mild. If the function of the component does not require a severe change in shape, designers should try to specify bends that are gentle enough so that forming can be done cold.

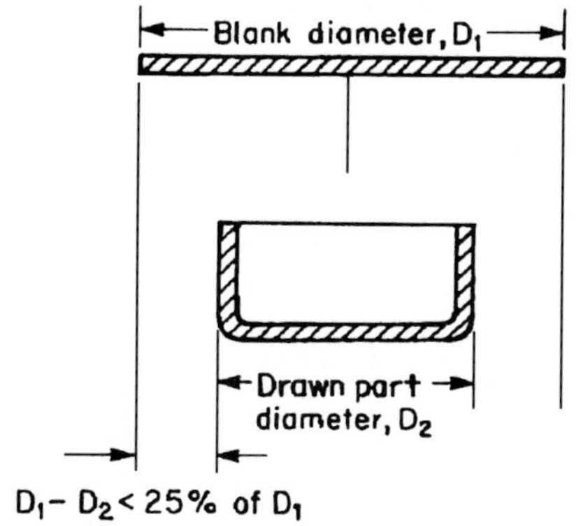

FIGURE 2.3.5 For magnesium alloy AZ31A-0 (annealed), the maximum percent reduction in cold drawing is 25 percent. For other anneal magnesium alloys, the maximum ranges from 15 to 25 percent.

For somewhat more severe deformation, the use of annealed sheet allows for sharper cold bends and more stock deformation than are possible with hard-rolled stock. The cold-forming limit of annealed AZ31A-0 alloy (M11310) is 25 percent. (See Fig. 2.3.5.) For other alloys, it is normally less.

Because of magnesium's low density, thicker sheet gauges can be used without a weight penalty. This can eliminate the need for stiffening ribs or other reinforcing members, which would add to the cost of the part. Designers therefore should consider the use of thicker but unribbed magnesium sheet instead of ribbed sheets of thinner but denser materials. Deep-drawn magnesium parts may exhibit some wrinkling at the corners. Cylindrical parts or generously rounded rectangular parts are therefore preferable to boxlike designs. In blanking stock thicker than 1.5 mm (0.060 in) or 3.0 mm (0.125 in) if annealed, some flaking of material at the edge where the blank breaks away from the sheet stock can be expected. If this is not permissible, allow for fineblanking or shaving. When bending magnesium tubing, bend radii should be equal or greater than the minimum values of Table 2.3.17.

TABLE 2.3.17 Minimum-Bend Radii for Round Magnesium Tubing Bent at Room Temperature

	Minimum-bend radius*		
Wall thickness	$\frac{1}{17}$ *D*	$\frac{1}{8}$ *D*	$\frac{1}{3}$ *D*
AZ61A (M11610) unfilled	5 *D*	2½ *D*	2½ *D*
AZ61A (M11610) filled†	2½ *D*	2½ *D*	2 *D*
AZ31B (M11311) unfilled	6 *D*	4 *D*	3 *D*
AZ31B (M11311) filled†	3 *D*	2 *D*	2 *D*
MIA (M15100) unfilled	6 *D*	6 *D*	2½ *D*
MIA (M15100) filled†	6 *D*	3 *D*	2½ *D*

*Centerline bend radius as a ratio of tube outside diameter *D*.

†Filled with low-melting-temperature alloy.

Source: Adapted from *Metals Handbook,* 9th ed., American Society for Metals, Metals Park, Ohio, 1979.

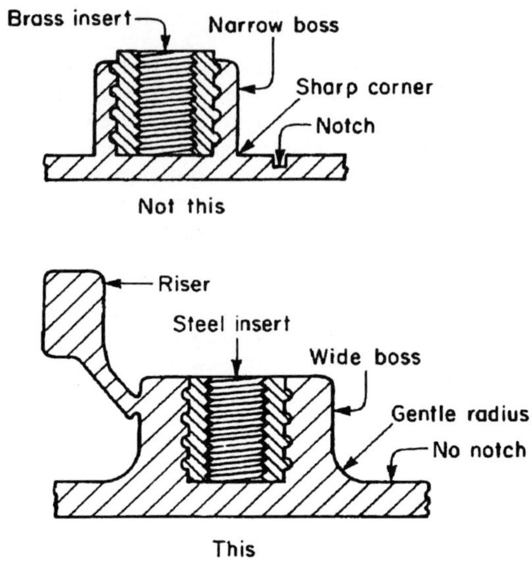

FIGURE 2.3.6 Magnesium sand-mold-casting practice with respect to inserts.

Sand-Mold Castings

Design recommendations for magnesium parallel those for other metals. Wall-thickness limitations are about the same as those for other nonferrous castings. Thin-walled sections should be kept small in area; 8 mm ($^5/_{16}$ in) is a desirable minimum wall thickness, but 5 mm ($^3/_{16}$ in) may be used in limited areas. (See Fig. 2.3.6.) Sharp corners, gouges, and notches should be avoided in areas of high stress. Generous fillets around bosses and other heavy sections are advisable.

Cast-in inserts are troublesome and should be avoided if other design approaches are feasible. For metallurgical reasons, the inserts, if used, should be made of steel rather than brass, bronze, or other nonferrous metals. Ample material must be allowed around cast-in inserts, but care must be taken so that the large casting mass in that area does not cause draws or cracks. Such bosses and other heavy sections should be located so that they can be risered or chilled. See Fig. 2.3.6 for an illustration of some of these points. When designing parts containing inserts, measures for prevention of galvanic corrosion must be part of the design. Shrinkage for magnesium sand castings ranges from 1 to 1.5 percent ($^1/_8$ to $^1/_{16}$ in/ft).

Die Castings

For design recommendations for magnesium die castings, see Chap. 5.4.

Weldments and Other Assemblies

Butt and fillet joints are preferred to lap joints for arc-welding magnesium. They are easier to weld and provide more satisfactory results in service. The minimum stock thickness for arc-welding magnesium is 0.8 mm (0.032 in). Thick plates up to the thickest available can be welded satisfactorily.

Spot and seam welds are of limited fatigue strength but are satisfactory for lower-stress applications. Riveted and bonded assemblies are often more satisfactory in joints subject to vibration or other high intermittent or cyclic stress. Spot welds in magnesium sheets of unequal thickness with a thickness ratio greater than $2\frac{1}{2}$:1 are not recommended.

Adhesive joints for magnesium should be lap joints with a large bonded area. Thin-sheet components can be bonded effectively, and bonding makes a more satisfactory assembly with thin stock than can be made by riveting.

Riveted joints of magnesium sheets or plates should utilize rivets of aluminum to avoid galvanic corrosion. Quarter-hard 5056 alloy is commonly used. Press fits in magnesium require considerable interference between the press-fit component and the base component. Table 2.3.18 provides information on normally recommended interference dimensions.

TABLE 2.3.18 Recommended Diametral Interference for Press Fit and Shrink Fit of Steel Inserts in Magnesium

		Operating temperature			
Outer diameter of insert, mm (in)	Magnesium wall thickness, mm (in)	21°C (70°F)	38°C (100°F)	93°C (200°F)	148°C (300°F)
12.7 (0.5)	1.5 (0.062)	0.010 mm (0.0004 in)	0.012 mm (0.0005 in)	0.023 mm (0.0009 in)	0.033 mm (0.0013 in)
25 (1.0)	2.4 (0.094)	0.015 mm (0.0006 in)	0.025 mm (0.0010 in)	0.043 mm (0.0017 in)	0.063 mm (0.0025 in)
50 (2.0)	3.2 (0.125)	0.025 mm (0.0010 in)	0.038 mm (0.0015 in)	0.079 mm (0.0031 in)	0.12 mm (0.0047 in)
75 (3.0)	4.0 (0.156)	0.038 mm (0.0015 in)	0.061 mm (0.0024 in)	0.12 mm (0.0047 in)	0.18 mm (0.0072 in)
100 (4.0)	4.8 (0.188)	0.053 mm (0.0021 in)	0.086 mm (0.0034 in)	0.16 mm (0.0062 in)	0.24 mm (0.0096 in)
125 (5.0)	5.5 (0.218)	0.071 mm (0.0028 in)	0.11 mm (0.0044 in)	0.20 mm (0.0080 in)	0.31 mm (0.0121 in)
150 (6.0)	6.3 (0.250)	0.091 mm (0.0036 in)	0.13 mm (0.0053 in)	0.25 mm (0.0098 in)	0.37 mm (0.0147 in)

Source: *Metals Handbook,* 9th ed., American Society for Metals, Metals Park, Ohio.

STANDARD STOCK TOLERANCES

Standard thickness tolerances for magnesium plate and sheet are shown in Table 2.3.19. Cross-sectional tolerances for other magnesium products are shown in Table 2.3.20.

TABLE 2.3.19 Thickness Tolerances for Magnesium Flat and Coiled Sheet and Plate

	Thickness tolerance, in;, plus and minus Specified widths, in			
Specified thickness, in	Up to 18 inclusive	Over 18 through 48	Over 48 through 60	Over 60 through 72
0.016–0.036	0.002	0.0025	0.005	0.005
0.037–0.068	0.0025	0.004	0.006	0.006
0.069–0.096	0.0035	0.004	0.006	0.006
0.097–0.125	0.0045	0.005	0.007	0.007
0.126–0.172	0.006	0.008	0.009	0.014
0.173–0.249	0.009	0.011	0.013	0.018
0.250–0.438	0.019	0.019	0.020	0.023
0.439–0.875	0.030	0.030	0.030	0.030
0.876–1.375	0.040	0.040	0.040	0.040
1.376–1.875	0.052	0.052	0.052	0.052
1.876–2.750	0.075	0.075	0.075	0.075
2.751–4.000	0.110	0.110	0.110	0.110
4.001–6.000	0.135	0.135	0.135	0.135

Source: Condensed from data supplied by the Dow Chemical Company.

TABLE 2.3.20 Standard Tolerances for Extruded Magnesium Bars, Rods, and Shapes (Cross-Sectional Dimensions)

	Tolerance, in, plus and minus	
Specified dimension, in	Allowable deviation from specified dimension when 75% or more of the dimension is metal	Wall thickness
	Circumscribing circle sizes less than 10 in in diameter	
0.124 and under	0.006	
0.125–0.249	0.007	
0.025–0.499	0.008	10% of specified dimension; 0.600 maximum,. 0.010 minimum
0.500–0.749	0.009	
750–0.999	0.010	
1.000–1.499	0.012	
1.500–1.999	0.014	
2.000–3.999	0.024	
4.000–5.999	0.034	
6.000–7.999	0.044	
8.000–9.999	0.054	
	Circumscribing circle sizes 10 in in diameter and over	
0.124 and under	0.014	
0.125–0.249	0.015	
0.250–0.499	0.016	
0.500–0.749	0.017	
0.750–0.999	0.018	
1.000–1.499	0.019	
1.500–1.999	0.024	
2.000–3.999	0.034	15% of specified dimension; 0.090 maximum, 0.015 minimum
4.000–5.999	0.044	
6.000–7.999	0.054	
8.000–9.999	0.064	
10.000–11.999	0.074	
12.000–13.999	0.084	
14.000–15.999	0.094	
16.000–17.999	0.104	
18.000–19.999	0.114	
20.000–21.999	0.124	
22.000–23.999	0.134	

Note: Dimensions enclosing hollow spaces may require 2 to 4 times the tolerances shown above. The tolerances applicable to a dimension composed of two or more component dimensions is the sum of the tolerances of the component dimensions if all the component dimensions are indicated.

Source: Courtesy the Dow Chemical Company. These tolerances are applicable to the average shape; wider tolerances may be required for some shapes, and closer tolerances may be possible for others.

PART 4
OTHER NONFERROUS METALS

The variety of nonferrous metals available to the design engineer is very broad. In fact, every known metallic element has some industrial use. The metals covered by this chapter, however, are those which have significant use in applications for which they are the major alloying ingredient and for which their use is more or less structural.

Metals whose prime applications are as part of chemical compounds, as coatings or as secondary alloying elements, are not covered in this chapter. Instead, they are covered to the degree appropriate in other chapters, e.g., as electroplated coatings, stainless steel alloying ingredients, dip coatings, etc.

These other nonferrous metals are specified when they have properties superior to those obtainable from the more common ferrous and nonferrous metals. One common advantage found in many of them is superior corrosion resistance. Another property necessary for some applications is higher-operating-temperature capability. Some of these metals have superior wear resistance; others have better electrical properties. Some are specified because of their appearance advantages. Most of them are more expensive than steel, aluminum, copper alloys, or magnesium. They are used in applications for which their support properties are worth the extra cost.

One exception in the group is the use of zinc for die castings. In this case, the prime advantage is manufacturing economy rather than special physical properties.

ZINC AND ITS ALLOYS

Zinc is a blue to gray metal that has a specific gravity of 7.1 and melts at 419°C (786°F). Its tensile strength in the wrought state ranges from 110 to 300 MPa (16,000 to 43,000 lbf/in^2). Zinc ranks third in worldwide consumption of nonferrous alloys (after aluminum and copper).

Zinc possesses inherent ductility and malleability. The superplastic alloy (Zn–22 Al) is very workable, especially at an elevated temperature of 260°C (500°F). Zinc also possesses fairly good corrosion resistance. Its solubility in copper accounts for the wide availability of brass alloys. Zinc's high place in the electromotive series accounts for its use as a galvanic protective coating for steel.

For structural applications, zinc is seldom used in the unalloyed state. However, the major use of zinc is nonstructural: 42 percent of current production is for galvanic coatings; 32 percent, for die-casting alloys; 16 percent, for brass; 6 percent, for rolled zinc; and 4 percent, for other applications.

A major use of wrought zinc is in the construction field: Roofing, flashing, gutters, leaders, downspouts, termite shields, and weather stripping are all made from sheet zinc. Other uses of sheet zinc are eyelets, meter cases, dry-battery cups, flashlight reflectors, fruit-jar caps, radio shielding, gaskets, nameplates, and parts for lamps. Wrought zinc is also used for photoengraving plates. As noted in Chap. 5.4, zinc is the basis for numerous die-cast parts: handles, gears, levers, pawls, etc.

Wrought zinc is made by hot rolling cast slabs. Extrusion and drawing are other methods employed for shapes other than sheets. Zinc foil is made by electroplating zinc on an aluminum drum and then stripping it off.

The available forms of wrought zinc are strip and sheet, foil, plate, rod, wire, and blanks for forging, extrusion, and impact extrusion. Foil thicknesses range from 0.025 to 0.15 mm (0.001 to 0.006 in). Strip thickness is from 0.075 to 6 mm (0.003 to 0.250 in), and widths are 3 to 900 mm (0.125 to 36 in).

The prime zinc die-casting alloys, as mentioned in Chap. 5.4, are ASTM AG40A (XXIII, SAE 903, or UNS Z33520) and ASTM AC41A (XXV, SAE 925, or UNS Z35531). For slush casting, zinc-based slush-casting alloys, 4.75 percent Al (Z34510) and 5.5 percent Al (Z30500), are recommended. For forming, the following are most suitable:

ASTM B69: Zn + 1.0 percent Cu + 0.01 percent Mg (Z45330)

Commercial rolled zinc (0.08 percent Pb) (Z41320 and Z44330)

Zn - Pb 0.05–0.10 percent, Fe 0.012 percent maximum, Cd 0.005 percent maximum, Cu 0.001 percent maximum (Z21210)

Zn - Pb 0.05–0.10 percent, Fe 0.012 percent, CD 0.06 percent, Cu 0.005 percent maximum (Z21220)

Bends in regular commercial rolled zinc should be at right angles to the grain or rolling direction and should have a minimum-bend radius of one material thickness. (See Fig. 2.3.7.) For forging and extrusion, use Zn-Mg alloys with up to 25 percent manganese. Commercial wrought-zinc alloys are easily machined, soldered, and spot welded by conventional methods.

LEAD AND ITS ALLOYS

Lead is a soft, heavy, very malleable, and very flexible metal. It has a specific gravity of 11.3 and a melting point of 327°C (621°F). Its tensile strength is low: 25 MPa (3600 lbf/in^2) after cold rolling.

Lead forms surface oxides easily but thereafter is very stable and resistant to corrosion. It also possesses good lubricity. Lead is easily fused and has a low strength, a low elastic limit, and low electrical conductivity. It is less expensive than other low-melting-temperature metals.

The following are significant commercial uses for lead and its alloys: storage batteries, electrical-cable covering, solder, babbitt bearings, ammunition, flashings and gutters for buildings, soundproofing and deadening, radiation shielding, chemical tank liners, piping, weights, electrodes, caulking material, and gaskets. For solders, tin is the predominant alloying material. Babbitt alloys contain 75 to 90 percent lead plus antimony and tin. Roofing sheet is 6 to 7 percent antimony. Health concerns and technological advances have reduced the use of lead in recent years in some applications.

Lead is commercially available in the following forms: lead wool for caulking, foil down to 0.0125 mm (0.0005 in) in thickness, and sheet 2.4 m (8 ft) wide in almost any thickness. Standard "3-lb" lead roofing is 1.2-mm ($^3/_{64}$-in) thick and weighs approximately 3 lb/ft^2.

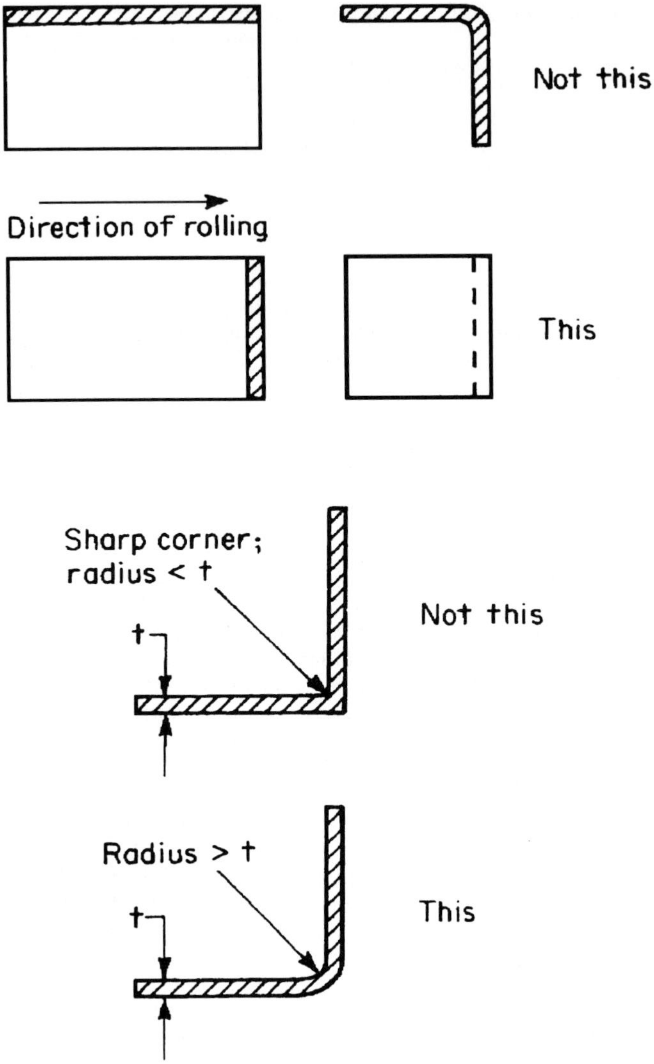

FIGURE 2.3.7 Design rules for bends in commercial zinc sheet.

TIN AND ITS ALLOYS

John Casteras
Alpha Metals, Inc.
Jersey City, New Jersey

Tin is a soft, silvery-white lustrous metal with a bluish tinge. It has a specific gravity of 7.3 and a melting point of 232°C (450°F). Its tensile strength is 14 to 28 MPa (2000 to 4000 lbf/in^2). Tin is a ductile, corrosion-resistant metal that is slightly harder than lead. It is highly reflective when polished, solders easily, and does not contaminate food.

Most tin is consumed as an alloying element in solder and bronze or as a coating for other metals. Coating applications include food containers (i.e., "tin cans"), electrolytic tin plate (i.e., tin-plated steel), tin-dipped-steel sheet, or terne-plated steel (terne is an alloy of lead and tin). Applications involving tin as the principal constituent are tinfoil (now replaced by aluminum foil except for electronic capacitors and bottle-cap liners), collapsible tubes, pewterware, costume jewelry, bearing linings, cooking vessels, and tubing and piping for food processing. Table 2.3.21 lists the forms of tin alloys that are available commercially.

Tin alloys are fully die-castable. Commonly cast alloys are white metal (92 percent Sn and 8 percent Sb), die-casting alloy (82 percent Sn, 13 percent Sb, and 5 percent Cu)(UNS L13820), tin babbitt alloy No. 1 (L13913), and tin babbitt alloy No. 5 (L13650).

Gravity casting is carried out with white metal or pewter. (Traditionally, pewter has been an alloy of about 88 percent tin and 12 percent lead; nontarnishing pewter or britannia metal is 91 percent tin, 7 percent antimony, and 2 percent copper.)

Spinning and various press-forming operations are easy with most tin alloys, notably pewter. Impact extrusion is another common process for which tin alloys are utilized. Pure tin, various tin-lead alloys, tin-lead-aluminum alloys, and "hard tin" are used.

TABLE 2.3.21 Commercially Available Mill Forms of Tin and Tin Alloys

Material	UNS or other designation	Mill forms/applications
	Tin and tin alloys	
Grade A tin	QQ-T-371 grade A, straights tin	Sheet, pipe and tube, foil, castings, powder
Hard tin	(99.6% Sn, 0.4% Cu)	Pipe and tube, foil
White metal	(92% Sn, 8% Sb)	Sheet, castings, jewelry
Pewter	L13912 L13911	Sheet Castings
Tin babbit alloy 1	L13913 CY 44A ASTM B23, alloy 1	Precision inserts of babbit-lined bearing shells, ingots, die castings
Tin babbit alloy 2	ASTM B23, alloy 2	
Tin babbit alloy 3	ASTM B23, alloy 3	
Die casting alloy	L13820 YC135A	Ingots, die castings
Tin babbit alloy 5	L13650 PY1815A	Ingots, die castings
	Tin-lead-antimony alloys	
Tin babbit alloy 8	L53565	Small ingots and bars, bearings
8(YT 155A)	L53560	Ingots and bars, die castings
Y10A	L53340	Ingots and bars, die castings
Tin babbit alloy 15	L53620	Ingots and bars, bearings
	Solders	
Antimonial tin solder	L13940 ASTM B-32 grade Sb 5	Bar, wire for soldering copper plumbing tubing
Tin-silver solder	(95% Sn, 5% Ag) ASTM B-32 grade Sn 95	Bar, wire for oxygen, air-condition- ing, food-handling equipment
Eutectic solder	L13630 ASTM B 32 grade 63	Bar, wire, paste for electronic circuit boards
60-40 solder	L13600 ASTM B-32 grade 60	Bar, wire, paste for electronic circuit boards
62-36-2 solder	(62% 5n, 36% Pb, 2% Ag	Bar, wire, paste for electronic circuit boards where thermal stress resistance is important

Note: For additional solders, see Chap. 7.4.

NICKEL AND ITS ALLOYS

Russ W. Kirchner
High Performance Alloys, Inc.
Tipton, Indiana

Nickel is a ductile, tough, magnetic, silver-white metal. It has a specific gravity of 8.8 and a melting point of 1452°C (2646°F). Nickel has a high resistance to corrosion in many media and is also abrasion-resistant. It is similar in some respects to iron, for example, in being strong, tough, and magnetic. Nickel also has good thermal conductivity and takes and retains a high polish. The tensile strength of nickel is approximately 480 MPa (70,000 lbf/in^2) when annealed and 790 MPa (115,000 lbf/in^2) when cold reduced.

Applications

The principal use of nickel is as an alloying agent with steel, to which it imparts corrosion resistance and strength. A sizable portion of current production is used for electroplating. About one-fourth of current production is devoted to alloys of which nickel is the principal ingredient. These alloys are used in applications demanding corrosion resistance (e.g., to caustic alkalies) and, often, the ability to maintain their properties at elevated temperatures.

The following are typical applications of nickel alloys: aircraft engine and aircraft structural parts, coinage, dies, cast propellers, valve seats, hard surfacing (from powder), corrosion-resistant chemical-processing equipment and containers, resistance heating wire, magnet parts, and food-handling equipment.

The following are specific applications for certain particular alloys: Incoloys 800 (N08800), 800H (N08810), and 800HT (N08811) are used for high-temperature furnace applications. Inconels are used for heat-treatment containers and furnace components, as well as for gas-turbine-engine parts subject to high heat, e.g., manifolds and certain springs. They are also used for dairy and food equipment, including milk-pasturizing apparatus and distilling equipment for spirits. Inconel X-750 (N07750) is used for gas-turbine blades and other high-temperature components. Monel alloys 400 (N04400), 405 (N04405), and K-500 (N05500) are used in marine applications such as propellers and propeller shafts and in food-service and food-processing applications. They are also used in refinery equipment, salt plants, steel-pickling equipment, and laundry and dry-cleaning machines. Hastelloys containing chromium, molybdenum, and sometimes tungsten are used for high-temperature and corrosive-environment applications. Hastelloy C-276 (N10276) is a "universal" corrosion-resistant alloy usable in both oxidizing and reducing environments. Hastelloy B-2 (N10665), nickel base, 28% molybdenum, is used in reducing corrosive environments. Nickel-chromium alloys such as ASTM B344-59T (80 percent Ni, 20 percent Cr) (N06003) are used for resistance heating wires and corrosion-resistant heating equipment.

Mill Process

Wrought-nickel alloy bar, plate, sheet, and strip are produced by casting, forging, and hot and cold rolling.

Grades for Further Processing

Most high-nickel alloys are best fabricated by forming and welding. They are generally difficult to cast, and machinability ratings usually range around 20 percent. The following is a list of alloys commonly used or particularly suitable for various manufacturing processes.

Sand-mold casting: Hastelloy B (N10001), Hastelloy C (N10002), Hastelloy C-4 (N06455), Inconel 625 (N066250, Monel 400 (N04400), and S Monel (N24030)

Forging: Monel 400 (N04400), Monel 405 (N04405), Monel K-500 (N05500), Hastelloy C-276 (N10276), Hastelloy C-4 (N06455), Hastelloy S (N06635), Inconel 600 (N06600), Inconel 625 (N26625), Incoloy 800H (N08810), Incoloy 800HT (N08811), and nickel chromium alloy, 80 percent Ni, 20 percent Cr

Machining: Hastelloy C (N10002), Monel R-405 (N04405), 400 (N04400), and K-500 (N05500)

Forming: Nickel 200 (N02200), Monel K-500 (N05500), K Monel (N04405), Hastelloy C-276 (N10276), Hastelloy B-2 (N10665), Hastelloy C-4 (N06455), Hastelloy C (N10002), and Inconel 625 (N06625)

Arc welding: Nickel 200 (N02200), Monel 400 (N04400), Monel 405 (N04405), Hastelloy B-2 (N10665), Hastelloy C-4 (N06455), Hastelloy S (N06635), Hastelloy X (N06002), and Hastelloy W (N10004)

Brazing: Monel, K Monel (N04405); for silver solder, Inconel

Resistance welding: Inconel 600 (N06600), Inconel 625 (N06625), Hastelloy B (N10001), Hastelloy C (N10002), Hastelloy B-2 (N10665), Hastelloy C-276 (N10276), Hastelloy C-4 (N06455), Hastelloy S (N06635), Hastelloy W (N10004), and Hastelloy X (N06002)

Gas welding: Not recommended for Hastelloy alloys

Soldering: Monel, K Monel (N04405), and Inconel

Available Shapes and Sizes

Nickel alloys are currently available in the form of plates, bars, sheets, strips, rods, wires, tubing, flats, powder, and wire cloth. Table 2.3.22 lists available sizes of mill forms and alloy availability in various forms.

TABLE 2.3.22 Thickness, Size Range, and Availability of Various Nickel-Alloy Mill Forms

	Thickness and size range of mill forms	
Form	Thickness range, in	Size range, in
Strip	0.001–0.125	Coils, 14
Plate	0.1875–4.00	10–150, width
Bar		
Square	$\frac{3}{8}$–$2\frac{1}{4}$	$\frac{3}{8}$–$2\frac{1}{4} \times 360$
Hexagonal	$\frac{3}{8}$–$2\frac{1}{2}$	$\frac{3}{8}$–$2\frac{1}{2} \times 360$
Square forged	$2\frac{1}{2}$–6	$2\frac{1}{2}$–6×72
Rod		
Cold drawn	$\frac{1}{16}$–4.00	456 maximum
Hot finished	$\frac{1}{4}$–4.50	288 maximum
Forged billets	12–25	
Wire, round		
Hot rolled	$\frac{1}{4}$–$\frac{7}{8}$	Coils
Cold drawn	0.001–0.875	Coils
Tube		
Cold drawn	0.012–8.00	0.002–0.500
Extruded	$2\frac{1}{2}$–$9\frac{1}{4}$	$\frac{1}{4}$–1.000

Available mill forms										
Material	UNS designation	Strip	Sheet	Plate	Bar and rod	Shapes	Wire	Pipe and tube	Forgings	Billets
Nickel and alloys										
Nickel 200	N02200	x	x	x	x	x		x		
Nickel 201	N02201	x	x	x	x	x		x		
Monel 400	N04400	x	x	x	x	x		x	x	x
Monel K-500	N05500				x				x	x
Monel R-405					x	x				
Nickel-base superalloys										
Inconel 600	N06600	x	x	x	x	x	x	x	x	x
Inconel 601	N06601	x	x	x	x	x	x	x	x	x
Inconel 625	N06625	x	x	x	x	x	x	x	x	x
Inconel X-750	N07750	x	x	x	x		x			
Hastelloy B-2	N10665	x	x	x	x		x	x	x	x
Hastelloy C-4	N06455	x	x	x	x		x	x	x	x
Hastelloy C-276	N10276	x	x	x	x		x	x	x	x
Hastelloy S	N06635		x		x				x	
Hastelloy X	N06002		x	x	x		x	x	x	
Unitemp HX	N06002		x	x	x		x	x	x	
Inconel 718	N07718	x	x	x	x				x	x
Udimet 500	N07500	x	x	x	x					x
Waspaloy	N07001	x	x	x	x		x			x
Nicrotung				Only as castings						
Rene 41, R-41*	N07041	x	x	x	x		x		x	
Inconel 713	N07713			Only as castings						
IN-100	N13100			Only as castings						
Low-expansion nickel alloys										
NI 36	K94610	x	x	x	x		x	x	x	
Ni 42	K94100	x	x	x	x		x	x	x	
Ni-Span C-902	N09902	x			x		x			

*Also as castings.

Source: Updated from AMCP 706-100.

COBALT AND ITS ALLOYS

Russ W. Kirchner
High Performance Alloys, Inc.
Tipton, Indiana

Cobalt is a white metal that resembles nickel but is much rarer and more expensive. It has a specific gravity of 8.8 and a melting point of 1495°C (2723°F). When used as an alloying ingredient, cobalt imparts high-temperature strength and corrosion resistance. The cobalt-based superalloys have high-temperature properties of high strength, shock resistance, low creep, and wear and corrosion resistance.

Common applications of cobalt alloys are jet-engine parts and other engine components, dental and surgical implants, burnishing and cutting tools, wear strips, permanent magnets, antifriction bearings, other high-wear machine parts, and agricultural equipment.

Particular applications of some of the notable cobalt alloys are as follows: Stellite 6 and 6B (UNS designation R30006) (cobalt, chromium, tungsten) have superior high-temperature properties and are used for fuel nozzles and bearings of gas turbine engines, ordnance parts (gun barrels), cutting tools, and knives. HS188 (UNS R30188) is used in burner cans of military gas turbines. HS25 (R30605), also designated L605, is also used in gas-turbine parts, bearings, and other high-temperature applications. N-155 (R30155), also designated Multimet, and Haynes 556 (R30556) are used in land-based turbine engine components. F90 (R30605), also designated L605, MP35N (R30035), and CCM (R30075), also designated F799 or F75, are all used for surgical implants and dental applications. Stellite 32 (R30002), also designated alloy F, and Eatonite (N06005) (cobalt, nickel, tungsten) are used for valve facings in internal-combustion engines. Invar (K93600) has zero thermal coefficient of expansion. Alnico (aluminum, nickel, cobalt) alloys are used for magnets.

Cobalt alloys are not easily machined. Stellite, for example, has a machinability rating of 15 (1112 steel = 100). The normal manufacturing process for the hard cobalt-chromimum-tungsten-molybdenum wear-resistant alloys is casting to as close to finished size as possible and then grinding. These alloys exhibit good fluidity even at low carbon levels. Cobalt alloys are particularly suitable for investment casting. Welding of cobalt alloys in plate and sheet form is feasible by GTAW, GMAW, and SMAW methods. Commercially available forms of cobalt alloys are shown in Table 2.3.23.

TABLE 2.3.23 Commercially Available Mill Forms of Cobalt and Cobalt Alloys

Material	UNS no.	Sheet	Plate	Bar and rod	Wire	Forgings	Billets
Cobalt		Roundelle powders					
HS25 (L605 or F90) R30605		x	x	x	x	x	x
HS188	R30188	x	x	x	x	x	x
6B,6BH	R30006	x	x	x	x	x	x
MP35N	R30035			x	x		
CCM (F799)	R30075			x	x		
F75	R30075			x	x		
Kovar	K94610	x	x	x	x	x	x

Note: Alloys 3 (R30001), 6 (R30006), 21 (R30021), 25 (R30605), and 31 (R30031) are supplied as ingots for casting.

TITANIUM AND ITS ALLOYS

Charles W. Hatmaker

Titanium and Alloys Corp.
Warren, Michigan

Titanium is a light metal (43 percent lighter than steel, 67 percent heavier than aluminum; specific gravity 4.5) with excellent corrosion resistance. It melts at 1668°C (3035°F) and has a tensile yield strength of 828 MPa (120,000 lbf/in^2) or higher, about equal to that of steel. Its strength-to-weight ratio is the highest of any structural metal, 30 percent greater than that of steel or aluminum. The outstanding strength of titanium is maintained at temperatures from −250 to 540°C (−420 to 1000°F). Titanium has low thermal and electrical conductivity and a low thermal coefficient of expansion. Its wider use has been hampered by its high cost (about double that of stainless steel) and its tendency toward contamination during some processing operations with hydrogen, nitrogen, or oxygen, all of which cause embrittlement.

Applications

Titanium is an important material in the aerospace industry. It is used for gas-turbine compressors and other jet-engine components, airframe parts, airborne equipment, and missile fuel tanks. Titanium is also used for chemical-processing equipment including heat exchangers, compressors, and valve bodies. Other applications are surgical implants and instruments, marine hardware, paper-pulp equipment, and portable machine tools.

Castings of titanium and some of its alloys can be produced in sizes up to 60 in in diameter. A vacuum-casting process is used with graphite or investment molds.

Grades for Further Processing

Special precautions are necessary in processing titanium at elevated temperatures because of its susceptibility to embrittlement from absorption of hydrogen, oxygen, and nitrogen. Protective atmospheres are required.

The grades of titanium most suitable for common manufacturing operations are as follows:

Casting: Titanium castings are not common. Ti-5A1-2.5Sn (R54521) is one alloy that is cast.

Forging: Commercially pure titanium is preferred to alloys. Among the latter, alpha-beta alloys are more easily forgeable than alpha alloys.

Machining and grinding: There is no free-machining titanium alloy. Commercially pure titanium is more machinable than the alloys and has a machinability about equal to that of 18-8 stainless steel.

Bending and forming: Commercially pure titanium has the best formability, followed by 99.0 percent pure titanium. Ti-3A1-2.5V (R56320) has good formability. Other suitable alloys are Ti-5A1-2.5Sn (R54521), Ti-6A1-4V (R56401), and Ti-3A1-13V-11Cr (R59010).

Welding: The lower-yield-strength grades (under 480 MPa, or 70,000 lbf/in^2) are generally most easily welded. Commercially pure unalloyed titanium is best. The alpha alloy Ti-5A1-2.5Sn (R54521) is good. Alloys Ti6A1-4V (R56401) and Ti-5A1-2.5Sn (R54521) also are satisfactory.

Heat treatment: Alloy Ti-6A1-4V (R56401) can be heat-treated to 1170 MPa (170,000 lbf/in^2) tensile yield strength. The alloy Ti-6A1-6V-2Sn (R56620) is heat-treatable to 1310 MPa (190,000 lbf/in^2). Beta alloys are heat-treatable to over 1380 MPa (200,000 lbf/in^2).

Available Shapes and Sizes

Titanium and its alloys are commercially available in a variety of wrought-mill forms such as plates, sheets, rods, wire, and tubing. Table 2.3.24 lists standard sizes, and Table 2.3.25 indicates which grades are available in each form.

Design Recommendations

When bending titanium sheet, generous bend radii are advisable. The minimum-bend radii for commercially pure titanium are 1.5 T at a forming temperature of 290°C (550°F) and 2 T at room temperature, where T = stock thickness.

TABLE 2.3.24 Thickness, Size Range, and Availability of Titanium and Titanium Alloys

Form	Thickness and size range of mill forms	
	Thickness range, mm (in)	Size range, m (in or ft)
Sheet	>0.4 (>0.0015)	Maximum width 1.2 (48 in)
Plate	3–150 (0.125–6)	2.4 × 8.9 (96 × 350 in)
Rod and bar	To 900 (36)	Maximum length 5.5 (18 ft)
Wire	2.5 min. dia. (0.001 min. dia.)	Coils or random length to 4 or 5 (14–16 ft)

TABLE 2.3.25 Available Mill Forms of Titanium Alloys

Material	UNS no.	Strip	Sheet	Plate	Bar and rod	Wire	Tube	Forgings	Billets
Unalloyed*†	R50250, R50550, R52250	x	x	x	x	x	x	x	x
5A1-2.5Sn*	R54521	x	x	x	x	x		x	x
5A1-5Sn-5Zr	R54550	x	x	x	x	x		x	x
8A1-1Mo-1V	R54810	x	x	x	x	x		x	x
7A1-4Mo*	R56740	x	x	x	x	x		x	x
6A1-6V-2Sn*	R56620	x	x	x	x	x		x	x
6A1-4V*	R56401	x	x	x	x	x	x	x	x
8Mn1	R56080	x	x	x					
13V-11Cr-3A1*	R59010	x	x	x	x	x		x	x

*Also as foil.

†Also extruded forms.

When designing forgings, the following limits are recommended for average-size nonprecision parts:

Minimum draft angle: 5 to 7°

Minimum web thickness (closed-die forgings): 16 mm (0.63 in)

Minimum rib width: 10 mm (0.39 in)

Maximum rib height: 4 times rib width

Minimum fillet radius of ribs: 25 percent of rib height

Note: These limits vary with the size of the forging and can be bettered at extra cost in precision forgings. See Fig. 2.3.8 for an illustration of these limits.

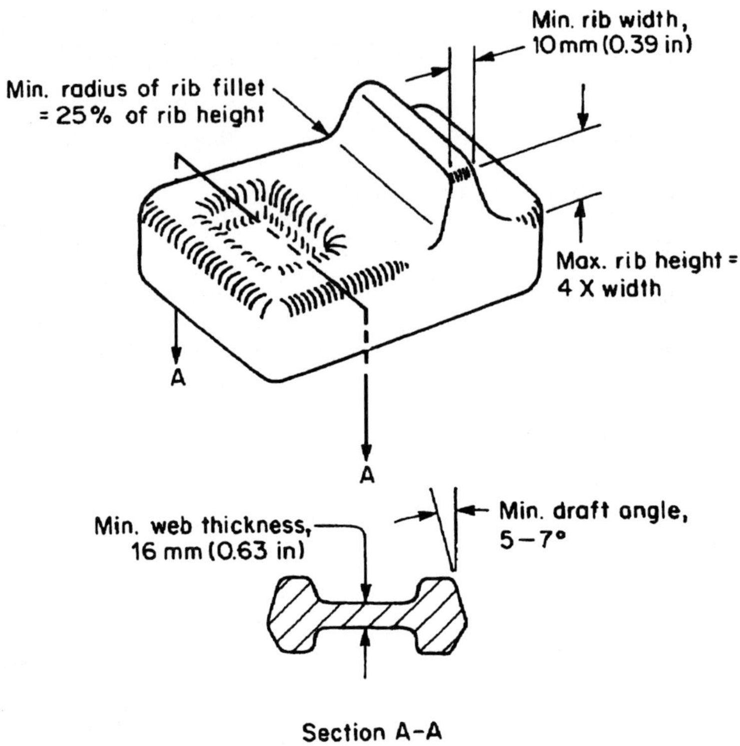

FIGURE 2.3.8 Design rules for titanium forgings.

REFRACTORY METALS AND THEIR ALLOYS

Jan Carlén
Boris Bryskin

Rhenium Alloys, Inc.
Elyria, Ohio

Metals are classified as *refractory* when they have an extremely high melting point, higher than the range of iron, nickel, and cobalt. Some metals in this category are chromium, niobium, molybdenum, tantalum, rhenium, and tungsten. Of these, tungsten, molybdenum, and tantalum are most commonly utilized unalloyed or as the major alloying ingredient.

Tungsten

Tungsten is a heavy, hard, ductile, and strong metal. Its specific gravity is 19.6, and its melting point, 3410°C (6170°F), the highest of any known metal. Tungsten is corrosion-resistant even at high temperatures. Its tensile strength is in the range of 345 to 1380 MPa (50,000 to 200,000 lbf/in^2) but can be as high as 3450 MPa (500,000 lbf/in^2).

Tungsten, in addition to its use as an alloying agent to increase the hardness of other metals and its role as a major constituent of tungsten carbide, has a number of very important uses. Many of these are for electrical applications. Tungsten is used for filament wires for incandescent lights and electron tubes. It is employed to make spark-plug electrodes, ignition breaker points for internal-combustion engines, circuit breakers, and lead-in wires for power electron tubes. Tungsten is employed for rocket nozzles and other aerospace components, magnets, vibration-damping devices, containers and shielding for radioactive environments, and reinforcing filament for metals, ceramics, and plastics.

Filament wire may be only 10 to 15 μm (0.0004 to 0.0006 in) in diameter and in some cases is as fine as 5 μm (0.0002 in).

Tungsten is processed at 1600 to 1800°C (2900 to 3300°F). The mill process for producing wrought shapes involves powder-metallurgy processes, followed by forging or rolling at elevated temperatures.

Table 2.3.26 lists the normally available forms of tungsten, tantalum, and molybdenum.

Molybdenum

Molybdenum is a silvery-white, ductile but hard metal with good resistance to corrosion. Its specific gravity is 10.2, and its melting point is 2620°C (4750°F). Molybdenum has greater structural strength over 870°C (1600°F) than any other com-

TABLE 2.3.26 Commercially Available Mill Forms of Tantalum, Tungsten, and Molybdenum Alloys

Material Common designation	Sheet	Plate	Bar and rod	Wire	Pipe and tube	Billets
Tantalum*	x		x	x	x	
Tungsten*	x	x	x	x	x	
Molybdenum*	x	x	x	x	x	x
Rhenium†	x	x	x	x	x	
Mo-0.05Ti, TZM (0.05Ti-0.01Zr)‡	x	x	x		x	x
Tantalum 10-W	x		x			

*Also as foil and powder.

†Including rhenium alloys with tungsten and molybdenum.

‡Also as forgings.

Source: Updated from AMCP 706-100.

mercial metal. The room-temperature tensile strength of a typical alloy (Mo - 1.5Ti) is 900 MPa (130,000 lbf/in^2).

The major use of molybdenum is as an alloying ingredient for steel. Only a small amount (often 0.15 to 0.50 percent) of molybdenum improves the heat treatability and toughness of steel. In stainless steels and in nonferrous alloys, molybdenum increases corrosion resistance. It is also added to cast iron to increase strength.

However, alloys containing molybdenum as the principal constituent are important too. Applications of these alloys comprise electrical parts such as arc-resistant contacts, furnace heating elements, and support members for electron-tube and lightbulb-wire elements; high-temperature structural parts for missiles and jet engines; tools for hot-metal working; boring bars; and nuclear-reactor parts.

Like other refractory metals, molybdenum is difficult to work. Wrought-mill forms are made from powder-metal ingots and hot-rolling processes. Because of their tendency to work-harden, shearing, bending, and other press-forming operations must be carried out at an elevated temperature. Welding requires a closely controlled protective atmosphere. Molybdenum alloys are hardened by work hardening rather than by heat treatment.

Available mill forms include sheet from 0.025 mm (0.001 in) to 9.5 mm (0.375 in) and wire as fine as 0.10 mm (0.004 in). Table 2.3.26 summarizes the forms available.

Tantalum

A silver-gray metal that resembles platinum and niobium, tantalum is very ductile, hard, and corrosion-resistant. It has a specific gravity of 16.6 and a melting temperature of 2850°C (5160°F). The tensile strength of annealed sheet is 350 MPa (50,000 lbf/in^2), and that of drawn wire is 900 MPa (130,000 lbf/in^2). The corrosion resistance of the pure metal is better than that of tantalum alloys.

The largest uses of tantalum are in the electronics industry, for capacitors, sputtering targets, filaments, and getter devices. Tantalum also is used in applications for which corrosion resistance and high-temperature performance are important. Examples of applications are chemical equipment such as heat-exchange coils for acid baths, surgical implants, permanent surgical sutures, surgical instruments, incandescent lights, electrical rectifiers, and rocket-motor parts. Tantalum is also alloyed with steel for high-temperature applications and as a carbide for cutting tools and wear-resistant parts.

Tantalum is more easily processed than the other common refractory metals. Unalloyed tantalum has about the same machinability as cold-finished steel; alloys may be somewhat less machinable. Forming is straightforward owing to the outstanding low-temperature ductility of tantalum; Ta-10 percent W is an easily formed alloy. Tantalum can be forged cold. It can be brazed with copper alloys in a vacuum and can be resistance-welded and arc-welded with TIG, metal-inert-gas (MIG), or electron-beam welding. Mill forms of tantalum and its alloys are made by cold rolling or drawing billets made from metal powders. A number of forms are commercially available: sheets, foil, rods, bars, tubing, and wire. Foil is available down to 0.005 mm (0.0002 in) in thickness. Sheet is made in thicknesses up to 6.5 mm (0.250 in), but 0.33 to 0.75 mm (0.013 to 0.030 in) is more common, and the normal thickness for condensers and heat exchangers is 0.38 to 0.5 mm (0.015 to 0.020 in). Sheet sizes are 0.9 by 2.4 m (36 by 96 in) or smaller. Wire is available in a diameter range from 0.025 to 0.38 mm (0.001 to 0.015 in). Common diameters for welded tubing are 19, 25, and 38 mm ($\frac{3}{4}$, 1, and $1\frac{1}{2}$ in). Table 2.3.26 summarizes the forms commercially available.

Rhenium

Rhenium is a silvery-white metal with very high density (21.04 g/cm^2) and a high melting point (3453 K, 3180°C). It is the only metal with good ductility in the temperature range 0 to 2775° K (−273 to 2500°C). Although soft in the recrystallized-annealed condition, rhenium work-hardens very quickly, and many process annealings are usually required when rhenium is cold formed. Most rhenium metal products are used for applications where the combination of its high-temperature strength and low-temperature ductility can be utilized. Despite its high density, there are many space applications for rhenium. Rhenium is also used as an alloying element with tungsten and molybdenum to improve their low-temperature ductility. The most common alloys are W-25/26%Re and Mo-41/48%Re.

PRECIOUS METALS AND THEIR ALLOYS

Considered by most persons only in terms of their use for jewelry and coinage, the precious metals are actually more important for their industrial applications. Covered here are the most significant precious metals for such applications: gold, silver, and platinum.

Gold

Gold is a soft yellow-colored metal with a specific gravity of 19.3 and a melting point of 1061°C (1943°F). It is a corrosion-resistant material, unaffected by many corrosive substances. It has a tensile strength, when cast, of 140 MPa (20,000 lbf/in^2) and a high cost ($3500 per pound in 1998).

Aside from well-known jewelry, coinage, and electroplating applications, gold has a number of important commercial uses. Among them are dentistry (alloys), infrared and red light reflectors, electrical contacts, particularly those in which contact pressure is low and no oxide or sulfide layer is permissible, laboratory crucibles and dishes for use in handling acids, electrical terminals, eyeglass frames, potentiometer wire (with iron and palladium), gold leaf (0.00013 mm, or 0.000005 in, thick) for decorative use in signs, books, etc., other chemical apparatus (in alloy with palladium and platinum), and grid wires in electron tubes.

Gold mill forms are produced by cold rolling or drawing. Since gold work-hardens, annealing at 500°C (932°F) is sometimes necessary. Gold leaf is produced by a hammering process rather than by rolling.

Available forms of gold and other precious metals are shown in Table 2.3.27.

Silver

Silver is a white metal that is very ductile and malleable. Its specific gravity is 10.7, and its melting temperature is 961°C (1762°F). Silver has the best electrical and heat conductivity of any metal. Its hardness lies between that of gold and copper; its malleability is second only to gold's. Its tensile strength is 180 MPa (26,000 lbf/in^2). Fine silver is 99.9 percent pure, sterling silver is 92½ percent silver and 7½ percent copper,

TABLE 2.3.27 Commercially Available Mill Forms of Precious Metals*

Metal	Strip	Sheet	Bar and rod	Wire	Pipe and tube	Foil	Powder
Gold		X	X	X	X	X	X
Silver	X	X	X	X	X		X
Platinum		X		X	X	X	X
Palladium		X		X	X	X	X
Rhodium		X		X			X
Ruthenium*	X		X				X
Osmium		Cast or sintered parts					
Iridium		X	X	X	X		X

*Also as sintered parts.

Source: From AMCP 706-100.

and commercial silver is 90 percent silver and 10 percent copper. Annual production of silver in the United States is about 1300 tons, and its price was approximately $75 per pound in 1998.

Major applications of silver are tableware, jewelry, electrical contacts, solders, dentistry (in alloy with tin and mercury), chemical and food equipment (clad metal), bearings in aircraft and rockets, electronic circuits, coinage, batteries, and electroplating. A substantial amount of silver is also used in the manufacture of photographic film.

Silver is customarily worked cold and annealed as necessary at 500°C (932°F). Wire of less than 0.025 mm (0.001 in) in diameter is difficult to draw. Available forms are tabulated in Table 2.3.27.

Platinum

Platinum is a heavy grayish-white metal that is very soft in the unalloyed state. Its specific gravity is 21.5, and its melting temperature is 1769°C (3217°F). Platinum is very ductile, more so than gold, copper, or silver. It is noncorroding and nontarnishing even at elevated temperatures and is inert to common strong acids. Its tensile strength when annealed is 120 MPa (17,000 lb/in^2). Annual world production is about 10 tons. Platinum lends its name to the platinum group of metals, all of which have similar properties. Other metals in this group are palladium, rhodium, iridium, osmium, and ruthenium.

Major applications for platinum are catalysts, jewelry, electrical contacts, particularly those with low pressure, thermocouples, resistance wires, resistance thermometers, electrodes (insoluble anodes), chemical-industry-equipment components (as cladding and in solid components), extrusion dies for glass fibers, crucibles for melting glass, filaments, burner nozzles, dental applications, aircraft spark plugs (alloyed with tungsten), electron-tube grids, brazing metal for tungsten, and various electroplated coatings.

Mill forms of platinum are produced by hot or cold working. Annealing after work hardening takes place at 1000°C (1832°F).

Platinum is available in the form of gauze, for catalyst applications; foil, down to 0.005 mm (0.0002 in) in thickness; and wire, down to 0.01 mm (0.0004 in) in diameter. See also Table 2.3.27.

CHAPTER 2.4
NONMETALLIC MATERIALS

Charles A. Harper

Technology Seminars, Inc.
Lutherville, Maryland

CLASSIFICATION OF MATERIALS

Especially in view of the explosive growth in recent years in the availability and use of polymeric materials (plastics and synthetic rubbers), nonmetallic materials are major constituents of present-day manufactured goods. For many applications they offer superior properties or less costly fabricability compared with metals, and designers have taken advantage of these attributes.

Nonmetallic is a broad category. It comprises organic materials of natural origin such as wood, leather, and natural rubber. It includes materials such as plastics and paper that are manufactured at least in part from natural substances. It also includes inorganic (mineral) materials such as glass, ceramics, and concrete. Figure 2.4.1 illustrates the generic relationships of common nonmetallic materials.

GENERAL PROPERTIES OF NONMETALLIC MATERIALS

Nonmetallic materials have varied properties, and few characteristics are applicable to all of them. Two that are almost universal are low electrical and heat conductivity. With the exception of carbon, nonmetallic materials, when dry, are nonconductors.

Nonmetallic materials are usually less tough and less strong than metals, except that the inorganic materials normally have very high compressive strengths. The inorganic materials also have superior high-temperature properties. Resistance to corrosion is a common property with many nonmetallics. Ease of fabrication is a property shared by polymers, wood, and some other organic materials. Table 2.4.1 lists the maximum service temperatures of selected nonmetallics.

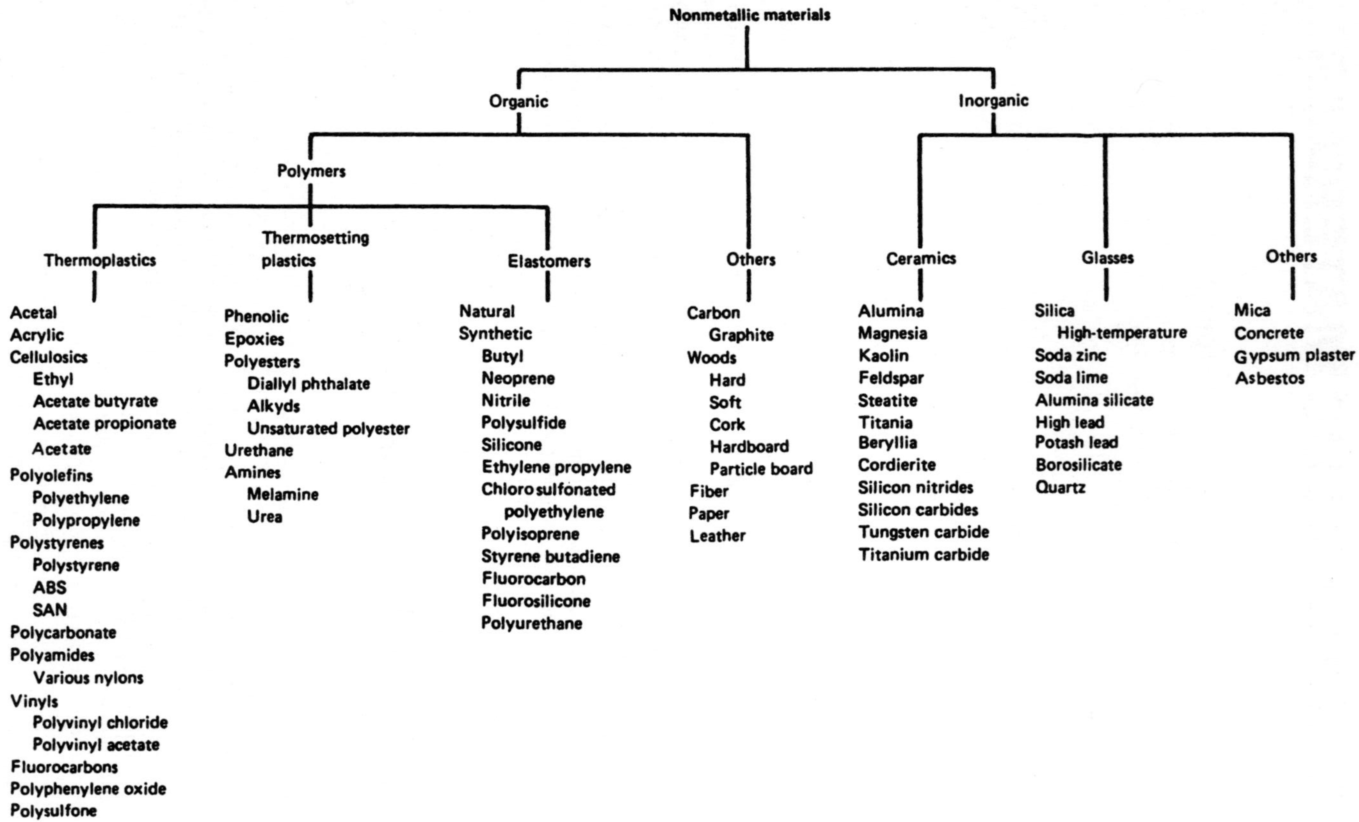

FIGURE 2.4.1 Generic relationships of common nonmetallic materials.

TABLE 2.4.1 Maximum Service Temperatures of Plastics and Other Nonmetallics, °C

Beryllia	2400	Polysulfone	150–175
Silicon carbide	2310	Nylon	80–150
Alumina	1950	Polycarbonate	95–135
Mullite	1760	Polypropylene	90–125
Cubic boron nitride	1600	Polyethylene	80–120
Porcelain enamel	370–820	Felt, rayon viscose	107
Silicones	260–320	Polyurethane	90–105
Polyesters	120–310	Acetal	85–105
Glass, soda-lime	290	Polystyrene	65–105
Epoxy	95–290	Cellulosic	50–105
Glass, borosilicate	260	ABS	60–100
Fluoroplastics	50–260	Acrylic	52–95
Phenolic	90–260	Natural rubber	82
Melamine	100–200	Vinyl	55–80

POLYMERS: PLASTICS AND ELASTOMERS

Practically stated and with the fine points of definition omitted, a plastic is an organic polymer available in some resin form or a form derived from the basic polymerized resin. These forms include liquids and pastes for embedding, coating, and adhesive bonding. They also encompass molded, laminated, or formed shapes, including sheet, film, and larger bulk shapes. While there are numerous minor classifications for polymers, depending on how one wishes to categorize them, nearly all can be placed in one of two major classifications. These two major plastic-material classes are thermosetting materials (or thermosets) and thermoplastic materials, as shown in Fig. 2.4.1. Although Fig. 2.4.1 shows elastomers separately (and they are a separate application group), elastomers, too, are either thermoplastic or thermosetting, depending on their chemical nature.

As the name implies, thermosetting plastics, or thermosets, are cured, set, or hardened into a permanent shape. This curing is an irreversible chemical reaction known as *cross-linking,* which usually occurs under heat. For some thermosetting materials, however, curing is initiated or completed at room temperature. Even then, however, it is often the heat of the reaction, or the exotherm, that actually cures the plastic material. Such is the case, for instance, with room-temperature-curing epoxy, polyester, or urethane compounds.

Thermoplastics differ from thermosets in that they do not cure or set under heat. They merely soften, when heated, to a flowable state in which under pressure they can be forced or transferred from a heated cavity into a cool mold. Upon cooling in a mold, thermoplastics harden and take the shape of the mold. Since thermoplastics do not cure or set, they can be remelted and rehardened by cooling many times. Thermal aging, brought about by repeated exposure to the high temperatures required for melting, causes eventual degradation of the material and so limits the number of reheat cycles.

The term *elastomers* includes the complete spectrum of elastic or rubberlike polymers that are sometimes randomly referred to as *rubbers, synthetic rubbers,* or *elastomers.* More properly, however, rubber is a natural material, and synthetic rubbers are polymers that have been synthesized to reproduce consistently the best properties of natural rubber. Since such a large number of rubberlike polymers exist, the broad term *elastomer* is most fitting and most commonly used.

TABLE 2.4.2 Commercially Available Shapes of Plastics Materials Other than Molding Compound Suitable for Further Fabrication

Material	Forms commonly available*	Typical applications
ABS	S,R	Decorative panels, vacuum forming
Acetal	S,P,R	Mechanical parts
Acrylic	S,P,R,T,E	Housewares, furniture, glazing
Cellulose acetate butyrate	S	Machine guards
Epoxy laminate	S,P,R,T	Aerospace structural parts, insulation
Melamine laminate	S,P,R,T	Switch boxes, circuit breakers, arc barriers
Nylon	S,P,R,T,E	Wear-resistant surfaces
Phenolic	S,P,R,T	Electrical and electronics parts
Polycarbonate	S,P,R,T,F,	Glazing
Polycarbonate, glass indent-reinforced	S,R	
Polyester	F	Membrane switches, graphic arts
Polyethylene, low-density	S,P,R,T,F	Tanks, containers
Polyethylene, high-density	S,P,R,T	Containers, coverings
Polypropylene	S,P,R,T	Tanks, ducts
Polyethylene, ultra-high molecular weight	S,P,R,E	Wear-resistant surfaces, machinery parts
Polystyrene	S	Thermoforming
Polyurethane	S,P,R,T,F	Seals, couplings, diaphragms, wheels, liners
Teflon	S,P,R,T,F	Low-friction surfaces
Viny, rigid	S,R,P	Vacuum forming
Vinyl, flexible	S,T,F	Low-pressure hose
Acrylic-PVC alloy	S	Thermoforming
Polyphenylene oxide	S,P,R	High-temperature applications
Polamide-imide reinforced	P,R	High-temperature seals and bearings

*T = tubing; S = sheet (0.50 mm [0.020 in] to 13 mm [0.5 in] thick); P = plate (13 mm [0.5 in] to 100 mm [4.0 in] thick); R = rod and round bar; E = square, rectangular, hexagonal, half-round, or other shapes; F = film (less than 0.50 mm [0.020 in] thick).

Applications of polymer materials, processing methods, and design recommendations are covered in Sec. 6 of this handbook. Materials for these processes are in granular or near-granular form, which is most convenient for molding operations. In addition, many bulk shapes and forms mentioned above are available commercially for machining and further fabrication. Table 2.4.2 lists the materials commonly available for such purposes.

OTHER ORGANIC MATERIALS

Carbon and Graphite

Carbon is a very common element and the key constituent of all organic materials. In the uncombined pure state, it exists as diamond or graphite. In a less pure state, it exists as charcoal, coal, or coke (amorphous carbon). Both amorphous carbon and graphite are produced in structural shapes when the particles are bonded together with elemental carbon.

Carbon and graphite exhibit properties similar to those of ceramics with two major exceptions. They are electrically and thermally conductive. The ceramiclike properties include greater compressive than tensile strength, a lack of malleability and ductility, and a resistance to high temperatures and corrosive environments. The usable temperature limits for carbon and graphite are on the order of 2400°C (4350°F) and even higher. Strength is actually higher at elevated than at room temperatures. Specific electrical resistance ranges from a low of 0.004 Ω · in (graphite) to 0.0022 Ω · in (carbon).

The production of carbon and graphite components involves two processes: (1) molding or extrusion followed by oven baking and (2) machining. High pressures and consequently significant die or mold costs are involved in the first method, which therefore is economically advantageous only for large-quantity production. Machining is more suitable for limited or moderate quantities.

Carbon and graphite components have mechanical, metallurgical, chemical, electrical, and nuclear applications. Typical uses include electrodes for the production of metals and chemicals in electric-arc furnaces, lighting electrodes, brushes for electric motors, electrodes in electrolytic cells, crucibles, molds for metal casting, resistance-furnace parts, and rocket components when high-temperature and thermal-shock resistance are important.

Carbon and graphite are available in round, square, and rectangular sections. Round bars commonly stocked range from 3 to 1100 mm ($\frac{1}{8}$ to 45 in) in diameter and from 300 to 2800 mm (12 to 112 in) in length. Rectangular bars range from 13 by 100 by 400 mm ($\frac{1}{2}$ by 4 by 16 in) long to 600 by 750 mm by 4.5 m (2 by $2\frac{1}{2}$ by 12 ft) long.

Graphite is favored over carbon for applications requiring extensive machining. Graphite machines fairly well, and tolerances comparable with those of rough machining of metals can be achieved. Carbon members are recommended if only cutoff or other minimum machining is required.

Wood

Wood has a number of desirable properties. It machines and fastens easily; it is attractive, it has a high strength-to-weight ratio; when dry, it has good electrical-, heat-, and noise-insulating properties; it is long-lasting in dry environments; and it accepts preservative treatment readily. On the negative side are its directional-strength characteristics (because of its grain), its large dimensional change and tendency to warp with changes in moisture content, its susceptibility to rot in moist environments, and its poor abrasion resistance. Table 2.4.3 summarizes the properties and uses of woods which grow in temperate climates.

CERAMICS AND GLASSES

Ceramics and glasses are nonorganic, nonmetallic materials made by fusing clays and other "earthy" materials that usually contain silicon and oxygen in various compositions with other materials.

TABLE 2.4.3 Properties and Uses of Wood

Species	Specific gravity	Yield strength, mPa (lbf/in^2)*	Workability with hand tools	Hardness†	Typical applications
Ash, black	0.49	50 (7200)	Poorer	520	Dowels, handles, lamp parts
Ash, white	0.6	61 (8900)	Poorer	1320	Handles, rollers, dowels
Birch	0.61	62 (9000)	Average	1210	Turnings, dowels, doors
Cherry	0.5	62 (9000)	Poorer	950	Furniture, caskets, blocks
Chestnut	0.43	42 (6100)	Best	540	Poles, railway ties, crates
Elm, American	0.5	52 (7600)	Poorer	830	Boxes, baskets, barrels
Fir, Douglas	0.44	52 (7400)	Poorer	600	Building lumber, crates
Hemlock	0.41	44 (6400)	Average	540	Building lumber, plywood
Hickory, bitternut	0.66	64 (9300)	Poorer	—	Tool handles, pallets
Maple, sugar	0.63	66 (9500)	Poorer	1450	Furniture, flooring, handles
Maple, silver	0.47	43 (6200)	Poorer	700	Building lumber, crossties
Oak, red	0.64	59 (8500)	Poorer	1310	Railroad ties, mine timbers
Oak, white	0.66	53 (7700)	Poorer	1320	Cooperage, fence posts
Pine, white	0.35	39 (5700)	Best	380	Sash, doors, paneling, boxes
Poplar, yellow	0.42	43 (6200)	Best	540	Furniture, plywood cores
Redwood	0.4	48 (6900)	Average	480	Outdoor furniture
Spruce	0.39	43 (6200)	Average	470	Building lumber, pulpwood

*Fiber stress at proportional limit in bending.

†Force in pounds required to embed a 0.444-in ball $^1/_2$ diameter.

Source: Prepared from material in *Wood Handbook,* U.S. Department of Agriculture.

Ceramics are hard, strong, brittle, and heat- and corrosion-resistant and are electrical insulators. They are used when these properties are important, particularly heat and corrosion resistance and electrical nonconductivity. Glass is used when transparency is important in addition to these other properties.

Also see Chap. 6.11 for information on the design of parts made from ceramic materials.

S • E • C • T • I • O • N • 3

FORMED-METAL COMPONENTS

CHAPTER 3.1
METAL EXTRUSIONS

THE PROCESS

Extrusion is a process in which a heated billet or ingot of metal is inserted in a chamber called a *container* and, by means of a hydraulically powered ram, is forced through a die hole of the desired shape. The material movement is in the same direction as the ram stroke. The metal emerges from the die in solidified form and closely conforms in cross section to the shape and dimensions of the die opening. Figure 3.1.1 illustrates the process schematically.

The ram stroke is usually horizontal. A lubricant is often employed to facilitate passage of the metal through the die. In the case of ferrous metals, a common method is the Sejournet process, in which molten glass is applied to the billet as a coating. The glass acts as both an insulator and a lubricant for the steel, which has been heated to the plastic or recrystallization temperature range. After extrusion, the section is stretch-straightened to remove twist and camber that can result from the process.

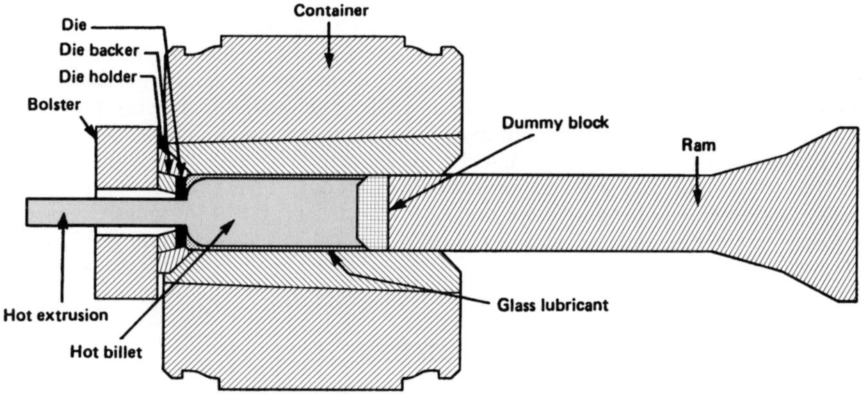

FIGURE 3.1.1 Sectional view of the extrusion process. (*Courtesy Babcock & Wilcox Co.*)

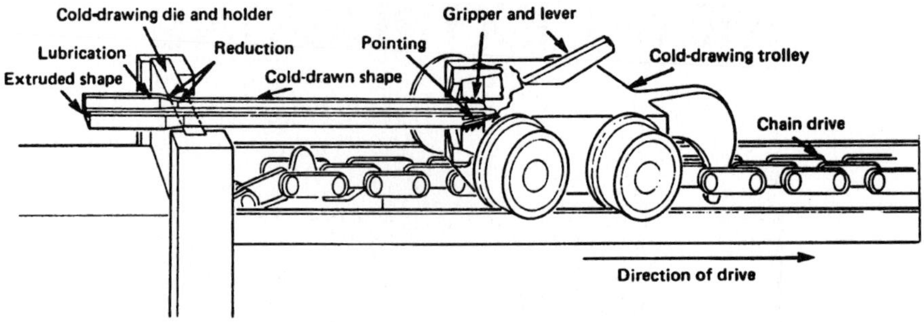

FIGURE 3.1.2 Cold drawing of an extrusion. (*Courtesy Babcock & Wilcox Co.*)

A common companion operation to extrusion is cold drawing of the extruded material. This secondary operation, illustrated by Fig. 3.1.2, tends to refine the molecular structure of the material and permit sharper corners and thinner walls in the extruded section. Cold drawing is more likely to be specified in the case of ferrous metals.

TYPICAL PARTS AND APPLICATIONS

Almost any part with a constant cross section may be suitable for extrusion. Typical extruded components have constant irregular or even intricate cross sections and often, but not necessarily, are long in comparison with their cross sections. The more intricate the cross section and the longer the part, the more advantageous extrusion will be as a manufacturing process. Cross sections that cannot be machined by normal machining processes are often economically produced by extrusion. Figure 3.1.3 illustrates typical extruded cross sections.

The maximum size of an extrudable cross section depends on the tonnage of the press available, the material to be extruded, and the complexity of the cross section. A standard method of measuring capacity is in terms of *circumscribing-circle diameter* (CCD). This is the size of circle into which the cross section will fit.

In aluminum, extrusions have been made with CCDs as small as 6.3 mm (0.25 in) and as large as 1.02 m (40 in). The bulk of aluminum-extrusion work, however, is done within a CCD of 250 mm (10 in). In steel, diameters are smaller because of the higher forces involved. The usual maximum CCD is about 150 mm (6 in). Extrusions of over 30 m (100 ft) in length and over 1 ton in weight can be made, but handling conditions usually limit piece length to 7.5 m (25 ft) or less.

Thin-walled extrusions are more difficult to produce than extrusions with heavy walls. Wall thickness for aluminum ranges from 1.0 mm (0.040 in) upward. For carbon steel, the minimum is 3.2 mm (0.125 in); for stainless alloys, 4.8 mm (0.187 in).

ECONOMIC PRODUCTION QUANTITIES

While commonly considered a high-production process, extrusion is surprisingly advantageous and should not be ruled out for short-run or even prototype production.

FIGURE 3.1.3 Typical extruded cross sections. (*Courtesy ITT Harper.*)

Conversely, for very high production levels and for shapes that can be produced by rolling, the latter process probably will prove more economical.

A prime advantage of the extrusion process is the low cost of tooling. Extrusion dies for the simpler shapes are low in cost. This is in contrast to rolling tooling, which usually involves a significant investment. Equipment for extrusion is applicable to any cross section within its tonnage capacity. Most such equipment is in the hands of vendors who specialize in the operation. Order charges cover extrusion die, setup, and operating costs. Operating costs are usually expressed on a per-unit-weight basis. There is customarily a minimum weight for each order of a special cross section. The minimum quantity may be 1000 kg for carbon steel, 500 kg for stainless, and 80 kg for aluminum. The net result is that the most economical operation takes place at moderately low production quantities or higher.

The major benefit of extrusion is the fact that machining work is often eliminated because irregular cross-sectional shapes can be incorporated in the extrusion dies. When applicable, this factor can outweigh other costs even for short runs.

A second factor of importance is the materials savings that can result, since no material is lost in the form of chips, as would be the case if the part were made by machining. The higher priced the raw material, the more important this factor becomes and the smaller the economical extrusion quantity also becomes.

Several other factors may make extrusions more economical for shorter-run production:

1. If the cross section is carried in stock by the supplier, there may be no minimum quantity. Some larger extrusion suppliers maintain stocks of common shapes such as angles, tees, ells, I beams, and flats.
2. If the die is standard or already in existence, there should be no tooling charge by the extruder.

GENERAL DESIGN RECOMMENDATIONS

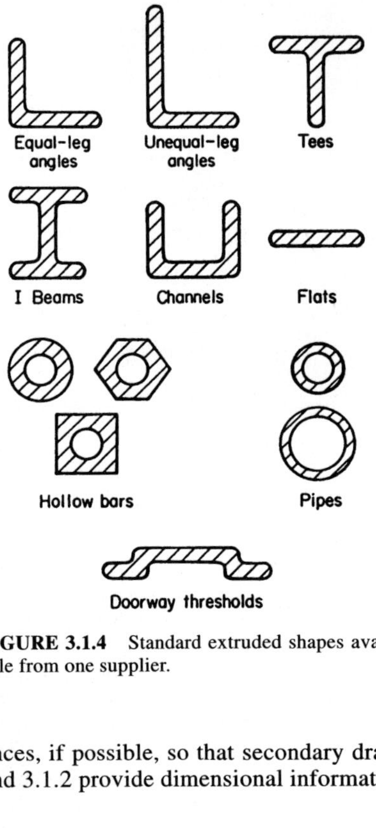

FIGURE 3.1.4 Standard extruded shapes available from one supplier.

Although extrusion is particularly advantageous when complicated shapes are required, it is still desirable to limit irregularities of shape as much as the function of the part permits. Metal flows less readily into narrow and irregular die sections, and distortion and other quality problems are more likely. Custom extruders charge higher prices for intricate cross sections.

Standard shapes may be available from extrusion shops or metals suppliers that will adequately serve the designer's purpose. Since these are always lower in cost than special shapes, the rule should always be "Use standard cross sections whenever possible, especially those carried in stock by the supplier." Figure 3.1.4 illustrates the standard cross sections carried in stock by one ferrous-metals extruder.

For best dimensional control (when component requirements are exacting), a secondary drawing operation is added after the extrusion of most metals. Although such an operation is entirely feasible, it does entail additional tooling, handling, and cost. Therefore, the designer should specify liberal enough tolerances, if possible, so that secondary drawing operations are not required. Tables 3.1.1 and 3.1.2 provide dimensional information to guide the designer in this respect.

SUITABLE MATERIALS FOR EXTRUSIONS

Quite a wide variety of metals is currently extruded commercially. The most common are the following (listed in order of extrudability): aluminum and aluminum alloys, copper and copper alloys, magnesium, low-carbon and medium-carbon steels, modified-carbon steels, low-alloy steels, and stainless steels.

The temperature required and the extrusion temperature range are two factors that

TABLE 3.1.1 Recommended Dimensional Tolerances for Ferrous-Metal Extrusions

Dimension	Tolerance
Cross-sectional dimensions	
0–40 mm (0–1.0 in)	±0.5 mm (0.020 in)
41–120 mm (1–3 in)	±0.8 mm (0.031 in)
121–160 mm (3–4 in)	±1.2 mm (0.047 in)
Over 160 mm (over 4 in)	±1.6 mm (0.063 in)
Cross-sectional dimensions if cold-drawn after extrusion (all dimensions)	
Angles	±2°
Surface finish	6.2 μm (250 μin)
Flatness (transverse)	
Dimensions to 25 mm	±0.25 mm (0.010 in)
Dimensions over 25 mm	±0.1 mm/cm (0.010 in/in)
Twist	
Widest dimension, 60 mm (1½ in)	2 mm/m (⅛ in/5 ft)
Widest dimension, 61–160 mm (1½–4 in)	3 mm/m (3/16 in/5 ft)
Widest dimension, 161 mm up (4 in up)	4 mm/m (¼ in/5 ft)
Camber	Maximum, 2 mm/m (⅛ in/5 ft)

affect the ease with which a metal can be extruded. The lower the required temperature and the broader the temperature range, the better.

The following is a discussion of the extrudability of various alloys.

Aluminum

Extrudability is in reverse order of strength. High-strength alloys require higher pressures and a slower extrusion rate. The most commonly extruded aluminum alloy is 6063, which combines very good extrudability with moderate strength, good formability, good weldability, and resistance to corrosion. Except for porcelain enameling, it finishes well.

Table 3.1.3 lists various aluminum alloys that are suitable for extrusion and gives information on suitable applications.

Copper and Copper Alloys

The most extrudable copper alloys are those with a copper content of 55 to 65 percent. As copper content increases beyond this range, extrudability decreases. The commercial coppers are extruded, but cross sections are limited to relatively simple shapes.

The following brasses are suitable for extrusion and are commonly used:

Architectural bronze (C38500): 57 Cu–40 Zn–3 Pb

Forging brass (C37700): 60 Cu–38 Zn–2 Pb

Free-cutting brass (C36000): 61 Cu–36 Zn–3 Pb

Aluminum bronzes, phosphor bronzes, and aluminum silicon bronzes also can be

TABLE 3.1.2 Recommended Dimensional Tolerances for Aluminum, Copper, and Brass Alloy Extrusions

Dimension	Tolerance
Cross-sectional dimensions*	
0–13 mm (0–0.50 in)	±0.25 mm (±0.010 in)
14–38 mm (0.51–1.50 in)	±0.30 mm (±0.012 in)
39–100 mm (1.51–4.00 in)	±0.50 mm (±0.020 in)
101–200 mm (4.01–8.00 in)	±1 mm (±0.040 in)
201–300 mm (8.01–12.00 in)	±1.6 mm (±0.065 in)
301–500 mm (12.01–20.00 in)	±2.5 mm (±0.100 in)
Angles	
Wall thickness, 0–5 mm (0–0.2 in)	±2°
Wall thickness, 5–19 mm (0.2–0.75 in)	±1½°
Wall thickness, 19 mm or more (over 0.75 in)	±1°

Surface finish (depth of surface defects)

Specified section thickness	Maximum depth
To 1.5 mm (0.06 in)	0.04 mm (0.0015 in)
1.5–6 mm (0.06–0.25 in)	0.06 mm (0.0025 in)
6–12 mm (0.25–0.5 in)	0.1 mm (0.004 in)
12 mm and over (0.5 in and over)	0.2 mm (0.008 in)

Twist

	Twist	
Widest dimension	Per length	Maximum
To 40 mm (1.5 in)	3°/m (1°/ft)	7°
40–75 mm (1.5–3 in)	1½°/m (½°/ft)	5°
75 mm and over (3 in and over)	¾°/m (¼°/ft)	3°

Camber (deviation from straightness)

Wall thickness	Camber
To 2.5 mm (to 0.094 in)	4 mm/m (0.05 in/ft)
Over 2.5 mm (over 0.094 in)	1 mm/m (0.0125 in/ft)

*Add 50% to these values if more than 25% of the dimension is over open space.

extruded. Nickel silvers are satisfactory for extrusion if the nickel content is not higher than 10 or 12 percent. Tin bronzes and cupronickel are difficult to extrude.

Steels

The extrusion of steels has been made possible by the Ugine Sejournet process noted above. High pressures (50 percent higher than for copper) and high temperatures are required with steel. High die wear is a complicating factor. However, stainless steel of various grades and carbon and alloy steels are extruded by commercial steel extruders.

TABLE 3.1.3 Characteristics of Some Aluminum Extrusion Alloys

Alloy	Extrudability	Corrosive resistance	Weldability	Remarks and applications
1100 (A91100)	150	A	A	Bright finish after anodizing
1350 (A91350)	150	A	A	Best electrical conductivity
3003 (A93003)	100	A	A	Tubing
6063 (A96063)	100	A	A	The most commonly extruded alloy; pipe, railing, furniture, architectural extrusions
6101 (A96101)	100	A	A	Good electrical conductivity
6463 (A96463)	100	A	A	Bright-appearance trim
6061 (A96061)	60	A	A	Heavy-duty structures, furniture, pipe lines
5086 (A95086)	25	A	A	
2014 (A92014)	20	C	B	Truck frames, aircraft structures
5083 (A95083)	20	A	A	
5454 (A95454)	60	A	A	For applications at temperatures above 65°C (150°F)
2024 (A92024)	15	C	B	Truck wheels, screw-machine parts, aircraft
7075 (A97075)	10	C	D	Aircraft parts, keys
7001 (A97001)	7	C	D	Used when higher strength is required

Source: From Aluminum Association data.

Magnesium

A number of magnesium alloys are suitable for extrusion: AZ31B (M11311), ZK60A (M16600), AZ80A (M11800), and AZ61A (M11610) are four recommended alloys. Extrusion is metallurgically desirable for magnesium because it refines the grain size.

Other Metals

Inconel and Monel are difficult to extrude, but simple cross-sectional shapes can be extruded. Titanium, molybdenum, and zirconium are all difficult to extrude and require the molten-glass lubricant that is used with steel. Zinc, tin, and lead are all generally suitable for extrusion.

DETAILED DESIGN RECOMMENDATIONS

Generous radii are advantageous for both internal and external corners of extruded cross sections. (See Fig. 3.1.5 for this and other recommendations.) It is possible to extrude sharp corners, but they cause the following problems:

1. Less smooth flow of material through the die, leading to increased dimensional variations and surface irregularities in the extruded part
2. Increased tool wear

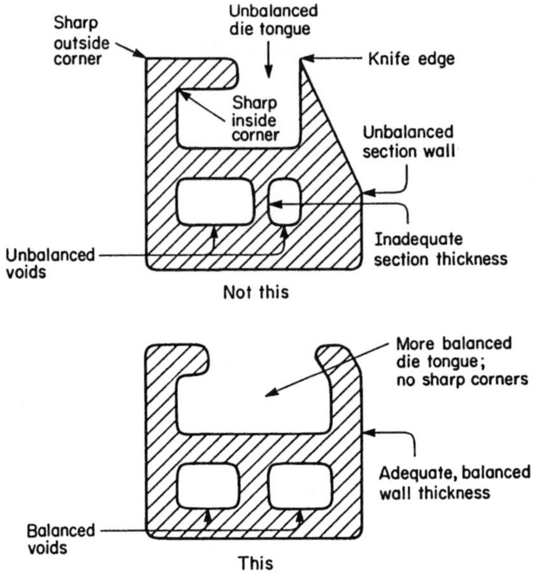

FIGURE 3.1.5 Good and bad practice in the design of cross sections to be extruded.

Table 3.1.4 Recommended Minimum Radii of Extruded Sections

Material	Corners, mm (in)	Fillets, mm (in)
Aluminum, magnesium, and copper alloys		
As extruded	0.75 (0.030)	0.75 (0.030)
After cold drawing	0.4 (0.015)	0.4 (01015)
Ferrous metals, titanium, and nickel alloys		
As extruded	1.5 (0.060)	3 (0.125)
After cold drawing	0.75 (0.030)	1.5(0.060)

3. Increased possibility of tool breakage
4. Less strength in the extruded part owing to stress concentrations

If a sharp internal corner is necessary, the included angle should be as large as possible and always more than 90°. The amount of radii necessary to avoid serious problems varies with the extrudability of the material used. Table 3.1.4 shows recommended minima.

Section walls should be balanced as much as the design function permits because die strain and extrusion distortion tend to occur with unbalanced cross sections. This is particularly true of hollow sections. (See Fig. 3.1.5.)

If long, thin sections have a critical flatness requirement, variations from flatness are reduced if ribs are added to the section. (See Fig. 3.1.6.)

Avoid knifelike edges of parts because they interfere with the smooth flow of material through the die. (See Fig. 3.1.5.)

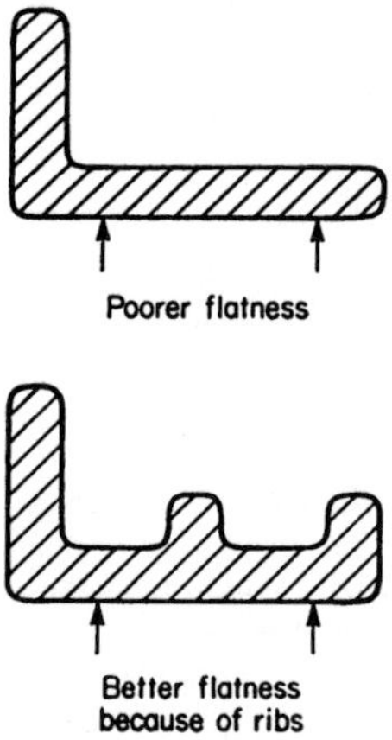

FIGURE 3.1.6 Variations from flatness of long sections are reduced if ribs are added to the sections.

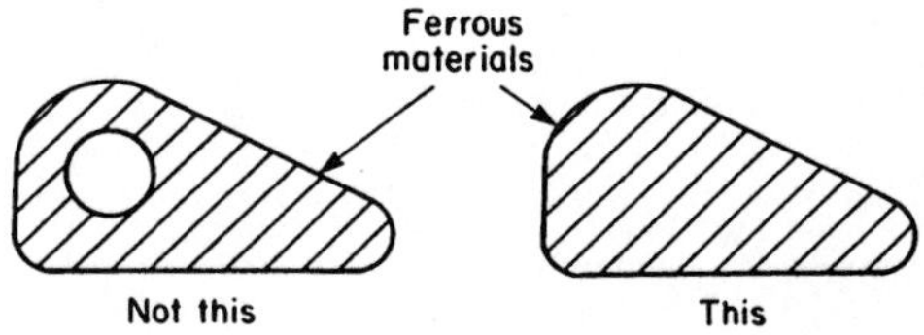

FIGURE 3.1.7 With steels and other less extrudable materials, holes in nonsymmetrical shapes should be avoided.

With steels and other less extrudable materials, holes in nonsymmetrical shapes should be avoided. (See Fig. 3.1.7.)

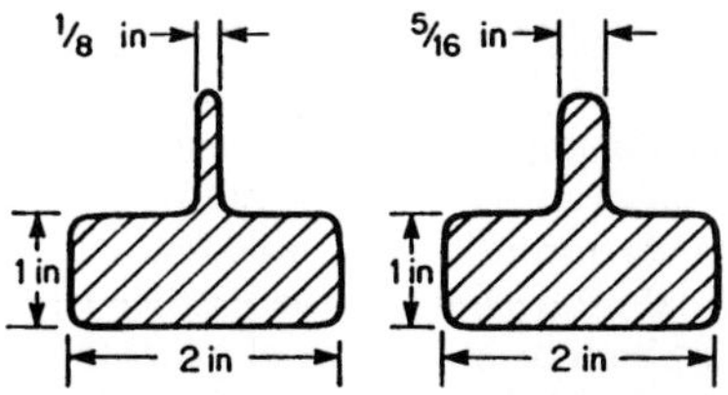

FIGURE 3.1.8 With less extrudable materials, avoid abrupt changes in section thickness.

With all metals, but particularly with steel and less easily extruded metals, avoid extreme changes in section thickness. Figure 3.1.8 illustrates limits for steel extrusions.

In steel extrusions, the depth of an indentation should be no greater than its width at its narrowest point. This is necessary to provide sufficient strength in the tongue portion of the extruding die. In copper alloys, magnesium, and aluminum, the depth of an indentation may be greater since extrusion pressures are lower. Figure 3.1.9 illustrates the limits for these materials.

With steel extrusions there are limitations on the cross-sectional length of any thin-walled segment. The ratio of length to thickness of any segment should not exceed 14:1. (See Fig. 3.1.10.) For magnesium, ratios of 20:1 can be used.

Symmetrical cross sections are preferable to nonsymmetrical designs to avoid unbalanced stresses and warpage. Sometimes it is advantageous, if the part is nonsymmetrical, to use a double section that is divided after extruding. Figure 3.1.11 illustrates an example.

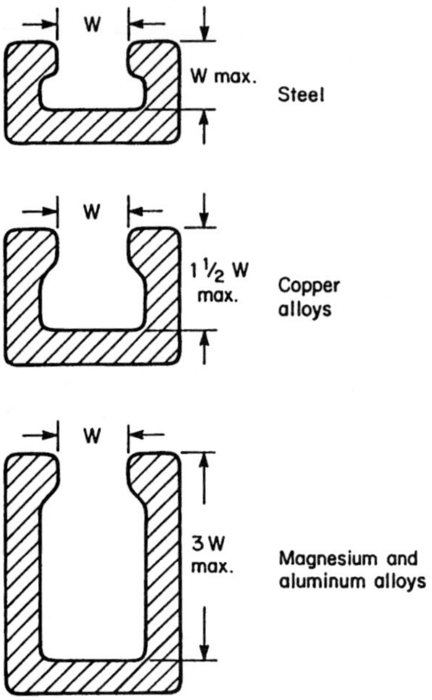

FIGURE 3.1.9 Design rules for indentations.

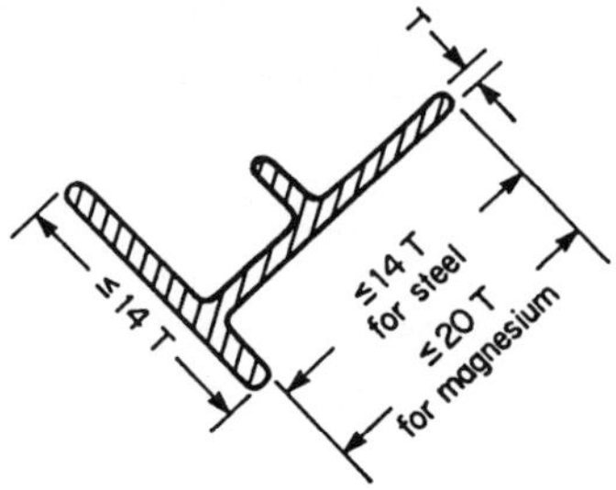

FIGURE 3.1.10 The length-to-thickness ratio of any section of an extrusion of steel or other difficult-to-extrude material should not exceed 14. For magnesium the limit is 20.

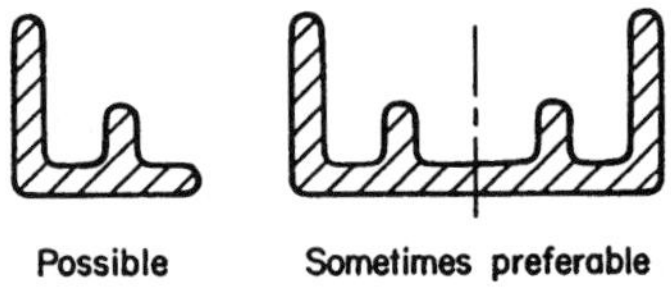

FIGURE 3.1.11 Sometimes it is preferable to produce a nonsymmetrical shape by extruding a symmetrical section and dividing it in two.

DIMENSIONAL FACTORS

Extrusions have more inherent piece-to-piece and drawing-to-piece dimensional variations than parts made with other processes (cold drawing, machining, etc.). The main reason for this discrepancy is the effect of heat. Extrusion is a hot process, and temperature and cooling-rate variations affect the final dimensions of the extruded parts. Another major factor is the wear of extrusion dies, both from passage of metal through the die and from polishing to remove adhered metal. When dimensions are critical, cold drawing the extruded section refines and smooths the surface as well as improving dimensional accuracy.

RECOMMENDED TOLERANCES

Tables 3.1.1 and 3.1.2 present recommended tolerances for extruded parts.

CHAPTER 3.2
METAL STAMPINGS

Conventional Stampings

John Stein

The Singer Company
Elizabeth, New Jersey

Short-Run Stampings

Federico Strasser

Santiago, Chile

STAMPING PROCESSES

Conventional Stamping

Simply put, *stamping* is a method of cold-working sheet metal to a prescribed size and shape by means of a die and a press. The die determines the size and shape of the completed workpiece. The press provides the force needed to effect the change.

Each die is specially constructed for the operation to be performed and is not suitable for other operations. The die is in two halves, between which the sheet metal is placed. When the two halves of the die are brought together, the operation is performed. Normally, the upper half of the die is the *punch* (the smaller member), and the lower half is the *die* (the larger member). When the two die halves are brought together, the punch enters the die.

The die, or matrix, has the desired opening cut into it by various methods. The punch has a shape that corresponds to that of the die but is smaller by an amount determined by the required *punch and die clearance,* which in turn is determined by the type and thickness of the material and the operation to be performed.

The two parts are mounted in a die set, or subpress, the die (in a simple blank-through die) being mounted on the base and the punch on the upper shoe. The use of a die set ensures proper alignment of the punch and die regardless of the condition of the press. The simplest dies are those for punching holes in a blank.

The machine used to consummate these changes of shape has a stationary bed, or bolster, on which the die portion is clamped. A guided slide, or ram, which has the

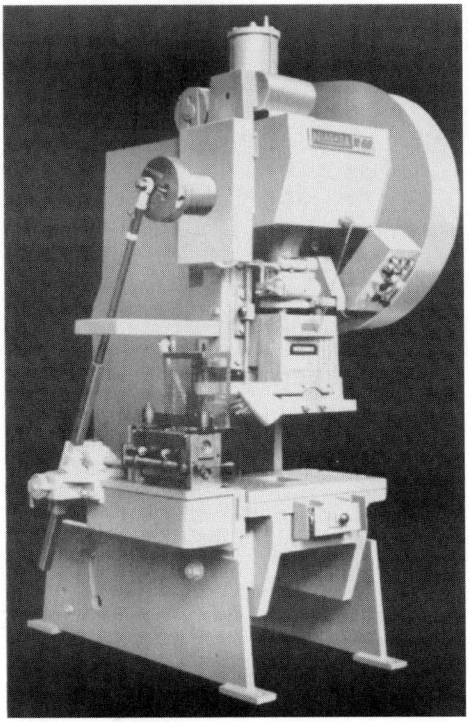

FIGURE 3.2.1 Typical crankshaft-type punch press. Coiled sheet metal is fed through the die automatically. Finished parts are ejected through the open back of the press. (*Courtesy Niagara Machine & Tool Works.*)

punch portion clamped to it, moves up and down perpendicularly to the bed. The motion and force of the ram are provided by a crankshaft, eccentric, or other mechanical means. Hydraulically actuated presses are also employed. Figure 3.2.1 shows a typical crankshaft-type press.

The stamping of sheet metal involves cutting or shearing, bending or forming, and drawing or deep-drawing operations. Cutting around the entire periphery of a part is called *blanking*. Cutting holes in a workpiece is called *punching* or *piercing*. A description of each category follows:

Blanking or Piercing. Blanking or piercing to a contour progresses through three stages (see Fig. 3.2.2): (1) plastic deformation, (2) penetration, and (3) fracture. In stage 1, the punch makes contact with the material, and pressure begins to be exerted until the elastic limit of the stock is exceeded and plastic deformation commences. In stage 2, the continuing pressure causes the punch to penetrate the stock, thereby displacing the blank or slug into the die opening, the displacement equaling the amount of penetration. In stage 3, the fracturing occurs. At this point, the blank or slug is separated from the parent stock.

Generally, the straight or cut band of the material will average approximately one-third of the stock thickness. This, of course, depends on the material's brittleness. The

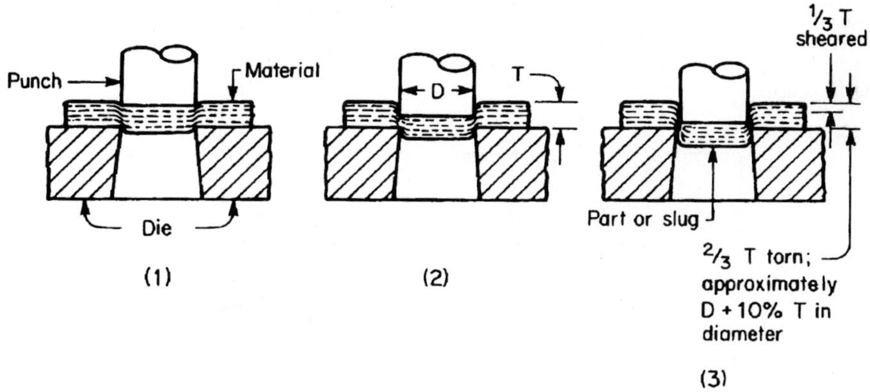

FIGURE 3.2.2 Blanking or piercing sequence.

punch could penetrate anywhere from 15 to 40 percent of the thickness before fracture occurs.

Forming. In forming, the operation produces one or more plane surfaces that are at an angle to the original or flat plane of the blank. Any change in the shape of the blank, no matter how small, is classified as forming.

The material to be formed must have the proper ductility to ensure its retention of deformation in tension without rupture. A schematic of the bending action in a pad die is shown in Fig. 3.2.3.

Drawing. When a part is designed so that no seams or other mechanical joints are permissible, it can be described as a "hollow" or "cup-shaped" body. The "body" is manufactured, by using a flat blank, with the drawing method. Drawing probably ranks second in importance to the cutting operations of press-metal working.

Generally speaking, drawing operations require the use of a triple-acting toggle joint or drawing press. Two or more draws (with necessary heat treatment between draws, depending on the material used) are required when the depth of the cup-shaped area is more than three-fourths of its diameter or width. In a simplified description, blanks are drawn by first confining them between the die and a pressure pad (this is where the triple-acting press comes in); then, as the press continues its cycle, the punch (which has the reverse shape of the die but is smaller by stock thickness) forces the material to flow inwardly from between the two confining surfaces.

The pressure applied between the die and the pressure pad during the entire press cycle must be such that the blank is kept from wrinkling. Wrinkles will prevent obtaining a smooth cup, and the possibility of tearing the cup increases. The pressure required is mainly the result of experience plus trial and error. For a schematic view of a typical drawing operation, see Fig. 3.2.4.

In addition to the three basic methods, conventional stamping includes shaving, trimming, embossing, coining, and swaging.

Shaving. Shaving is a secondary operation after blanking or piercing. It produces a smooth edge on the workpiece instead of the breakaway edge shown in Fig. 3.2.2. This is accomplished by removing only a small amount of stock from the edge of the

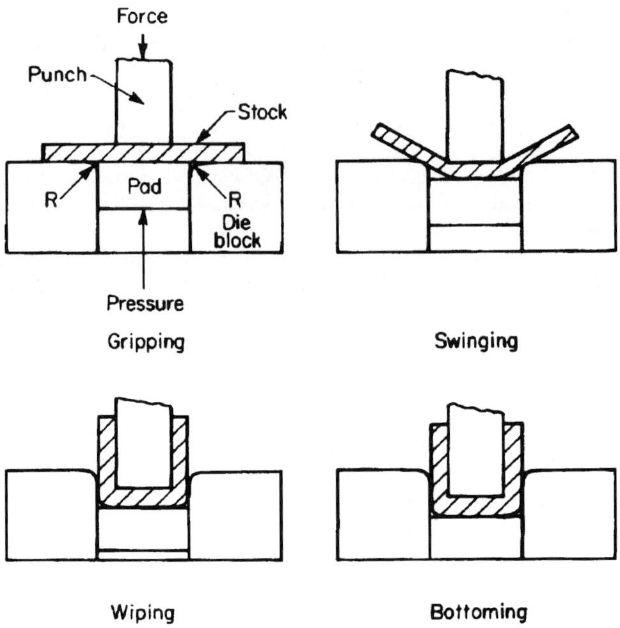

FIGURE 3.2.3 Schematic view of the bending action that occurs in a forming operation.

part. The deformed, fractured portion of the edge is removed, leaving chiplike scrap material and a relatively square and unfractured edge on the part.

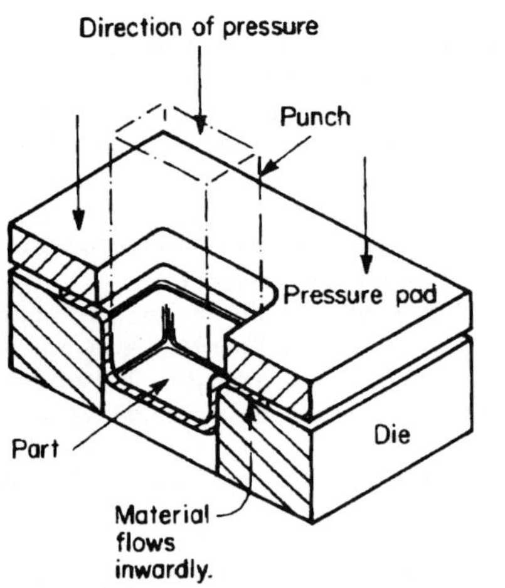

FIGURE 3.2.4 Schematic illustration of deep drawing.

Trimming. This method is similar to blanking except that it occurs after forming, drawing, or other operations when extra metal is left in the part for holding or locating purposes or as a stock allowance. The removal of this extra stock is called *trimming*.

Embossing. Embossing produces shallow surface designs from alternately raised and depressed areas with little or no change in material thickness. Raised areas in the punch correspond to depressed areas in the die. Nameplates and stiffening ribs are two applications. See Fig. 8.6.1, in Chap. 8.6, "Designing for Marking," for illustrations of embossed parts and simple embossing dies.

Coining. In this process, the blank is entirely captive within the die. The die squeezes the workpiece under very high pressure, causing the workpiece material to flow into every die depression. Dies of quite substantial proportions and special presses are often required because of the high tonnage involved.

Unlike embossing, coining permits different designs to be imparted on either side of a blank, as is the case with all coins.

Swaging. This process involves an "open" die, also of substantial construction. The part is also squeezed into a cavity, but in contrast to coining, the excess material is not contained but allowed to flow at will.

Knuckle-joint presses are customarily used for both coining and swaging because they combine high tonnage and a slow squeezing action to the workpiece rather than the sharp impact of a conventional press.

Short-Run Methods

General-purpose-equipment methods involve a variety of means for working sheet metal. They differ from conventional metal-stamping processes in utilizing tooling of universal or wide-range applicability rather than dies made for only one operation. They rely on the skill of the operator rather than on tooling for their accuracy. Their production rates are far lower than those of conventional stamping. Common general-purpose sheet-metal processes are the following:

Straight-Line Shearing. This process utilizes two blades. The lower blade is stationary; the upper is movable. Sheet material is placed between the blades. The movable blade is forced down into the stock, cutting and fracturing it with an action similar to that which occurs when sheet metal is blanked.

Notching. Similar to straight shearing except that it produces an angle cut in the workpiece, notching is used to remove excess stock prior to a forming, bending, or drawing operation.

Nibbling. This process involves cutting out a contoured or other shape by punching a series of overlapping round or square holes along the edge of the part.

Rotary Shearing. This process is a means for cutting sheet in a contoured shape or a straight line by passing it between two tapered wheel cutters.

Folding-Machine Bending. This process produces straight-line bends. The sheet is clamped between two beams, while a movable beam pivots upward, folding the sheet against the edge of the clamping beam. Sharp or rounded folds can be made. Hand-lever force is adequate for most metal thicknesses.

Press-Brake Bending. This process is performed in a long, narrow gap-frame press. Bends are normally made on the centerline of the ram in V-shaped dies. Universal tooling of a wide variety, however, can produce many different bend shapes. With successive operations, complex forms are possible.

Turret Punching. This process involves the use of a machine that has punches and dies mounted on synchronized indexing tables. As the tables are indexed, different sets of punches and dies come into operating position. Thus, as a sheet is moved on the

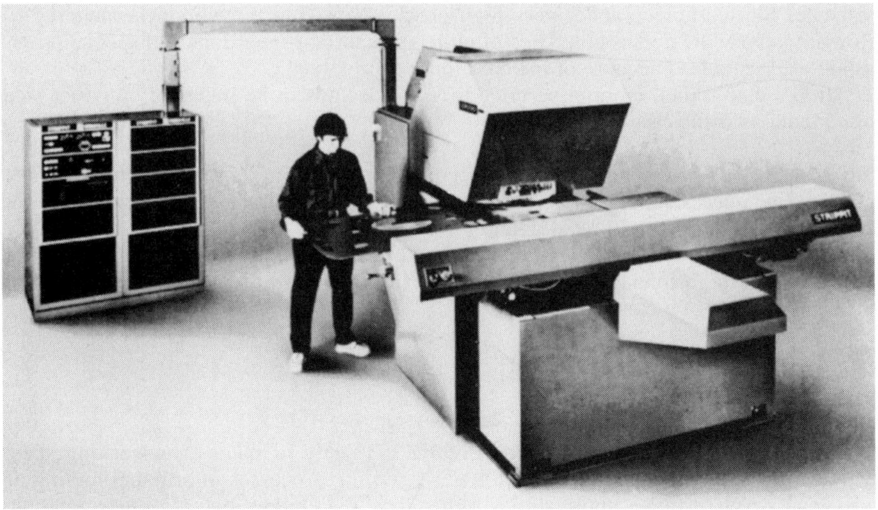

FIGURE 3.2.5 Turret punching machine equipped with computer control. (*Courtesy Strippit-Houdaille.*)

machine's work table and different dies are used, holes and notches of various sizes and shapes can be punched and various-shaped blanks can be produced by nibbling. Some turret punching machines rely on manual manipulation of the workpiece and manual indexing of the turret. Others provide for numerical or computer control of these elements. Fig. 3.2.5 illustrates a typical machine, whereas Fig. 3.2.6 illustrates typical blanks that it can produce. Although primarily used for blanking, with appropriate dies, these machines can produce embossed shapes, countersunk holes, welding projections, electrical breakaway holes, and louvers. Tapping heads can be added as accessories.

Steel-Rule Die Blanking. This process utilizes a die that employs a thin strip of hardened steel bent to the outline of the blank and held on its edge by a wooden base. The exposed edge of the steel band (or *steel rule*) is sharpened to a cutting edge (see Fig. 3.2.7). The workpiece is placed between this edge and a flat opposing surface. The cutting edge penetrates the workpiece and cuts it to the outline of the part. Normally, soft materials such as leather, paper, and rubber are blanked with this method. However, it is surprisingly effective for short runs of nonferrous metals and even unhardened steel. Steel-rule dies are quite inexpensive.

Nonstamping Blanking Methods. When quantities are small, it may be advantageous to produce the blank by nonstamping methods, even though subsequent piercing, forming, or other operations are produced by punch-press methods. Among these nonstamping methods are contour sawing (see Chap. 4.10), routing, laser-beam machining (see Chap. 4.21), other machining methods, flame cutting (see Chap. 4.11), and hand cutting and filing.

Master Die Set: Adjustable-Die Stamping. This is a useful method for secondary stamping operations (after blanking). Punching, notching, countersinking, and other operations can be performed. The system utilizes reusable punch-die combinations for each hole or other element stamped in the workpiece. These combinations are fastened

FIGURE 3.2.6 Typical turret-punched parts. (*Courtesy Wiedemann Division, The Warner & Swasey Co.*)

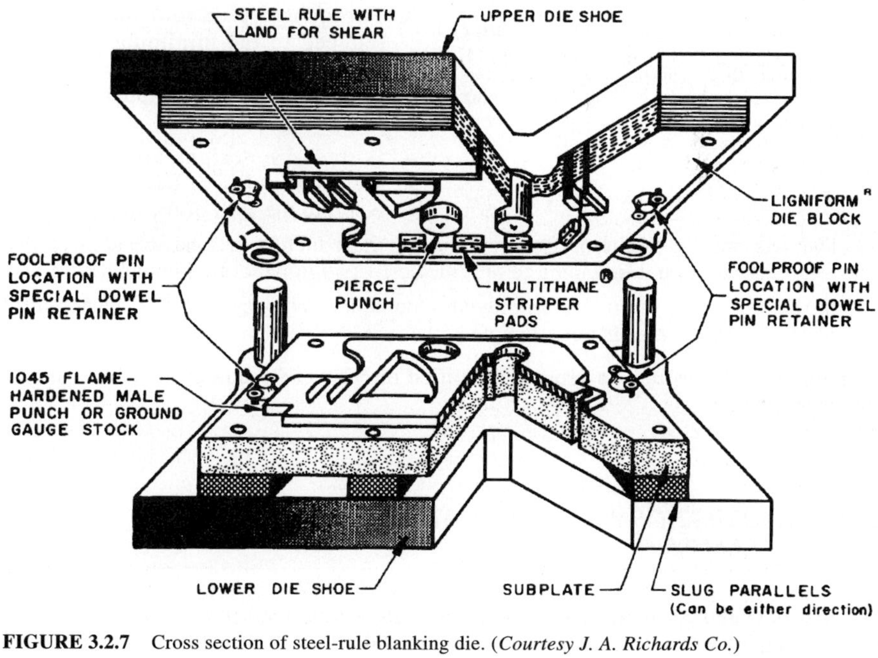

FIGURE 3.2.7 Cross section of steel-rule blanking die. (*Courtesy J. A. Richards Co.*)

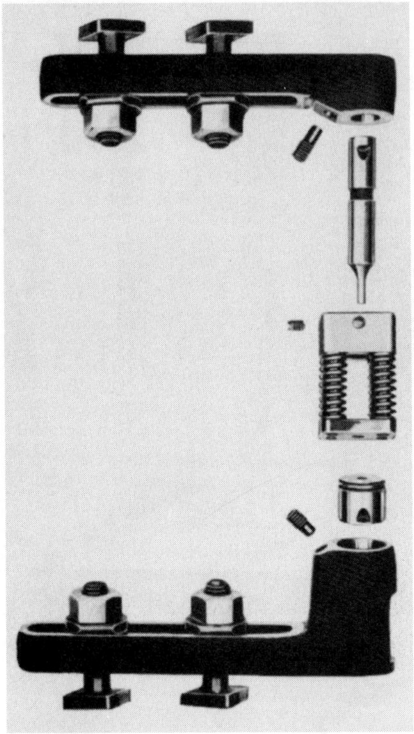

FIGURE 3.2.8 Exploded view of a punch-die combination used with a master reusable die set. The stripper is self-contained, and holes of different sizes and shapes can be produced by changing to the appropriate punch and die inserts. (*Courtesy S. B. Whistler & Sons, Inc.*)

to a master reusable die set. The number of punch-die sets used and their locations determine the configuration of the stamped workpiece. In each punch-die combination are incorporated stripper devices, which are bolted to the master die plates or held magnetically. Templates are often used to locate the punch-die combinations, particularly if the job is rerun periodically. Figure 3.2.8 shows a typical punch-die combination.

Special Short-Run Tooling Methods. These methods involve tooling that, although more or less conventional in principle, is constructed especially for short-run work. Differences between this kind of tooling and its operating method and that of high-production tooling can include some or all of the following:

1. Less expensive and more easily machinable materials can be employed.
2. Often no provision is made for regrinding. There is no need for regrinding if the whole lot can be run in one setup.
3. The quantity of tool members is reduced to the minimum possible. Often no die shoes, punch pads, guideposts, spring strippers, tripper stops, wear plates, etc., are used.
4. Formal die designs are omitted or greatly simplified. Die construction is left to the skill of the toolmaker.
5. Die sets may be omitted. The tightness of the punch press and the skill of the setup operator are relied upon to provide correct alignment of the punch and die.
6. Single-operation tooling is employed. Compound, progressive, and other multiple-operation tooling is not attempted.

Figure 3.2.9 illustrates a typical simplified die especially constructed for a short-run operation.

CHARACTERISTICS AND APPLICATIONS OF METAL STAMPINGS

If cleverly designed, metal stampings, like injection-molded parts and die castings, can contribute strongly to design for manufacturability (DFM). They can incorporate

FIGURE 3.2.9 The simplified die on the left is especially constructed for a short-run application with a conventional die for the same part shown to the right. (*Courtesy Ebway Corp.*)

diverse functions that otherwise would require additional parts. Springs and other flexible sections, snap-fit elements, folding tabs, and press-fitable designs can be incorporated and can eliminate the need for other parts, including separate screw fasteners or other fasteners. Projection welding and spot welding are easily provided for. Such innovations can reduce the number of parts in an assembly and make the stamped part itself easy to assemble.

Perhaps the major characteristic of stamped-metal parts is the fact that with a few exceptions, the wall thickness is essentially uniform throughout. Finished stampings are sometimes quite intricate in shape, with many tabs, arms, holes of various shapes, recesses, cavities, and raised sections. In all cases, however, the wall thickness is essentially uniform. Thick bosses of the type found in many castings are absent. For drawn shells, there is almost an infinite variety of shapes.

The wall thickness ranges from a low of about 0.025 mm (0.001 in) to about 20 mm (0.79 in), although pure bending or shearing operations are produced on even heavier stock, Most stamping, however, is performed in the range of about 1.3-mm (0.050-in) to 9.5-mm ($\frac{3}{8}$-in) stock thickness. The size of metal stampings ranges from the smallest parts used in wristwatches to large panels used in trucks or aircraft. The largest press brakes are as long as 9 m (30 ft).

The characteristic edge of a stamped part, as illustrated in Fig. 3.2.2., should be kept in mind by the design engineer, especially if bearing surfaces at the edges are involved or if, for appearance or other reasons, smooth edges are needed. The designer also should be aware of the burr on one side of stamped pieces and be careful to design parts so that burrs either are easily removed or do not interfere with subsequent operations or function.

Stampings can be machined after blanking and forming if dimensions that are more accurate than can be produced by stamping or shapes not feasible purely by stamping are required. Examples are reamed center holes of stamped pulleys or gears, surfaces ground for flatness, and grooves or relief areas requiring a change in thickness of the part.

A collection of typical stampings is shown in Fig. 3.2.10. Stampings produced by short-run methods are shown in Fig. 3.2.11.

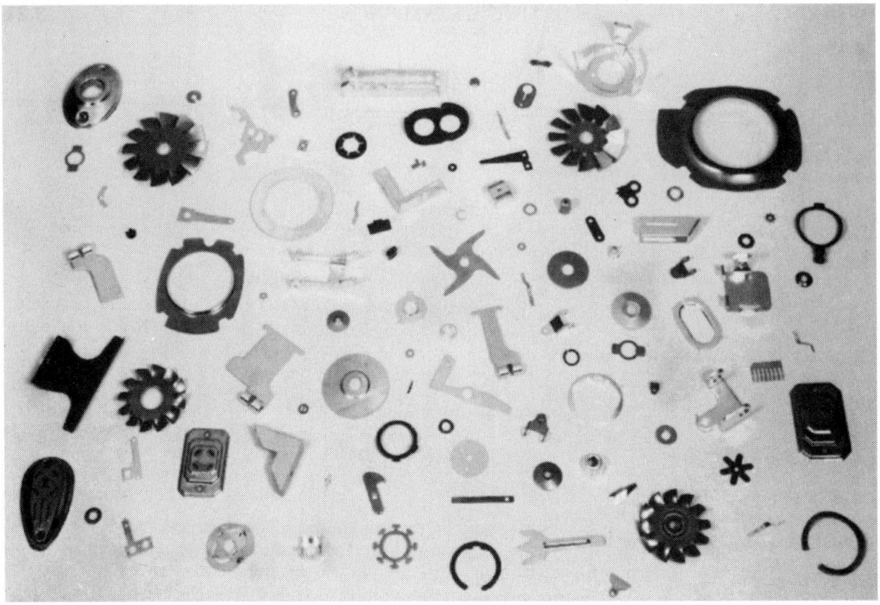

FIGURE 3.2.10 Collection of typical conventional stampings. (*Courtesy Torin Corporation.*)

FIGURE 3.2.11 Stampings produced by short-run methods. (*Courtesy Dayton Rogers Mfg. Co.*)

ECONOMIC PRODUCTION QUANTITIES

Conventional Stampings

Conventional stamping is a high-production process. Production is very fast, from 35 to 500 or more strokes per minute. If total production is sufficient to justify compound or progressive dies, all blanking and forming can be done in one press stroke. In these cases, parts can be produced complete at a rate of thousands per hour.

A progressive die for the production of parts similar to those illustrated in Fig. 3.2.10 requires high production rates (e.g., 250,000 pieces per year) to justify the investment. Conventional dies to produce such parts might consist of (1) a compound blanking and piercing die and (2) a pad-forming die, which together would cost only about half the price of the progressive die. As a general rule, a progressive die should not be considered unless two secondary press operations can be eliminated.

Forming dies vary greatly in cost, depending on their complexity and size. A simple conventional die to form one bend can be very inexpensive, whereas a complex forming or drawing die for a large part can require a major investment.

As a result of these often sizable tooling costs for metal stampings, coupled with low unit labor costs, even with multiple operations, conventional metal stamping is a high-production process.

Punch presses are relatively low in cost compared with other high-production equipment. However, press cost is not a significant factor in economic-lot-size calculations because presses are versatile. Almost any given press has a wide range of stamping-operation capabilities.

Short-Run Stampings

As a very rough approximation, it can be stated that for average conditions, the borderline between short-run and regular (middle-volume) production may be assumed to be between 5000 and 10,000 pieces per run or per lot. Probably more important, however, is the total quantity expected to be produced by the tooling over the life of the product. If this quantity is less than 20,000, then short-run methods will probably provide the lowest overall costs. When 10,000 to 20,000 total parts are required, it may be advantageous to have both short-run and conventional tooling quoted so as to enable a comparative-cost study to be made.

Another rule of thumb for differentiating between short-run and regular stampings is the following: When the cost of the dies themselves exceeds the cost of the parts produced, it is a short-run job.

Labor costs per piece are invariably higher per unit for short-run methods. The materials cost also may be somewhat higher in short-run jobs because of poorer stock utilization. Setup time is usually greater, and other secondary factors such as material handling, scrap, and work-in-process inventory are apt to be somewhat higher with short-run methods owing primarily to the greater number of operations involved.

The approach that yields the lowest costs for the total of all these factors over the life of the product is the one that should be used.

It should be noted that the total tooling cost of the separate dies used for short-run methods is always a fraction of that of a multistage die that incorporates all the necessary operations simultaneously. (The estimated cost ratio is as much as 1:6 and up to 1:8 or more.) Very significant is the fact that in a majority of cases short-run methods permit the use of exisiting universal-type dies, especially for hole punching and right-angle bending.

Another advantage of the short-run approach is the shorter lead time required for tooling the short-run variety. It should be noted, however, that the quality of stampings produced with permanent-type tooling is usually superior to that of those produced with temporary dies; thus interchangeability of parts is better.

Short-run stamping methods should be considered under any of the following conditions: (1) For experimental, prototype, or pilot-lot production, particularly when engineering changes are likely to occur and the use of permanent tooling would not yet be advisable. (2) For spare-parts production after the original tooling has been disposed of. (3) For products such as industrial, medical, or laboratory equipment or for other fields in which unit production levels are not large. (4) In cases in which faster delivery of a component is essential for the commercial success of a product (e.g., a seasonal article whose development has been started late or has taken too long). With shortened tooling lead time, production may start earlier. In these cases, the basic economic requirements (low cost and high productivity) must be disregarded in the temporary tooling. These considerations are duly taken into account when permanent-type tooling is developed later. (5) When standby tools are desired for high-production permanent-type dies whose working is essential for the uninterrupted manufacturing process. (6) Last but not least, for low-budget business when it is not considered convenient to invest a large sum in costly permanent tooling.

SUITABLE MATERIALS FOR STAMPINGS

One rule of thumb is that any material that can be produced in the form of sheet or strip can be press-worked. The exceptions, of course, are brittle nonmetallic materials such as glass and the somewhat similarly brittle, very-high-hardness metals, e.g., those of hardness above R_c 50.

Materials for stamping are classified in three groups: (1) ferrous metals, (2) nonferrous metals, (3) nonmetallics.

Ferrous Metals (See also Chap 2.2)

By far the most widely used metal for general stamping applications is cold-rolled steel in sheet, strip, and coil form having a carbon content of between 0.05 and 0.20 percent. The lower-carbon steels are the least expensive for stamping.

Steels with 0.10 percent or less carbon are most suitable for severe forming applications. However, steel with up to 0.15 percent carbon normally can be bent 180° on itself in any direction. If the carbon content is 0.15 to 0.25 percent, steel can be bent 180° over one thickness of material. Use of aluminum-killed steels is also advantageous when the design calls for bending or folding the material down on itself.

Comparatively few materials can be drawn successfully, since the material must have the required ductility and tensile strength. Aluminum-killed (sometimes called *special-killed*) steel is the preferred material for deep-drawing applications. Drawing-quality steel is superior to regular commercial-quality steel for these applications but is not as drawable as the aluminum-killed material. Cold-rolled steel is preferred over hot-rolled for drawing applications.

For coining applications, carbon or alloy steels of no more than 0.30 percent carbon content should be specified.

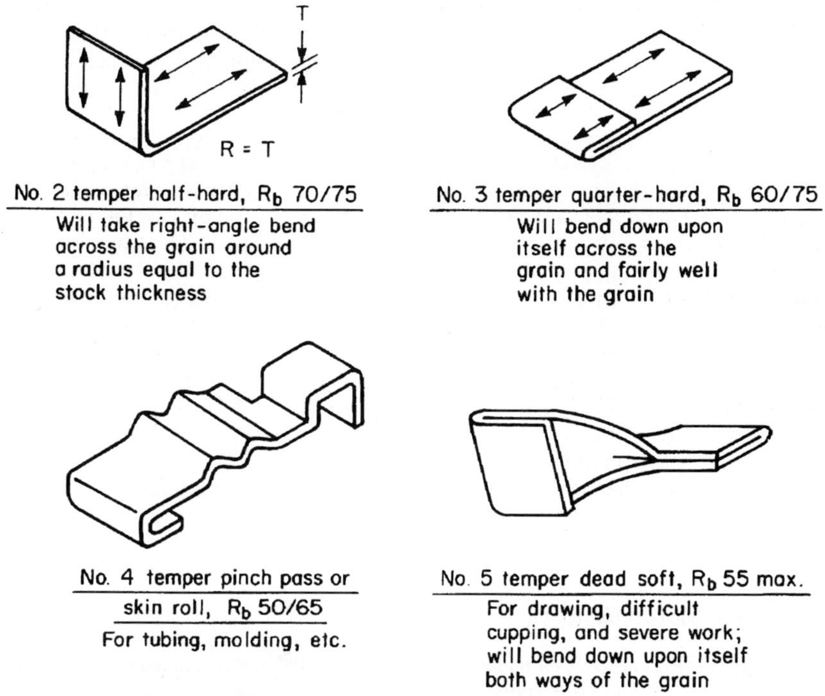

FIGURE 3.2.12 Temper ranges and forming capabilities of low-carbon open-hearth steel C-1010.

Of the stainless steels, Types 302 (S30200), 304 (S30400), and 305 (S30500) have maximum stamping and forming capabilities. However, with all stainless steels, higher tonnages are required and punch and die life suffers because of the tougher material. Since the stainless-steel groups and some steel-alloy groups such as the silicon steels are more difficult to work, especially when forming, liberal tolerances should be specified. For coining applications using stainless steel, the following grades are usable: 301 (S30100), 302 (S30200), 304 (S30400), 305 (S30500), 410 (S41000), and 430 (S30500) (305 is more costly than 301 or 302).

Temper is an important factor in the formability of both ferrous and other materials. Figure 3.2.12 illustrates the forming capabilities of four tempers of C-1010 open-hearth steel.

Galvanized and other coated steels are normally practical materials.

Nonferrous Metals (See also Chap. 2.3)

Aluminum and copper alloys are the two principal nonferrous materials used for stampings. Other stampable nonferrous metals are nickel alloys, zinc, magnesium (when heated), titanium, and many less common metals.

Copper and many of the brasses have excellent stamping characteristics. Cartridge brass is particularly suited for deep-drawing and severe forming applications. Yellow brass has similar stampability. Phosphor bronze and beryllium copper, even though less workable, are frequently used for electrical contacts and other applications for which spring action is required. The preferred alloys for coining are those with very high copper content: electrolytic-tough-pitch (ETP) copper, gilding, commercial bronze, and red brass.

Aluminum alloys 1100 (A91100) and 3003 (A93003) have excellent forming characteristics and are low in cost. If greater strength is required, 3004 (A93004), 5052 (A95052), 5154 (A95054), and 5036 (A95036) are suitable. If the finished stamping requires a smooth or polished surface, 5053 (A95053), 5252 (A95252), or 5457 (A95457) is recommended. Types 1100 (A91100), 3003 (A93003), and 5005 (A95005) are the aluminum alloys most easily deep-drawn.

Nickel 200 (NO2000) and Monel 400 (NO4400) are the most easily press-worked nickel alloys. If the amount of forming is not severe, they are often processed in the one-eighth-hard or quarter-hard tempers to promote better blanking and piercing.

Magnesium alloys are normally formed at temperatures between 120 and 430°C (250 and 800°F). However, large-radius bends can be made in most alloys at room temperature. Alloys AZ31B-0 (M11311-0) and LAI41A-0 (M14141-0) are the most easily formable alloys at room temperature.

Gold, silver, and platinum are all highly suitable for press working, including coining.

Nonmetallic Materials

Many nonmetallic materials in sheet form are processed with punch presses and tooling of the type used for sheet metal. Operations are normally limited to blanking and piercing, although with some materials bending, light forming, and embossing are possible. Another exception is ABS plastic, which can be deep-drawn. Some cup-shaped food containers of ABS are made by deep-drawing on metal-working presses.

Nonmetallic materials commonly blanked and pierced include fiberboard, paper, leather, rubber, cork, wood and wood-based composition board, and various plastics, especially laminated thermosets. Except for glass-reinforced laminates, these materials are ideally suited for blanking with steel-rule dies and other short-run tooling.

DESIGN RECOMMENDATIONS

Stock Utilization

Stampings should be designed for the economical use of material. Shapes that can be nested close together are better than those which must be more widely spaced on the stock material. An L-shaped part will nest better than a T shape. Other examples are illustrated in Fig. 3.2.13. Such improved stock utilization also requires close liaison between the designer and the die maker or, at the very least, the ability of the designer to visualize a strip layout such as the die maker would make.

Consideration also should be given to utilizing scrap portions to produce additional parts. In the case of large projects, many parts no doubt will require the same thickness and type of material. By designing a small part of such a shape that it can be made from a piece of stock left over from the blanking of a large part, the designer will save material. The designer should make a drawing notation in such cases to carry

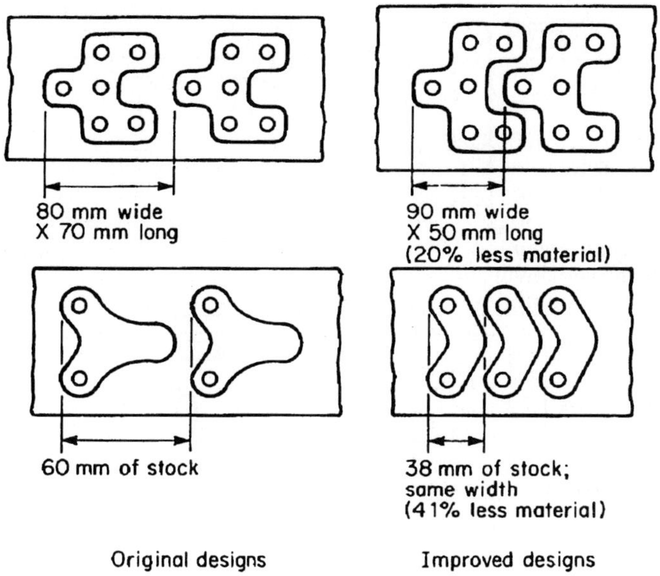

FIGURE 3.2.13 Two examples of parts that were redesigned to provide better nesting of blanks and thus improved material utilization.

the information to the manufacturing personnel. Figure 3.2.14 illustrates an example of this kind of step. Another common example is the typical motor lamination die in which the field and armature parts are made from the same strip with very little waste of material. (See Fig. 3.2.15.)

Holes

The diameter of pierced holes should be not less than stock thickness, as illustrated in Fig. 3.2.16. Special intermeshed punch-support sleeves or the fine-blanking process (Chap. 3.3) permits smaller holes, but with conventional stamping tooling, punch breakage becomes excessive if pierced holes below the prescribed minimum are attempted.

The spacing between holes should be a minimum of 2 times stock thickness; 3 times is preferable from a die-strength standpoint; i.e., if the wall thickness is too little, the die's ability to resist the pressure of piercing is seriously impaired.

The minimum distance from the edge of a hole to the adjacent edge of the blank should be at least stock thickness, but preferably it should be $1\frac{1}{2}$ to 2 times that (see Fig. 3.2.17). Too small a spacing will cause the part to bulge in the edge area adjacent to the hole.

Piercing a hole before forming is less costly than piercing or drilling as a secondary operation. The minimum distance between the lowest edge of the hole and the other surface should be $1\frac{1}{2}$ times stock thickness plus the radius of the bend, as illustrated in Fig. 3.2.18. Distortion of the hole will occur if this minimum distance is not observed.

The writers have employed the following method to eliminate or minimize distor-

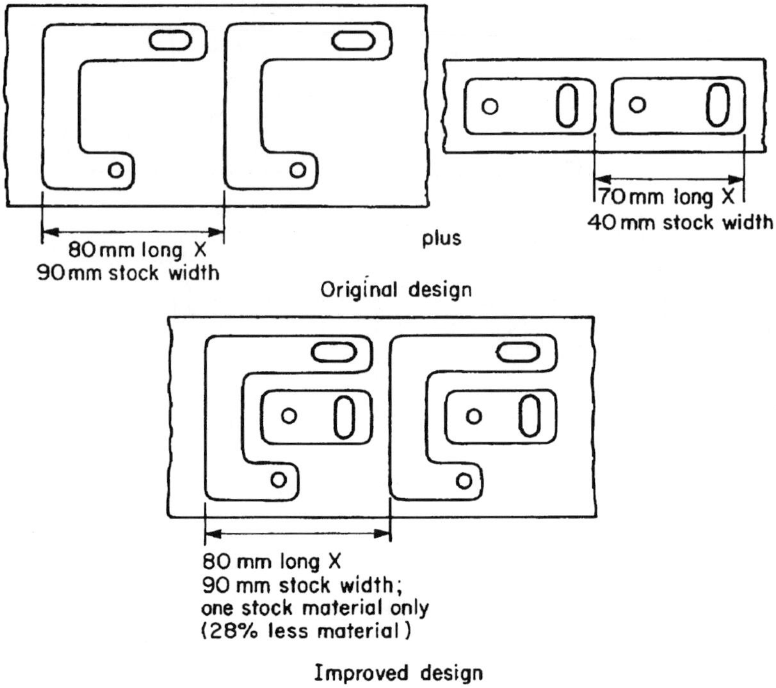

FIGURE 3.2.14 Example of how minor redesign allowed a part to be blanked from scrap material left over from the blanking of a companion part.

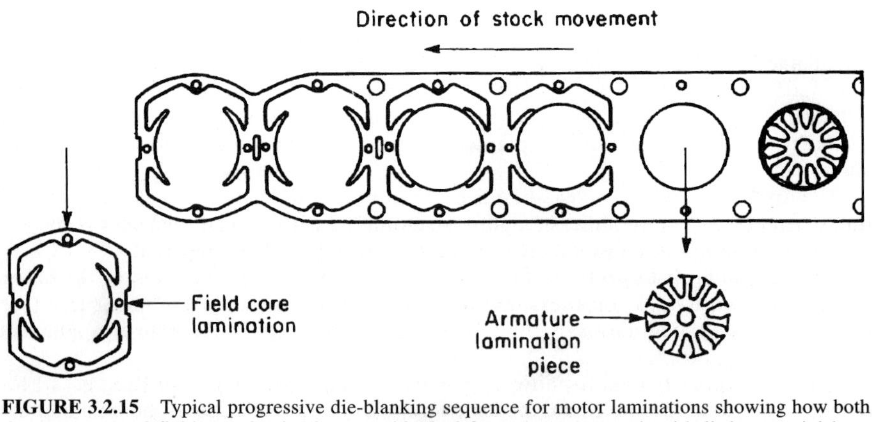

FIGURE 3.2.15 Typical progressive die-blanking sequence for motor laminations showing how both the armature and field core laminations are blanked from the same stock with little material loss. (*Courtesy The Singer Company.*)

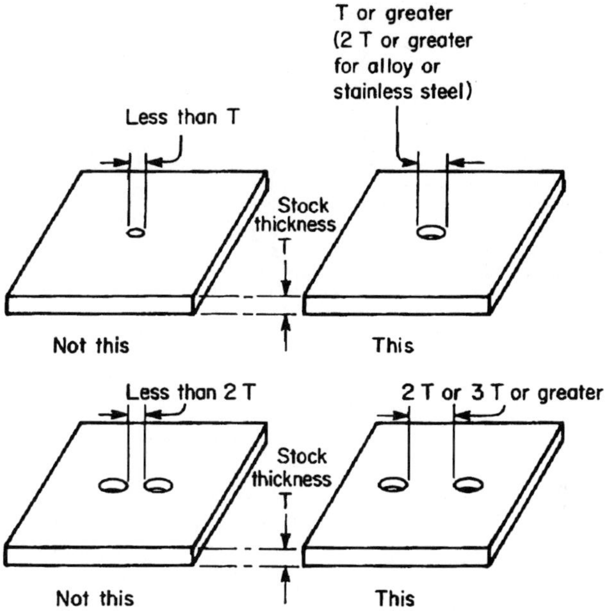

FIGURE 3.2.16 Design rules for size and spacing of holes.

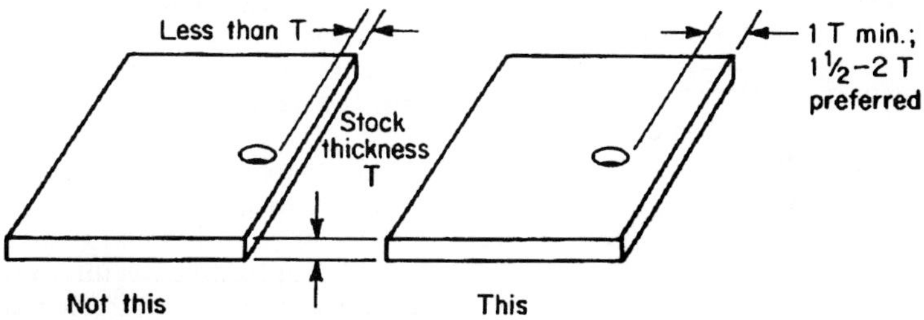

FIGURE 3.2.17 Pierced holes should not be located too close to the edge of the part.

tion when the design requires the lowest edge of the hole to be closer than the recommended minimum. A nonfunctional window, either square or rectangular, is pierced directly beneath the desired hole or holes. (A minimum wall thickness between the holes should, of course, be observed.) Thus, during forming, no (or at least very little) stress or distortion is transmitted to the hole. (See Fig. 3.2.19.) Trial and error and experience are necessary to determine stress.

Often it is desired to include two aligned holes in opposite legs of a U-bent part for holding a shaft or for some other purpose. Designers should realize that it is difficult to form such a part from a prepierced blank and have the holes aligned precisely. Several alternatives can be considered: (1) Pierce or drill the holes after forming. This is more expensive but provides excellent alignment. (2) Use broad tolerances on the holes, or make one a slot, i.e., allow for misalignment if the function of the part per-

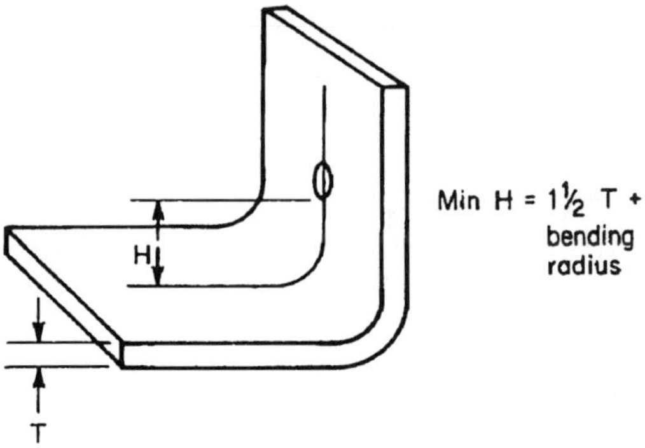

FIGURE 3.2.18 Minimum spacing between a pierced hole and a bend to avoid distortion of the hole.

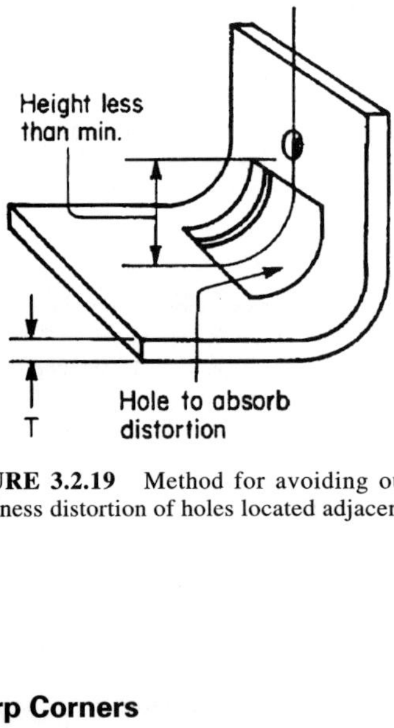

FIGURE 3.2.19 Method for avoiding out-of-roundness distortion of holes located adjacent to a bend.

mits. (3) Include a pilot hole in the bottom of the U bend. This hole is located over a pin in the pad of the forming die that will consistently position the blank. Another requirement, if truly close alignment is to be achieved with this method, is the use of stock of close thickness control. Although material of close thickness tolerance commands a premium price, the extra cost may be more than offset by the savings realized by not having to perform the second operation. (See Fig. 3.2.20.)

The designer of stampings should always try to specify round holes instead of holes of square, rectangular, or other shapes. Tooling costs for round-hole punches and dies are far below those for holes of other than round shapes.

Sharp Corners

Sharp corners, both internal and external, should be avoided whenever possible. Sharp external corners of punches or dies tend to break down prematurely, causing more pull-down, larger burrs, or rougher edges of the blanked part in the area of the corner. Correspondingly sharp interior corners of punches and dies are a stress-concentration point and can lead to cracking and failure from heat treatment or in use. A general rule of thumb is to allow a minimum corner radius of one-half stock thickness and never less than 0.8 mm ($^1/_{32}$ in). (See Fig. 3.2.21.) It should be remembered, however, that there will inevitably be a sharp corner wherever two edges produced by separate shearing, slotting, or blanking operations intersect at approximately a right angle or less. Such corners may have to be rounded by tumbling the part or performing some other secondary operation.

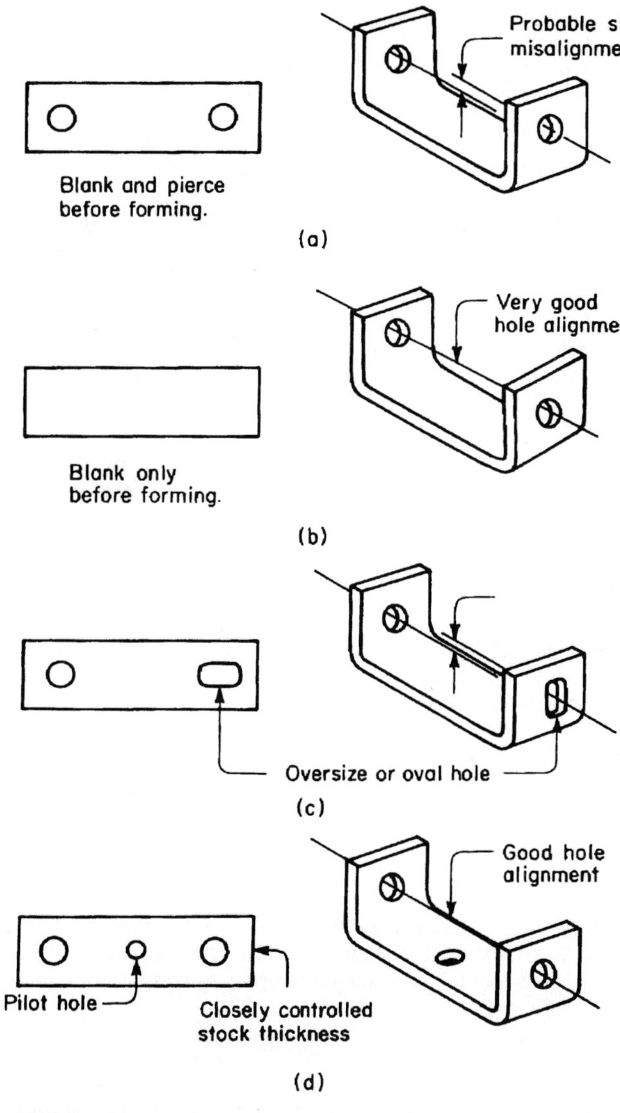

FIGURE 3.2.20 Problem of alignment of holes in opposite legs of a U-bent part. (*a*) Normal method. Not recommended if close hole alignment is required. (*b*) More accurate method. Pierce or drill holes after forming. (*c*) Oversize or oval hole allows for misalignment. (*d*) Pilot hole assures that blank is centered in forming die.

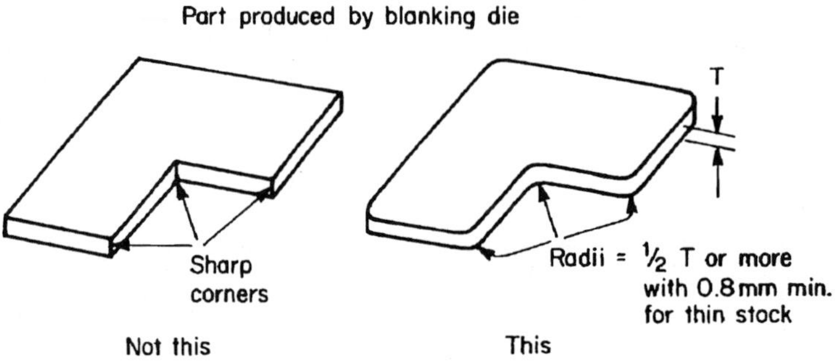

FIGURE 3.2.21 Design rules for fillets and radii of blanked parts.

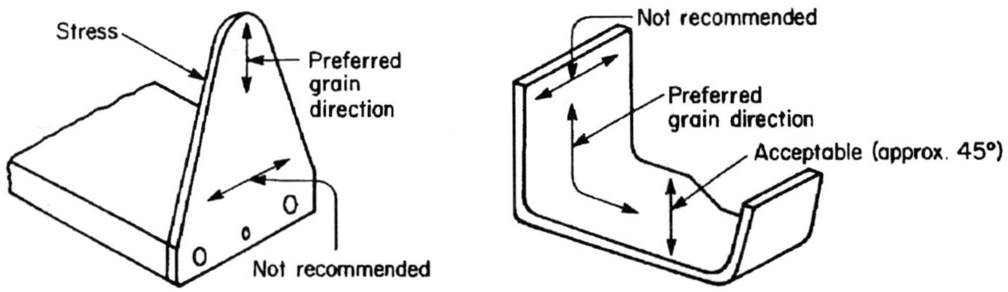

FIGURE 3.2.22 The grain direction of rolled sheet metal has a bearing on its strength and bendability.

Grain Direction

Whether a part remains flat or is eventually formed, the designer should consider its strength requirements relative to the grain direction of the material. If necessary, the desired grain direction should be indicated on the part drawing. (Grain direction usually runs lengthwise in the coil or strip.) See Fig. 3.2.22.

Strip Stock

Many times it is possible to design a part so that it can be cut off from strip stock rather than being blanked by a more expensive die, which also leaves a web of unused material. Granted, this method applies only when the part can have two sides parallel, as in simple brackets, but if a specific shape or extra close width tolerances are not required, it is nonetheless an inexpensive and simple approach. Some examples are shown in Fig. 3.2.23.

In designing a part when for economic reasons a cutoff or parting tool is used with strip stock, avoid featheredges. In Fig. 3.2.24, method A is preferred, method B is acceptable, and method C is to be avoided.

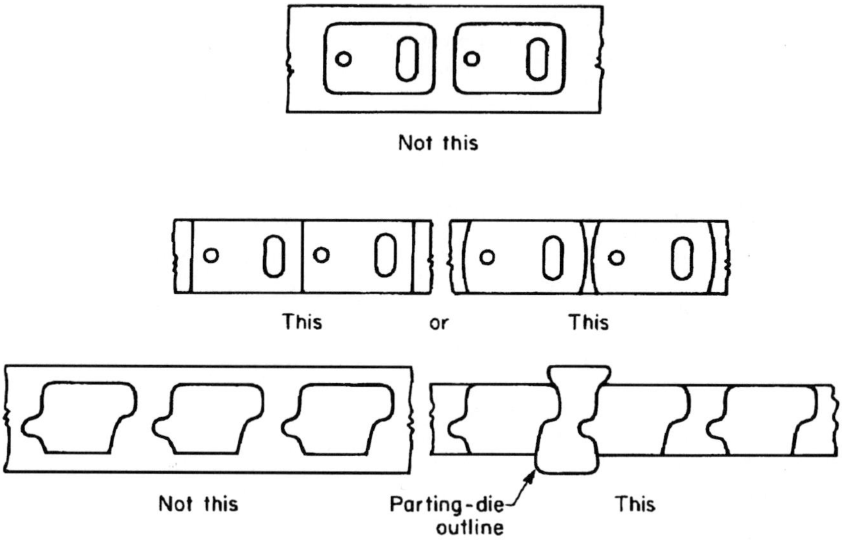

FIGURE 3.2.23 Use of strip stock of the width of the part with a parting die utilizes material better than wider strip stock with a full-periphery blanking die.

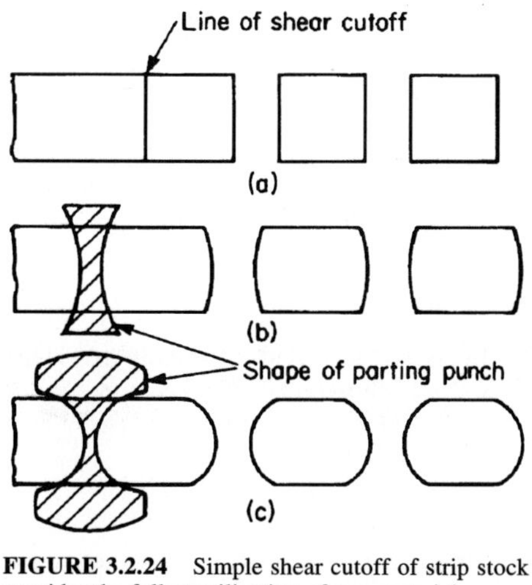

FIGURE 3.2.24 Simple shear cutoff of strip stock provides the fullest utilization of raw material.

Narrow Sections

Long, narrow projections should be avoided because they are subject to distortion and require thin, fragile punches. As a general rule, long sections should not be narrower than $1\frac{1}{2}$ times stock thickness. If the projection or web is relatively short, however, this precaution can be relaxed. Figure 3.2.25 illustrates some examples.

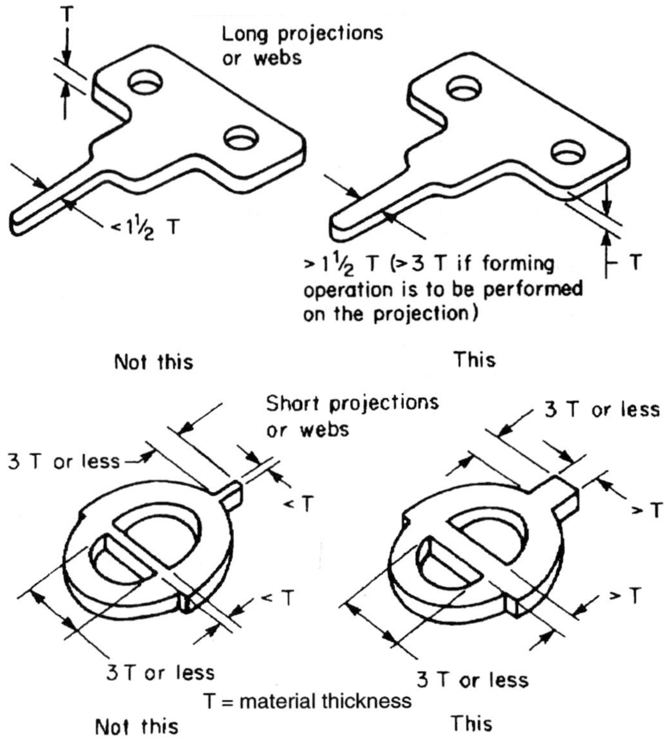

FIGURE 3.2.25 Narrow projections and webs cause die punches to be narrow and fragile. This should be avoided. Projections should be still wider if they undergo bending operations.

Shaving Allowances

If parts are shaved after blanking to provide a smooth edge, the recommended stock allowance for this operation is as shown in Table 3.2.1.

Reinforcing Ribs

When the formed section of a part requires extra resistance to flexing greater than that afforded by the strength, thickness, or temper of the material, stiffening ribs should be provided. Figure 3.2.26 illustrates a typical design.

Screw Threads

Sheet-metal stampings almost invariably are part of an assembly, and the methods for joining the parts into an assembly are many. One of the most widely employed and least expensive is the use of a screw fastener. An added advantage is the ease with which the components can be disassembled if necessary.

TABLE 3.2.1 Per-Side Shaving Allowance for Steel

Thickness, mm (in)	Allowance, mm (in)			
	No. 1 temper		No. 5 temper	
	First shaving	Second shaving	First shaving	Second shaving
1.2 (0.046)	0.08 (0.003)	0.04 (0.0015)	0.06 (0.0025)	0.03 (0.00125)
1.6 (0.062)	0.10 (0.004)	0.05 (0.002)	0.08 (0.003)	0.04 (0.0015)
2.0 (0.078)	0.13 (0.005)	0.06 (0.0025)	0.09 (0.0035)	0.044 (0.00175)
2.4 (0.093)	0.15 (0.006)	0.08 (0.003)	0.10 (0.004)	0.05 (0.002)
2.8 (0.109)	0.18 (0.007)	0.09 (0.0035)	0.13 (0.005)	0.06 (0.0025)
3.2 (0.125)	0.23 (0.009)	0.11 (0.0045)	0.18 (0.007)	0.08 (0.003)

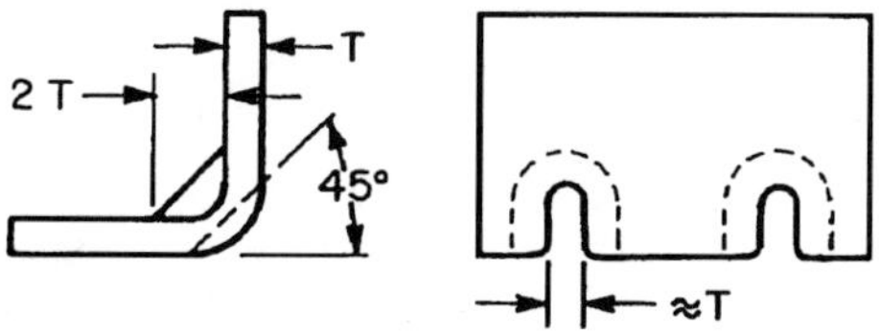

FIGURE 3.2.26 Stiffening ribs for a right-angle bend.

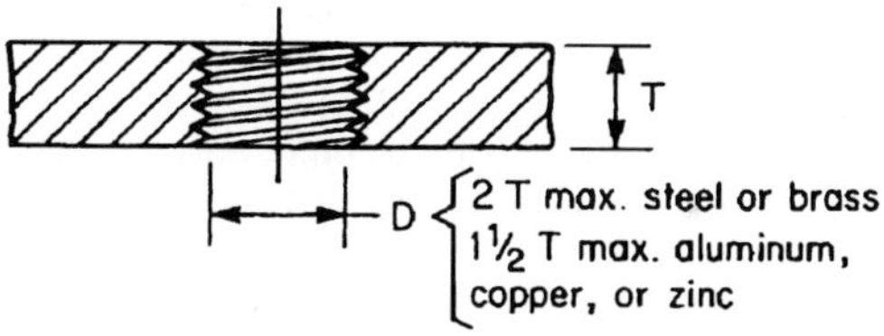

FIGURE 3.2.27 Design recommendations for screw threads in flat stock.

A disadvantage, in many cases, is the fact that the material thickness of a typical stamping does not permit an adequate number of threads for even minimum tightening ability. A rule of thumb for the minor thread diameter (tap-drill size) is that it not exceed twice the stock thickness for steel and brass and 1½ times the stock thickness for aluminum, copper, and zinc. (See Fig. 3.2.27.)

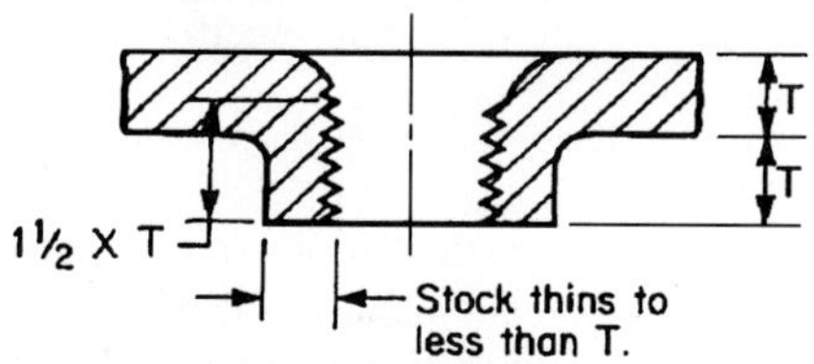

FIGURE 3.2.28 Design recommendations for screw threads in extruded holes.

One of the simplest and least expensive solutions to the problem of getting sufficient thread length is by use of extruded or flanged holes. Generally, the height of the extrusion is limited to the stock thickness; taking into account the unavoidable lead-in or pull-down radius, one can expect a full thread length approximately 1½ times the stock thickness. (See Fig. 3.2.28 for a typical

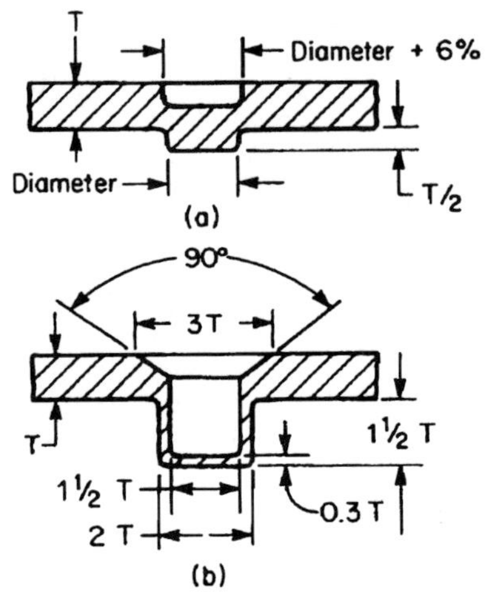

FIGURE 3.2.29 Design recommendations for set-outs when they are used to replace separate rivets, pins, cam followers, locators, etc.

design.) However, it is extremely difficult to define by formulation a standard for these extrusions. As a rule, most are developed in the die room and are based on trial and error.

Set-Outs

Set-outs can serve many purposes such as locators, rivets, cam followers, pins, etc. As such, they are economical in that separate components need not be purchased, handled, and assembled. Generally, to avoid fracture of the stock material, set-outs should be limited in height to one-half the stock thickness. The punch, of course, must be slightly larger in diameter than the set-out. (See Fig. 3.2.29*a.*) If the set-out is made hollow, it is possible to obtain a height of approximately 1½ times stock thickness (see Fig. 3.2.29*b*).

Burrs

Designers should bear in mind the difference between the two sides of a blanked or sheared sheet (as shown in Fig. 3.2.2) and decide which face of the finished part should have the sharp edge and which side should have the rounded or pulled-down edge. When the stamping must move against another part, normally the rounded edge should be on the side in contact. It is also advisable to locate the sharp or burred side so that it cannot cause injury when the part is in use. This may be preferable to the inclusion of blanket notes like "Remove all burrs" or "Break sharp edges," which are antieconomical because burr removal is expensive (see Chap. 4.23). Curled, folded, or sharply bent edges should be designed so that the burr side is on the interior of the bend. In blueprints of stampings, unless otherwise specified, it is generally understood that the burr side is the top one as drawn.

Formed Parts

For tooling economy, the designer should specify shapes that can be produced with standard existing, universal bending dies. This implies the following specific recommendations:

1. The inside bend angle preferably should be 90°.
2. In channel forming, the relationship between leg height and width (relationship between *W* and *H* in Fig. 3.2.30) should allow the use of a single standard 90° bending tool, especially if production quantities are limited. Normally, a ratio of channel width to leg height of at least 2:1 is required, but the actual minimum ratio depends on the width of the forming punch.

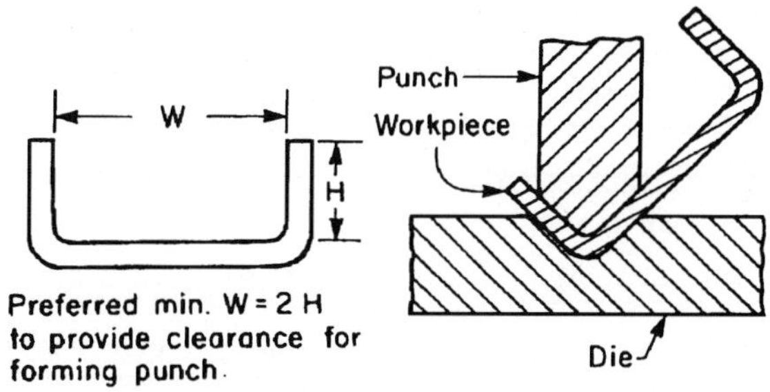

FIGURE 3.2.30 Preferred proportions for channel bends.

3. To avoid twisting and distortion, the width of the formed portion of the part should be at least 3 times stock thickness. (See Fig. 3.2.25.)

Drawn Parts

Drawing operations are very seldom employed for small lots because drawing requires more sophisticated, more expensive tooling and more development work and time than simply bent stampings. In short-run jobs, quite often drawn shells and boxes are changed to fabricated shells, which are easier to make and require simpler and less expensive tooling.

The only kind of drawing operation that is sometimes performed in small lots is very shallow drawing of round shells, which requires comparatively simple tools without blank-holding equipment. There are no hard-and-fast rules for the design of such workpieces. The limits of drawing without blank holders depend on the combination of comparatively thick stock, small shell height, and large shell diameter. Heights vary between 5 and 10 percent of shell diameter in thin- and medium-gauge stock, as shown in Fig. 3.2.31.

When it is imperative to make drawn shells of conventional design and shape (seamless shells), a few basic recommendations will help to keep down tooling and overall production costs:

1. Avoid tapered-wall shells and/or flanged shells. They are much more expensive than straight cylindrical shells.

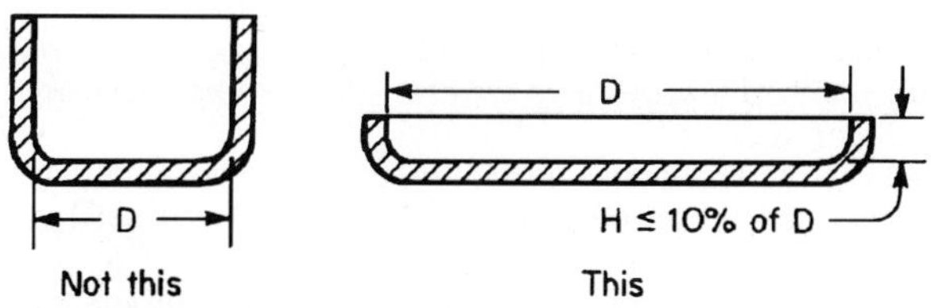

FIGURE 3.2.31 Recommended proportions of drawn parts produced in short-run quantities when simple tooling, without blank-holding features, can be used.

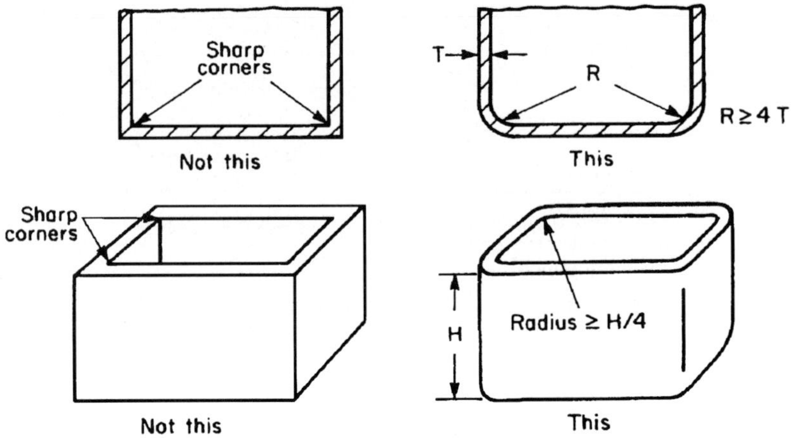

FIGURE 3.2.32 Design rules for deep-drawn stampings.

2. Do not specify both inside and outside diameters; only one of these dimensions can be controlled because of variations in wall thickness.
3. Avoid sharp corners in the bottoms of drawn shapes. The recommended minimum radius is 4 times stock thickness. (See Fig. 3.2.32.)
4. For rectangular boxes, specify corner radii at least 0.25 times the depth of draw. (See Fig. 3.2.32.)

Countersinks and Counterbores

Countersinking, counterboring, and deep chamfering, as illustrated in Fig. 3.2.33, can be expensive, especially if quantities are limited or machining is required. Designers should avoid specifying these features unless they are really necessary because of the additional tooling and operational costs they entail.

Other Recommendations: Short-Run Stampings

The best designs for short-run stampings are those which require the least tooling or even none not already available. Designs that can be produced by straight squaring shears, notching tooling, and standard punches and forming blocks are best. Regularly shaped parts with standard-size features are most likely to be producible with this kind of equipment and tooling.

Other than the need to economize on tooling costs, the design of short-run and conventional stampings should follow the general principles described above. It also should be borne in mind that even though large-volume production provides an opportunity to amortize high tooling costs, it is still preferable to minimize tooling costs in high-production applications. Consequently, in cases in which the outer contour of a blank is not functional (need not match mating parts), the designer should shape the blank so that its production will entail the least expense for tooling.

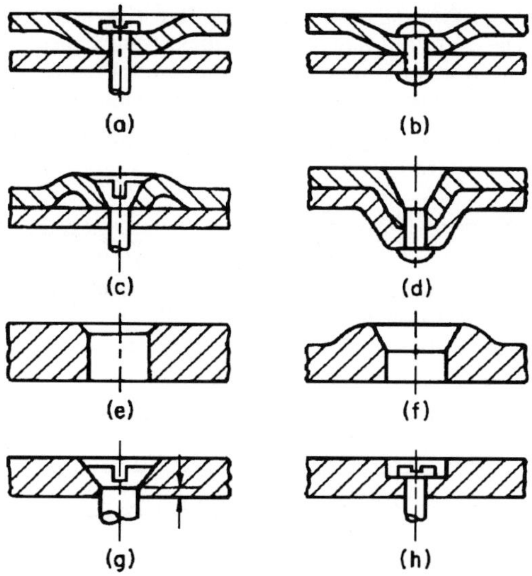

FIGURE 3.2.33 Various countersinks and counterbores used with metal stampings. *a, b, c,* and *d* represent preferred approaches which are relatively easy to produce. With properly designed dies (double-acting tools) and a favorable dimensional relationship between stock thickness and clearance hole size, the hole may be punched simultaneously with the forming of the depression. In *e* and *f,* the countersink is produced by coining. This requires soft material, and some development is frequently needed, so this approach is not so suitable for short runs. Unless the chamfer is small (as in *e*), a bulge will be raised around the hole as shown in *f*. *g* and *h* illustrate machined countersinks and counterbores, respectively. Because of the secondary machining operations, these are more costly to produce.

DIMENSIONAL FACTORS

Realistic, practical tolerances on metal stampings at large, whether in short-run, medium-run, or high-production lots, depend on several details. The chief ones are the design of the part, its function, the size of the part, the press operations to be performed, and the material to be employed (kind and thickness). Die accuracy, die wear, metal springback variations (due primarily to material temper and thickness variations), changes in die dimensions after sharpening, feed, and workpiece-placement variations in the die all affect final-part accuracy. This means that each specific part and each individual feature or dimension must have its own practical tolerance.

Short-run stampings may not have as high an accuracy as those produced from conventional high-production tooling for several reasons. In high-production tooling, the extra time and cost of building high accuracy into the tooling is more easily justified. Multiple-operation dies usually are more accurate than separate operations in which the part must be positioned manually for each separate step. Finally, some types of short-run tooling (e.g., steel-rule dies) inherently cannot be made to quite the high level of accuracy of conventional tooling. It also must be stated that the precision of stampings of any type and kind has its limitations. Below these practical limits it is necessary to incur some additional machining operation, such as milling, grinding, reaming, lapping, honing, burnishing, etc., if the part demands the highest precision.

RECOMMENDED TOLERANCES

Tables 3.2.2 and 3.2.3 show recommended tolerances for blanks. Forming-tolerance recommendations are contained in Table 3.2.4.

TABLE 3.2.2 Recommended Dimensional Tolerances for Sheet-Metal Blanks Produced with Short-Run Tooling

	Closest tolerance, mm (in)	Normal production tolerance, mm (in)
Width of slit cold-rolled sheet stock up to 230 mm (9 in) wide and 1.5 mm (0.060 in) thick	±0.05 (0.002)	±0.13 (0.005)
Length, width, and other dimensions of blanks produced by shear, parting-tool, and notching operations	±0.13 (0.005)	±0.25 (0.010)
Dimensions of parts produced by numerical-control turret punching machines	±0.05 (0.002)	±0.13 (0.005)
Diametral tolerance for holes less than 25 mm (1.0 in) in diameter		
Stock under 1.5 mm (0.060 in) thick	+0, −0.05 (0.002)	+0, −0.10 (0.004)
Stock thicker than 1.5 mm (0.060 in)	+0, −0.10 (0.004)	+0, −0.30 (0.012)

TABLE 3.2.3 Recommended Dimensional Tolerances for Sheet-Metal Blanks Produced with Blanking Dies

	Dimension		
	Up to 75 mm (3 in)	76 to 200 mm (3 to 8 in)	200 to 600 mm (8 to 24 in)
Material thickness, mm (in)	Tolerance, mm (in)		
To 1.5 (0.060)	0.08 (0.003)	0.15 (0.006)	0.25 (0.010)
Over 1.5 to 3.0 (0.125)	0.15 (0.006)	0.25 (0.010)	0.40 (0.016)
Over 3.0 to 6.3 (0.250)	0.30 (0.012)	0.45 (0.018)	0.80 (0.031)

Note: Tolerances are plus only for outside dimensions and minus only for inside dimensions. For holes produced with blanking dies, the diametral tolerances shown in Table 3.2.2. also apply.

TABLE 3.2.4 Recommended Tolerances for Dimensions of Formed Stampings

	Closest tolerance	Normal production tolerance
Angle of bend of formed legs		
Special dies	±½°	±1°
Press brake or other universal tooling	±1°	±2°
Bending brake	±2°	±3°
Height or length of bent leg (dimension *H*)	±0.2 mm (0.008 in)	±0.4 mm (0.016 in)
Height of extruded holes	±0.2 mm (0.008 in)	±0.3 mm (0.012 in)
Deviations from flatness	±0.3%	±0.5%

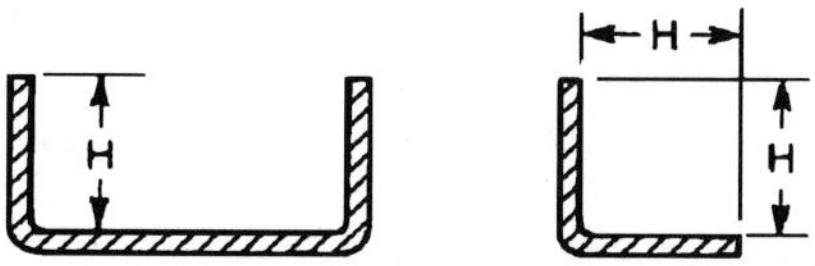

CHAPTER 3.3
FINE-BLANKED PARTS

Joe K. Fischlin
American Feintool, Inc.
White Plains, New York

The fine-blanking process is a press-working technique that utilizes a special press and precision tools and dies to produce parts that are more nearly complete and ready for assembly as they leave the fine-blanking press than parts blanked by conventional methods. Fine-blanking produces parts with cleanly sheared surfaces over the entire material thickness. By comparison, conventionally stamped parts generally exhibit a cleanly sheared edge over only approximately one-third of the material thickness, and the remainder shows fracture. With conventional stamping, when these surfaces are functional, they may require some form of secondary operation such as shaving, milling, reaming, broaching, hobbing, or grinding. Often, several of these operations are necessary to complete the part. Apart from the improved quality of the sheared surfaces when fine-blanking is used, high dimensional accuracy can be obtained, and the process also allows operations that cannot normally be accomplished in conventional stamping dies.

THE FINE-BLANKING PROCESS

Basic Press Requirements and Functions

To produce fine-blanked parts, a precise die with low punch and die clearance (approximately $\frac{1}{2}$ percent of material thickness) is required along with a triple-action press to clamp the material during the shearing operation. The three actions in the press provide the shearing pressure, the V-ring pressure, and the counterpressure. These must be held constant throughout the stroke to ensure good-quality parts. The press also must be capable of a fast approach stroke, a slow shearing speed between 4 and 15 mm/s (0.160 and 0.600 in/s), and a fast ram retraction to obtain good stroking rates.

In production, to obtain the maximum number of parts between punch and die regrinds, it is imperative that the punch does not enter the die. This is ensured through a ram stroke, or shut-height setting, accurate within 0.01 mm (0.0004 in).

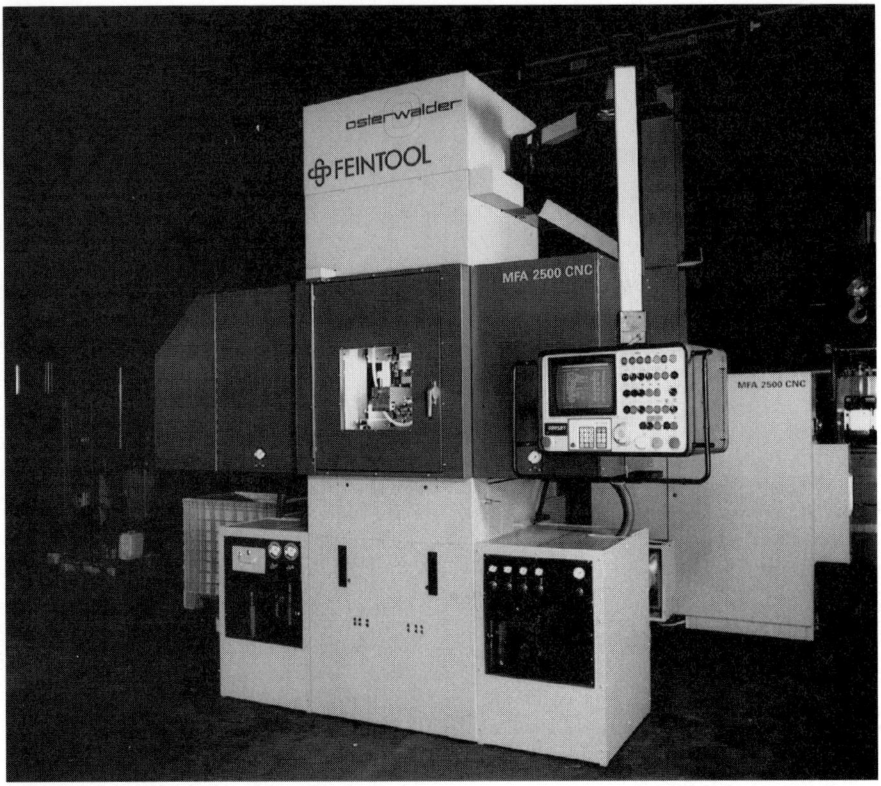

FIGURE 3.3.1 A 250-ton fine-blanking press. (*Courtesy American Feintool, Inc.*)

Press Capacities

Currently, fine-blanking presses from 40 to 1400 tons total capacity are in use on a production basis. Figure 3.3.1 illustrates a typical fully automated fine-blanking press with a total capacity of 250 tons.

The Press Cycle

The sequence of operations during one cycle of the fine-blanking press, shown in Fig. 3.3.2, is as follows: (1) The material is fed into the die. (2) The upstroking ram lifts the lower table and die set. This raises the material to the die face. (3) As the tool closes, the V ring is embedded in the material. The material is clamped between the V-ring (or "stinger") plate and the die plate outside the shear periphery. The counterpunch (which is under pressure) clamps the material against the blanking-punch face inside the shear periphery. (4) While the V-ring pressure and the counterpressure are held constant, the punch continues upstroking, cleanly shearing the part into the die and the inner-form slug into the punch. At the top dead-center position, all pressures are shut off. (5) The ram retracts, and the die opens. (6) Almost immediately after the tool

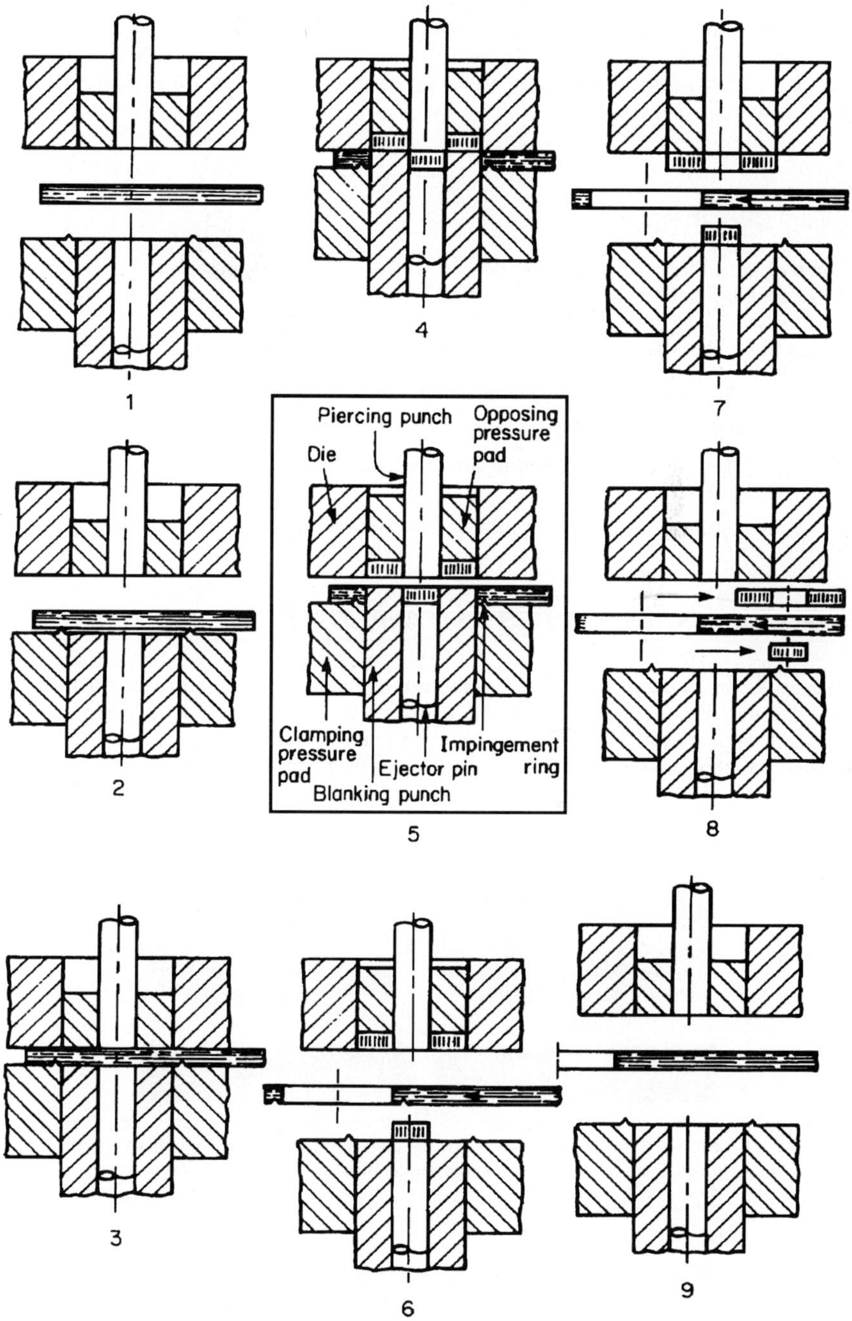

FIGURE 3.3.2 One cycle of a fine-blanking press.

opens, the V-ring pressure is reapplied. This strips the punch from the skeleton material and pushes the inner-form slugs up out of the punch. The material feed begins. (7) The counterpressure is reimposed, ejecting the part from the die. (8) The part and slugs are removed from the die area either by an air blast or by a removal arm. (9) The cycle is complete and ready to recommence.

Tooling: Dies, Progressive Dies

Whenever possible, a fine-blanked part is produced with a compound die so that it is produced completely in a single operation. Fine-blanking progressive dies have been in use for several years.

Operations that can be carried out in progressive fine-blanking dies include

- Chamfering (internal and external forms)
- Bending (legs, offsets, protrusions, etc.)
- Coining for counterbores, thinning material on hammer-type components, flanges, and coins or tokens
- Forming (shallow "draws" are possible in certain parts and materials)

TYPICAL CHARACTERISTICS AND APPLICATIONS

The reasons for considering fine-blanking include the need for improved surface finish and squareness of the sheared edges, higher dimensional accuracy, and superior flatness or appearance compared with conventional stampings. Figure 3.3.3 compares the sheared-edge condition from a conventionally stamped part with a fine-blanked part.

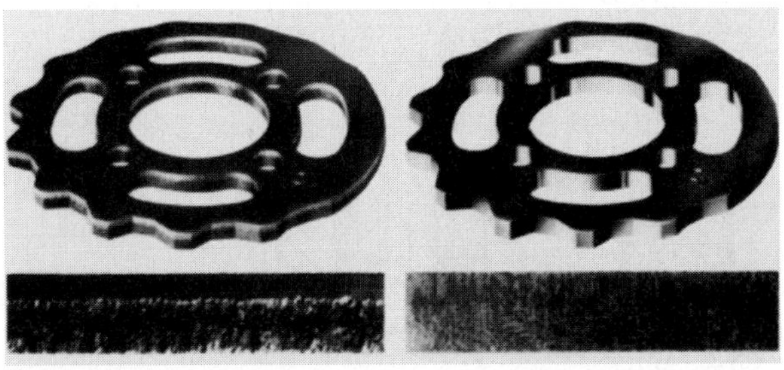

FIGURE 3.3.3 Comparison of the edge surface produced by conventional stamping (left) and fine-blanking (right). (*Courtesy American Feintool, Inc.*)

Such features as cam tracks, locating pins, rivets, and depressions can all be incorporated into the fine-blanked part. Figure 3.3.4 shows a semipierced form used in an automotive seat-back-recliner assembly. The production of gears, gear segments, ratchets, and racks is a major field for fine-blanking application.

FIGURE 3.3.4 This automotive seat-back-adjustment assembly uses semi-pierced main plates, coined double gear, spacer, and eccentric plate, all produced by fine-blanking, to eliminate separate parts that otherwise would be necessary. (*Courtesy American Feintool, Inc.*)

Surface Finish and Edge Squareness

An average surface finish for fine-blanked parts is 0.45 μm (32 μin). It is possible to obtain a superfinish of 0.1 to 0.2 μm (8 to 16 μin) by using carbide dies for applications such as special cams for which a polished surface is required. Because of die wear and materials in use, the 0.45-μm (32-μin) figure may be exceeded after a certain number of parts have been produced.

The perpendicularity of sheared edges will seldom be exactly 90° but will not fluctuate more than approximately 40 to 50 min. The quality of material sheared and the condition of the fine-blanking tooling are most influential for these two features.

Part and Material Sizes

As with most stampings, the vast majority of fine-blanked parts are made from coil or strip stock. On occasion, profile-rolled or -extruded bars are fine-blanked when the part configuration dictates, but this is not common. However, even when working with flat stock, the finished part can have coined sections, offsets, bends, embossed areas, etc. These latter examples may be obtained with progressive dies.

The size range for fine-blanked parts is very broad. Watch parts with overall lengths of about 2.5 mm (0.100 in) and smaller are being produced regularly. Conversely, large machinery plates about 760 mm (30 in) across have also been made. The majority of parts do fall in the range of 25 by 25 mm to 250 by 250 mm (1 by 1 in to 10 by 10 in).

Thickness, as with overall part size, also covers a broad range. Parts as thin as 0.13 mm (0.005 in) and as thick as 13 mm (0.5 in) are produced on a regular production basis. The most widely produced parts fall within a range of 0.8- to 8-mm (0.030- to 0.300-in) material thickness.

Die Roll (Pull-Down)

Through fine-blanking, a certain deformation of the piece part occurs on the die side of the part. (See Fig. 3.3.5.) This deformation, which is termed *die roll* or *pull-down,*

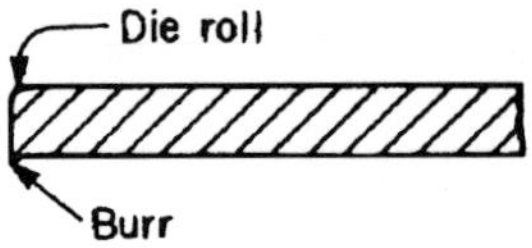

FIGURE 3.3.5 Die roll and edge burr that may occur on fine-blanked parts.

is less marked on fine-blanked parts than on conventionally stamped parts. This is due to the retention of the material during blanking through the V-ring pressure and counterpressure.

The sharper the angle on the outer form, the smaller the radius on these corners, and the thicker the material, the greater will be the die roll, which on certain gear-wheel parts may be as much as 30 percent of the material thickness. Less die roll is experienced on hard than on soft materials.

Burrs

Owing to inevitable wear on the shearing punch, blunting occurs and causes burrs to form on the piece part after a certain number of parts have been produced. (This is also shown in Fig. 3.3.5.) The degree of burr produced is related to the type of material worked and the number of parts produced between resharpenings of the punches.

If the burr is objectionable, it can be removed by tumbling, vibratory finishing, or belt sanding. All three methods are acceptable and must be selected according to the type of part and finished condition required.

Outside Chamfers

In certain cases, outside chamfers on a part also can be produced. This is accomplished by making a V-ring impression on the die side of the part in a preceding station in the die, exactly at the location where the outer edge will be. The part is then blanked out, parting the V half and producing a chamfer on the die-roll side of the part. This procedure has been carried out successfully to a chamfer depth of about 30 percent of material thickness.

Offsets

Using specially formed elements, it is possible to produce fine-blanked piece parts with offsets to a maximum of 75° in a single operation. Until the present time, materials of maximum 6-mm (0.24-in) thickness have been worked by this method. Figure 3.3.6 shows a typical offset part producible in one press stroke.

Under certain circumstances, a normal fine-blanked edge finish will not be achieved in the offset zone, but for most piece parts of this type this finish is not a requirement, since functional surfaces seldom lie in these areas. The dimensions of both inner and outer forms relative to one another on both sides of the offset zone can be held far more accurately by this method of production than is possible when producing the offset in a secondary operation.

Semipiercing

Semipiercing in fine-blanking is shown in Fig. 3.3.7. The material has been partially pierced or sheared, but no fracture or division of the material has been made in the

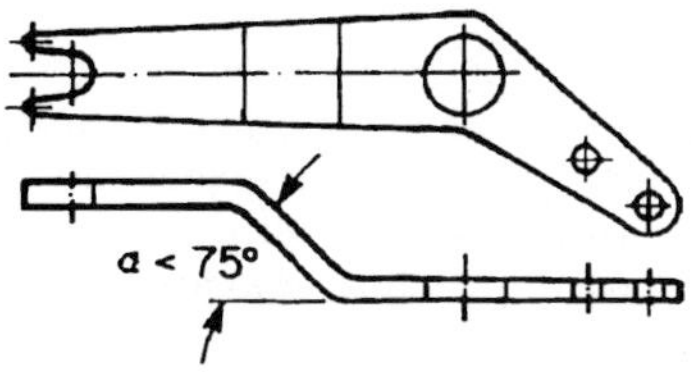

FIGURE 3.3.6 Offset part that can be produced in a single fine-blanking operation.

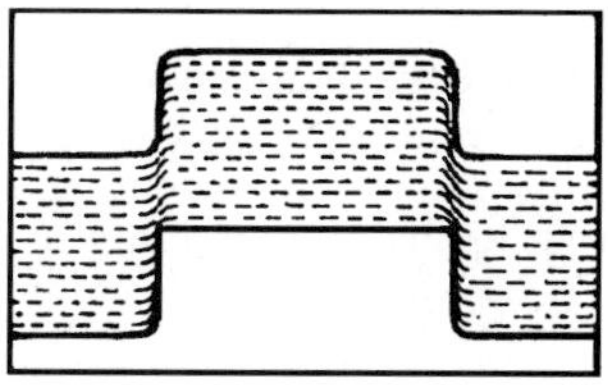

FIGURE 3.3.7 Semipiercing.

shear zone. The picture shows that the material structure has been displaced at 90° in the shear zone. The "cold forming" involved causes the material tensile strength in the shear zone to increase, disproving the often propagated opinion that such sections are weak and may not be placed under load or strain.

Semipierced faces may be used on both the internal and the external forms as dowel pins, rivets, cam tracks, or gear forms. (See also Fig. 3.3.4.)

Additional examples of fine-blanked parts are shown in Fig. 3.3.8.

FIGURE 3.3.8 A collection of typical fine-blanked parts. (*Courtesy American Feintool, Inc.*)

ECONOMIC PRODUCTION QUANTITIES

Fine-blanking generally can be classified as a high-volume production process, since tool quality and cost normally require a reasonable quantity of parts to justify the expenditure. A comparison of various production methods will show true justification, since tooling costs are not the only factor to be taken into account. In some cases, quantities of 1000 to 5000 pieces can completely amortize tooling costs. This is the case when an expensive secondary operation such as numerical-control contouring, grinding, or a difficult broaching is eliminated by designing the part to utilize fine-blanking. As a general guideline, however, minimum quantities can be considered to be between 10,000 and 20,000 parts.

Production-Time Comparisons

The operating cycle is normally somewhat slower for fine-blanking than for conventional stamping. A press rate of 45 strokes per minute would be a fair average for fine-blanking operations. In comparing fine-blanking with conventional stamping, total costs for all operations necessary to complete the part, including secondary machining, must be considered.

MATERIALS SUITABLE FOR FINE-BLANKED PARTS

Unless they are to be drawn or severely formed, conventionally blanked parts are generally made with material in a relatively hard condition to obtain a good shear-break characteristic and minimize burrs. In contrast, fine-blanking processing requires materials that exhibit a good cold-forming quality.

Steel: Structural Requirements and Shearing Characteristics

Approximately 90 percent of all fine-blanked parts are produced from steel. With regard to the above-mentioned hardness and cold-forming characteristics, it must be mentioned that in steels the required features generally relate directly to the material structure. As the carbon content increases, the structure of the steel becomes increasingly important. A good dispersion of small cementite spheres within a soft ferrite mass (spheroidize-annealed) is best. This enables the cutting elements to shear the part cleanly rather than induce a shear-break characteristic in the unevenly dispersed structure of a nonannealed material. The quality of the material used will influence not only the appearance of the piece parts but also the die and punch life between resharpening operations.

Provided that it is pickled and oiled (is scale-free), hot-rolled material can be fine-blanked. Here again, the structure of the steel is probably of the greatest importance. Generally, cold-rolled material will provide the best results, but overall economics related to the function of the part, etc., must be considered. Best results in steel can be obtained with material having a tensile strength from approximately 260 to 520 MPa (38,000 to 76,000 lbf/in^2) and a carbon content of 0.08 to 0.35 percent. However, it is possible to produce parts in materials with a tensile strength up to 685 MPa (100,000 lbf/in^2) and a carbon content of 0.7 to 0.9 percent. In this carbon range and above, the quality of the steel used and the type of part configuration to be produced are more critical and must be carefully evaluated in each case.

Nonferrous Materials

Nonferrous materials such as brass, copper, and aluminum also can be fine-blanked. When parts are produced from brass, there should be no lead content. While lead is required in chip-forming operations, it will cause tearing on the sheared surfaces of fine-blanked parts. Copper may tend to "cold-weld" to the sides of the punch and die and cause a buildup on these surfaces. This generally can be overcome to an acceptable degree by using a suitable cutting lubricant. Soft and half-hard aluminum can be fine-blanked providing cleanly sheared surfaces. Hard grades of aluminum (e.g., T5 and T6) can be fine-blanked but will show edge fracture because of the material structure and hardness.

In certain cases, materials with a high level of hardness (not generally recommended for fine-blanking processing) have been sheared successfully on a production basis. Each case must be evaluated on its merits, and generally some trial or testing is required. An overall evaluation of material suitability is provided in Table 3.3.1.

DESIGN RECOMMENDATIONS FOR PIECE PARTS

Corner Radii

On fine-blanked parts, corners must be rounded. If a radius is not introduced or is too small, the edge near the corner will show tears in the material over the shear zones. The main punch and inner-form punches are likely to chip in these areas after only a short period of use.

The corner angle, material thickness, and type of material determine the minimum radius required. In general, the following values can be applied:

Obtuse angles: radius 5 to 10 percent of material thickness

Right angles: radius 10 to 15 percent of material thickness

Acute angles: radius 25 to 30 percent of material thickness

These data relate to external forms. Inner corners can be radiused with two-thirds of these values. In general, the maximum possible radius should be specified in all cases, as this provision can only be beneficial to tool life. (See Fig. 3.3.9.)

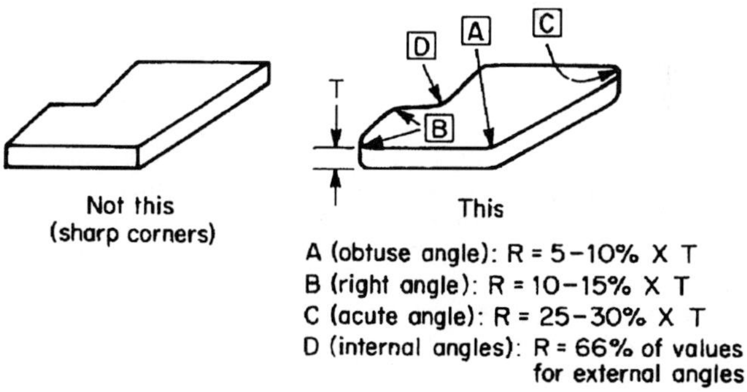

FIGURE 3.3.9 Recommended corner radii for fine-blanked parts.

Holes and Slots

Holes in materials 1 to 4 mm (0.039 to 0.157 in) thick generally can be blanked with the width of sections from inner to outer form corresponding to approximately 60 to 65 percent of the material thickness. When working thicker materials, the specific shear pressure becomes higher in such critical areas. Also, the heat-up rate of the cut-

TABLE 3.3.1 Suitability of Various Materials for Fine-Blanking Based on Hardness or Temper Condition

Material	Excellent	Average	Poor	Not recommended
Aluminum				
1100 (A91100)	O, H2, H4	H6	H8	
3003 (A93003)	O, H2	H4	H6, H8	
5052 (A95052)	O	H2, H4	H6, H8	
Heat-treatable aluminum				
2014 (A92014)	O		T4	T6
6061 (A96061)	O		T3, T4	T6, T8
7075 (A97075)	O			T6
Brass*	Quarter- and half-hard	Three-quarter-hard	Hard, spring	
Bronze	Quarter- and half-hard	Three-quarter-hard	Hard, spring	
Beryllium copper	Soft R_B45–78		Half-hard R_B88–96	Hard R_B96–102
Phosphor bronze				Spring R_B89–96
Monel metal	Quarter- and half-hard			
Nickel silver, 65% Cu, 18% Ni	Quarter-hard R_B50–75 Half-hard R_B68–82	Hard R_B80–90	Extra-hard R_B87–94	Spring R_B89–96
Carbon steel				
Low-carbon C1010–C1020 (G10100–G10200)	R_B50–75; temper nos. 3, 4, 5	R_B70–85; temper no. 2	R_B84–90; temper no. 1; poor to not recommended depending on part	
Medium carbon C1021–C1035 (G10210–G10350)		Fully annealed		
High carbon C1036–C1070 (G10360–G10700)		Under 1/8 in thick Sphereodize-annealed	Over 1/8 in thick Sphereodize-annealed	

Alloy steels 8615 (G86150), 8620 (G86200), etc., 4130 (G41300), 4140 (G41400), 4620 (G46200)	Fully annealed Under 1/8 in thick Spherodize-annealed	Over 1/8 in thick	
Stainless steel 200, 300 Series	Fully annealed; no thickness limit		
400 Series, heat-treatable	Fully annealed; under 1/16 in thick		Over 1/16 in thick

Note: Silicon steel, up to $2\frac{1}{2}\%$ silicon content, is suitable for fine-blanking only if hot-rolled-annealed, after rolling to R_B85–90, pickled, and oxalate-coated. Tool steel, AISI 52 (R_B80–90), is feasible for fine-blanking. However, stock thickness should not exceed 1/16 in, and configuration of part must be "ideal." Low die life must be taken into consideration.

*Brass usually can be blanked successfully only if lead-free. Cartridge brass C26000 (SAE70 or ASTME) is the most commonly used.

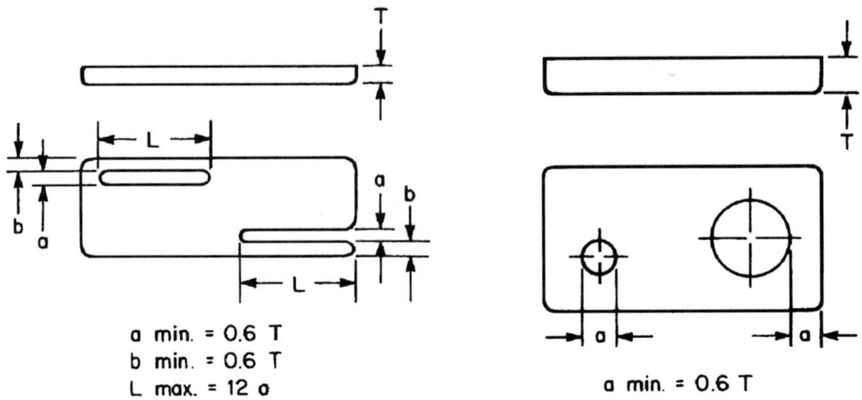

FIGURE 3.3.10 Design rules for slots in fine-blanked parts.

ting elements increases. Practice has shown that the sections mentioned above must be increased for materials over 4 mm (0.157 in) thick.

Basically, the same data apply for both hole diameters and slot widths. Recommendations for actual applications can be determined from the data supplied in Fig. 3.3.10*a* and *b*.

Tooth Forms

Tooth forms for gears, ratchets, etc., may be produced by fine-blanking when the width of the tooth on the pitch-circle radius is 60 percent of material thickness or more. In certain cases, width can be reduced to 40 percent of stock thickness. Governing factors are the tooth form and, once again, the material being worked.

It should be specially noted that the crests of the teeth and the tooth-root form must be radiused. In practically all cases this is allowable as the tooth form in these areas generally has no actual function.

Lettering and Marking

Relatively deep marking and lettering, for appearance or identification purposes, often can be incorporated in the piece part in the blanking operation without extra cost. It should be noted that, when possible, markings should be laid off on the die side of the piece part so that they can be imprinted through the ejector. If this is not possible for any reason, the marking can be made through a punch insert on the piece-part burr side.

The wider the markings and the deeper the coined depth into the material, the greater the danger of bruising on the reverse side of the part. The example shown in Fig. 3.3.11 is for a telephone coin-insert plate. Here the maximum possible coining depth has been reached.

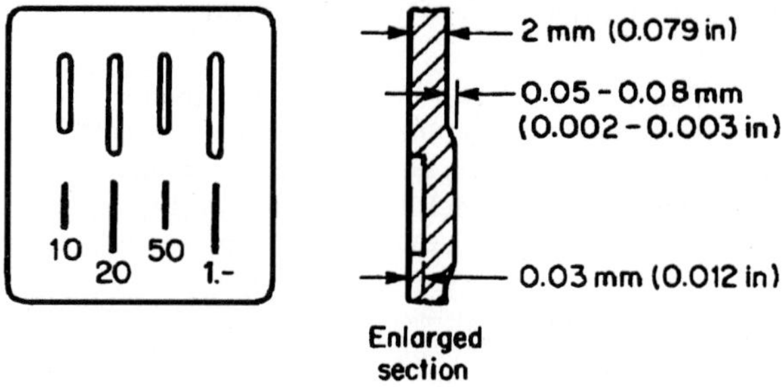

FIGURE 3.3.11 Coining of a pay telephone money-insert plate showing the maximum depth of coining without bruising the opposite face of the part.

Semipiercings

Generally, the maximum depth to which offsets or semipiercings (see Fig. 3.3.7) are recommended to be pierced is 70 percent of material thickness.

Countersinks and Chamfers

If the countersink of an inner form lies on the die side of the part, it can be made in one stroke, i.e., together with the actual blanking of the part in one operation. Countersinks of 90° can be introduced to depths of one-third material thickness without disagreeable material deformation becoming noticeable. This is valid for material thicknesses up to approximately 3 mm (0.125 in). Naturally, the angle of the countersink may be increased or decreased. In such cases, the volume of material to be compressed should not exceed the volume of one-third material thickness at 90°. (See Fig. 3.3.12.)

When the countersink lies on the burr side (punch side) of the piece part, it must be made in a preceding station in the tool. In the operation in which the piece part is finally blanked out, the countersink hole may be used for location or pilot purposes.

Should countersinks be required on both sides of the hole, these must be introduced in one or two preceding stations, depending on the depths necessary.

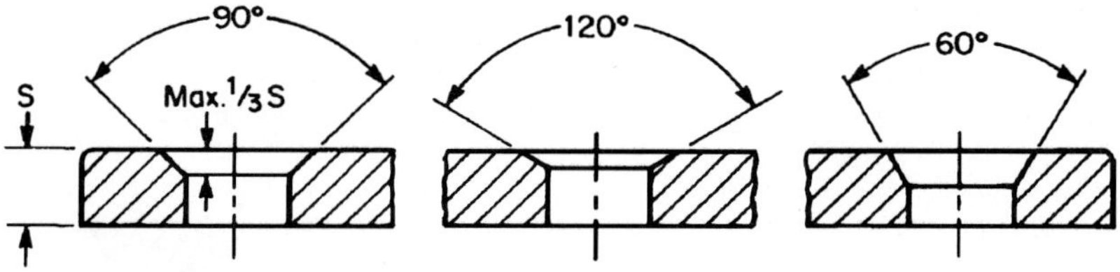

FIGURE 3.3.12 Countersink variations producible by fine-blanking without significant stock deformation. Countersinks of 90° usually can be made to depths of one-third material thickness without disagreeable deformation of the adjacent surface.

DIMENSIONAL FACTORS AND TOLERANCES

The major factors influencing piece-part quality are the stock material and the accuracy of the press tool. The actual configuration of the piece part is also of importance.

Since there is no taper on the punch or the die in fine-blanking tools, resharpening the tool allows the size of the original die apertures to be maintained. This factor ensures piece-part repeatability and maintains part accuracy for the entire life of the die.

Piece-part flatness obtainable in production varies widely in relation to part size, part configuration and quality, and thickness of material being worked. Generally, the piece part will remain as flat as the stock material introduced to the die.

Table 3.3.2 indicates typical tolerances that can be maintained for various dimensional characteristics of fine-blanked parts.

TABLE 3.3.2 Recommended Tolerances for Fine-blanked Parts

Material thickness, mm (in)	Tolerance of inner form and outer configuration and dimensions, mm (in)	Tolerance of hole spacing, mm (in)
Material tensile strength to 410 MPa (60,000 lbf/in^2)		
0.5–1.0 (0.020–0.040)	±0.008/0.013 (0.0003/0.0005)	±0.013 (0.0005)
1.0–3.0 (0.040–0.120)	±0.013 (0.0005)	±0.025 (0.001)
3.0–5.0 (0.120–0.200)	±0.025 (0.001)	±0.038 (0.0015)
5.0–6.3 (0.200–0.250)	±0.039 (0.0015)	±0.038 (0.0015)
Material tensile strength over 410 MPa (60,000 lbf/in^2)		
0.5–1.0 (0.020–0.040)	±0.013/0.025 (0.0005/0.001)	±0.013 (0.0005)
1.0–3.0 (0.040–0.120)	±0.025 (0.001)	±0.025 (0.001)
3.0–5.0 (0.120–0.200)	±0.038 (0.0015)	±0.038 (0.0015)
5.0–6.3 (0.200–0.250)	±0.050 (0.002)	±0.038 (0.0015)

Note: The above tolerances are average indicative values. For parts between 3 and 6 mm (⅛ to ¼ in) in thickness and up to 150 mm (6 in) in length, indicative flatness is within 0.1%.

CHAPTER 3.4
FOUR-SLIDE PARTS

Kenneth Langlois
Torin Machinery
Torrington, Connecticut

INTRODUCTION

Four-slide parts are metal stampings, usually of complex configuration, produced on high-speed, multiple-ram, highly versatile machines. A number of operations can be performed before a part is ejected. Parts often can be produced completely in one four-slide operation.

Four-Slide Process

The major functional elements and basic process sequences of four-slide machines are shown in Fig. 3.4.1 for a horizontal machine. (There are also vertical machines with similar capabilities but with somewhat improved ability to perform some secondary operations.) Metal strip material moves from left to right from a continuous coil on a strip reel through a straightener, a feed mechanism, and one or more presses in the press area and, finally, into the forming area, from which complete parts are ejected.

Standard press operations can be performed on strip material in the press area as long as the total tonnage and stroke required are within the capacity of the machine. One or more press units are self-contained and controlled from a camshaft. Press operations include piercing, slotting, trimming, notching, dimpling, coining, lancing, extrusion, blanking, forming, embossing, and drawing (shallow).

In the forming area, four forming slides operate at 90° angles around the center form. Each slide is independently operated by a cam on one of the machine's interlocking camshafts. The forming slides advance (and dwell) in sequence as necessary to form the part. In some machines, a second forming action is available. Forming in more than one stage on the center form permits complex forming of a single part in more than one machine cycle without any reduction in production rate. In all forming positions, the stroke, timing, and velocity characteristics of each forming tool can be independently set and adjusted as necessary. After all forming and other secondary operations have been completed, the part is ejected from the machine.

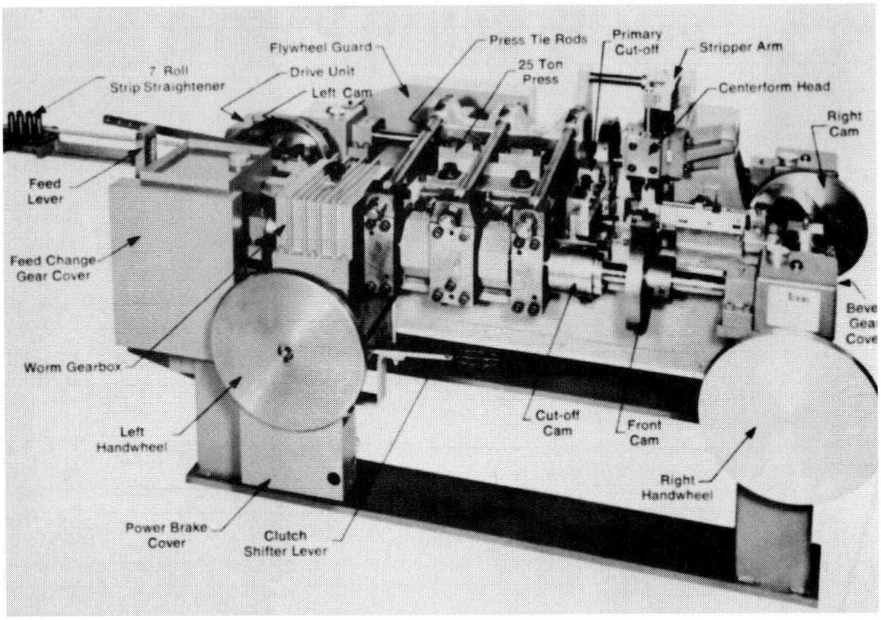

FIGURE 3.4.1 Typical vertical four-slide machine. (*Torin V-81 Verti-Slide.*)

Straight or shaped ends on parts can be produced by a variety of cutoff methods. Usually, a standard cutoff punch and fixed die blocks are used. However, the forming slides can all be tooled for cutoff as desired.

As many as eight or more parts can be produced in each machine cycle with multiple tooling, substantially increasing the machine's productivity. The duplicate parts may be formed side by side across the strip width or following each other along the strip. Or two or more parts may be produced simultaneously from separate strips fed in together in parallel. This kind of multiple processing is usually limited to relatively simple parts.

Eliminating Secondary Operations

The structure and operating characteristics of four-slides allow these machines to perform at one time many operations that are customarily carried out as secondaries in conventional metal stamping. These operations include drawing (limited by the stroke of the press portion of the four-slide machines, but performed in one or more die stations), reaming (normally performed by an air-operated reaming head), tapping (performed by an auxiliary tapping head), welding (both spot and butt welds can be produced), assembly (simple assembly equipment can be integrated into the machines), and packaging (automatic packaging apparatus can be added at the ejection point of a vertical machine).

TYPICAL PART CHARACTERISTICS AND APPLICATIONS

Compared with parts produced on conventional punch presses, four-slide parts are more complex, involve a higher degree of forming, and usually have the basic strip shape and edges integrated into their design. In contrast, the usually larger capacity of a conventional power press, in both die size and tonnage, is superior whenever fabrication of the part is mostly a matter of blanking. Parts that require forming to more than shallow angles out of the strip plane take fullest advantage of the four to six slide tools in a four-slide's forming area. The slide motions available can bend strip in almost any conceivable way: it can be turned back against itself or bent in several directions.

The ends of four-slide parts can be overlapped, butted, and interlocked as necessary to form closed configurations. Ends also can be welded together to produce a part that otherwise might be made at a higher cost on other machines, either out of metal tubing or as a drawn cup with the bottom blanked out. Welding is also used to secure such pieces as contacts, screws, or nuts to strip parts as subassemblies before ejection from the machine.

Since four-slide parts are normally produced from strip stock, the edge of the part, its thickness (unless coined or drawn), and its surface finish are normally those of the unprocessed material.

The size range of four-slide parts does not extend to the large sizes routinely produced on conventional metal-stamping presses. The largest standard four-slide machines currently available from one manufacturer can produce parts up to 350 mm (14 in) long, 100 mm (4 in) wide, and 2.1 mm (0.084 in) thick. However, most four-slide parts are made from stock narrower than 25 mm (1 in) and thinner than 1.5 mm (0.060 in).

Other limitations of four-slide parts, in comparison with press-produced stampings, are a shallower depth of draw and considerably less severe coining. It is also not feasible to stamp parts on the bias, should this be necessary to ensure favorable stress development in hardening.

Figure 3.4.2 shows a collection of typical four-slide parts.

ECONOMICS OF FOUR-SLIDE PROJECTION

Parts making on four-slide equipment is a high-production process. This is a result of three prime factors: (1) production rate, (2) tooling cost, and (3) setup time.

The production rate on vertical four-slide machines generally ranges between 100 and 300 parts per minute, depending on the complexity of the parts and assuming that the rate is not limited by a secondary operation. Horizontal four-slides are usually somewhat slower than vertical machines but are less expensive for a given strip-size capacity. More intricate parts requiring multiple forming are made more rapidly on either type of four-slide machine than on a punch press. Secondary operations such as drawing, reaming, tapping, etc. normally require a considerable reduction in machine speed to keep operations synchronized, but output is still considerably faster than it would be if operations were performed separately. The life expectancy of high-production four-slide tooling is in the millions, with 200,000 to 300,000 parts being made between tool sharpenings.

The press and forming tools for making less precise parts on four-slides may be of

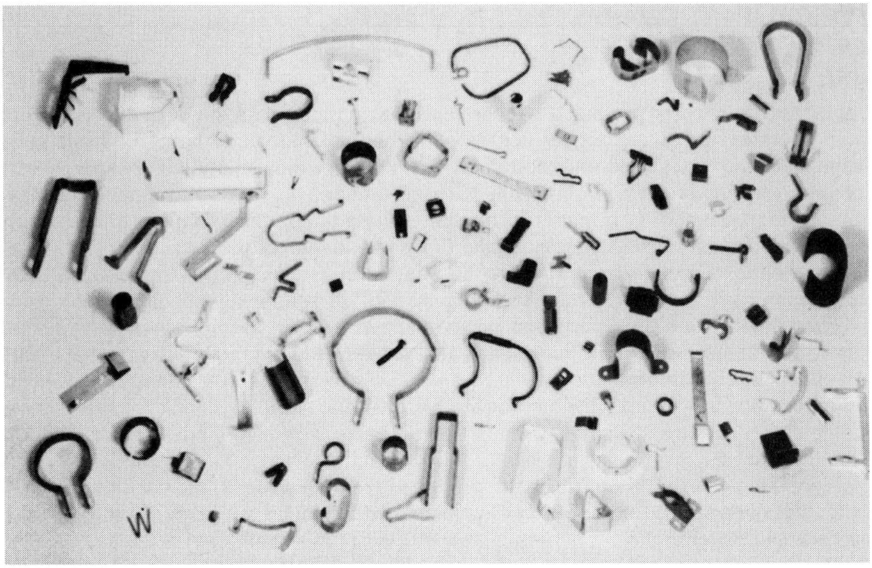

FIGURE 3.4.2 Parts that involve relatively complex forming can usually be more economically produced on four-slides. (*Courtesy Torin Corporation.*)

the type made in job shops for short runs, in which the appearance and life expectancy of tools are not as important. The life expectancy of this kind of forming tool may be no more than 25,000 to 50,000 parts. Rail die construction is sometimes used in the press area.

In general, tooling cost for a given high-precision component can be expected to be lower on four-slides than on power presses no matter what the quality and complexity of the tools.

Initial setup is usually considered part of tooling cost, and on four-slide production it can be significant. Setup is longer than for conventional punch presses because there are more individual tools (as well as cams) to mount in the machine and adjust. Rerun setup tends to be longer for the same reason. Greater skill is also required. However, the ability to adjust four-slide tools individually provides an advantage of flexibility, for example, in making small adjustments to compensate for variations in materials characteristics or tool wear. Setup time for four-slide machines when running parts of average complexity is from 2 to 6 h.

Four-slide machines, particularly the vertical type, are more likely to be used for long runs. However, the "minimum" length of run for four-slides tends to vary widely from company to company. And even with higher setup time, a need for precise dimensional control and/or elimination of secondary operations may lead to using four-slides for runs well below the minimum. A rule of thumb used by many manufacturing engineers says that a four-slide is likely to be the more economical method whenever its running time is equal to or greater than its setup time. This occurs usually for quantities in the range of tens of thousands per run.

SUITABLE MATERIALS*

Tables 3.4.1 through 3.4.6 give the major forming characteristics of the six families of strip material most commonly used in four-slide production. Smaller bend radii than those shown in the tables may be possible if the bend is made in more than one step or if V bending is used instead of tangential bending.

High-Formability Steels

Low-carbon cold-rolled strip steel (Table 3.4.1) and spheroidize-annealed spring steel (Table 3.4.2) are the most common high-formability materials used in four-slide production. Cold-rolled strip steel (available in five tempers) is the basic ferrous material for four-slide parts other than springs. AISI 1075 (G10750) spheroidize-annealed spring steel is the main material for spring parts; AISI 1095 (G10950) is for spring parts in thinner sections. These latter two materials are hardened and tempered after forming to attain desired properties.

Silicon-killed steel, which is used mainly in core hardening, carburizing, and other special applications, is not usually specified for four-slide forming. Aluminum-killed steel should always be specified when fabrication of the part in the four-slide is to include drawing.

The austenitic (or fracture) grain size, which is established during the hardening process, very much affects the formability of steel. The larger the grains, the less formable is the material and so the more likely that it will develop cracks in forming. The process of cold rolling stretches the grains along the length of the strip. As a result, the material is more formable across the strip than along the strip.

Tempered Spring Steels

Behavior of the tempered spring steels (Table 3.4.3) is difficult to predict in forming. To control the mechanical properties, the strip mill should be required to fulfill two of three critical specifications: hardness, tensile strength, and minimum inside-bend radius without fracture. In addition, allowable limits on reproducibility of mechanical properties should be established with the mill.

Stainless Steels

Type 302 (S30200) is the most common stainless steel specified for four-slide parts (Table 3.4.4). Other stainless steels that may be used (austenitic types, such as the 200 and 300 Series) have similar mechanical properties, and the data in the table are also applicable to them. Stainless steel does not have nearly the degree of springback of

*Much of the remaining parts of Chap. 3.4, including Tables 3.4.1 through 3.4.6, was prepared by James. J. Skelsky, Jr., and has been updated for this edition.

TABLE 3.4.1 Formability of Low-Carbon, Cold-Rolled Strip Steels

			Relative formability			
				Recommended min. forming radius for 90° bends*		
Temper	Tensile strength, 1000 lbf/in^2	Hardness, Rockwell B	Bending characteristics	Perpendicular bend†	45° bend†	Parallel bend†
No. 1 (hard)	80–100	84 min.‡ 90 min.§	Recommended for flat blanking only	2.0	3.0	Limited formability
No. 2 (half-hard)	55–75	70–85	Easy blanking; will bend to sharp right angle across grain and rounding bend along grain	1.0	1.5	2.0
No. 3 (quarter-hard)	45–65	60–75	Will bend flat on itself across grain and to sharp right angle along grain	Sharp	0.5	1.0
No. 4 (planished¶)	42–54	65 max.	Will bend flat on itself across or along grain	Sharp	Sharp	Sharp
No. 5 (dead-soft)	38–50	55 max.	For deep drawing; will bend flat on itself across or along grain; tends to kink when unwound from coil	Sharp	Sharp	Sharp

*Ratio, inside-bend radius to stock thickness.

†Compared with rolling direction.

‡Thickness 0.070 in and greater.

§Thickness less than 0.070 in.

¶Also called *skin-rolled.*

Source: Machine Design.

TABLE 3.4.2 Formability of Spheroidize-Annealed, Cold-Rolled Spring Steel

Grade	Tensile strength, 1000 lbf/in^2	Thickness, in	Hardness, Rockwell	Relative formability			
					Recommended min. forming radius for 90° bends*		
				Bending characteristics	Perpendicular bend†	45° bend†	Parallel bend†
AISI 1095 (0.90–1.05% C) (G10950)	75–95	Under 0.015	15T 86–90	Flat on itself both ways	0.5	1.0	1.5
		0.015–0.024	30T 70–76	Flat across; one thickness with grain	0.8	1.2	1.8
		0.025–0.034	45T 53–62	Flat across; one thickness with grain	1.0	1.5	2.0
		0.035–0.062	B 80–90	One thickness across; two thicknesses with grain	1.2	1.8	2.5
AISI 1075 (0.70–0.80% C) (G10750)	65–85	Under 0.015	15T 85–88	Flat on itself both ways	Sharp	0.8	1.2
		0.015–0.024	30T 67–74	Flat on itself both ways	0.5	1.0	1.5
		0.025–0.034	45T 48–58	Flat across; one thickness with grain	0.8	1.2	1.8
		0.035–0.062	B 75–85	Flat across; one thickness with grain	1.0	1.5	2.0

*Ratio, inside-bend radius to stock thickness.

†Compared with rolling direction.

Source: Machine Design.

TABLE 3.4.3 Formability of Tempered Spring Steels

Grade	Temper	Hardness, Rockwell, and tensile strength, 1000 lbf/in^2				Relative formability; min. radius for 180° bends perpendicular to rolling direction*			
		Under 0.015 in thick	0.015–0.024 in thick	0.025–0.034 in thick	0.035 in thick and over	Under 0.015 in thick	0.015–0.02 4 in thick	0.025–0.034 in thick	0.035 in thick and over
SAE 1095 (0.90–1.05% C) (G10950)	Min. ductility: not recommended for forming; best spring properties	15N 86.5–88.5 275–330	30N 70–73 260–275	A 75–77 245–260	C 48–51 245 max.	†	†	†	†
	Average ductility: reasonable formability; good spring properties	15N 84–86 260–320	30N 65–68 240–260	A 73–75 230–240	C 44–47 230 max.	7.5	6.0	5.2	4.8
	Max. ductility: for blanking and forming; moderate spring properties	15N 81.5–83.5 235–270	30N 61–64 220–235	A 70–72 210–220	C 39–42 210 max.	6.5	5.5	4.8	4.4
SAE 1074 (0.70–0.80% C) (G10740)	Min. ductility: not recommended for forming; best spring properties	15N 86.5–88.5 270–320	30N 70–73 250–270	A 75–77 240–250	C 48–51 240 max.	†	†	†	†
	Average ductility: reasonable formability; good spring properties	15N 84–86 260–315	30N 65–68 240–260	A 73–75 230–240	C 44–47 230 max.	6.5	5.8	5.0	4.5
	Max. ductility: for blanking and forming; moderate spring properties	15N 81.5–83.5 220–270	30N 61–64 205–220	A 70–72 195–205	C 39–42 195 max.	5.5	4.5	4.0	3.6

*Ratio, inside-bend radius to stock thickness. Because of springback this is not an indication of the actual size of the radius formed.

†Extremely limited formability.

Source: Machine Design.

TABLE 3.4.4 Formability of Stainless Steels

Alloy	Temper	Tensile strength, 1000 lbf/in^2	Hardness		Relative formability, recommended min. forming radius for 90° bends*		
			Rockwell B or C	Rockwell superficial	Perpendicular bend†	45° bend†	Parallel bend†
Type 302 stainless (S30200)	Annealed	75–100	B 88 max.	30T 75 max.	Sharp	Sharp	Sharp
	Quarter-hard	125 min.	C 25 max.	30N 46 min.	Sharp	0.5	1.0
	Half-hard	150 min.	C 32 min.	30N 52 min.	0.5	1.0	1.5
	Three-quarter-hard	175 min.	C 37 min.	30N 56.5 min.	1.0	1.5	2.0
	Hard	185 min.	C 40 min.	30N 60 min	1.5	2.0	3.0
17-7PH stainless (S17700)	Condition A	150 max.	B 92 max.	30T 75 max.	0.5	1.0	1.5
	Condition B	200 min.	C 41 min	30N 61 min.	2.5	3.0	4.0

*Ratio, inside-bend radius to stock thickness.

†Compared with rolling direction.

Source: Machine Design.

TABLE 3.4.5 Formability of Copper Alloys

Alloy	Temper	Hardness			Electrical conductivity, % IACS	Relative formability; recommended min. forming radius for 90° bends*		
		Tensile strength, 1000 lbf/in^2	Rockwell B	Rockwell superficial, 30T		Perpendicular bend†‡	45° bend†	Parallel bend†
Brass (C26000)	Annealed	45–53	10–45	20–50	26–28	Sharp	Sharp	Sharp
alloy no. 6	Quarter-hard	49–59	40–65	43–60	(annealed)	Sharp	Sharp	Sharp
0.7 Cu–0.3 Zn	Half-hard	57–67	60–77	56–68		Sharp	Sharp	Sharp
ASTM B 36	Three-quarter-hard	64–74	72–82	65–72	Sharp	Sharp	Sharp	Sharp
	Hard	71–81	79–86	70–74		Sharp	Sharp	0.5
	Extra hard	83–92	85–91	74–77		Sharp	0.5	2.0
	Spring	91–100	89–93	76–78		Sharp	1.0	3.0
	Extra spring	95–104	91–95	77–79		0.5	1.5	3.5
Nickel silver	Annealed	49–70	29–73	35–65	5	Sharp	Sharp	Sharp
(C75200)	Quarter-hard	69–87	70–88	63–75	(annealed)	Sharp	Sharp	Sharp
alloy no. 4	Half-hard	78–95	81–92	71–78		Sharp	Sharp	Sharp
18% Ni,	Hard	92–107	90–96	76–80		Sharp	Sharp	0.5
ASTM B 122	Extra hard	102–115	95–99	79–82		Sharp	Sharp	1.5
	Spring	108–120	97–100	80–83		Sharp	0.8	2.5
	Extra spring	112–127	98–101	81–84		0.5	1.0	3.0
Phosphor bronze	Annealed	40–55	0–45	16–46	18	Sharp	Sharp	Sharp
(C51000)	Half-hard	55–70	53–78	52–71	(annealed)	Sharp	Sharp	Sharp
Grade A	Hard	72–87	80–88	69–75		Sharp	Sharp	1.0
0.95 Cu–0.05 Sn	Extra hard	84–99	86–92	73–78		Sharp	0.5	2.5
ASTM B 103	Spring	91–105	88–94	75–79		Sharp	1.0	3.0
	Extra spring	96–109	89–95	76–80		0.5	1.5	5.5

Phosphor bronze	Annealed	53–67	20–66	27–62	18	Sharp	Sharp	Sharp
(C52100)	Half-hard	69–84	69–88	63–75	(annealed)	Sharp	Sharp	0.5
Grade C	Hard	85–100	89–95	73–80		Sharp	0.5	2.0
0.92 Cu–0.08 Sn	Extra hard	97–112	93–98	77–82		Sharp	2.0	4.0
ASTM B 103	Spring	105–119	95–100	78–83		Sharp	3.5	5.5
	Extra spring	110–122	96–101	79–83		0.5	4.0	7.5

*Ratio, inside bend radius to stock thickness.

†Compared with rolling direction.

‡Material under 0.010 in: use 50 to 75 percent of above ratios.

Source: Machine Design.

TABLE 3.4.6 Formability of Beryllium Copper Alloys

Alloy	Temper*	Precipitation-hardening treatment	Tensile strength, 1000 lbf/in^2	Hardness: Rockwell B or C	Hardness: Rockwell superficial	Electrical conductivity, % IACS	Relative formability; recommended min. forming radius for 90° bends†: Perpendicular bend‡§	45° bend‡	Parallel bend‡
C17200							Sharp	Sharp	Sharp
	A	Heat-treatable	60–78	B 45–78	30T 46–67	17–19			
	¼H		75–88	B 68–90	30T 62–75	16–18	Sharp	Sharp	Sharp
	½H		85–100	B 88–96	30T 74–79	15–17	0.5	0.7	1.1
	H		100–120	B 96–102	30T 79–83	15–17	1.0	2.0	2.5
	AT	3 h at 600°F	165–190	C 36–41	30N 56–61	22–25			
	¼HT	2 h at 600°F	175–200	C 38–42	30N 58–63	22–25			
	½HT	2 h at 600°F	185–210	C 39–44	30N 59–65	22–25	Not formed after aging		
	HT	2 h at 600°F	190–215	C 40–45	30N 60–66	22–25			
	AM	Mill-hardened	100–110	C 18–23	30N 37–42	22–32	1.0	1.0	1.0
	¼HM		110–120	C 21–26	30N 42–46	22–32	1.0	2.0	2.0
	½HM		120–135	C 25–30	30N 46–50	22–32	3.0	3.0	3.0
	HM		135–150	C 30–35	30N 50–54	22–32	6.0	6.0	6.0
	XHM		170–185	C 32–38	30N 55–58	22–32	¶	¶	¶
C17000	A	Heat-treatable	60–78	B 45–78	30T 46–67	17–19	Sharp	Sharp	Sharp
	¼H		75–88	B 68–90	30T 62–75	16–18	Sharp	Sharp	Sharp
	½H		85–100	B 88–96	30T 74–79	15–17	Sharp	0.7	1.0
	H		100–120	B 96–102	30T 79–83	15–17	1.3	2.4	2.9
	AT	3 h at 600°F	150–180	C 33–38	30N 53–58	22–25			
	¼HT	2 h at 600°F	160–185	C 35–39	30N 55–59	22–25	Not formed after aging		
	½HT	2 h at 600°F	170–195	C 37–40	30N 56–60	22–25			
	HT	2 h at 600°F	180–200	C 39–41	30N 58–61	22–25			

C17000	AM	Mill-hardened	100–110	C 18–23	30N 37–42	25–33	1.0	1.0	1.0
	¼HM		110–120	C 21–26	30N 42–46	25–33	2.0	2.0	2.0
	½HM		120–135	C 25–30	30N 46–50	25–33	3.0	3.0	3.0
	HM		135–150	C 30–35	30N 50–54	25–33	6.0	6.0	6.0
	XHM		160–175	C 32–38	30N 55–58	25–33	§	§	§
C17500	A	Heat-treatable	35–55	B 20–45	30T 20–45	25–30	Sharp	Sharp	Sharp
	H		70–85	B 70–80	30T 69–75	22–27	0.5	0.5	0.6
	AT	3 h at 900°F	100–110	B 92–100	30T 77–82	48–55	1.9	1.9	1.9
	HT	2 h at 900°F	110–125	B 96–102	30T 79–82	48–52	2.4	2.4	2.6

*A = solution-heat-treated (annealed); AM, HM, and XHM = special mill processing and precipitation treatment; H = hard (except for tempers with M designation); T = precipitation-heat-treated.

†Ratio, inside-bend radius to stock thickness.

‡Compared with rolling direction.

§For material under 0.010 in: use 50 to 75 percent of above ratios.

¶Extremely limited formability.

Source: Machine Design.

tempered steels. Series 400 stainless steels have properties similar to AISI 1075 spheroidize-annealed, cold-rolled spring steel.

Nonferrous Strip Materials

The nonferrous alloys (Tables 3.4.5 and 3.4.6) offer a wide selection of mechanical properties. However, there may be large differences in hardness and tensile strength among various alloys that have roughly equivalent formability. As in the preceding tables, the radius-to-thickness ratios given are based on production experience; smaller ratios may be attained in particular part configurations. Work-hardened copper alloys normally have seven or eight degreees of hardness. Beryllium copper alloys are available in four heat-treatable conditions or in up to five mill-hardened conditions; parts produced on four-slides are almost always made from annealed beryllium copper.

DESIGN RECOMMENDATIONS

Design freedoms and limitations for producing metal-strip parts in four-slides are similar to those for power presses in that they are determined by material characteristics or part shape. But there are also special differences. Design recommendations for external contours of strip parts are given in Fig. 3.4.3. Recommendations for internal contours, tabs and slots and lugs, bridges, and curls are given in Fig. 3.4.4. These recommendations will help in producing four-slide parts of optimal quality at maximum speed and lowest cost.

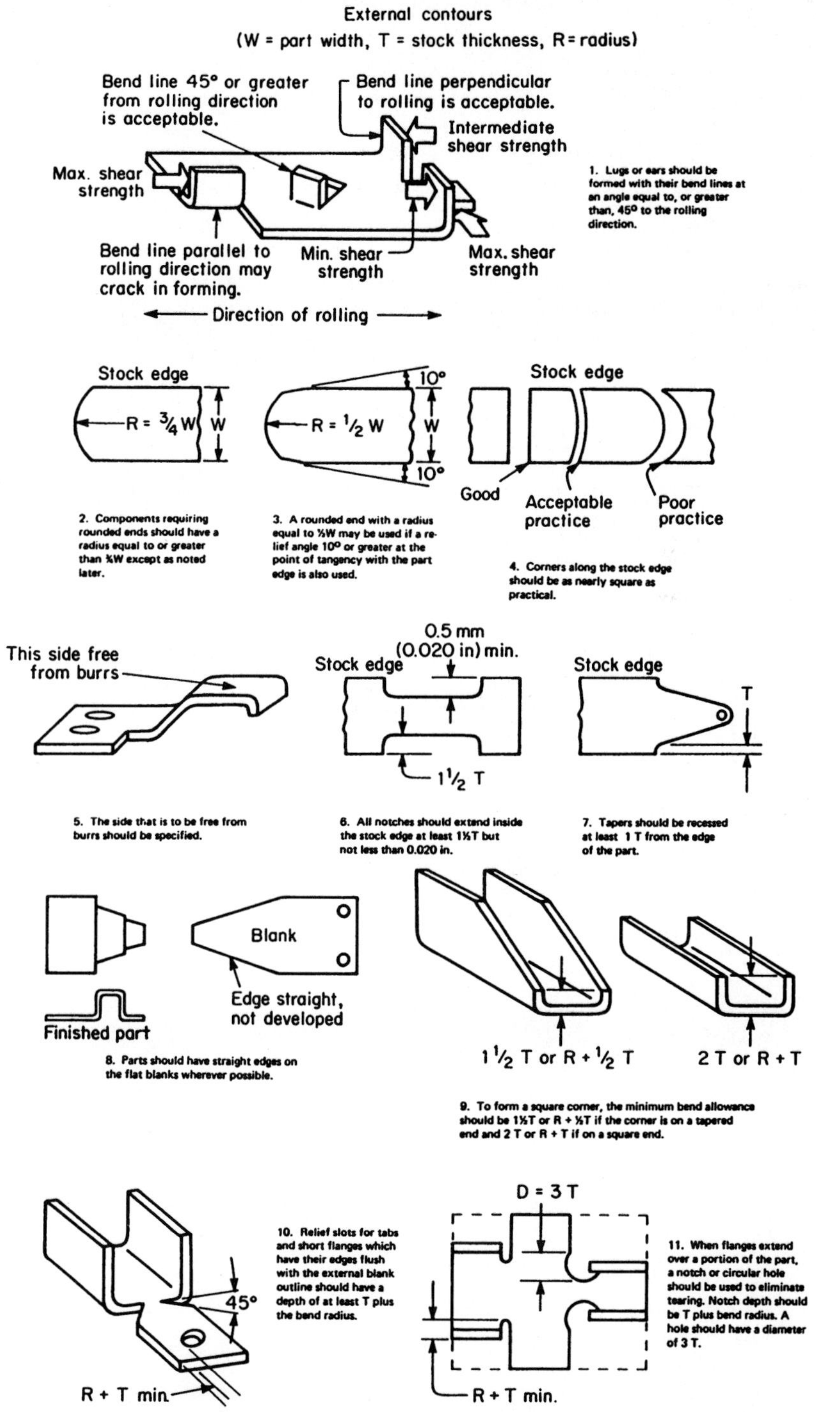

FIGURE 3.4.3 Design recommendations for external contours in strip parts. (*Machine Design.*)

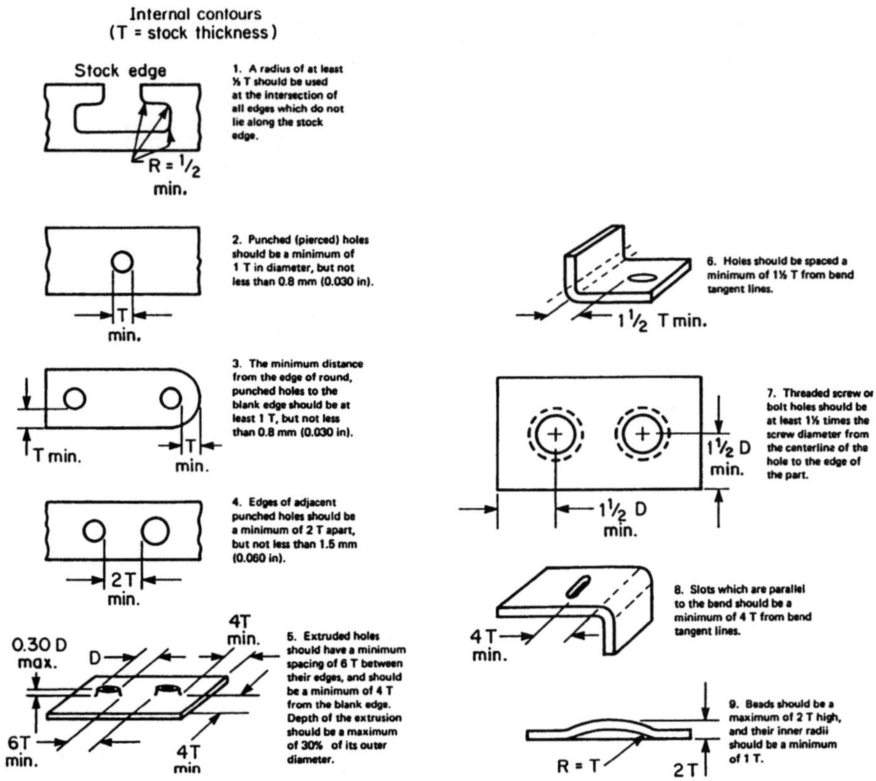

FIGURE 3.4.4 Design recommendations for internal contours, tabs and slots, and lugs, bridges, and curls in strip parts. (*Machine Design.*)

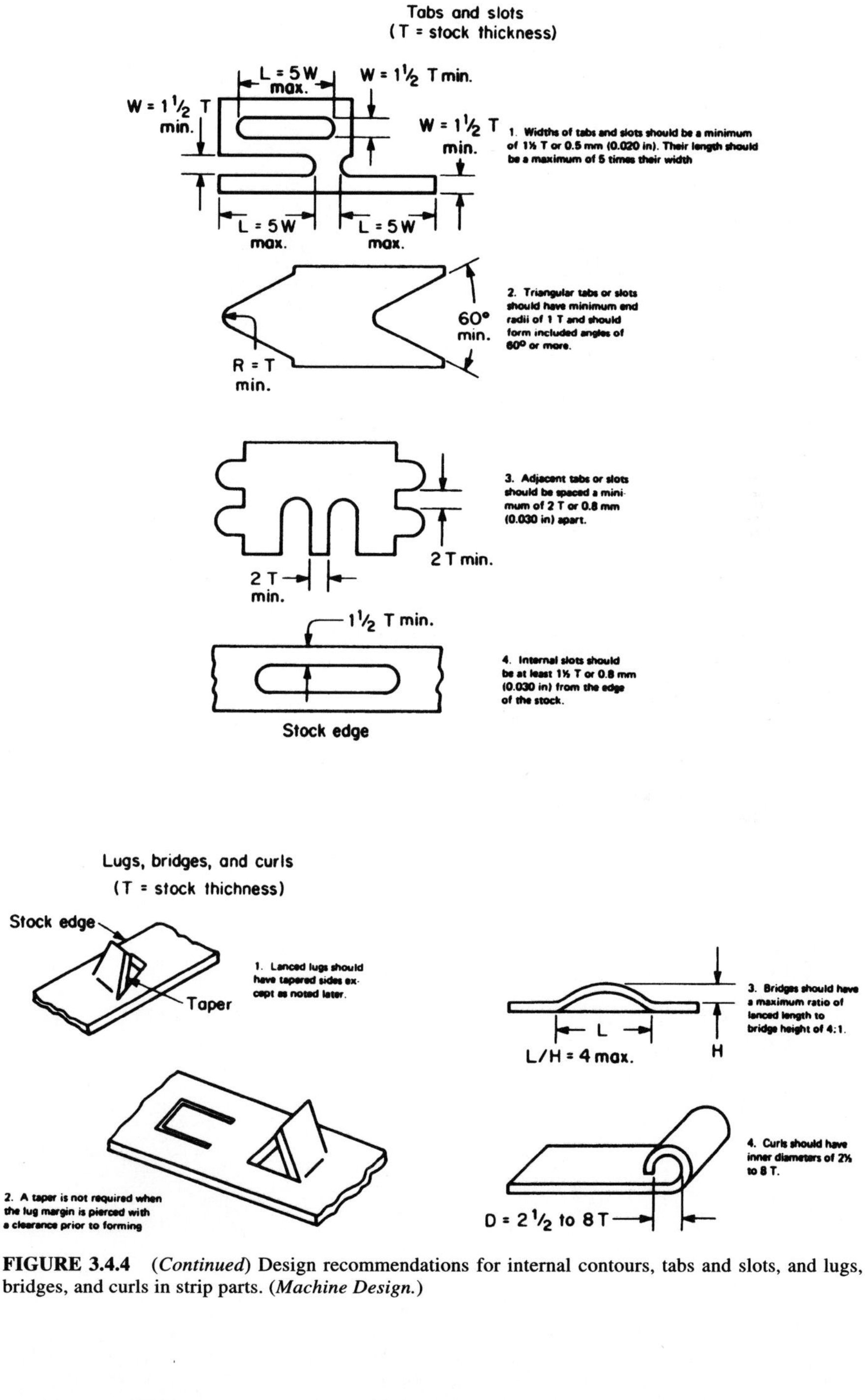

FIGURE 3.4.4 (*Continued*) Design recommendations for internal contours, tabs and slots, and lugs, bridges, and curls in strip parts. (*Machine Design.*)

CHAPTER 3.5
SPRINGS AND WIRE FORMS*

Technology Committee
Spring Manufacturers Institute
Wheeling, Illinois

Springs and wire forms are made in an almost limitless variety of shapes, sizes, and materials. A large proportion of springs are not formed out of round wire, do not have the familiar helical-coil shape, and even appear not to obey Hooke's law. Wire forms can function as springs or as rigid structural members, or as both, but all are made of metal wire, round or otherwise.

These two types of metal components are often made of the same wire materials and perform the same functions. They are frequently made on the same types of machines and so are subject to the same forming limitations and economic considerations.

Figure 3.5.1 lists the various types of spring elements, typical configurations, and the spring action that each exerts.

TYPES OF SPRINGS

Compression Springs

Compression springs are open-coil helical springs that offer resistance to a compressive force applied axially. They are usually coiled as constant-diameter cylinders. Other common forms of compression springs, such as conical, tapered, concave, or convex springs or various combinations of these, are used as required by the application. While square, rectangular, or special-section wire may have to be required, round wire is predominant in compression springs because it is readily available and adaptable to standard tooling.

*This chapter by the Spring Manufacturers Institute, Inc. (SMI), contains advisory information only. SMI is not a standards-development organization. It does not develop and publish industry standards. It does not provide interpretations of any standards with regard to a manufacturer's specific product. No person has authority in the name of SMI to issue an interpretation of any standard or an interpretation of the technical information and comments published in this chapter that relate to any manufacturer's specific product.

Type	Configuration	Action
Helical compression Round and rectangular wire	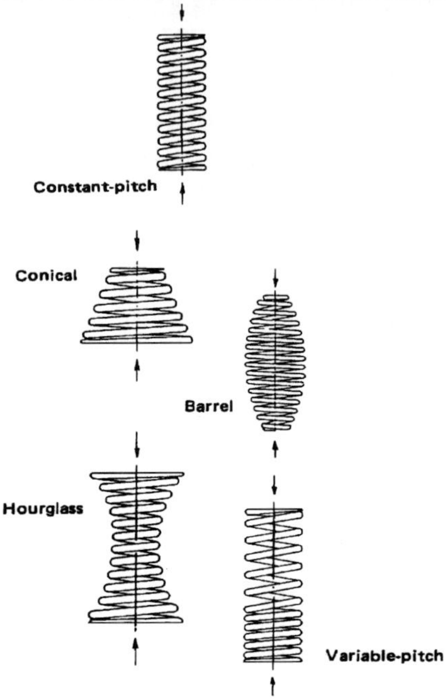	Push; wide load and deflection range; constant rate. Push; wide load and deflection range. Conical spring can be made with minimum solid height and with constant or increasing rate. Barrel, hourglass, and variable-pitch springs used to minimize resonant surging and vibration.
Helical extension	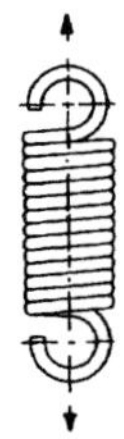	Pull; wide load and deflection range; constant rate.
Torsion	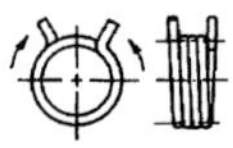	Twist; constant rate.
Flat springs Cantilever	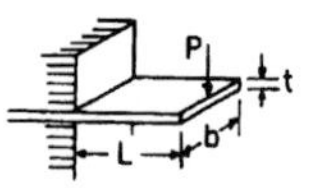	Push or pull; wide range of loads; low deflection range.

FIGURE 3.5.1 Various types of spring elements, typical configurations, and the spring action that each exerts.

Type	Configuration	Action
Simple beam	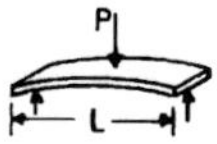	
Other springs		
Torsion bar	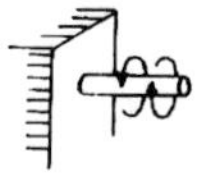	**Twist.**
Constant force	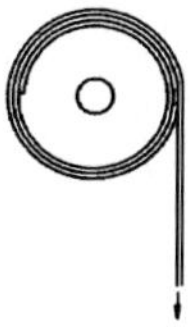	**Pull; very long deflection at constant load or low rate.**
	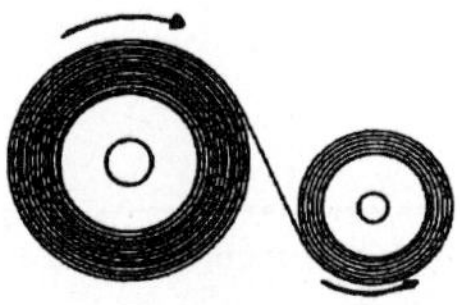	**Twist; exerts close-to-constant torque over many turns.**
Spiral		**Twist.**
Hairspring		
Brush		**Twist or push.**
Motor, power, or clock	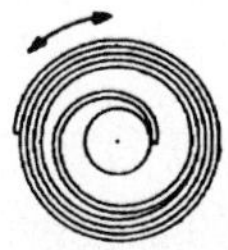	**Twist; exerts torque over many turns.**
Spring washer Belleville		**Push; high loads, low deflections; choice of rates (constant, increasing, or decreasing).**
Wave		**Push; light loads, low deflection; limited radial space.**

FIGURE 3.5.1 (*Continued*) Various types of spring elements, typical configurations, and the spring action that each exerts.

Type	Configuration	Action
Slotted		Push; higher deflections than Bellevilles.
		Push; for axial loading of bearings, etc.
Curve	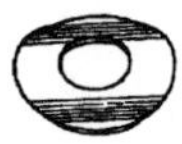	Push; used to absorb axial end play.
Volute	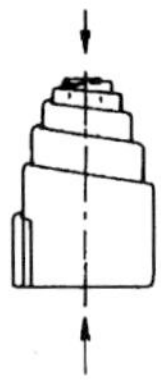	Push; may have an inherently high friction damping.

FIGURE 3.5.1 (*Continued*) Various types of spring elements, typical configurations, and the spring action that each exerts.

Extension Springs

These springs absorb and store energy by offering resistance to a pulling force. Various types of ends are used to attach an extension spring to the source of the force. Simple loops and hooks are the most frequently used ends. These springs may be wound with little or no initial tension, sometimes are wound with clearance between coils, but most often are close-wound with initial tension.

Torsion Springs

Torsion springs, whose ends are rotated in angular deflection, offer resistance to externally applied torque. The wire itself is subjected to bending stresses rather than torsional stresses, as might be expected from the name. Springs of this type are usually close-wound and reduce in coil diameter and increase in body length as they are deflected. The designer must consider the effects of friction and arm deflection on the torque.

Flat Springs

The two most common types of flat spring are cantilever and simple beam springs. Both are fabricated from flat strip material and store energy on being deflected by an external load. Only a small portion of a complex stamping actually may be functioning as a spring. For purposes of design, that portion often may be considered as an independent simple spring form, while the rest of the part is temporarily ignored.

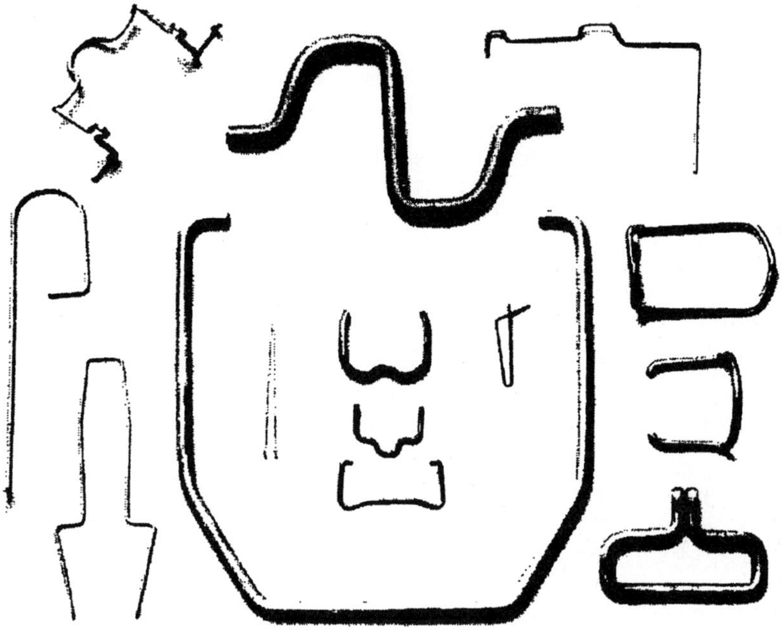

FIGURE 3.5.2 Typical wire forms.

TYPES OF WIRE FORMS

Wire forms (Fig. 3.5.2) are either spring elements, structural elements, or both. Springs usually are designed in one piece, mainly because it is difficult to join most spring materials. Structural wire forms, which are not deflected in application, often consist of two or more parts that have been welded or otherwise joined. They may be specified for the same types of applications as stampings, and a cost comparison may be called for.

Wire forms, particularly structural types, often have special characteristics other than the basic shapes produced on wire coilers or four-slides. For example, a large variety of wire ends or terminations is available to match application needs. These include swaged, chamfered, turned, ground, headed, pierced, and grooved ends. Many of these can be produced automatically, while others require secondary operations.

Various types of wire eyes or loops can be produced on wire-forming machines automatically, depending on the complexity of the other forming requirements. Flat and shaped wire cross sections may be specified, too, as needed and with the same types of tooling.

Various welding methods—projection, T, butt, cross-wire—may be used to fabricate assemblies with structural wire forms. While classified as wire forms, these assemblies often include stampings and other types of parts. All standard cut, rolled, or wood-screw threads also may be specified as secondary operations on wire forms.

MANUFACTURING PROCESSES

Helical compression, extension, and torsion springs are formed from continuous coils of wire on spring coilers or torsion winders. Flat springs are produced from continuous coils of strip material on either four-slide machines (Chap. 3.4) or power presses (Chap. 3.2). Wire forms are made on four-slides and sometimes on torsion winders if one or more helical coils are formed as part of their configuration. Fabricating wire forms on a four-slide involves the forming tools used in making strip parts but rarely requires the machine's press tools.

There are two types of heat treatment of springs: low-temperature heat treatment of prestrengthened materials and high-temperature heat treatment to strengthen annealed materials. Low-temperature heat treating, sometimes called *stress relieving,* is by far the more common process. Its purposes are to stabilize the spring dimensionally, relieve excessive residual stresses, and increase the yield strength of cold-drawn materials. Stress relief is accomplished in certain beryllium copper and nickel-base alloys with a heat-treatment process called *age hardening.*

SUITABLE MATERIALS

Wire

There are five major metallurgical classifications of wire spring materials: (1) high-carbon steel, (2) alloy steel, (3) stainless steel, (4) nickel-base alloys, and (5) copper-base alloys. In selecting spring materials, the designer must consider such factors as temperature and corrosion resistance, conductivity, physical properties, formability, and availability. When there are a large number of stress cycles and a long fatigue life is therefore required, the quality of the material must be tightly controlled. The wire material must have a uniform internal structure, and its surface must be free of pits, seams, scratches, and any other flaws that can impair its fatigue life. The most common long-service materials are high-carbon steels or alloy steels, since both can be obtained free of seams and of aircraft quality (such as is specified for valve springs).

The costs of spring wire materials vary widely and are constantly changing. In general, the common spring materials, in order of relative cost from lowest to highest, are hard-drawn steel, oil-tempered steel, alloy steel, music wire, stainless steel (average of several), phosphor bronze, and beryllium copper.

Flat Stock

The most commonly used flat spring materials are high-carbon spring steels, either annealed or tempered. Commonly specified grades are 1050 (G10500), 1075 (G10750), and 1095 (G10950). Types 301 (S30100), 302 (S30200), and 1.7-7 PH (S17700) stainless steel are often specified when corrosion resistance is needed, and phosphor bronze and beryllium copper alloys are the most common for high electrical conductivity. The characteristics of flat spring materials are summarized in Chap. 3.4. The costs of strip materials for flat springs such as spring wire vary widely and are constantly changing. In general, the following are the common strip materials, in order of relative cost from lowest to highest: high-carbon steel (annealed), alloy steel, high-carbon steel (tempered), stainless steel, copper-base alloys (average of several), and beryllium copper.

DESIGN RECOMMENDATIONS

General

Dimensional and load requirements must be specified in designing any type of spring for any application. Depending on the type of application, there may be these additional considerations: operating environment, spring rate, dynamic load characteristics, space occupied, and operating life.

Specifying dimensional and load tolerances closer than needed may add unnecessarily to the cost of a spring. No tolerances should be specified at all unless they are required by the spring's function. Dimensional tolerances that do not exceed commercial standards permit the spring to be fabricated by ordinary production methods.

Compression Springs

A form similar to that in Fig. 3.5.3 is recommended as a guide in specifying compression springs. The functional design characteristics of the spring should be given as mandatory specifications. Among the most important specifications is *spring load,* which is defined as the force that is applied to a spring and causes deflection to a finite position. *Spring rate* is linear in only about the central 60 percent of its deflection range. Secondary characteristics, which may be useful for reference, should be identified as *advisory data.* This practice controls the essential requirements while providing as much design flexibility as possible to the spring manufacturer in meeting these requirements.

The *solid height* of a compression spring is defined as the length of the spring when it is under sufficient load to bring all coils into contact with adjacent coils and additional load causes no further deflection. Solid height should be specified by the user as a maximum allowable within the available space. The actual number of coils in the spring is determined by the spring manufacturer.

There are four basic types of compression-spring ends, as shown in Fig. 3.5.3. The particular type of end specified affects the pitch, solid height, number of active and total coils, free length, and seating characteristics of the spring. Since grinding is a separate operation, compression springs with unground ends are more economical. Open ends, ground or unground, may involve additional cost in assembly because the springs are then more susceptible to tangling.

Extension Springs

A form similar to that in Fig. 3.5.4 is recommended as a guide in specifying extension springs. Most extension springs are specified with initial tension, which is an internal force that holds the coils tightly together. Unlike a compression spring, which has zero load at zero deflection, an extension spring can have a preload at zero deflection.

The two most common loops or hooks for extension springs are the twist loop and

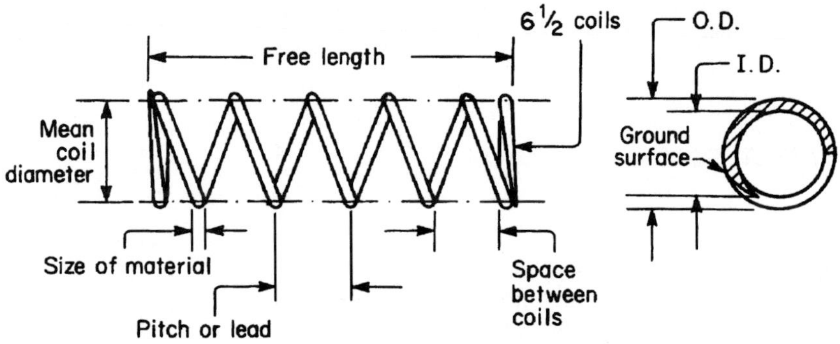

MANDATORY SPECIFICATIONS
(Fill in only those required)

1. FREE LENGTH
   a._____in max.,_____in min. or
   b._____in ± _____ in
2. OUTSIDE DIAMETER
   a. _____in max. or
   b. _____in ± _____in
3. INSIDE DIAMETER
   a._____ in min. or
   b._____ in ± _____in
4. Load_____lb ± _____ lb @ _____in
   Load_____ lb ± _____ lb @ _____ in
   Rate _____ lb per in ± _____ lb per in
   between lengths of _____in and _____ in
5. Maximum solid height _____ in
6. Direction of helix (L, R or optional) __________
7. Type of ends ____________________

ADVISORY DATA

1. Wire diameter_____ in
2. Mean coil diameter _____ in
3. No. of active coils ________
4. Total no. of coils ________

SPECIAL INFORMATION

1. Type of material ____________________
2. Finish ____________________
3. Squareness (free): within _____°
4. Frequency of compression, _____ cycles/s, and working range, _____ in to _____ in of length.
5. Operating temp. _____°F
6. End use or application ____________________
7. Other ____________________

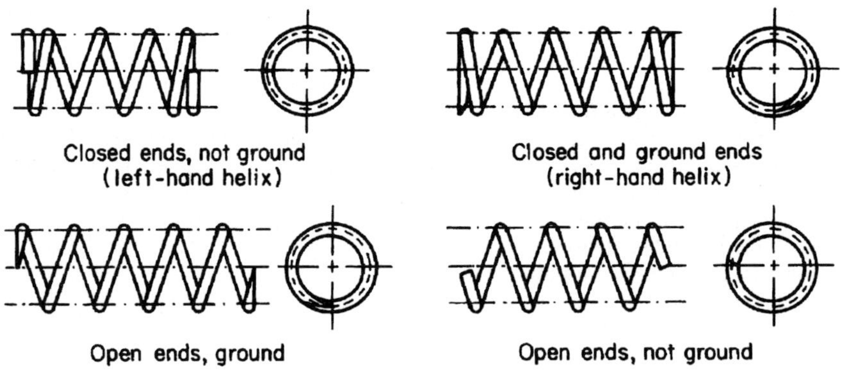

FIGURE 3.5.3 Representative form for use in specifying compression springs.

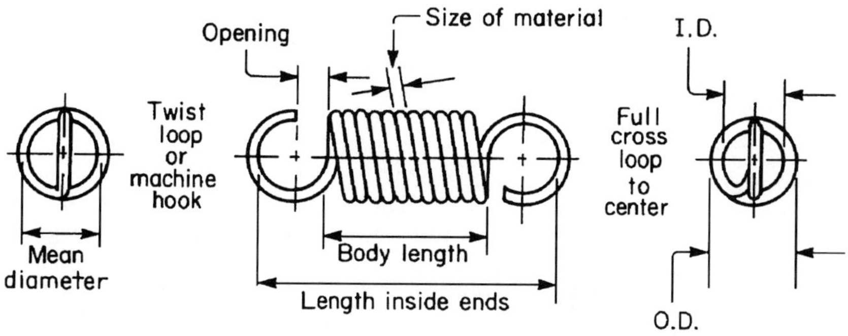

MANDATORY SPECIFICATIONS
(Fill in only those required)

1. LENGTH INSIDE ENDS
   a. ____ in max., ____ in min. or
   b. ____ in ± ____ in or
   c. approx. ____ in
2. OUTSIDE DIAMETER
   a. ____ in max. or
   b. ____ in ± ____ in
3. INSIDE DIAMETER
   a. ____ in. min. or
   b. ____ in ± ____ in
4. Load ____ lb ± ____ lb @ ____ in
   Load ____ lb ± ____ lb @ ____ in
   Rate ____ lb per in ± ____ lb per in
   between ____ in and ____ in
5. Maximum extended length (inside ends) without set ____ in
6. Relative loop position, ____ ° max. separation of loop planes.
7. Direction of helix (L, R, or optional) ____________
8. Type of ends ____________

ADVISORY DATA

1. Wire diameter ____ in
2. Mean coil diameter ____ in
3. No. of active coils ________
4. Body length ____ in
5. Initial tension ____ lbs.

SPECIAL INFORMATION

1. Type of material ____________________
2. Finish ____________________
3. Frequency of extension, ____ cycles/s, and working range, ____ in to ____ in of length.
4. Operating temp. ____ °F
5. End use or application ____________________
6. Other ____________________

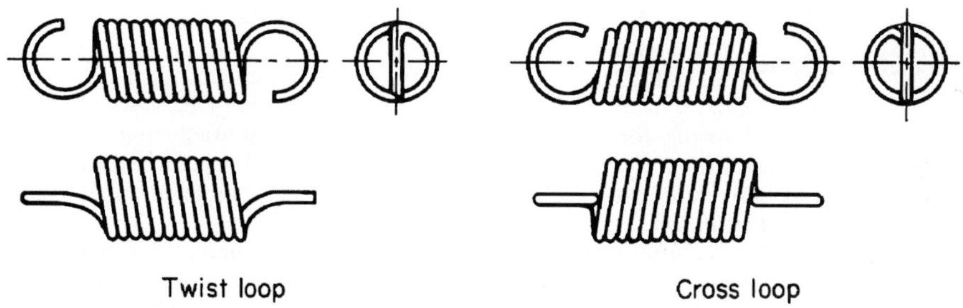

FIGURE 3.5.4 Representative form for use in specifying extension springs.

the cross loop shown in Fig. 3.5.4. These loops are made with standard looping tools and so should be specified whenever possible to minimize cost. If required by the application, many special types of extension-spring loops are offered by the spring industry at additional cost.

Torsion Springs

A form similar to that in Fig. 3.5.5 is recommended as a guide in specifying torsion springs. It may be necessary also to provide the spring manufacturer with a separate drawing that details the end or arm configuration.

Torsion springs with coils that are close-wound (touching) are most economical. Wire-diameter tolerances, which are usually quite close, then govern the dimensional variations in body length and therefore in the spacing of the spring arms. If the spring is close-wound (or the intercoil space in open-wound coils is smaller than the wire diameter), the springs do not tangle when shipped in bulk. Thus the cost of special packing is avoided. Machine setup time is lower with torsion springs that are close-wound than with those which have small intercoil separation because tool adjustment must be more precise with the latter to avoid adjacent coils touching and affecting spring torque.

The types of ends on torsion springs should be considered carefully. While there is a good deal of flexibility in specifying special ends, the cost may be increased and a tool charge incurred. The longer and more extensively formed the ends, the higher is the cost for tooling and secondary operations. Therefore, short, straight ends should be specified whenever possible. Minimum end lengths are usually 3 times wire diameter on one end and at least tangent to the coil outside diameter on the other end.

Flat Springs

Design considerations for flat springs are comparable with those for parts covered by Chaps. 3.2 and 3.4. The formability of the spring material used is an important factor.

DIMENSIONAL FACTORS

The very nature of spring forms, materials, and standard manufacturing processes causes inherent variations. Such spring characteristics as load, mean coil diameter, free length, and the relationship of ends or hooks therefore change to some degree throughout a production run.

The amount of manufacturing variation in particular spring characteristics depends in large part on changes in other spring characteristics, primarily spring index but also wire diameter, number of coils, free length, deflection, and ratio of deflection to free length. Therefore, normal or average tolerances on performance and dimensional characteristics differ for each spring design. For this reason, block tolerances, which may be preprinted on drawing forms, cannot realistically be applied to springs.

If the designer specifies spring tolerances that are equal to or wider than normal manufacturing variations, no special production methods or inspection procedures will be necessary. If the tolerances are narrower, on the other hand, the cost of the springs

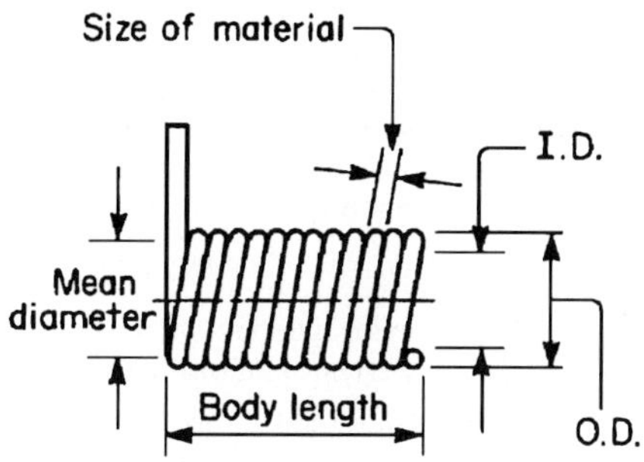

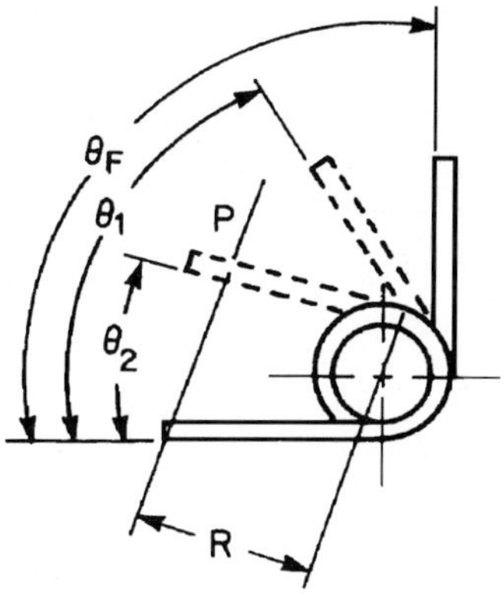

MANDATORY SPECIFICATIONS
(Fill in only those required)

1. To work over____ in dia. shaft.
2. OUTSIDE DIAMETER
   a. ____ in max. or
   b. ____ in ± ____ in
3. INSIDE DIAMETER
   a. ____ in min. or
   b. ____ in ± ____ in
4. Torque ____ in-lbs. ± ____ in-lbs.
   at θ_1 = ____°.
   Torque ____ in-lbs. ± ____ in-lbs.
   at θ_2 = ____°.
5. Length of space available ____ in
6. Maximum wound position ____ turns or
   θ_F = ____° from free position.
7. Length of moment arm (R) ____ in
8. Direction of helix (L, R, or optional) ____________
9. Type of ends ____________________

ADVISORY DATA

1. Wire diameter ____ in
2. Mean coil diameter ____ in
3. No. of coils ____________
4. Rate ____ in-lbs. per turn (360°).
5. θ_F = ____° free angle reference.

SPECIAL INFORMATION

1. Type of material ____________________
2. Finish ____________________
3. Frequency of rotation, ____ cycles/s, and
   working range, θ = ____° to θ = ____° deflection.
4. Operating temp. ____°F
5. End use or application ____________________
   ____
   ____
6. Other ____________________
   ____
   ____

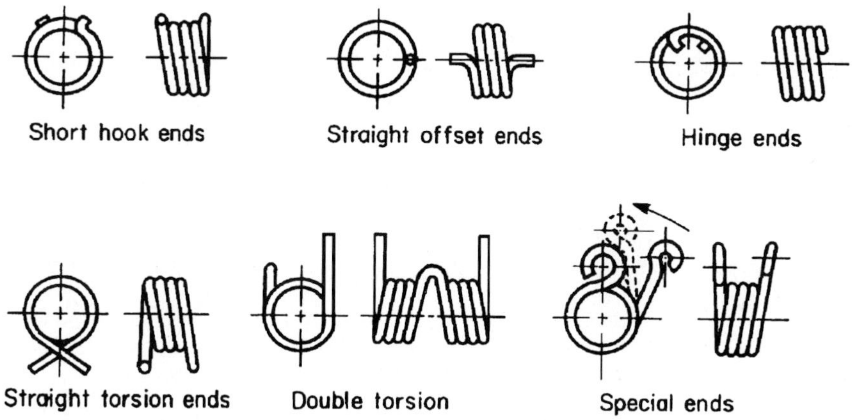

FIGURE 3.5.5 Representative form for use in specifying torsion springs.

may be expected to be higher. If spring tolerances are narrow but are based entirely on application need, of course, the higher spring cost cannot be avoided.

Tables 3.5.1 through 3.5.7 give tolerances on major spring dimensions based on normal manufacturing variations. It is recommended that these tables be used as guides in establishing tolerances, particularly in estimating whether or not application requirements may increase spring cost.

Note: The Spring Manufacturers Institute would like to acknowledge with its thanks the contribution of illustrations by The Barnes Group, Bristol, Connecticut, and Industrial Components Division, Forestville, Connecticut.

TABLE 3.5.1 Normal Free-Length Tolerances for Compression Springs, Squared and Ground, ±in/in of Free Length

Number of active coils per inch	Spring index, *D/d**						
	4	6	8	10	12	14	16
0.5	0.010	0.011	0.012	0.013	0.015	0.016	0.016
1	0.011	0.013	0.015	0.016	0.017	0.018	0.019
2	0.013	0.015	0.017	0.019	0.020	0.022	0.023
4	0.016	0.018	0.021	0.023	0.024	0.026	0.027
8	0.019	0.022	0.024	0.026	0.028	0.030	0.032
12	0.021	0.024	0.027	0.030	0.032	0.034	0.036
16	0.022	0.026	0.029	0.032	0.034	0.036	0.038
20	0.023	0.027	0.031	0.034	0.036	0.038	0.040

Note: For springs less than ½ in long, use the tolerances for ½ in. For closed ends not ground, multiply above values by 1.7.

*Mean coil diameter = *D*; wire diameter = *d*.

TABLE 3.5.2 Normal Load Tolerances for Compression Springs ±Percent of Load

Length	Deflection from free length to load, in														
tolerance, ±in	0.05	0.10	0.15	0.20	0.25	0.30	0.40	0.50	0.75	1.00	1.50	2.00	3.00	4.00	6.00
0.005	12	7	6	5											
0.010		12	8.5	7	6.5	5.5	5								
0.020		22	15.5	12	10	8.5	7	6	5						
0.030			22	17	14	12	9.5	8	6	5					
0.040				22	18	15.5	12	10	7.5	6	5				
0.050					22	19	14.5	12	9	7	5.5				
0.060					25	22	17	14	10	8	6	5			
0.070						25	19.5	16	11	9	6.5	5.5	5		
0.080							22	18	12.5	10	7.5	6	5		
0.090							25	20	14	11	8	6	6		
0.100								22	15.5	12	8.5	7	5.5		
0.200										22	15.5	12	8.5	7	5.5
0.300											22	17	12	9.5	7
0.400												21	15	12	8.5
0.500												25	18.8	14.5	10.5

TABLE 3.5.3 Coil-Diameter Tolerances for Compression and Extension Springs, ±in

	Spring index, D/d*						
Wire diameter, in	4	6	8	10	12	14	16
0.015	0.002	0.002	0.003	0.004	0.005	0.006	0.007
0.023	0.002	0.003	0.004	0.006	0.007	0.008	0.010
0.035	0.002	0.004	0.006	0.007	0.009	0.011	0.013
0.051	0.003	0.005	0.007	0.010	0.012	0.015	0.017
0.076	0.004	0.007	0.010	0.013	0.016	0.019	0.022
0.114	0.006	0.009	0.013	0.018	0.021	0.025	0.029
0.171	0.008	0.012	0.017	0.023	0.028	0.033	0.038
0.250	0.011	0.015	0.021	0.028	0.035	0.042	0.049
0.375	0.016	0.020	0.026	0.037	0.046	0.054	0.064
0.500	0.021	0.030	0.040	0.062	0.080	0.100	0.125

TABLE 3.5.4 Normal Free-Length Tolerances for Extension Springs

Spring free length (inside ends), in	Tolerance, ±in
½ or less	0.020
Over ½ to 1 inclusive	0.030
Over 1 to 2 inclusive	0.040
Over 2 to 4 inclusive	0.060
Over 4 to 8 inclusive	0.093
Over 8 to 16 inclusive	0.156
Over 16 to 24 inclusive	0.218

TABLE 3.5.5 Normal Load Tolerances for Extension Springs, ±Percent of Load

Index D/d	Free length, deflection L/F	Wire diameter, in 0.015	0.022	0.032	0.044	0.062	0.092	0.125	0.187	0.250	0.375	0.437
	12	20.0	18.5	17.6	16.9	16.2	15.5	15.0	14.3	1 3.8	13.0	12.6
	8	18.5	17.5	16.7	15.8	15.0	14.5	14.0	13.2	12 .5	11.5	11.0
	6	16.8	16.1	15.5	14.7	13.8	13.2	12.7	11.8	11 .2	9.9	9.4
4	4.5	15.0	14.7	14.1	13.5	12.6	12.0	11.5	10.3	9.7	8.4	7.9
	2.5	13.1	12.4	12.1	11.8	10.6	10.0	9.1	8.5	8. 0	6.8	6.2
	1.5	10.2	9.9	9.3	8.9	8.0	7.5	7.0	6.5	6.1	5.3	4.8
	0.5	6.2	5.4	4.8	4.6	4.3	4.1	4.0	3.8	3.6	3 .3	3.2
	12	17.0	15.5	14.6	14.1	13.5	13.1	12.7	12.0	11.5	11.2	10.7
	8	16.2	14.7	13.9	13.4	12.6	12.2	11.7	11.0	10 .5	10.0	9.5
	6	15.2	14.0	12.9	12.3	11.6	10.9	10.7	10.0	9. 4	8.8	8.3
6	4.5	13.7	12.4	11.5	11.0	10.5	10.0	9.6	9.0	8 .3	7.6	7.1
	2.5	11.9	10.8	10.2	9.8	9.4	9.0	8.5	7.9	7.2	6.2	6.0
	1.5	9.9	9.0	8.3	7.7	7.3	7.0	6.7	6.4	6.0	4 .9	4.7
	0.5	6.3	5.5	4.9	4.7	4.5	4.3	4.1	4.0	3.7	3 .5	3.4
	12	15.8	14.3	13.1	13.0	12.1	12.0	11.5	10.8	10.2	10.0	9.5
	8	15.0	13.7	12.5	12.1	11.4	11.0	10.6	10.1	9. 4	9.0	8.6
	6	14.2	13.0	11.7	11.2	10.6	10.0	9.7	9.3	8.6	8.1	7.6
8	4.5	12.8	11.7	10.7	10.1	9.7	9.0	8.7	8.3	7.8	7.2	6.6
	2.5	11.2	10.2	9.5	8.8	8.3	7.9	7.7	7.4	6.9	6.1	5.6
	1.5	9.5	8.6	7.8	7.1	6.9	6.7	6.5	6.2	5.8	4 .9	4.5
	0.5	6.3	5.6	5.0	4.8	4.5	4.4	4.2	4.1	3.9	3 .6	3.5
	12	14.8	13.3	12.0	11.9	11.1	10.9	10.5	9.9	9. 3	9.2	8.8
	8	14.2	12.8	11.6	11.2	10.5	10.2	9.7	9.2	8.6	8.3	8.0
	6	13.4	12.1	10.8	10.5	9.8	9.3	8.9	8.6	8.0	7.6	7.2
10	4.5	12.3	10.8	10.0	9.5	9.0	8.5	8.1	7.8	7.3	6.8	6.4
	2.5	10.8	9.6	9.0	8.4	8.0	7.7	7.3	7.0	6.5	5.9	5.5
	1.5	9.2	8.3	7.5	6.9	6.7	6.5	6.3	6.0	5.6	5 .0	4.6
	0.5	6.4	5.7	5.1	4.9	4.7	4.5	4.3	4.2	4.0	3 .8	3.7
	12	14.0	12.3	11.1	10.8	10.1	9.8	9.5	9.0	8.5	8.2	7.9
	8	13.2	11.8	10.7	10.2	9.6	9.3	8.9	8.4	7.9	7.5	7.2
	6	12.6	11.2	10.2	9.7	9.0	8.5	8.2	7.9	7.4	6.9	6.4
12	4.5	11.7	10.2	9.4	9.0	8.4	8.0	7.6	7.2	6.8	6.3	5.8
	2.5	10.5	9.2	8.5	8.0	7.8	7.4	7.0	6.6	6.1	5.6	5.2
	1.5	8.9	8.0	7.2	6.8	6.5	6.3	6.1	5.7	5.4	4 .8	4.5
	0.5	6.5	5.8	5.3	5.1	4.9	4.7	4.5	4.3	4.2	4 .0	3.3
	12	13.1	11.3	10.2	9.7	9.1	8.8	8.4	8.1	7.6	7.2	7.0
	8	12.4	10.9	9.8	9.2	8.7	8.3	8.0	7.6	7.2	6 .8	6.4
	6	11.8	10.4	9.3	8.8	8.3	7.7	7.5	7.2	6.8	6 .3	5.9
14	4.5	11.1	9.7	8.7	8.2	7.8	7.2	7.0	6.7	6.3	5.8	5.4
	2.5	10.1	8.8	8.1	7.6	7.1	6.7	6.5	6.2	5.7	5.2	5.0
	1.5	8.6	7.7	7.0	6.7	6.3	6.0	5.8	5.5	5.2	4.7	4.5
	0.5	6.6	5.9	5.4	5.2	5.0	4.8	4.6	4.4	4.3	4 .2	4.0
	12	12.3	10.3	9.2	8.6	8.1	7.7	7.4	7.2	6.8	6.3	6.1
	8	11.7	10.0	8.9	8.3	7.8	7.4	7.2	6.8	6.5	6 .0	5.7
	6	11.0	9.6	8.5	8.0	7.5	7.1	6.9	6.5	6.2	5. 7	5.4
16	4.5	10.5	9.1	8.1	7.5	7.2	6.8	6.5	6.2	5.8	5.3	5.1
	2.5	9.7	8.4	7.6	7.0	6.7	6.3	6.1	5.7	5.4	4 .9	4.7
	1.5	8.3	7.4	6.6	6.2	6.0	5.8	5.6	5.3	5.1	4 .6	4.4
	0.5	6.7	5.9	5.5	5.3	5.1	5.0	4.8	4.6	4.5	4 .3	4.1

TABLE 3.5.6 Coil-Diameter Tolerances of Torsion Springs, ±in

Wire diameter, in	Spring index, D/d* 4	6	8	10	12	14	16
0.015	0.002	0.002	0.002	0.002	0.003	0.003	0.004
0.023	0.002	0.002	0.002	0.003	0.004	0.005	0.006
0.035	0.002	0.002	0.003	0.004	0.006	0.007	0.009
0.051	0.002	0.003	0.005	0.007	0.008	0.010	0.012
0.076	0.003	0.005	0.007	0.009	0.012	0.015	0.018
0.114	0.004	0.007	0.010	0.013	0.018	0.022	0.028
0.172	0.006	0.010	0.013	0.020	0.027	0.034	0.042
0.250	0.008	0.014	0.022	0.030	0.040	0.050	0.060

*Mean coil diameter = D; wire diameter = d

TABLE 3.5.7 Free-Angle Tolerances of Torsion Springs, ±°

Number of coils, N	Spring index, D/d* 4	6	8	10	12	14	16	18	20
1	2	3	3.5	4	4.5	5	5.5	5.5	6
2	4	5	6	7	8	8.5	9	9.5	10
3	5.5	7	8	9.5	10.5	11	12	13	14
4	7	9	10	12	14	15	16	16.5	17
5	8	10	12	14	16	18	20	20.5	21
6	9.5	12	14.5	16	19	20.5	21	22.5	24
8	12	15	18	20.5	23	25	27	28	29
10	14	19	21	24	27	29	31.5	32.5	34
15	20	25	28	31	34	36	38	40	42
20	25	30	34	37	41	44	47	49	51
25	29	35	40	44	48	52	56	60	63
30	32	38	44	50	55	60	65	68	70
50	45	55	63	70	77	84	90	95	100

*Mean coil diamer = D; wire diameter = d.

CHAPTER 3.6
SPUN-METAL PARTS

Harry Saperstein
Livingston, New Jersey

THE PROCESS

The process of metal spinning involves the forming of sheet metal to specific circular shapes on a lathe. The sheet-metal blank rotates and is pressed against a form block of the desired shape, called a *chuck,* by means of suitably shaped spinning tools. These tools are manipulated manually (conventional spinning) or automatically (power spinning). Figure 3.6.1 illustrates conventional spinning schematically.

Although automatic conventional-spinning equipment is available, power spinning usually differs from conventional manual spinning in that the metal is worked more severely and is reduced in thickness. *Flow turning, flow forming, hydrospinning,* and *displacing spinning* are terms for this process. With power spinning of this type, the shapes that can be produced are more limited than with the conventional process,

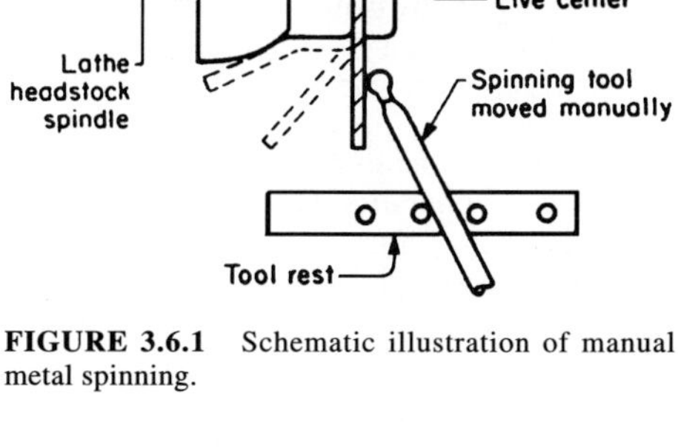

FIGURE 3.6.1 Schematic illustration of manual metal spinning.

FIGURE 3.6.2 Computer-controlled power-spinning machine for heavy-walled parts. (*Courtesy Autospin, Inc.*)

which has few shape limitations. The power-spinning process also requires extremely high constant pressure. A typical power-spinning machine is shown in Fig. 3.6.2.

It is common for the hand spinner to use an additional hand tool, a *back stick,* in the left hand to keep the metal from wrinkling as it is pressed against the chuck by the spinning tool. A lubricant, brown laundry soap, Albany grease or cup grease, beeswax, petroleum jelly, etc., may be used. The chuck can be machined from hard maple wood, cast iron, brass, steel, or a fiber-composition block.

For parts with reentrant contours, i.e., shapes that are narrower at the open end than at the closed end or in the middle (e.g., a light-bulb shape), it is necessary to use a segmented chuck for small-lot production. After the part has been spun, the chuck segments, called *splits,* are removed piece by piece. If the chuck were in one piece, it would be trapped inside the spun part.

For high-production applications, some limited reentrant shapes can be made through the use of an off-center form roll instead of a segmented chuck. The roll is withdrawn from the workpiece easily after spinning has been completed.

With some forms, particularly deep forms and cylindrical shapes, and some materials, to avoid excessive stretching or stress of metals, it is necessary to use a succession of spinning operations and possibly an intermediate annealing operation before the part receives the final spinning. Intermediate-form chucks are called *breakdown chucks.*

Sometimes parts of considerable depth can be produced very economically by performing a preliminary press-draw operation on a standard set of drawing dies and then making the final shape on a spinning form. This preliminary press operation replaces the breakdown spinning operation and provides better control over metal thickness.

FIGURE 3.6.3 Typical parts produced by metal spinning. (*Courtesy Autospin, Inc.*)

TYPICAL CHARACTERISTICS AND APPLICATIONS: SPUN-METAL PARTS

To be adaptable to spinning, an article must be symmetrical about its axis. It may be a cylinder with one or more diameters, a hemisphere, a cone, or some variation of these shapes. For practical purposes, spinnings can be as small as 6 mm ($\frac{1}{4}$ in) and as large as 4 m (12 ft) in diameter. The thickness of the metal to be spun can vary from about 0.1 mm (0.004 in) to 120 mm (4 or 5 in) on special machines and with hot material. The most common thicknesses, however, are 0.6 to 1.3 mm (0.024 to 0.050 in). Maximum thickness and size are limited only by the size of the equipment and the power available to make the metal flow.

Typical components produced by metal spinning are lamp bases, reflectors, hollowware (pitchers, tankards, vases, candlesticks, etc.), platters, pots, pans, bowls, nose cones, round trays, housings, surgical instruments, cooking utensils, aircraft parts, and components for electronic equipment. Figure 3.6.3 shows typical components which are commonly produced by metal spinning.

ECONOMIC PRODUCTION QUANTITIES

A major advantage of spinning is its extremely low tooling cost compared with the major competitive process, punch-press drawing. Since spinning tooling is simple, lead time for tooling is short, and when necessary, production on a new part can start within hours. Typically, maple chunks for parts without reentrant shapes cost only one-fiftieth as much as draw and trim dies for the same part.

Even though the unit labor cost for spun parts is much higher than that of comparable parts made on a punch press, the avoidance of a higher tooling investment makes the process attractive when low quantities are involved. Spinning is also useful in making experimental parts and preproduction samples at low cost. In an industry in which periodic styling changes are important, in the long run it may be economical to spin the pieces and make frequent changes in chucks rather than make expensive changes to stamping dies. Steel chucks are good for unlimited amounts of pieces. For precision work, extrathick metal pieces may be spun and then machined to final dimensions.

Generally, spinning is always advantageous when quantities are low or uncertain. It may be preferable for quantities as high as 5000 or more if draw dies for a part are particularly expensive, if lead time is critical, or if power-spinning equipment is available.

SUITABLE MATERIALS FOR SPINNING

The suitability of metals for spinning parallels their suitability for other cold-forming operations. Ductility is the prime prerequisite. Another desirable property is a minimum tendency to work-harden. When a material lacks these properties, annealing between operations is necessary. A popular material for spun parts is 2S aluminum because it possesses both properties. Soft copper is another preferred choice for the same reasons.

For convenience of reference, Table 3.6.1 indicates the relative spinnability of various commercially available metals. A rating of 100 indicates maximum suitability for the type of spinning indicated, while lower rating values indicate proportionally less ease of forming with spinning methods. Note that in some cases (e.g., brass) an average value for a number of alloys has been indicated. Specific alloys and specific parts designs and spinning conditions may involve values different from those shown.

DESIGN RECOMMENDATIONS

In spinning, the conical shape is the easiest to form and the most economical. The metal is not subjected to such severe strain when worked down to its extreme depth because the angle at which the chuck meets the metal is small and allows better control of the metal during the spinning operation. The hemispherical shape is more difficult to spin because the angle grows increasingly sharper as the metal is forced farther back on the chuck. In spinning a cylinder, the metal is exposed to greater strain because of the sharp angle. This operation requires more time and skill. (See. Fig. 3.6.4 for an illustration of these basic shapes.)

Manual metal spinning is a versatile process, and spun parts have few shape restrictions except for the basic limitations of circular cross section. The following design recommendations, therefore, are not mandatory rules but rather suggestions for promoting ease of manufacture:

1. Blended radii and fillets are preferable to sharp corners for ease of spinning. Sharp corners tend to cause thinning of stock and, in the case of external corners, breakage of wood or masonite chucks. A desirable minimum is 6 mm ($\frac{1}{4}$ in), although 3 mm ($\frac{1}{8}$ in) usually causes no problems. (See Fig. 3.6.5.)
2. If the part has cylindrical sides and a wood chuck is used, allow a taper of 2° or more, if possible, to facilitate removal of the part from the chuck. With steel chucks, less taper is required; as little as $\frac{1}{4}$° will be satisfactory.
3. If desired, a stiffening bead can be rolled on the edge of the spun shell. It is

TABLE 3.6.1 Spinnability of Various Materials

		Manual spinning[a]		Power spinning[b]		
Metal	UNS designation	Shallow or flat shapes[c]	Severe or deep shapes[d]	Shallow, 40–90°	Medium, 25–40°	Severe, 11–25°
Group I: Aluminum and alloys						
Non-heat-treatable						
1100-0	A91100[h]	100	100			
3003-0	A93003[h]	100	99			
3004-0	A93004[h]	100	90	100	75	50
5052-0	A95052[h]	90	65			
5086-0	A95086[h]	88	75			
Heat-treatable						
2014-0	A92014[h]	75	60[e]			
2024-0	A92024[h]	65	45	100	95	65
6061-0	A96061[h]	95	85			
Group II: Copper and alloys						
Copper	Various	100	92	100	90	80
Brass	Various	95	75	100	80	75
Bronze	Various	90	60	90	70	45
Group III: Steels						
Low-carbon 1010 to 1020	G10100 to G10200	100	95	100	95	90
SAE 4130	G41300	80	50[e]	90	75	NA
Galvanized (electrolytic coatings)		100	80	100	90	NA
4340	G43400	75	45[e]	85	70	NA
AISI-8630	G86300	70[e]	40[e]	NA	NA	NA
Tricent 4405 (modified 4340)	G43406	75	45[e]	85	70	NA
Group IV: Stainless steels						
Unstabilized types						
202	S20200	100	78	100	92	85
302	S30200	98	80			
304	S30400	98	90	100	95	90
305	S30500	100	100			
Stabilized types						
316 (molybdenum)	S31600	90	60			
321 (titanium)	S32100	85	50	98	90	80
347 (columbium)	S34700	90	50			
17-7 PH	S17700	80[e]	60[e]	70[e]	NA	NA
Straight chromium types						
405	S40500	90	70	90	80	70
410	S41000	90	70			
430	S43000	90	50			
446	S44600	50[e]	35[g]			
Magnetic shielding alloys, Mumetal, 1950, etc.		95	70	100	85	65

TABLE 3.6.1 Spinnability of Various Materials (*Continued*)

		Manual spinning[a]		Power spinning[b]		
Metal	UNS designation	Shallow or flat shapes[c]	Severe or deep shapes[d]	Shallow, 40–90°	Medium, 25–40°	Severe, 11–25°
		Group V: Nickel and nickel alloys				
Monel	Various	100	88	100	92	NA
Inconel 800	N08800	90	70	90	80	NA
Inconel 702	N07702	85	60	NA	NA	NA
Incoloy 801	N08801	80	55	NA	NA	NA
Inconel X	N07750	90	50	NA	NA	NA
A-286	S66286	70[e]	60[e]	NA	NA	NA
Nickel	Various	100	92	NA	NA	NA
HastelloyB	N10001	75	35	90	75	60
HastelloyC	N10002	50	10	90	75	60
HastelloyX	N06002	NA	NA	90	75	60
		Group VI: Miscellaneous				
Lead		96	90	NA	NA	NA
Pewter	L13912 L13911	100	99	NA	NA	NA
Zinc		100	100	NA	NA	NA
Tantalum		86	45	NA	NA	NA
Magnesium		80[e]	70[g]	90[e]	80[e]	70[e]
Molybdenum		75[e]	35[e]	NA	NA	NA
Titanium[e,g]		60	25	90	85	75

Note: NA, not available. Much experimental and development work is in progress, but definite ratings are not yet available.

[a]Manual or conventional spinning using hand or lever tools.

[b]Power spinning is basically for conical shapes in which the spinning-roll power is applied through mechanical or hydraulic setups. Angles shown are from sidewall to centerline (one-half the included angle).

[c]Shallow or flat shapes are relatively shallow contours or simple flanges, dish shapes, etc.

[d]Severe or deep shapes are parts that have extreme deformation of blank, tending toward cylindrical contour.

[e]Spinnings are made at elevated temperatures.

[f]Nature of this alloy necessitates very frequent intermediate annealing.

[g]No lubricant is needed when hot-spinning titanium at 1150 to 1250°F.

[h]0-temper (annealed).

Source: Updated from Frank W. Wilson (ed. in chief), *Tool Engineers Handbook,* copyright © 1959 by American Society of Tool Engineers. Used with the permission of the McGraw-Hill Book Company.

preferable, from a production standpoint, to have the bead face outside rather than inside the part. Otherwise, rechucking in a hollow external chuck will be necessary because both inside and outside beads are formed "in air" with no chuck support; bead dimensions should be loosely toleranced. (See Fig. 3.6.6.)

4. Use as shallow a part design as possible; i.e., avoid deep cylindrical designs, which require repeated operations and annealing. A spinning ratio (depth-to-diameter ratio) of less than 1:4 is preferable. Spinning ratios are classified as follows:

Shallow: less than 1:4 Moderate: 1:4 to 3:4 Deep: 3:4 to 5:4

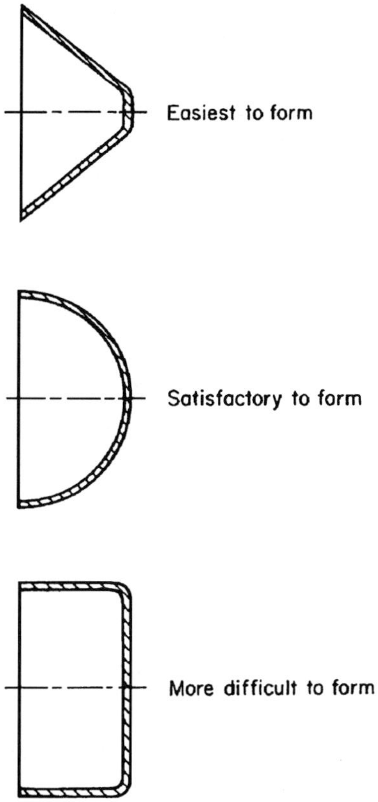

FIGURE 3.6.4 Comparison of basic spun shapes arranged in order of ease of spinning.

5. Internal flanges and other configurations of reentrant shapes, as mentioned earlier, are more costly to produce because they require special, more complex chucks or spinning without backup support for the work. (See Fig. 3.6.7.)
6. Dimension parts to surfaces adjacent to the chuck (usually inside dimensions). This allows the chuck maker to apply these dimensions directly to the chuck, and it avoids variations in diameter or length caused by variations in material thickness. (See Fig. 3.6.8.)
7. Since some thinning of material is normal in conventional metal spinning, this should be allowed for if final wall thickness is critical. Specifying material 25 or 30 percent thicker than the finished-part thickness is usually sufficient to allow for such reduction in wall thickness.
8. However, material too thick for easy spinning should not be specified. Both extra-thick and extra-thin materials make spinning more difficult. See Table 3.6.2 for recommended stock thicknesses for best spinnability.
9. Avoid reverse-form designs, if possible, because they require additional operations and can cause considerable thinning of stock. (See Fig. 3.6.9.)

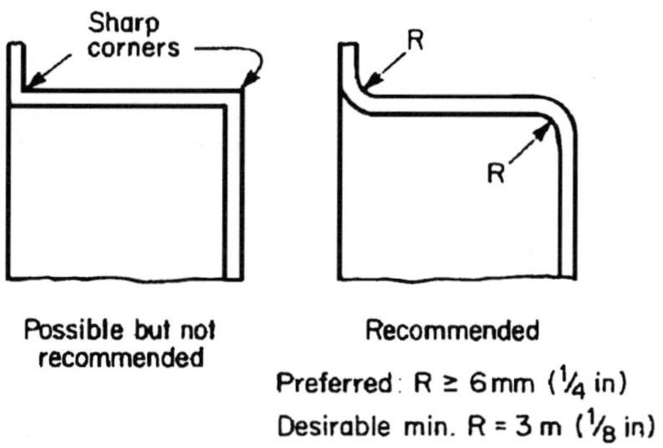

FIGURE 3.6.5 Sharp corners in spun parts are possible but should be avoided.

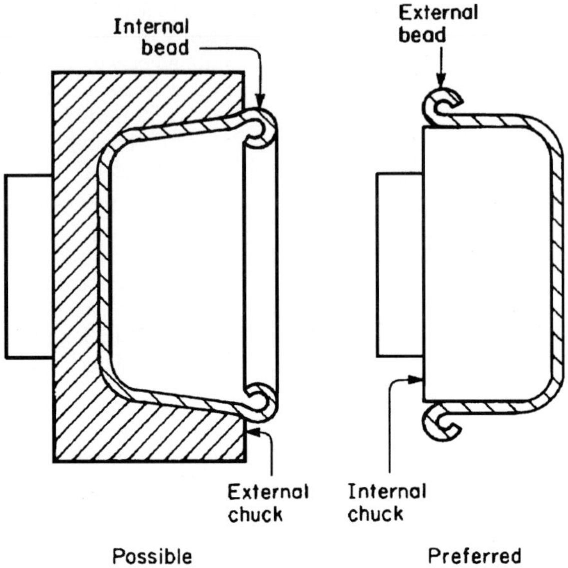

FIGURE 3.6.6 Beads placed on the outside of a part are more economically spun than beads on the inside.

10. The flatness of a flat-bottomed part will not be improved by the spinning operation. If bottom rigidity is important (freedom from "oil canning") and flatness is not required, a cone angle of 5° will provide rigidity and is easy to spin.

TOLERANCES

Table 3.6.3 presents recommended minimum dimensional tolerances for conventionally spun parts. Table 3.6.4 provides similar data for power-spun parts.

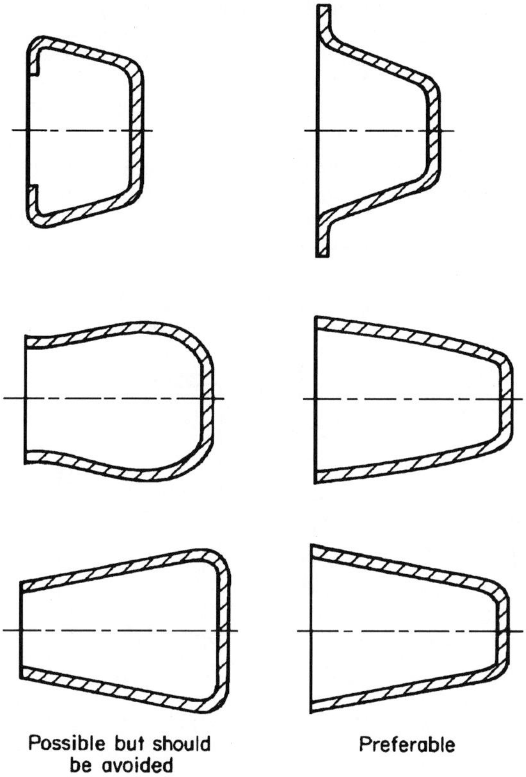

FIGURE 3.6.7 Avoid reentrant shapes, if the design purpose permits, because they add to the part cost.

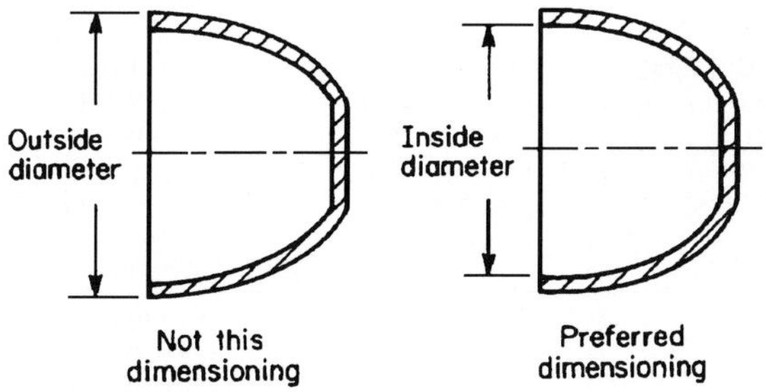

FIGURE 3.6.8 Spun shapes should normally be dimensioned to the surface that touches the chuck.

TABLE 3.6.2 Recommended Stock Thickness for Easy Spinning

	Recommended stock thickness, mm (in)			
	Aluminum		Steel	
Blank diameter, mm (in)	Minimum	Maximum	Minimum	Maximum
Less than 100 (less than 4)	0.6 (0.025)	1.2 (0.045)	0.5 (0.020)	1.0 (0.040)
Over 100 to 500 (over 4 to 20)	0.8 (0.032)	2.0 (0.080)	0.6 (0.025)	1.5 (0.063)
Over 500 (over 20)	1.0 (0.040)	3.0 (0.125)	0.8 (0.032)	2.0 (0.080)

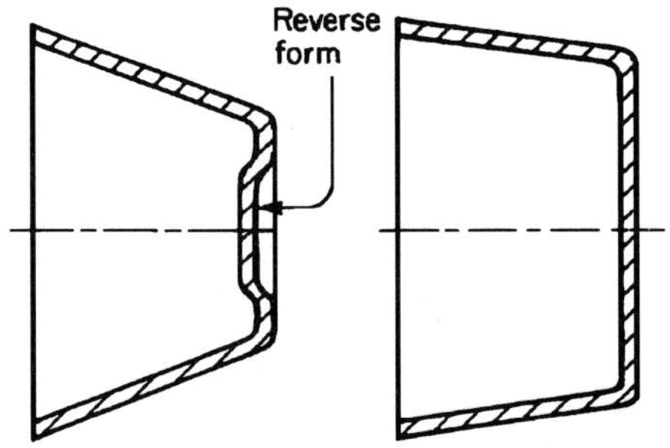

FIGURE 3.6.9 Avoid reverse bends, if possible, because they necessitate multiple spinning operations with separate chucks.

TABLE 3.6.3 Recommended Tolerances for Conventional Spinning

Nominal dimension	Diameters, mm (in)	Depths, mm (in)	Angles, °
Under 40 mm (under 1.6 in)	±0.25 (0.010)	±0.4 (0.016)	±1
40–125 mm (1.6–5.0 in)	±0.4 (0.016)	±0.8 (0.030)	±3
125–500 mm (5–20 in)	±0.8 (0.030)	±0.8 (0.030)	±3
500 mm–1 m (20–40 in)	±1.5 (0.060)	±1.1 (0.044)	±5
1–2 m (40–80 in)	±3 (0.120)	±1.5 (0.060)	±5

Wall thickness, mm (in)	Recommended tolerances, mm (in)
Under 1.3 (0.050)	+0.13 (0.005), −0.38 (0.015)
Under 2.5 (0.100)	+0.25 (0.010), −0.63 (0.025)
Over 2.5 (0.100)	+0.25 (0.010), −1.3 (0.050)

TABLE 3.6.4 Recommended Tolerances for Power-Spun Parts

Dimension	Recommended tolerances, mm (in)	
	Tight	Normal
Diameter, 0–150 mm (0–6 in)	±0.05 (0.002)	±0.13 (0.005)
Diameter, over 150 mm (6 in)	±0.08 (0.003)	±0.20 (0.008)
Length	±0.13 (0.005)	±0.25 (0.010)
Wall thickness	±0.05 (0.002)	±0.13 (0.005)

CHAPTER 3.7
COLD-HEADED PARTS

Charles Wick
Consultant

THE PROCESS

In cold heading or upsetting, force applied by one or more blows of a tool displaces metal in a portion or all of a slug, wire, or rod blank. This plastic flow produces a section or part of different contour, larger in cross section than the original material. Force is generally applied by a moving punch to form that portion of the blank protruding from a stationary die. As seen in Fig. 3.7.1, some parts are upset in the punch,

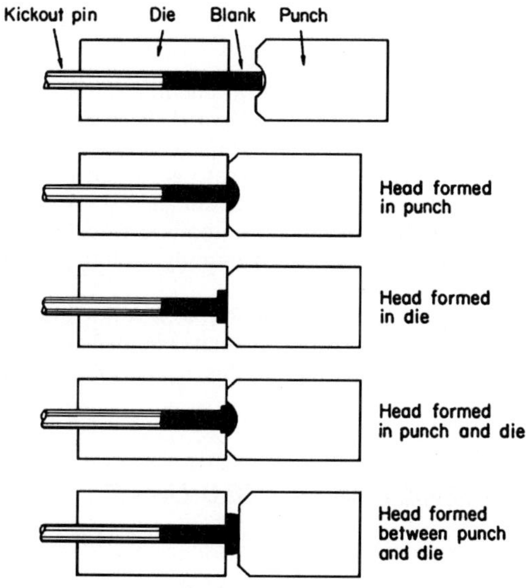

FIGURE 3.7.1 Upsetting can be done in the moving punch, in the stationary die, in both punch and die, or between the punch and the die. (*National Machinery Co.*)

some in the die, some in both punch and die, and others between the punch and the die.

Related Processes

Upsetting is often combined with forward or backward extrusion, particularly in cold-forming large-headed, small-shanked parts. Warm heading is applied for processing difficult-to-head metals such as medium-to high-carbon steels, austenitic stainless steels, and other materials that work-harden rapidly. In this compromise between cold and hot heading, the metal is preheated to between 150 and 540°C (300 and 1000°F) or more (depending on the particular metal, but below its recrystallization temperature) to increase its plasticity and reduce the force required for heading.

Rotary forging is a related process in which deformation is induced progressively rather than by a single direct squeeze. The workpiece is subjected to a combined rolling and squeezing action to produce progressive upsetting. This method is particularly suitable for thin, intricate flange sections.

Secondary Operations

Many secondary operations are performed on cold-upset parts, depending on the design and application of the part. These include grooving, shaving, pointing, flattening, bending, piercing, drilling, tapping, threading (often done with a thread-rolling machine made an integral part of the header), heat treating, and plating.

Cold-Heading Machines

Most cold-heading machines used in high production are fed coil stock. A machine may include a straightener, draw die, feed rolls, and shear. Sheared slugs are automatically transferred to a stationary die, and the projecting end of each slug is struck axially by the upsetting tool (punch). In some cases, as in nail making, the stock can be upset first and then sheared to length. Multistation and progressive headers, as well as bolt-making machines, are equipped with automatic-transfer mechanisms to move parts to successive stations and may have trimming, pointing, and thread-rolling stations.

Headers and forming machines are available in a wide variety of sizes and types. They are frequently rated by maximum-diameter-cutoff capacity and feed lengths (generally up to 14 times rated diameter, but this capacity can vary).

TYPICAL CHARACTERISTICS OF COLD-UPSET PARTS

Cold upsetting and heading are best suited to the high-volume production of relatively small parts. Producing heads on a wide variety of fasteners is still the greatest single use of the process. However, when combined with other metal-forming operations such as extruding, piercing, trimming, and threading, the process is being increasingly applied to the production of many different parts. Raw materials range up to 50 mm (2 in) or more in diameter. The size of upset parts varies from balls ½ mm (0.020 in) in diameter to large shafts, stem pinions, and other parts weighing 3.2 kg (7 lb) or more. The longest pieces headed on commonly used equipment are about 230 mm (9 in) long, but special machines will handle longer parts. Materials used range from aluminum alloys to medium-carbon steels. Stainless steels as well as copper and nickel alloys are also cold-formed.

Cold-headed parts have improved physical properties, with metal-grain flow lines following the contours of the parts. They have smooth surfaces and an absence of stress-inducing tool marks. On the other hand, there are limitations on the intricacy of the parts that can be produced. Undercuts, reentrant shapes, and thin walls normally are not feasible. Shank ends of parts are usually irregular, and secondary operations are necessary if the ends must be fully square. Screw machines are often preferred for shorter production runs and for parts requiring intricate forms, sharp angles, and closer tolerances.

Typical applications of cold-upset parts include, in addition to conventional fasteners, the following: valves, knobs, shelving spacers, electrical contacts, cams, terminal posts, rollers, shafts, balls, bearing races, electrical terminals, transistor bases, wrist pins, cable connections, spark-plug bodies, ball-joint housings or sockets, spindles and shafts, gear blanks, pistons, hose fittings, switch housings, pipe plugs and fittings, wheel hubs, and stub axles. Figures 3.7.2 and 3.7.3 illustrate typical examples.

ECONOMIC PRODUCTION QUANTITIES

Cold-upsetting or -heading processes are most suitable for producing large quantities of relatively small metal parts. Long production runs are desirable because the relatively high cost of tools and the substantial setup time can be distributed over a large number of workpieces, thus reducing the unit cost. Also, initial investment in capital equipment (large progressive headers and part formers) is expensive, but write-offs can be rapid with sufficient production requirements.

Minimum-quantity levels for economic production generally range from 25,000 to 50,000 parts, but quantities as low as 1000 have been produced. The exact minimum run for economic production depends on many factors, including the size and shape of the workpiece, the material from which it is made, tooling cost, setup time, and the costs of competitive processes.

Cost of the raw material can be a significant factor affecting the competitiveness of the process. As the material cost increases, the quantity of parts required to justify the process decreases because of material savings (little or no waste). However, parts made of high-carbon and alloy steels as well as other less ductile materials are generally uneconomical to produce in this way because of the higher pressures required (necessitating more costly equipment) and the possible need for several operations with intermediate annealing treatments.

Many parts previously made on screw and automatic bar machines are now being produced on headers and formers. One rule of thumb is that cold heading will be more economical than machining if shank length is more than 3 times the shank diameter of the part and production requirements are sufficiently high.

FIGURE 3.7.2 One die and two punches form these double-blow cold-headed parts. (*National Machinery Co.*)

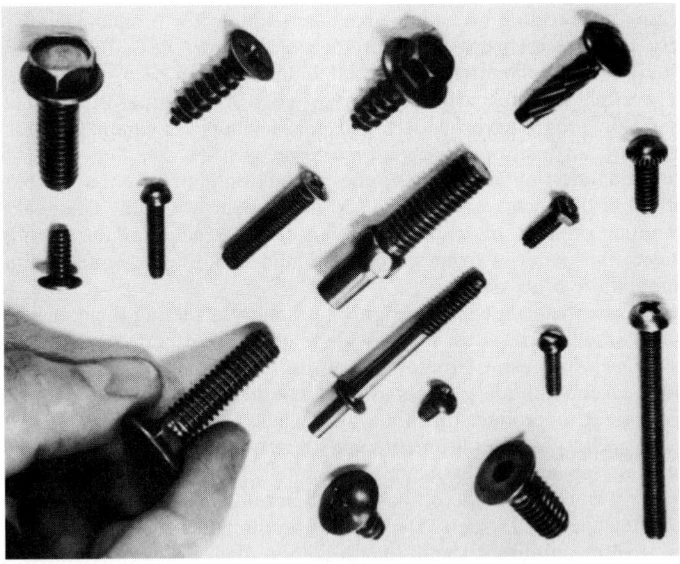

FIGURE 3.7.3 Combination cold-heading and thread-rolling machines form these completed fasteners. (*National Machinery Co.*)

GENERAL DESIGN CONSIDERATIONS

All parts to be upset should be designed for the smallest stock diameter possible, minimum amount of upset, smallest-diameter head (or other upset portion), and fewest strokes necessary to satisfy requirements. Thin, large-diameter heads (in comparison with the original wire diameter) are more difficult to upset than thicker ones. Symmetrical parts are the easiest to cold-form, but unsymmetrical parts can be produced. No draft allowance is required, but parts must be designed so that they can be removed from the tooling.

Designers should try to visualize optimal metal flow in the design of parts to be upset. For example, generous fillets and rounded corners facilitate metal flow.

Close cooperation between designer and manufacturing engineer often can improve a part and reduce its cost. Parts originally designed to be made by some other process may require modifications to permit upsetting at low cost. Sometimes two or more functional components can be combined into a single upset part, thereby eliminating assembly costs.

SUITABLE MATERIALS

Any material that is malleable when cold can be cold-formed. Materials having the lowest yield strength and the greatest range between yield and ultimate strengths are the most malleable. Since strain or work hardening during cold working decreases formability, the rate of strain hardening of the material is an important consideration. Hardness of the material is also critical, since too hard a material is impracticable to cold-work.

Prior operations, sometimes required to prepare material for upsetting, can include heat treating, drawing, descaling, and lubricating. The upsetting properties of most steels are improved by annealing or spheroidizing.

Whenever possible, the designer should avoid specifying exact material analyses, since this limits the upsetter's flexibility and may increase production costs. Several different analyses of a basic material often can satisfy service requirements, but they may vary in upsettability, availability, and cost.

Steels

Low-carbon steels such as 1006 (G10060), 1008 (G10080), 1010 (G10100), 1012 (G10120), 1015 (G10150), and 1018 (G10180) are the easiest to upset. Such steels have a carbon content too low to permit a marked response to heat treatment, but they have a minimum tensile strength of 380 MPa (55,000 lbf/in^2), and physical properties can be substantially improved by cold working. If wear-resistant surfaces are required, they can be case-hardened. When subsequent through-hardening is required, medium low-carbon steels such as 1019 (G10190), 1022 (G10220), and 1025 (G10250) are used.

Medium high-carbon steels such as 1038 (G10380) are also fairly easy to cold-work but more difficult than low-carbon steels, and their headability decreases as their carbon content increases. Such steels are used for parts requiring a minimum tensile strength of 830 MPa (120,000 lbf/in^2).

High-carbon steels containing up to about 0.62 percent carbon are cold-upset, but they have limited workability, and maximum carbon contents of about 0.50 percent are generally recommended. Rapid work hardening of higher-carbon steels restricts the amount of cold working possible without annealing.

The chemical composition of carbon steels also can affect their headability. A silicon content of more than 0.35 percent can increase the rate of work hardening and resistance to plastic flow. A manganese content above 1.00 percent also may increase pressure requirements. Sulfur content should be limited to about 0.05 percent. Resulfurized steels generally are not recommended but may be necessary when subsequent machining is required. Phosphorus contents greater than 0.04 percent also can cause higher pressure requirements.

Alloy steels are more difficult to cold upset, but parts are produced from such materials. Alloy steels such as 4037-H (H40370), 4118 (G41180), 8620 (G86200), and 8640 (G86400) are suitable for cold upsetting and have strengths suitable for many fasteners and other cold-headed parts. The nickel content of alloy steels should be limited to about 1.00 percent, chromium to about 0.90 percent, and molybdenum to about 0.40 percent for best results. Limitations on manganese content depend on the carbon content of the alloy steel, from as much as 1.35 percent manganese in steels containing up to about 0.12 percent carbon to 0.90 percent manganese in steels containing up to 0.30 percent carbon. In some cases, small amounts of boron in the steels can achieve hardness requirements while allowing the use of more easily formed lower-carbon materials.

Rimmed steels are ideal for heading operations when large surface expansions are required, but such steels generally are not satisfactory for cold extrusion. As a result, killed steels are usually preferred for cold-forming operations combining upsetting and extruding. Vacuum degassing of steel at the mills is recommended by some upsetters. Heat treatment, preferably spheroidize-annealing, to establish the microstructure of the material is an important consideration. The hardness of the steel that is best suited for upsetting generally varies from about R_b 72 for low-carbon steels to about R_b 85 for higher-carbon and alloy steels.

Cold heading or upsetting is carried out with either hot-rolled or cold-drawn steel. The surface quality of the raw material is also important. The material should be free of surface imperfections such as scratches, seams, folds, and blisters, which can cause cracking during deformation and mar the appearance of the finished part. Cold-heading-quality steel, which has sufficient surface stock removed to eliminate deep seams, is available from mills.

Internal material defects such as porosity and nonmetallic inclusions also can cause failures in upsetting, but they are even more critical in extruding. When parts are to be cold-formed by upsetting and extruding, cold-extrusion-quality steels (having better surface quality and uniform internal structure) are often specified.

Stainless Steels

High initial strength and rapid work-hardening characteristics make stainless steels more difficult to cold-upset, but some grades work-harden less rapidly and are more formable than others. Stainless steels most widely cold-headed include the austenitic types 305 (S30500) and 316 (S31600) and the ferritic and martensitic types 410 (S41000) and 430 (S43000). Type 410 is one of the most widely used hardenable corrosion-resistant steels for upsetting, but type 430 (a ductile, nonhardenable type) is more corrosion resistant. Type 384 (S38400), which was developed for cold heading, provides a good combination of headability and corrosion resistance.

Stainless steels that have higher work-hardening rates and less headability include types 302 (S30200), 304 (S30400), 309 (S30900), 310 (S31000), 316 (S31600), 431 (S43100), and 440C (S44004). Most of these require less severe cold heading, and some cannot be considered for producing recessed-head-type fasteners.

Nickel Alloys

Cold-heading properties of nickel-alloy materials range from good to poor. Monel metal has fairly good cold workability, and Nichrome is also readily cold-worked. Inconel also is cold-headed, but this is more difficult.

Aluminum Alloys

Aluminum alloys are among the most easily headed materials. The most widely used alloy is 2024 (A92024), which is generally upset in the H13 temper and then heat-treated to the T4 condition. The most easily headed aluminum alloys are 1100-0 (A9110-0), 1100-H14 (A91100-H14), 6053-H13 (A96053-H13), 3003-0 (A93003-0), 3003-H14 (A93003-H14), 6061-0 (A96061-0), and 6061-H13 (A96061-H13). The next most easily formable are 2117-0 (A92117-0), 2117-H13 (A92117-H13), 2117-H15 (A92117-H15), 2017-O (A92017-0), 2017-H13 (A92017-H13), 2024-0 (A92024-0), 2024-H13 (A92024-H13), 5005-0 (A95005-0), 5005-H32 (A95005-H32), 5052-H32 (A95052-H32), 5056-0 (A95056-0), and 5056-H32 (A95056-H32). Somewhat less headable are 7575-0 (A97075-0), 7075-H13 (A97075-H13), and 7178-H13 (A97178-H13). Aluminum alloy 6061-T913 (A96061-T913), used only for nail wire, is more difficult to head.

Copper and Copper Alloys

Copper and alloys such as brass and bronze also possess excellent heading properties, and their hardness and strength can be improved considerably by cold working. Brass containing 65 to 70 percent copper and 30 to 35 percent zinc is widely used and easily formed into intricate shapes. Commercial bronze (90 percent copper, 10 percent zinc) has better cold-working properties than the brass just described. The electrolytic-tough-pitch (ETP) type is the copper most widely cold-headed, but oxygen-free and tellurium coppers can be formed just as easily. Care is required in upsetting leaded copper alloys because of the possible danger of splitting the material.

Other Materials

Titanium, beryllium, and refractory metals are less formable at room temperatures, and they are likely to crack when cold-headed. As a result, they are often processed warm.

DETAILED DESIGN RECOMMENDATIONS

Maximum Upset Diameters

When low-carbon steel is upset, there are limitations on the ability of the material to deform without cracking or splitting. Figure 3.7.4 shows generally accepted maximum upset diameter limits. Free upsets between die and punch do not contain the metal at the sides of the upset, and under these conditions, the maximum upset diameter is about $2\frac{1}{2}$ times the blank diameter. If the metal is contained in or controlled by punch surfaces, the maximum upset diameter is about 3 times the blank diameter. The exact maximum diameter depends on the material quality, and there are examples of upsetting carried out at 4 times the blank diameter.

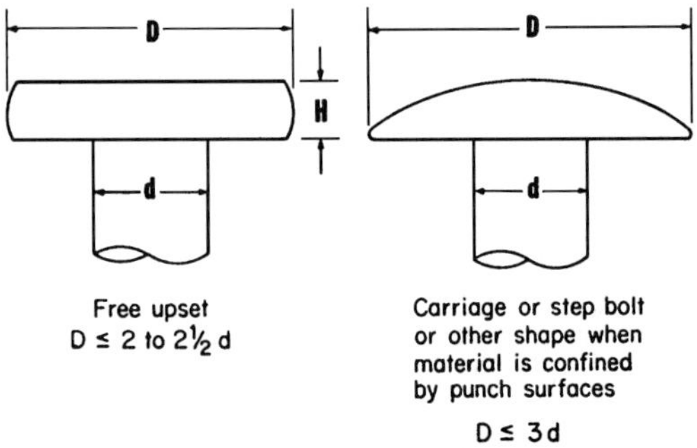

FIGURE 3.7.4 Generally accepted maximum upsetting diameters.

Lengths

The length of a headed part is considered to be the distance from the extreme end of a given piece to the nearest point at which the diameter of the upset portion is at maximum. As a general rule, the length for cold heading should be between one diameter of the stock and 150 mm (6 in) for the most economical operation. If the length is shorter than one diameter, handling becomes difficult and usually requires special tooling. For parts longer than 225 mm (9 in), special equipment such as rod headers may be required.

Amount of Upset

The maximum length that can be upset without unwanted bending is expressed in multiples of the blank diameter. A general rule is that up to $2\frac{1}{4}$ to $2\frac{1}{2}$ diameters in length can be upset in one blow. If this volume is not enough to form the required upset, two blows must be used. Up to about $4\frac{1}{2}$ diameters generally can be upset in two blows and 6 to 8 diameters with a triple blow. For economy, parts should be designed to require the smallest number of blows.

Part Symmetry

Parts symmetrical about their axes are the easiest to form because they permit even metal flow and balanced forces against the tool. Although asymmetrical parts can be produced, they may exhibit interrupted grain flow and entail more complex tooling and shorter tool life.

Head Shape

Unless flats are required on heads for wrenches or some other contour is necessary, circular heads should be specified for maximum economy. Square, rectangular, hexagonal, and other shapes of heads are commonly produced, but if sharp edges are required, trimming is generally necessary, thus adding to the cost. Tighter tolerances can be maintained when an indentation on the head of a part is placed as closely as possible to the body or shank axis.

Recesses and Slots

Hollow upsets should be avoided when possible because they require more die maintenance and there is a possibility of cracks forming around the edges of the recesses. However, many fasteners with hexagonal-socket heads and other special recesses are being produced successfully on cold headers. Recesses in such heads should not extend into the bodies of the formed parts.

Screwdriver slots can be cold-formed in the heads. In flat-headed parts, the slot must have a curved (instead of a flat) bottom to eliminate contact between the slotting punch and the dies. In round-headed parts, the slots can be straight. (See Fig. 3.7.5.)

FIGURE 3.7.5 Cold-formed screwdriver slots.

Collars or Double Heads

When a second upset is produced some distance from the end of the part, it must be small enough to drop from the dies as they are released and cannot generally be more than one-third to one-half larger than the diameter of the wire being used. Upset sections also should be at least one wire diameter apart. (See Fig. 3.7.6.) Double heads or collars that exceed these limits usually must be made by a secondary upsetting or machining operation.

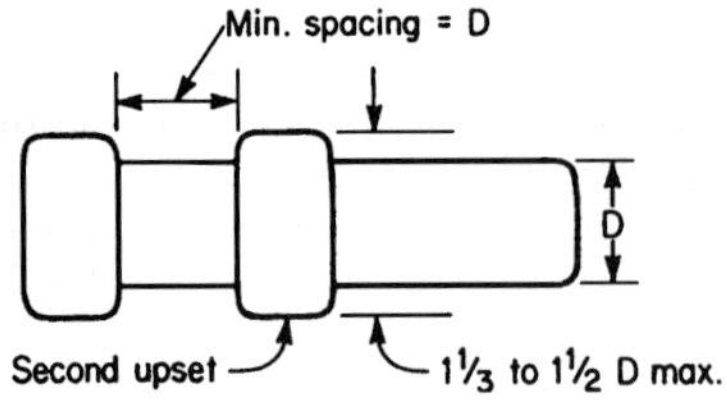

FIGURE 3.7.6 Design rules for second upsets.

Round, square, and hexagonal collars are most common. Such collars are easier and more economical to form if the diameter above the collar exceeds the diameter below the collar. Thin, large-diameter collars, as in the case of heads, are harder to produce than thick ones. For maximum economy, they should be in the range of $1\frac{1}{2}$ to $1\frac{2}{3}$ times the wire diameter. (See Fig. 3.7.7.)

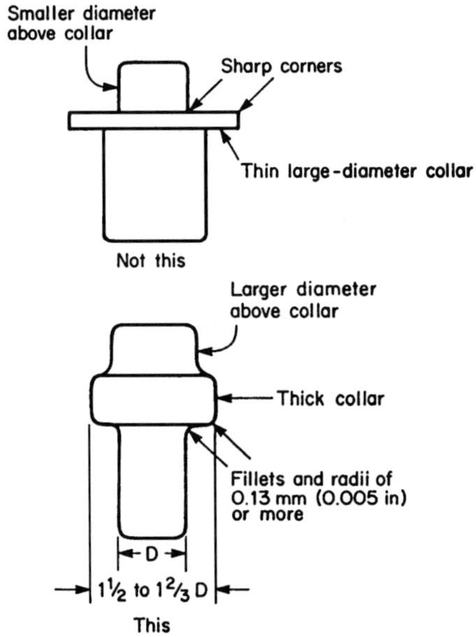

FIGURE 3.7.7 Collars are easier and more economical to form if the diameter above the collar exceeds that below. Thin, large-diameter collars are harder to produce than thick ones. Collar diameters should be in the range of $1\frac{1}{2}$ to $1\frac{2}{3}$ times the wire diameter for maximum economy in upsetting. Normal flow of material produces slightly rounded edges on collars. Radii and/or fillets make cold-upset parts more economical and easier to produce.

Collars often require sharp corners, but economical cold heading requires forming with slightly rounded edges. A typical collar rounding with normal flow of material during heading is also shown in Fig. 3.7.7. If straight sides and sharp corners are required, a trimming or secondary shaving operation is necessary.

Corners

As in all cases of plastic flow, sharp corners should be avoided because they interfere with smooth metal flow, make it difficult to fill, reduce tooling strength, and can cause stress concentrations. Parts with radii and fillets are more economical and easier to produce. A minimum radius of about 0.13 mm (0.005 in) and even more for stainless steels and nickel alloys, depending on part size and location of the radii, is recommended. Axial undercuts, such as the ones shown on the part in Fig. 3.7.8, can be formed in cold heading, and they are almost always less expensive than producing square corners.

Shanks

When a shank must have some portion that is noncircular in section, the length of that portion should be kept to a minimum. Also, when a shank requires multidiameters, a generous radius or taper should be provided between diameters.

Locking devices to prevent fasteners from rotating (such as square shanks below the heads) can be cold-formed, but tooling is simplified and costs reduced if two or four lugs are specified instead of a square shank (Fig. 3.7.9).

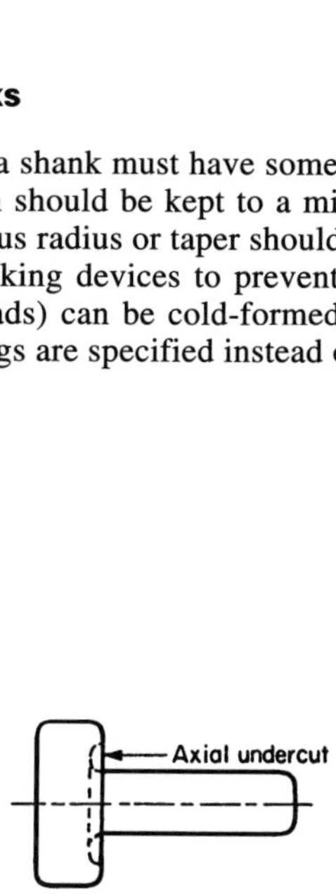

FIGURE 3.7.8 Undercut beneath the head of this stainless-steel plunger is formed during upsetting and eliminates the need for producing square corners.

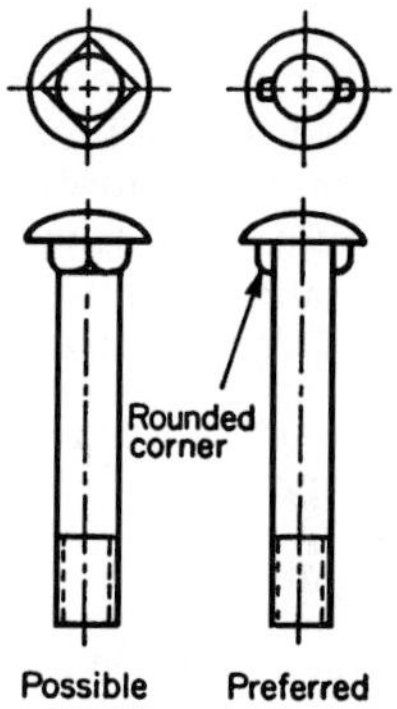

FIGURE 3.7.9 Tooling is simplified if two or four lugs are used instead of a square shank.

Serrations, Ribs, and Knurls

Only straight knurls can be produced in heading, and they normally require a draft or taper of about 2° to facilitate removing the part from the punch and die. Sharp diamond-pattern knurls or straight knurls without taper must be rolled after upsetting and are therefore less economical.

Headed knurls are not usually as sharply defined as those produced by trimming or rolling and are therefore not recommended for press fits, close tolerances, or cases in which maximum holding power is recommended.

Lugs, Fins, Projections, and Dimples

Lugs form best on cold-headed parts when they are designed with a radius on their bottom and do not extend to the edge of a head, as shown in Fig. 3.7.9. When lugs are extended to the edge of a head, they will not usually fill out sharply, and a somewhat irregular configuration can be expected. If a projection or dimple (as for welding) is needed near the edge of a head and it is not joined to the body or shank of the part, it should be placed at least one projection diameter from the body and at least 0.8 mm ($\frac{1}{32}$ in) from the perimeter of the head.

Points or Chamfers

Forming of points or chamfers on cold-headed parts is generally limited to parts having a length 10 to 12 times the smallest diameter at the pointed end when an unsupported knockout pin is used and up to 20 diameters with a supported pin (Fig. 3.7.10). For parts beyond these limitations, the point usually must be machined in a subsequent operation.

Embossing

Forming of numbers, letters, or other identification marks on parts in a header may add slightly to the cost, but it is generally cheaper than processing the parts through a secondary operation. Depressed characters in the part are usually cleaner and more legible. However, tooling tends to wash off in service, and it therefore may be expensive in the long run. Depression in the punch, producing raised lettering in the part, allow longer tool life, but character legibility is not as good.

Threading

On parts for which thread length must be as close to the head as practicable, maximum allowable distance from the underside of the head to the first complete thread should be specified. As shown in Fig. 3.7.11, this should be $2\frac{1}{2}$ threads for sizes up to 25 mm (1 in) in diameter and $3\frac{1}{2}$ threads for larger sizes.

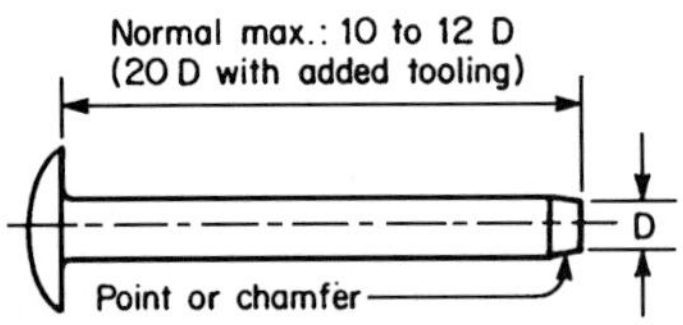

FIGURE 3.7.10 Length of parts having formed points or chamfers is generally limited to 10 to 12*D* with unsupported knockout pins and up to 20*D* with supported pins.

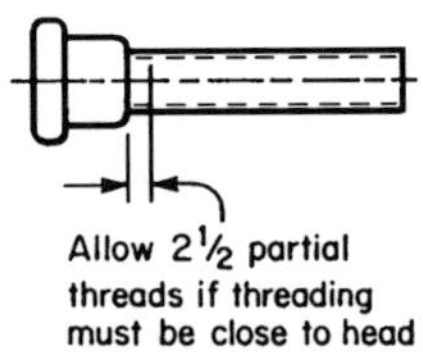

FIGURE 3.7.11 Recommended maximum distance from the underside of the head or shoulder to the first complete thread is $2\frac{1}{2}$ threads for sizes up to 1 in in diameter.

Combining Parts

Designers should give major consideration to the possibilities of combining two or more components into a single cold-formed part. Advantages of such designs include reduced costs and increased part strength. Examples include fasteners combined with spacers, washers, or special heads

DIMENSIONAL FACTORS AND TOLERANCES

Tolerances that can be maintained in cold upsetting vary with the style (shape and volume) of the upset, length-to-diameter ratios, material being formed, part size, quality of the machine and tooling, and other factors. Also, after prolonged use, tooling wears and parts lose size. Tolerances are primarily a function of tool-manufacturing practices and wear, and smaller parts generally can be held to closer tolerances.

As with any manufacturing process, part costs increase when closer tolerances are specified. Care should be exercised not to specify tolerances closer than those actually required for the specific application of the part. Tolerances should be as liberal as possible for increased economy, longer production runs (due to prolonged tool life), and reduced scrap.

Diameters

Diameter tolerances can easily be maintained within those specified for standard fasteners. Tolerance ranges for the diameters of upset heads are given in Table 3.7.1,

TABLE 3.7.1 Tolerance Ranges for Upset-Head Heights and Diameters

Nominal shank diameter, in	Tolerance ranges*	
	For upset height, in	For upset diameter, in
1/16	0.008	0.014
1/8	0.010	0.018
3/16	0.012	0.021
1/4	0.014	0.025
5/16	0.015	0.029
3/8	0.017	0.032
7/16	0.019	0.036
1/2	0.021	0.040
5/8	0.025	0.048
3/4	0.028	0.055
7/8	0.032	0.062
1	0.036	0.070

*Tolerance range generally applied bilaterally. For example, a range of 0.014 would be ± 0.007 in.

Source: Industrial Fasteners Institute.

TABLE 3.7.2 Tolerance Range on Shoulder Diameters (Round, Square, and Hexagonal), in

Body size	Economical	Normal	Expensive
0.029–0.099	±0.0025	±0.0015	
0.100–0.179	±0.003	±0.0015	±0.001
0.180–0.312	±0.003	±0.002	±0.0015
0.313–0.500	±0.0035	±0.0025	±0.0015
Tolerance range on shoulder lengths, in			
0.0029–0.500	±0.0075	±0.0055	±0.0025

Source: Elco Industries, Inc.

TABLE 3.7.3 Tolerance Range on Hexagonal-, Round-, and Square-Collar Diameters, in

Body size	Economical	Normal	Expensive
0.070–0.125	±0.004	±0.0025	±0.0015
0.126–0.250	±0.005	±0.003	±0.002
0.251–0.375	±0.006	±0.004	±0.0025
0.376–0.625	±0.007	±0.005	±0.003
0.626–1⅛	±0.008	±0.006	±0.0035

Source: Elco Industries, Inc.

TABLE 3.7.4 Tolerance Range for Unthreaded-Body Diameter (below Shoulder or above and below Collars), in

Body size	Economical	Normal	Expensive
0.029–0.099	±0.002	±0.001	±0.0005
0.100–0.179	±0.002	±0.001	±0.00075
0.180–0.312	±0.002	±0.0015	±0.001
0.313–0.500	±0.0025	±0.002	±0.0015
0.501–0.562	±0.0025	±0.002	±0.0015

Note: The shape and volume of the material in the head determine the allowable tolerances to be held on an unthreaded-body section.

Source: Elco Industries, Inc.

shoulders in Table 3.7.2, collars in Table 3.7.3, and bodies or shanks in Table 3.7.4. These tolerances are not necessarily limitations but can be used as a guide. The tolerance ranges listed as economical are the most easily held. Although those listed as normal are more costly, they do not generally require secondary operations, but special tooling may be needed. Tolerance ranges listed as expensive may require secondary operations.

Lengths

Tolerance ranges for the heights or lengths of upset heads are given in Table 3.7.1, shoulder lengths in Table 3.7.2, collar thickness in Table 3.7.5, and overall lengths in Table 3.7.6.

TABLE 3.7.5 Tolerance Range for Collar Thickness, in

Body size	Economical	Normal	Expensive
0.030–0.062	±0.004	±0.0025	±0.0015
0.063–0.125	±0.005	±0.003	±0.002
0.126–0.250	±0.006	±0.0035	±0.0025
0.251–0.312	±0.007	±0.004	±0.003
0.313–0.500	±0.008	±0.0045	±0.0035

Source: Elco Industries, Inc.

TABLE 3.7.6 Length Tolerances for Headed Parts

For lengths, in	Tolerance range, in
For diameters up to and including ¾ in	
Up to 1	1/32
Over 1 to 2	1/16
Over 2 to 6	3/32
Over 6	3/16
For diameters over ¾ in	
Up to 1	1/16
Over 1 to 2	1/8
Over 2	3/16

Source: Industrial Fasteners Institute.

Other Dimensions

As wide a tolerance as possible should be allowed on lugs, fins, tapers and projections because metal-flow behavior is sometimes unpredictable in forming these shapes. Radii generally can be formed to tolerances of ±0.13 mm (0.005 in).

Geometry

Concentricity variations of 0.025 to 0.08 mm (0.001 to 0.003 in) TIR are not uncommon in blank sizes up to 6 mm (¼ in) in diameter if the difference between the two referenced diameters is not too great. Greater concentricity tolerance is generally needed for larger-size parts and larger differences in diameters. If closer tolerances are needed, secondary operations such as machining, shaving, or grinding are generally required.

Straightness of cold-upset parts can generally be held to about 0.4 percent of length.

Surface Finishes

A finish of 0.4 to 0.8 μm (16 to 32 μin) usually can be obtained economically. Upsetting of heads, however, sometimes can decrease the quality of surface finish unless high pressures are exerted on the sidewalls of the containing die. High-quality finishes generally also require the use of high-quality clean wire.

Surface finish also can vary among different parts or different areas of the same part, depending on the surface of the wire or bar being headed, the amount of cold working performed in various areas, the lubricant used, and the condition of the tools. The best finish on any part is usually achieved when direct contact has been made with the tools.

CHAPTER 3.8
IMPACT- OR COLD-EXTRUDED PARTS

John L. Everhart, P.E.
Metallurgical Engineer
Westfield, New Jersey

THE PROCESS

The process is generally called *impact extrusion* by the producers of nonferrous parts and *cold extrusion* by those producing ferrous parts, but no distinction will be made between these terms in this text. Impact extrusion is a chipless method that involves forcing a metal to flow through an orifice by compressive stressing and thereby forming a part of the desired shape. The operation is shown schematically in Figs. 3.8.1 and 3.8.2. Figure 3.8.1 represents backward extrusion; Fig. 3.8.2, forward extrusion. For the production of certain parts, combined backward and forward extrusion is used.

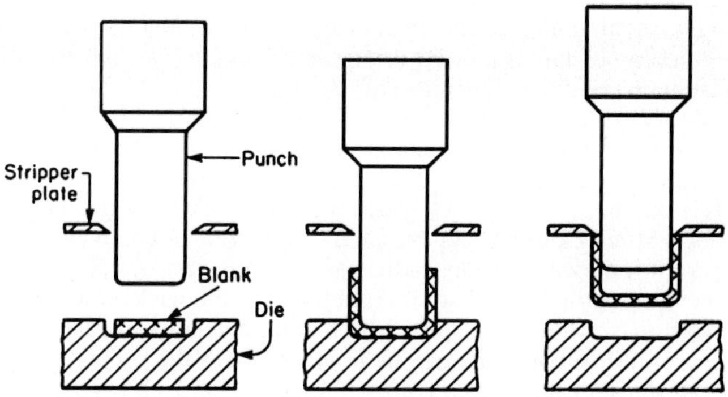

FIGURE 3.8.1 Backward extrusion. Blank in die chamber (*left*); extrusion of shell (*center*); stripping shell (*right*).

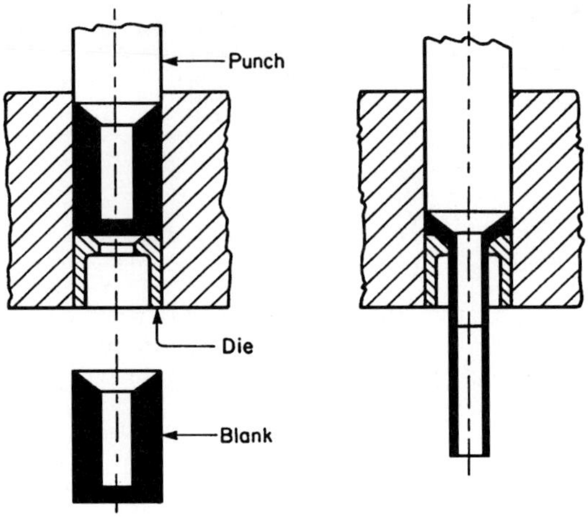

FIGURE 3.8.2 Forward extrusion. Blank (*lower left*); blank in die chamber (*upper left*); extrusion of shell (*right*).

Backward Extrusion

In backward extrusion, a blank of metal to be extruded is placed in the cavity of a die and compressed by a punch. The metal is forced to flow through the annular opening between punch and die, and it rises along the punch. The thickness of the base is determined by the press stroke and is independent of the wall thickness.

Forward Extrusion

In forward extrusion, a blank is placed in the cavity of a die having a restriction. The punch fits so closely that metal cannot flow between it and the die walls. As the punch descends, it compresses the blank and causes it to extrude through the orifice formed by the projection on the die and the nose of the punch.

Blanks for forward extrusion are generally preformed to the contour necessary to obtain the desired configuration of the extruded part. Forward extrusion requires less power than backward extrusion for the production of parts of equivalent size, and smaller presses can be used. Because part length is practically independent of punch length, a short, stocky punch can be used. However, secondary tooling may be required to permit removal of the extruded part from the die, and this requirement will increase the cost of tooling over that required for a simple backward extrusion.

With forward extrusion, the shape of the blank plays a role in shaping the part. In some cases, several different parts can be produced with the same set of tools merely by changing the shape of the blank.

Combined Extrusion

Backward and forward extrusion can be used simultaneously to form parts. This procedure is followed frequently.

TYPICAL CHARACTERISTICS OF IMPACT-EXTRUDED PARTS

Parts produced by impact extrusion often require only trimming to length after forming. In other cases, operations of a secondary nature are reduced from those required with several competing processes. In addition, certain components that cannot be made by other methods can be formed.

The process can be applied to produce parts from a wide variety of metallic materials. In commercial production are parts made from aluminum, copper, lead, magnesium, tin, zinc, and their alloys and from various steels.

The use of impact extrusion should be considered when

1. The part length is greater than 2 or 3 times the diameter.
2. The part is hollow, with one end closed or partially closed.
3. The part would require considerable machining if produced by another method.
4. The base must be thicker than the sidewalls.
5. Zero draft is desired in the sidewalls.
6. Redesign will permit use of a one-piece part instead of an assembly.
7. Large-quantity production is planned.

Impact extrusion should probably not be considered if

1. The part can be readily made on a screw machine.
2. Only a few parts are required.
3. The part is asymmetrical, with a length greater than 3 to 4 times the diameter.
4. The bottom must be flat from edge to edge.

Backward extrusion can be used to produce a wide variety of shells with circular, oval, square, rectangular, and other polygonal cross sections. In general, the sidewalls must be perpendicular to the base to permit flow during extrusion, but vertical ribs can be formed on the internal and external walls. Multiwall shells also can be formed.

Advantages of backward extrusion include simplicity of tooling and ease of removal of the shell after forming. However, the process has limitations. The punch must be long enough to accommodate the entire length of the shell, plus an allowance for trimming, and a long-stroke press is frequently required. There is therefore a limit to the length of part that can be formed by this procedure. Further, in the extrusion of long shells the punch has a tendency to float, and this tendency is increased by play in the moving head of the press. As a result, there is some difficulty in maintaining uniform wall thicknesses in long extrusions.

Forward extrusion can be used to produce open- or closed-end tubes. The bottom of a closed-end tube can be of the same thickness as the sidewalls or heavier. Sidewalls can be parallel or tapered, and parts can be produced with fillets. (*Note:* Forward extrusion in this chapter applies to the production of discrete pairs that normally have one end closed. Chapter 3.1, "Metal Extrusions," covers the production of long pieces of constant cross section made with a closely related process.)

Since a true alignment of punch and die is ensured, tolerances can be held more closely with forward extrusion than is possible with backward extrusion of long shells. However, the shapes that can be produced by forward extrusion are more limited than those which can be formed by backward extrusion. In particular, the base of the forward-extruded part must be plain because there is no forging action against the die bottom. In general, the forward-extruded parts produced are solid rods or tubular forms. The tubes can be open or closed and have parallel or tapered walls, and the wall thickness of the tapered end can be varied.

Figure 3.8.3 illustrates typical impact-extruded parts.

FIGURE 3.8.3 A variety of typical impact-extruded parts.

Combined Extrusions

Most of the shapes produced by backward extrusion can have forward-extruded rods or tubes formed at the same time. (See Fig. 3.8.4.) The two sections of such parts can have the same or different cross-sectional areas and different wall thicknesses that can have the same or different shapes. A cup having a square cross section can be backward-extruded at the same time as a tube having a circular cross section is being forward-extruded. Tubes can be extruded in both directions from a flanged portion. (See Fig. 3.8.5.) Combined extrusion often can be used to produce a part that can replace an assembly.

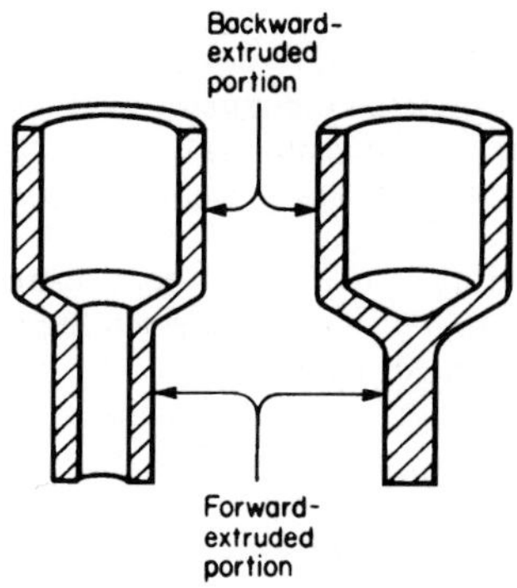

FIGURE 3.8.4 Combined extrusions.

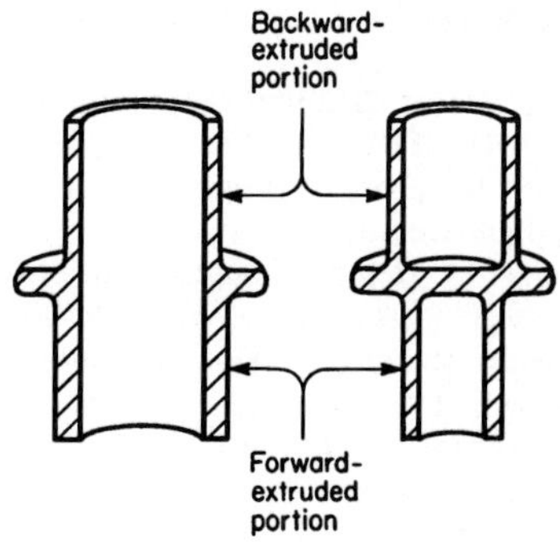

FIGURE 3.8.5 Combined extrusions from a flange.

ECONOMIC PRODUCTION QUANTITIES

Generally, impact extrusion is a mass-production operation. For parts of small size, 1000 pieces per month might be uneconomical, while 10,000 or more pieces would be desirable. With larger parts, 1000 pieces per month might be sufficient for economical operations. Going to the other extreme, a producer of aluminum parts states that 50,000 is an economical lot size for automatic production. The same producer, however, notes that if tooling is available, any lot size is practical. For steel, suggested minimum economic production quantities are

10,000 for parts weighing 1 to 2 g (0.035 to 0.07 oz)
5000 for parts weighing 2 to 500 g (0.07 oz to 1.1 lb)
3000 for parts weighing over 500 g (1.1 lb)

SUITABLE MATERIALS

Components are produced commercially from aluminum, copper, lead, magnesium, tin, zinc, and their alloys and from various steels. Certain less common materials (e.g., zirconium alloys for nuclear applications) are employed for specialty products.

In the nonferrous-alloy systems, the pure metal requires a lower extrusion pressure than the alloys, and therefore, greater reductions of area can be made with the pure metals than with the alloys. For example, the practical minimum wall thickness for backward extrusion of a 1-in-diameter shell is 0.010 in using commercial aluminum and 0.040 in for the high-strength 7075 (A97075) alloy. In the steels, the carbon steels permit lower extrusion pressures than the alloy steels, but structure also plays a significant role.

Aluminum

The selection of a particular alloy depends on the properties required. Alloys such as 1100 (A91100) and 3003 (A93003), which are strain-hardenable and cannot be heat-treated for property improvement, are used to meet moderate-strength requirements. The heat-treatable alloys 6061 (A96061), 6066 (A96066), and 6070 (A96070) are used when greater strength or improved machinability is required. The heat-treatable alloys 2014 (A92014), 7001 (A97001), and 7075 (A97075) are used when the highest possible strengths obtainable in aluminum alloys are required.

Copper

Because of its excellent ductility and high electrical conductivity, oxygen-free copper is the most widely used of the copper-base materials, but ETP copper and zirconium copper are also used for components requiring high electrical or thermal conductivity. Brasses, nickel silvers, and copper-nickel alloys are also used for impact-extruded parts, their selection being based usually on strength requirements and corrosion resistance. These materials are all strain-hardenable, with the exception of zirconium copper, and cannot be improved by heat treatment.

Lead and Tin

These materials are considered together because their major field of application is specialized. They are used mainly for collapsible tubes. Lead and tin are the most readily impact-extruded metals. Unalloyed lead and tin, lead-antimony to 2 percent antimony, and tin-copper to 1 percent copper are used.

Magnesium

The impact extrusion of magnesium differs from that of aluminum in one respect. Because the structure of magnesium is such that it cannot be worked readily in the cold, the blanks must be warmed. Generally, temperatures in the range of 230 to 370°C (450 to 700°F) are used. As a result, the parts are not strain-hardened. Typical alloys used for impact-extruded parts are AZ31B (M11311), AZ61A (M11611), AZ80A (M11800), and ZK60A (M16600).

Zinc

Pure and commercial zincs have been impact-extruded for quite a few years, and a few alloys such as zinc-cadmium (0.6 percent Cd) and zinc-copper (up to 1 percent Cu) are also employed. Generally, the blanks are warmed, and there is no strain hardening as the result of extrusion.

Steels

Low-carbon steels can be extruded more readily than steels of higher carbon content, and the alloying elements generally must be held to relatively low values, although some high-alloy steels have been extruded successfully. Low-carbon and some low-alloy steels can be processed in the stress-relieved normalized condition for some shapes. The higher-carbon and steel-alloy steels require complete spheroidization for successful cold extrusion, particularly for backward extrusion. Recently, boron-containing steels have been developed for cold-forming operations. They do not increase the extrusion pressure, and they permit the use of relatively low-carbon steels to achieve properties that can be met with non-boron-containing steels only at higher carbon contents.

Steels being used commercially for cold extrusion include

Carbon steels 1006–1050 (GI0060–G10500)
Manganese steels 1330–1345 (GI3300–G13450)
Molybdenum steels 4012–4047 (G40120–G40470)
Chromium-molybdenum steels 4118–4140 (G41180–G41400)
Manganese-molybdenum steels 4615–4626 (G46150–G46260)
Chromium steels 5115–5140 (G51150–G51400)
Nickel-chromium-molybdenum steels 8615–8630 (G86150–G86300)

DESIGN RECOMMENDATIONS FOR BACKWARD EXTRUSION

Symmetry

To minimize tooling deflection, it is desirable to aim for symmetry in the design of parts unless they are quite short in relation to their diameter. Figure 3.8.6 shows examples of good and poor design. Parts on the right can be produced with close tolerances and with a minimum of lateral movement of the punch. In producing parts such as those shown on the left, the forces tend to be off center and lateral movement can occur. One way to minimize floating of the punch is to include a small recess in the part design (see Fig. 3.8.7); the point on the punch required to produce this recess aids in maintaining alignment.

Flow

In backward extrusion, the metal is forced to flow through an orifice to form the sidewalls. Parts must therefore be designed to prevent restrictions to flow:

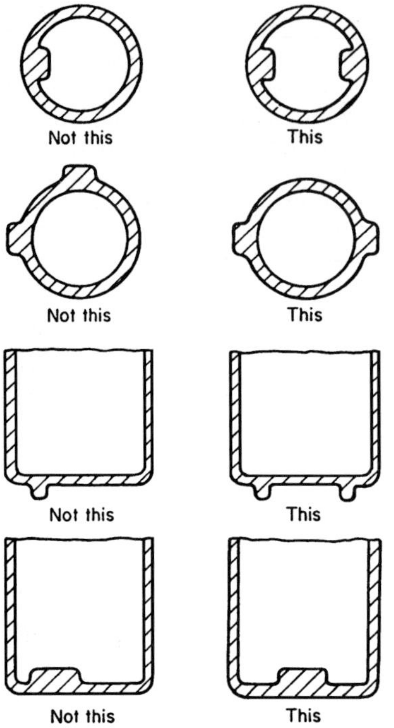

FIGURE 3.8.6 Impact-extruded parts should be symmetrical to avoid lateral movement of tooling punches and unequal wall thickness.

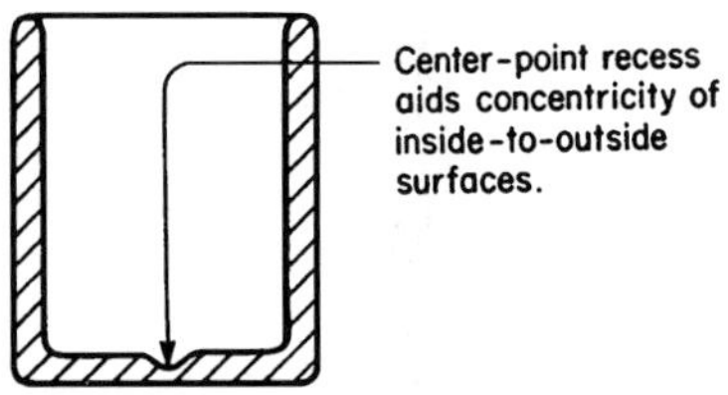

FIGURE 3.8.7 A small recess in the bottom of the part minimizes floating of the punch and thereby helps ensure uniform wall thickness.

1. Assist flow by incorporating suitable inclined bottom surfaces near the sidewalls and by rounding the outside corners of the part as shown in Figs. 3.8.8*b* and 3.8.9. Note that the inside bottom radius should be as small as possible.
2. To ensure minimum restriction to flow, whenever possible, use a design that includes a bottom angle approaching 120°. (See Figs. 3.8.8*d* and *e*.)

Bottom Design

The bottom of a backward-extruded part can be thicker than the walls or of equal thickness. However, a thicker bottom can aid in metal flow, and some authorities recommend that the bottom-wall thickness should be at least 15 percent greater than that of the sidewalls. (See Fig, 3.8.9.) The thickest area should be at the base of the sidewalls, but it can be tapered to be thinner at the center.

The inside bottom surface of a part should not be completely flat but can be flat for approximately 80 percent of the inside diameter. (See Fig. 3.8.9.) Spherical bottoms should be avoided, especially in long shapes; use an angular bottom instead, as shown in Fig. 3.8.8*d*. Bottom cross sections can be circular, oval, square, rectangular, or of any other regular polygonal shape. They also can be irregular provided they are symmetrical. However, circular cross sections are preferable. When a nozzlelike projec-

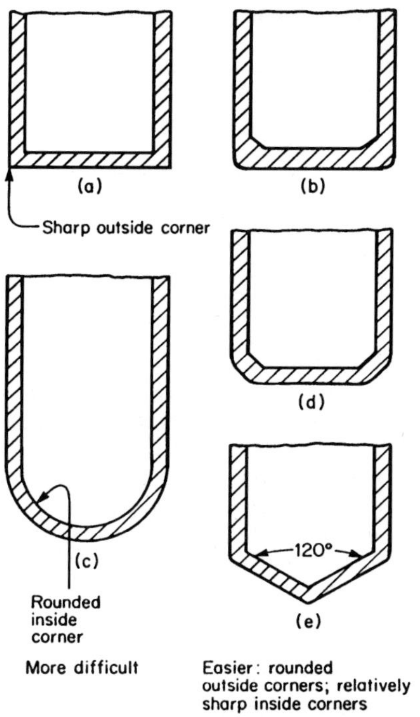

FIGURE 3.8.8 Inclined bottom surfaces and rounded outside corners assist metal flow in backward extrusion.

tion is required on the bottom, ample taper should be allowed on both inside and outside. This strengthens the punch extension that forms the hole inside the nozzle and helps strip the extruded part from the punch.

Length-Diameter Ratio

The length of parts that can be formed by backward extrusion is limited by the column strength of the punch and also depends on the material being extruded. For commercial operations in general, the length of an aluminum part should not be greater than 6 to 10 times the diameter; for steel parts, the length should be more limited, probably 3 to 4 times the diameter. (See Table 3.8.1.) The greater the length, the greater is the tendency of the punch to float and consequently the greater the variation in wall thickness to be expected.

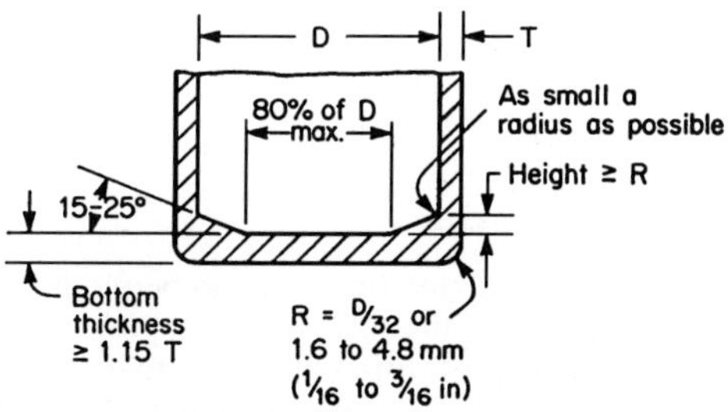

FIGURE 3.8.9 Details of the recommended bottom design for backward extrusion.

Draft

No draft is required on the walls of backward extrusions.

TABLE 3.8.1 Practical Length-Diameter Ratios for Backward Extrusion

Aluminum and softer aluminum alloys	8:1 to 10:1
Harder aluminum alloys	5:1 to 7:1
Copper and copper alloys	3:1 to 5:1
Magnesium and magnesium alloys	6:1 to 9:1
Zinc and zinc alloys	4:1 to 6:1
Carbon steels	3:1 to 4:1
Low-alloy steels	2:1 to 3:1

Ribs

Longitudinal ribs can be produced on either the outside or the inside of the sidewalls. External ribs need not be spaced equally but should be arranged in a symmetrical design.

External ribs can have heights up to 2 times the wall thickness, but 1 times wall thickness is preferable. (See Fig. 3.8.10.) Internal ribs can run the full length of the shell or be designed to terminate at a desired distance from the closed end. The maximum recommended height of internal ribs is 3 times wall thickness. (See Fig. 3.8.11.)

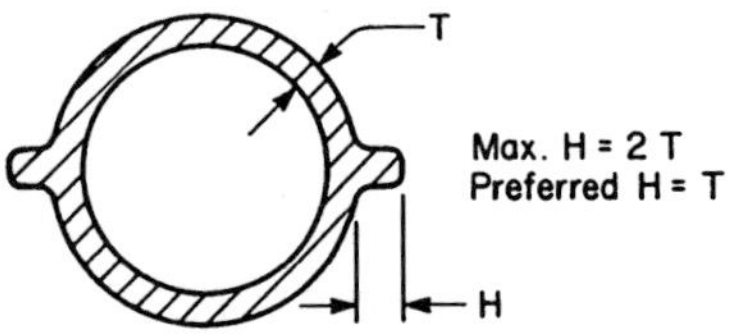

FIGURE 3.8.10 The maximum recommended height of external ribs is 2 times wall thickness, but 1 times wall thickness is preferred.

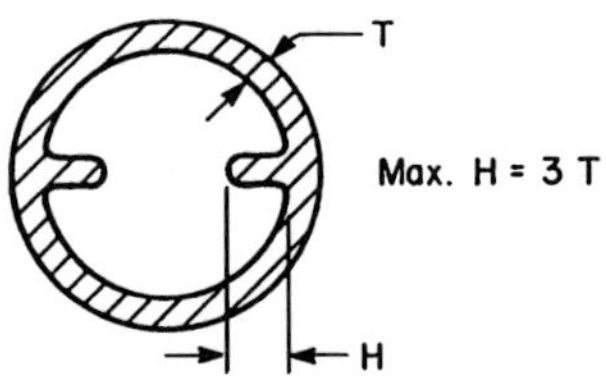

FIGURE 3.8.11 The maximum recommended height of internal ribs is 3 times wall thickness.

Bosses

External and internal bosses in the axial direction can be produced as integral parts of the design. They can be produced on the same shell, and they can be of the same or of different diameters and lengths and of different cross sections. External bosses can be produced in any size up to the diameter of the shell and can be solid or hollow. Solid bosses can be of almost any symmetrical shape. Internal bosses can be longer than the sidewalls. Cross sections can be round, oval, square, or of other regular configurations.

The diameter of an inside boss should be not more than one-fourth the diameter of the shell. The preferred location for bosses is on the axis of a part. If not coaxial, bosses should be in a symmetrical arrangement; otherwise, metal will tend to flow unevenly. Avoid sharp edges and corners on bosses, for they make it difficult for the metal to flow properly.

TABLE 3.8.2 Practical Minimum Wall Thicknesses for Backward-Extruded Aluminum Parts

Outside diameter, mm (in)	Alloy and wall thickness, mm (in)			
	1100 (A91100)	6061 (A96061)	2014 (A92014)	7075 (A97075)
25 (1)	0.25 (0.010)	0.38 (0.015)	0.90 (0.035)	1.0 (0.040)
50 (2)	0.50 (0.020)	0.76(0.030)	1.8 (0.070)	2.0 (0.080)
75 (3)	0.76 (0.030)	1.1 (0.045)	2.7 (0.105)	3.0 (0.120)
100 (4)	1.0 (0.040)	1.5 (0.060)	3.6 (0.140)	4.0 (0.160)
125 (5)	1.3 (0.050)	1.9 (0.075)	4.4 (0.175)	5.0 (0.200)
150 (6)	1.5 (0.060)	2.3 (0.090)	5.3 (0.210)	6.0 (0.240)
175 (7)	1.9 (0.075)	2.8 (0.110)	6.2 (0.245)	7.0 (0.280)
200 (8)	2.5 (0.100)	3.3 (0.130)	7.0 (0.280)	8.0 (0.320)
225 (9)	2.8 (0.110)	3.7 (0.145)	8.0 (0.320)	9.0 (0.360)
250 (10)	3.2 (0.125)	4.2 (0.165)	8.9 (0.350)	10.0 (0.400)

Source: From *Impacts Design Manual,* Aluminum Association, Inc., New York.

Wall Thickness

The minimum wall thickness that can be produced commercially varies with the diameter of the part and the material being extruded. Table 3.8.2 gives the minimum wall thickness for aluminum-based materials.

Threads and Undercuts

Threads and undercuts must be produced in secondary operations.

Double-Wall Shells

Double-wall shells can be produced without difficulty by incorporating a wide-diameter center tube. This center tube can be longer than the sidewalls and be open or closed at the top or bottom. Multiwall extruded parts also can be produced.

Asymmetrical Parts

Although it is preferable to design symmetrical parts, especially for the production of parts having large ratios of length to diameter, asymmetrical parts can be produced if the height is not too great. To produce such parts, however, it is necessary to secure a uniform flow of metal to all edges of the die by adjusting the tool angles. This process is strictly empirical and may be quite time-consuming.

Secondary Operations

Many parts produced by backward extrusion require only trimming to length. However, certain operations that may be required to finish a part cannot be performed during extrusion. These include heading, flanging, flaring the open end, ironing, drilling, threading, and forming undercuts. If possible, designs requiring these operations should be avoided.

DESIGN RECOMMENDATIONS FOR FORWARD EXTRUSION

Symmetry

Symmetrical shapes are preferred for forward extrusion.

Length-Diameter Ratio

The length of a part produced by forward extrusion is limited only by the size of the blank that can be handled and the ease of removal of the part from the die. Forward extrusions can be made with length-diameter ratios of 100:1 or greater.

Draft

It is possible to design for zero draft on the sidewalls.

Flanges

Flanges can be formed during forward extrusion. They can be heavy, stepped, or formed in a variety of shapes. The following suggestions will aid in attaining the easiest flow of the metal during extrusion:

1. The bottom corners of the part can be square and do not require a fillet where bottom and sidewall join. There should be a fillet, however, where the extrusion and the flange join. (See Fig. 3.8.12.)

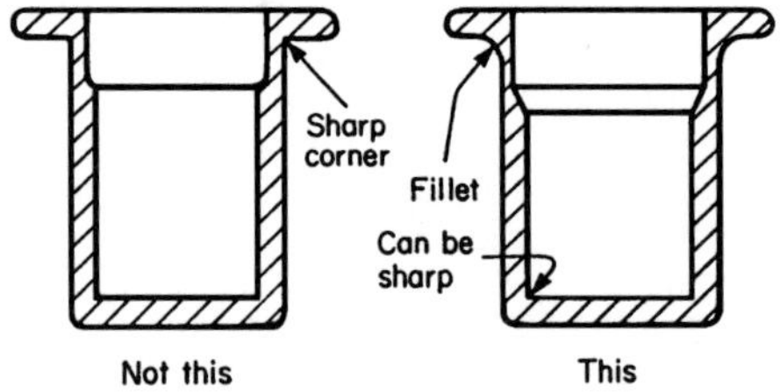

FIGURE 3.8.12 Forward extrusions should have a radius at the junction of a flange and the sidewall. Bottom corners, however, can be sharp.

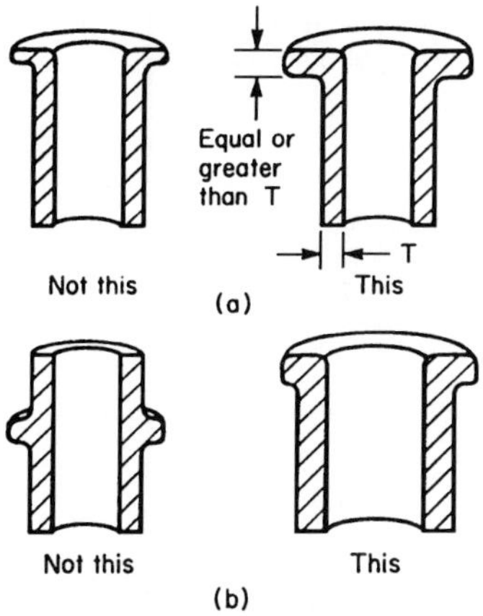

FIGURE 3.8.13 Flanges should be at least as thick as the sidewall (*a*) and be located at or near the ends of the forward extrusion (*b*).

2. Flanges should be symmetrical.
3. They should have thicknesses equal to or greater than the sidewall. (See Fig. 3.8.13*a*.)
4. They should be placed at the end of the part or close to the end. (See Fig. 3.8.13*b*.)
5. Diameters of flanges should be as small as possible.

Threads and Undercuts

As with backward extrusion, threads and undercuts must be produced by secondary operations.

DIMENSIONAL FACTORS AND TOLERANCES

Tolerances attainable in backward extrusion depend on such factors as tool wear, regularity of metal flow, size and shape of the part, wall thickness, and material being extruded. When thin-walled tubes are extruded, for example, tool wear will cause increases in the wall thickness of the shell and a corresponding reduction in its length. The thickness on the base is influenced by the lengthening of the punch by thermal expansion, and allowance must be made for this effect in setting tolerance limits. Tolerances attainable commercially also vary with the alloy.

Aluminum

Aluminum has been backward-extruded for many years, and tolerances that can be attained have been quite well established. Representative tolerances for round, square, and rectangular shells are indicated in Table 3.8.3.

Copper

Tolerances that can be maintained on small parts are

Wall thickness: ±0.08 mm (0.003 in)

Diameter: ±0.08 mm (0.003 in)

TABLE 3.8.3 Typical Tolerances for Regular Shapes in Aluminum Alloys in Backward Extrusion

			Tolerances, ±mm (in)			
Outside dimensions, mm (in)	Wall thickness, mm (in)	Bottom thickness, mm (in)	Outside diameter	Inside diameter	Bottom	Length
Round shells						
6 (¼)	0.25–0.50 (0.010–0.020)	0.50–0.76 (0.020–0.030)	0.13 (0.005)	0.15 (0.006)	0.18 (0.007)	0.8 (1/32)
25 (1)	0.38–0.63 (0.015–0.025)	0.63–0.88 (0.025–0.035)	0.13 (0.005)	0.15 (0.006)	0.25 (0.010)	0.8 (1/32)
50 (2)	0.50–0.76 (0.020–0.030)	0.81–1.1 (0.032–0.042)	0.18 (0.007)	0.20 (0.008)	0.30 (0.012)	0.8 (1/32)
75 (3)	0.63–1.3 (0.025–0.050)	1.0–1.7 (0.040–0.065)	0.20 (0.008)	0.25 (0.010)	0.38 (0.015)	0.8 (1/32)
Square shells			Side	Side		
44×44 (1¾ × 1¾)	0.63 (0.025)	0.88 (0.035)	0.2 (0.008)	0.25 (0.010)	0.38 (0.015)	0.8 (1/32)
50×50 (2 × 2)	0.63 (0.025)	0.88 (0.035)	0.2 (0.008)	0.25 (0.010)	0.38 (0.015)	0.8 (1/32)
56×56 (2¼ × 2¼)	1.6 (0.062)	1.9 (0.075)	0.2 (0.008)	0.25 (0.010)	0.38 (0.015)	0.8 (1/32)
Rectangular shells			Side	Side		
50×25 (2 ×1)	0.63 (0.025)	0.88 (0.035)	0.2 (0.008)	0.25 (0.010)	0.38 (0.015)	0.8 (1/32)
64×56 (2½ × 2¼)	0.78 (0.025)	1.3 (0.050)	0.2 (0.008)	0.25 (0.010)	0.38 (0.015)	0.8 (1/32)

TABLE 3.8.4 Typical Tolerances for Backward-Extruded Magnesium Parts

Dimension	Tolerances, ±
Diameter	0.2% of diameter
Bottom thickness	0.13 mm (0.005 in) for all sizes
Wall thickness, mm (in)	
0.50–0.75 (0.020–0.029)	0.05 mm (0.002 in)
0.76–1.12 (0.030–0.044)	0.08 mm (0.003 in)
1.13–1.50 (0.045–0.059)	0.1 mm (0.004 in)
1.51–2.50 (0.060–0.100)	0.13 mm (0.005 in)

Magnesium

Typical tolerances attainable in commercial production are given in Table 3.8.4.

Zinc

Tolerances that can be held on small parts are

Inside diameter: +0.00, −0.15 mm (+0.000, −0.006 in)

Outside diameter: ±0.08 to ±0.13 mm (±0.003 to ±0.005 in)

Wall thickness, 0.25 to 0.5 mm thick: ±0.08 mm (0.010 to 0.020 in thick: ±0.003 in)

Steels

Tolerances that can be met in commercial production are

Backward Extrusion

	Outside diameter, mm (in)	Inside diameter, mm (in)	Length, mm (in)
Small parts	±0.025 (0.001)	±0.1 (0.004)	±3 (0.12)
Large parts	± 0.4 (0.016)	±0.35 (0.014)	±6 (0.24)
Deviation of outside diameter from inside diameter: 0.5–1.2% of nominal OD			
Out-of-roundness: 0.2 to 0.6% of nominal OD			

Forward-Extruded Solid Parts

	Diameter, mm (in)
Small parts	±0.05 (0.002)
Large parts	± 0.13 (0.005)

Surface Finish

The surface finish that can be attained on a backward-extruded part depends on the condition of the tools and the alloy being extruded. If tools are properly maintained, finishes as low as 0.8 μm (32 μ in) rms can be achieved. Finishes that are attained commercially on several materials are:

Aluminum	
1100	0.8–1.6 μm (32–63 μin)
6001	1.6 μm (63–250 μin)
7075	3.1–6.3 μm (125–250 μin)
Copper	1.6 μm (63 μin)
Steels	1.5–5.0 μm (60–200 μin)

As in backward extrusion, the surface finish attainable in forward extrusion depends on the condition of the tools. Surface finishes of 1 to 5 μm (40 to 200 μin) have been attained on aluminum in commercial operations; 1.5 to 5 μm (60 to 200 μin), on steel.

CHAPTER 3.9

ROTARY-SWAGED PARTS

THE PROCESS

Rotary swaging is a process primarily applicable to cylindrical parts made from rods, tubes, and wires. It alters the diameter or shape of such components by means of a large number of controlled-impact blows applied radially by one or more pairs of opposed dies. The dies are shaped to give the part the required form.

In the common rotary-swaging machine, as shown schematically in Fig. 3.9.1, the spindle rotates, throwing dies and hammers outward away from the work and against a roller cage that surrounds the spindle. Each time that the hammer blocks pass directly against opposing rollers, they are driven inward. This forces the die sections to close and strike the workpiece. As the hammer blocks pass out from in between opposing rollers, the dies are thrown open and are ready for the next forming stroke. For most machines, there will be upwards of 1000 such die strokes per minute.

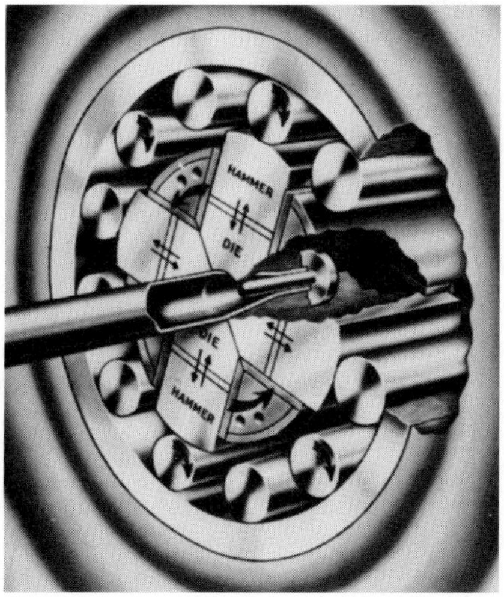

FIGURE 3.9.1 Die motions in a conventional rotary-swaging machine. (*Courtesy Fenn Mfg. Co.*)

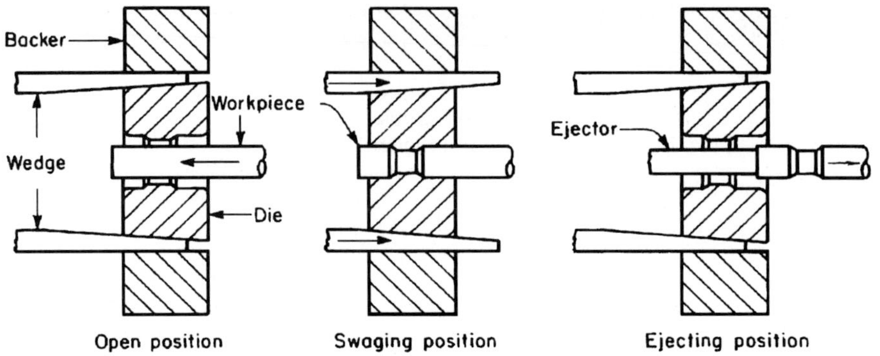

FIGURE 3.9.2 Key elements of a die-closing swaging machine. (*From* Metals Handbook, *American Society for Metals.*)

A variation of the rotary-swaging process is *stationary-die swaging.* In this process the spindle remains stationary, but the roller cage rotates. The advantage of the stationary-die process is that it permits nonround cross sections to be produced.

Some swagers are equipped with wedged-shaped members between the hammer and the die blocks. When these wedges are retracted, there is a larger opening to retract the workpiece. Hence reductions in diameter are possible at locations other than the end of the workpiece, which then can still be withdrawn from the machine. Figure 3.9.2 illustrates the operation of this machine, which is called a *die-closing swager.*

Owing to the resistance to the entry of work into the die (because of the taper of the die opening), for some work it is necessary to have a mechanical feed or a mechanically assisted feed for the workpiece. This may consist of a rack-and-pinion, roll-feed, or hydraulic system.

Mandrels are often used to support tubing during swaging, particularly if the tubing has thin walls.

Flat swaging is a punch-press or header operation related only in name to rotary swaging. It involves a reduction in thickness of all or a portion of a sheet-metal workpiece.

TYPICAL PARTS AND APPLICATIONS

Swaging invariably involves a workpiece of rodlike or tubular shape. Most swaged parts have a round cross section. The swaging operation reduces the outside diameter of the work and may change the outside shape (or the inside shape in the case of tubing).

The size range of parts processible by rotary swaging extends from a minimum practicable finished diameter of about 0.5 mm (0.020 in) to a normal maximum starting diameter for tubing of 150 mm (6 in). The maximum diameter processible, however, is a function of the size of the swaging machine and theoretically has no limit if a large enough machine were constructed. Machines have been built to handle tubing with a 350-mm (14-in) diameter, but tonnage requirements limit the maximum diameter of solid parts swaged to much smaller sizes.

For fine tubing swaged over a mandrel, flexible wire is used as a mandrel, and music wire as fine as 0.08 mm (0.003 in) in diameter has been used.

Through-feed tooling permits the processing of indefinitely long workpieces. Long lengths of straight or coiled tubing or wire are reduced in diameter by rotary swaging.

The following parts have been produced at least in part by rotary swaging: (1) Pointed tubing and bar ends prepared for cold drawing. (2) Tapered tubing and solid bars for such items as furniture legs, golf clubs, and fishing rods. (3) A reduction in diameter of a tube, bar, or wire for its entire length. (4) Pins, punches, needles (including crochet and sewing machine needles), buttonhooks, bicycle spokes, dental-tool handles, and other parts having tapers or changes in diameter. (5) Automotive parts such as torque tubes, steering posts, forged axle housings, drag links, drive shafts, and exhaust pipes. (6) Phillips and conventional screwdriver blades, soldering-iron tips, square ends of round bars, and other irregular shapes producible on stationary-die machines. (7) Tubular parts of irregular cross section made by swaging tubing over a shaped mandrel. Examples are wrench sockets, rifle barrels, internal rachets, and splines. (8) The assembly of fittings to cable, wire, and hose. (9) Parts that have diameter reductions between shoulders. Convoluted flexible tubing is one example.

An additional application is the diameter or wall reduction of items too fragile for the application of other processes such as cold drawing. An example is the size reduction of soft solder wire. Another is the final sizing of thin-walled lipstick cases.

Figure 3.9.3 shows a selection of parts and assemblies produced by rotary swaging.

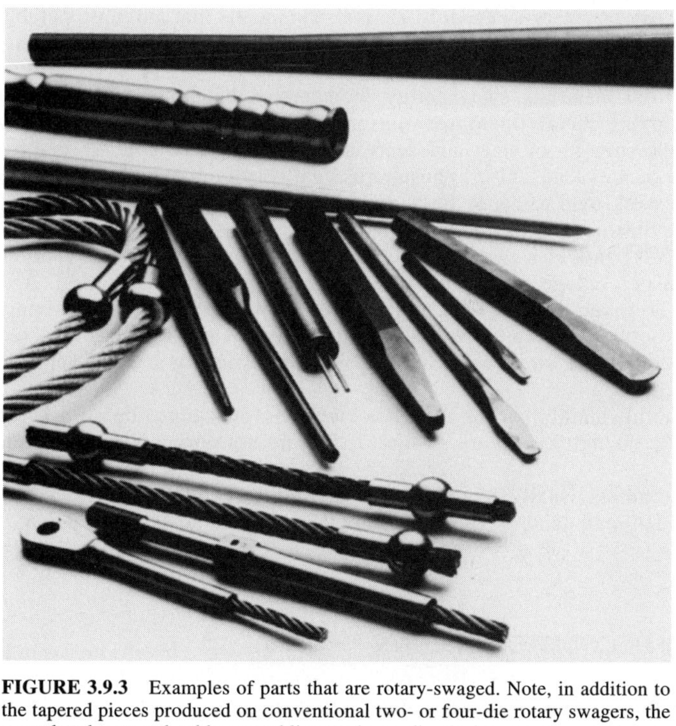

FIGURE 3.9.3 Examples of parts that are rotary-swaged. Note, in addition to the tapered pieces produced on conventional two- or four-die rotary swagers, the convoluted part and cable assemblies made on die-closing machines and parts with rectangular cross sections made on stationary die machines. (*Courtesy Fenn Mfg. Co.*)

ECONOMIC PRODUCTION QUANTITIES

Rotary swaging is a process suitable for moderate to high levels of production. Its advantages are low tooling and only moderate equipment cost, fairly high output rates, and little or no materials loss.

Simple two-piece dies for smaller parts are low in cost. Four-piece dies and dies for long tapers, large diameters, or more complex forms can run much higher. Mandrels are similarly reasonably priced except when splines, flutes, or other intricate shapes are involved.

The low tooling cost and fast setup (usually less than an hour) make swaging economical for production runs as small as the low hundreds of pieces. The high output rates (400 pieces per hour is an easily attained rate for a simple operation) and materials savings often make it also suitable for the highest production applications.

Competitive processes for lower production levels are spinning and machining, and for high production levels, punch-press drawing, cold heading, and impact extrusion.

SUITABLE MATERIALS

Most metals, provided there is a sufficient difference between their yield and ultimate strengths, can be rotary-swaged. In general, the metals that are most suitable for other metal-forming processes are also the most readily swageable. Best results are obtained with low-carbon steels and ductible nonferrous metals.

For ferrous materials, swageability decreases as the carbon content or the percentage of alloying metals increases. Chromium, nickel, tungsten, and manganese all increase the strength of steel and decrease its ability to be cold-worked by swaging. Additives such as lead, sulfur, phosphorus, and beryllium, which increase the machinability of steel, tend to cause fractures during swaging and make the resulting alloy unsuitable for this process.

Table 3.9.1 provides an index of swageability of various metals. The index is the maximum percentage reduction that can be made in one operation. All materials included, however, can be swaged by a greater amount than indicated simply by stress relieving between successive swaging operations. In cases of severe reduction of cross-sectional area, stress relieving may not be sufficient, and full annealing may be required.

In general, metals harder than Rockwell B 100 cannot be cold-swaged; those between R_b 90 and R_b 100 are swageable but are not normally recommended for this operation.

Welded tubing usually does not swage as well as seamless tubing because the metal at the welded seam tends to flow less smoothly than the other material.

DESIGN RECOMMENDATIONS

There are limits to the steepness of taper angle attainable with swaging. With conventional hand-feed workpieces, the maximum included taper angle is 6 to 8°. Larger angles result in severe feeding difficulties. With power feeds, steeper taper angles, up to 14° (included angle), are attainable. Smaller tubular components of softer metals

TABLE 3.9.1 Maximum Reduction in Cross-Sectional Area Obtainable by Swaging

Work metal		
Common designation	UNS designation	Reduction in area, %
Plain carbon steels		
Up to 1020	Up to G10200	60
1020–1050	G10200–G10500	50
1050–10955	G10500–G10950	40
Alloy steels		
0.20% C		50
0.40% C		40
0.60% C		20
High-speed tool steels*		
All grades		20
Stainless steels		
300 Series		50
400 Series		
Low-carbon		40
High-carbon		10
Aluminum alloys		
1100-0	A91100†	70
2024-0	A92024†	20
3003-0	A93003†	70
5050-0	A95050†	70
5052-0	A95052†	70
6061-0	A96061†	70
7075-0	A97075†	15
Other alloys		
Copper alloys‡	Various	60–70
A-286	S66286	60
Inconel X-750	N07750	60
Kovar	K94610	80

*Spheroidize-annealed.

†0-temper (annealed).

‡Annealed.

Source: Adapted from *Metals Handbook,* vol. 4: *Forming,* 8th ed., American Society for Metals, Metals Park, Ohio, 1969.

such as copper can be swaged, in some cases with included angles as high as 30°. For ease of fabrication, however, the taper angle should be specified to be as shallow as possible. (See Fig. 3.9.4.)

It is not possible with conventional swaging machines to have a shoulder steeper than the maximum taper angle permitted by the workpiece and the setup. This is true because any portion of the part of reduced diameter is produced only by feeding the part axially into tapered dies. With die-closing machines, however, the die can feed radially against the work and thereby produce shoulders much closer to the perpendicular. To allow for metal flow and die withdrawal, a perfectly perpendicular wall is not feasible, and the shoulder should have a 15° slope compared with the perpendicular. (See Fig. 3.9.5.)

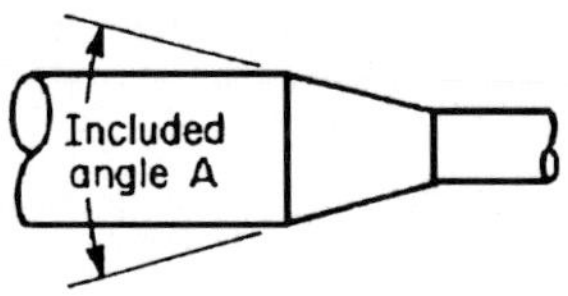

FIGURE 3.9.4 Taper angles of swaged parts should be as shallow as possible. For hand-feed applications (small parts), *A* should not exceed 6 to 8°. For most power-feed applications, *A* should not exceed 14°. For small diameters of copper or aluminum tubing with power feed, *A* can be as high as 30°.

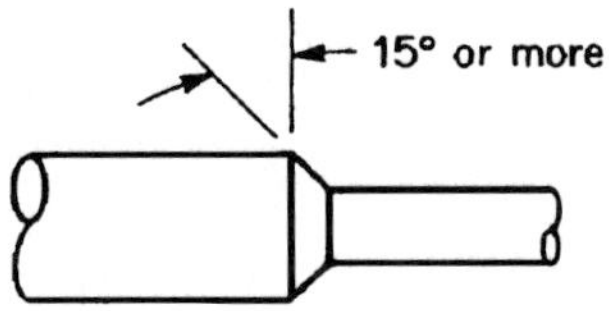

FIGURE 3.9.5 The maximum steepness of shoulders from a die-closing machine is 15° from the perpendicular.

Extreme reductions in cross-sectional area should be avoided because they increase cycle time and die wear, and should the specified reduction exceed that attainable in one pass, stress relieving and an additional swaging operation will be required. See Table 3.9.1 for the normal maximum reduction per pass.

When tubing is swaged without a mandrel, the wall will thicken in proportion to the amount of diameter reduction, and design specifications should allow for this. For example, tubing of 50-mm (2-in) diameter and 1.3-mm (0.050-in) wall thickness, when swaged to 25-mm (1-in) diameter, will show a wall thickness increase to 2.5 mm (0.100 in). There is also some lengthening of the stock, although this is less proportionally unless a mandrel is used.

It is generally preferable, for minimum tool and unit costs, to swage tubing without a mandrel. However, this practice is generally limited to heavier tubing whose diameter is no more than 25 times wall thickness. This limit is generally necessary to avoid wrinkling of stock, although with very low feed rates and 6° taper angles, tubing diameters up to 70 times wall thickness have been swaged without a mandrel.

It is best to limit the length of tapered sections to the maximum that can be swaged in one pass by the available machine. It is possible by successive operations to swage a continuous taper of any length. However, the areas where successive passes overlap are subject to some demarcation no matter how carefully the swaging dies are made. Commercially available long-taper machines will handle 375-mm (15-in) tapers in tubing up to 56 mm ($2\frac{1}{4}$ in) in diameter.

Where a part with a shoulder is intended to be swaged, 6 mm ($\frac{1}{4}$ in) should be allowed between the shoulder and the start of the tapered section for die clearance. (See Fig. 3.9.6.)

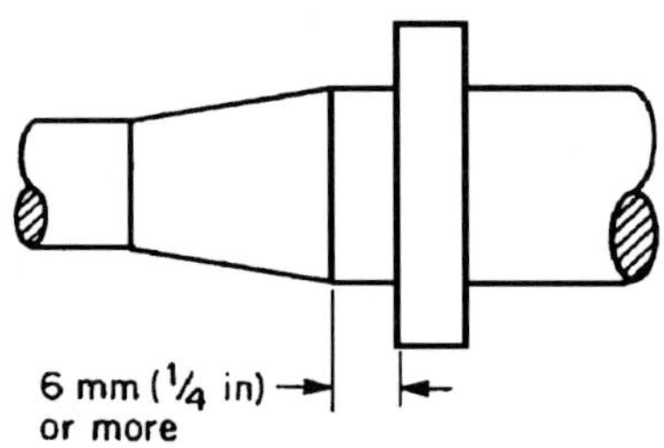

FIGURE 3.9.6 Minimum clearance between shoulders and swaged areas.

Internal splines or other irregular internal shapes require a very slight draft to facilitate removal of the mandrel from the swaged part. Inside-diameter tolerances must allow for this. A common amount of draft is 0.004 in/ft of length.

With die-closing swaging machines, because of die and wedge configurations, the maximum reduction in area per pass is 25 percent. With center reductions, this is the maximum reduction that can be made regardless of the

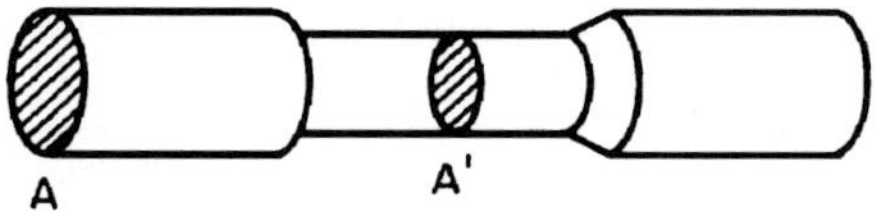

FIGURE 3.9.7 Die-closing swagers are limited by machine and die configurations to a maximum reduction of 25 percent in the cross-sectional area in parts swaged. A = original cross-sectional area; A' = swaged cross-sectional area.

number of passes. Otherwise, the part could not be removed from the die. (See Fig. 3.9.7.)

When swaging with shaped mandrels, the maximum special angle of flutes or grooves from the longitudinal axis is 30°.

DIMENSIONAL FACTORS AND TOLERANCES

The closeness of dimensional control of swaged parts is a function of a number of factors, the most important of which are the accuracy and condition of the dies and the machine and the size of the workpiece. Dimensions vary more at the ends of a swaged area than in the center portion because of the effect of axial flow of material. As in all metal-forming operations, springback is a factor that affects the finished dimensions; more ductile materials are easier to control dimensionally. Dimensions and surface finish are better at low feed rates and lesser area reductions per pass.

In swaging tubular parts, the use of a mandrel is valuable if dimensions must be accurately controlled. This applies also when fine surface finishes are required. If the inside diameter of a swaged tubular part is critical, a mandrel must be used.

Recommended tolerances for diametral dimensions are shown in Table 3.9.2. Surface-finish tolerances are shown in Table 3.9.3.

TABLE 3.9.2 Tolerances for Diameters of Rotary-Swaged Parts*

	Tolerance, mm (in)	
Finished diameter, mm (in)	Normal	Tightest
0 to 3 (0 to ⅛)	±0.05 (0.002)	±0.03 (0.001)
Over 3 to 6 (⅛ to ¼)	±0.08 (0.003)	±0.04 (0.0015)
Over 6 to 12 (¼ to ½)	±0.10 (0.004)	±0.05 (0.002)
Over 12 to 25 (½ to 1)	±0.13 (0.005)	±0.08 (0.003)
Over 25 to 50 (1 to 2)	±0.18 (0.007)	±0.10 (0.004)
Over 50 to 100 (2 to 4)	±0.25 (0.010)	±0.13 (0.005)
Over 100 (4)	±0.38 (0.015)	±0.18 (0.007)

*They also apply to outside diameters and cross-sectional dimensions of shapes swaged with stationary-die machines. They apply to inside diameters of tubular parts if swaged over a mandrel after reaming.

TABLE 3.9.3 Surface Finish of Parts after Rotary Swaging, μm AA (μin)

	After swaging	
Before swaging	Normal	Finest
6.3 (250)	1.6 (63)	0.8 (32)
3.2 (125)	0.8 (32)	0.4 (16)
1.6 (63)	0.4 (16)	0.21 (10)
0.8 (32)	0.2 (10)	0.05 (5)

Note: For aluminum workpieces, apply values in the next higher line.

CHAPTER 3.10
TUBE AND SECTION BENDS

THE PROCESSES AND THEIR APPLICATIONS

Tubing and piping are bent for many reasons. A frequent reason is the need to convey fluids around or to provide clearance from machine or structural elements. Another is to provide for expansion and contraction of tubing systems. Heat-transfer coils and tubular boiler members require bending. Tubular parts are often used as structural components in vehicles and machines and in furniture, railings, handles, etc. Bends in such members (while preserving the walls from collapse) are frequent requirements.

Nontubular parts such as channels, angles, I beams, etc., require the same techniques and have similar structural applications.

The common bending methods are as follows:

Compression Bending

The workpiece is clamped and bent around a stationary form with the aid of a follower block or roller. There is somewhat more compressive force than elongation on the workpiece (although there is still elongation on the outer portion of the bend), and the method's name is derived from this fact. See Fig. 3.10.1 for an illustration of the process.

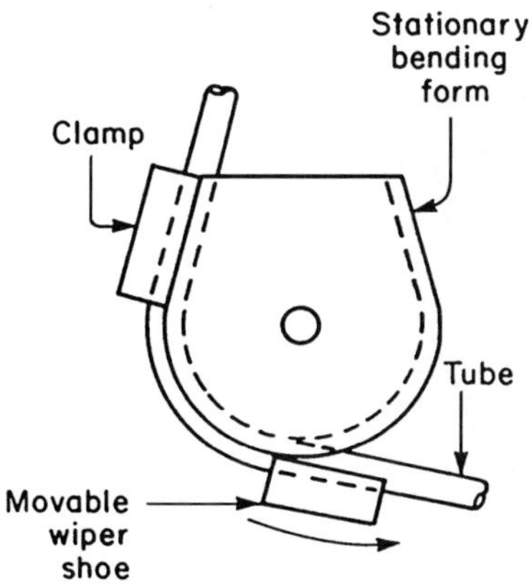

FIGURE 3.10.1 Compression bending. The follower block wipes the tubing around the stationary bending form.

Compression bending is a common method, often performed by hand, on tubing or other sections of heavier wall thickness and larger bending radius. Thin-walled tubing usually is not bent by this method.

The normal minimum centerline radius for compression bends is 4 times the tubing diameter. With thinner-walled tubing with good support, bend radii as small as $2\frac{1}{2}$ times diameter can be used. Bend angles range up to 170° per bend.

Since there is little stretching of the outer surface, plated and painted tubing can be bent with this method.

Draw Bending

In this method, the workpiece is clamped against a bending form as in compression bending, but the bending form rotates, drawing the workpiece through a pressure die and, in many cases, over a mandrel. This method is suitable for thin-walled tubing, especially when bent to a small radius. It provides closer control over the workpiece than other bending methods.

Draw bends are made when dimensional requirements are stringent (e.g., in the aircraft industry) or when tight bends of thin-walled tubing are required. Tubing 12 to 250 mm ($\frac{1}{2}$ to 10 in) in diameter can be processed by draw bending. While bend radii as close as one diameter can be achieved, they require considerable extra care, a close-fitting internal mandrel, and external dies and shoes. Draw bending is more common than compression bending when powered equipment is employed. Bends up to 180° can be made. Ultrasonic vibration, sometimes applied to the tools or workpiece, greatly diminishes the friction between the tooling and workpiece. This enables tubing of difficult-to-bend materials to be processed more easily and bent to a tighter radii.

Ram-and-Press Bending

With this method, a workpiece is placed between two supports, and a rounded-form die (ram or punch) is pressed against it. The two supports pivot as the ram moves forward, maintaining support of the workpiece.

This method, though it provides less control over metal flow, is nonetheless rapid. It is used in production applications on heavy tubing or pipe and rolled or extruded sections whenever some distortion of the workpiece section is permissible and rapid production is important. Figure 3.10.2 illustrates the process. With currently available machines, ram-and-press bending is applied to pipe and tubing from 10 to 350 mm ($\frac{3}{8}$ to 14 in) in diameter. The method is suitable for bends of up to 165°. Extremely heavy sections can be bent.

The minimum centerline bending radius for ram bending is 3 times diameter unless deformation or collapsing of the bent section is permissible (as in some structural applications); 4 to 6 times diameter is a preferred radius.

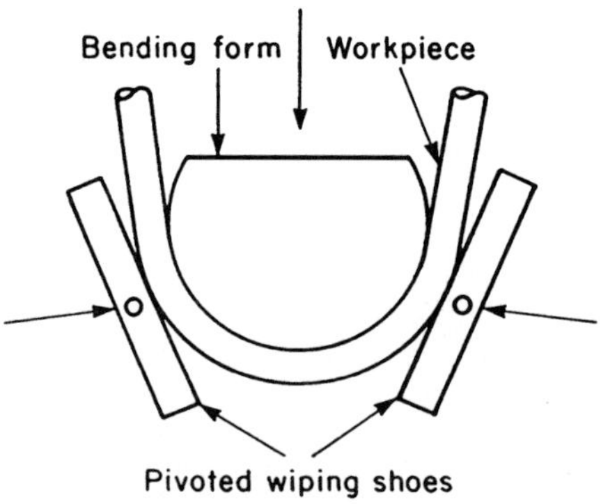

FIGURE 3.10.2 Ram-and-press bending. (*Courtesy Teledyne Pines.*)

Roll Bending

This method employs three or more grooved rollers arranged so that the workpiece passing between them is bent to a curved shape. The process can be used to form rings or continuous coils but is limited to heavy-walled workpieces.

With currently available equipment, roll bending is applicable to pipe up to 200 mm (8 in) in diameter. The minimum-bend radius is normally about 4 times tubing diameter. Greater values (6 times or more) are preferred. Full circles and coils are easily made with this approach, but straight sections must be allowed for on both ends of the bend.

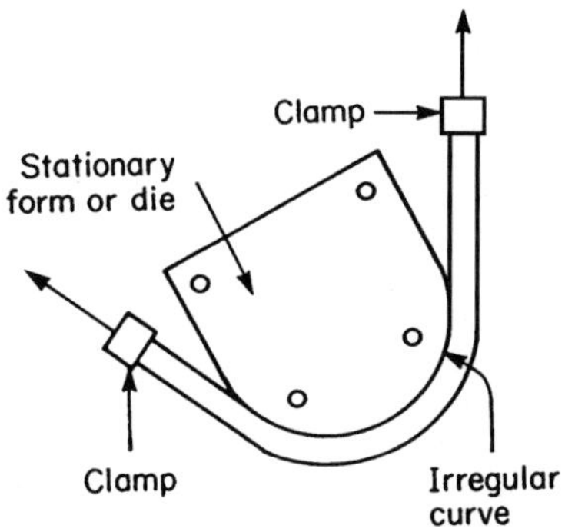

FIGURE 3.10.3 Stretch bending.

Stretch or Tension Bending

With this method, the workpiece is stretched longitudinally to the yield point and is then wrapped around a form or die. (See Fig. 3.10.3.) The method is not rapid but has less springback than other methods and can produce bends of nonuniform radius. A mandrel is not required, and the method is useful for nontubular cross sections.

Wrinkle Bending

In this method, a series of wrinkles is deliberately made on one side of the tubing. The wrinkles cause a shortening of this side, which then becomes the inside of the bend. Wrinkling is induced by heating a localized area by gas torch while applying an axial compressive force to the tube.

This type of bend can be applied to heavy-wall and large pipe. It is particularly adapted to field conditions in which bending equipment is not available. Pipe of 50 to 650 mm (2 to 26 in) except wrought- and cast-iron pipe can be bent by this method. The minimum centerline bend radius is about 2 times diameter.

Roll-Extrusion Bending

This process creates a large-radius bend in large, heavy-walled pipe by internally swaging the pipe wall on one side. This causes the wall to elongate and the pipe to bend. Bends are distortion-free. The minimum centerline bend radius is 3 times diameter. A straight section of 50 mm (2 in) or more for a 125-mm- (5-in-) diameter pipe must be allowed between bends. Successive bends in different planes are feasible. Only pipe and tubing can be processed with this method. Current commercial equipment has diameter capacities of from 125 to 300 mm (5 to 12 in).

ECONOMIC PRODUCTION QUANTITIES

Tube- and section-bend specifications are seldom dictated by production-quantity economics. Rather, functional requirements of the workpiece determine whether bending operations will be specified. Except for simple bends in solid sections, a certain amount of fitted form blocks, follower blocks, and wiper shoes are required regardless of the production quantity. If the weight or size of the workpiece is sufficient, power bending equipment is required.

If close-radius, tight-tolerance bends are required in thin-walled tubing on an experimental or small-quantity order, bending forms and blocks may be made from hardwood or fiber-reinforced thermoplastics. Sand, rosin, or low-melting-temperature alloys are used for internal support in place of a mandrel. The identical part, if produced on a higher-production basis, would probably be bent on a powered draw bender with hardened and chromium-plated bending blocks, pressure dies, and a ball-type mandrel.

Press-and-ram benders are used for the highest-production applications, especially when bend radii are generous and some distortion of the bent part is permissible. Automobile muffler piping is an example. Production rates of 1000 bends per hour are feasible with hydraulic presses; higher rates are possible with mechanical presses.

Production draw-bending machines have production-rate capacities that vary with the size of machine. Machines rated for 50-mm- (2-in-) diameter tubing, for example, can produce 200 bends per hour; those rated for 25-mm (1-in) tubing can produce 600 bends per hour. Production compression-bending machines may be somewhat faster, up to 900 bends per hour.

Stretch bending is considerably slower than the above processes and generally is not used in high-production conditions. Roll-extrusion bending of extra large, heavy pipe is a slow process (1 h per bend) but nevertheless is considerably faster than the hot-bending methods that otherwise would be necessary.

SUITABLE MATERIALS

Generally, the more ductile materials and those in the fully annealed state bend most satisfactorily. (Sometimes, however, some deviation from the dead-soft condition is preferred to help avoid collapse of the outer wall.) Ductility is particularly important for small-radius bends when the percentage elongation of the outer surface is very large.

Table 3.10.1 lists various common metals in the order of increasing ductility (percentage elongation). Those capable of higher elongation are more suitable for small-radius bends.

Compression bending does not require as much ductility as draw bending because there is less elongation on the outer surface of the bend. Wrinkle bending is suited to all ferrous pipe except cast and wrought iron. Aluminum, copper, and brass materials are also applicable to the process.

Comments on the suitability of certain common tubing materials for bending follow.

1. *Steel tubing.* This tubing bends easily. Welded tubing with internal flashing may cause problems if mandrel bending is required. Seamless steel tubing may give problems if the wall thickness is not uniform.

TABLE 3.10.1 Percentage Elongation (Small-Radius Bendability) of Various Metals

Percent elongation	Material*
10–15	High-carbon steel as rolled; aluminum alloys 2014 (A92014), 2011 (A92011), and 2024 (A92024); magnesium alloys AZ61A (M11610) and AZ31B (M11311)
15–25	High-carbon steels, chromium steels, nickel steels, aluminum 1100 (A91100)
25–30	Low-carbon steels as rolled, chromium-vanadium steels, aluminum 1060 (A91060)
30–35	Stainless steels 332 (N08800) and 904L (N08904)
35–40	Low-carbon steel; stainless steels 302 (S30200), 304 (S30200), 314 (S31400), and 316 (S31600)
40–50	Oxygen-free copper (C10200) and phosphorous-deoxidized copper (C12200)
50–65	Red brass (C23000), low-lead brass (C33000), silicon bronze (C65100), cartridge brass (C26000), inhibited admiralty metal (C44400 and C44500)

*Unless otherwise noted, all materials listed are in the annealed condition.

2. *Steel pipe.* This pipe is bent easily but, for best results, it should be free from internal and external scale.
3. *Galvanized and aluminized tubing.* These are suitable for bending, but the former is limited to bend radii larger than 4 times diameter.
4. *Copper and copper-alloy tubing.* Especially in the annealed condition these are very easily bent. The copper-nickel alloys are not as bendable and have greater springback. (Other copper alloys, if annealed, have very little springback.)
5. *Aluminum tubing.* This tubing is easily bendable but may tear or collapse if it is in the soft condition. Anodized coatings generally bend without damage.
6. *Stainless-steel tubing.* This tubing is frequently bent and is generally more suitable for bending than low-carbon steel. The 300 Series is the most common variety for bent-tubing applications.

DESIGN RECOMMENDATIONS

When specifying bends in tubing and other sections, the designer should bear in mind the following points:

1. Even when the proper material has been specified and good tooling and equipment are available, tight bends (small radii) and large-angle bends are more troublesome and more costly than large-radius bends with less severe angles. The designer should specify the gentlest and shallowest bend that serves the function of the part.
2. If possible, design with large enough bend radii so that a mandrel is not required, since mandrel bending is slower and more expensive. Table 3.10.2 lists minimum radii for bending steel tubing without a mandrel.

TABLE 3.10.2 Minimum Centerline Radii for Bending Steel Tubing without a Mandrel

	Minimum centerline radius, mm (in)					
	Wall thickness, mm (in)					
Tubing outside diameter, mm (in)	0.9 (0.035)	1.25 (0.049)	1.65 (0.065)	2.1 (0.083)	2.4 (0.093)	3 (0.120)
5 (3/16)	8 (5/16)	6 (1/4)	5 (3/16)			
6 (1/4)	13 (1/2)	10 (3/8)	8 (5/16)			
8 (5/16)	22 (7/8)	19 (3/4)	16 (5/8)			
10 (3/8)	38 (1 1/2)	32 (1 1/4)	28 (1 1/8)	25 (1)		
13 (1/2)	57(2 1/4)	50 (2)	44 (1 3/4)	38 (1 1/2)		
19 (3/4)	100 (4)	76 (3)	64 (2 1/2)	50 (2)		
25 (1)	200 (8)	150 (6)	100 (4)	76 (3)	50 (2)	50 (2)
38 (1 1/2)			300 (12)	250 (10)	200 (8)	150 (6)
50 (2)				600 (24)	500 (20)	400 (16)
64 (2 1/2)					600 (24)	500 (20)
76 (3)						630 (25)

Source: Based on data from *Metals Handbook,* vol. 4: *Forming,* 8th ed., American Society for Metals, Metals Park, Ohio.

3. In all cases, the minimum practicable bend radius for the bending process employed must be considered. The minimum radii for the processes covered in this chapter are summarized in Table 3.10.3. Table 3.10.4 provides recommended minimum radius values for draw-bending seamless steel tubing. These values, from one manufacturer, provide for easiest bendability by using draw-bending equipment. For magnesium tubing, see Table 2.3.17.

4. It should be borne in mind that tight bends (with a small centerline radius) are easier if the bend angle is small. (For example it is easier to make a tight bend if the part is bent 45° than it is if the part is bent 120°.)

5. If a part has multiple bends in more than one plane, a straight length should be allowed between bends. The amount of this length varies for different bending methods but generally should be one or two diameters. (See Fig. 3.10.4.)

TABLE 3.10.3 Minimum Bend Radii for Various Methods: Annealed Low-Carbon Steel Tubing Bent 45° or More

Method	Tooling	Minimum radius, times tubing outside diameter
Compression	Forming roller	4
	Follower block	2 1/2
Draw		See Table 3.10.4
Ram-and-press	Between centers	6
	With pressure or wing dies	3
Roll	Three rolls	6
Wrinkle		2
Roll-extrusion		3

TABLE 3.10.4 Recommended Minimum Bend Radii: Draw Bending Easily Bent Tubing

Outside diameter, mm (in)	Wall thickness mm	Wall thickness in	Approximate minimum centerline radius (multiple of tubing outside diameter): Without mandrel	Ball mandrel	Plain mandrel	Shoe-and-ball mandrel
13–24 (1/2–7/8)	0.9	0.035	6 1/2	2 1/2	3	1 1/2
	1.25	0.049	5 1/2	2	2 1/2	1 3/4
	1.65	0.065	4	1 1/2	1 3/4	1
25–39 (1–1 1/2)	0.9	0.035	9	3	4 1/2	2
	1.25	0.049	7 1/2	2 1/2	3	1 3/4
	1.65	0.065	6	2	2 1/2	1 1/2
40–54 (1 5/8–2 1/8)	1.25	0.049	8 1/2	3 1/2	4 1/2	2 1/4
	1.65	0.065	7	3	3 1/2	1 3/4
	2.10	0.083	6	2 1/2	3	1 1/2
55–79 (2 1/4–3)	1.65	0.065	9	3 1/2	4	2 1/2
	2.10	0.083	8	3	3 1/2	2 1/4
	2.80	0.109	7	2 1/2	3	2
80–100 (3 1/4–4)	2.10	0.083	9	3 1/2	4 1/2	3
	2.80	0.109	8 1/2	3	4	2 1/2

Source: Courtesy Wallace Supplies Mfg. Co. Welded steel or fully annealed seamless mechanical tubing.

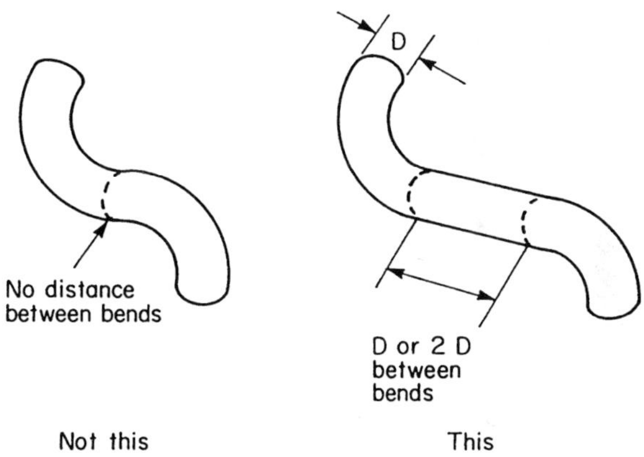

FIGURE 3.10.4 Allow a straight length between bends.

DIMENSIONAL FACTORS AND TOLERANCES

Springback is the major factor influencing the accuracy of the bend angle. This, in turn, is a function of material hardness, consistency of wall thickness, and section size, as well as of the bending methods. Tables 3.10.5 and 3.10.6 provide recommended bend-angle and bent-workpiece envelope tolerances.

TABLE 3.10.5 Recommended Bend-Angle Tolerances for Bends in Tubing and Other Sections

Method of bending	Bend-an gle tolerance, °	
	Normal tolerance	Tightest tolerance
Compression bending	±2	±1/4
Draw bending	±2	±1/4
Ram-and-press bending	±3	±1/2
Roll bending	±4	±2
Stretch bending	±2	±1/4
Wrinkle bending	±4	±2
Roll-extrusion bending	±2	±1/4

TABLE 3.10.6 Recommended Envelope Tolerances for Compound Tube and Section Bends

Bending method	Workpiece length, m (ft)	Normal tolerance, mm (in)	Tight tolerance, mm (in)
Compression bending	1 (3) or less	±13 (0.5)	±0.8 (0.030)
Draw bending	1 (3) or less	±13 (0.5)	±0.8 (0.030)
Ram-and-press bending	1 (3) or less	±25 (1.0)	±6 (0.25)
Roll bending	1 (3) or less	±25 (1.0)	±6 (0.25)
Stretch bending	1 (3) or less	±13 (0.5)	±0.8 (0.030)
Wrinkle bending	2.4 (8) or less	±50 (2.0)	±13 (0.5)
Roll-extrusion bending	2.4 (8) or less	±50(2.0)	±13 (0.5)

CHAPTER 3.11
ROLL-FORMED SECTIONS

THE PROCESS

Roll forming is a process in which strip or coiled sheet metal is fed continuously through a series of contoured rolls arranged in tandem. As the stock passes through the rolls, it is gradually formed into a shape with a desired uniform cross section. Only bending takes place; the metal gauge remains essentially constant.

The number of pairs of rolls (called *stands*) required to form a particular part depends on the material, the shape to be formed, and the precision required. There may be as few as 3 or 4 roll stands for a simple shape or as many as 28 or 30 if the cross section is complex.

Other operations such as notching, piercing, slotting, embossing, straightening, welding, marking, and longitudinal bending can be incorporated into the operation. Most often, this is done by arranging punch presses or other equipment in line with the roll former. The most common operation thus incorporated is a "flying" cutoff operation after forming operations have been completed. Punch-press operations are normally located ahead of the roll forming.

Straightening guides or rolls are often located at the exit end of the machine to correct for twist distortion. It is also possible to install cutters for the purpose of trimming the formed part to the exact width; this, however, is not a common practice.

Figure 3.11.1 shows the cross section of one part as it would appear if cut off after each of 14 stands of progressively contoured rolls.

TYPICAL PARTS AND APPLICATIONS

Roll forming has its best application in large quantities of parts with constant complex cross-sectional shapes. Parts often are of long length. As with extrusions, however, short parts can be made by cutting the formed strip to length.

Sheet stock as thin as 0.13 mm (0.005 in) and as thick as 25 mm (1 in) can be processed. However, except for heavy-wall welded steel pipe and structural members made by this process, normal commercial limits are 0.25 to 4 mm (0.010 to 0.156 in or 32 to 9 gauge).

Processible widths range from 25 mm to $2\frac{1}{2}$ m (1 to 100 in). However, 1 m (40 in) is a common commercial maximum, and 400 mm (16 in) is the widest stock processible on the most commonly available machines. Deep forms are difficult; 100 mm (4 in) is a practical depth limit for easily made cross-sectional shapes.

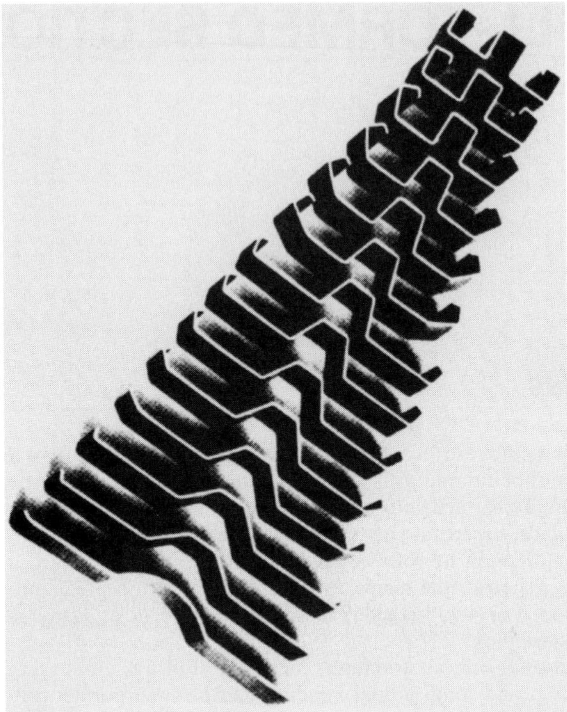

FIGURE 3.11.1 Evolution of a roll-formed section. (*Based on a photograph that was made by taking sections from a length of strips as it was stopped during its passage through 14 stands of rolls. Courtesy the Lockformer Co.*)

Roll-formed components have many applications, especially when mass production is involved. In the building industry, components manufactured by this process include roof and siding panels, purlins, joists and studs, window frames, door-trim ridge rolls, downspouts, architectural trim, and copper electrical conductors.

Appliance parts made by roll forming include panels for stoves, refrigerators, and other appliances, particularly when wraparound forms are involved, refrigerator shelves, and lighting-fixture parts. Curtain rods and tracks for sliding doors and draperies, drawer handles, and metal picture-frame members are also made by roll forming. When curling (longitudinal-bending) equipment is included in the forming line, such parts as bicycle fenders and wheels, Quonset-hut frames and roofing, barrel hoops, drum rings, and bezels are formed.

Roll forming can be carried out with plated, galvanized, lithographed, vinyl-coated, and painted stock (if the paint is baked and the forming is not too severe or if a bendable-paint formulation is used). Dual-layered components also can be made at one time. These may include bimetallic parts such as those made when a thin stainless-steel facing sheet and a carbon steel support sheet are formed together or when fiber, paper, asbestos, or rubber inserts are placed in a metallic form.

Long pieces such as railroad-car and truck-trailer trim members can be made with this process. The maximum length of part is dictated only by application requirements

FIGURE 3.11.2 Representative cross-sectional shapes that can be made by roll forming. (*Courtesy the Lockformer Co.*)

and handling conditions. Figure 3.11.2 illustrates some typical cross sections that can be produced by this process.

ECONOMIC PRODUCTION QUANTITIES

Roll forming is a mass-production process. Production rates are rapid, but tooling and setup costs are high. Generally, production requirements in excess of 100,000 ft/year are necessary before the process becomes economical. In some cases, depending on the configuration to be produced and the efficiency of alternative methods, 500,000 ft of annual production may be necessary to justify the process. In others, a run of only 10,000 ft may be justifiable.

Setup times are lengthy because of the large number of tooling elements that are interrelated. Once operative, however, roll forming is an extremely fast cycle operation. Forming speeds can run as high as 300 ft/min for some shapes and average from 50 to 100 ft/min for most applications. The low labor cost inherent in such a production rate and the savings in materials cost (compared with extrusion or machined parts) provide very low cost production if quantities are large enough to amortize the initial tooling and equipment charges.

SUITABLE MATERIALS

Any metals suitable for other bending or forming processes are processible on cold-roll-forming equipment. Since only bending action takes place compared with the cold flow involved in spinning, swaging, or drawing operations, for example, maximum ductility is not required of materials, and all normally formable materials can be processed by roll forming.

For best results, however, it is always preferable to use the more ductile alloys that are available commercially. One factor in favor of this approach is the fact that in roll forming almost all bends are made parallel to the grain lines of the stock. Fractures are more likely when bends are in this direction than at right angles to the grain of the workpiece material.

Low-carbon steels are the most frequently processed roll-forming materials. Others are aluminum (preferably the nonaging alloys), brass and bronze, copper (both rigid and flexible), and zinc. Stainless steels and high-alloy steels are utilized in roll forming and can be processed satisfactorily. Clad, plated, galvanized, anodized, and organically finished metals are suitable for the process. Prepainted stock, if coated with a durable, nonbrittle baked finish, is fully processible. The relatively gentle forming action of roll forming makes it particularly suitable for the forming of polished, embossed, or otherwise prefinished stock.

DESIGN RECOMMENDATIONS

The first guide rule is the universal one of having the parts designer understand the process. The designer should be in close contact with the engineer who designs the forming rolls or someone who thoroughly understands roll-forming operation, equipment, and tooling. Just as in the case of progressive-stamping dies, the best results are obtained when the forming sequence is known by the product designer. Reviewing the sequence carefully can avoid unnecessarily high tooling costs, high tool wear, and forming problems.

Bending Radii

Radii for both inside and outside corners should be as generous as possible. The minimum practical radius is a radius equal to one stock thickness. However, 2 times stock thickness is preferable. (See Fig. 3.11.3.) Some alloys, such as aluminum 2014 (A92014) and 2024 (A92024), require an even larger radius, except in the 0 (annealed) temper. Bending radii smaller than one stock thickness are actually possible, but they require sharp corners on the forming rolls, and roll life is shortened significantly.

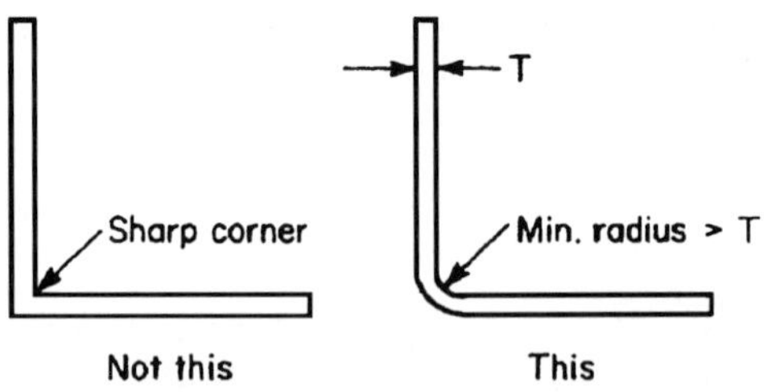

FIGURE 3.11.3 The minimum-bend radius for roll-formed components is one stock thickness, but 2 times stock thickness is preferable.

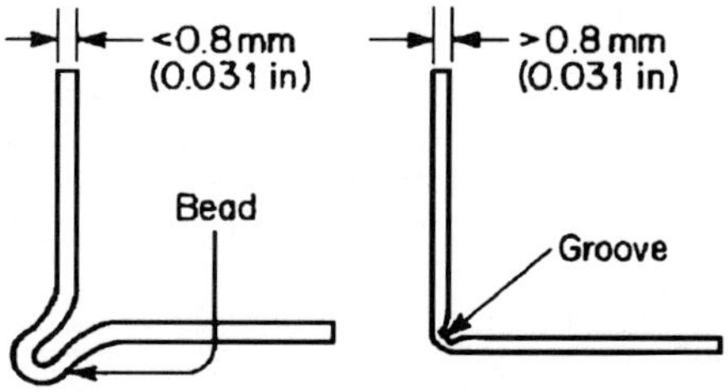

FIGURE 3.11.4 Bending sharp corners in roll-formed parts can be achieved by beading (thin stock) or grooving (heavy stock).

If sharp bends are essential, they can be achieved through one of two approaches. For stock up to 0.8 mm (0.031 in) in thickness, an external bead can be used. For stock over 0.8 mm thick, a formed groove one-third to one-half the stock thickness facilitates bending. See Fig. 3.11.4 for an illustration of both approaches.

Part Length

From the standpoint of ease of forming, it is preferable to form the stock in long lengths and to cut to length after forming. The normal process is to accomplish this as one continuous automatic operation. If this is not possible, there are minimum-length limitations for the roll-formed pieces. There is an entrance and exit flare distortion at each end. For one class of machines, the flare amounts to about 1.3 mm (0.050 in) and extends 75 mm (3 in) from each end.

As a rule of thumb, consider that parts shorter than 3 times the centerline spacing of rolls of the machine employed will not feed or form satisfactorily. [Centerline spacing for machines processing average-length stock is 200 to 250 mm (8 to 10 in).] Therefore, precut parts should normally be 600 to 750 mm (24 to 30 in) long or longer.

Depth of Form

Deep profiles require correspondingly larger-diameter forming rolls, which are more costly and require larger machines. More important, however, is the fact that different diameters of forming rolls have different peripheral speeds and exert different amounts of pull on the stock. Of necessity, slippage occurs with its resultant frictional heat, roll wear, and possibility of scoring the stock. Minimizing the depth of form reduces these roll-diameter differences and their attendant problems. (See Fig. 3.11.5.) One rule-of-thumb maximum form depth that has been established for average-size forming machines is 100 mm (4 in).

Note. Some forming machines are equipped with differential gearing that allows certain large-diameter rolls to be run at a slower number of revolutions per minute than other rolls and thereby more closely match their peripheral speed to the stock speed. Such machines can form deeper sections. If considering a deeply formed section, the designer should ascertain if such equipment is on hand.

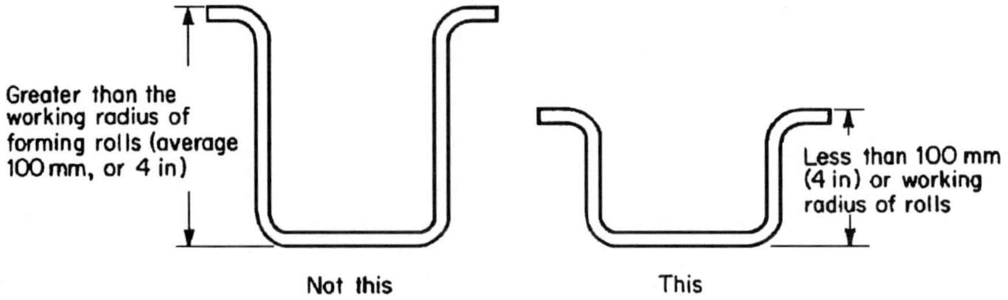

FIGURE 3.11.5 The depth of a roll-formed section should be kept as small as possible to simplify tooling, reduce tooling wear, and preserve stock finish.

The depth of form is much more critical if the stock is prefinished because the coated surface can be damaged by the sliding contact of deep forming rolls.

Wide Sections

Thin, wide parts may have some tendency to show waviness unless unformed areas are less than 125 mm (5 in) wide. If a longitudinal stiffening rib is formed in the wide area, these irregularities can be avoided and the component will be more rigid. Similarly, waviness or unevenness at the edges can be avoided by incorporating a flange, hem, or rib near or at the edge. (See Fig. 3.11.6.)

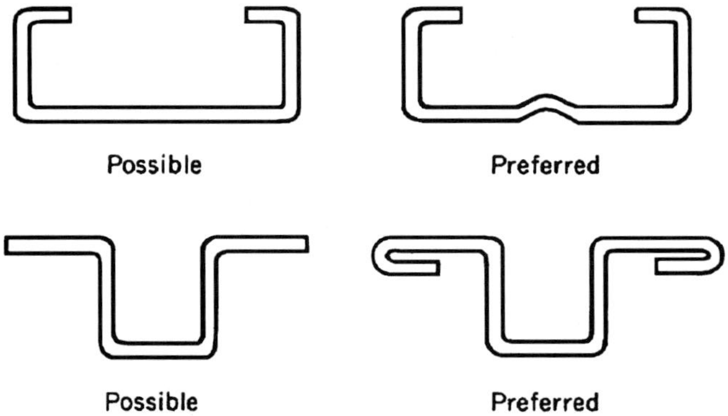

FIGURE 3.11.6 Stiffening bends help avoid any tendency toward waviness or other irregularities in wide areas and near edges.

Symmetrical or Balanced Forms

Shapes that are symmetrical about a vertical centerline are the most satisfactory for roll forming. If symmetry is not possible, it is desirable to have approximately the same amount of bending on each side of the centerline. Completely nonsymmetrical forms also are entirely possible, but they are more apt to require straightening as part of the forming process.

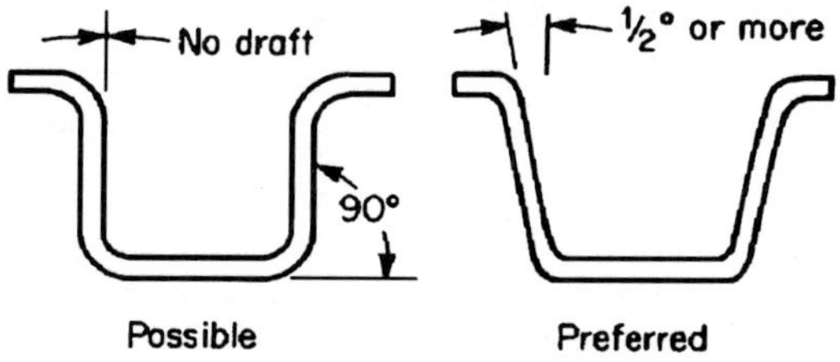

FIGURE 3.11.7 Avoid exactly vertical sidewalls.

Vertical Sidewalls

Exactly vertical sidewalls can be troublesome to the roll former. They cause excessive roll wear and scoring of the workpiece. A draft angle of ½° or more is preferred. (See Fig. 3.11.7.)

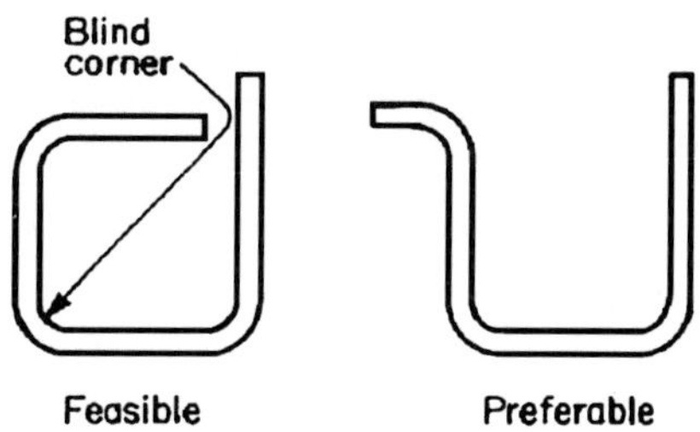

FIGURE 3.11.8 Blind corners can be formed, but they are less accurate than corners formed with rollers in contact with both sides of the stock.

Blind Corners

Blind corners and radii, as illustrated in Fig. 3.11.8, are feasible but should be avoided if precise bends are needed. Accurate forming is facilitated when both sides of the stock can be controlled by the rolls.

Minimum Leg Length

The minimum practical length for formed legs is 3 times stock thickness. (See Fig. 3.11.9.)

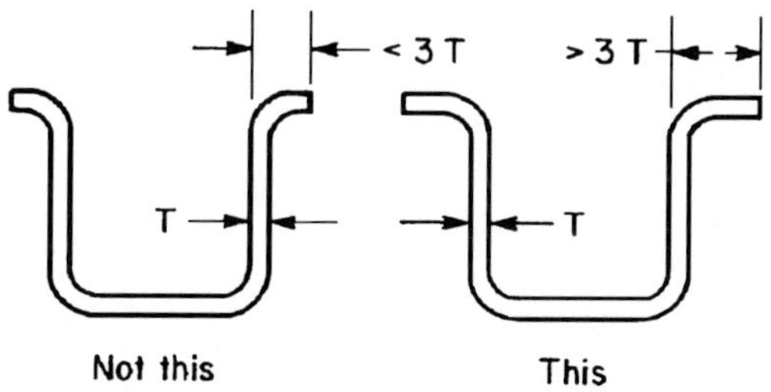

FIGURE 3.11.9 Normal minimum leg length is 3 times stock thickness.

DIMENSIONAL FACTORS AND TOLERANCES

Dimensional variations of roll-formed parts are caused by the same factors that affect other sheet-metal-forming processes. They are springback; variations in hardness, thickness, and yield point of material; tooling deviations and tooling wear; machine deflection; and variations in setup and adjustment of tooling. Closer dimensional tolerances usually can be held with thinner stock and with smaller parts. Table 3.11.1 provides recommended dimensional tolerances for roll-formed components.

TABLE 3.11.1 Recommended Tolerances for Roll-Formed Sections

Dimension	Tolerance
Section or leg width or height dimensions up to 50 mm (2 in) or critical dimensions over 50 mm	±0.25 mm (0.010 in)
Section or leg width or height dimensions over 50 mm (2 in) or any noncritical dimension	±0.4–0.8 mm (0.015–0.030 in)
Angle	±1°
Twist	3.5 mm/m (⅛ in/3 ft)
Camber	0.8 mm/m (1/32 in/3 ft)

Note: Tighter tolerances than these can be held in many cases at extra cost.

CHAPTER 3.12
POWDER METALLURGY PARTS*

Originally prepared by

B. H. Swan

Industrial Powder Met (Consultants), Ltd.
London, England

Updated by

Robert Powell

Consultant
Bordentown, N.J.

THE PROCESS

Powder metallurgy (P/M) parts are formed by compressing metal powders in a die and then sintering the piece thus formed. The basic process includes as a first step the selection of pure metals or alloys in suitable powder form and the weighing and mixing of them with or without nonmetals. Normally, dry-powder lubricants such as stereates and waxes are added to reduce the friction between the powder particles and the particles and die walls.

The mixed powder is accurately metered, by volume, into suitable dies contained within an automatic high-speed compacting press. The powder is molded to the required shape, normally at room temperature, by the application of high-tonnage compacting pressure. (Recent developments have shown advantages in warm compacting.) No binder or adhesive material is used in this operation.

Following compaction, the *green compacts* are heat-treated by the process known as *sintering* to induce optimal strength. Specialized sintering furnaces, utilizing accu-

*Note that the standard terminology of the Metal Powder Industries Federation for the process of making parts made from metal powders is *powder metallurgy* (*P/M*). Forgings from metal powders are referred to as *powder metallurgy forgings,* or *P/F.* These terms are used in this chapter.

rately controlled atmospheric conditions to suit the particular alloy being produced, are used for this operation. For certain products, manufacture is now complete, and the components are ready for use without further processing.

Should dimensional tolerances of extreme accuracy by required, the components can be subjected to a repressing or calibrating operation, carried out in high-speed presses, utilizing tools manufactured to a maximum degree of accuracy. Certain sintered alloys, owing to their extremely high strength and capability of resisting deformation under load, cannot be repressed or calibrated. These must be accepted within the tolerances achieved at the sintering stage of manufacture unless other secondary operations are performed.

Figure 3.12.1 shows the action of typical compacting dies. Metal powders can be abrasive to the die, and it is probable that the die will be manufactured from tungsten carbide.

Split dies are not acceptable to the process for a variety of technical reasons. The die also must be designed to allow ejection of the component by simple pressure on the underside of the component by the lower punch, following removal of the upper punch.

Secondary Operations

Machining. Following sintering, most alloys may be readily machined, if necessary, by the normal techniques of turning, milling, drilling, tapping, thread rolling, grinding, etc. In operations for which cutting tools are used, the effect of porosity (see below) must be taken into consideration. Cutting speeds, feeds, tool angles, etc., should conform to the recommendations of the sintered-component producer.

Surface Finishes. A wide variety of protective surface finishes can be applied to the finished components. These include

Electroplating of zinc, nickel, copper, chromium, etc. (Care must be taken to overcome the problem of electrolyte entrapment in the porous structure.)

Stream treatment, a blue oxidizing process for increased wear resistance, also used as a pore-filling process prior to electroplating.

Oil and chemical blackening.

Phosphate coatings as an antirust treatment.

Varnishing as an antirust treatment.

Heat Treatment. Most iron-based sintered alloys are suitable for conventional heat treatment. With correct control of the combined carbon content during the sintering operation, alloys that are suitable for through hardening by heating and quenching can be produced. Such alloys are usually suitable for induction hardening or other localized hardening treatments.

Most iron-based alloys can be surface-hardened; the preferred route is by carbonitriding. In this process, the rate and depth of carburization will be affected by the residual porosity of the component.

Liquid-cyanide hardening is not suitable for the treatment of sintered components owing to absorption of the liquid salts into the porous structure of the material. Prolonged washing in water following hardening does not remove entrapped salts, which subsequently corrode the component both internally and externally.

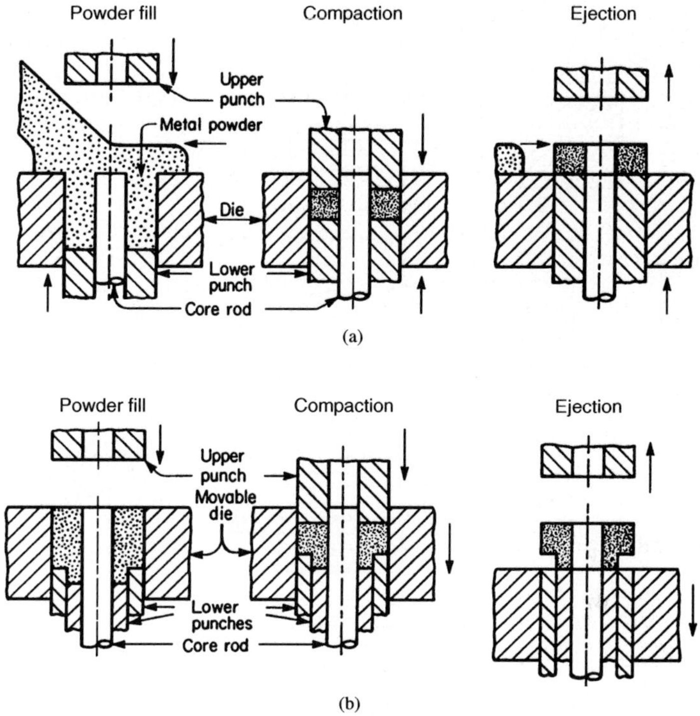

FIGURE 3.12.1 (*a*) Fixed-die pressing system. The die is fixed in the press table. The lower punch is withdrawn when the feed shoe (not shown) is over the die cavity. Suction fill occurs. The feed shoe is withdrawn. The upper punch enters the die. The upper and lower punches move simultaneously. Powder is compacted. The upper punch is now withdrawn. The lower punch continues to rise, ejecting the new component from the die. The feed shoe moves across the face of the die, pushing the part to a collecting station. The cycle is repeated. (*b*) Withdrawal-die pressing system. The main lower punch is fixed; the die is movable. The die is filled. The upper punch enters the die. The die is withdrawn at half the speed of travel of the upper punch. When two lower punches are employed, one lower punch is fixed and one is movable. When the movable punch has completed its compaction motion, it is allowed to move to enable compaction of the second level to occur. When compaction has been completed, the upper punch is raised. The die is further withdrawn to effect ejection by stripping the die from the component. The feed shoe moves across the face of the die, pushing the part to the collection station. The die rises to allow it to fill from the feed shoe. The cycle is repeated.

TYPICAL CHARACTERISTICS AND APPLICATIONS

General Features

P/M parts up to 15 cm (6 in) are commonplace. They may be intricate in shape but usually have straight, parallel sides. They possess a high degree of dimensional accuracy. Generally, they are ready for use without subsequent operations. P/M components can be self-lubricating and can be produced from unusual alloys if necessary. A

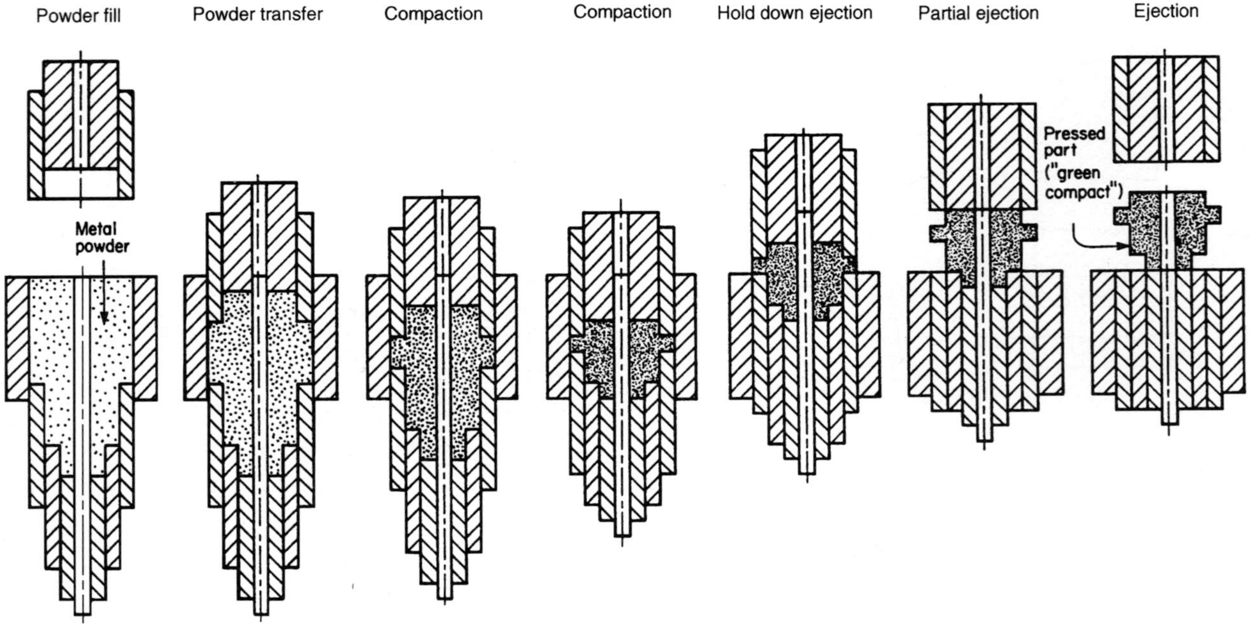

FIGURE 3.12.1 (*Continued*) (*c*) Typical tooling action for a complex component.

FIGURE 3.12.2 Typical P/M parts.

very high degree of surface polish can be achieved on sintered components, particularly when such components are repressed in highly polished dies manufactured from tungsten carbide.

Although complex part shapes are feasible and even most advantageous in many cases, there are some limitations on the shape of P/M parts. Undercuts, cross holes, and screw threads, for example, are best provided with secondary operations. P/M parts usually have low resistance to shock loading and generally lower physical properties than wrought metals.

Applications include machine parts of various kinds, especially when other production processes would involve considerable or difficult machining. Cams, slide blocks, levers, gears, bushings, ratchets, guides, spacers, splined parts, sprockets, pawls, and connecting rods are typical parts. Business machines, sewing machines, automobiles, and firearms are typical products that utilize powder metallurgy. Figure 3.12.2 illustrates typical P/M parts. Figure 3.12.3 shows P/M parts used in automobiles and other vehicles.

Porosity

When metal-powder mixtures are being compacted within a die, extremely high compacting tonnages are required to achieve a product. By careful control of the applied compacting load, the volume or density of the component being compacted can be controlled within very accurate limits. The fact that under normal compacting loads, 100 percent dense material has not been achieved results in a material that is porous. This porosity exists as interconnected pores of extremely small size. The capillary-size pores, being interconnected, can absorb liquids and may be readily impregnated with oils, greases, waxes, etc. As density is increased by increasing compaction load, the number and size of the pores will decrease and the ability to retain impregnated fluids will be reduced. This characteristic of sintered material is the basis of the well-known "self-lubricating" range of bearings and bushings.

FIGURE 3.12.3 A collection of P/M parts for automotive and other vehicle applications. (*Courtesy Cincinnati Incorporated.*)

Such porous materials that have applications in cams, gears, etc., in addition to bearings and bushings, are impregnated with lubricating oil following the sintering or repressing operation. Under operating conditions, the lubricant is withdrawn from the porous mass by capillary action, produced by movement of the component or a mating surface. When the component or mating surface comes to rest, the lubricant is reabsorbed into the porous structure by capillary action, the porous material acting as a wick.

General Size Parameters

The maximum size of components that can be processed is controlled by the compacting pressure that can be applied by the available press. The required compacting tonnage can be estimated by calculating the face area of the upper surface of the component and multiplying this by the tonnage per unit area required to obtain the desired density. Densities range from 5.8 to 7.2 g/cm^3.

For iron and its alloys, the tonnage requirement will be about 4 to 8 $tons/cm^2$ (25 to 50 $tons/in^2$) of face area. For copper and its alloys, 2 to $3\frac{1}{2}$ $tons/cm^2$ (10 to 20 $tons/in^2$) of face area is required. Presses of up to 500 tons capacity are in common usage, and those above 1000 tons are extremely rare.

POWDER METALLURGY FORGINGS (P/F)

While the production of porous sintered components fulfills the needs of many engineering applications, the lower strengths and lack of ductility inherent in porous materials have led to a demand for nearly fully dense high-strength material to be manufactured by P/M methods. This requirement is being met by powder metallurgy forging (P/F).

In this process, a compacted, sintered preform, usually coated with a protective envelope of graphite, is heated to forging temperature and forged to final shape and tolerance in closed precision dies. Limited flash or waste is produced, and the precision powder forging is usually ready for use.

Since considerable deformation can occur to the preform during forging, it is frequently possible to use a preform of very simple shape to produce complicated forged components. An example is the production of an angled spur gear for an automobile differential. This gear is forged from a simple cylindrical preform containing a hole of suitable size. The teeth of the gear are "thrown" during the forging operation.

In designing powder metallurgy forgings, the general design principles for porous sintered components should be observed. P/F tools may be made parallel, and no requirement for draft is usually encountered.

ECONOMIC PRODUCTION QUANTITIES

The P/M process provides high rates of production, low process costs, low scrap rates, and full utilization of raw materials. Because of the use of automatic presses and expensive tooling, the process is more suitable to large-quantity production runs. Quantities of 20,000 pieces and larger are usually considered as acceptable for production. However, the main advantage of low-cost production is usually achieved when quantities of 100,000 pieces or more are produced at one setup of the press.

It is not unusual to manufacture 1 or 2 years' requirements at one setup of a press in order to take advantage of the economy offered. The capability of this process to manufacture components of very complex shape to extremely close tolerances has, however, led to a demand for small-quantity production despite the premium that must be paid for the smaller quantities. Some manufacturers will entertain production of 2000 to 5000 pieces, providing a premium price is agreed on. Extremely large quantities on the order of 1 million pieces per month can be accommodated; extremely high-speed presses are available for special applications.

MATERIALS FOR P/M PARTS

The list of materials used in P/M components includes ferrous and nonferrous metals and nonmetallics. Iron, in combination with copper and/or nickel, and often with carbon, is common. Bronze is a major material in bushings and bearings. Other materials are aluminum and aluminum alloys, beryllium, brass, cobalt, copper, copper-nickel, iron-nickel, molybdenum, nickel and nickel alloys, steel of various carbon contents, stainless steel, titanium, tungsten, and zirconium.

The most commonly used materials are iron with 2 to 5 percent copper and 0.60 percent carbon. This material sinters well, has little size change during sintering, and is of reasonably low cost. The pure iron and iron-graphite materials are of lowest cost and are fabricated easily.

The alloying materials to be avoided whenever possible, purely on a cost basis, are molybdenum, nickel, tin, and similar materials of high purchase price. Copper in increased proportions also adds to the cost of a powdered-iron mixture. It may be essential, however, to use one or more of these materials if parts of extremely high strength are required.

Table 3.12.1 presents information on physical properties of conventional P/M parts made from freely available materials. Table 3.12.2 presents similar information for powder forgings.

TABLE 3.12.1 Typical Properties of Common Ferrous P/M Structural Materials

Material	Density (g/cm^3)	Tensile strength 10^3 psi	Tensile strength MPa	Yield strength 10^3 psi	Yield strength MPa	Elongation (%)	Hardness (Rockwell)
Carbon steel	6.6	43	290	35	240	<1.0	60HRB
(MPIF F-0008)	7.0	57	390	40	275	1.0	70HRB
Carbon steel	6.6	75	515	*		<0.5	28HRC
(MPIF F-0008 heat-treated)	7.1	95	655	*		<0.5	35HRC
Copper steel	6.7	60	415	55	380	<1.0	73HRB
(MPIF FC-0208)	7.2	75	515	65	450	<1.0	84HRB
Copper steel	6.8	90	620	*		<0.5	35HRC
(MPIF FC-0208 heat-treated)	7.1	105	725	*		<0.5	43HRC
Nickel steel	6.9	50	345	30	205	2.5	59HRB
(MPIF FN-0205)	7.2	60	415	35	240	4.0	69HRB
Nickel steel	6.9	120	825	*		<0.5	36HRC
(MPIF FN-0205 heat-treated)	7.1	145	1000	*		<0.5	40HRC
Infiltrated steel (MPIF FX-1005)	7.3	77	530	50	345	4.0	82HRB
Infiltrated steel (MPIF FX-1005 heat-treated)	7.3	120	825	*		<0.5	38HRC
Low-alloy steel (MPIF FL-4605 heat-treated)	6.95	130	895	*		<0.5	60HRC
Stainless steel							
(MPIF SS-316 N2)	6.4	60	415	39	270	10.0	62HRB
(MPIF SS410)	6.5	105	725	*		<0.5	23HRC

*Yield and ultimate tensile strength are approximately the same for heat-treated materials.

Source: Material designations and data from Metal Powder Industries Federation Standard 35, *Materials Standards for P/M Structural Parts.*

TABLE 3.12.2 Typical Mechanical Properties of Powder Metallurgy Forgings from Ferrous Structural Materials (ASTM B848)

MPIF alloy designation*	Heat treatment†	Hardness (Rockwell)	Tensile strength psi	Tensile strength MPa	Yield strength psi	Yield strength MPa	Elongation %	Impact energy Charpy V-notch (ft·lbf)
P/F-1040	N	80HRB	75,000	515	45,000	310	27	4
P/F-1040	Q	30HRC	140,000	965	120,000	825	12	15
P/F-1060	N	80HRB	85,000	585	50,000	345	22	2
P/F-1060	Q	40HRC	195,000	1345	175,000	120	58	10
P/F-10C40	N	97HRB	100,000	690	70,000	480	15	3
P/F-10C60	N	23HRC	115,000	790	100,000	670	11	2
P/F-11C60	N	28HRC	130,000	895	90,000	620	11	3
P/F-4220	N	84HRB	75,000	515	55,000	380	25	25
P/F4240	N	93HRB	90,000	620	60,000	415	18	12
P/F4260	N	22HRC	110,000	760	75,000	515	15	5
P/F-4620	N	96HRB	80,000	550	60,000	415	20	25
P/F4620	Q	28HRC	140,000	965	130,000	895	24	60
P/F4620	Q	38HRC	190,000	1310	155,000	1070	20	35

*From Metal Powder Industries Federation Standard 35, *Materials Standards for P/M Structural Parts.*
†N = normalized; Q = quenched-hardened and tempered to the hardness listed.

DESIGN RECOMMENDATIONS

In compaction of a part, pressure is applied equally from above and below to the powder being compacted. The powder is contained within a die, and owing to the irregular shape of the powder particles employed, the powders will interlock mechanically and become a coherent mass. Because of this interlocking characteristic, powder will not "flow" under pressure as plastic material is known to flow during molding. It is essential, therefore, that the component be designed so that the powder can be conformed to a required shape during the filling of the die and before pressing commences.

Draft

Draft is not desirable and in production usually will produce problems. Lack of draft is an advantage because die walls can be absolutely parallel to each other, enabling component faces to be parallel and of close tolerance. An exception is the sidewalls of recesses formed by a punch entering the top side of a part. In these cases, a draft of 2° or more is advisable, as shown in Fig. 3.12.4.

Wall Thickness

While uniform wall thickness is of minor advantage in avoiding some warpage during the sintering operation, it is by no means essential. The main economic advantages of the process are usually achieved with components of extremely complicated shape in which wall thickness varies over a very wide range. The minimum recommended wall thickness is 1.5 mm (0.060 in). This stricture also applies to the wall between a hole

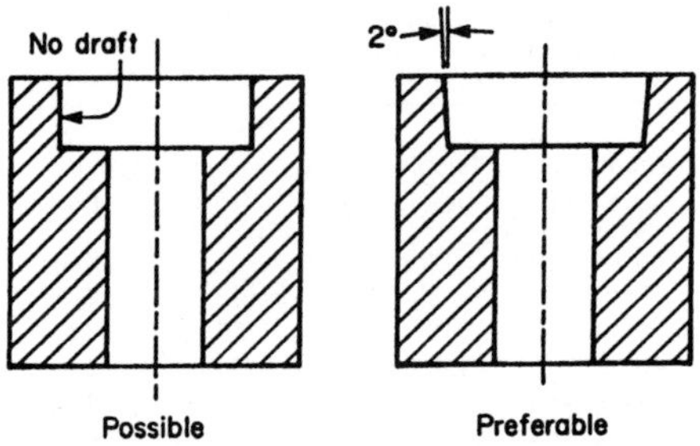

FIGURE 3.12.4 When recesses occur in the top of a part, 2° minimum draft assists removal of the upper punch.

and the outside surface of the part and the wall between two holes. Smaller wall thicknesses, down to 1.0 mm (0.040 in), can be produced for special applications. However, this may lead to problems in powder filling the die cavity, resulting in localized low-density areas. There also may be problems of low punch strength under compression and ejection problems, which all increase the cost of production. Very thin sections or projections also may be liable to damage during handling between the press and furnace, since compacted components, prior to sintering, are fragile. Figure 3.12.5 illustrates the recommended minimums.

The normal maximum ratio of wall thickness to length is 18:1. With special tooling, the ratio can be increased to 30:1. The normal maximum length of P/M parts is 100 mm (4 in). (See Fig. 3.12.6.)

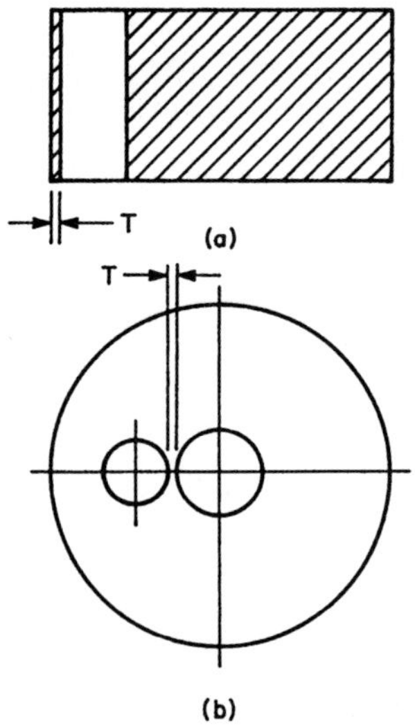

FIGURE 3.12.5 Minimum wall thickness *T* under normal conditions is 1.5 mm (0.060 in). With special tooling, minimum $T = 1.0$ mm (0.040 in).

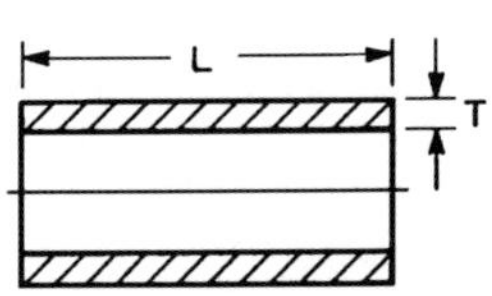

FIGURE 3.12.6 Rules for the maximum length of thin-walled parts.

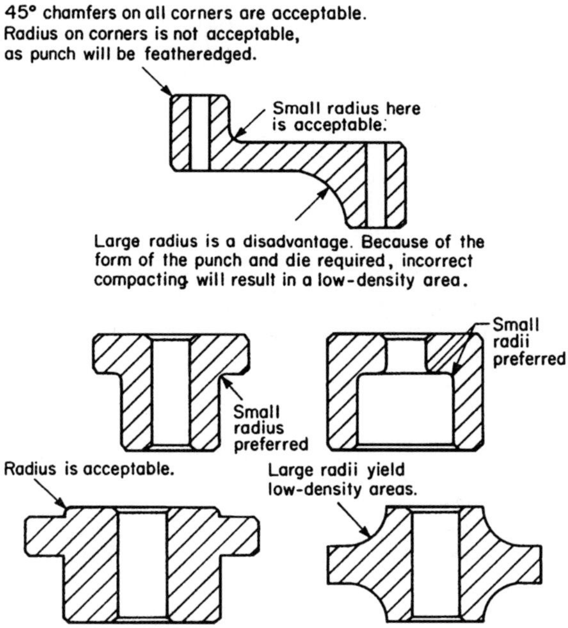

FIGURE 3.12.7 Radii in P/M components.

Radii

Small radii at component corners, both internal and external, can be advantageous in extending tool life. Extremely large radii in certain directions can induce problems of density variation. (See Fig. 3.12.7.) However, large radii in horizontal cross sections are not a problem, whereas sharp corners reduce tool strength and may cause filling problems. Figure 3.12.8 illustrates good and bad practice.

Care must be exercised when curved surfaces adjoin one another, since the necessary punches may have fragile featheredges. Figure 3.12.9 shows one such case that was corrected by redesign of the part.

Holes

Holes in the direction of pressing are freely accepted. They can be molded in complicated shapes at close tolerances.

The minimum diameter of holes is 1.5 mm (0.060 in). Smaller holes lead to problems of premature fracture of core rods during ejection. In addition, very slender core rods may bend during compaction, giving rise to the condition known as "banana bores."

Holes at right angles to the direction of pressing cannot be achieved and must be machined after sintering. They should therefore be avoided if possible.

Blind holes are frequently pressed. The quality of such holes is improved by putting a drill-point angle on the top of the core rod to assist in spreading powder during pressing. Blind holes smaller than 6.3 mm (0.25 in) in diameter are not recom-

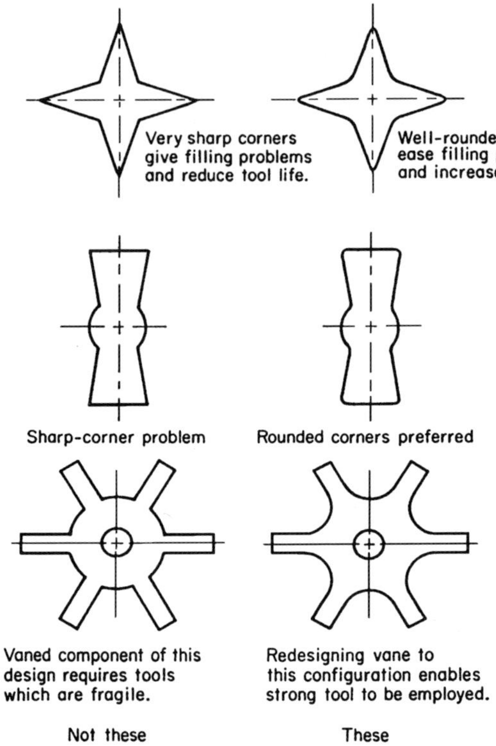

FIGURE 3.12.8 Advantages of rounded corners over sharp corners in horizontal cross sections of P/M parts.

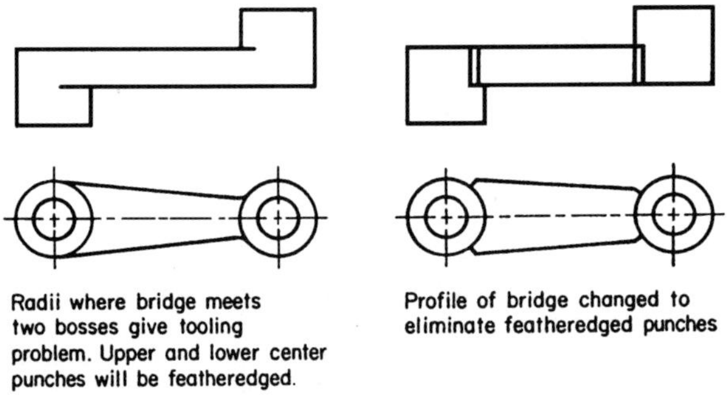

FIGURE 3.12.9 Avoidance of featheredged punches.

mended unless they are very shallow. The rigidity of the core rod during pressing is of major importance. Figure 3.12.10 shows typical blind holes.

To increase the depth of deep blind holes, use steps, thus enabling a stronger core rod to be employed. Figure 3.12.11 illustrates design recommendations for stepped holes.

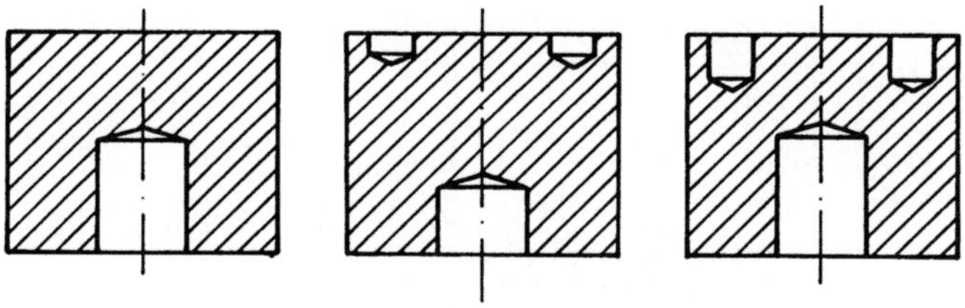

These blind holes are all suitable for production. Blind holes from above should be shallow; blind holes from below may be deep.

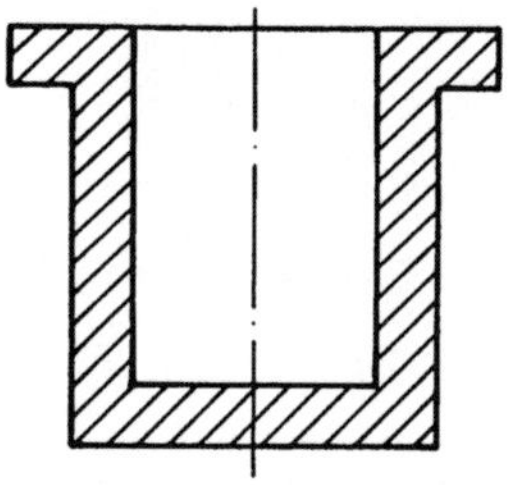

This type of deep blind hole is possible, but tooling is expensive. Powder must be transferred from fill position to ready-for-pressing position before any compacting commences.

FIGURE 3.12.10 Blind holes.

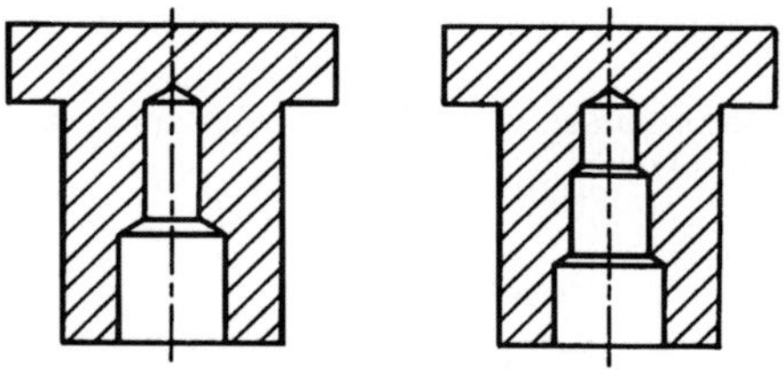

Acceptable stepped holes. They will increase core-rod rigidity when a small diameter is very slender. Plain-diameter core rods, however, of larger diameter, are preferred wherever possible.

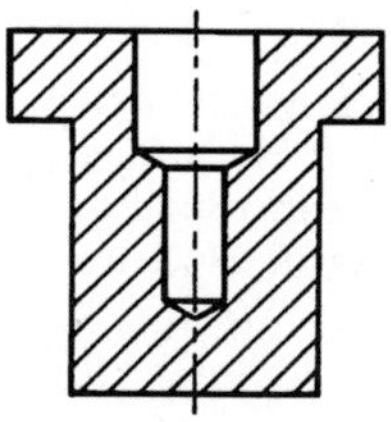

This type of stepped hole from the top is not recommended unless the total depth of the hole is shallow.

FIGURE 3.12.11 Stepped blind holes.

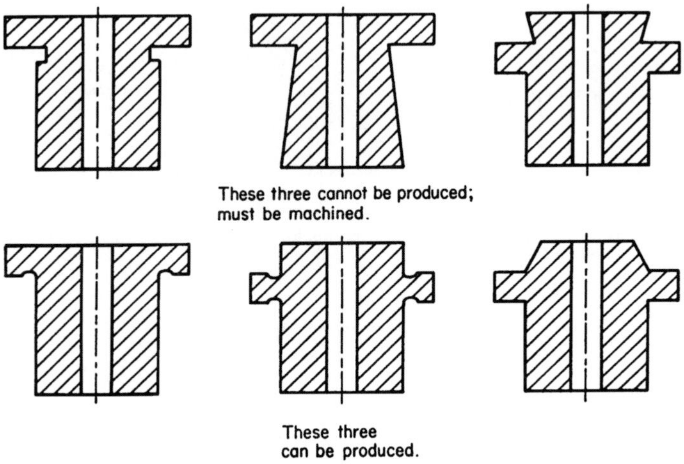

FIGURE 3.12.12 Undercuts: problems and some solutions.

Undercuts

Undercuts are not permissible because of problems in ejecting the component from the die. Figure 3.12.12 illustrates undercut problems and possible alternatives.

Screw Threads

Screw threads cannot be produced in the P/M process and are obtained, if needed, by machining.

Gears

Spur, helical, and crown gears can be produced to close tolerances by powder metallurgy. Figure 3.12.13 presents design recommendations.

Inserts

The molding of inserts into the compact, though not impossible, is not usual and normally is not recommended. It adversely affects the high production rates normally obtained with the process. One area in which these problems have been successfully overcome, however, is the molding of electrical wire "tails" into electrical contact brushes.

Angled Sidewalls

These can cause tooling problems (fragile, knife-edged punches) if the angle is too small or the intersection of the angled wall with other surfaces is a single edge. The

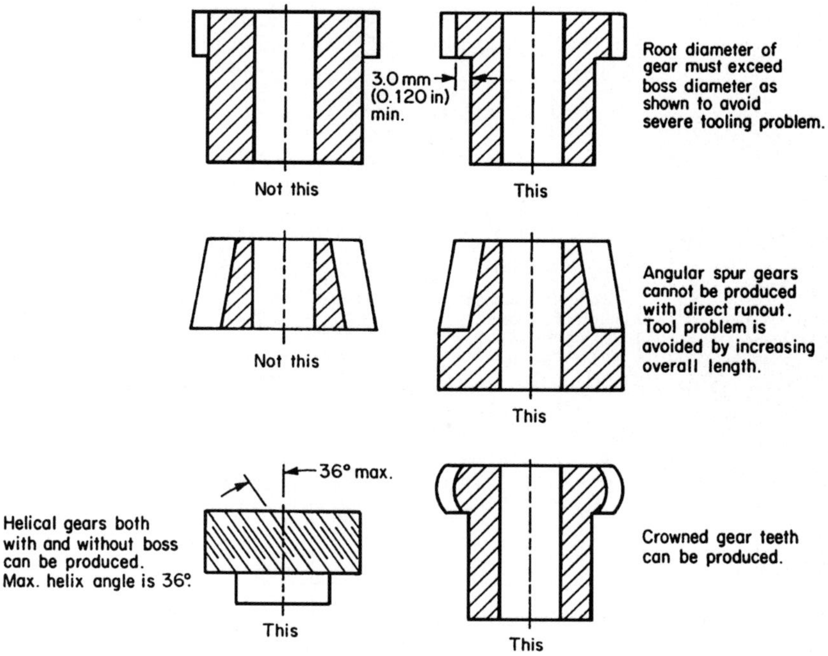

FIGURE 3.12.13 Recommendations for the design of P/M gears.

use of narrow, flat areas where angled surfaces intersect enables stronger punches to be used and prevents die damage due to variations of punch stroke. Figure 3.12.14 illustrates design recommendations.

Spherical Surfaces

These have molding characteristics similar to those of tapered surfaces. Die damage is avoided if a narrow flat is provided at the parting line, as shown in Fig. 3.12.15.

Steps and Recesses

It is advisable to minimize the number of steps in a part and the depth of recesses in order to avoid making tooling unduly complicated.

Knurls and Serrations

These can be produced only in the vertical direction.

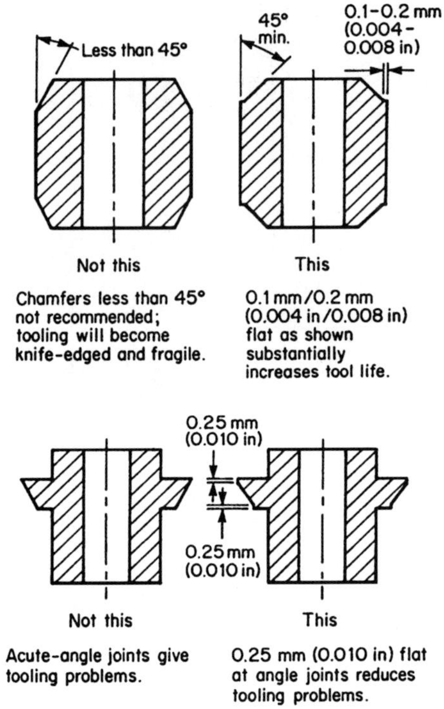

FIGURE 3.12.14 Angled sidewalls.

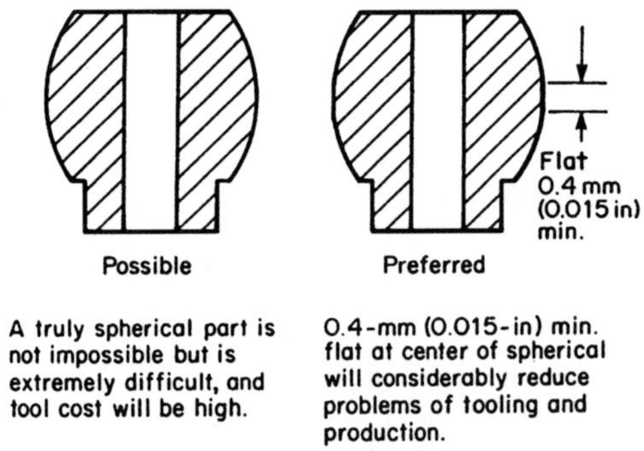

FIGURE 3.12.15 Spherical surfaces.

Lettering or Surface Marking

This presents no problems but can be produced only on the upper and lower faces. For clarity, the letters or marks should be sunk into the surface of the component.

DIMENSIONAL TOLERANCES

The dimensional tolerances that can be maintained are one of the most attractive features of this process. It should be clearly understood, however, that the tolerances provided by the process fall into two distinct patterns:

1. Tolerances that can be maintained at the end of the sintering operation
2. Tolerances that can be maintained only after a repressing or calibrating operation

It will be obvious that components in the as-sintered condition are less expensive than those requiring the repressing operation.

Table 3.12.3 (with Fig. 3.12.16) illustrates typical dimensional tolerances for P/M parts in the as-sintered and repressed or calibrated condition and for powder metallurgy forgings.

TABLE 3.12.3 Typical Dimensional Tolerances for P/M Parts*

	±Tolerances, mm (in)		
Typical dimensions, mm (in)	As-sintered	After repressing	Powder-metallurgy forgings
Outside cross sections			
A, 13 (0.5)	0.04 (0.0015)	0.013 (0.0005)	†
B, 9 (0.35)	0.04 (0.0015)	0.013 (0.0005)	†
C, 50 (2)	0.075 (0.003)	0.025 (0.001)	0.18 (0.007)
D, 40 (1.6)	0.075 (0.003)	0.025 (0.001)	0.15 (0.006)
E, 100 (4)	0.13 (0.005)	0.04 (0.0015)	0.25 (0.010)
Inside cross sections and holes			
F, 5 (0.2)	0.025(0.001)	0.006 (0.00025)	†
G, 10 (0.4)	0.05 (0.002)	0.006 (0.00025)	0.075 (0.003)
H, 7.5 (0.3)	0.04 (0.0015)	0.006 (0.00025)	0.06 (0.0025)
I, 13 (0.5)	0.05 (0.002)	0.013 (0.0005)	0.075 (0.003)
Hole, center to center			
J, 25 (1)	0.05 (0.002)	0.025 (0.001)	0.2 (0.008)
K, 75 (3)	0.10 (0.004)	0.04 (0.0015)	0.25 (0.010)
Flange thickness			
L, 4 (0.15)	0.075 (0.003)	0.05 (0.002)	†
M, 6.3 (0.25)	0.075 (0.003)	0.05 (0.002)	0.18 (0.007)
Overall height			
N, 31 (1.25)	0.10 (0.004)	0.10 (0.004)	0.25 (0.010)
O, 50 (2)	0.13 (0.005)	0.13 (0.005)	0.4 (0.015)
Other vertical dimensions			
P, 7.5 (0.3)	0.075 (0.003)	0.05 (0.002)	†
Q, 38 (1.5)	0.13 (0.005)	0.13 (0.005)	0.3 (0.012)

*See Fig. 3.12.16.

†Powder metallurgy forgings are not usually made this small.

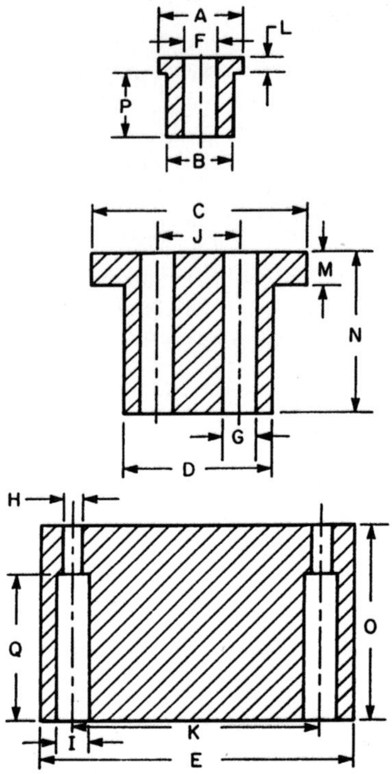

FIGURE 3.12.16 Drawing of P/M parts for use with Table 3.12.3 (typical tolerances for dimensions of P/M parts).

CHAPTER 3.13
FORGINGS

Originally prepared by

Paul M. Heilman

Bridgeport Brass Company
Norwalk, Connecticut

Updated with the assistance of

Philip J. Guichelaar

Western Michigan University
Kalamazoo, Michigan

FORGING PROCESSES

Forging is the controlled plastic deformation of a piece of metal into a useful shape, usually at an elevated temperature. Pressure or repeated press strokes may be used. It resembles such mill operations as hot rolling, blooming, cogging, and hot extrusion except that total deformation is usually much greater and final part geometry is more complex. All these processes refine grain structure and improve physical properties of the metal. The wrought metal in a forging, though, is shaped to become a specific, individual part.

The machinery used for forging must be capable of applying large compressive forces. Traditionally, drop hammers, both gravity and power-assisted, were used and continue to be used. Crank presses and screw presses have become more prevalent because part dimensions are more consistent, and automation is facilitated. Power-driven rolls that produce a shape in one revolution may be used as a preliminary forming operation to develop a rudimentary part shape prior to press or hammer forging. Hydraulic presses are widely used for forging large, simple shapes.

Open-die and hand forgings are made by forming metal between parallel, flat dies. In between successive blows, the operator manipulates the workpiece to gradually form a gross outline of the final machined part. This process is useful for making very large parts and prototype parts.

Impression-die forging uses a matched set of dies with contoured impression in each half to form the part in two or three steps. Depending on the forgeability of the

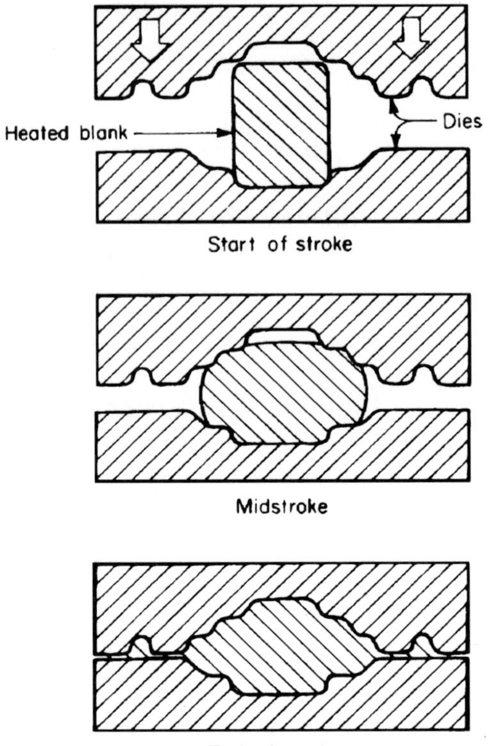

FIGURE 3.13.1 Conventional impression-die forging sequence. (*Based on* American Machinist, *July 1978, Special Report 705, Fig. 3.*)

metal involved and the geometric complexity of the final part, an impression-die forging may be made directly from a heated metal billet or after roll forging to redistribute mass. (Roll forging involves a relatively long, thin workpiece fed into power-driven rolls, which are shaped to vary the section of the bar along its length.) Figure 3.13.1 illustrates conventional impression-die forging.

Precision forgings and low-draft or no-draft forgings are refinements of impression-die forgings. These represent closer approaches to final part shape, with the last operation performed with dies of high accuracy. Conversely, blocker-type forgings are less precise, made with larger fillet radii and thicker cross sections. While extensive final machining is required with blocker forgings, tooling costs are lower, and the process is thus attractive for low production quantities.

Ring rolling forms an axisymmetrical shape from a hollow cylindrical preform made by open-die forging. The preform is placed between two rollers, and as they rotate, the preform wall thickness is reduced and the ring diameter increases. Depending on the shape of the rolls, rectangular or contoured cross sections are produced.

Secondary shaping operations such as twisting, bending, and coining may be performed on forged parts to develop final part geometry. Forgings also may be heat-treated in a separate operation. Most forgings are shot-blasted after forging, and their surfaces may be chemically cleaned.

Cold heading and impact extrusion are cold-forging techniques covered in this handbook in Chaps. 3.7 and 3.8, respectively.

TYPICAL CHARACTERISTICS AND APPLICATIONS

Controlled grain structure is the primary benefit of the forging process. With proper design, it is possible to align grain flow with directions of the principal stresses that will occur when the part is loaded in service. *Grain flow* is the directional pattern that metal crystals assume during plastic deformation. Strength, ductility, and impact resistance along the grain are significantly higher than they would be in the randomly oriented crystals of cast metal or weld metal. Because hot working refines grain structure, physical properties are also improved across the grain. (See Fig. 3.13.2.)

Forging helps ensure structural integrity from piece to piece. Internal pockets, voids, inclusions, laps, and similar flaws are easier to avoid by good forging quality control than they are in castings.

In most cases, forgings must be machined before use. Open-die forgings and parts made of difficult-to-forge metals often have several times as much metal to machine away as will be left in the finished part. In any case, enough metal must be left on each machined face so that the part will clean up. An allowance is added for this purpose during initial design of the part.

Because of high-strength and light-weight requirements, makers of aircraft engines and structures, along with other aerospace manufacturers, are the most significant users of forgings on a value basis. However, their use of forgings is exceeded in numbers and variety by the makers of different kinds of land-based vehicles and portable equipment.

Moving parts are forged to reduce inertia forces, and parts that must be supported by other structures are forged to reduce overall weight and complexity. Hand tools that people lift and handle are forged to reduce weight. Parts whose failure would cause injury or expensive damage are forged for safety. Forgings also seldom have internal flaws to create blemishes after machining or cause leakage when pressure tightness is a requirement.

Decorative parts, even when stressed very lightly, may be produced from forgings to reduce scrap losses and ensure a plateable surface, since forged or machined surfaces of forgings can be polished and plated without revealing blemishes or other internal flaws.

Size limits of forgings depend more on the facilities available than on any inherent

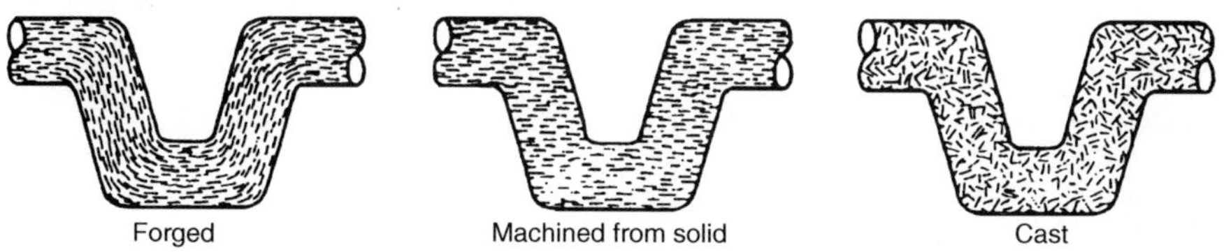

FIGURE 3.13.2 Forging produces parts with an unbroken grain flow following the contour of the part.

FIGURE 3.13.3 A collection of typical forgings. Note that some still have flashing attached. (*Courtesy National Machinery Company.*)

limits of the process itself. Small, highly stressed, high-speed-machine components may weigh less than an ounce. Large, open-die forgings may weigh several tons.

Some typical forging applications are the following: landing gear parts for aircraft, automotive connecting rods, universal joints, crankshafts, off-highway and farm equipment parts, plumbing valves, tees, elbows, ordinance components, railroad wheels, axles, gears, oil-field machinery components, turbine disks and blades, and bearing assemblies. Figure 3.13.3 illustrates a collection of typical forgings.

Forging Nomenclature

Geometric shapes on an impression-die forging are named for the direction in which metal must flow to fill the die impressions. Any wall filled by flow parallel to die motion is a *rib,* and a projection is called a *boss* when it is filled parallel to die motion. The wall filled by generally horizontal flow, perpendicular to die motion and parallel to the parting plane, is the *web.* A *recess* is a small web area surrounded by thicker metal. Figure 3.13.4 illustrates these terms.

It is not practical to forge a through hole in a web. When a hole through a web or a boss will be needed, a recess may be forged in one or both sides. The thin web remaining is punched out later, and the hole subsequently may be cleaned up by machining.

Flash

To be sure that the die cavities will fill completely, excess metal is usually provided. As the die halves come together, the excess is extruded into a gutter at the parting line,

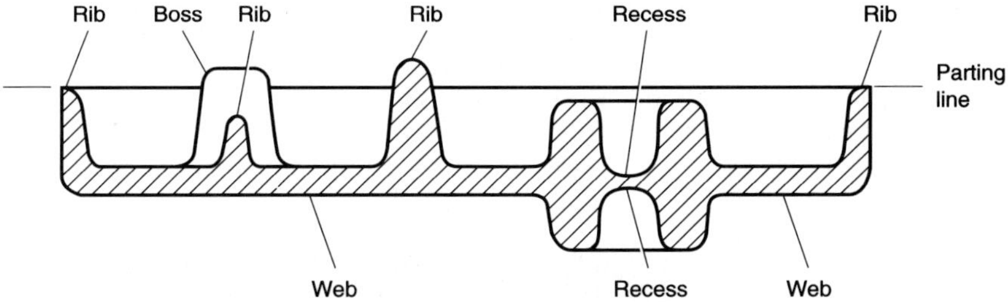

FIGURE 3.13.4 Forging nomenclature relates all features of a part to the direction of die motion.

producing a part with a fringe of flash metal around it. This flash is trimmed off in a separate operation.

ECONOMIC PRODUCTION QUANTITIES

Forging cost is the sum of a number of factors: metal cost including scrap, labor cost, and overhead expenses of the production facility. The one-time cost that does not vary in proportion to number of pieces is the die cost. Even though a forging-die set may require repair or replacement over the course of a long production run, the cost of this is treated as an operating expense according to established industry practice. It is initial die cost, then, that controls the economics of lot size.

By its nature, forging is a high-production process. Its costs grow more attractive as die cost becomes a smaller fraction of piece cost. This is especially true for small parts. A minimum economical lot size for 100-g ($\frac{1}{4}$-lb) forgings usually ranges around 5000 pieces. Very large forgings weighing from 50 kg (110 lb) to $\frac{1}{2}$ ton may be economical in lots as small as 2 or 3 pieces. This approximation assumes ordinary conventional forgings of readily forgeable alloys. However, when only a few pieces are needed, the forging buyer can reduce die costs by utilizing a supplier who does prototype work and has the skills and equipment for such jobs.

One approach is to eliminate die costs entirely by ordering hand forgings made with open general-purpose dies. The time of a highly skilled person and the overhead of specialized facilities will then be major cost items. Much more machining also may be required. However, the parts will have the advantage of the forged controlled-grain structure, freedom from porosity, and minimal nonmetallic inclusion content.

Forgings made in blocker-die impressions come closer to conventional forged-part shapes. They, too, may require extensive machining, but the cost of finishing dies will be avoided. It is common practice to use open-die forgings and blocker forgings in prototype models with the anticipation that conventional or even precision forgings will be justified at a later date in production.

SUITABLE MATERIALS FOR FORGING

Most metals and alloys can be forged at elevated temperatures. However, the ease with which they deform plastically varies widely. Some alloys remain very strong even when heated almost to their melting temperatures. Some have high coefficients of friction at forging temperature, and it is difficult to make them slide along die surfaces. Some are susceptible to metallurgical degradation or to the formation of mechanical flaws in the course of hot working. Differences of this nature are summed up in Table 3.13.1, which ranks metals and their alloys by forgeability.

For preliminary planning and decision-making purposes, the alloys of aluminum, magnesium, and copper along with mild steels, may be regarded as readily forgeable. There are differences among them, but some of these differences tend to balance out. For example, aluminum can be forged at lower temperatures than steel, but it flows less readily and requires higher pressures. There are differences among alloys in each group also. Nevertheless, concessions from conventional design practice are seldom necessary when a material from one of the widely used families of alloys is selected. Steel forgings are usually heat-treated after finish machining to develop the static and dynamic strength properties needed in service.

The stainless steels are somewhat more resistant to plastic flow, but fully formed conventional forgings are produced routinely in these alloys. Superalloy forgings, though, are usually produced only as simpler shapes. Great care is exercised to establish grain-flow patterns that will suit the parts for their intended service. If any part will require elaborate contours or drastic section changes, however, these features must be provided in subsequent operations.

Some metals require atmospheric control during forging. At the extreme, beryllium is sometimes forged by first sealing metal powder in an evacuated welded-steel jacket. The part is formed; then the jacket is removed. Vacuum-hot-pressed beryllium billets also can be forged in impression dies if suitable precautions are taken.

DESIGN RECOMMENDATIONS

Forging Drawings

Good practice dictates that a forging drawing be prepared. Shapes and dimensions of a part as it will be forged, before any machining is done, are shown on this drawing. Die design and processing requirements are dictated by the way in which the part is drawn. Grain flow must be aligned with the direction of highest principal stress. An experienced designer usually can visualize metal flow from bar stock to final forging and the resulting grain-flow pattern. A forging manufacturing engineer may have concerns, however, about potential laps and locally excessive die wear and recommend changes to the forging drawing that may affect grain flow and the exact shape of the final part. Flash is not customarily indicated on the forging drawing.

For a number of reasons, forging design should be developed in partnership between the forging user and the forging producer. To neglect technical contributions that either partner can make is to risk a needless waste of money and performance.

It is often advisable to use metal-flow simulation software to study blocker and finisher shapes for forgings. The simulation software shows how a metal bar changes shape under the action of the forging press or hammer, predicts total forging loads and tooling stresses, indicates where laps and other defects may form, and describes grain-flow patterns. A three-dimensional simulation takes a few hours if the part shape is available in a computer-aided design (CAD) file. In contrast, the dies for a prototype forging take several weeks to machine and then may have to be modified after forging trials are completed.

TABLE 3.13.1 Relative Forgeability*

Base metal	Alloy	UNS designation
Aluminum	2104	A92104
	2024	A92024
	6061	A96061
	7075	A97075
Magnesium	AZ31B	M11311
	AZ61A	M11610
	ZK60A	M16600
	HM21A	M13210
	EK31A	M13310
Copper, brass, and bronze	CA377	C37700
	CA464	C46400
	CA485	C48500
	CA642	C64200
	CA673	C67300
	CA675	C67500
	CA102	C10200
	CA110	C11000
	CA147	C14700
	CA150	C15000
	CA694	C69400
Steel	1010–1030	G10100–G10300
	1050–1095	G10500–G10950
	4140	G41400
	4340	G43400
	8740	G87400
	1045	G10450
Martensitic stainless steel	430	S43000
	405	S40500
	450	S45000
	431	S43100
	410	S41000
	455	S45500
	420	S42000
	440C	S4404
Maraging steel	250	
	200	
Austenitic stainless steel	316	S31600
	317	S31700
	302	S30200
	304	S30400
	310	S31000
	321	S32100
	347	S34700
	303	S30300
Nickel	Nickel 200	N02200
	Monel 400	N04400

TABLE 3.13.1 Relative Forgeability* (*Continued*)

Base metal	Alloy	UNS designation
Precipitation-hardening stainless steel	17-7PH	S17700
	AM 355	S35500
	17-4PH	S17400
Titanium	Ti-6Al-6V-2Sn	R56620
	Ti-6Al-4V	R56401
	Ti-7Al-4Mo	R56740
	Ti-5Al-2.5Sn	R54521
	Ti-13V-11Cr-3Al	R59010
Iron-base superalloy	16-25-6	
	19-9DL	
Cobalt-base superalloy	S-816	
	L-605	R30605
	V-36	R30036
Niobium	Unalloyed	
	Nb-1Zr	R05261
Tantalum	Unalloyed	
	Ta-10W	R05255
Molybdenum	Unalloyed	
	Mo-0.5Ti-0.1Zr	R03630
	Mo-0.5Ti	R03620
Nickel-base superalloy	Hastelloy X	N06002
	Inconel 718	N07718
	Waspaloy	N07001
	Incoloy 901	N09901
	Inconel X-750	N07750
	M-252	N07252
Tungsten	Unalloyed	
Beryllium	Unalloyed	

*These listings, in order of decreasing forgeability, are approximate. There are too many variables for a precise summation that would fit every case. Also, the least forgeable alloys of a basic metal may be more difficult to forge than the more forgeable alloys of the next metal listed. Any tentative choice based on this preliminary guide should be confirmed for the specific forging operation under consideration.

More forgings are produced in two-part impression dies. The design of such forgings is the topic of the following discussions.

Parting Line

As the die halves come together and confine metal in their cavities, their mating surfaces define a parting line around the edges of the forging. The parting line is indicated on the forging drawing, and determining its location is a critical step in forging design.

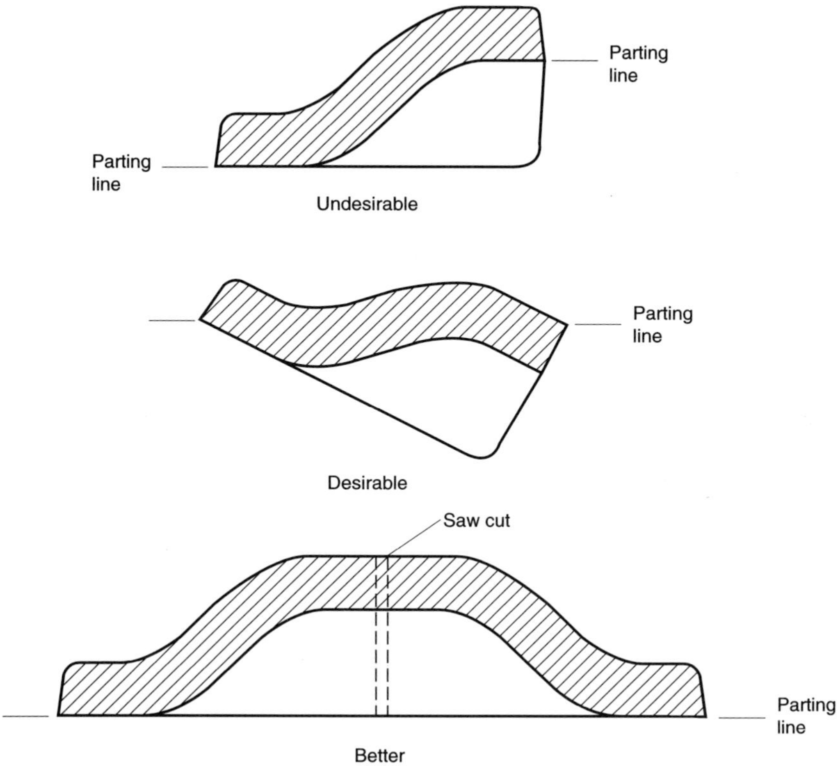

FIGURE 3.13.5 Ideally, the parting line should lie in one plane, perpendicular to the direction of die motion. Jogs in parting lines impose side-thrust forces on the die halves. It is expensive to absorb these forces with die counterlocks. Often, symmetry can be achieved by forging pieces as pairs. If they are right- and left-handed mates, this approach has extra merit.

Ideally, the parting line will lie in one plane perpendicular to the axis of die motion, as shown in Fig. 3.13.5. Sometimes it can be located so that one die half will be completely flat, and the line will surround the largest projected area of the piece. (See Fig. 3.13.6.)

If the parting line cannot lie in one plane, it is desirable to preserve symmetry so as to prevent high side-thrust forces on the dies and the press. Such forces can be countered, at extra die cost, if they are unavoidable. No portion of the parting line should incline more than 75° from the principal parting plane, and much shallower angles are desirable.

An obvious essential is to select a parting line that will not entail any undercuts in either die impression, since the forging must come out of the die after it is made.

Because metal flow at the parting line is outward into the flash gutter, grain flow in the forging has a corresponding pattern. Depending on the way in which the part will be loaded, it may be desirable to change parting-line location to control grain flow. (See Fig. 3.13.7.)

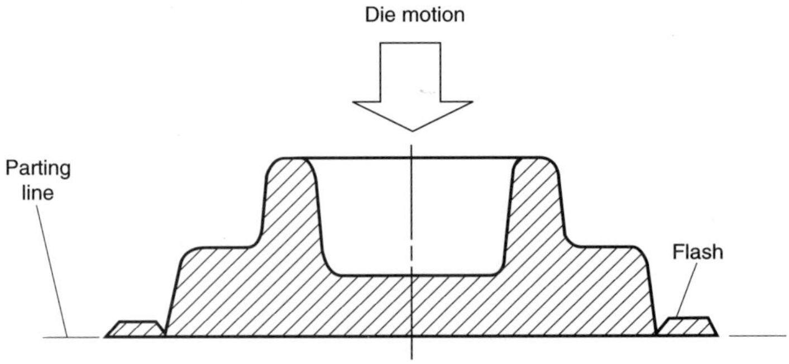

FIGURE 3.13.6 The parting line here is in one plane perpendicular to die motion, and the impression is entirely in one die half. This is usually the most economical tooling arrangement for two-part impression dies.

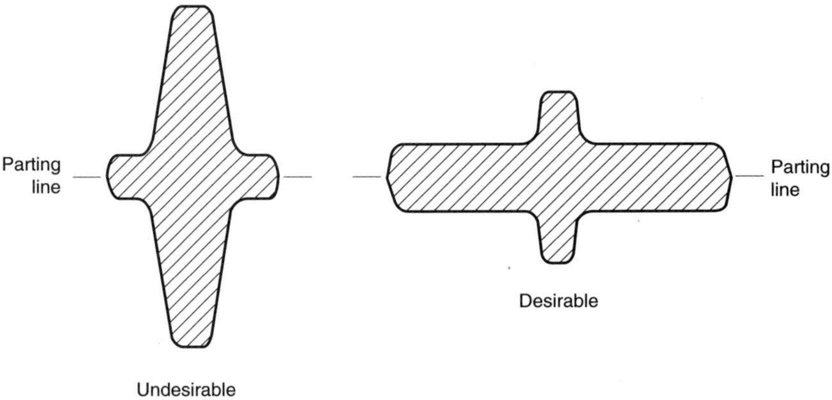

FIGURE 3.13.7 (*a*) When there is a choice, locate the parting line so that metal will flow horizontally, parallel to the parting line.

Draft

Die impressions are tapered so that forgings can be removed from their dies, and forged surfaces that lie generally parallel to die motion are correspondingly tapered. This taper, called *draft,* also promotes flow into relatively deep die cavities.

Draft is specified as an angle with respect to the die-motion axis. Conventionally, a standard draft angle will be specified for all affected surfaces on a forging, which simplifies tooling for die sinking. It is also conventional to call for matching draft on both die halves to make surfaces of unequal depth meet at the parting line. Table 3.13.2 shows typical standard draft angle ranges for finished forgings in the various alloy families.

Sometimes, a parting-line location presents tapered surfaces automatically because of a part's shape. For example, a cylinder lying parallel to the parting plane has such natural draft except for small bands next to the parting line. The draft needed there

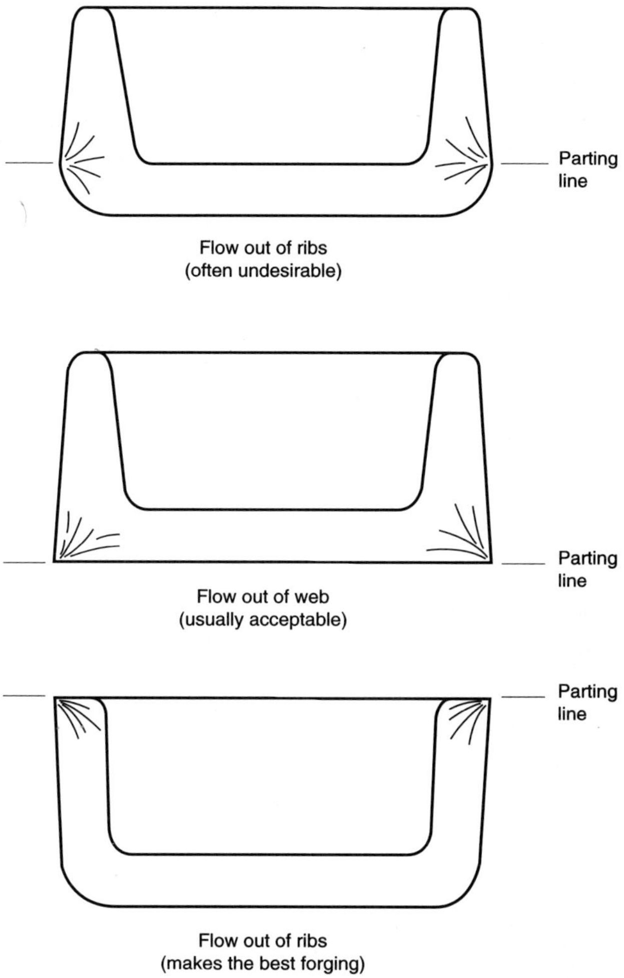

FIGURE 3.13.7 (*Continued*) (*b*) The parting line location governs when the constricted grain flow associated with flash will occur on the part. The designer can locate the parting line to achieve the objectives of each part's function.

TABLE 3.13.2 Typical Draft Angles

Alloy family	Draft angle, °
Aluminum	0–2
Magnesium	0–2
Brass and copper	0–3
Steel	5–7
Stainless steel	5–8
Titanium	5–6

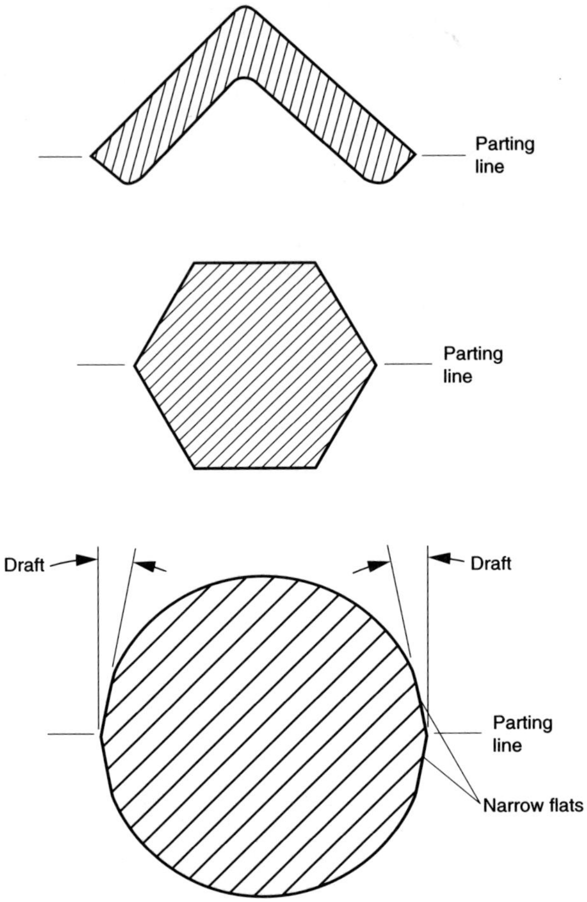

FIGURE 3.13.8 These shapes have natural draft. Usually, the cylindrical section will be modified slightly to provide draft in the narrow region next to the parting line.

will be provided by narrow tapered tangents, but they need not be indicated on the forging drawing. (See Fig. 3.13.8.)

Low-draft and no-draft forgings can be produced in some metals, such as aluminum and brass. This usually applies to selected surfaces for which reduction or elimination of draft yields significant benefits.

Ribs, Bosses, Webs, and Recesses

Metal flow is relatively easy to manage when ribs and bosses are not too high and narrow, and it is easiest when the web is relatively thick and uniform in thickness. (See Fig. 3.13.9.)

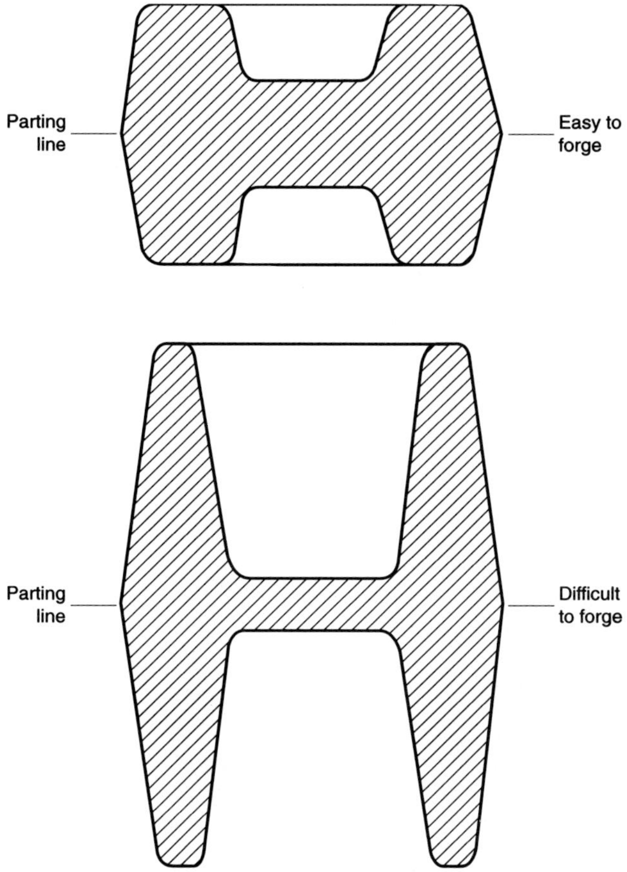

FIGURE 3.13.9 As the web becomes thinner and the ribs become deeper, forging difficulty increases.

Correspondingly, forging becomes more difficult when large amounts of metal must be moved out of relatively thin webs into such projections as deep ribs and high bosses. It is helpful to taper such webs toward the ribs and bosses. Deep recesses also are easier to forge if they have spherical bottoms. When successive forging operations are entailed, it can be advantageous to design for a fairly large punch-out hole in the thin-web section. During finish forging, after the hole has been punched, flash flows inward at its edges and helps to relieve excessive die forces.

Surface textures, designs, and lettering on forged surfaces are simply very small ribs and recesses. Locate these features on surfaces that are as nearly perpendicular to die motion as possible, and locate them away from zones of wiping metal flow. Call for raised lettering and numbers, which can be produced by milling recesses in the die. It is more difficult to achieve die projections that will form recessed symbols on the forging.

Radii

Forgings are designed with radii on all their external corners except at the parting line. It would require a sharp internal angle in the die to form a sharp corner on the forging. This is a vulnerable stress raiser; also, excessive pressure would be required to fill sharp corners. Both considerations suggest generous corner radii. A common practice is to call for full radii at the edges of all ribs and the same radius on each corner of a boss, web, or other shape.

Fillet radii on a forging correspond to corners in die impressions that metal must round to fill ribs and bosses. If metal flows past a sharp corner and then doubles back, the forging may be flawed with a lap or cold shut, and the die may not fill completely. This is more likely if the sharp die corner or sharp fillet in the forging is near the edge of the piece.

While all radii should be ample for easy forging, they can be made smaller in readily forgeable metals whenever there is a good reason for doing so. Adding forging costs should be justified by the benefits gained.

Table 3.13.3 shows typical radii in terms of forging proportions. The deeper the impression, the larger the radius should be, both at the fillet around which metal must flow and at the corner that must fill with metal.

TABLE 3.13.3 Typical Minimum Radii for Forgings

	Minimum radius, mm (in)	
Depth of rib or boss, mm (in)	Corner	Fillet
13 (½)	1.6 (1/16)	5 (3/16)
25 (1)	3 (⅛)	6.3 (¼)
50 (2)	5 (3/16)	10 (⅜)
100 (4)	6.3 (¼)	10 (⅜)
200 (8)	16 (⅝)	25 (1)
400 (16)	22 (⅞)	50 (2)

Machining Allowance

Design features that promote easy forging add to the metal that must be machined away. Ample draft angles, large radii, and generous tolerances can all have this effect. The machining allowance should allow for the worst-case buildup of draft, radii, and all tolerances. If a part is forged with locating pads on it for setup reference, calculate tolerance buildup from those points. See Table 3.13.4 for typical allowances for machined surfaces. Extra metal is sometimes provided to keep critical machined surfaces away from the grain-flow pattern that occurs in the flash region near the parting line.

Machining allowances or finishing allowances are added to external dimensions and subtracted from internal dimensions.

Other Forging Processes

Several of the limitations laid down for impression forgings made in two-part dies can be bypassed and eliminated by upset- or cored-forging techniques. Shapes that would constitute undercuts for two-part dies can be forged. Sharp external corners are feasi-

TABLE 3.13.4 Typical Machining Allowances for Forging

Alloy family	Forging size: projected area at parting line, mm (in) To 640 cm^2 (100 in^2)	To 2600 cm^2 (400 in^2)	Over 2600 cm^2 (400 in^2)
Aluminum	0.5–1.5 (0.020–0.060)	1.0–2.0 (0.040–0.080)	1.5–3.0 (0.060–0.120)
Magnesium	0.5–1.5 (0.020–0.060)	1.0–2.0 (0.040–0.080)	1.5–3.0 (0.060–0.120)
Brass	0.5–1.5 (0.020–0.060)	1.0–2.0 (0.040–0.080)	1.5–3.0 (0.060–0.120)
Steel	0.5–1.5 (0.020–0.060)	1.5–3.0 (0.060–0.120)	3.0–6.0 (0.120–0.240)
Stainless steel	0.5–1.5 (0.020–0.060)	1.5–2.5 (0.060–0.100)	1.5–5.0 (0.060–0.200)
Titanium	0.8–1.5 (0.030–0.060)	1.5–3.0 (0.060–0.120)	2.0–6.0 (0.080–0.240)
Niobium	0.8–2.5 (0.030–0.100)		
Tantalum	0.8–2.5 (0.030–0.100)		
Molybdenum	0.8–2.0 (0.030–0.080)	2.0–3.0 (0.080–0.120)	

ble. Draft can be reduced, and no draft at all may be possible on some surfaces. Deep recesses also can be forged.

TOLERANCES

The tolerances summarized below should be regarded as guidelines rather than as absolutes. Adjustments can be made from these values when it is advisable for reasons of either manufacturing economy or the component's function. They apply to impression die forgings made in two-part die sets. Aluminum forgings are commonly made to higher precision than listed here.

Length and Width Tolerances

Dimensions generally parallel to the parting plane and perpendicular to die motion are subject to length and width tolerances. When a forged projection extends more than 150 mm (6 in) from the parting plane, dimensions to its extremities, measured parallel to die motion, are also subject to these tolerances. Length and width tolerances are commonly specified at +0.3 percent of each dimension, rounded off to the next higher $\frac{1}{2}$ mm or $\frac{1}{64}$ in.

Die-Wear Tolerances

These tolerances apply only to dimensions generally parallel to the parting plane and perpendicular to die motion. The corresponding variations parallel to die motion are included in die-closure tolerances.

Die-wear tolerances are plus variations of external dimensions and minus variations of internal dimensions. They do not affect center-to-center dimensions. Thus they allow for erosion of die metal and corresponding enlargement of the forged parts.

TABLE 3.13.5 Typical Die-Wear Tolerances*

Alloy family	%
Aluminum, 2014 (UNS-A92014)	0.4
Aluminum, 7075 (UNS-A97075)	0.7
Magnesium	0.6
Brass and copper	0.2
Mild steel	0.4
Alloy steel	0.5
Martensitic stainless steel	0.6
Austenitic stainless steel	0.7
Titanium	0.9
Superalloys	0.8
Refractory alloys	1.2

*Plus variations of external dimensions and minus variations of internal dimensions.

While this tolerance is applied routinely to all horizontal dimensions, as a practical matter, dies are subject to severe wear only in the zones of harsh metal flow.

Table 3.13.5 shows typical tolerances. Multiply each horizontal dimension by the appropriate factor, and round off the tolerance to the next higher $\frac{1}{2}$ mm or $\frac{1}{64}$ in. These are tolerances on the dimensions themselves. They are based on die wear at each surface of half these values.

Die-Closure Tolerances

Dimensions parallel to die motion between opposite sides of a forging are affected by failure of the two die halves to close precisely. The plus tolerances on such dimensions are shown in Table 3.13.6. There is no minus tolerance in this category. Effects of die wear on these vertical dimensions are included in the die-closure tolerances. An added tolerance of 0.3 percent applies to any projection that extends more than 150 mm (6 in) from the parting plane.

Match Tolerances

A lateral shift of one die half with respect to the other moves all features on opposite sides of the forging correspondingly. Table 3.13.7 shows typical tolerances in terms of piece weight and material.

Straightness Tolerances

For relatively long, thin parts, a typical straightness tolerance is 0.3 percent of length. When this aspect of forging accuracy is critical, forged parts are often straightened in secondary cold operations.

TABLE 3.13.6 Typical Die-Closure Tolerances, mm (in)

Alloy family	Forging size: projected area at parting line, mm (in)						
	To 65 cm^2 (10 in^2)	To 135 cm^2 (30 in^2)	To 320 cm^2 (50 in^2)	To 650 cm^2 (100 in^2)	To 3200 cm^2 (500 in^2)	To 6450 cm^2 (1000 in^2)	Over 6450 cm^2 (1000 in^2)
Aluminum, magnesium, and brass	0.8 (1/32)	0.8 (1/32)	1.6 (1/16)	2.4 (3/32)	3.2 (1/8)	4.8 (3/16)	6.3 (1/4)
Steel	0.8 (1/32)	1.6 (1/16)	2.4 (3/32)	3.2 (1/8)	4.0 (5/32)	4.8 (3/16)	6.3 (1/4)
Martensitic stainless steel	0.8 (1/32)	1.6 (1/16)	2.4 (3/32)	3.2 (1/8)	4.8 (3/16)	6.3 (1/4)	8.0 (5/16)
Austenitic stainless steel	1.6 (1/16)	2.4 (3/32)	3.2 (1/8)	4.0 (5/32)	4.8 (3/16)	6.3 (1/4)	8.0 (5/16)
Titanium	1.6 (1/16)	2.4 (3/32)	3.2 (1/8)	4.8 (3/16)	6.3 (1/4)	8.0 (5/16)	9.5 (3/8)
Super alloys	1.6 (1/16)	2.4 (3/32)	3.2 (1/8)	4.8 (3/16)	6.3 (1/4)	8.0 (5/16)	9.5 (3/8)
Refractory alloys	2.4 (3/32)	3.2 (1/8)	4.0 (5/32)	4.8 (3/16)	6.3 (1/4)	8.0 (5/16)	9.5 (3/8)

TABLE 3.13.7 Typical Match Tolerances, mm (in)

Alloy family	Forging size: weight after trimming, mm (in)							
	To 2.3 kg (5 lb)	To 11 kg (25 lb)	To 23 kg (50 lb)	To 45 kg (100 lb)	To 90 kg (200 lb)	To 225 kg (500 lb)	To 450 kg (1000 lb)	Over 450 kg (1000 lb)
Aluminum, magnesium, brass, and steel	0.4 (1/64)	0.8 (1/32)	1.2 (3/64)	1.6 (1/16)	2.4 (3/32)	3.2 (1/8)	4.0 (5/32)	4.8 (3/16)
Stainless steel, titanium, and superalloys	0.8 (1/32)	1.2 (3/64)	1.6 (1/16)	2.4 (3/32)	3.2 (1/8)	4.0 (5/32)	4.8 (3/16)	6.3 (1/4)
Refractory alloys	1.6 (1/16)	2.4 (3/32)	3.2 (1/8)	4.0 (5/32)	4.8 (3/16)	6.3 (1/4)	8.0 (5/16)	9.5 (3/8)

TABLE 3.13.8 Typical Flash-Extension Tolerances, mm (in)

Alloy family	Forging size: weight after trimming, mm (in)							
	To 5 kg (10 lb)	To 10 kg) (25 lb	To 25 kg (50 lb)	To 50 kg (100 lb)	To 100 kg (200 lb)	To 250 kg (500 lb)	To 500 kg (1000 lb)	Over 500 kg (1000 lb)
Aluminum, magnesium, brass, and steel	0.8 (1/32)	1.6 (1/16)	2.4 (3/32)	3.2 (1/8)	4.8 (3/16)	6.3 (1/4)	8.0 (5/16)	9.5 (3/8)
Stainless steel, titanium, and superalloys	1.6 (1/16)	2.4 (3/32)	3.2 (1/8)	4.8 (3/16)	6.3 (1/4)	8.0 (5/16)	9.5 (3/8)	12.7 (1/2)
Refractory alloys	3.2 (1/8)	4.8 (3/16)	6.3 (1/4)	8.0 (5/16)	9.5 (3/8)	12.7 (1/2)	15.8 (5/8)	19 (3/4)

Flash-Extension Tolerances

Although there are many other possibilities, the most common flash-removal method is by a punching operation in contoured dies. This may produce clean, trimmed edges, but a small bead of flash is allowed. Conventional flash-extension tolerances shown in Table 3.13.8 are appropriate when this procedure is acceptable.

Draft Angles

Common tolerances on draft angles are $+2°$ and $-1°$.

Radii

The conventional tolerance on all corner and fillet radii is plus or minus one-half the radius. On any corner where metal will be removed later, the plus radius tolerance governs how much metal will be left for producing a sharp corner on the finished part. The minus radius tolerance, which would only limit sharpness of the forged corner, is not enforced.

Total Tolerances

When a forging drawing is being prepared, the tolerances, plus and minus, for each dimension are arithmetic sums of all individual tolerances that apply to the surfaces involved. Other tolerances that may appear as notes, such as those on draft angles, radii, mismatch, die wear, and straightness, are also additive because they affect these surfaces.

A forging should be dimensioned so that enough metal will be available on every surface to satisfy all functional requirements of the finished part.

CHAPTER 3.14
ELECTROFORMED PARTS

E. N. Castellano

General Manager
Liqwacon Corporation
Thomaston, Connecticut

ELECTROFORMING PROCESS

Electroforming is the production of articles by electrodeposition. It is similar to electroplating, except that the latter involves only a resurfacing of an existing article, whereas electroforming involves the fabricating of a new object which did not previously exist. It differs in the extra process time, the greater thickness of the coating, and the extra controls and process steps required to ensure that it is uniform, smooth, stress-free. There are three steps in the process: (1) Prepare a mandrel of the appropriate size, shape, and finish; (2) electroplate it to the thickness required (much thicker than in electroplating for surface coating); and (3) separate the mandrel from the electroplated material. (Sometimes this is not done, or only part of the mandrel is removed.)

TYPICAL CHARACTERISTICS AND APPLICATIONS OF ELECTROFORMED PARTS

Electroforming is versatile. Parts fabricated by electroforming can be simple or complex and can have walls as thin as 0.025 mm (0.001 in) and as thick as 13 mm ($\frac{1}{2}$ in) or more. They can range from a very small size, e.g., 10 mm ($\frac{3}{8}$ in) in length, to waveguides that weigh up to 220 to 270 kg (500 to 600 lb) and are 1.5 to 1.8 m (5 to 6 ft) in length.

An outstanding characteristic of electroforms is their near-perfect reproduction of surface details. This property accounts for a long-standing application of electroforming: the production of compact audio and video disc stamper molds. High accuracy of shapes and dimensions is another property of electroforms.

Holes as small as 0.013 mm (0.0005 in) can be made by attaching nonconductive filaments to the electroforming mandrel. Larger holes of various shapes can be made by application of nonconductive materials to the mandrel. Bosses can be incorporated, and if collapsible, dissolvable, or flexible mandrels are used, undercuts, reentrant angles, and reverse tapers are possible. Dissimilar metals can be laminated and cold-welded, and components such as plates, pins, tubes, etc., can be incorporated as inserts in the electroform.

Major applications of electroformed parts are found in the aerospace, electronics, and electrooptics industries. Common electroformed parts include duplicating plates, molds, paint masks, surface-finish standards, waveguides, reflectors, rocket thrust chambers, venturi nozzles, missile nose cones, medical and prosthetic devices, and holographic masters.

Designers should be aware that most electroformed parts have some degree of stress that may cause distortion of the part after it is removed from the mandrel. Special attention to the chemical and electrolytic aspects of the process can minimize this effect.

Figures 3.14.1 through 3.14.4 illustrate typical parts.

FIGURE 3.14.1 Large, heavy-walled ($^1/_4$-in nickel thickness) waveguide with "grow-on" stainless-steel flanges. (*Photograph used with permission of GAR Electroforming Division, Mite Corp., Danbury, Conn.*)

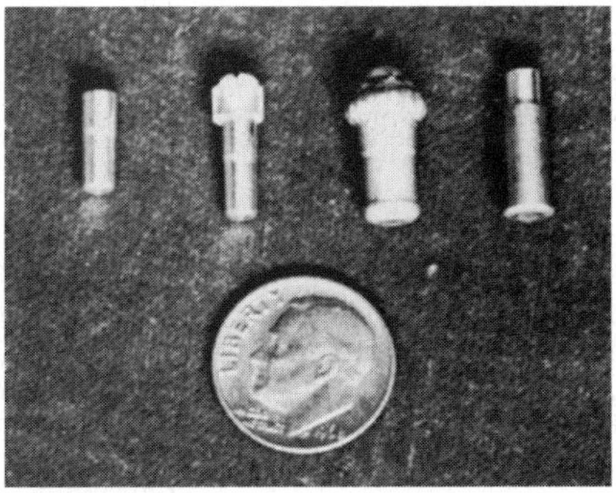

FIGURE 3.14.2 Miniature precision electroformed components. (*Photograph used with permission of GAR Electroforming Division, Mite Corp., Danbury, Conn.*)

ECONOMICS OF ELECTROFORMING

Electroforming is applicable to production levels ranging from one of a kind to mass production. In many cases in which prototypes or short runs are involved, mandrels can be provided with short lead times. The production time in such cases, however, may not be short, particularly if large, thick-walled parts are to be produced. Several hours or even several days may be required for the formation of heavy-walled electroforms.

However, the process is employed for high production levels when multiple mandrels are used. In such cases, one operator can tend as many as 500 parts in process at

FIGURE 3.14.3 Large electroformed nickel vacuum manifold jacket for a space shuttle. (*Photograph used with permission of GAR Electroforming Division, Mite Corp., Danbury, Conn.*)

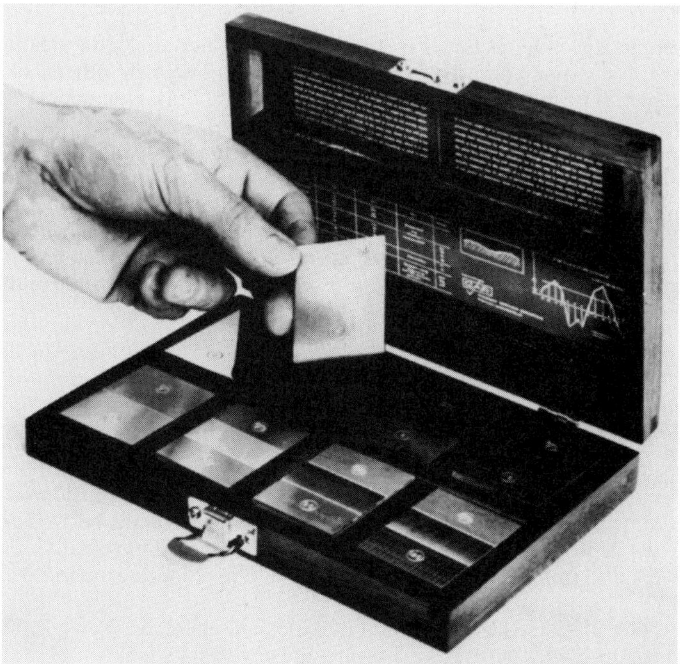

FIGURE 3.14.4 Electroformed microfinish comparator blocks. (*Photograph used with permission of GAR Electroforming Division, Mite Corp., Danbury, Conn.*)

one time. Unit labor costs, therefore, are low. In other cases in which fewer multiple mandrels are used, labor costs are more substantial.

Electroform production involves skilled labor. At one electroforming company, it takes an average of 7 years' apprenticeship to train an electroforming technician: 5 years as an apprentice machinist–tool-and-die maker and 2 years' electrochemical training.

Little or no material wastage occurs in electroforming. However, the most commonly used materials are somewhat costly compared with metals generally used in other forming and fabricating operations.

Tooling costs also may be high if the workpiece is complex or large. Tooling costs for electroformed components can run from practically negligible amounts for simple parts to tens of thousands of dollars for steel molds to cast complex mandrels for aerospace projects. The small components illustrated in Fig. 3.14.2 can be produced on mandrels made by lathes or screw machines.

The total effect of these factors is a relatively high cost level for electroformed parts. However, when the special attributes of electroforms—precise surface reproduction, accurate dimensions, and complex seamless shapes—are involved, no other method of fabrication is as satisfactory, and the cost of electroforming becomes a favorable factor.

SUITABLE MATERIALS

Basically, the prospective customer is limited to two materials for fabricating, copper and nickel, with possible alternatives of gold and silver. There are others, such as iron, lead, and cobalt, but for practical consideration there are only two, copper and nickel, which can be used very successfully in most applications.

Copper, the most frequently used electroforming material, is inexpensive, plates easily, and has low residual stresses after plating. Its ultimate tensile strength is 200 to 325 MPa (30,000 to 55,000 lbf/in^2). Nickel has high tensile strength, 350 to 650 MPa (60,000 to 110,000 lbf/in^2), and is the most widely used material when structural strength is important. It is also used when corrosion resistance is required. Iron is low in cost but tends to be brittle and highly stressed when electrodeposited and is subject to corrosion.

DESIGN RECOMMENDATIONS

One important design parameter of electroforming that must be considered is inside-corner weakness, as shown in Fig. 3.14.5. Inside corners should be well rounded to ensure an even deposit of metal. (If sharp inside corners are unavoidable, they can be made appreciably stronger by the intermediate added soldering operation shown in Fig. 3.14.6.)

Outside sharp corners cause a buildup of deposited metal, as shown in Fig. 3.14.7. When this is undesirable, outside corners should be well rounded.

Although complex shapes are well within the capabilities of electroforming, economy is served by minimizing the number and depths of holes and grooves and by keeping bosses low and wide. Otherwise, mandrels are more complicated, and shaped

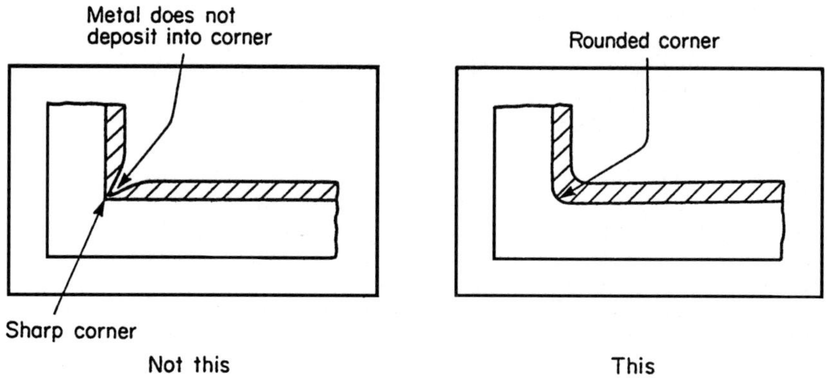

FIGURE 3.14.5 Avoid sharp internal corners on electroformed parts.

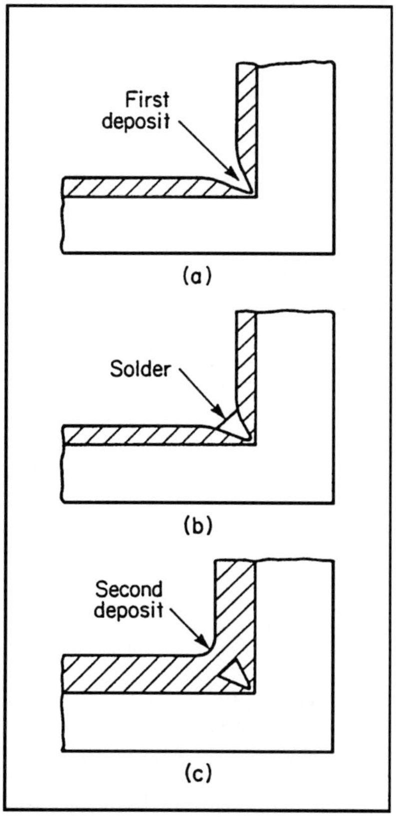

FIGURE 3.14.6 Using solder to augment metal coverage in a sharp corner. (*From Douglas C. Greenwood,* Engineering Data for Product Design, *McGraw-Hill, New York, 1961.*)

anodes may be required for uniform metal thickness.

Undercuts, reverse tapers, and reentrant angles especially should be avoided, if possible, because they necessitate a flexible or dissolvable mandrel that is less accurate and more costly. Included angles should be as wide as possible to promote uniform metal coverage. Holes and grooves should be as wide as possible, at least $1\frac{1}{2}$ times their depth. (See Fig. 3.14.8.)

Figure 3.14.9 shows the normal thickness distribution of electroformed nickel, copper, and silver in angles, grooves, and holes. Hole depth and diameter affect the ratio. The depth of the angle and the radius at the apex of the angle also affect the ratio. The values shown are approximate and suitable for design purposes only. The location of the anode in the electroforming tank and the precise configuration of the electroform itself may affect the values shown.

Wall thickness should be designed to be essentially uniform. Although it can be controlled by masking, anode shaping, and anode placement, precise control is difficult. Sudden or large-magnitude wall-thickness changes should particularly be avoided.

Electroforms made on permanent metal mandrels should have some internal draft or taper to facilitate mandrel removal after forming. Taper of 0.1 percent (0.001 in/in) is usually adequate.

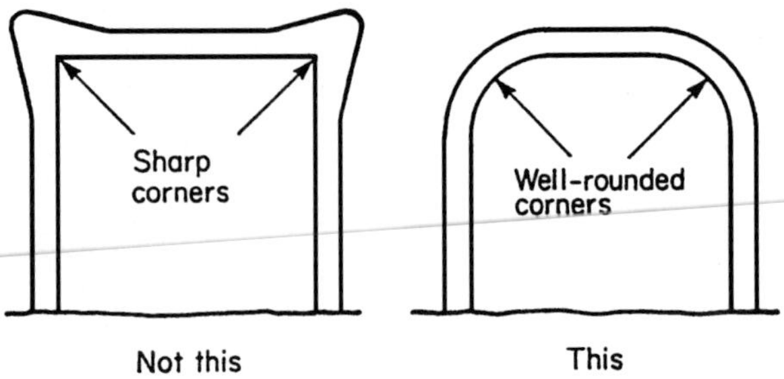

FIGURE 3.14.7 Metal builds up at outside corners. Round such corners as much as possible.

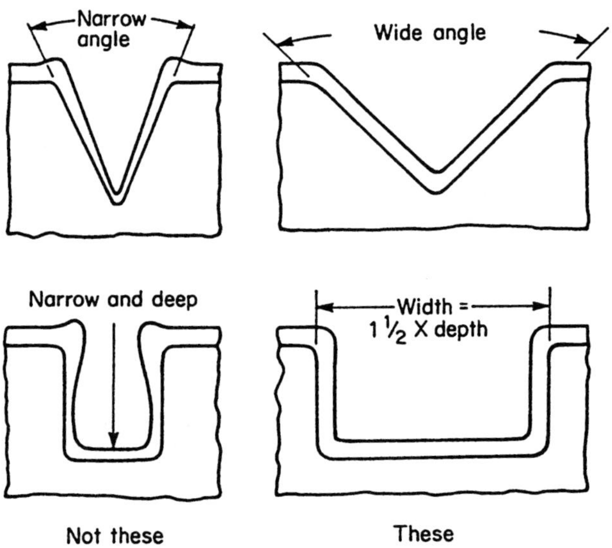

FIGURE 3.14.8 Narrow angled surfaces and narrow, deep grooves lack an adequate metal deposit and should be avoided.

TOLERANCES

Tolerances that can be held on electroformed parts depend primarily on the accuracy of the mandrel. This in turn depends on the skill of the toolmaker and the capabilities of the equipment and tools. If the mandrel is machined accurately enough, a tolerance as close as ±0.0025 mm (0.0001 in) can be held. However, such a tolerance would be expensive. A more realistic allowance for normal everyday production with metal mandrels would be ±0.025 mm (0.001 in).

Surface finish tolerances also depend on the finish of the mandrel, and values as low as 0.05 μm (2 μin) have been held. Recommended tolerances for production conditions are shown in Table 3.14.1.

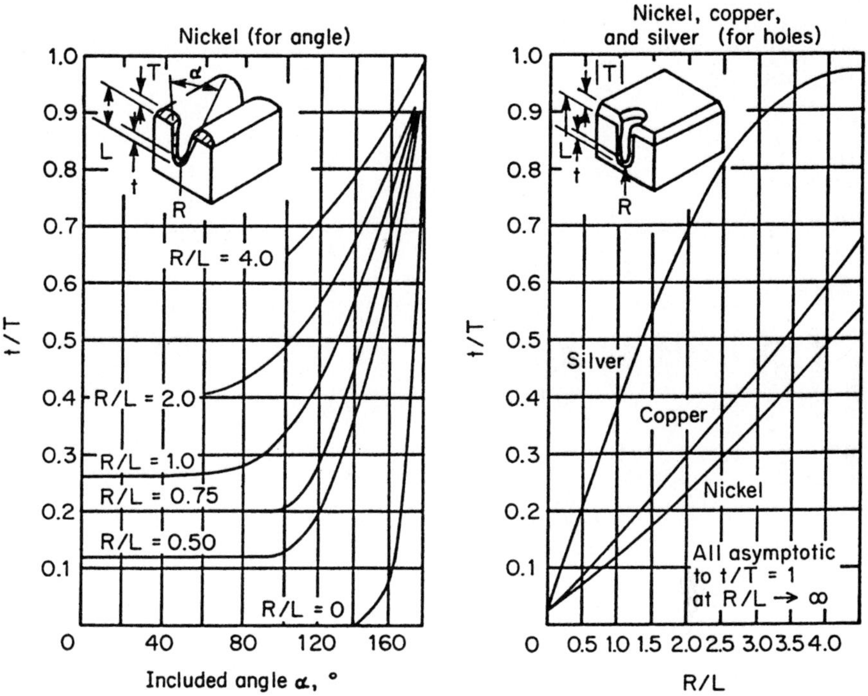

FIGURE 3.14.9 These diagrams show the distribution of metal deposit in holes, angles, and grooves. *t/T* is the ratio of metal thickness in deep areas to that in prominent, exposed areas. *R/L* is the ratio of bottom radius to the groove or hole depth. (*From Douglas C. Greenwood,* Engineering Data for Product Design, *McGraw-Hill, New York, 1961.*)

TABLE 3.14.1 Surface-Finish Tolerance Recommendations for Production Conditions of Electroformed Parts

Nominal surface roughness	Tolerance as a percentage of rated value
0.05–0.10 μm (2–4 μin)	+25, −35
0.2 μm (8 μin)	+25, −30
0.4 μm (16 μin)	+15, −25
0.8 μm (32 μin) and above	+15, −20

Source: ASA B-46.1 1962, *Surface Texture Standard,* American Society of Mechanical Engineers, New York.

Wall-thickness variations within a part can be substantial if the part shape is complex, as indicated by Fig. 3.14.9. However, repeatability of wall thickness from part to part at corresponding points can be maintained within ±0.025 mm (0.001 in).

CHAPTER 3.15

PARTS PRODUCED BY SPECIALIZED FORMING METHODS

Dieter E. A. Tannenberg

Senior Vice President, AM International, Inc.
President, Multigraphics Division
Mount Prospect, Illinois

STRETCH-FORMED PARTS

The Process

In *stretch forming,* sheet-metal workpieces are formed over a block to the required shape while being held in tension. To retain the contour of the form block, the workpiece is stretched just beyond its yield point (usually 2 to 4 percent total elongation). Bars and rolled or extruded sections are also processible by this method.

In the form-block method, a single male die is used. Each end of the blank is held by a gripper. Movement of either the block or the grippers stretches the blank over the form block and gives the workpiece its shape. Figure 3.15.1 illustrates this method. It

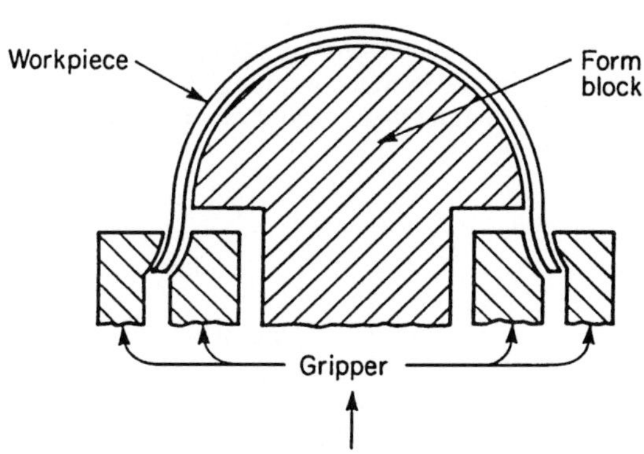

FIGURE 3.15.1 Stretch forming.

shows a linear motion between the form block and the stretched blank. However, rotary motion also can be employed.

The mating-die method uses a two-piece die mounted in a single-action press. The workpiece material is placed between the die halves, which come together and form the part.

Typical Characteristics and Applications

Many shapes that can be produced by other sheet-forming methods also can be produced by stretch forming. Stretch forming is best in the production of shallow or nearly flat contours. Applications include aerospace parts (some of which are impossible or difficult to produce by other methods), automotive body panels and frame parts, both interior and exterior, and architectural forms that call for compound curves, reverse bends, twists, and bends in two or more planes.

Stretch forming is limited in its ability to form sharp contours and reentrant angles. It is also not possible to apply deep forming in the direction of the free (ungripped) edges. Best results are achieved by the use of rectangular blanks rather than blanks of other shapes. It is difficult to improve surface finish through stretch forming because tool contact with the surface is incidental. In stretch forming, some springback is always involved, but it can be greatly reduced and controlled by overforming. Large workpieces, e.g., automotive roof panels, are within the economical size range of the process.

Economic Production Quantities

Stretch forming is a sheet-metal-working method that does not require high production quantities. In fact, it can be used advantageously in many cases with prototype and job-shop-quantity production levels. The reasons for this adaptability are as follows: (1) Form blocks can be made of low-cost materials such as wood, plastic, cast iron, or low-carbon steel for about one-third of the cost of conventional forming dies; (2) less press investment is required because about 40 percent less tonnage is required than for comparable forming by conventional methods; and (3) the handling and changing of setups are very simple because only one form block and two sets of grippers are involved.

Stretch forming is seldom suited to progressive or transfer operations. However, automatic material-handling equipment can be adapted to the machine for production runs. When such equipment is used and the gripping and stretching cycle also is mechanized, stretch forming with mating dies is more rapid and approaches the output rate of conventional press drawing.

Suitable Materials

Steel, nickel, aluminum, and titanium alloys and heat-resisting and refractory metals are used for stretch-formed components. Three commonly used suitable steels are 1008 (G10080), 1008 aluminum–killed, and 1010 (G10100). However, low-carbon steel that does not have uniform mechanical properties throughout the sheet shows stretch-strain marks. Titanium with high yield and tensile strength requires automatic equipment to determine the amount of strain for uniform results.

Design Recommendations

Designers of parts to be stretch-formed should bear in mind the following:

1. Sharply defined contours and reentrant angles should be avoided; stretch forming is more suitable to parts with shallow, gentle bends.
2. Parts should be formed from blanks that are essentially rectangular rather than round, triangular, trapezoidal, etc.
3. Designs that have deep forming in the direction of the free edges (as in a deep-drawn part) are not feasible with stretch forming.

Tolerances

Dimensions of formed elements of less than 500 mm (20 in) normally can be held with care to ±0.25 mm (0.010 in). Larger or less controlled dimensions may require a tolerance of ±0.8 mm ($\frac{1}{32}$ in). With springback allowance, bend angles can be controlled to ½°.

EXPLOSIVELY FORMED PARTS

The Process

In *explosive forming,* one of the high-velocity-forming (HVF) methods, a workpiece is shaped by an instantaneous high pressure that results from the detonation of an explosive charge. There are a number of variations of the process, including *contact-explosive operation,* in which the charge is affixed to the workpiece, and *standoff operation,* in which the charge is separated from the workpiece. In the latter case, a fluid medium, usually water, transmits the shock wave. Standoff operations rarely have shock pressures exceeding 520 MPa (75,000 lbf/in^2), and pressures are often less than 340 MPa (50,000 lbf/in^2). However, in contact operations, the explosion pressure can exceed 2070 MPa (300,000 lbf/in^2).

Figures 3.15.2 and 3.15.3 illustrate two different techniques of the standoff method: the unconfined or free-forming technique and the confined system. In both confined and unconfined systems, a tank, a crane, a vacuum pump, a detonator, a blank holder, solid and composite dies, ports and seals, and liquid or solid transmission media constitute the primary or auxiliary equipment. Dies up to about 10 tons are used in explosive-forming operations. Although water is usually the preferred forming medium, oil, rubber sheets, cast plastics, molten metals, sand, and glass beads are used for certain applications.

Typical Characteristics and Applications

Explosive forming is used to produce parts made of sheet metal, metal plate, tubing, welded sheet metal, and heated blanks. Complex shapes and metals with special properties (such as work-hardening stainless steels) are formed through explosive-forming operations. Explosive forming can produce extremely large items that would be impracticable to form by conventional methods. Part size is limited only by the size of

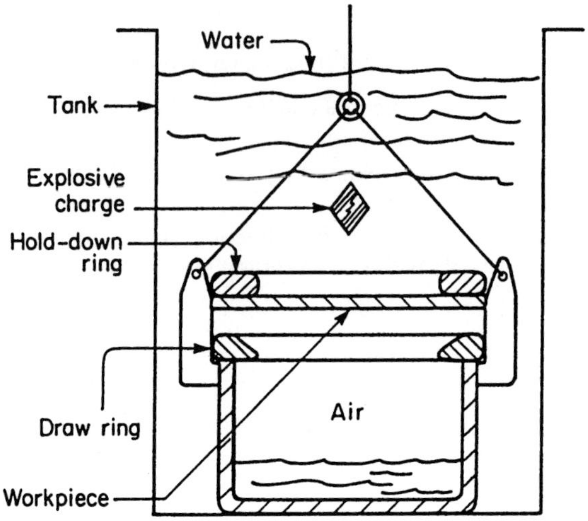

FIGURE 3.15.2 Explosive forming system by free forming.

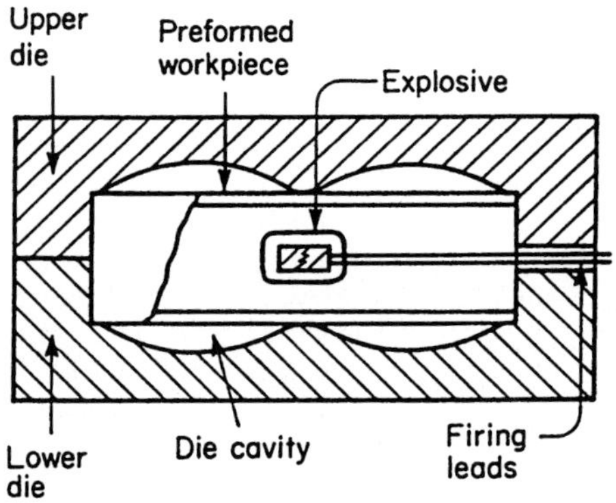

FIGURE 3.15.3 Explosive forming (confined system).

blanks available, currently about 4 m (13 ft) in diameter. The process is most advantageous for those parts that would require costly tooling and equipment if other forming processes had to be employed instead.

Typical applications include aircraft and jet engine components, ducts, panels, and housings. Tubing normally can be bulged to up to 35 percent over the original diameter if several forming steps with intermediate annealing are used.

Economic Production Quantities

Explosive forming is best suited to low-quantity production. Tooling and equipment costs are lower than for conventional forming processes, and lead time is shorter, but output rates are much lower. On occasion, explosive forming may prove economical for production runs of several thousand, but this is the exception rather than the rule. The process is also usually uneconomical for forming simple shapes of smaller parts. These should be produced by conventional methods.

Suitable Materials

About 40 metals and many more metal combinations are suitable for explosive forming. These include aluminum, beryllium, cobalt, copper, molybdenum, niobium, nickel, tantalum, tungsten, zirconium, and various steels.

Ductility and toughness are the major characteristics that determine formability. In forming a part, it is general practice not to exceed the elongation determined by tension testing. The construction of the water tank, the design of the blank holder and die, and the characteristics of the transmission medium all influence the apparent formability of the material. Through modified tooling design and increase of temperature, formability can be increased.

In Table 3.15.1, the formability of some metals, using annealed-aluminum alloy 1100-0 (G91100, 0 temper) as a basis, is compared.

Design Recommendations

Concentric shapes are more easily formed by explosive forming than nonconcentric shapes. Tooling is less expensive, and the shaping of the explosive charge is far sim-

TABLE 3.15.1 Relative Explosive Formability of Selected Metals

Designation			
Conventional	UNS	Condition	Relative formability
1100 aluminum	(A91100)	0 temper	100
Tantalum			88
Copper			80
1010 steel	(G10100)		78
6061 aluminum	(A96061)	T6 temper	76
20 niobium stainless steel			66
5% chromium steel			62
321 stainless steel	(S32100)		59
347 stainless steel	(S34700)		55
Inconel X-750	(N07750)		50
René 41	(N07041)		48
Hastelloy X	(N06002)		40
PH15-7 molybdenum stainless steel	(S15700)		40
4130 steel	(G41300)	Normalized	32
Titanium 6Al-4V	(R56401)		12
301 stainless steel	(S30100)	Full-hard	8

Source: Adapted from data in *Metals Handbook,* vol. 14: *Forming,* 9th ed., American Society for Metals, Metals Park, Ohio, 1988.

pler. Nonsymmetrical and nonconcentric shapes require more development of process conditions and, if possible, should be avoided.

Explosive forming is often a practical process for producing complex shapes in smaller parts, but for large parts it is advisable to keep shapes as simple as possible. Corners, especially inside corners, if sharp, produce a stress concentration in the forming die, resulting in shortened die life. Therefore, it is preferable to design parts with as gentle contours as possible.

Tolerances

Die accuracy, die wear, explosion pressure, the choice of pressure-transmission medium, and setup variations all have considerable influence on the dimensional accuracy of the workpiece. These factors make it advisable to specify as liberal a tolerance as possible, even though dimensional control within ±0.025 mm (0.001 in) is possible with small parts if all variables are carefully controlled. Table 3.15.2 presents recommended tolerances for production conditions.

ELECTROMAGNETICALLY FORMED PARTS

The Process

Electromagnetic forming (EMF), another HVF process, is a technique in which a high-intensity magnetic field is used to shape metallic workpieces. By the passage of a pulse of electric current through a forming coil, the workpiece is formed without mechanical contact. The electric current produces a very intense magnetic field that lasts only a few microseconds. The resulting eddy currents induced in a conducive metal part interact with the magnetic field to cause repulsion between the workpiece and the forming coil. With the help of this force of repulsion and a die or a mandrel, the work metal is stressed far beyond its yield strength. The forming coil must be stiff and strong enough to withstand such forces while the workpiece is being formed.

Two variations of electromagnetic forming are illustrated in Figs. 3.15.4 and 3.15.5. In all cases, a basic circuit that consists of a forming coil, an energy-storage capacitor, switches, and a power supply of nearly constant current to charge the capacitor is needed. In most electromagnetic processes, the pulse lasts between 10 and 100 μs. By the use of field shapers, the shaping force, which forms the workpiece, is concentrated in certain regions of the workpiece. This technique lengthens the life of the forming coil by preventing high pressures on weaker parts of the coil.

TABLE 3.15.2 Recommended Tolerances for Explosively Formed Parts

Class of part	Recommended tolerance, mm (in)
Sheet and plate components, formed in dies	±0.25 (0.010)
Parts free-formed from plate (no die)	±6.3 (0.250)
Formed welded assemblies (with die)	±0.8 (0.032)
Parts formed from tubing (with die)	±0.25 (0.010)

FIGURE 3.15.4 Compression forming of tubing. (*From* Metals Handbook, *American Society for Metals.*)

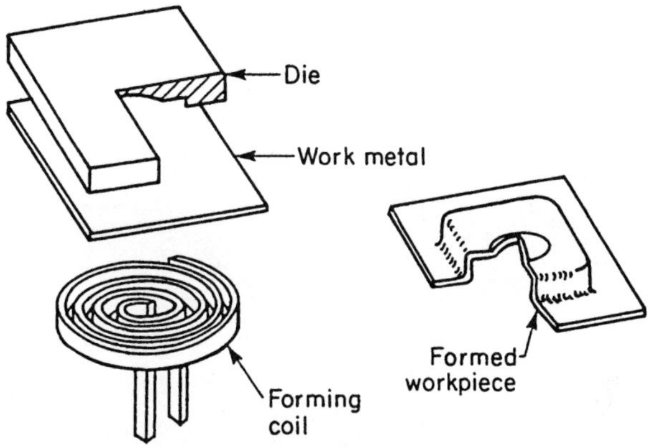

FIGURE 3.15.5 Contour forming of sheet. (*From* Metals Handbook, *American Society for Metals.*)

Typical Characteristics and Applications

Electromagnetic forming is primarily an assembly operation. The most common application is the crimping of tubular parts onto fittings or other components. Forming without assembly also can take place. Tubular parts can be expanded, compressed, or formed. Occasionally, flat sheet material also can be formed by this method. The method is also used for piercing or shearing. Several forming and assembly operations are often combined into a single step.

Advantages of electromagnetic crimping are as follows: (1) No tool marks are produced on the surfaces of parts; (2) no lubricant is required; (3) metal parts can be assembled to ceramic, phenolic, or other nonmetallic fittings; (4) by accurately controlling the magnetic field, swivel-joint assemblies can be produced; (5) parts contained in a vacuum or inside a sterile plastic bag can be assembled because magnetic lines of force pass freely through container walls; and (6) metallic bands can be compressed over rubber and other soft materials without excessive deformation of the materials.

The following definite design limitations should be considered: (1) Not all workpiece contours can be easily shaped by the magnetic field. It is not always possible to apply high pressure in one area and low pressure in an adjacent area. (2) Nonsymmetrical parts are not suitable for the magnetic-forming process. (3) Small-diameter expansion coils have limited magnetic-forming force, so small tubing cannot normally be expanded. (4) EMF is also not so feasible with heavy-gauge sheet or tubing.

The automotive industry uses the electromagnetic-forming method to assemble air-conditioner components, high-pressure hoses, shock-absorber dust covers, rubber boots on ball joints, oil-cooler heat exchangers, and accessory motor packages. Electromagnetic forming is used to form universal-joint yokes, drive linkages, wheels, and various other fittings or linkages.

Electrical equipment manufactured or joined by electromagnetic forming includes components of high-voltage fuses, insulators, and lighting fixtures. Aircraft parts include torque shaft assemblies, control rods and linkages, assembly and forming of cooling-system ducts, and sizing of tubing.

Recreation equipment such as aluminum baseball bats, anodized-aluminum boat hooks, and plastic (ABS) or stainless-steel fishing gaffs are produced by electromagnetic forming. Figure 3.15.6*a* to *f* illustrates typical components produced with the aid of electromagnetic forming.

(a)

(b)

(c)

FIGURE 3.15.6 Typical components produced with the aid of electromagnetic forming. (*a*) Lightning-reflector facets. (*b*) Automotive cruise-control assembly. (*c*) Aircraft ducting.

(d)

(e)

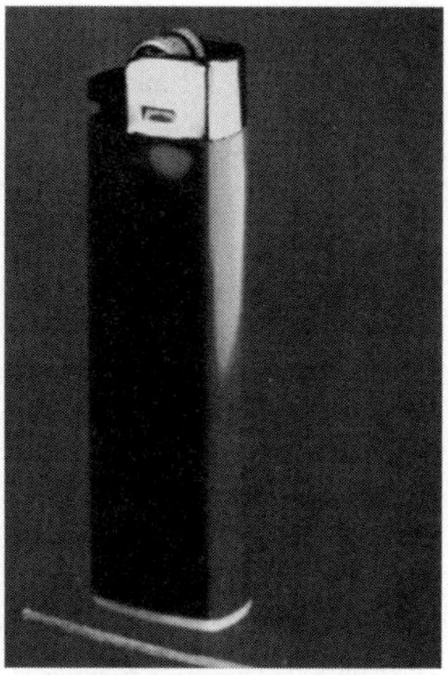

(f)

FIGURE 3.15.6 (*Continued.*) Typical components produced with the aid of electromagnetic forming. (*d*) Automotive fuel-pump assembly. (*e*) Aircraft oil filter. (*f*) Flow-control valve for pocket cigarette lighter. (*Courtesy Maxwell Laboratories, Inc.*)

Economic Production Quantities

Electromagnetic forming is most advantageous in medium- and high-production applications. The operation is fairly rapid, with output rates of 400 pieces per hour being common and outputs into the thousands per hour feasible with automatic handling. No special operator skills are required with the usual high-volume operation. Labor costs are therefore quite modest. Tooling costs also may be modest but can rise when a special coil and field shapers are required. Coil costs vary with the energy level, forming force, operating ratio, size, and other factors. Equipment, including an energy-storage bank, may require a sizable investment.

Suitable Materials

Good electrical conductors such as silver, copper, aluminum, gold, mild steel, and many other metals and their alloys can be formed easily. Stainless steel, because of its low conductivity, is difficult to form by this process. Metals to be electromagnetically formed should have a resistivity not greater than 15 $\mu\ \Omega$/cm.

Nonconductive materials can be formed by placing a conductive material called a *driver* between this coil and the workpiece.

Design Recommendations

The workpiece should not have slots or other cutouts in the area to be formed, since the process demands a continuous electrical path in this area. Small perforations and other minor irregulatories do not seriously hamper current flow and thus can usually be permitted. (See Fig. 3.15.7.)

Figure 3.15.8 illustrates design recommendations for one type of crimped joint.

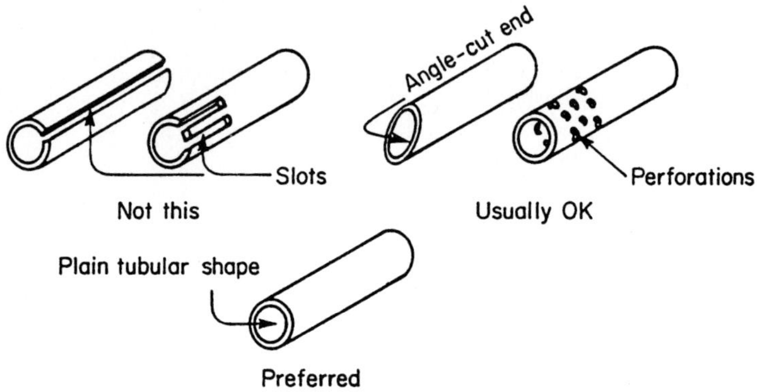

FIGURE 3.15.7 Avoid slots and other openings in parts to be electromagnetically formed.

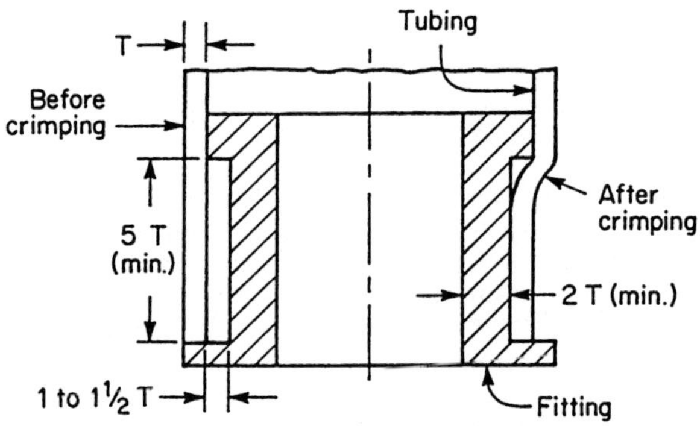

FIGURE 3.15.8 Design recommendations for axially loaded joints made by electromagnetic forming.

CHAPTER 3.16

METAL INJECTION-MOLDED PARTS

Andrew D. Hanson

Product Group Vice President
PCC—Advanced Forming Technology
Longmont, Colorado

THE METAL INJECTION-MOLDING PROCESS

Metal injection molding (sometimes referred to by the initials MIM) is a powder metallurgy process. It differs from conventional powder metallurgy in that the metal powder—with binders—is injected under high pressure into the die or mold cavity rather than flowing by gravity. The metal powder is finer than conventional powder. Particles are typically one-tenth the size and measure less than 20 μm in diameter. A much higher percentage of binders and flow agents are part of the powder mix. Binders are typically proprietary but generally include various waxes and thermoplastics. The compounded material is injected with conventional plastic injection-molding machines, which are slightly modified, into metal dies, normally of one to eight cavities. Injection cycle times are similar to those with plastic molding, ranging from 15 to 60 seconds.

The molded component is called a *green part.* As in conventional powder metallurgy, it is not yet ready for use. It is strong enough only for in-plant handling. The binders are first driven off either by heating, by solvent extraction, or by a combination of both methods. Approximately 80 percent of the binders are removed. Heating is typically at about 260°C (500°F) for from 8 to 24 hours. Solvent methods, involving the use of heated vapors, reduce the process time to 1 to 4 hours. Debinding times are directly correlated with wall thickness. After debinding, the part, now called a *brown part,* is porous and fragile but remains the same size as when molded.

The brown parts are then sintered as in conventional powder metallurgy. Vacuum or inert-atmosphere furnaces are used, and temperatures approximate 1300°C (2400°F) but depend on the particular material being processed. However, sintering temperatures are well below the melting point of the material. The sintering cycle, like binder removal, may be a batch or continuous process. It normally requires about 24 to 36 hours for the entire cycle. During sintering, metal particles diffusion bond, and the part becomes dense and strong enough for use. Density increases from approximately 60 percent in the brown stage to 95 to 99 percent of theoretical.

Shrinkage is considerable in sintering. The amount of shrinkage depends on the percentage of binders in the original powder blend and the final density of the part. Between 19 and 25 percent is the typical linear shrinkage. Fortunately, shrinkage tends to be nearly equal in all directions, so the final dimensions of the finished part can be controlled. Gravity and subsequent drag associated with densification and shrinkage must be addressed, however, especially on components of greater mass.

If the necessary dimensional tolerances of the part are very close, a final sizing operation, usually by press forming, can be employed. Machining operations also may be performed.

TYPICAL CHARACTERISTICS AND APPLICATIONS OF METAL INJECTION-MOLDED PARTS

Typical metal injection-molded parts are small and complex. The more complex the part, the more likely it is that metal injection molding will be advantageous. High precision is possible. These parts differ from those made by conventional powder metallurgy processes in that the sidewalls do not have to be parallel.

The process is most applicable to parts that are smaller than a golf ball in volume, weigh between 0.5 and 150 g (0.2 and 2.2 oz), and have a wall thickness less than 6.3 mm (¼ in). The typical maximum dimension is below 100 mm (4 in). Multifunctional parts with thin walls, sharp corners, undercuts, screw threads, cross holes, contours, gear segments, and similar features are feasible without secondary operations. Any shape that can be injection molded is feasible, and sometimes more complex shapes can be produced by sintering two injection-molded parts together. Densities are normally greater than those of conventional powder metallurgy parts, running between 95 and 99 percent of theoretical.

Typical applications have been rotors for stepper motors, sear housings for police pistols, hammers and links for printing machines, stainless-steel orthodontic retainers, small hermetically sealable housings for electronic components, shotgun trigger guards, and medical, camera, computer peripheral, automotive fuel-injector, steering column, seat, and firearm components. Figure 3.16.1 shows some typical metal injection-molded parts.

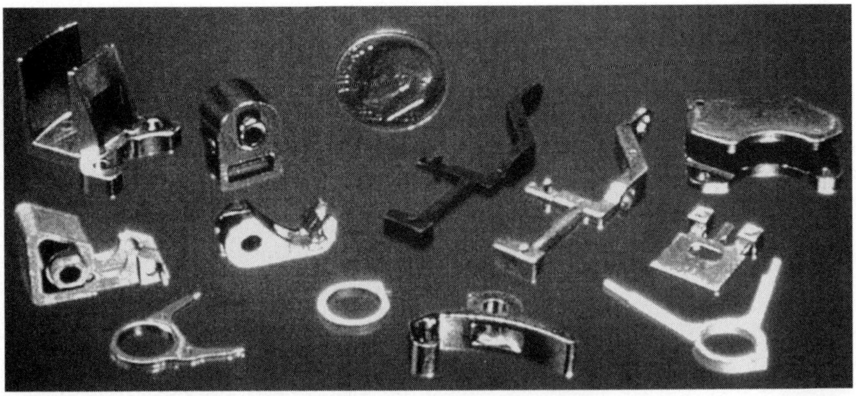

FIGURE 3.16.1 A collection of typical metal injection molded parts. (*Photo courtesy of Precision Castparts Corporation.*)

ECONOMIC PRODUCTION QUANTITIES

Metal injection molding is a mass-production process like the injection molding of plastics. The major factor is the amortization of the costs of designing and fabricating a mold. Other costs include the development of the molding, debinding, sintering, and if applicable, the sizing process. The net effect of all these factors is that a production quantity of 10,000 pieces per year is normally a minimum for economical production.

MATERIALS SUITABLE FOR METAL INJECTION-MOLDED PARTS

Metal injection-molded parts can be made from a wide variety of materials. Low-carbon and alloy steels, stainless steels, nickel alloys, and copper alloys are examples. The process also has been used for high-speed steel, titanium and tungsten alloys, cemented carbides, aluminide intermetallics, and ceramics.

Materials that are easily die cast, such as zinc, aluminum, and magnesium, could be processed by metal injection molding, but die casting is usually much more economical for such materials. Table 3.16.1 lists some common metals and alloys proces-

TABLE 3.16.1 Common Materials Suitable for Metal Injection-Molded Parts

Material	Remarks
Low-alloy steels	
2%–8% Ni-steel	High strength/heat-treatable
MIM 4600 Series MIM 4630 MIM 4650	Similar to high-strength, low-alloy wrought materials with excellent toughness and heat-treatability
Austenitic stainless steels	
MIM 316L MIM 304L	Excellent corrosion resistance and ductility
Ferritic stainless steel	
MIM 430	Good soft magnetic characteristics
Martensitic stainless steel	
MIM 420	High strength/heat-treatable
MIM 440	High hardness and corrosion resistance
Precipitation-hardening stainless steel	
MIM 17-4 PH	High strength and good corrosion resistance in an as-sintered or heat-treated condition
Duplex structure stainless steel	
MIM 316 (duplex)	Stronger and harder than MIM 316L; slightly magnetic
Iron-cobalt, nickel-based and other alloys	
MIM (ASTM) F15	Medical device MIM alternative
2%–3% Fe-Si	Soft magnetic applications
MIM nickel alloys	Controlled expansion for metal-to-glass sealing
Copper-based	
MIM Cu-100%	High electrical conductivity
MIM Cu-Ni-Sn	High-strength bronze, corrosion resistant
MIM Cu-Ni 84/16 (Spinodal)	High strength, conductivity

TABLE 3.16.2 Specifications of Low-Alloy and Stainless Steels for Metal Injection

Material designation (condition)	Minimum values			Typical values					
	Tensile properties			Tensile properties				Hardness (Rockwell)	
	Ultimate strength, 10^3 psi	Yield strength (0.2%), 10^3 psi	Elongation %	Ultimate strength, 10^3 psi	Yield strength, (0.2%), 10^3 psi	Elongation %	Density gm/cm^3	Apparent (direct)	Micro (converted)
Low-alloy steels (MIM material properties)									
MIM-4600 as-sintered	37	16	20.0	42	18	40	7.6	45 HRB	
MIM-4650 as-sintered	55	25	11.0	64	30	15	7.5	62 HRB	
MIM-4650 quenched and tempered	215	190	<1.0	240	215	2	7.5	48 HRC	55 HRC
MIM-2700 as-sintered	55	30	20.0	60	37	26	7.6	69 HRB	
Stainless steels (MIM material properties)									
MIM-316L as-sintered	65	20	40.0	75	25	50	7.6	67 HRB	
MIM-duplex as-sintered	68	26	33.0	78	33	43	7.6	84 HRB	
MIM-17-4 PH as-sintered	115	94	4.0	130	106	6	7.5	27 HRC	
MIM-17-4 PH solution treated and aged	155	140	4.0	172	158	6	7.5	33 HRC	42 HRC

Source: Reprinted with permission from Standard 35, *Materials Standards for Metal Injection-Molded Parts, 1993–1994*, Metal Powder Industries Federation, Princeton, N.J., 1994.

sible by metal injection molding. Table 3.16.2 provides composition and yield-strength data for suitable materials for the process.

DESIGN RECOMMENDATIONS

Many of the design principles and recommendations applicable to injection molding of plastics and summarized in Chap. 6.2 also apply to metal injection-molded parts. Some specific design recommendations for metal injection-molded parts are as follows.

1. Uniform wall thickness is very advantageous. When this is not possible, one rule of thumb specifies that the thicker wall should not be more than 2 times the thinner wall. The transition from thick to thin walls also should be as gradual as possible. Figure 3.16.2 illustrates this point.

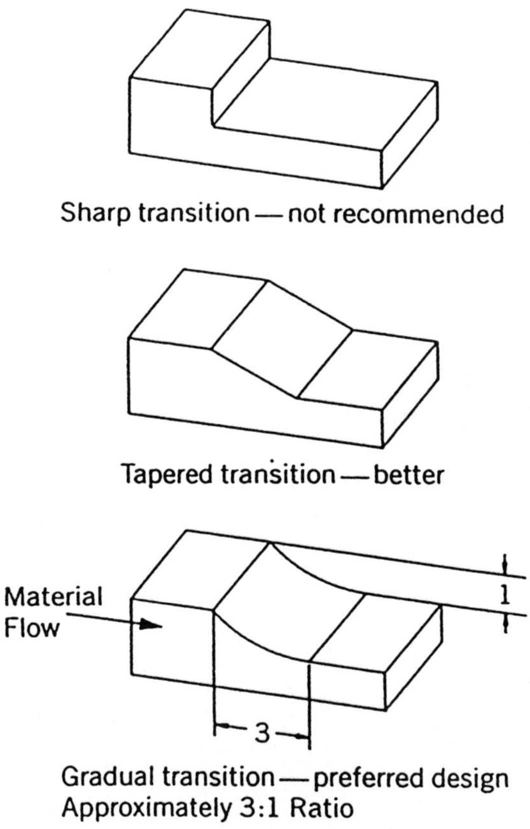

FIGURE 3.16.2 When wall thickness cannot be uniform, make the most gradual transition possible from thick to thin sections.

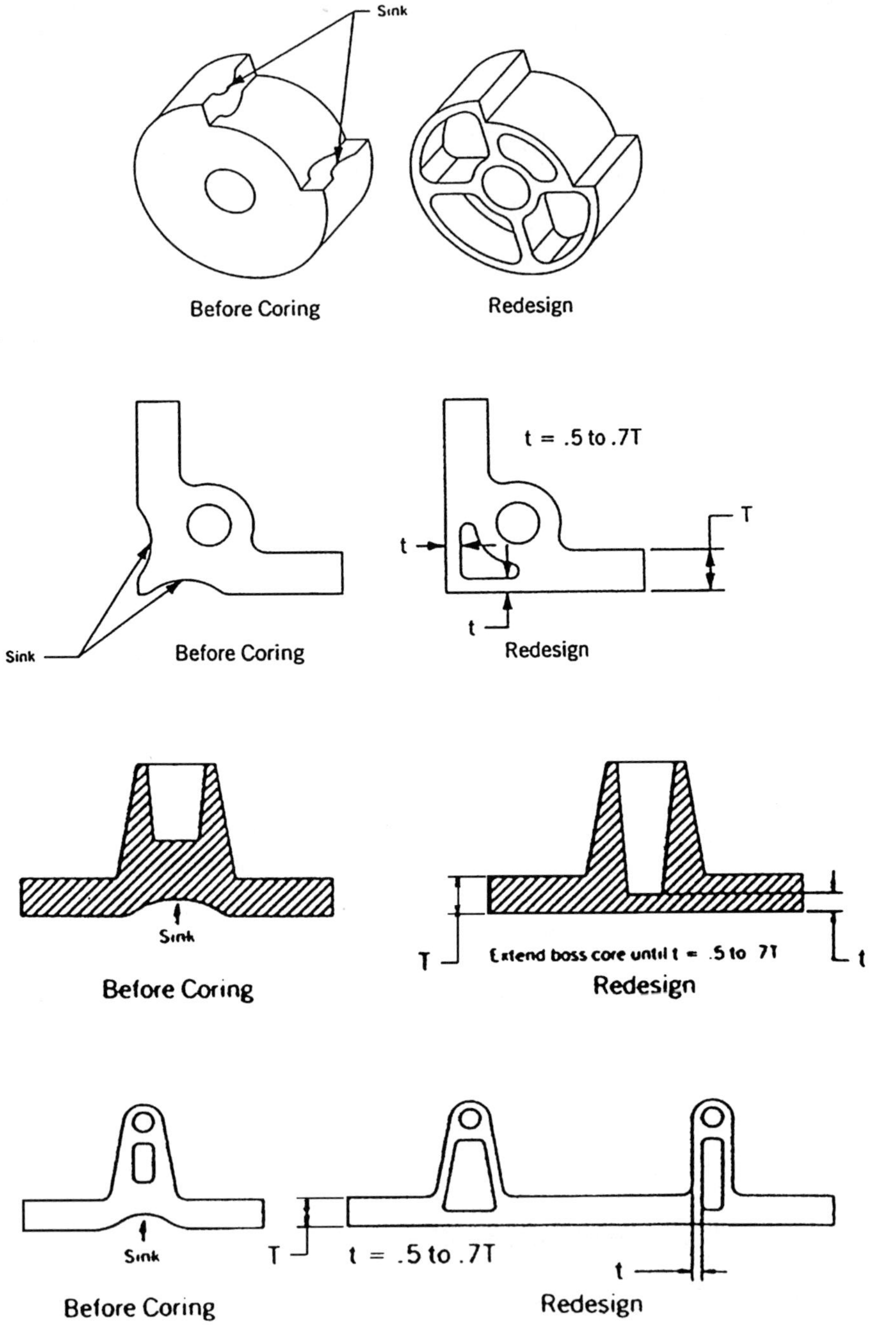

FIGURE 3.16.3 Examples of design changes in metal injection molded parts to reduce excessive wall thickness.

2. Walls also should not be too thick. Thick cross sections result in slower binder removal. They also increase the injection-molding cycle and can cause surface depressions or "sinks" opposite the heavy section due to material shrinkage. A rule of thumb is to keep wall thickness to 6.35 mm (¼ in) or less. If the part has sections that are heavier, cores should be added to reduce the cross section. Figure 3.16.3 illustrates some methods of reducing wall thickness of heavier sections.

3. Parts should have a draft angle on sidewalls to ensure easy part removal from the mold. Angles of ½ to 2° per side are recommended for both inside and outside walls. Larger draft angles should be used for deep sections, complex configurations, and when an inside core or die section is used, because the part will tend to shrink around it.

4. Generous radii and filets should be used as much as possible at all corners. Rounded corners aid material flow in molds and reduce stress concentrations in molds or in the part. A suggested rule is to use a corner radius of one-half the adjacent wall thickness and no less than 0.4 mm (0.015 in).

5. Bosses and ribs are feasible and often desirable in MIM part design. Ribs enable a part to be designed to be strong and rigid even when wall thickness and mass must be reduced. Since these details can be produced by the tooling, they can be incorporated into the design at minimal unit cost. Figure 3.16.4 shows suggested designs for bosses, and Fig. 3.16.5 illustrates recommended dimensions for reinforcing ribs.

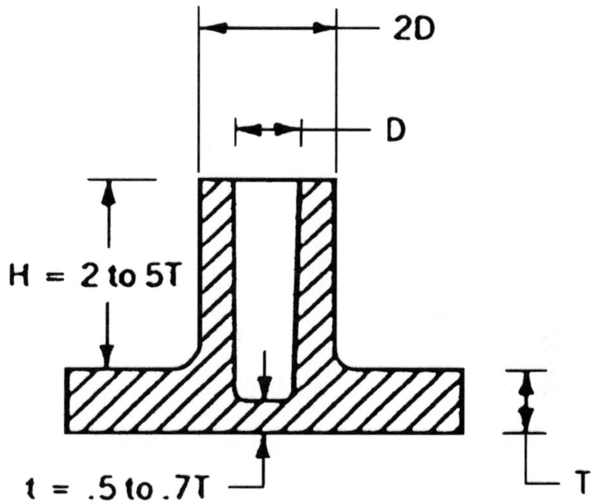

FIGURE 3.16.4 Suggested design for bosses in metal injection molded parts.

TOLERANCE RECOMMENDATIONS

As a general rule, dimensions of MIM parts can be held to tolerances of ±0.08 mm (±0.003 in) to ±0.13 mm (±0.005 in). Dimensions of over 25 mm (1 in) should be allowed plus or minus 0.5 percent of the nominal dimension, or ±0.005 in/in. If closer

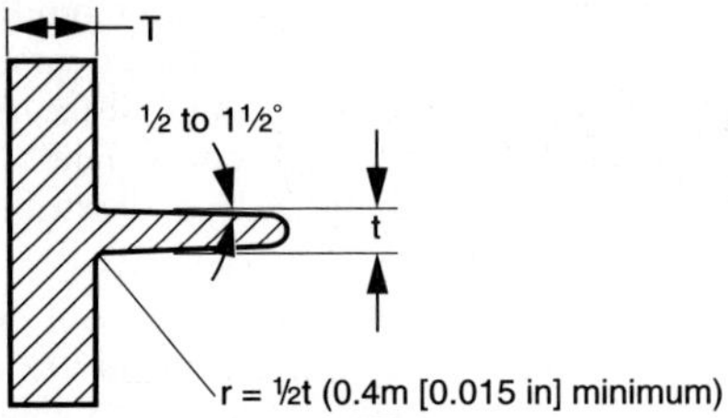

FIGURE 3.16.5 Suggested design for ribs in metal injection molded parts.

dimensional control is needed, secondary operations should be performed. Non critical dimensions should be allowed at least ±0.13 mm (±0.005 in) or 1 percent, whichever is larger.

As-sintered surface finishes can be in the range of 0.8 m to 1.1 μm (32 to 45 μin). Surface finishes can be enhanced by standard surface finishing or polishing processes.

S • E • C • T • I • O • N • 4

MACHINED COMPONENTS

CHAPTER 4.1
DESIGNING FOR MACHINING: GENERAL GUIDELINES

THE MACHINING PROCESS

All machining, whether heavy single-point planing or turning, form-tool turning or milling, grinding, honing, or lapping, involves essentially the same process at the point where the cutting tool meets the work. Figure 4.1.1 illustrates this process.

Material lying in front of the cutting tool is compressed as the tool advances and fail in shear in a narrow zone extending at an angle from the cutting edge to the surface of the workpiece ahead of the tool. For practical purposes in single-point cutting this shear zone can be considered a plane. As the cutting tool advances into the work, the shear plane also constantly moves forward. The material that passes through the shear plane is deformed. This material comprises the chip. In the case of ductile materials, it is apt to consist of a continuous ribbon of deformed and heated metal moving away from the workpiece along the face of the cutting tool. In the case of nonductile or brittle materials, the shear action periodically causes fracture, and the chips consist of discrete pieces rather than a continuous ribbon of material.

Since the energy expended in cutting is manifested as heat, the chip, the cutting tool, and even the workpiece experience a considerable rise in temperature. This temperature rise can be reduced when a fluid coolant is applied to the cutting tool. In addition to reducing temperature, the coolant lubricates the tool in its movement against the workpiece and, more important, the movement of chips against the face of the tool.

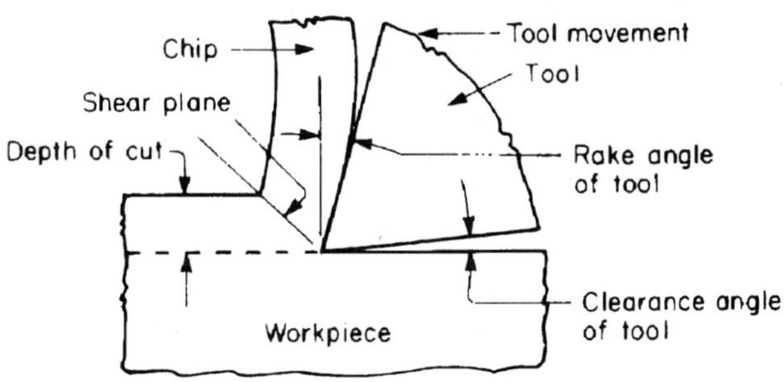

FIGURE 4.1.1 Action of a metal-cutting tool.

Grinding operations including honing and lapping exhibit the same basic interaction between workpiece and cutter. However, the cutter in such abrasive-machining operations is an abrasive particle, which may be very small. The shape of the abrasive particle also may vary considerably from that of the metal-cutting tool shown in Fig. 4.1.1.

TYPICAL MACHINED PARTS

Machined parts are universally used in industrial and consumer products of every description. They are found in applications for which precision is required. If high dimensional accuracy is not necessary, stampings, castings, or stock shapes or molded parts used as is will be more economical. However, if surface finish, flatness, roundness, circularity, parallelism, or close fit is involved, some machining of the workpiece will practically always be involved.

Almost invariably, if the part is in motion, is in contact with a part that is in motion, or fits precisely with another part, machining operations will be employed in its manufacture. For most interchangeable parts, machining is a probable step in the manufacturing sequence.

Of course, wonders are worked with stampings and molded components and with new precision techniques such as powder metallurgy, fine-blanking, and investment casting, but these processes usually only reduce rather than eliminate the need for machining if the part has a truly precision application.

Machined parts can be as small as the miniature screws, shafts, gears, and other parts found in wristwatches and small precision instruments. They can be as large as the huge turbines, turbine housings, and valves found in hydroelectric power stations.

Metals in a broad variety, both ferrous and nonferrous, are the normal materials used for machined components. However, plastics (with or without reinforcement), hard rubber, carbon, graphite, wood, and ceramics are also employed.

RECOMMENDED MATERIALS FOR MACHINABILITY

Table 1.4.14 in Chap. 1.4 contains a summary of the machinability of common metals, including those generally suitable for a broad cross section of machining operations. The other chapters in this section cover parts produced by specific machining operations and provide additional and more specific materials recommendations. Chapters 2.2 to 2.4 in Sec. 2 include additional recommendations of materials that can be machined advantageously. Table 4.1.1 summarizes how changes in certain materials properties affect machinability.

TABLE 4.1.1 Effects of Material Properties

Material factor	Probable effect of decrease in material factor on		
	Machinability[a]	Finishability	Tool life
Strength/hardness[b]	Improves	None	Improves
Ductility[c]	Improves	Improves	Improves
Strain hardenability[b]	Improves	Improves	Improves
Coefficient of friction[d]	Improves	Improves	Improves
Heat conductivity[e]	None	None	Reduces
Heat capacity[e]	None	None	Reduces
Chemical reactivity[f]	None	Improves	Improves
Grain size	Improves	Improves	Reduces
Abrasive insolubles	Improves	Improves	Improves
Free-machining additions	Decreases	Decreases	Decreases

[a]*Machinability* refers to ease of chip removal.

[b]Tensile strength and hardness are the simplest, but not always reliable, guides to machinability. High-temperature alloys, for example, are difficult to machine in spite of their low room-temperature hardness and strengths. High strain hardenability and reactiveness to tool materials are the reasons.

[c]While lower ductility seems to help machining, inadequate ductility (like that of molybdenum and tungsten) can cause spalling at exit cuts or on clamped edges.

[d]Low frictional resistance is desirable; hence the use of cutting fluids is recommended.

[e]Low heat conductivity (especially if combined with low heat capacity, as in titanium) contributes to high tool temperature and local high workpiece temperature.

[f]Chemical reactivity of certain metals (such as titanium) can cause galling, smearing, and welding of machined metal to the tool.

Source: *Machine Design.*

DESIGN RECOMMENDATIONS: MACHINED PARTS

1. If possible, avoid machining operations. If the surface or feature desired can be produced by casting or forming, the cost is almost always lower. (See Fig. 4.1.2.)
2. Specify the most liberal surface finish and dimensional tolerances possible, consistent with the function of the surface, to simplify the prime machining operation and to avoid costly secondary operations like grinding, reaming, lapping, etc. (See Fig. 4.1.2.)
3. Design the part for easy fixturing and secure holding during machining operations (see Fig. 4.7.4). A large, solid mounting surface with parallel clamping surfaces should be provided to assure a secure setup.
4. Avoid designs that require sharp corners and sharp points in cutting tools because these make the tools more subject to breakage.
5. Use stock dimensions whenever possible if so doing will eliminate a machining operation or the need for machining an additional surface. (See Fig. 4.1.3.)
6. It is preferable in all single-point machining operations to avoid interrupted cuts, if possible, because they tend to shorten tool life or prevent the use of faster-cutting carbide or ceramic tools.

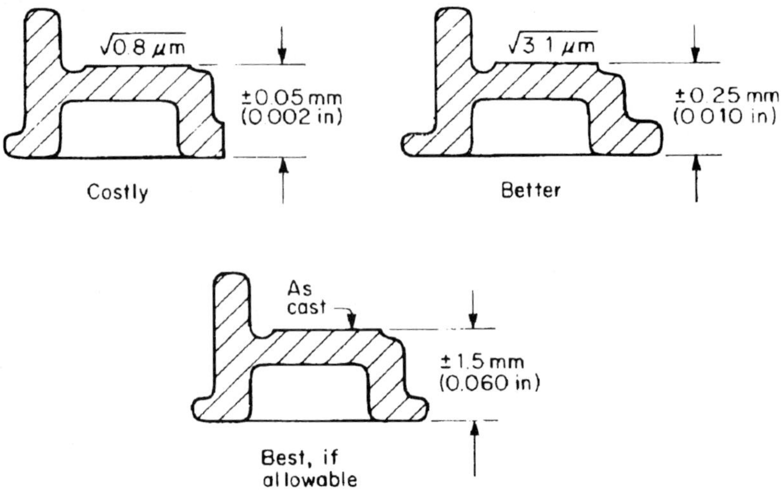

FIGURE 4.1.2 Avoid tolerances that necessitate machining operations if the as-cast, as-forged, or as-formed dimensions and surface finishes would be satisfactory for the parts' function.

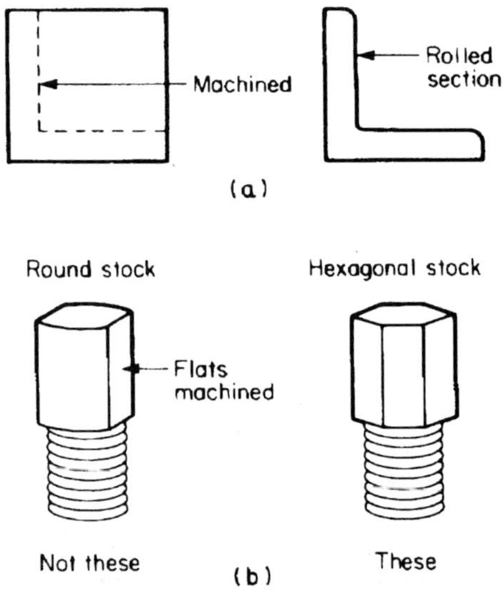

FIGURE 4.1.3 Use stock dimensions whenever possible, and minimize the amount of machining.

7. Design the part to be rigid enough to withstand the forces of clamping and machining without distortion. The forces exerted by a cutter against a workpiece can be severe, as can the clamping forces necessary to hold the workpiece securely. Parts that may be troublesome in this respect are those with thin walls, thin webs, or deep pockets and deep holes that require machining. Also design the part

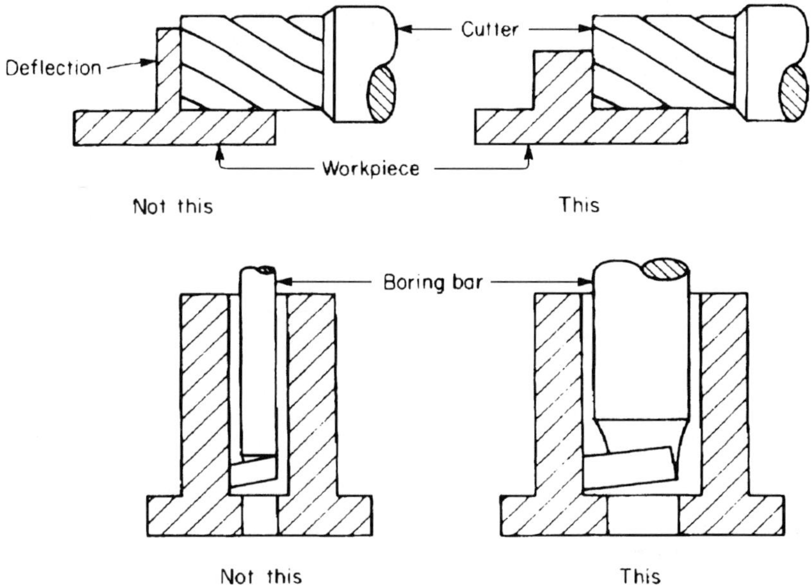

FIGURE 4.1.4 Design the part to be rigid enough so that it will withstand cutting and clamping forces without significant deflection and so that cutting tools and toolholders also do not deflect.

so that a rigid cutter can be employed while still permitting access to the surface. (Figure 4.1.4 illustrates this point. See also Fig. 4.4.8.)

8. Avoid tapers and contours as much as possible in favor of rectangular shapes, which permit simple tooling and setups.
9. Reduce the number and the size of shoulders because they usually require extra operational steps and additional material.
10. Avoid undercuts, if possible, because they usually involve separate operations of specially ground tools. (See Fig. 4.1.5.)
11. Consider the possibility of substituting a stamping for the machined component. If tooling is available, or if quantities are sufficient to amortize the tooling cost, a stamped-sheet-metal part invariably will be lower in cost than one made by machining, provided of course that the dimensional accuracy and surface finish are adequate for the component's function. (Figure 4.1.6 illustrates one such example.)
12. Avoid the use of hardened or difficult-to-machine materials unless their special functional properties are essential for the part being machined.
13. For thin, flat pieces that require surface machining, allow sufficient stock for both rough and finish machining. In some cases, stress relieving between rough and finish cuts also may be advisable. Rough and finish machining on both sides is sometimes necessary. Allow about 0.4 mm (0.015 in) stock for finish machining.
14. It is preferable to put machined surfaces in the same plane or, if they are cylindrical, with the same diameter to reduce the number of operations required. When surfaces cannot be in the same plane, they should be located, if possible, so that they all can be machined from one side or from the same setup.

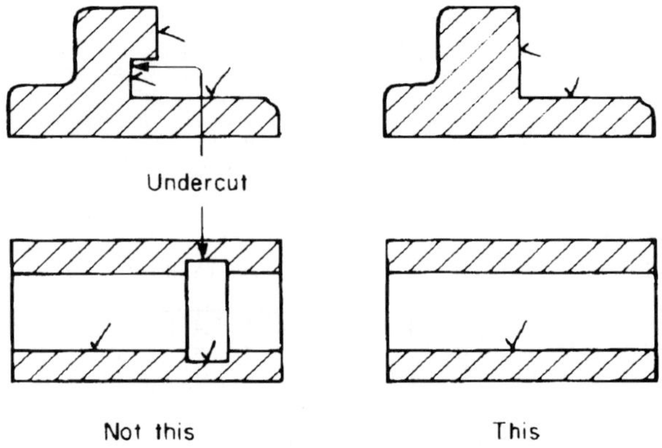

FIGURE 4.1.5 Avoid undercuts as much as possible because they require extra machining operations, which may be costly.

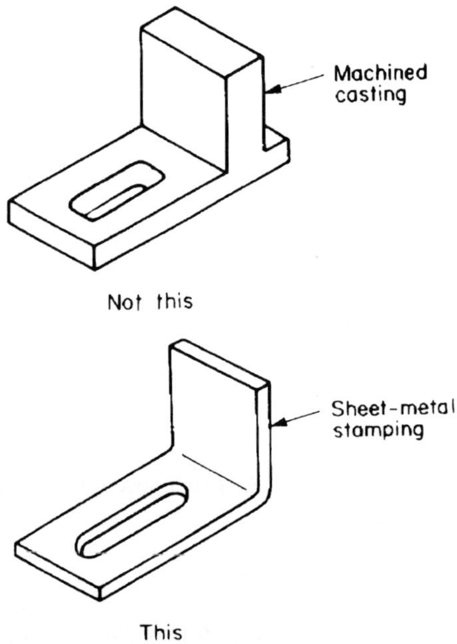

FIGURE 4.1.6 Stampings are often less costly than machined castings.

15. Provide access room for cutters, bushings, and fixture elements.

16. Design workpieces so that standard cutters can be used instead of cutters that must be ground to a special form. (See Fig. 4.1.7.)

17. Avoid having parting lines or draft surfaces serve as clamping or locating surfaces. Provide alternative clamping and locating surfaces if possible.

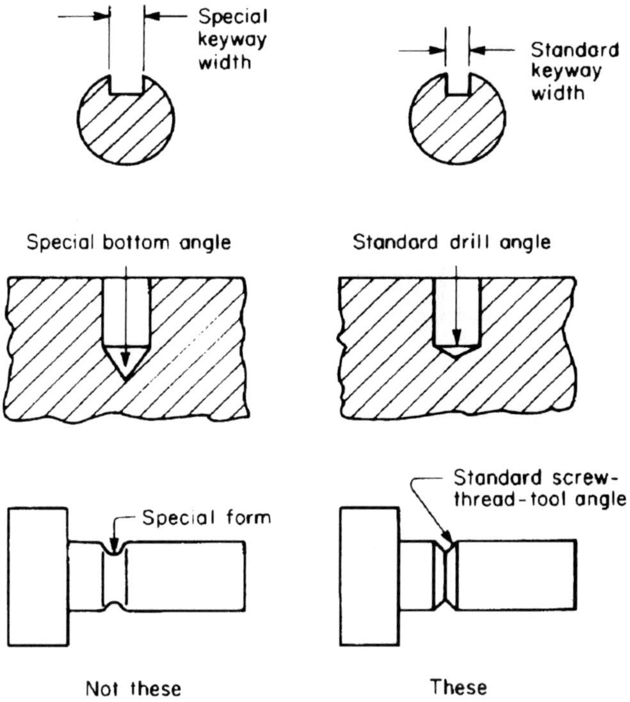

FIGURE 4.1.7 Design parts so that standard cutting tools can be used.

18. Avoid projections, shoulders, etc., which interfere with the overrun of a cutter. Instead, provide clearance space at the end of the cut. The space can be cast or formed to minimize machining. This also can provide a noncritical space for burrs.

19. Burr formation is an inherent result of machining operations. The designer should expect burrs, provide relief space for them, if possible, and furnish means for easy burr removal. (See Chap. 4.23.)

CHAPTER 4.2
PARTS CUT TO LENGTH

Ted Slezak
Armstrong-Blum Mfg. Co.
Chicago, Illinois

Cutoff is a common manufacturing operation. It occurs whenever a long workpiece is divided into pieces of shorter, more useful length. Very often it is part of a multiple operation as, for example, in the functioning of an automatic screw machine or progressive-stamping die. Less frequently but still often, cutoff is employed as a separate operation. It is with this last case that this chapter is concerned.

PROCESSES FOR CUTTING TO LENGTH

The following cutoff processes are the ones most frequently employed: band-saw, hacksaw, circular-saw, abrasive-wheel, friction-saw, shear, lathe-type-machine, thin-wall-tubing cutters (shear type), flame cutting, and other methods such as electron-beam and laser-beam, EDM wire cutting, etc. (See Figs. 4.2.1 through 4.2.5.)

Cutoff machines have three major components:

1. The stock-feeding mechanism, which manually or automatically feeds uncut lengths of stock to the cutoff tool
2. The actual cutting mechanism—saw, single-point tool, abrasive wheel, etc.
3. The discharge table or chute for cut lengths

For manual cutting, the first and third components may be simple worktable surfaces to hold the uncut stock and cut parts. For automatic equipment, hoppers, conveyors, and sophisticated control apparatus may be involved.

TYPICAL APPLICATIONS

Often, cutoff operations supply a component in the form of a finished part or nearly so. Examples are structural members, spacers, and simple pins. More frequently, cutting to length is the first of a series of machining, forming, and finishing operations.

FIGURE 4.2.1 Hacksaw cutoff of bar stock. (*Courtesy Armstrong-Blum Mfg. Co.*)

Some of the situations in which it may be advisable to provide a workpiece of a specified length with little or no other operations to be performed at the same sequence are as follows:

1. When the workpiece does not require any further processing except welding or assembly
2. When additional operations required are most economically performed at a subsequent operation
3. When one blank can be made into a variety of end pieces and it is preferable to defer the second operation as part of a finish-to-order procedure

As a result of these factors, the parts included in this section can be made of a number of materials, metallic and nonmetallic; in a number of cross sections, regular, irregular, solid, or hollow; and, of course, in a variety of lengths.

The minimum length for each process depends on the economics of materials usage, labor costs, and process capability factors such as squareness and surface finish of the cut. Most cutoff processes will provide cut lengths down to about 1.3 mm (0.050 in).

See Table 4.2.1 for common applications and parameters of the more frequently used methods.

FIGURE 4.2.2 Circular-saw cutoff of aluminum billets. (*Courtesy Hill Acme Co.*)

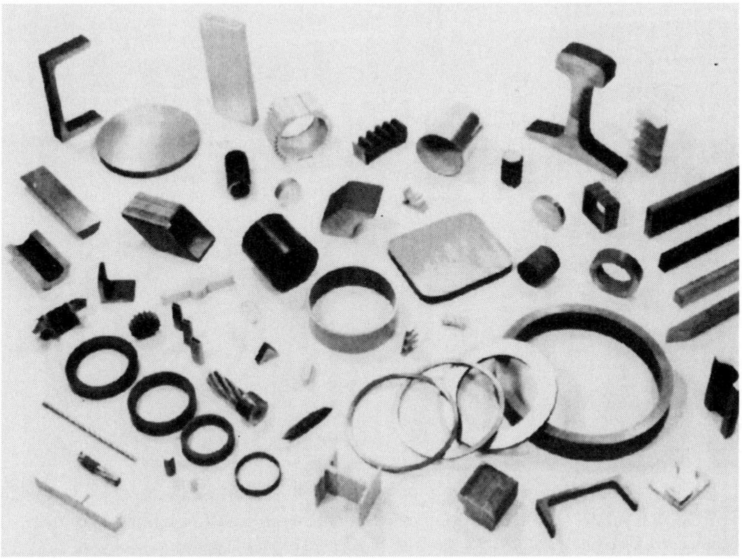

FIGURE 4.2.3 Typical parts and shapes cut by abrasive saws. (*Courtesy W. J. Savage Co.*)

FIGURE 4.2.4 Lathe-type cutoff machine. (*Courtesy Modern Machine Tool Co.*)

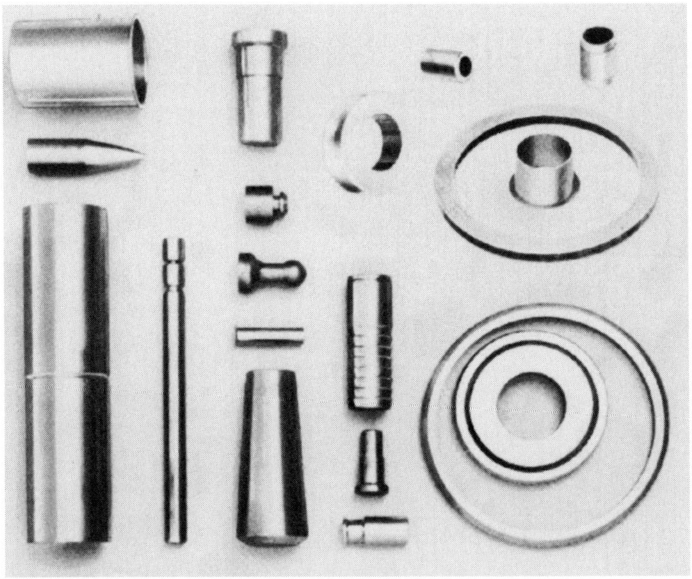

FIGURE 4.2.5 Typical parts cut off with lathe-type cutoff machines. (*Courtesy Modern Machine Tool Co.*)

TABLE 4.2.1 Characteristics and Applications of Various Cutoff Methods

Method	Kerf width	Typical cutting rate, cm^2/min (in^2/min)	Suitable materials	Applications and remarks
Band saw	1.5 mm (1/16 in)	25–190 (4–30)	Machinable materials to R_c 45	Broad application; good for fragile shapes like thin-walled extrusions.
Hacksaw	3 mm (1/8 in)	19–39 (3–6)	Machinable materials to R_c 45	Slower than band saw, but saw blades and equipment are less costly. Both have automatic cutting.
Circular saw	5.6–8.4 mm (0.220–0.330 in)	45–230 (7–36)	Machinable materials to R_c 45 mm (12 in) thick.	Accurate, smooth-cut surface; automat-fic; best for high-volume cutting of structural pieces, ingots, etc., to 300
Abrasive disk	Average 3.8 mm (0.150 in)	40–190 (6–30)	Conventional materials and those not normally machinable	For hardened steel, ceramic, glass, etc.; or fragile sections, metal-braided hose. Wheel cost is high, but equipment is inexpensive. Good for job shop, short production work.
Friction	2.5–9.5 mm (0.100–0.375 in)	Rapid	Conventional materials and those which are hard or work-harden; not good a for copper, brass, or aluminum	Structural shapes and billets, etc.; noisy. Cutting accuracy and surface finish are poor.
Shear	None	Most rapid	Machinable materials to R_c 45	Best for sheet metals but used for billets to 200 mm (8 in) thick; causes edge distortion.
Lathe	1.5–6 mm (1/16–1/4 in)	100 to 2000 short pieces	Metals to R_c 45	Round bar or tubular parts to 200 mm (8 in) in diameter; fully automatic. Used for machine shafts, rollers, spacers, pins, coupling, and nipple blanks. Equipment is similar to that of screw machines.

TABLE 4.2.1 Characteristics and Applications of Various Cutoff Methods (*Continued*)

Method	Kerf width	Typical cutting rate, cm^2/min (in^2/min)	Suitable materials	Applications and remarks
Thin-wall tubing (shear-type machines)	None	100 to 1700 cut pieces of tubing per hour	Used primarily for more ductile materials since cut tubing is usually bent, swaged, or flared	Several types of equipment are available. Suitable for tubing to 1.6-mm (0.065-in) wall thickness and diameters to 1500 mm (6 in).
Flame cut		See Chap. 4.11.		Heavy structural shapes and plate.
Electron and laser beam		See Chap. 4.21.		
Electrical discharge		See Chap. 4.18.		

ECONOMIC PRODUCTION QUANTITIES

Automatic cutoff equipment requires a relatively short setup time. Therefore, a relatively moderate lot size (100 to 1000 pieces) will be economical. For manually advanced stock, of course, a lot quantity as small as 1 piece is economical.

At the opposite extreme, high production, there are no real limits except for truly mass-production quantities (upward of 100,000 per year) of parts requiring secondary operations. In this case, it usually is justifiable to utilize a special machine to combine cutoff with the secondary operations, and pure cutoff is not used.

DESIGN CONSIDERATIONS

Cut-to-length parts are inherently simple, and there is little that designers need to do to tailor their designs to facilitate cutoff operations. Their best approach is to ensure that the tolerances and other specifications are compatible with the capabilities and limitations of the most economical methods available to them.

Thus designers should not specify a smooth, burr-free cutoff surface if the part is a structural member that will be welded at the ends; they should not demand a perfectly square end face if the application of the part does not require it; they should expect and allow for distortion if the part is one that can be cut off with a shear. Designers should consider the burrs, kerf, and surface effects (hardness, discoloration, and smoothness) of the cutoff process they will use. If the part is used in high production, and hopper feeding of the uncut lengths of stock would be advisable, the part should be round rather than hexagonal, square, or some other shape. This is desirable for easy feeding.

DIMENSIONAL FACTORS AND TOLERANCES

Length variations of parts cut to length may be caused by the following: (1) variations in operator skill, (2) machine wear, (3) variations in machine adjustment, settings, and automatic feed, (4) variations in teeth set (for saws) or wheel width (for abrasive cutoff), and (5) blade "wander." The last-named may be a significant factor for hacksaw, band-saw, and sometimes, abrasive-wheel cutoff. Blade wander occurs when the blade or wheel skids to one side at the start of a cut. It occurs with band-saw and hacksaw cutoff when heavy cutting pressure and inadequate blade tension cause the blade to flex. This condition, of course, affects the squareness of the cut and therefore the length accuracy.

Cutoff machines (except circular saws designed for carbide) are less rigid than other machine tools and therefore are subject to producing parts with greater surface and dimensional variations.

Table 4.2.2 summarizes the length, squareness, and surface-finish tolerances recommended to be specified for parts cut to length with common cutoff methods.

TABLE 4.2.2 Recommended Tolerances for Parts Cut to Length

Process	Length tolerance		Squareness tolerance, ±%	Surface finish	
	Normal	Close		Normal	Close
Band saw	±0.25 mm (0.010 in)	±0.13 mm (0.005 in)	0.2	5.0–7.5 μm (200–300 μin)	2.5–5.0 μm (100–200 μin)
Hacksaw	±0.25 mm (0.010 in)	±0.13 mm (0.005 in)	0.2	5.0–7.5 μm (200–300 μin)	2.5–5.0 μm (100–200 μin)
Abrasive wheel	±0.25 mm (0.010 in)	±0.13 mm (0.005 in)	0.2	3.2 μm (125 μin)	1.6 μm (63 μin)
Circular saw	±0.20 mm (0.008 in)	±0.10 mm (0.004 in)	0.2	3.2 μm (125 μin)	1.6 μm (63 μin)
Friction saw	±3.0 mm (0.125 in)	±0.50 (0.020 in)	1.0	25 μm (1000 μin)	12.5 μm (500 μin)
Thin-wall tubing	±0.13 mm (0.005 in)		0.2	3.2 μm (125 μin)	1.6 μm (63 μin)
Shear	±0.25 mm (0.010 in)*	±0.13 mm (0.005 in)*	0.2		
Lathe	±0.13 mm (0.005 in)		0.2	3.2 μm (125 μin)	1.6 μm (63 μin)
Flame	±3.0 mm (0.125 in) (plates over 150 mm, or 3 in, thick)	±1.5 mm (0.060 in) (plates to 150 mm, or 3 in, thick)	0.5	25 μm (1000 μin)	12.5 μm (500 μin)

*Applies to sheared area only. Area of "breakaway" can be 3 mm (1/8 in) longer or shorter than sheared area and, in the case of bars over 100 mm (4 in) in diameter, can vary by as much as 10 mm (3/8 in).

CHAPTER 4.3
SCREW-MACHINE PRODUCTS

Fred W. Lewis
Standard Locknut and Lockwasher, Inc.
Carmel, Indiana

THE PROCESS

The production of screw-machine products can use one of three basic machine types and many different combinations of metal-removing or -displacing tools. These tools operate on the internal and external surfaces of the workpiece in a predetermined, automatic sequence. The three basic machine types are Swiss, single-spindle, and multiple-spindle.

The Swiss-type machine incorporates five radially located tools with two to three end-working or axial tools (used for end drilling, tapping, or chamfering) as optional accessories. The incorporation of a sliding headstock permits the bar material, usually ground to $\pm$ 0.025 mm (0.001 in) or less, to be either advanced or retracted past the radial tools, which cut in a predetermined sequence.

The single-spindle machine can be equipped with from two to four radial tools and has a six- to eight-hole turret. The turret can contain tools that cut both internally and externally. The stock can be fed out more than once, depending on the work to be done. Figure 4.3.1 illustrates a typical single-spindle screw machine.

The multiple-spindle machines are built with 4, 5, 6, or 8 spindles (one design contains 12). At each spindle there is a radial-tool slide and an axial-tool holder. All cross-slide tool holders and the end tool holders advance to the workpiece simultaneously to produce one completely machined component with each indexing of the spindle carrier.

A combination of any of some 32 different cutting or forming operations can be incorporated into the tooling for any of the machines described above. The amount of work done is limited only by the number of tool positions available and the tool layout engineer's ingenuity.

These operations include single-point and form turning, shaving, skiving, drilling, reaming, counterboring, boring, recessing, tapping, die-head and single-point threading, roll threading, knurling, and burnishing. Special attachments enable the following additional operations to be performed: slotting, milling, broaching, punching, staking, peening, and impression marking.

Swiss-type and single-spindle machines can be equipped with a pickoff attachment that provides a means of holding the workpiece stationary after cutoff while an auxil-

FIGURE 4.3.1 Typical single-spindle screw machine. (*Courtesy Wickman Corporation.*)

iary rotating cutter machines a screwdriver slot, flats, cross holes, or other features not possible when the stock is rotating.

Multiple-spindle machines usually have one spindle that can be stopped so that operations of this kind can be performed by a rotating cutter in an auxiliary tool-holding spindle.

TYPICAL CHARACTERISTICS

Automatic-screw-machine products can be recognized (but not always) by the fact that they are cylindrical in shape and can have several outside diameters as well as a hexagonal or square-surfaced portion. They may or may not be threaded on one or both ends and may have an internal axial hole with more than one diameter. The hole may be chamfered and tapped.

Threads may be of different size and different pitch at each end of the component and may be both external and internal.

Knurls are specified on many screw-machine products and are used for finger grips on cylindrical nuts and knobs, for holding an insert in plastic or die-cast material, or for an interference fit within a mating piece.

Diameters of screw-machine parts range from those of the smallest watch parts to about 200 mm (8 in). Lengths can be as short as 1 mm (0.040 in) and as long as 1 m (3.3 ft).

Features that usually can be incorporated in screw-machine products without secondary operations include internal and external steps and tapers, axial holes with countersinks, counterbores, and internal threads, external threads, intricate external forms (from form tools), knurls, cross holes, slots, wrench flats, recesses, undercuts, and trepanned holes. Eccentric surfaces can be turned if suitable chuck jaws or collets are used. Parts can be turned from tubular as well as from solid stock. Circumferential markings can be rolled into the workpiece.

The following features usually (but not always) require secondary operations: nonround holes, marking on ends or flats, reamed or tapped cross holes, axial slots or grooves, gear teeth or splines, heat treatments, and ground surfaces.

Roller-burnished surfaces can be produced in screw machines, but electroplating and other surface treatments require secondary operations.

TYPICAL APPLICATIONS

Automatic-screw-machine parts are used in numerous products common in modern life. Swiss screw machines produce shafts and pinions for instruments and the holder for the ball in a ball-point pen. Contacts for electrical devices, pins, valves, and other small components are produced on these machines. Single- and multiple-spindle machines are used to produce a variety of close-tolerance components, including rivets, nuts, screws, bolts, and other fasteners; shafts; spacers; washers; hose, tubing, and pipe fittings; valve stems; pulleys; bushings; spools; gear blanks; push rods; rollers; inserts for plastics and die castings; and blanks for cutting tools and plug gauges. Figure 4.3.2 illustrates some of these typical screw-machine products.

ECONOMIC PRODUCTION QUANTITIES

Automatic screw machines have been used to produce fewer than 100 pieces of a particular part. Skilled screw-machine setup persons often prefer to use these machines to produce such small lots, especially if a part is similar to others normally produced or if it uses the same tooling. Normally, however, automatic screw machines are used for high-quantity production when hundreds of thousands or even millions of pieces are required.

Output rates vary with the size of the part, the amount of stock to be removed, and the machinability of the workpiece material. They range from a few seconds to about 5 min per piece for single-spindle machines. Multiple-spindle machines produce at higher rates because several pieces are machined simultaneously. Operator labor time per piece is considerably less than machine time because of multiple machine assignments. One operator can tend 4 or more single-spindle machines and may sometimes be assigned 10 or more. Normally, 2 or more multiple-spindle machines, as available,

FIGURE 4.3.2 Screw-machine parts like these are used in many products in everyday use. (*Courtesy National Screw Machine Products Association.*)

are assigned per operator. Labor costs per unit, therefore, are quite low for screw-machine products.

Setup times run from under 1 h to 8 h, depending on the complexity of the part, which determines the number and kind of tools required, and the need to alter available feed cams.

Tooling consists of the various cutting tools required, often including form tools or special counterbores, feed cams, collet pads (if a special bar shape is used), and checking gauges. Compared with tooling costs for other mass-production methods, those for screw machines are low.

Multiple-spindle machines require more tooling than single-spindle machines, but the extra cost is recovered when production quantities are sufficiently large.

SUITABLE MATERIALS

Screw-machine products can be produced from any machinable material that is available in bar form. Metals, both ferrous and nonferrous, plastics, and other organic materials have been processed.

Materials specified for automatic-screw-machine production should be stable at normal conditions of temperature and humidity. In steels, the free-machining grades in the 1100 and 1200 Series plus the leaded grades are preferred. Cold-drawn material is preferable to hot-rolled steel. For copper parts, half-hard or harder annealed alloys are

TABLE 4.3.1 Recommended Free-Machining Materials for Screw-Machine Products*

Material	Common designation	UNS designation	Machinability rating
Magnesium	ASTM AZ-61A	M11610	500
Magnesium	ASTM AZ-91A	M11910	500
Aluminum	2017-T4	A92017 (T4 temper)	300
Aluminum	2011-T3	A92011 (T3 temper)	300
Brass	High-leaded (342)	C34200	220
Zinc	ASTM AG40A	Z33520	200
Brass	Medium-leaded (340)	C34000	180
Phosphor bronze	FC 544	C54400	180
Carbon steel	12L14	G12144	105–195
Carbon steel	1212	G12120	100
Carbon steel	1119	G11190	90
Carbon steel	1114	G11140	75

*All materials listed have a surface-finish rating of "excellent."

preferred, while with aluminum, the best finishes are produced on the harder-tempered alloys.

Machinability is an important factor in the cost of screw-machine products. Minimum costs are usually associated with the most machinable grade of a particular base metal or alloy even if the more machinable grade carries a price premium. Data on machinability and various other properties of metallic and other materials can be found in Sec. 2 of this *Handbook.* Machinability data shown there are applicable to screw-machine products. For a summary of common high-machinability materials used for screw-machine work, see Table 4.3.1.

DESIGN RECOMMENDATIONS

Stock Size and Shape

Bar stock sizes and tolerances often can be used to advantage when designing for screw-machine production. The largest diameter of a component should, if possible, be the diameter of the bar stock used in order to conserve material and reduce machining. Standard sizes and shapes of bar stock should be used in preference to special diameters and shapes.

Basic Part Shape and Complexity

Although complex form tools for external surfaces and counterbores for internal surfaces are a little more costly than simpler tools, it nevertheless is advisable, insofar as possible, to keep the design of screw-machine parts as simple as possible. The fewer the tool stations required and the fewer the characteristics to be machined and gauged, the lower is the cost of the part in the long run.

Standard tools should be used as much as possible. The designer should specify standard, common sizes for holes, screw threads, slots, knurls, etc., so that readily available cutting tools and gauges can be used instead of those which must be specially fabricated.

Avoiding Secondary Operations

Screw machine products should be designed, as much as possible, to be complete on cutoff from the bar material. If secondary operations cannot be avoided, they should be minimized. If slots, flats, or other surfaces not generated by the rotation of the bar are required, they should be designed so that they can be machined when the part is held in a pickoff attachment. Such surfaces should be small so that heavy cuts are not required. As many of the other features of the part as possible, especially internal surfaces and screw threads, should be located on one end of the part only so that all machining can be performed on the screw machine before cutoff. (Fig. 4.3.3 illustrates some examples.)

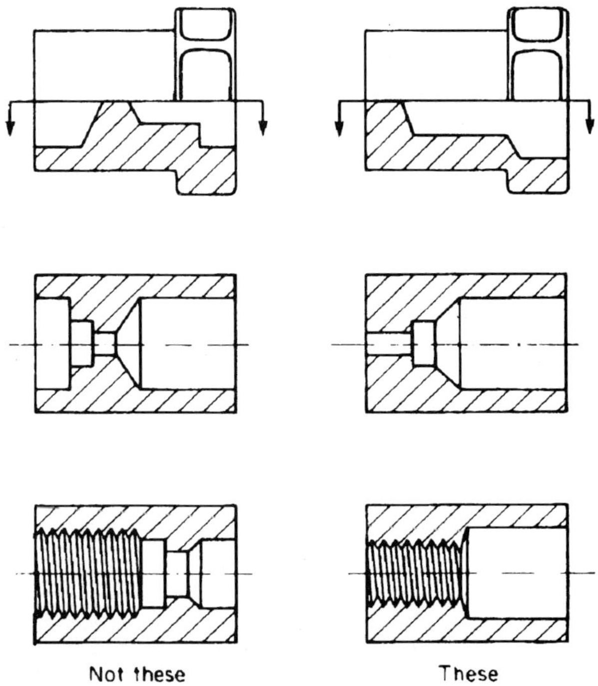

FIGURE 4.3.3 Screw-machine parts should be designed so that, if possible, they can be produced completely on the screw machine without subsequent machining. All parts on the left would have to be rechucked and machined on the left end.

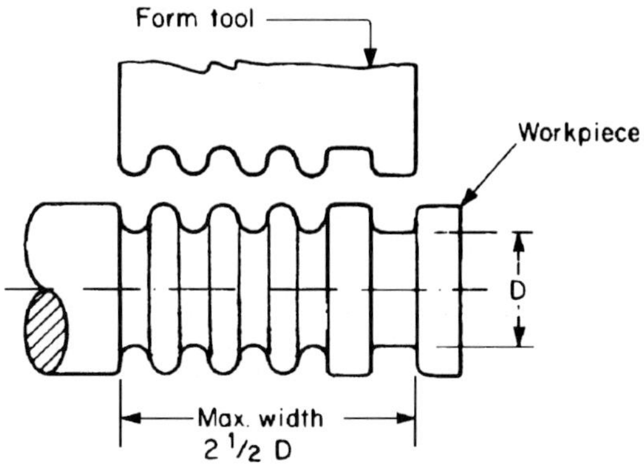

FIGURE 4.3.4 Form-tool width limitation.

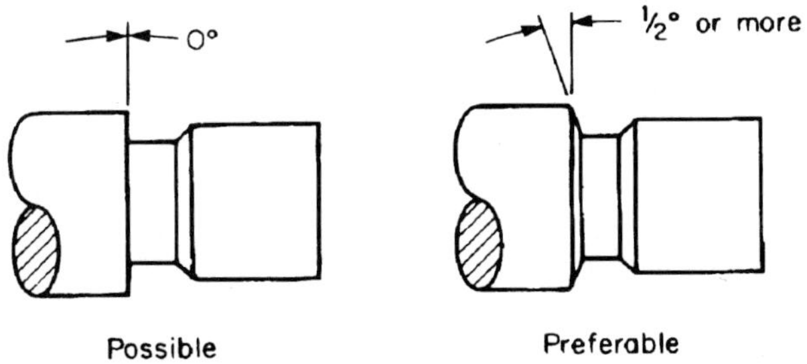

FIGURE 4.3.5 Sidewalls and faces at a slight angle from the perpendicular prevent tool marks when the tool is withdrawn.

External Forms

There are limits to the length of formed areas that are practicable to machine with one form tool. As shown in Fig. 4.3.4, the length of such areas should normally not exceed $2\frac{1}{2}$ times the minimum workpiece diameter.

Sidewalls of grooves and other surfaces that are perpendicular to the axis of the workpiece should have a slight draft. This will prevent tool marks on the machined surface as the tool is withdrawn. The recommended draft is $\frac{1}{2}°$ or more. (See Fig. 4.3.5.)

When the part is produced from square or hexagonal stock, beware of turned diameters that are of the same dimensions as the distance between the opposite flats of the stock. Small variations in stock size or concentricity will result in a surface that is not fully cylindrical. The best solution is to design the turned surface to be about 0.25 mm (0.010 in) or more smaller than the stock size.

The part also should be designed so that the form tool used to produce it is not fragile. For example, avoid deep, narrow grooves, sharp corners, etc.

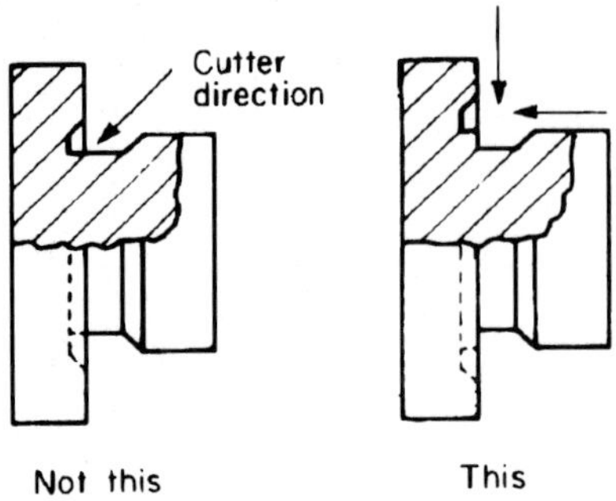

FIGURE 4.3.6 Use undercuts obtainable with transverse or axial tool movement (or both) rather than angular movement. (*Based on Fig. 12, Rupert Le Grand,* Manufacturing Engineers' Manual, *McGraw-Hill, New York, 1971, p. 128.*)

Undercuts

Because standard tool slides in screw machines operate at right angles to the axis of the workpiece, an angular undercut, internal or external, is difficult to produce and should be avoided whenever possible. (See Fig. 4.3.6.)

Annular grooves are easier to incorporate in the external surface of the part than as an internal recess. External grooves can be incorporated into form tools. Internal recesses require tools with both axial and transverse motion. If there is a choice, design parts to contain external grooves, as shown in Fig. 4.3.7.

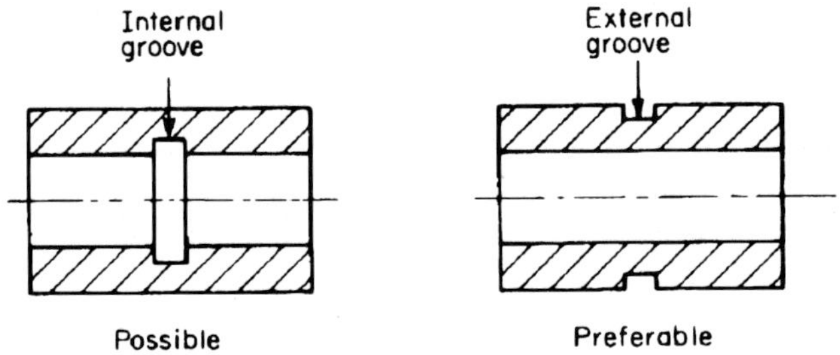

FIGURE 4.3.7 External grooves can be incorporated in screw-machine parts more economically than internal recesses.

Holes

The bottom shape of blind holes should be that made by a standard drill point, as shown in Fig. 4.3.8*c*. When flat-bottom holes are required, allow a drill point as shown in Fig. 4.3.8*b*, preferably 3 mm (0.120 in) deep.

Although deep, narrow holes can be provided if necessary, it is better to limit the depth of blind holes to 3 to 4 times the diameter.

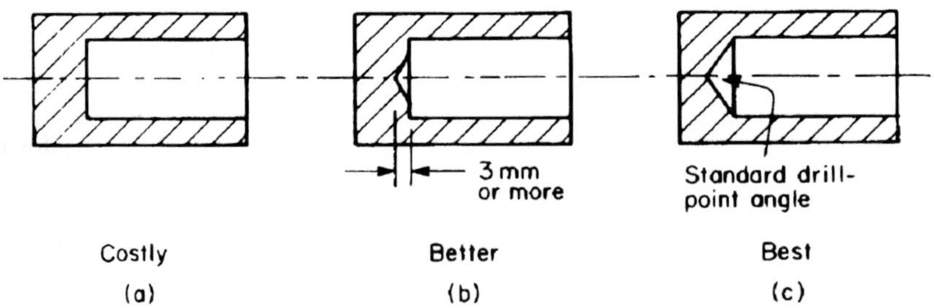

FIGURE 4.3.8 Hole bottoms are most economical if they use standard drill-point angles. If flat bottoms are required, some drill-point depression in the center should be allowed.

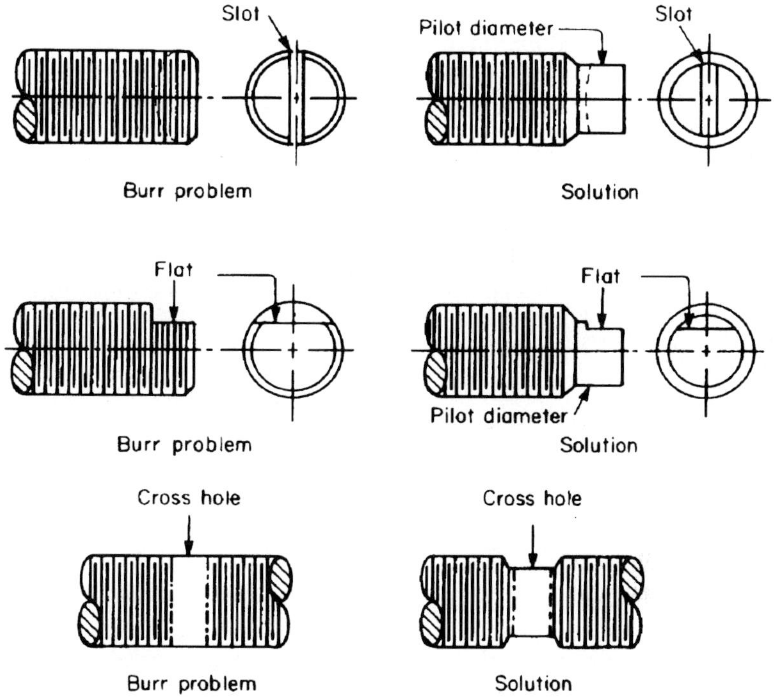

FIGURE 4.3.9 Means for avoiding burrs in threaded parts.

Screw Threads

Internal and external screw threads are very common features of screw-machine products. Design recommendations for screw threads are found in Chap. 4.8. Rolled threads are preferable to cut threads in screw-machine products, and designs should be made accordingly. Some design recommendations for avoiding burrs where screw threads intersect other surfaces are shown in Fig. 4.3.9.

Knurls

A knurled area should be kept narrow. Its width should not exceed its diameter.

An exact number of teeth cannot be produced by knurling, so a typical drawing note specifies an approximate number of teeth per inch or the general size of the knurl (coarse, medium, fine), type of knurl (straight, diagonal, diamond), and use for the knurl (finger grip, appearance, press fit) to guide the producer.

Sharp Corners

Avoid sharp corners in the design of screw-machine parts. Sharp corners, both internal and external, cause weakness or more costly fabrication of form tools.

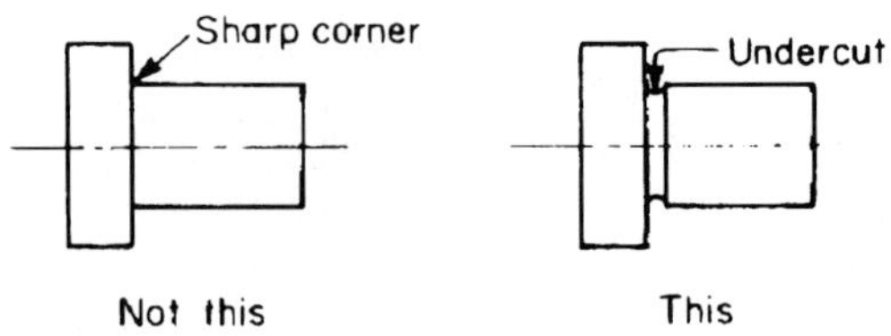

FIGURE 4.3.10 Undercuts supply corner space, if needed, with fewer problems than sharp corners.

When a sharp corner is required, it can be produced within specified limits, which are usually close enough to meet the functional requirements of the part. When unwanted sharp corners occur, they can be given a commercial corner break of 0.4 mm ($^1/_{64}$ in) by 45°.

An inside corner can be made "sharp" by providing an undercut at the corner. This would eliminate the inevitable radius produced by a form at this intersection. (See Fig. 4.3.10.)

Spherical Ends

When a spherical end is required on a screw-machine part, it is better to design the radius of the spherical end to be larger than the radius of the adjoining cylindrical portion. This procedure eliminates the need for a blending of two surfaces that may not be perfectly concentric and which usually are formed by separate tools. (See Fig. 4.3.11.)

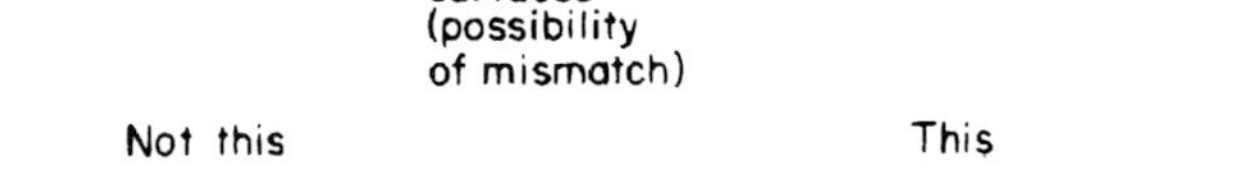

FIGURE 4.3.11 Make spherical ends with a large radius to avoid problems of blending surfaces formed by separate cutters.

Slots and Flats

These can be produced with a concave surface at the bottom or end, its radius of curvature being the same as that of the milling cutter used. By transverse feeding of the cutter, the bottom or end can be made flat. If the function of the part does not require a flat surface, allow a concave surface, which is easier to produce. (See Fig. 4.3.12.)

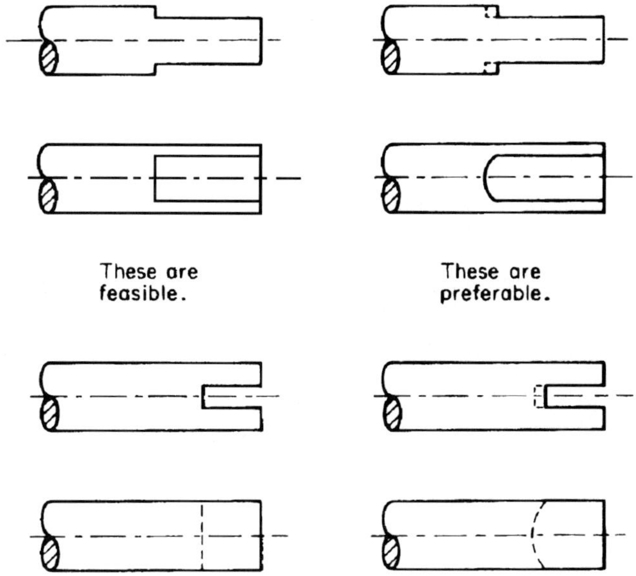

FIGURE 4.3.12 Permit curved bottoms of slots and flats if possible.

Marking

Impression markings can be made on screw-machine products as part of the basic screw-machine operation if rollers marking as shown in Fig. 4.3.13*c* is used. Other positions of the mark require secondary operations.

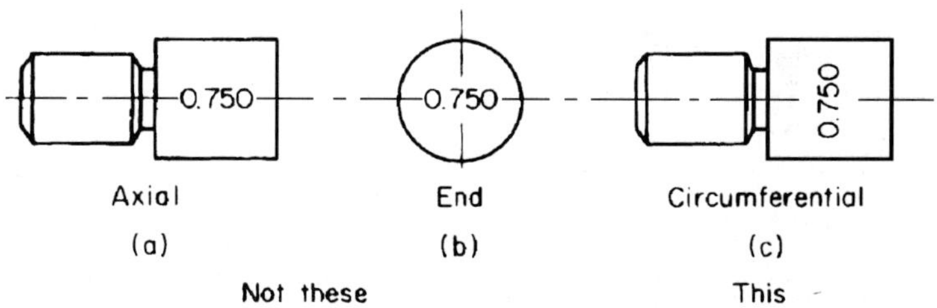

FIGURE 4.3.13 Position impression markings so that roller marking tools can be used as part of the basic operation.

Special Cross Sections

When production quantities are large enough and the configuration of the part to be produced is appropriate, designers should consider the use of specially shaped bar stock. This stock can be rolled, drawn, or extruded to a shape suitable for the shape of the final part. If necessary, the automatic screw machine can be equipped with special collets to accept the stock's cross section.

DRAFTING RECOMMENDATIONS

Dimensions

When dimensioning a drawing, use a common reference line or surface for all dimensions except diameters. Be sure that all distances are clearly dimensioned. Show dimensional relationships for concentricity, eccentricity (if intended), squareness, angularity, and surface finish. A measurable surface roughness should be designated. If a surface is to be used as a "seat," it should be so noted.

Keyway depths should be specified from the side of the piece opposite the keyway because this is easily measured (see Fig. 4.3.14). The same approach is used in specifying the location of a flat.

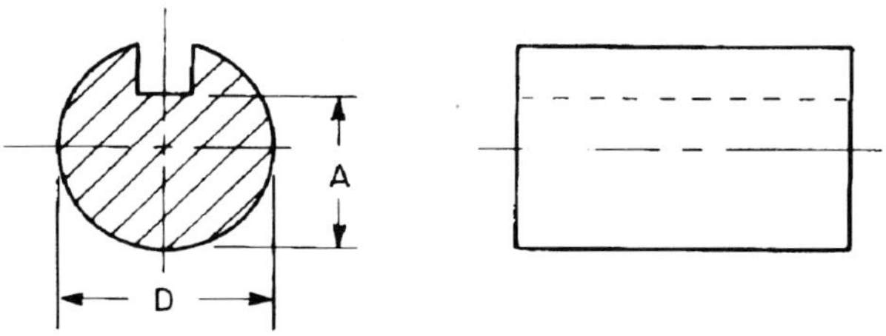

FIGURE 4.3.14 *A* is the preferred dimension to show the depth of the keyway. The depth cannot be measured from the top center of the part because the material has been removed.

When a specific concentricity is required for a threaded surface, be sure that the concentricity limits take into consideration the tolerance limits of the threads.

Internal recesses are often dimensioned incorrectly. The use of the common axis of the part as a datum line is the key to correct dimensioning. (See Fig. 4.3.15.)

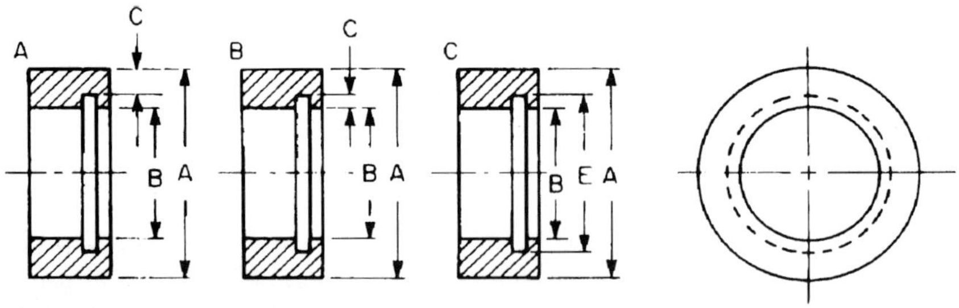

FIGURE 4.3.15 Preferred dimensioning of recesses is shown in *C*. *A* and *B* views show dimensions that are difficult to measure.

Tapers

The preferred method for dimensioning both inside and outside tapers is to specify the end diameters and the length and not the angle of the taper. However, when the tapered surface extends to the end of the part, the edge may be chamfered or radiused to remove a sharp edge. This produces a measuring problem, since the diameter to be measured will have been removed.

TOLERANCE RECOMMENDATIONS

Recommended dimensional tolerances for screw-machine products are shown in Table 4.3.2.

TABLE 4.3.2 Recommended Tolerances for Screw-Machine Products

Operation	Commercial tolerance, mm (in)	Commercial surface finish, μm (μin)
Drilling (including cross drilling and back-end drilling)	±0.075 (±0.003)	1.6–3.1 (63–125)
Reaming (including cross reaming)	±0.025 (±0.001)	0.8–1.6 (32–63)
Counterboring	±0.10 (±0.004)	1.6–3.1 (63–125)
Recessing	±0.25 (±0.010)	1.6–3.1 (63–125)
Tapping (including roll tapping)	Class 2	
Boring	±0.025 (±0.001)	1.6–3.1 (63–125)
Broaching	±0.15 (±0.006)	0.8–1.6 (32–63)
Rough turning	±0.13 (±0.005)	1.6–3.1 (63–125)
Finish turning	±0.05–0.13 (±0.002–0.005)	0.4–1.6 (16–63)
Form-tool turning, rough	±0.13 (±0.005)	1.6–3.1 (63–125)
Form-tool turning, finish	±0.025–0.05 (±0.001–0.002)	0.4–1.6 (16–63)
Roller shaving	±0.025 (±0.001)	0.4–1.6 (16–63)
Skiving	±0.025–0.05 (±0.001–0.002)	0.2–0.8 (8–32)
Die-head threading	Classes 2 and 3	
Single-point threading	Class 3	
Roll threading	Classes 2 and 3	
Slotting	±0.13–0.25 (±0.005–0.010)	1.6–3.1 (63–125)
Other milling	±0.13 (±0.005)	1.6–3.1 (63–125)
Burnishing	±0.013 (±0.0005)	0.15–0.25 (6–10)

CHAPTER 4.4
OTHER TURNED PARTS

Theodore W. Judson
GMI Engineering and Management Institute
Flint, Michigan

This chapter covers two general classifications of parts: (1) parts identical or similar to the screw-machine products covered by Chap. 4.3 when those parts are made on equipment other than automatic screw machines and (2) all other components requiring turning or other operations performed on lathes and lathe-type machines.

TURNING PROCESSES

Turning is a conventional material-removal operation that produces surfaces of rotation on the workpiece. Like other machining operations, turning removes material by the shear process to produce the desired shape, size, and surface finish. It is generally accomplished by causing cutting tools, varying in geometry and working separately or simultaneously, to move in a precise path with respect to a rotating workpiece. Workpieces, generally made of metal in the form of bars, tubes, castings, and forgings, are held by means of chucks, collets, and centers. The workpiece rotates about its center axis and the centerline of the machine spindle.

Common cutting-tool modes on turning equipment are shown in Fig. 4.4.1. They include facing, straight turning, taper turning, grooving, cutoff, threading (single-point threading is shown in Fig. 4.4.1; other thread-cutting operations described in Chap. 4.8 for both internal and external threads are commonly carried out on turning equipment), tracer turning (form turning, as described in Chap 4.3, is also common), drilling, reaming, and boring.

In tracer turning, the single-point cutting tool, as it traverses the work, moves inward and outward in response to the motion of a stylus that bears against a template or master part. After one or more passes, the shape of the template is reproduced in the workpiece. (See Fig. 4.4.2.)

A wide variety of turning machines is available. The simplest and most basic is the engine lathe, shown in Fig. 4.4.3. Tool action is parallel and transverse to the axis of rotation, in most cases, cutting tools are mounted individually or in a tool turret on a cross slide and are used successively as work progresses. Stock material in the form of

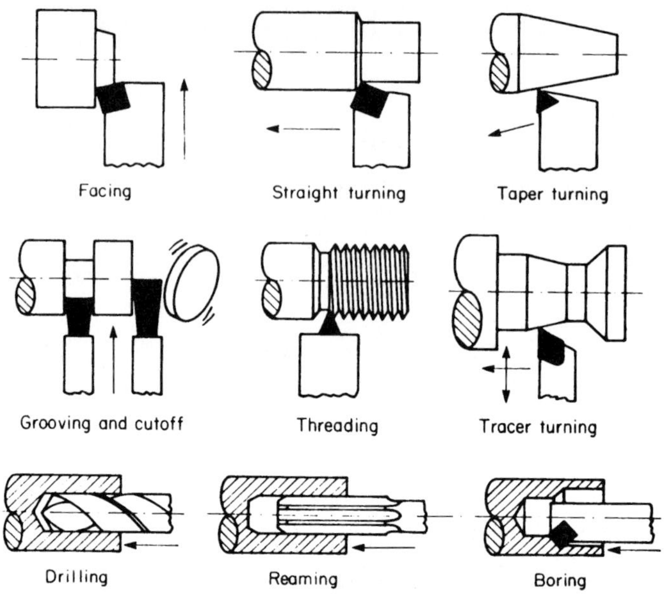

FIGURE 4.4.1 Basic operations performed on turning equipment.

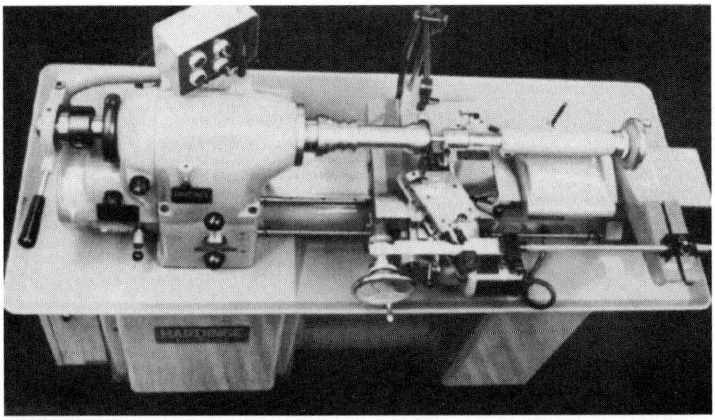

FIGURE 4.4.2 Mechanical-tracing attachment. (*Courtesy Hardinge Brothers, Inc.*)

tubes or bars can be extended through a hollow spindle and held by a chuck or centers. Other irregular work such as forgings and castings is usually chucked. Engine lathes range in size from swing capacities of 150 mm (6 in) to 1½ m (60 in) and in bed lengths from 1 m (40 in) to as much as 15 m (50 ft) in length. Machines have automatic motion of both carriage and cross slide. The bench lathe is a smaller variety of the engine lathe. It is considered more suitable for the smaller and lighter range of work-

FIGURE 4.4.3 Engine lathe. (*Courtesy DoAll Company.*)

pieces. A toolroom or toolmaker's lathe is similar to an engine lathe but is built to more accurate specifications. It usually is equipped with a wider speed and feed range.

In a turret lathe, often called a *hand-screw machine,* cutting tools can be selected, preset, and mounted in a hexagonal turret attached to a saddle and mounted on the lathe bed. This saddle also may be power-driven and includes a mechanism for indexing the turret. The cross slide contains a four-position quick-indexing turret, and a tool holder is mounted on the rear side of the cross slide. Positive stops are available for all cutting tools, but for machines without automatic, computer, or numerical control, an operator is required to manipulate the tools.

Collet chuck capacity ranges from 4.8 to 110 mm ($\frac{3}{16}$ to $4\frac{1}{2}$ in). Maximum chuck capacity ranges from 250 to 380 mm (10 to 15 in). Bar-feeding attachments are common. In addition to manual-control models, automatic control of machine functions is available.

Computer-controlled and numerically controlled lathes provide automatic continuous-path control of tool motion. The machines can be engine, bench, or turret lathes, the last-named sometimes combining full contouring capabilities with two six-station turrets. A side-mounted turret replaces the cross-slide construction of a standard turret lathe. The computer-controlled or numerically controlled machine finds its advantage in automatic repeatability of the machining operation. Full process instructions are stored for later use. Quick modification of the part specifications is possible if required. Figure 4.4.4 illustrates a computer numerically controlled lathe.

Chucking machines, or chuckers, are similar to automatic screw machines except

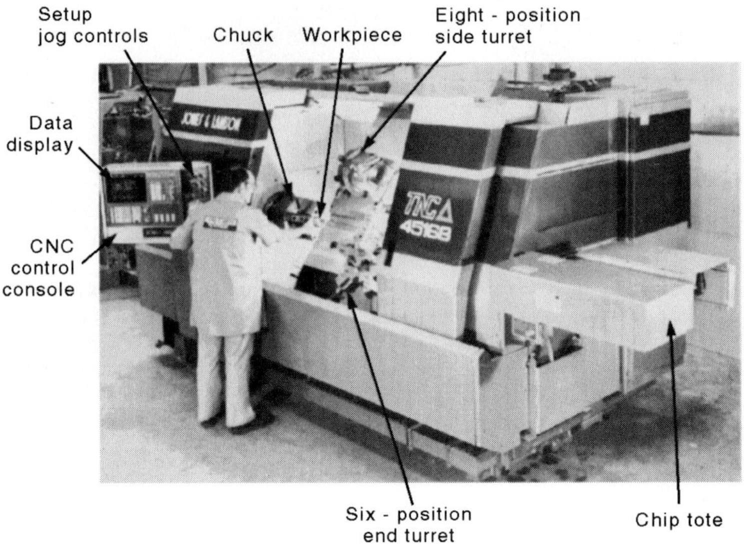

FIGURE 4.4.4 Computer numerically controlled lathe. (*Courtesy Jones & Lamson, Waterbury Farrel Division of Textron, Inc.*)

that the machining is performed on a discrete workpiece instead of a length of bar stock. Workpieces are inserted in the chuck (or chucks if the machine is a multiple-spindle type), and cross-slide and end-turret tools perform a variety of operations automatically. Chucking machines may be either horizontal or vertical as illustrated by the machine in Fig. 4.4.5.

FIGURE 4.4.5 Vertical double-spindle chucking machine. (*Courtesy The Bullard Company.*)

TYPICAL CHARACTERISTICS AND APPLICATIONS

Turned parts all share one common characteristic: They have curved machined surfaces that are produced during the rotation of the part about its axis. Aside from this common characteristic, shape, size, and material can vary widely. Workpieces vary in size from minute needle-valve components weighing a small fraction of an ounce to huge turned rolls weighing many tons.

Except for the lower quantities normally involved, parts made from bar stock on bench, engine, and turret lathes have the same characteristics as the screw-machine products described in Chap. 4.3.

Turning is a major machining operation. Aside from screw-machine products, the most common application of turning is in secondary operations performed on workpieces produced by other processes. (Secondary operations on screw-machine parts are also very common.) If precision surfaces are required when some portion has a round shape, turning is likely to be employed. This is most common when the surface in question contacts a moving part or is used for sealing. Turning also may be employed when the primary operation cannot produce the required configuration, for example, when undercuts or annular grooves are required in cold-headed or powder metallurgy parts.

Among the classes of parts that frequently require turning operations are castings of various types, forgings, extrusions, welded and brazed assemblies, and molded nonmetallic parts. Although internal and external surfaces of revolution are the norm, flat surfaces are produced on many parts by facing. One limitation of turning is that the part must be rotated during the operation. This may impose some limitations on size, weight, and shape.

Some common turned parts produced on various lathes and chuckers are brake drums and brake disks, valve parts, nozzles, pipe and tube fittings including tees and elbows, rollers of various kinds, pistons, turbine rotors, hubs, crankshafts and many other shafts, pins and axles, gun barrels, and handles. Figure 4.4.6 illustrates typical parts machined on chucking machines.

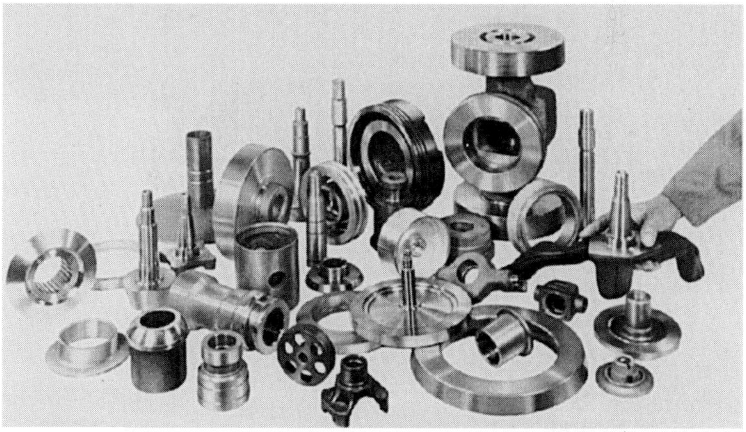

FIGURE 4.4.6 Typical parts made by chucking machines. (*Courtesy The Bullard Company.*)

ECONOMIC QUANTITIES FOR TURNING

Turning operations are employed for production quantities ranging from one piece to many millions. When production quantities warrant or part specifications dictate, turning operations can be varied in their degree of sophistication from manual, numerical, computer, or completely automatic mechanical control. Further, total system automation including conveyors, load-unload devices, in-process inspection, verifying probes, and memory devices can be integrated with and supplement primary turning machine tools.

Production quantity is a key factor in the choice of turning machines. The higher the quantity and the longer the production run, the more capital funds can be allocated to increased automatic operation. Greater use of automatic systems usually means a reduction in labor cost and floor space and increased production output. This can be translated into lower cost even though the system cost is initially higher than for a more manually controlled machining system. A cost analysis, individualized for each situation, will aid in determining the degree of automatic operation best suited to the machining condition.

Tooling costs for parts machined on bench and engine lathes are very modest because general-purpose chucks, cutting tools, and measuring equipment are normally used.

When production schedules require multiple units on the order of 10 to 25 or more, turret lathes begin to be advantageous. Tooling costs for turret lathes will be almost as low as for engine lathes if general-purpose tooling is used but higher if form tools and special reamers and counterbores are used. Even with the most thorough specialized tooling and gauging, tooling costs for turret lathes are still modest compared with those of other processes and lower than for screw machines. (For a comparison of costs and setup times between screw machines and turret lathes for one part, see Chap. 1.2.)

Tracer operation and numerical and computer control, though they provide automatic operation of turning equipment, actually reduce tooling costs. Single-point cutting tools, guided to produce special contours, replace special form cutters. (See Fig. 4.4.7.

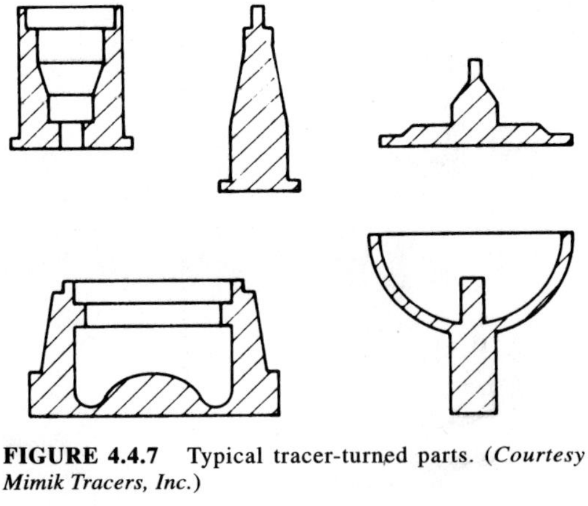

FIGURE 4.4.7 Typical tracer-turned parts. (*Courtesy Mimik Tracers, Inc.*)

In general, turning-machine operations are used more frequently in the lower ranges of production quantities. The most economic range for each process can be summarized as follows:

Engine and bench lathes: very low to low quantities

Turret lathes: low to medium quantities

Tracer lathes: low to medium quantities

Numerically controlled and computer-controlled lathes: low to medium quantities

Single-spindle chuckers: medium to high quantities

Multiple-spindle chuckers: high to very high quantities

SUITABLE MATERIALS FOR TURNING

Materials listed as suitable for screw machines in Chap. 4.3 are equally applicable for use on other turning equipment. Operations that can be performed are identical, although there is a greater preponderance of single-point cutting with engine and bench lathes and even with turret lathes and a more extensive use of form tools in screw machines.

In addition to the bar materials used in automatic screw machines, castings of various kinds, forgings, extrusions, powder-metal parts, and other forms of material are processed on these other turning machines. Properties that make these materials machinable closely parallel those for bar materials.

Table 4.4.1 gives machinability ratings for a selection of materials used in chuckers and other turning machines.

DESIGN RECOMMENDATIONS

In addition to the recommendations listed below, see the following chapters, which have design recommendations especially applicable to turned parts: Chap. 4.3, "Screw-Machine Products," Chap. 4.5, "Machined Round Holes," and Chap. 4.8, "Screw Threads." The following are additional design recommendations not covered in related chapters but applicable to turned parts.

1. The design should incorporate standard tool geometry at diameter transitions, exterior shoulders, grooves, and chamfer areas.
2. The design should minimize unsupported, delicate small-diameter work when possible to reduce work deflection from the cutting tool. Keeping parts as short as possible will help in this regard. Short, stubby parts are easier to machine than long, thin parts, which require tailstock or steady-rest support. (See Fig. 4.4.8.)
3. A product design that requires an irregular and interrupted cutting action should be avoided when possible. Hole intersections, curved or slant surface drilling, and hole or slotting operations before turning are illustrations of this condition.
4. When castings or forgings are designed with large shoulders or other areas to be faced, the surface should be 2 to 3° from the plane normal to the axis of the part. Such an incline provides edge relief for cutting tools. (See Fig. 4.4.9).

TABLE 4.4.1 Machinability Ratings for Various Materials When Turned with Single-Point Cutting Tools*

Material	Designation	Condition and remarks	Brinell hardness number	Rating
Steel	1212 (G12120)	Cold-drawn resulfurized	150–200	100
Steel	12L13 (G12134)	Leaded, cold-drawn	150–200	105
Steel	1010 (G10100)	Cold-drawn	125–175	65
Steel	1040 (G10400)	Cold-drawn	125–175	60
Steel	1070 (G10700)	Cold-drawn	175–225	50
Stainless steel	43OF (S43020)	Free-machining, ferritic	135–185	55
Stainless steel	31.6 (S316000)	Austenitic	135–185	40
Stainless steel	17-7PH (S17700)	Precipitation hardening, wrought	150–200	40
Cast iron	A 48, class 20 (F11401)	Ferritic, annealed	120–150	80
Cast iron	A 48, class 30 F121010)	Pearlitic	190–220	45
Cast iron	60-40-18 (F32800)	Ductile, ferritic, annealed	140–190	80
Aluminum	2024 (A92024)	Wrought, solution-treated	75–150	325
Aluminum	208.0 (A92080)	As sand- or permanent-mold-cast	40–100	405
Brass	280 (C28000)	Wrought, annealed	10–70 R_b	150
Brass	934 (C93400)	High-leaded tin bronze	40–150	215
Magnesium	AZ92A (MI 1920)	As cast	40–90	430
Nickel	200 (NO2200	Annealed or cold-drawn	80–170	60
Zinc	AG40A (Z33520)	Die-cast	80–100	175

*Recommended cutting speed with high-speed steel tool and heavy cut, compared with cutting speed for 1212 cold-drawn steel. Rating expressed as a percentage.

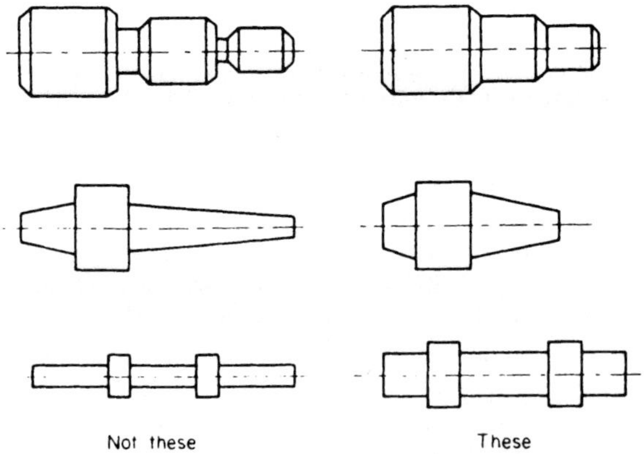

FIGURE 4.4.8 Keep parts short and stocky to minimize deflection.

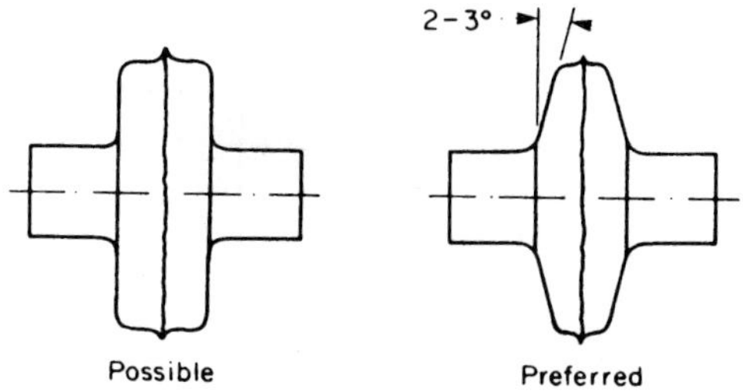

FIGURE 4.4.9 Cast or forged-in relief on surfaces to be faced provides tool clearance.

5. Radii, unless critical for the part's function, should be large and conform to standard tool nose-radius specifications. Often, the radius can be left to manufacturing preference. (See Fig. 4.4.10.)
6. Specify a break of sharp corners where sharpness or burrs may be hazardous or disadvantageous to the function of the part. The product design should be specific in this regard; i.e., do not specify "Break all corners" unless this is really necessary, because such operations are quite costly. Sharp corners and burrs can be minimized if chamfers or curved surfaces are placed at the intersection of other surfaces, as shown in Fig. 4.4.11. Often such curves or chamfers can be included in a casting or forging at no extra cost before machining.

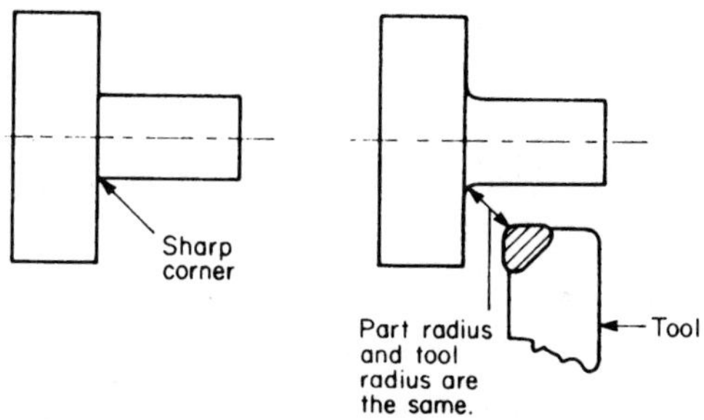

FIGURE 4.4.10 Avoid sharp corners. If possible, leave radius dimensions to the discretion of the manufacturer.

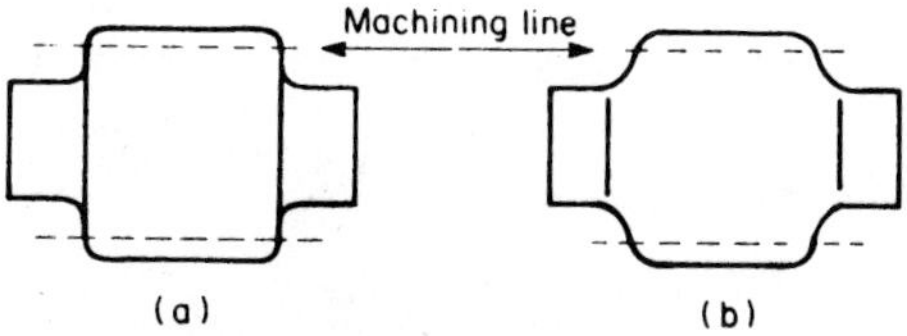

FIGURE 4.4.11 Sharp corners and burrs can be minimized by including chamfers or curved surfaces in the part before machining. Design *a* results in sharp corners after machining. The shape of *b* eliminates the sharp corners and reduces the possibility of burrs.

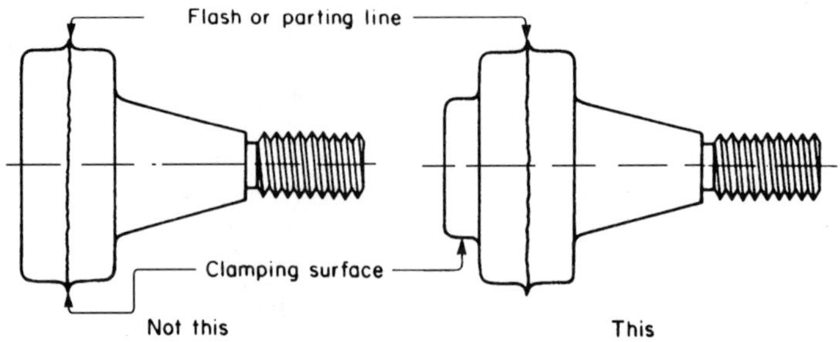

FIGURE 4.4.12 Avoid designs that require clamping on parting lines or flash areas.

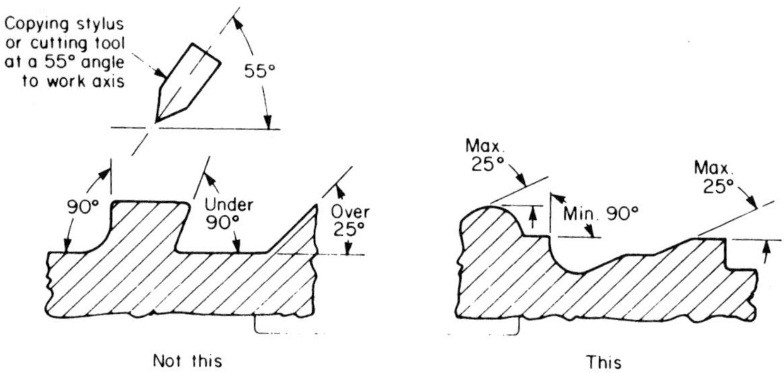

FIGURE 4.4.13 Design recommendations for tracer-turning operations when the cutting tool is angled 55° from the workpiece axis.

7. The design of the product must be such that parting lines, draft angles, and forging flash are excluded from surfaces used in the clamping or locating of the part. (See Fig. 4.4.12.)
8. When a part is to be tracer-turned, the turned contour should be such that easy tracing is possible with a minimum number of changes of stylus and cutting tool. Grooves with parallel or steep sidewalls are not feasible in one operation, and undercuts should also be avoided. Figure 4.4.13 illustrates feasible contours for a cutting tool (and stylus) inclined at an angle of 55° from the axis of the part. Figure 4.4.14 illustrates feasible contours when they are perpendicular to the axis.

DIMENSIONAL CONTROL IN TURNING

As a general rule, close dimensional limits for turning are inversely related to workpiece size and length. The larger these dimensions, the greater is the possible variation after machining.

Machine construction and maintenance also play an important part in the control of variation from basic dimensions for different turning operations. Machine design and

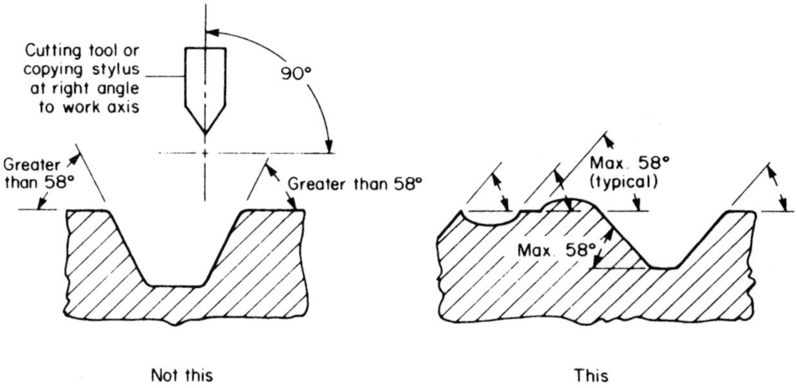

FIGURE 4.4.14 Design recommendations for tracer-turning operations when the tool is at right angles to the workpiece axis.

construction must provide control over operating disturbances that may account for piece-to-piece variation in the work done by a machine. Control of operational disturbances such as vibration, deflection, thermal distortion, and wear of functional parts of the machine are among the critical factors.

Part deflection, tool wear, measuring-tool accuracy, and operator skill are other factors. Another is the fact that turning tools cut on the part's radius. This causes deviations in the diameter of the part to be twice as large as deviations in cutting-tool position.

Surface finish is an important characteristic of all turned workpieces. The quality of finish is related to the factors mentioned above. In addition, surface finish is directly related to feed rate. The finer the feed rate, the better the surface finish. Tool sharpness, tool geometry, and tool material are also factors. Workpiece material is another important consideration.

In general, well-maintained, vibration-free machines with the proper cutting tools and high-machinability materials produce the most accurate size control and the best surface finishes. Typical machining tolerances are specified in Table 4.4.2 for well-maintained turning machines. The machining tolerances shown in the table are average figures for different-diameter work. Closer attention to machine construction and condition and to variations in cams and cutting tools and in setup can substantially improve on these commercially acceptable tolerance variations.

TABLE 4.4.2 Recommended Tolerances for Parts Made on Engine Lathes, Turret Lathes, and Chucking Machines, mm (in)

Nominal dimension	To 6.3 (0.250)	Over 6.3 to 12.7 (to 0.500)	Over 12.7 to 19 (to 0.750)	Over 19 to 25 (to 1.000)	Over 25 to 50 (2.000)	Over 50 to 100 (4.000)
Turned diameters	±0.025 (0.001)	±0.025 (0.001)	±0.025 (0.001)	±0.025 (0.001)	±0.05 (0.002)	±0.08 (0.003)
Internal dimensions	See Chap. 4.5, "Machined Round Holes"					
Shoulder location (turning)	±0.13 (0.005)	±0.13 (0.005)	±0.13 (0.005)	±0.13 (0.005)	±0.13 (0.005)	±0.13 (0.005)
Shoulder location (form tools)	See Chap. 4.3, "Screw-Machine Products"					
Overall length	±0.13 (0.005)	±0.13 (0.005)	±0.13 (0.005)	±0.13 (0.005)	±0.13 (0.005)	±0.13 (0.005)
Surface finish	6.3–0.4 μm (250–16 μin)					

CHAPTER 4.5
MACHINED ROUND HOLES

HOLE-MAKING PROCESSES

There are a variety of machining processes by which holes are produced. This chapter deals with the conventional, traditional methods—drilling, reaming, boring, etc.—and the design of workpieces to facilitate such operations.

Drilling is a machining process in which a round hole is produced or enlarged by means of an end-cutting rotating too, a drill. *Reaming* is a related process in which an existing round hole is enlarged to accurate size (diameter and straightness) and smooth finish by means of an end-cutting rotating tool, a reamer. Figure 4.5.1 illustrates some typical drills and a typical reamer.

With drills, the width of the cut from each flute usually is the full radius of the tool; in reaming, only 0.13 to 0.38 mm (0.005 to 0.015 in) is removed from each side of the hole. Reamers are run at about two-thirds the rotary speed of drills of the same diameter but are fed about 50 percent more per revolution.

Drill fixtures or jigs are employed in production drilling to locate the hole accurately in the work. A bushing serves to guide the drill. Reamers may be fed through a fixed bushing to ensure the most accurate hole location, but in many cases they are simply allowed to float and follow the location and direction of the drilled hole.

Boring is a machining operation that enlarges an existing hole by removing metal with a rigidly mounted single-point cutting tool. Either the workpiece of the tool rotates on the centerline of the hole or the tool feeds into the work parallel to the axis of rotation. The rigidity of the spindle and the boring tool rather than the guiding effect of a drill bushing or the machined hole determine the accuracy of the cut.

Other processes that are used to machine round holes are trepanning and gun drilling. Countersinking and counterboring are secondary operations for existing holes. These processes and their applications are described in greater detail below. Still other hole-making processes such as internal grinding, flame cutting, EDM, chemical machining, and other advanced processes are covered in subsequent chapters. Chapter 4.21 describes the hole-drilling capabilities of abrasive-jet, ultrasonic, electron-beam, and laser-beam machining.

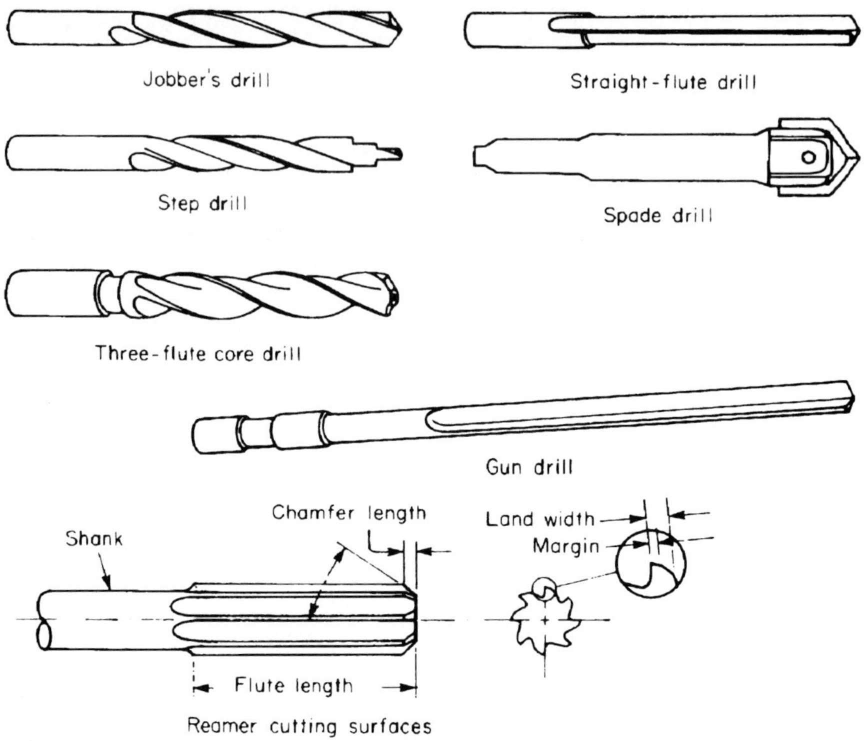

FIGURE 4.5.1 Typical drills and a typical reamer.

APPLICATIONS OF HOLE-MACHINING OPERATIONS

Holes are machined in workpieces whenever the primary production method, e.g., casting, forging, molding, extruding, stamping, etc., either does not produce holes or does not produce them with the necessary size, accuracy, straightness, surface finish, or other characteristics required by the function of the part. Machined holes usually have some mechanical function to perform as, for example, (1) serving as the bearing for a shaft, pin, or axle, with or without an inserted bushing or mechanical bearing, or (2) providing the location and support for a holding pin, screw, or bolt. The hole may be threaded for this purpose.

Clearance, vent, or lubrication holes may be drilled, but they are more apt to be cast, molded, or punched if the process permits, since their dimensions are less critical.

The following summarizes the normal dimensional range of holes machined by conventional drilling (or drilling and reaming):

Usual diameter range: 1.5–38 mm (0.060–1.5 in)

Minimum diameter (spade-type microdrills): 0.025 mm (0.001 in)

Minimum diameter (reaming): 0.3 mm (0.013 in)

Maximum diameter: 150 mm (6 in)

Usual maximum depth: 3 times diameter

Practical maximum depth: 8 times diameter

Hardness of drilled material: usually less than R_c 30

Maximum hardness: R_c 50, rarely to R_c 60

Bored holes are found in components when particular accuracy of diameter, location, straightness, or direction is required. They are more common when production quantities are low, making drill fixtures unjustifiable, and when material is of poor machinability, making single-point carbide-tip machining more necessary. Typical parts with bored holes include gearboxes, sewing-machine arms, bearings, hydraulic cylinders, pump housings, die sets for metal stamping, machine-tool frames, and many other machine components.

The following summarizes the normal dimensional limits for bored holes:

Minimum diameter: 2.5 mm (0.100 in), with fishtail-type solid cutting tool

Maximum diameter: 1.2 m (48 in) or more on vertical boring mills; limited only by the maximum size of machine available

Maximum hardness of material: R_c 60

Maximum depth of conventional boring bars: 5 times diameter

Maximum depth of solid carbide boring bars: 8 times diameter

ECONOMIC PRODUCTION QUANTITIES

Drilling is the most common machining operation and probably the least costly in terms of volume of metal removed per monetary unit. There is no lower limit to economic production quantity because manual drill presses are low-priced, versatile machines and the cutting tools (drills) are readily available. For the lowest-quantity work, the traditional method of laying out, scribing, and center-punching hole locations may be used. If quantities total about 100 or more, or if greater location accuracy is required, it may be economical to construct a drill fixture with bushings. The use of drill fixtures is the common production method even for the highest levels of production.

For high and very high levels of production, of, say, 10,000-unit quantities or greater, multiple-spindle drill presses may be used. Figure 4.5.2 shows a typical multiple-spindle drill for this level of production. The various spindles often have universal-joint drive, which allows them to be moved to new positions when the setup is changed.

Low-quantity work may be carried out on a vertical milling machine, with which hole locations are made by moving the machine table by an exact amount. This approach is often used in fabricating jigs and fixtures when a boring operation follows drilling. The process can be automated through the use of numerically controlled or computer-controlled machines and, as such, is economical for one-of-a-kind to moderate quantity production.

Boring operations take place at all levels of production. The precision jig boring machine is designed for one-of-a-kind tooling work. At the other extreme, high-production boring machines, often specially designed for a particular product, are used.

FIGURE 4.5.2 Multiple-spindle drill for high-production work. (*Courtesy National Automatic Tool Co., Inc.*)

The most notable example of such an operation is the boring of engine cylinders in the automotive industry.

SUITABLE MATERIALS

The easily machined materials noted in accompanying chapters are also advantageous for drilling and other hole-making operations. Machinability is not quite so important for drilling operations as it is for operations such as milling, turning, gear cutting, etc., because drilling is rapid. Often handling and fixture elements consume more time than actual drilling. However, when a large number of holes are to be drilled and finished with other operations such as reaming, countersinking, and tapping, easy machinability may be vital.

Because it is easier to use carbide, ceramic, and diamond-tipped single-point cutting tools when a hole is bored, the processing of very hard or difficult-to-machine materials is not so severe a handicap as it is with drilling or reaming, although, with such materials, machine speeds and feeds must be reduced commensurately with the reduced machinability.

DESIGN RECOMMENDATIONS

As with other machining operations, the best way to minimize the cost of hole-machining operations is simply to avoid them. If it is possible to cast, mold, or pierce a hole with sufficient accuracy to fulfill the functional requirement of the part, this is preferable.

If a machined hole is necessary, although drilling is really a roughing operation, do not ream or bore unless the requirements of the part truly demand the better finish and closer dimensions that these secondary operations provide.

Drilling

The following are recommended design practices for drilled holes:

1. The drill entry surface should be perpendicular to the drill bit to avoid starting problems and to help ensure that the hole is in the proper location. (See Fig. 4.5.3.)
2. The exit surface of the drill also should be perpendicular to the axis of the drill to avoid breakage problems as the drill leaves the work. Figure 4.5.3 also illustrates this.
3. If straightness of the finished hole is particularly critical, it is best to avoid interrupted cuts unless a guide bushing can be placed at each reentry surface. If the drill intersects another opening on one side, some deflection will occur. Even when straightness is not critical, it is important that the center point of the drill remain in the material throughout the cut to avoid extreme deflection and possible drill breakage. Figure 4.5.4 illustrates the alternatives.
4. It is best to use standard drill sizes whenever possible to avoid the added cost of special drill grinding.
5. Through holes are preferable to blind holes because of easier clearance for tools and chips, especially when secondary operations such as reaming, tapping, or honing are required.
6. When blind holes are specified, they should not have flat bottoms. The preferred drill bit generates a pointed hole, and if other bottom shapes are specified, secondary operations are required. Square-bottomed holes also cause reaming prob-

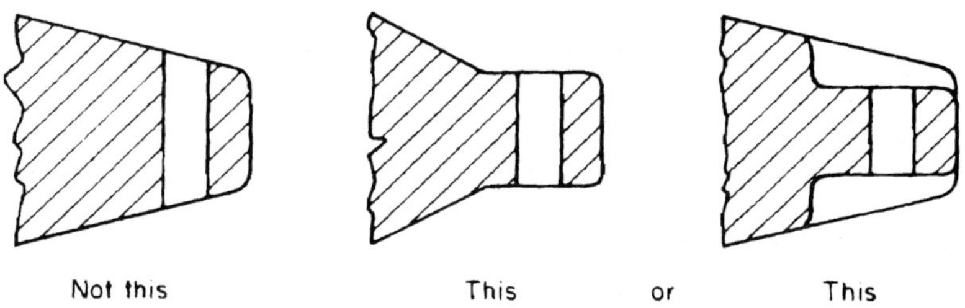

FIGURE 4.5.3 The entrance and exit surface should be perpendicular to the drill bit.

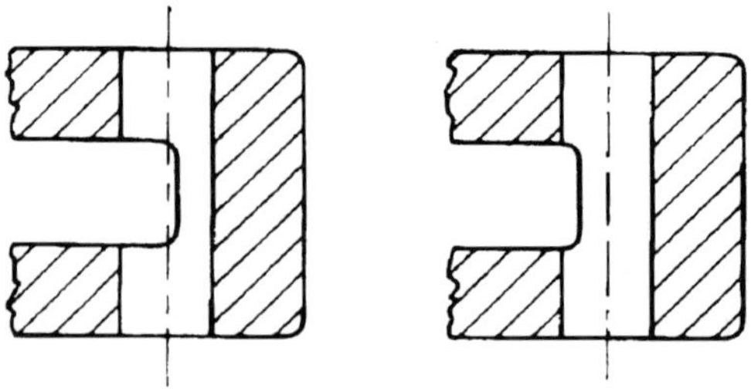

FIGURE 4.5.4 If holes with intersecting openings are unavoidable, it is important that the center point of the drill remain in the work throughout the cut.

lems because reamers have tapered ends and require room for chips to fall if tool wear and breakage are to be avoided. (See Fig. 4.3.8.)

7. Avoid deep holes (over 3 times diameter) because of chip-clearance problems and the possibility of deviations from straightness (see Fig. 4.5.10). Note that while extremely deep holes (e.g., gun barrels) can be drilled, they require different tooling, equipment, and techniques (see "Gun Drilling and Gun Reaming" below). They are not necessarily costly if gun-drilling is available but nevertheless constitute a special operation and for this reason should be avoided if possible.
8. Avoid designing parts with very small holes if the small size is not truly necessary, because small drills are more susceptible to breakage. About 3-mm ($\frac{1}{8}$-in) diameter is a desirable minimum for convenient production.
9. If large finished holes are required, it is desirable to have cored (cast-in) holes in the workpiece prior to the drilling operations. This saves material and reduces the power required for drilling.
10. If the part requires several drilled holes, dimension them from the same surface to simplify fixturing. (See Fig. 4.5.5.)
11. Rectangular rather than angular coordinates should be used to designate the location of drilled, reamed, and bored holes. They are easier and more nearly foolproof for the machinist to use in laying out the part or a drill fixture. (See Fig. 4.5.6.)
12. Insofar as possible, design parts so that all holes can be drilled from one side or from the fewest number of sides. This simplifies tooling and minimizes handling time.
13. Design a part so that there is room for a drill bushing near the surface where the drilled hole is started. (See Fig. 4.5.7.)
14. Standardize the size of holes, fasteners, and other screw threads as much as possible so that the number of drill spindles and drill changes can be minimized.
15. When production quantities are large enough to justify multiple-drilling arrangements, the designer should bear in mind that there are limitations as to how closely two simultaneously drilled holes can be spaced. Smaller holes can be spaced more closely than larger ones because gearing, chucks, and bushings can be smaller. As a rule of thumb for small holes of 6-mm ($\frac{1}{4}$-in) diameter or less, spacing should not be less than 19 mm ($\frac{3}{4}$ in), center to center, although in some cases 13 mm ($\frac{1}{2}$ in) is possible.

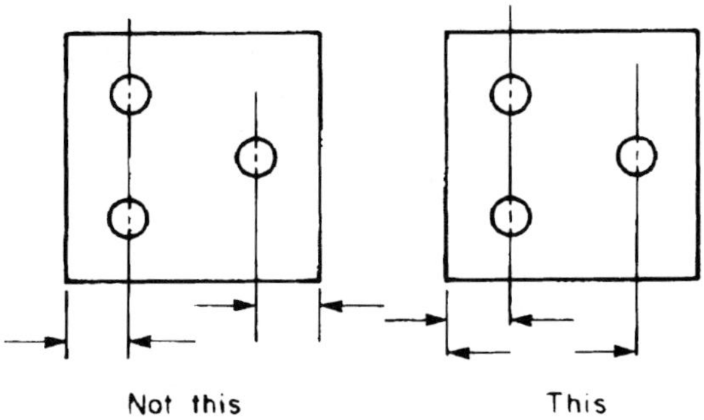

FIGURE 4.5.5 Locate all holes from one surface insofar as possible.

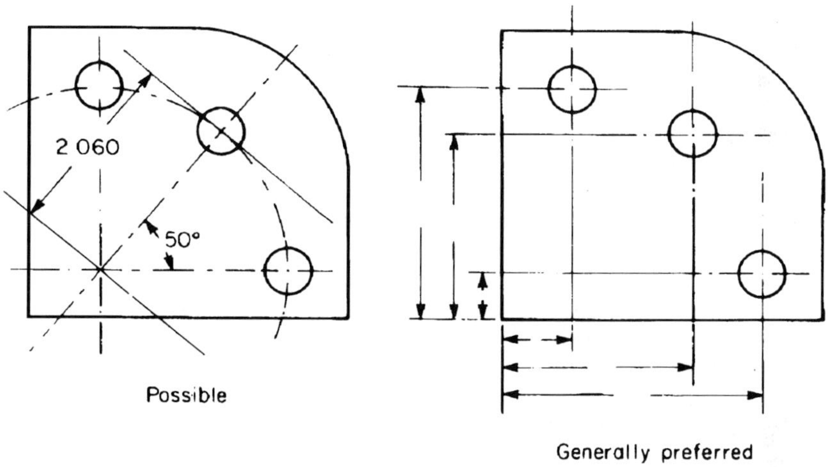

FIGURE 4.5.6 Rectangular coordinates are preferable to angular coordinates for showing hole locations in drawings.

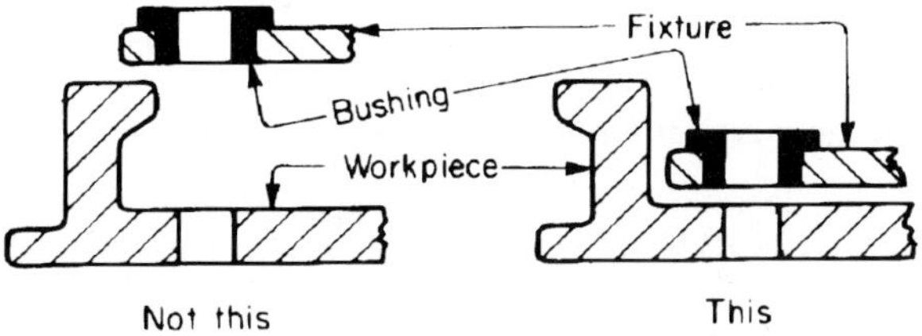

FIGURE 4.5.7 Allow room for drill bushings close to the workpiece surface to be drilled.

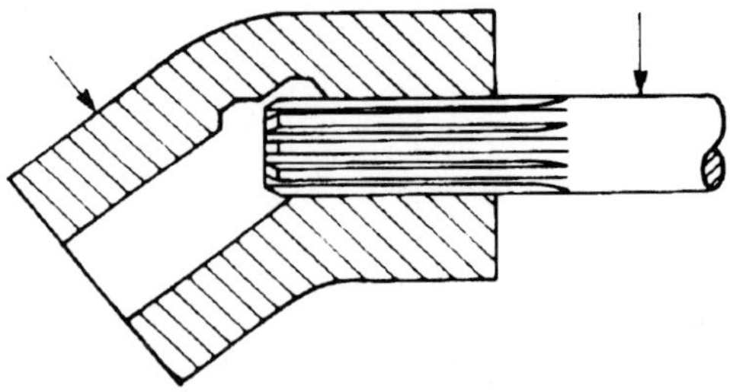

FIGURE 4.5.8 Avoid intersecting drilled and reamed holes if at all possible.

Reaming

The following are good design practices for reaming:

1. Even though it is good practice to ream with a guide bushing when the hole location or alignment is critical, do not depend on reaming to correct location or alignment discrepancies unless the discrepancies are very small.
2. Avoid intersecting drilled and reamed holes if possible to prevent tool breakage and burr-removal problems. (See Fig. 4.5.8.)
3. If a blind hole requires reaming, good practice calls for extra drilled depth to provide room for chips. (See Fig. 4.5.9.)

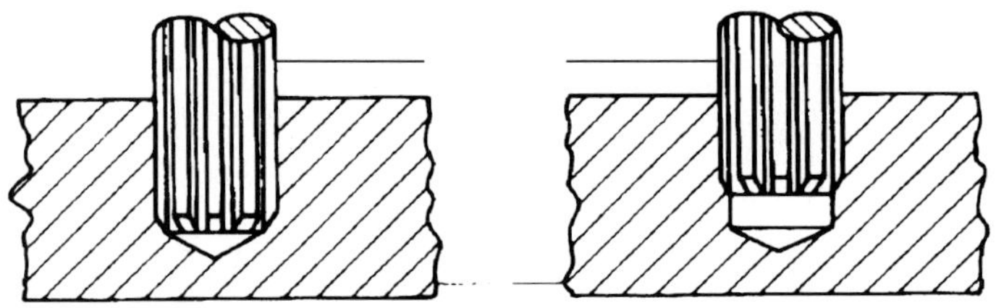

FIGURE 4.5.9 Provide extra hole depth if blind holes are to be reamed.

Boring

Recommended design practices for boring follow:

1. Even when boring operations are employed, avoid designing holes with interrupted surfaces. Interrupted cuts tend to throw holes out of round and cause vibration and tool wear.
2. Avoid designing holes with a depth-to-diameter ratio of over 4 or 5:1; otherwise, accuracy may be lost owing to boring-bar deflection. (If carbide boring bars are available, depth-to-diameter ratios of 8:1 are feasible.) If deep holes are unavoidable, consider the use of stepped diameters to limit the depth of the bored surface. (See Fig. 4.5.10.)

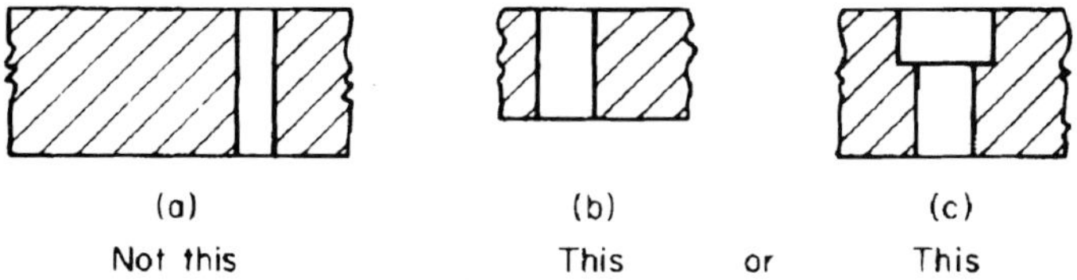

FIGURE 4.5.10 Avoid deep, narrow holes as in *a*. Drilled holes should not be deeper than 3 times diameter; bored holes can be as deep as 5 times diameter. Deeper holes are possible but can cause tool deflection and breakage problems. For deep, narrow holes, consider stepped diameters as in *c*.

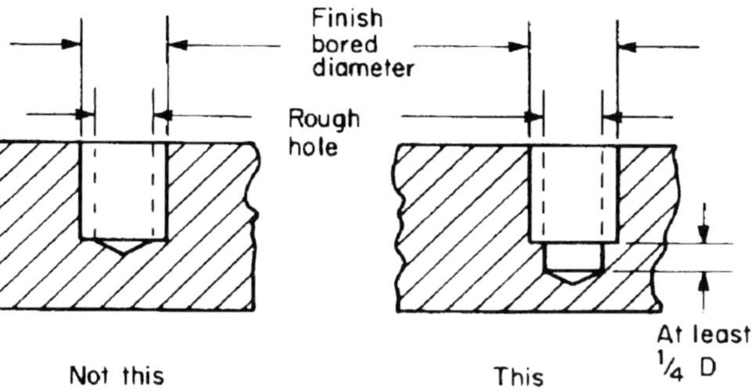

FIGURE 4.5.11 Blind holes to be bored should be one-fourth diameter deeper than the final bored hole to allow space for chips.

3. Use through holes whenever possible. If the hole must be blind, allow the rough hole to be deeper than the bored portion by an amount equal to at least one-fourth the hole diameter. (See Fig. 4.5.11.)
4. Remember that except for small quantities of special-diameter holes, boring is more expensive than drilling and reaming. Equipment is more costly, and the operation is slower. Use boring only when the accuracy requirements demand it. Do not specify bored-hole tolerances unless really necessary.
5. With boring as with other precision machining operations, the part must be rigid so that deflection or vibration as a result of cutting forces is avoided. Care also must be taken in the workpiece and fixture design to avoid deflection of the part when it is clamped in the fixture, for if this occurs, machined surfaces will be off location when the part springs back from its clamped position.

DIMENSIONAL FACTORS

The correctness of drill sharpening is probably the major factor affecting the accuracy of both the diameter and the straightness of drilled holes. A drill can be made to cut oversize or undersize simply as a result of the way it is sharpened.

The play and lack of rigidity in the typical drill spindle constitute another important factor. Whether or not a drill bushing is used also affects the accuracy of a drilled hole. Bushings tend to reduce the bell-mouthing of holes and the tendency of drills to cut oversize.

Even though commercial twist drills have a diameter slightly smaller than nominal size, the usual variation in diameter of drilled holes is in the oversize direction. Plus tolerances for drilled holes are greater than minus tolerances.

Thermal expansion of the material to be drilled can be a factor in the diameter of close-tolerance holes. High-coefficient-of-expansion materials such as aluminum can expand from frictional heat during the drilling operation and, on cooling, contract to a diameter less than that of the drill.

In reaming, the location and direction of the hole are affected by the previous drilling operation even if a bushing is used for reaming.

When precision boring, e.g., jig boring, is employed, a number of factors become important, not because boring per se is involved but because of the high precision usually required. Temperature changes of the workpiece due to heat generated from cutting or from other causes can affect the accuracy of bored holes. Workpiece distortion from clamping also can be a problem. Machine condition and rigidity and boring-bar rigidity are especially important in jig boring work.

RECOMMENDED TOLERANCES

Tables 4.5.1 through 4.5.4 provide recommended dimensional tolerances for drilled, reamed, and bored holes. Data for surface-finish tolerances can be found in Chap. 1.4.

TABLE 4.5.1 Recommended Tolerances for Diameters of Drilled Holes

Hole diameter, mm (in)	Recommended tolerance, mm (in)
0 to 3 (0–⅛)	+0.08, −0.025 (+0.003, −0.001)
Over 3–6 (⅛–¼)	+0.1, −0.025 (+0.004, −0.001)
Over 6–13 (¼–½)	+0.15, −0.025 (+0.006, −0.001)
Over 13–25 (½–1)	+0.2, −0.05 (+0.008, −0.002)
Over 25–50 (1–2)	+0.25, −0.08 (++0.010, −0.003)
Over 50–100 (2–4)	+0.3, −0.1 (+0.012, −0.004)

TABLE 4.5.2 Recommended Tolerances for Diameters of Reamed Holes

Hole diameter, mm (in)	Recommended tolerances, mm (in)
0–13 (0–½)	±0.013 to ±0.025 (0.0005 to 0.001)
Over 13–25 (½–1)	±0.025 (0.001)
Over 25–50 (1–2)	±0.05 (0.002)
Over 50–100 (2–4)	±0.08 (0.003)

TABLE 4.5.3 Recommended Tolerances for Diameters and Depths of Bored Holes

Hole diameter, mm (in)	Normal tolerance, mm (in)	Close tolerance, mm (in)
	Precision production boring	
0–19 (0–3/4)	±0.025 (0.001)	±0.005 (0.0002)
Over 19–25 (3/4–1)	±0.04 (0.0015)	±0.005 (0.0002)
Over 25–50 (1–2)	±0.05 (0.002)	±0.01 (0.0004)
Over 50–100 (2–4)	±0.08 (0.003)	±0.02 (0.0008)
Over 100–150 (4–6)	±0.1 (0.004)	±0.025 (0.001)
Over 150–300 (6–12)	±0.13 (0.005)	±0.05 (0.002)
	Jig boring	
0–25 (0–1)	±0.013 (0.0005)	±0.0025 (0.0001)
Over 25 (1)	±0.025 (0.001)	±0.005 (0.0002)
	Depth (blind or partially bored holes)	
Precision production boring	±0.08 (0.003)	±0.025 (0.001)
Jig boring	±0.025 (0.001)	±0.013 (0.0005)

TABLE 4.5.4 Recommended Location Tolerances for Machined Holes

	±Distance from true position	
	Normal tolerance	Close tolerance
1. Using manual layout, center punch, and drill	0.5 (0.020)	0.25 (0.010)
2. Using drill fixture with bushing; part located in fixture from existing surface	0.25 (0.010)	0.13 (0.005)
3. Using precision milling or numerical-control (NC) machine; part located in fixture from existing surface	0.2 (0.008)	0.1 (0.004)
4. Using precision milling or NC machine; part located by indicating, optical measurement, or from previous hole	0.05 (0.002)	0.025 (0.001)
5. Jig boring machine; part located as in no. 4	0.025 (0.001)	0.005 (0.0002)

COUNTERSINKING

Countersinking is a drilling-related operation by which a chamfer is put on the edge of a hole. The surface cut by the tool is concentric with the hole and at an angle of less than 90° to the axis of the hole.

Countersinking is a fast operation usually performed to remove the burrs caused by drilling. It also may be used to provide clearance for a tapered screw head or other object inserted in the hole.

To facilitate burr removal by countersinking, sufficient flat area around the hole and perpendicular to its axis should be provided.

COUNTERBORING

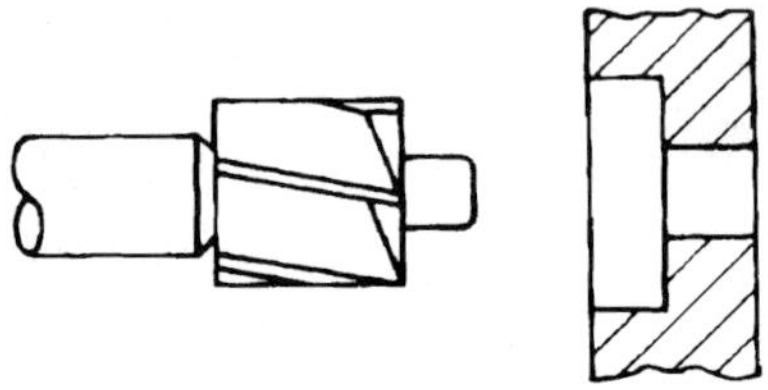

FIGURE 4.5.12 Counterboring tool and the counterbored surface that it produces.

Counterboring is another drilling-related operation. Its purpose is to enlarge a hole for part of its depth, the enlargement being approximately concentric with the original hole. Figure 4.5.12 illustrates a counterbored hole and a typical counterboring tool.

Holes are counterbored to allow them to accept some multidiameter mating part, usually a cap screw whose head is to be recessed when the screw is inserted.

Drilled-hole diameter tolerances can be applied also to the diameter of the counterbored holes. A common tolerance for counterbored depths is $\pm$ 0.010 in, although tighter tolerances usually can be held if necessary.

GUN DRILLING AND GUN REAMING

Gun drilling is a special drilling operation that was originally developed for the manufacture of gun barrels. It utilizes a single-flute drill, usually carbide-tipped, with an oil-coolant passage throughout its length. Oil is forced through the passage under high controlled pressure. It clears the chips from the hole, and as the tool cuts, it cools the tip and provides lubrication and support for the drill body. The tool is guided by a bushing at the start but thereafter is self-piloting in the drilled hole. Hence deep, straight holes are possible. Either the tool or the workpiece can be rotated, but usually it is the workpiece. Figure 4.5.13 schematically shows the gun-drilling process.

Gun reaming (sometimes called gun boring) is a similar process in its use of an oil-coolant passage and a single-flute cutter that is guided by a bushing and the machined hole. It is used to refine the diameter, straightness, or surface finish of existing holes.

Gun drilling and reaming are typically employed for deep holes, those with a depth of from 5 to over 300 times diameter. They can be employed for shallow holes as well and are advantageous if the hole is partial or overlaps or intersects another surface at an angle or if the workpiece material is hard. Any materials suitable for machining with carbide cutters can be gun-drilled. Figure 4.5.14 illustrates some typical gun-drilled parts.

Gun drilling can be employed for hole diameters from 3 to 50 mm ($\frac{1}{8}$ to 2 in). Because special equipment and tooling are required, gun drilling is normally most advantageous for moderate and higher levels of production.

There are few special design considerations for parts that will receive this operation. The major one, perhaps, is the need to provide access for the placement of accurate bushings for the entry surface and at each point where the hole is interrupted. Gun drills and gun reamers depend on bushings for location and direction of the hole. Accurate bushing and fixture construction is essential if the hole is to be accurate.

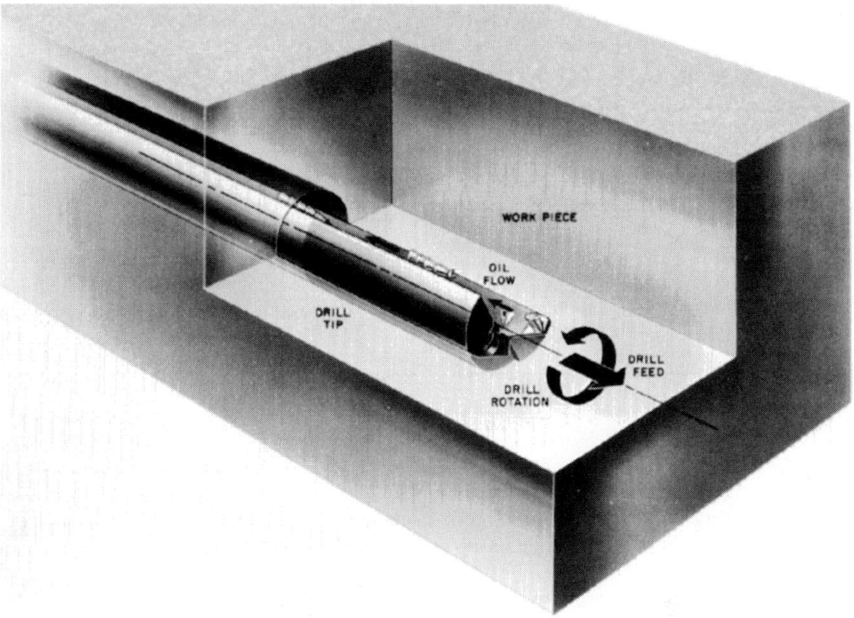

FIGURE 4.5.13 Phantom view of gun drilling. (*Courtesy Eldorado Tool & Mfg. Corp.*)

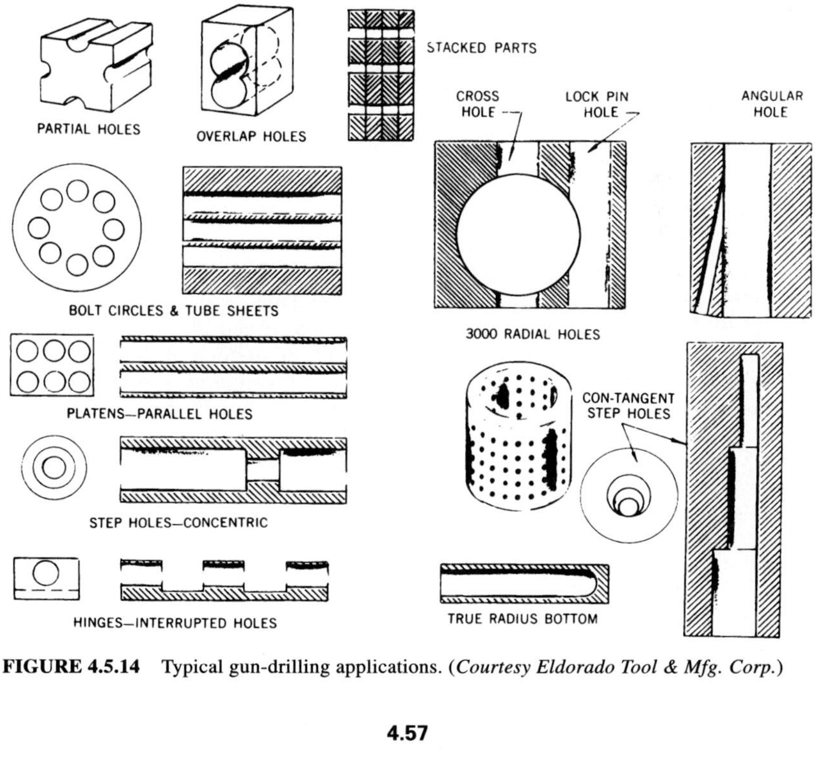

FIGURE 4.5.14 Typical gun-drilling applications. (*Courtesy Eldorado Tool & Mfg. Corp.*)

TABLE 4.5.5 Recommended Tolerances for Gun-Drilled and Gun-Reamed Holes

Hole diameter	Normal tolerance*	Close tolerance*
Under 16 mm (⅝ in)	±0.04 mm (0.0015 in)	±0.025 mm (0.001 in)
Over 16 mm (⅝ in)	±0.05 mm (0.002 in)	±0.04 mm (0.0015 in)
	Surface finish	
Gun-drilled	2.5 μm (100 μin)	0.8 μm (32 μin)
Gun-reamed	1.6 μm (63 μin)	0.5 μm (20 μin)
Straightness	0.13 mm TIR/150 mm (0.005 in/6 in)	0.05 mm TIR/150 mm (0.002 in/6 in)

*TIR = total indicator runout.

Another desirable requirement is to have a part small enough in diameter to be chucked and rotated. The process is more satisfactory if the part (rather than the drill) can be rotated. This avoids the necessity for controlling the whipping motion that can affect long-rotating gun drills.

The recommended tolerances applicable to gun-drilled holes are shown on Table 4.5.5.

TREPANNING

Trepanning is a machining operation for producing holes or circular grooves by the action of one or more cutters revolving around a center. It has at least four applications:

1. Making disks from flat stock when production quantities are small. Disks up to ¼ in in thickness and 6 in in diameter can be made by this method.
2. Making large, shallow through holes (of diameter equal to or greater than 5 times stock thickness).
3. Machining circular grooves, such as would be used, for example, to retain O rings.
4. Machining deep holes of 2 in or more in diameter. This application is a variation of gun drilling in that the cutter is self-piloting, and forced lubrication and special equipment are used. It differs from conventional gun drilling in that a solid-center core is produced as the hole is drilled. Diameter and straightness tolerances comparable with those applied to gun-drilled holes also apply to gun-trepanned holes.

CHAPTER 4.6
PARTS PRODUCED ON MILLING MACHINES

Theodore W. Judson
GMI Technical and Management Institute
Flint, Michigan

MILLING-MACHINE PROCESSES

The material-removal process intended to produce plane, form, or profile surface geometry by a rotating cutter is termed *milling*. Normally, the rotating spindle and cutter remain stationary, and the workpiece moves in the *X, Y,* or *Z* axis.

Process operations such as broaching, planing, grinding, and shaping are competitors of milling because these too can produce plane and form geometric surfaces. Milling cutters can be classified by two principal cutting actions: those which remove material by cutting on the side and those whose cutting action may be described as end cutting in addition to side cutting. Common types of milling cutters and typical operations are shown in Fig. 4.6.1. Figure 4.6.2 illustrates the development of a milling machine into a present-day computer-controlled machining center. Figure 4.6.3 shows a production machine convertible to either vertical or horizontal operation.

CHARACTERISTICS AND APPLICATIONS OF PARTS PRODUCED ON MILLING MACHINES

Milling is an effective means of removing large amounts of material and an efficient method of producing highly precise contours and shapes. It is a very versatile process. Principally, however, milling is used to produce flat surfaces and slots. When close fitting is required between moving parts, when sealing is involved, or simply when an accurate fit is desired, milling-machine operations probably will be advantageous. Slots, grooves, and keyways including screwdriver slots are other commonly milled features. Contoured surfaces include rack and circular gears, helical forms, ratchets, sprockets, cams, and surfaces given a contour to improve their appearance.

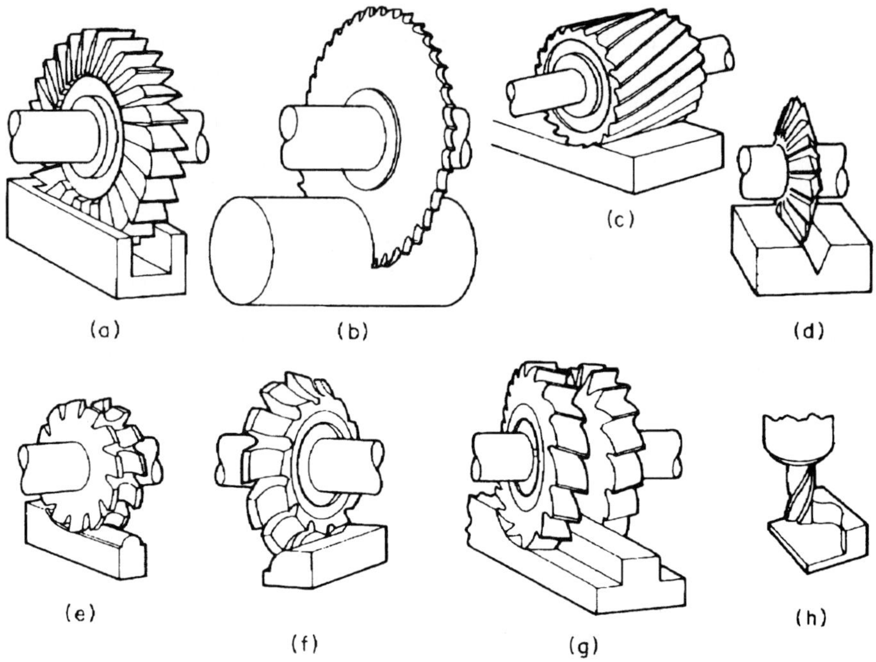

FIGURE 4.6.1 Types of cutters and typical milling operations. (*a*) Slotting with a straight-tooth side-milling cutter. (*b*) Slotting with a concave slitting saw. (*c*) Plain milling. (*d*) Angle milling. (*e*) and (*f*) Form milling. (*g*) Straddle milling. (*h*) Profile milling with an end mill. (*Courtesy Niagara Cutter, Inc.*)

Surface configurations produced by milling also can be produced by casting, forging, extruding, or cold forming but not to the degree of precision provided by milling machining. Most parts receiving milling operations also are processed with other machining operations, e.g., drilling, reaming, boring, turning, grinding, etc. Milling-machine operations are normally part of a full manufacturing sequence.

Common components produced with the aid of milling-machine operations include automobile engine blocks and automobile cylinder heads, open-end wrenches and other hand tools, gearboxes, brackets, ribs, fittings, flanges, spars, beams, pumps, printing presses, machine-tool components, aircraft parts, office machines and computer peripheral equipment, and compressors. In fact, virtually all machines, particularly if they entail heavy duty or precision, employ milling-machine operations in their manufacturing sequence. Another common application is the milling of surfaces prior to surface grinding. Figure 4.6.4 illustrates a number of surfaces produced by milling-machine operations.

ECONOMICS OF MILLING-MACHINE OPERATIONS

Tooling costs for milling-machine operations can be very modest if general-purpose equipment and cutters are used. If necessary, standard support blocks and work clamps

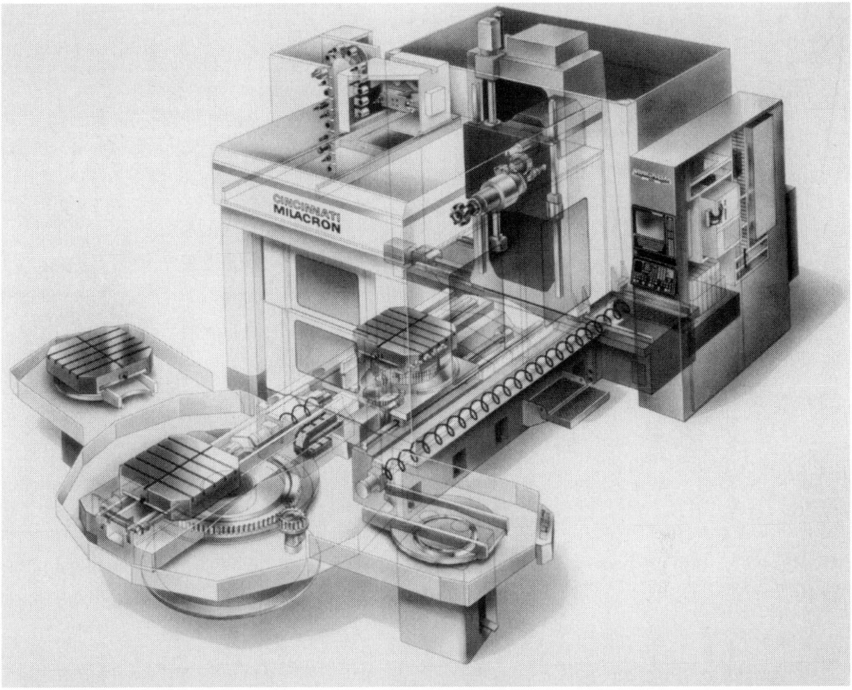

FIGURE 4.6.2 The modern development of the milling machine. This computer numerically controlled machining center can perform a variety of milling and boring operations automatically on a series of identical or different parts. Automatic tool and workpiece pallet changes as well as the cutting parameters (feed, speed, depth of cut, etc.) are all under computer control. (*Courtesy Cincinnati Milacron, Inc.*)

can be adapted to virtually all workpieces. Therefore, milling operations can be undertaken with little or no investment (provided general-purpose machining facilities are available). Consequently, milling usually plays a large part in tool and die work, prototype making, and limited-quantity manufacturing. Low production dictates the use of a mill that features flexibility and adaptability. Such a mill can be changed easily in tooling and fixturing for a variety of workpiece shapes and sizes.

At the other end of the production scale, high-level mass-production, special-purpose machines are used; these may combine drilling, boring, tapping, and other machining operations with milling. Index tables or transfer equipment may be employed with one or more machining heads at each station. Such sophisticated equipment may cost millions of dollars but is justifiable when handling labor is virtually eliminated and machining accuracy and repeatability are ensured.

For moderately high production levels, general-purpose production machines such as those illustrated in Figs. 4.6.2 and 4.6.3 are used with holding fixtures specially designed for the workpiece and often with specially ground cutting tools. These production mills, especially the bed type, offer increased rigidity and are more capable of production accuracy on a continuing-production basis without frequent adjustment. Tooling costs in such instances can range from very modest to quite high levels, depending on the sophistication and complexity of the fixtures, cutters, and gauges used.

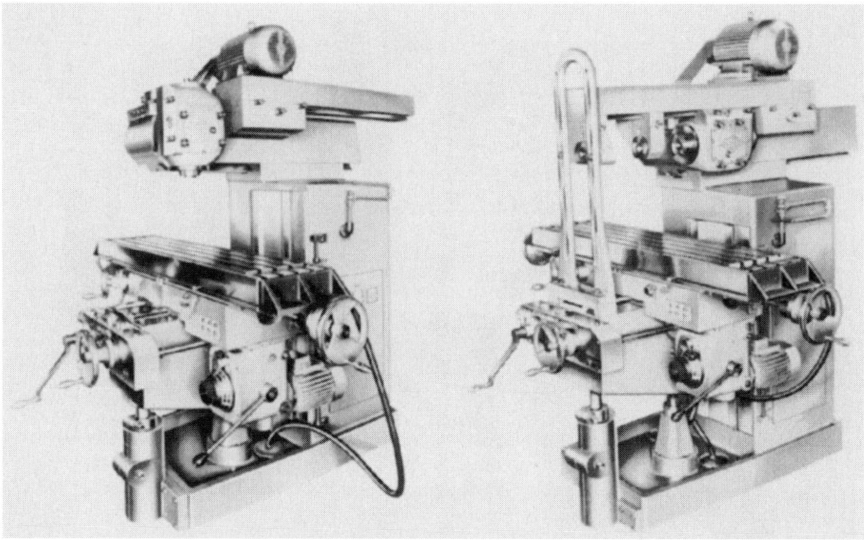

FIGURE 4.6.3 Milling machine convertible for either vertical operation (*left*) or horizontal operation (*right*). (*Courtesy DoAll Company.*)

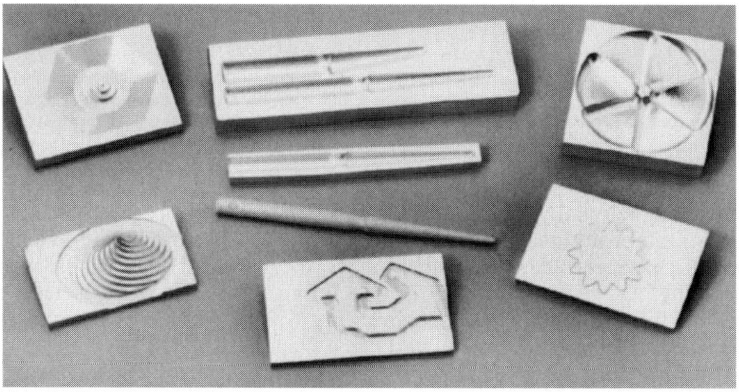

FIGURE 4.6.4 Complexity of surfaces that can be generated with present-day milling machines. (*Courtesy DoAll Company.*)

Output rates from milling operations vary tremendously, depending on the amount of stock to be removed, the workpiece material, the type of cutter used, the amount of workpiece handling involved, and several other factors. Feed rates for milling cutters range from a few millimeters to over 3 m (118 in)/min.

In addition to depreciation and amortization, cost factors involved in milling include workpiece material, labor, energy, cutter sharpening and replacement, equipment maintenance, and scrap and rework allowances. For many higher-production applications, direct labor can be supplied by relatively unskilled workers because operators' duties involve simply loading, unloading, and gauging. For low-quantity production, operators customarily perform their own setup and adjustments, and much higher skill levels and higher direct labor wage rates are required.

Even though milling is a relatively rapid machining method, it is more costly than casting, cold forming, extruding, etc., if these processes can provide the flatness, surface finish, and dimensional accuracy required for the part. This and the factors given above point to a somewhat greater use and economy of milling at low levels of production than at mass-production volumes.

SUITABLE MATERIALS FOR MILLING

Castings, forgings, hot- and cold-rolled or drawn shapes, and extrusions of both ferrous and nonferrous material are commonly processed on milling machines. Perhaps less common but also frequently milled in secondary operations are powder-metal and cold-formed components including cold-headed parts, impact extrusions, and metal stampings. Often, however, materials most suitable for these forming processes are not optimal for milling-machine operations.

Nevertheless, a wide variety of metallic (and nonmetallic) materials is feasible for milling machinery. Best results are usually obtained, however, with materials having high general machinability ratings. In addition to these, soft materials and those with hardness up to about R_c 45 are fully processible, but at higher cost.

Table 4.6.1 provides data on the average amount of material removable per horsepower for milling operations on various classes of commonly milled materials. Table 4.6.2 provides machinability ratings for some selected specific materials milled with high-speed steel side-milling cutters.

Materials of higher hardness require reduced cutting speeds and reduced feed per tooth. Such a relationship holds true for both ferrous and nonferrous metals. Higher physical-mechanical properties are inversely related to usable cutting speeds and feeds in milling.

TABLE 4.6.1 Horsepower Required per Unit Volume of Metal Removed per Minute for Various Classes of Milling-Machined Materials

Material	Hardness (Brinell hardness number)	Average volume removed per minute per horsepower, cm^3 (in^3)
Malleable cast iron	Soft or hard	21 (1.3)
Cast steel	Soft or hard	13 (0.8)
Wrought steel	100–150	16 (1.0)
Wrought steel	150–250	13 (0.8)
Wrought steel	250–350	11 (0.7)
Wrought steel	350–450	9.8 (0.6)
Cast iron, soft	150–180	29 (1.8)
Cast iron, medium	180–225	21 (1.3)
Cast iron, hard	225–350	16 (1.0)
Bronze	Average	21 (1.3)
Brass	Average	26 (1.5)
Aluminum alloy	Average	66 (4.0)

TABLE 4.6.2 Machinability Ratings for Various Materials When Machined with Side-Milling Cutters

Material	Common designation	UNS designation	Condition and remarks	Brinell hardness number	Rating
Steel	1212	G12120	Cold-drawn, resulfurized	150–200	100
Steel	12LI4	G12144	Cold-drawn, leaded	150–200	105
Steel	1117	G11170	Cold-drawn, resulfurized	150–200	85
Steel	1010	G10100	Cold-drawn, low-carbon	125–176	55
Steel	1040	G10400	Cold-drawn, medium-carbon	125–175	50
Steel	1070	G10700	Cold-drawn, high-carbon	175–225	35
Alloy steel	4140	G41400	Cold-drawn, medium-carbon	175–225	30
Stainless steel	43OF	S43020	Free-machining, ferritic	135–185	80
Stainless steel	303	S30300	Free-machining, austenitic	135–185	65
Stainless steel	17-7PH	S17700	Precipitation hardening, wrought	150–200	25
Cast iron	A 48, class 20	F11401	Ferritic, annealed	120–150	100
Cast iron	A 48, class 30	F12101	Pearlitic, as cast	190–220	55
Cast iron	60-40-18	F32800	Ductile, annealed	140–190	85
Aluminum	2024	A92024	Wrought, solution-treated	75–150	65
Aluminum	208.0	A02080	As-sand- or permanent-mold-cast	40–100	400
Brass	934	C93400	High-leaded tin bronze as cast	40–150	135
Brass	340	C34000	Medium-leaded, cold-drawn	60–100 R_b	165
Magnesium	AZ92A	MI1920	As cast, annealed	40–90	400
Nickel	200	N02200	Annealed or cold-drawn	80–170	45
Zinc	AG40A	Z33520	Die-cast	80–100	165

Source: Compiled from data in the *Machining Data Handbook,* 3d ed., by permission of the Machinability Data Center; © 1980 by Metcut Research Associates, Inc. Ratings are based on recommended cutting speed with cutters of high-speed steel compared with that recommended for 1212 material.

DESIGN RECOMMENDATIONS

Despite the fact that milling is a rapid-removal process, milling operations, when necessary, do add to product cost. If surfaces produced by prior processes such as casting, extruding, forging, drawing, etc., provide a surface of sufficient flatness and proper configuration so that milling is not required, costs will be reduced. Designers therefore should strive to provide a product design that does not require the accuracy provided by milling. There are numerous commercial products, particularly those which are mass produced, for which processes like fine-blanking, powder metallurgy, and precision casting are utilized to produce configurations that, in less clever designs, would have to be produced by milling operations.

A number of product design rules common to other machining operations also apply to milling operations. For example, sharp inside and outside corners should be avoided. The part should be easily clamped, machined surfaces should be accessible, easily machined materials should be specified, and the design should be as simple as possible.

There are a number of additional design recommendations, particularly applicable to milling, that will facilitate the manufacture of a part when these operations are required. These recommendations, listed below, can result in shorter setup time, more economical cutters, improved processing, and lower cost.

1. The product design should permit the use of standard cutter shapes and sizes rather than special, nonstandard cutter designs. Slot widths, radii, chamfers, corner shapes, and overall forms should conform to those of cutters available off the shelf rather than those which require special fabrication. Specialized form-relieved cutters are costly and difficult to maintain. (See Fig. 4.6.5.)
2. The product design should permit manufacturing preference as much as possible to determine the radius where two milled surfaces intersect or where profile milling is involved. This will permit the use of standard available or most easily ground cutters. (See Fig. 4.6.6.)
3. When a small, flat surface is required, as for a bearing surface or a bolt-head seat perpendicular to a hole, the product design should permit the use of spotfacing, which is quicker and more economical than face milling. (See Fig. 4.6.7.)
4. When spotfaces or other small milled surfaces are specified for castings, it is good practice to design a low boss for the surface to be machined, as shown in Fig. 4.6.8. This simplifies machining and paint removal and usually results in a less sharp edge.
5. When outside surfaces intersect and a sharp corner is not desirable, the product design should allow a bevel or chamfer rather than rounding. Bevels and chamfers may be created by face mills, whereas rounding requires a form-relieved cutter and a more precise setup, both of which are most costly to maintain. (See Fig. 4.6.9.)
6. Similarly, when form-milling or machining rails, it is best not to attempt to blend the formed surface to an existing milled surface because exact blending is difficult to achieve. The approaches illustrated in Fig. 4.6.10 are preferred.
7. Keyway design should permit the keyway cutter to travel parallel to the center axis of the shaft and form its own radius at the end. A standard cutter should be able to form both the width and end radii of the keyway slot. (See Fig. 4.6.11.) This principle also applies to other slots, saw cuts, and shell and face milling.

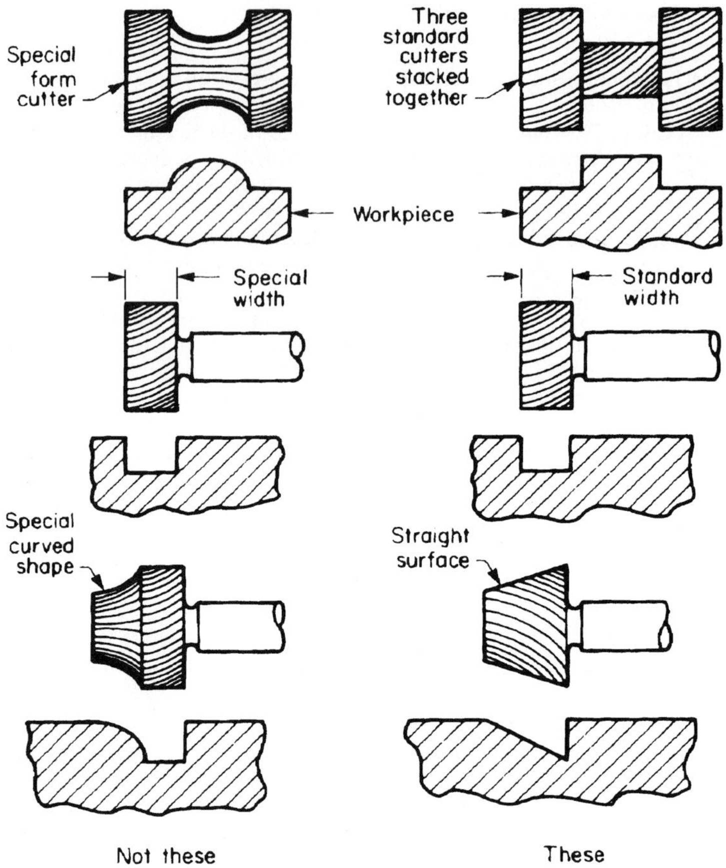

FIGURE 4.6.5 Product design should permit the use of standard cutter shapes and sizes rather than special nonstandard cutter designs.

8. A design that requires the milling of surfaces adjacent to a shoulder or flange should provide clearance for the cutter path. Small steps or radii or inclined flange or shoulder surfaces as shown in Fig. 4.6.12 should be used.
9. A product design that avoids the necessity of milling at parting lines, flash areas, and weldments will generally extend cutter life.
10. As with other surface-machining processes, the most economical designs are those which require the fewest separate operations. Surfaces in the same plane or at least in the same direction and in parallel planes are preferred.
11. Milling operations often can be performed more economically if the product design lends itself to stacking so that a milled surface can be incorporated into a number of parts in one gang-milling operation. This also can occur if parts can be "sliced" from a long workpiece after the milling operation. (See Fig. 4.6.13.) Even if the parts do not have flat adjoining surfaces as shown, provided they are designed so that they can nest together, gang milling may then be employed.

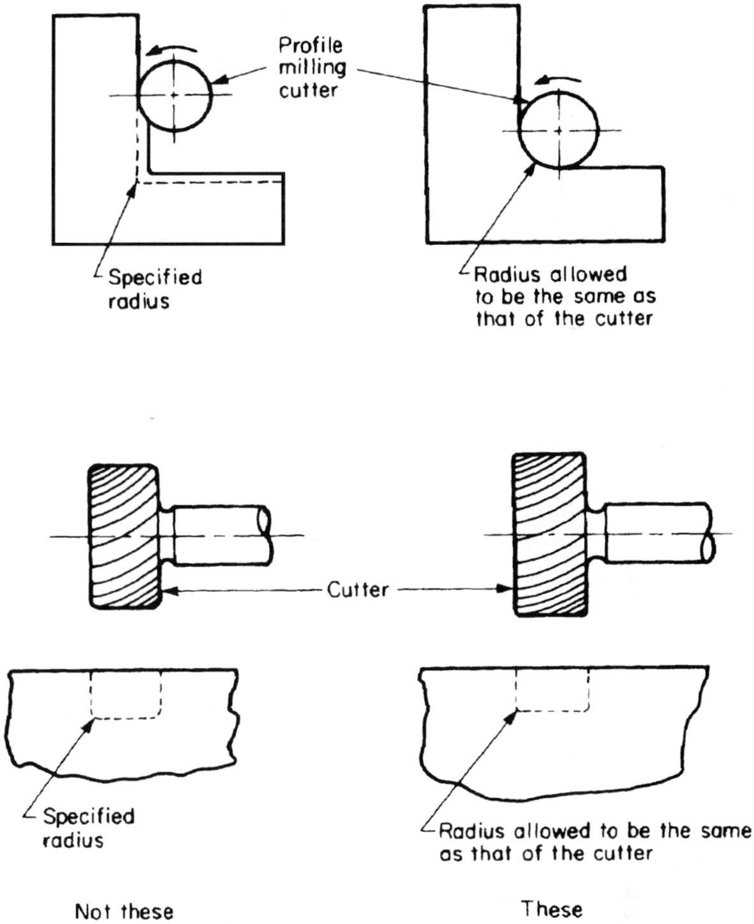

FIGURE 4.6.6 Product design should permit the use of the radii provided by the cutting tool.

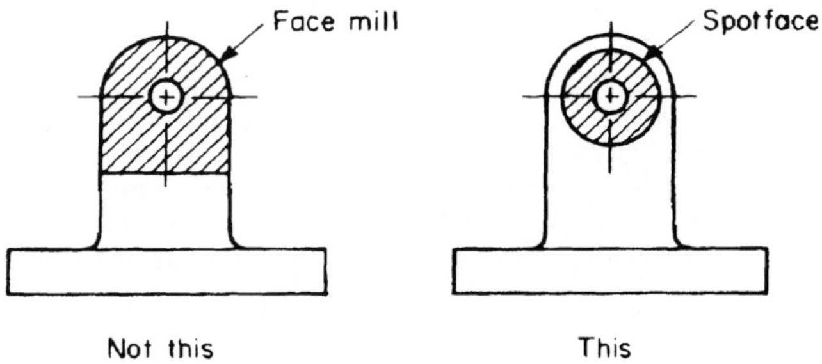

FIGURE 4.6.7 Spotfacing is quicker and more economical than face milling for small, flat surfaces.

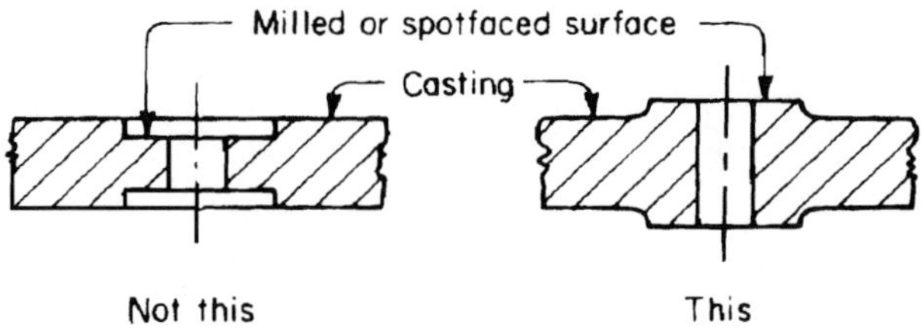

FIGURE 4.6.8 A low boss simplifies the work of machining a flat surface.

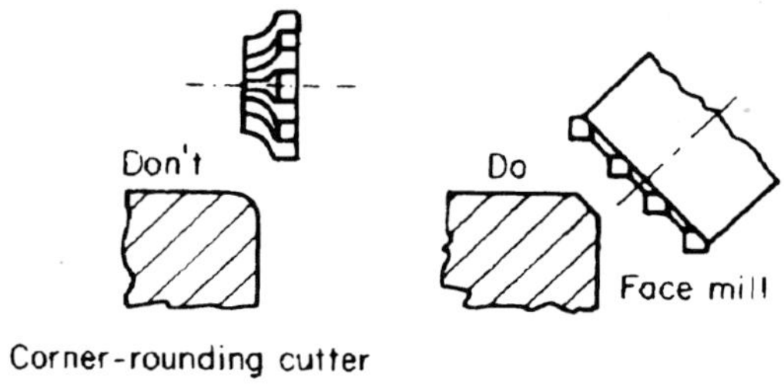

FIGURE 4.6.9 Allowing a beveled rather than a rounded corner can provide more economical machining.

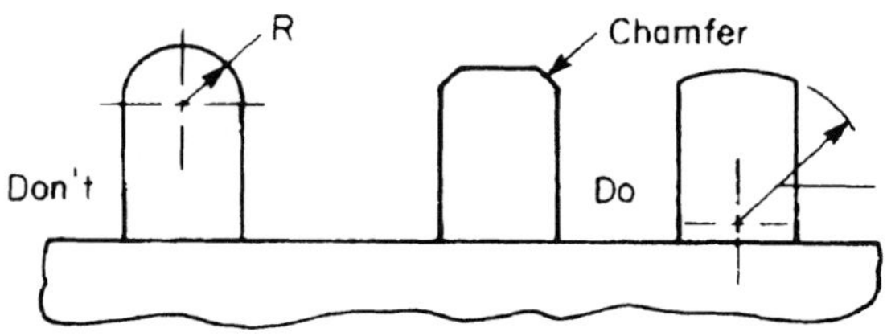

FIGURE 4.6.10 It is better not to specify a blended radius on machined rails.

12. The product design should provide clearance to allow the use of larger-size cutters rather than small cutters in order to permit high material-removal rates, more efficient use of machine horsepower, and lower dynamic operating conditions when machining. Smaller-sized cutters are less rugged and require higher operating speeds to machine effectively. They are more subject to vibration, chatter, and deflection of tool and machine components. Large clearances also facilitate the use of carbide-insert cutters. These provide high production rates, minimal cutter maintenance, and less frequent downtime for tool changes.

13. In end-milling slots in mild steel, the depth should not exceed the diameter of the cutter. (See Fig. 4.6.14.)

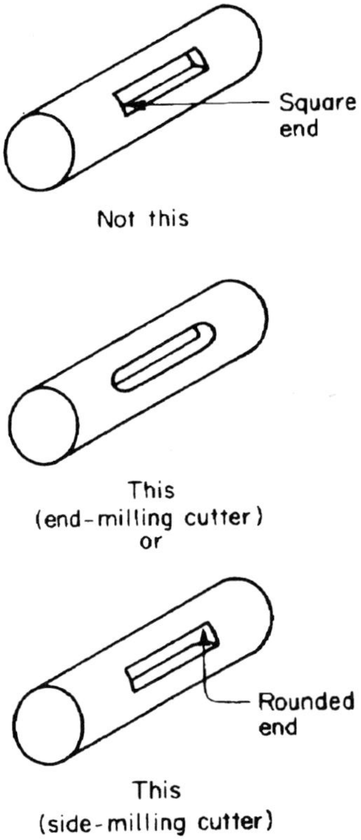

FIGURE 4.6.11 Keyways should be designed so that a standard cutter can produce both sides and ends in one operation.

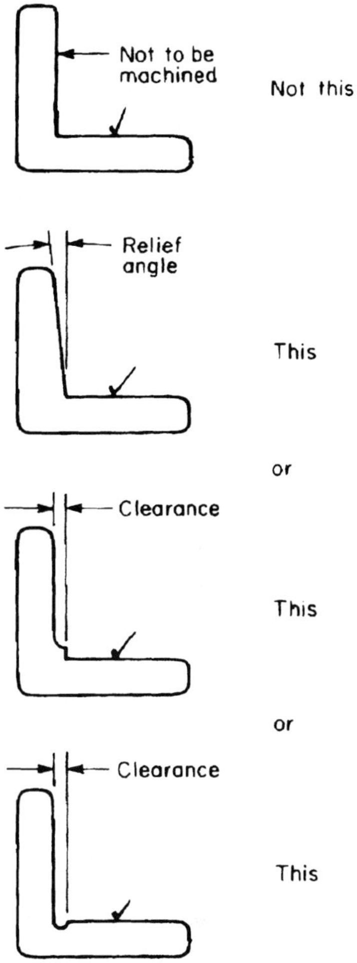

FIGURE 4.6.12 Provide clearances for the milling cutter.

DIMENSIONAL FACTORS AND TOLERANCES

Recommended dimensional tolerances for various milling operations are shown in Table 4.6.3. It should be noted that data in this table are given for production work and properly maintained machines. The tolerance-holding capabilities of the operation are influenced by more than just the condition of the cutter, machine, and work-holding device. Operational disturbances such as tool wear, machine wear, deflection, and vibration and the rigidity and stability of the workpiece itself figure importantly in tolerance-holding capabilities. Milling, by nature, produces high intermittent forces conducive to vibration and deflection. Workpiece materials affect tolerance control.

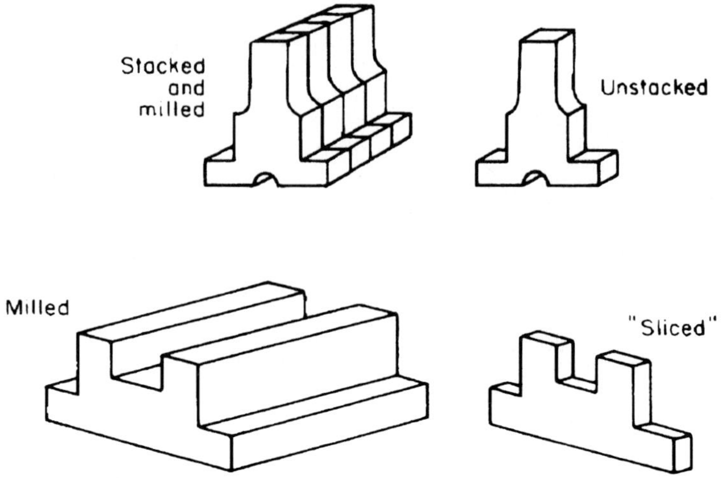

FIGURE 4.6.13 Designs that permit stacking or "slicing" with form milling often provide an economical approach.

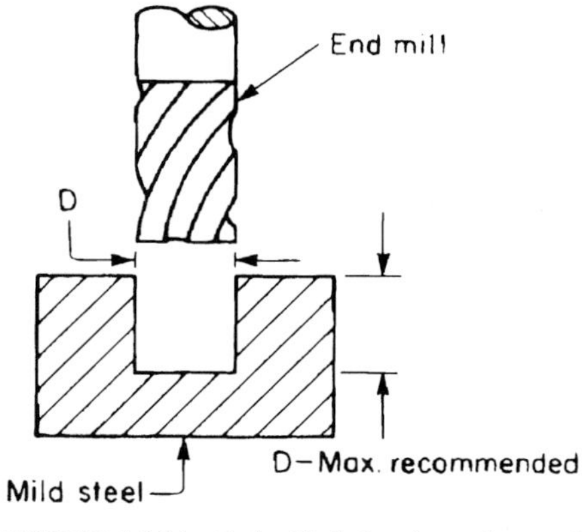

FIGURE 4.6.14 End-milled slots in steel normally should not be deeper than the cutter diameter.

Materials of good machinability, perhaps with fine grain structures and reasonable hardness, machine more precisely than very hard or very soft large-grained materials.

The number of parts to be produced before cutter or insert replacement is a major factor in controlling the quality of the surface finish. A low feed per tooth results in a smoother surface finish. Those factors which reduce friction, such as cutting fluid, high cutting speed, and workpiece materials of good machinability, all tend to produce desired surface smoothness. Accordingly, surface finishes as low as 5 μin have been obtained by milling, while conditions sometimes make it nearly impossible to produce surface finishes less than 500 μin. The surface finish that can be expected on production runs of free-machining irons, steels, and nonferrous materials falls in the range of 1.5 to 3.8 μm (60 to 150 μin).

TABLE 4.6.3 Recommended Dimensional Tolerances for Milling-Machine Operations (Variations from Basic Dimension)

Operation	Basic dimension, mm (in)					
	To 6.3 (0.250)	To 12.7 (0.500)	To 19 (0.750)	To 25 (1.000)	To 50 (2.000)	To 100 (4.000)
Straddle milling	±0.5 (0.020)	±0.5 (0.020)	±0.5 (0.020)	±0.5 (0.020)	±0.5 (0.020)	±0.5 (0.020)
Slot width	±0.04 (0.0015)	±0.04 (0.0015)	±0.05 (0.002)	±0.05 (0.002)	±0.05 (0.002)	±0.06 (0.0025)
Slot width (end mill)	±0.05 (0.002)	±0.06 (0.0025)	±0.06 (0.0025)	±0.06 (0.0025)		
Face milling	±0.05 (0.002)	±0.05 (0.002)	±0.05 (0.002)	±0.05 (0.002)	±0.05 (0.002)	±0.05 (0.002)
Hollow milling		±1.5 (0.060)	±2.0 (0.080)	±2.5 (0.100)		

Source: *General Motors Engineering Standards.*

CHAPTER 4.7

PARTS PRODUCED BY PLANING, SHAPING, AND SLOTTING

THE PROCESSES

Planing, shaping, and slotting are machining processes that provide a cutting action as a result of a straight-line reciprocating motion between the tool and the work.

Planing

This process is utilized for large components. The workpiece is fixed to a table that moves back and forth against single-point cutting tools. The tools are stationary except for feeding between strokes of the machine. Tools can be fed across the workpiece in a horizontal, vertical, or angular direction. As many as four tool holders, each of which can produce a separate machined surface, may be employed simultaneously.

Shaping

Shaping is a process similar to planing that is used for smaller workpieces or smaller surfaces. The workpiece is stationary (except for feeding between strokes), while the single-point cutting tool moves. The tool is supported by a ram that reciprocates with a linear motion. In the conventional shaper this motion is horizontal. Feeding of the workpiece between strokes is normally automatic and is in the horizontal direction across the stroke of the ram. However, the tool may be fed vertically, horizontally, or at an angle, as desired. Figure 4.7.1 illustrates a standard horizontal shaper.

Slotting

This process is identical to shaping except that the motion of the ram is vertical instead of horizontal. This factor sometimes provides a more convenient means of holding certain workpieces, most commonly those for which a rotary rather than linear feed is used.

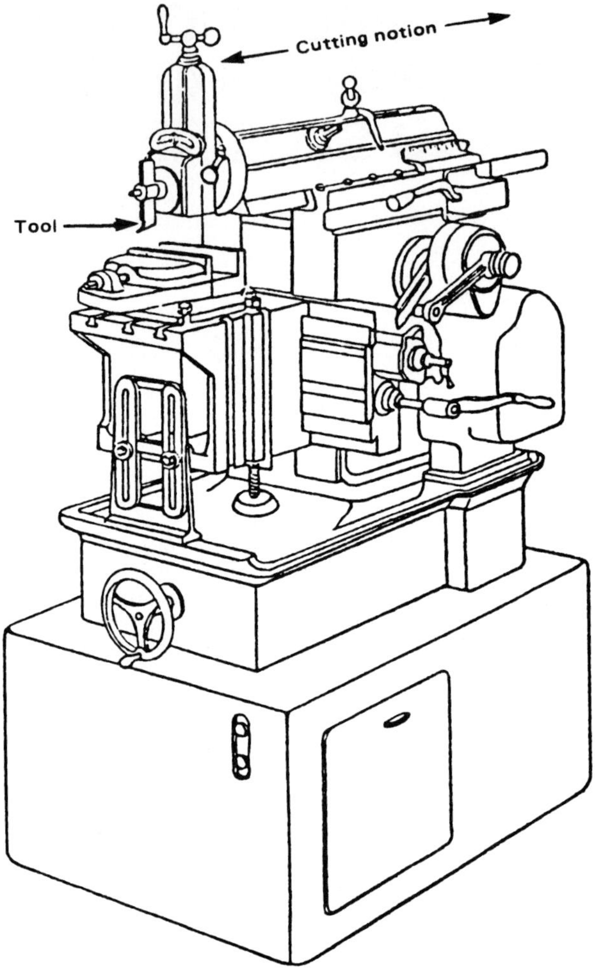

FIGURE 4.7.1 Standard horizontal shaper. (*Courtesy General Motors Engineering Standards.*)

CHARACTERISTICS AND APPLICATIONS OF PLANER-MACHINED PARTS

Planers are used most frequently to machine large, flat surfaces. Machinery bases, diesel engine blocks, and locomotive and ship parts commonly undergo planer operations. Large, rough castings and welded assemblies that require some afterweld machining are candidates for planing. A typical planer will have a maximum workpiece capacity of 1 by 1 by 4 m (3 by 3 by 12 ft). However, machines have been built with part-length capacities as long as 15 m (50 ft) or more. Surfaces shorter than 300 mm (12 in) usually are not practical to machine on a planer except as part of a gang-machining operation.

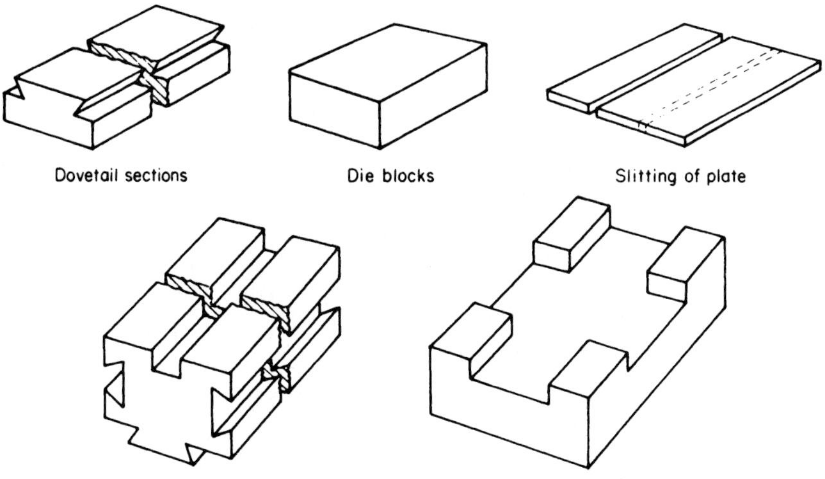

FIGURE 4.7.2 Typical parts and surfaces machined on a planer.

Dovetail and V-groove machine-tool surfaces are frequently machined on a planer. (Planed surfaces lend themselves better than milled surfaces to subsequent hand-scraping operations.) Another common application is the machining of large die blocks, since difficult-to-machine alloys and tool steels up to a hardness of about R_c 46 can be processed.

Although primarily used for the production of flat surfaces, planing also can be employed to produce contoured surfaces. This can be accomplished by hand manipulation of the tool height (usually in following a scribed line of the workpiece). It also can be accomplished through the use of a tracer attachment that automatically changes the tool position from stroke to stroke of the machine. With a special work holder to rotate the work during cutting, helical surfaces or slots can be machined. Form tools also will machine curved and irregular surfaces of limited size. Figure 4.7.2 illustrates typical parts and surfaces machined on a planer.

CHARACTERISTICS AND APPLICATIONS OF SHAPER-MACHINED PARTS

Shaper parts are smaller than those run on planers. Usually, they are small enough to be moved easily by hand and clamped in a vise-type work holder on the machine table. Surfaces machined with shapers are usually, but not necessarily, flat and can be in the horizontal, vertical, or angular plane.

Shapers seldom have a stroke longer than 900 mm (36 in), which is therefore the longest surface machinable. A minimum surface is less than 13 mm ($\frac{1}{2}$ in), with the usual case about 150 to 400 mm (6 to 16 in). The width capacity of a 36-in shaper normally is about 20 in, and that of a 12-in shaper is about 10 in.

Although some shapers are equipped with tracer attachments that permit them to reproduce accurately surface contours both at right angles and parallel to the machine

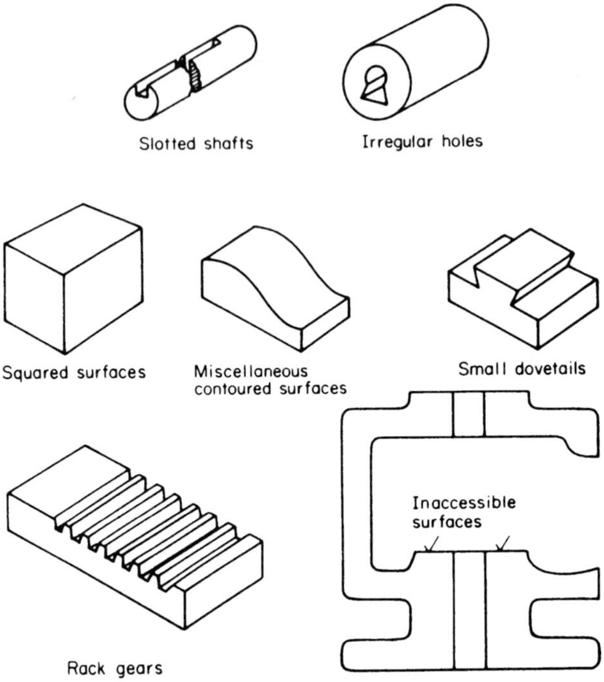

FIGURE 4.7.3 Typical parts and surfaces machined on a shaper.

stroke, most such contours are made by manually controlling the height of the cutting tool during each stroke. Such surfaces therefore are only roughly machined. A third method involves the use of a template mounted below the shaper table to raise and lower it during cross feeding and thereby to produce gentle contours at right angles to the machine stroke.

The shaper is a very versatile machine, even more versatile than the planer. Some inaccessible surfaces that are not feasible to machine with other types of equipment can be produced by shaping. Examples are deep internal slots and contours in blind holes or inaccessible places.

Although single-point tools are normally used, shapers can cut with form tools, especially when a repeated form is required as, for example, in the machining of rack gears.

Shapers are frequently used for slot, keyway, and spline cutting. Such operations may require the use of a dividing head mounted on the shaper table. Figure 4.7.3 illustrates typical shaper-machined parts.

ECONOMIC PRODUCTION QUANTITIES

With some exceptions, planing, shaping, and slotting are most suitable for unit or low-quantity production. They are all very versatile processes with very low tool costs.

The machines frequently can be found in maintenance and model shops and toolrooms but seldom in quantity production. (The milling machine has supplemented the shaper and the planer for production applications because of its faster metal-removal rate, which is at least twice that of the latter machines.)

Since only inexpensive single-point cutting tools and standard work holders and clamps are required and setup times are usually short, planers, shapers, and slotters are especially suited for one-of-a-kind or emergency production.

MATERIALS FOR PLANING, SHAPING, AND SLOTTING

Although, as indicated above, planers, shapers, and slotters can machine difficult alloys and high hardnesses (to approximately R_c 46), it still is advantageous, when these machines are used, to specify as machinable a material as the functional requirements of the workpiece will allow.

The machinability table in Chap. 4.1 provides general data on ferrous, nonferrous, and nonmetallic materials. These data are also applicable to planer, shaper, and slotter operations. Table 4.7.1 also provides machinability information for the high-speed steel-cutting-tool machining of materials commonly machined on shapers.

DESIGN RECOMMENDATIONS

Because of the flexible capability of planers, shapers, and slotters, there are very few restrictions on the design of parts to be machined with them. There are some rules, however, that should be adhered to, either for economy of the operation or for dimensional control. They are as follows:

1. Since the cutting forces in planing and shaping may be abrupt and rather large, design parts so that they can be easily clamped to the work table and are sturdy enough to withstand deflection during machining. (See Fig. 4.7.4.)
2. It is preferable to put machined surfaces in the same plane to reduce the number of operations required. (This stricture does not apply if a multitooled planer can machine both surfaces simultaneously.)
3. Avoid multiple surfaces that are not parallel in the direction of reciprocating motion of the cutting tool because this would necessitate additional setups.
4. Avoid contoured surfaces unless a tracer attachment is available, and then specify gentle contours and generous radii as much as possible.
5. With shapers and slotters it is possible to cut to within 6 mm ($\frac{1}{4}$ in) of an obstruction or the end of a blind hole. (See Fig. 4.7.5.) If possible, allow a relieved portion at the end of the machined surface.
6. For thin, flat pieces that require surface machining, allow sufficient stock for a stress-relieving operation between rough and finish machining, or, if possible, rough-machine equal amounts from both sides. Allow about 0.4 mm (0.015 in) for finish machining. Then finish-machine on both sides.

TABLE 4.7.1 Relative Machinability of Various Materials Used on Planers and Shapers (Based on Recommended Cutting Speed with High-Speed-Steel Tools)

Material	Designation	Condition and remarks	Brindell hardness number	Rating
Steel	1212 (G12120)	Cold-drawn	150–200	100*
Steel	1117 (G11170)	Cold-drawn	150–200	85
Steel	1137 (G11370)	Cold-drawn	175–225	80
Steel	12L13 (G12134)	Cold-drawn, leaded	100–150	100
Steel	Plain low-carbon	Cold-drawn	85–125	90
Steel	Plain medium-carbon	Cold-drawn	175–225	70
Steel	Plain high-carbon	Cold-drawn	175–225	65
Steel	Alloy, medium-carbon	Cold-drawn	175–225	65
Stainless steel	430F (S43020)	Ferritic, annealed	135–185	40
Stainless steel	304 (S30400)	Austenitic, annealed	135–185	30
Stainless steel	17-7 PH (S17700)	Precipitation-hardened and annealed	150–200	35
Cast steel	Plain low-carbon	Annealed and normalized	100–150	80
Cast iron	Gray	Ferritic, annealed	120–150	100
Cast iron	Gray	Pearlitic, as cast	190–220	70
Cast iron	Ductile	Ferritic, annealed	140–190	100
Cast iron	Malleable	Ferritic, malleabilized	110–160	145
Cast aluminum alloy		As cast in sand or permanent mold	40–120	280

*A rating of 100 is equivalent to a cutting speed of 30 m (100 ft)/min. Other ratings are proportional.

Source: This table is compiled from data in the *Machining Data Handbook,* 3d ed., by permission of the Machinability Data Center; © 1980 by Metcut Research Associates, Inc.

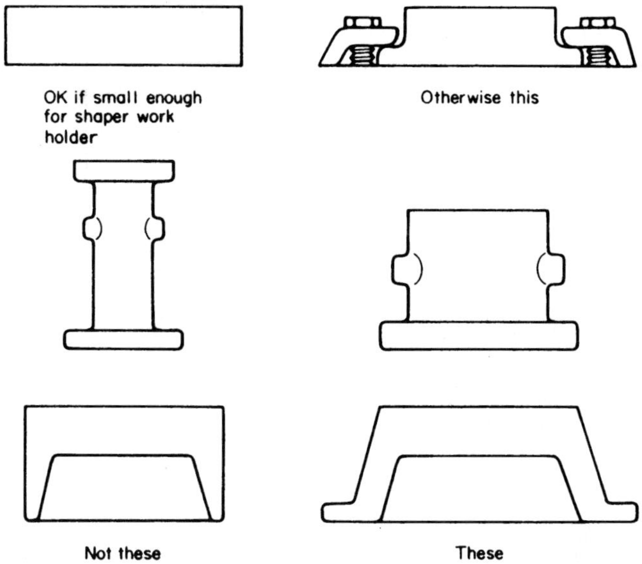

FIGURE 4.7.4 Design planer- and shaper-machined parts to be sturdy enough to withstand cutting-tool forces and to be solidly clamped.

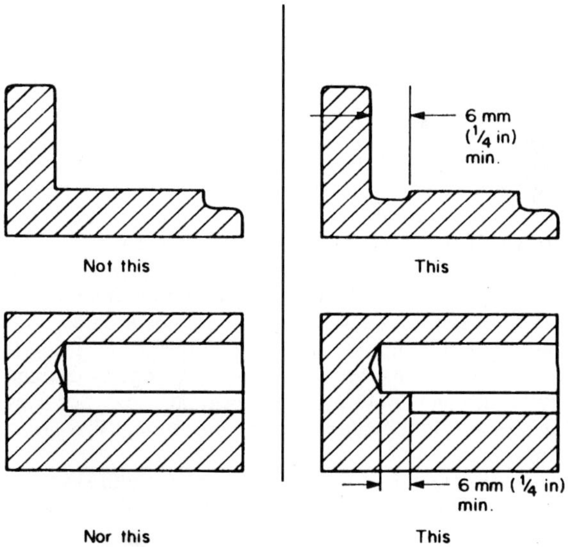

FIGURE 4.7.5 Avoid machined surfaces too close to an obstruction at the end of the cut.

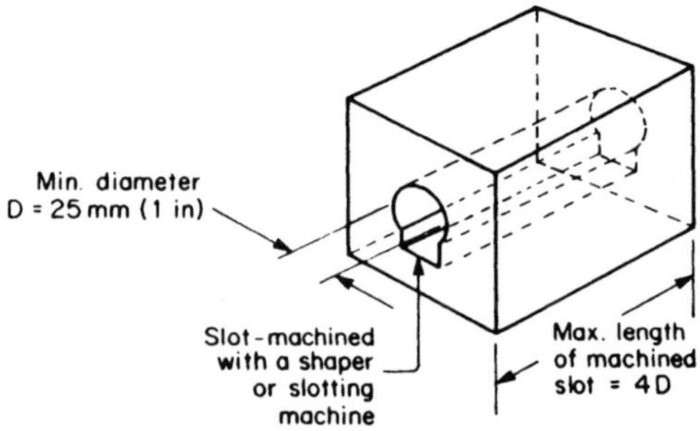

FIGURE 4.7.6 The minimum-size hole in which a keyway, slot, or other contour can be shaper-machined is about 1 in. Slots and contours should not be longer than 4 times the largest dimension of the opening or the hole diameter.

7. The minimum size of holes in which a keyway or a slot can be machined with a slotter or shaper is about 1 in. (See Fig. 4.7.6.)
8. Because of a lack of rigidity of long cutting-tool extensions, it is not normally feasible to machine a slot longer than 4 times the hole diameter (or the largest dimension of the opening). (See Fig. 4.7.6.)

DIMENSIONAL FACTORS

Planers, shapers, and slotters are rugged, precision machines. If they are maintained in proper operating condition, any major dimensional variations will occur from human factors, the design and condition of the part itself, and the clamping method.

The squareness and flatness of the clamping surface of the workpiece are of paramount importance. A piece that is distorted in clamping will spring back after machining and will not have a true surface. This is especially true of parts machined on planers when the cutting forces are very large and when particularly strong clamping is required.

The solidity of clamps, supports, and stops is quite important, especially in planers. Movement of the workpiece can cause tool and equipment damage as well as dimensional errors. Similarly, deflection of the part as a result of cutting forces can cause significant dimensional variations. Workpieces must be rigidly designed and well supported if accuracy requirements are high. Warpage also can occur as a result of the release of internal stresses in the material during machining. Flat pieces are particularly susceptible to this condition.

Tool rigidity is another factor of importance, especially in slotting operations or in the shaping of internal surfaces when there is substantial overhang of the tool or toolholder.

Slower cutting speeds, lighter cuts with finer feeds, and the use of lubricants all tend to improve accuracy when tool or workpiece deflection factors are operating. Similarly, sharp tools, correctly ground, and fine feeds facilitate smooth surface finishes.

TABLE 4.7.2 Recommended Dimensional Tolerances for Parts Produced by Planing, Shaping, and Slotting

	Preferred tolerance for manufacturing, mm (in)	Average tolerance, mm (in)	Closest tolerance, mm (in)
Flatness			
In 0.5 m^2 (5 ft^2) (planers)	±0.13 (0.005)	±0.05 (0.002)	±0.013 (0.0005)
In 0.1 m^2 (1 ft^2) (planers or shapers)	±0.08 (0.003)	±0.025 (0.001)	±0.013 (0.0005)
In 65 cm^2 (10 in^2) (shapers)	±0.05 (0.002)	±0.025 (0.001)	±0.013 (0.0005)
Surface finish			
Cast iron	3 μm (125 μin)	1.5 μm (60 μin)	0.8 μm (32 μin)
Steel	6 μm (250 μin)	3 μm (125 μin)	0.8 μm (32 μin)
Surface location			
Small surfaces	±0.25 (0.010)	±0.13 (0.005)	±0.025 (0.001)
Large surfaces, over 0.1 m^2 (1 ft^2)	±0.4 (0.015)	±0.2 (0.008)	±0.05 (0.002)
Contour to specified position			
Manual feed	±1.0 (0.040)	±0.7 (0.028)	±0.4 (0.015)
Tracer feed	±0.13 (0.005)	±0.05 (0.002)	±0.025 (0.001)

Table 4.7.2 presents recommended tolerances for dimensions and surfaces produced on planers, shapers, and slotters.

CHAPTER 4.8
SCREW THREADS

Engineering Staff
Teledyne Landis Machine
Waynesboro, Pennsylvania

THREAD SYSTEMS

Figures 4.8.1 through 4.8.6 illustrate common thread forms.

The *unified national screw thread* (see Fig. 4.8.1) was adopted in 1948 as the preferred system for fasteners in the United States, Great Britain, and Canada. It is very similar to the earlier American standard system. Common designations are UNC (coarse), UNF (fine), UNEF (extra fine), and UNS (special).

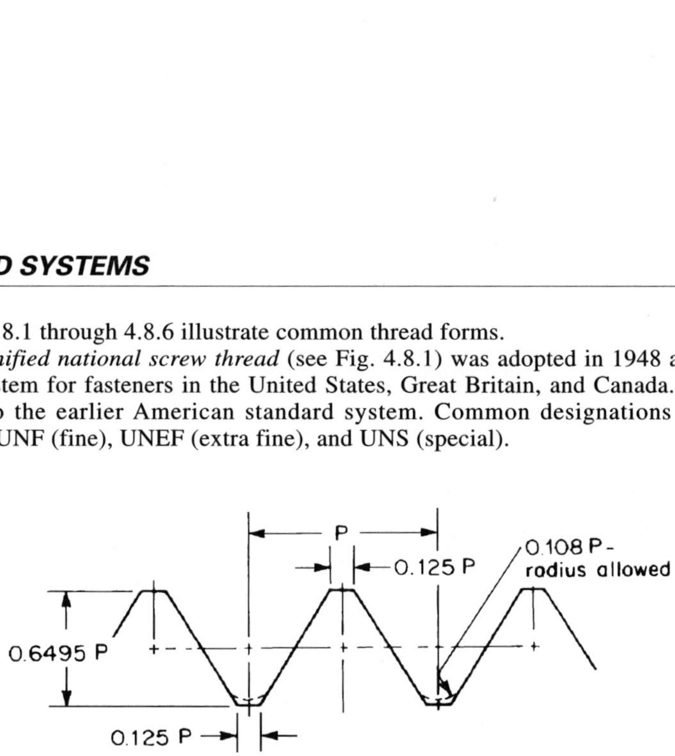

FIGURE 4.8.1 Unified national thread.

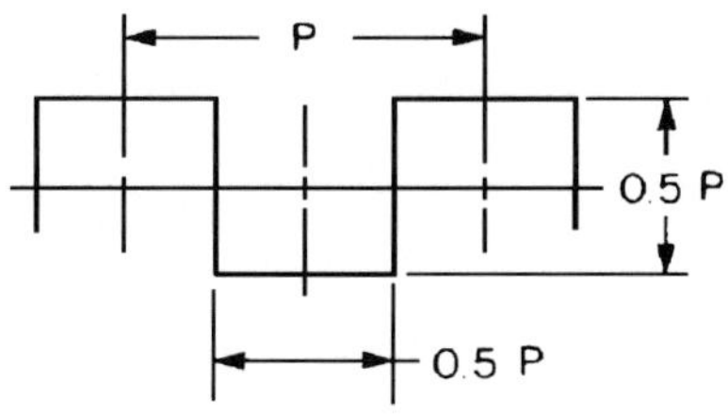

FIGURE 4.8.2 Square thread.

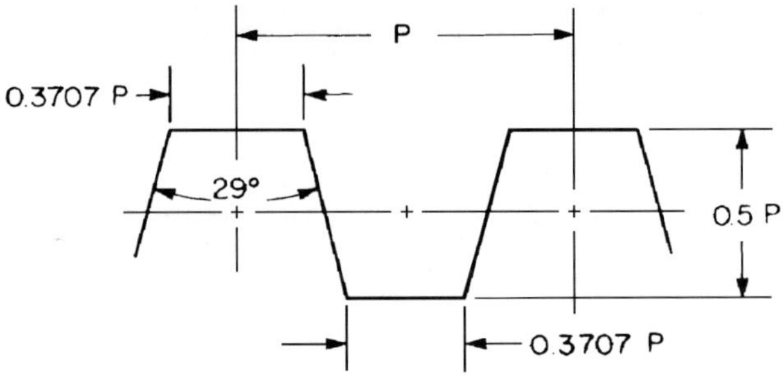

FIGURE 4.8.3 General-purpose Acme thread.

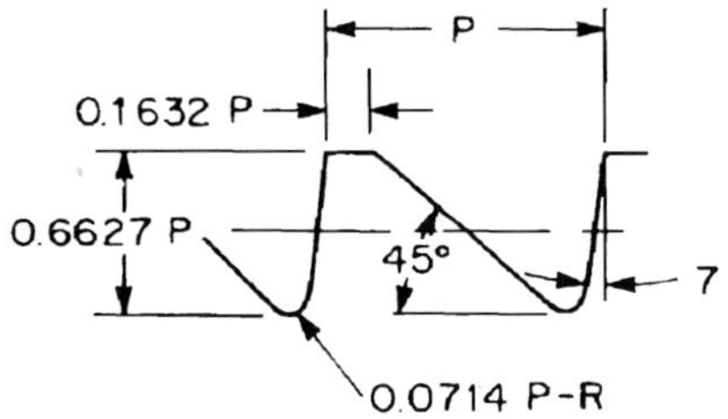

FIGURE 4.8.4 National buttress thread.

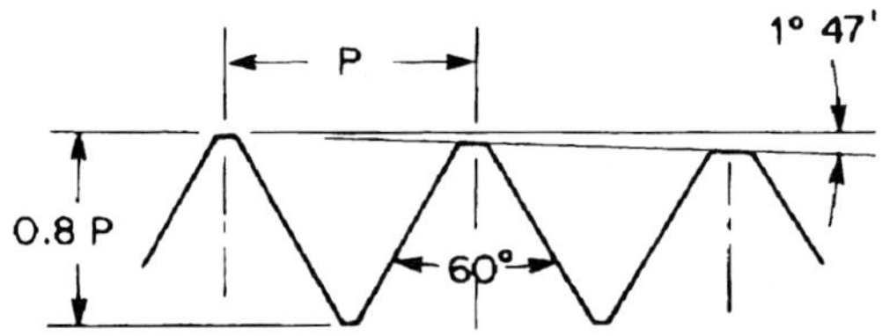

FIGURE 4.8.5 NPT pipe thread.

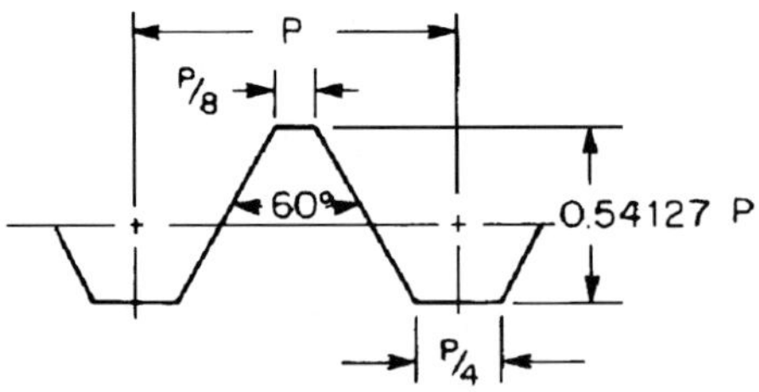

FIGURE 4.8.6 ISO metric thread.

There are three common thread classes in the unified system. Class 1 has the loosest fit and the broadest dimensional tolerances, Class 2 is the most common class for fasteners with closer fits and tolerances, and Class 3 is for more precise or critical applications. The letter *A* designates external threads and the letter *B* internal threads.

The standard method for designating a screw thread is to specify in sequence the nominal size, number of threads per inch, thread-series symbol, and thread-class symbol, supplemented optionally by pitch diameter and its tolerance. An example of an external thread designation and what it means is

¼–20–UNC–3A
thread-class designation
thread-series designation
number of threads per inch (pitch)
nominal size (in)

The *square thread form* (see Fig. 4.8.2) is the most efficient form for the transmission of power. However, it is more expensive to produce than other forms and has been largely superseded by the Acme thread form.

Acme threads (see Fig. 4.8.3) are also used for power transmission and are easier to manufacture than square threads, but their power-transmission capabilities are slightly lower. Some valve stems and many lead screws use this thread form.

Buttress threads (see Fig. 4.8.4) transmit power in one direction with virtually the full efficiency of a square thread but are relatively easily produced because of the tapered backside of the tooth form. They are used in military applications and when tubular members are screwed together.

American standard taper pipe thread (NPT), with the form shown in Fig. 4.8.5, is the standard thread for piping in the United States. *Straight* (nontapered) *pipe threads* and *dry-seal pipe threads* have similar forms.

The *ISO* (International Organization for Standardization) *metric screw thread* (see Fig. 4.8.6) is the prime metric screw thread for fasteners.

Standard nomenclature for thread forms is illustrated by Fig. 4.8.7. Coarse threads are suitable for general use, particularly in machines and other fastener applications in which quick and easy assembly is important. Fine threads are used when the design requires increased strength or reduced weight.

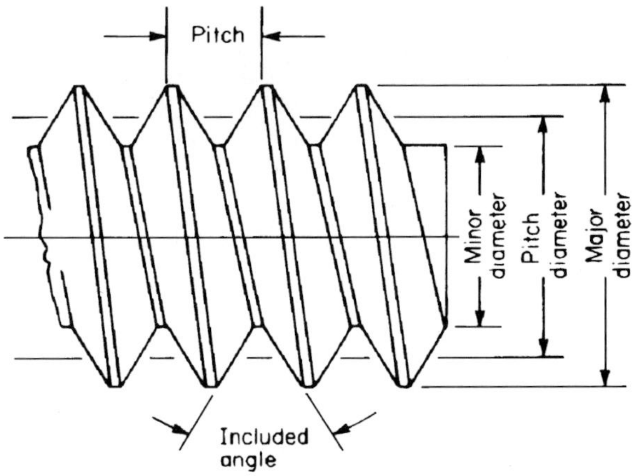

FIGURE 4.8.7 Standard nomenclature for screw-thread elements.

In addition to fasteners—bolts, machine screws, setscrews, cap screws, and studs—and power applications (vises, clamps, fixtures, and screw jacks), screw threads are used to control position accurately as in machine lead screws and vehicle-steering mechanisms, to feed materials, and to change rotary to linear motion.

The size range of commercial screw threads is vast. Screw threads as small as 0.3 mm (0.012 in) in diameter and 140 threads per centimeter (360 per inch) are used in watches. At the other extreme, 600-mm (14-in) pipe is threaded with 8.5-mm-pitch (two threads per inch) pipe thread.

Self-tapping screws are used for wood, sheet metal, fiberboard, and other softer materials. The thread form differs from that used in machine screws and the shank is normally tapered. Figure 7.1.31 illustrates some typical self-tapping screws.

THREAD-MAKING PROCESSES: THEIR APPLICATIONS AND ECONOMICS

Hand Dies

An acorn or button die for external threads must be employed by hand; it is the least desirable of the methods that can be used to cut external threads. (See Fig. 4.8.8.) However, such dies can be used to advantage when a limited number of small-to medium-sized threads are to be cut and when accuracy of the thread lead in relation to the thread axis is not essential. Compared with other thread-making tooling, they are relatively inexpensive and easy to use.

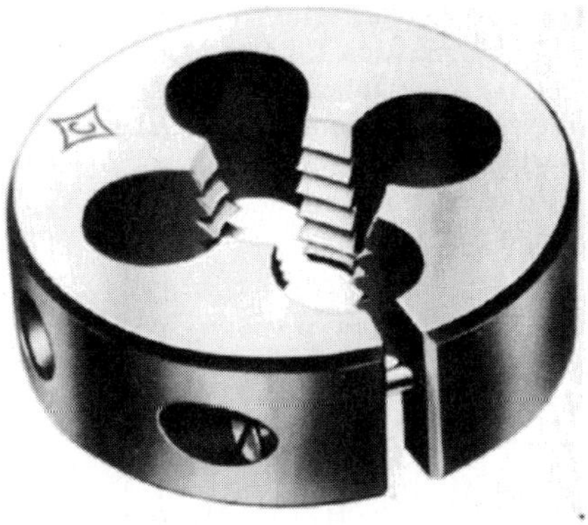

FIGURE 4.8.8 Button die for external threads. (*Courtesy Cleveland Twist Drill Co., subsidiary of Acme-Cleveland Corp.*)

Single-Point Threading

With this method, a single-point tool having a profile corresponding to the profile of the thread is used as a means of generating the thread. Internal or external threads can be produced by this method. A lathe is used, and its carriage is moved longitudinally along the part by a lead screw that is gear-driven from the spindle. The lead screw

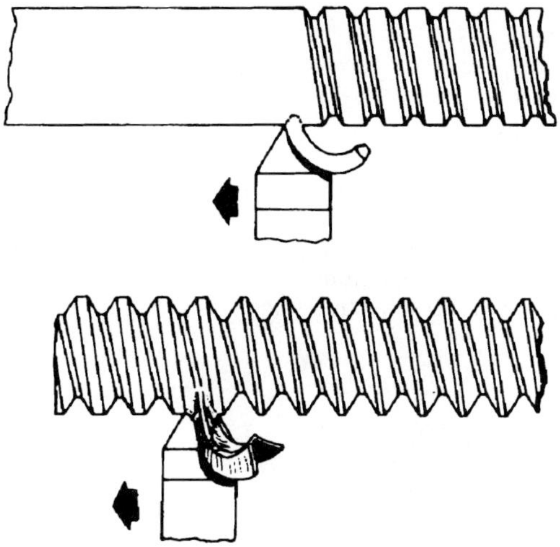

FIGURE 4.8.9 Single-point screw-thread cutting. (*Courtesy Teledyne Landis Machine.*)

moves the carriage and hence the tool at a rate exactly equal to the lead or pitch of the thread being produced. Generally, the thread is produced by making successive multiple passes. (See Fig. 4.8.9.)

Single-point threading is used more often when the workpiece is too large in diameter, the pitch too coarse, the material too difficult to machine, or the quantity too small to warrant using a die head. Holes as small as 8 mm ($^5/_{16}$ in) can be threaded by this method.

Thread-Cutting Die Heads

Die heads (not to be confused with thread-rolling heads) are an efficient and popular means of threading. They are versatile, have relatively wide ranges, and are made in a variety of models and sizes for application to many types of machines, including lathes, chuckers, multiple-spindle screw and threading machines, drill presses, and other types.

Die heads have four or five insert form cutters. When the head is fed axially from the end of the work, the threads are cut. Once engaged, the head is self-feeding at the rate of the thread lead. Cutter inserts can be removed for resharpening. Figure 4.8.10 illustrates a stationary, self-opening die head used with production lathes, chuckers, and screw machines.

Die heads can be used economically from low to moderate to high production levels depending on the circumstances. Compared with rolling, the blank need not have its diameter controlled as accurately because a certain amount of oversize can be trimmed away by the throat section of the chaser. Die-head chasers cost less than thread rollers and usually can be salvaged if partially damaged. Setup for die-head cutting is also usually faster than for thread rolling.

Pipe-thread cutting is a common application for die heads.

FIGURE 4.8.10 Stationary self-opening die head. (*Courtesy Teledyne Landis Machine.*)

Thread Milling

This process involves the use of a form-milling cutter that machines the thread form as the workpiece revolves. The most common type is the multiple-rib or multiple-form type, as shown in Fig. 4.8.11. Single-rib cutters are also used. With these, the workpiece must make as many revolutions as there are threads on the work.

Thread milling can be applied internally and externally and can be used to produce most thread forms regardless of whether they are straight or tapered. Minimum inter-

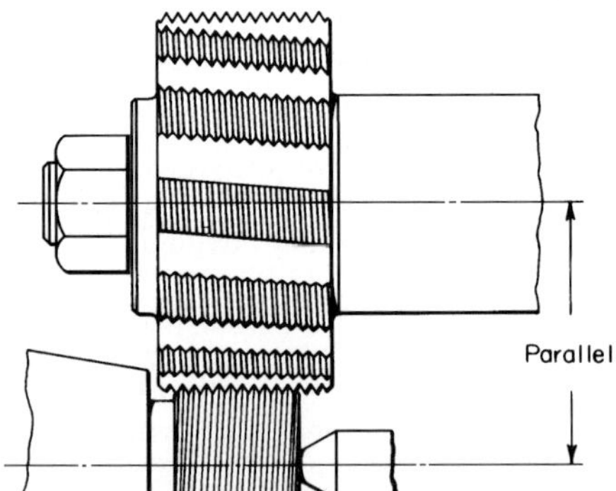

FIGURE 4.8.11 Thread milling. (*Courtesy Teledyne Landis Machine.*)

nal thread size is determined by the diameter of the cutter. Since interference is more pronounced because the cutter does not clear itself, the cutter normally should not exceed one-third of the hole diameter.

Thread forms that have flanks approaching 90° (to axis) are impossible to mill because the cutter cannot enter the cut without shaving the flank.

Some very coarse threads that are to be ground are rough-milled and then finished by grinding, possibly after a heat treatment.

Although thread milling is slower than die cutting, it is often necessary that a thread be milled because of coarse pitch, large or odd-shaped parts, a high helix angle, extremely long thread lengths, workpiece geometry, poor machinability of the workpiece material, or other considerations. As such, a single part or 10,000 pieces might be an economical production quantity.

Tapping

FIGURE 4.8.12 Solid tap for cutting internal threads. (*Courtesy Teledyne Landis Machine.*)

This process involves the use of a cylindrical form cutter, a tap, that has multiple cutting edges. The tap rotates and is fed axially into the work to produce internal threads. Both solid (Fig. 4.8.12) and collapsible (Fig. 4.8.13) taps are used. The operation can be carried out by hand or with drill presses, lathes, automatic screw machines, or special tapping machines.

Solid taps are used mainly to thread diameters ranging from 1.2 mm (0.047 in) to 150 mm (6 in). While the collapsible tap is limited by design factors on the low side to around 32-mm ($1\frac{1}{4}$-in) diameters, it is supplied for diameters as large as 600 mm (24 in). Solid taps are most economical in the 1.5-mm ($\frac{1}{16}$-in) to 50-mm (2-in) range. Although it is necessary to reverse the tap to back it out, in many cases the reversing operation can be done at much higher speed to reduce backout time.

Thread Grinding

Center-type grinding and centerless cylindrical grinding as described in Chaps. 4.13 and 4.14 are used in the production of some screw threads. Single- or multiple-rib-form wheels are employed with center-type grinding, while multiple-rib wheels are employed with centerless grinding. There is axial motion between the work and the wheel as the work rotates. Figures 4.8.14 and 4.8.15 illustrate the processes.

With center-type grinding, regardless of whether a single- or a multiple-rib wheel is used, the material specifications and the form, length, and quality of thread will determine the number of passes required to complete it. The number of passes can vary from one to five or six. With centerless grinding, the part is normally finished in

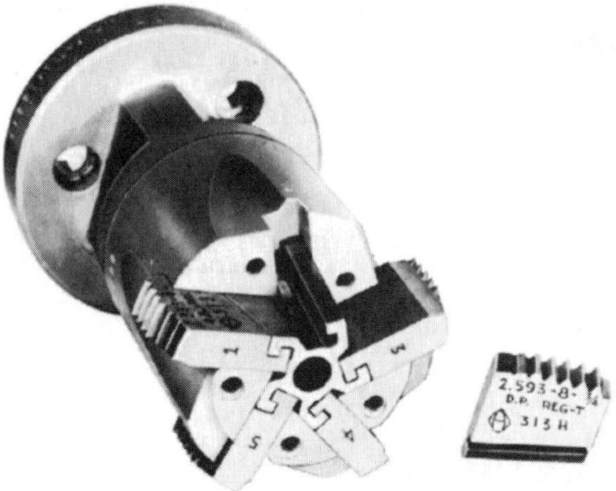

FIGURE 4.8.13 Collapsible tap. (*Courtesy Teledyne Landis Machine.*)

FIGURE 4.8.14 Center-type thread grinding. (*Courtesy Teledyne Landis Machine.*)

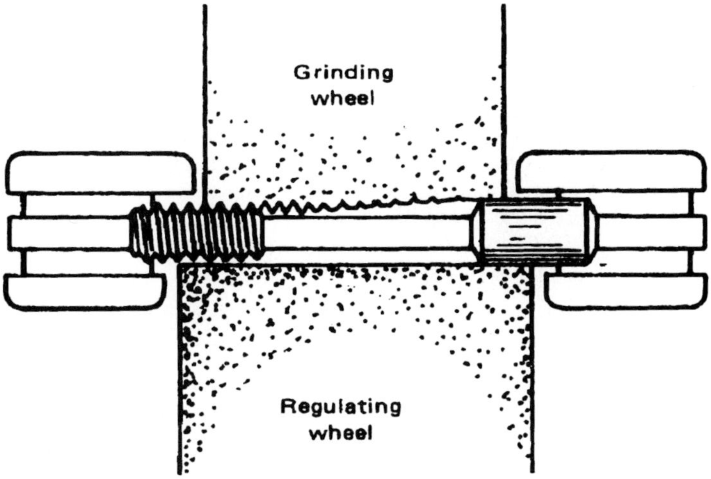

FIGURE 4.8.15 Centerless thread grinding. (*Courtesy Teledyne Landis Machine.*)

one pass through the machine. As the work moves across the wheel, first it is sized to the correct diameter and then the threads are formed.

Threaded parts that are ground include those which are too hard to cut, mill, or roll, when a fine finish is required or when precision form, lead, and pitch requirements must be held before and, most particularly, after hardening. Forms that are produced include API, NPT, and other taper pipe threads, 60° unified and metric, 55° Whitworth, 29° and 40° worm, 47°30″ British Association, 53°8″ Lowenhertz, Buttress, and others.

Centerless-ground threaded parts include continuous threaded parts such as setscrews, studs, threaded bushings, threaded size-adjusting bushings for boring heads, thread gauges, worm gears, powdered-iron screws, and self-threading insert bushings.

Materials that can be thread-ground include hardened and annealed screw stock, the alloyed high-speed tool and stainless steels, and sintered iron. The last-named is used widely for continuously threaded screws.

Center-type thread grinders are applicable to short as well as long production runs. Single-rib grinding wheels are more applicable to low production quantities and multi-rib wheels to mass production.

Setup times for hand operation range from $\frac{1}{2}$ to 1 h and, for automatic operation, from $1\frac{1}{2}$ to 2 h.

Centerless thread grinding is used for high production quantities. On diameters 10 mm ($\frac{3}{8}$ in) and larger, moderate quantities of 10 to 15,000 make for economical setups and reasonable runs. When grinding diameters smaller than 10 mm, particularly when diameter and length are the same or the length is up to $1\frac{1}{2}$ times the diameter, the quantity should exceed 15,000. On the smaller diameters, the setup is proportionally harder and takes longer. However, the production rate is higher than on the larger diameters. It is possible to run a $\frac{1}{4}$-in, 20-pitch, $1\frac{1}{4}$-mm-long setscrew at 7500 pieces per hour.

Thread Rolling

Thread and form rolling is accomplished by having hardened-steel dies penetrate round blanks. The exertion of adequate force displaces the material into the voids and produces a form the reverse of that on the die. Figure 4.8.16 shows the process schematically when flat dies are used. The main advantages of rolled threads over threads produced by other manufacturing processes are that they have improved physical characteristics, greater accuracy, and a high degree of surface finish. Another advantage is that there is no waste, since no material is removed in the formation of the thread.

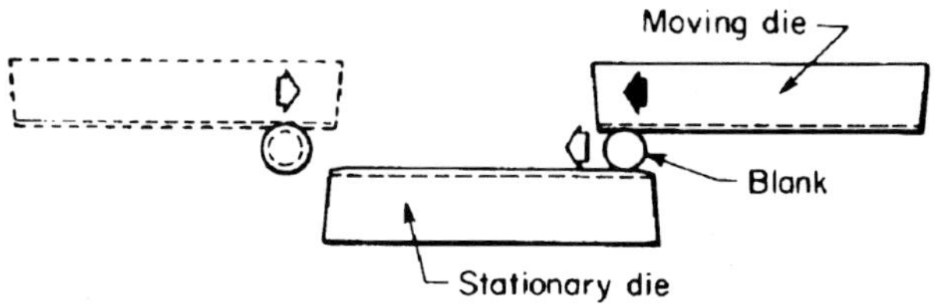

FIGURE 4.8.16 Thread rolling with flat dies. (*Courtesy Teledyne Landis Machine.*)

With regard to the physical characteristics of a rolled thread, there is a substantial increase in the tensile and shear strengths and resistance to fatigue. When a thread is produced by other manufacturing processes, the grain fibers of the metal are severed in the formation of the thread. However, when a thread is rolled, the grain fibers are made to flow in continuous unbroken lines following the contour of the thread. This is shown in Fig. 4.8.17.

Thread rolling is accomplished with reciprocating flat-die rolling machines, machines of the cylindrical-die type, thread-rolling heads, thread-rolling attachments, single-bump rolling equipment, and planetary rolling machines. Because the workpiece diameter before thread rolling should be controlled accurately, centerless grinding sometimes precedes the operation.

In addition to straight and taper threads, such forms as oil grooves and worm and gear forms are produced routinely by cold forming on thread-rolling machines. Sometimes, however, the geometry of the part is not conducive to thread-rolling applications. Figure 4.8.18 presents illustrations of typical parts that have had threads and other forms produced by the rolling process.

In comparison with a cutting tool that is less expensive, tends to wear quickly, but can be easily reground, the roll set is more expensive, may or may not be regrindable, and must produce more parts to justify its cost. Therefore, the rolling process generally must be matched to longer runs or to cases in which extra tool cost can be amortized. In some cases, when the necessary rolling die is available and can be substituted for a thread-cutting head, rolling may be economical for moderate-sized lots. Thread-rolling dies have a long life (from tens of thousands to millions of pieces), and resharpening during the life of the die is not necessary.

FIGURE 4.8.17 Grain structure of cut threads (*a*) contrasted with the stronger, plastically deformed structure of rolled threads (*b*). (*Courtesy Teledyne Landis Machine.*)

Cold-Form Tapping

Forming taps produce internal screw threads by plastic flow of material near the hole walls rather than by metal removal, as with conventional cutting taps. Figure 4.8.19 illustrates a typical forming tap.

The method has the advantages that no chips are formed, that the threads are strong, and that tapping speeds are higher. However, only a limited range of soft, ductile materials is suitable for cold-form tapping, and the percentage of thread is best held to 65 percent or less to avoid overfilling at the minor diameter. Torque requirements for tapping are also higher than for cutting taps. The characteristic thread form has a small groove at the crest of the thread, the width of which decreases with higher percentages of thread.

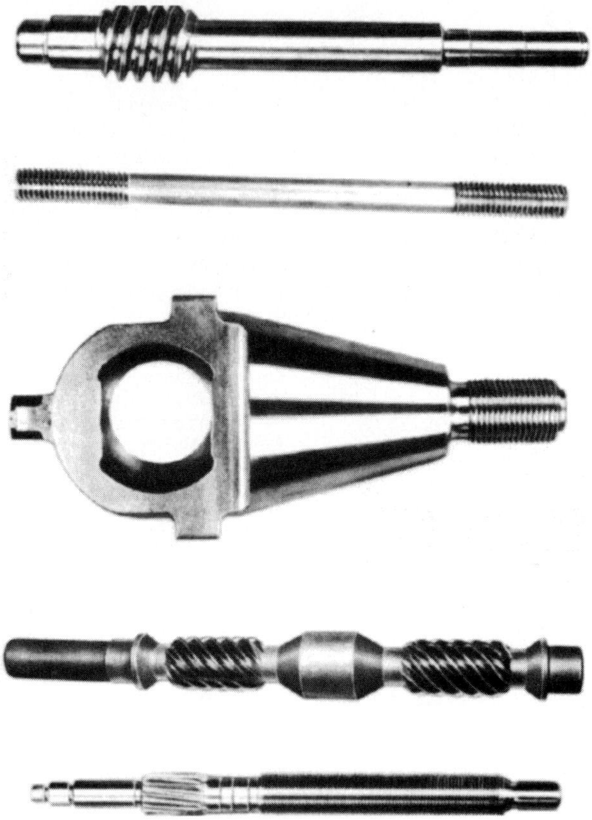

FIGURE 4.8.18 Typical parts with rolled threads. (*Courtesy Teledyne Landis Machine.*)

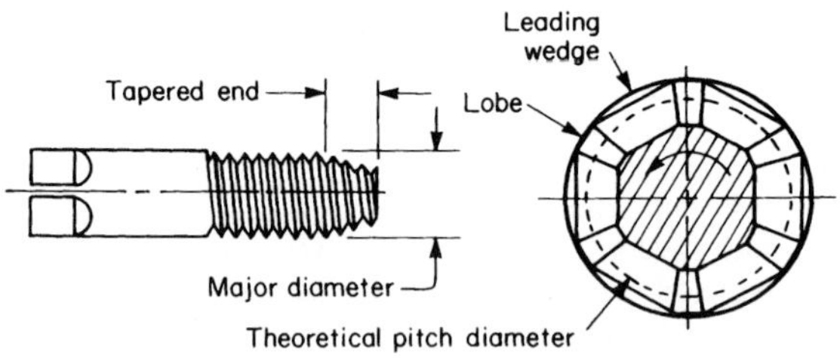

FIGURE 4.8.19 Cold-forming tap. (*From American Machinist.*)

SUITABLE SCREW-THREAD MATERIALS

Cut Threads

Often, the end use of the workpiece or considerations other than the threading operation dictate the selection of the material. However, when a choice is possible, selecting one of the free-cutting grades of material will give a more accurate thread of smoother finish. Compared with threading non-free-machining grades, producing a thread on free-cutting material will result in higher production at lower machining and tool costs. Soft, non-free-machining metals are especially difficult to thread, for they produce stringy chips that weld to the cutting edge.

In many cases metals selected on the basis of cost are more expensive in the end. A proportionately greater amount of time is spent obtaining a satisfactory thread. Also, higher tool cost is involved, tool life is poorer, and more downtime is required for tool changes.

Materials suitable for threading follow generally those suitable for most machining operations. Brasses and bronzes cut better and at higher speed than steels, free-machining steels cut better than unleaded or non-free-machining grades, and as carbon content increases and/or additives such as chromium or molybdenum are introduced, machinability drops quite rapidly. Aluminum, in bar stock, is generally quite good, but cast aluminum can be quite abrasive and cause excessive tool wear. Cast iron is brittle and presents a problem of maintaining a good form on the crest of the thread. Low-carbon steels, such as the 1010 and 1020 grades, while soft enough for easy machining, tend to tear, and it is difficult to obtain a good finish.

In steels, it is difficult to cut good threads when the Brinell hardness is below 160. This is due mainly to the difficulty in breaking the chip in such soft steel. In harder materials, the chip can be broken more easily. Easier breaking causes less interference at the cutting face of the tool and allows freer cutting and a smoother finish. Materials above R_c 34 are usually not suitable for die chasers and taps, which, generally speaking, are manufactured from high-speed steels. The single-point process using carbide is better suited for materials above R_c 34.

Difficult-to-machine materials sometimes can be more advantageously threaded by thread milling. In setting speeds and feeds, consideration must be given to the workpiece hardness and cutter material, taking into account the fact that 60° thread forms do not make for a cutter with strong teeth.

Ground Threads

Materials generally suitable for other form-grinding operations are satisfactory also for ground threads. The most suitable materials are the hardened steels and any metals that will be heat-treated above R_c 33 before threading. Aluminum and comparable soft materials are the most difficult to grind because they tend to load the wheel and cause burning.

Formed Threads

Different properties are required for thread forming than for cutting, and materials that can be cut may not be suitable for thread rolling or cold-form tapping. Factors that promote thread formability are low hardness, a low yield point, elongation of 12 percent or more, a fine-grained microstructure, and freedom from work hardening.

Leaded and sulfurized steel and leaded brasses do not work out well for thread rolling and should be considered only for cut threads. The use of thread rolling is also generally not recommended for materials that exceed R_c 32 hardness. With materials harder than this, die life is substantially reduced. Table 4.8.1 indicates the rollability and expectable die life for commonly thread-rolled metals.

TABLE 4.8.1 Rollability of Materials: Rolled-Thread Finish and Proportional Die Life*

Standard designation	UNS designation	Thread finish	Soft	R_c 5–24	R_c 25–32	R_c 33 and over	Remarks
			Proportional die life				
Carbon and alloy steels							
1008–1095	G10080–G10950	E	H	H–M	M	L	Excellent rollability.
1108–1151	G11080–G11510	G	H	H–M	M		These are free-machining steels with high sulfur content. The highest sulfer materials, 1119, 1144, and 1200 Series, should be avoided when possible.
1211–1215	G12110–G12150	F	M				
1111–1113	G11110–G11130	F	H				
1330–1345	G13300–G13450	E	H	M	M–L	L	These are medium-alloy steels with alloys such as manganese, molybdenum, chromium, and nickel. Work hardening of material requires higher pressures, and some reduction in roll life over the 1000–1200 Series will be experienced.
4118–4161	G41180–G41610	E	H–M	M	M–L	L	
4320–4340	G43200–G43400	E	H–M	M	M–L	L	
4815–4820	G48150–G48200	E	H–M	M	M–L	L	
5115–5160	G51150–G51600	E	H-M	M	M-L	L	
6118–6150	G61180–G61500	E	H–M	M	M–L	L	
8720–8740	G87200–G87400	E	H–M	M	M–L	L	
9255–9260	G92550–G92600	E	H–M	M	M–L	L	
Stainless steels							
302–304	S30200–S30400	E		M–L	L		These are nonhardenable austenitic steels containing higher quantities of nickel and chromium. High work hardening occurs with percent alloy. They are also nonmagnetic. Material does not seam.
305, 321	S30500, S32100	E		M	L		
347, 348	S34700, S34800	E		M	L		
329, 430F	S32900, S43020	E		M–L	L		Nonhardenable ferritic chromium stainless, but magnetic.
446	S44600	E		M–L	L		

430–443	S43000–S44300	E	M	L	Lower work hardening but higher pressures required owing to carbon.
414, 420F	S41400, S42020	E	M–L	L	Hardenable martensitic chromium steels; magnetic. They are the best suited for rolling of the stainless grades; low work hardening.
440F	S44020	E	H–M	L	
410, 431	S41000, S43100	E	H–M	L	
440C	S44004	E	H–M	L	

Wrought copper and copper alloys

Material designation	Alloy name	Maximum hardness	Finish	Die life	Remarks
C10200	Oxygen-free copper	RF40	E	H	More than 90% copper; excellent rollability.
C11000	Electrolytic copper (ETP)	RF40	E	H	
C21000	Gilding, 95%	RB40	E	H	Copper-zinc alloys basically good for rolling except when zinc exceeds 30%. This tends to produce poor finish as indicated in C27000–C28000.
C23000	Red brass, 85%	RB55	E	H	
C26000	Cartridge brass, 70%	RB60	E	H	
C27000	Yellow brass, 65%	RB55	F	H	
C31400	Leaded commercial bronze	RB65	P	M-H	Copper-zinc alloys with lead added for improved machining characteristics; poor to fair for rolling. Higher lead produces poorer thread finish and is not recommended for rolling
C33500	Low-leaded brass	RB60	G	M-H	
C34200	High-leaded brass, 64½%	RB55	P	M	
C34500	Thread-rolling brass	RB75	G	M-H	
C36000	Free-cutting brass	RB70	P	L	
C37000	Free-cutting Muntz metal	RB70	P	L	
C38500	Architectural bronze	RB65	P	L	
C44300–C44500	Admiralty brasses	RB75	E	H	Copper-zinc alloy with 1.0% tin; excellent rollability.
C46400, C46500	Naval brass	RB75	P-F	M	Copper-zinc alloy with lead and tin not conducive to good rolling characteristics. Alternative material should be used.
C48500	Naval brass, high-leaded	RB80	P	L	
C50200	Phosphor bronze E	RB50	G	H	Copper-tin alloy generally good for rolling, but increasing tin content reduces rollability. C54400 contains some lead and zinc, thereby reducing rollability
C51000	Phosphor bronze A	RB65	G	H	
C54400	Phosphor bronze B-2	RB70	P	L	

TABLE 4.8.1 Rollability of Materials: Rolled-Thread Finish and Proportional Die Life* (*Continued*)

Wrought copper and copper alloys

Material designation	Alloy name	Maximum hardness	Finish	Die Life	Remarks
C65100	Low-silicon bronze B	RB70	E	M	Copper with silicon as basic alloy; average rollability
C65500	High-silicon bronze A	RB75	FG	M	
C67500	Manganese bronze A	RB70	P	M	High-zinc alloy; alternative material should be used.
C70600	Copper-nickel 10%	RB70	G	M–H	High-nickel alloy; reduce rollability proportionally.
C74500	Nickel-silver, 65-10	RB70	E	H	Copper with zinc and nickel as alloy; rollability good to excellent. As alloy increases, rollability \ decreases.
C75200	Nickel-silver, 65-18	RB70	G–E	H	

Annealed copper casting alloys

With regard to the rollability of copper casting alloys in the annealed condition, most are rated as having poor rollability and poor die life. Copper alloys with basic quantities of tin, zinc, or silicon rate slightly better in die life, with poor to fair finish. It is recommended that these materials be avoided when possible and be considered only for low production quantities.

Wrought aluminum and aluminum alloys

Material designation: SAE no. & temper	UNS no.	Condition	Maximum hardness	% Elongation	Finish	Die life	Remarks
1100–0	A91100	Annealed	RB23	45	E	H	99% aluminum recommended for rolling. Work-hardens very slowly; cannot be heat-treated. Major alloy is silicon.
1100–HI4	A91100	Half hard	RB32	20	G–E	H	
1100–HI8	A91100	Full hard	RB44	15	FG	M	
2011–T3	A92011	Heat-treated and cold-worked	RB95	15	FG	M–H	Lower-quality finish is a result of lead and bismuth alloys; not generally recommended for rolling.
2011–T6	A92011	Heat-treated and aged	RB97	17	F	M–H	

2014–0	A92014	Annealed	RB45	18	G	M–H	Copper, silicon, and manganese major alloys; higher strength requires greater roll pressure.
2014–T4	A92014	Heat-treated and aged	RB105	20	G–E	M–H	
2017–0	A92017	Annealed	RB45	22	E	H	Good rollability; most commonly used for rolling.
2017–T4	A92017	Heat-treated and aged	RB105	22	E	H	
2024–0	A92024	Annealed	RB47	22	E	H	
2024–T3	A92024	Heat-treated and cold-worked	RB120	18	E	H	
3003–0	A93003	Annealed	RB28	40	E	H	99% aluminum recommended for rolling; work-hardens very slowly; cannot be heat-treated. Major alloy is manganese.
3003–H14	A93003	Half hard	RB40	16	G	H	
3003–H18	A93003	Full hard	RB55	10	P–F	L–M	
5052–0	A95052	Annealed	RB47	30	E	H	Fair to good rollability in the lower-hardness condition; major alloy manganese with chromium.
5052–H34	A95052	Half hard	RB68	14	F	M	
5056–0	A95056	Annealed	RB65	35	E	H	Major alloy magnesium; recommend rolling in annealed condition only.
5056–H18	A95056	Strain-hardened	RB105	10	P	L–M	
061–0	A96061	Annealed	RB30	30	E	H	Good to excellent rollability in conditions.
6061–T4	A96061	Heat-treated and aged	RB25	65	G–E	H	
7075–0	A97075	Annealed	RB60	16	F	H	Generally not recommended for rolling.
7075–T6	A97075	Heat-treated and aged	RB150	11	P	M	
Wrought nickel and nickel alloys							
The nickel alloys in general can be produced with a good to excellent thread finish. The Inconel and Hastelloy series result in a poor to fair finish. The higher tensile strength of nickel alloys requires high roll pressures, and therefore, medium to low die life can be expected. It is recommended that annealed material be used whenever possible.							

*Letter designations for finish: E, excellent; G, good; F, fair; P, poor. Letter designations for die life: H, high; M, medium; L, low. Elongation factor: generally acceptable results can be achieved when percent elongation equals 12 or more.

Cold-form tapping requires even greater cold workability than external-thread rolling. Cold-workable grades of brass, copper, and aluminum and low-carbon steel are the most commonly used materials.

DESIGN RECOMMENDATIONS FOR SCREW THREADS

External threads made by all processes should not terminate too close to a shoulder or other larger diameter. Space must be provided for the thread-cutting tool. In fact, there should be an area of thread relief or undercut where the diameter of the workpiece is less than the minor thread diameter. (See Fig. 4.8.20.) This allows room for the throat

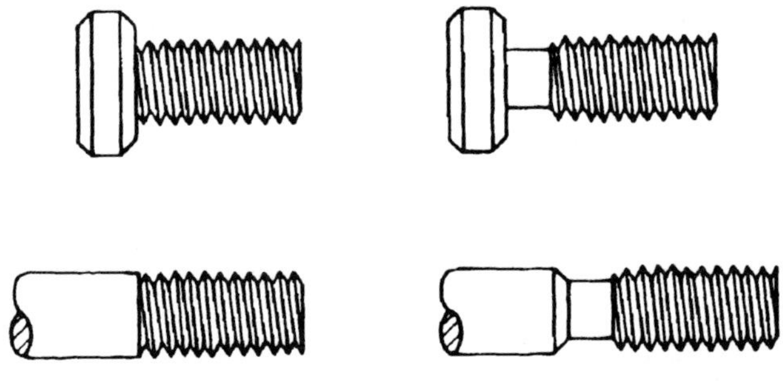

FIGURE 4.8.20 Allow thread relief at the end of the threaded length.

angle of the thread cutter, which would otherwise produce an incomplete thread at the end. It also reduces the chance of tool breakage. The width of this relief depends on the size of the part, coarseness of the thread, and throat angle of the threading tool. From 1.5 mm ($^1/_{16}$ in) to 19 mm ($^3/_4$ in) or more should be allowed. When possible, the width of the relief should be increased to allow use of chasers having the maximum length of the throat or chamfer. This will provide maximum efficiency of the operation and maximum tool life.

Internal threads should have a similar relief or undercut even though, for blind holes, it necessitates an added recessing operation before threading. Blind holes, even if not provided with an undercut, require some unthreaded length at the bottom for chip clearance. Best and most economical of all is the through hole, which provides both chip clearance and relief if the threads extend to the opposite surface. Figure 4.8.21 illustrates these alternatives.

In many applications no more than 60 or 65 percent of the thread height is required for adequate thread strength. Threads in this range machine more easily, requiring only 75 percent of the torque needed for conventional threads. If high strength is not required, consider the use of a reduced-height thread form. (See Fig. 4.8.22.)

Similarly, the length threaded should be kept as short as possible consistent with the functional requirements of the part. Shorter threads machine more quickly and provide longer tool life. For internal threads, where tap breakage may be a problem, limit the depth of the threaded portion to two diameters.

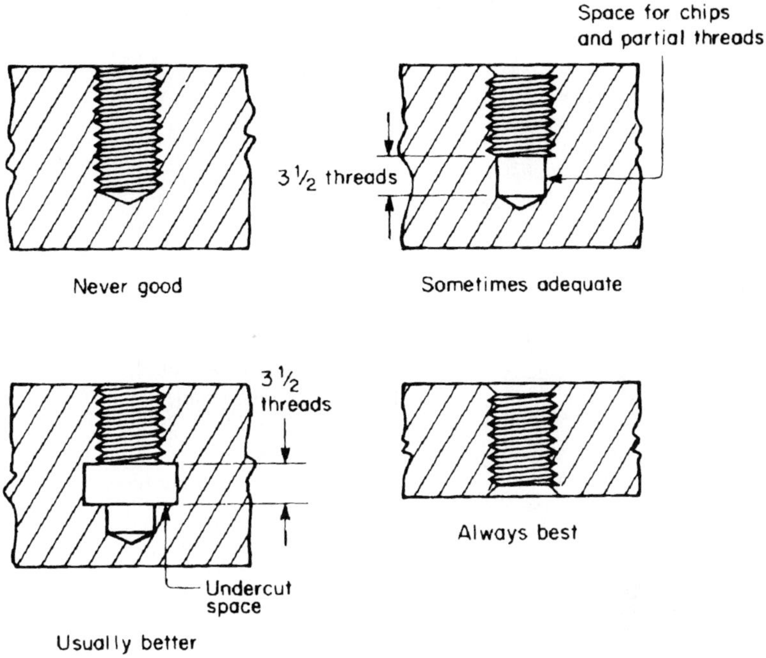

FIGURE 4.8.21 Allow chip clearance with internal threads.

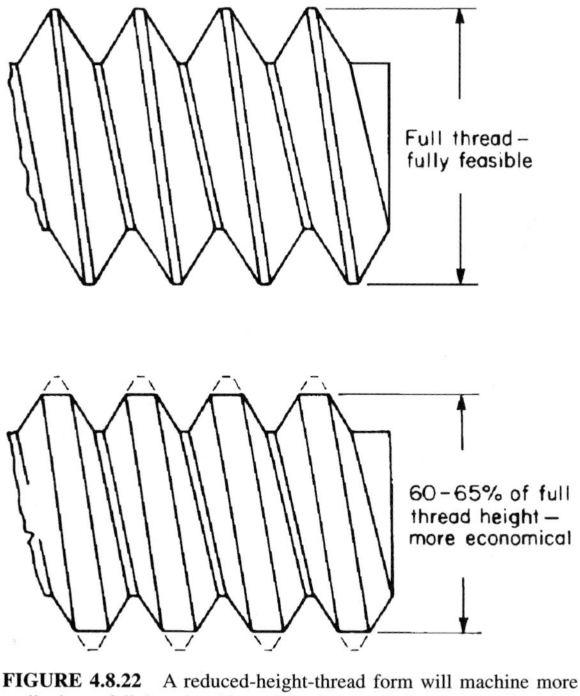

FIGURE 4.8.22 A reduced-height-thread form will machine more easily than a full thread and has adequate strength for most applications.

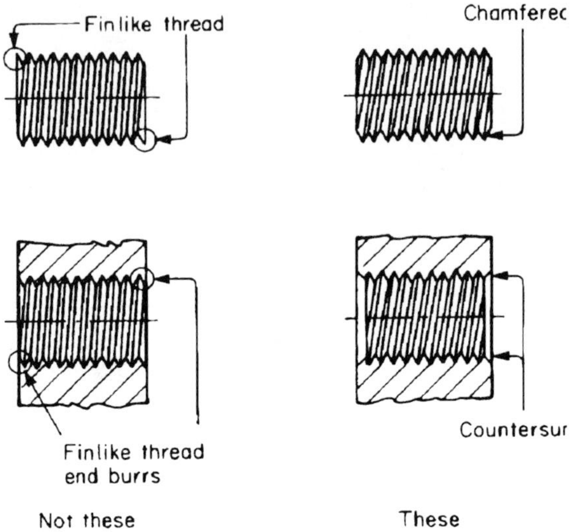

FIGURE 4.8.23 Specify chamfers and countersinks at the ends of threaded sections.

The design of threaded products should include a chamfer at the ends of the external threads and a countersink at the ends of the internal threads. These inclined surfaces prevent the formation of finlike threads at the ends, help to minimize burrs, and assist the threading tool in starting to cut or form the threads. (See Fig. 4.8.23.)

Aside from chamfers and countersinks, the surface at the starting end of the screw thread should be flat and square with the thread's center axis. Otherwise, proper starting of the thread-making tool may be difficult.

Slots, cross holes, and flats should not be placed where they intersect screw threads. Most thread-making processes are adversely affected by surface interruptions, and burrs are almost inevitable where the surfaces intersect. Burrs on thread surfaces are especially costly to remove. When cross holes are unavoidable, they should be countersunk.

Standard thread forms and sizes with off-the-shelf threading tools are always more economical than threads made with special tools.

Tubular parts must have a wall heavy enough to withstand the pressure of the cutting or forming action. This stricture applies to both internal and external threads. Castings and forgings of odd shapes should not have thin sections at a portion of the thread's circumference. Otherwise, out-of-roundness will occur.

Tolerances closer than required for the given function should not be specified. Class 2 threads are usually satisfactory for most work.

Threads to be ground should not be specified to have sharp corners at the root. Normally, a radius of 0.08 mm (0.003 in) is the very minimum that can be expected, and much larger radii on the order of 0.25 mm (0.010 in) are preferable. (See Fig. 4.8.24.)

Centerless-ground threads should have a length-to-diameter ratio of at least 1:1, but preferably the length should be longer than the diameter. Parts to be centerless-thread-ground also should not have large burrs, be flattened or egg-shaped from shearing, or be bent or crooked. Taper and flatness also should be avoided because they will not be removed by the thread-grinding operation.

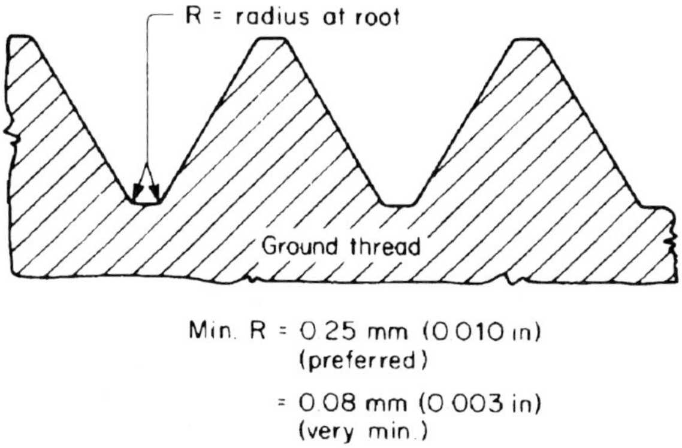

FIGURE 4.8.24 Ground threads require a generous root radius.

Parts for thread rolling have similar requirements of roundness, straightness, and freedom from taper and burrs. Uniformity of hardness is also important for thread rolling, as is accurate blank diameter.

Except for the largest sizes, coarse threads are slightly more economical to produce than fine threads and should be specified in preference to fine threads if the part's function permits. Coarse threads also assemble more rapidly.

DIMENSIONAL FACTORS AND TOLERANCES

The same factors induce dimensional variations in screw threads as affect the dimensions of other types of surfaces produced by thread-making equipment. The accuracy and conditions of tooling and equipment are key factors for all thread-making processes. So are the skill of the worker, the suitability of the material, and the feed rate of the threading tool.

When conditions are optimal and extra care and extra time (sometimes considerable) are taken, Class 4 and 5 threads can be produced by all the methods covered by this chapter. Suitably accurate measuring equipment is also required to guide and control the accuracy of the final results.

Hand dies are normally not capable of the highest precision and are primarily applicable to Class 1 and 2 threads only. Other thread-cutting methods can be used for Classes 1 through 5, with costs increasing sharply at the higher precision levels. Best lead accuracy occurs when the advance of the thread-cutting tool is controlled by lead screw rather than by the tap or die. Surface finishes smoother than 1.6 μm (63 μin) are not normally attainable by thread cutting.

With thread milling, by using a careful setup and moderate feed, the outside, pitch, and root diameters can be held to ±0.025 mm (0.001 in). Lead depends on the accuracy of the lead screw of the machine and can be held as close as 0.001 mm/cm (0.0001 in/in). Surface finish can be held to 1.6 μm (63 μin), with finer finishes sometimes possible with finer feed. Thread milling is a suitable method for accurate classes of thread, especially if the workpiece material has some machinability limitations.

At one time, thread grinding was essential to achieve Class 4 and 5 threads. This circumstance has changed, and Class 5 threads are now both rolled and cut. Grinding is used most frequently when the hardness of workpiece precludes other methods. If finishes and accuracy greater than those specified for Class 3 threads are required, reduced production rates should be expected. Center-type grinders can hold flank angles of thread forms to $\pm\frac{1}{4}°$ and lead accuracy to within ±0.002 mm/cm (0.0002 in/in) and the cumulative error to not more than 0.06 mm in 300 mm (0.0024 in/ft).

All classes of thread can be rolled. The piece-to-piece accuracy of rolled threads depends on various factors, particularly the consistency of the blank diameter and the uniformity of material and structure from piece to piece. Tolerances cannot be met if there are variations in these factors. Centerless grinding is a common preliminary operation to thread rolling to ensure an accurate blank diameter. The surface finish of rolled threads is superior to that of cut threads or about as smooth as the surface of the rolling dies. Generally, a limit of 0.8 μm (32 μin) can be specified if a smooth surface is required.

Tables 4.8.2 and 4.8.3 provide information on the approximate dimensions and tolerances of standard screw threads. Additional data on screw-thread dimensions and tolerances can be found in the publication *Screw Thread Standards for Federal Services,* Handbook H28, published by the U.S. Department of Commerce. A similar publication is *Unified Inch Screw Threads* (ANSI B1.1-1974), published by the American National Standards Institute.

TABLE 4.8.2 Dimensions and Tolerances for Unified and National Threads

			Tolerances, in												
			Major diameter external threads†			Pitch diameter‡								Minor diameter,§ internal threads	
Sizes*	Basic major diameter, in, D	Threads per inch, n	Class 1A	Classes 2A and 3A	Classes 2 and 3	Class 1A	Class 1B	Class 2A	Class 2B	Class 3A	Class 3B	Class 2	Class 3	Classes 1B, 2B, and 3B	Classes 2 and 3
Fine-pitch threads															
0 (0.060)	0.0600	80		0.0032	0.0034			0.0018	0.0023			0.0017	0.0013	0.0049	0.0049
1 (0.073)	0.0730	72		0.0035	0.0036			0.0019	0.0025			0.0018	0.0013	0.0055	0.0054
2 (0.086)	0.0860	64		0.0038	0.0038			0.0020	0.0027			0.0019	0.0014	0.0062	0.0055
4 (0.112)	0.1120	48		0.0045	0.0044			0.0024	0.0031			0.0022	0.0016	0.0074	0.0066
6 (0.138)	0.1380	40		0.0051	0.0048			0.0026	0.0034			0.0024	0.0017	0.0077	0.0070
8 (0.164)	0.1640	36		0.0055	0.0050			0.0028	0.0036			0.0025	0.0018	0.0077	0.0063
10 (0.190)	0.1900	32		0.0060	0.0054			0.0030	0.0039			0.0027	0.0019	0.0079	0.0062
12 (0.216)	0.2160	28		0.0065	0.0062			0.0032	0.0042			0.0031	0.0022	0.0084	0.0062
¼	0.2500	28	0.0098	0.0065	0.0062	0.0050	0.0065	0.0033	0.0043	0.0025	0.0032	0.0031	0.0022	0.0077	0.0060
⅜	0.3750	24	0.0108	0.0072	0.0066	0.0057	0.0074	0.0038	0.0049	0.0029	0.0037	0.0033	0.0024	0.0073	0.0065
½	0.5000	20	0.0122	0.0081	0.0072	0.0064	0.0084	0.0043	0.0056	0.0032	0.0042	0.0036	0.0026	0.0078	0.0072
¾	0.7500	16	0.0142	0.0094	0.0090	0.0775	0.0098	0.0050	0.0065	0.0038	0.0049	0.0045	0.0032	0.0085	0.0080
1	1.0000	12	0.0172	0.0114	0.0112	0.0088	0.0114	0.0059	0.0076	0.0044	0.0057	0.0056	0.0040	0.0100	0.0090
1¼	1.2500	12	0.0172	0.0114	0.0112	0.0092	0.0120	0.0062	0.0080	0.0046	0.0060	0.0056	0.0040	0.0100	0.0090
1½	1.5000	12	0.0172	0.0114	0.0112	0.0096	0.0125	0.0064	0.0083	0.0048	0.0063	0.0056	0.0040	0.0100	0.0090
Coarse-pitch threads															
1 (0.073)	0.0730	64		0.0038	0.0038			0.0020	0.0026			0.0019	0.0014	0.0062	0.0062
2 (0.086)	0.0860	56		0.0041	0.0040			0.0021	0.0028			0.0020	0.0015	0.0070	0.0070
4 (0.112)	0.1120	40		0.0051	0.0048			0.0025	0.0033			0.0024	0.0017	0.0090	0.0089
6 (0.138)	0.1380	32		0.0060	0.0054			0.0028	0.0037			0.0027	0.0019	0.0098	0.0103
8 (0.164)	0.1640	32		0.0060	0.0054			0.0029	0.0038			0.0027	0.0019	0.0087	0.0082
10 (0.190)	0.1900	24		0.0072	0.0066			0.0033	0.0043			0.0033	0.0024	0.0106	0.0110
12 (0.216)	0.2160	24		0.0072	0.0066			0.0034	0.0044			0.0033	0.0024	0.0098	0.0092
¼	0.2500	20	0.0122	0.0081	0.0072	0.0056	0.0073	0.0037	0.0048	0.0028	0.0036	0.0036	0.0026	0.0108	0.0101

TABLE 4.8.2 Dimensions and Tolerances for Unified and National Threads (*Continued*)

			Tolerances, in												
			Major diameter external threads†			Pitch diameter‡								Minor diameter,§ internal threads	
Sizes*	Basic major diameter, in, D	Threads per inch, n	Class 1A	Classes 2A and 3A	Classes 2 and 3	Class 1A	Class 1B	Class 2A	Class 2B	Class 3A	Class 3B	Class 2	Class 3	Classes 1B, 2B, and 3B	Classes 2 and 3
Coarse-pitch threads (*Continued*)															
⅜	0.3750	16	0.0142	0.0094	0.0090	0.0065	0.0085	0.0044	0.0057	0.0033	0.0043	0.0045	0.0032	0.0109	0.0111
½	0.5000	13	0.0163	0.0109	0.0104	0.0074	0.0097	0.0050	0.0065	0.0037	0.0048	0.0052	0.0037	0.0117	0.0123
½	0.5000	12	0.0172	0.0114		0.0077	0.0100	0.0051	0.0066	0.0038	0.0050			0.0125	
¾	0.7500	10	0.0194	0.0129	0.0128	0.0088	0.0115	0.0059	0.0077	0.0044	0.0057	0.0064	0.0045	0.0128	0.0136
1	1.0000	8	0.0225	0.0150	0.0152	0.0101	0.0132	0.0068	0.0088	0.0051	0.0066	0.0076	0.0054	0.0150	0.0148
1¼	1.2500	7	0.0246	0.0164	0.0170	0.0111	0.0144	0.0074	0.0096	0.0055	0.0072	0.0085	0.0059	0.0171	0.0154
1½	1.5000	6	0.0273	0.0182	0.0202	0.0121	0.0158	0.0081	0.0105	0.0061	0.0079	0.0101	0.0071	0.0200	0.0180
2	2.0000	4½	0.0330	0.0220	0.0254	0.0143	0.0186	0.0095	0.0124	0.0071	0.0093	0.0127	0.0089	0.0267	0.241
2½	2.5000	4	0.0357	0.0238	0.0280	0.0155	0.0202	0.0104	0.0135	0.0078	0.0101	0.0140	0.0097	0.0300	0.0270
3	3.0000	4	0.0357	0.0238	0.0280	0.0161	0.0209	0.0107	0.0139	0.0080	0.0104	0.0140	0.0097	0.0300	0.0270
3½	3.5000	4	0.0357	0.0238	0.0280	0.0166	0.0215	0.0110	0.0143	0.0083	0.0108	0.0140	0.0097	0.0300	0.0270
4	4.0000	4	0.0357	0.0238	0.0280	0.0170	0.0221	0.0113	0.0147	0.0085	0.0111	0.0140	0.0097	0.0300	0.0270

*Listings in parentheses indicate unified threads.

†Major diameter of internal threads may extend to a $p/24$ flat.

‡British: effective diameter.

§Minor diameter of external threads may extend to a $p/8$ flat.

Source: Data from American Standard ASA B1.1-1949, published by the American Society of Mechanical Engineers, New York. Values are based on a length of engagement equal to the nominal diameter.

TABLE 4.8.3 Dimensions and Tolerances for Metric Screw Threads

	External threads						Internal threads				
	Major diameter		Pitch diameter				Minor diameter		Pitch diameter		
Basic thread designation	Maximum	Minimum	Maximum	Minimum	Minor diameteR (flat root), maximum	Minor diameter (rounded root), minimum*	Minimum	Maximum	Minimum	Maximum	Major diameter, minimum
M1.6 × 0.35	1.581	1.496	1.354	1.291	1.202	1.075	1.221	1.321	1.373	1.458	1.600
M2 × 0.4	1.981	1.886	1.721	1.654	1.548	1.408	1.567	1.679	1.740	1.830	2.000
M3 × 0.5	2.980	2.874	2.655	2.580	2.439	2.272	2.459	2.599	2.675	2.775	3.000
M4 × 0.7	3.978	3.838	3.523	3.433	3.220	3.002	3.242	3.422	3.545	3.663	4.000
M5 × 0.8	4.976	4.826	4.456	4.361	4.110	3.869	4.134	4.334	4.480	4.605	5.000
M6 × 1	5.974	5.794	5.324	5.212	4.891	4.596	4.917	5.153	5.350	5.500	6.000
M8 × 1.25	7.972	7.760	7.160	7.042	6.619	6.272	6.647	6.912	7.188	7.348	8.000
M8 × 1	7.974	7.794	7.324	7.212	6.891	6.596	6.917	7.153	7.350	7.500	8.000
M10 × 1.5	9.968	9.732	8.994	8.862	8.344	7.938	8.376	8.676	9.026	9.206	10.000
M10 × 1.25	9.972	9.760	9.160	9.042	8.619	8.272	8.647	8.912	9.188	9.348	10.000
M12 × 1.75	11.966	11.701	10.829	10.679	10.072	9.601	10.106	10.441	10.863	11.063	12.000
M12 × 1.25	11.972	11.760	11.160	11.028	10.619	10.258	10.647	10.912	11.188	11.368	12.000
M16 × 2	15.962	15.682	14.663	14.503	13.797	13.271	13.835	14.210	14.701	14.913	16.000
M16 × 1.5	15.968	15.732	14.994	14.854	14.344	13.930	14.376	14.676	15.026	15.216	16.000
M20 × 2.5	19.958	19.623	18.334	18.164	17.252	16.624	17.294	17.744	18.376	18.600	20.000
M20 × 1.5	19.968	19.732	18.994	18.854	18.344	17.930	18.376	18.676	19.026	19.216	20.000
M24 × 3	23.952	23.577	22.003	21.803	20.704	19.955	20.752	21.252	22.051	22.316	24.000
M24 × 2	23.962	23.682	22.663	22.493	21.797	21.261	21.835	22.210	22.701	22.925	24.000
M30 × 3.5	29.947	29.522	27.674	27.462	26.158	25.306	26.211	26.771	27.727	28.007	30.000
M30 × 2	29.962	29.682	28.663	28.493	27.797	27.261	27.835	28.210	28.701	28.925	30.000
M36 × 4	35.940	35.465	33.342	33.118	31.610	30.654	31.670	32.270	33.402	33.702	36.000
M36 × 2	35.962	35.683	34.663	34.493	33.797	33.261	33.835	34.210	34.701	34.925	36.000

*For reference.

Source: From American National Standard ANSI B1.13-1979, published by the American Society of Mechanical Engineers.

CHAPTER 4.9
BROACHED PARTS

Robert Roseliep
General Manager
General Broach and Engineering Co., Inc.
Chesterfield, Michigan

THE PROCESS

Broaching is the cutting of a machinable material by passing a cutter with a series of progressively stepped teeth over or through it. These teeth travel in a plane parallel to the surface being cut and, by removing a predetermined amount of stock, produce precision contours and finishes.

Most often, all the cutting teeth will be contained in one tool, known as a *broach,* to rough-out and finish-cut the part completely in a single machine stroke. (See Fig. 4.9.1.) When excessive stock prevents a one-stroke application, additional strokes can be employed, utilizing the same tool, if possible, or a series of tools. The location of the tool in relation to the workpiece must be changed by an amount equal to the stock removed on each stroke if the same tool is used for multiple strokes. The broach can be pulled or pushed and can be vertical or horizontal.

External Broaching

This method involves machining an external surface of the part. The workpiece is usually clamped in a holding fixture and the tool secured in a broach holder. Either the broach holder or the fixture is attached to the powered slide of a broaching machine, and the other is held in a fixed position relative to the surface to be broached. *Pot broaching* and *straddle broaching* are two processes whereby the workpiece is surrounded by broaches exerting balanced cutting pressures, which eliminate the need for clamping the workpiece.

Internal Broaching

This method requires a hole or opening in the workpiece to allow the broaching tool to be inserted and pulled or pushed through the part. When the cut is balanced, as with a

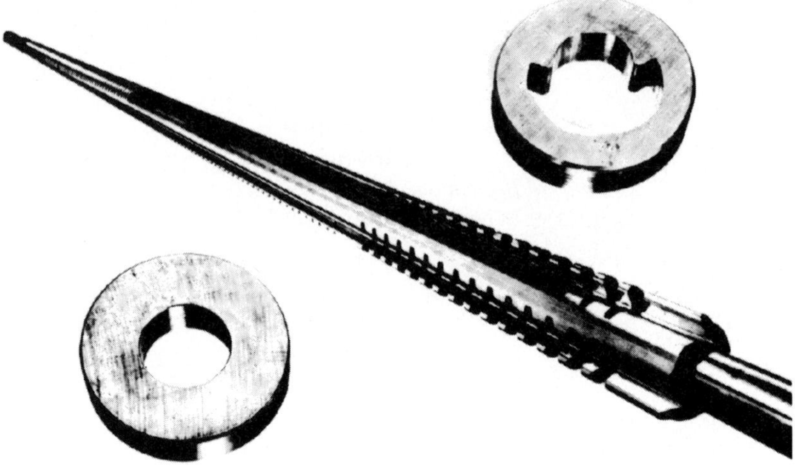

FIGURE 4.9.1 Complex-form pull broach showing the part blank before broaching and the finished configuration after pulling the broach through the part. (*Apex Broach & Machine Co.*)

round or splined hole, the broach is usually allowed to position the workpiece centrally to the cut. Unbalanced cuts require guiding the broach. Guiding usually is accomplished in the design of the tool or by fixturing with guides above and below or through the part. *Blind-hole broaching,* in which the tool cannot pass completely through the part, is performed by using a single tool or a set of tools with a limited number of cutting teeth. The tool is pushed into the workpiece until the teeth are past the surface being cut and is then retracted.

Machines specifically designed and built for broaching are used for the process. They may be completely automatic, semiautomatic, or basically manual; with a manual machine, the operator handles the workpiece and the tool and directs the motion of the machine. The machines are manufactured in a variety of standard and special types and sizes. Most are hydraulically powered, with some being electromechanically, gear, screw, or chain driven.

TYPICAL CHARACTERISTICS OF BROACHED PARTS

Virtually any part that is made by chip-forming machining could be a candidate for broaching. Some parts have no practical alternative method of manufacturing and must be broached. Typical parts are those with square, circular, or irregular holes, the key slot in lock cylinders (Fig. 4.9.2), splines and matching holes with straight-sided, involute, cycloidal, or specially shaped teeth, cam forms, gears, ratchets, and other complex forms requiring tight tolerances and precision finishes. Both helical and straight shapes are feasible. Sizes range from very small parts to parts weighing several tons, such as the stationary steam-turbine-rotor forms illustrated in Fig. 4.9.3.

Two exceptional characteristics, extremely high speed of production and outstand-

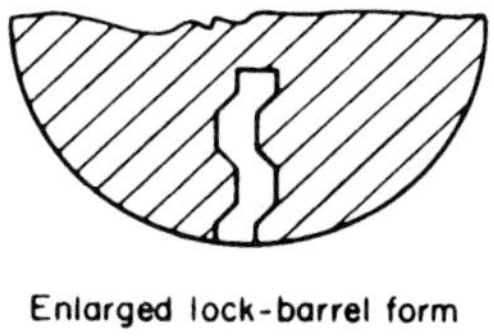

FIGURE 4.9.2 Key slots in lock cylinders are normally broached.

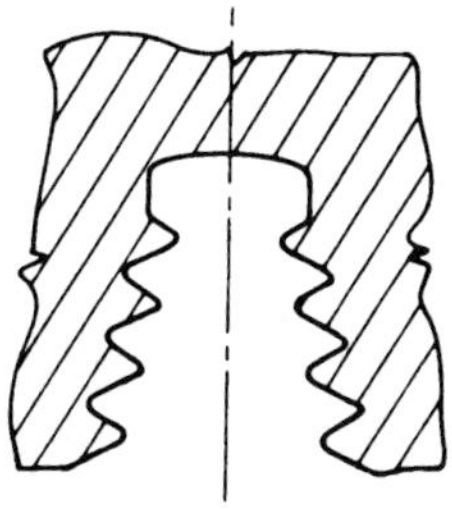

FIGURE 4.9.3 Mounting openings for steam-turbine blades are commonly broached.

ing repetitive accuracy, promote broaching over conventional machining processes. Broaching is often chosen to replace milling, planing, shaping, hobbing, slotting, boring, reaming, and grinding. A fine surface finish is produced because of the shaving action of the final teeth. Usually, no additional surface-refining operations are required. Tool marks are axial rather than circumferential.

ECONOMIC PRODUCTION QUANTITIES

Broaching usually requires high-volume production. Small production quantities often cannot justify the initial cost of the broaching tool and other tooling and the need for a special machine. The exceptions are cases in which there is no practical alternative method of machining or when standard broaching tools, already on hand, can be employed.

Broaches range in cost from the low hundreds of dollars for a simpler tool to several thousands of dollars for complex form-cutting tools, such as that required for the turbine-rotor form shown in Fig. 4.9.3. Production rates usually will range from 15 to more than 100 times higher than with alternative machining methods. For example, pot broaching 24 gear teeth in an SAE 1144 steel blank, 22 mm ($^7/_8$ in) thick, at 1000 pieces per hour, using fully automated tooling on one broaching machine, replaced 16 hobbing machines and four operators. A minimum production requirement of 1 million parts was needed to justify the tooling and machine costs.

Tooling a machine for more than one part or for a group of similar parts with the same machined surface can often make broaching an economical operation on small-lot quantities.

SUITABLE MATERIALS FOR BROACHING

Most of the known metals and alloys, especially steels, cast irons, bronze, brass, and aluminum, some plastics, hard rubber, wood graphite, asbestos, and other composites, have been broached. In all cases, machinability of the material is the key factor. Broaching of porous forgings and castings having nonuniform densities creates the problems of unacceptable surface finish and poor size control and should be avoided.

Controlling material hardness is important to prevent excessive part-to-part varia-

TABLE 4.9.1 Typical Machining Results with Commonly Broached Materials

Material		Brinell			Tolerance ±	
Common designation	UNS designation	hardness number	Finish μm	Finish μin	mm/surface	in/surface
SAE 1008/1010	G10080/G10100	60–70 R_b	1.1/1.6	45–65	0.025	0.001
SAE 1020/1023	G10200/G10230	70–74 R_b	1.1/1.5	45–60	0.025	0.001
SAE 1040	G10400	80–86 R_b	0.8/1.1	30–45	0.025	0.001
SAE 1064/1065	G10640/G10650	18–20 R_c	0.6/1.3	25–50	0.038	0.0015
SAE 1069/1070	G10690/G10700	18–20 R_c	0.6/1.3	25–50	0.038	0.0015
SAE 1095	G10950	23–25 R_c	0.9/1.5	35–60	0.046	0.0018
SAE 1144	G11440	93–97 R_b	0.8/1.5	30–45	0.020	0.0008
SAE 4027	G40270	91–95 R_b	0.6/1.3	25–50	0.013	0.0005
SAE 4140/4142	G41400/G41420	92–94 R_b	1.3/2.0	50–80	0.038	0.0015
SAE 4145	G41450	92–94 R_b	1.3/2.0	50–80	0.038	0.0015
SAE 4340 cast	G43400	31–33 R_c	2.0/3.0	80–120	0.064	0.0025
SAE 5140	G5140	12–16 R_c	1.5/2.0	60–80	0.051	0.002
SAE 52100	G52986	18–25 R_c	1.1/1.5	45–60	0.020	0.0008
SAE 6150	G61500	18–22 R_c	1.1/1.5	45–60	0.020	0.0008
SAE 8620	G86200	18–27 R_c	1.0/1.5	40–60	0.020	0.0008
SAE 8640	G86400	18–25 R_c	1.3/2.0	50–80	0.025	0.001
SAE 8642	G86420	18–25 R_c	1.3/2.0	50–80	0.025	0.001
SAE 8645	G86450	18–25 R_c	1.3/2.0	50–80	0.025	0.001
Gray cast iron		88–94 R_b	2.0/2.5	80–100	0.063	0.0025
Pearlitic malleable iron		90–96 R_b	1.1/1.5	45–60	0.013	0.0005
303 stainless steel	S30300	85–90 R_b	0.8/1.1	30–45	0.038	0.0015
17-4PH stainless steel	S17700	34–40 R_c	0.5/0.8	20–30	0.030	0.0012
403 stainless steel	S40300	27–32 R_c	0.6/0.9	25–35	0.030	0.0012
410 stainless steel	S41000	84–92 R_b	0.4/0.8	15–30	0.025	0.001
416 stainless steel	S41000	18–22 R_c	0.4/0.8	15–30	0.025	0.001
M-2 high-speed steel	T11303	24–28 R_c	1.1/1.5	45–60	0.038	0.0015
Inconel 750X	N07750	29–32 R_c	0.8/1.1	30–45	0.025	0.001
Greek Ascoloy	G41800	32–38 R_c	0.9/1.1	35–45	0.025	0.001
Rene 41	N07041	40–42 R_c	0.8/1.0	30–40	0.051	0.002
Stellite 31	R30031	30–32 R_c	2.0/3.0	80–120	0.051	0.002
2014-T6 Aluminum	A92014	70–80 R_b	0.8/1.1	32–45	0.058	0.0023
Copper	C12200	45–85 R_f	1.1/1.5	45–60	0.038	0.0015
Tellurium copper	C14500	45–50 R_f	0.8/1.1	30–45	0.025	0.001
Free-cutting brass	C36000	70–75 R_b	0.5/0.9	20–35	0.025	0.001
Navy M bronze	C92200	65–70 R_b	0.8/1.1	30–45	0.038	0.0015
Magnesium		60–75 R_b	0.2/0.4	8–15	0.038	0.0015
Aluminum bronze (8%)	C95600	75–83 R_b	1.4/2.0	55–80	0.063	0.0025

tions. Wide variations can result in poor or inconsistent surface finish and are a major factor in tool life and size control. The ideal hardness range of ferrous parts is between R_c 25 and R_c 32 with a tolerance of 3 to 5 points.

Table 4.9.1 lists some of the more commonly broached materials and the surface finishes and tolerances that may be achieved under normal conditions.

DESIGN RECOMMENDATIONS

Entrance and Exit Surfaces

A part to be broached should be designed so that it can be easily located and held in the proper attitude. Surfaces contiguous to the area to be cut should be square and relatively flat. Care in selecting the location of parting lines and gates to prevent poor support during machining is important. The designer should visualize how the part is to be retained and supported and avoid the possibility of uneven or inconsistent surfaces in these areas.

This is especially true in internal broaching, in which the tool is not retained or guided by the machine or fixturing. Uneven or inclined surfaces can cause side-thrust pressures to the tool, which can result in inaccuracies in the finished hole and possible tool failure.

External broaching is not so demanding, provided the part is so designed that the holding fixture can retain and support it rigidly during the cutting stroke. However, it is still advisable to design the part with square supporting faces whenever possible.

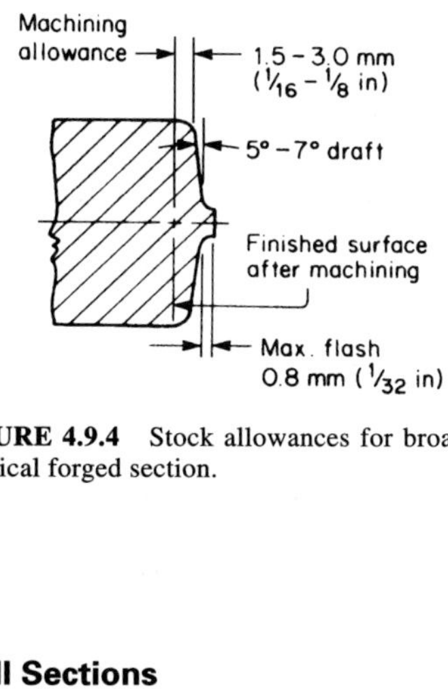

FIGURE 4.9.4 Stock allowances for broaching a typical forged section.

Stock Allowances

When forgings are planned for broaching, they should be held to as close dimensions as possible, allowing only minimum stock for finishing. Figure 4.9.4 shows a typical forged section with stock allowances recommended to avoid overloading the broach tool during production runs.

Castings require a greater stock allowance to ensure that inclusions, scale, and hard spots are removed and clean surfaces are produced in machining. Cold-punched or pierced holes present much the same problems as castings, and stock allowances should be ample enough to allow for blanking breakout.

Wall Sections

It is advisable to avoid frail or thin wall sections and to maintain a uniform thickness for any wall that will be subjected to machining forces. Sections should be at least sufficient to withstand fixture-retaining pressures and to minimize deflections caused by cutting forces.

Families of Parts

If all of a group of parts will require a similar broaching operation, the designer should attempt to design the parts so that all use the same broaching tool and, if possible, the same holding fixture. For example, a number of levers of different sizes or shapes could be designed with the same square hole in one end.

Round Holes

Starting holes may be cored, punched, bored, drilled, flame-cut, or hot-pierced. When the starting hole is drilled or bored, 0.8-mm ($\frac{1}{32}$-in) stock on the diameter of holes up to 38 mm ($1\frac{1}{2}$ in) in diameter and 1.6-mm (1/16-in) stock on larger holes are usually sufficient for cleanup. When cored holes are planned, draft angles, surface texture, and size variations must be taken into consideration in determining core size so as to assure cleanup.

Long holes should be chambered as shown in Fig. 4.9.5 to improve accuracy as well as to reduce costs. Table 4.9.2 shows the recommended maximum total length of the total hole surface for various diameters.

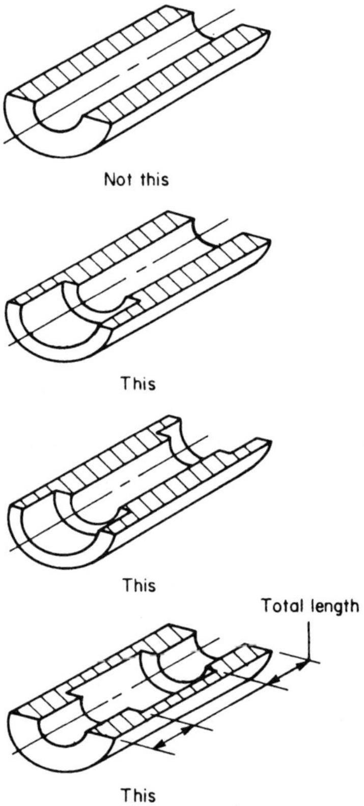

FIGURE 4.9.5 Long holes should be chambered.

Internal Forms

Symmetrically shaped internal forms are usually broached by starting from round holes, and the guidelines under "Round Holes" apply to them. Irregularly shaped internal forms may be started from round holes as shown in Fig. 4.9.6 or from cored, punched, pierced, or machined irregular holes. Sometimes the cost of removing excess stock prior to broaching may prove to be higher than that of broaching from the round hole, and the product designer should investigate this cost before finalizing the part-blank design. If this is not practicable, it is suggested that optional constructions for the part-blank starting hole be specified.

Whenever stock allowance for broaching can be controlled by the method used to form the blank, e.g., casting, stamping, or forging, it is always advisable to leave a minimum amount of stock for cleanup plus draft, mismatch, or out-of-round tolerances.

TABLE 4.9.2 Recommended Maximum Total Length of Hole for Various Diameters

Hole diameter		Maximum total length	
mm	in	mm	in
1.4–1.5	0.055–0.060	3.2	0.125
1.5–1.9	0.060–0.075	4.8	0.188
1.9–2.3	0.075–0.090	5.5	0.218
2.3–2.8	0.090–0.110	6.3	0.250
2.8–3.8	0.110–0.150	9.5	0.375
3.8–5.1	0.150–0.200	12.7	0.500
5.1–6.3	0.200–0.250	15.9	0.625
6.3–7.6	0.250–0.300	19.0	0.750
7.6–8.9	0.300–0.350	22.2	0.875
8.9–10.2	0.350–0.400	25.4	1.000
10.2–12.1	0.400–0.475	31.8	1.250
12.1–14.0	0.475–0.550	34.9	1.375
14.0–16.5	0.550–0.650	41.3	1.625
16.5–20.3	0.650–0.800	50.8	2.000
20.3–25.4	0.800–1.000	63.5	2.500
25.4–31.8	1.000–1.250	82.5	3.250
31.8–38.1	1.250–1.500	102	4.000
38.1–41.3	1.500–1.625	121	4.750
41.3–46.0	1.625–1.812	131	5.500
46.0–50.8	1.812–2.000	152	6.000
50.8–53.9	2.000–2.125	178	7.000
53.9–57.2	2.125–2.250	203	8.000
57.2–63.5	2.250–2.500	267	10.500
63.5–76.2	2.500–3.000	305	12.000
76.2–88.9	3.000–3.500	457	18.000

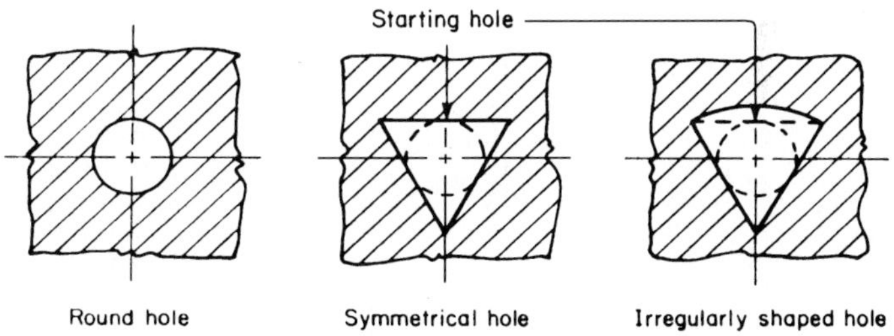

FIGURE 4.9.6 Irregularly shaped broached holes are started from round holes.

Internal Keyways

Whenever possible, it is advisable to design keyways to ASA specifications as shown in Table 4.9.3. In doing so, standard keyway broaches, available from some manufacturers as off-the-shelf tools, can be used. Many subcontract broaching sources stock these standard keyway broaches and solicit the broaching of any quantity of parts.

TABLE 4.9.3 Standard Keyway Sizes, in

$$G = \frac{A}{2} + \frac{0.010}{0.012}$$

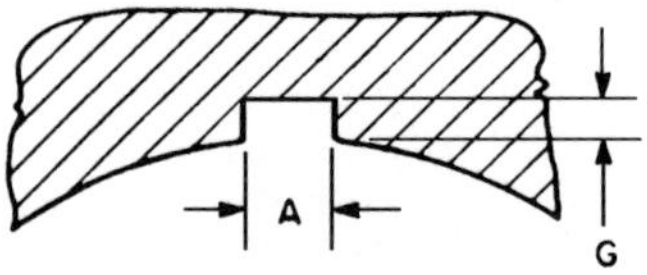

A, nominal dimension	Minimum hole size	Minimum length cut*	Maximum length cut	*G*	No. of cuts
1/16	3/8	3/8	1 1/4	0.042	1
3/32	7/16	1/2	1 1/2	0.058	1
3/32	5/8	5/8	2 1/2	0.058	1
1/8	1/2	1/2	1 1/2	0.074	1
1/8	7/8	5/8	2 1/2	0.074	1
5/32	19/32	1/2	1 1/2	0.089	1
5/32	23/32	5/8	2 1/2	0.089	1
3/16	11/16	5/8	2 1/2	0.105	1
3/16	15/16	11/16	3 1/2	0.105	1
7/32	11/16	5/8	2 1/2	0.120	1
7/32	15/16	11/16	3 1/2	0.120	1
1/4	11/16	5/8	2 1/2	0.136	1
1/4	1	11/16	4	0.136	1
1/4	1 7/16	7/8	6	0.136	1
9/32	7/8	11/16	4	0.152	1
9/32	1 1/4	7/8	6	0.152	1
5/16	1	11/16	4	0.167	1
5/16	1 5/16	7/8	6	0.167	1
3/8	1 1/16	11/16	4	0.199	1
3/8	1 5/16	7/8	6	0.199	1
7/16	1 9/16	11/16	4	0.230	1
7/16	2	1	8	0.230	2
1/2	1 1/2	11/16	4	0.261	1
1/2	1 1/2	1	8	0.261	2
9/16	1 3/4	11/16	4	0.292	1
9/16	1 5/8	1	8	0.292	2
9/16	2 1/4	1 1/8	12	0.292	2
5/8	1 7/8	11/16	4	0.324	1
5/8	2 1/2	1	8	0.324	2
5/8	2 1/4	1 1/8	12	0.324	2
3/4	1 7/8	11/16	4	0.386	1
3/4	2	1	8	0.386	2
3/4	2 1/4	1 1/8	12	0.386	3
7/8	2 1/4	11/16	4	0.449	1
7/8	2 1/4	1	8	0.449	2
7/8	2 1/4	1 1/8	12	0.449	3
1	2 1/4	5/8	2 1/2	0.511	1
1	2 1/4	7/8	6	0.511	2
1	2 1/4	1 1/8	12	0.511	3

*Minimum length of part recommended to prevent the part from dropping between the teeth of the broach.

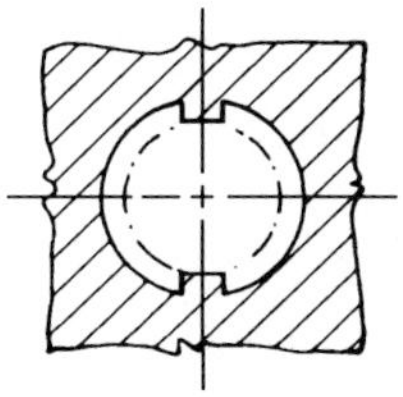

FIGURE 4.9.7 A balanced shaped hole is preferable to prevent the broach from drifting to one side.

Internal Keys

Pilot holes for internal keys should be on the same centerline as the finished hole. Balanced designs having more than one key equally spaced are advantageous to prevent the broach from drifting and should be used when hole location is critical (see Fig. 4.9.7). However, two keys require two keyways or other space on the mating part, and this also must be considered.

Straight-Splined Holes

1. Parallel or straight-sided splined holes should be designed to SAE standards.
2. Involute splines should be designed to SAE, DIN, or AGMA standards. Fine diametral pitches and stub-tooth forms are advisable since shallow-depth splines reduce the length of broach required (Fig. 4.9.8).

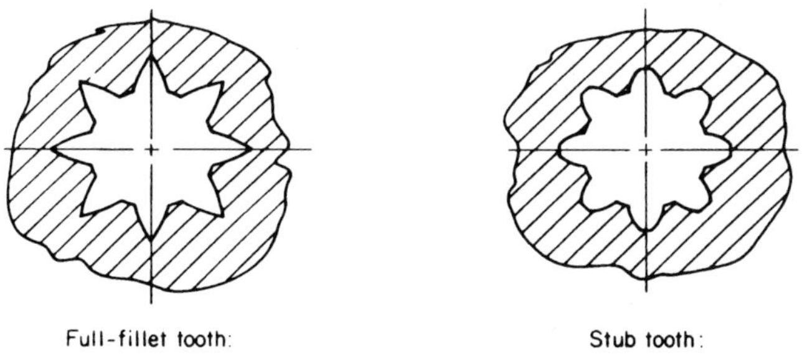

FIGURE 4.9.8 Stub-tooth forms are preferred.

3. Long holes should be chambered or relieved similarly to round holes, as shown in Fig. 4.9.6. If only tooth areas are broached, the reduction of circumferential area being cut allows the total hole length given in Table 4.9.2 to be increased by 33 percent for splined holes. This does not apply if the entire hole surface including the root of the spline is broached.

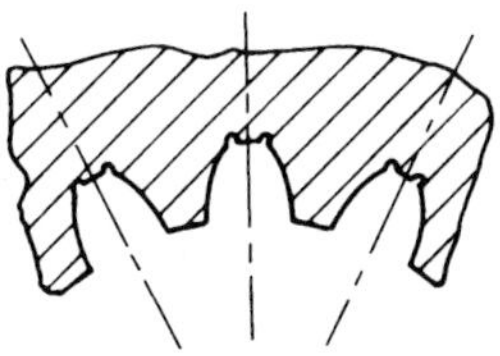

FIGURE 4.9.9 This shape allows room for the upset burr on a cold-rolled spline.

4. Broaching allows the designer to modify the spline profile to suit product requirements. For example, Fig. 4.9.9 illustrates a method to provide clearance for the upset burr of a cold-rolled spline shaft. This undercut procedure does increase the cost of the tool, but it can economically eliminate assembly problems.

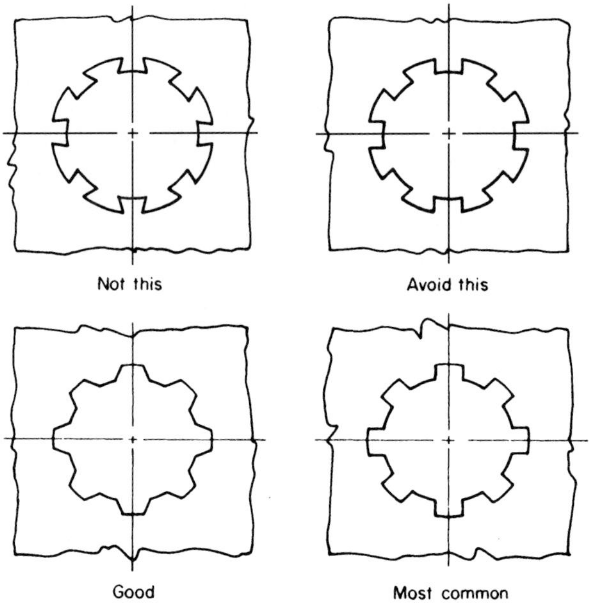

FIGURE 4.9.10 Avoid dovetail or inverted-angle splines.

5. Dovetail or inverted-angle splines should be avoided whenever possible (see Fig. 4.9.10).

Spiral Splines

The guidelines for straight splined holes also apply to spiral splines:

1. Spiral splines with helix angles greater than 40° cannot be broached by using conventional methods. It is advisable to use the lowest helix angle possible.
2. Splines with helix angles greater than 10° usually will require the broach to be driven rotationally during its travel through the part. The designer should provide some means of retaining the part to prevent rotation. This can be accomplished by an irregular contour of some prominence such as a projection, notch, hole, or indentation in the face of the part.

Tapered Splines

Tapered splines should be avoided (see Fig. 4.9.11):

1. Splines that taper on tooth thickness usually cannot be broached.
2. Splines that taper with the bore should be avoided. Single keyways are an exception, provided the bore is large enough to permit insertion of a mandrel having a tapered slot to guide the broach.

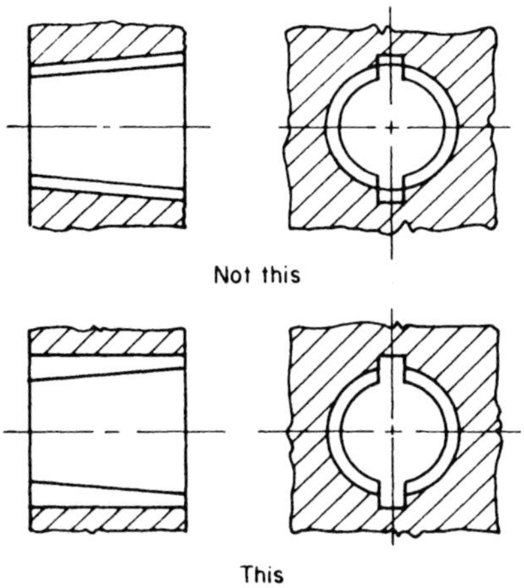

FIGURE 4.9.11 Avoid tapered splines.

3. Shallow tapers, such as those used for steering (Pitman) arms to provide a solid lock-up, are often broached to the smallest through-hole size and swaged to size in a secondary operation. In rare instances, a combination push broach and swage tool is inserted to depth and retracted.

Square and Hexagonal Holes

It is advantageous to use a slightly oversized starting hole, particularly for square holes. This method is optional, but it will reduce the cost of broaching considerably (see Fig. 4.9.12).

Avoiding sharp corners at the major diameter is recommended to reduce broach costs. This is best accomplished by specifying a slightly smaller major diameter. If corner radii are a design requirement, they will add to the cost of the broach (see Fig. 4.9.13).

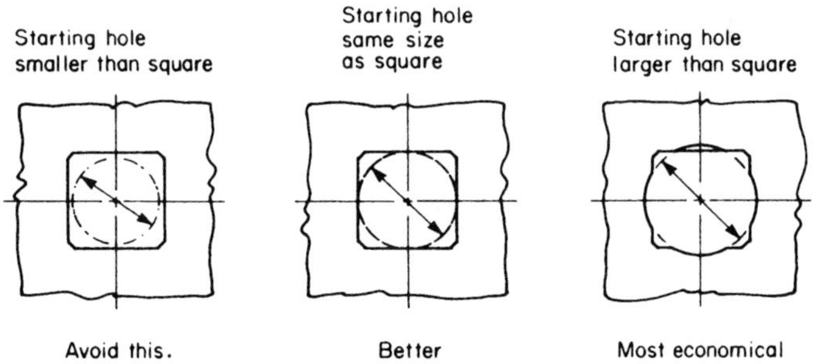

FIGURE 4.9.12 Use a slightly oversize starting hole.

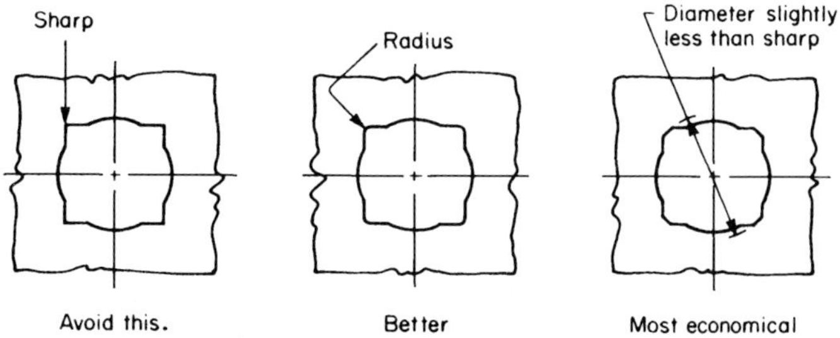

FIGURE 4.9.13 Avoid sharp corners on a major diameter.

Saw-Cut or Split Splined Holes

When the part will have an intersecting cut into the splined hole, such as is used to provide a clamping method or to allow expansion for the mating part, the splined hole should be designed with an omitted space as shown in Fig. 4.9.14. This allows room for the burr produced by the saw cut.

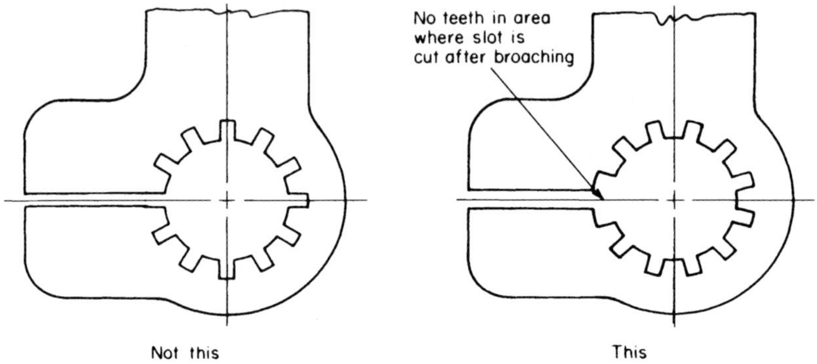

FIGURE 4.9.14 Allow room for the burr produced by the saw cut by eliminating one tooth.

Blind Holes

Broaching blind holes should be avoided if at all possible. If splines or similar shapes are necessary in blind holes, there should be a relief at the bottom of the broached area to allow the chip to break off. This relief area should be as wide as possible to retain the material removed by broaching and to allow for the fact that the broach will be shorter as it is sharpened. Grooves that produce a tapered heel for the broached area are recommended (see Fig. 4.9.15).

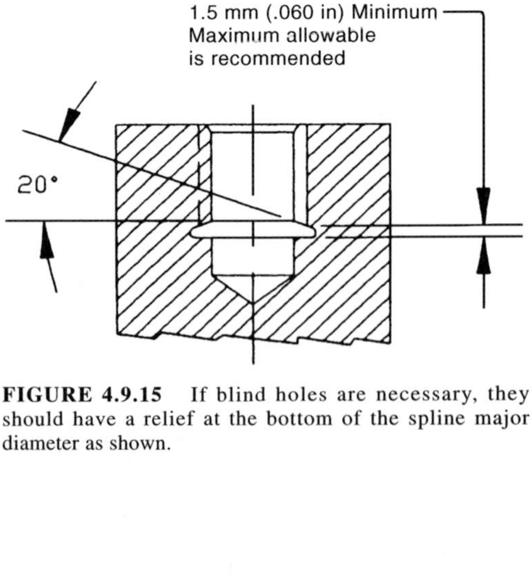

FIGURE 4.9.15 If blind holes are necessary, they should have a relief at the bottom of the spline major diameter as shown.

Gear Teeth

Internal gear teeth should be given the same consideration as internal involute splines

Chamfers and Corner Radii

In all situations in which corners must be broken by machining, chamfers are preferred over radii.

1. Sharp internal corners should be avoided to eliminate stress points and minimize tooth-edge wear. Chamfers are preferred to simplify manufacture, but radii may be specified (see Fig. 4.9.16).

FIGURE 4.9.16 Internal-corner design.

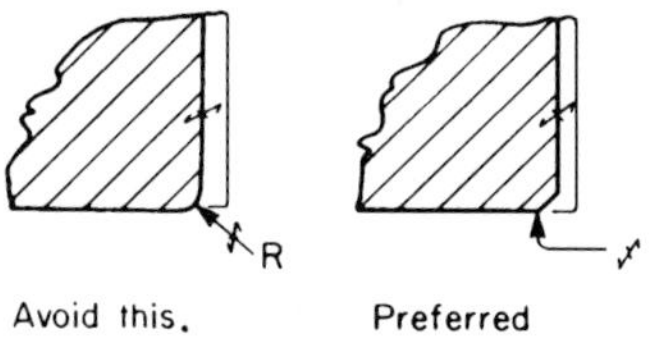

FIGURE 4.9.17 Chamfer outer corners rather than rounding them.

2. Outer corners or edges that must be machined should be chamfered rather than rounded, also for ease in manufacturing (see Fig. 4.9.17).
3. Sharp corners or edges of intersecting outer broached surfaces should be avoided whenever possible. Castings, forgings, and extrusions should be designed with a corner break that does not require machining.

External Surfaces

Whenever possible, external machined surfaces should be relieved to reduce the area that must be broached.

1. Reliefs of undercuts in the corners will simplify the broaching operation (see Fig. 4.9.18).
2. Large surfaces should be broken into a series of bosses whenever possible (see Fig. 4.9.19).

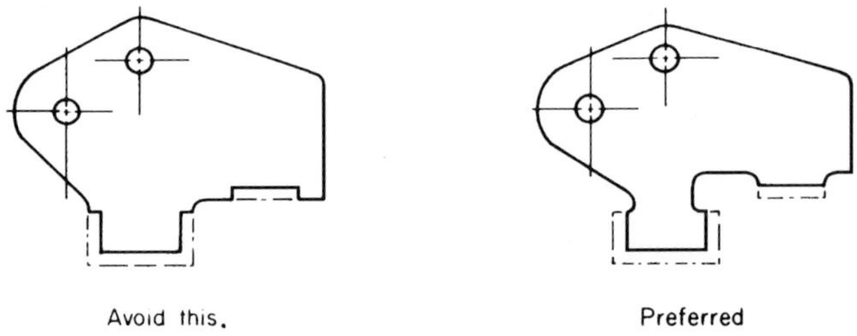

FIGURE 4.9.18 Reliefs or undercuts in the corners simplify broaching of external surfaces.

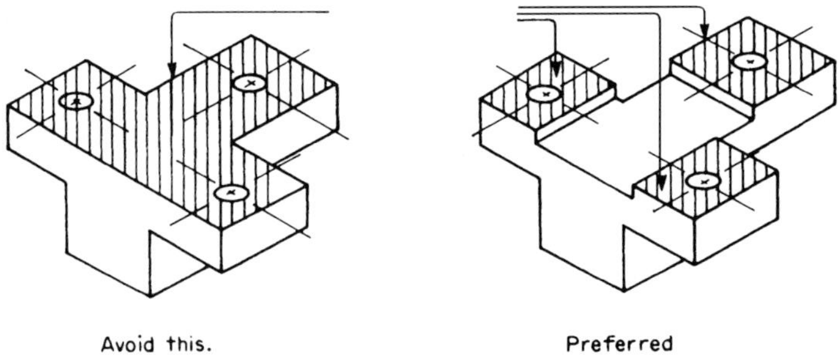

FIGURE 4.9.19 Break large surfaces into series of bosses.

Undercuts

Machined undercuts, such as when grinding may be required after broaching or when parts are relieved for mating-part fits, should be as shallow as possible Avoid sharp or narrow undercut configurations.

Burrs

Owing to the predetermined and controlled cut per tooth, burrs from broaching are generally smaller than burrs produced by other methods of machining. Chamfers or reliefs on the exit edge of the surface to be broached are recommended to contain the burr produced and eliminate a deburring operation.

Unbalanced Cuts

Unbalanced stock conditions caused by cross holes or other interruptions that could engender tool deflections should be avoided.

DIMENSIONAL FACTORS

Internal broaching, especially single-pass operations, usually can be held to closer tolerances than external applications because accuracy is primarily dependent on the tool itself. External broaching or internal applications that require guiding the tool or multiple passes are subject to residual factors. These comprise clearances of machine and fixture slides, accuracy in the repetition of multiple passes, rigidity of work-holding fixtures, and part alignment to the tool assembly.

Uniformity of material, consistency of datum faces (part-support faces, locating points, and clamping areas), and strength of the part are important to guarantee piece-to-piece tolerance control. Other factors that will affect tolerances and size control are effectiveness of tool maintenance and resharpening, machinability of the material, condition of the broaching machine, proper type and use of coolant, cutting speed, and proper design of the tool. Proper tool design is the most important factor in controlling size and finish.

RECOMMENDED TOLERANCES

Surface Finish

The surface finish produced by broaching is generally of high quality. While it does not match a grinding finish, it will be superior to the finish produced by most other manufacturing methods. By employing good tool design and proper coolant oils, finishes of a burnished quality can be obtained in good-machinability-rated materials.

Table 4.9.1 shows the surface finish that may be expected under normal conditions. Smoother surface finishes can be obtained in most materials but should not be specified unless absolutely necessary.

Flatness

Parts of uniform section and sufficient strength to withstand cutting pressures can be expected to be broached within 0.013 mm (0.0005 in) TIR (total indicator reading). An exception may be found on the exit edge of soft or gummy materials such as aluminum or stainless steels, where the metal extrudes during the cut and snaps back. A flatness of 0.025 mm (0.001 in) is a safe assumption for most broached parts.

Parallelism

Parallelism of surfaces machined in the same cutting stroke should be within 0.025 mm (0.001 in) TIR on good- to fair-machinability-rated materials.

Squareness

For parts that can be fixtured and retained on true surfaces, a squareness of 0.025 mm (0.001 in) TIR is possible, and tolerances of 0.08 mm (0.003 in) can be obtained consistently under controlled conditions in good-machinability-rated materials.

Concentricity

Broaches usually will follow the pilot hole, and concentricity errors due to broach drift should not exceed 0.025 to 0.05 mm (0.001 to 0.002 in) for round or similarly shaped holes in good- to fair-machinability-rated materials. Free-cutting materials such as brass allow the broach greater freedom for drifting during the cut. A special broach design and selection of the proper broaching machine often can solve this problem for the product designer.

Chamfers and Radii

Tolerances on chamfers and radii should be as liberal as possible. Radii under 0.8 mm (0.030 in) should have a minimum tolerance of 0.13 mm (0.005 in); 0.25 mm (0.010 in) should be allowable on larger sizes. Generous tolerances reduce broach manufacturing and maintenance costs.

Basic Dimensions

Apply the values in Table 4.9.1.

CHAPTER 4.10
CONTOUR-SAWED PARTS

THE PROCESS

Contour Sawing

Contour sawing is a band-sawing process. A multiple-tooth cutter, made from an endless band of sheet steel with cutting teeth on one edge, is power-driven across the workpiece. The workpiece rests on a horizontal table, and the band moves vertically in a downward direction. The work is fed either manually or with power assistance against the blade, and depending on how the work is guided, a straight or a curved cut is made as the blade advances into the work. For both conventional manually fed models and machines equipped with a servofeed attachment for heavier work, curved cuts are controlled visually as the operator follows a line on the workpiece. CNC and optical tracer control are also sometimes employed.

Machines employ an open-yoke C-shaped frame. The table may be stationary or, for larger work, movable. A third type of machine for large work is of radial-arm design; the entire C frame pivots and moves forward and backward. See Fig. 4.10.1 for the range of contour-sawing operations.

Saw bands are of either carbon or high-speed steel. Carbide-tooth inserts also can be used; the teeth are offset to provide clearance for the band, especially when curves are being cut.

When a quantity of thin parts is to be cut, it may be advisable to stack a number of blank sheets and cut the parts simultaneously.

Friction Contour Sawing

By operating a band saw at extremely high speeds [1600 to 5000 m/min (6000 to 15000 ft/min)], the work is melted (or softened) at the cutting edge by frictional heat, and the saw teeth remove the softened metal. Cuts similar to those made by conventional contour sawing can be made, and heat-treated steel and other material too hard for conventional band sawing can be processed. The blade, which is somewhat heavier than a conventional blade and not as sharp, remains sufficiently cool because any one part of it is in contact with the work for only a brief instant.

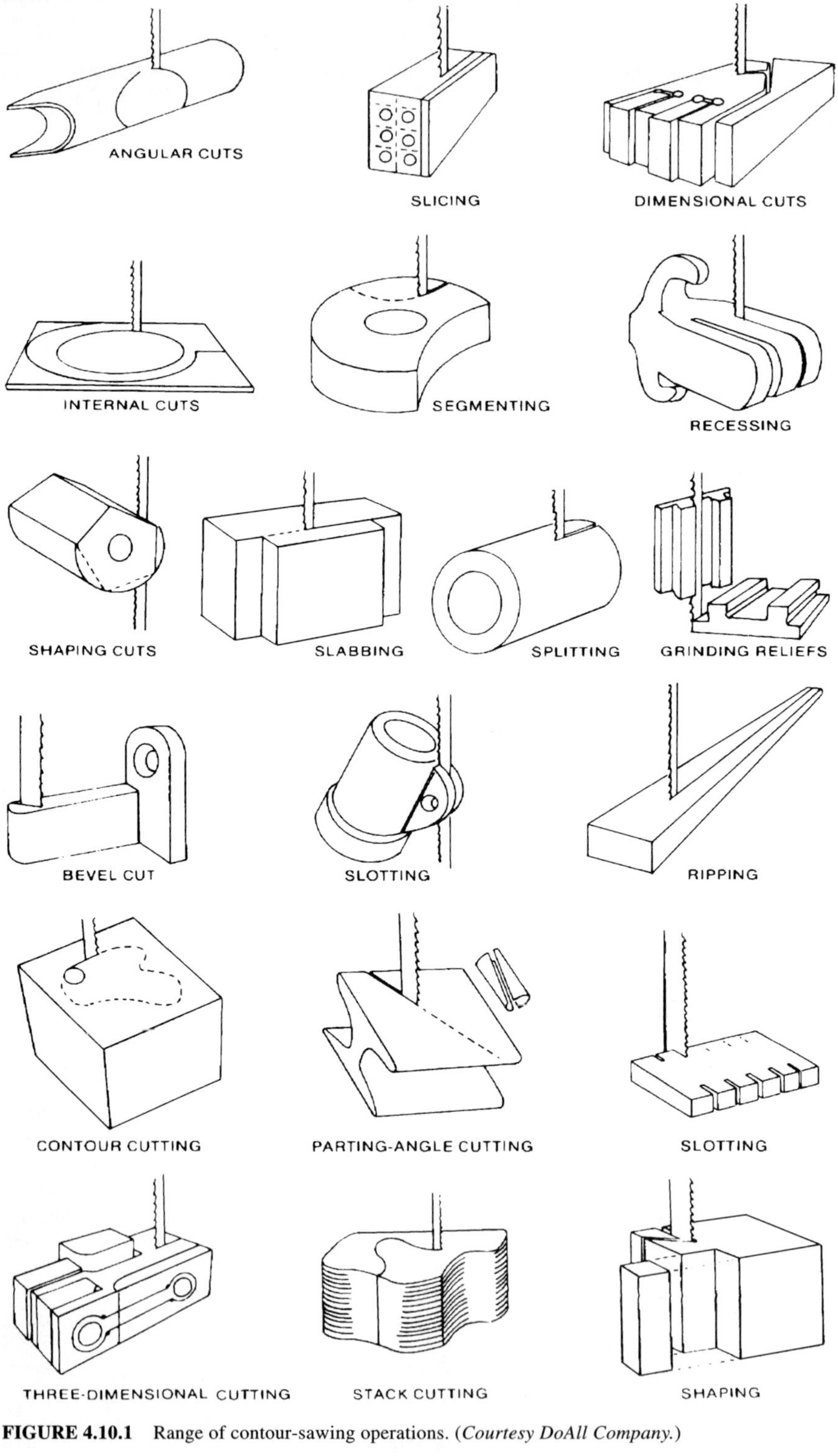

FIGURE 4.10.1 Range of contour-sawing operations. (*Courtesy DoAll Company.*)

Diamond-Edge Sawing

Another method employed in contour-sawing difficult-to-machine materials is diamond-edge or aluminum oxide–edge sawing. These cutting materials require a coolant; the method is not rapid but is suitable for hard or particularly tough materials.

TYPICAL CHARACTERISTICS AND APPLICATIONS

Parts cut from plate or sheet are perhaps the most typical parts produced by contour sawing. Parts that would otherwise be produced by punch-press blanking or flame cutting are good candidates for contour sawing. The process, of course, can be used for three-dimensional parts as well as for parts cut from sheet or plate. However, since the blade is straight where it passes through the work, the cut surface must be straight in at least one dimension. Angular cut surfaces to 45° are obtainable by tilting the machine table.

Compared with milling, contour sawing is rapid because only a narrow slit of material is converted to chips. However, it is less accurate than conventional machining and produces a rougher surface. Hence it is more suitable for rough machining, for which dimensional-accuracy requirements are not severe.

Conventional contour sawing is applicable to all machinable materials up to a hardness of R_c 45. Friction and abrasive-edge methods are used for harder materials, the former being most common for ferrous materials up to 12 mm ($\frac{1}{2}$ in) thick.

The maximum thickness or maximum workpiece dimension cuttable with conventional or abrasive methods depends only on the C-frame opening and load-carrying capability of the saw used. On larger commercially available machines, material as thick as 1.4 m (55 in) can be cut.

Thin sheet pieces can be cut by all contour-sawing methods, especially when the sheets are stacked. For adequate blade life, however, the material must be thick enough so that two or more cutting teeth are in contact with the work at any one instant. This means that for the finest-pitch blades available (32 teeth per inch), the minimum stack or sheet thickness for unstacked sheets is 1.6 mm (0.063 in).

Profile shapes cuttable by contour sawing are practically limitless, except that the sharpness of internal corners (or external corners cut in one pass) is limited by the blade width. (See "Design Recommendations" below.) The smallest normal radius with the narrowest commercial blades is 3 mm ($\frac{1}{8}$ in).

Among typical parts produced by contour sawing are parts for machinery of various kinds (farm machines, trucks, machine tools, special machines, and scientific apparatus), tanks, jigs and fixtures, boats and ships, aircraft, railway equipment, and structural parts. See Fig. 4.10.2 for typical applications of the process.

ECONOMIC PRODUCTION QUANTITIES

It can be seen from the preceding list of typical parts that the contour-sawing process is most applicable to parts with low to moderate levels of production. Since contour-sawing equipment is general-purpose, there are few or no fixture costs, and setup times are very short. There are virtually no one-time costs to amortize. This method, therefore, is not costly for one-of-a-kind or low-level production.

FIGURE 4.10.2 Collection of typical contour-sawed parts. (*Courtesy DoAll Company.*)

Processing time for contour-sawed parts is rapid compared with other machining methods. This advantage adds to the attractiveness of the process at various levels of production. Cutting speeds range from 0.08 to 0.5 cm^2/min (0.5 to 3.2 in^2/min). See Table 4.10.1 for more information on cutting feed rates for some metals.

As a rule of thumb, contour sawing is usually economically advantageous for quantities under 1000. Above that amount, punch-press or other methods usually will entail a lower cost per production lot. Larger quantities may be advantageous in cases for which saw cutting provides more rapid metal removal than other rough-machining methods or for which the cut scrap is usable for some other part. Normally, however, contour-sawed parts are most frequently produced when samples, prototypes, and other low quantities are required.

TABLE 4.10.1 Feed Rate for Contour-Sawing Various Materials, 12.7-mm (½-in) Stock Thickness

Workpiece material	Carbon steel blade	High-speed steel blade
Low-carbon steel; low-carbon, low-alloy steel	10 cm/min (4 in/min)	27 cm/min (10.8 in/min)
High-carbon alloy steel	2.5 cm/min (1 in/min)	12 cm/min (4.7 in/min)
Gray cast iron	19 cm/min (7.5 in/min)	40 cm/min (16 in/min)
Tool steel	5.3 cm/min (2.1 in/min)	15 cm/min (6.1 in/min)
Titanium		3.8 cm/min (1.5 in/min)

SUITABLE MATERIALS

Contour-Sawed Parts

The suitability of materials for use in conventionally contour-sawed parts closely parallels their suitability for other machining operations. Brass, bronze, copper, magnesium, aluminum, cast iron, steel, and unreinforced plastics are contour-cut without difficulty.

Factors that influence band sawability are hardness (as indicated, R_c 45 is a practical limit for conventional sawing), work-hardening tendency, "stringiness" or "gumminess," and uniformity of structure.

Table 4.10.2 lists some commonly contour-sawed ferrous and nonferrous metals. Relative machinability for contour-sawing operations is shown for certain metallurgical conditions and hardness levels of each metal.

Friction-Sawed Parts

Materials too hard for conventional contour sawing usually can be processed advantageously by friction contour sawing. Both nonferrous and ferrous metals and nonmetallic materials can be cut. Tool steels, armor plate, and alloys like stainless steel, Hastelloy, and Tantung G are processed by this method. Hard ferrous materials are the most frequently processed and the most suitable for friction-contour-sawed parts. Ferrous materials tend to retain the frictional heat at the cutting point better than some other metals and have the advantage of being oxidized by a jet of air at the point of cutting. The air cools the blade but actually helps the cutting by oxidizing the heated steel in the same way as the phenomenon which occurs when steel is flame-cut.

Greater hardness aids the process by increasing friction and causing heat to build more rapidly. Soft materials like aluminum, copper, and brass do not work as well, producing a poor edge, although thin sheet aluminum is frequently cut with the friction method and can give satisfactory results.

Reinforced-plastics materials that pose problems with conventional contour sawing are good candidates for the friction method.

Diamond-Edge-Sawed Parts

Hard materials not suitable for friction sawing can be contour-cut with diamond- or aluminum oxide–edged blades. Common materials suitable for this method are tungsten carbide, glass, various ceramics materials, stone, and hardened steel. The last-named material cuts more rapidly with friction sawing but more accurately and with a better finish with abrasive-edge methods.

Brittle materials, such as ceramics, often can be cut with friction contour saws, but edge cracking and shattering can occur with this method, making abrasive-edge sawing more practical. High-melting-temperature hard materials are best suited for the abrasive-edge method.

TABLE 4.10.2 Relative Machinability of Materials Suitable for Contour-Sawed Parts

Material	Designation	Condition*	Brinell hardness no.	Rating
Free-machining carbon steel	1212 (G12120), 1213 (G12130), 1215 (G12150)	CD	150–200	100
Low-carbon steel	1008 (G10080), 1010 (G10100), 1015 (G10150), 1018 (G10180), 1020 (G10200), 1025 (G10250)	HR, N, A, or CD	125–175	87
Medium-carbon steel	1030 (G10300), 1035 (G10350), 1040 (G10400), 1050 (G10500)	HR, N, A, or CD	125–175	84
High-carbon steel	1060 (G10600), 1070 (G10700), 1080 (G10800), 1090 (G10900), 1095 (G10950)	HR, N, A, or CD	175–225	60
Medium-carbon alloy steel	4140 (G41400), 5060 (G50600), 8640 (G86400), 9260 (G92600)	HR, A, CD	175–225	63
Wrought aluminum	2024 (A92024), 6066 (A96066)	CD	30–80	285
Copper	102 (C10200), 110 (C11000)	CD	R_b 60–100†	75
Brass	330 (C33000), 332 (C33200), 342 (C34200), 360 (C36000), 370 (C37000)	CD	R_b 60–100†	120
Wrought magnesium	AZ92A (M11920), AZ80A (M11800), ZK61A (M16610)	A, CD	40–90	385

*CD = cold-drawn; N = normalized; A = annealed; HR = hot-rolled.

†R_b = Rockwell B hardness.

Source: Compiled from data in the *Machining Data Handbook,* 3d ed., by permission of the Machinability Data Center; © 1980 by Metcut Research Associates, Inc. Based on cutting speed with high-speed-steel saw blades.

DESIGN RECOMMENDATIONS

Design restrictions with contour sawing are few. The following are the major rules to be kept in mind by the design engineer:

1. Remember that the process provides contours in two dimensions only; the saw's cutting edge is in a straight line, and the contours that can be produced are those

which can be generated by a straight line. Spherical or radiused surfaces must be obtained in secondary operations.

2. Radii of contours should be as generous as possible. Although internal radii as sharp as 3 mm (⅛ in) can be made with a 3-mm-wide blade, a blade this narrow is much slower in cutting than a wide blade, is susceptible to breakage, and increases cutting costs. The widest blade possible for the radius required should be used to permit a heavier cutting force to be applied. The effect is more important as workpiece thickness increases. External radii can be made sharp by cutting past the corner in two passes. This requires extra time, however, and reduces the usability of the surplus material surrounding the contoured part; it should be avoided if possible. Figure 4.10.3 illustrates the relationship between blade width and minimum radius for conventional sawing. Friction and abrasive-edge sawing have similar limitations.

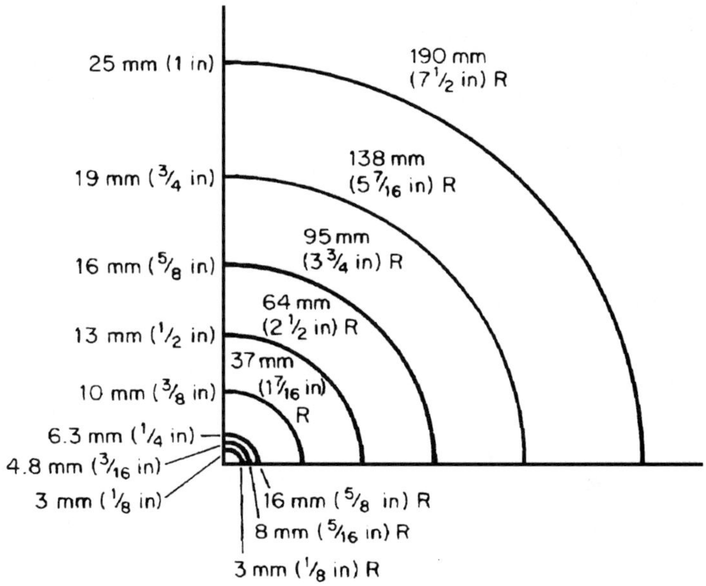

FIGURE 4.10.3 The minimum internal radius of contour-sawed surfaces depends on the blade width.

3. As in all sawing processes, there is kerf loss in contour band sawing. This must be allowed for by the designer and manufacturing engineer. Kerf widths range from 0.8 mm (1/32 in) to about 4 mm (5/32 in), depending on the cutting process, sawtooth set, speed, and other factors.

4. Contour-sawed holes should be avoided if possible. Since normal band-sawing practice involves an endless blade, it is necessary when sawing such shapes to predrill a hole, thread the blade through the hole, and weld the blade. Although these steps are perfectly feasible and not uncommon, they do require extra operations and cost and should be avoided. It is better to design the part with an access slot from one edge. (See Fig. 4.10.4 for an illustration.)

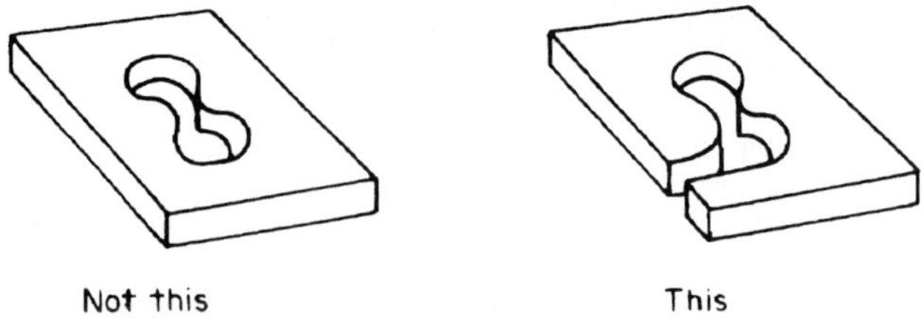

FIGURE 4.10.4 The part on the left requires cutting and rewelding of the band-saw blade.

5. Since contour sawing is essentially a rough-machining process, be sure to allow sufficient stock for finish machining should the function or fit of the part require finish machining. (If at all possible, finish machining should be avoided because it is far more costly than contour sawing.) To allow for surface variations and any work hardening that may have occurred (more likely if friction sawing is used), sawed dimensions should be increased by 1 mm (0.040 in) or more.

RECOMMENDED TOLERANCES

Table 4.10.3 presents tolerances consistent with the capabilities of contour-sawing processes. The closest dimensional values usually require CNC, photoelectric servo equipment, or extreme care and the aid of a magnifying glass if the control is manual. The close surface-finish values in Table 4.10.3 occur only with a fine-pitch blade, low feed, and high cutting speeds.

TABLE 4.10.3 Recommended Tolerances for Contour-Sawed Parts

	Normal	Closest
Closeness of cut contour to prescribed dimension	±0.8 mm (±0.030 in)	±0.25 mm (±0.010 in)
Squareness, flatness, and straightness	0.04 mm/cm (0.004 in/in)	0.02 mm/cm (0.002 in/in)
Surface finish	5–7.5 μm (200–300 μin)	1.5–5 μm (60–200 μin)

CHAPTER 4.11
FLAME-CUT PARTS

Paul Sopko
Airco Welding Products
Murray Hill, New Jersey

THE PROCESSES

Oxygen (Flame) Cutting

This procedure is defined as a group of cutting processes through which the severing or removing of metals is effected by means of the chemical reaction of oxygen with the base metal at elevated temperatures. With oxygen cutting, a small area of the metal is preheated with oxygen and a fuel gas to the ignition temperature of the workpiece material. Then a stream of pure oxygen is directed onto the heated area. The oxygen rapidly oxidizes the workpiece material in a narrow section (the *kerf*) as the molten oxide and metal are removed by the kinetic energy of the oxygen stream. The thermal energy generated by the oxidation is a significant factor in the propagation of the process.

The process uses a torch with a tip whose functions are (1) to mix the fuel gas and the oxygen in the right proportion to produce the initial heating and continuous preheating and (2) to supply a uniformly concentrated stream of high-purity oxygen to the reaction zone for the purpose of oxidizing and removing the molten materials. The torch unit is then moved (manually or by machine) across the material to be cut at a controlled speed sufficient to produce a continuous cutting action. The fuel gases most commonly used for oxygen cutting are acetylene, natural gas, propane, and MAPP gas (a proprietary formulation).

The process is also known as *gas cutting* or *oxyfuel gas cutting*.

Cutting machines range from small, portable units that cut straight lines, bevels, and circles to large, 24-torch computer-controlled flame-cutting systems. Within this range are many small- to medium-sized general-purpose machines as well as machines designed for specific applications. Figure 4.11.1 shows a multiple-torch computer-controlled machine in operation.

FIGURE 4.11.1 Multiple cutting with a numerically controlled flame-cutting machine. (*Courtesy Airco Welding Products Division, The BOC Group, Inc.*)

Plasma Cutting

Newer than oxygen cutting, plasma cutting is a metal-removal process that uses equipment similar to that used for the earlier process. Plasma is the state of matter produced when a gas is subjected to intense electrical and/or thermal forces that cause the molecules to be broken down to ions. When a high-voltage electric arc is used, the arc heats the gas until some of its atoms momentarily lose one or more electrons (this process is called *ionization*). As the ionized gas is expelled from the torch nozzle, the atoms regain their missing electrons and release energy previously absorbed by the ionization process. This recombination energy is added to the energy of the electric arc to produce an intensely hot flame. The plasma flame from the torch tip hits the workpiece in a thin, high-energy jet. Metal in the path of the plasma jet is melted or vaporized and washed through the kerf. A plasma-cutting machine can make fast, square, clean cuts in all types of electrically conductive materials. Its primary use is for nonferrous materials.

Guidance of the Torch

The key to successful application of the flame-cutting process is smooth, accurate guidance of the cutting tool, the torch-tip combination. This led to the introduction and widespread use of the cutting machine.

A *tracing template* is a full-size pattern used to guide the cutting of required shapes from metal plate. It can be compared to a tailor's pattern used to cut fabric. In electronic tracing, a line or edge template drawing provides a path for an electrooptical tracer to follow that, through a servo system, causes the machine to move the torches along a path of the same shape. The resulting cut part will be identical to the template if the effect of the diameter (kerf) of the cutting-oxygen stream is neglected. For this reason, *kerf compensation* is used to attain the correct dimension.

Computer and Numerical Control

Computer- and numerical-controlled flame cutters use neither template nor drawings but, instead, utilize perforated or magnetic tape, magnetic disks, or integrated memory circuits that carry, in digital form, the description of the part to be cut. The memory device also has "commands" relating to cutting speeds, burner ignition, register points, and other ancillary functions. The computer can develop an optional layout of cut parts on the plate stock to achieve maximum utilization of the material. Its display can prompt the operator to perform the correct sequence of steps in setting up and operating the flame cutter.

CNC units can store libraries of preprogrammed shapes that can be used to simplify the programming of new parts. Setup is greatly accelerated.

TYPICAL CHARACTERISTICS AND APPLICATIONS

Flame-cut parts are normally made from flat plate. Shapes that can be oxygen- or plasma-cut vary from simple rectangles and circles to complex curves and contours. Straight-line segments can be of any length. The machine components carrying the cutting torches ride on tracks, and track extensions can be added as necessary.

Contoured shapes may be simple arcs or circles or may employ compound, complex curves. Round and irregular-shaped holes are also feasible. Figure 4.11.2 illustrates typical examples of flame-cut shapes.

The flame-cut edge of a workpiece is normally perpendicular to the plate surface. It also may be beveled, with or without lands, or double-beveled (beveled top and bottom). Beveled edges are normally ready for welding immediately after flame cutting. Beveling with oxyfuel torches requires special steps, but beveling with plasma is routine; in fact, bevel cuts made with plasma are often superior to comparable perpendicular plasma cuts.

FIGURE 4.11.2 Typical flame-cut parts. (*Courtesy Airco Welding Products Division, The BOC Group, Inc.*)

Plasma cutting of stainless steel gives favorable results in edge quality. The face of the cut is smooth and clean, the edges are sharp, and there is almost no slag. For carbon steel, however, the quality of the edge may not be as good with plasma as with oxyfuel gas.

Plate thicknesses of commercial significance are generally within the range of 3 to 300 mm ($\frac{1}{8}$ to 12 in), but cuts in stock as thick as 2.4 m (94 in) are possible. Sheets less than 3 mm ($\frac{1}{8}$ in) thick are usually flame-cut only in stacks.

Plates up to 4 m (12 ft) wide and of any practical length can be flame-cut with most equipment. In optical, tracer-controlled machines, maximum sizes usually range from 1.2 to 4 m (48 to 144 in). Machines of 18-m (60-ft) width capacity are also available and are guided by numerical control. The minimum size for flame-cut parts depends on practical and economic factors. Parts with a major dimension less than 25 or 50 mm (1 or 2 in) are usually more economically made by other methods.

Conventional oxygen cutting is limited to ferrous metals and titanium, while plasma cutting is applicable to any metallic material. Plasma-cut parts may not be as thick as parts cut with oxyfuel. The principal benefits of plasma cutting are realized on thicknesses of 32 mm ($1\frac{1}{4}$ in) or less. Plasma cutting of thicknesses over 64 mm ($2\frac{1}{2}$ in) entails a reduction in edge quality and an increased energy requirement that may offset the increased speed of the process. For thicknesses over 75 mm (3 in), plasma cutting is generally slower than oxyfuel cutting; the maximum thickness is 115 mm ($4\frac{1}{2}$ in).

With oxygen cutting, a large quantity of heat is liberated in the kerf. Much of this thermal energy is transferred to the area adjacent to the kerf at a temperature above the critical temperature of steel. Since the torch is constantly moving forward, the source of heat quickly moves on and the mass of cold metal near the kerf acts as a quenching medium, rapidly cooling the heated metal. The steel hardens to a degree that depends on the amount of carbon and alloying elements present as well as on the thickness of the material being cut. With the more conductive alloy steels, quench cracks may appear at the surface.

The depth of the heat-affected zone ranges from 0.8 to 6 mm ($\frac{1}{32}$ to $\frac{1}{4}$ in). There is also a shallower zone of measurably increased hardness. From 0.4 to 1.5 mm ($\frac{1}{64}$ to $\frac{1}{16}$ in) deep, it exhibits an increased hardness of 30 to 50 points on the Rockwell C scale. This hardness can be removed by annealing but not by stress relieving. With plasma cutting these heat-affected zones are narrower.

Typical applications of flame-cut parts include shipbuilding, building construction and equipment components, pressure and storage vessel parts, blanks for gears, sprockets, handwheels, and clevises. The most common applications probably are heavy-walled parts to be welded as part of some frame or structure.

Specialized flame-cutting machines are used in the steel industry to sever billets, blooms, slabs, or rounds (with either hot or cold cutting). As in beveling, both plasma and oxyfuel gas also can be used for gouging and grooving, but plasma is simpler. Gouging is used extensively to remove deep defects in steel revealed by scarfing or by radiographic, magnetic, ultrasonic, and other inspection methods. Among other gouging applications are removing tack welds, defective welds, blowholes, or sand inclusions in castings, welds in temporary brackets or supports, flanges from piping and heads, and old tubes from boilers. Gouging also is used in demolition work and in the preparation of plate edges for welding.

ECONOMIC PRODUCTION QUANTITIES

Flame cutting is generally more economical per inch of cut than other machine operations used to generate comparable shapes in steel plates. It is most advantageous for lower-quantity production. A run long enough to justify a substantial tooling investment would give casting, forging, or stamping a competitive advantage in many cases. The prime advantages of flame cutting are greatly reduced cost of tooling, minimum setup time, flexibility and simplicity of operation, and short lead time.

A not uncommon application of flame cutting is in maintenance work or one-of-a-kind production, in which the operator lays out the part outlines on the plate to be cut and then operates the cutting torch freehand or perhaps with the aid of a straightedge or compass.

Even in repetitive production, the special tooling is quite simple and inexpensive. It consists of a drawing (optical tracing), a template, or, with a computer-controlled machine, a stored program. The cost of these items is negligible compared with blanking or forging dies and is even far below that of a simple pattern for sand casting.

This low cost of tooling and the fast cutting action of the flame or plasma method ensure its economic advantage for low-quantity production. With multiple-torch machines (up to 16 torches) and stack cutting, the advantage extends to medium and, in some cases, to high levels of production.

Typical cutting rates are 20 to 30 in/min and range up to 120 in/min for oxyfuel cutting. For plasma cutting, rates are typically 50 to 125 in/min but can be as high as 300 in/min. Rates vary inversely with stock thickness. Use of multiple cutting torches effectively multiplies these rates when a series of parts is to be cut.

MATERIALS SUITABLE FOR FLAME CUTTING

Low-carbon steel and wrought iron are the best materials for flame cutting. Medium-carbon and low-alloy, low-carbon steels also are usually satisfactory. Processibility decreases as the carbon and alloy content increases. Nonferrous metals are normally not suitable for flame (oxygen) cutting because they do not oxidize readily. Galvanized steel, especially in thin gauges, also is not well adapted to flame cutting.

Conventional flame cutting is largely an oxidation process. As the amount and number of alloying materials increase, the oxidation rate decreases from that of pure iron. For ferrous metals with a high alloy content, such as cast iron and stainless steel, variations in normal oxygen-cutting methods must be used. These include preheating, torch oscillation, flux addition, and iron-powder addition to the flame.

Plasma cutting can be used on almost any metal, including metals difficult to flame-cut. Stainless steel and aluminum are the materials most frequently cut by the plasma process. Plasma cutting also makes fast, clean cuts in the following electrically conductive materials: brass, bronze, nickel, tungsten, mild steel, alloy steel, tool steel, Hastelloy, copper, cast iron, molybdenum, Monel, and Inconel.

Table 4.11.1 summarizes applicable cutting methods for the most commonly processed materials.

TABLE 4.11.1 Applicable Cutting Methods for Commonly Flame-Cut Materials

Material	Method
Mild steel	Oxyfuel or plasma
Low-alloy steel	Oxyfuel or plasma
Titanium	Oxyfuel or plasma
High-alloy steel	Plasma
Stainless steel	Plasma
Copper	Plasma
Aluminum	Plasma

DESIGN RECOMMENDATIONS

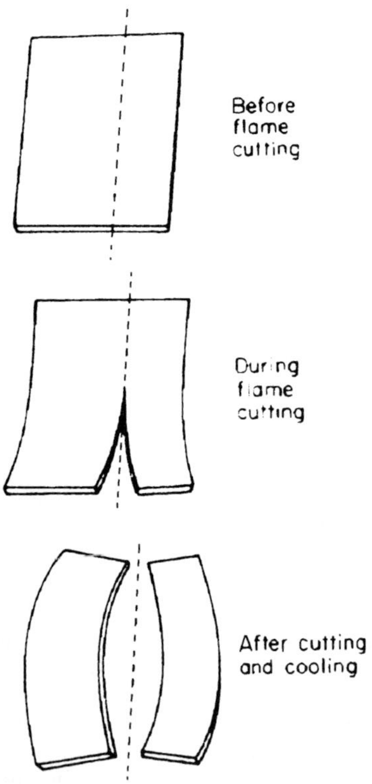

FIGURE 4.11.3 Deformation that occurs in flame cutting shown greatly exaggerated. Wide pieces (such as the left-hand piece) deform less than narrow pieces (such as the right-hand piece) because the larger unheated area resists deformation.

Heat Effects

It must be recognized that the flame-cutting process involves a chemical reaction between iron and oxygen with a resulting heat effect. With this condition existing, narrow sections may pick up extra heat and not be capable of resisting warpage; thin plates may buckle, or inadequate stock may not provide sufficient material for the reaction (see Fig. 4.11.3). However, water spray at the cut can eliminate some distortion problems on thin plate.

Usually distortion is severe enough to cause problems only in plate thinner than 8 mm ($\frac{5}{16}$ in). It is also more severe when long, narrow sections are cut. Therefore, it is advisable to design narrow sections to be as short as possible. (See Fig. 4.11.4.) Parts designed for balanced cutting heat by cutting on both sides will also have less distortion.

Kerf

It is important to allow for kerf when designing parts for flame cutting. Kerf widths range from approximately 1.3 mm (0.050 in) for 19-mm- ($\frac{3}{4}$-in-) thick material to 4 mm (0.150 in) for 200-mm- (8-in-) thick material. When templates are used, the size must be adjusted to allow for the kerf width.

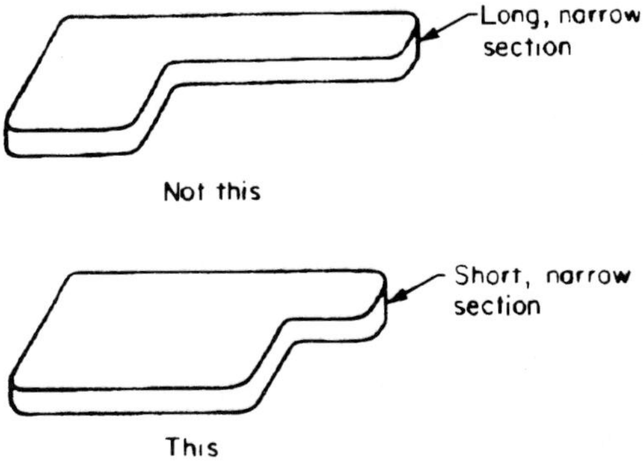

FIGURE 4.11.4 Design narrow sections to be as short as possible to avoid distortion.

Minimum Radii

Small radii, especially internal radii, are difficult to flame-cut. External corners can be sharp if the cutting torch moves past the corner, but this may reduce the usability of the plate material from which the workpiece is cut. Generous internal and external radii should be allowed whenever possible. Figure 4.11.5 illustrates normal minimum allowable radius dimensions for various plate thicknesses.

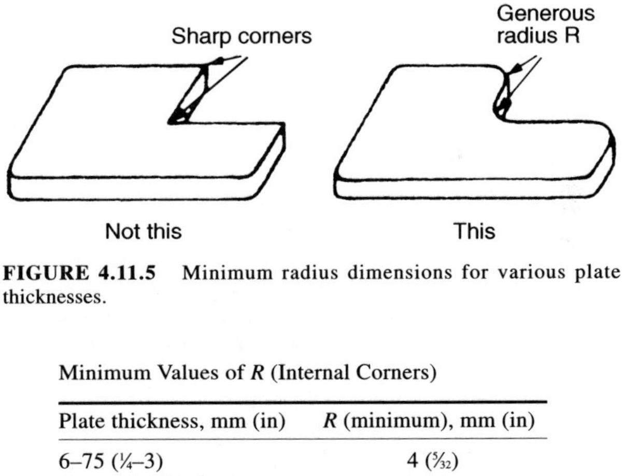

FIGURE 4.11.5 Minimum radius dimensions for various plate thicknesses.

Minimum Values of *R* (Internal Corners)

Plate thickness, mm (in)	*R* (minimum), mm (in)
6–75 (1/4–3)	4 (5/32)
Over 75–150 (3–6)	5 3/16)
Over 150–200 (6–8)	6 (1/4)

Sweeping Curves

Generous fillets and sweeping curves produce a pleasing design effect, but if they are not required for structural strength, their omission may result in considerable cost savings. A curved edge entails a large scrap loss and extensive cutting from even the most ideally sized rectangular plate. Rectangular shapes normally will result in a lower overall cost.

Machining Allowance

When flame-cutting tolerances do not meet the requirements of the finished part, flame cutting can be used to remove the major portion of the material, leaving enough stock for milling, drilling, boring, grinding, or other machining operations. When using flame cutting for rough stock removal and machining for finishing to final tolerances, allow at least 1.5 mm ($\frac{1}{16}$ in) of stock for finishing to minimize cutter damage due to the hardened material at the flame-cut edge.

Nested Parts

Designing parts so that they can be nested on the plate and employ common edges will make one cut do the work of two. It also will improve utilization of the plate material. This principle is the same as that which applies to metal stampings and is illustrated in Fig. 3.2.13.

Minimum Holes Size and Slot Widths

Holes and slots can be produced by flame cutting, but because of the kerf, there is a minimum size for such openings. This size varies with the plate thickness and is summarized in Table 4.11.2.

Specifying Quality of Cut

Not all flame cutting must be of the same quality. For workpieces for which either dimensional accuracy or appearance can be sacrificed, higher speeds and longer drags can be used. A shop can set up two or even three standards for quality, and designers can specify the quality required for each cut. The highest-quality level could specify

TABLE 4.11.2 Minimum Slot and Hole Sizes for Flame-Cut Parts

Plate thickness	Minimum slot size, width × length	Minimum hole diameter
6–35 mm ($\frac{1}{4}$–$1\frac{3}{8}$ in)	8 × 19 mm ($\frac{5}{16}$ × $\frac{3}{4}$ in)	16 mm ($\frac{5}{8}$ in)
Over 35–75 mm ($1\frac{3}{8}$–3 in)	9.5 × 22 mm ($\frac{3}{8}$ × $\frac{7}{8}$ in)	22 mm ($\frac{7}{8}$ in)
Over 75–100 mm (3–4 in)	13 × 25 mm ($\frac{1}{2}$ × 1 in)	32 mm ($1\frac{1}{4}$ in)
Over 100–127 mm (4–5 in)	19 × 32 mm ($\frac{3}{4}$ × $1\frac{1}{4}$ in)	41 mm ($1\frac{5}{8}$ in)
Over 127–165 mm (5–$6\frac{1}{2}$ in)	25 × 44 mm (1 × $1\frac{3}{4}$ in)	54 mm ($2\frac{1}{8}$ in)
Over 165–200 mm ($6\frac{1}{2}$–8 in)	38 × 64 mm ($1\frac{1}{2}$ × $2\frac{1}{2}$ in)	64 mm ($2\frac{1}{2}$ in)

Source: From *Tooling and Production,* March 1974.

flame-cut parts with square, smooth edges cut to close limits. This grade would apply to parts to be mated without further machine work or to parts to be joined by automatic welding. The second grade could call for good appearance but with lower standards of dimensional accuracy. A proper use for this grade of cut quality would be for lightening holes, the external edges of the equipment, or edges on which manual welding is to be performed. When further machine work is to be done on the cut surfaces, a third-grade cut might be used. In such cases, a mere severance cut might be sufficient.

When more detailed means of specifying cut quality are required, the American Welding Society (P.O. Box 351040, Miami, Fla. 33135) has a series of specifications that cover flatness, angularity, roughness, top-edge rounding, and notches. In any case, it is advisable that the design of the part to be flame-cut indicate the quality level required. Only the quality level needed for the function of the part should be specified. Specifying overly high quality levels unnecessarily increases product cost.

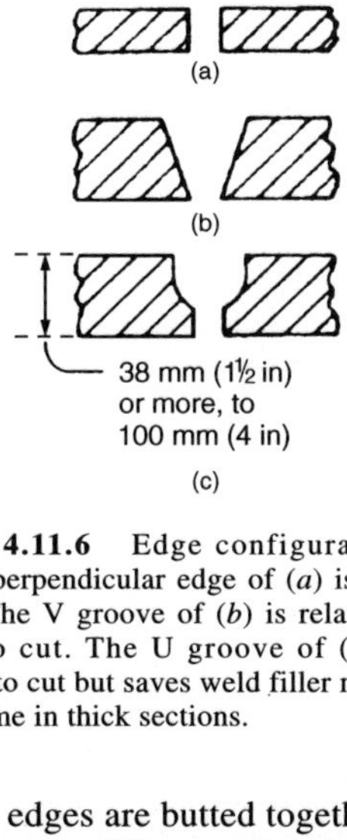

FIGURE 4.11.6 Edge configurations. The square or perpendicular edge of (*a*) is more economical. The V groove of (*b*) is relatively inexpensive to cut. The U groove of (*c*) is more expensive to cut but saves weld filler material and welding time in thick sections.

Edge Design

Square edges as shown in Fig. 4.11.6*a* are most economical, but some rather complex edge designs can be produced in one or two passes by using multiple torches. Edge design for plate to be welded depends on the thickness of the weld, the required access for the welding gun to get root penetration, and the welding process to be used. Proper edge design reduces both filler-metal usage and welding labor.

Heavy plates over 38 mm ($1\frac{1}{2}$ in) thick are usually prepared for welding with chamfered or grooved edges. V grooves as shown in Fig. 4.11.6*b* are easy and inexpensive to cut but require more filler metal. J grooves prepared on each plate produce U grooves (Fig. 4.11.6*c*) when the edges are butted together. These contoured grooves are more costly to flame-cut but use less filler metal and are less expensive to weld.

Curved edges with bevels or grooves require special equipment or additional operator skill levels. They are therefore more expensive than perpendicular cuts and should be avoided if they are not necessary.

TOLERANCE RECOMMENDATIONS

Table 4.11.3 presents recommended dimensional tolerances for normal production conditions for various plate thicknesses. Closer tolerances can be obtained if conditions are optimal and extra care and time can be taken. However, the values shown should be used for most applications.

TABLE 4.11.3 Recommended Dimensional Tolerances for Flame-Cut Parts

Plate thickness	Length, width, or hole or slot location	Hole or slot dimension
6–35 mm (1/4–1 3/8 in)	±1.5 mm (±0.060 in)	±0.8 mm (±0.030 in)
Over 35–75 mm (1 3/8–3 in)	±1.5 mm (±0.060 in)	±1.2 mm (±0.045 in)
Over 75–100 mm (3–4 in)	±2.4 mm (±0.090 in)	±1.5 mm (±0.060 in)
Over 100–127 mm (4–5 in)	±2.4 mm (±0.090 in)	±2.4 mm (±0.090 in)
Over 127–165 mm (5–6 1/2 in)	±2.4 mm (±0.090 in)	±3.0 mm (±0.125 in)
Over 165–200 mm (6 1/2–8 in)	±3.0 mm (±0.125 in)	±3.0 mm (±0.125 in)

Note: Under the most precise conditions of template or computer-controlled cutting with a highly accurate template and no significant heat distortion, dimensions of ±0.6 mm ±0.024 in) can be held on plate 75 mm (3.0 in) thick or less.

CHAPTER 4.12
INTERNALLY GROUND PARTS

Roald Cann
Bryant Grinder Corporation
Springfield, Vermont

THE PROCESS

Essentially, internal grinding removes material from the inside of a hole. Generally, the part is caused to revolve about the desired axis of the hole, and a grinding wheel that revolves about its own axis is placed inside the hole and fed radially to contact the part. If size or external shape makes rotation of the part about the hole axis impractical, the part is kept stationary, and the grinding wheel is given a planetary motion about the hole axis in addition to its rotation about its own axis, as in a jig grinder. Figure 4.12.1 illustrates a typical internal grinding machine.

Hole-size control is often accomplished by *diamond sizing,* feeding the wheel at a fixed distance from where it was when it was last dressed by the relatively fixed diamond.

Profiles are usually dressed directly into the wheel and then reproduced in the part by a purely radial feed. A true spherical profile is usually generated directly with the corner of a cylindrical wheel whose axis intersects the axis of revolution of the part. Profiles that are longer than a practical wheel length are produced by moving a shorter barrel-shaped wheel along the required path by means of a template or a CNC program. Profiles are dressed into wheels by profiled diamond rolls or by mechanisms that use pivots, templates, or a CNC program to impart the required path to a single-point diamond.

TYPICAL CHARACTERISTICS OF INTERNALLY GROUND PARTS

Characteristics Likely to Require Internal Grinding

The most frequent reason for internal grinding is that the part material is too hard to be machined economically by a nonabrasive process and must be machined after hardening to correct heat-treatment distortion and to remove scale or decarburized material.

FIGURE 4.12.1 Internal grinding machine with an outside-diameter clamping diaphragm chuck and two wheel spindles. (*Courtesy Bryant Grinder Corporation.*)

Other frequent material-related reasons are that materials are too fragile for other processes, are smeared by other processes (electric-motor or synchro lamination stacks, porous bearings), or are too abrasive for economical tool life (some filled plastics.)

Thin walls either all around or partway around parts or interruptions such as splines or cross holes often dictate an abrasive process rather than a cutting process. A complex surface form with a tight form tolerance often can be ground most economically because of the ease with which the grinding-wheel form can be restored by the dresser as compared with the difficulty of maintaining the form with either a form tool or a template-guided tool. Because a grinding wheel has its cutting surface continually renewed by dressing, and because it can remove very small and closely controlled amounts of stock, internal grinding can be more economical than precision boring when tolerances fall below 0.013 mm (0.0005 in) on size, 0.0025 mm (0.0001 in) on roundness, and 0.5 μm (20 μin) on surface finish.

Characteristics that Influence the Choice between Internal Grinding and Honing

Some of the above-mentioned characteristics also make honing a suitable or even a preferred process. (For more details, see Chap. 4.16.) A brief review of the strengths and weaknesses of both processes follows. Internal grinding becomes difficult when hole length exceeds 1200 mm (48 in) or 6 times the hole diameter, whereas such lengths present no problem to honing. Internal production grinding is practical for hole diameters from 2 m (80 in) to 1 mm (0.040 in), although toolroom work occasionally has gone a little smaller; honing normally is not used over 1-m (40-in) or under 6-mm (0.25-in) diameter. Both processes are used with interruptions such as splines or cross holes, but honing has less tendency to round over their corners or to remove too much material in their vicinity. Both processes are suitable for blind holes, but internal grinding can reach closer to the blind end. Honing requires a 6- to 9-mm- ($\frac{1}{4}$- to $\frac{3}{8}$-in-) long relief, whereas grinding is practical with 0.8 mm (0.030-in) relief or can produce a 0.25-mm (0.010-in) corner radius, although a 3-mm- ($\frac{1}{8}$-in-) long relief is preferred.

Although both processes can produce fine surface finishes, grinding is generally practical only down to finishes of 0.13 to 0.2 μm (5 to 8 μin) AA, whereas honing continues to be practical down to 0.05 μm (2 μin) or less. However, only grinding can produce precise location and orientation of the hole axis with respect to other surfaces, although both processes can produce excellent roundness and straightness.

Typical Parts and Materials

Internally ground holes may be cylindrical or conical, have faces or shoulders inside the hole, and have a noncircular cross section as in a vane pump ring, or have an axial profile as in a gun chamber or ball-bearing track.

The following list shows the range of reasonably common applications of internal grinding; inner-ring bores and outer-ring tracks of ball and roller bearings, bores and inside faces of universal-joint needle-bearing cups, bores of automotive hydraulic-valve-lifter bodies, bores and tapered seats of fuel-injection-nozzle bodies, bores and external faces of gear or gear clusters, bores of connecting rods, bores of valve rocker arms, noncircular bores of hydraulic-vane pumps or refrigeration compressors, internal and external spheres of ball joints and axial-piston-pump slippers, internal spheres of constant-velocity universal-joint cages and outer races, bores of drill bushings, bearing sleeves, bores and faces of machine-tool spindles and metal or ceramic magnets or cores, contoured bores of gun chambers and wire-drawing dies, and tapered-joint bores of glass plumbing. Most parts fall within a hole-diameter range of 1 to 100 mm (0.040 to 4 in), but many fall between 100 and 500 mm (4 and 20 in), and some reach 80 in in diameter. Materials are usually through-hardened or case-hardened wrought steels but include cast steels, nitrided steels, stainless steels, cast iron, soft magnetic-lamination iron, bronze, aluminum, the various tungsten and titanium carbides, titanium, Alnico, ferrite, graphite, ceramics, glass, and glass- or mineral-filled plastics. Figures 4.12.2 and 4.12.3 illustrate typical internally ground parts.

ECONOMIC PRODUCTION QUANTITIES

Since there usually are no large initial costs for such items as dies or molds, there usually is no minimum lot size below which internal grinding would be uneconomical. Therefore, quantities as small as one piece are suitable for internal grinding. At the other extreme, production of thousands of parts per hour for many years also may be economic. Lot size and desired production rate do influence the decision between acquiring machinery and subcontracting to the possessor of machinery or between acquiring manually operated machinery and acquiring fully automatic machinery. Exceptions, in which initial costs might be significant, are cases requiring a formed diamond-dressing roll, an axial-profile template, or a cross-section-contour cam. If

FIGURE 4.12.2 Typical internally ground parts. The internal surfaces of these parts are routinely ground despite thin walls, interruptions, and special contoured cross sections. (*Courtesy Bryant Grinder Corporation.*)

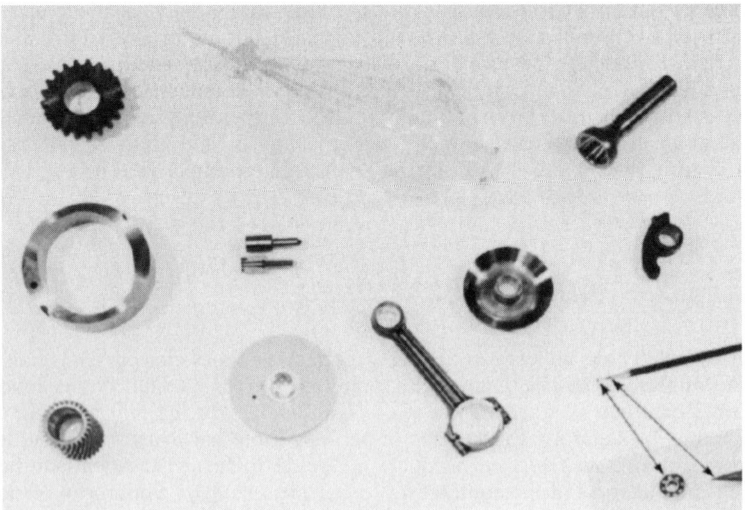

FIGURE 4.12.3 Typical internally ground parts. (*Courtesy Bryant Grinder Corporation.*)

production quantities justify it, even usual jobs will benefit from initial investment in optimal part-holding and -locating tooling and perhaps in a costly diamond or cubic boron nitride grinding wheel if the material is difficult to grind.

For truly high production quantities, as with automotive-bearing parts, handling is fully automatic at rates between 100 and 600 parts per hour per machine on machines requiring one operator for every 3 to 7 machines and one setup person for every 10 to 20 machines. On such machines, changeover tooling is required for each additional part within a family; changeover time ranges between 10 and 60 min if the tooling is preset and grinding-cycle adjustments are made to prerecorded settings.

SUITABLE MATERIALS

A combination of materials such as inserts, welds, solders, or adhesives interspersed with the main material may add to grinding time by requiring frequent dressing of the wheel if there is no compromise wheel that neither wears too rapidly when in contact with the harder material nor clogs too rapidly when in contact with the softer material. Chromium plating combined with hardened steel is a fine choice. Copper or cadmium plating with hardened steel can cause problems, as can soft-steel inserts in or a decarburized layer on hardened steel and some plastic fillers in motor lamination stack slots. For similar reasons, part design should allow inexpensive removal of such materials as soft black scale and paint from internal surfaces that are to be ground.

Sometimes compromise is necessary; e.g., air-hardening steels have good dimensional stability but are more resistant to stock removal than the less stable oil- or water-hardening steels.

Hardened alloy steels become increasingly difficult to grind with increasing hardness of the carbide or the carbide-forming alloying element, i.e., from chromium to molybdenum to tungsten to vanadium. An increasing percentage of carbide and therefore of alloying elements (if sufficient carbon is assumed to be present) also means increasing difficulty. Within any one alloy, increasing R_c hardness also increases grinding difficulty. However, increasing nickel content only slightly increases grinding difficulty. Soft and ductile materials may require very little force, but they do require a wheel that wears rapidly in order to prevent a buildup of part material in the pores of the wheel.

Some soft materials, e.g., austenitic stainless steels, are tricky to grind because they act soft during heavy cuts but work-harden severely during light cuts. The free-machining characteristics of resulfurized steels also tend to benefit grinding, but such benefits are less clearly indicated with leaded steels, which may even have such a nonuniform lead distribution as to cause localized excess lead concentrations to load the wheel. Through-hardened steels distort less than surface-hardened steels, though they grow more, but growth is not a problem because the amount is small and somewhat predictable. Tempering does provide some stress relieving or equalization and, combined with proper quenching techniques and fixtures, can do much to minimize distortion.

Specific grindability values for materials suitable for internal grinding can be found in Chap. 4.15, "Flat-Ground Surfaces," since flat surface and internal grindability are closely parallel.

DESIGN RECOMMENDATIONS

Standardization

When a product has a number of internally ground components, the designer should strive for maximum part quantity by standardizing on one part design in place of two or more different designs whenever this would achieve an economy while still satisfying all functional requirements. When parts cannot be identical, much can be gained by designing them to be members of a family with as many common characteristics as possible. This procedure usually will permit consideration of the total family volume when justifying investment in some degree of automation of the process, and it will minimize the amount of setup or changeover time.

Holding, Locating, and Driving Surfaces

Parts that are to be held in a chuck must present surfaces to the gripping portions of the chuck and to the locating portions of either the chuck or the chuck-loading fixture. Thin-walled parts will show a greater influence of outside-diameter quality on inside-diameter quality. Face-clamping chucks are more forgiving of outside-diameter quality but require better face flatness to prevent a misshapen inside diameter unless either the inside diameter is very short or the part is very rigid.

If the part is magnetically driven for internal centerless grinding, the end face of the part must provide sufficient area for good magnetic holding. An area equal to 25 percent of the area to be ground internally usually will be sufficient.

Length of Hole to Be Ground

Deep, narrow holes necessitate the use of less rigid spindles or quills. These increase grinding time and the possibility of waviness and chatter. A rule of thumb is to avoid hole length-to-diameter ratios in excess of 6. The problem is less severe as diameter increases and more severe as material grindabiity decreases. The rule applies to the reach distance to the hole as well as to the ground portion, as shown in Fig. 4.12.4.

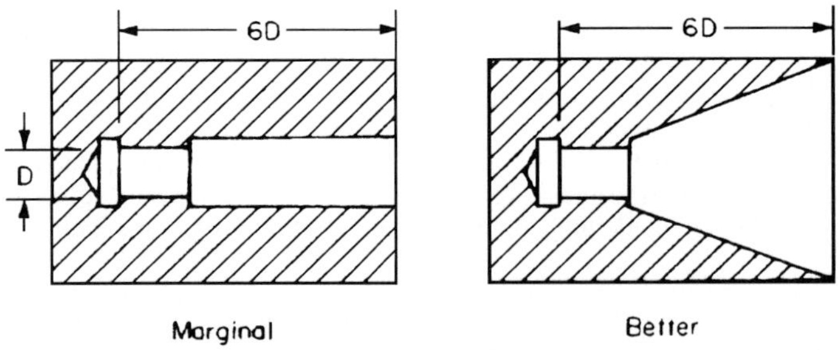

FIGURE 4.12.4 Holes deeper than 6 times diameter and with overly long-reach distances to the ground hole should be avoided unless the area is wide enough to provide rigid support for the wheel spindle.

Blind Holes

Grinding may take longer for a blind hole because the flow of coolant in the region of contact between the wheel and the part may be marginal, forcing a reduction in the infeed rate, and the flow may be nonuniform, making it more difficult to keep straightness and taper within tolerance. When necessary, however, a blind hole is practical, provided a sharp corner is not required at the blind end. While a corner radius of 0.25 mm (0.010 in) is possible, a previously machined corner relief or neck is better. If the desired profile is a straight line, a relief of at least 3-mm ($\frac{1}{8}$-in) axial length will minimize straightness and taper problems by permitting sufficient reciprocation of the wheel to correct minor deviations. (See Fig. 4.12.5.)

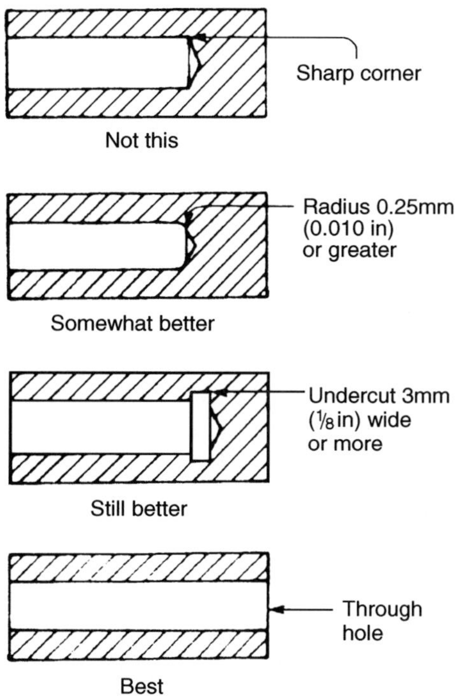

FIGURE 4.12.5 Treatment of blind holes.

Circumferential Interruptions

These interruptions should be avoided if possible. The less rigid the quill or spindle, the greater the tendency of the wheel to remove more stock in the vicinity of a cross hole or to round the corners of a keyway or spline slightly. When such deviations would exceed the tolerance, they usually can be reduced at the cost of longer grinding time by reducing the infeed rate and lengthening spark-out time.

Axial Interruptions

Similarly to the circumferential case, a reciprocating wheel has a tendency to remove more stock near an interruption in the axial line of contact such as a circumferential valve groove or a relief. In contrast to outside-diameter grinding, it is usually best not to put a relief in the middle of a long inside diameter even if only the ends of the inside diameter are to be used. If the part material shows reasonable grindability with the wheel being used, axial interruptions will incur a grinding-time penalty only when the hole diameter is less than 2 in and at the same time the hole length-to-diameter ratio exceeds 3. This effect is usually negligible with plunge grinding (no reciprocation).

Amount of Stock

Although not directly specified by the designer, the amount of stock is often the result of design decisions. Table 4.12.1 presents suggested stock allowances for internal grinding. Since many variables, such as type of material and heat treatment, quality of previous operations, and rigidity of the internal-grinding setup, will affect the minimum amount of stock required, some experimenting is worthwhile if the production quantity is high. Table 4.12.1 can be used to find a starting point. Careful design and process planning may permit leaving less stock than shown in the table.

Use of a Liner or Insert

In some cases, problems related to material, locating surfaces, or length of reach can be eliminated by grinding a liner that is later pressed into or otherwise made integral with the less-suitable actual part. The advantages must be weighed against the cost of an extra part and the effects of tolerance buildup. Chromium plating and flame or plasma spraying are other approaches to be considered.

Internal Faces

If possible, leave a relief (the larger the better) in the center of an internal face. One reason for having the relief is to facilitate coolant flow into the contact zone. Another is to permit the use of a cupped wheel rather than a wheel with nearly full end-face contact and to permit this cupped wheel to have a diameter greater than one-half the part-face diameter. Nearly full wheel–end-face contact does not grind well because of poor coolant access and low wheel surface speed near the wheel center. A cupped wheel grinds much better but will leave an unground region in the center of the face unless the rim of the cup passes through the center. (See Fig. 4.12.6.)

Hole with Confined Entrance

The entrance must be large enough to pass a reasonable wheel size and avoid interference with the quill or spindle even after the wheel has worn somewhat. (See Fig. 4.12.7.)

TABLE 4.12.1 Suggested Stock to Be Left for Internal Grinding*

Diameter of hole, in	Length of hole, in													
	5/16	1/2	3/4	1	1 1/2	2	2 1/2	3	3 1/2	4	5	6	7	8
1/8	0.004	0.004												
	0.005	0.005												
1/4	0.005	0.005	0.006	0.006										
	0.006	0.006	0.008	0.008										
1/2	0.005	0.005	0.006	0.006	0.008	0.008	0.006	0.006	0.008	0.008	0.010	0.010		
3/4	0.006	0.006	0.008	0.008	0.010	0.010	0.010	0.010						
	0.008	0.008	0.010	0.010	0.012	0.012	0.012	0.012						
1	0.008	0.008	0.008	0.008	0.010	0.010	0.010	0.010	0.010	0.010				
	0.010	0.010	0.010	0.010	0.012	0.012	0.012	0.012	0.012	0.012				
1 1/2	0.008	0.008	0.010	0.010	0.010	0.012	0.012	0.012	0.012	0.012	0.012			
	0.010	0.010	0.012	0.012	0.012	0.015	0.015	0.015	0.015	0.015	0.015			
2	0.010	0.010	0.010	0.012	0.012	0.012	0.015	0.015	0.015	0.015	0.015	0.015	0.015	
	0.012	0.012	0.012	0.015	0.015	0.015	0.018	0.018	0.018	0.018	0.018	0.018	0.018	
2 1/2	0.012	0.012	0.012	0.012	0.015	0.015	0.015	0.015	0.018	0.018	0.018	0.018	0.018	0.018
	0.015	0.015	0.015	0.015	0.018	0.018	0.018	0.018	0.020	0.020	0.020	0.020	0.020	0.020
3	0.012	0.012	0.012	0.015	0.015	0.015	0.015	0.018	0.018	0.018	0.018	0.018	0.018	0.018
	0.015	0.015	0.015	0.018	0.018	0.018	0.018	0.020	0.020	0.020	0.020	0.020	0.020	0.020
4	0.015	0.015	0.015	0.015	0.015	0.018	0.018	0.018	0.018	0.020	0.020	0.020	0.020	0.020
	0.018	0.018	0.018	0.018	0.018	0.020	0.020	0.020	0.020	0.025	0.025	0.025	0.025	0.025

TABLE 4.12.1 Suggested Stock to Be Left for Internal Grinding* (*Continued*)

Diameter of hole, in	Length of hole, in													
	5/16	1/2	3/4	1	1 1/2	2	2 1/2	3	3 1/2	4	5	6	7	8
5	0.018 0.020	0.018 0.020	0.018 0.020	0.018 0.020	0.018 0.020	0.018 0.020	0.020 0.025	0.020 0.025	0.020 0.025	0.020 0.025	0.020 0.025	0.025 0.030	0.025 0.030	0.025 0.030
6	0.020 0.025	0.020 0.025	0.020 0.025	0.020 0.025	0.020 0.025	0.020 0.025	0.020 0.025	0.025 0.030	0.025 0.030	0.025 0.030	0.025 0.030	0.025 0.030	0.030 0.035	0.030 0.035
7	0.025 0.030	0.025 0.030	0.025 0.030	0.025 0.030	0.025 0.030	0.025 0.030	0.025 0.030	0.025 0.030	0.025 0.030	0.030 0.035	0.030 0.035	0.030 0.035	0.030 0.035	0.030 0.035
8	0.025 0.030	0.025 0.030	0.025 0.030	0.025 0.030	0.025 0.030	0.025 0.030	0.025 0.030	0.025 0.030	0.025 0.030	0.030 0.035	0.030 0.035	0.030 0.035	0.030 0.035	0.030 0.035

*Under average conditions and rigid wall thickness; for hardened steel. Stock is given for diameter of hole and excludes runout of hole due to improper centralization of work in fixture.

Source: From *The ABC of Internal Grinding,* Norton Company, Worcester, Mass., 1961.

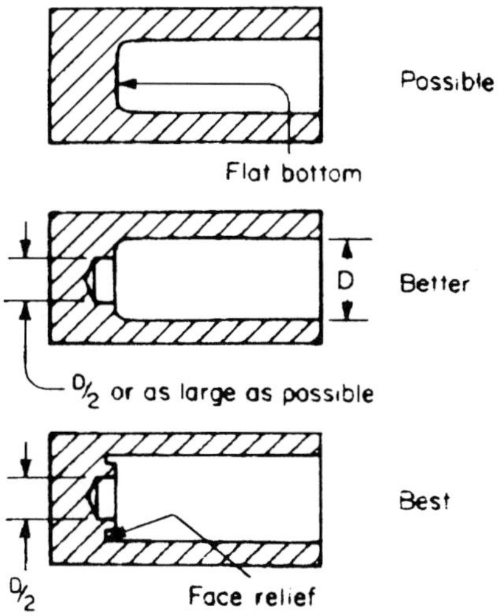

FIGURE 4.12.6 Flat-ground bottoms should be avoided. Center relief of one-half the hole diameter or more avoids the need to grind in the center area, where the grinding wheel's surface speed is low. Relief at the outer corner also helps provide flatness.

Access for Coolant

Because a rapidly spinning internal-grinding wheel is a very effective "flinger" type of centrifugal seal, coolant is unlikely to work its way into the grinding contact zone against the centrifugal action of the wheel. Either an axial through flow or a nozzle aimed at that portion of the wheel outside diameter which is about to enter the contact zone must be allowed by the part design. (On sufficiently large blind parts, a through flow has been accomplished by introducing the coolant through a hollow wheel spindle.)

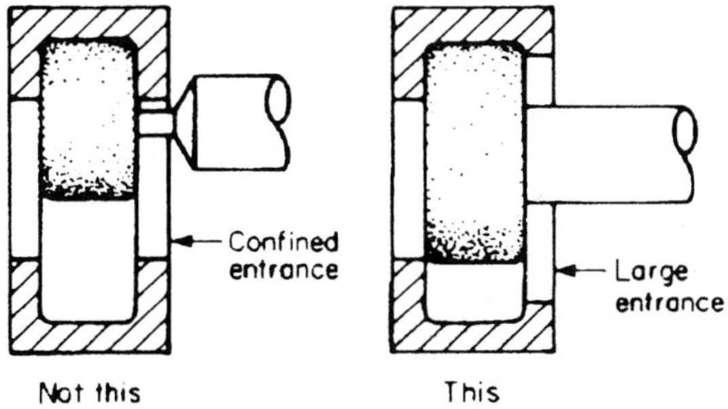

FIGURE 4.12.7 Hole with confined entrance.

DIMENSIONAL FACTORS

There are definite limitations to the degrees of perfection that are practical for the various dimensional attributes of internally ground surfaces. Important factors are whether or not the machine is in good condition, whether the ways are straight, and whether the axis of the wheel, the axis of the part, and the contact point of the dresser are coplanar.

Any one internally ground surface will be subject to tolerances on some of the following attributes: roundness, straightness, taper (these three sometimes combined as cylindricity), concentricity, squareness, size, location in various directions, surface finish, surface waviness, axial profile, and cross-sectional contour, all of which affect cost.

Since the grinding wheel and the quill or spindle of an internal grinder must be slender and long enough to reach into the hole, their rigidity can be limited. This limitation distinguishes internal grinding from all other grinding processes. Severely diminished rigidity causes the wheel to tend to follow the existing surface rather than to generate a new, true surface. This makes it more difficult to control hole size and taper. In extreme cases, it can cause an unacceptable amount of waviness or chatter in the ground surface as a result of wheel unbalance or self-excited vibration between the wheel and the part. Another factor is the limited number of abrasive grains in contact with the work when a small internal-grinding wheel is used. This causes greater wheel wear and necessitates more frequent dressing. Both of these conditions are accentuated if materials of low grindability are used.

The roundness of centerless internally ground surfaces depends largely on the roundness of the outside diameter of the workpiece. The accuracy of form internal grinding depends on the accuracy of the form of the wheel as well as on the other factors indicated above. Similarly, the profile accuracy of nonround holes depends on the accuracy of the mechanism which controls the position of the grinding wheel.

Factors affecting surface finish in internal grinding are parallel to those involved in other grinding methods. (See Chap. 4.15.)

TOLERANCE RECOMMENDATIONS

Recommendations in this subsection are based in part on standards of the antifriction-bearing industry. Table 4.12.2 presents recommendations for tolerances on hole diameters. These tolerances include allowances for the effect of taper and roundness deviations.

Surface finishes between 0.4 and 0.8 μm (15 and 30 μin) AA are normal for internal grinding. Allowing a surface finish coarser than 0.8 μm (30 μin) AA rarely will produce cost savings. A surface finish below 0.4 μm (15 μin) AA is likely to lengthen grinding time significantly, although finishes as fine as 0.13 μm (5 μin) often can be obtained.

Squareness of bores and faces or shoulders ground in one placement of the workpiece can be held within 0.00025 to 0.00063 mm (0.000010 to 0.000025 in) TIR (total indicator runout). If operations are performed separately, a squareness tolerance of 0.005 mm (0.0002 in) TIR is recommended. The same tolerances can be applied to parallelism specifications.

TABLE 4.12.2 Recommended Tolerances for Internally Ground Holes

Hole diameter, mm (in)	Normal tolerance, mm (in)	Tightest tolerance, mm (in)
0–25 (0–1)	±0.004 (±0.00015)	±0.0013 (±0.00005)
25–50 (1–2)	±0.005 (±0.0002)	±0.0013 (±0.00005)
50–100 (2–4)	±0.008 (±0.0003)	±0.0025 (±0.0001)
100–200 (4–8)	±0.013 (±0.0005)	±0.0032 (±0.00013)
200–400 (8–16)	±0.020 (±0.0008)	±0.005 (±0.0002)

Concentricity tolerances with chucking grinders can be as low as 0.0025 mm (0.0001 in) TIR, but 0.013 mm (0.0005 in) TIR is preferable if it can be tolerated for both chucking and centerless machines.

Roundness deviations normally can be held within 0.0025 mm (0.0001 in) on radius, and 0.00025 mm (0.000010 in) is not unusual.

Location in the axial direction can readily be held within ± 0.013 mm (0.0005 in) and with extra care and favorable conditions to 0.0013 mm (0.00005 in).

Cross-sectional contours are customarily allowed to deviate by 0.025 mm (0.001 in) on either side of the perfect contour, but some are held to 0.005 mm (0.0002 in).

CHAPTER 4.13
PARTS CYLINDRICALLY GROUND ON CENTER-TYPE MACHINES

Wes Mowry
Norton Company
Worcester, Massachusetts

THE PROCESSES

Two basic processes are used for grinding external cylindrical surfaces: center-type grinding, in which the workpiece is clamped to the machine spindle, and centerless grinding, in which the workpiece rotates freely between two opposing grinding wheels.

In center-type grinding, the workpiece normally is held between pointed "centers" as in a lathe. However, it also may be held by a chuck or be mounted on a faceplate. The grinding wheel (except for tapered surfaces) rotates on an axis parallel to that of the work; it is fed perpendicularly to and from the work. When necessary, transverse motion between the wheel and the work also is provided by movement of the headstock-tailstock table or, on some large machines, by movement of the wheel. The headstock-tailstock table also can swivel to provide grinding of tapered surfaces. The work and the wheel are driven by separate motors so that they move in opposite directions at the point of contact. When necessary, steady rests are used to support the work during grinding.

TYPICAL APPLICATIONS

Shafts and pins are commonly ground on center-type grinding machines. Parts with steps (multiple diameters), tapers, and ground forms are suitable for the process. Diameters as large as 1.8 m (72 in) can be ground on commercially available equipment. Workpiece deflection limits the minimum diameter that is practical to process. (Centerless grinding is more suitable for thin and long pins.) A practical minimum diameter for short cylindrical elements is perhaps 3 mm ($\frac{1}{8}$ in).Cast and forged workpieces with cylindrical projections and other bulky and irregular parts not grindable on centerless equipment are ground on center-type equipment.

Typical parts ground on center-type machines are crankshaft bearings; bearing rings; machine arbors, spindles, and pins, particularly those with tapers; shafts; bushings; axles; rolls; and parts with interrupted cylindrical surfaces, e.g., splined and slotted shafts or components with holes. (See Fig. 4.13.1.)

Plunge cylindrical grinding is limited to ground surfaces shorter than the width of the grinding wheel.

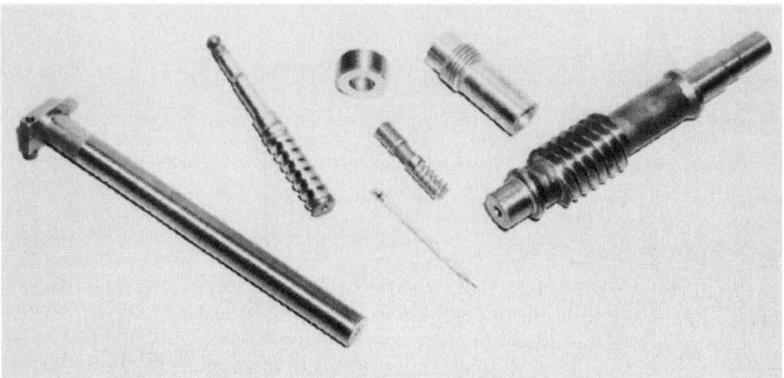

FIGURE 4.13.1 Thread grinding, performed on the parts shown, is typical of the kind of operations that can be performed on center-type grinding machines. (*Courtesy Jones & Lamson, Waterbury Farrel Division of Textron, Inc.*)

COSTS AND ECONOMIC PRODUCTION QUANTITIES

Center-type cylindrical grinding is most economical for short and moderate production runs. Setup times for center-type machines are shorter than for centerless grinders, but process time is significantly longer, especially when the time for machining centers in workpieces is considered.

Center-type machines have the advantage for toolroom, experimental, and low-quantity job-shop production. When quantities exceed about 100 pieces, however, if centerless equipment can be used, it is usually advantageous from a cost standpoint.

Most operations for center-type grinding can be set up in 30 min or less, especially if the operation is manual and the workpiece-holding method remains the same for successive jobs. More time is required for form grinding or automatic operation. Typical output rates range from 10 to 130 pieces per hour, with about 60 pieces per hour being a fair average for operations involving a single surface or a single cut.

SUITABLE MATERIALS

Materials suitability for center-type cylindrical grinding parallels that for other grinding processes. See Chap. 4.15, "Flat-Ground Surfaces," for a discussion of grindable materials.

DESIGN RECOMMENDATIONS

When designing components for center-type grinding, observe the following suggestions:

1. Keep the parts as well balanced as possible for better finish and accuracy.
2. Avoid long small-diameter parts because their tendency to deflect from the pressure of the grinding wheel hampers control of the finished-diameter dimension, straightness, roundness, and finish. The use of steady-rest supports (which must be adjusted as the diameter is reduced by the grinding) is helpful but not fully satisfactory. Parts with a length more than 20 times their diameter are particularly troublesome. Parts shorter than 8 times their diameter are best.
3. Profile shapes can be plunge-ground and sometimes can be made quite intricate. However, profiles are better kept as simple as possible. The following workpiece features cause difficulties in or increase the complexity of trueing the grinding wheel and should be avoided if possible:
   (***a***) *Tangents to radii.* Nontangent radii or angular reliefs (wheel chamfers) are preferable. (See Fig. 4.13.2*a*.)
   (***b***) *Grooves in parts.* These should be eliminated or made as shallow and wide as possible. (See Fig. 4.13.2*b*.)
   (***c***) *Angular shapes and tapers.* (See Fig. 4.13.2*c*.)
   (***d***) *Component radii.* (See Fig. 4.13.2*d*.)
4. Interrupted surfaces present grinding problems and should be avoided or specified with an expectation of somewhat less accurate and more costly grinding. The surface adjacent to interruptions tends to be ground more deeply than nearby continuous surfaces. (See Fig. 4.13.3.)
5. Undercuts on facing surfaces are not normally possible on regular cylindrical grinding machines except to a shallow degree when the wheel is turned at an angle and specially trued or with small wheels normally used for internal grinding. If possible, undercuts are best avoided in either case. (See Fig. 4.13.4.)
6. If fillets are used, make them as large as possible. As in the case of other grinding processes, however, the most preferable procedure is to machine or cast a relief in the workpiece at the junction of two ground surfaces. (See Fig. 4.13.5.)
7. Center holes on workpieces held between centers should have an exact 60° angle and uniformity of shape for accurate cylindrical grinding. If precision of the grinding is critical, lapping the center holes may be required after heat-treating the workpiece.
8. Tubular parts, especially if thin-walled, can be deformed slightly when held in a three-jaw chuck. If the exterior cylindrically ground surface must be held to precise limits, it may be necessary to specify a heavier wall or to use a center-mounting method.
9. In cylindrical center-type grinding, as with other grinding processes, it is advisable to minimize the stock removed by grinding. Pregrinding machining should be as accurate as possible.

DIMENSIONAL FACTORS

Center-type cylindrical grinding is a process of high precision. As in other machining processes, the accuracy of final dimensions reflects the condition of the equipment and the skill of the operator. Worn bearings, centers, or machine ways, poor coolant action, incorrect grinding wheels, and deflection of the workpiece all can adversely

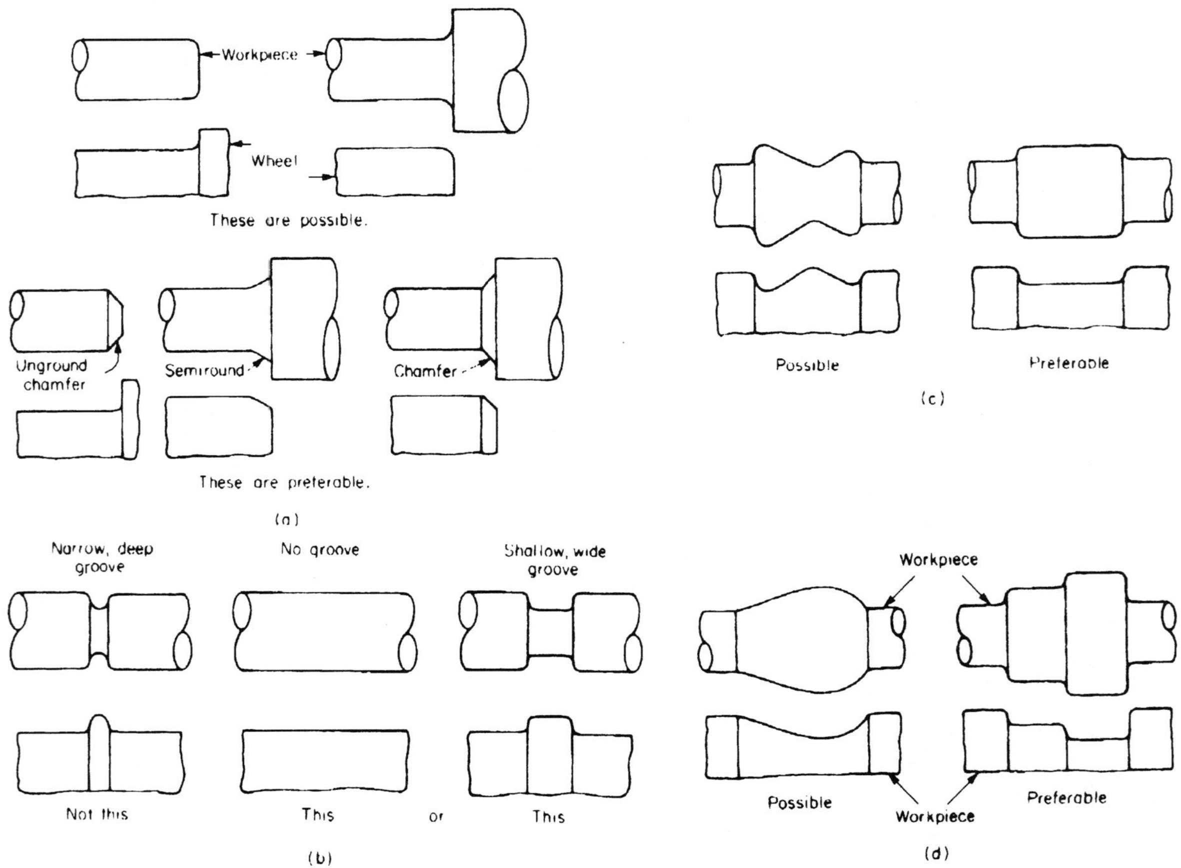

FIGURE 4.13.2 Plunge-ground shapes that make wheel dressing more difficult are (*a*) tangent radii and straight surfaces; (*b*) grooves, especially if deep and narrow; (*c*) tapers and angular surfaces; and (*d*) special flowing curves and large radii.

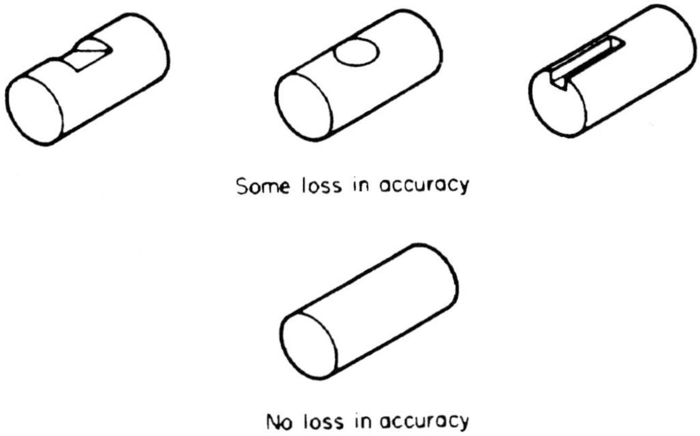

FIGURE 4.13.3 Interrupted surfaces reduce the accuracy of cylindrically ground parts.

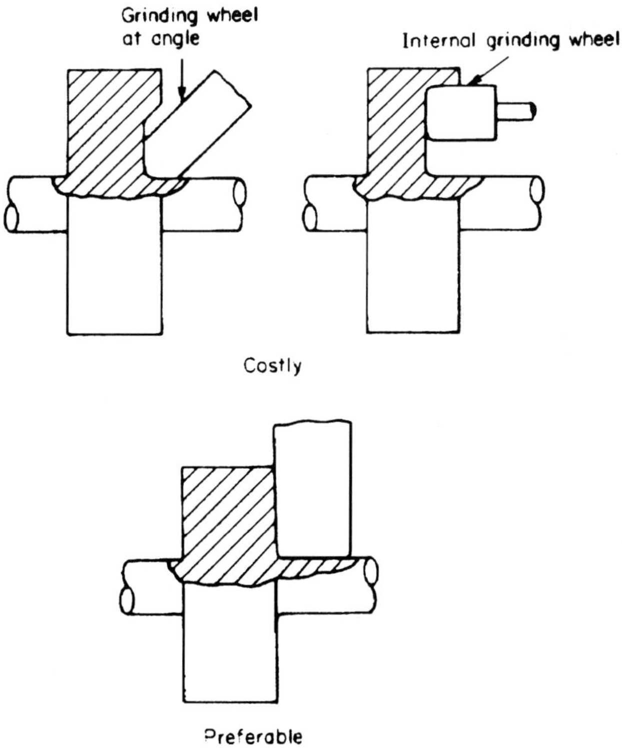

FIGURE 4.13.4 Ground undercuts on facing surfaces are costly and should be avoided.

affect the precision of the finished surface and dimensions. Maintenance of the correct position of steady-rest supports for long work is particularly important. Another vital factor is the accuracy, especially the roundness, of center holes in workpieces. Use of the correct grinding wheel and correct feed and speed also are important.

Table 4.13.1 provides recommended dimensional tolerances for cylindrically ground parts under normal production conditions.

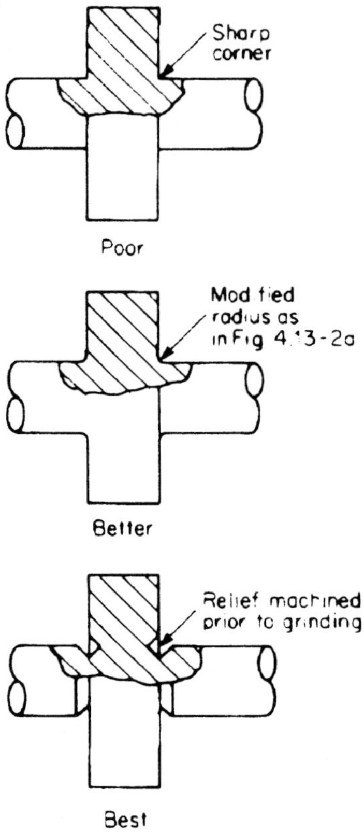

FIGURE 4.13.5 The best practice is to machine or cast a relief at the junction of two surfaces before grinding.

TABLE 4.13.1 Recommended Dimensional Tolerances for Center-Ground Parts

	Recommended tolerance	
Dimension	Normal	Tight
Diameter	±0.0125 mm (0.0005 in)	±0.0025 mm (0.0001 in)
Paralellism	±0.0125 mm (0.0005 in)	±0.0050 mm (0.0002 in)
Roundness	±0.0125 mm (0.0005 in)	±0.0025 mm (0.0001 in)
Surface finish	0.2 μm (8 μin)	0.05μm (2 μin)

CHAPTER 4.14
CENTERLESS-GROUND PARTS

L. J. Piccinino
Norton Company
Worcester, Massachusetts

THE PROCESS

Centerless grinding is an abrasive method for finish-machining cylindrical surfaces. Workpieces are unmounted but pass between two opposed grinding wheels. There are several variations of the process:

In *through-feed grinding* the part passes axially between the main grinding wheel and the regulating wheel; a metal blade supports the workpiece at the correct height. This method is limited to cylindrical parts without heads, shoulders, or other projections that would prevent through movement of the part between the wheels. It is illustrated schematically in Fig. 4.14.1.

Infeed grinding differs from through-feed grinding in that the workpiece does not move axially during grinding and in that the grinding wheel may be dressed to the shape to which the part is to be ground. The process is comparable with plunge or form grinding on a cylindrical grinder. The grinding wheel is fed into the work and retracts for workpiece removal. The infeed method is not quite as rapid as through-feed grinding. (See Fig. 4.14.2.)

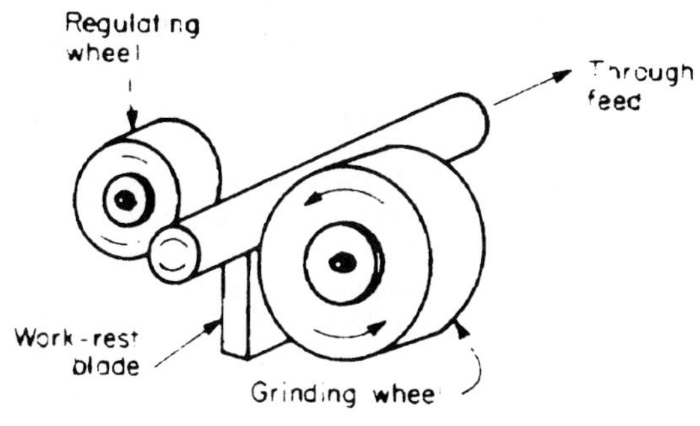

FIGURE 4.14.1 Through-feed centerless grinding.

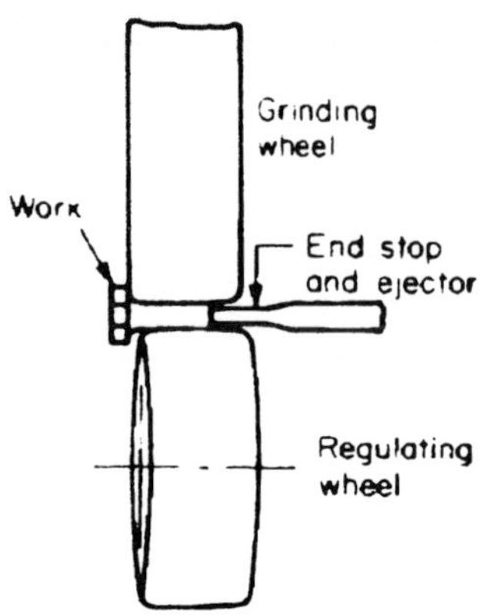

FIGURE 4.14.2 Infeed centerless grinding. (*Courtesy Cincinnati Milacron.*)

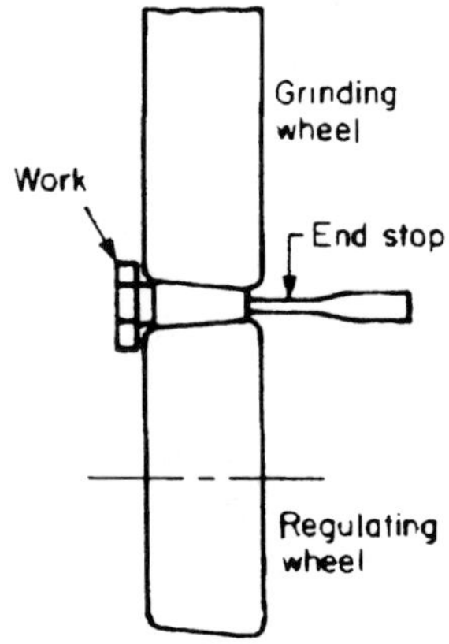

FIGURE 4.14.3 End-feed centerless grinding. (*Courtesy Cincinnati Milacron.*)

End-feed grinding is used for tapered work. Either the grinding wheel or the regulating wheel or both are dressed to the proper taper. The workpiece is fed axially to a fixed end stop. Figure 4.14.3 illustrates the process.

Internal centerless grinding is used for ring- and sleeve-shaped parts. The workpiece is supported between three rolls, which establish its location and cause it to rotate. An internal grinding wheel removes stock. Since the part is located from its outside surface, the ground inner surface is nearly perfectly concentric with the outside.

TYPICAL APPLICATIONS

The centerless-grinding method can be applied to solid parts with diameters as small as 0.1 mm (0.004 in) and as large as 175 mm (7 in). Rings and tubing somewhat larger—250 mm (10 in)—can be centerless-ground with equipment currently available. Parts as short as 10 mm (0.4 in) and as long as 5 m (16 ft) are centerless-ground.

Centerless grinding produces accurate cylindrical surfaces. The method is ideally suited for pins, shafts, and rings when close-tolerance outside diameters, precise roundness, and smooth surfaces or all three are required. Long, slender parts that would be subject to deflection in conventional cylindrical grinding can be ground accurately by the centerless method. Screw threads also can be ground with this process.

For through-feed grinding, the surface to be ground must be a straight cylinder. If the part has two or more diameters, only the largest can be ground with the through-feed method. Figure 4.14.4 shows a variety of parts ground with the through-feed centerless method.

The infeed method is slightly slower but possesses the advantage of having the capability of producing multidiameter or formed surfaces. Tapered parts or parts with steps are ground with this method. Valve tappets, arbors, yoke pins, shackle bolts, and distributor shafts are typical examples.

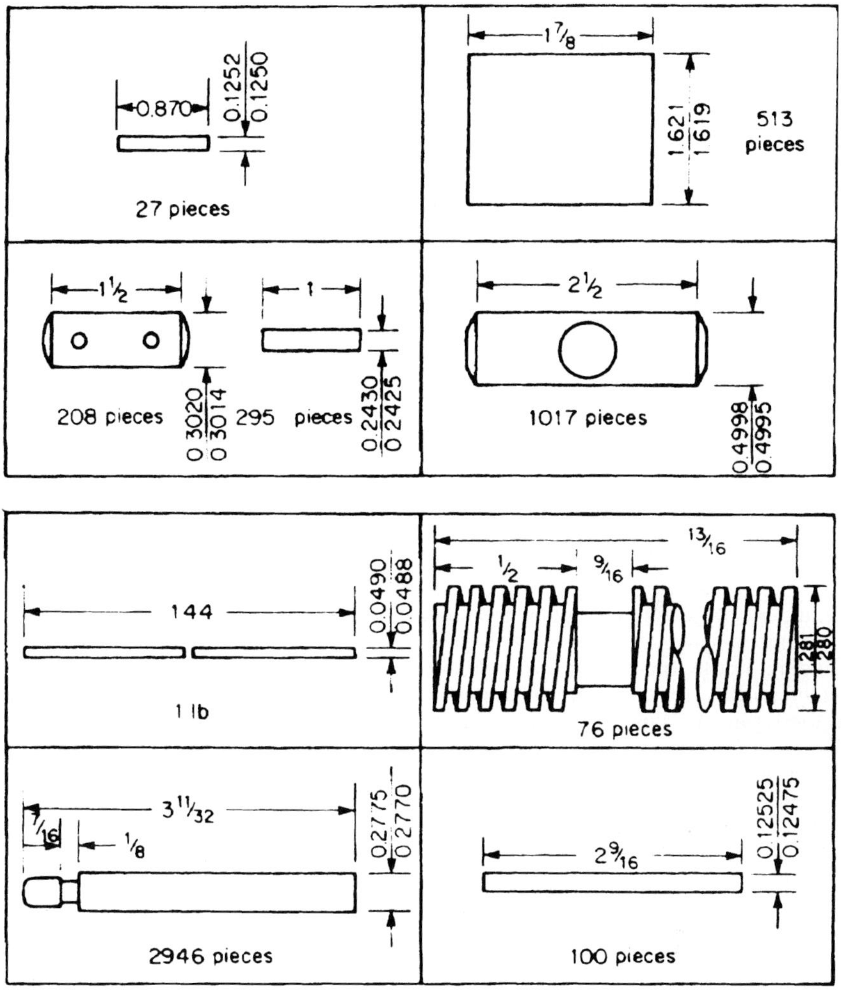

FIGURE 4.14.4 A variety of parts produced with through-feed centerless grinding. (*Courtesy Cincinnati Milacron.*)

ECONOMIC PRODUCTION QUANTITIES

Centerless grinding is primarily a mass-production process. It combines high output levels with the precision of finished dimensions.

Short parts processed by the through-feed method and using an automatic magazine or hopper feed can be centerless-ground at rates as high as 6000 pieces per hour. Times may be considerably longer for end-feed or infeed processed parts, particularly the latter when larger diameters, greater stock removals, or multiple diameters are involved. For infeed grinding, a rate of 30 to 240 pieces per hour is more typical. For through-feed grinding, production rates are more easily stated in terms of length processed per minute. Rates range from 1 to 4 m/min (3 to 12 ft/min) per pass to as much as 9 m/min (30 ft/min) per pass. From one to six passes may be involved.

Centerless grinding is more rapid than conventional cylindrical grinding because loading and unloading time is much shorter. There is also no need for drilling center holes on the ends of the workpiece.

Setup times for centerless grinding range from a few minutes to as much as 4 h, depending on the changes required and the machine used. Sometimes a change in the material being ground may require more setup time than a workpiece change because it may entail changes in wheels and coolant.

Job shops can group lots of parts of like material and nearly the same diameter and economically process lots of 100 pieces or less. Justification for centerless grinding in comparison with center-type cylindrical grinding depends on the availability of equipment and wheels and other tooling and the complexity of the parts to be ground. Single-diameter parts can be ground as easily with the centerless method as with center-type grinding if quantities are 10 to 25 or more. Complex parts may require quantities of 1000 or more to justify the centerless approach.

SUITABLE MATERIALS

Centerless grinding has the same broad range capability for processing various materials as the other grinding processes have. However, there is one bonus with centerless grinding: brittle, fragile, and easily distorted parts and materials are more suitable for centerless than for conventional center-type cylindrical grinding. With the centerless method, parts are supported along all or most of their length, and there is no end pressure or deflection from wheel pressure. This makes the grinding of glass, porcelain, rubber, plastics, cork, and other brittle or easily distorted materials more practical on centerless equipment.

For comments on the suitability of most materials to grinding methods in general, see Chap. 4.15, "Flat-Ground Surfaces."

DESIGN RECOMMENDATIONS

The following suggestions should be kept in mind by the designers wishing to take maximum advantage of the centerless-grinding approach.

1. If possible, the ground surface of the workpiece should be its largest diameter to permit the use of through-feed grinding. (See Fig. 4.14.5.)
2. Short pieces are more susceptible to having unspecified taper or concave or barrel-shaped surfaces. To help avoid this problem, keep ground surfaces at least one diameter in length if possible. (See Fig. 4.14.6.)
3. Parts with irregular shapes cannot have a ground surface longer than the grinding-wheel width unless the shape permits a combination of infeed and through-feed grinding. (See Fig. 4.14.7.)
4. It is preferable to avoid grinding the ends of infeed centerless-ground parts. This stricture includes radii on the ends of parts. If the end must be finished, avoid

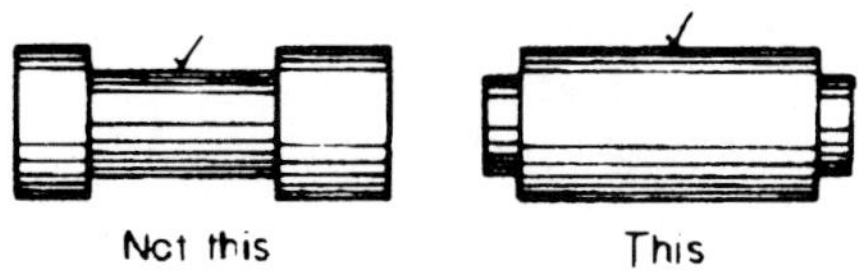

FIGURE 4.14.5 Only the largest diameter of the workpiece can be through-feed centerless-ground. If a smaller diameter must be centerless-ground, as in the left-hand part, the slower plunge centerless-grinding method must be used.

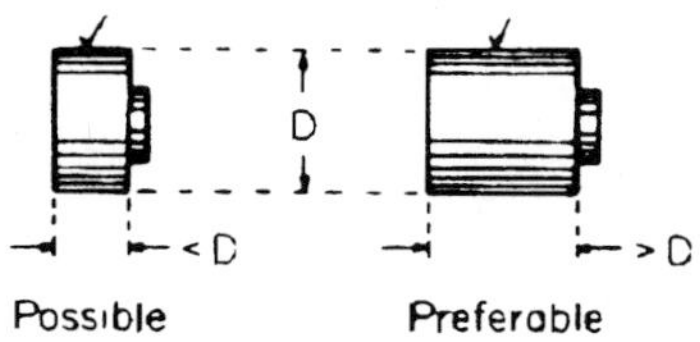

FIGURE 4.14.6 The width of a centerless-ground part should be at least as large as its diameter.

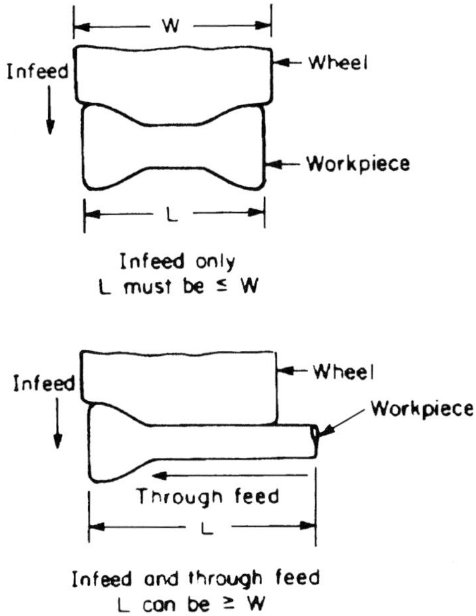

FIGURE 4.14.7 Parts with irregular surfaces cannot be longer than the width of the grinding wheel unless both infeed and through-feed are used and the part is stepped in one direction as shown.

square, nearly square, or round ends. The included angle of a pointed end should be 120° or less. (See Fig. 4.14.8.)

5. As with cylindrical grinding, it is best to avoid fillets and radii and instead use undercut or relief surfaces. This eliminates difficult wheel dressing when there is a fillet. (See Figs. 4.13.2*a* and 4.13.5.) When radii or fillets are used, they should be as large as possible, and on any one part all of them should be the same size to simplify wheel dressing.
6. When a part is designed for form centerless grinding (infeed method), the form should be as simple as possible to reduce wheel dressing and other costs.
7. If accuracy is critical, avoid keyways, flats, holes, and other interruptions to the surface to be ground or make them as small as possible. (See Fig. 4.13.3.)

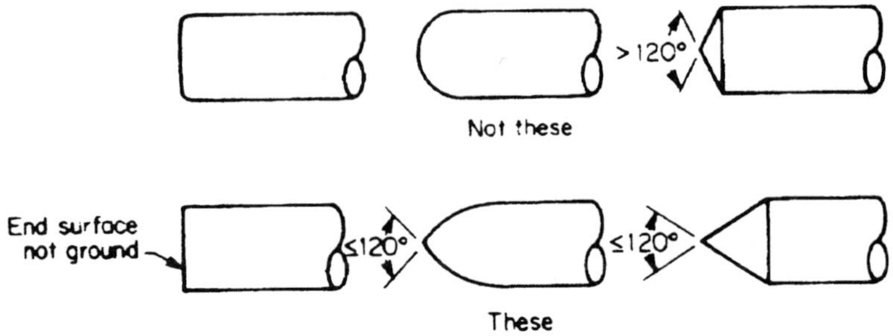

FIGURE 4.14.8 Do not design the ends of centerless-ground parts to have ground surfaces unless infeed grinding is used and the end surface has an included angle of less than 120°.

8. If a flat is necessary at the end of a shaft to provide a surface against which a setscrew is to be tightened or for some other reason, and if tolerances are tight, it is preferable to put flats on opposite sides of the part. This prevents the tendency for a high spot to develop opposite the flat. Another remedy is to retain a full cylindrical section at the end of the shaft. (See Fig. 4.14.9.)

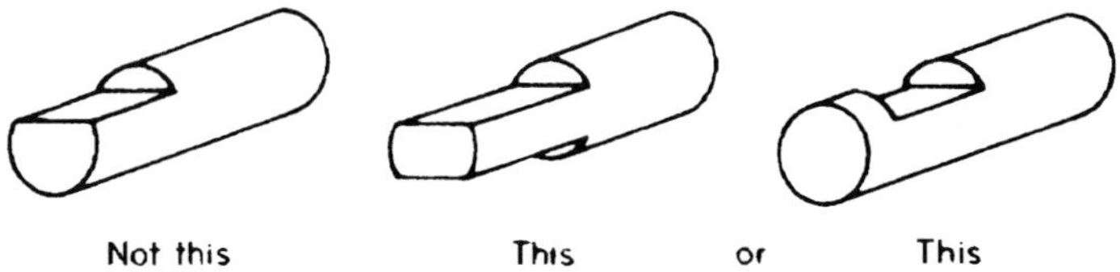

FIGURE 4.14.9 When interruptions are unavoidable, balance them if possible or provide full cylindrical surfaces on both sides.

DIMENSIONAL FACTORS

Like other grinding methods, centerless grinding can be extremely accurate if all conditions are correct. These conditions include the condition of the equipment, particularly wheel-spindle bearings, the use of the proper wheel and coolant, and the evenness of the temperature of the workpiece, machine, and coolant.

Roundness control in centerless grinding is remarkably good even though the part does not rotate about a fixed center. The reason for this is the geometry of the setup of grinding wheel, regulating wheel, and workpiece support, which, if correct, systematically cause the grinding action to remove the high spots of a part as it is rotated against the wheels.

Another favorable factor in centerless grinding is the fact that the setting of the grinding wheel affects the *diameter* of the workpiece rather than the *radius* from its center point, as in conventional cylindrical grinding. Thus a factor of 2 is involved in the accuracy of wheel settings.

Table 4.14.1 presents recommendations for dimensional tolerances for production centerless grinding.

TABLE 4.14.1 Recommended Dimensional Tolerances for Centerless Ground Parts

Dimension	Recommended tolerance	
	Normal	Tight
Diameter	±0.0125 mm (±0.0005 in)	±0.0025 mm (±0.0001 in)
Parallelism	±0.0125 mm (±0.0005 in)	±0.0025 mm (±0.0001 in)
Surface finish	0.20 μm (8 μin)	0.05 μm (2 μin)

CHAPTER 4.15
FLAT-GROUND SURFACES

R. Bruce MacLeod
Vice President
Taft-Peirce Supfina
Cumberland, Rhode Island

SURFACE-GRINDING PROCESS

Surface grinding is a process that moves a workpiece in a horizontal plane so that it passes under a grinding wheel in such a way that a precise amount of material is removed and a flat surface is produced. (Note that the term *surface grinding* has come to designate the grinding of surfaces that are basically flat rather than cylindrical.) Various types of surface-grinding machines are illustrated schematically in Fig. 4.15.1. The most common variety, with a horizontal wheel spindle and a reciprocating table, is shown in Fig. 4.15.2.

With horizontal-spindle machines, the work table or the wheel advances a short distance with each reciprocating stroke of the table. The wheel thus generates a flat surface. With vertical-spindle machines, wheels are normally cup-shaped, and flat surfaces can be generated from one pass of the workpiece beneath the wheel.

Grinders are normally equipped with magnetic chucks to hold the work. When nonmagnetic materials are ground, vises, other holding fixtures, vacuum chucks, or adhesive methods are employed.

Metal removal in grinding results from the cutting action of small, irregularly shaped abrasive particles that form the wheel structure along with a bonding agent. Common abrasives are aluminum oxide, silicon carbide, diamond, and cubic boron nitride. Common bonding agents are vitrified materials (fused clays), silicates, various plastic or rubber materials, and metals.

As the wheel cuts, it wears, smoothing some abrasive grains but causing others to fracture or break away from the wheel. In the latter two cases, new sharp edges of abrasive are exposed. These continue the cutting action of the wheel.

Periodically, it is advisable to dress or true the wheel. Usually performed with a diamond-tipped tool, this procedure fractures or removes abrasive grains that have worn smooth and restores the cutting surface of the wheel to the correct straightness or form.

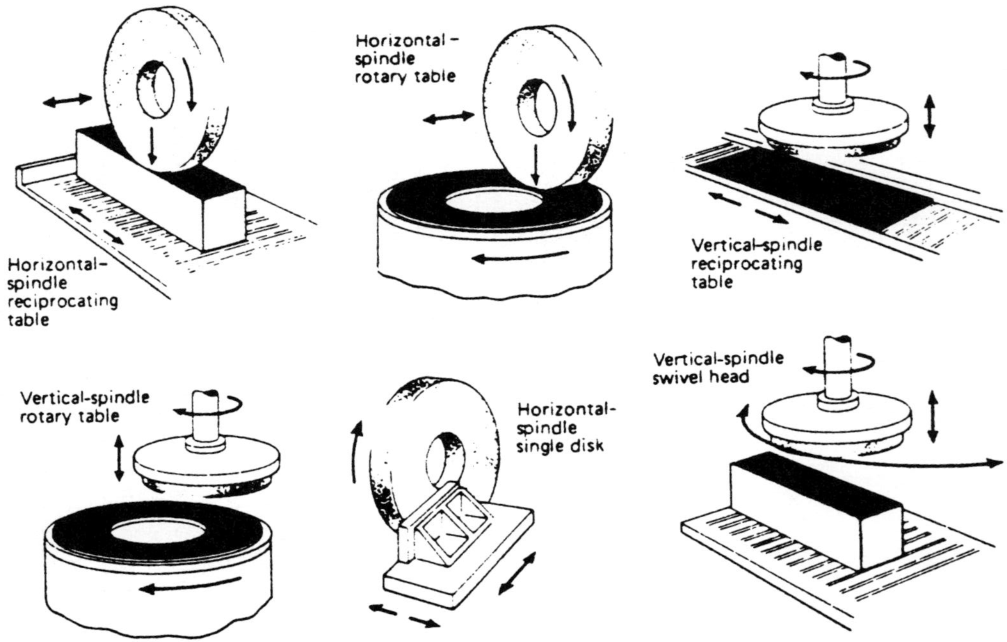

FIGURE 4.15.1 Surface-grinding methods. (*From American Machinist.*)

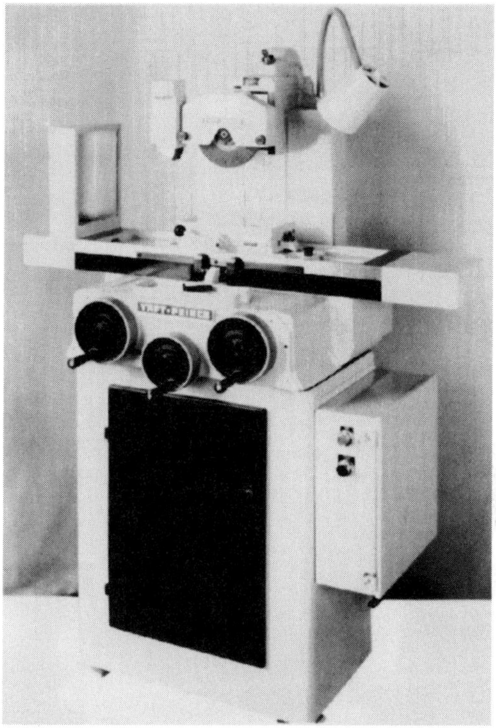

FIGURE 4.15.2 Typical horizontal-spindle reciprocating-table surface grinder. (*Courtesy Taft-Peirce Manufacturing Co.*)

TYPICAL CHARACTERISTICS OF SURFACE-GROUND PARTS

Parts are surface-ground for several reasons: (1) to produce a flat surface, (2) to produce a desired dimension, usually a highly accurate one, (3) to produce a smooth surface finish, (4) to produce a profiled surface (this is a less common application), and (5) to sharpen a cutting tool. The process was originally intended for removing stock on hardened pieces, fine finishing, and working to close tolerances. It has now been broadened to include *abrasive machining,* a process that can compete with milling for fast metal removal.

Each type of grinder is suited to a particular part or process. Parts such as machine ways, precision parallels, molds, and dies (see Figs. 4.15.3 and 4.15.4) are ground on a horizontal-spindle reciprocating-table type of grinder. Large blocks, tables, and rough castings can be machined on a vertical-spindle rotary-table type of grinder. These parts are often ground without prior operations.

There are few part-size limitations for surface grinding. Reciprocating-table surface grinders are available with table sizes ranging from 125 by 250 mm (5 by 10 in) up to 1.2 by 6 m (48 by 240 in). Rotary-table grinders have table diameters ranging from 150 to 4400 mm (6 to 174 in). In addition, there are not many material limita-

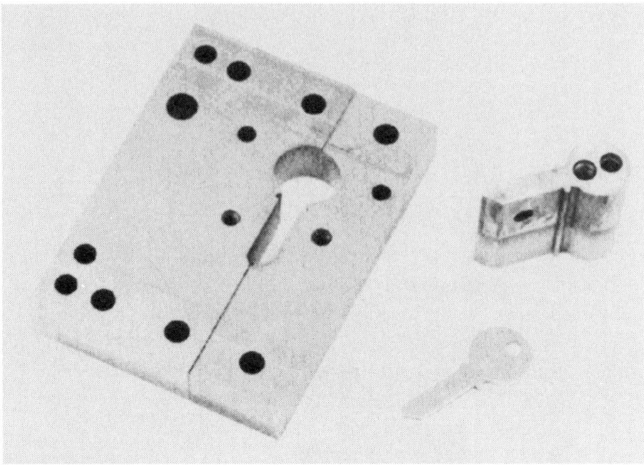

FIGURE 4.15.3 These elements of a blanking die (for a key blank) are typical of the kind of component which is surface-ground. (*Courtesy Taft-Peirce Manufacturing Co.*)

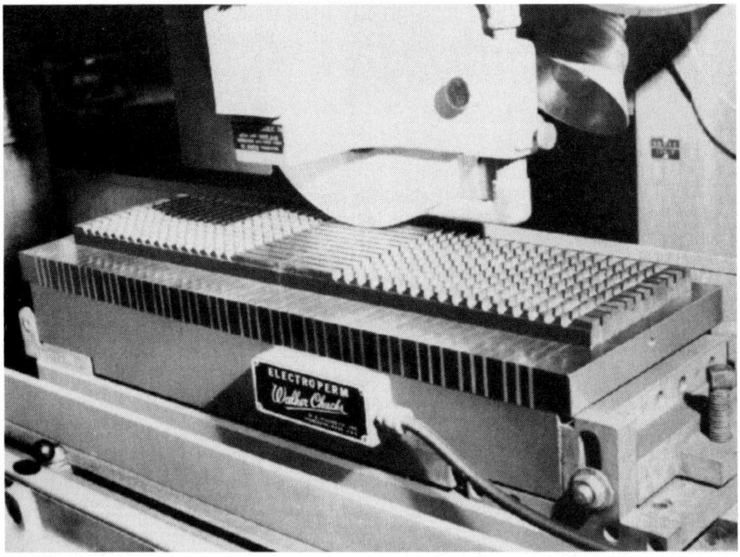

FIGURE 4.15.4 Series of broaching-tool elements being surface-ground. (*Courtesy O. S. Walker Co., Inc.*)

tions on parts that can be ground. The most commonly ground materials are ferrous and can be held with magnetic chucks. Most surface-ground ferrous parts are also hardened. Nonferrous materials include brass, plastic, and glass.

ECONOMIC PRODUCTION QUANTITIES

Much less setup time is involved in normal surface-grinding operations than with other types of metal-removal equipment. Most parts ground on horizontal-spindle reciprocating-table surface grinders are hard and magnetic. Therefore, the same wheel and the same magnetic chuck often can be used for different part configurations. Tooling costs are also usually very low because wheels and chucks are invariably general-purpose items applicable to a variety of parts.

Horizontal-spindle grinders use a wheel that presents only a line contact with the work. For this reason, if much stock is to be removed, several passes of the part under the wheel are required before the correct size is reached. This means that the horizontal-spindle grinder is not the most efficient type for high-volume production unless the work is not suited to a vertical-spindle machine.

All these facts point to the most advantageous use of horizontal-spindle reciprocating-table machines for low-quantity production. Small hand-operated toolroom surface grinders that have no stops to set or feeds to adjust are ideally suited for single parts such as dies, molds, gauges, and cutting tools.

Large-volume production on surface-grinding equipment is most efficiently carried out on the vertical-spindle rotary-table type of machine. Parts can be loaded automatically onto a chuck or into a special fixture, rotated under the wheel, which is positioned automatically to remove the proper amount of stock, and then unloaded automatically as the table moves out from under the wheel. The vertical-spindle type of grinder can be supplied with up to 300 hp, which, combined with the large wheel-contact area, means that the required amount of stock often can be removed in one pass. If more than one pass is required, two spindles can be used, with the first head doing the roughing and the second head the finishing.

SUITABLE MATERIALS FOR GRINDING

Virtually any material can be shaped by grinding. In fact, grinding is the only practical method for surface machining many hard or brittle materials like glass, ceramics, and tungsten carbide. Although there are differences in how easily various materials can be ground, these differences are not reflected in recommended grinding speeds nearly as much as differences in machinability are reflected in recommended feed and speeds in metal-cutting operations. The major effect of different grinding properties lies in requirements for a proper grinding wheel.

Steel in both carbon and alloy grades and other high-tensile-strength metals, both hardened and unhardened, are ground with aluminum oxide wheels. (Aluminum oxide is the most common wheel material.) For hard, brittle materials like carbide, stone, and ceramics, diamond is the preferred abrasive, although silicon carbide is also used. Cast iron is ground with silicon carbide wheels, as are materials of low tensile strength such as soft brasses, bronzes, aluminum, copper, plastics, and other nonmetallic materials. Hardened tool-and-die steels, which are difficult to grind, are processed with grinding wheels that use cubic boron nitride as an abrasive.

Grindability, or ease of grinding, is most commonly expressed in terms of the *grinding ratio,* which is defined as the ratio of the volume of material removed from the workpiece to the volume of material lost from the wheel because of wear. The higher the grinding ratio, the easier the material is to grind. Unfortunately, a grinding ratio for a particular workpiece material is not a constant for that material. It applies to only one particular set of conditions: one grinding method, wheel, feed, speed, etc. However, it does provide an indication of relative grindability in that a more grindable material will always show a higher grinding ratio than a less grindable material if wheel and operating conditions are correctly set.

Grindability is rarely a factor in materials selection. Functional factors and processibility for other primary operations are usually more critical. However, when grind-

TABLE 4.15.1 Grinding Ratios for Stainless Steels*

Steel designation				
Common	UNS	Condition	Rockwell hardness	Grinding ratio
		Group 1		
AM-355	S35580	Solution-annealed	B95	7
310	S31000	Hot-rolled	B92	8
AM-355	S25580	Precipitation-hardened	C42	9
304	S30400	Hot-rolled	B77	9
17-4PH	S17400	Precipitation-hardened	C45	9
D3	*T30403*	*Hardened tool steel*	*C60*	*9*
		Group 2		
302	S30200	Hot-rolled	B81	15
316	S31600	Hot-rolled	B76	17
440C	S44004	Hardened	C55	19
430	S43000	Hot-rolled	B91	21
440A	S44002	Hardened	C53	23
		Group 3		
440A	S44002	Annealed	B91	40
440C	S44004	Annealed	C22	40
202	S20200	Hot-rolled	B83	50
410	S41000	Annealed	B98	55
314	S31400	Hot-rolled	B81	65
1020	*G10200*	*Hot-rolled carbon steel*	*B67*	*65*
1019	*G10190*	*Cold-drawn carbon steel*	*B85*	*85*
410	S41000	Hardened	C41	85
		Group 4		
303†	S30300	Annealed	B85	110
303†	S30300	Cold-drawn	B98	110
303	S30300	Hot-rolled	B78	220
430F	S43020	Hot-rolled	B81	240
416	S41600	Hardened	C42	550
416	S41600	Hot-rolled	B89	600

*One tool steel and two plain-carbon steels are listed (in italics) for comparison.

†Low sulfur content.

Source: Adapted from *Metals Handbook,* 8th ed., vol. 3: *Machining,* American Society for Metals, Metals Park, Ohio.

TABLE 4.15.2 Grinding Ratios for Several Heat-Resistant Alloys in One Test

Alloy	Condition*	Grinding ratio
S-590	A	13
S-816	A	9
16-25-6	A	5.5
Nimonic 80	A	2.3
HS-21	B	240
A-286	B	32
J-1570	B	7.7

*Condition A: A-46-J8-V grinding wheel down-fed 0.002 in per pass. H4 grinding fluid (containing fatty materials and synthetic soaps) in 4% concentration. Condition B: A-60-J8-V grinding wheel down-fed 0.002 in per pass. G2 grinding fluid (dark sulfochlorinated oil containing fats) in 100% concentration.

Source: Data from *Metals Handbook,* 8th ed., vol. 3: *Machining,* American Society for Metals, Metals Park, Ohio.

TABLE 4.15.3 Relative Grindability of Tool Steels

Grindability	Steels
Low	A7 (T30107), D7 (T30407), M3 (class 1) (T11323), M4 (T11304), M43 (T11343), M44 (T11344), T15 (T12015)
Medium	A2 to A6 (T30102 to T30106), A8 to A10 (T30108 to T30110), D2 to D5 (T30402 to T30405), H steels (T208XX), M1 (T11301), M2 (T11303), M7 (T11307), M10 (T11310), T1 (T12001), T4 (T12004), F steels (T606XX), P4 (T51604)
High	O steels (T315XX), W steels (T723XX), S steels (T419XX), L steels (T612XX), P steels, except P4 (T516XX)

Source: Adapted from *Metals Handbook,* 8th ed., vol. 3: *Machining,* American Society for Metals, Metals Park, Ohio.

ability is important in the selection of a material, the following points should be kept in mind: (1) Nonhardened materials usually grind more rapidly than hardened materials. (2) Very soft materials also are often difficult to grind because they tend to clog the grinding wheel and do not yield as smooth and as bright a finish as harder alloys. This is particularly true of the softer nonferrous metals such as aluminum, copper, etc. (3) In general, the grindability of materials closely parallels their machinability. For example, resulfurized steels with high machinability ratings also show high grinding ratios.

Tables 4.15.1 (stainless steels), 4.15.2 (heat-resistant alloys), 4.15.3 (tool steels), and 4.15.4 (refractory metals) give relative grindability information for materials that normally are relatively costly to grind.

TABLE 4.15.4 Grinding Ratios for Surface Grinding Refractory and Other Alloys*

Workpiece metal: Common designation	Workpiece metal: UNS designation	Hardness†	Wheel classification	Wheel speed, ft/min	Grinding fluid	Table speed, ft/min	Downfeed, in	Grinding ratio‡
Steels								
D-6ac		R_c 57	A_s46-H8-V	6000	Sulfurized oil	40	0.001	75
302	S30200	R_b 86	A_s60-H8-V	6000	Sulfurized oil	60	0.001	20
302	S30200	R_b 86	A_s60-H8-V	6000	Water-soluble oil	60	0.001	8
Heat-resistant alloy								
René 41	N07041	365 BHN	A_s-46-J8-V	4000	KNO_2 solution	40	0.001	10
Refractory metals								
Niobium		225 BHN	A_s-46-K8-V	4000	Chlorinated oil	60	0.0005	4.0
D-31		225 BHN	A_s-46-K8-V	2000	KNO_2 solution	40	0.0005	7.5
Ta-10W	R05255	240 BHN	A_s-46-J8-V	2000	KNO_2 solution	40	0.001	4.8
TZM		232 BHN	A_s-46-N5-V	2000	KNO_2 solution	40	0.001	25
Tungsten		R_c 26	A_s-46-N5-V	4000	KNO_2 solution	40	0.001	4.3

*Cross feeds were 0.05 in per pass in all tests except for Ta-10W, which was ground at 0.025 in per pass. Designations of grinding wheels are based on American Standard Marking Chart B5.17.

†BHN = Brinell hardness number.

‡Ratio of volume of metal removed in grinding to the volume of wheel wear. The higher the index, the easier the metal is to grind.

Source: Adapted from *Metals Handbook,* 8th ed., vol. 3: *Machining,* American Society for Metals, Metals Park, Ohio.

DESIGN RECOMMENDATIONS

A part usually is not designed specifically for surface grinding. Often surface grinding is a necessary step in a series of machining operations intended to produce a part. Since surface grinding may be a necessary step in finishing a part, consideration must be given to the design so that the grinding operation can be performed easily. The quantity of parts to be run is an important factor in design. If production is to be of a very high volume, the part, if possible, should be designed so that it can be ground on a vertical-spindle surface grinder. This means that it should not have any surface higher than the surface to be finished (see Fig. 4.15.5). It also means that the part should be magnetic or be easily clamped in an automatic fixture. A high-volume part that is to be form-ground or has projections above the surface to be finished should be designed for a horizontal-spindle-powered table surface grinder. As with the rotary-table machine, a magnetic material is preferable, although more complicated holding fixtures can be used.

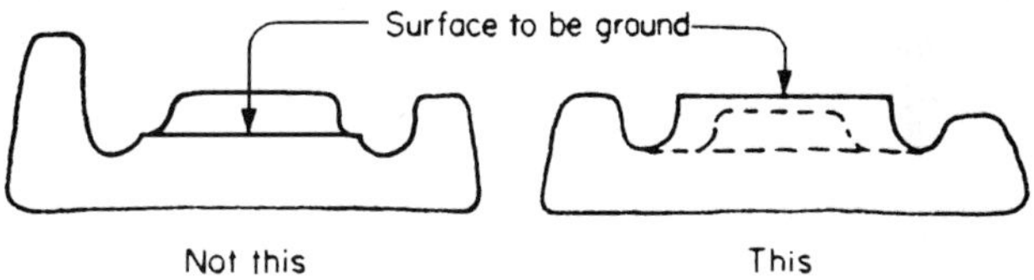

FIGURE 4.15.5 Avoid designs that include surfaces higher than the surface to be ground because these obstruct the motion of the grinding wheel.

On a hand-operated machine, there is very little need to consider part design beyond allowing enough room to permit clearing the recommended-diameter wheel. Surfaces to be ground should be easily accessible.

If the requirements of the part permit the wider tolerances and less smooth surface finish of milling or another machining operation, it is generally more economical to avoid surface grinding. The part drawings should specify as liberal tolerances and finishes as possible so that the manufacturing engineer is not bound to specify grinding but instead can utilize whatever metal-removal process is most economical.

Other design recommendations for surface-ground parts follow:

1. Design the part to be easily held by magnetic chucks. Large, flat locating surfaces are the best.
2. Do not specify a better ground finish than necessary. Material can be removed faster with coarser-grit wheels that do not provide as smooth a finish as a fine-grit, slower-cutting wheel. Plunge grinding does not produce as fine a finish as when a traversing wheel motion is used. If plunge grinding is feasible for the part, specify a liberal surface finish.
3. Design the part so that, as much as possible, surfaces to be ground are all in the same plane. (See Fig. 4.15.6.)
4. If very flat surfaces are required, avoid openings in the surfaces because the grinding wheel tends to cut slightly deeper at the edge of an interrupted surface. Unsupported surfaces that deflect from grinding-wheel pressure also should be avoided. (See Fig. 4.15.7.)

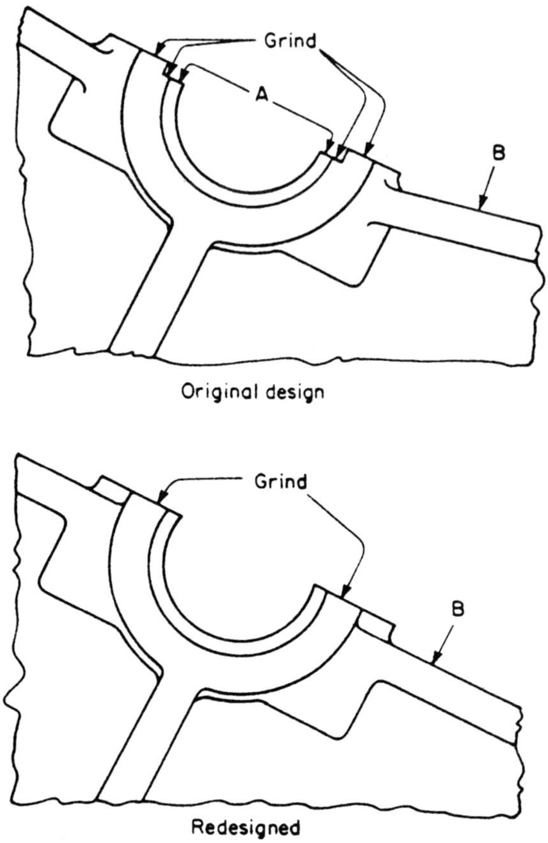

FIGURE 4.15.6 Redesign of a crankshaft bearing bracket. Economy was achieved in making the part suitable for grinding in a single setup. Shoulders at *A* were eliminated, putting all ground surfaces in one plane. Surface *B* was modified to eliminate access interference. (*Courtesy Machine Design.*)

5. When grooves or other forms are ground on a surface grinder, the same rule for corners as in cylindrical grinding applies: Corner relief is preferable to sharp or radiused corners.
6. Avoid blind cuts: designs that force the wheel to be stopped during the cut or reversed with too little clearance provided.
7. Designs should provide for minimum stock removal by grinding, especially if horizontal-spindle equipment will be used.
8. Avoid extremely thin sections; burning or warping can occur.
9. When possible, avoid dissimilar materials. Wheel loading and growth differences can cause problems.
10. Indicate clearly on drawings the permitted straightness and parallelism required.

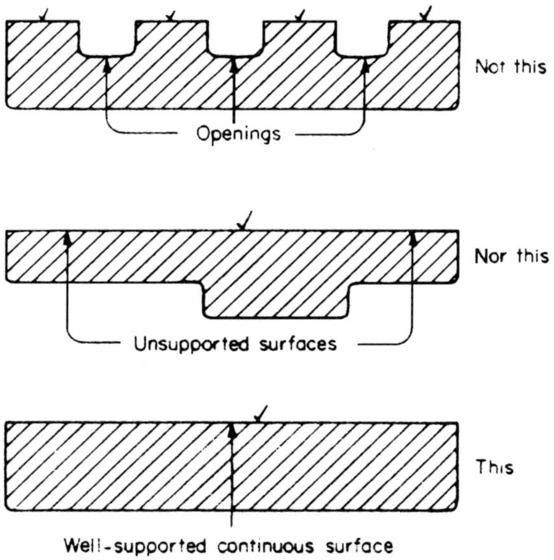

FIGURE 4.15.7 If flatness is very critical, avoid openings in the ground surface and poorly supported areas that will deflect under grinding-wheel pressure.

DIMENSIONS AND FINISHES

It is possible to meet thickness tolerances of less than ±0.0025 mm (0.0001 in) and flatness of less than one light band on surface grinders. However, in production, the more liberal tolerances shown in Table 4.15.5 are recommended as a guide.

Dimensional variations in flat-ground surfaces are affected by the same factors that affect cylindrical grinding: machine condition, accuracy and cleanliness of the chuck or fixture, suitability of the wheel and coolant used, wheel speed, depth of cut, and traverse rate, grindability of the workpiece material, uniformity of temperature, freedom of the workpiece from internal stresses, and other factors.

Straightness and flatness are strongly affected by operator technique, but if these are critical, the designer must help by calling for a stress-relief operation prior to finish grinding. Also, straightness should be related to length, e.g., "straight within 0.0001 in/ft" or "no more than 0.005-mm deviation from a straight line in any 60-cm section."

Surface finish is often a function of process time. The better the finish required, the longer it will take to grind a given area. Decreased feeds and slower traverse rates favor improved finishes, as does increased spark-out time—an important factor. Other factors that provide smoother surface finishes are the use of fine-grit abrasive wheels, the use of grinding fluids, slow traverse, diamond dressing of the wheel, the maintenance of good wheel balance, and the use of a hardened but high-grindability workpiece material.

TABLE 4.15.5 Recommended Tolerances for Flat-Ground Parts

Production	Tolerance
Vertical-spindle machines: rotary or reciprocating table	
Normal production	
Size	±0.025 mm (0.001 in)
Finish	0.8 μm (32 μin) and up
Tightest feasible	
Size	±0.0025 mm (0.0001 in)
Finish	0.05 μm (2 μin) and up
Horizontal-spindle machines	
Normal production	
Size	±0.025 mm (0.001 in)
Finish	0.5 μm (20 μin) and up
Short run	
Size	±0.0025 mm (0.0001 in)
Finish	0.2 μm (8 μin) and up
Toolroom	
Size	±0.0013 mm (0.000050 in)
Finish	0.05 μm (2 μin) and up

ABRASIVE-BELT GRINDING

Abrasive-belt grinding or *abrasive-belt machining,* as it is sometimes called, employs an abrasive-coated cloth belt instead of a solid grinding wheel. The belt passes over several pulleys, one of which, called the *contact wheel,* provides opposing pressure to the workpiece at the point of contact. Belts as much as 3 m wide allow the entire surface to be covered in one pass. Surface speeds of 1070 to 2280 m/min (3500 to 7500 ft/min) are used with high contact pressures.

Abrasive-belt equipment for surface grinding, cylindrical grinding (between centers), and centerless grinding is available. Internal grinding by belt is not feasible.

Abrasive-belt grinding is particularly useful for removing substantial amounts of metal from large surfaces. It is competitive and often superior to milling, broaching, planing, and shaping in such situations. Typical examples are removal of slag from steel-plate components used in the fabrication of construction equipment, finish machining of flat surfaces on diesel engine castings, and remachining of large, heavy foundry plates (which had become warped after years of service) to restore surface flatness.

Workpieces being abrasive-belt-ground tend to remain cool because the belt is effective in carrying away heat. Heat damage, therefore, is minimized. Burrs are also minimal. Form grinding is not feasible.

Metal-removal rates of 100 cm^3/min (6 in^3/min) are not uncommon, and with large machines and workpieces, 600 cm^3/min (37 in^3/min) is quite feasible. This rate is greater than that normally achieved with conventional milling even with large cutters and heavy cuts. The fastest metal removal is achieved with coarse-abrasive belts. Finer abrasives remove stock more slowly but provide a better surface finish. Setup and workpiece-handling times for abrasive-belt grinders are equivalent to those for conventional grinders and other machine tools. Belt-changing time is not a significant cost factor, but the cost of abrasive belts is higher per unit of metal removed than the cost of grinding wheels.

All grindable materials are also processible by the abrasive-belt process. Of the commonly ground materials, cast iron machines most rapidly.

Design recommendations for surface grinding with abrasive wheels also apply to the operation when performed with an abrasive belt. Differences, which stem from the special characteristics of belt grinding, are as follows:

1. Surfaces to be flat-ground can be cored out or relieved for much of the surface because interruptions in the surface actually aid the belt-grinding process. Consequently, this provides the designer with an opportunity to save material in the workpiece. When the size of the surface to be ground is reduced in this way, remaining sections to be ground should be at least 6 mm (¼ in) wide. (See Fig. 4.15.8.)
2. The configuration of the workpiece to be ground should be such that no projections, lips, or shoulders intersect the surface to be ground in such a way that the edge of the belt is used. The belt edge breaks down rapidly, preventing a proper grinding operation from continuing. (See Fig. 4.15.9.)

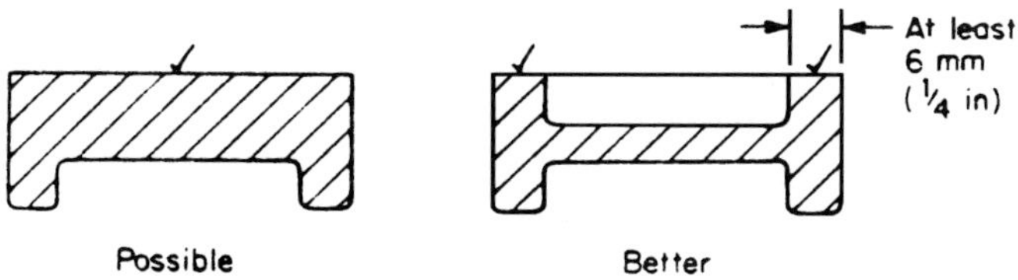

FIGURE 4.15.8 Redesign of a cast machine base to reduce weight and abrasive-belt-grinding time. Note that this may not be suitable if flatness limits are extremely close.

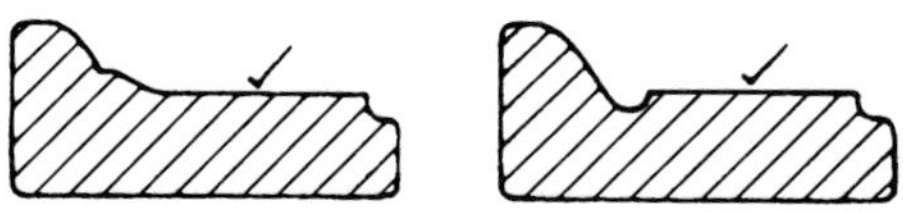

FIGURE 4.15.9 With abrasive-belt grinding, avoid projections, lips, or shoulders that would require the use of the edge of the belt.

Surface finish and dimensional accuracy of abrasive-belt grinding are improved if a fine-grit belt is used, but cutting speed and belt life are proportionally poorer. Coarse abrasives can provide dimensional control within ±0.05 mm (0.002 in), and fine belts can work to as fine as ±0.013 mm (0.0005 in). However, a practical minimum tolerance is ± 0.025 mm (0.001 in). Practical surface-finish tolerances if fine-grit belts can be used are 0.25 μm (10 μin) or 1.0 μm (40 μin) with a coarse-grit belt.

LOW-STRESS GRINDING

This is a variation of conventional grinding in which conditions are closely controlled to minimize residual stresses in the workpiece surface. Coarse wheels frequently dressed with low surface speeds and low infeed rates are used. This method is employed with heat-sensitive materials and when the component being ground is subjected to high stresses. Design considerations are identical to those for conventionally ground parts. Dimensional-tolerance and surface-finish specifications can also be the same as those for regularly ground components.

CHAPTER 4.16

HONED, LAPPED, AND SUPERFINISHED PARTS

R. W. Militzer, P.E.

Consulting Engineer
Fenton, Michigan

Honing, lapping, and superfinishing are all low-velocity abrasive processes. Each constitutes a final machining process whose purpose is to provide a refinement of geometry or finish of surfaces produced by another machining process.

HONED PARTS

Honing Process

Of the three chief low-velocity abrasive machining processes, the most common is honing. The honing process involves the application of one or more abrasive stones or "sticks" to a workpiece with area contact (instead of essentially line contact, as with grinding processes), precisely controlled pressure, and relatively slow motion in several directions simultaneously. The purpose is to generate an accurate surface configuration and an improved finish.

The process utilizes bonded abrasives, either silicon carbide, aluminum oxide, cubic boron nitride, or diamonds, as a cutting medium. In honing cylindrical or spherical figures, these abrasive elements are applied by using a combination of rotating and reciprocating motions.

The process can be performed on either automatic or manual machines. On automatic machines, it can be fully mechanized, including parts handling, machine actuation, size control, and workpiece qualification and segregation in and out of the honing machine. As a manual operation, honing is applied on parts small enough to be handheld, with the machine supplying reciprocating and rotating motions to the abrasive, along with the radial pressure required to force the abrasive against the workpiece (see Fig. 4.16.1).

Because honing will not generate true positioning of the bore but rather will follow the previously generated location, a self-alignment feature is provided in a honing fix-

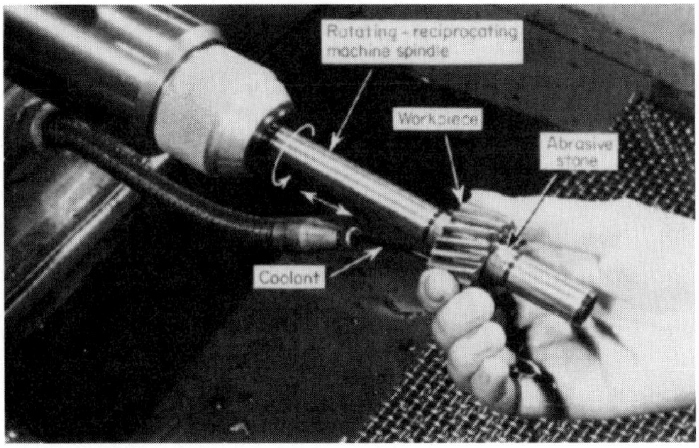

FIGURE 4.16.1 Honing the bore of a pinion gear to accurate dimensions. (*Courtesy XLO Micromatic.*)

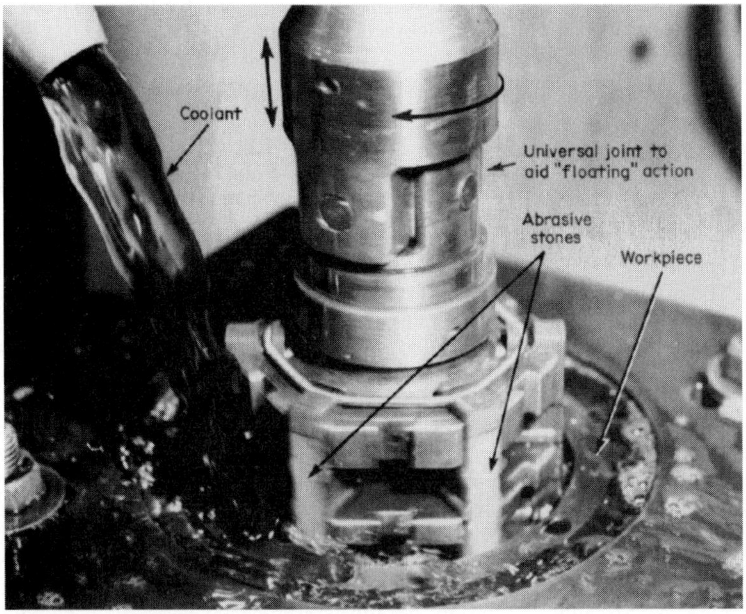

FIGURE 4.16.2 Honing a diesel-engine cylinder liner. (*Courtesy XLO Micromatic.*)

ture that retains the workpiece or in the driving element of the honing tool, as shown in Fig. 4.16.2.

Workpiece Characteristics

The honing process can be applied to any geometric figure, but it is used primarily for inside diameters and flat surfaces. Other shapes generated by the honing process are spherical, toroidal, elliptical, arcuate, and cylindrical. Bores as small as 2.4 mm ($^3/_{32}$ in) in diameter by 1.6 mm ($^1/_{16}$ in) long up to 1.2 m (48 in) in diameter by 9.1 m (30 ft) long can be finished by the honing process. In honing external cylindrical surfaces, the common area of activity includes parts ranging from 6.3 mm ($^1/_4$ in) in diameter by 12.7 mm ($^1/_2$ in) long to 450 mm (18 in) in diameter by 9.1 m (30 ft) long. The application of the honing process to flat surfaces is usually confined to areas of less than 645 cm^2 (100 in^2), although this limitation is not inflexible. Spherical diameters (see Fig. 4.16.3) ranging from 3 mm ($^1/_8$ in) to 300 mm (12 in) can be honed. Both internal and external spherical surfaces are processed by the honing operation.

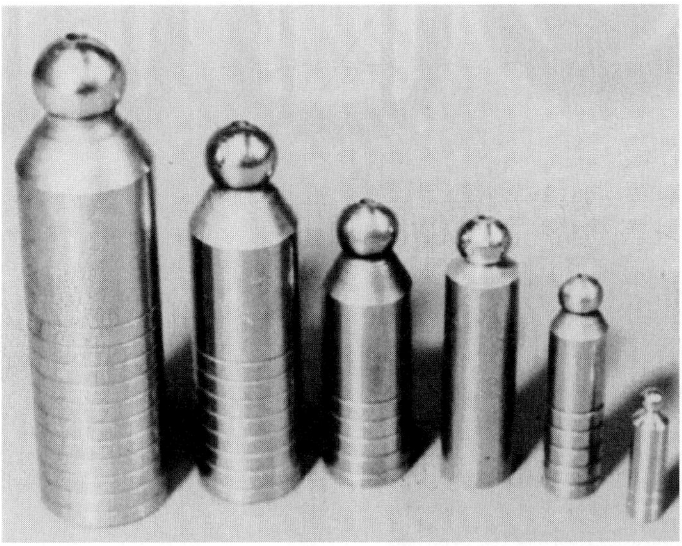

FIGURE 4.16.3 Ball-joint members and other spherical surfaces are given an accurate finish by honing. (*Courtesy XLO Micromatic.*)

The honing operation, because it is an area-abrading process, can correct previously generated errors such as out-of-roundness, waviness, and taper, as shown in Fig. 4.16.4.

Because of the multidirectional action of the abrading elements, the typical surface has a crosshatch pattern as shown in Fig. 4.16.5. This type of surface finish is very efficient in a moving bearing application such as an automotive-cylinder bore or a valve-body bore, since the peaks carry the load of the mating part and the valleys in the surface structure provide reservoirs for lubricating oil.

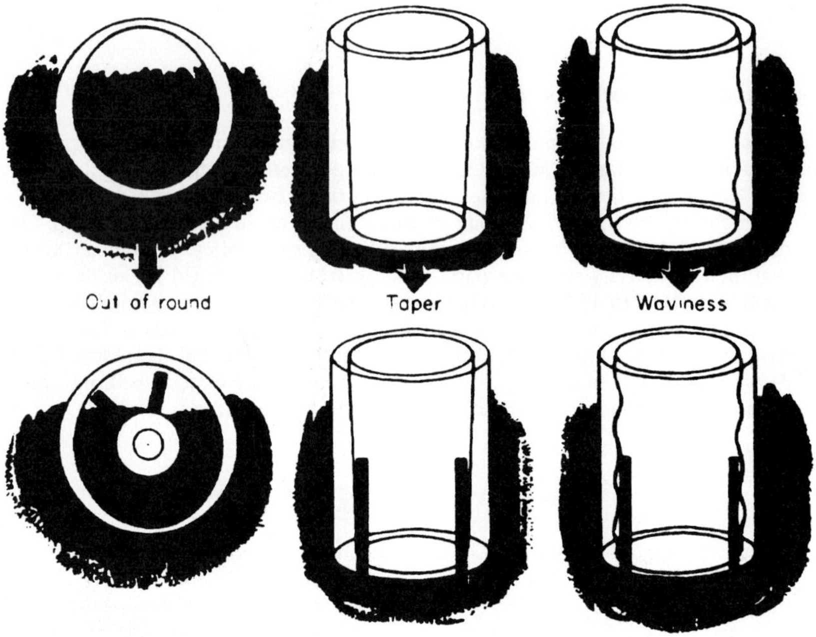

FIGURE 4.16.4 The abrasive stones of a hone create a true cylindrical surface by abrading off high spots and other inaccuracies. (*Courtesy XLO Micromatic.*)

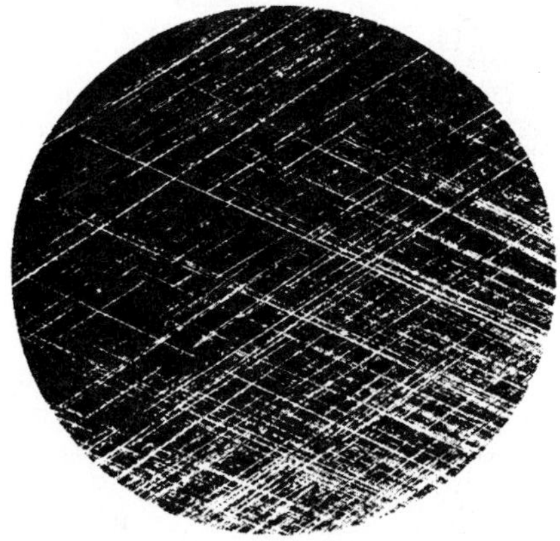

FIGURE 4.16.5 Typical crosshatch pattern of the scratch marks on a honed surface. (*Courtesy XLO Micromatic.*)

Another advantage of honing is the freedom from heat effects. The low-velocity abrasive action avoids surface damage due to heat, as can occur with normal grinding operations.

Gear teeth, races for ball and roller bearings, crankpin bores, drill bushings, gun barrels, piston pins, and hydraulic cylinders are often processed with honing operations.

Production Quantities

Honing can be economical for all levels of production. Under conditions of one of a kind or very limited quantities, manually controlled drill presses and lathes can be used in combination with standard honing tools. Setup time is very short.

For the better-equipped shop, general-purpose horizontal or vertical honing machines are employed. These provide both reciprocating and rotary motion and also may include built-in measuring apparatus. Setup time for such honing equipment is also very short. Short-run applications include the bores of hydraulic-valve bodies, machine-tool spindle bearings, and aircraft engine components. A typical small-lot job honing setup may have fewer than 10 valve bodies of one size and be changed over to other sizes as many as 6 or 8 times in an 8-h working shift.

When honing is applied to mass-production conditions, specialized equipment such as that described under "Honing Process" is employed. Typical volume runs of honed automotive engine blocks run as high as 2 million per year. Production runs of connecting rods, transmission gears, bearing bores, etc., also run into the multimillions on an annual basis.

Materials for Honing

Any material that can be machined in conventional cutting-tool operations can be honed. The most common materials to be honed are steel, cast iron, and aluminum. Each of these materials, even steel in its most extreme cases of heat-treated hardness, is easily honable. Other materials such as bronze, stainless steel, plastics, etc. can be processed by low-velocity abrading.

Design Recommendations

On the surface of a workpiece being honed, allowance must be made for the multidirectional application of the abrading members. Accurate geometric characteristics can be generated only when the abrading elements can be applied uniformly and repetitiously over the entire area of the surface to be honed. For example, projections such as shoulders, bosses, etc., must be avoided when designing a workpiece for the honing operation. The same general rule applies to the honing of other geometric figures such as spherical surfaces, flat surfaces, and outside diameters. The area adjacent to and beyond the surface to be honed must be free of interfering projections.

When honing an inside diameter, the abrading elements must overrun the ends of the bore by an amount equal to one-fourth to one-half of the length of the abrasive used (see Fig. 4.16.6).

Keyways, ports, undercuts, and other surface interruptions frequently present problems on the surface to be honed. Because an abrading element has a tendency to "overcut" whenever an edge surface is passed over by the abrasive, some degree of "washout," or depression of a surface, can be expected around the edge of the surface interruption. When they are essential to the functional design

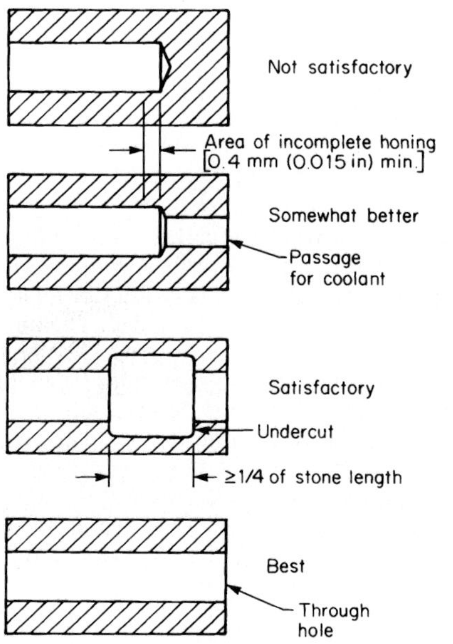

FIGURE 4.16.6 Design recommendations for internal cylindrical surfaces which are honed.

of the workpiece, interruptions such as keyways or ports should be kept as small as the limits of good design will permit so that the abrading elements can pass over these interruptions with minimal effect.

In designing workpieces that require application of the honing process, it is important that the part have easily identifiable locating surfaces. Also, the part must have convenient clamping pads that will not cause workpiece distortion during application of the process.

Recommended Tolerances

In general, the honing process can generate as close geometric and surface-finish conditions as any metal-working process. It is possible and practical to generate cylinder bores in automotive engine blocks on a high-volume-production basis to geometric tolerances of roundness, straightness, and size control within 0.008 mm (0.0003 in). On hardened-steel bores under 25 mm (1 in) in diameter, the honing process can develop bore or outside-diameter characteristics of 0.0008-mm (0.000030-in) accuracy on a continual-production basis but with some sacrifice of production rate. With additional care and cost, still closer values can be obtained. For normal production, however, the values in Table 4.16.1 are recommended.

Hardened-steel parts can be honed to any desired degree of surface finish from 0.05 μm (2 μin), since this is primarily a factor of the grit size of the abrading element, the coolant used, the surface speeds applied, and the forces with which the abrasives are applied to the surface of the work. The degree of surface-finish refinement obtainable on other materials is limited by the molecular and granular structure of the material. Very fine surface finishes are more costly. However, finishes of 0.25 to 0.4 μm (10 to 15 μin) can be obtained easily; they constitute a valid range for drawing specifications.

LAPPED PARTS

Lapping Process

The lapping process involves the application of a master form called a *lap* and an abrasive that is moved continuously in a random pattern over the surface of the workpiece at a low velocity and lightly applied pressure. The abrasive is usually in loose-powder or paste form but may be a solid block or an abrasive cloth or paper. The purpose of the operation is to refine the finish and dimensions of the surface to a finer degree than would be obtainable in any metal-cutting or -grinding operation. Peaks in the surface are removed by a continuous rubbing contact until the desired geometry, dimension, and surface finish are obtained. The amount of material removed is very small. (See Fig. 4.16.7.)

With loose abrasives, the lap material is always softer than the workpiece. Cast iron, steel, soft nonferrous metals, wood, and leather can be used. Abrasive particles become embedded in the soft lap and cut the workpiece when the lap is moved across it. The process differs from honing in that motion is usually in one direction, light pressure is employed, and the abrasive is generally in loose form. Lapping is more limited than honing in stock-removal capabilities.

The process has evolved from a purely manual operation to one that may be either

TABLE 4.16.1 Recommended Dimensional Tolerances for Honed Parts

Nominal dimension	0 to 6.3 mm (0.25 in)	To 12.7 mm (0.50 in)	To 19 mm (0.75 in)	To 25 mm (1.0 in)	To 50 mm (2.0 in)	To 100 mm (4.0 in)
Normal tolerance	+0.013, −0.000 mm (+0.0005, −0.0000 in)	+0.013, −0.000 mm (+0.0005, −0.0000 in)	+0.013, −0.000 mm (+0.0005, −0.0000 in)	+0.013, −0.000 mm (+0.0005, −0.0000 in)	+ 0.020, −0.000 mm (+0.0008, −0.0000 in)	+0.025, −0.000 mm (+0.0010, −0.0000 in)
Tight tolerance	+0.005, −0.000 mm (+0.0002, −0.0000 in)	+0.008, −0.000 mm (+0.0003, −0.0000 in)	+0.008, −0.000 mm (+0.0003, −0.0000 in)	+0.010, −0.000 mm (+0.0004, −0.0000 in)	+0.013, −0.000 mm (+0.0005, −0.0000 in)	+0.002, −0.000 mm (+0.0008, −0.0000 in)

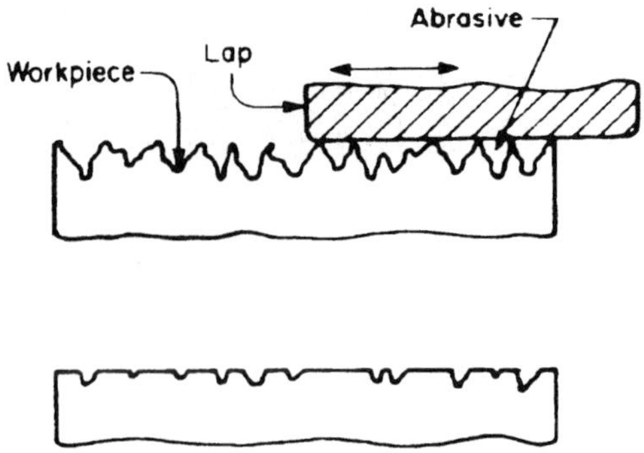

FIGURE 4.16.7 The lapping process.

FIGURE 4.16.8 A production lapping machine for flat surfaces. Parts can be unloaded and loaded on one table while lapping takes place on the other table. (*Courtesy Lapmaster Division, Crane Packing Co.*)

manual or completely automatic. Figure 4.16.8 illustrates a production machine in which the lapping action is automatic.

Workpiece Characteristics

The types of parts most frequently benefited by lapping operations are those which require one or more of the following: (1) an improved surface finish, (2) extreme accuracy, (3) correction of minor imperfections in shape, and (4) precise fit between mating surfaces. Many components have mating metal surfaces that must be sealed together to contain a pressure differential. Examples are valve spools, fuel-injector plungers, seal rings, piston rods, valve stems, cylinder heads, and spherical valve seats. When the function of a part is primarily to maintain a seal in a fluid or gas chamber, the surface of the sealing members must be so highly refined that normal metal-cutting operations will not meet the requirements, and a lapping operation is used to obtain a final surface finish and geometry.

The lapping operation is most frequently applied to cylindrical, flat, or spherical surfaces. Both male and female cylindrical, conical, and spherical surfaces can be lapped, as can gear teeth, cams, and closely fitting machine parts. Sizes range from holes or pins as small as 0.8 mm (0.030 in) to as large as 250 to 300 mm (10 to 12 in) in diameter, with length proportional to the diameter as required for good bearing conditions.

Flat surfaces requiring the lapping operation generally are found on small workpieces ranging from 6 to 80 cm^2 (1 to 12 in^2), but in some cases parts as large as 1300 cm^2 (200 in^2) can be lapped successfully.

Other parts that are lapped are gauge blocks, plug and ring gauges, thread gauges, drill-bushing inside diameters, gears, cam rolls, and bearings. Figure 4.16.9 illustrates typical flat lapped parts.

FIGURE 4.16.9 Collection of typical flat lapped parts. (*Courtesy of Lapmaster Division, Crane Packing Co.*)

Production Quantities

Lapping operations, on the basis of an economical lot size, span the range from one-of-a-kind fabrication to high-level mass production. At low volumes of production, hand lapping is employed. Since laps are relatively inexpensive and may be usable for a variety of parts, tooling costs are quite low. Equipment costs are also nil or minimal because the lapping motion is provided manually or with the aid of a drill or lathe spindle.

When hand lapping is considered part of the manufacturing process, component interchangeability generally is sacrificed. Parts that are hand lapped are usually mated for life, although certain precision measuring techniques make "lap to print" a feasible process. Production rates for hand-lapped parts, of course, may be quite low.

In recent years, high-volume automatic lapping machinery has been introduced as a final operation for certain workpieces with extremely high production requirements. Centerless lapping machines are available that can lap as many as 1000 pieces per hour of parts of the general configuration of piston pins, providing consistent repeatability of precision on an automatic basis. Machines are also available to perform automatic lapping of flat surfaces at production rates as high as 2000 pieces per hour.

It should be noted that metal removal by lapping is generally considered to be expensive. In all cases this operation should be considered only as a final step in the manufacturing process to provide refinements not possible with machining and grinding operations.

Materials for Lapping

Almost any solid material used in industrial processes can be lapped. Steel and cast iron are the most frequently employed materials for parts requiring lapping operations, but other materials such as glass, aluminum, bronze, magnesium, plastics, and ceramics can be finished by this process. Soft materials that can retain loose abrasive are less satisfactory for lapping than hard materials because trapped abrasive can later cause wear on a part that mates with the lapped part. One solution is to use a self-destructing type of abrasive, i.e., one that breaks down quickly and loses its cutting power.

Design Recommendations

When designing a part to be lapped, several important factors must be considered. A clear, uninterrupted area free of shoulders and projections is most easily lapped. Any shoulders, projections, or other interruptions require the lap to be applied so as to avoid the projected interference when the lap is moved back and forth across the work surface; this is difficult. Figures 4.16.6 and 4.16.7 apply equally to lapping and to honing, except that coolant passages are less necessary with lapping. Clearance recesses that permit overrun of a lap to obtain flatness or roundness and surface finish are required.

When two opposite sides of a workpiece must be machine lapped to a highly refined parallelism, the two surfaces should extend beyond other surfaces of the workpiece so that the lap and machine table can make unobstructed contact with the surfaces to be lapped.

Recommended Tolerances

Extremely close tolerances and fine surface finishes can be obtained with proper utilization of a lapping operation. Surface finishes as fine as 1 μin AA can be obtained by lapping. A common approach to inspecting flat surfaces that have been lapped is to

TABLE 4.16.2 Recommended Dimensional Tolerances for Lapped Surfaces

	Normal tolerance	Tight tolerance
Diameter or other dimension	±0.0006 mm (0.000025 in)	±0.0004 mm (0.000015 in)
Flatness, roundness, or straightness	±0.0006 mm (0.000025 in)	±0.0003 mm (0.000012 in)
Surface finish	0.1–0.4 μm (4–16 μin)	0.025–0.1 μm (1–4 μin)

use a reflective light-band instrument. It is not uncommon to find flat surfaces lapped to one-light-band flatness tolerance (0.000012 in). For cylindrical workpieces, lapping can, if necessary, provide roundness within 0.00013 mm (0.000005 in). Minimum size tolerances obtainable by lapping are related directly to the surface characteristics and geometric accuracy of the preceding operation. The limited stock-removal capability of lapping indicates the importance of obtaining accurate boundary dimensions prior to the lapping operations and utilizing the lap for the purpose for which it was intended, that of refining final dimensions and surface finish.

Table 4.16.2 presents recommendations for tolerances of lapped parts under normal production conditions.

SUPERFINISHED PARTS

Superfinishing Process

This process is generally considered as falling between honing and lapping. It involves the application of an area-contacting abrasive block to the workpiece with controlled bidirectional movement between the two. The abrasive is loosely bonded so that it wears to the contour of the workpiece. Little or no dressing is required. The primary motion is at a rate of 3.7 to 4.6 m/min (12 to 15 ft/min), with a side-to-side oscillation of about 5 mm ($^3/_{16}$ in) at a frequency of up to 800 cycles/min. An oil-based cutting fluid is used. From 0.005 to 0.025 mm (0.0002 to 0.001 in) of stock is removed. A production superfinishing operation is illustrated in Fig. 4.16.10.

Workpiece Characteristics

Superfinishing is not considered a stock-removal process. It is utilized almost wholly as a surface-refining process with additional benefits available in improvements to the geometric characteristics of the workpiece.

The superfinishing process has been applied to many surface configurations, e.g., inside and outside diameters, conical surfaces, spherical surfaces, flats, flutes, keyways, and recesses. It is most frequently applied when the workpiece requires a highly refined surface finish (often to the degree frequently referred to as *mirror finish*) for sealing or bearing surfaces. Results obtained from the superfinishing process can be very similar to those achieved by lapping.

Typical parts finished by superfinishing are pistons, piston rods, shaft-sealing surfaces, crank pins, valve seats, bearing races, and steel-mill rolls.

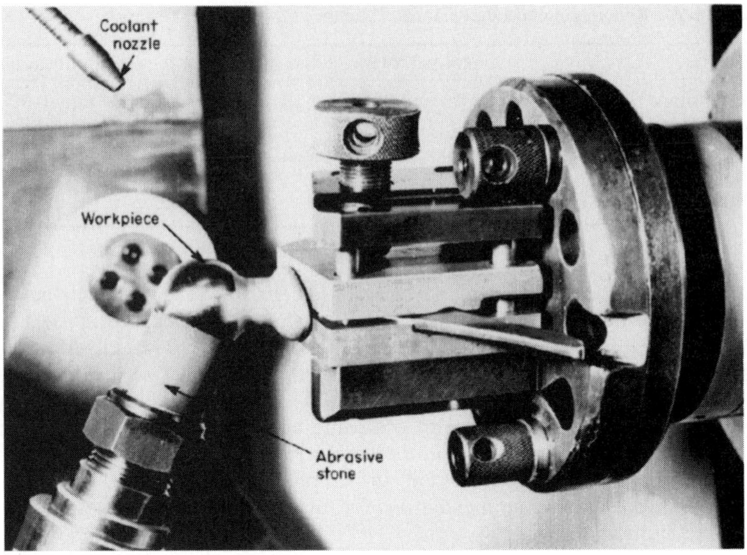

FIGURE 4.16.10 Superfinishing of a ball-joint component. The abrasive stone assumes a shape to fit the ball. Motion between the ball and the stone removes surface roughness and geometric imperfections from the ball surface. (*Courtesy Giddings & Lewis, Inc.*)

Production Quantities

Superfinishing is most often applied to high-volume production. It is frequently used in conjunction with completely automatic parts-handling systems. Production rates on an operation of this type can range as high as 240 pieces per hour.

General-purpose superfinishing equipment is also available. A typical use of larger machines of this type is the finishing of rolls for steel mills for which production rates may range from 4 to 6 parts per hour.

Materials for Superfinishing

It is possible to apply this process to any metallic parts, both ferrous and nonferrous. Superfinishing is most frequently applied to hardened and ground alloyed-steel components.

Design Recommendations

The same needs for accessibility of the workpiece surface and for relief areas at the ends of finished surfaces exist for parts to be superfinished as for those to be honed or lapped. Keyways, holes, and grooves or other interruptions in surfaces to be superfinished should be avoided.

Recommended Tolerances

The degree of surface-finish refinement attainable is related to the grit size of the abrasive used in the process, the applied force, relative surface speeds, the consistency and cleanliness of the coolant application, and the length of time during which the process is applied. As a general rule, the longer the application, the finer the abrasive grit size, and the lighter the force applied, the higher the degree of surface refinement attainable.

Surface finishes of 0.025 to 0.075 μm (1 to 3 μin) are not uncommon in the application of superfinishing. Under precisely controlled conditions with the highest-quality hardened-steel workpiece material, there are instances of obtaining surface finishes to less than 0.025 μm (1 μin). Recommended drawing specifications for most production applications, however, are 0.1 to 0.2 μm (4 to 8 μin), with tighter limits being specified only when essential and when preceding operations provide a relatively smooth surface.

With regard to refinement of geometric tolerances and control of workpiece size, results are limited by the relatively low stock-removal capability of the process.

CHAPTER 4.17
ROLLER-BURNISHED PARTS

C. Richard Liu
School of Industrial Engineering
Purdue University
West Lafayette, Indiana

THE PROCESS

Roller burnishing is a method of surface refinement by pressure rolling without removing metal. The roller-burnishing tool consists of a series of tapered, highly polished, and hardened rolls positioned in slots within a retaining cage. The tool is sized so that the rolls develop within the workpiece a pressure that exceeds the yield point of the softer workpiece. As the rolls rotate, they compress the peaks of the workpiece surface pattern into the valleys. They work-harden and slightly compact the surface by producing localized plastic deformation. Figure 4.17.1 illustrates the process.

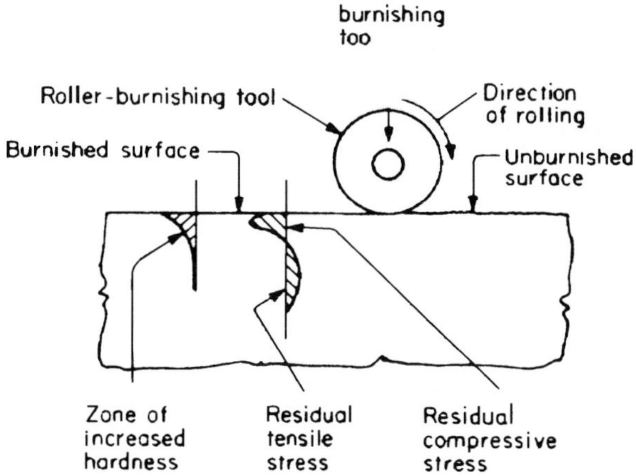

FIGURE 4.17.1 Schematic illustration of the roller-burnishing process.

In a variation of this process, the rollers ride on a cam instead of a cone or cylinder. Called *bearingizing,* this combines a peening action with the regular rolling.

Results very similar to regular roller burnishing are obtained on straight bores by *ball burnishing,* or *ballizing.* A smooth, round ball slightly larger than the bore is pushed through the hole, leaving a closely controlled finish.

Roller burnishing is a finishing process used in conjunction with or in replacement of reaming, honing, lapping, and/or grinding.

TYPICAL CHARACTERISTICS AND APPLICATIONS

The major application of this process is to improve the geometric and mechanical properties of surfaces of revolution. The process cannot be applied to surfaces with complicated geometry. Inside or outside cylindrical surfaces for hydraulic application, tapered bores, flat seats for seals and valves, conical spindles, and fillets are typical surfaces to which roller burnishing is applied (see Fig. 4.17.2).

The process cannot be applied to parts with as complicated geometry as shot peening can handle (see Chap. 8.7). It also requires uniformity in section thickness or very thick sections.

Geometric and Mechanical Properties

The process is usually employed to improve the following aspects of a mechanical component: (1) dimensional accuracy (possibly up to 0.0066 mm, or 0.00025 in), (2) surface finish, (3) surface hardness, (4) wear resistance (excellent for porous material), and (5) fatigue and corrosion resistance.

Dimensional Accuracy

Parts machined by other processes can be brought to final size by roller burnishing. Improving the dimensional consistency of parts to be press-fit is a common application. There are, however, limitations on the degree of geometric improvement possible. To end straight and circular after burnishing, a hole must be straight and circular before burnishing.

Surface Finish

Cylindrical holes have been roller-burnished in a single pass to a finish of from 0.05 μm (2 μin) to 0.35 μm (14 μin), depending on the initial finish and the workpiece material. High-quality surface finish (under 0.25 μm, or 10 μin) without embedded abrasive is essential for seals that control leakage of air, oil, or water.

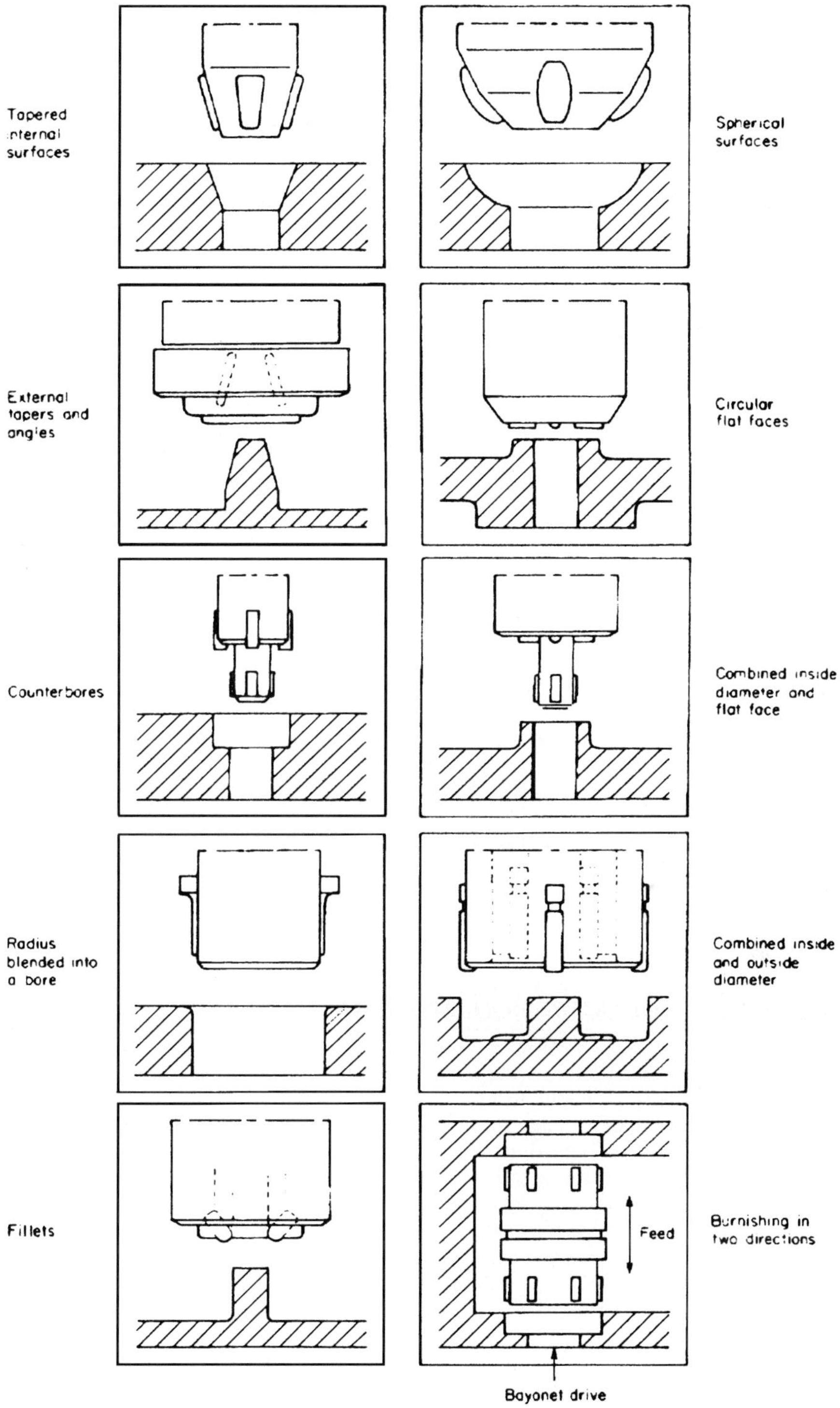

FIGURE 4.17.2 Typical roller-burnished-part configurations and the burnishing tool used. (*Courtesy Sandvik, Inc.*)

Surface Hardness

The rolled surface is plastically deformed and work-hardened. Surface hardness (BHN) increases from 5 to 25 percent with penetration from 0.13 mm (0.005 in) to 12.5 mm (0.5 in), depending on the size and pressure of the rolls. The increase often eliminates the need for heat treating or other surface treatment as a means of improving wear resistance.

Fatigue and Corrosion Resistance

Surface pits, scratches, tool marks, and openings in the grain structure are greatly reduced by roller burnishing. The chance of holding reactive substances or contaminants by surface pits is reduced, and corrosion resistance is increased. The rolling process also closes cracks and introduces compressive residual stress beneath the rolled surface, most of the time more deeply than shot peening; fatigue resistance is therefore improved. Roller burnishing is especially effective in improving the fatigue resistance of fillets. The compressive surface residual stresses of roller burnishing are invariably balanced by internal tensile stresses. This may result in a lowered elastic limit or creep resistance.

Part Size

Tools are available from one supplier's stock to size holes as small as 4.8 mm ($^3/_{16}$ in) and as large as 380 mm (15 in) and from 3.2 mm ($^1/_8$ in) to 150 mm (6 in) for outside-diameter rolling. The maximum length of the rolled surface depends on the maximum tool deflection allowable during rolling. For example, a particular toolmaker's specification for maximum work length for holes 13 mm ($^1/_2$ in) or smaller is 120 mm (4.56 in) and for larger inside diameters is 230 mm (9.06 in). It is claimed that the effective work length is practically unlimited for inside diameters larger than 59 mm (2.3 in).

ECONOMIC PRODUCTION QUANTITIES

The process is applicable for a wide range of production quantities. Tooling costs are modest, and normally available drill presses, lathes, screw machines, and other general-purpose equipment are suitable for the operation. Production rates are high.

The tool life of roller-burnishing tools is much longer than that of abrasive-finishing tools. The wear parts in a roller-burnishing tool are the mandrel and rolls and occasionally a roll cage, which are inexpensive. Roller burnishing can be carried out with less skilled labor and at higher production rates than abrasive processes.

Economic benefit has been reported for moderate and mass-production quantities. Economic justification for very small quantities depends on the availability of burnishing tools of the necessary size.

SUITABLE MATERIALS

Any ductile or malleable material can be roller-burnished. The following materials are documented with at least some degree of success: wrought steel at hardness levels ranging up to R_c 45, gray malleable and nodular cast iron, stainless steel, cast and wrought aluminum, copper, brass, and cast and wrought magnesium. (This list is not intended to be exclusive of other materials.)

Successful application also has been reported for rolling very hard material.* The process also has good potential for powdered metal, which is hard to grind. Roller burnishing does not close the pores to restrict the flow of lubricant in the burnished surface.

DESIGN RECOMMENDATIONS

Wall Thickness

The walls of burnished holes should be uniform or thick enough to withstand the pressure of the rollers with only local plastic deformation at the contact line, without producing general yielding of the entire wall. Thin-walled sections may need to be supported in a sleeve to prevent general yielding. Workpieces with walls as thin as 1.6 mm (0.060 in) can be burnished successfully with caution.

Cutouts and Interrupted Holes

With interrupted holes, it is troublesome to get and maintain tool support. The burnishing tool must have many small rolls. Usually there is no problem if intersecting holes, keyways, and other interruptions account for less than 10 percent of the circumference. Otherwise, burnishing should be sequenced before machining. This alternative also may solve the problem presented by thin-walled workpieces.

Blind Holes

Through holes are preferable to blind holes if roller burnishing is to be performed. With blind holes, a length of about 1.5 mm (0.060 in) from the bottom cannot be reached by the tool. If it is desired to roll the hole closer to the bottom, enough clearance should be given in the part's design. (See Fig. 4.17.3.)

*S. E. Kalen and J. T. Black, "The Anatomy of a Roller Burnished Surface, " *Proceedings of the International Conference on surface Technology*, May 1973.

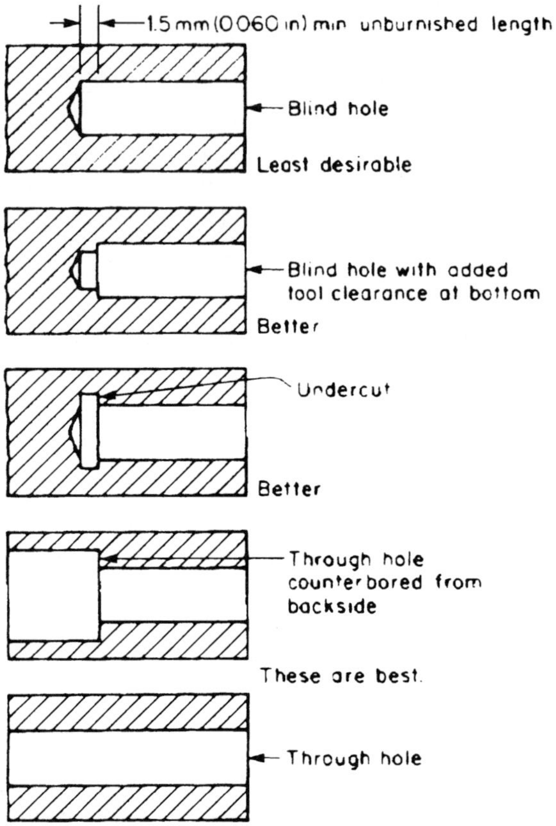

FIGURE 4.17.3 Through holes are preferable to blind holes.

Stock Allowance for Roller Burnishing

The process changes the dimension of the burnished surface slightly, and allowance should be made for this fact. The diameter of holes normally increases by between 1 part per 1000 on larger holes and 2 parts per 1000 on smaller holes. However, the stock allowance required depends on the surface profile produced by the prior machining process. A surface profile with a high peak-to-valley pattern but with uniformity otherwise is excellent for burnishing. A turned outside cylindrical surface is of this type, and a reduction from 5 μm (200 μin) to 0.15 μm (6 μin) in surface finish is possible. For the inside of a cylindrical surface, a conventional reamed finish is not a good starting point. The reamer must be modified according to the roller-burnishing-tool manufacturer's recommendations. A fine-ground surface leaves little material to be rolled down by burnishing. Such a part can be burnished only lightly insofar as sizing is concerned, but its mechanical properties may still be improved. Typical stock allowances for burnishing are shown in Table 4.17.1. The exact allowance for a given job should be determined by experimenting with the actual part.

TABLE 4.17.1 Stock Allowance–Surface Finish Chart

		Internal surfaces			External surfaces	
		Surface finish, μin			Surface finish, μin	
Workpiece size range, in	Stock allowance, in	Machined	Roller-burnished	Stock allowance, in	Machined	Roller-burnished
			High ductility			
0.125–0.484	0.0004	80	8	0.0004	80	8
	0.0007	125	8	0.0006	100	8
0.500–1.000	0.0007	60	8	0.0005	60	8
	0.0015	125	8	0.001	180	8
1.031–2.000	0.001	60	8	0.0007	100	8
	0.002	125	8	0.001	180	8
2.031–6.500	0.0015	60	8	0.001	125	8
	0.002	125	8	0.0015	300	8
	0.003	200	8	0.002	500	8
			Low ductility			
0.125–0.484	0.0004	80	18	0.0003	60	18
	0.0007	100	18	0.0005	90	18
0.500–1.000	0.0007	90	18	0.0005	100	18
	0.001	125	18	0.0007	140	20
1.031–2.000	0.001	125	18	0.0005	100	18
	0.0015	180	20	0.001	180	20
2.031–6.500	0.0015	120	18	0.001	125	18
	0.002	200	24	0.0015	200	20

Note: The table is a guide only, derived from experiments. Under particular conditions, the results may be slightly different. The table indicates that in the 0.500- to 1.000-range, a hole machined in a high-ductility material to 125 μin and 0.0015 in smaller than the burnishing tool size will be burnished to 8 μin. If the hole is finished to 60 μin before burnishing, only 0.0007 of stock need be left for burnishing to 8 μin.

High-ductility materials have more than 18% elongation and less than R_c 32. They include annealed steel, stainless steel, aluminum, brass, bronze, and malleable iron. Low-ductility materials have less than 18% elongation and a maximum hardness of R_c 40. They include gray cast iron, modular iron, heat-treated steel, magnesium alloys, and hard copper alloys.

Source: Courtesy Sandvik, Inc.

Rolled Fillets

When a fillet is to be rolled, the radius machined should be smaller than the radius called for in the part print. For example, if the drawing for the finished part calls for a 1.5-mm radius on the fillet, the prerolled fillet should be machined to 1.35 mm, or 10 percent less. When the fillet is rolled to size, uniform stressing will be produced.

Tapers

Tapers of 15° or less can be rolled successfully with proper control. A narrow land on the taper is more amenable to rolling than a wide one. The wall thickness surrounding the burnished surface should be uniform to avoid creating an egg shape.

DIMENSIONAL FACTORS AND TOLERANCES

The accuracy and quality of finish obtained by roller burnishing depends on (1) workpiece hardness, ductility, and porosity; (2) the accuracy and finish generated by the preceding operations, and (3) the amount of movable metal left on the surface to be roller-finished. Any diameter difference before burnishing will vary the surface finish from workpiece to workpiece.

In a cylindrical hole, a tolerance of ±0.015 mm (±0.0006) can be obtained if the hole is machined before roller burnishing to within ±0.05 mm (±0.002 in). Closer tolerances can be held under closely controlled conditions.

CHAPTER 4.18

PARTS PRODUCED BY ELECTRICAL-DISCHARGE MACHINING

Originally prepared by

Stuart Haley

Colt Industries
Davidson, North Carolina

Updated by

Glenn Coggins

Agie U.S.A., Ltd.
Davidson, North Carolina

THE EDM PROCESS

Electrical-discharge machining (EDM) is a material-removal process that uses a series of fine, controlled electrical sparks or discharges to erode the workpiece material. These discharges, which are generated many thousands of times per second, pass from the electrode to the workpiece. The spark energy vaporizes a minute, very localized area of the workpiece. The vaporized material is immediately resolidified into small particles by the cooling effect of the dielectric fluid that is present in the gap between workpiece and electrode. The small particles are then flushed from the gap, allowing the process to be repeated. As the process is repeated, a cavity is produced that mirrors the electrode shape.

The EDM process is applied in two primary systems: die-sinker (or ram) systems and wire (or wire-cut) systems. While there is a process of EDM grinding, it is not in common use.

Die-Sinker EDM

These systems utilize a shaped electrode that is fed into the workpiece. There may be single (Z) axis servo-driven systems that erode only in the vertical direction or more sophisticated CNC systems that allow orbiting and contouring capability in multiple

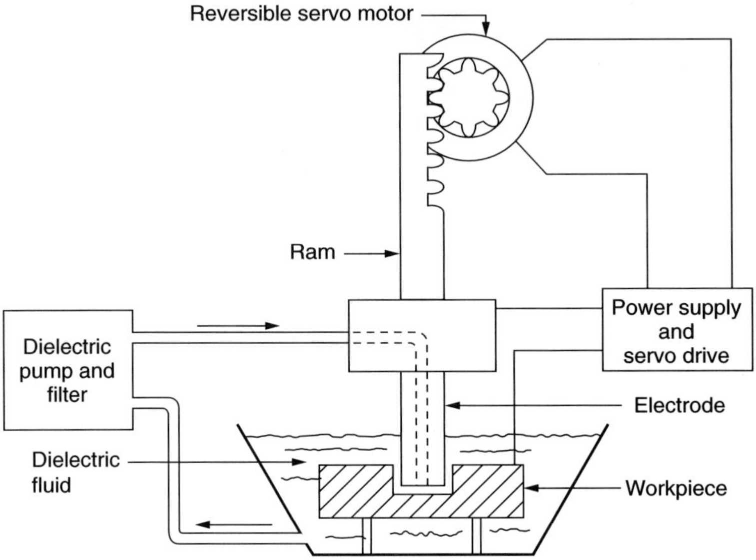

FIGURE 4.18.1 Basic elements of die-sinking EDM. (*Courtesy American Machinist.*)

axes, thus distributing electrode wear or creating special shapes. Figure 4.18.1 illustrates the die-sinker EDM process.

Electrodes are made of a number of electrically conductive materials, the most common being graphite and copper. Other materials in use are copper-tungsten, silver-tungsten, brass, steel, carbide, and tungsten. The electrode is made undersize by an amount appropriate to the discharge power being employed.

Wire EDM

This is a variation of the process in which a constantly moving strand of wire replaces the shaped electrode. The wire of 0.025 to 3 mm (0.001 to 0.013 in) diameter passes through the workpiece in the vertical axis and cuts a contour in the horizontal plane. The operation thus is somewhat like contour band sawing, except that spark discharges from the wire rather than cutting teeth remove the workpiece metal. This process is illustrated in Fig. 4.18.2.

In both processes, multiple cuts are used when close tolerances and fine finishes are needed.

CHARACTERISTICS OF EDM'D PARTS

Parts produced by the EDM process generally have one or more of the following characteristics:

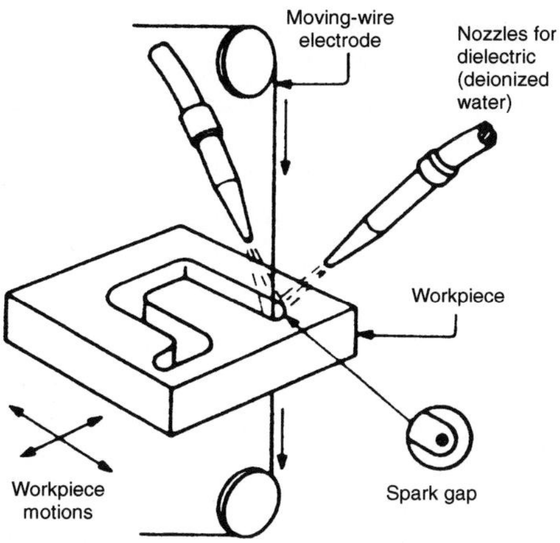

FIGURE 4.18.2 Wire EDM produces the desired shape of cut by making a contoured slit through the workpiece material. (*Courtesy American Machinist.*)

1. They include intricate shapes, usually internal shapes, that are difficult or impossible to machine conventionally.
2. The part is made of hardened metals or other materials that are difficult or are not feasible to machine by conventional machining methods.
3. They are of such value that the relatively slow material-removal rate of EDM is allowable.
4. They are made of electrically conductive materials.
5. They are delicate and must be shaped with a process that presents very low cutting pressure.
6. Deep, narrow slots or holes are required, especially if they are not round or have shallow entrance angles.
7. Hardening is required, but heat treatment after machining could cause severe distortion or other problems.
8. They are burr-free, eliminating secondary operations that would be required with other processes.

The limitations of parts produced by EDM are as follows:

1. The workpiece must be electrically conductive.
2. There may be a thin—0.0013 to 0.13 mm (0.00005 to 0.005 in)—layer of highly stressed and minutely cracked metal on the cut surface. This is more prominent on older die-sinking EDM equipment. Newer generators and energy-monitoring equipment, especially with wire EDM machines, can reduce this layer greatly.
3. The finish of EDM'd surfaces produced on die-sinking machines at production metal-removal rates may be quite rough, ranging as high as 13 μm (500 μin). However, with very low metal removal rates and advanced electronic controls, the surface of a final finishing cut can be as fine as 0.15 μm (6 μin). Wire EDM can produce somewhat better surfaces: as high as 6 μm (120 μin) with roughing cuts

and as fine as 0.10 μm (4 μin) under ideal conditions, although 0.25 μm (10 μin) may be a more realistic minimum for most equipment.

4. With ram EDM, cavity and part-to-part dimensional variations result from electrode wear during metal removal, and cavities may be slightly tapered from the point of electrode entry. Additional electrodes may be required for finishing cuts.

Figure 4.18.3 illustrates some typical die-sinker cuts.

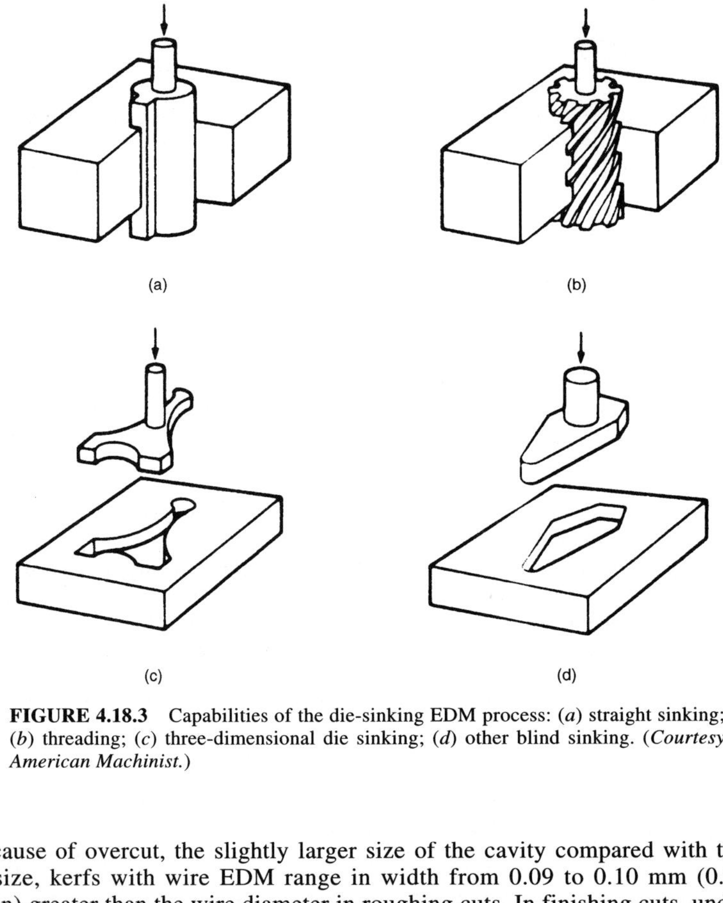

FIGURE 4.18.3 Capabilities of the die-sinking EDM process: (*a*) straight sinking; (*b*) threading; (*c*) three-dimensional die sinking; (*d*) other blind sinking. (*Courtesy American Machinist.*)

Because of overcut, the slightly larger size of the cavity compared with the electrode size, kerfs with wire EDM range in width from 0.09 to 0.10 mm (0.0035 to 0.004 in) greater than the wire diameter in roughing cuts. In finishing cuts, under ideal conditions, they are as small as 0.0025 to 0.005 mm (0.0001 or 0.0002 in) over the wire diameter. With ram EDM, the overcut can range between 0.5 and 0.013 mm (0.020 and 0.0005 in) per side but may be as much as 3 mm ($\frac{1}{8}$ in) in roughing cuts.

Mold and die making are the most common applications of EDM, but many other toolroom and special-production operations are performed. Stamping, extruding, wire drawing, die casting, and forging dies and plastics molds are toolroom applications, as are form tools, especially of carbide. Short-run production with wire EDM can eliminate the need for blanking dies. Wire EDM systems can cut parts up to 25 cm (10 in) thick, and systems have been built for parts 50 cm (20 in) thick. However, dimensional tolerances must be relaxed for thick parts. Holes as fine as 0.15 mm (0.006 in) can be drilled in thin stock by EDM.

Wire EDM systems are capable of four-axis contouring. It is possible to cut shapes that are different on the top from on the bottom. This is advantageous in intricate extrusion dies, mold applications, and some prototype or production applications.

Table 4.18.1 presents some examples of production parts made by EDM.

ECONOMICS OF EDM

The majority of EDM equipment in existence is used for toolroom applications. It is used for a wide variety of parts in one-of-a-kind or job-lot quantities. Dominance in toolroom versus production applications is generally due to the low metal removal rates of EDM.

A distinction must be made between die-sinker and wire systems. Die-sinker EDM is generally employed when conventional machining methods cannot be used. While some EDM operations can be performed at metal removal rates of 200 to 250 cm^3/h (12 to 15 in^3/h), an undesirable surface finish results. Most die-sinker EDMs operate at about 8 cm^3/h ($\frac{1}{2}$ in^3/h) and often less. Additionally, there is the cost of the electrodes and the need to replace them periodically due to wear.

The low cutting rate is compensated for somewhat by the greater ease of machining a male electrode from an easily machined material than a female cavity in a workpiece material of very low machinability. Also, for production applications, electrodes can sometimes be cast, molded, or extruded at a lower cost than if they were fully machined.

Wire EDM systems currently cut at speeds of 130 to 140 cm^2/h (20 to 22 in^2/h) in thicker material (e.g., 5 cm or 2 in thick) with good flushing conditions. Since electrodes do not need to be fabricated and much wire equipment can run unattended even for extended periods, the process can be economical in a variety of situations. However, EDM equipment is expensive—several times as costly as conventional machining equipment. Although cutting forces are relatively low, the machine must still be rigid, and there is the extra cost of electric-spark components not present in regular machine tools. Wire-cut EDM machines are more complex and costly than the ram-type machines.

SUITABLE MATERIALS FOR EDM

The EDM process will machine any electrically conductive material regardless of hardness. Hardness does not determine a material's suitability for cutting with EDM, but increasing hardness makes conventional machining more costly and EDM relatively more attractive. Hardened steel and carbide are the most commonly EDM'd materials. Others are nonhardened steel, refractory metals, Alnico, brass, copper, graphite, and aluminum. Polycrystalline diamond, used in form tools and other cutting tools, is increasingly being cut by EDM. Cast iron is normally processed by other machining methods because it may contain nonconductive impurities.

TABLE 4.18.1 Examples of EDM Production Applications

Product	Description of EDM operation	Workpiece material	Characteristic of machine cut	Cutting time per piece	Estimated yearly production
Carburetor	Drill idle bleed holes	Zinc, aluminum	Precise; burr-free hole	7 s	2,500,000
Automobile grille	Trim die-casting flash	Aluminum	Delicate geometry; burr-free	10 s	3,000,000
Electrical contact point	Trim contact point	Udimet	Burr-free; close tolerance	$\frac{1}{2}$ s	12,000,000
Gas-turbine blade	Cut platform periphery	René 41	Sharp corner; burr-free	12 min	6,000
Turbine shroud	Pierce airfoil shapes	18-8 stainless	Airfoil geometry; repetitive; burr-free	14 min	2,000,000 holes
Turbine blade	Pierce small holes at acute angle	Udimet	Burr-free holes	21 min	100,000 holes
Heat exchanger (regenerator)	Trim to size	0.002-in-thick stainless honeycomb	Delicate workpiece	10 min	10,000
Phonograph needle	Trim featheredge	Zirconium	Burr-free; delicate part	2 s	8,000,000
Tube, fuel metering	Slot thin wall	Brass	Burr-free; delicate part	18 s	4,000,000

DESIGN RECOMMENDATIONS

The following strategies can be employed by designers to enhance the processibility of parts with EDM:

1. Relax surface-finish requirements for the part, if feasible. This allows the manufacturer to produce the part with fewer passes, at a higher current level and a higher metal-removal rate.
2. Design the part so that the amount of stock that must be removed by EDM is relatively small. If it can be designed so that the bulk of the stock can be roughed out by traditional machining techniques or other processes with the finishing operations performed by EDM, a significant reduction in the amount of time required for the EDM operations can be achieved. (See Fig. 4.18.4.)
3. Design the part so that (*a*) several parts can be stacked and machined simultaneously or (*b*) a single part can have several EDM operations performed simultaneously.

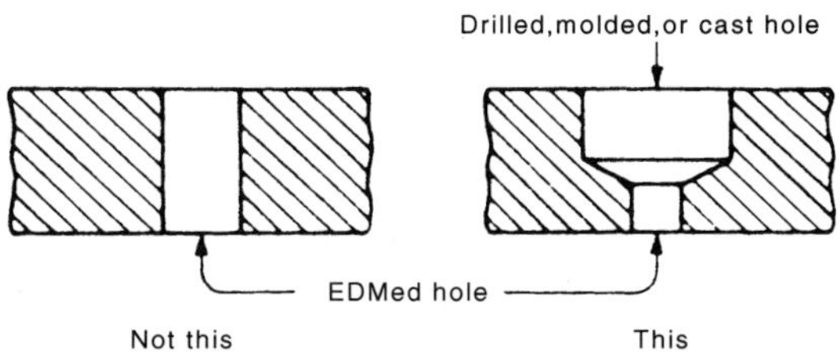

FIGURE 4.18.4 Design the part so that as much of the machining as possible can be performed by conventional machining or other manufacturing process rather than by the slower EDM.

When die-sinking EDM is used, designers have the responsibility to design a part so that the electrodes required can be produced at low cost. This means that the electrode shape (the cavity shape in the EDM'd part) should be kept as simple as possible. The shape of the electrode should be compatible with the capabilities of the machining or other process used to produce it. Excessively thin, fragile electrode sections should be avoided, especially if graphite is employed as the electrode material. Narrow openings and deep slots may have dielectric-flow problems. They tend to cause overheating and warping of metal electrodes. Sometimes, several simple electrode shapes can be combined to machine a complex cavity.

Designers also must be aware of overcut. Because of this effect, dead-sharp internal corners in cavities are not possible even if the electrode has a sharp edge. The minimum radius obtainable will be equal to the amount of overcut. A desirable minimum specified radius is 0.1 mm (0.004 in). (See Fig. 4.18.5.) With ram EDM, it is also advisable to specify cavity tolerances that allow a taper angle of 2 to 20 minutes per side.

When existing holes are to be enlarged or reshaped by EDM, through holes are preferred to blind holes because they permit easier flow of dielectric fluid past the area being machined.

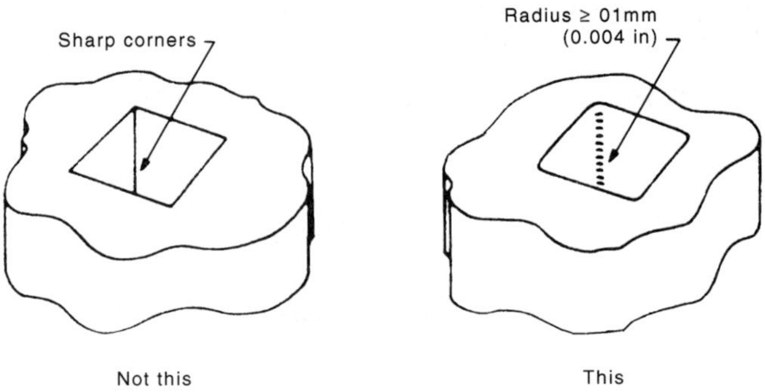

FIGURE 4.18.5 Dead-sharp internal corners will not be produced even if the electrode has sharp corners. Allow for an internal radius.

DIMENSIONAL FACTORS AND TOLERANCE RECOMMENDATIONS

The surface finish of EDM'd parts is a function of time and spark energy. Finer finishes require a finishing cut with low current, short spark duration, and a high discharge frequency. These parameters dictate low metal-removal rates. The finer the required finish, the longer it takes to achieve. In most cases with die-sinker EDM, a finish of 0.8 μm (30 μin) is adequate. Better finishes and better machining times may be achieved with orbiting or rotating electrodes. Finishes of 0.4 μm (15 μin) can be specified. With wire EDM, a finish of 1.0 μm (40 μin) can be specified if it is desired to avoid the need for finishing cuts; 0.25 μm (10 μin) is a feasible tolerance with multiple cuts.

Dimensional accuracy in die-sinker EDM is a function of electrode accuracy, stability of the tooling, quality of the generator/power supply, electrode wear, and preexisting stress in the workpiece material. Skill of the operator also may be a factor. Tolerances of ±0.05 to 0.13 mm (±0.002 to 0.005 in) are feasible with a single cut and to ±0.005 mm (±0.0002 in) with multiple cuts.

Wire EDM is inherently more accurate than die-sinker EDM. The factors that affect the accuracy are the CNC program, the quality of the wire being used, and the correctness of operation of the generator parameters, as well as preexisting stresses in the workpiece material. A dimensional tolerance of ±0.0025 mm (±0.0001 in) is possible under favorable circumstances, but a recommended tolerance for noncritical dimensions would be ±0.05 mm (±0.002 in).

CHAPTER 4.19
ELECTROCHEMICALLY MACHINED PARTS

James W. Throop
GMI Engineering and Management Institute
Flint, Michigan

ELECTROCHEMICAL MACHINING

The Process

Electrochemical machining (ECM) is the reverse of electroplating. It involves the controlled removal of metal from the workpiece by anodic dissolution in an electrolytic cell. The workpiece is the anode, and a specially shaped tool is the cathode. The electrolyte flows through a small gap between the tool and the workpiece, normally about 0.25 mm (0.010 in) but as small as 0.025 mm (0.001 in). The rate of metal removal is proportional to the direct-current density, and the current density is large: about 16 to 300 A/cm^2 (100 to 2000 A/in^3). The tool is normally fed into the workpiece at a rate of 0.25 to 20 mm/min (0.010 to 0.75 in/min). The electrode is usually sodium chloride or another salt. Suspended solids are removed from the electrolyte by various methods. For drilling small, deep holes, however, several processes use an acid electrolyte. Metal removal forms a cavity in the workpiece that is very close in shape to that of the electrode. Hence intricate shapes can be cut in one operation. Figure 4.19.1 illustrates the process schematically. Cleaning of the workpiece is usually necessary after electrochemical machining.

Applications and Capabilities

ECM is most suitable for materials and shapes that are difficult to machine by conventional methods. The production of small and irregular, deep holes is a common application. The minimum hole size is about 0.25 mm (0.010 in), although 0.8 to 3 mm (0.030 to 0.125 in) is more practical. Hole depth can be 50 or more times diameter. Since the hardness of the workpiece does not reduce the cutting rate, heat-treated or work-hardened materials can be handled easily. Single or multiple cavities can be sunk simultaneously. Depending on the type of equipment used, other operations are

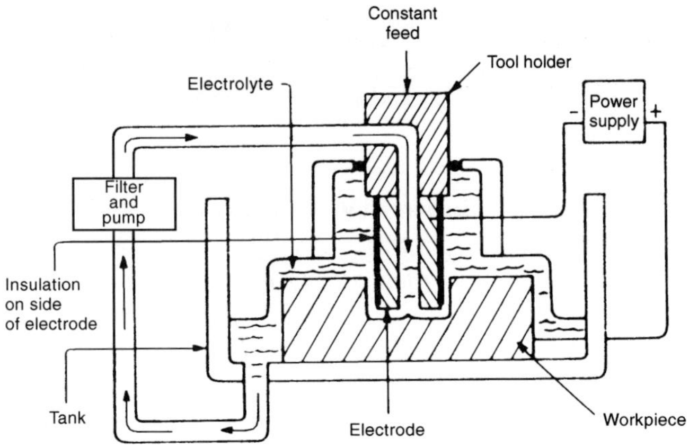

FIGURE 4.19.1 Electrochemical machining (ECM).

turning, facing, trepanning, profiling, contouring, slotting, "embossing," deburring, etching, and marking. Another application is the production of internal undercut surfaces of various contours. Selective masking of tooling electrodes and control of machining time can produce internal contours that would be extremely difficult to make by other methods.

ECM is used for jet engine parts, nozzles, cams, forging dies, burner plates, and other contoured shapes. The aircraft industry has the largest number of current applications.

Surfaces produced by ECM are burr-free with smooth corners. The process also provides machining without inducing additional stresses in the workpiece. A third advantage is freedom from hydrogen surface embrittlement.

Economic Production Quantities

ECM is best suited to quantity production. The tools used, though very close to the shape of the surface being cut, are not exactly the same, and a certain amount of tool development is required. Thus the initial cost of tooling is much higher than that for electrical-discharge machining. However, once the tooling has been developed and satisfactory parts have been produced, the tooling can last almost indefinitely and requires little maintenance. This fact, plus the high initial investment required for ECM equipment and the more rapid metal-removal rates for ECM compared with the competitive process EDM (25 times as fast), makes ECM most suitable for higher volumes of production.

However, owing to high power requirements and other costs, the process is not competitive with conventional machining for normal materials and contours. A generalization that can be made is that ECM is most competitive for hard workpieces, complex shapes, and moderate to high levels of production. Table 4.19.1 compares various characteristics of ECM versus EDM.

TABLE 4.19.1 ECM versus EDM: Various Characteristics

	ECM	EDM
Equipment cost	Very high	High
Tooling cost	High	Moderate
Metal-removal rate	High	Low
Usual production quantities	Medium to high	Low to medium
Suitability for difficult-to-machine materials	Excellent	Excellent
Power consumption per unit of metal removed	High	Moderate

Suitable Materials

ECM requires materials that are electrically conductive. Therefore, plastics, ceramics, glass, and rubber are not suitable. However, all metals, including cast iron, steel, copper and its alloys, aluminum, and many exotic alloys, can be processed. In fact, it is the difficult-to-machine or normally nonmachinable metals for which ECM is most advantageous. Typical among these are hardened steel, including alloy and tool steels; nickel alloys; cobalt alloys; tungsten, molybdenum, zirconium, and other refractory metals; and titanium and various high-strength alloys. Some of these metals may be less easily electrochemically machined than others, but whenever conventional machining is a problem because of poor machinability, ECM is probably a favorable method.

Design Recommendations

Holes and cavities produced by ECM are easier to machine if a taper is allowed on the walls. A normal taper is about 0.1 percent, or 0.001 in/in. This can be eliminated, if necessary, by insulating the sides of the electrode, a procedure that is fully feasible but presents an added complication in making the electrode. (See Fig. 4.19.2.)

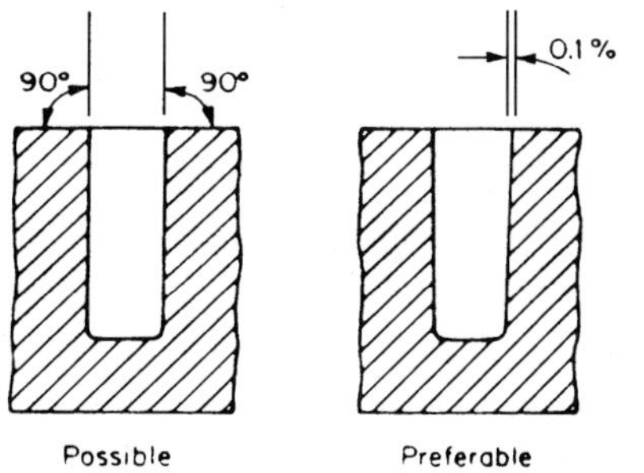

FIGURE 4.19.2 A taper of 0.1 percent on sidewalls facilitates the electrochemical machining of cavities.

The minimum internal radius for ECM cavities is about 0.18 mm (0.007 in), but designers should allow more than this, preferably at least 0.4 mm (0.015 in). (See Fig. 4.19.3.) It is also advisable to avoid specifying too sharp an external radius. The minimum achievable with ECM is 0.05 mm (0.002 in). (See Fig. 4.19.4.)

When specifying irregular cavities, the designer should allow for ample deviation from the nominal shape if at all possible because this will minimize one of the costs of ECM, the trial-and-error development of the exact electrode shape required for the particular cavity shape.

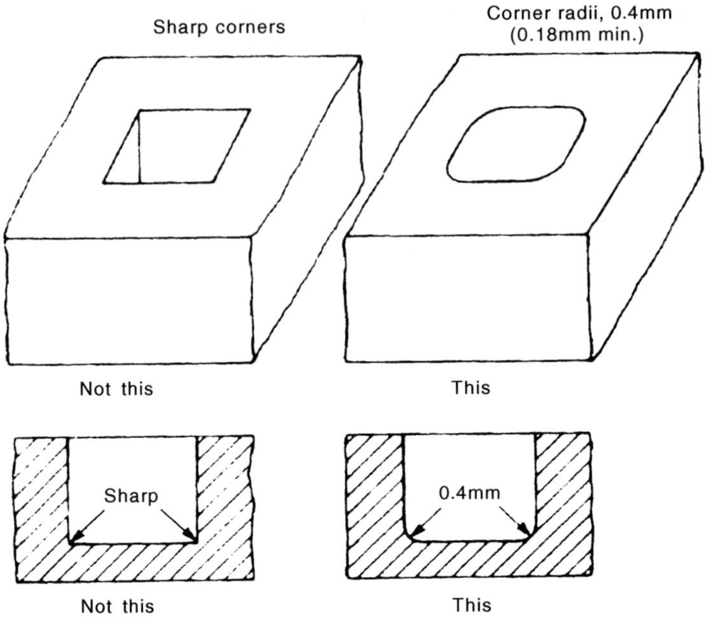

FIGURE 4.19.3 Minimum internal radii in ECM'd cavities.

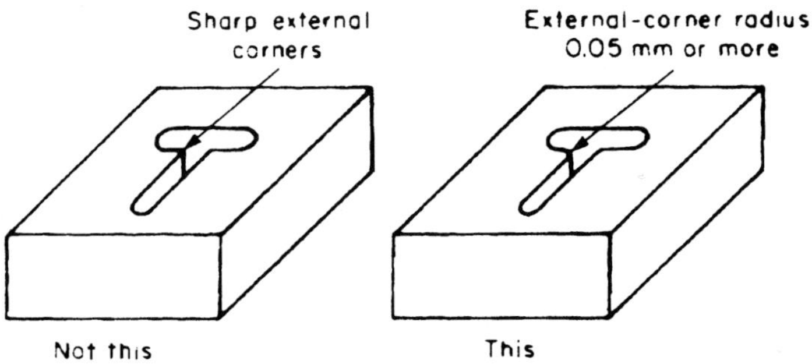

FIGURE 4.19.4 The recommended minimum external-corner radius with ECM is 0.05 mm (0.002 in).

Dimensional Factors

Overcut variations are the largest source of dimensional deviations in surfaces produced by ECM. Aside from the differences between the size and shape of the machined surface and those of the electrode, other nonuniformities contribute to the dimensional variations of the part. These include current density, voltage, electrolyte flow, electrolyte concentration, electrolyte temperature, and electrode feed-rate variations. Other factors are machine deflection (from electrolyte pressure) and deviations in precision of the electrode itself. Despite these variations, however, close dimensional tolerances are achievable with ECM. These are summarized in Table 4.19.2.

TABLE 4.19.2 Recommended Dimensional Tolerances for Electrochemically Machined (ECM) Surfaces

	Normal tolerance	Most exact tolerance
Specific dimensions	±0.13 mm (±0.005 in)	±0.025 mm (±0.001 in)
Contours	±0.25 mm (±0.010 in)	±0.05 mm (±0.002 in)
Surface finish (at end of electrode)	1.6 μm (63 μin)	0.4 μm (16 μin)
Surface finish (sidewall)	3 μm (125 μin)	1.6 μm (63 μin)

ELECTROCHEMICAL GRINDING

The Process

Electrochemical grinding (ECG) is very similar to electrochemical machining. Both processes involve reverse electroplating, the electrolytic dissolution of metal from the workpiece by means of a cathodic electrode tool, an electrolyte, and a direct electric current. The process differs in that ECM relies on high flow rates of electrolyte to remove the metallic salts that form, whereas ECG uses the grinding-wheel electrode rotating at a high speed (1200 to 1800 surface m/min, or 4000 to 6000 surface ft/min) to provide abrasive scrubbing action.

In ECG, the grinding wheel is electrically conductive (except for the abrasive grains that protrude about 0.025 mm, or 0.001 in, from the wheel's surface). The electrolyte is flooded on the workpiece. That which flows in the narrow gap between the protruding abrasive grains is most highly ionized and performs the deplating action. The abrasive grains prevent the wheel from shorting out to the workpiece.

The process is not exclusively electrolytic. About 5 to 10 percent of the metal (and sometimes more) is removed by abrasive action.

The direct-current voltage between the wheel and the work ranges from 4 to 8 V. The current density is 80 to 300 A/cm^2 (500 to 2000 A/in^2). In contrast to normal grinding, the wheel advances along the work slowly; the depth of cut is equal to or almost equal to the full amount of material to be removed.

Applications and Capabilities

ECG is used primarily to machine materials that are difficult to machine conventionally: steels harder than R_c 60, stainless steels, high-nickel alloys, tungsten carbide, fragile components, those which are heat sensitive, or those which must be free of machining stresses or burrs. A major application is the sharpening of carbide cutting tools. In this application, ECG provides much faster grinding than conventional methods. The ground cutters are without burrs and are free of mechanically or heat-induced stresses. Wheel cost is only about 20 percent of that of the regular diamond-wheel grinding of carbide because wheel wear with ECG is very low.

The low stress-inducing characteristics of the ECG process make it suitable for other components that are fragile or susceptible to heat damage or to distortion from normal grinding stresses. Examples are aircraft honeycomb materials, surgical needles, thin-walled tubes, laminated materials, and various parts thinner than 1.5 mm

FIGURE 4.19.5 The narrow, closely spaced grooves in these small blocks illustrate the fine machining on delicate parts that can be achieved with electrochemical grinding. (*Courtesy Anocut, Inc.*)

($\frac{1}{16}$ in) that have tight flatness tolerances and a tendency to distort after grinding. Figure 4.19.5 illustrates such a part.

The process is suitable for form-ground, flat, or cylindrical surfaces. Because arcing damage can occur, the process is avoided for highly stressed parts.

Economic Production Quantities

Electrolytic grinding machines tend to be expensive because of the cost of corrosion protection from the electrolyte solution and the cost of the electrical circuitry. Once this investment has been made, the process is fairly easily justifiable, although setup times are considerably greater than with conventional grinding. With moderate and higher levels of production, the labor and grinding-wheel savings of ECG are more fully realized. Typical metal-removal rates with ECG are 1.6 cm^3/min/1000 A (0.1 in^3/min/1000 A). The rate varies with different metals. However, for materials harder than R_c 60, it averages 5 to 10 times the rate attainable with conventional grinding.

Suitable Materials

Like ECM, electrochemical grinding requires that the workpiece material be electrically conductive. Carbides, hardened steels, stainless steels, and jet-engine alloys are commonly ground by ECG. Table 4.19.3 compares ECG with milling and grinding for a variety of common materials.

Design Recommendations

Generous inside radii are necessary with ECG. With extreme care and frequent wheel dressing, inside radii can be held to 0.25 to 0.38 mm (0.010 to 0.015 in). For practical purposes, however, it is better to allow 0.75 to 1.0 mm (0.030 to 0.040 in). Specify more liberal tolerances if the groove is deep.

Outside corners can be specified to be sharp if the direction of the corner is such that the wheel passes across it when entering or leaving the workpiece. Outside corners parallel to the plane of the grinding wheel should not be expected to be sharp. Electrochemical action concentrated at the corner will round the edge to about a 0.05-mm (0.002-in) radius in most cases. If a sharp edge is not required, designers should allow outside-corner radii up to 0.13 mm (0.005 in) to cover all normal process variations. (See Fig. 4.19.6.)

Electrolytic action also will remove some material from the surfaces on either side of the wheel as well as from outside corners. If flatness requirements are stringent and the ground surface is wider than the wheel, a final nonelectrolytic-grinding pass over the full surface may be necessary. If possible, in such cases the part to be ground

TABLE 4.19.3 ECG Compared with Milling and Grinding

Material	UNS designation	Stock removal (milling)	Stock removal (grinding)	Tooling costs and replacement	Size control	Production of fragile parts	Potential of heat damage	Quality of surface finish	Problems with burrs
Machinery steel		−	=	−	−	+	=	=	+
Tool steel, soft		−	=	−	−	+	+	=	+
Tool steel, hard		+	+	−	−	+	+	=	+
Cast iron		−	−	−	−	+	=	−	=
Copper		−	+	−	=	+	=	+	+
Brass		−	+	−	=	+	=	=	+
Aluminum		−	−	−	−	+	=	−	+
Tungsten		−	+	−	=	+	+	+	+
Tungsten carbide		+	+	+	=	+	+	+	=
Beryllium		−	=	−	−	=	=	−	+
300 stainless		−	+	+	=	+	+	+	+
400 stainless		−	+	+	=	+	+	+	+
Titanium		−	=	=	=	+	+	=	+
A-236	G10450	−	+	+	=	+	+	=	+
Waspaloy	N07001	+	+	+	=	+	+	+	+
Inconel 718	N07718	+	+	+	=	+	+	+	+
Inconel X 750	N07750	+	+	+	=	+	+	+	+
René 41	N07041	+	+	+	=	+	+	+	+
Haynes Stellite 21	R30021	+	+	+	+	+	+	=	=
Hastelloy X	N06002	+	+	+	=	+	+	=	+
Udimet 500	N07500	+	+	+	=	+	+	+	+

Note: = , processes about equal; −, ECG inferior to conventional machining; +, ECG superior to conventional machining. ECG compares favorably with precision machining on many work materials and is generally superior in several aspects (metal-removal rate, heat, burrs).

Source: Updated from *American Machinist,* p. 48.

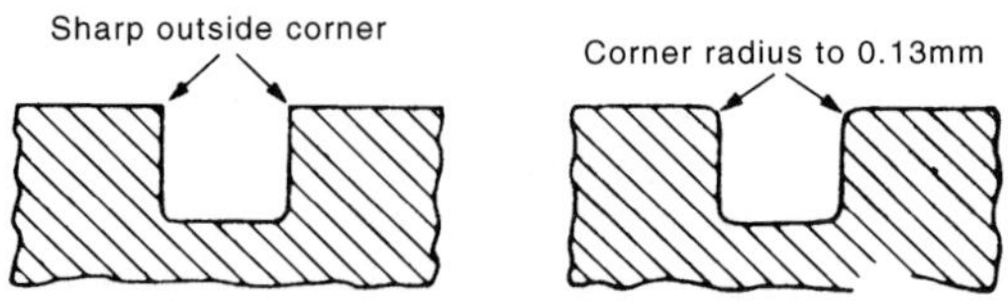

FIGURE 4.19.6 Outside sharp corners in ECG'd parts should be avoided. The groove design at left requires a secondary grinding operation across the groove; the design on the right shows the normal result of ECG of grooves.

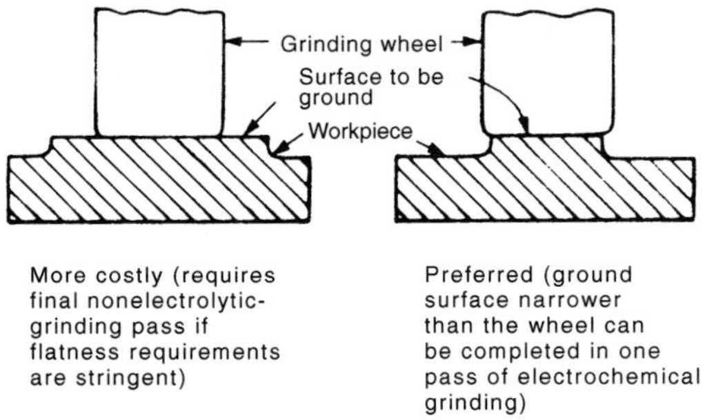

FIGURE 4.19.7 The ground surface should be narrower than the wheel.

should be designed so that the ground surface is narrower than the wheel. Figure 4.19.7 illustrates this.

Dimensional Factors

Causes of workpiece dimensional variations in ECG are very similar to those of ECM. One difference is the fact that in ECG the electrode (the grinding wheel) also physically removes some material from the workpiece. Whenever the wheel contacts the work, the variable of overcut (always present in ECM) is virtually eliminated.

Table 4.19.4 lists recommended dimensional tolerances for electrochemically ground surfaces.

TABLE 4.19.4 Recommended Tolerances for Electrochemically Ground (ECG) Surfaces

	Normal tolerance	Most exact tolerance
Specific dimensions	±0.025 mm (±0.001 in)	±0.013 mm (±0.0005 in)
Contours	±0.13 mm (±0.005 in)	±0.05 mm (±0.002 in)
Surface finish, plunge-grinding carbide	0.25 μm (10 μin)	0.13 μm (5 μin)
Surface finish, traverse-grinding carbide	0.40 μm (16 μin)	0.25 μm (10 μin)
Surface finish, steel	0.75 μm (30 μin)	0.40 μm (16 μin)

ELECTROCHEMICAL HONING

Electrochemical honing (ECH) is very similar to electrochemical grinding in that electrolytic metal removal is combined with mechanical abrasion. It incorporates the normal elements of honing, particularly a spindle with both reciprocating and rotary motion, a tool that holds a number of abrasive stones, and a means for keeping the stones in contact with the work surface. In addition, the ECH machine includes a direct-current power supply (6 to 30 V) and a means for pumping, filtering, containing, and temperature-controlling the electrolyte. The tool that holds the abrasive stones is made to a precise size for the bore involved. It functions as the cathode in the electrolytic circuit. The workpiece is the anode. The normal starting gap between tool and work is 0.08 to 0.13 mm (0.003 to 0.005 in). This increases to about 0.5 mm (0.020 in) as the operation proceeds, but the stones, which are about half as long as the bore depth, remain in contact with the bore surface. They remove metallic salts and metal from the surface, especially from the high spots, allowing the electrolytic action to proceed. The electrolyte flows through the hollow tool and through radial holes in the tool to flood the gap between the tool and the work uniformly. Current density ranges from 20 to 45 A/cm^2 (120 to 300 A/in^2). The honing action may continue for a few seconds for each piece after the current has been shut off to provide a final mechanical sizing of the bore. Sometimes, however, the reverse takes place. The mechanical action stops before the electrolytic action so that any stressed surface molecules in the bore are removed electrolytically.

Electrochemical honing is presently used only for finishing internal cylindrical surfaces. Holes from 10 to 300 mm ($\frac{3}{8}$ to 12 in) in diameter and up to 300 mm (12 in) in depth can be electrochemically honed. In addition to the faster cycle time inherent in the process, it has other advantages: surfaces are free from burrs and heat damage, little surface stress is induced, and the life of abrasive stones is greatly extended. Such parts as the bores of hardened gears and pumps are currently being finished by ECH.

Electrochemical-honing equipment is considerably more costly than the conventional type because of its electrical and fluid-handling elements and the need for corrosion protection. This factor, plus the more costly nature of ECH tooling and longer setup time compared with conventional honing, makes the process more economical for longer production runs than for toolroom or job-shop conditions. Where ECH finds economical application, however, it demonstrates metal-removal rates 3 to 5 times faster than regular honing.

Blind holes can be electrochemically honed, but they are not as easily handled as through holes. With blind holes, sidewall parallelism and diametral accuracy are facilitated if an undercut is machined below the honed area. A drain hole for the electrolyte also aids the operation. These parts are essentially the same as for conventional honing, illustrated in Fig. 4.16.6.

Tolerances of $\pm$0.0025 mm ($\pm$0.0001 in) on electrochemically honed diameters are possible, but $\pm$0.0060 to 0.0125 mm ($\pm$0.00025 to 0.0005 in) is recommended for more usual conditions. Surface finish can be as fine as 0.05 μm (2 μin), with the normal range being 0.1 to 0.8 μm (4 to 32 μin).

ELECTROCHEMICAL-DISCHARGE GRINDING

Electrochemical-discharge grinding (ECDG), sometimes known as *electrochemical-discharge machining* (ECDM), is a process that combines the attributes of two others:

electrochemical grinding and electrical-discharge grinding. Almost all metal removal, however, is a result of one action, the anodic dissolution of the workpiece metal. This is also the basis of electrochemical grinding and the other electrochemical processes. With ECDG, there are differences in process parameters: current voltage, electrolyte composition, gap, etc. Nevertheless, metal removal follows the same process: Negatively charged ions in the electrolyte solution combine with the workpiece metal to form removable oxides. Electrical discharges break up the oxide film, allowing it to be swept away by electrolyte flow.

The grinding wheel is graphite and contains no abrasive. It forms the cathode of the electrolyte circuit; the workpiece is the anode. The graphite wheel rotates to provide a surface speed of 1200 to 1800 m/min (4000 to 6000 ft/min) to circulate the electrolyte. Both flat and contoured surfaces are machined in one pass.

Electrochemical-discharge grinding is used primarily for the sharpening of carbide cutting tools, including form tools and thread-chasing-die inserts. It is used for hardened-steel or nickel-alloy parts and for fragile parts like honeycombs, thin-walled tubing, and surgical needles. Surfaces ground by ECDG are free of burrs and stresses. Only electrically conductive workpiece materials can be used.

ECDG is about 5 times faster than the nearest competitive process, electrical-discharge grinding. Its graphite-wheel tooling is also cheaper than that required by EDG. (It is, however, less accurate than EDG and requires considerably more electrical current.) As with the other electrochemical processes, equipment costs are higher than for conventional equipment. Cutting rates are typically 0.1 cm^3/min (0.006 in^3/min) for carbide and 0.25 cm^3/min (0.015 in^3/min) for steel.

Design limitations are similar to those of mechanical-grinding processes. The easy machinability of graphite wheels makes intricate form grinding relatively easy with ECDG.

Recommended dimensional tolerances for ECDG are $\pm$0.025 mm ($\pm$0.001 in) with $\pm$0.013 mm ($\pm$0.0005 in) being possible under conditions of close control. Surface-finish ranges are 0.13 to 0.40 μm (5 to 16 μin) for carbide and 0.40 to 0.75 μm (16 to 30 μin) for steel.

CHAPTER 4.20
CHEMICALLY MACHINED PARTS

Welsford J. Bryan
INA Bearing Co., Inc.
Cheraw, South Carolina

CHEMICAL MACHINING PROCESS IN GENERAL

Chemical milling, photochemical blanking and milling, chemical engraving, and chemical deburring are closely related processes. In all of them, metal is removed by the etching action of an acid or alkaline solution working on the exposed surfaces of the workpiece. This chemical action replaces the mechanical-cutting operation of conventional machining.

All these processes require a sequence of operations, some of which are common to all of them. These include

1. *Cleaning and rinsing.* The parts to be processed must be free of oils, dirt, heat-treatment scale, and polishing or lapping paste. If all traces of foreign matter are not removed, metal removal will be nonuniform. The standard cleaning procedures—vapor degreasing, solvent wiping, or alkaline cleaning—are usually employed.
2. *Masking.* Areas not to be chemically machined may be masked. Several methods, depending on which chemical machining process is used, may be employed.
3. *Etching (chemical machining).* This process is usually carried out by immersing the workpiece in the chemical solution.
4. *Mask removal, final cleaning, finishing, and drying, or antirust treatment.* These are common subsequent steps.

SPECIFIC CHEMICAL MACHINING PROCESSES AND THEIR APPLICATIONS

Chemical Milling

Parts can be chemically milled over their entire surfaces, or a surface can be selectively machined by masking areas not requiring machining. The preponderance of applications are in the aerospace industry, mainly for parts having large surface areas requiring small depths of metal removal. Parts such as aircraft skin and fuselage sections, airframe extrusions, and other parts needing weight reduction and maximum strength-to-weight ratios can have shallow pockets or cavities produced by chemical machining.

When selective machining is involved and the scribe-and-peel method is used, masking material is applied to the workpiece by spraying, dipping, or brushing. The dry maskant is scribed from a pattern and peeled from the workpiece to expose the areas that are to be chemically milled. Maskant also can be applied selectively by the silk-screening process or by photographic methods.

Machining takes place when the workpiece, contained in a work basket or held by a rack, is immersed in the chemical solution. Agitation of the work or circulation of the solution is necessary to ensure uniform rates of metal removal from all exposed surfaces. The maskant material is removed from the workpiece by hand or with the aid of a demasking solution.

When scribe-and-peel maskants are used, cuts as deep as 13 mm (0.5 in) can be made. With the thinner and less chemical-resistant silk-screened masks, the depth of cut is limited to 1.5 mm (0.060 in), but a more accurate and detailed cavity is possible. Photoresists provide still more detail and accuracy, but the cut is limited to a depth of 1.3 mm (0.050 in). Sheets and plates are tapered in thickness by immersing and withdrawing the workpiece from the etching bath at a controlled rate. Steps are produced by repeated cycles of etching with different areas masked.

The advantages of chemical milling are that (1) metal can be removed from one or more sides, (2) brittle or hardened metals with thin cross sections can be machined, (3) burrs are not produced, (4) shapes that cannot be conventionally machined or cast or extruded can sometimes be chemically machined, and (5) internal stresses are not introduced as in conventional machining. However, there are disadvantages to the process: (1) Pollution-control equipment is required. Safeguards are needed for handling chemicals. (2) The depth of cut is limited to about 13 mm ($\frac{1}{2}$ in) in plate materials and to less than this with forgings, castings, and extrusions. (3) Interior surfaces of cylinders or other shapes that limit circulation do not yield uniform removal rates. (4) Materials should be of homogeneous structure and composition. Welded and brazed areas and porous castings often yield unacceptable results. (5) Scratches, dents, surface waviness, and other irregularities tend to be retained in the surface of the chemically milled part (except with magnesium, for which such defects tend to "wash out").

Chemical milling is most frequently used to reduce the weight or thickness of parts after other operations. Webs and ribs that are too thin for extruding, casting, forging, or machining can be brought to the desired degree of thinness after being produced initially to a heavier dimension. (See Fig. 4.20.1.)

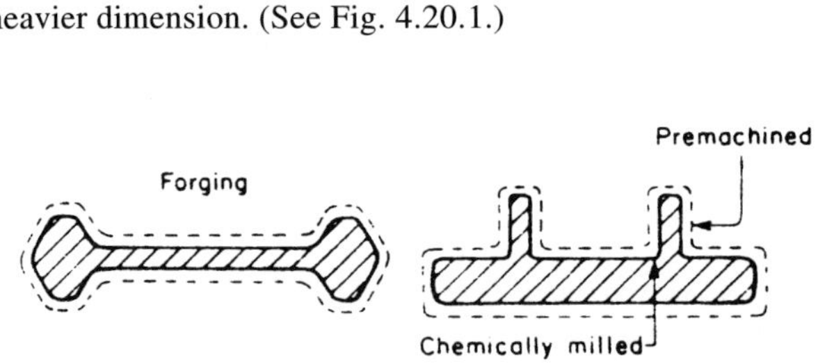

FIGURE 4.20.1 Chemical machining can be used to reduce the thickness of ribs and webs to a greater degree than is possible with mechanical metal-removal or metal-forming processes. (*From* Nontraditional Machining Processes, *American Society of Tool and Manufacturing Engineers, Dearborn, Mich., 1967.*)

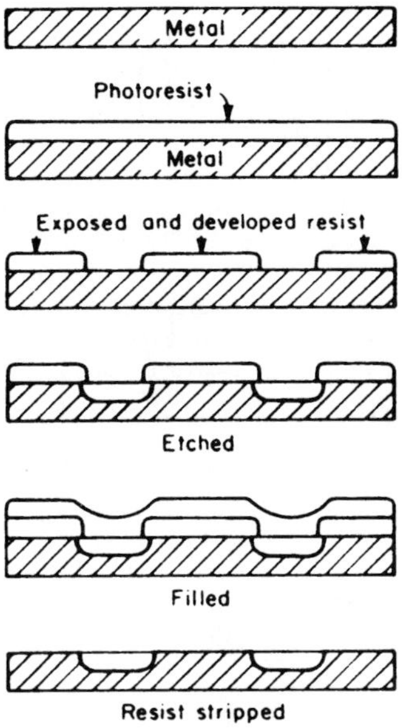

FIGURE 4.20.2 Major steps of chemical engraving. Film thicknesses and engraving depth are exaggerated for clarity. (*From* Non-traditional Machining Processes, *American Society of Tool and Manufacturing Engineers, Dearborn, Mich., 1967.*)

Chemical Engraving

This is a closely related process in which chemical machining techniques are used to remove metal from selected areas of nameplates or other components to produce the lettering, figures, or other nomenclature required. Chemical engraving is a substitute for mechanical pantograph engraving. Lettering can be either depressed or raised.

Normally, a photoresist maskant is used, especially if the engraving is to have fine detail. The maskant also can be applied by silk screening. The etched area may be filled with paint or other material of a color contrasting with that of the unetched part. Sometimes, aluminum parts are anodized in the etched area in a color contrasting with that of the rest of the part. Figure 4.20.2 illustrates the major steps of chemical engraving.

The process is generally used when the images must be more durable than those obtainable by printing methods. The most common metals employed are aluminum, brass, copper, and stainless steel. Typical parts produced by chemical engraving are instrument panels, nameplates, printing plates, signs, and pictures. Parts that require engraving with fine detail are especially suited to chemical engraving. Hard materials or those of low machinability are also more easily engraved chemically than by pantograph engraving.

Photochemical Blanking

This process produces blank shapes from sheet metal by covering the sheet with a precisely shaped mask made with photographic techniques and then removing unmasked metal with chemical-machining techniques. The process is illustrated schematically in Fig. 4.20.3. The photoresist material is applied by dipping, spraying, flow coating, or roller coating. Two photoresist masks may be placed opposite each other, one on either side of the sheet, to minimize the undercut.

The major application of photochemical blanking is the production of very intricately shaped parts from thin sheet metal. The finished parts are burr-free. Typical parts are electric-motor laminations, shadow masks for color television, fine screens, printed-circuit cards, slotted-disk springs, and gaskets.

Photochemically blanked parts can range in thickness from 0.0013 to 3 mm (0.00005 to $\frac{1}{8}$ in) but are more commonly in the range of 0.0025 to 0.8 mm (0.0001 to $\frac{1}{32}$ in). They can reach 60 cm (24 in) in length. Parts are normally flat, but they can be curved in one dimension. Figure 4.20.4 shows a collection of typical parts.

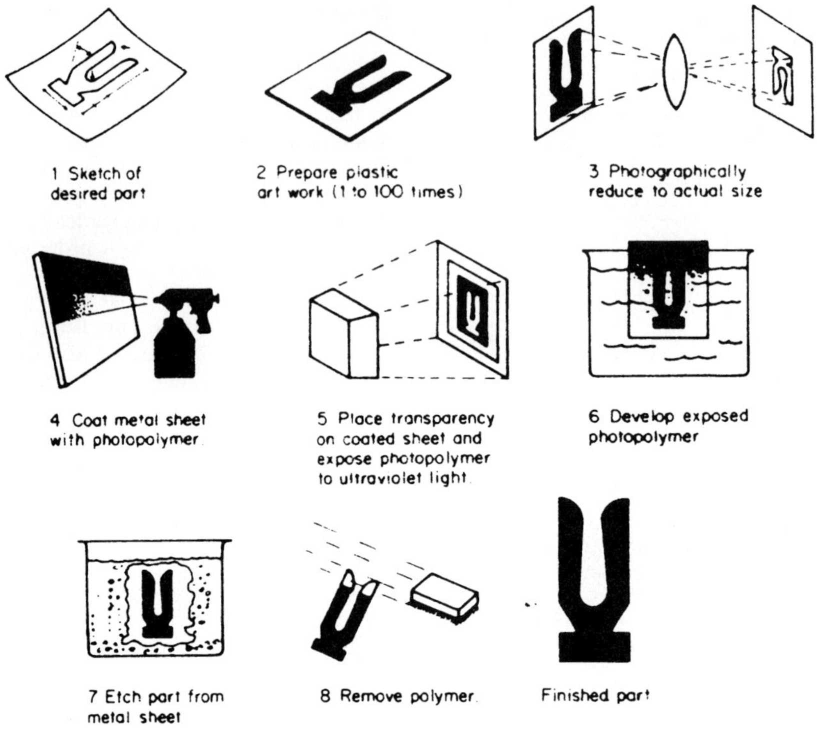

FIGURE 4.20.3 Major steps of photochemical blanking. (*Courtesy* Mechanical Engineering.)

The advantages of photochemical blanking include the ability to handle hard, very thin, or brittle materials that are difficult to blank conventionally, the ability to make extremely complex shapes without blanking dies, and the burr-free nature of the blanked parts. Disadvantages include the need for photographic facilities, the high degree of skill required by the operator, the inability of the process to produce sharp corners, and the fact that the etching chemicals require safety and anticorrosion precautions.

Chemical Deburring

Chemical machining techniques have proved to be very effective in burr removal. Parts to be deburred are placed in a rotating horizontal barrel or racked. During the operation, the parts or the deburring solution, or both, are kept in motion.

The process is designed to remove or reduce burrs developed during machining, metal stamping, or other manufacturing processes. If burrs are excessively thick, the parts may still be processed in the deburring bath to reduce the burrs to manageable sizes for other deburring processes.

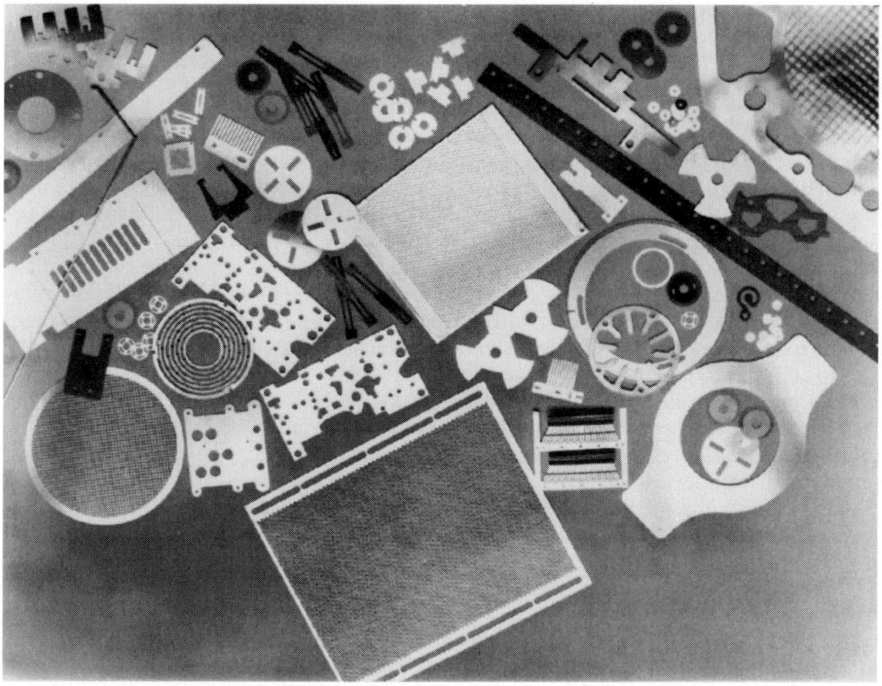

FIGURE 4.20.4 Typical photochemically blanked parts. (*Courtesy Conard Corp.*)

Chemical deburring is suitable for both ferrous and nonferrous materials. Ferrous materials may be in the annealed, normalized, or hardened condition, but prior to the process, all heat-treatment scale must be removed. This is normally accomplished with a pickling operation. The workpiece is not normally distorted during processing.

Small parts are processed most readily because they can be loaded into a rotating barrel, which provides good contact between the deburring liquid and the parts. Larger parts are racked, and parts with as great a length as the tank size accepts can be chemically deburred.

Since the chemical deburring solution removes stock from all surfaces, not only from the burrs, it is sometimes necessary to protect surfaces against metal loss. This is done by coating the area to be protected with a maskant.

The system also can be used to good advantage as an economical method of rework or as a means of salvaging work that would otherwise have to be scrapped. Examples are oversized screws or parts with undersize cavities or holes.

Chemical deburring has the advantage of being economical because individual part handling is usually eliminated. It provides controlled, reproducible results with an improved surface finish of the part. Large burrs can be removed, but in the process large amounts of metal are removed from other unmasked surfaces, and this may be undesirable because it causes changes in basic dimensions. Chemical deburring, like the other chemical machining processes, also necessitates the use of pollution-control equipment.

ECONOMICS OF CHEMICAL MACHINING

Quality and process-capability factors are more important than economic factors in choosing chemical milling for weight reduction of aerospace components. When economic factors are important, chemical machining processes normally are most advantageous at low levels of production. Exceptions are chemical deburring, which can be competitive with other deburring methods at mass-production levels, and chemical engraving when fine detail is involved. With fine detail, other engraving methods may not be as suitable even if quantities are large.

Normally, however, the low equipment and tooling costs attendant on chemical machining processes make them attractive for modest quantities. Conversely, the relatively low metal-removal rates and high unit costs for these processes make them unattractive when quantities are large enough to amortize the tooling required for conventional processes.

Equipment for chemical machining is less costly than conventional machine tools despite the need for corrosion protection and antipollution apparatus. Photographic or other mask-making equipment also may be required. Tooling, consisting of masks, photographic masters, or silk screens for producing masks, racks, and holding fixtures, is also invariably less costly than conventional machining fixtures or dies. When photochemical blanking is used, the artwork and production of transparencies for the etching mask can require from 2 to 10 labor-hours.

Metal-removal rates range normally from 0.0025 to 0.02 mm/min (0.0001 to 0.0008 in/min) for steel alloys to as much as 0.08 mm/min (0.003 in/min) for aluminum alloys. In chemical deburring, rates tend to be low, especially if size control is to be maintained. Chemical blanking and milling can be in the higher range of these figures. Table 4.20.1 gives removal rates and applications for chemical deburring.

Other factors affecting unit cost are the cost of chemicals, the cost of maskant material, if used, the labor cost of applying them, and the cost of controlling and disposing of the etchant solutions. Unless the process is mechanized, the mask preparation, masking, and chemical machining may have a high labor content.

Chemical machining methods are well suited to prototype work as well as to limited-quantity production.

SUITABLE MATERIALS

In general, any material that can be chemically etched or dissolved can be chemically machined. Many metals, both ferrous and nonferrous, have been processed.

TABLE 4.20.1 Metal-Removal Rates in Chemical Deburring

Removal rate	Application
0.0025–0.0038 mm/min (0.00010–0.00015 in/min)	Honinglike metal removal, polishing of surfaces, very close tolerance work where thread size must be controlled
0.0046–0.0064 mm/min (0.00018–0.00025 in/min)	General-purpose chemical deburring
0.0090–0.0115 mm/min (0.00035–0.00045 in/min)	Aggressive metal removal where metal-removal rate is more important than size control

Nonmetallics such as plastics are less suitable because the chemical solution used for etching the workpiece also may attack the maskant material. The process was originally developed for aluminum and magnesium alloys but was soon extended to carbon, alloys, and stainless steels, brass, copper, nickel alloys, and other metals.

Aluminum alloys are the simplest to chemically machine, followed by magnesium and titanium. Low-alloy and stainless steels are next. The superalloys are the most difficult. Some alloys like René 41 (N07041), Waspaloy (N07001), and Inconel X750 (N07750) require extremely active etchants that are too strong for use with photoresists.

The quality of the material chosen for chemical processing is as important as the basic choice of materials. Factors to be considered include grain size, rolling direction, hardness, internal cleanliness (freedom from inclusions), and surface quality (freedom from oxides or other effects of corrosion).

Materials should be of homogeneous structure and composition. Chemical machining in a weld seam may produce pits and uneven metal removal. Aluminum castings also may produce uneven results because of the porosity and nonuniformity of the cast metal. Brazed joints also may yield unacceptable results.

The most common materials for chemical engraving are aluminum, brass, copper, and stainless steel. If a matte finish in the etched area is desired (which helps the adhesion of filler material), it is relatively difficult to achieve with brass and copper, which normally produce only a smooth finish in the etched areas.

Chemical deburring is applicable to both ferrous and nonferrous metals. Parts can be annealed, normalized, carburized, carbonitrided, or through-hardened. However, prior to the process, all heat-treatment scale must be removed. Usually, low-carbon steels are preferred to high-carbon varieties because they do not smut, but if high-carbon steels are used, the smut can be removed by pickling.

DESIGN RECOMMENDATIONS FOR CHEMICALLY MACHINED PARTS

When designing a part to be chemically machined, the designer must remember that the solution will work on all surfaces of the workpiece with which it comes in contact. Any surface that does not require chemical machining must be protected by a maskant, or the loss of material from the surface must be allowed or compensated for. Usually, parts that are to be chemically deburred are not selectively coated with a maskant but are designed slightly oversize to allow for metal removal from all areas. In other cases the designer may allow for metal removal simply by specifying a liberal undersize tolerance.

When designing parts to be chemically deburred, it is important to know what manufacturing processes will be used and the type and size of burr that can be expected. Then the designer should calculate the maximum amount of metal that will be lost and compensate for it, bearing in mind that the edges of the part will break down at a rate approximately twice that of metal removal on other surfaces. The metal-removal allowance should be based on what is required to remove the worst burr. Sharp corners normally can be rounded to a radius of about 0.7 to 1.0 mm (0.003 to 0.004 in).

The metal-removal allowance also must allow for the fact that etching takes place horizontally as well as vertically when a cavity is chemically machined or a part is chemically blanked. This effect is illustrated by Fig. 4.20.5. Etching produces an undercut beneath the edge of any masks used. Normally, the undercut immediately below the mask will be approximately as large as the depth of cut.

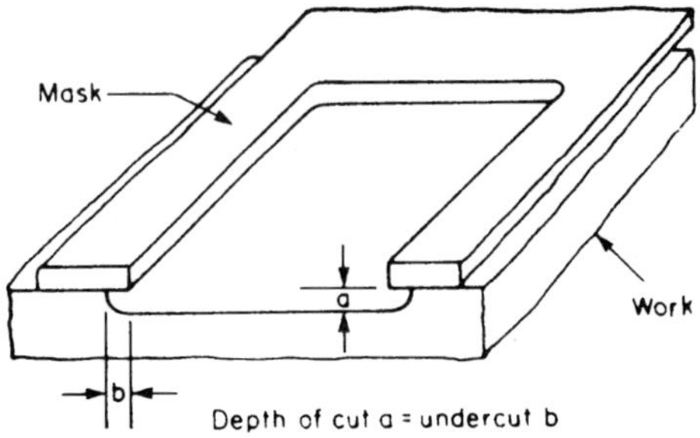

FIGURE 4.20.5 Undercut is produced by the effect of horizontal etching. (*From B. W. Niebel and A. Draper,* Product Design and Process Engineering; © *1974 by McGraw-Hill, Inc. Used with permission of McGraw-Hill Book Company.*)

There is also a tapered effect to this undercut because of the longer exposure to the etching solution of vertical surfaces nearer the mask. Designers of chemically blanked parts must allow for the fact that because of this effect the edges of such parts will not be perfectly square. The normal taper for a part chemically blanked from one side is one-tenth the stock thickness. If etching takes place on both sides, the taper is half as great, $T/20$.

The part should be designed so as to prevent any entrapped chemical solutions or gases. Porous materials and part designs that have deep, narrow cavities or folded-metal seams should be avoided.

Parts that will have masks produced by photographic or silk-screen methods should have surfaces as flat as possible. Tracing templates for scribing spray- or dip-applied masks are also more economical if the chemically machined surface is flat or curved in only one direction.

When aluminum is to be chemically milled, the grain direction should be noted on the drawing. Etched areas should be laid out so as to minimize cuts across the grain or be laid out at a 45° angle to the grain direction.

Good design practice for chemically milled parts generally excludes deep and narrow cuts, holes, narrow lands, and severe tapers. The minimum land width between chemically machined cavities depends on the masking method. With scribed masks, designers should specify a land width twice the depth of the cut and in no case less than 3 mm ($\frac{1}{8}$ in). The land need not be wider than 25 mm (1 in). With silk-screened masks, much narrower lands are possible but with much shallower cuts. Even narrow lands (with still shallower cuts) can be obtained with photoresists. (See Fig. 4.20.6.)

The minimum width of a chemically milled cavity should be twice the depth of the cut plus 1.5 mm (0.060 in) for cavities up to 3 mm (0.120 in) deep and twice the depth of the cut plus 3 mm (0.120 in) for deeper cuts. (See Fig. 4.20.7.) When photoresists are used, the recommended minimum width for slots, holes, or other cavities is simply twice the depth of the cut. However, if a particularly narrow hole or slot is required in a photochemically blanked part, the following minimum widths, as a function of the stock thickness T, can be obtained by etching from both sides: aluminum workpieces, $1.4T$; copper alloys, $0.7T$; stainless steel, $1.4T$; and steel, $1.0T$.

When only shallow etching is required, minimum practicable widths of etched areas are as follows: with scribe-and-peel maskants, 1.5 mm (0.060 in); with silk-screened masks, 0.25 mm (0.010 in); and with photoresists, 0.13 mm (0.005 in).

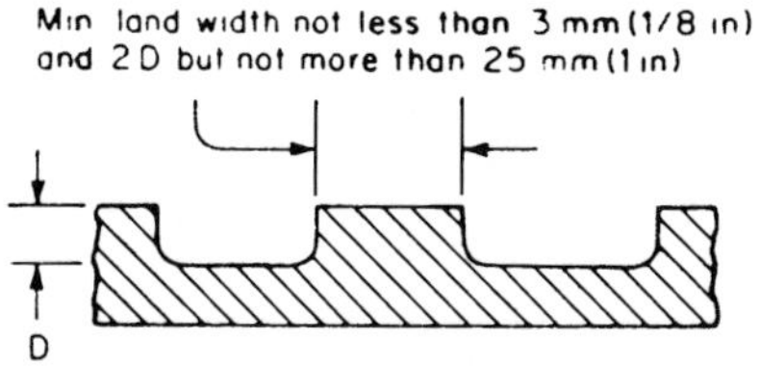

FIGURE 4.20.6 Minimum land widths in chemically milled parts.

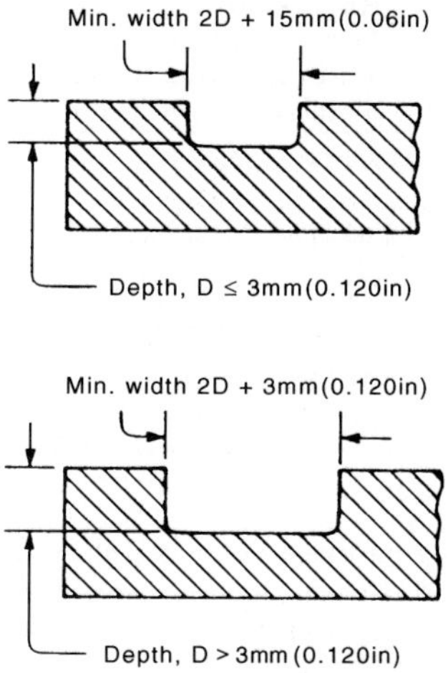

FIGURE 4.20.7 Minimum slot, hole, and cavity widths in chemically milled parts.

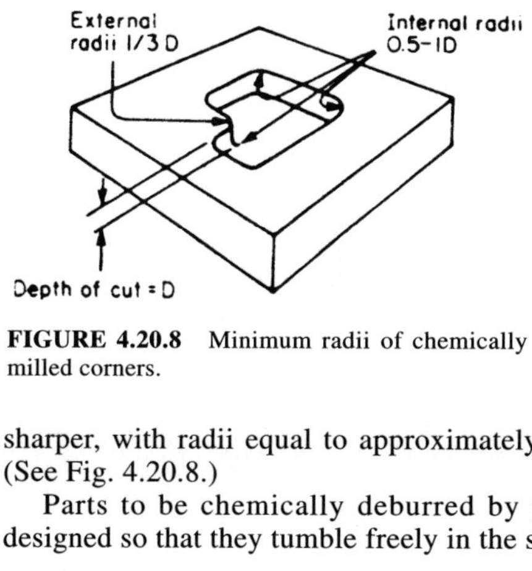

FIGURE 4.20.8 Minimum radii of chemically milled corners.

When a part is to be chemically milled from sheet or plate stock, it often is preferable to provide excess material at the periphery of the part for trimming after chemical milling. The trimming operation may be more economical than masking the edges of the stock, which can be costly and time-consuming.

Sharp internal corners are not produced by chemical machining. A radius of 0.5 to 1 times the etched depth should be expected and allowed for in drawing specifications. External corners can be sharper, with radii equal to approximately one-third of the depth of cut if necessary. (See Fig. 4.20.8.)

Parts to be chemically deburred by processing in a rotating barrel should be designed so that they tumble freely in the solution without interlocking or tangling.

DIMENSIONAL FACTORS AND TOLERANCE RECOMMENDATIONS

Quite close tolerances can be held with chemical machining operations. The major dimensional variable is the accuracy of the part's dimensions before chemical machining. Other factors are the accuracy of the artwork or template used to produce the mask employed, the accuracy of the mask itself (a result of the accuracy of scribing, if the mask is produced manually, or of the silk-screening or photographic process if it is

TABLE 4.20.2 Recommended Tolerances on the Depth of Cut for Chemical Milling

Depth of cut, mm (in)	Recommended tolerance, mm (in)
Up to 2.15 (0.085)	±0.025 (0.001)
2.16–3.05 (0.086–0.120)	±0.038 (0.0015)
3.06–3.94 (0.121–0.155)	±0.050 (0.002)
3.95–4.84 (0.156–0.190)	±0.064 (0.0025)
4.85–5.84 (0.191–0.230)	±0.076 (0.003)
5.85–7.10 (0.231–0.280)	±0.089 (0.0035)
7.11–8.65 (0.281–0.340)	±0.102 (0.004)
8.66–10.16 (0.341–0.400)	±0.114 (0.0045)

produced by these methods), the correctness of the undercut allowance, and the uniformity of etching and other process conditions.

Table 4.20.2 provides tolerance recommendations for the depth of chemical milling cuts. Tolerances on the length and width of chemically milled cavities are as follows:

With scribe-and-peel maskant and cuts to 1.3 mm (0.050 in) deep: ±0.38 mm (0.015 in).

With scribe-and-peel maskant and cuts over 1.3 mm (0.050 in) deep: ±0.64 mm (0.025 in).

With silk-screened maskant: ±0.25 mm (0.010 in).

With photoresist: Use Table 4.20.3.

Table 4.20.3 provides recommended tolerances for photochemical blanking of various metals.

Tolerances for chemically engraved parts are customarily specified at ±0.13 mm (0.005 in), but tighter values can be held if necessary, especially for letter-size dimensions and for the depth of etching. Tolerances for chemically deburred parts should be the same as prior to deburring, although, as indicated above, basic dimensions will change slightly because of metal removed during the deburring process.

TABLE 4.20.3 Recommended Length and Width Tolerances for Photochemically Blanked Parts

Stock thickness, mm (in)	Tolerance, mm (in)				
	Aluminum alloys	Copper and alloys	Nickel	Stainless alloys	Low-alloy steel
0.050 (0.002)	0.050 (0.002)	0.025 (0.001)	0.025 (0.001)	0.025 (0.001)	0.025 (0.001)
0.127 (0.005)	0.076 (0.003)	0.050 (0.002)	0.076 (0.003)	0.050 (0.002)	0.050 (0.002)
0.25 (0.010)	0.102 (0.004)	0.076 (0.003)	0.127 (0.005)	0.076 (0.003)	0.102 (0.004)
0.50 (0.020)	0.15 (0.006)	0.13 (0.005)	0.25 (0.010)	0.13 (0.005)	0.15 (0.006)
1.0 (0.040)	0.20 (0.008)	0.15 (0.006)		0.25 (0.010)	0.25 (0.010)
1.5 (0.060)	0.30 (0.012)	0.18 (0.007)		0.35 (0.014)	0.30 (0.012)

Note: The values apply to normal production conditions and to dimensions of 50 mm (2 in) or less. Larger dimensions require proportionally greater tolerances. Under highly controlled conditions and at higher cost, values 50 percent tighter than those shown can be obtained.

TABLE 4.20.4 Normal Surface-Finish Ranges after Chemical Machining

Metal	Normal surface finish	
	Range	Average
Aluminum		
Cuts to 6.3 mm (0.250 in) deep	1.75–3.13μm (70–125 μin)	2.25 μm (90 μin)
Cuts deeper than 6.3 mm (0.250 in)	2.00–4.13 μm (80–165 μin)	2.88 μm (115 μin)
Magnesium	0.75–1.75 μm (30–70 μin)	1.25 μm (50 μin)
Titanium	0.75–1.25 μm (30–50 μin)	0.63 μm (25 μin)
Steel	0.75–6.25 μm (30–250 μin)	1.75 μm (70 μin)

Surface finish for chemically machined parts depends on the initial surface finish of the workpiece before chemical machining, the etchant used, the workpiece material, the depth of the cut, and the heat-treatment condition of the workpiece. The smoother the surface finish before chemical machining, the better the results after the operation. Table 4.20.4 lists expected surface-finish ranges with commonly chemically machined metals.

The ideal surface finish for chemically engraved surfaces is 0.5 to 1.3μm (20 to 50 μin). This level of finish is obtainable with the proper etchant and the shallow cuts typical of chemical engraving. For additional information, see military specification MIL-C-81760, "Chemical Milling of Metals."

CHAPTER 4.21
PARTS PRODUCED BY OTHER ADVANCED MACHINING PROCESSES

ABRASIVE-JET MACHINING

The Process

Abrasive-jet machining (AJM) achieves its cutting effect from the action of fine-powdered abrasive impinged on the workpiece by a high-velocity stream containing the abrasive in a gas carrier. The stream is focused through a nozzle opening of 0.13 to 0.81 mm (0.005 to 0.032 in) and travels at 150 to 300 m/s (500 to 1000 ft/s). The carrier gas is normally air but may be carbon dioxide or nitrogen and is at a pressure of 200 to 830 kPa (30 to 120 lbf/in^2). The abrasive is aluminum oxide, silicon carbide, or glass, and the nozzle is tungsten carbide or sapphire.

For precision cutting, the nozzle is mounted on apparatus that provides accurate positioning; for rough cutting, deburring, or stripping, the nozzle is usually handheld. In some cases, rubber, glass, or copper masks are used to confine the abrasive action to a certain portion of the workpiece surface.

Applications and Characteristics

Cutting, drilling, slotting, trimming, etching, cleaning, deburring, carving, and stripping can be performed by abrasive-jet machining. Some current applications are trimming resistors to precise values; stripping varnish from wires; cutting patterns and shapes in silicon semiconductors; abrading and frosting glass; drilling, cutting, and trimming thin sheets of tungsten or hardened steel; etching trade names and numbers on parts; removing plating or other surface coatings, particularly in a portion only of the workpiece surface; removing broken tools from holes; and making final adjustments or minor modifications in hardened-steel molds. Other products machined by AJM are dental devices, jewelry, hard-alloy-steel thrust bearings, and laminated optical filters. (See Fig. 4.21.1.)

Very little heat is generated during abrasive-jet machining; there is no heat damage to the workpiece. Therefore, the process enjoys a major advantage, applicability to the machining of heat-sensitive components. The minimum slot width machinable with AJM is 0.13 mm (0.005 in). Taper in the walls of AJM cuts is inherent; it increases as nozzle-to-work spacing increases.

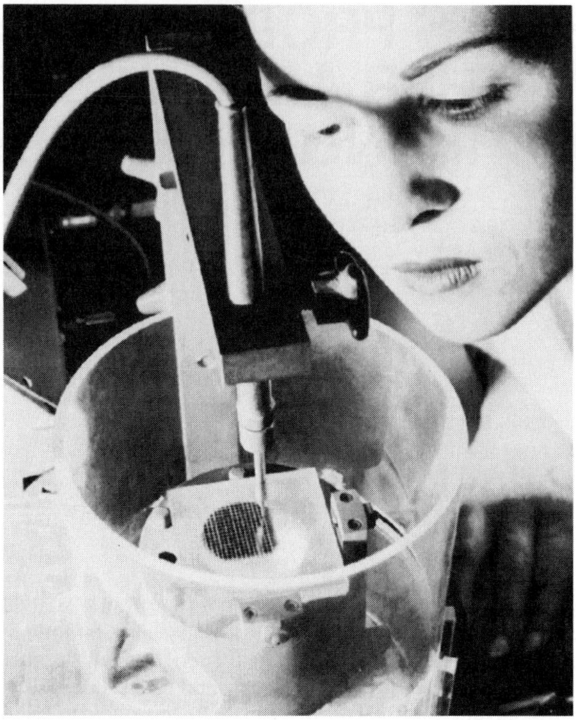

FIGURE 4.21.1 Abrasive jet linked to a pantograph cuts intricate patterns in silicon semiconductors. (*Courtesy S. S. White Industrial Products.*)

Costs and Economic Production Quantities

Manual abrasive-machining equipment is inexpensive. Tooling costs also are low, making the process economical for small quantities. In any case, the choice of the AJM process does not depend on production quantity. The process is slow—0.016 cm^3 (0.001 in^3) of material is normally removed per minute—and is used only when more conventional and more rapid metal-removal processes cannot be employed because of the nature of the workpiece material. The process is also used primarily for finishing and light cutting operations when the slow cutting rate is not too serious an adverse factor.

Suitable Materials

Abrasive-jet machining is used most advantageously for hard, fragile, heat-sensitive materials. Porcelain, glass, ceramic, sapphire, quartz, tungsten, chromium-nickel alloys, hardened metals, and semiconductors such as germanium, silicon, and gallium are suitable for the process.

Design Recommendations

Designers specifying abrasive-jet machined parts should make allowance in their designs for the following:

1. The taper of the sidewalls of holes, slots, and other cuts should be at least 0.05 mm/cm (0.005 in/in) of depth.
2. The part configuration should allow access room for the abrasive-jet nozzle; i.e., cuts should not be specified immediately alongside steps or bosses.
3. Severing cuts should allow for kerf, at least 0.13 mm (0.005 in) but preferably 0.45 mm (0.018 in).
4. Corners cannot be sharp. As a minimum, allow a radius of 0.1 mm (0.004 in).

These design recommendations are illustrated in Fig. 4.21.2.

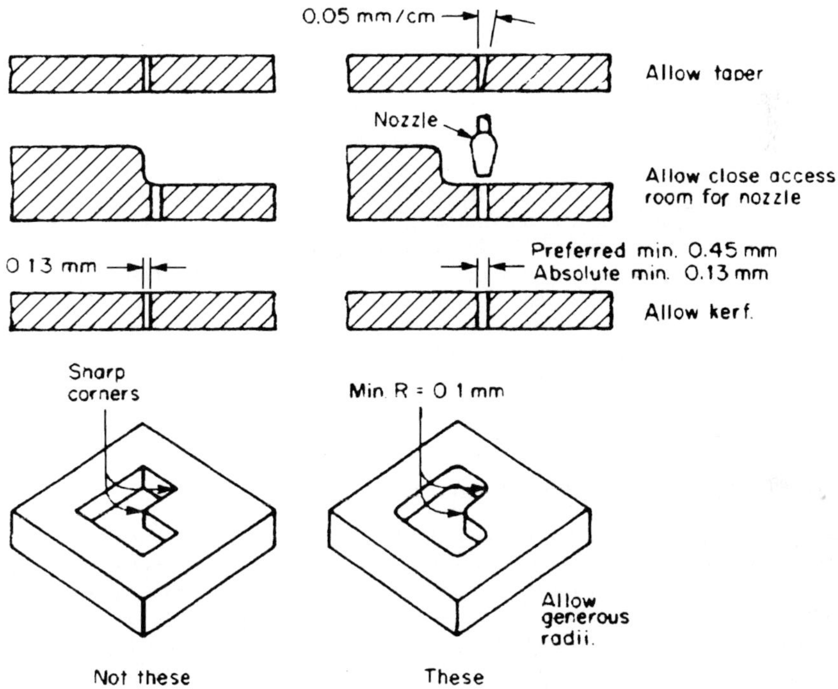

FIGURE 4.21.2 Abrasive-jet machining design recommendations.

Recommended Tolerances

The normal tolerance for the dimensions of machined areas is ±0.13 mm (0.005 in), although ±0.05 mm (0.002 in) is possible if extra care is taken. A desirable surface-finish tolerance is 1.3 μm (50 μin), which would allow the use of larger, faster-cutting abrasives. If necessary, surface finish can be held to 0.25 μm (10 μin) through the use of fine, slower-cutting abrasives.

ABRASIVE-FLOW MACHINING

Abrasive-flow machining (AFM) involves the use of a viscous semisolid medium rather than the gas-abrasive mix of AJM. In AFM, the workpiece is clamped between two cylinders. The putty-like medium is pumped hydraulically from one cylinder to the other and extruded through or over the workpiece. Abrasive grains in the medium rub against the surfaces of the workpiece, particularly where there are restrictions, sharp corners, or burrs, and gently remove material. One to hundreds of reversals of direction of the medium compound may take place. Pressure of the medium is 690 to 20,000 kPa (100 to 3000 lbf/in^2). Medium compounds are proprietary and contain thickening agents and lubricants as well as the abrasive particles. Particle size ranges from No. 8 to No. 500. Abrasive materials may be aluminum oxide, silicon carbide, boron carbide, or diamond. Fixtures used to hold the workpiece should be abrasive-resistant in areas contacted by the medium; they are made of hardened steel, ceramic, or urethane.

Applications and Characteristics

AFM is primarily a deburring method for burrs in locations not easily accessible with conventional methods. (If burrs can be removed by vibratory or barrel tumbling, that method is usually more economical than AFM.) Other uses of AFM are the radiusing of sharp corners, particularly at the intersection of internal machined surfaces, polishing, and minor surface removal. One example of surface removal is the removal of the recast layer from holes produced by EDM or laser-beam machining. Holes as small as 0.4 mm (0.016 in) in diameter can be processed on AFM equipment. Although it may not be practical to remove large burrs by AFM because of its low rate of material removal, results are normally more uniform with AFM than with manual methods. Figure 4.21.3 illustrates examples of abrasive-flow-machined parts.

Costs and Economic Production Quantities

Equipment of moderately high cost and the need for fixturing and for development of proper operating parameters dictate the need for middle-sized or higher production quantities of most parts to justify the AFM process economically. Given a moderate production quantity, one-time costs can be recovered from the significant reduction in each-piece production times with the process. However, for particularly inaccessible surfaces, AFM may be the only practical finishing method, and small production lots or even prototype quantities may be processed advantageously.

Suitable Materials

All materials that can be machined by abrasive methods are suitable for abrasive-flow machining. Softer materials, of course, are processed most quickly, but AFM is best applied to aerospace and similar components with complex machined surfaces and hard, tough materials of poor machinability.

FIGURE 4.21.3 Examples of parts processed with abrasive-flow machining. (*a*) Deburring and polishing ratchet teeth. (*b*) Polishing turbine blades. (*Courtesy Extrude Hone Corp.*)

Design Recommendations

Designers should follow these recommendations:

1. Avoid blind holes if AFM is to be used. The process requires a through flow of abrasive compound.
2. Allow the same degree of out-of-roundness in holes after AFM as existed before the operation; AFM will not improve the roundness of holes.
3. Wide holes and slots (over about 12 mm, or 0.5 in) suffer from some loss of AFM efficiency because of boundary effects. For maximum process effectiveness, keep holes and slots narrow, down to about 0.6 mm (0.024 in). (The minimum practical hole diameter, as noted, is 0.4 mm, or 0.016 in.)
4. If AFM efficiency is important, make sure that parts are not too fragile. Cutting is most efficient when high pressures and more viscous media are used.
5. Corners that are to be rounded should have a radius of 0.013 to 2.0 mm (0.0005 to 0.080 in).

Tolerance Recommendations

When product function permits, maximum production economy results when $\pm$ 25 percent is allowed on the radii of rounded corners and on the amount of stock to be removed by AJM. If necessary, however, radii and surface-stock removal can be controlled within $\pm$ 10 percent. The latter limit requires extra care and production time and more extensive testing on the establishment of process parameters.

Surface-finish improvement normally results in roughness readings 50 percent finer than those which existed before the AJM operation. In extreme cases, roughness can be reduced by 90 percent.

ULTRASONIC MACHINING

The Process

Ultrasonic machining (USM) actually involves two different processes: *ultrasonic impact grinding* and *rotary ultrasonic machining.* Ultrasonic impact grinding (USIG) involves the rapid oscillation of a shaped tool immersed in a slurry of abrasive that is also in contact with the workpiece. This oscillation drives abrasive particles against the workpiece and cuts in it a cavity that has the same shape as the tool. The oscillating frequency of the tool is from 19,000 to 25,000 Hz, and its amplitude is only 0.013 to 0.063 mm (0.0005 to 0.0025 in). The gap between the tool and the workpiece is small (0.025 to 0.1 mm, or 0.001 to 0.004 in), and the abrasive slurry is pumped through this gap. The tool is normally of low-carbon or stainless steel and is fastened to an ultrasonic generator through a "horn" of Monel metal. The abrasive particles may be aluminum oxide, silicon carbide, or boron carbide.

Rotary ultrasonic machining is similar to conventional drilling or milling of glass or other nonmetallics except that the rotating diamond-coated tool is also vibrated at an ultrasonic frequency (20 kHz). There is no abrasive slurry as in ultrasonic impact grinding, but there is a coolant (usually water) that flushes away the removed material.

The ultrasonic vibration reduces the pressure on the cutting tool and the friction at the point of cutting. It provides better coolant flow and better flushing of removed material. These factors result in faster cutting action.

Typical Applications and Characteristics

Ultrasonic impact grinding is most advantageous when applied to the machining of irregular holes in thin sections or shallow, irregular cavities. Materials not suitable for other processes can be machined by USM (see subsection "Suitable Materials"). Fragile parts and materials like honeycomb are processible without undue difficulty. Drilling, cutting, deburring, etching, polishing, cleaning, and machining to produce a coined-like or embossed-like surface are normal applications. Holes as small as 0.08 mm (0.003 in) across, round or nonround, can be drilled. The maximum hole size with currently available 2.4-kW machines is about 90 mm (3.5 in) in diameter, although larger holes can be made by trepanning or by feeding the cutter in a transverse direction. The normal maximum hole depth is 25 to 50 mm (1 to 2 in). Multiple holes machined in one pass, cavities with curved axes, and screw threads can be produced by the process.

Rotary ultrasonic machining is used primarily to drill glass and other hard nonmetallic materials. Milling and grinding of these materials are also accomplished. The rotary ultrasonic method provides faster material removal and less danger of cracking or chipping these brittle materials than can be achieved with regular diamond-tool machining. The ultrasonic assist enables holes to be drilled straighter, deeper, and closer to edges than otherwise would be feasible. Holes up to about 50 mm (2 in) in diameter can be core drilled.

Surfaces cut by both processes have low surface stresses and no heat effects and are free from burrs. With ultrasonic impact grinding, there is an overcut on the diameter or width of the machined cavity of twice the average particle size. Most holes, especially deep ones, have a sidewall taper due to cutting on the sides of the tool. Figure 4.21.4 illustrates typical parts machined ultrasonically.

Costs and Economic Production Quantities

USIG is a relatively slow cutting method. Costs of tooling and equipment are moderate. However, choice of the process is not based particularly on production quantity. USIG is chosen when it is the most suitable method for the workpiece and material involved. When other conventional processes are usable, they are generally more economical. Cutting rates with USIG vary greatly with different materials, ranging from 0.03 to 4 cm^3/min (0.002 to 0.25 in^3/min). Tool wear, which ranges in ratio from 1:1 to 1:200 with workpiece-material removal, is another adverse cost factor. Rotary ultrasonic machining of hard nonmetallic materials, however, is faster than conventional machining of these materials.

Suitable Materials

USIG is most advantageous for hard, brittle, nonconductive materials. Actually, all materials can be cut by USIG, but processing materials softer than R_c 45 is not recommended; materials harder than R_c 64 are best suited to the process. EDM is a compet-

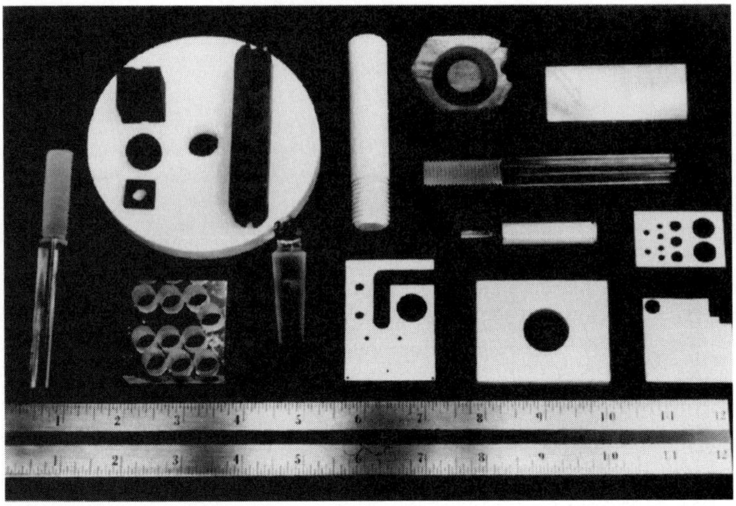

FIGURE 4.21.4 Alumina, glass, and ferrite parts machined ultrasonically. (*Courtesy Branson Sonic Power Co.*)

ing process and provides more rapid cutting of metals and other conductive materials; USIG is used primarily for nonmetallic materials not suitable to EDM. Materials commonly machined ultrasonically, listed in order of decreasing ultrasonic machinability, are glass, mother-of-pearl, ferrite, glass-bonded mica, germanium, carbon and graphite, quartz, ceramic, synthetic ruby, boron carbide, tungsten carbide, and tool steel. Table 4.21.1 lists metal-removal rates for some of these materials. Materials for rotary ultrasonic machining are similar: glass, quartz, ferrite, ruby, sapphire, and various ceramics. Softer, ductile materials are not so suitable; they tend to clog the diamond tool.

TABLE 4.21.1 Typical Material-Removal Rates for Materials Machined by Ultrasonic Impact Grinding

	Volume of material removed per minute, cm^3 (in^3)	Tool feed rates per minute, mm (in)
Glass	3.87 (0.236)	3.8 (0.150)
Ferrite	3.21 (0.196)	3.2 (0.125)
Mica, glass-bonded	3.21 (0.196)	3.2 (0.125)
Germanium	2.18 (0.133)	2.2 (0.085)
Graphite	2.05 (0.125)	2.0 (0.080)
Quartz	1.67 (0.102)	1.7 (0.065)
Ceramic	1.54 (0.094)	1.5 (0.060)
Boron carbide	0.39 (0.024)	0.38 (0.015)
Tungsten carbide	0.36 (0.022)	0.36 (0.014)
Tool steel	0.26 (0.016)	0.25 (0.010)

Design Recommendations

Designers should follow these recommendations:

1. Shallow holes and cavities are more suitable for USIG than deep ones. Holes should not be deeper than $2\frac{1}{2}$ times diameter.
2. Through holes or holes with through passages for abrasive slurry are preferred to blind holes. (See Fig. 4.21.5.)
3. If the workpiece material is brittle and a through hole is machined, the part should be designed so that a backup plate can be cemented or clamped to the exit surface. This will prevent chipping of the workpiece at the exit surface. (See Fig. 4.21.6.)
4. Allow for taper in holes machined by USIG, especially deep holes. Taper averages 0.05 mm/cm (0.005 in/in), as illustrated in Fig. 4.21.7. If necessary, taper can be reduced by using two successive machining operations.
5. Allow large radii at the bottom of blind holes, especially with USIG, because tool wear is concentrated at the corners of tools. For the same reason, do not specify sharp detail at the bottom of blind holes. (See Fig. 4.21.8.)

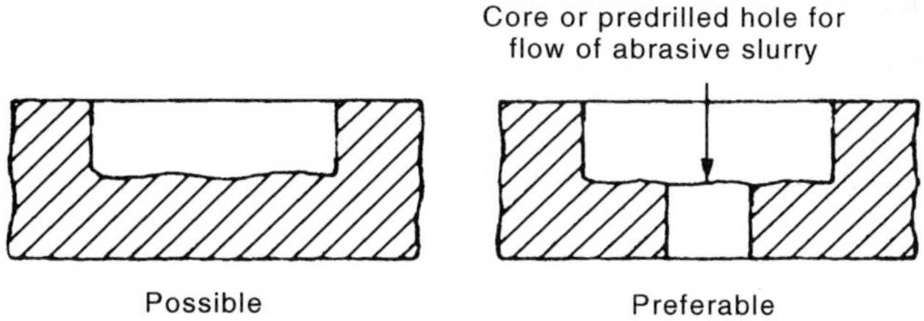

FIGURE 4.21.5 Through holes or holes with through passages for the USIG abrasive slurry are preferable.

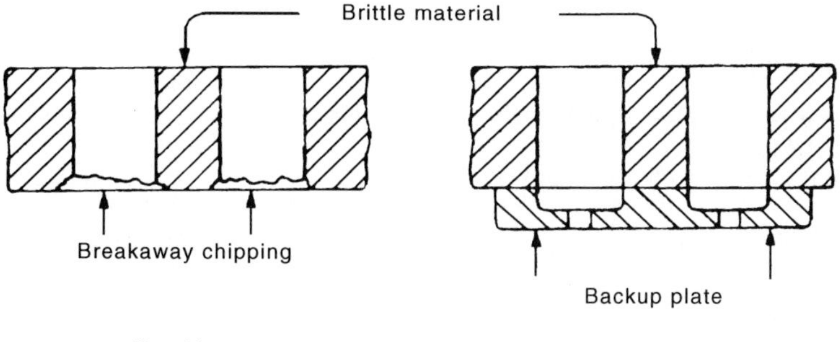

FIGURE 4.21.6 Parts that are ultrasonically machined from brittle materials should have a backup plate attached to the exit surface to prevent chipping at the edge.

Dimensional Factors and Recommended Tolerances

Overcut and tool wear are the two primary factors affecting the accuracy of ultrasonically impact-ground surfaces. Other factors are the rigidity of the setup, the size of the

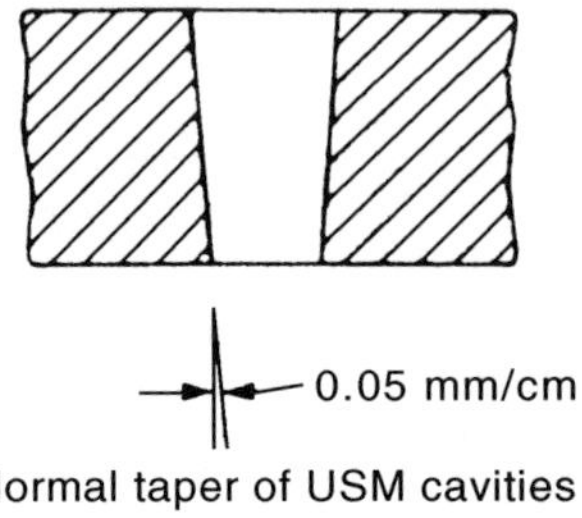

FIGURE 4.21.7 Normal taper of sidewalls of cavities made by USIG.

abrasive, the temperature of the slurry, and the design of the cutting tool. Nevertheless, very good accuracies are attainable. The recommended dimensional tolerance for USIG surfaces is ±0.025 mm (±0.001 in). However, ±0.013 mm (±0.0005 in) can be held if necessary. A surface-finish tolerance of 1 mm (40 μin) should be allowed, but slower-cutting, finer abrasive can produce surfaces as fine as 0.25 μm (10 μin).

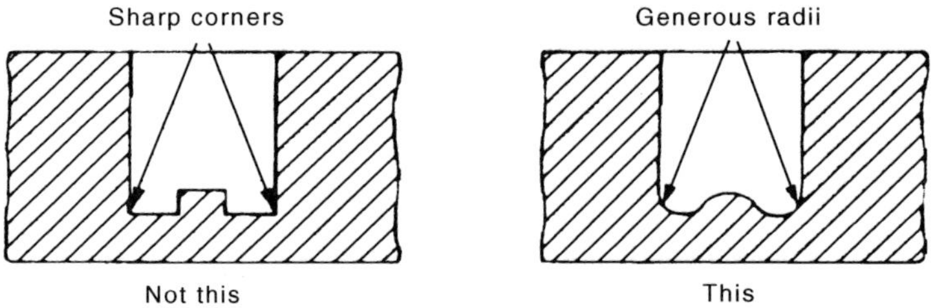

FIGURE 4.21.8 Do not specify sharp corners at the bottom of cavities machined ultrasonically.

HYDRODYNAMIC MACHINING

Hydrodynamic machining, sometimes called *water-jet machining,* uses a high-velocity narrow jet of liquid as a cutting agent. The jet of water, sometimes with polyethylene oxide or another long-chain-polymer additive, travels at about 600 m/s (2000 ft/s), or twice the speed of sound. Material is removed from the workpiece by the impingement of this jet. Pressures of 69 to 415 MPa (10,000 to 60,000 lbf/in^2) drive the liquid through a fine sapphire nozzle orifice. The resulting jet is 0.05 to 1.0 mm (0.002 to 0.040 in) wide.

The process is presently applicable commercially to soft nonmetallic materials, though very thin, soft metals also can be cut. Gypsum board, urethane, and polystyrene foam, $\frac{1}{8}$-in plywood, rubber, various thermoplastics, and fiberglass-reinforced plastics are suited to hydrodynamic cutting. Shoe soles, asbestos brake-shoe linings, and furniture parts made from laminated paperboard are examples of production parts cut from other materials.

The process is used most often for cutout operations on material in sheet form. It has a number of advantages for such work: The kerf is narrow, only 0.025 mm (0.001 in) wider than the nozzle orifice; the dwell of the jet does not widen the kerf; there is no heat effect to the cut edge; and little or no dust is created. One drawback is the high noise level that can accompany the process.

Cutting rates can be rapid with some materials but vary considerably from one material to another. For cutout operations, tooling is minimal. Equipment is normally designed and fabricated for a specific application.

Tolerances for cutout pieces depend primarily on the accuracy of the mechanism that provides movement of the nozzle with respect to the material to be cut. Tolerances of ±0.25 mm (0.010 in) are normally achievable.

When abrasive particles are added to the water jet, the process is referred to as *abrasive water-jet machining,* and metals and other hard materials can be cut with the process. Abrasive particles are fed into the high-speed steam of water, travel with the water through an orifice, and impinge on the workpiece. The abrasive particles cut the workpiece material as they strike it. The jet diameter is typically 0.5 to 2.5 mm (0.020 to 0.100 in). The abrasive slurry is captured after it leaves the workpiece.

The process is used to machine metals, ceramics, concrete, glass, and composites. Almost all materials can be machined by this method. Reinforced plastics, honeycombed materials, and others difficult to machine by other methods are particularly suitable. It has the advantage over laser and electron-beam machining in not producing any significant heat-affected zone. It is also superior to laser-beam machining for metal sheets thicker than 13 mm (0.5 in). Blind-hole and small-hole drilling generally are not feasible. Kerfs are wider than with laser-beam machining or electron-beam machining, typically 0.75 to 2.3 mm (0.030 to 0.090 in). Except in soft, ductile materials, burrs are not produced. Metal plates up to about 200 mm (8 in) can be cut by the process.

Equipment for abrasive water-jet machining is lower in cost than that required for laser-beam or electron-beam machining, but capital costs are still a major item. Another is the cost of the abrasive used, most commonly garnet. Cutting speeds for materials thin enough for laser-beam machining are somewhat slower than with laser-beam machining. The process is also considerably slower than flame cutting but produces a cleaner edge. In general, it is most economical for those materials for which a heat-affected zone is a problem. As with laser-beam, electron-beam, and flame cutting, no cutting tools are required.

The surface finish of the cut edge typically ranges from 1.6 to 6.3 μm (63 to 250 μin). Dimensions for cut parts should have a tolerance of ±0.25 mm (±0.010 in), though closer values are possible, if necessary, with precision equipment.

ELECTRON-BEAM MACHINING

The Process

Electron-beam machining (EBM) is, with minor differences, the same process as used for electron-beam welding and described in Chap. 7.1. A high-velocity beam of electrons, focused on a small point of the workpiece, intensely heats the workpiece material at that point so that it melts and vaporizes. Whereas in electron-beam welding the objective is to melt the workpiece material so that it flows together and fuses, in EBM the objective is to cut completely through the workpiece. Higher power levels and higher beam velocities are utilized in EBM compared with electron-beam welding. Another difference is the greater need for a full vacuum with EBM. In EBM, the beam impinges on an area of 0.32 to 0.64 mm^2 (0.0005 to 0.001 in^2) and has a power density of 15 million W/mm^2 (10 billion W/in^2).

Applications and Characteristics

EBM is most suitable for fine cuts in relatively thin workpieces. Drilling fine holes in metals is the major application, but any material can be machined by the process.

Holes and slots only a few thousandths of an inch wide and very precise contoured cuts are quite feasible. The process is particularly suitable for cuts that are too fine for EDM or ECM.

EBM is used for drilling metering holes such as are used for diesel fuel injection, gas orifices for pressure-differential devices, wire-drawing dies, spinnerets, sleeve-valve holes, scribing thin films, and removing broken taps of small diameter. Holes as small as 0.013 mm (0.0005 in) in diameter are practicable in 0.025-mm- (0.001-in-) thick material, as are slots as narrow as 0.025 mm (0.001 in). Length-to-diameter ratios of holes up to 10 or even 15 in in some cases can be achieved. However, 6.4 mm (0.25 in) is a practical maximum depth of cut. If the workpiece material is over 0.13 mm (0.005 in) thick, a 1 to 2° taper in the sidewalls of the through cut should be expected.

Cratering usually occurs on the workpiece surface adjacent to the hole entrance. There also may be some spatter on the same surface, but this is easily removed. The edges of holes and slots tend to show nonuniform surfaces. There is a heat-affected zone about 0.25 mm (0.010 in) deep adjacent to the cut. Otherwise, the workpiece is distortion-free, since there is no pressure or contact between the workpiece and any cutter. Figure 4.21.9 illustrates characteristics of a hole machined by an electron beam.

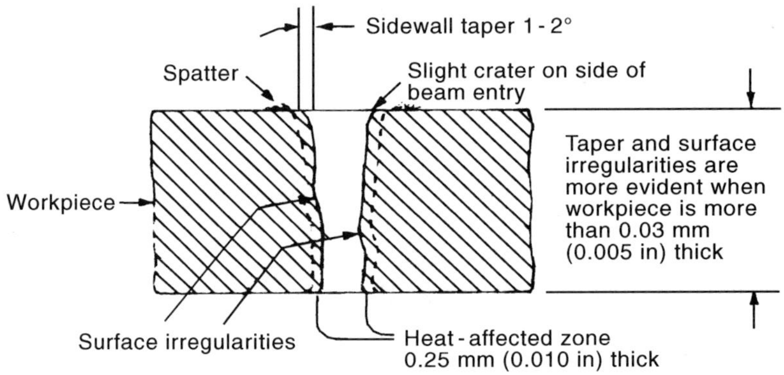

FIGURE 4.21.9 Typical characteristic of an electron-beam-machined hole.

Costs and Economic Production Quantities

Cutting rates are rapid for thin materials. For example, producing 0.1-mm- (0.004-in-) diameter holes in 0.5-mm (0.020-in) stock requires less than $^1/_{10}$ s. Slots 0.05 mm (0.002 in) wide in 0.25-mm- (0.010-in-) thick material can be machined at 65 to 150 mm/min (2.5 to 6 in/min). Nevertheless, the volume of metal removal with EBM is actually low compared with conventional methods, averaging about 0.8 to 2 mm^3/min (0.00005 to 0.00012 in^3/min). The time required to evacuate the vacuum chamber for each machine load is a factor that increases production time. The high cost of equipment and the need for skilled operators are further adverse cost factors.

Suitable Materials

As indicated above, any material can be machined by EBM. Metals, ceramics, plastics, and composites all are easily machined, although the cutting rate is slower for materials with high melting and vaporization temperatures. Hardened steel, stainless steel, molybdenum, nickel, cobalt, titanium, tungsten, and their alloys; quartz, ceramics, and synthetic sapphire have all been successfully machined by EBM.

Design Recommendations

Designers should follow these recommendations:

1. Workpieces to be machined by the electron-beam process should be kept as small as possible so that a larger number of pieces will fit into the vacuum chamber. It may even be advisable to create an assembly of several parts rather than having one bulkier part if only a portion of the workpiece requires EBM.
2. The normal minimum radius for internal corners is 0.25 mm (0.010 in). Sharper corners should not be specified.
3. For best results with through cuts, the workpiece should be as thin as possible. A practical maximum is 6.3 mm (0.25 in), but workpieces considerably thinner than this machine more rapidly with less sidewall taper.
4. Designers should allow for the machined surface effects of EBM, which for some applications may be undesirable. In these cases, stock should be allowed for secondary operations. Figure 4.21.9 illustrates these surface effects.

Recommended Tolerances

A tolerance of $\pm$ 10 percent should be allowed on hole diameters and slot widths. The normal surface-finish specification should be 2.5 μm (100 μin), although surfaces as fine as 0.5 μm (20 μin) can be produced under optimal conditions.

LASER-BEAM MACHINING

Like EBM, laser-beam machining (LBM) utilizes a process that is applicable to welding as well as to machining. The basic process is described in Chap. 7.1 and illustrated in Fig. 7.1.4. When used for machining, the process operates at somewhat higher energy levels than for welding. The narrow, highly focused beam of coherent light of extremely high intensity melts and vaporizes material at the point where it strikes the workpiece. The beam may consist of intermittent pulses or of a continuous beam. A typical focused spot on the workpiece is 25 to 50 μm (0.001 to 0.002 in) wide. The density of energy at this spot is extremely high, amounting to millions of watts per square centimeter. When this level of energy strikes the workpiece, material at the point of impingement is vaporized or melts and is swept away from the beam. In some cases, a jet of gas coaxial with the laser beam is directed against the workpiece. Oxygen is the most common gas for rapid cutting, while inert gases are best to improve the edge-surface finish.

Applications and Typical Characteristics

Laser-beam machining is most commonly used for high-precision machining or micromachining of thin parts that are difficult to machine by conventional methods. Holes smaller than 1.3 mm (0.050 in) and stock thinner than 5 mm (0.200 in) give the best results. Larger holes can be trepanned. The drilling of very small holes with a large depth-to-diameter ratio is a particularly advantageous application. Increasingly, the process is being used to cut out blank parts from sheet metals and other materials up to about 13 mm (0.5 in) in thickness. Slitting, trimming, and perforating are other applications.

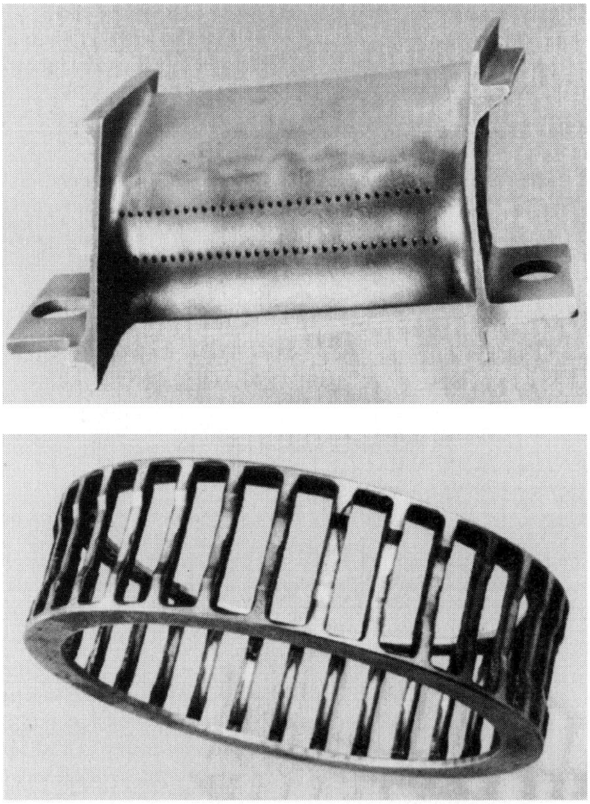

FIGURE 4.21.10 Examples of LBM: (*a*) Cooling holes in turbine-engine blades. (*b*) Slots in high-strength steel for sprag-type clutch. (*Courtesy Apollo Lasers, Inc.*)

Additionally, by using two intersecting lasers, machining operations equivalent to milling and turning can be performed. However, these are normally limited to materials not feasible to machine by conventional methods.

Common examples of LBM are drilling 0.1-mm (0.004-in) holes in glass contact lenses, drilling 0.13-mm- (0.005-in-) diameter holes in plastic aerosol-spray nozzles, trimming thick and thin film resistors to precise values, scribing silicon and ceramic substrates for electronic microcircuits, and making contoured cuts in cloth for garment components. Figure 4.21.10 illustrates additional typical examples.

The LBM process has the advantage of being usable for machining inaccessible places. It can operate through transparent materials and various atmospheres. There is no tool contact with the work. This and the fact that the heating effect is very much localized permit the machining of brittle, heat-shock-sensitive, and fragile workpieces.

The minimum hole diameter is about 0.005 mm (0.0002 in), but 0.13 mm (0.005 in) is more common. Length-to-diameter ratios of up to 50:1 are feasible with 0.13-mm- (0.005-in-) diameter holes. Holes with an angle to the surface as shallow as 15° are possible. Kerf widths for slits and profile cuts as narrow as 0.1 mm (0.004 in) are used, but 0.4 mm (0.015 in) is a better normal value for most applications.

LBM holes deeper than 0.5 mm (0.020 in) exhibit noticeable taper. There is also considerable nonuniformity of hole diameter. Both taper and other irregularities become more pronounced as the depth of the cut increases. Holes normally also are not perfectly round. There is a tendency toward cratering at the entrance surface of the cut, and a narrow heat-affected zone of about 0.13 mm (0.005 in) borders the machined surface.

Costs and Economic Production Quantities

Actual drilling time for small holes in thin stock is only hundredths of a second. Linear cutting of sheet materials also may be rapid for most materials, particularly with oxygen assist. For example:

Titanium alloys from 0.5 to 10 mm (0.020 to 0.400 in) thick: cut at 2.5 to 12 m/min (100 to 500 in/min)

Mild steel 5 mm (0.200 in) thick: cut at 2.5 m/min (100 in/min)

Plastics 0.8 mm to 3 mm (0.030 to 0.125 in) thick: cut at 2.5 to 7.5 m/min (100 to 300 in/min)

However, equipment costs are high for LBM (though lower than those for EBM, which requires a vacuum chamber). Operational costs may be high due to the complexity of the operation. Higher production quantities and the use of hard or difficult-to-machine materials favor the process. Laser-equipped turret machines have found a niche in the job-shop production of sheet-metal parts.

Suitable Materials

All materials are laser-beam machinable, but not necessarily with the same technique or process. The most practical materials for LBM are those which are difficult or impossible to cut by conventional methods. Ceramics, glass, carbides, and some aerospace alloys fall into this category. Copper, aluminum, gold, and silver are not so suitable because of their high thermal conductivity. Aluminum also tends to form a dross at the cut edge that requires a secondary operation for removal. Mild steel and titanium are commonly processed by laser beam. Other materials that have been machined successfully are plastics, rubber, beryllium, zirconium, stainless steel, tungsten, cast iron, brass, molybdenum, cloth, cardboard, wood, boron- and graphite-reinforced epoxy composites, and various laminated materials. Laminates and composites may require special steps because the constituent materials behave differently when exposed to the laser energy.

Design Recommendations

Since LBM is a process that uses energy in the form of light, best results are achieved when the workpiece surface absorbs rather than reflects the beam energy. Surfaces should be dull and unpolished until after LBM has been completed.

Other design recommendations closely parallel those for EBM. Workpieces should be thin in areas through-cut by LBM because machining time is faster and taper and irregularities are minimized with thinner materials. Corner radii allowances should be 0.25 mm (0.010 in) or greater. Allowances should be made for taper averaging 3° per side and for a heat-affected zone of about 0.13 mm (0.005 in) deep, for cratering on the entrance surface and for spatter and other residue on machined surfaces.

Recommended Tolerances

Hole-diameter and slot-width tolerances should be ±0.025 mm (±0.001 in). When a reactive gas assist is used, allowances should be increased to ± 0.1 mm (±0.004 in). Dimensions of parts cut by laser from sheet stock should have a tolerance of ±0.13 mm (±0.005 in).

CHAPTER 4.22
GEARS

J. Franklin Jones
Springfield, Vermont

DEFINITION

Gears are machine elements that transmit angular motion and power by the successive engagement of teeth on their periphery. They constitute an economical method for such transmission, particularly if power levels or accuracy requirements are high.

APPLICATIONS

Gears are useful when the following kinds of power or motion transmission are required: (1) a change in speed of rotation, (2) a multiplication or division of torque or magnitude of rotation, (3) a change in direction of rotation, (4) conversion from rotational to linear motion or vice versa (rack gears), (5) a change in the angular orientation of the rotational motion (bevel gears), and (6) an offset or change in location of the rotating motion.

Power-transmission gears are normally of coarse pitch and can be very large. Power-plant gears as large as 7 m (24 ft) have been made. They can transmit tens of thousands of horsepower. Contrasted with these are fine-pitch miniature instrument and watch gears as small as 2 mm (0.080 in) in pitch diameter. Normally, these small gears transmit motion only; the power they transmit is negligible.

Gears of 20 diametral pitch (see "Gear Elements") or coarser are classified as coarse-pitch gears. Fine-pitch gears are those with a diametral pitch greater than 20. The usual maximum fineness is 120 diametral pitch. However, involute-tooth gears can be fabricated with diametral pitches as fine as 200 and cycloidal-tooth gears with diametral pitches to 350.

Other typical gear applications are automotive transmissions, differentials, and steering mechanisms; small appliances such as electric hand mixers, drills, and other power tools; machine tools; aircraft engines; sewing machines; toys; clocks; military fire-control devices; inertial-navigation apparatuses; conveyor drives; winches; countermechanisms; elevators; agricultural equipment; and instruments of various kinds.

Figures 4.22.1 and 4.22.2 illustrate typical gears.

FIGURE 4.22.1 This helical ring gear for an ore-grinding mill is 24 ft in diameter and over 3 ft wide; it weighs 70 tons. (*Courtesy The Falk Corporation, subsidiary of Sundstand Corporation.*)

FIGURE 4.22.2 Collection of miniature zinc die-cast gears. These are used in such products as timing devices, automobile speedometers, fishing-rod reels, etc. (*Courtesy Fisher Gauge Limited.*)

KINDS OF GEARS

Various kinds of gears are illustrated in Fig. 4.22.3 and can be described as follows:

Spur gears are the most common variety and the most economical to manufacture. Their overall shape is cylindrical, and teeth are parallel to the axis of rotation. Axes of mating gears also are parallel.

Helical gears also have a cylindrical overall shape, but the teeth are set at an angle to the axis. This provides smoother and quieter action but somewhat reduces the straightforwardness of manufacturing operations. Helical gears may be run on non-parallel axes. When the axes are at right angles, the gears are known as *crossed helical gears.*

Herringbone gears are double-helical gears. Both right-hand and left-hand helix angles exist side by side across the face of the gear. Thus axial thrust from helical teeth is neutralized.

Internal gears have teeth on the inside surface of a hollow cylinder.

Rack gears have teeth on a flat surface instead of on a curved or cylindrical surface. They provide straight-line instead of rotary motion.

Bevel gears have teeth on a conical surface and operate on axes that intersect, usually at right angles. Because the teeth are tapered, bevel gears are more difficult to produce (at least by machining methods) than spur, helical, or rack gears.

Miter gears are mating bevel gears with equal numbers of teeth and with axes at right angles.

Spiral bevel gears are bevel gears that have teeth that are curved and oblique to any plane passing through the axis. They have the same advantages of quietness and smoothness as helical gears. Spiral bevel gears are more difficult to manufacture than straight bevel gears. The negative draft of the back side of gear teeth complicates molding and forming processes.

Hypoid gears are spiral bevel gears with offset nonintersecting axes.

Zerol bevel gears are spiral bevel gears with zero spiral angles. Teeth, however, are curved and lie in the same general direction as straight teeth.

Skew bevel gears are similar to spiral bevel gears in that the teeth are oblique. However, the teeth are straight rather than curved.

Worm gears are crossed-axis helical gears in which the helix angle of one of the gears (the *worm*) has a high helix angle so that it resembles a screw. The worm can drive the mating gear (the *worm gear*) but not vice versa. The axes of the two gears are normally at right angles.

An *hourglass worm* has a smaller diameter in the middle than at the ends and thus conforms more closely to the shape of the cylindrical mating worm gear.

A *pinion* is a gear with a small number of teeth. In a set of two gears running together, the one with the smaller number of teeth is called the *pinion.*

A *crown gear* is a bevel gear in the form of a disk. The pitch surface is a plane perpendicular to the axis.

Face gears consist of a pinion gear in combination with a mating gear of disk form. Axes are usually at right angles and may or may not be intersecting.

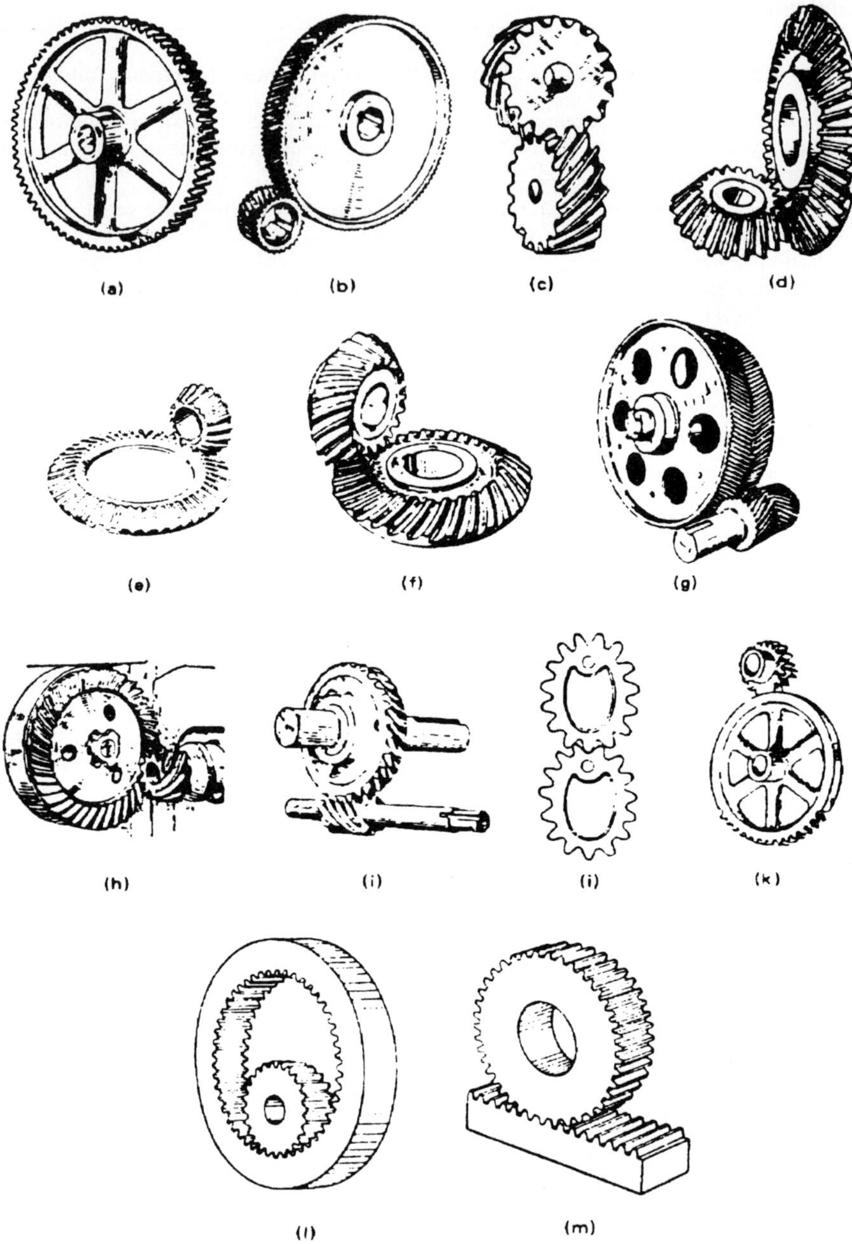

FIGURE 4.22.3 Kinds of gears; (*a*) spur gear; (*b*) parallel helical gears; (*c*) crossed helical gears; (*d*) straight bevel gears; (*e*) zerol bevel gears; (*f*) spiral bevel gears; (*g*) herringbone gears; (*h*) hypoid gears; (*i*) worm gears; (*j*) elliptical gears; (*k*) intermittent gears; (*l*) internal gears; (*m*) rack and pinion. (*a through k from* The New American Machinist's Handbook, *McGraw-Hill, New York, 1955; l and m from* Maintenance Engineers' Handbook.)

GEAR ELEMENTS

Figure 4.22.4 illustrates common elements and standard dimensions of gears. Elements and dimensions bearing further discussion are the following:

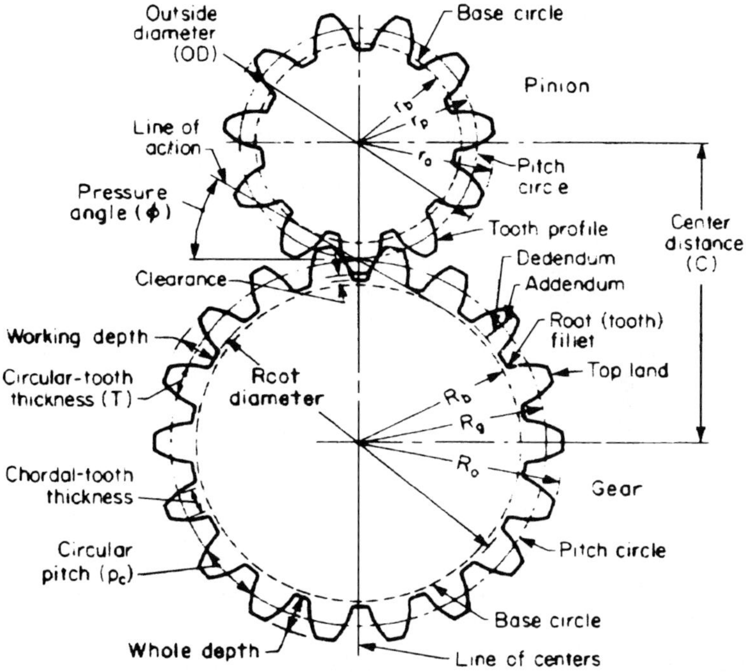

FIGURE 4.22.4 Gear elements and key dimensions. (*From G. W. Michalec,* Precision Gearing, *Wiley, New York, 1966.*)

Diametral pitch is a measure of the coarseness of the gear. It is equal to π divided by the circular pitch in inches. For any gear it is the ratio of the number of teeth to its pitch diameter (in inches). *Circular pitch,* as shown in Fig. 4.22.4, is the distance along the pitch circle from tooth to tooth. The *pitch diameter* is the diameter of the *pitch circle,* the imaginary circle that rolls without slipping with the pitch circle of the mating gear. The *module* is the reciprocal of the diametral pitch.

The *pressure angle* of a gear is the angle between the tooth profile and a line perpendicular to the pitch circle, usually at the point where the pitch circle and the tooth profile intersect.

The most common and preferred tooth profile for parallel-axis gears is the *involute-tooth form.* This tooth profile provides smooth gear operation through a wide range of center distances. It facilitates tooling standardization because fewer cutters are needed for a range of gear sizes.

Backlash is the amount by which tooth spacing at the point of contact exceeds tooth width.

The *helix angle* is the angle between the tooth face and a line parallel to the axis of the gear.

The *gear ratio* is the ratio of the number of teeth in the larger gear to the number in the pinion or smaller gear.

GEAR-MANUFACTURING PROCESSES AND THEIR APPLICATIONS

Machining

The variety of machining processes used in the manufacture of gears includes milling, hobbing, shaping, broaching, shear cutting, and several specialized processes as primary machining methods. Shaving, grinding, honing, lapping, and burnishing are used as refining operations to improve accuracy and surface finish.

Milling. A milling cutter with teeth ground to the shape of the gear-tooth spacing is fed across the gear blank. After each tooth space is cut, the cutter is returned to its starting position, the blank is indexed, and the cycle is repeated. Tooth spaces therefore are machined one at a time. The blank is stationary during the cutting cycle. Indexing of the blank between cutting cycles may be manual or automatic, depending on the degree of sophistication of the milling machine.

Gear milling can apply to both roughing and finishing operations, although the latter is less common because of accuracy limitations. For the same reasons and because other methods are more rapid, gear milling is normally confined to replacement-gear making or low-quantity production. Milling cutters are less costly than hobs and other types of cutters.

Spur, helical, and straight bevel gears are machined by gear milling. Bevel gears require two passes because of the tooth taper. Spiral bevel gears are not feasible for gear milling. Internal gears can, in some cases, be cut by milling machines.

Hobbing. Hobbing is a generating process. A spindle carrying a gear blank is geared to a second spindle carrying a rotating hob (cutting tool) at approximately right angles to the blank. The hob is similar in appearance to a worm gear but has gashes to form cutting edges. The hob, if it has a single pitch, makes one revolution for each tooth of the workpiece, and as the two rotate continuously, the hob is fed parallel to the face of the gear tooth.

The hob teeth are helical, and when a spur gear is machined, the hob axis is inclined from the perpendicular position by the helix angle of the hob. (See Fig. 4.22.5.) The generation action of the hob cutting teeth is shown in Fig. 4.22.6.

Hobbing is used to produce spur, helical, and worm gears and worms but not bevel or internal gears. Production rates are high. Though hobbing is most economical at medium and high rates of production, its accuracy, versatility, and ease of setup make it adaptable to low-quantity production also. Though more expensive than milling and shape cutters, gear hobs are cheaper than shear cutters and gear broaches.

Shaping. Shaping is also a generating method. A cutting tool resembling a gear is mounted on a spindle parallel to the axis of the gear to be cut. The blank and the cutter are geared together, and as they slowly rotate, the cutter reciprocates in an axial direction and generates the teeth of the gear in successive cuts. The gearlike cutter is slightly tapered to allow a clearance angle on the sides of the cutting teeth. Cutting occurs on the downward stroke. On the upward (return) stroke, the cutter and the work are

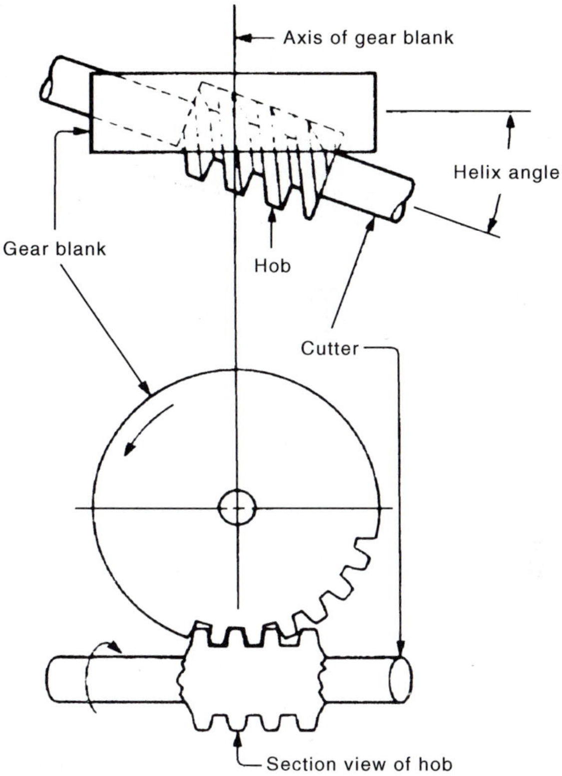

FIGURE 4.22.5 Gear hobbing. Rotation of the hob and gear blank is synchronized and continuous. When a spur gear is hobbed, the axis of the hob is positioned at an angle to the plane of the gear blank. This angle is the same as the helix angle of the hob.

moved apart by approximately 0.5 mm (0.020 in) to prevent the cutter from rubbing against the work.

When helical gears are to be cut, a guide that produces a helical motion in the cutter spindle is provided. A separate guide is required for each helix angle, and the amount of the helix is limited.

Shaping can be used to produce both internal and external spur and helical gears. Bevel gears are not produced by this method. However, herringbone and face gears are machinable by shaping. Shaping is also well suited to the machining of gears that are located close to obstructing surfaces or in a cluster because the clearance required for cutter overtravel is much less than with hobbing or milling.

Since tooling is relatively inexpensive, shaping can be used for low-quantity production. However, because the process can be applied to certain gears not machinable by other methods, and because it can be made automatic, it is used at moderate and high production levels as well, even though it is not as rapid as hobbing.

Broaching. As described in Chap. 4.9, broaching is a useful machining method for gears, especially when production runs are large. Both internal and external spur and

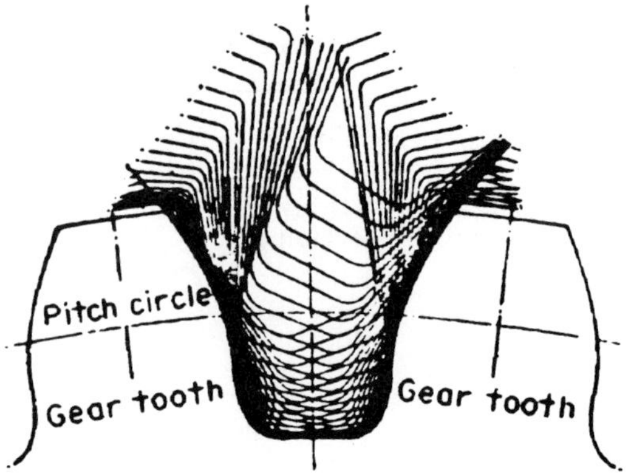

FIGURE 4.22.6 Generating action of hob cutting teeth. (*From D. W. Dudley,* Gear Handbook, *McGraw-Hill, New York, 1962.*)

helical gears can be broached, although internal gears are the more common application. External gears, especially if large, require bulky and expensive broaching tools.

Fine finishes and high accuracy levels are possible with broaching. Gears are usually completely machined in one pass, although separate roughing and finishing operations are sometimes employed. If the gear is helical, the tooth or work rotates as the broach advances. Either the work or the tool is pushed or pulled.

Broaches are expensive and therefore normally require high-volume production if tool costs are to be amortized. Broaching's short cycle time adds to its attractiveness for mass-production applications.

Shear Cutting. This process could be considered a cross between broaching and gear shaping. Like the gear broach, the formed cutters simultaneously remove material from all tooth spaces of the gear. Like the gear shaper, the cutting head has a reciprocating motion. The cutters all advance slightly with each stroke of the cutting head until the full form of the gear is machined.

Shear cutting is applicable to spur gears of diameters up to 500 mm (20 in) and face widths up to 150 mm (6 in). While helical gears are not machinable with the process, internal spur gears can be shear-cut.

The process is most advantageous with high production. Shear-cutting cycles are short, often less than 1 min for gears smaller than 150 mm (6 in) in diameter, but tool costs are rather high.

Bevel-Gear-Machining Processes. These processes are complex, with numerous variations that depend on the size, pitch, gear ratio, bevel angle, and spiral angle (if any) involved. The gears are machined with processes that use both reciprocating cutters and circular cutters. While both generating and nongenerating processes are employed, generating procedures are more important because nongenerating methods are limited in applicability to only one of the two mating bevel gears.

Straight-bevel-gear methods include straight-bevel-gear planing, two-tool planing, milling (see above), dual-rotating-cutter machining (Coniflex), and Revacycle. *Spiral-bevel-gear methods* include planing and face-mill-cutter machining. Face-mill-cutter machines include both generating and form-cutting types, machines that cut one gear-tooth side at a time and those which cut two sides simultaneously.

Straight-Bevel-Gear Planing. This is a variation of gear shaping adapted to straight bevel gears. With this method, the gear blank is held stationary (except for indexing), and the cutting tool reciprocates across the gear blank. To this extent, the process is similar to gear shaping. It differs in that cutting strokes are not parallel but are radial from one pivot to produce the taper required in bevel-gear teeth. Templates are used to control the location of the tool slide and thus enable the reciprocating cutter to trace the tooth profile.

The straight-bevel-gear planing method is used primarily for large, coarse-pitch gearing. It is economical for repair or short-run production. While it is not a rapid method, cutting-tool costs are low.

Two-Tool Planer-Type Machines. These machines have separate tools on separate tool slides for each side of the gear tooth. The tool guides are mounted on a cradle that provides a rolling motion in relation to the gear blank as the latter rotates and the cutting tools reciprocate across the blank. This rolling motion enables the cutter surfaces to simulate the teeth of an imaginary crown gear and to generate the necessary profile shape of the tooth.

The two-tool planer method is used for straight bevel gears of both fine and coarse pitch. For gears with diametral pitches of 24 or more, both a rough and a finishing cutter are incorporated in one tool holder. For gears with diametral pitches coarser than 24, separate roughing and finishing operations are required. Tooling costs are low, but cutting cycles are somewhat long. This method therefore is best suited to low production quantities.

Dual Rotating Cutters. This method involves circular-milling-type cutters that have interlocking cutting teeth and cut both sides of the gear-tooth space simultaneously. The cutter spindles are tilted to provide the necessary taper of the gear teeth and the desired tooth-pressure angle. During cutting, the spindles change their position to generate the tooth profile. After each tooth is cut, the gear blank is indexed and the cutter spindles roll back to their starting position. The process is applicable to bevel gears smaller than 215-mm ($8\frac{1}{2}$-in) pitch diameter, 32-mm ($1\frac{3}{8}$-in) face width, and 3 or finer diametral pitch.

The dual-rotating-cutter method is faster than the two-tool planer method but not as rapid as the Revacycle method described below. Tooling and equipment are more costly than those required for two-tool planing. Medium production quantities are most advantageous.

Revacycle Process. This process is a rapid high-production method for straight-bevel-gear machining. A fairly large circular cutter passes across the face of the gear blank, which is stationary during each cut. The cutter is actually a circular broach in that each successive cutting tooth is larger than the one that precedes it along the cutter's circumference. The cutter makes only one revolution per tooth space. The increasing size of the cutting teeth and the movement of the cutter across the face of the gear produce the required tooth taper. A gap in the teeth provides time and space for the gear blank to be indexed between cuts. Commercial-quality pinions and gears can be finish-machined in one operation. Such cutters and other Revacycle tooling are more expensive than the tooling required for other bevel-gear-cutting methods. The 5- to 6-h setup time is also significant. However, the short cutting cycle, normally less than 3 s per tooth and sometimes 2 s or less, permits a low total cost of production when high volume is involved.

Spiral-Gear Planing-Generator Cutting. This process is an extension of straight-bevel-gear planing. Generation of the tooth profile normally takes place with a single

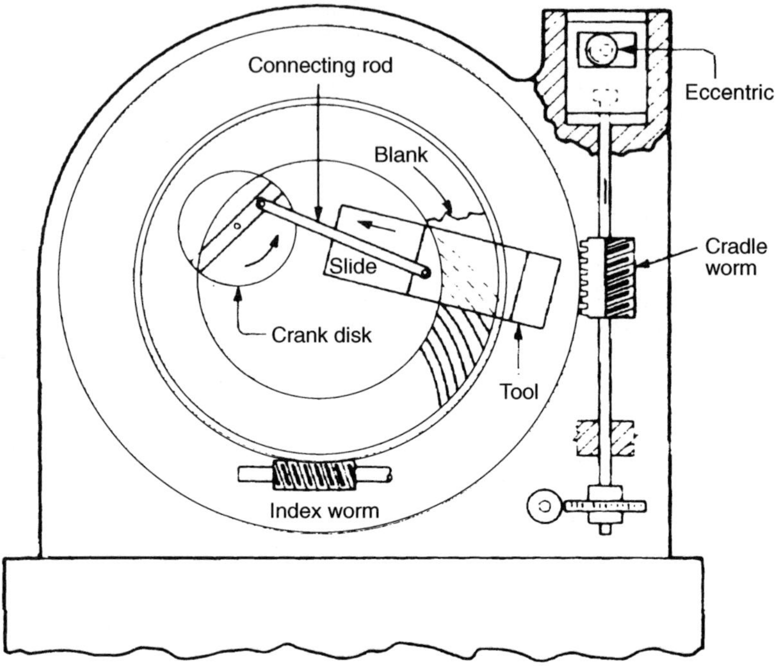

FIGURE 4.22.7 Spiral-gear planing-generator cutting. (*From D. W. Dudley,* Gear Handbook, *McGraw-Hill, New York, 1962.*)

cutter. The spiral path of the cutter vis-à-vis the gear blank is achieved by rotating the blank by a fixed, controlled amount during the cutting stroke. The process is illustrated schematically in Fig. 4.22.7. The gear blank's timed rotation is continuous so that each stroke of the cutter removes material from successive teeth spaces. The tooth form is generated during successive rotations of the blank. The blank cradle also has a rolling, oscillating motion to provide tooth generation. One side of all teeth is machined before the machine is reset to cut the other side of all teeth. Sometimes a roughing-cut cycle precedes the finishing-cut cycles.

The process is generally accurate and is used for spiral-bevel, zerol, and hypoid gears that are too large for available face-mill-cutting machines. Gears up to approximately 2.5-m (100-in) pitch diameter, 30-mm (1.2-in) face width, and one diametral pitch are produced by this method. Cutting times are fairly lengthy, but since tooling costs are low, the process is economical for the modest quantities normally involved in large-gear production.

Face-Mill-Cutting Methods. These include both form-cutting and generating types. The *face-mill-form-cutting method* is faster and is suitable for the gear member of a spiral-bevel-gear set if the ratio between pinion and gear is 3:1 or higher. It is not suitable for machining the pinion-gear member. Figure 4.22.8 illustrates the action of a form cutter contrasted with that of the generating type. Nongenerated machining is best suited for high-production applications like automotive differentials. Spiral-bevel, hypoid, and zerol-bevel gears are cut by these methods. Two well-known proprietary nongenerating processes for these gears are the Formate and Hexiform methods.

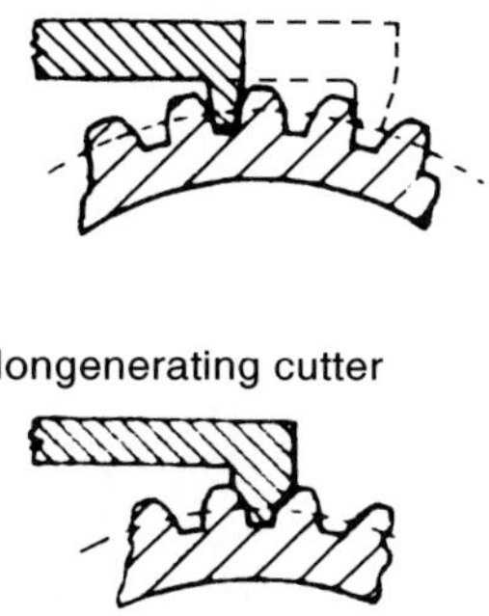

FIGURE 4.22.8 The action of a form cutter as contrasted with a generating-gear tooth cutter. (*From D. W. Dudley,* Gear Handbook, *McGraw-Hill, New York, 1962.*)

Formate cutting is similar to Revacycle cutting in that the cutter is, in effect, a circular broach. By placing the cutting teeth on the face of the cutter, however, and by properly orienting the cutter spindle, a curved or spiral configuration of the gear teeth is produced.

With *face-mill-generating cutting methods,* the cutter spindle is mounted so that it has a rolling motion in a timed relationship to the cutter and workpiece spindles. As shown in Fig. 4.22.8, at any instant cutting occurs at only one point, but the motions of the cutter with respect to the workpiece cause the tooth form to be generated by successive teeth and successive cuts. Sometimes the cutter produces only one side of the gear teeth in one setup; in other cases, both sides of the teeth are machined in the same operation. In most cases, separate roughing and finishing operations are used.

These generating methods are suitable for almost all spiral-bevel gears. However, they are used most often for small and medium-sized gears and for small and moderate quantities. Gears up to about 850 mm (33 in) in pitch diameter can be machined on presently available equipment. (Larger spiral-bevel gears are machined on planing-type machines.) As mentioned above, spiral-bevel pinion gears are invariably machined by face-cutter-generating methods.

Worm-Gear Sets. Worms can be machined on lathes or by hobbing or thread milling. Lathe-turning and thread-milling methods are the same as those used for screw threads and described in Chap. 4.4, except that in some cases the tooth form has a convex (helicoidal) shape instead of being straight-sided.

Enveloping worm gears are normally machined by hobbing, although conventional milling machines are sometimes used for limited quantities. Nonenveloping worm gears are manufactured by the methods used for other helical gears.

Gear-Finish-Machining Methods

These methods include shaving, grinding, honing, lapping, and burnishing. They all have the same or similar objective: to provide more accurate and smoother gear-tooth surfaces and thereby to provide smoother, quieter, and more uniform gear action. They remove variations that may be inherent from the initial machining operation, distortions due to relieved stresses or heat treatment, and nicks, burrs, and other surface irregularities. These finish-machining operations are not by any means essential to gear machining; in most cases they are not necessary. For precision-gearing applications and when noise reduction is important, however, they are quite common.

Shaving. The most common gear-finish-machining operation, shaving, involves the use of a highly accurate cutting tool that is gear-shaped and can mesh with the gear to be shaved. The teeth of the cutter conform precisely to the shape of the final gear teeth, but each cutter tooth is slotted or gashed at several points across its width to

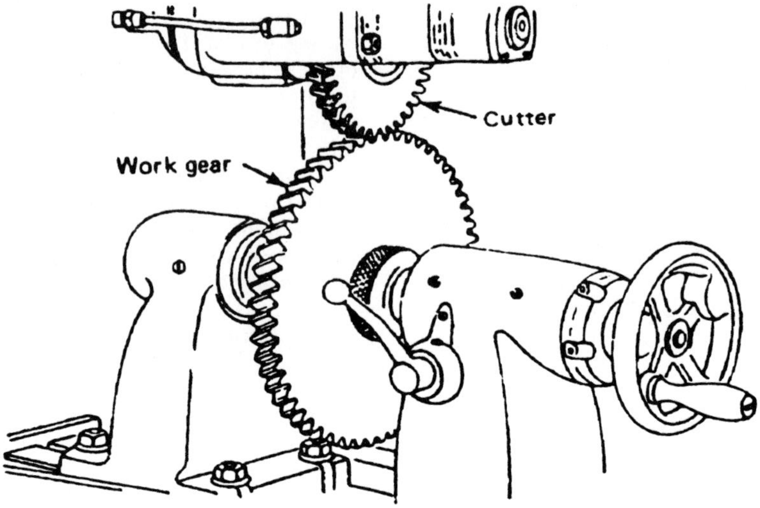

FIGURE 4.22.9 Gear shaving. (*From D. B. Dallas,* Tool and Manufacturing Engineers Handbook, *3d ed., McGraw-Hill, New York, 1976.*)

provide multiple cutting edges. The cutter's tooth arrangement is helical even if the cutter is to be used on straight spur gears. In use, the cutter engages the workpiece gear with the axes of gear and cutter oriented about 15° differently from one another. As the gears rotate, there is an axial sliding motion of the teeth, and the multiple cutting edges of the shaver cutter remove minute amounts of material from the surfaces of the gear. A high degree of dimensional accuracy and a smooth surface finish result. Figure 4.22.9 illustrates the process. Shaving is applicable to gears with diametral pitches of 2 to 180 and pitch diameters of 4 mm to 5.5 m (0.15 to 220 in). Materials should not be harder than R_c 40.

Shaving tools are expensive, much more costly than hobs. The need for special shaving machines adds to costs, and setup must be careful. These factors necessitate long production runs if costs are to be kept to a minimum.

Grinding. Although sometimes used to machine fine-pitch gears from solid stock, grinding is normally a finishing method for gears machined by other methods. The usual application is for post-heat-treatment machining to remove the distortion that occurs during heat treatment and otherwise improve tooth spacing, form, and surface finish.

There are numerous gear-grinding methods. Some use form-dressed wheels to form the finished gear teeth fully, while others generate the tooth shape through relative motions between the gear and the grinding-wheel spindle. Some generating methods employ multiple or ribbed wheels. Some grinding methods are quite similar to gear-machining methods except that a grinding wheel takes the place of the cutter.

Almost any gear that can be machined also can be finish ground. Exceptions are internal gears smaller than 63 mm ($2\frac{1}{2}$ in) in diameter and very large gears 2 m (about 6 ft) in diameter when sufficiently large grinding machines are not available. Grindability is best when gear-material hardness ranges from R_c 40 to 60.

Gear grinding tends to be costly, but it is used at all levels of production when the application of the gear demands heat treatment and high dimensional accuracy.

Honing. This process bears some similarity to gear shaving. The honing tool is a plastic gear impregnated with abrasive. As in a gear-shaving tool, the teeth are helical so that, when engaged with the workpiece gear, the axes of gear and tool are not parallel. The honing action takes place when the honing tool drives the workpiece gear for a time in each direction and there is transverse motion of the tool across the face of the gear.

Nicks, burrs, and surface irregularities are reduced or eliminated by the abrasive action of the particles embedded in the tool. Gear honing is normally performed after heat treatment, and if distortion from heat treatment is minimal, it can be substituted for grinding. It is far faster than grinding. Some errors in tooth-to-tooth spacing and in tooth form can be corrected by honing, but the process is primarily one of surface improvement. Honing is used in the manufacture of spur and helical gears (internal and external) with diametral pitches of 32 or coarser and is most commonly employed in high-production situations.

Lapping. A similar method for finishing gears after heat treatment, lapping involves an external abrasive compound and either a gear-shaped lapping tool or two mating gears. The gear pair or the gear-lapping-tool pair are run together with the compound. This procedure corrects minute errors of profile, tooth spacing, concentricity, and helix angle and improves the gear-tooth surface.

If a lapping tool is used, it is usually made of cast iron or bronze to provide an open grain that retains the abrasive. As in gear honing, the tool has a helix angle to provide nonparallel axes and a sliding motion at the point of tooth contact. Normally, additional across-the-tooth reciprocating motion is imparted during the process to ensure abrasive action across the full width of the gear.

Lapping is applicable to all classes of gears, including bevel, spiral-bevel, and hypoid gears. The last-named gears are normally run in pairs rather than with a lapping tool. Lapping tools, when required, are reasonable in cost. Cycle time is usually from $\frac{1}{2}$ to 2 min. The operation is applicable to all levels of production, being particularly suitable for small-quantity situations if mating gears are lapped together.

Burnishing. This process involves engagement of the workpiece gear with a burnishing gear hardened with accurately ground and polished teeth. Sometimes three burnishing gears, spaced 120° apart, are used. The pressure of the burnishing tool or tools improves the surface finish of the gear teeth and can roll down the burrs or nicks that occur during handling. However, it cannot significantly improve tooth position, tooth form, or concentricity. The operation is rapid and involves rolling the unhardened workpiece gear in both directions in the presence of a lubricant. Burnishing can apply to all levels of production, but at high production volumes, investment in honing equipment and tooling becomes justifiable, and results from honing normally will be superior to those obtained by burnishing.

Casting Methods

Gears can be cast to the semifinished state by all casting processes, but process limitations, particularly dimensional limitations, restrict the widespread use of all except die casting. (Casting of gear blanks in steel, bronze, and other materials is, of course, quite common.) For a description of these casting methods, see Chaps. 5.1, 5.2, 5.3, and 5.4.

Sand-Mold Casting. The sand-mold casting of gears provides only the lowest levels of accuracy. Large mill gears produced in underdeveloped countries are one application. In such cases, gear-tooth profiles and surfaces are improved to some degree by hand filing. Other applications include slow-moving mechanisms when irregularities of gear action and backlash are not objectionable. Normally, however, gears cast in sand molds are used only when machining facilities are not available.

Plaster-Mold, Permanent-Mold, and Investment-Cast Methods. These casting methods are more accurate than sand-mold casting and are usable in many commercial applications, generally those for which loads and velocities are not high and tooth-profile and positional errors are tolerable. These methods, though far more accurate than sand and shell molding, still do not provide the accuracy obtainable by machining. Of the methods, permanent-mold casting is better suited to simpler configurations, while investment casting is applicable to the most complex shapes, although normally only in the smaller sizes. With investment casting, helical and spiral-bevel gears can be cast.

Die Casting. This process is widely used for the production of gears employed in appliances, business machines, instruments, cameras, etc. When quantities are large, loads are light or moderate, and commercial tolerances are sufficient, die-cast gears may be called for. Spur, helical, worm, bevel, and shouldered and stepped gears can be produced. The approximate normal maximum pitch diameter is 150 mm (6 in).

If face width exceeds 6 mm ($\frac{1}{4}$ in), draft is required to permit the workpiece to be removed from the die. Trimming operations are necessary after the part has been removed from the die, and in some cases (e.g., when there is draft in the teeth) trimming involves shaving or broaching the teeth. Trimming die costs may be very high in such cases.

Gear-Forming Methods

A number of forming methods are available for gear making, either for forming the teeth in material that is processed further to provide the finished gear or for making the gear completely. These processes include extrusion, cold rolling and drawing, stamping, powder metallurgy, and forging. In many cases, these methods are fully as accurate as machining. Often they also are applicable to high-strength ferrous metals, which is not the case with die casting and some other casting methods.

Extrusion. Typical cross sections of extruded gears are shown in Fig. 4.22.10. Best results are obtained with pinions of coarse pitch.

The process involves forward extrusion as described in Chap. 3.1. To get the accuracy required for gears, a secondary drawing operation is required after extrusion. The stock is then cut off, turned, and bored as necessary on lathes or screw machines equipped with special collets.

Gears made from extruded stock are used in various commercial applications such as clocks, instruments, and appliances. The process is limited to straight spur gears and is most economical when quantities are moderate or greater.

Cold Drawing. The cold drawing of gears goes beyond extrusion to involve repeated passes of bar material through dies of progressively smaller openings, each closer to the final gear shape than the one preceding it. (The process is described in greater detail in Chap. 2.2, Part 2.) As with extruded-gear stock, the material is cut off and

	Number of teeth	Outside diameter, in	Pitch diameter, in	Roof diameter, in	Pitch	Tooth thickness, in
	18	1.241	....	0.972	16	0.098
	7	1.487	1.062	0.775	6	0.261
	8	0.985	0.800	0.544	10	0.157
	7	1.430	....	0.712	6	0.261
	15	2.020	....	1.588	8	0.1963
	9	1.325	....	0.920	14	0.196
	16	1.285	....	0.985	15	0.112
	11*	1.8	....	1.265 hole drawn to 1.215		
	7	Segment for Easy washer, body 0.720, radii 1/32 in				
	12*	5.25	4.438	3.625	Tolerance = ±0.034	

FIGURE 4.22.10 Typical cross sections of extruded gear stock. (*Data courtesy The American Brass Co., except for final gear, from Reynolds Metals Company.*)

machined to produce all surfaces except the gear teeth themselves. The process produces gears with teeth of fine surface finish and high density. Accuracy is approximately equivalent to that achieved by gear-machining methods. For high-precision or high-load applications, however, drawn gears may be inferior to machined gears because of built-in stresses from the drawing operations.

The cold-drawing method is applicable to spur pinion gears in various drawable materials including steel. Involute and cycloidal tooth forms can be produced, as can other special shapes such as noncircular gears and ratchets. Maximum outside diameters currently available are on the order of 50 mm (2 in). Quantity production is necessary to amortize tooling costs.

Cold-drawn gears are used in business machines, motion-picture projectors, switches, timers, clocks, watches, cameras, scale mechanisms, and similar equipment.

Powder Metallurgy. The methods of powder metallurgy described in Chap. 3.12 are being used increasingly for the fabrication of gears. Spur gears, helical gears to a 35° helix angle, bevel gears, and face gears can be made with this process. Coarse spur gears involve the most straightforward processing. Pitches as fine as 64 can be produced. The size of gears producible depends on the press size available, but gears of diameters up to 90 mm ($3\frac{1}{2}$ in) are routinely made, and diameters to 150 mm (6 in) are possible.

To achieve the accuracy required for most applications, powder-metallurgy gears are subjected to repressing or coining after sintering. In such cases, accuracies to AGMA No. 9 are possible, whereas without repressing accuracies are on the order of AGMA No. 6.

Repressed gears have an excellent surface finish. The surface of nonrepressed gears is sufficiently porous that they can be vacuum-impregnated with oils for long-life lubrication.

An advantage of powder-metallurgy processes for gear manufacture is that projections, notches, collars, bosses, keyways, and other irregular shapes can be incorporated into the part without additional operations. Tooling costs tend to be high. Therefore, large quantities of powder-metallurgy gears are required if the full manufacturing economy of this method is to be realized.

Iron, brass, and steel powder-metallurgy gears are used in appliances, instruments, electrical tools, farm and garden equipment, business machines, and automobiles.

Stamping. The stamping processes described in Chaps. 3.2 and 3.3 can be used as an economical means of producing spur gears of good accuracy. Best results require accurate tooling to blank all teeth and pierce the center hole in one press stroke. Short-run stamping methods that utilize less elaborate tooling and multiple press strokes are less apt to be sufficiently accurate for the usual applications. Hence this method is economical only for large-quantity production.

Metal stamping is a sheet-metal operation. The thickness of the stock that can be stamped for gears depends somewhat on the coarseness of the teeth. Normally, it ranges from 0.25 to 2.5 mm (0.010 to 0.100 in). When greater face widths are required, a number of individual blanked pieces are laminated together. These are fastened by riveting, press fitting, or welding.

Although the edges of stamped gears exhibit the same areas of drawdown, shear, and breakaway evidenced by the edges of other stampings, the sheared portion can exhibit quite high levels of accuracy. AGMA accuracy numbers as high as Q9 are attainable.

When fine-blanking is used, virtually the entire edge is sheared and is of this quality. The fine-blanking process, because of its accuracy and the smooth, square edges it produces, is well suited to gear blanking.

Diametral pitches range from 20 to 120. Slots, tabs, extra holes, and special shapes are producible without extra operations if the necessary elements can be incorporated in the blanking die. Applications include electric and water meters, clocks, instruments, counters, and appliances.

Stamping rates range from 35 to 200 pieces per minute. Secondary operations are tumble deburring and (sometimes) plating.

Forging. Forging (see Chap. 3.13) is normally used only to produce gear blanks, but there has been limited production of forged gears in recent years. The process involved has additional steps compared with the most common commercial forging processes; it includes accurate machining of the forging blank and both rough and finish forging. The process is most applicable to straight-bevel and face gears. Spur gears are possible, but die life is short. Spiral-bevel gears have been produced experimentally.

Forged gears are not as accurate as machined gears (under controlled conditions, tolerances almost equivalent to those from rough machining can be maintained) but are superior in fatigue strength. Large-quantity production is required to amortize tooling and process-development costs.

Plastics-Gear Molding

Injection Molding. Injection molding (see Chap. 6.2) is the least expensive method of producing gears in large quantities. Except for the careful mold making and close process controls necessitated by the high dimensional accuracy required by gears, the molding process is the same as for any other component.

Plastics gears are used in many applications when power requirements are relatively low and quiet operation, light weight, corrosion resistance, and the inherent lubricity of plastics are advantageous. All gear configurations including helical, bevel, and spiral-bevel gears can be molded easily. In fact, gear clusters and combination parts using cams, bosses, lips, and holes are easily made in one piece by injection molding. Diametral pitches range from 20 to 120.

Plastics gears are less satisfactory if temperatures, loads, and velocities are high. They have a much higher thermal coefficient of expansion than metals and are thus less satisfactory if they are to be used in a wide temperature range. Nylons and some other materials are subject to moisture absorption and expansion so that care must be exercised in adjusting the mesh of the gear train if humid environments will be encountered. The mating of plastics and powder-metallurgy gears is not recommended because the porous surface of the powder-metallurgy gear can cause the plastics gear to wear rapidly. The resiliency of most thermoplastics helps provide smooth and quiet operation and can minimize backlash if the gear set is adjusted to have full engagement of mating teeth. Thermoplastics have less friction, light weight, freedom from the need for lubrication, and good impact and corrosion resistance.

Injection molding, of course, is a high-production process. Production rates are rapid, ranging from 60 to 80 pieces per hour per mold cavity. The cost of injection molds is substantial enough to require quantity production for the amortization of costs.

Compression Molding and Other Thermosetting-Plastics Processes. These processes are used less frequently than injection molding for gears but are practicable for many applications. The processes described in Chap. 6.1 are used with a variety of thermoset materials. Thermosetting-plastics gears have most of the advantages and limitations of injection-molded thermoplastics gears. One advantage is better applicability to moderately high temperatures. The thermoset processes usually are also better suited to low- and medium-production quantities.

SUITABLE MATERIALS FOR GEARS

Sufficient strength to transmit the power involved is a first requisite for any gear material. Machinability is also important for machined gears, for two reasons: A considerable amount of metal removal is involved when gears are machined, and it is easier to achieve precision of machining and smooth surface finishes (which are important in gears) when the metal used has favorable machinability ratings.

Other properties that are almost always desirable and may be necessary for certain applications are corrosion resistance, dimensional stability, impact strength, light weight, high temperature resistance, heat treatability, wear resistance, natural lubricity or compatibility with lubricants, noise-damping properties (or limited noise-generating properties), and (lest we forget) low cost.

Machined Gears

Machined gears are most frequently made from steel because of its high strength and favorable cost. Carbon steel is low in cost, is satisfactory for machining and case hardening, and is very commonly used in commercial gears. Alloy steels offer heat-treating, strength, and corrosion- and wear-resistance advantages but may be less machinable than plain carbon steels. Their cost also is higher. Leaded and resulfurized (free-machining) steels should be considered for machined gears whenever possible. However, they do have reduced impact strength and are less suitable for high-power applications. Stainless steels are used when corrosion resistance is essential but should be avoided if this factor is not required. They are more costly and much more difficult to machine, are not particularly good for wear resistance, and in the 300 Series, are not heat-treatable.

Cast irons have some desirable properties for gearing. They have good machinability, are low in cost, and have noise-damping characteristics. However, except for ductile and malleable cast irons, they have low shock resistance. They can be heat-treated for greater strength. Cast steels have better physical properties than cast irons but are more costly and less machinable and lack good noise-damping properties.

Table 4.22.1 lists the common ferrous materials used for machined gears. Table 4.22.2 indicates gear steels suitable for case and local surface hardening.

Bronze is a superior gear material, especially for applications in which the bronze gear mates with a steel gear and in which sliding friction is involved, as with worm gears. Most gear bronzes have excellent machinability and wear and corrosion resistance. Initial material cost, however, is high.

Aluminum is a good material for lightly loaded gears. It machines easily and provides a good surface finish. It also is advantageous when light weight and corrosion resistance are important.

Table 4.22.3 lists specific alloys of bronze, brass, and aluminum suitable for machined gears.

TABLE 4.22.1 Ferrous Metals for Machined Gears

Material	Common designation	UNS designation	Machin-ability	Yield strength,* MPa (lbf/in²)	Remarks
Low-carbon steel	1020	G10200	Good	207 (30,000)	For commercial gears of low or moderate power rating; can be case-carburized
Medium-carbon steel	1040	G10400	Fairly good	275 (40,000)	Can be induction- or flame-hardened to BHN 500
High-carbon steel	1060	G10600	Fair	370 (54,000)	For higher power ratings
Resulfurized carbon steel	1117	G11170	Excellent	235 (34,000)	Same as for 1020
Chrome-molybdenum-alloy steel	4140	G41400	Fairly good	415 (60,000)	Suitable for nitriding and flame hardening
Nickel-chrome-molybdenum-alloy steel	4340	G43400	Fair	450 (65,000)	Suitable for nitriding and induction hardening
Stainless steel	303	S30300	Fair	240 (35,000)	Best corrosion resistance; nonhardenable; nonmagnetic; for lower power applications
Stainless steel	416	S41600	Fair	275 (40,000)	Can be heat-treated; for low- to medium-power applications
Cast iron	G2500	F10005	Very good	170 (25,000)	Sound damping properties; for moderate power, low-shock applications; larger size
Cast iron	G4000	F10008	Good	275 (40,000)	Same as G2500 but with better mechanical properties
Ductile iron	80-60-3	F34100	Good	415 (60,000)	Better impact and fatigue strength than cast iron

*Annealed condition.

Die-Cast Gears

Die-cast gears can be made from zinc, aluminum, brass, or magnesium. Zinc is the preferred metal because of its easier, lower-temperature castability. Table 4.22.4 summarizes pertinent information about the more commonly used alloys for gears.

Formed Gears

Formed gears made by cold drawing can be manufactured from plain carbon steel, alloy steel, stainless steel, malleable iron, brass, and aluminum. Cold-formable varieties of these metals should be used rather than free-machining varieties. In the case of steel and brass, sulfur or lead lowers ductibility. Steel blanks should be no harder than R_c 28.

TABLE 4.22.2 Typical Gear Steels Used for Various Case-Hardening and Local Hardening Processes*

Carburizing	Nitriding	Carbonitriding	Cyaniding	Induction hardening	Flame hardening
1015 (G10150)	4130 (G41300)	1010 (G10100)	1020 (G10200)	1040 (G10400)	1040 (G10400)
1019 (G10190)	4140 (G41400)	1117 (G11170)	5132 (G51320)	1050 (G10500)	1050 (G10500)
1117 (G11170)	4340 (G43400)	5130 (G51300)		4340 (G43400)	4140 (G41400)
4320 (G43200)	8630 (G86300)	8620 (G86200)			6150 (G61500)
4615 (G46150)	Nitrolloy N				
4320 (G43200)	Nitralloy 135				
4615 (G46150)	Nitralloy 135 (modified)				
4815 (G48150)					
8620 (G86200)					
8720 (G87200)					
9310 (G93100)					

*Updated from G. W. Michalec, *Precision Gearing: Theory and Practice*, Wiley, New York.

TABLE 4.22.3 Nonferrous Metals for Machined Gears

Material	Designation		Machinability	Yield strength, MPa (klbf/in^2)	Remarks*
	Common	UNS			
Nickel-aluminum-bronze 10%	642	C64200	60	241–469 (35–68)	Strength, corrosion resistance, good antigalling and antiseizing properties
Gunmetal	SAE 62	C90500	30	124–145 (18–21)	Cast alloy used for worm wheels, good resistance to wear; casting methods: S, PM, I, P
Manganese bronze	SAE 420	C86400	80	221–240 (32–42)	Cast alloy with high mechanical strength, good corrosion resistance, and castability; casting methods; S, PM, I, P
Nickel-tin-bronze	947	C94700	20–30	345–414 (50–60)	Cast alloy with good castability, corrosion and wear resistance, heat treatable; casting methods: S, PM, I, P
Free-cutting brass	360	C3600	100	124–359 (18–52)	Low cost, noncorrosive; best machinability of copper alloys
Aluminum alloy	2024-T4	A92024, T4 temper	Very good	325 (47)	Most widely used aluminum alloy; surface finish excellent; for light-duty applications
Aluminum alloy	7075-T6	A97075, T6 temper	Very good	330 (48)	Best mechanical properties of aluminum gear alloys

*S = sand mold casting; PM = permanent mold; I = investment; P = plaster mold.

TABLE 4.22.4 Metals for Die-Cast Gears

Material	Designation		Tensile strength, lbf/in^2	Remarks
	Common	UNS		
Zinc	ASTM AG-40A (XXIII)	Z33520	41,000	Fastest cycle; best die life; best surface finish and dimensional accuracy; good impact strength
Zinc	ASTM AC-41A (XXV)	Z35531	47,000	Slightly diminished castability, finish, and accuracy but better physical properties than AG-40A
Aluminum alloy	413	A04130	36,000	Light weight; corrosion resistance
Aluminum alloy	380	A03800	43,000	Light weight; corrosion resistance
Magnesium alloy	ASTM AZ91A	M11910	33,000	Poor corrosion resistance; very light; for low-load applications
Die-casting yellow brass	858	C85800	55,000	Good machinability; short die life; corrosion resistance for water service applications

Extruded-Gear Sections

These sections are made from metals recommended for other extrusion applications in Chap. 3.1. Aluminum, brass, and magnesium are common examples.

Molded Gears

Plastics materials for molded gears include nylon, polyacetal, polycarbonate, polyurethane, phenolic laminated with fabric or paper, and on occasion, a wide range of other thermoplastics. Other materials such as polystyrene, polyethylene, polypropylene, and cellulose acetate are used for toys and other inexpensive products with very low load applications.

Real economy is achieved when plastics-gear materials are injection-molded, but for small-quantity work they normally are machined. Some plastics, especially glass-filled varieties, are not easy to machine. Sharp cutting tools are required to minimize burrs. Temperature increases from cutting should be avoided in order to control dimensions. Table 4.22.5 lists common nonmetallic gear materials.

Stamped Gears

Stamped gears are made from all low- or medium-carbon steels, all brass alloys including bronze, and half- and full-hard aluminum. Machinable grades of these materials usually also work easily with gear-blanking dies.

TABLE 4.22.5 Nonmetallic Gear Materials

Material	Fillers	Tensile strength, lbf/in^2	Fabrication method	Remarks
Phenolic laminates	Kraft paper	17,000-21,000	Machined from compression-molded stock or blank	Not easy to machine; poor surface finish; used for heavy-duty service
	Cotton-canvas fabric	9,000–14,500		
Nylon 66		11,200	Injection molding	High wear resistance; subject to hydroscopic expansion; can be filled with graphite or molybdenum disulfide for better lubricity
	Glass fiber	19,000–31,000		
Polyacetal		10,000	Injection molding	Less creep and better dimensional stability than nylon and more precise dimensions
	Glass fiber	8,500–13,500		
Polycarbonate		9,000–10,500	Injection molding	
Polyurethane	Glass fiber	5,000–10,000	Injection molding	Especially quiet; molding control more critical

Powder-Metallurgy Gears

These gears are made with ferrous and copper alloys, the ferrous alloys being by far the more common. Case hardening of steel- and iron-based materials is possible. Table 4.22.6 lists some common powder metals for gear applications. Table 4.22.7 presents hardness levels for steel and iron gears.

DESIGN RECOMMENDATIONS

The closely dimensioned and standardized configuration inherent in the function of gears severely limits the latitude of the designer in selecting a low-cost alternative design. Nevertheless, there are choices that the designer can make that will have a significant bearing on the cost and performance of gear components. Some points for consideration, applicable to machined, molded, cast, formed, or stamped gears, are the following:

1. Normally, the coarsest-pitch gear system that performs the required function will be the most economical to produce. The designer, if given a choice between fine-pitch gearing and coarse-pitch gearing and provided operating requirements permit, should choose the coarser-pitch system.
2. Helical, spiral, and hypoid systems are more difficult and costly to manufacture than straight-tooth designs. Straight-tooth systems should be specified unless noise or other considerations necessitate a helical configuration.
3. Dimensional tolerances, as controlled by AGMA or DIN gear numbers and permissible tooth-to-tooth and total cumulative variations and surface finishes, should be as liberal as the function of the gears permits. Gears, like other manufactured components, are subject to geometrically increased costs as tolerances are reduced.

See Fig. 4.22.11 for an illustration of these points. The following are design recommendations for gears produced by specific manufacturing processes:

Machined Gears

1. Shoulders, flanges, or other portions of the workpiece larger than the root diameter of the gear should not be located close to the gear teeth; otherwise, there will not be sufficient clearance for the gear-cutting tool. When gears are hobbed, the space between the teeth and the shoulder should be greater than one-half of the hob diameter. Gear shapers require less clearance, but a good general rule is to avoid all such obstructions if possible and, if not possible, to keep the spacing liberal.
2. Other gear configurations also require clearance areas or undercuts to provide room for cutting tools. Herringbone gears should have a groove between halves; internal gears in blind holes require an undercut groove or other recessed space for cutter overtravel. (See Fig. 4.22.12.)
3. If steel gears are to be heat-treated for increased strength, consideration should be given to using non-heat-treated gears of larger size instead. The increased cost of larger gears may be less than the cost of heat treating and of grinding or lapping after heat treating to correct heat-treatment distortion.

TABLE 4.22.6 Powdered Metals Suitable for Gears

Material	Composition, %	Designation (MPIF)*	Ultimate tensile strength, MPa (lbf/in^2)†	Remarks
Copper steel	Fe-2Cu-0.8C	FC-0208	469 (68,000)	Admixture of iron and copper powders; an older formulation, still used; has good hardenability with consistent hardness, especially for larger parts
Nickel steel	Fe-2Ni-0.8C	FN-0208	448 (65,000)	Admixture of iron and nickel powders; produces durable and strong gears
Nickel steel	Fe-0.85Mo-55C	FL-4405	428 (62,000)	Prealloyed powder; favorable cost; excellent compressibility for a prealloy makes higher density more practical and aids consistency
Nickel-steel mixture	Fe-2Ni-0.55Mo-55C	FLN2-4405	428 (62,000)	Prealloyed powder with admixed nickel added to increase fatigue life
Bronze	Cu-10Sn	CT-1000 (repressed)	150 (22,000)	Density 7.2 for this strength; can be self-lubricating at lower density (but with lower strength)
316 Stainless steel		SS-316 N2	410 (60,000)	Density 6.5 for this strength; corrosion resistant; nitrogen alloyed

*Metal Powder Industries Federation.

†At a density of 7.0 g/cm^3. Higher densities provide greater ultimate tensile strength and better fatigue life.

TABLE 4.22.7 Relative Hardness Levels for Steel or Iron Gears

Hardness			
Brinell	Rockwell C	Machinability	Comments
150 200		Very easy	Very low hardness; minimum load-carrying capacity
200 250	24	Easy	Low hardness; moderate load-carrying capacity; widely used in industrial gear work
250 300	24 32	Moderately hard to cut	Medium hardness; good load capacity; widely used in industrial work
300 350	32 38	Hard to cut; often considered the limit of machinability	High hardness; excellent load capacity; used in lightweight, high-performance jobs
350 400	38 43	Very hard to cut; many shops unable to handle	High hardness; load capacity excellent provided heat treatment develops proper structure
500 550	51 55	Requires grinding to finish	Very high hardness; wear capacity good; may lack beam strength
587	58 63	Requires grinding	Full hardness; usually obtained as a surface hardness by case carburizing; very high load capacity for aircraft gears, automobile gears, trucks, tanks, etc.
	65 70	May be surface-hardened after final machining	Superhardness; generally obtained by nitriding; very high load capacity

Source: Taken from Darle W. Dudley (ed.), *Gear Handbook,* McGraw-Hill, New York.

4. Heat-treated gears should be of uniform cross section to minimize heat-treatment distortion. Nitriding is a preferred heat treatment because it minimizes distortion and the consequent probability that grinding will be necessary after heat treatment.

5. In many cases, it is best to machine gears as separate parts and to assemble them afterward to shafts or other machine components. Some examples of this are found when (*a*) gears are clustered together, (*b*) broached internal gears must be placed in a blind hole, and (*c*) the gear or shaft component would not be rigid enough for machining.

6. The greater the helix of helical gears, the more difficult it is to machine to high levels of accuracy. Use as shallow a helix angle as possible. (See Fig. 4.22.13.)

7. Wide-faced gears are more difficult to machine to a given tolerance than narrow-faced gears. Small, long center holes are also more costly and are subject to loss of squareness. Extremes of both cases should be avoided. (See Fig. 4.22.14.)

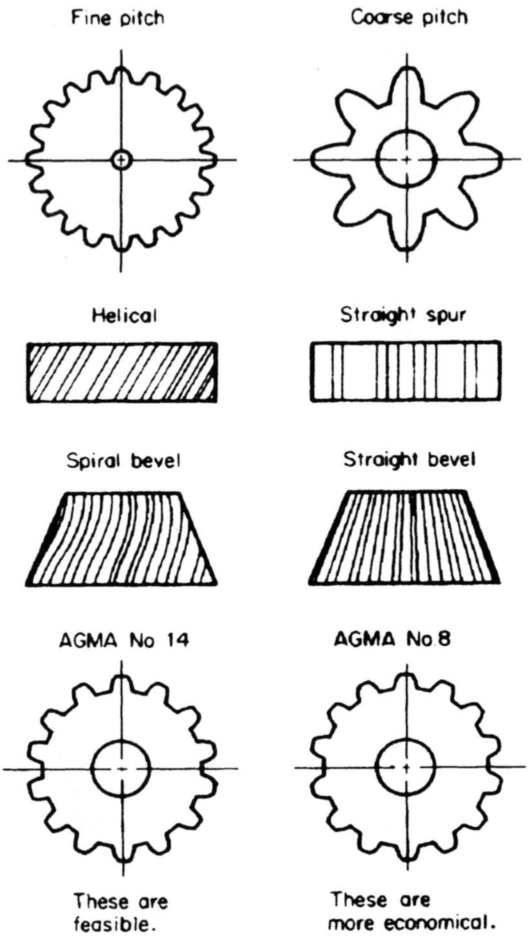

FIGURE 4.22.11 The more economical gear designs are shown on the right.

8. Gear blanks should be designed so that they can be clamped securely and without distortion during machining from the cutting tool or clamping device. (See Fig. 4.22.15.)

9. When gears must be press-fitted to a shaft or other component, the fitting surface should not be too close to the teeth. The gear section must be heavy enough so that the pressure of the fit does not overstress the gear or change its dimension.

10. Standard pitches should be used as much as possible to minimize tooling costs and tooling inventories and to ensure better availability of cutting tools or stocked gears. Preferred pitches recommended by Michalec* are shown in Table 4.22.8.

*George W. Michalec, *Precision Gearing: Theory and Practice,* Wiley, New York, 1966.

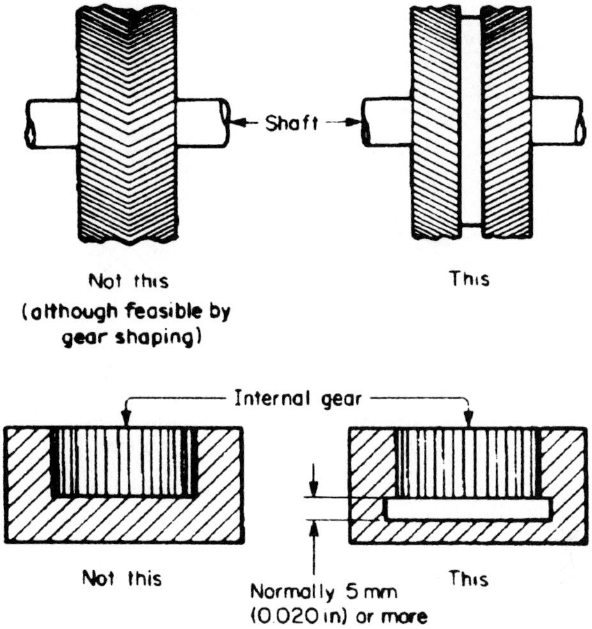

FIGURE 4.22.12 Clearance or undercuts at the end of gear teeth are advisable to provide room for cutting tools.

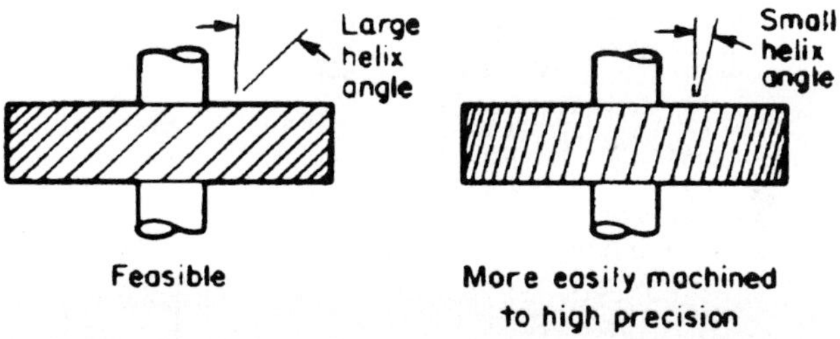

FIGURE 4.22.13 Helical gears of low helix angle are easier to machine.

11. Correct design of the gear blank is important. Especially with bevel gears, the specified dimensions of the blank should be closely maintained, and the blank should be capable of being held securely in a rigid setup for machining the teeth.

12. When gears are to be subjected to a secondary operation for improved tooth accuracy and finish, it is important to specify the proper stock allowance. Too great an allowance reduces production and increases costs. Since finish-machining operations have low material-removal rates, too little stock allowance may not permit the proper dimension or surface finish to be realized. Table 4.22.9 lists recommended allowances for shaving, grinding, lapping, and honing for various diametral pitches.

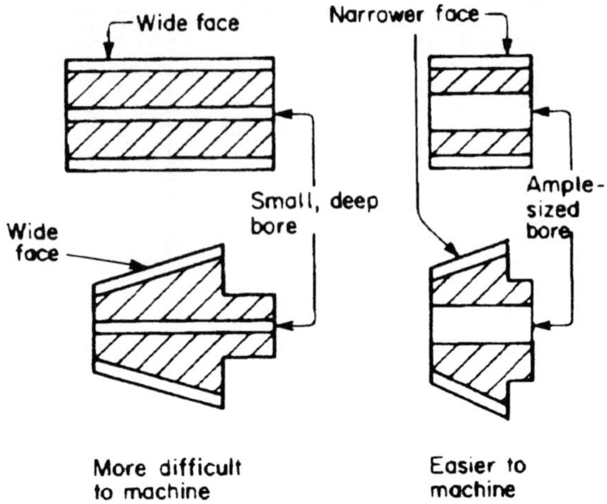

FIGURE 4.22.14 Narrower-face large-bore gears are generally easier to machine.

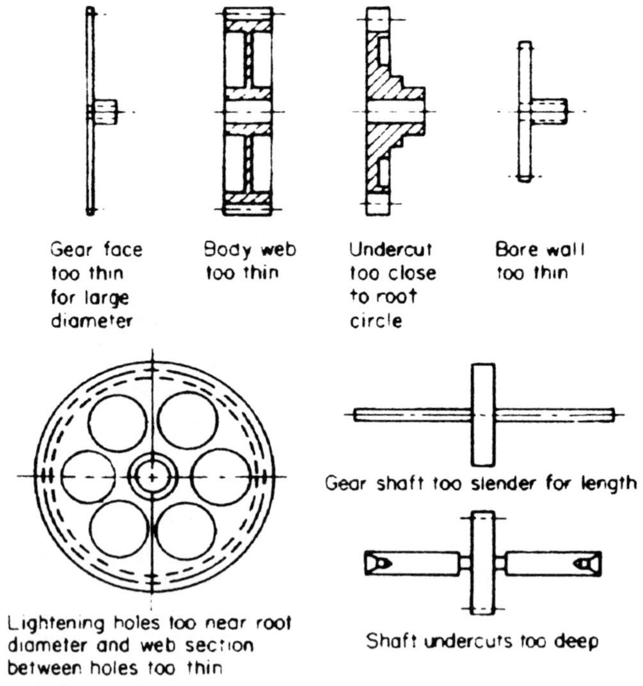

FIGURE 4.22.15 Gear blanks should be designed so that they can be clamped securely without distortion and can withstand the forces of the machining cutters. (*From G. W. Michalec,* Precision Gearing, *Wiley, New York, 1966.*)

TABLE 4.22.8 Preferred Pitches for Gears

Pitch class	Preferred diametral pitch
Coarse	½
	1
	2
	4
	6
	8
	10
Medium coarse	12
	14
	16
	18
Fine	20
	24
	32
	48
	64
	72
	80
	96
	120
	128
Ultrafine	150
	180
	200

Source: From George W. Michalec, *Precision Gearing: Theory and Practice,* Wiley, New York, 1966.

TABLE 4.22.9 Recommended Stock Allowances for Gear-Finishing Operations

	Stock allowances, mm (in)		
Diametral pitch	Shaving	Grinding or honing	Lapping
2.5–4.5	0.075–0.1 (0.003–0.004)	0.020 (0.0008)	0.015 (0.0006)
5–6	0.06–0.09 (0.0025–0.0035)	0.018 (0.0007)	0.013 (0.0005)
7–10	0.05–0.075 (0.002–0.003)	0.015 (0.0006)	0.010 (0.0004)
11–14	0.04–0.06 (0.0015–0.0025)	0.013 (0.0005)	0.008 (0.0003)
16–18	0.025–0.05 (0.001–0.002)	0.010 (0.0004)	0.008 (0.0003)
20–48	0.013–0.04 (0.0005–0.0015)	0.008 (0.0003)	0.005 (0.0002)
Over 48	0.008–0.02 (0.0003–0.0007)	0.005 (0.0002)	0.005 (0.0002)

13. The involute form of tooth is easy to machine. It works at any center distance and is used perhaps 99 percent of the time. It should be the standard specified tooth form for all normal gearing.

Formed, Cast, and Molded Gears

1. Cold drawn gears are an exception to the rule that coarse-pitch gears are more economical than fine-pitch gears to manufacture. Because less metal needs to be displaced, a finer pitch is preferable; 18 or more teeth should be specified per full gear. Pressure angles of 20° or more should be specified. Fillets should be 0.13 mm (0.005 in) at the very minimum and preferably should be greater than 0.25 mm (0.010 in). Figure 4.22.16 illustrates these points.

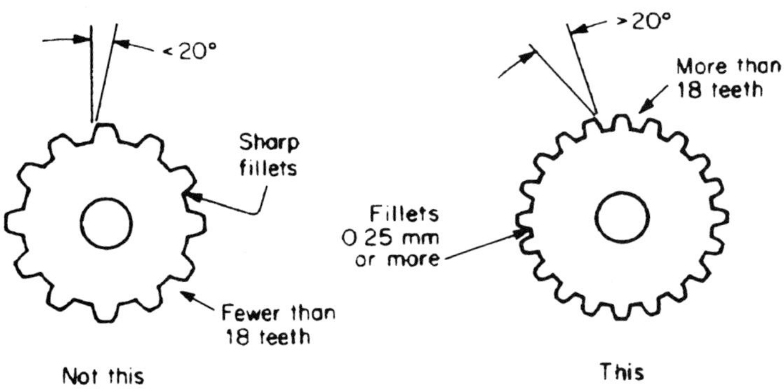

FIGURE 4.22.16 Design recommendations for cold drawn gear stock.

2. The designer should consider the use of molded and cast gears when bevel gears are required because the cost advantages of molded and cast gears are even greater in this case.

3. With plastics and die-cast gears, the use of a metal insert for the shaft or shaft hole should be considered. An insert normally provides a more accurate bore than the bore achievable by molding or casting, reduces shrinkage distortion, and provides a stronger means of fastening the gear.

4. If load requirements are severe and the load-carrying ability of the gear material is limited, as is the case with molded plastic or die-cast gears, for example, consider the use of a tooth form with a 20° or 25° pressure angle instead of 14.5°.

DIMENSIONAL FACTORS

Dimensional variations in gears result in noise, vibration, operational problems, reduced load-carrying ability, and reduced life. These problems are compounded at higher gear-operating speeds. Dimensions critical to precision gear operation are

TABLE 4.22.10 Comparison of Gear-Classification Systems

AGMA quality no.*	Approximate DIN quality no.	Quality classification	Typical application
(1)			
(2)			
3			Appliances, hand tools, pumps,
4	12	Commercial	clocks, farm machines, fishing
5	11		reels, hoists, slow-speed machinery
6	10		
7			
8	9–10		Instruments, aircraft engines,
9	8–9		turbines, professional cameras,
10	7–8	Precision	machine-tool speed drives,
11	6–7		automotive transmissions, high-
12	5–6		speed machinery
13	4–5		
14	3–4		Military navigation, precision
15	2–3	Ultraprecision	instruments, computer
16	2		equipment
(17)	1		
(18)			
(19)			

*Nos. 1, 2, 17, 18, and 19 are derived by extrapolating AGMA tolerance curves for quality Nos. 3–16.

pitch, concentricity, tooth profile, tooth thickness, and tooth-surface finish. Two indices that combine the effect of several of these dimensions are tooth-to-tooth composite error (TTCE) and total composite error (TCE). *Tooth-to-tooth composite error* is the combined effect of pitch, profile, and tooth-thickness variations. *Total composite error* is the combined effect of these tooth-to-tooth errors plus runout.

The American Gear Manufacturers Association (AGMA) has incorporated these factors into a comprehensive accuracy-classification system. Dimensional-accuracy levels are designated by quality numbers that apply to gears of varying size, diametral pitch, and type. There are 14 levels of accuracy. Quality No. 3 applies to gears of low-commercial quality. Quality No. 16 applies to gears of the highest precision. The complete specification is detailed in AGMA Standard 390.03, *Gear Classification Manual.* The German DIN Standard 3962 (867) uses a similar system.

Table 4.22.10 compares AGMA and DIN quality numbers for typical applications. Table 4.22.11 summarizes dimensional tolerances represented by the range of quality numbers for spur, helical, and herringbone gears.

Gear Accuracy versus Cost

There is a definite and direct relationship between gear accuracy and manufacturing costs. A machined gear of ordinary quality is approximately AGMA 7 to 8. Making a better quality normally will call for better-than-average processing. Allowing wider tolerances generally will permit methods less refined than one normally finds in present-day shops.

TABLE 4.22.11 Examples of Tooth-to-Tooth Composite Tolerance and Total Composite Tolerance for Various AGMA Quality Numbers: Spur, Helical, and Herringbone Gears

AGMA quality number	Normal diametral pitch	Pitch diameter, in	Total composite tolerance, in	Tooth-to-tooth composite tolerance, in
5	20	1.0	0.0074	0.0036
5	64	16.0	0.0070	0.0025
6	24	0.25	0.0049	0.0033
6	80	10.0	0.0044	0.0017
7	2	4.0	0.0124	0.0039
7	100	1.0	0.0022	0.0012
8	20	2.5	0.0030	0.0012
8	50	6.3	0.0025	0.0010
9	1	100	0.0149	0.0018
9	200	0.10	0.0007	0.00053
10	30	0.63	0.0011	0.00061
10	64	6.3	0.0012	0.00047
11	8	2.5	0.0016	0.00059
11	60	0.16	0.00065	0.00043
12	12	1.0	0.00086	0.00043
12	64	16.0	0.00067	0.00024
13	24	0.25	0.00048	0.00031
13	80	10.0	0.00041	0.00016
14	2	4.0	0.00044	0.00037
14	100	1.0	0.00021	0.00011
15	20	2.5	0.00029	0.00012
15	50	6.3	0.00024	0.00009

Note: For complete tolerance data for AGMA gear quality numbers, see AGMA Standard 390.03, *Gear Classification, Materials and Measuring Methods for Unassembled Gears,* from which these examples were taken.

Consider as an example a gear of 152-mm (6-in) pitch diameter and 8 diametral pitch. AGMA tolerances for involute-tooth profile for such a gear range from 0.12 mm (0.046 in) for quality No. 4 to 0.0015 mm (0.00006 in) for quality No. 17 (AGMA curves extrapolated). Figure 4.22.17 plots the cost of holding these dimensions within the indicated tolerances.

There is a limit to the degree of accuracy attainable in machining the tooth profile. This is represented by the vertical line of 0.00005 in on the graph. Even by allowing a large tolerance, there is a minimum cost that cannot be bettered with any machining method. This is represented by the low horizontal line on the graph.

The graph shows the relation of cost and accuracy and contains several warnings:

1. At high levels of accuracy a very great increase in the amount of effort (or money) spent will produce only a slight improvement in accuracy.
2. When accuracy is waived in favor of saving money, the additional savings resulting from the large increase in allowable error becomes slight.

Control of accuracy to the high AGMA quality-number levels requires control of the environment and all manufacturing conditions and necessitates secondary machining operations like grinding, shaving, lapping, etc.

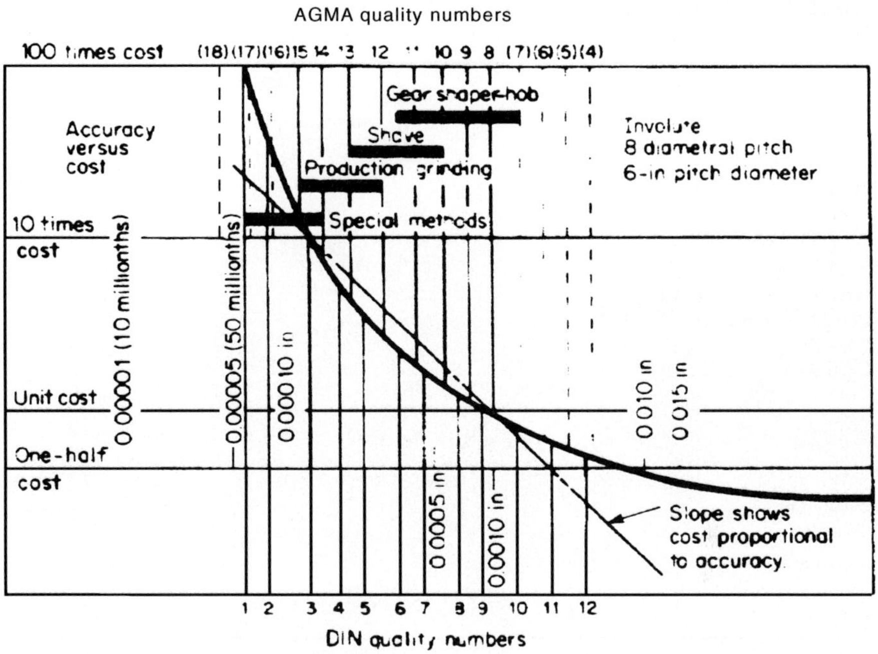

FIGURE 4.22.17 Graphical comparison of gear quality and manufacturing cost. (*Courtesy* Manufacturing Engineering, *Society of Manufacturing Engineers.*)

TOLERANCE RECOMMENDATIONS

Table 4.22.12 shows recommended tolerances in terms of AGMA quality numbers for various gear-manufacturing processes.

TABLE 4.22.12 Recommended Tolerances in Terms of AGMA Quality Numbers for Various Gear-Making Processes

Process	Highest quality number	
	Normal	With extra care
Sand-mold casting	1	3
Plaster-mold casting	3	5
Permanent-mold casting	3	5
Investment casting	4	6
Die casting	5	8
Injection and compression molding	4	8
Powder metallurgy	6	9
Stamping	6	9
Extrusion	4	6
Cold drawing	6	9
Milling	6	9
Hobbing	8	13
Shaping	8	13
Broaching	7	12
Grinding	10	15
Shaving	9	14
Honing	10	15
Lapping	10	15

CHAPTER 4.23
DESIGNING PARTS FOR ECONOMICAL DEBURRING

LaRoux K. Gillespie
Allied Signal Aerospace
Kansas City, Missouri

Burrs form on the edges of all machined and stamped parts. Flash, which has properties similar to those of burrs, occurs on every cast part. In the majority of cases, these protrusions must be removed to prevent interference fits or short circuits, to improve fatigue life, or to prevent cut hands.

DEBURRING PROCESSES

The deburring processes in relatively common use in industry are as follows:

Barrel Tumbling

A large group of parts having burrs are placed in a rotating barrel with a small, pebble-like medium, a fine abrasive powder, and water. The barrel and its contents are then slowly rotated until the burrs are worn off. Typical deburring times are from 4 to 12 h. (Also see Chap. 8.2.)

Centrifugal Barrel Tumbling

This process is similar to barrel tumbling except that the barrel is placed at the end of a rotating arm. Rotation of the arm adds a centrifugal force of up to $25g$ to the weight of the parts in the barrel, resulting in 25 to 50 times faster deburring than with conventional barrel tumbling.

Spindle Finishing

This process also is similar to barrel tumbling except that the workpiece is fastened to the end of a rotating shaft and then placed in a barrel rotating in the opposite direction.

Although each part must be handled individually, deburring times are only 1 to 2 min per part.

Vibratory Deburring

This process is similar to barrel tumbling except that the mixture of parts and medium is vibrated rather than rotated.

Abrasive-Jet Deburring

A high-velocity stream of small abrasive particles or miniature glass beads is sprayed at the burrs. The combination of impact, abrasion, and peening actions breaks or wears away the burrs. Deburring times vary from 30 s to 5 min per part, depending on workpiece size and complexity. (Also see Chap. 4.21.)

Water-Jet Deburring

A 0.25-mm- (0.010-in-) diameter jet of water at very high velocity cuts burrs and flash from the workpiece. In nonmetals, this process can deflash part contours at a rate of 250 lineal mm/s (600 lineal in/min).

Brush Deburring

Motorized rotating brushes abrade burrs from parts. Deburring time ranges from 10 s to 5 min per part.

Sanding

Belt sanders are widely used to deburr flat parts, while disks and flap wheels are used on contoured parts. It is relatively easy to deburr 600 stamped parts in an hour with little significant automation. While sanding removes heavy burrs, it also produces a very small burr itself.

Mechanical Deburring

A variety of specialized machines mechanically cut burrs off. They utilize chamfering tools, knives, or grinding wheels. Automotive gears can be deburred at a rate of 400 gears per hour with an automated unit.

Abrasive-Flow Deburring

Hydraulic cylinders force an abrasive-laden putty-like material over edges having burrs. The properties of the putty concentrate the abrading action at the edges of a part. In a typical application, 30 parts per hour can be deburred. With automation, it is possible with some parts to produce 400 parts per hour. Some dimensional changes occur on surfaces contacting the putty-like medium. (Also see Chap. 4.21.)

Liquid Hone Deburring

A 60-grit abrasive is suspended in water and then forced over burr-laden edges. Very fine burrs can be removed in a 5-min cycle. There is very little edge radiusing with this process.

Chemical Deburring

Buffered acids dissolve burrs, and large groups of small parts can be chemically deburred in 5 to 30 min. Because the acids attack all surfaces, some dimensional changes occur on all features. (Also see Chap. 4.20.)

Ultrasonic Deburring

A combination of buffered acids and a fine abrasive medium is ultrasonically agitated to wear and etch away minute burrs. Burrs produced by honing are removed by this process.

Electropolish Deburring

Burrs are removed by electrolysis in a mild acid solution. This "reverse" electroplating operation removes stock from all surfaces and produces excellent surface finishes while deburring. (Also see Chap. 8.2.)

Electrochemical Deburring

This process is similar to electropolish deburring except that a salt solution and a shaped electrode are required. Stock removal occurs only at the edges to be deburred, although some light etching occurs at other places on the part. A surface residue left on the part later must be brushed or wiped off. Deburring-cycle time is typically 2 min per part without automation. Actual deburring time is only 15 s.

Thermal-Energy Deburring

The high-temperature wavefront produced by igniting natural gas in a closed container vaporizes and burns off burrs. The short duration of the wavefront resulting from ignition produces temperatures of only 95°C (200°F) on components, although burrs may see 3300°C (6000°F) temperatures. Up to 800 parts per hour can be deburred by this process.

Manual Deburring

Workers with special knives, files, scrapers, and other tools cut burrs from parts without any mechanization.

CNC Machining Centers

CNC machining centers can remove burrs by brushing parts using a simple brush and arbor in a toolholder. When appropriate, the machine changes tooling and begins a brushing path instead of a cutting path.

Robotic Deburring

A computer-controlled robotic arm with a compliant brush or other deburring tool maintains tool contact with the edge to be deburred. Since compliant tools do not require exact edge location, programming is simple.

Laser Deburring

This process uses CNC equipment and laser energy to remove burrs.

Plasma Glow Deburring

This process uses a low-temperature plasma with argon and oxygen to remove very thin burrs and hairlike projections on plastics.

Edge Rolling

Slitting lines for sheet metal often use a roller pushing on the edge of the strip to literally push the metal burr back into the parent metal.

Skiving

Some slitting lines for sheet metal use a cutting tool insert to scrape off burrs as the strip passes the tool.

Burnish Deburring

Sheet-metal strips can be deburred merely by running the edge of the strip over a smooth metal surface. The contact force rubs or burnishes the edge such that the burr is removed.

Edge Coining

Press tooling is designed such that the punch coins the edge and pushes the burr back into the parent metal.

TYPICAL BURR CHARACTERISTICS

The only exceptions to the universality of burrs are parts produced by chemical or electrochemical machining. Such nontraditional processes as electrical discharge and laser machining typically produce edges with small resolidified nodules. These recast particles are frequently as objectionable as burrs or flash.

Burrs can be small, loose, or flexible projections, or they can be short humps firmly adhering to the workpiece edge. The majority of burrs have a triangular cross section. The more pronounced burrs also have a rectangular, flexible portion extending out of the triangular portion (see Fig. 4.23.1). On most carefully machined parts, burr thickness measured at the base of the burr is in the order of 0.8 mm (0.030 in). Height varies from 0.025 mm (0.001 in) to 0.13 mm (0.005 in), depending upon the conditions which produced the burr. Burr properties are not very repeatable. On stamped parts burr height is usually 0.13 mm (0.005 in) or shorter. Under "gentle" profile-milling conditions, burr thickness in stainless steel will vary by ±0.8 mm (0.030 in) from nominal thickness.

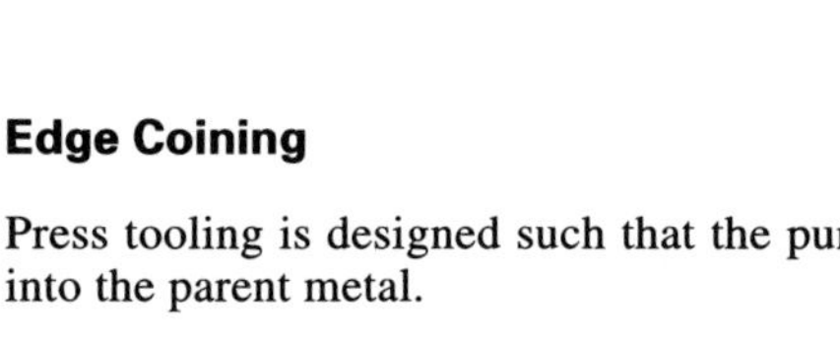

Typical shape of most common burrs

(a)

Typical shape of more pronounced burrs

(b)

FIGURE 4.23.1 Typical burr shapes. (*a*) Typical shape of most common burrs. (*b*) Typical shape of more pronounced burrs.

Since burrs and related protrusions can be formed by several basic mechanisms, it is possible to have two or more types of burrs on a single part. Deburring consistency is affected because burr properties vary widely.

Because most materials are basically work-hardening in nature, the burrs will be harder than the parent metal. For most processes, burr thickness is the most important variable in removal. Except for small burrs, burr height usually has little to do with the economics of the removal process.

ECONOMICS OF DEBURRING

For machined parts, the usual cost of removing burrs ranges from 1 to 50 percent of the basic machining cost. Costs for commercial parts are in the low portion of this range, i.e., from 1 to 5 percent of total fabrication costs, while deburring costs for precision aerospace parts are often at the high extreme. For die-cast parts, flash and gate removal and concurrent surface-finishing operations constitute from 25 to 35 percent of a part's cost.

Most deburring and deflashing methods are economic over a wide range of production quantities from prototype lots to mass production. Only six of the deburring processes described above require any tooling, and the cost of tools, when they are required, is seldom significant. Highly automated processes, however, cost more when several tool sets and specialized handling equipment are involved.

Brushing, barrel tumbling, and vibratory deburring are generally the most economical processes when burrs are on exposed edges. In large quantities, deburring costs for these processes are in the order of 0.1 to 6 cents per part.

SUITABLE MATERIALS

While all materials can be deburred, materials that produce the smallest burrs are obviously more desirable. Low ductility is the most significant property for minimizing burr size. On many cast irons, for example, one cannot detect a visible burr. A material with 0.5 percent elongation may, however, form large burrs if it is heat-sensitive. When cutting tools rub on such materials, they change the microstructure at the edges of the cut so that the material at the edges is plastic and easily forms burrs. On heat-treatable materials, it may be desirable to machine the workpiece in a nonductile condition and then heat-treat it to provide the ductility required in service. When it is not possible to machine the workpiece in a low-ductility condition, a noticeable burr will form. Removal of the burr can sometimes be hastened by a hardening operation. While the part and burr become harder, the burr may become noticeably more brittle and thus easier to remove.

Table 4.23.1 indicates burr-formation tendencies for certain materials, expressed in terms of a strain-hardening exponent n. High values of n (.50, for example) generally indicate that thick burrs will occur.

DESIGN RECOMMENDATIONS

Minimizing and controlling burr size through part and process design are preferred to developing methods of removing undesirable burrs. Additionally, a key to minimizing deburring costs lies in being able to answer the two groups of questions in Tables 4.23.2 and 4.23.3.

Design to Allow Burrs

Many components can be designed so that burrs are allowable. Figures 4.23.2 and 4.23.3 present two examples in which burr-free conditions are not required. The cutoff

TABLE 4.23.1 Burr-Forming Tendencies of Some Materials Expressed in Terms of a Strain-Hardening Exponent *n*

(Higher values of n indicate thicker burrs.)

Material	Treatment	*n*
1100 aluminum (A91100)	900°F, 1 h, annealed	.20
2024 aluminum (A92024)	T-4	.15
Copper	1000°F, 1 h, annealed	.55
Copper	1250°F, 1 h, annealed	.50
Copper	1500°F, 1 h, annealed	.48
70-30 leaded brass (C85310)	1250°F, 1 h, annealed	.50
70-30 brass (C26000)	1000°F, 1 h, annealed	.56
70-30 brass (C26000)	1200°F, 1 h, annealed	.52
1005 steel (G10050)	Annealed	.32
1018 steel (G10180)	Annealed	.25
1020 steel (G10100)	Hot-rolled	.22
1212 steel (G12120)	Hot-rolled	.24
1045 steel (G10450)	Hot-rolled	.14
1144 steel (G11440)	Annealed	.14
4340 steel (G43400)	Hot-rolled	.09
52100 steel (G52986)	Spherodize-annealed	.18
52100 steel (G52986)	1500°F, annealed	.07
18-8 stainless	1600°F, 1 h, annealed	.51
18-8 stainless	1800°F, 1 h, annealed	.53
304 stainless (S30400)	Annealed	.45
303 stainless (S30300)	Annealed	.51
202 stainless (S20200)	1900°F, 1 h, annealed	.30
17-4 PH stainless (S17400)	1100°F, aged	.01
17-4 PH stainless (S17400)	Annealed	.05
Molybdenum	Extruded annealed	.13
Haynes Stellite 25 (R30605)	Solution heat-treated	.50
Vanadium	Annealed	.35

Source: Adapted and updated from Joseph Datsko, *Material Properties and Manufacturing Processes*, Wiley, New York.

TABLE 4.23.2 Is a Burr Allowable?

Would it cause an electrical short circuit?
Would it jam a mechanism?
Would it cause interference fits?
Would it cause misalignment?
Would it be a safety hazard? (Would it cut someone's finger during assembly or in use?)
Would it cause unallowable stress concentrations?
Would it accelerate wear beyond allowable limits?

TABLE 4.23.3 Why This Edge Quality?

Why is burr-free condition required?
Why are _____ maximum edge radii required?
Where is burr-free condition required?
Where are _____ maximum edge radii required?
How is burr-free condition measured?
How is edge break condition measured?
What happens if part is not burr-free?
What happens if part does not have _____ maximum edge radii?
How can part be redesigned to minimize the burr?

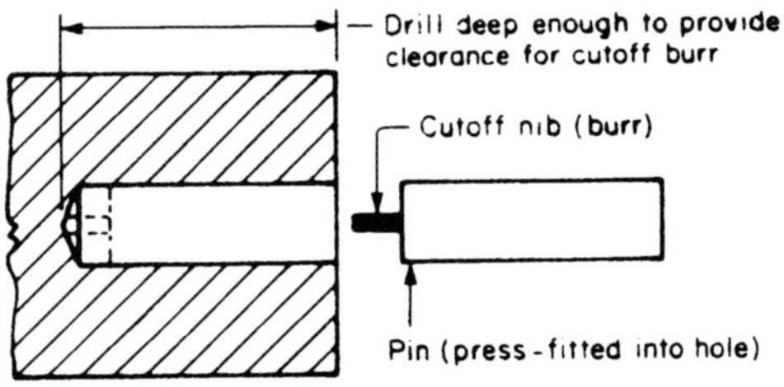

FIGURE 4.23.2 Design to accommodate cutoff burr.

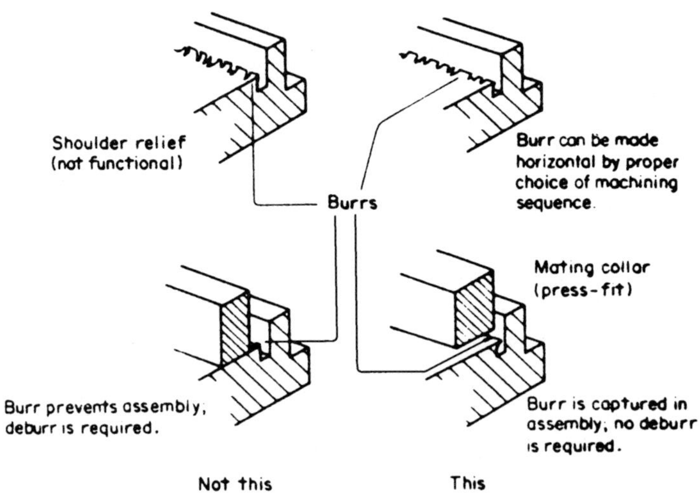

FIGURE 4.23.3 Specify burr direction.

burr in Fig. 4.23.2 need not be removed if the mating hole in the housing is drilled deep enough to accommodate it. The location of the burr in Fig. 4.23.3 determines whether or not deburring is required. If the burr is directed toward the shoulder rather than perpendicularly to the horizontal surface, the mating part will easily slide flush with the shoulder without deburring. If the burr comes loose, it is trapped within the groove and cannot interfere with the assembly.

Burrs can be an asset on thin parts. Watch manufacturers, for example, have used the burr formed by drilling and tapping thin flanges to provide extra screw-holding torque. Recent studies on aerospace parts indicate that removing burrs on some drilled holes *decreases* fatigue life. Burrs on precision miniature electron-beam-welded parts act as fill metal for the weld. In each of these cases, it is essential that the designer indicate that burrs are allowable.

The addition of small undercuts as shown in Fig. 4.23.4 can eliminate the need for deburring. In this case the burr produced by the cutoff tool does not project past the inner diameter and thus is allowable.

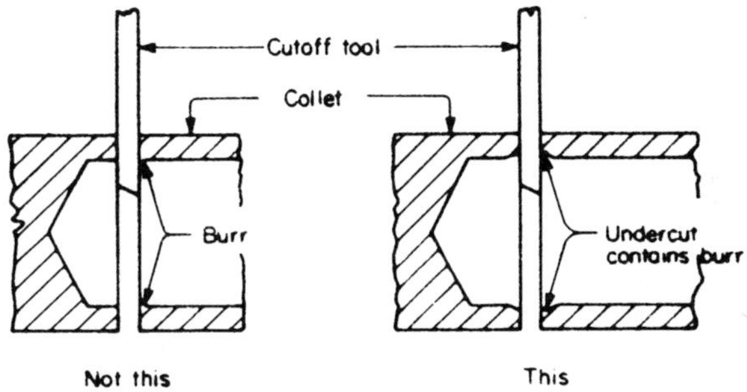

FIGURE 4.23.4 Recess minimizes burr and eliminates the need to deburr.

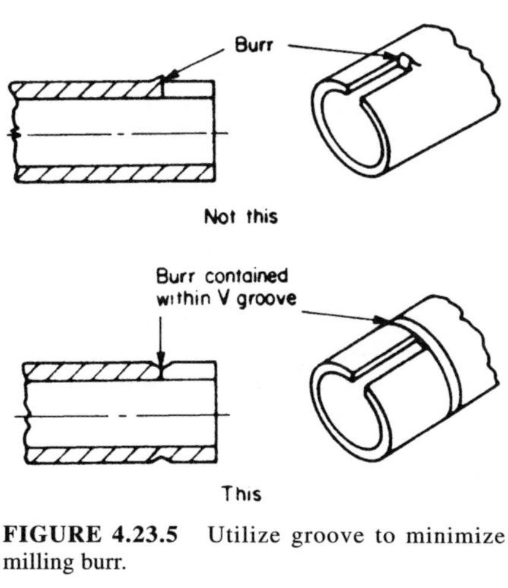

FIGURE 4.23.5 Utilize groove to minimize milling burr.

Chamfers added to sharp edges often will eliminate the need for deburring. The chamfering tool removes the large burrs formed by facing, turning, and boring and produces a relief for mating parts. Although chamfering also produces small burrs, they need not be removed in many cases, and if removal is necessary, it is easy.

If a small burr is allowable on the outside diameter of a slotted part, offer an optional V groove at the bottom of the slot, as shown in Fig. 4.23.5. Although a small burr also forms at the sides of the slots, it may not be large enough to require removal, and if it does, it is much easier to remove than the burr at the bottom of the slot.

Some of the tough vinyl coatings used

to enhance appearance can be applied over burrs and sharp edges, thus eliminating the need for burr-free conditions. In the case of an electroplated coating, however, burrs may accelerate failure of the coating because they can act as sources of high stress concentration.

Design to Minimize Deburring Problems

When a product's function dictates that burrs cannot be allowed, a designer can utilize two other approaches to minimize the cost of deburring. The designer can change the shape of the part to minimize burr size or change the geometry to put the burr in an area more accessible to the deburring medium or tool. On some parts, the relocation of a boss or shoulder or the allowance of a shallow counterbore greatly reduces the deburring effort. Sheet-metal parts, for example, should have large radii rather than sharp corners to minimize burrs. Sharp corners encourage punch and die wear with resultant large burrs. When sharp corners are truly necessary, it is sometimes possible to provide them by using more expensive progressive dies.

Burrs formed by machining through threads are extremely difficult to remove. Chapter 4.3, "Screw-Machine Products," offers a number of design recommendations to avoid such burrs and facilitate their removal.

The angle between a machined surface and intersecting surfaces greatly influences burr size. When the included angle is 150° or larger, little or no burr typically forms. As this angle decreases, burrs become much thicker and longer and hence require considerable deburring time.

The amount of radius that can economically be produced after removing the burr is also a function of the angle between intersecting surfaces (see Fig. 4.23.6). Large radii can be produced relatively quickly when the included angle is large. Finishing times are 20 times longer when the included angle is 30° than when it is 120°. However, precision edge-radius tolerances are harder to maintain when large angles are present. When a component has features involving several different edge angles, edge radii will vary significantly. Designers must recognize this when assigning tolerances to edge radii if they want to eliminate the extra costs required to produce equal radii.

Undercuts of the type shown in Fig. 4.23.7 should be avoided because it is difficult to reach burrs under the ledge and in the corners. If an undercut occurs on only one side of a part, the manufacturing engineer can prevent the occurrence of heavy burrs on these edges by ensuring that the cutter enters the workpiece at these edges. (See Fig. 4.23.8.)

Design for Easy Flash Removal

Since flash on die-cast, forged, or molded parts has many of the same characteristics as burrs, many of the preceding suggestions apply to flash, fins, and gates. Two additional rules, however, need to be observed for parts that will have flash on them. Both part designers and tool designers must be aware of these rules:

1. For appearance and ease of removal, parts should be designed so that flash occurs at edges rather than on surfaces.
2. Gates should be designed to facilitate their removal.

Figure 4.23.9 illustrates the first of these rules. Other design recommendations

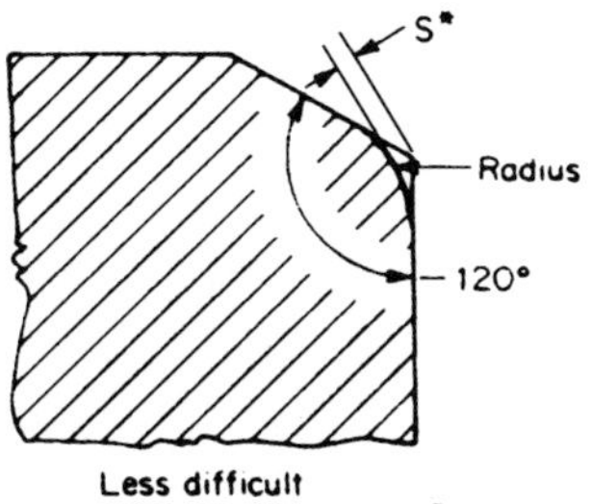

Radius		S	
in	μm	in	μm
0.002	50.8	0.0003	7.6
0.003	76.2	0.0004	10.2
0.005	127.0	0.0007	17.8
0.010	254.0	0.0015	38.1
0.015	381.0	0.0022	55.9

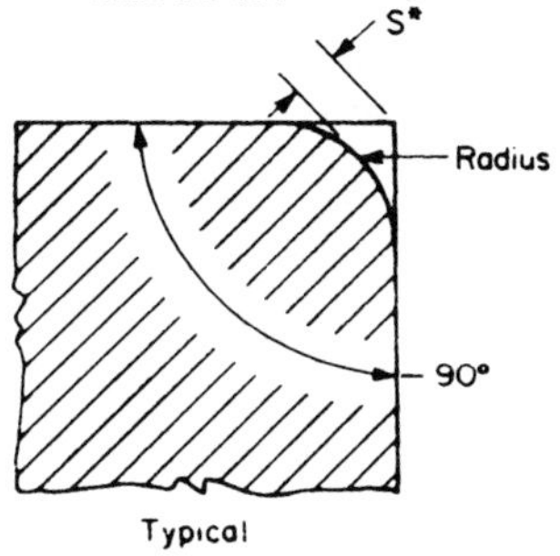

Radius		S	
in	μm	in	μm
0.002	50.8	0.0008	20.3
0.003	76.2	0.0012	30.5
0.005	127.0	0.0021	53.3
0.010	254.0	0.0041	104.1
0.015	381.0	0.0062	157.5

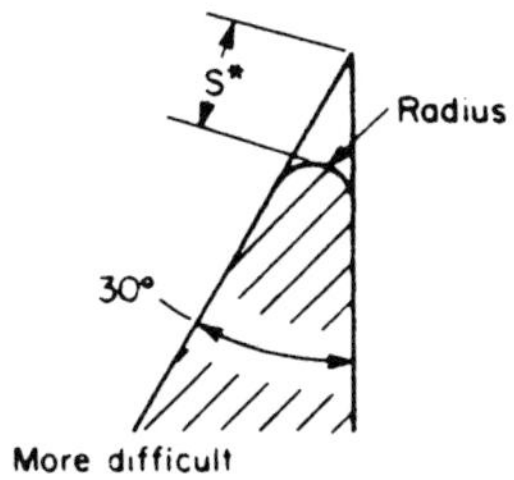

Radius		S	
in	μm	in	μm
0.002	50.8	0.0077	195.6
0.003	76.2	0.0116	294.6
0.005	127.0	0.0194	492.8
0.010	254.0	0.0387	983.0
0.015	381.0	0.0581	1475.7

FIGURE 4.23.6 Effect of geometry on edge radiusing. S = stock removal to produce indicated radius. (*From L. K. Gillespie, "The Effect of Edge Angle on the Radius Produced by Vibratory Finishing," Bendix Corporation (Kansas City Division), Report BDX-613-1279, February 1975.*)

involving flash and gates can be found in the sections of this *Handbook* covering cast, forged, and molded parts.

Define Allowable Conditions

The single most significant factor in minimizing deburring costs is knowing what edge-condition is actually required on each edge of a part. The second most significant factor is defining these conditions in such a manner that manufacturing personnel know exactly what is allowable. Table 4.23.4 illustrates the approaches that can be used to define allowable burrs or edge conditions. International standards are now available to help users define appropriate edge requirements.

Although chamfering produces a small burr, it is generally smaller than the burrs produced by other processes, and thus chamfering may represent all the deburring required. Either drawing notes or an in-plant standard should be used to indicate

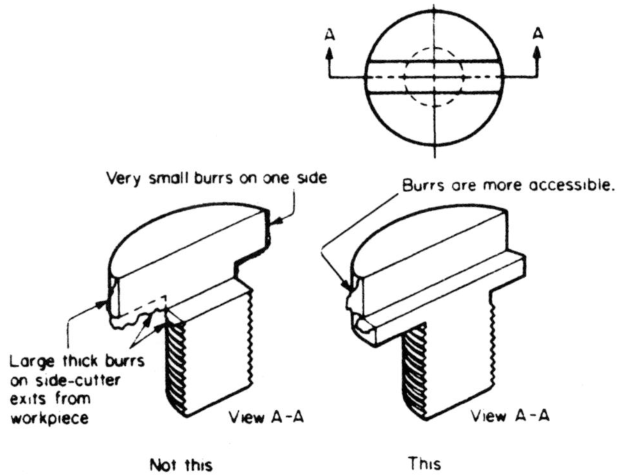

FIGURE 4.23.7 Slotting through flanges makes deburring difficult.

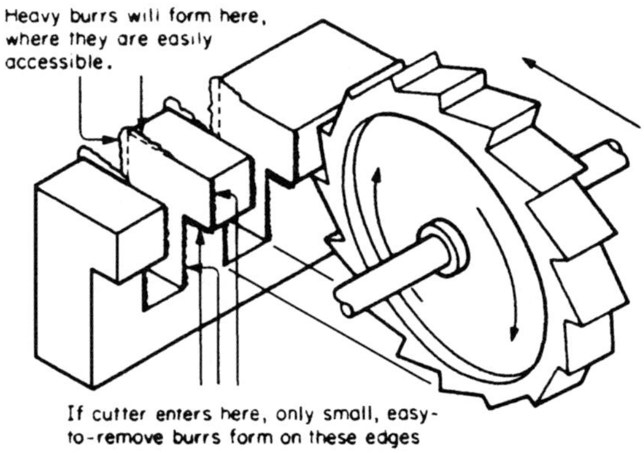

FIGURE 4.23.8 Design so that burrs occur on accessible edges.

whether or not chamfering represents adequate deburring. When a smooth blend is required, it should be specified as a radius. Edge breaks (chamfers) should be so specified that either a chamfered or a radiused condition is allowable. This allows the manufacturing engineer to determine whether a machining or a deburring process will provide the most economical edge condition. A typical corner break is 0.4 mm ($^1/_{64}$ in) by 45°. Radii should not be specified larger than 0.50 mm (0.020 in) or smaller than 0.08 mm (0.003 in).

The direction a burr faces is sometimes more critical than its actual size. In these cases, the orientation of the part should be noted on the drawing. With symmetrical threaded parts, it is helpful to the manufacturer if the designer indicates from which end of the part the screw is started. This precaution eliminates the need to deburr both ends.

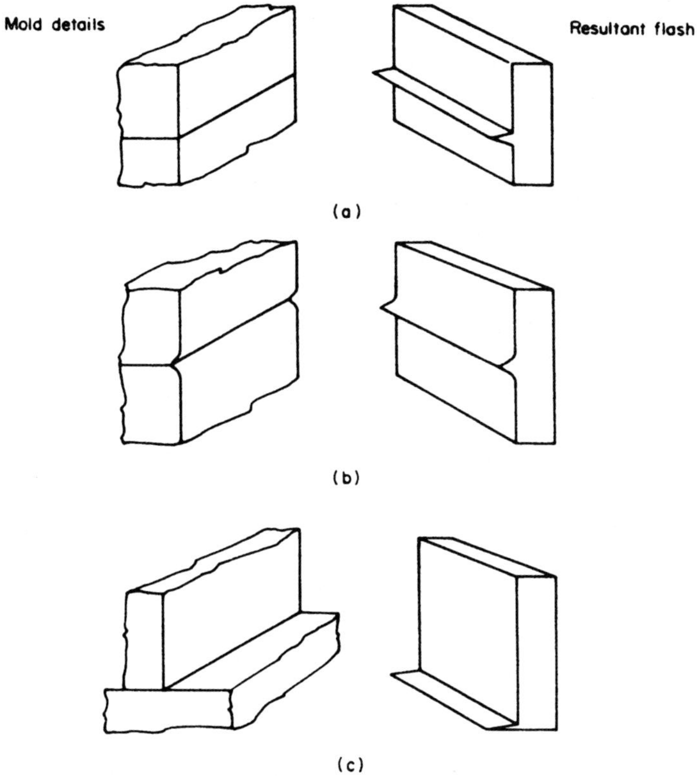

FIGURE 4.23.9 Effect of flash on ease of removal. (*a*) Difficult to remove all flash. (*b*) Rib allows access to flash and tends to obscure incomplete flash removal in slight offset in mold halves. (*c*) Easiest to remove flash.

TABLE 4.23.4 Methods of Defining an Allowable Burr or Edge Condition

Define it on the print, process-engineering specification (manufacturing specification), production traveler (routing sheets), or inspection traveler.
Define it by an interpretive memorandum. Such a memorandum could include sketches, photographs, measuring techniques, etc.
Define it with photographs of acceptable and unacceptable conditions.
Define it by the use of comparative masters (the master is given a tool or gauge number or a visual aid or visual standard number).
Define it by go–no go. If the burr fits the gauge, it is acceptable.
Define it by taking specific exception to general skill specifications.
Define it by special specifications.
Define it by such phrases as "Firmly adhered burrs or raised metal is allowable in this area provided a microtool 90° hook will not dislodge them."

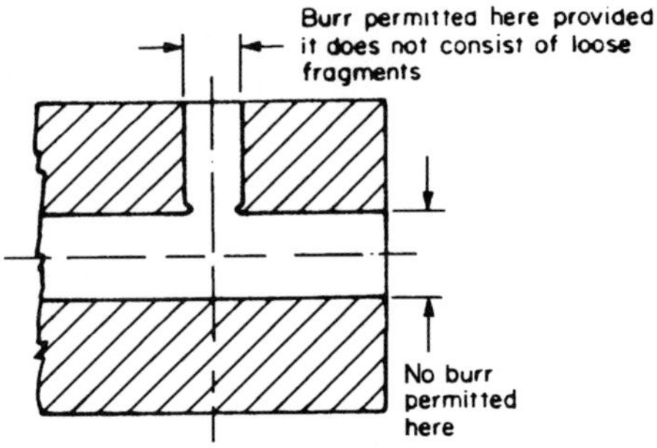

FIGURE 4.23.10 Define allowable burr size and location.

A burr always forms at the intersection of two holes. If a burr cannot be tolerated in one hole but can in the other, this must be noted (see Fig. 4.23.10). Defining where burrs can exist on formed parts may eliminate the need to deburr the sheet stock.

On many parts, the only significant edge requirement is that all sharp edges be removed. In this case, beating over burrs and dulling edges is adequate.

DIMENSIONAL FACTORS

Most deburring processes remove material from exposed surfaces while removing burrs. As a result, dimensional changes will occur during most deburring processes. These changes can be closely controlled if burr sizes are consistent from lot to lot. Unfortunately, extensive studies of burr properties have shown that they will vary significantly even on a single-piece part. For this reason, it is important to allow some tolerance on part dimensions and edge radii for the deburring operation. The values in Table 4.23.5 indicate realistic allowances that should be provided for deburring operations. For high-volume or critical parts, it is often possible to develop approaches which minimize the undesirable side effects and hence reduce the tolerances shown in this table.

The data in Table 4.23.5 are based on removing a burr that is 0.08 mm (0.003 in) thick at its base. This is a typical burr thickness. The loose or flexible portion of most burrs can be removed in a shorter time. The data in Table 4.23.5 are based on typical deburring cycles. Better surface finishes can be obtained in many of these processes by extending finishing time slightly and adding burnishing compounds.

TABLE 4.23.5 Allowances Recommended for Deburring Processes[a]

Process	Edge radius, mm (in)	Stock loss, mm (in)	Stock-loss location A	B	C	Surface finish, μm AA (μin AA)[b]
Barrel tumbling	0.08–0.5 (0.003–0.020)	0–0.025 (0–0.001)	x			1.5–0.5 (60–20)
Vibratory deburring	0.08–0.5 (0.003–0.020)	0–0.025 (0–0.001)	x			1.8–0.9 (70–35)
Centrifugal barrel tumbling	0.08–0.5 (0.003–0.020)	0–0.025 (0–0.001)	x			1.8–0.5 (70–20)
Spindle finishing	0.08–0.5 (0.003–0.020)	0–0.025 (0–0.001)	x			1.8–0.5 (70–20)
Abrasive-jet deburring	0.08–0.25 (0.003–0.010)	0–0.05 (0–0.002)[c]	x			0.8–1.3 (30–50)
Water-jet deburring	0–0.13 (0–0.005) (p)	0 (p)			x	
Liquid hone deburring	0–0.13 (0–0.005)	0–0.013 (0–0.0005)[d]			x	
Abrasive-flow deburring	0.025–0.5 (0.001–0.020)	0.025–0.13 (0.001–0.005)[d]	x			1.8–0.5 (70–20)
Chemical deburring	0–0.05 (0–0.002)	0–0.025 (0–0.001)		x		1.3–0.5 (50–20)
Ultrasonic deburring	0–0.05 (0–0.002)	0–0.025 (0–0.001)		x		0.5–0.4 (20–15)
Electrochemical deburring	0.05–0.25 (0.002–0.010)	0.025–0.08 (0.001–0.003)[e]		x		
Electropolish deburring	0–0.25 (0–0.010)	0.025–0.08 (0.001–0.003)	x			0.8–0.4 (30–15)
Thermal-energy deburring	0.05–0.5 (0.002–0.020)	0			x	1.5–1.3 (p) (60–50)
Power brushing	0.08–0.5 (0.003–0.020)	0–0.013 (0–0.0005)			x	
Power sanding	0.08–0.8 (0.003–0.030)[f]	0.013–0.08 (0.0005–0.003)	x			1.0–0.8 (40–30)
Mechanical deburring	0.08–1.5 (0.003–0.060)				x	
Manual deburring	0.05–0.4 (0.002–0.015)[g]			x		

[a]Based on a burr 0.08 mm (0.003 in) thick and 0.13 mm (0.005 in) high in steel. Thinner burrs can generally be removed much more rapidly. Values shown are typical. Stock-loss values are for overall thickness or diameter. Location A implies that loss occurs over external surfaces, B implies that loss occurs over all surfaces, and C indicates that loss occurs only near edge. (p) indicates best estimate.

[b]Values shown indicate typical before and after measurements in a deburring cycle.

[c]Abrasive is assumed to contact all surfaces.

[d]Stock loss occurs only at surfaces over which medium flows.

[e]Some additional stray etching occurs on some surfaces.

[f]Flat sanding produces a small burr and no radius.

[g]Chamfer is generally produced with a small burr.

S • E • C • T • I • O • N • 5

CASTINGS

CHAPTER 5.1
CASTINGS MADE IN SAND MOLDS

Edward C. Zuppann

Meehanite Worldwide Division
Meehanite Metal Corporation
White Plains, New York

THE PROCESS

The making of a casting starts with the making of a pattern of the part to be cast. A refractory mold is then made from this pattern. The pattern is removed from the mold for reuse, and the cavity left by the pattern is filled with molten metal. When solid, the metal is the shape of the part. Figure 5.1.1 illustrates a mold as it would appear in a section view just after the metal has been poured.

The pattern is usually of wood for low production quantities, of aluminum for intermediate quantities, and of hard ferrous metals for high production. The "sand" refractory is usually a mixture of high purity: silica sand, bentonite clay, organic addi-

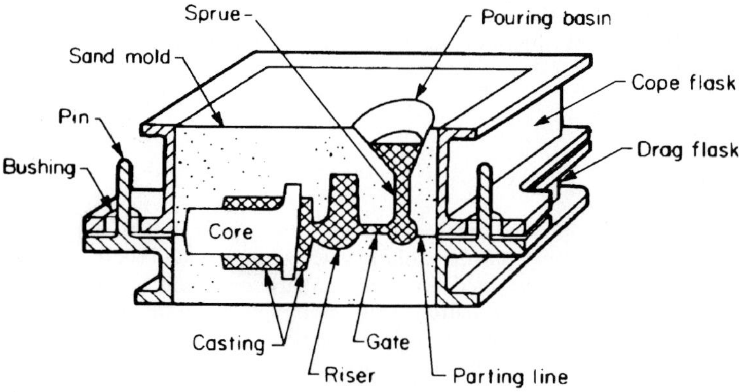

FIGURE 5.1.1 Typical mold arrangement for a sand-mold casting.

tives, and water. This mixture is formed around the pattern by ramming and squeezing. Larger molds are dried to some degree before pouring. After the refractory has been compacted or chemically hardened, the mold is opened at a prearranged parting location and the pattern removed. Often the mold halves are made separately by using part of the pattern for each half. Accurate, strong sand components, called *cores,* are placed in the mold cavity to produce holes and internal cavities in the casting. The two halves of the mold are placed together by using pins and bushings for location. Metal is poured into the cavity through a previously prepared opening.

Green-Sand Molding

In this process, a moist, plastic, rammable refractory mixture is used. After ramming, removal of the pattern, and finishing, the mold is filled with metal while still in the damp, or green, state. This process can utilize the simplest hand ramming for short runs but usually is mechanized for intermediate quantities and is fully automatic for high production rates. Usually, the green-sand process is the most economical and is applicable to all but the largest castings.

Dry-Sand Molding

This process is similar to green-sand molding except that the molds are baked or dried before pouring. It is usually employed for multiton castings for which core and metal weights are great.

Cold-Cure Molding

This process uses chemical bonding of the sand by various organic and inorganic binders. These binders are blended into the sand immediately before it is placed on the pattern. The speed of chemical hardening can be regulated. The pattern is removed after hardening, and the mold is quite true to the pattern. Cold-cure molding is used on all sizes but predominantly on larger castings. Cores of all sizes, but especially large cores, also are made by this process. Costs and accuracy are greater than for green-sand molding.

Shell Molding

Shell molding is accomplished by coating a hot ferrous pattern with sand that has been mixed with a thermosetting plastic. This plastic, when heated, bonds the particles of sand together. At a predetermined time, any unhardened sand is removed, leaving a "shell" of bonded sand on the pattern to be completely cured. After hardening, the shell, 10 to 20 mm (0.4 to 0.8 in) thick, is ejected from the pattern. Two half-shell molds are mated together and glued. The mold can be filled while resting either horizontally without backup material or vertically with a packed shot or sand backing. Complicated casting shapes increase the strength and therefore the accuracy of the mold. This process is the most accurate of the sand-casting methods and is generally used for small, complicated parts for which accuracy is important. Accurate cores for all types of molding are made by this process if quantities or the desire for accuracy are great enough.

Lost-Foam (Evaporative Pattern) Molding

This is a process in which a polystyrene foam pattern, embedded in green sand, is not removed prior to pouring of the molten metal. Instead, the pattern vaporizes from the heat of the molten metal. The metal replaces the foam and takes the shape of the mold cavity.

The pattern is coated with a permeable refractory coating, which is dried before the pattern is placed in the mold flask. Loose sand is then introduced to the flask. Vibration is normally used to compact the sand around the pattern, and a vacuum may be applied to the mold during pouring.

The foam pattern is made by introducing partly expanded pellets of polystyrene material into a metal mold. Steam heat expands and softens the pellets so that they fill the mold and take its shape. The mold is cooled, and the part is ejected. Gating and multiple patterns may be glued together prior to the casting process.

Lost-foam molding has several advantages: (1) The process does not require two mold halves (there are no parting-line problems), (2) the part can be more intricate and have a better surface finish than that producible in the normal sand-molding process, (3) cores are not required, (4) cleaning is greatly reduced, and (5) there is less labor content in the process.

TYPICAL CHARACTERISTICS OF SAND-CAST PARTS

Complex castings can be made with casting processes using sand molds. Sand-mold casting has particular advantages when complicated shapes and differing section sizes are encountered. Intricate shapes with undercuts, reentrant angles, and complex contours, which would be very difficult to machine, are practicable to cast by sand-mold methods.

Another advantage of this method is that through stress-analysis techniques the designer can reassess his or her product. Usually, metal can be removed in areas of low stress and added in areas of high stress with relatively simple alterations to the pattern. This is especially true when prototypes are being made and when wood patterns are used.

The sizes of sand castings vary from about 30 g (1 oz) to 200 tons or more. However, foundries usually specialize in a particular size range. The section size depends to some extent on the metal.

Almost any metal that can be melted can be cast. Mold-metal reactions are limiting factors for a few materials. Table 5.1.1 lists metals commonly cast in sand molds.

Sand-mold castings have somewhat irregular, grainy surfaces. Dimensional variations should be expected in as-cast surfaces, and machining is normally required if moving contact with other parts is involved or if seals are required. Most sand-cast metals (e.g., iron, aluminum, magnesium, and brass) are freely machinable.

TABLE 5.1.1 Metals Commonly Cast in Sand Molds

Designation		Tensile strength,	
Common	UNS	MPa (ksi)*	Remarks
		Cast iron	
G1800	F10004	124 (18)	Used where high strength is not a requirement; best machinability, damping properties, and resistance to thermal stress
G2500	F10005	172 (25)	Used for small cylinder blocks, pistons, gear boxes, clutch plates, and light-duty brake drums
G4000	F10008	275 (40)	Used for diesel engine cylinders, camshafts, pistons, and liners; best strength and wear resistance of these three cast irons
		Ductile iron	
60-40-18	F32800	410 (60) 275 (40) (Y)	Used for auto crankshafts, hubs, parts requiring shock resistance
		Magnesium	
AZ63A	M11630	200 (29) 130 (19) (Y)	Good castability; general casting alloy having good strength, ductility, and toughness
AZ92A	M11920	170 (25) 130 (19) (Y)	Pressure-tight components with good tensile and yield strength
		Copper alloys	
Leaded semired	C84400	235 (34) 103 (15) (Y)	For low-pressure valves and fittings, hardware parts, brass plumbing fixtures; machinability rating: 90% of C36000
Leaded red brass	C83600	255 (37) 117 (17) (Y)	Good general-purpose casting alloy; used for fire-equipment fittings, small gears, small pump parts; machinability: 84% of C36000
High-lead tin bronze	C92200	275 (40) 138 (20) (Y)	Valves, flanges, pump parts, gears, bushings
		Aluminum	
C355.0	A33550	248 (36) 172 (25) (Y)	Crankcases, gear housings air compressors, fittings
A356.0	A13560	234 (34) 165 (24) (Y)	Auto oil pans, transmission cases, real axle housings
B443.0	A24430	117 (17) 41 (6) (Y)	Marine fittings, pipe fittings, cooking utensils; very good castability
		Stainless steel	
CF-8M	J92900	550 (80) 290 (42) (Y)	Similar to wrought 316; used for aircraft parts, chemical processing, electronics, nuclear equipment, food processing, mining, fertilizer equipment, missiles

Cast components are usually stable, rigid, and strong compared with parts made by other processes. Cast iron has excellent sound- and vibration-damping properties.

Sand-mold castings are typically used for machine-tool bases, structures, and other components, pump housings, internal-combustion-engine blocks and cylinder heads,

TABLE 5.1.1 Metals Commonly Cast in Sand Molds (*Continued*)

Designation		Tensile strength,	
Common	UNS	MPa (ksi)*	Remarks
		Stainless steel (*Continued*)	
CF-8	J92600	531 (27) 255 (37) (Y)	Similar to wrought 304; used for aerospace castings, beverage, nuclear power, pharmaceutical, and chemical equipment
CA-15	J91150	793 (115) 690 (100) (Y)	Similar to wrought 410; used for jet engine parts, gears, bushings, liners, furnace burners, ship propellers, valve bodies
		Nickel alloys	
Nickel CZ-100 alloy		345 (50) 124 (18) (Y)	Standard grade nickel casting alloy with excellent castability; used for pressure tight components, pumps, valves, equipment for processing caustics at elevated temperatures
M-35-1 (Monel)		448 (65) 172 (25) (Y)	Pumps, valves, fittings; used with wrought nickel-copper alloys
		Carbon and alloy steel	
Carbon and low-alloy steels available in wrought form also can be sand-mold cast. Applications include those where greater strength, toughness, or wear resistance are required compared with that providable with cast irons. Tensile strengths range from 60 to 100 ksi for carbon steels and from 70 to 200 ksi for low-alloy steels. The bulk of casting output consists of medium-carbon steels. Uses for steel castings include parts for trucks, trailers, agricultural and earth-moving machinery, and machine tools and equipment for process industries and mining. Some representative steel casting alloys are the following: SC1020, SC1045, SC4140, SC8620, SC8630, and ASTM Spec. A356/A356M grades 1 and 2.			

*(Y) = yield point.

transmission cases, gear blanks, crankshafts, connecting rods, and many other machine components.

Figures 5.1.2 and 5.1.3 illustrate typical industrial castings.

ECONOMIC PRODUCTION QUANTITIES

Depending on the mold-making method used, sand-mold casting can be an economical means of production of applicable parts at all quantity levels. The processes used to produce sand molds for castings vary from simple hand ramming to completely automatic methods. In between, various amounts of mechanical assistance and semiautomatic equipment are used. If only a few pieces are required, a foundry with a minimum of mechanization often can give the lowest price. Automotive-type quantities require highly automated systems.

The cost of a casting includes the amortization of the pattern cost together with the manufacturing labor, metal, and overhead costs. With short runs, the pattern may be made of wood inexpensively. Even this low cost, however, spread over only a few

FIGURE 5.1.2 This turbine manifold is an example of the complexity of part that can be produced by sand-mold casting. (*Courtesy Meehanite Worldwide Division of Meehanite Metal Corporation.*)

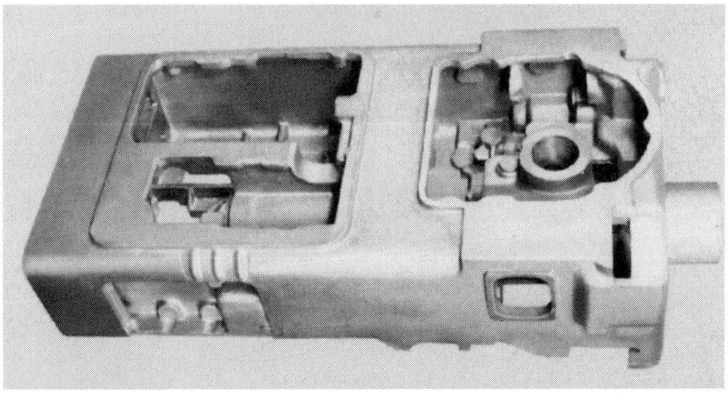

FIGURE 5.1.3 Cast-iron machine housing. (*Courtesy Meehanite Worldwide Division of Meehanite Metal Corporation.*)

pieces, might be quite high on a per-unit basis. High production runs require more expensive metal patterns to withstand the wear of repeated use. Such patterns, however, would entail a low cost when divided among many pieces.

Frequent, small orders require setup time. Handling of small orders is expensive. Making a few prototype castings can easily involve a unit cost 10 times as high as that of the same casting made in substantial quantities. Larger quantities per order should permit a lower price.

A light-sectioned, rangy casting will always cost more per unit of weight than the same weight in a chunk. The difference can be as much as 4:1.

When obtaining quotations for castings, the designer should accurately estimate the weight. This weight should be used by all quoting foundries. The foundry should quote a piece price with a differential (metal cost only) for the difference between the actual and the estimated weight.

DESIGN CONSIDERATIONS AND RECOMMENDATIONS

Almost any shape that is designed can be cast. The problem is one of economics. A small consideration in design may mean substantial differences in costs. Consult the foundry early in the design stage when changes are easy. If the designer has several alternatives, discuss these. The foundry representative usually can determine which is the cheapest for the foundry to produce. The representative often is able to suggest changes that mean little in the design but which cut the casting cost or increase reliability.

Shrinkage

A factor affecting sand-mold and other castings is the natural shrinkage of cast metal as it cools and solidifies. As well as reducing workpiece dimensions compared with the size of the mold cavity, it can cause induced stresses and distortion. The amount of shrinkage varies with different metals, but it is predictable and can be compensated for by making patterns slightly oversize. Table 5.1.2 lists normal shrinkage for metals frequently sand-cast.

TABLE 5.1.2 Shrinkage Allowance for Metals Commonly Cast in Sand Molds

Metal	Percent
Gray cast iron	0.83–1.3
White cast iron	2.1
Ductile cast iron	0.83–1.0
Malleable cast iron	0.78–1.0
Aluminum alloys	1.3
Magnesium alloys	1.3
Yellow brass	1.3–1.6
Gunmetal bronze	1.0–1.6
Phosphor bronze	1.0–1.6
Aluminum bronze	2.1
Manganese bronze	2.1
Open-hearth steel	1.6
Electric steel	2.1
High-manganese steel	2.6

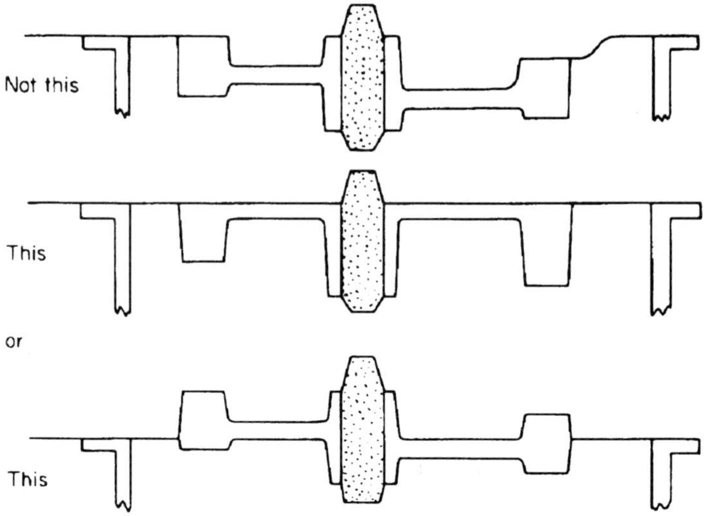

FIGURE 5.1.4 Straight parting lines are more economical than stepped parting lines.

Parting Line

The *parting line* is a continuous line around a part that separates the two halves of the mold. It is more economical and more accurate if the parting line can be on a flat plane, as shown in Fig. 5.1.4. Contoured parting lines usually result in fewer parts per mold, more costly patterns, less accuracy, more difficult "debugging," higher losses, and a need for more skilled molders, all of which increase costs. The greater the degree of contouring, the greater are the problems and costs.

Figure 5.1.5 illustrates an example in which a designer was able to redesign a part to eliminate an undercut. This permitted the use of a simpler pattern with the parting line coincident to a flat plane of the part instead of across its center. A flatter, more economical flask also could be used.

Draft

Each pattern must be easily removable from the weak, brittle molding sand. To facilitate removal, the pattern must have some degree of taper, or draft. With little or no draft, the pattern tends to tear the mold rather than slipping out smoothly. (See Fig. 5.1.6.) Castings having no draft or with undercuts can be made only by incorporating in the mold separately made, added cores at additional expense.

The amount of draft needed is related to the method of molding and drawing of the pattern, the material the pattern is made from, the degree of precision, and the surface smoothness of the pattern. In high-production work, the draw mechanism, being quite accurate, allows the pattern to be drawn if it is of high quality even if it has little draft.

The less the draft, the higher is the quality of the pattern needed and the greater is the cost. In high production, the higher pattern cost often can be easily saved in metal

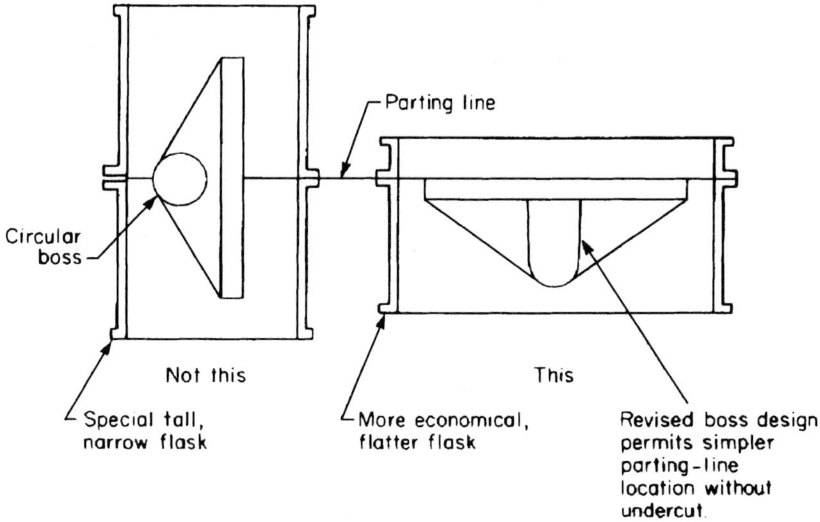

FIGURE 5.1.5 Elimination of an undercut on this casting permitted the use of a standard flask and a more straightforward parting-line location.

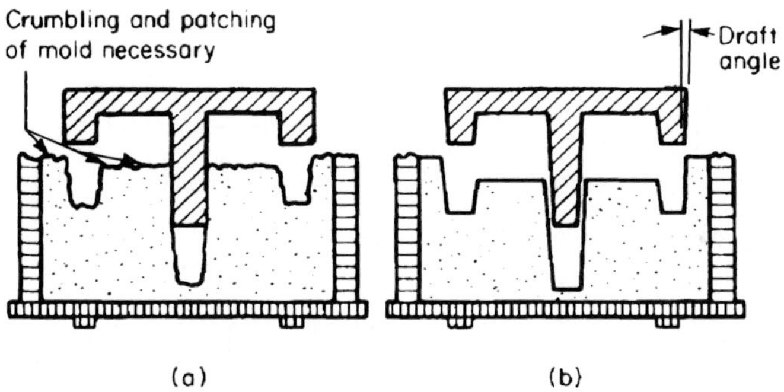

FIGURE 5.1.6 (*a*) Poor stripping from the mold results when no allowance is made for draft. (*b*) Ample draft permits easy and safe stripping.

weight. Table 5.1.3 presents recommended draft angles for sand-molded castings (and patterns) under various conditions.

Casting Soundness

With most casting metals, we must consider liquid-to-solid shrinkage. This shrinkage can leave a void at the last point to solidify. Near that point, usually in a heavy section, the foundry operator adds a riser (reservoir) designed to contain metal hotter than

TABLE 5.1.3 Draft Angles for Outside Surfaces of Sand-Molded Castings*

	Pattern material					
	Wood		Aluminum		Ferrous	
	Pattern-quality level					
Ramming method	Normal	High	Normal	High	Normal	High
Hand	5°	3°	4°	3°		
Squeezer	3°	2°	3°	2°		
Automatic			2°	1°	1½°	½°
Shell molding					1°	¼°
Cold cure	3°	2°	2°	1°		

*The draft on inside surfaces should be twice that on the outside.

the casting. The riser supplies metal to the casting during solidification and contains the final void. The riser is removed later, leaving a solid casting.

By considering risering in the original design, the engineer can make the part easier to produce and reduce rejects and, therefore, costs. Risers are attached to the heaviest sections. All surrounding sections should be thinner. The thinnest sections are farthest from the riser and solidify first. Solidification progresses toward the heaviest section and then the riser. The designer should visualize the direction of solidification and taper sections as necessary to ensure that solidification proceeds toward the direction of risers, thus minimizing the chance for voids to occur. Figure 5.1.7 illustrates good and bad practice.

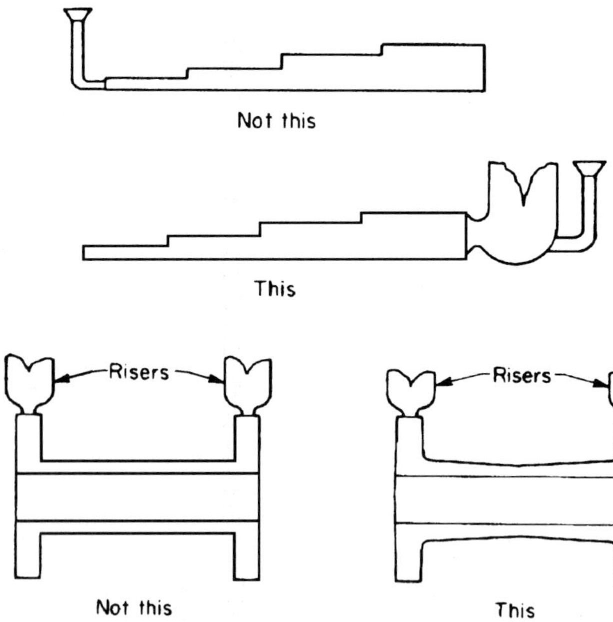

FIGURE 5.1.7 Incorrect and correct designs of castings and riser location to facilitate void-free filling of molds.

When walls cannot be graduated with respect to a riser location, the next best approach is to have even wall thickness and allow the foundry operator to control the heat gradient with the metal flow.

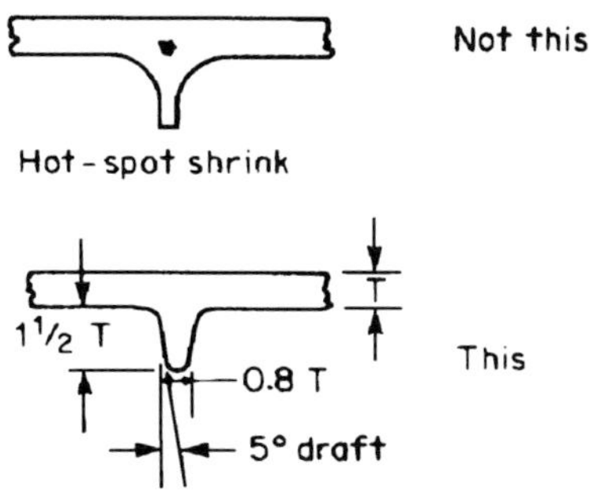

FIGURE 5.1.8 Incorrect and correct casting-rib design.

Ribs and Webs

These are effective in providing increased stiffness to a component with a minimum increase in weight. Figure 5.1.8 illustrates desirable proportions for stiffening ribs. Heavier sections where the rib intersects the casting wall can cause hot-spot shrinks as shown.

The number of ribs intersecting at one point should be minimized to avoid hot-spot effects. (See Fig. 5.1.9.) When it is necessary to bring a number of ribs or other members together at one point, a cored hole at the point of intersection will help to speed solidification. This will prevent shrink voids and structural weakness and distortion. A preferable alternative is to use a circular web to connect the ribs, thus avoiding the need for a core piece. Figure 5.1.10 illustrates these solutions.

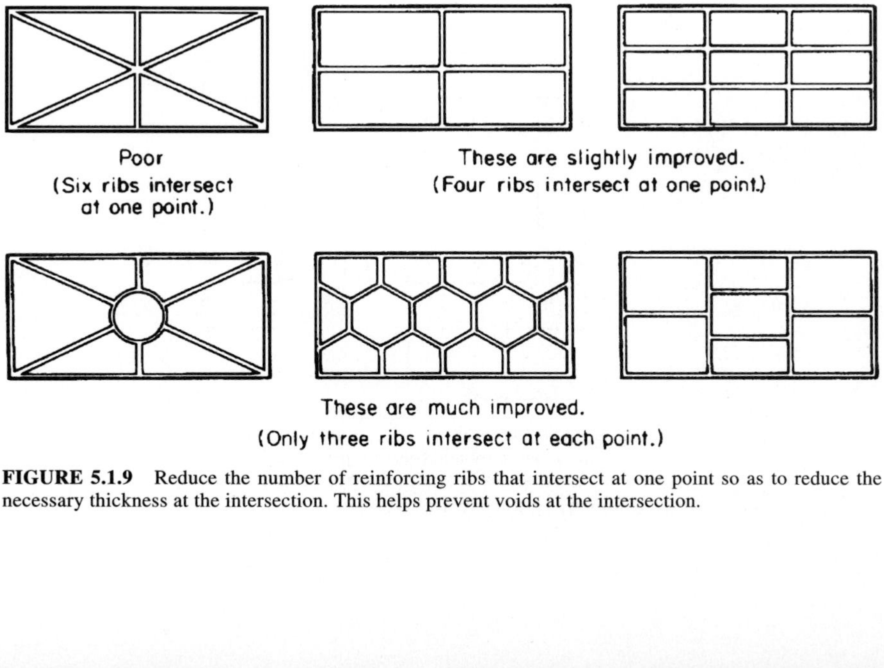

FIGURE 5.1.9 Reduce the number of reinforcing ribs that intersect at one point so as to reduce the necessary thickness at the intersection. This helps prevent voids at the intersection.

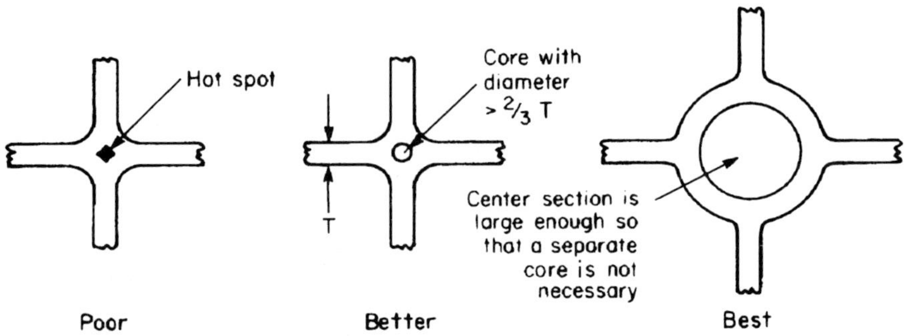

FIGURE 5.1.10 Design alternatives to prevent hot-spot voids at rib and casting wall intersections.

Corners and Angles

Hot spots are the most common defect in casting design. They cause weak zones and points of high stress. There are many variations of this problem.

The crystal growth of solidifying metal progresses inward from each surface. Outside corners radiate heat in two directions and cool quickly. Inside corners heat the sand in the corner from two directions, creating a hot spot that retards solidification. Rounding of corners of all types avoids local structural weakness, as shown in Fig. 5.1.11. Rounding the inside corners decreases the severity of the hot spot and lessens the stress concentration. However, too much rounding will promote a shrink defect in a corner. Rounding both the inside and the outside of the corner and using the same center for the radii are better.

In a T section, as illustrated in Fig. 5.1.12, a dished contour opposite the intersect-

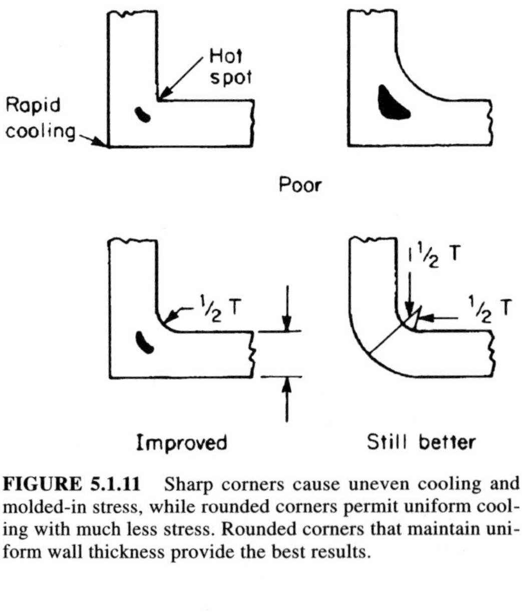

FIGURE 5.1.11 Sharp corners cause uneven cooling and molded-in stress, while rounded corners permit uniform cooling with much less stress. Rounded corners that maintain uniform wall thickness provide the best results.

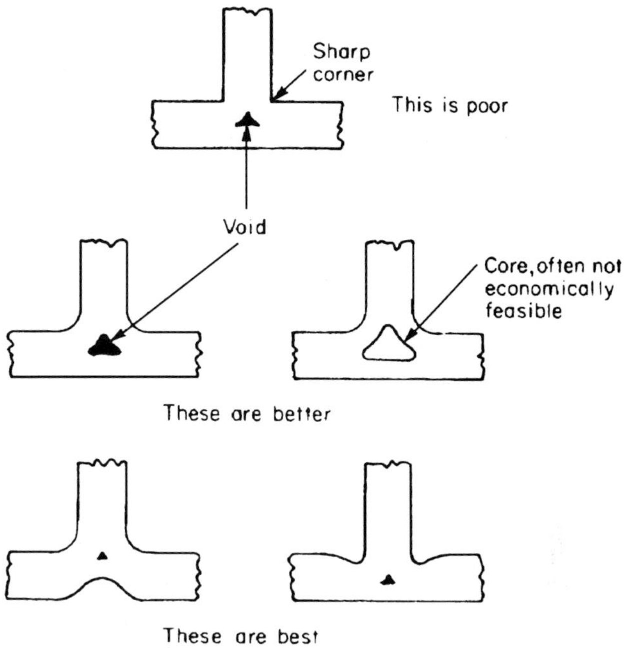

FIGURE 5.1.12 Incorrect and correct design of T joints in castings.

ing member minimizes the shrink so that larger inside radii can be used to minimize stress concentration and hot spots. Two dished contours, one on each side of the center leg, are also effective.

The ultimate in removing the shrink area is to add a separately made core to the intersection. Except for "x-ray quality" work, however, this is more trouble and expense than it is worth. When the voids are minimized, they retreat to a zone of almost zero stress and are thus insignificant.

The more acute the angle, the greater is the problem, for both the rapid-cooling area and the hot-spot effect are increased. Figure 5.1.13 illustrates the cooling effect. The intersection of two walls of a casting should be at a right angle, if possible, to minimize heat concentration. Figure 5.1.14 illustrates an example.

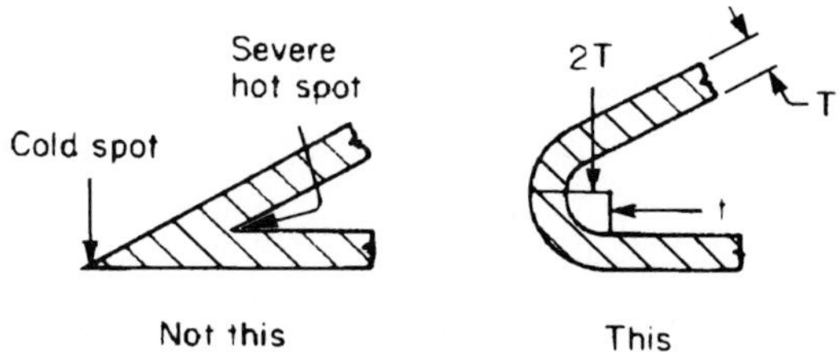

FIGURE 5.1.13 Sharp-corner acute angles cause areas of uneven cooling and should be avoided.

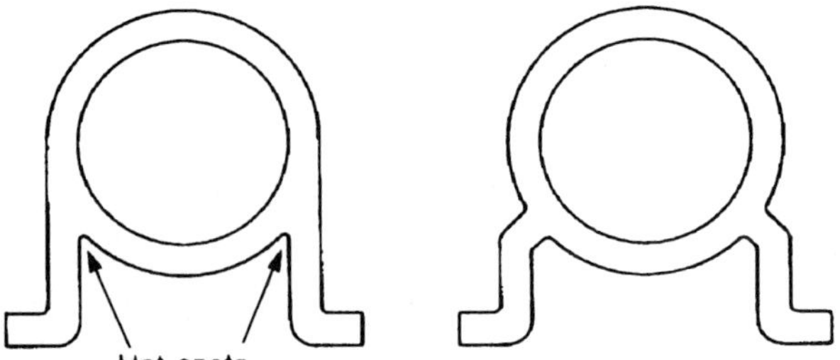

FIGURE 5.1.14 The intersection of two walls of a casting should be at right angles to avoid sharp-corner hot spots.

Wall Thickness

Table 5.1.4 shows the normal minimum economical wall thickness for various metals. In general, problems increase rapidly if sections are too small [under 6 mm (¼ in)] in all metals. It is often cheaper to increase the section size than to pay an increased price required to cover foundry scrap losses. The farther the metal must flow in the mold, the heavier the section must be. Therefore, there is a limit to the savings that are gained by reducing the weight of a casting. Heavy sections also can cause problems, especially if they are isolated, as described above.

TABLE 5.1.4 Minimum Economical Section Thickness for Green-Sand Castings*

Section length	To 300 mm (12 in)	To 1.2 m (6 ft)	To 3.6 m (12 ft)
Aluminum	3–5 mm (⅛–3/16 in)	8 mm (5/16 in)	16 mm (⅝ in)
Brass and bronze	2.4–3 mm (3/32–⅛ in)	8 mm (5/16 in)	16 mm (⅝ in)
Ductile iron	5 mm (3/16 in)	13 mm (½ in)	19 mm (¾ in)
Gray iron, low strength	3 mm (⅛ in)		
Gray iron, 138-MPa (20,000-lbf/in^2) tensile strength	4 mm (5/32 in)	10 mm (⅜ in)	
Gray iron, 207-MPa (30,000-lbf/in^2) tensile strength	5 mm (3/16 in)	10 mm (⅜ in)	19 mm (¾ in)
Gray iron, 276-MPa (40,000-lbf/in^2) tensile strength	6 mm (¼ in)	13 mm (½ in)	25 mm (1 in)
Gray iron, 345-MPa (50,000-lbf/in^2) tensile strength	10 mm (⅜ in)	16 mm (⅝ in)	25 mm (1 in)
Magnesium alloys	4 mm (5/32 in)	8 mm (5/16 in)	16 mm (⅝ in)
Malleable iron	3 mm (⅛ in)	6 mm (¼ in)	
Steel	8 mm (5/16 in)	13 mm (½ in)	25 mm (1 in)
White iron	3 mm (⅛ in)	13 mm (½ in)	19 mm (¾ in)

*Thinner walls are economical in shell, dry-sand, or cold-cure molded castings. Some designs of castings lend themselves to thinner walls without problems.

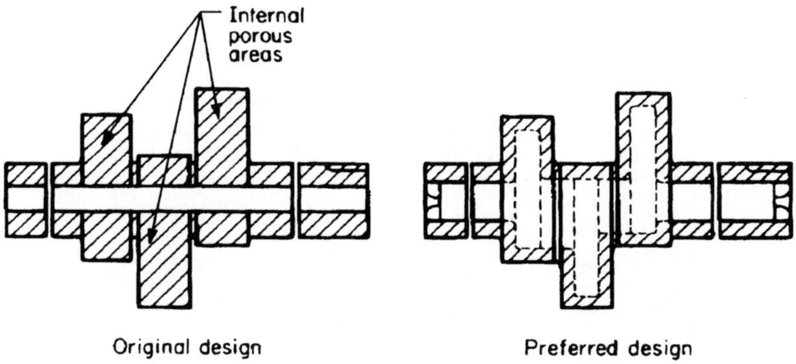

FIGURE 5.1.15 Keeping wall thicknesses uniform promotes sounder castings because heavy sections are more prone to internal porosity.

It is best to have sections and walls as uniform as possible in thickness. This avoids problems with voids and porosity in heavy sections and added stresses or distortion due to uneven cooling. Figure 5.1.15 illustrates a typical example.

Section Changes

Abrupt changes in sections should be avoided. Fillets and tapers are preferable to sharp steps. Normally, a difference of greater than 2:1 in relative thickness of adjoining sections should be avoided. If a section change of over 2:1 is unavoidable, the designer has two alternatives: (1) Design the section as two separate castings to be bolted together later, or (2) use a wedge form between the unequal sections. The taper of the wedged area should not exceed 1:4. Figure 5.1.16 illustrates these points.

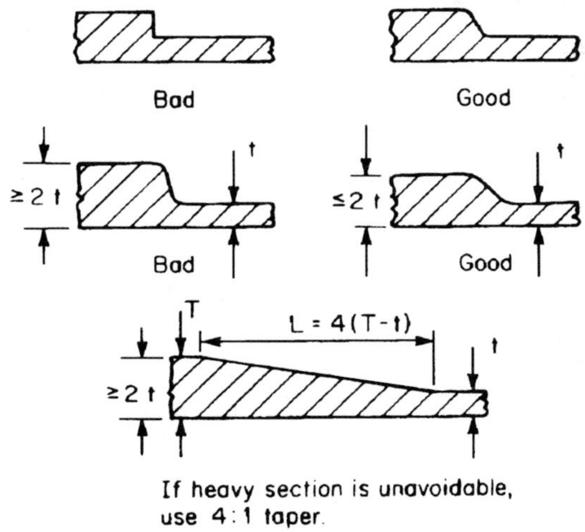

FIGURE 5.1.16 Design rules for areas where section thickness must change.

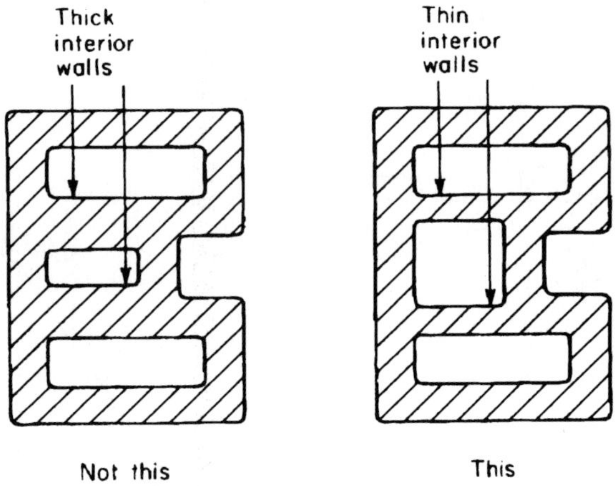

FIGURE 5.1.17 Interior walls should be 20 percent thinner than exterior walls since they cool more slowly.

Interior Walls and Sections

These walls and sections should be 20 percent thinner than the outside members because they cool more slowly than the external walls. In this way, thermal and residual stresses are reduced and metallurgical changes are minimized (Fig. 5.1.17).

Lightener Holes

To reduce weight in low-stressed areas, lightener holes can be added. These in themselves can be stress raisers if they have sharp corners.

Holes and Pockets

The incorporation of through holes and pockets (cavities) in sand-mold castings is straightforward and economical. If these are located so that they can be produced by the mold rather than by an extra core, they may reduce the cost of the casting by saving material.

With green-sand molding, the so-called green-sand core is very weak. It will break if touched by the pattern during the draw. The rule, therefore, is that the draft on the inside of a pocket must be twice as much as on the surrounding outside surfaces. This outside acts as a guide during the drawing of the pattern. Another rule is that the depth of a hole or pocket should not be more than $1\frac{1}{2}$ times its narrowest dimension if it is in the drag (bottom) half of mold; this depth should be no more than the narrowest dimension if the hole or pocket is in the cope (top) half of the mold. Figure 5.1.18 illustrates this rule.

Small holes are generally not economical to mold or core. It is usually cheaper and more satisfactory to drill them. The break-even point is between 13- and 25-mm ($\frac{1}{2}$- and 1-in) diameter. A drill dimple can be cast in the part to facilitate offhand drilling if applicable, as shown in Fig. 5.1.19.

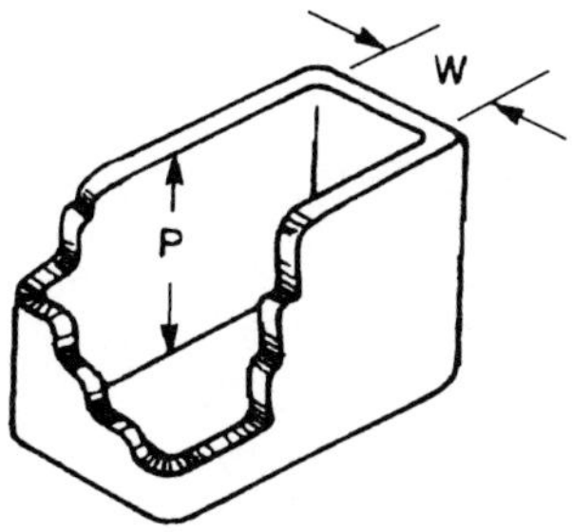

FIGURE 5.1.18 Design rules for the correct proportions of rectangular pockets or holes in castings: $P \leq 1\frac{1}{2}\ W$ if in drag; $P \leq W$ if in cope.

With shell molding, pockets that are much deeper than their width can be drawn with high-quality pattern equipment. This can justify the more expensive shell process.

Holes are generally stress raisers, and they should have extra metal around them to compensate. This metal should be carefully blended to minimize hot spots and stress-raising corners. (See Fig. 5.1.20.)

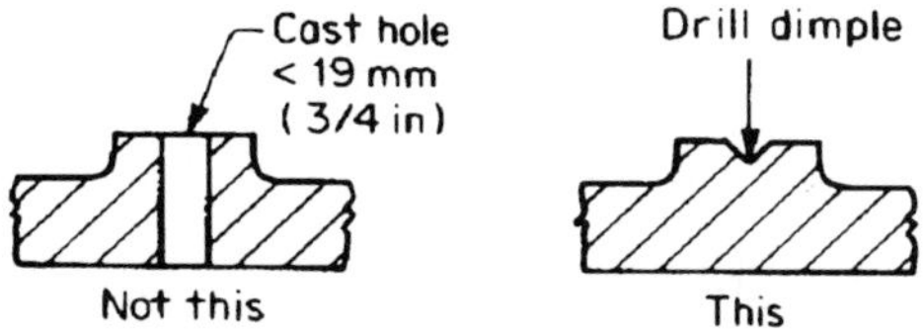

FIGURE 5.1.19 Holes less than 19 mm ($\frac{3}{4}$ in) in diameter are cheaper and better if drilled after casting.

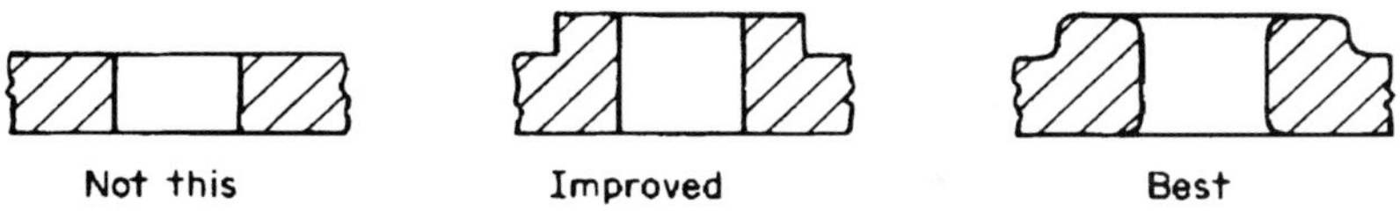

FIGURE 5.1.20 When a hole is placed in a highly stressed section, add extra material around the hole as a reinforcement.

Bosses and Pads

These small raised areas are usually necessitated to minimize surface machining. If too large, they create voids and hot spots. Because of these problems, bosses, pads, and lugs should be minimized. Figure 5.1.21 illustrates workable alternatives.

Cores

Cores are subject to heat and high floating pressures. Their usual organic binders break down, releasing gases that must be vented to prevent bubbles in the metal. The decomposed sand must be removable by cleaning processes. More significantly, cores are expensive to make, handle, set in the mold, and clean out later. For these reasons they should not be used if at all avoidable. Figure 5.1.22 illustrates some economical alternatives.

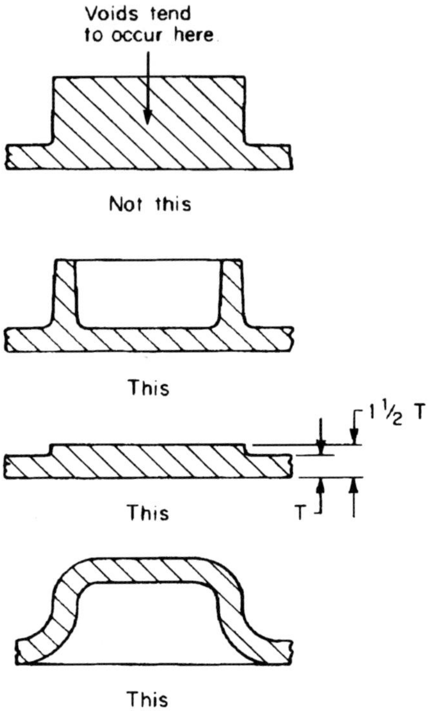

FIGURE 5.1.21 Design suggestions for minimizing material thickness at bosses.

Small cores in heavy sections are extremely hard to remove. If small, round holes are required, it is almost always cheaper and more satisfactory to drill them. The core should have a diameter at least equal to the surrounding wall thickness and preferably twice the wall thickness or more. (See Fig. 5.1.23.)

Side bosses and undercuts normally necessitate the use of cores as part of the mold. Since cores add significantly to the cost of castings, it behooves the designer to eliminate side bosses and undercuts from casting designs. Figure 5.1.24 illustrates examples of various components for which this was done.

Internal cores require adequate provision for removing the gases generated when the core comes in contact with the molten metal. Figure 5.1.25 illustrates the addition of venting holes to a particular casting for this purpose and to facilitate removal of the core sand.

Gears, Pulleys, and Wheels

These apparently simple objects are often improperly designed. Residual casting stresses involved can markedly reduce their strength, resulting in premature breakage. To minimize the stresses, a balance between the section size of the rim, spokes, and hub must be attempted. (See Fig. 5.1.26.) It is also desirable to have an odd number of spokes. Curved spokes as shown in Fig. 5.1.27 dissipate additional stress. Solid

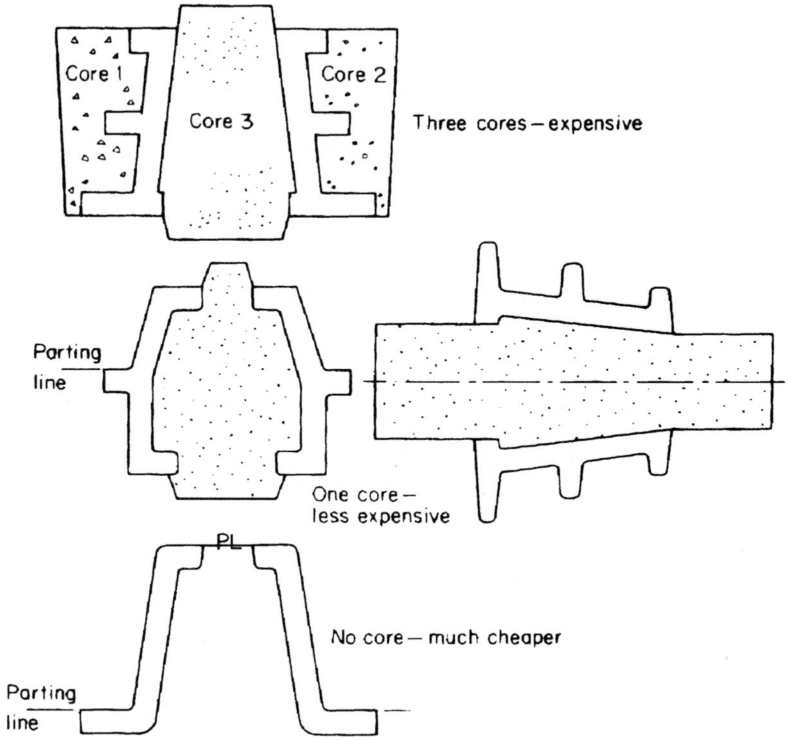

FIGURE 5.1.22 Minimize the need for cores as much as possible by eliminating undercuts.

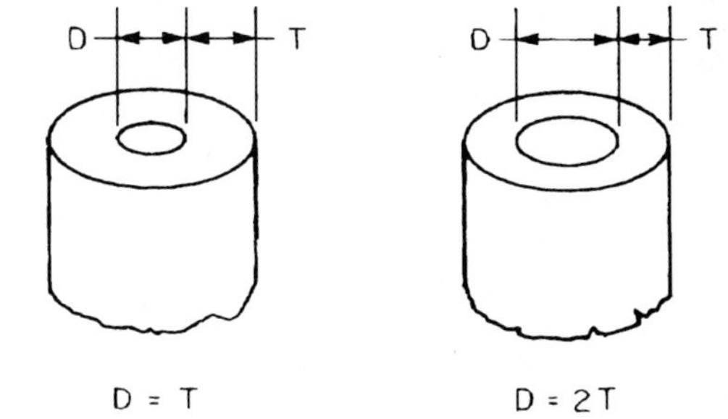

FIGURE 5.1.23 Small cored holes should be avoided since removal of core material after casting is costly.

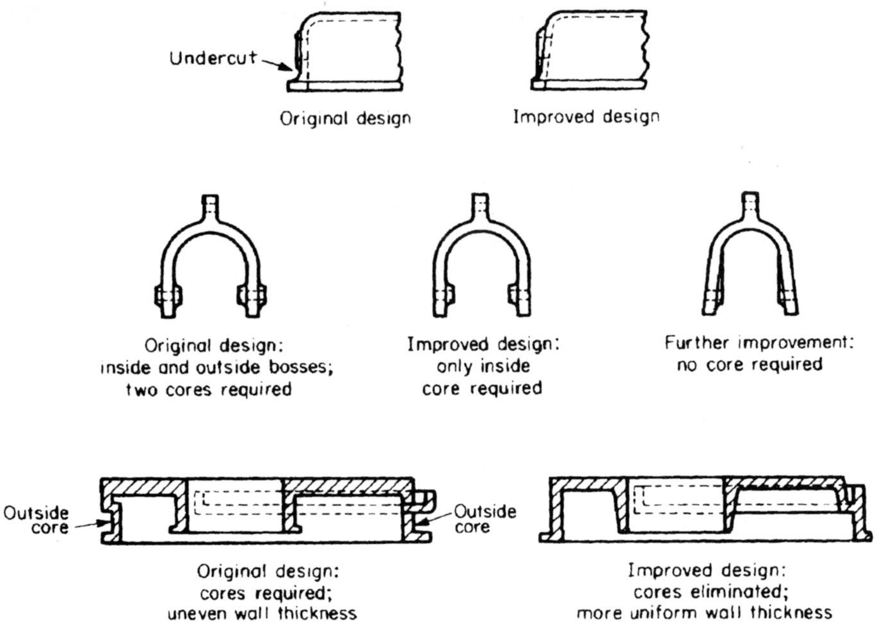

FIGURE 5.1.24 External undercuts frequently necessitate the use of separate mold core pieces. Illustrated are three parts that were redesigned to eliminate the need for such cores without reducing the function of the parts.

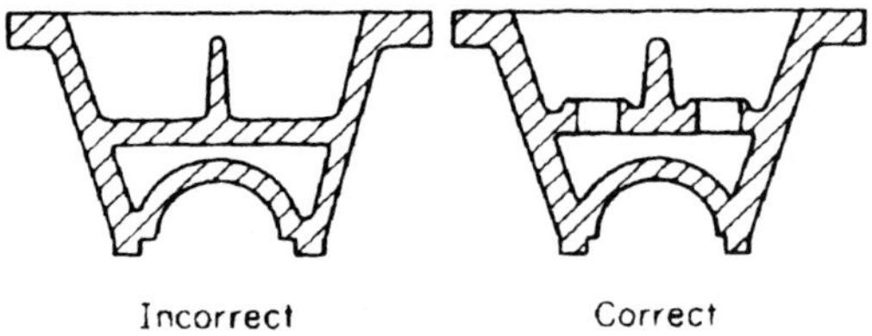

FIGURE 5.1.25 Internal pockets in castings may require cored vent holes to allow gases released during pouting to escape and to facilitate cleaning after casting.

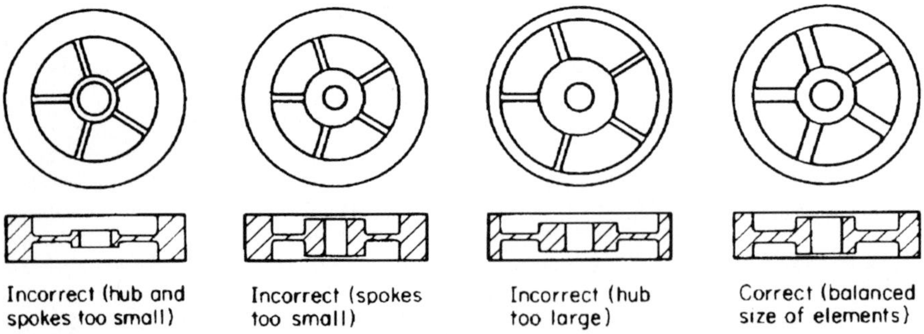

FIGURE 5.1.26 Incorrect and correct proportions of elements of pulleys and gear blanks.

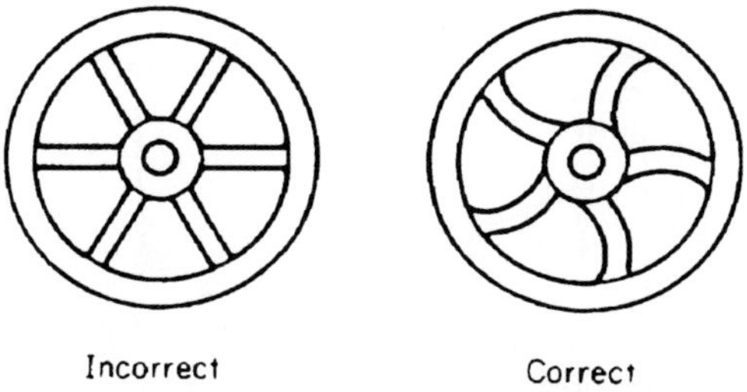

FIGURE 5.1.27 An odd number of curved wheel spokes helps dissipate cast-in stresses.

webbed wheels can have dished surfaces that allow the stress to be reduced. In this case, excessive section variation is to be avoided (see Fig. 5.1.28).

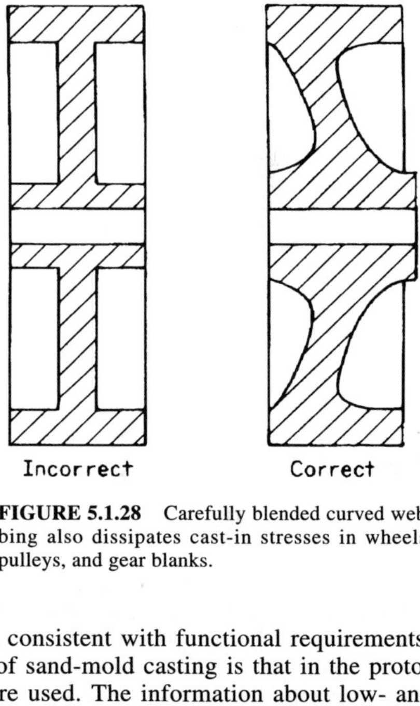

FIGURE 5.1.28 Carefully blended curved webbing also dissipates cast-in stresses in wheels, pulleys, and gear blanks.

Lettering and Other Data

Any lettering should be parallel to the parting plane so that the mold pockets for the letters will draw. When placed in any other position, the lettering would require an expensive core.

Such data as part numbers, date of casting, foundry trademark, cavity number, and sometimes serial numbers, etc. all need a place on the casting. These must be in a position where they will not interfere with machining. They can be either sunken or raised above the surface.

Weight Reduction

Economies of production are realized when, consistent with functional requirements, casting weight is minimized. An advantage of sand-mold casting is that in the prototype state easily changed wooden patterns are used. The information about low- and high-stressed areas can be determined by stress analysis of prototype castings, and metal can be removed from low-stressed areas and added to high-stressed areas by way of an inexpensive pattern change, often of wax. Many examples may be found for cutting weight 20 to 50 percent while increasing strength 50 to 100 percent through this technique. (See Fig. 5.1.29.)

Inserts of Different Metals

It is sometimes desirable to incorporate a section of a different metal, either harder or softer than the base metal, in a casting. This can be done and may save money. An example is the cast-aluminum aircraft brake of Fig. 5.1.30 with a cast-iron wear surface included as an insert. Various metal combinations can be used. Most are held in the casting more by mechanical interlocking than by metallic fusion. The inserts must

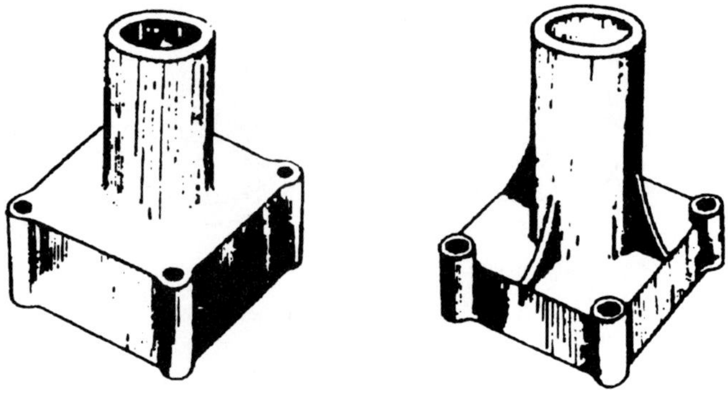

FIGURE 5.1.29 A hydraulic-ram head was reduced in cost when the casting was redesigned to require less material.

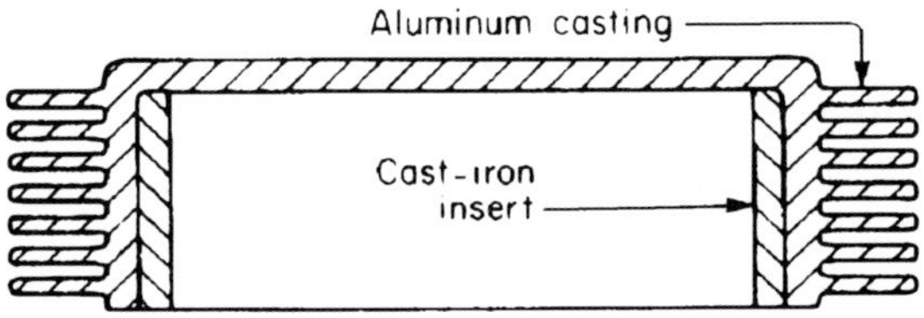

FIGURE 5.1.30 A cast-iron wear-surface insert in an aluminum aircraft-brake casting.

have projections to prevent turning or lateral movement. These inserts are not generally pressure-tight unless sealed after casting.

Design to Facilitate Machining

Castings designed with sharp angles or edges are a common source of machine-shop trouble. This is caused by the faster cooling of the corner section at the joint and is accentuated by the presence of a fin or flash.

By rounding edges and corners sufficiently to eliminate chilling, the risk of hard corners or edges is avoided, and, in turn, machine-shop costs are reduced. When this is done, the parting should be made to permit all corners and edges to be rounded. (See Fig. 5.1.31.)

Machining Allowances

A stock allowance must be added to surfaces that are to be machined. The allowance provides for dimensional and surface variations in the as-cast workpiece. The amount of stock allowance depends on the size of the surface to be machined and also somewhat on the machining method and the final accuracy required.

If only flatness is desired and some unmachined areas are not objectionable, a minimum of additional metal is needed. If a fully machined surface without imperfections is necessary, more metal must be removed. Normal full allowances are shown in Table 5.1.5.

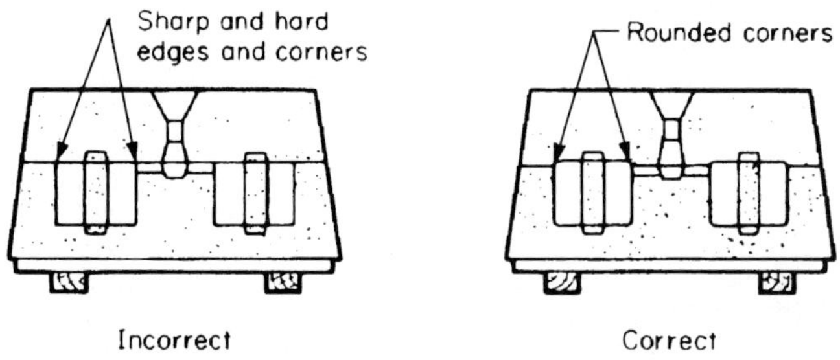

FIGURE 5.1.31 Avoid sharp corners and fins in areas to be machined after casting.

TABLE 5.1.5 Allowance for Machining: Each Side

		Allowance, mm (in)	
	Casting size, mm (in)*	Drag and sides	Cope surface
Gray iron	Up to 150 (up to 6)	2.3 (3/32)	3 (1/8)
	150–300 (6–12)	3 (1/8)	4 (5/32)
	300–600 (12–24)	5 (3/16)	6 (1/4)
	600–900 (24–36)	6 (1/4)	8 (5/16)
	900–1500 (36–60)	8 (5/16)	10 (3/8)
	1500–2100 (60–84)	10 (3/8)	13 (1/2)
	2100–3000 (84–120)	11 (7/16)	16 (5/8)
Cast steel	Up to 150 (up to 6)	3 (1/8)	6 (1/4)
	150–300 (6–12)	5 (3/16)	6 (1/4)
	300–600 (12–24)	6 (1/4)	8 (5/16)
	600–900 (24–36)	8 (5/16)	10 (3/8)
	900–1500 (36–60)	10 (3/8)	13 (1/2)
	1500–2100 (60–84)	11 (7/16)	14 (9/16)
	2100–3000 (84–120)	13 (1/2)	19 (3/4)
Malleable iron	Up to 75 (up to 3)	1.5 (1/16)	2.3 (3/32)
	75–300 (3–12)	2.3 (3/32)	3 (1/8)
	300–450 (12–18)	3 (1/8)	4 (5/32)
	450–600 (18–24)	4 (5/32)	5 (3/16)
Ductile iron	Up to 150 (up to 6)	2.3 (3/32)	6 (1/4)
	150–300 (6–12)	3 (1/8)	10 (3/8)
	300–600 (12–24)	5 (3/16)	19 (3/4)
	600–900 (24–36)	6 (1/4)	19 (3/4)
	900–1500 (36–60)	8 (5/16)	25 (1)
	1500–2100 (60–84)	10 (3/8)	28 (1 1/8)
	2100–300 (84–120)	11 (7/16)	32 (1 1/4)
Nonferrous metals	Up to 150 (up to 6)	1.6 (1/16)	2.3 (3/32)
	150–300 (6–12)	2.3 (3/32)	3 (1/8)
	300–600 (12–24)	3 (1/8)	4 (5/32)
	600–900 (24–36)	4 (5/32)	5 (3/16)

**Casting size* refers to the overall length of the casting and not to the length of a particular measurement.

TABLE 5.1.6 Allowance for Machining Cylinders; Machining Located from the Bore*

Diameter, mm (in)	Range of allowance; total of both sides, mm (in)
0–25 (0–1)	3–4.5 (0.12–0.18)
25–100 (1–4)	3–5 (0.12–0.20)
100–200 (4–8)	4.5–6 (0.18–0.24)
200–300 (8–12)	5–8 (0.20–0.32)
300–600 (12–24)	6–10 (0.24–0.40)

Note: Use as a rough guide only.

*When machining location is other than the bore, additional stock may be necessary to compensate for allowable tolerance between locating points and bore.

Casting shrinkage across a bore is less predictable than that for other dimensions. For this reason, it is difficult to make general recommendations for machining allowances for cylinder bores. The more bores to be machined per casting, the more variables there are, and therefore, more stock is allowed. The larger the casting, the more unpredictable is its size, and again, more allowance is made. Table 5.1.6 presents ranges of recommended allowances for cylinder-bore machining.

DIMENSIONAL FACTORS AND TOLERANCE RECOMMENDATIONS

Many factors influence the degree to which any particular cast piece differs in dimensions from those specified. These include differences in methods, equipment, metal, and sand from foundry to foundry or from lot to lot; variations in metal temperature, cooling time, and cooling conditions; and differences in mold hardness, pattern inaccuracies, shrinkage variations, internal stresses, and many more. Because of the effect of these factors in normal production, it is best not to specify tolerances any tighter than absolutely necessary. Tight tolerances require more precise pattern making, a statistical checking program, reworking the pattern, and often 100 percent gauging with resulting scrap. The purchaser pays for all of this sooner or later.

When castings are dimensioned in inches, the general practice is to specify noncritical or nominal dimensions in fractions of inches or one- or two-place decimals. Critical dimensions are usually given as three-place decimals with a tolerance range.

Recommended tolerances are shown in Table 5.1.7. These are for average conditions. Under specific conditions, the ranges can be decreased

TABLE 5.1.7 Typical Tolerances for Green-Sand and Oil-Sand Castings*

Location	Dimension	Tolerance
One side of parting line	0–25 mm (0–1 in)	±0.6 mm (±0.023 in)
	25–75 mm (1–3 in)	±0.8 mm (±0.030 in)
	75–150 mm (3–6 in)	±1.2 mm (±0.045 in)
	150–230 mm (6–9 in)	±1.5 mm (±0.060 in)
	230–300 mm (9–12 in)	±2.3 mm (±0.090 in)
	300–400 mm (12–16 in)	±2.6 mm (±0.102 in)
	400–500 mm (16–20 in)	±2.9 mm (±0.114 in)
	500–600 mm (20–24 in)	±3.2 mm (±0.125 in)
	600–760 mm (24–30 in)	±3.5 mm (±0.138 in)
	760–900 mm (30–36 in)	±3.8 mm (±0.150 in)
	Area at parting line	
Additional tolerance for dimensions across parting line (tolerance to be added to that above)	6–65 cm^2 (1–10 in^2)	±0.5 mm (±0.020 in)
	65–320 cm^2 (10–50 in^2)	±0.9 mm (±0.035 in)
	320–650 cm^2 (50–100 in^2)	±1.0 mm (±0.040 in)
	650–1600 cm^2 (100–250 in^2)	±1.3 mm (±0.050 in)
	1600–4000 cm^2 (250–600 in^2)	±1.5 mm (±0.060 in)
	4000–6500 cm^2 (600–1000 in^2)†	±2.0 mm (±0.080 in)
	Dimension	
Between two cores	0–75 mm (0–3 in)	±0.8 mm (±0.030 in)
	75–150 mm (3–6 in)	±1.5 mm (±0.060 in)
	150–230 mm (6–9 in)	±2.3 mm (±0.090 in)
	230–600 mm (9–24 in)	±3.0 mm (±0.120 in)
	600–1500 mm (24–60 in)	±4.5 mm (±0.180 in)
	Over 1500 mm (over 60 in)	±6.3 mm (±0.250 in)
Cores: shell, hot-box, cold-cure, etc. (one side of core box)‡	0–25 mm (0–1 in)	±0.15 mm (±0.006 in)
	25–50 mm (1–2 in)	±0.30 mm (±0.012 in)
	50–75 mm (2–3 in)	±0.45 mm (±0.018 in)
	75–150 mm (3–6 in)	±0.75 mm (±0.030 in)
	150–230 mm (6–9 in)	±1.0 mm (±0.040 in)
	230–300 mm (9–12 in)	±1.3 mm (±0.050 in)
	Over 300 mm (over 12 in)	±1.3 mm plus 0.2% (±0.050 in plus 0.2%)
Shift, mold or core; largest casting dimension *A* greater than smallest *B*	0–200 mm (0–8 in)	±2 mm (±0.080 in)
	200–450 mm (8–18 in)	±3 mm (±0.120 in)
	450–900 mm (18–36 in)	±5 mm (±0.190 in)
	900–1500 mm (36–90 in)	±6 mm (±0.250 in)

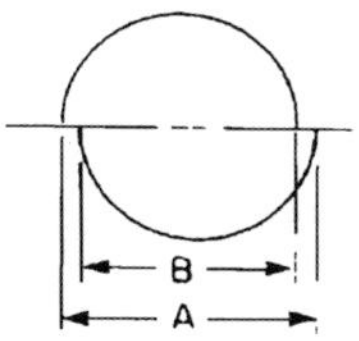

TABLE 5.1.7 Typical Tolerances for Green-Sand and Oil-Sand Castings* (*Continued*)

Location	Angle	Tolerance
Draft angles		
Outside	1–3°	±½°
Cope (top)	3–7°	±½°
Drag (bottom)	2–6°	±½°
Oil-sand cores	2–5°	±½°
Shell-sand cores	½–3°	±¼°
	Dimension	
Wall thickness	3 mm (⅛ in)	±15%
	4–14 mm (3/32–9/16 in)	12½%
	15–25 mm (19/32–1 in)	±10%

*This table is primarily for gray iron but is applicable to other metals. Aluminum castings should be capable of holding 33% narrower ranges, while steel and copper alloys usually need 50% wider ranges. Molds made by cold-cure processes can be expected to hold tighter tolerances, up to 50% of the tolerances on one side of the parting line. However, they require more tolerance across the parting line.

†Over 900 mm, add 0.2%; over 36 in, add 0.002 in/in.

‡Add ±0.4 mm (±0.015 in) to these values if the dimension is across the core-box parting line; applicable to molds by these processes also.

CHAPTER 5.2
OTHER CASTINGS

B. W. Niebel

Professor Emeritus of Industrial Engineering
The Pennsylvania State University
University Park, Pennsylvania

PERMANENT-MOLD CASTINGS

The Process

Permanent-mold casting makes use of reusable metal molds that are filled with metal by gravity as in sand-mold casting. Cores of metal or sand may be used. Molds are generally made of cast iron and frequently have a thick [up to 0.75-mm (0.030-in)] coating of sodium silicate and clay or other insulating materials over the cast-iron surface.

Permanent molds are preheated to 150 to 200°C before pouring and are given a graphite dusting every three or four shots. Thermal balance is very important, and auxiliary water cooling or radiation pins are used to cool heavy sections. Proper venting of the cavity is very important in order to avoid misruns. The process is illustrated by Fig. 5.2.1.

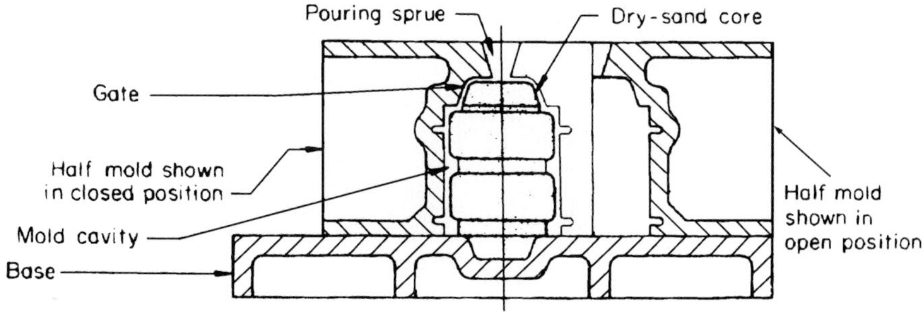

FIGURE 5.2.1 Permanent-mold casting. (*Courtesy General Motors Engineering Standards.*)

Low-Pressure Permanent-Mold Casting

In this type of casting, the liquid metal (usually aluminum) is forced by low air pressure from a silicon carbide crucible into the die cavity. If a proper heat balance is maintained, risers are eliminated and there is a high casting yield.

Typical Characteristics and Applications

The permanent-mold casting process is especially applicable for producing small- and medium-sized components of aluminum-base, magnesium-base, and bronze alloys. Seldom do permanent-mold castings weigh 100 kg (220 lb) or more.

Metal-mold castings have some distinct advantages over typical sand-mold castings. These include closer dimensional tolerances, better surface finish, greater strength, and more economical production in larger quantities. Disadvantages are the possibility of hot tearing, which results from the inability of the metallic mold to accommodate the contraction forces of the solidifying metal, and difficulty in removing the casting from the mold because the mold cannot be broken up. For these and other reasons, permanent-mold castings are apt to be relatively simple in shape, less intricate than many die castings. The ceramic-mold coating may cause a somewhat poorer casting surface finish and wider tolerance than are found in a typical aluminum or zinc-base die casting. Permanent-mold castings are superior to die castings from the standpoint of strength, density, pressure tightness, and mold cost. Improved density is the reason that automotive pistons are cast in permanent molds rather than die-cast.

Typical permanent-mold castings include gears, splines, hydraulic cylinders, motor parts, bushings, air-cooled cylinder heads, wheels, gear housings, washing-machine agitators, pipe fittings, and the like. Figure 5.2.2 illustrates a fuel-injector body produced by permanent-mold casting.

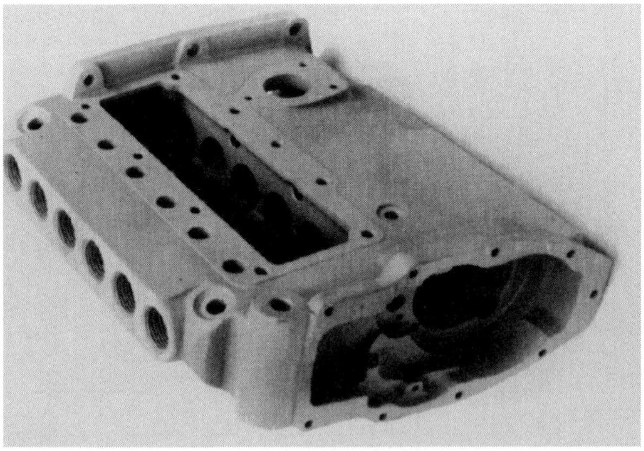

FIGURE 5.2.2 Fuel-injector body produced by the permanent-mold casting process. (*Courtesy Oberdorfer Foundries, Inc.*)

Economic Production Quantities

Both low-pressure permanent molding and permanent molding represent competitive methods of manufacture for quantities up to approximately 40,000 pieces. Molds for permanent-mold castings are more expensive than patterns for sand-mold castings but less costly than die-casting dies. The labor content of a typical permanent-mold casting, on the other hand, is less than that of an equivalent sand-mold casting but higher than that of an equivalent die casting.

Figure 5.2.3 illustrates the relative cost of various casting methods for various quantities of one particular part. Note that the break-even point for permanent-mold castings when compared with die castings is approximately 20,000 pieces. The break-even point of low-pressure permanent-mold castings when compared with die castings is larger.

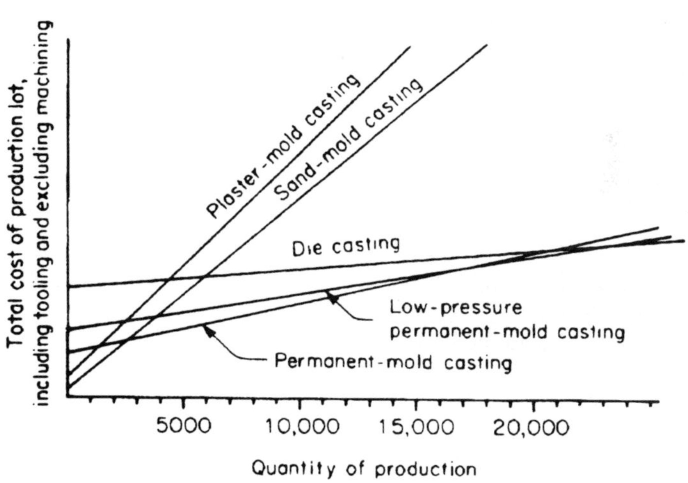

FIGURE 5.2.3 Relative cost of different casting process for one particular aluminum alloy part at various production levels.

Quantities in the range of 2500 pieces or more allow the process to compete with sand casting. High rates of production are possible when multiple-cavity dies are used. Die life is relatively long, being up to 50,000 parts per cavity when casting small gray-iron parts if protected with ceramics and considerably more in connection with the casting of nonferrous alloys. Quantities of 100,000 parts per cavity are not unusual when casting magnesium.

Suitable Materials

Permanent-mold casting can be used to produce components of aluminum, magnesium, copper-base alloys, and gray iron. Its widest application has been in connection with aluminum alloys. See Table 5.2.1 for a summary of suitable aluminum alloys.

In magnesium, alloys AZ91E (M11918), AZ63A (M11630), and AZ92A (M11920)

TABLE 5.2.1 Common Aluminum Alloys for Permanent-Mold Casting

Aluminum Association designation	UNS designation	Fluidity	Shrinkage tendency	Resistance to hot cracking	Pressure tightness	Machinability	Corrosion resistance	Weldability
359.0	A03590	A	A	A	A	C	B	A
356.0	A03560	A	A	A	A	C	B	B
357.0	A03570	A	A	A	A	C	B	B
443.0	A04430	B	A	A	A	E	B	A
332.0	A03320	A	B	A	B	D	C	B
213.0	A02130	B	C	C	C	B	D	B
771.0	A07710	C	C	D	D	A	B	—
201.0	A02010	C	D	D	C	A	D	B
711.0	A07110	E	D	E	D	A	C	C
240.0	A02400	C	D	D	D	C	D	D

Source: Data from the Aluminum Association.

are commonly used. For copper-base alloys, aluminum bronze 89-1-10 (C95300) is a good choice.

Design Recommendations

Castings to be produced by permanent-mold methods should be simple in design with fairly uniform wall sections and without undercuts and complicated coring. If a specific design requires undercuts or relatively complicated coring, semipermanent molds should be considered. Here sand cores are used with the metal mold.

Holes. Holes are best kept in the direction of parting of the mold halves so that separate core pieces are not required. (See Fig. 5.2.4.) Note also that the minimum diameter of cored holes is 6 to 9 mm (0.25 to 0.35 in), depending on the hole depth. The greater the hole depth, the greater the minimum diameter of the core. The hole depth should not be greater than six diameters; otherwise, cores will not be sufficiently rigid.

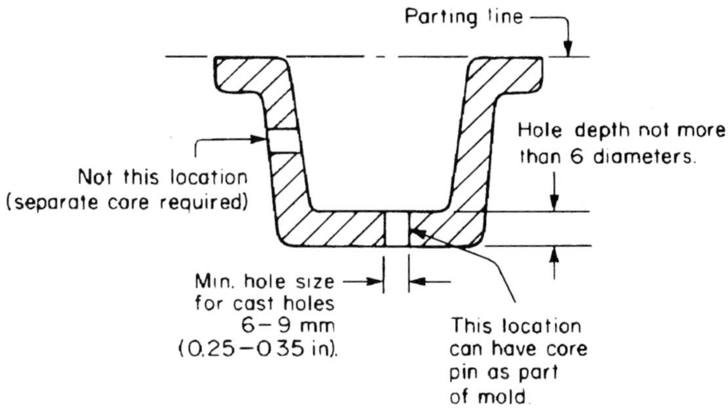

FIGURE 5.2.4 Rules for cored holes.

Draft. Draft is required on the walls of permanent-mold castings to permit easy withdrawal of the casting from the mold. More draft is required on inside sections, since the metal shrinks away from the mold surface but tightens around cores during solidification. The recommended minimum draft for exterior surfaces is 1 to 3° and for interior surfaces is 2 to 5° with greater draft angles being required for lower walls. (See Table 5.2.2.)

Fillets and Radii. Permanent-mold castings should be designed with generous fillets and radii to facilitate metal flow, promote more uniform cooling, and prevent stress concentrations. Fillets should have a radius at least equal to one wall thickness, while outer radii should be 3 times wall thickness. (See Fig. 5.2.5.)

Wall Thickness. The minimum wall thickness of permanent-mold castings is approximately 3 mm ($\frac{1}{8}$ in), and this stricture applies to small castings. Table 5.2.2 lists average minimum values for a range of casting sizes.

TABLE 5.2.2 Recommended Dimensions for Permanent-Mold Castings

Minimum wall thickness			
Casting size	Under 75 mm (3 in)	75–150 mm (3–6 in)	Over 150 mm (6 in)
Minimum wall thickness	3 mm (1/8 in)	4 mm (5/32 in)	5 mm (3/16 in)

Corner radii
Inside radius = t (t = average wall thickness)
Outside radius = $3t$ (blend tangent to walls)

Draft angles					
	Outside surfaces	Inside surfaces	Short cores	Average cores	Recesses
Minimum angle	1°	2°	2°	2°	2°
Preferable angle	3°	5°	3°	3°	5°

Machining allowances			
Casting size	Under 250 mm (10 in)	Over 250 mm (10 in)	Sand-core surfaces
Minimum allowance	0.8 mm (1/32 in)	1.2 mm (3/64 in)	1.5 mm (1/16 in)
Preferred allowance	1.2 mm (3/64 in)	1.5 mm (1/16 in)	2.0 mm (5/64 in)

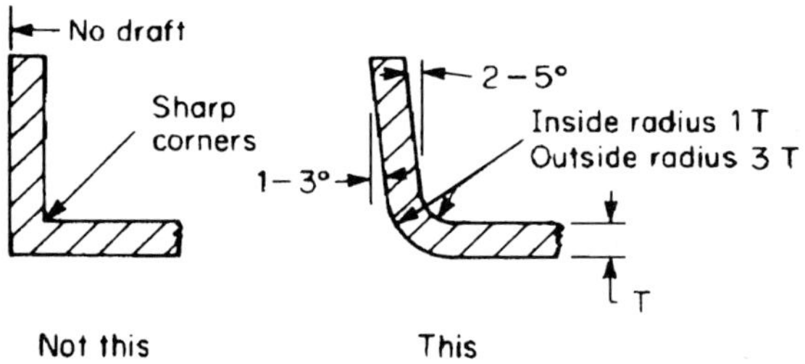

FIGURE 5.2.5 Avoid sharp corners in permanent-mold castings.

Ribs. Ribs for permanent-mold castings should follow the guidelines outlined in Chap. 5.1 for sand-mold castings.

Inserts. Inserts of metals other than the casting metal can be incorporated readily in permanent-mold castings. Normally some locking features—grooves, knurls, slots, holes, or lugs—are required to ensure that the insert is secured in the casting. Full wall thickness should be provided around the inserts to avoid overstressing the casting.

Markings. Lettering and other identifying markings can be cast into permanent-mold castings with no additional cost except that required in constructing the mold. Design recommendations for markings in die castings (Chap. 5.4) apply to permanent-mold castings as well.

Parting Line. As in the case of sand-mold and other castings, a flat parting surface between mold halves is preferable. The casting should be designed to permit this and to have a convenient parting-plane location.

Variations in Wall Thickness. These cause casting problems and workpiece inaccuracies because of uneven cooling. If section changes are unavoidable, however, the transition from a heavy to a thin section should be as gradual as possible.

Machining Allowances. Although permanent-mold castings are more accurate than sand-mold castings, there are still many instances in which they must be machined before castings are ready for functional use. The amount of material added to permanent-mold castings for machining is less than that required for sand-mold castings. Normal allowances, summarized in Table 5.2.2, range from 0.8 mm ($^1/_{32}$ in) to 2 mm ($^5/_{64}$ in).

Recommended Tolerances

Figure 5.2.6 presents typical dimensional tolerances for average permanent-mold castings.

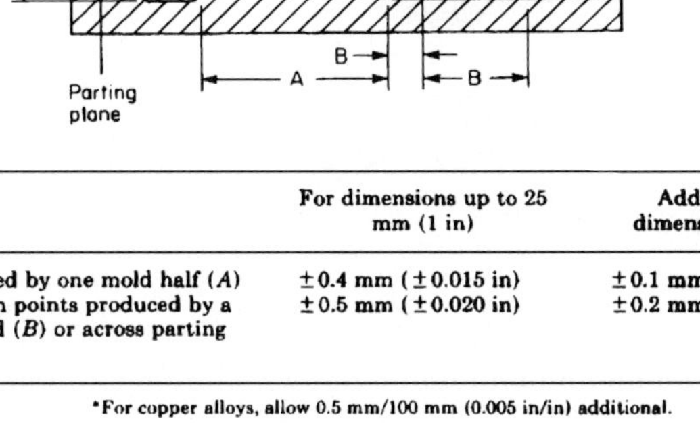

	For dimensions up to 25 mm (1 in)	Additional tolerances for dimensions over 25 mm (1 in)
Dimensions produced by one mold half (*A*)	±0.4 mm (±0.015 in)	±0.1 mm/100 mm (±0.001 in/in)*
Dimensions between points produced by a core and the mold (*B*) or across parting plane (*C*)	±0.5 mm (±0.020 in)	±0.2 mm/100 mm (±0.002 in/in)*

*For copper alloys, allow 0.5 mm/100 mm (0.005 in/in) additional.

Surface finish		
Low-pressure process	0.5–3.2 μm (20–125 μin)	
Conventional process	3.8–13 μm (150–500 μin)	

FIGURE 5.2.6 Recommended tolerances for permanent-mold castings.

CENTRIFUGAL CASTINGS

The Process

Centrifugal casting is a specialized casting process that provides a cost-effective method for producing hollow cylindrical parts and circular plates as well as intricate parts. The most common centrifugal process, true centrifugal casting, is used for pipe liners, steel tubing, and other hollow, symmetrical objects. It involves the use of centrifugal force to hold the metal against the outer walls of the rotating tubular mold while the volume of metal poured determines the wall thickness and internal diameter. No core is used, and the inner surface tends to collect the lighter dross and impurities.

Molds are divided into two classes, permanent and lined. In the lined mold, a flask with a baked or green-sand lining is placed horizontally on a casting machine, where it is rotated by a system of rollers actuated by an electric motor. Molten metal is fed into one end of the mold cavity and is carried to the walls of the cavity by centrifugal force. Rotative speeds sufficient to produce a force of up to 150 times the force of gravity are feasible. High forces are necessary for casting exceptionally thick walls.

Permanent molds are made of steel, iron, or graphite. Graphite molds are used principally in the casting of titanium. Usually, a thin refractory wash is sprayed on the inside surface to increase mold life and permit faster solidification.

Horizontal molds are generally used to cast long cylinders and pipes, whereas vertically spinning molds are generally used in the production of short castings of variable diameters and different shapes.

In one process variation, semicentrifugal casting, the process is used for forming

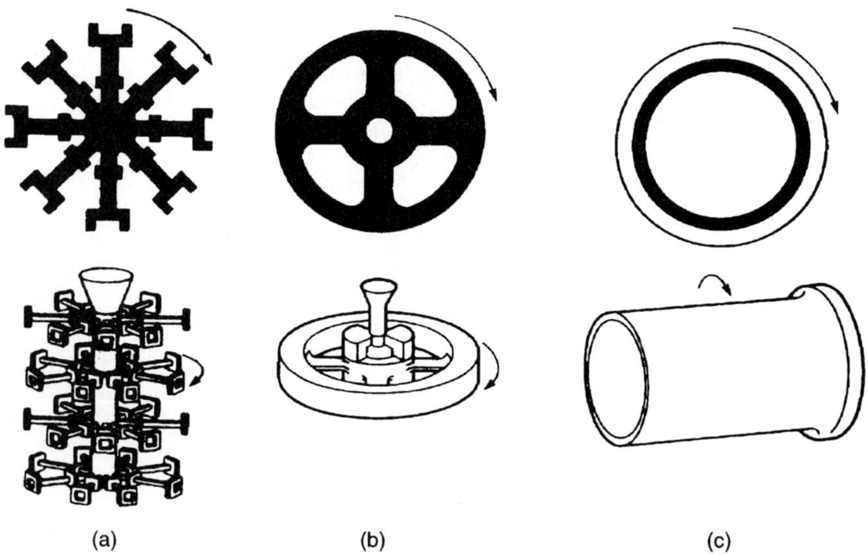

FIGURE 5.2.7 Three types of centrifugal castings are: (*a*) centrifuging; (*b*) semicentrifugal; (*c*) true centrifugal. (*From David C. Ekey and Wesley P. Winter,* Introduction to Foundry Technology, *McGraw-Hill, New York, 1958.*)

symmetrical shapes such as wheels, nozzles, and similar parts about the rotative vertical axis. The molten metal is introduced through a gate, which is placed on the axis, and flows outward to the extremities of the mold cavity.

In centrifuging, several molds are located radially about a central riser or sprue, and the entire mold is rotated, with the central sprue acting as the axis of rotation. Centrifugal force provides a means for obtaining greater pouring pressures at every point within the mold cavity. This type of centrifugal casting is best used for small, intricate parts where feeding problems might otherwise be encountered.

These three varieties of centrifugal casting are illustrated in Fig. 5.2.7.

Typical Applications

Even though cylinders as small as 13 mm ($\frac{1}{2}$ in) in diameter can be centrifugally cast, from an economic standpoint, 125 mm (5 in) is considered about the minimum diameter. Cylinders up to 3 m (10 ft) in outside diameter with lengths up to 16 m (52 ft) have been cast successfully by centrifugal methods. Typical centrifugal castings include large rolls, gas and water pipe, engine-cylinder liners, wheels, nozzles, bushings, sheaves, gears, ship-shaft liners, bearing rings, and flanged shapes, as well as more intricately shaped components.

Economic Production Quantities

Centrifugal-casting methods have application for very small quantities (down to one), for which sand is used as the mold material, up to large production quantities. Graphite molds are used for small or medium-sized runs. Usually, at least 50 pieces are required to justify a graphite mold. With medium to large runs, it is usually most economical to use a metal mold. Here 400 pieces is representative of the typical break-even point.

Design Considerations

As compared with permanent-mold and sand casting, for which a liberal amount of gating and risering is required, the filling of the mold in all types of centrifugal casting is enhanced by the centrifugal force in back of the metal. There are no gates in true centrifugal castings, and fewer, relatively shorter nubs are needed for centrifugal castings than for typical sand-mold castings. Although the need for radii at adjoining flanges and other design features is not as pronounced as in sand castings, a radius should always be specified so as to avoid stress concentrations.

True centrifugal castings may be cast with a range of wall thickness of 6 to 125 mm ($\frac{1}{4}$ to 5 in). Usually a wall thickness of at least 9 mm (0.35 in) is preferred. The minimum size of cored holes or cored control bores is 25 mm (1 in). If smaller holes are desired, they should be drilled after casting.

Sand molds for true centrifugal castings do not require draft, but draft should be applied for centrifuged parts in accordance with usual sand-casting practice. When using cast-iron molds, draft is applied. As the diameter decreases, the amount of draft needed increases.

TABLE 5.2.3 Typical Tolerances for Centrifugal Castings

	Tolerances	
Metal molds	Normal tolerance, ±	Minimum tolerance, ±
Dimensions to 13 mm (½ in)	0.5 mm (0.020 in)	0.13 mm (0.005 in)
Dimensions over 13 mm to 130 mm =(½ in to 5 in)	0.5 mm (0.020 in)	0.25 mm (0.010 in)
Dimensions over 130 mm (5 in)	0.5 mm (0.020 in)	0.38 mm (0.015 in)
Sand molds	Normally ±0.8 mm (0.030 in) for dimensions to 80 mm (3 in); for values for larger dimensions, see Table 5.1.6.	
Internal diameters	(Determined by centrifugal force) ±1.5 mm (1/16 in)	
Surface finish		
Metal molds	2.5–13 μm (100–500 μin)	
Sand-faced molds	6.3–25 μm (250–1000 μin)	

Suitable Materials

In general, any alloys that can be cast in sand molds can be centrifugally cast. Those metals which are cast regularly on a production basis include carbon steels, alloy steels, brasses, bronzes, aluminum alloys, nickel alloys, cast iron, and copper.

Those alloys having short freezing ranges are particularly desirable for centrifugal casting. Usually, high-phosphorus alloys should be avoided because this element results in a "wetting" characteristic and subsequent removal from the mold is difficult. Also to be avoided are high-lead alloys, which tend to separate.

Recommended Tolerances

Castings produced by any of the three centrifugal methods have tolerances typical of those experienced in sand casting. Table 5.2.3 lists the tolerances that should be readily maintained by the process.

Centrifugal castings made in metal molds will have surface finishes ranging from 4.4 to 6.3 μm (175 to 250 μin). The surface roughness of parts formed against sand will typically range from 12 to 20 μm (300 to 500 μin).

PLASTER-MOLD CASTINGS

Plaster-mold casting is very similar to sand-mold casting except that the mold and cores are made of plaster (or a combination of plaster and sand) instead of packed sand. The mold is usually made in a wood or metal frame containing the pattern. After the plaster sets, the frame and pattern are removed, and the mold halves and cores, if any, are assembled and are normally baked to remove moisture and improve permeability. A new mold is required for each casting as in sand-mold casting. In practice, the mold is poured hot, and a vacuum system continually removes the water vapors produced from the loss of water of hydration. Figure 5.2.8 shows a plaster mold after pouring.

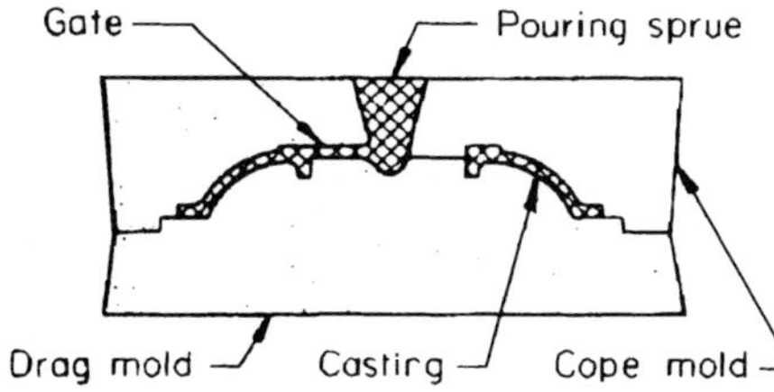

FIGURE 5.2.8 Plaster mold after pouring. (*Courtesy General Motors Engineering Standards.*)

Typical Characteristics

The process is useful for producing small nonferrous parts. It permits the production of smooth surfaces, fine detail, thin sections, and dimensionally accurate geometries in the as-cast condition.

Because of their excellent surface finish and good dimensional accuracy, components frequently can be produced to requirements without secondary machining and finishing operations. Plaster molds permit accurate control of shrinkage, distortion or warpage being negligible.

Generally, the surfaces of plaster-mold castings have a satin texture. However, the special surface textures of the pattern can be reproduced in the mold. Plaster-mold castings can be plated easily without special preparation. Surfaces are easily polished and buffed. Castings weighing more than 10 kg are seldom made, although some as large as 100 kg have been cast. The typical casting weighs about 125 to 250 g. Castings as small as 1 g have been made. Typical parts produced include valves, fittings, clutches, sprockets, compressor wheels, gears, impellers, lock components, pistons, cylinder heads, etc.

Economic Production Quantities

The process is relatively high in cost because of the long time required for the plaster to set and the fact that the production rate is low. Also, molds must be poured hot and over a vacuum to remove water vapors. Cooling time is long because plaster is a good insulator. Thus, the cycle time is long.

Either small or large quantities can be run economically by placing the pattern in the line when the proper metal is being poured. One pattern typically can produce 150 to 250 pieces per week. Multiple patterns or plates will increase production accordingly.

The cost of the patterns and core boxes employed in the plaster-mold process is approximately 50 percent higher than the corresponding equipment used in sand casting.

General Design Considerations

Principal design considerations for plaster-mold castings include the following:

1. Keep overall dimensions as small as possible. The process is most suitable for small parts. Mold and materials economies result when castings are small.

2. Keep sections as thin as possible (see limits below), with due regard for strength, stiffness, and allowance for machining.
3. Endeavor to keep sections uniform unless differences are essential to meet service requirements. If sections cannot be uniform, make variations in thickness as gradual as conditions permit.
4. Use cores when savings in metal or machining can be effected or to promote sounder castings. However, avoid slender or fragile cores and those not readily supported by the mold.
5. Avoid undercuts that will prevent removal of the pattern from the mold.
6. Design castings to minimize secondary operations.
7. Employ fillets and rounded corners to strengthen the casting and make it easier to remove the pattern from the mold.

Suitable Materials

Aluminum alloys, zinc alloys, yellow brass, aluminum bronze, manganese bronze, silicon-aluminum bronze, nickel brass, and other bronze alloys with less than 1.5 percent lead all can be cast in plaster molds. The material of the mold can stand a maximum pouring temperature of about 1200°C (2200°F). Consequently, ferrous metals and others with melting points greater than 1100°C (2000°F) cannot be cast. Magnesium is not suitable for plaster casting because of potential chemical reactions between the molten magnesium and the mold or any water remaining in it.

Aluminum alloys form the bulk of plaster-cast parts. Alloys 355.0 (A03550), 356.0 (A03560), 357.0 (A03570), and 443.0 (A04430) are suitable to the process.

Detailed Design Recommendations

Wall Thickness. Although plaster-mold casting can produce thin and sound walls, the following minimum wall thicknesses for given areas should be observed when designing castings:

Walls with projecting areas up to 650 mm^2 (1 in^2): 1-mm (0.040-in) minimum thickness

Walls with projected areas above 650 mm^2 to 1950 mm^2 (1 to 3 in^2): 1.5-mm (0.060-in) minimum thickness

Walls with projected areas above 1950 mm^2 to 9750 mm^2 (3 to 15 in^2): 2.4-mm (0.090-in) minimum thickness

Inserts. Inserts can be used with plaster-mold castings. Due regard should be given to potential corrosion problems, and steps must be taken to locate the insert properly in the mold. Fastening methods as described above under "Permanent-Mold Castings" also apply to plaster-mold castings.

Markings. These can easily be incorporated in plaster-mold castings and are clearly reproduced. Lettering should be depressed in the casting rather than raised for minimum cost and, as in other types of castings, should be on a surface parallel to the mold-parting plane for easy pattern removal.

Draft. For outside surfaces draft should be ½° or more. For inside surfaces, it should be at least 1 to 3°.

Holes. Holes can be cored easily with good accuracy and finish. They should be perpendicular to the mold-parting plane to avoid the need for extra core pieces. Holes less than 13 mm (½ in) in diameter are generally more economically drilled. Drill spots are accurate and can eliminate jigs.

Machining Allowance. Stock allowances for machining should be 0.8 mm (0.030 in). On small parts the machining allowance may be less than 0.8 mm for holes and broaching.

Figure 5.2.9 illustrates the preceding recommendations.

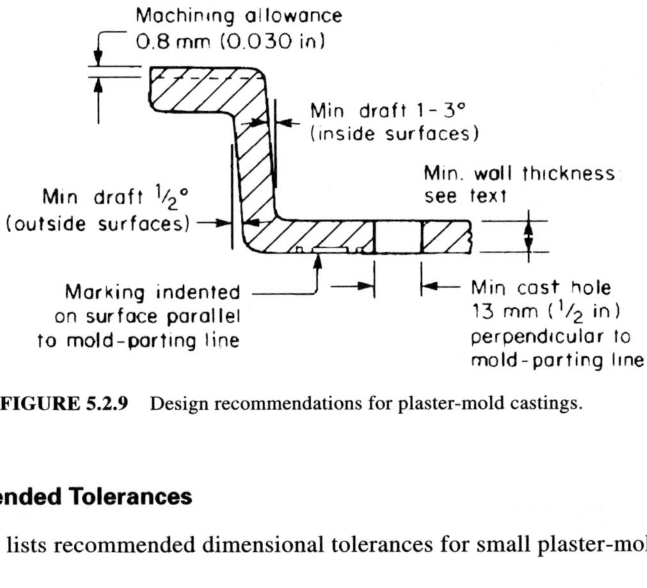

FIGURE 5.2.9 Design recommendations for plaster-mold castings.

Recommended Tolerances

Table 5.2.4 lists recommended dimensional tolerances for small plaster-mold castings.

TABLE 5.2.4 Recommended Tolerances for Plaster-Mold Castings

	Tolerance
Dimensions wholly in one-half of the mold	0.05 mm/cm (0.005 in/in)
Dimensions across mold-parting line	0.10 mm/cm (0.010 in/in)
Dimensions subject to horizontal shift between cope and drag of mold	0.10 mm/cm (0.010 in/in)
Flatness	0.05 mm/cm (0.005 in/in) in any direction
Surface finish	0.8–1.3 μm (30–50 μin)

CERAMIC-MOLD CASTINGS

The Process

Ceramic-mold casting is frequently referred to as *cope-and-drag investment casting.* The major distinction between ceramic-mold casting and investment casting is that the former relies on precision-machined metal patterns rather than the expendable wax patterns used in conventional investment casting. Ceramic-mold casting is similar to plaster-mold casting except that the mold materials are more refractory, require higher preheat, and are suitable for most castable alloys, particularly ferrous alloys. The refractory slurry consists of fine-grained zircon and calcined high-alumina mullites or, in some cases, fused silica.

In ceramic-mold casting, a thick slurry of the mold material is poured over the reusable split and gated metal pattern, which is usually mounted on a match plate. A flask on the match plate contains the slurry, which gels before setting completely. The mold is removed during the time when the gel is firm to prevent bonding to the pattern.

The Shaw process and the Unicast process are two proprietary variations of the ceramic-mold casting method. Before pouring, the molds are usually preheated in a furnace to reduce the temperature difference between the mold and the molten metal and to maximize the permeability available through the microcrazed mold structure.

Typical Applications

Ceramic-mold casting has particular application in the casting of ferrous and other high-melting-temperature alloys in which the geometry is complex. Parts as heavy as 700 kg (1500 lb) are feasible. Even on such large castings the process has excellent accuracy and reproducibility. Fine detail and smooth surfaces are feasible.

Typical products cast by this process include stainless steel and bronze pump parts such as impellers and impeller cores, molds for plastic molded components, core boxes of cast iron, beryllium copper molds for the blow molding of thermoplastic resins, tool-steel molds for the molding of rubber components, tool-steel milling cutters, and press-working dies and die-casting dies of H-13 steel.

Economic Production Quantities

For complex parts, this process can be advantageous when the production quantity is as small as only one part. Savings are found in the machining that is eliminated. The process may be economic at higher production levels for parts that are too large to be advantageously investment cast but which require the desirable characteristics of investment castings: complex shape, dimensional accuracy, the use of alloys with a high melting temperature.

General Design Considerations

In this process, mechanical properties suffer the loss of the chill effect because the mold materials are such good insulators that a coarse-grained structure results. The designer can compensate for such a reduction in strength by making critically stressed sections somewhat heavier, or the metallurgist can add grain refiners to the molten metal before pouring. Areas subjected to major stress can be chilled by incorporating suitable chills in the mold.

Design considerations and recommendations that apply to plaster casting also apply to ceramic-mold casting.

Suitable Materials

Most castable materials can be ceramic-mold cast. Typical materials include stainless steel, bronze, beryllium copper, gray iron, and tool steel.

Recommended Tolerances

An as-cast surface finish of 3 μm (120 μin) or better can readily be achieved. Dimensional tolerances are ±0.1 mm (0.004 in) for the first 25 mm (1 in), with incremental tolerances of ±0.025 mm/cm (0.0025 in/in) for larger dimensions. Across the parting line, an additional tolerance of ±0.3 mm (0.012 in) should be provided.

CHAPTER 5.3
INVESTMENT CASTINGS

Robert J. Spinosa
Cooper Industries
Monroe, North Carolina

THE PROCESS

Investment casting (sometimes called the *lost-wax process*) utilizes a one-piece ceramic mold. The mold is made by surrounding a wax or plastic replica of the part with ceramic material. After the ceramic material solidifies, the wax replica is melted out, and metal is poured into the resulting cavity.

There are two basic mold-making variations: (1) the *monolithic method,* which utilizes a solid mold contained by a stainless steel or Inconel flask, and (2) the *shell method,* which utilizes a thin-walled (about 6-mm- or ¼-in-thick) mold and no flask. (See Fig. 5.3.1.)

The complete sequence for making investment castings with the shell method is as follows:

1. Wax or plastic patterns are injection-molded. When wax is used, the molds can be inexpensive, being made from a low-temperature alloy sprayed or cast around a master-part pattern. The pattern is made with an allowance for shrinkage.
2. Patterns with gates and runners are assembled to form a "tree."
3. The tree is dipped into a slurry of fine ceramic material.
4. Refractory grain is sifted onto the coated pattern with a "rainfall coater" or by fluidized bed.
5. The coat of slurry and grain is dried with air. Typically, this requires 2 h in an environment with considerable air movement and relative humidity of 50 percent or less.
6. Steps 3, 4, and 5 are repeated with successively coarser refractory material to a thickness of 5 to 8 mm ($^3/_{16}$ to $^1/_3$ in) depending on the size of the casting tree. A final dip in the slurry material provides a sealing coating.
7. The coated tree is thoroughly dried. The pattern, gate, and sprue material is melted out of the mold by inverting (sprue opening down) and heating it in an autoclave or oven for about 5 to 10 min.

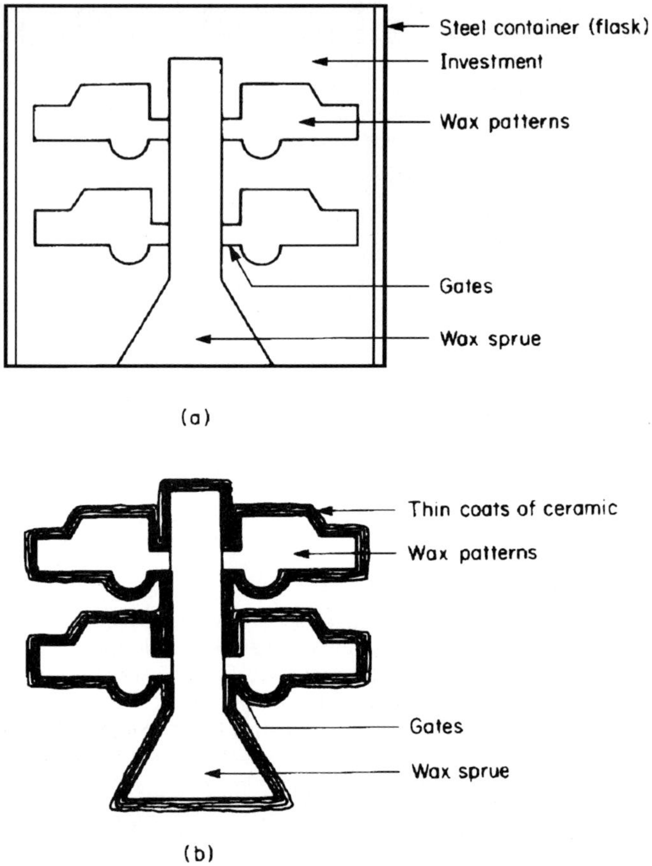

FIGURE 5.3.1 Molds for investment casting: (*a*) monolithic or solid mold before wax is removed; (*b*) shell mold before wax is removed.

8. The mold is burned out to remove all traces of wax or plastic and to cure the investment. The burnout temperature is typically 980°C (1800°F) for a minimum of 2 h.
9. Hot molds are filled with molten metal by gravity, pressure, vacuum-assist, or centrifugal force.
10. Shake out: After the mold cools and the cast metal solidifies, the mold material is removed from the tree, typically with an air-powered hammer or vibratory machine.
11. The tree is descaled. This is achieved by immersing the tree for 10 to 15 min in a 600°C (1100°F) caustic salt bath followed by an immediate cold-water dip and then neutralizing and cleaning rinses.
12. Parts are removed from the tree by bandsaw or abrasive wheel.
13. Gates are removed from the parts by means of a grinding wheel, abrasive belt, milling machine, or lathe.

Current practice often involves robotic handling of the shells and automation of the various steps. This provides advantages of cost, consistency, and uniformity of the coatings. However, where applicable, manual methods are still used.

The process for monolithic (solid) molds is very similar, the difference being that the mold is first coated and then inserted in a flask, which is then filled with investment material.

The shell process is better adapted to larger parts and has a shorter cycle time. It also has the advantage of less tendency of surface decarburization of the cast part. On the other hand, mold cracking is more likely, especially with plastic patterns. In addition, detail definition at the part surface may be poorer because of more rapid cooling of the melted metal.

TYPICAL CHARACTERISTICS AND APPLICATIONS

Historically, investment castings most likely were specified when the following part characteristics were involved: intricate shape, close tolerances, small size, and high-strength alloys. Costs of the investment material and other items inhibited its use for larger castings. However, advances in the process involving greater use of the shell method and with robotic assist have reduced the cost differential with other casting processes, especially for larger sizes. Investment castings currently range in size from 1 g (0.035 oz) to upwards of 90 kg (200 lb).

Investment castings are most likely to be used when the shape of the part involves contoured surfaces, undercuts, and other intricacies that make machining difficult or unfeasible. They are used for mechanical components in such products as sewing machines, business machines, firearms, and surgical and dental devices. Turbine blades, valve bodies, ratchets, cams, pawls, gears, hose fittings, cranks, levers, vanes, connectors, support rings, impellers, manifolds, hand tools, golf club heads, and radar waveguides are typical parts. Figure 5.3.2 shows a collection of typical investment-cast parts.

ECONOMIC PRODUCTION QUANTITIES

Investment casting is best suited to moderately low and medium production levels. In many cases in which machining would be extensive, however, it is applied successfully to high-production applications as well.

The process also can be used for prototype quantities by having the wax patterns made by stereolithography or another rapid prototyping process (with the necessary shrinkage allowances). Prototype parts can then be cast from these patterns using the normal investment casting process.

Tooling costs for wax patterns can be low, especially if the mold is made from a master pattern by using the spray or casting methods. The lowest-cost molds are made by spraying low-temperature alloys over the pattern. Such molds are used for wax patterns only and have a life of 800 to 1200 pieces. More nearly permanent soft-metal molds made by casting will have a life of up to 12,000 pieces.

For still higher production levels, mold cavities are machined from aluminum or steel. Aluminum machined molds (for wax patterns) are quite inexpensive. With steel

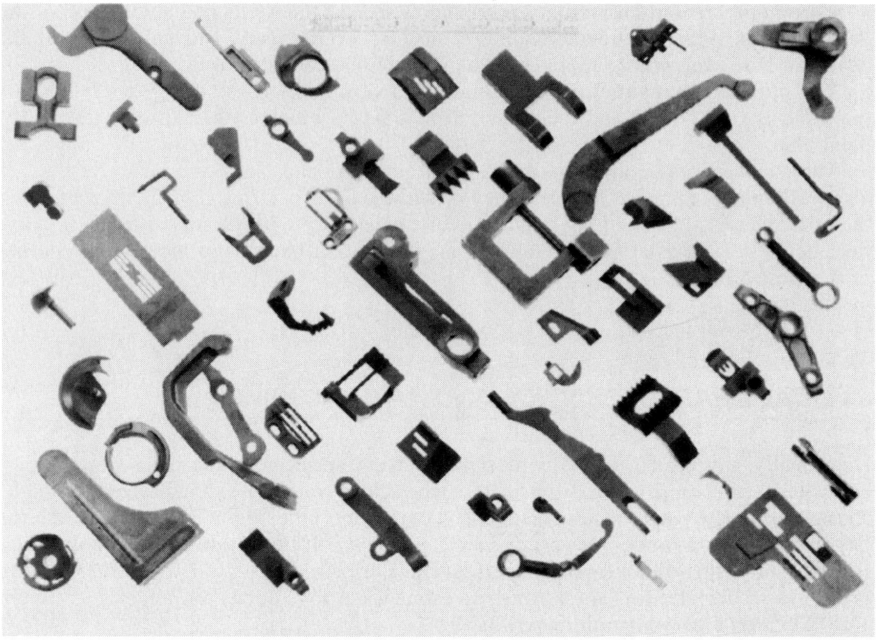

FIGURE 5.3.2 Collection of investment-cast parts. Note their complexity and size, as indicated by the 6-in ruler. (*Courtesy The Singer Company.*)

mold material and with multiple cavities, mold costs are the same as those for other injection-molded plastic parts.

Because of a high labor and indirect-materials content, the unit cost of investment castings tends to be higher than that of powder-metal, die-cast, or other parts made from alternative high-production methods. There are exceptions, when a part is particularly complex or of a configuration or material that is not suitable for other processes, but generally, economical quantities for investment castings do not extend to mass-production situations.

GENERAL DESIGN CONSIDERATIONS

Perhaps the greatest advantage of investment castings is the high degree of design freedom that they permit. Complex shapes that would be too costly to machine can be produced quickly and economically as investment castings. In many cases, what would otherwise be two or more separate parts can be designed as one integral casting, eliminating assembly operations. The process also allows designers to select from a wide range of alloys. Ideally, the time for deciding if a part is to be made by the investment-casting process is when it is on the drawing board. When designing for investment casting, designers should keep the toolmaker and investment caster in mind. They should remember that the pattern is injection-molded and therefore should observe good practices that apply to injection-molded parts. These include the use of a well-located, straight parting line, adequate draft, and avoidance of undercuts. Designers should use generous fillets and radii whenever possible. This not only makes a better-looking part but also produces a stronger one as well.

SUITABLE MATERIALS

A wide variety of metals, both ferrous and nonferrous, can be used to make investment castings. Any metal that is meltable in standard induction or gas furnaces can be used. Vacuum melting and pouring can be incorporated in the process if the metal used and its application require this step.

Materials that are not easily machinable are good candidates for some investment-casting applications because the process can provide dimensional precision and intricate shapes that would otherwise require machining. Table 5.3.1 lists a number of alloys with high-castability ratings.

DETAILED DESIGN RECOMMENDATIONS

Minimum Wall Thickness

The minimum thickness of casting walls is determined primarily by the fluidity of the metal to be cast. Another factor is the length of the section involved. If the section is long, a heavier wall may be required. Table 5.3.2 provides recommended minimum wall thicknesses for various investment-castable metals.

Flatness and Straightness

Deviations from flatness and straightness can be minimized if ribs and gussets are incorporated in the part. Because of pattern shrinkage, investment shrinkage, and metal shrinkage during solidification, there is always a tendency for an investment-cast part to "dish" (develop concave surfaces where flat surfaces are specified). This condition takes place in areas of thick cross section. As illustrated in Fig. 5.3.3, dishing can be minimized by designing parts with uniformly thin walls.

Radii

Although sharp corners are achievable, generous radii are preferred whenever possible. Ample fillets and radii facilitate die filling for the pattern and mold filling for the cast part and also tend to produce better-quality, more accurate parts. Although it is possible to cast sharper corners, a minimum fillet radius of 0.75 mm (0.030 in) should be specified, and even 1.5 to 3.0 mm (0.06 to 0.125 in) is preferable. (See Fig. 5.3.4.)

TABLE 5.3.1 Suitable Materials for Investment Casting

Material	Fluidity	Shrinkage	Resistance to hot tearing	Castability rating
Carbon steels				
1040 (G10400)	B	B	B	B+
1050 (G10500)	B	B	B	B+
1060 (G10600)	B	A	B	B+
Alloy steels				
2345 (G23450)	B	B	B	A−
4130 (G41300)	B	B	B	A−
4140 (G41400)	B	B	B	A−
4150 (G41500)	B	B	B	A−
4340 (G43400)	B	B	B	A−
4640 (G46400)	B	B	B	A−
6150 (G61500)	B	B	B	A−
8640 (G86400)	B	B	B	A−
8645 (G86450)	B	B	B	A−
Austenitic stainless steels				
302 (S30200)	A	A	A	A+
303 (S30300)	A	A	B	A
304 (S30400)	A	A	A	A+
316 (S31600)	A	A	A	A+
CF-8M (V92900)	A	A	A	A+
Ferritic stainless steels				
410 (S41000)	A	C	B	A
416 (S41600)	A	C	B	B+
430 (S43000)	A	C	B	A−
431 (S43100)	A	C	B	A−
Precipitation-hardening stainless steel				
17-4 PH (S17700)	A	C	B	B+
Nickel alloys				
Monel (QQ-N-288-A) (N04020)	A	B	B	B+
Inconel 600 (AMS 5665) (N06600)	A	B	B	B+
Cobalt alloys				
Cobalt 21 (R30021)	A	A	B	A−
Cobalt 31 (R30031)	A	A	B	A−
Aluminum alloys				
A 356 (A13560)	A	A	A	A+
C 355 (A33550)	A	A	A	A
Tool steels				
A-2 (T30102)	B	B	B	B+
H-13 (T20813)	B	B	B	B+
S-1 (T41901)	B	B	B	A−
S-2 (T41902)	B	B	B	A−
S-4 (T41904)	B	B	B	A−
S-5 (T41905)	B	B	B	A−
Copper alloys				
Gunmetal (C90500)	A	C	A	B+
Naval brass (C46400)	B	B	A	B+
Phosphor bronze (C94400)	B	B	A	B+
Red brass (C23000)	A	A	A	A−
Silicon brass (C87800)	A	A	A	A+
Beryllium copper 10C (C82000)	A	A	A	A
Beryllium copper 20C (C82500)	A	A	A	A+
Beryllium copper 30C (C82200)	A	A	A	A

Note: A = excellent; B = good; C = poor.

TABLE 5.3.2 Suggested Minimum Wall Thickness for Various Investment-Casting Metals

Metal	Minimum wall thickness, mm (in)
Ferrous metals	
Low-carbon steel	1.8 (0.070)
High-carbon steel	1.5 (0.060)
Low-alloy steel	1.5 (0.060)
Stainless steel, 300 Series	1.0 (0.040)
Stainless steel, 400 Series	1.5 (0.060)
Cobalt-base alloys	0.75 (0.030)
Nonferrous metals	
Aluminum	1.0 (0.040)
Beryllium copper	0.75 (0.030)
Brass	1.0 (0.040)
Bronze	1.5 (0.060)

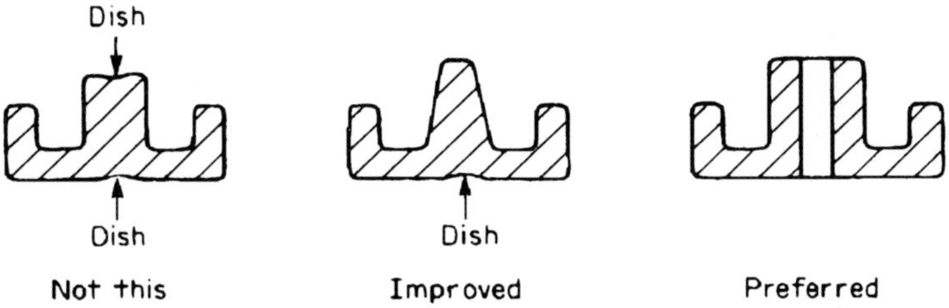

FIGURE 5.3.3 Keep walls uniformly thin to avoid "dishing."

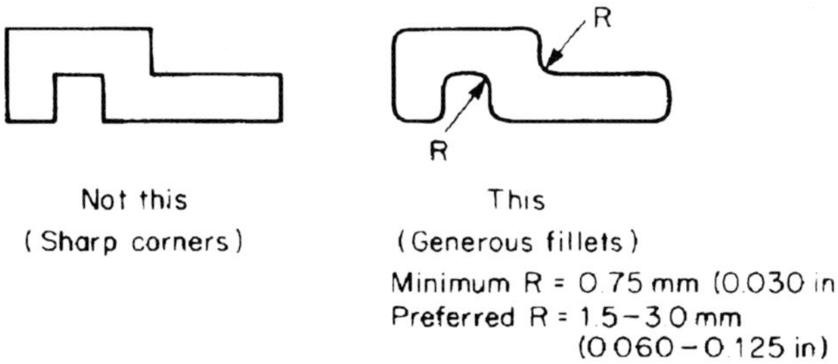

FIGURE 5.3.4 Use generous fillets and radii.

Curved Surfaces

Both convex and concave surfaces can be produced easily. However, concave surfaces can be cast to somewhat greater accuracy than convex surfaces because the cast metal, when solidifying and shrinking, hugs the mold instead of shrinking away from it.

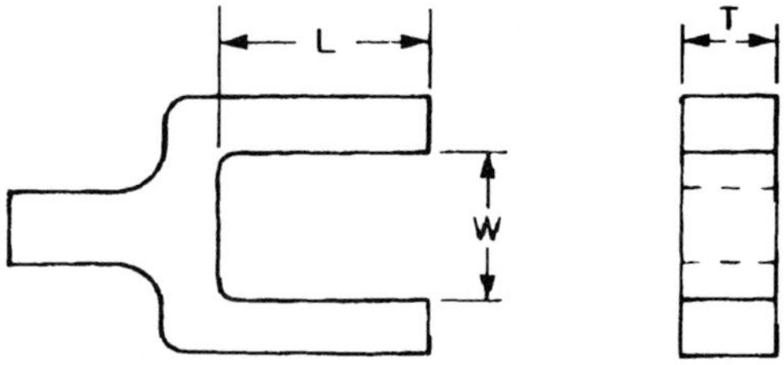

Thickness T, mm (in)	Minimum W, mm (in)
1.5 (1/16)	1.5 (1/16)
6 (1/4)	2.4 (3/32)
10 (3/8)	3 (1/8)
13 (1/2)	5 (3/16)
25 (1)	6 (1/4)

FIGURE 5.3.5 Forks and yokes have minimum practical slot widths.

Parallel Sections

Parallel sections can be classified into two basic geometric types: (1) forks or yokes and (2) clamps or pinch collars. They each offer different challenges to the caster. See Fig. 5.3.5, where T represents the thickness, L the length of the tine of the fork, and W the inside width of the fork. As the thickness T increases, the width of the opening W also must increase so that the mold ceramic will fill the slot.

For clamps or pinch collars, as illustrated in Fig. 5.3.6, the minimum slot width W for ferrous metals is 1.5 mm (0.060 in) and for nonferrous metals is 1.0 mm (0.040 in).

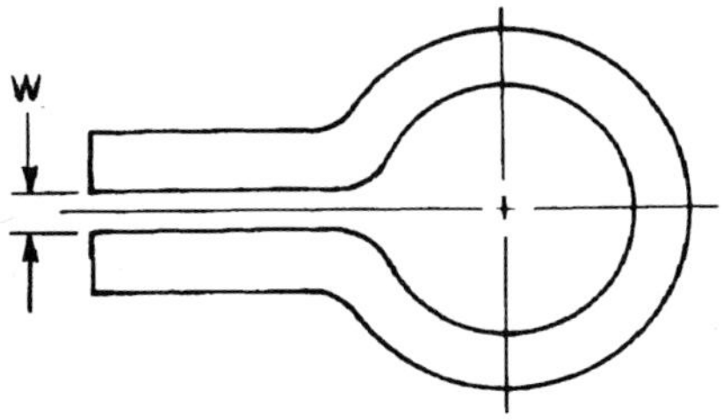

FIGURE 5.3.6 Minimum slot widths W for clamps and pinch collars are 1.5 mm (0.060 in) for ferrous metals and 1.0 mm (0.040 in) for nonferrous metals.

Keys and Keyways

When keys and keyways are required, the desirable ratio of width to depth for ease of casting is 1.0 or more (see Fig. 5.3.7). The minimum castable key width is 2.3 mm (0.090 in) for ferrous metals and 1.5 mm (0.060 in) for nonferrous alloys.

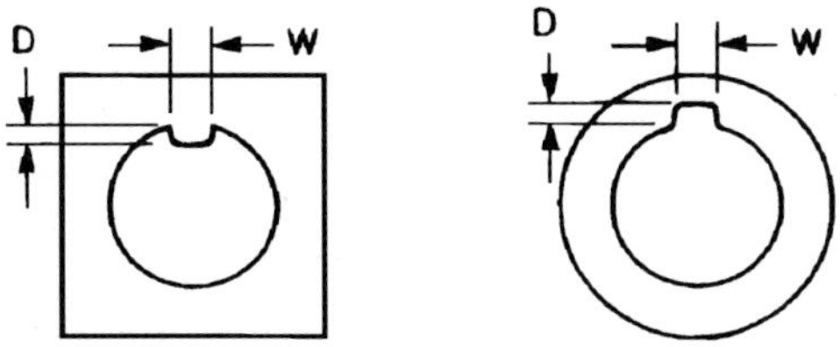

FIGURE 5.3.7 Keys and keyways should have a width-to-depth ratio of 1:1 or more. Keys should not be narrower than 2.3 mm (0.090 in) for ferrous metals and 1.5 mm (0.060 in) for nonferrous metals.

Holes

Holes can be incorporated into investment castings when required. However, it becomes impractical to cast when holes are less than 1.5 mm (0.060 in) in diameter for nonferrous alloys or less than 2.2 mm (0.087 in) for ferrous alloys.

Blind Holes

Blind holes usually are not recommended to be cast, since in many cases if the operator is not careful in precoating and investing the mold, air pockets that would cause globs of metal in the hole could be formed. However, if a blind hole is required, the length of depth should not exceed the diameter.

Through Holes

In most cases a through hole may be cast more economically than it can be produced by drilling after casting. Also, in many cases, by incorporating a hole in the casting, sinks or cavitation can be minimized, since the wall thickness is more uniform. The same tolerance applies for through holes as for blind holes. There is a definite ratio between the hole diameter D and the length of the hole L in through holes. This L/D ratio should not be greater than 4:1 in ferrous metals or 5:1 in copper and aluminum alloys.

Ceramic Cores

The use of ceramic cores makes intricate internal configurations of investment castings possible. This is particularly important for intricate blades or internal threads in materials that are not easily machinable. Once the metal has been cast, the cores are leached out. In designing pattern molds for a part to be made with a ceramic core, provision must be made in the mold to support the ceramic core when the wax is injected into the mold.

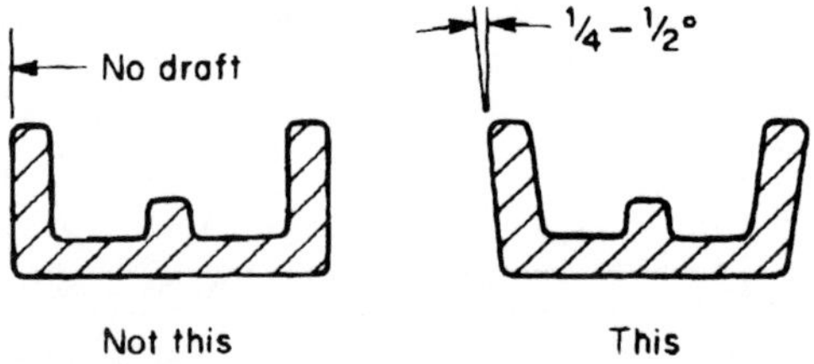

FIGURE 5.3.8 A small amount of draft is necessary for the removal of patterns from the molding die.

Draft

A small amount of draft is necessary to be able to remove wax patterns from the molding die; $\frac{1}{4}$ to $\frac{1}{8}$° is adequate. (See Fig. 5.3.8.)

Screw Threads

Screw threads, both internal and external, can be investment-cast. In most cases, however, it is more satisfactory to machine them than to cast them. The major reasons for this are

1. Cast threads usually require a secondary chasing operation after casting to refine the surface and dimensions.
2. Threaded sections in wax or plastic patterns are a complicating factor in pattern making. If internal threads are cast, the threaded hole must be a through hole to allow removal of the investment material from the casting.

Undercuts

Undercuts pose no problem to the casting operation, but they do complicate the pattern-molding operation and therefore should be avoided if possible.

DIMENSIONAL FACTORS

By far the most significant factor influencing the dimensional accuracy of investment castings is the shrinkage of the materials used as they change from the molten to solid state. The wax or plastic pattern, the investment material, and the cast metal all exhibit this characteristic to some degree. Although these shrinkages are predictable, there are some variations in the amount of shrinkage and its effect on any part that result in some loss of dimensional accuracy. The effect of shrinkage is more evident in heavy-wall or unequal-wall sections. The following are normal shrinkage allowances to accommodate differences between master-pattern and final-part dimensions:

Shell-mold method: 1.6–1.7 percent

Flask-mold method: 1.1 percent

Other factors adversely affecting dimensional control are

- Temperature variations in pouring
- Fluidity of the metal used
- Pattern and pattern-mold dimensional variations
- Pattern distortion during handling
- Investment-mold cracks and other variations
- Wax shrinkage and sag over a period of time

RECOMMENDED TOLERANCES

Tolerance recommendations for angles, general dimensions, roundness, and flatness are shown in Fig. 5.3.9 and Tables 5.3.3 through 5.3.5.

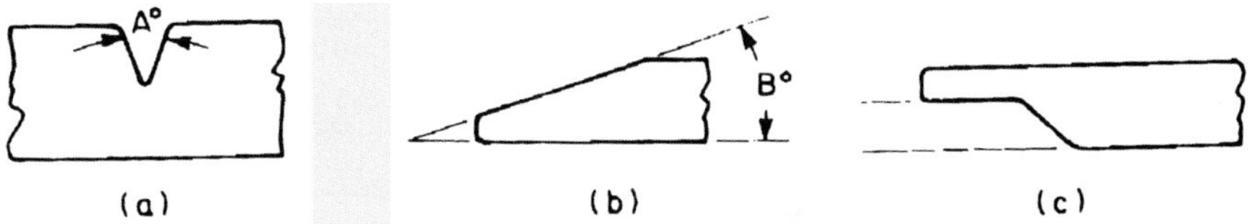

FIGURE 5.3.9 Recommended tolerances for angles: (*a*) angular openings as *A* (±1.5°); (*b*) angular shapes as *B* (±0.5°); (*c*) parallelism (±2.0°).

TABLE 5.3.3 General Dimensional Tolerances

	Recommended tolerance, mm (in)	
Dimension, mm (in)	Normal	Tight
Up to 6 (up to 0.25)	±0.4 (±0.015)	±0.08 (±0.003)
6–13 (0.25–0.50)	±0.4 (±0.015)	±0.10 (±0.004)
13–25 (0.50–1.0)	±0.4 (±0.015)	±0.13 (±0.005)
25–50 (1.0–2.0)	±0.4 (±0.015)	±0.18 (±0.007)
50–100 (2.0–4.0)	±0.8 (±0.030)	±0.40 (±0.015)
100–150 (4.0–6.0)	±1.1 (±0.045)	±0.60 (±0.025)
Over 150 (over 6.0)	±1.5 (±0.060)	±0.80 (+0.030)

TABLE 5.3.4 Recommended Allowance for Out-of-Roundness: Round and Tubular Sections

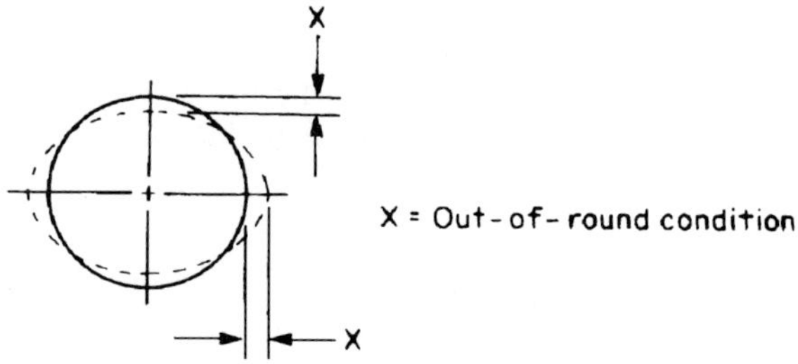

Solid sections	
Nominal diameter, mm (in)	Recommended out-of-roundness allowance, mm (in)
Under 13 (under 0.5)	0.15 (0.006)
13–25 (0.5–1.0)	0.25 (0.010)
25–38 (1.0–1.5)	0.33 (0.013)
38–50 (1.5–2.0)	0.38 (0.015)
Hollow-tube sections	
Outside diameter, mm (in)	Recommended out-of-roundness allowance, mm (in)
Up to 25 (up to 1.0)	0.25 (0.010)
25–38 (1.0–1.5)	0.33 (0.013)
38–50 (1.5–2.0)	0.38 (0.015)
50–75 (2.0–3.0)	0.63 (0.025)

TABLE 5.3.5 Recommended Flatness and Straightness Tolerances

Length of section, mm (in)	Allowable deviation, mm (in)
Up to 6 (up to 0.25)	±0.05 (±0.002)
6–13 (0.25–0.50)	±0.10 (±0.004)
13–25 (0.50–1.00)	±0.20 (±0.008)
25–50 (1.00–2.00)	±0.40 (±0.016)
50–100 (2.00–4.00)	±0.70 (±0.028)
100–150 (4.00–6.00)	±1.00 (±0.040)

CHAPTER 5.4
DIE CASTINGS

John L. MacLaren
Vice President, Marketing
Western Die Casting Company
Emeryville, California

THE PROCESS

Die casting involves the injection, under high pressure and with high velocity, of molten metal into a split metal die. Zinc, lead, and tin alloys, with melting points below 390°C (730°F), are cast in a "hot-chamber" die-casting machine, as shown in Fig. 5.4.1, whose injection pump can be immersed continuously in the molten material. Because molten aluminum dissolves ferrous parts, aluminum alloy is cast in a "cold-chamber" machine, whose design avoids continuous molten-metal contact by using a ladle to introduce molten metal to the machine.

Dies, even those for castings of simple shape, are complex mechanisms with many moving parts. The die must be rugged enough to withstand metal injection pressures of up to 69 MPa (10,000 lbf/in^2), yet it must also reproduce fine surface detail in the casting. Larger dies are usually cooled by channeling water behind the heavier casting sections and inside cores.

Casting removal is accomplished by ejector pins bearing on the casting's face. The pins are advanced after the casting has solidified sufficiently.

Production die castings are almost always die-trimmed, using a special hardened-steel tool shaped to match the actual casting to remove peripheral parting-line flash, in-gates, gates to overflow pads, and, as the design requires, flash in cored holes and openings. If pressure tightness is required, the castings may be impregnated with sodium silicate or organic compounds.

CHARACTERISTICS

Process Advantages

1. High rate of production.
2. High accuracy in sustaining dimensions part to part.

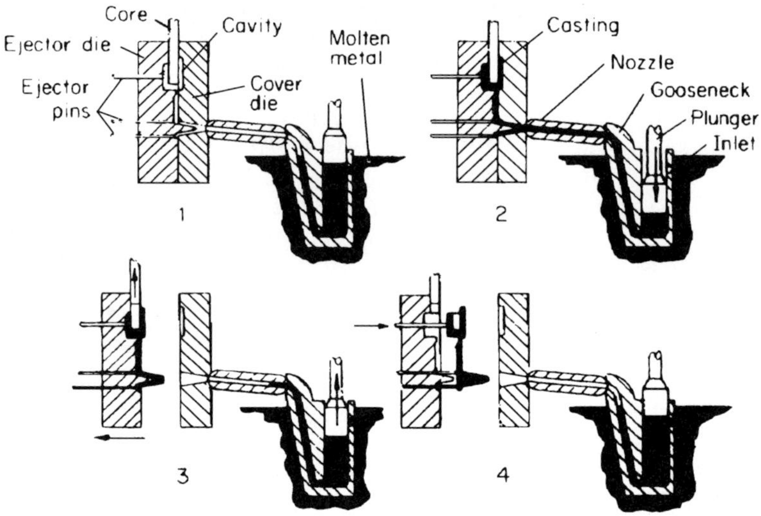

FIGURE 5.4.1 Operating sequence of the hot-chamber die-casting machine: (1) Die is closed, and hot chamber (gooseneck) is filled with molten metal. (2) Plunger injects molten metal through gooseneck and nozzle and into the die cavity. Metal is held under pressure until it solidifies. (3) Die opens, and core slides, if any, retract. Casting stays in the ejector half of the die. Plunger returns, pulling molten metal back through nozzle and gooseneck. (4) Ejector pins push casting out of ejector half of the die. As plunger uncovers the inlet, molten metal refills gooseneck. (*Courtesy Aluminum Industrie Vaasen.*)

3. Smooth surface finishes for minimum mechanical finishing or surfaces simulating a wide variety of textures.
4. Ability to incorporate such cast-in details as holes, openings, slots, trademarks, numbers, etc.
5. Intricate shapes and details at no greater cost than if they were omitted (once the die has been made), thus often making it attractive for the designer to incorporate two or more components in one die casting.
6. Thinner walls than can be produced with other casting processes.
7. The process uses a range of alloys that fit many design requirements: zinc for intricate form and platability; aluminum for higher structural strength, rigidity, and light weight; brass for still higher strength and corrosion resistance; and magnesium for good strength with extra lightness and superior machinability.
8. Ability to cast in inserts such as pins, studs, shafts, linings, bushings, fasteners, strengtheners, and heating elements.
9. Ability to cast pressure-tight parts.

Process Limitations

1. Dies are complicated and expensive, particularly if they have moving elements for coring details, so that lower casting prices usually can pay for the tooling only at high rates of usage.

2. Microporosity is common in die castings because the die is filled rather violently and solidification begins in less than $\frac{1}{5}$ s; air and vaporized die lubricant can become trapped in the cavity. With a well-designed part and good die design, porosity can be reduced to a harmless level, but highly stressed parts having heavy sections and bosses may give trouble nevertheless. Designers should allow for lower tensile strengths in such areas. Also, because pockets of porosity can produce surface blisters, heat treatment of parts or use in environments above 200°C (400°F) should be avoided. (Also see below under "Wall Thickness.")
3. Undercuts cannot be incorporated in simple two-piece dies. When they are essential, they often can be added by using core slides (moving die elements), but they add to the die's cost and also may increase the casting price by slowing production rates.
4. Owing to the very high metal pressures, sand cores or collapsing metal cores cannot normally be used, so some "hollow" shapes are not readily die-castable. In special situations for which the added cost is justified, zinc core pieces of the appropriate shape, incorporating undercuts, can be used and subsequently melted out by heating the casting to 390°C (730°F).
5. Size is limited at the upper level by equipment availability. Today, this is about 3000-ton locking pressure, for a top limit of about 30 kg (65 lb) for zinc, 45 kg (100 lb) for aluminum, and 16 kg (35 lb) for magnesium die castings, and a maximum effective area (less openings) of about 0.77 m^2 (1200 in^2) including the metal-distribution system to the cavity.
6. The alloys having the most attractive properties for the designer are often those with the least castability, so the list of alloys that can be processed with maximum economy is restricted. Alloys with high melting temperatures are not practicable for die castings.
7. Flash is always present except in very small zinc die castings; its removal may present an economic burden.
8. To get the part out of the die, taper, or draft, is needed. Thus walls and other details cannot normally be made perpendicular to the parting line except through use of expensive core slides.

APPLICATIONS

Die casting is a preferred process for making nonferrous-metal parts of intricate shape for such mass-production items as automobiles, appliances, outboard motors, hand tools, builders' hardware, electronics parts, electric switchgear, computer peripherals, business machines, optical and photographic equipment, and toys. Figures 5.4.2 and 5.4.3 illustrate typical die-cast components.

ECONOMIC PRODUCTION QUANTITIES

In large-volume situations, tooling cost is subordinate to piece price, and it is in these situations that die casting is most advantageous. At the other end of the scale, where die cost must be amortized over a small number of parts, it is desirable to evaluate

FIGURE 5.4.2 This one-piece die-cast chassis for a high-speed serial electronic printer illustrates the possibility of combining a number of parts to eliminate machining and assembly costs. The die casting replaces 82 separate components and fasteners in the forerunner assembly. Drilling and tapping have been eliminated by incorporating cored holes, most of which receive thread-forming screws for fastening. The part is used as die-trimmed and deburred, secondary machining having been completely eliminated. (*Courtesy Diablo Systems, Inc., a Xerox Company.*)

FIGURE 5.4.3*a* These aluminum die castings for a motorcycle engine show how good design can provide ribs where needed for strength and cored holes for fastening while maintaining fairly constant wall thickness. (*Courtesy Aluminum Industrie Vaasen.*)

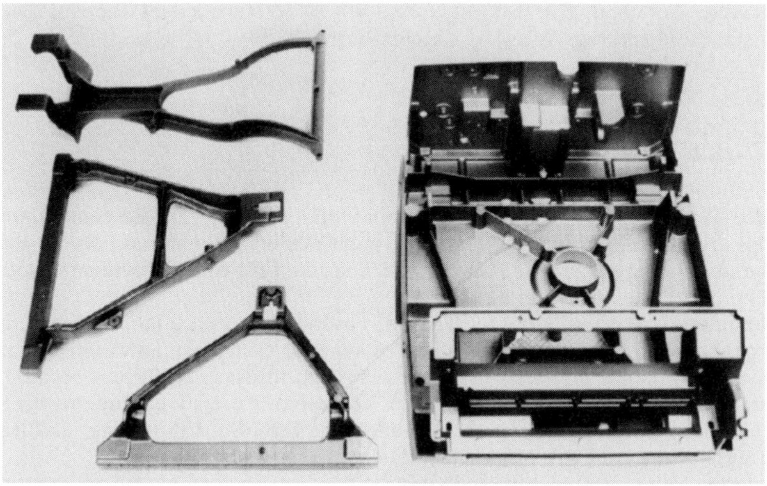

FIGURE 5.4.3*b* These die-cast aluminum components for a weight scale demonstrate intelligent use of ribbing to promote strength and stiffness. (*Courtesy Aluminum Industrie Vaasen.*)

closely overall costs, including machining, in relation to costs that would be incurred with other manufacturing methods. Often a total quantity in the range of 5000 to 10,000 pieces can justify the tooling cost.

Die-casting dies are relatively expensive; the greater the dimensional accuracy and detail specified in the drawing, the more the die will cost. A simple insert die for a small straightforward part can be very modest in cost, whereas a self-contained die for a complex part such as an automotive automatic-transmission housing will involve an expenditure well into hundreds of thousands of dollars.

Die life is a significant aspect of the economics of die casting because of the die's cost. Typically, a die for aluminum or magnesium parts will last for about 125,000 shots, a zinc die is good for 1 million or more shots, and a brass die will last only for 5000 to 50,000 shots, depending on the weight of the casting and/or die material.

Die-casting production rates are high. Typically, a 0.5-kg (1-lb) aluminum die casting having a projected area of 320 cm^2 (50 in^2) can be produced at the rate of about 100 per hour, while a large, complex 9-kg (20-lb) part might run 10 or 15 pieces per hour. Casting rates for magnesium are somewhat faster than for aluminum. Zinc-casting rates are more than double the rates for aluminum on parts of similar size because zinc need not be ladled in the hot-chamber machine, runs at a lower temperature, and uses readily automated equipment.

Flash removal adds to the cost of die casting. Die trimming is a fairly inexpensive means of removal, but a trimming die must be built, increasing the cost of the tooling package. There is an economic break-even point, often at the level of 10,000 to 20,000 castings over the life of the die, below which it is preferable to remove flash by hand sanding or filing, or both.

The designer should make adequate allowance, in planning a product-introduction date, for the fairly long lead times needed for die design and construction. A simple insert die usually can be built and tried out, the samples subjected to dimensional analysis, and the die corrected within 8 weeks. A complex die will take 30 weeks or more for the same procedures. And there is no way in which these intervals can be appreciably shortened: Die making is on a 40- to 50-h workweek schedule, not normally subject to speedup by use of a second shift.

SUITABLE MATERIALS

Because of limitations in the hot-work tool steels available for die construction, die castings are made from the lower-melting-point nonferrous materials. By far the most popular are aluminum and zinc alloys, which account for over 90 percent of total production.

Aluminum is predominant for larger die castings because of its lower cost. Zinc is preferred for smaller castings, for which its superior castability outweighs the materials-cost factor. Other materials are more restricted in use: magnesium because of its relatively high cost and other factors, brass because of its relatively high melting point [910°C (1670°F)] and resulting shortening of the life of costly tooling, and lead and tin because of material cost and relatively low mechanical properties.

In recent years, ferrous-metal die casting has been carried out on an experimental and limited-production basis. High melting temperatures of about 1700°C (3100°F) are involved, and these necessitate the use of special refractory metals for dies and a number of special procedures. The process is most advantageous for difficult-to-machine tool steels, alloy steels, and stainless steels.

Table 5.4.1 lists the most suitable alloys for die casting with information about their typical applications, melting points, and tensile strength.

GENERAL DESIGN CONSIDERATIONS

The designer is urged to study the overall function of the product and consider the possibility that several functions can be incorporated into one die casting, with integral features for attachment and assembly, as typified by the chassis illustrated in Fig. 5.4.2. Full advantage also should be taken of opportunities for the reduction of machining that die casting affords.

Dies, after being machined, must finally be hardened by heat treating, which makes subsequent alterations difficult. It is therefore important for the product designer to finalize the best design for the die-cast part and reach agreement with the die caster on the producibility of the design before the die itself is designed and construction begins.

The designer also should consider ejector-pin locations early in production design, preferably in consultation with the die caster. If the impressions left by the pins are not tolerable or cannot be hidden in a cored-out zone, alternatives such as ring or sleeve ejection are available, but they are more costly to incorporate into the die than pins.

Abrupt section changes, sharp corners, and walls at an acute angle to one another disturb the continuity of metal flow and promote porosity and surface irregularities. Therefore, radii should be as generous as possible, with differing sections blending into one another. Blind recesses in the die, such as are needed to form bosses, tend to cause subsurface porosity because of trapped air. This can cause drills to wander and taps to break in secondary machining and should therefore be avoided.

Simulation software, in conjunction with computer-aided design (CAD) programs, is increasingly becoming available to designers of die castings. It can aid in optimiz-

TABLE 5.4.1 Alloys Recommended for Die Casting

Material	Alloy designation		Melting point, °C (°F)	Ultimate tensile strength, kPa (lbf/in^2)*	Castability	Remarks
	Commercial	UNS				
Aluminum	380	A03800	593 (1100)	317,000 (46,000)	Excellent	Most popular aluminum alloy; best combination of properties and ease of use
	383	A03830	582 (1080)	310,000 (45,000)	Excellent	Higher-silicon version of 380; used when thin walls and pressure tightness are required
	360	A03600	596 (1105)	303,000 (44,000)	Good	Used when better corrosion resistance and ductility are required
Zinc	AG40A (Zamak No. 3)	Z33520	387 (729)	283,000 (41,000)	Excellent	Used for the majority of commercial applications for thin walls and good platability
Magnesium	AZ91B	M11913	596 (1105)	234,000 (34,000)	Excellent	Used for the majority of commercial applications for lightness with strength
Brass	858	C85800	899 (1650)	379,000 (55,000)	Fair	Used when high strength, elongation, and corrosion resistance are required

*The typical values indicated are for separately die-cast test bars and do not represent values for specimens cut from die castings.

ing materials flow and cooling of the casting in the die. Sometimes, casting design changes result from such a simulation, providing an improvement in the part's castability and a reduction in production problems.

SPECIFIC DESIGN RECOMMENDATIONS

Wall Thickness

The easiest die casting to make and the soundest in terms of minimum porosity is one that has uniform wall thickness. Sharp changes in sectional area or heavy sections over 6 mm ($\frac{1}{4}$ in) thick should be avoided if possible. When a heavy section seems to be indicated, its underside should be cored out.

The skin of a die casting is its strongest part. The injected liquid metal chills rapidly on contact with die cavity surfaces, resulting in a fine-grained, dense structure generally devoid of porosity. This skin measures between 0.38 and 0.63 mm (0.015 and 0.025 in) thick, depending on casting size. Thus the strength-to-weight ratio of die-cast sections improves quickly as wall thickness is reduced, to the point where it is maximized when the wall is "all skin," i.e., from 0.75 to 1.3 mm (0.030 to 0.050 in) thick.

Table 5.4.2 indicates the minimum-wall-thickness ranges, by alloy classification, for varying single-surface areas that can be achieved consistently and economically.

Heavy bosses behind the surface can cause visible "sinks" on decorative parts having flat, expansive areas. These marks are caused by shrinkage of the relatively large mass of metal in the boss during cooling after the adjacent walls have been frozen, which draws the wall in toward the boss. The effect is magnified with thinner walls. Shrinks become a problem when the part is plated or painted, since these treatments accentuate surface irregularities. When a boss must be so located, the shrinkage problem may be mitigated by moving the boss away from the wall and connecting it to the wall with a short rib of the same thickness, as shown in Fig. 5.4.4.

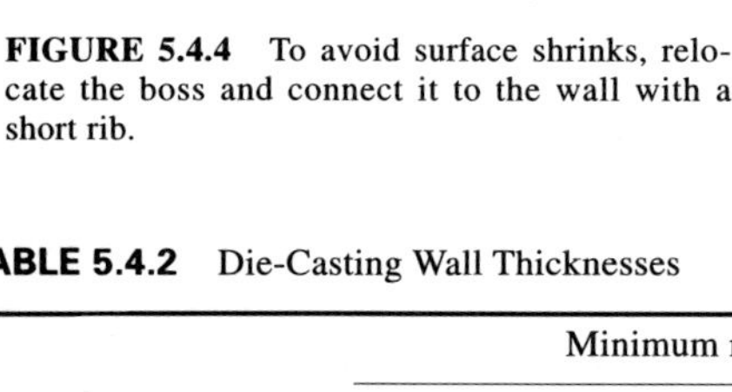

FIGURE 5.4.4 To avoid surface shrinks, relocate the boss and connect it to the wall with a short rib.

TABLE 5.4.2 Die-Casting Wall Thicknesses

Surface area, cm^2 (in^2)*	Minimum recommended wall thickness, mm (in)		
	Zinc alloys	Aluminum and magnesium alloys	Copper alloys
Up to 25 (4)	0.38–0.75 (0.015–0.030)	0.75–1.3 (0.030–0.050)	1.5–2.0 (0.060–0.080)
25–100 (4–15)	0.75–1.3 (0.030–0.050)	1.3–1.8 (0.050–0.070)	2.0–2.5 (0.080–0.100)
100–500 (15–80)	1.3–1.8 (0.050–0.070)	1.8–2.2 (0.070–0.090)	2.5–3.0 (0.100–0.120)
500–2000 (80–300)	1.8–2.2 (0.070–0.090)	2.2–2.8 (0.090–0.110)	
2000–5000 (300–800)	2.2–4.6 (0.090–0.180)	2.8–6.3 (0.110–0.250)	

*The area of a single, separately identifiable plane to be cast at the stated wall thickness.

Zinc die-casting technology has recently been developed to reduce minimum wall thickness considerably below Table 5.4.2 values, to as low as 0.38 mm (0.015 in) in limited zones for parts in the 25- to 100-cm^2 (4- to 15-in^2) size range. This specialized procedure has appeal mainly to high-volume industries because the much more sophisticated and therefore more expensive tooling and production processing that are involved must be offset by materials-cost savings in large-scale production. The designer should confer with the die caster before tackling "thin-walled zinc" part design.

Ribs and Fillets

Ribs are mainly incorporated into a die casting to reinforce it structurally, replacing heavy sections that would be otherwise necessary. (See Figs. 5.4.5 and 5.4.6.) They should be perpendicular to the parting line to allow for casting removal from the die, although external ribs running parallel to the parting line can be incorporated by using core slides.

To avoid sinks, ribs should be no wider than the thickness of the casting wall and no higher than 4 times their width for complete filling and ease of removal from the die. The minimum distance between two adjacent ribs should be the sum of their heights. Ample draft (at least 2° per side) also helps with ejection (see Fig. 5.4.7).

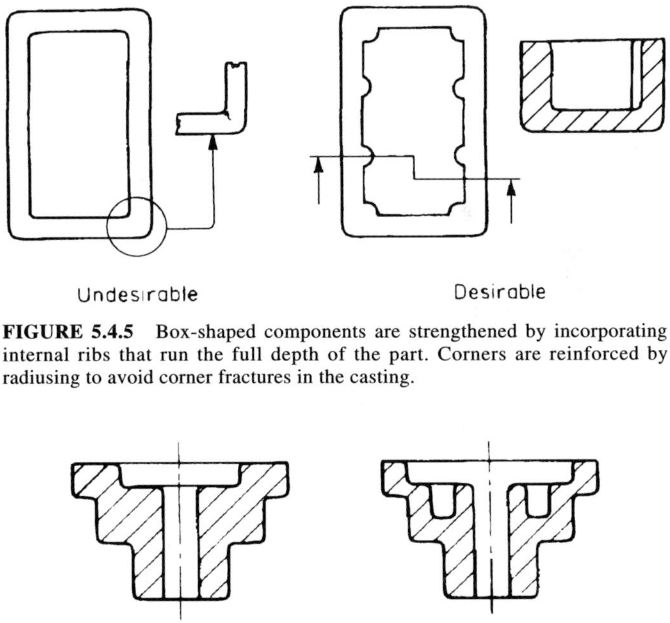

FIGURE 5.4.5 Box-shaped components are strengthened by incorporating internal ribs that run the full depth of the part. Corners are reinforced by radiusing to avoid corner fractures in the casting.

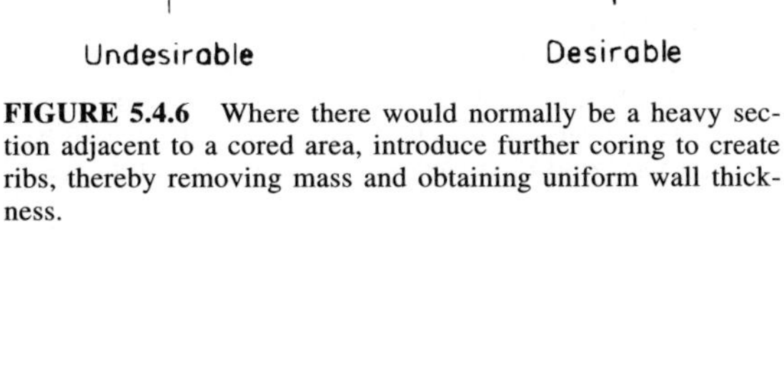

FIGURE 5.4.6 Where there would normally be a heavy section adjacent to a cored area, introduce further coring to create ribs, thereby removing mass and obtaining uniform wall thickness.

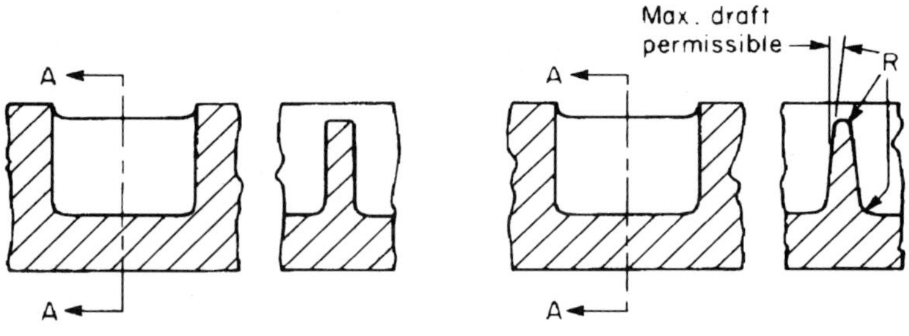

FIGURE 5.4.7 Provision of maximum draft eases ejection of deep ribs from the die.

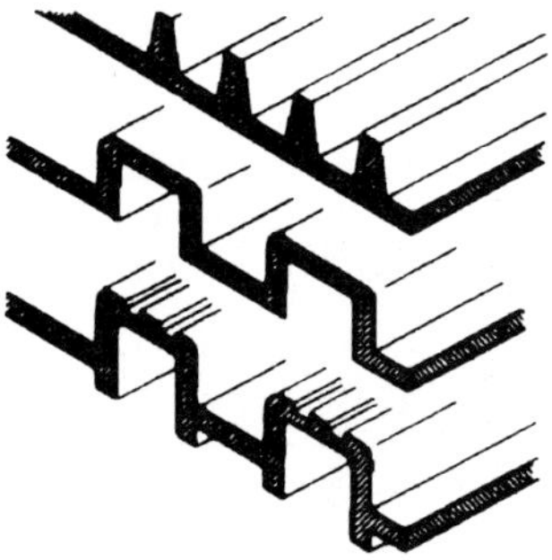

FIGURE 5.4.8 Ribbing of various forms used to stiffen the wall of a die casting.

Ribs and fillets are also used to improve the rigidity and strength of standing bosses. Ribs need not be designed all on one side of a wall to strengthen it; they can be formed in both halves of the die, as shown in the lower two illustrations of Fig. 5.4.8. The rib is then part of the wall, strengthening it without thickening it locally.

If ribs are designed to cross, they should do so at right angles. Acute-angle intersections cause the die to overheat in the area between the ribs. Multiple intersections of radial ribs should be avoided; otherwise, the intersection will contain porosity. (These rules are identical to those for other casting processes and are illustrated in Figs. 5.1.9 and 5.1.10.)

Ribs are fairly easy to incorporate into an existing hardened die by using electrical-discharge machining (EDM) techniques. Thus the designer may underdesign initially, test sample castings, then add strength if necessary by removal of die steel until the optimum combination of mechanical properties and casting-material conservation is reached. This is preferable to overdesigning and having to lighten the die casting later by welding the die, which is a costly, life-limiting procedure.

Draft

The sidewalls of die castings and other features perpendicular to the parting line must be tapered, or drafted, as much as possible to facilitate removal from the die. When the draft angle is abnormally small, even the slightest depression in the drafted surface of the die will prevent ejection without causing drag marks in the surface of the casting. Draft angle depends on the alloy and varies inversely with wall depth, as shown in Table 5.4.3. Although this table indicates that the draft required on outside walls is one-half the draft for inside walls, the designer should, if possible, specify the inside angle for the outside as well and be as liberal as possible in the selection of angle.

TABLE 5.4.3 Draft Requirements for Die Castings

	Draft angle for inside wall,°*			
Depth of wall, mm (in)	Copper alloys	Aluminum alloys	Magnesium alloys	Zinc alloys
0.25–0.50 (0.01–0.02)	18	16	13	10
0.50–1.0 (0.02–0.04)	14	12	10	7
1.0–2.0 (0.04–0.08)	10	8	7	5
2.0–3.8 (0.08–0.15)	7	6	5	3.5
3.8–7.5 (0.15–0.3)	5	4	3.5	2.5
7.5–15 (0.3–0.6)	3.5	3	2.5	1.8
15–25 (0.6–1.0)	2.5	2.2	2	1.5
25–50 (1.0–2.0)	2	1.5	1.5	1
50–100 (2.0–4.0)	1.3	1	1	0.7
100–175 (4.0–7.0)	1	1	0.7	0.5
175–250 (7.0–10.0)	1	0.7	0.6	0.5

*For outside walls, one-half of the above angles may be used.

Source: Data taken from ADCI-E4-65, American Die Casting Institute.

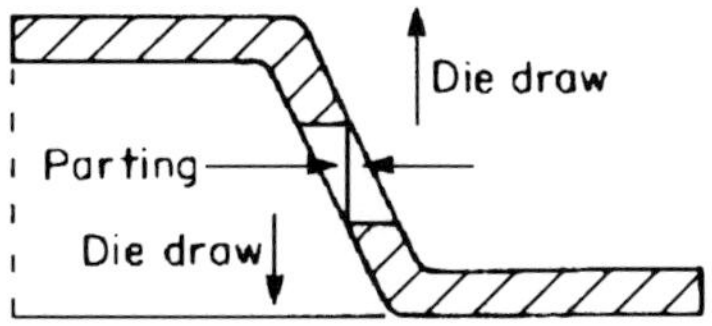

FIGURE 5.4.9 When a die casting's wall slopes sufficiently, through holes can sometimes be formed by using "kissing cores" built into opposite die halves.

The designer may use draft in another way, as shown in Fig. 5.4.9, to incorporate a side hole without resorting to a core slide in the die. When details in opposite die halves come together or "shut off" at an angle to form an opening, as in this illustration, the angle should be no less than 5° from the vertical to avoid galling at the shutoff zone.

Radii

Sharp internal corners in a die casting are to be avoided for several reasons. As the alloy shrinks on the core, induced stresses are concentrated at the corner instead of being distributed through the surrounding mass, thus weakening the part and perhaps even causing it to fracture. The abrupt change in metal-flow direction during injection also can produce subsurface porosity at the corner to weaken this area further. Sharp edges on cores are difficult to maintain because they are points of heat concentration, with resulting premature erosion of the die material.

Sharp external corners are undesirable because they become a localized point of heat and stress buildup in the die steel that can cause die cracking and early failure. Therefore, radii and fillets should be as generous as possible, preferably at least $1\frac{1}{2}$ times wall thickness for both inside and outside radii. (See Figs. 5.4.10 and 5.4.11.)

Holes

The die-casting process can accommodate the coring in of holes into the body of the casting at right angles to the parting line. However, there are core-length limits,

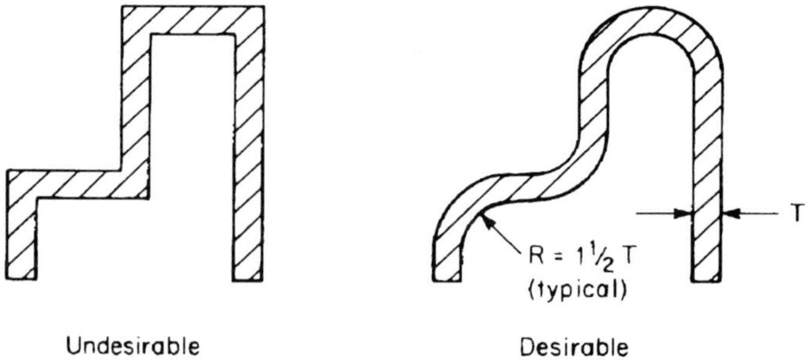

FIGURE 5.4.10 Allow generous radii at internal and external corners.

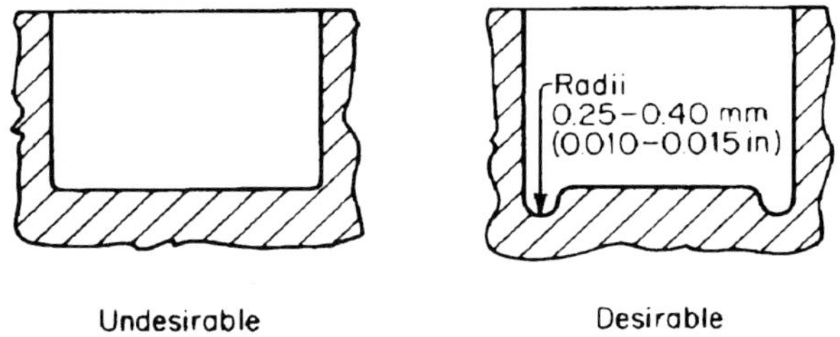

FIGURE 5.4.11 Sharp corners in the die cavity concentrate die stresses and shorten die life.

depending on diameter, that should not be exceeded. (See Table 5.4.4.) Long, slender cores lead to excessive core breakage. Cores also must be drafted adequately to ensure their longevity. (See Table 5.4.5.) A lower limit on core diameter of 3 mm (0.120 in) for aluminum and magnesium and 1.5 mm (0.060 in) for zinc should be observed, since smaller cores are prone to frequent breakage and erosion from heat buildup.

When small-diameter cores are needed several inches apart in a casting face, consideration should be given to incorporating design features such as cored-out recesses that will support casting shrinkage in the spans between cores. Otherwise the delicate cores will have to absorb these forces, with possible resulting bending and breakage.

TABLE 5.4.4 Optimum Greatest Depth of Cored Holes as Related to Diameter*

	Diameter of hole, mm (in)								
	3 (1/8)	4 (5/32)	5 (3/16)	6 (1/4)	10 (3/8)	13 (1/2)	16 (5/8)	19 (3/4)	25 (1)
Alloy	Maximum depth, mm (in)								
Zinc†	10 (3/8)	14 (9/16)	19 (3/4)	25 (1)	38 (1½)	50 (2)	80 (3⅛)	115 (4½)	150 (6)
Aluminum†	8 (5/16)	13 (1/2)	16 (5/8)	25 (1)	38 (1½)	50 (2)	80 (3⅛)	115 (4½)	150 (6)
Magnesium†	8 (5/16)	25 (1)	16 (5/8)	25 (1)	38 (1½)	50 (2)	80 (3⅛)	115 (4½)	150 (6)
Copper				13 (1/2)	25 (1)	32 (1¼)	50 (2)	90 (3½)	125 (5)

*Requirements for tapped holes are shown in Table 5.4.6.

†For cores larger than 25 mm (1 in) in diameter the diameter depth shall be 1:6.

TABLE 5.4.5 Draft Requirements for Cored Holes in Die Castings

Depth of hole, mm (in)	Draft,°			
	Copper alloys	Aluminum alloys	Magnesium alloys	Zinc alloys
1.5–2.5 (0.06–0.1)	11	10	8	6
2.5–5 (0.1–0.2)	9	8	7	4.5
5–10 (0.2–0.4)	7	6	5	3
10–20 (0.4–0.8)	5	4	3.5	2.5
20–38 (0.8–1.5)	3.5	3	2.5	1.8
38–75 (1.5–3)	3	2	1.8	1.3
75–150 (3–6)	2	1.5	1	0.9

Source: From ADCI-E7-65, American Die Casting Institute.

Table 5.4.6 provides recommended diameters for die-cast holes to be tapped without prior drilling.

Since cores incorporated into moving slides are also denied the benefit of an ejection system to move the casting off them, they need as much draft as possible to minimize casting distortion when the slide is withdrawn.

Cores are fitted through the die block and should not be located closer than about 3 mm ($\frac{1}{8}$ in) to a cavity wall. Otherwise, local weakening in the steel can result in die cracking and premature failure.

In designing for consistent walls and weight reduction, it is best to avoid a situation requiring sliding cores. If possible, the part should be designed so that it can be made in a simple two-piece die. (See Fig. 5.4.12.)

Through-wall cored holes for tapping should be countersunk on both sides, as shown by Fig. 5.4.13.

Concentricity requirements are important in designing for cored details. If a central core and a circle of holes are to be concentric, they should be designed so that they can all be cored from the same die half.

Designers are urged to take advantage of the economies offered through use of thread-forming screws. Such fasteners roll their own thread, avoiding the need to tap into cored holes and, unlike thread-cutting (self-tapping) screws, generate no chips. Core diameters are somewhat greater than for tapped holes, so the manufacturer's recommendations should be followed.

Holes and openings in sidewalls parallel to the parting line usually require a slide to carry the core, which adds appreciably to the die's cost and slows casting-cycle time. Two designs that handle this situation without a slide are shown in Figs. 5.4.9 and 5.4.14.

Core Slides

It is always preferable to locate a core slide at the parting line even if it means stepping the parting line to accommodate the core's centerline. Flash that inevitably forms around the slide is then ejected with the casting, whereas with cores that are "submerged" below the parting line, the flash may shear off in place when the casting is ejected to cause heavy wear in the slide's ways and possible seizure. A further drawback of submerged cores is the difficulty of getting a die-release agent to them.

TABLE 5.4.6 Cored Holes for Tapping

When required, cored holes in zinc, magnesium, and aluminum die castings may be tapped without removing the draft by drilling. Recommended sizes for tapping are based upon allowing 75% of full depth of thread at bottom or small end of the cored hole and 60% at the top or large end of the cored hole.

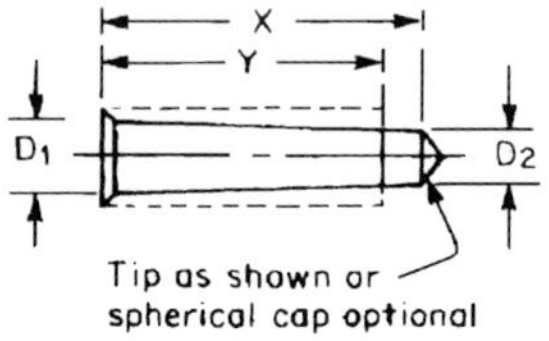

Tap size†		Dimensions, in			
		Hole diameters‡		Maximum threaded depth of hole Y	Maximum depth of cored hole X
		D_1	D_2		
0.138	6–40 NF	0.119	0.114	0.208	0.270
0.138	6–32 NC	0.113	0.107	0.162	0.240
0.164§	8–36 NF	0.142	0.134	0.321	0.390
0.164§	8–32 NC	0.140	0.133	0.307	0.385
0.190	10–32 NF	0.165	0.159	0.380	0.458
0.190	10–24 NC	0.158	0.150	0.380	0.482
0.216	12–28 NF	0.188	0.181	0.432	0.519
0.216	12–24 NC	0.184	0.176	0.432	0.534
	¼–28 NF	0.222	0.215	0.500	0.587
	¼–20 NC	0.211	0.201	0.500	0.625
	5/16–24 NF	0.280	0.272	0.625	0.727
	5/16–18 NC	0.269	0.259	0.625	0.762
	⅜–24 NF	0.343	0.334	0.750	0.852
	⅜–16 NC	0.326	0.314	0.750	0.906
	7/16–20 NF	0.398	0.388	0.875	1.000
	7/16–14 NC	0.382	0.368	0.875	1.053
	½–20 NF	0.461	0.451	1.000	1.125
	½–13 NC	0.440	0.425	1.000	1.192
	9/16–18 NF	0.519	0.508	1.125	1.262
	9/16–12 NC	0.497	0.481	1.125	1.333
	⅝–18 NF	0.582	0.571	1.250	1.387
	⅝–11 NC	0.554	0.536	1.250	1.497
	¾–16 NF	0.701	0.689	1.500	1.656
	¾–10 NC	0.672	0.652	1.500	1.750
	⅞–14 NF	0.819	0.804	1.750	1.928
	⅞–9 NC	0.789	0.767	1.750	2.027
	1–14 NF	0.944	0.929	2.000	2.178
	1–8 NC	0.903	0.878	2.000	2.312

*From ADCI Standard E8-65, American Die Casting Institute.

†NF = national fine; NC = national coarse.

‡These dimensions are subject to ±0.002-in tolerance for ⅝-in or smaller diameters and ±0.003-in tolerance for larger diameters.

§Minimum sizes recommended for aluminum or magnesium alloys.

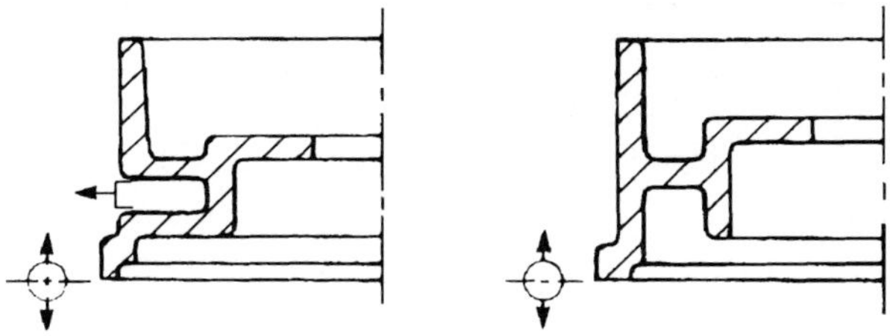

FIGURE 5.4.12 Avoid external-wall undercuts and the need to incorporate core slides in designing for weight reduction.

FIGURE 5.4.13 Countersink on both sides any through holes that are to be tapped in order to avoid a deburring operation where the tap emerges.

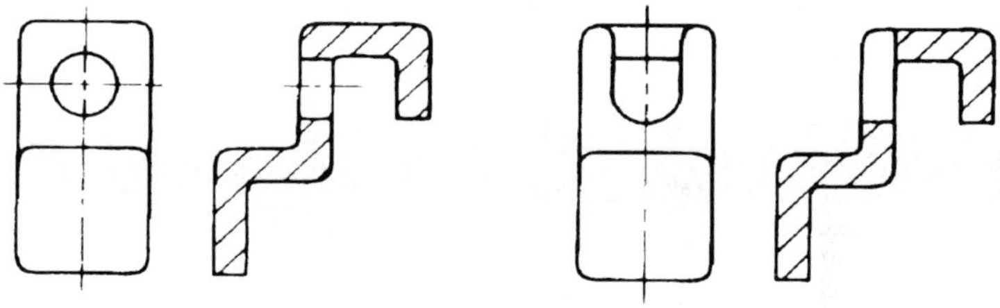

FIGURE 5.4.14 Core slides can be avoided by using this hole design.

A sliding core should never be designed to intersect the opposite die half, since imperfect die closure (possibly the result of flash at the parting plane not being fully removed) would result in a damaged die. It is preferable, in the situation of Fig. 5.4.15, to core through the outside wall only and plan on drilling the inside holes.

Threads

External screw threads can be formed on die castings, but when a precision fit is required, it is recommended that the threads be machined. In less demanding situations, the die-casting process can produce acceptable external threads in several ways, the most practical of which is to cut the thread between the die halves and eject the

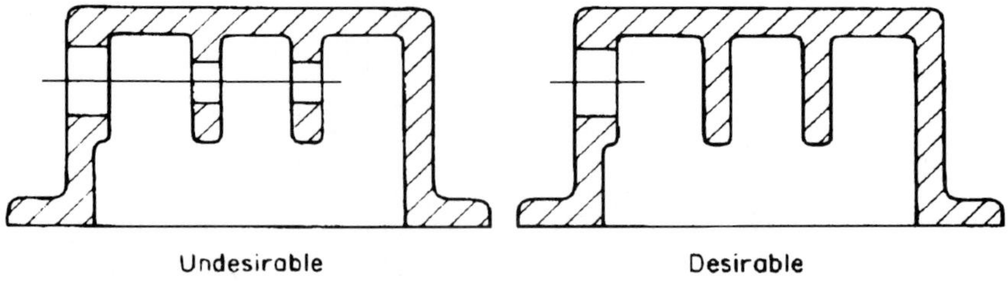

FIGURE 5.4.15 Avoid designing for core slides that must fit accurately through both die halves. Drill holes in internal ribs.

part in the normal manner. Accuracy will be affected to the extent that the die halves may be misaligned, and for this reason pitches finer than 24 threads per inch for aluminum and magnesium and 32 threads per inch for zinc are not recommended.

Die trimming such a part by using tooling shaped to the thread profile seldom produces a good finish, again because of the slight die mismatch customarily encountered. Figure 5.4.16 shows another thread form having flats at each side that can be readily trimmed along a straight line. Although the die cavity becomes more expensive, the extra cost is usually offset by a more acceptable product that is easier to deflash.

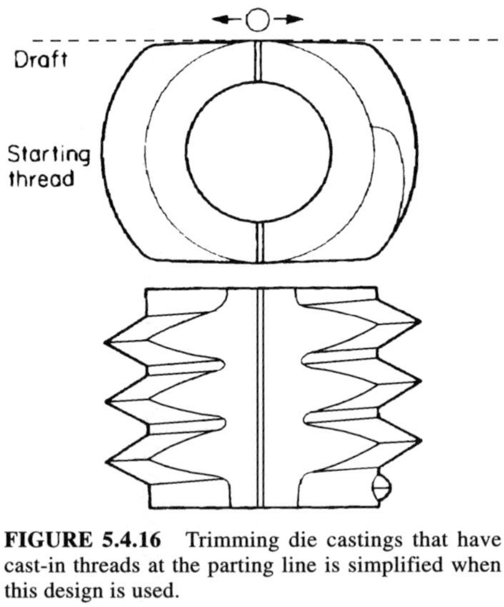

FIGURE 5.4.16 Trimming die castings that have cast-in threads at the parting line is simplified when this design is used.

It is possible to die-cast internal threads by using techniques that involve either unscrewing the casting from a threaded core or rotating the core out of the casting during ejection. Such a design should be confined to zinc parts, since the other alloys shrink more tightly onto cores and the draft needed for release affects thread function.

Inserts

Inserts can be incorporated in die-cast parts where necessary, though with increased cycle time because of the time required to load inserts into the die. The most common type of insert is the threaded stud used for assembly. It must be designed so that material will shrink onto its shank with sufficient force to prevent movement in use. Figure 5.4.17 illustrates various forms that will resist both axial and rotational forces. The insert must have a very precise fit into its retaining recess in the die block, and the insert's threads must be kept away from the casting face; otherwise, they will load up with metal. Positive location within the die must be provided to prevent insert movement during the casting cycle.

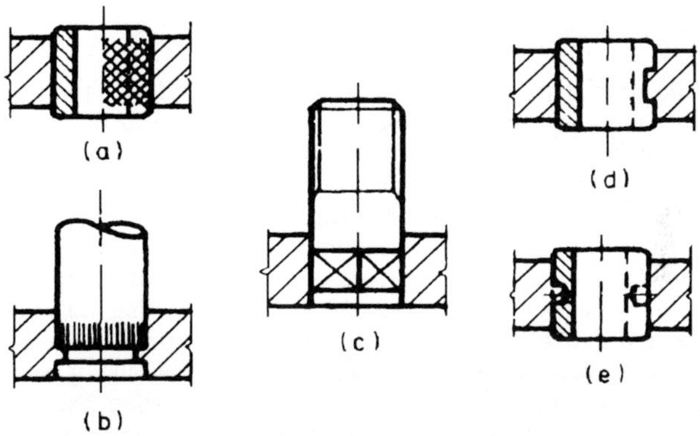

FIGURE 5.4.17 Insert designs to prevent rotation and pullout: (*a*) knurled circumferences; (*b*) recess and longitudinal grooves; (*c*) machined and undercut square; (*d*) localized machined flats; (*e*) drilled anchor holes.

Machining Allowance

When a die casting requires a machining operation, the added material (machining allowance) to be removed should not exceed 0.5 mm (0.020 in), which is about the average thickness of the dense, fine-grained skin of the casting. (See Fig. 5.4.18.) Deeper cuts could open up unsightly subsurface porosity and possibly affect function. The machining allowance should not be less than 0.25 mm (0.010 in) to avoid excessive tool wear.

Holes for tapping should be cored rather than drilled. A drilling operation is eliminated, and the tap will cut into dense material for a higher-quality thread. Tapping into a porous substructure laid open by drilling can result in tap breakage.

If an area to be machined covers ejector-pin locations, their impressions should be left standing to 0.4 mm (0.015 in)—the usual specification is 0 to 0.4 mm (0.015 in) depressed—so that they are removed in machining.

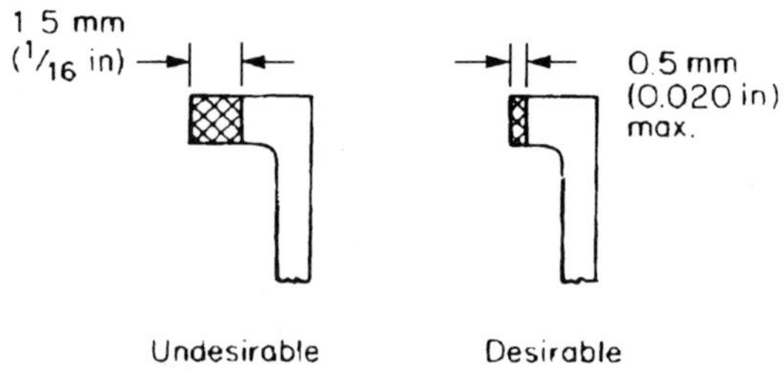

FIGURE 5.4.18 Limit machining allowance to 0.25 to 0.50 mm (0.010 to 0.020 in).

Flash and Gate Removal

Designers should be realistic about the degree of flash and gate removal they specify. With complex castings having massive core slides, the cost of complete removal by a combination of die trimming and hand operations can be as much as, or even considerably more than, the cost of the raw casting.

A well-built trimming die will remove flash almost to the most extremely drafted point of the casting wall. But because the cutting edges wear, an allowance must be made. Commercial die castings are considered to be adequately trimmed if flash and gates are removed to within 0.38 mm (0.015 in) of the casting wall or, in the case of very heavy gates more than 2.2 mm (0.090 in) thick, within 0.75 mm (0.030 in).

Designers should avoid an angled junction of an external wall with the parting line. It is preferable to add a minimum-draft shoulder at the parting line, as shown in Fig. 5.4.19 so that most of the gate material will come away in trimming.

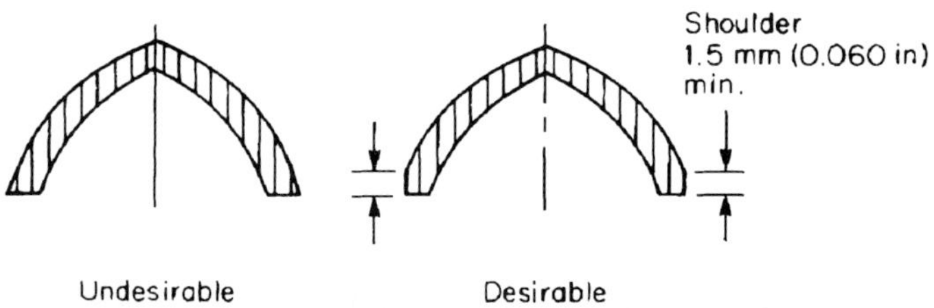

FIGURE 5.4.19 Allow a 1.5-mm (0.060-in) minimum height shoulder at parting line for more nearly complete removal of in-gates by the trimming die.

If the sidewalls of the part are intricately configured, a trimming tool to match would be expensive and hard to maintain. In such situations, it is advisable to add a shoulder between wall detail and parting line to allow for use of a single trimming die or a lathe-turning operation to remove gates and flash. (See Fig. 5.4.20.)

When sidewalls are formed in both halves of the die, mismatch will complicate flash and gate removal. If possible, the part should be redesigned to a straight parting line. Flash is then at the bottom of the walls, where it is more readily trimmable, and an unsightly seam line on the sidewalls is avoided. A further benefit is elimination of the costly stepped parting line in the die.

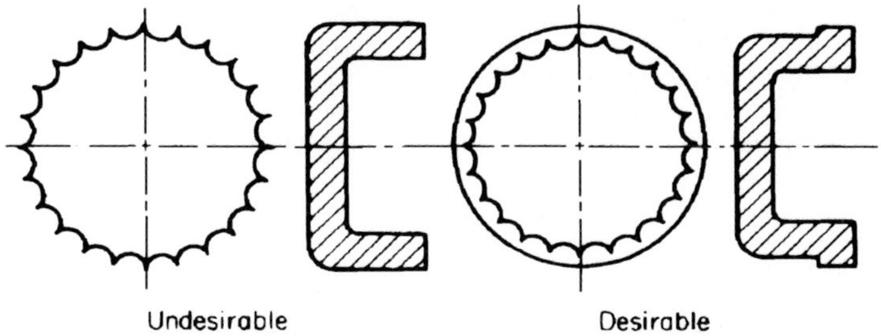

FIGURE 5.4.20 Provide a shoulder between parting line and scallop to simplify gating and trimming.

Die-Sinking Economics

In machining cavities, the die builder is working basically with a circular cutter. If the casting is designed with convex features in outside walls, as in Fig. 5.4.21, it is a straightforward job to mill the corresponding concavities into the steel. Likewise, a nonsymmetrical rib end can be put into the steel, though at considerable cost, but symmetrical features are easy.

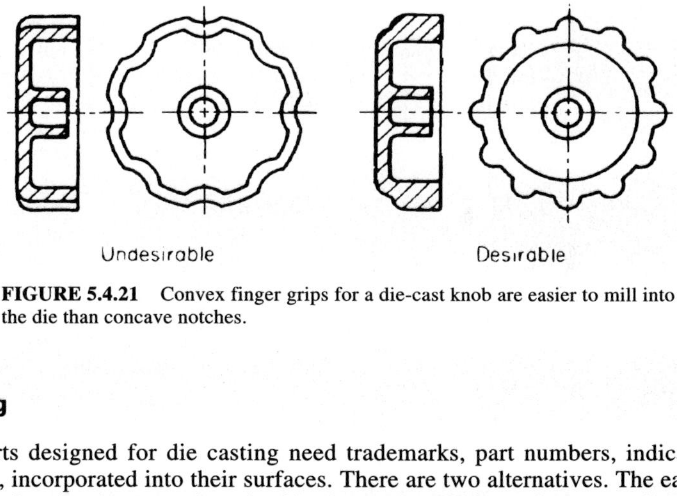

FIGURE 5.4.21 Convex finger grips for a die-cast knob are easier to mill into the die than concave notches.

Lettering

Many parts designed for die casting need trademarks, part numbers, indications for dials, etc., incorporated into their surfaces. There are two alternatives. The easy way is to specify that the characters be raised in the casting. This can be accomplished by relatively inexpensive engraving of the die. If the characters are to be depressed into the casting, however, all the background steel on that face of the die must be painstakingly removed around the characters.

If the designer wants the economy of raised characters but does not wish them to project above the surrounding surface, a raised pad can be incorporated into the die to form a depressed area in the casting. Then when the pad is engraved, the lettering will come out flush.

There are some basic rules to follow for lettering: minimum character width, 0.25 mm (0.010 in), to allow for filling; character height, 0.25 to 0.5 mm (0.010 to 0.020 in), also to allow for filling; and at least 10° draft for clean ejection. Lettering cannot be located on sidewalls in two-part dies, as it would constitute an undercut. It is restricted to features parallel to the parting line or on sidewalls formed by core slides.

Surface Design

Large, plain areas on a die casting are vulnerable cosmetically because any slight casting imperfection will be readily visible. The surface may be sanded or polished, but a less expensive means of resolving the problem is to mask irregularities by designing in ribs, serrations, other details, or mold texturing that "breaks up" the surface of the part. Sidewalls can be textured by using some of the shallower options, provided they are drafted sufficiently to prevent the texture forming an undercut.

Integral Means of Assembly

Die casting offers the designer wide latitude in providing appendages that make assembly to other parts quick and simple. These features are of two general types: cast-in studs to receive spring clips and cast-in lugs, rivets, and lips that are deformed to effect closure. All the alloys can be used with the spring-retainer approach, but zinc alloy, because of its ductility and ready formability, is favored when deformation is the means of assembly. These fastening methods are discussed and illustrated in Chap. 7.5.

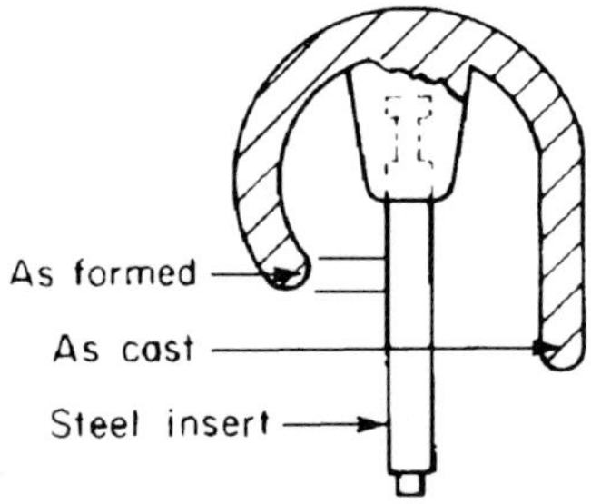

FIGURE 5.4.22 A small knob cast onto a steel insert and formed to final shape by spinning.

Spinning is a handy procedure for permanently assembling a zinc die casting to another component. This technique also may be used to form a variety of noncastable but very useful shapes like the one shown in Fig. 5.4.22.

DIMENSIONS AND TOLERANCES

The as-cast dimensional variations of a die casting, as with all casting processes, depend in part on the size of the casting. This dependency is largely due to thermal expansion and contraction of both the die and the casting. The die expands at operating temperatures, and the casting shrinks after it leaves the die. Both of these vary linearly. Variations in die operating temperatures and the temperature of the molten metal entering the die add to the need for a design tolerance on die-casting dimensions.

In designing the die for a particular part, the die builder enlarges the dimensions by a "shrink factor," usually 0.6 percent, to account for die expansion and metal contraction (the former being about half the latter in most situations). This works out well for a dimension between points that are not restrained during contraction, but when the metal shrinks onto a constraining feature such as a massive core, shrinkage will be less.

Tolerances across the parting line and between core slides and main die blocks must be greater because of the clearances that are incorporated into these features in the die to enable them to function at elevated temperatures. Recommended tolerances also allow for gradual wear in die components over the life of tooling. This is a significant factor. Another is warpage. In production, in the time taken for a casting to cool to room temperature, varying rates of contraction, a function of part design, can produce warpage. Such distortion is common in all castings, particularly large-area, thin shapes.

If dimensional limits in the as-designed part must be exceptionally close, the impact on the costs of die development and casting production needs to be weighed carefully to determine whether the close-tolerance requirement might be handled more economically in a secondary machining operation.

Because of the guesswork involved in developing a die design, the ultimate die casting will invariably have some dimensions that fall outside normal commercial tolerances. When such dimensions do not impact form, fit, or function, "buy-off" and corresponding print changes are recommended. Unnecessary corrections are costly; they adversely affect die life, and they will only delay production.

Table 5.4.7 presents recommended tolerances for dimensions determined by cavity dimensions in either half of the die. Table 5.4.8 provides additional tolerances to be applied if the dimension crosses the die-parting line. Table 5.4.9 provides additional tolerances to be applied if the dimension is affected by moving die cores.

TABLE 5.4.7 Recommended Tolerances for Die-Casting Dimensions Determined by Cavity Dimensions in Either Half of the Die, mm (in)

	Die-casting alloy			
	Zinc	Aluminum	Magnesium	Copper
	For critical dimensions			
Dimensions to 25 mm (1 in)	±0.08 (±0.003)	±0.10 (±0.004)	±0.10 (±0.004)	±0.18 (±0.007)
Each additional 25 mm over 25 to 300 mm (each additional in over 1 in to 12 in)	±0.025 (±0.001)	±0.038 (±0.0015)	±0.038 (±0.0015)	±0.05 (±0.002)
Each additional 25 mm over 300 mm (each additional in over 12 in)	±0.025 (0.001)	±0.025 (±0.001)	±0.025 (±0.001)	
	For noncritical dimensions			
Dimensions to 25 mm (1 in)	±0.25 (±0.010)	±0.25 (±0.010)	±0.25 (±0.010)	±0.35 (±0.014)
Each additional 25 mm over 25 to 300 mm (each additional in over 1 in to 12 in)	±0.038 (±0.0015)	±0.05 (±0.002)	±0.05 (±0.002)	±0.08 (±0.003)
Each additional 25 mm over 300 mm (each additional in over 12 in)	±0.025 (±0.001)	±0.025 (±0.001)	±0.025 (±0.001)	

TABLE 5.4.8 Parting-Line Tolerances, in Addition to Linear-Dimension Tolerances (Based on Single-Cavity Die)*

Projected area of die casting, cm^2 (in^2)	Additional tolerances, die-casting alloy, mm (in)			
	Zinc	Aluminum	Magnesium	Copper
Up to 300 (50)	±0.10 (± 0.004)	±0.13 (±0.005)	±0.13 (±0.005)	±0.13 (±0.005)
300–600 (50–100)	±0.15 (±0.006)	±0.20 (±0.008)	±0.20 (±0.008)	
600–1200 (100–200)	±0.20 (±0.008)	±0.30 (±0.012)	±0.30 (±0.012)	
1200–1800 (200–300)	±0.30 (±0.012)	±0.40 (±0.015)	±0.40 (±0.015)	

*Parting-line tolerances, in addition to linear-dimension tolerances, must be provided when the parting line affects a linear dimension. The above tolerances are to be added to linear tolerances worked out for a dimension as provided in Table 5.4.7.

TABLE 5.4.9 Moving-Die-Part Tolerances, in Addition to Linear-Dimension Tolerances*

Projected area of die casting portion, (cm^2 in^2)	Additional tolerances, die-casting alloy, mm (in)			
	Zinc	Aluminum	Magnesium	Copper
Up to 60 (10)	±0.10 (± 0.004)	±0.13 (±0.005)	±0.13 (± 0.005)	±0.25 (±0.010)
60–120 (10–20)	±0.15 (±0.006)	±0.20 (±0.008)	±0.20 (±0.008)	
120–300 (20–50)	±0.20 (±0.008)	±0.30 (±0.012)	±0.30 (±0.012)	
300–600 (50–100)	±0.30 (±0.012)	±0.40 (±0.015)	±0.40 (±0.015)	

*Moving-die-part tolerances, in addition to linear-dimension tolerances and parting-line tolerances, must be provided when a moving die part affects a linear dimension.

Concentricity tolerance recommendations for various conditions are shown in Table 5.4.10. Commercial flatness tolerances prescribe that a die casting be flat within 0.003 in/in of the maximum dimension measured corner to corner, with a minimum standard of 0.008 in for parts 3 in or less in dimension, as measured by a feeler gauge on a surface plate. A die caster agreeing to meet the tolerance will mechanically straighten the part to this standard if necessary.

Tighter-than-commercial tolerances often can be accommodated, but the procedures required cannot be established precisely until sample castings are subjected to trials. Extremely close tolerances may involve stress relieving and special straightening fixtures. The designer is cautioned against specifying close-tolerance straightness in a design unless it is absolutely necessary, for it is costly to obtain.

MINIATURE DIE CASTINGS*

Die castings under 86 g (3 oz) are classified by some authorities as miniature die castings. Although most design considerations are identical, and although designers will

*This section was contributed by Fred H. Jay, Manager, Sales and Applications Engineering, Zinc Components Division, Fisher Gauge Limited, Petersborough, Ontario.

TABLE 5.4.10 Recommended Concentricity Tolerances

1. Surfaces in fixed relationship in one die section	Diameter of largest surface *A*	Recommended tolerance, TIR
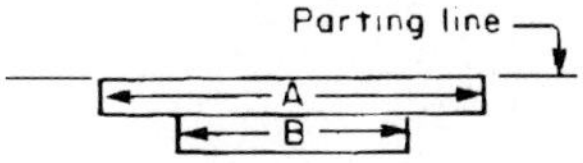	75 mm or less (3 in or less) Over 75 mm (over 3 in)	0.1 mm (0.004 in) 0.1 mm plus 0.0015 mm/additional mm (0.004 in plus 0.0015 in/additional in)
2. Surfaces formed by opposite sections of die, based on single-cavity die	**Projected area of die casting, e.g., area of *C***	**Additional tolerance (added to basic tolerance)**
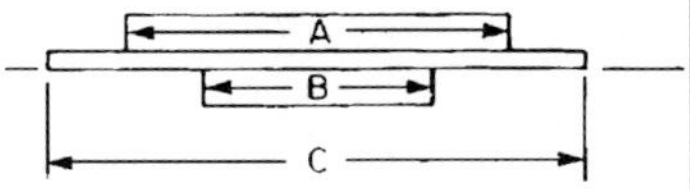	Less than 320 cm^2 (50 in^2) 320–650 cm^2 (50–100 in^2) 650–1300 cm^2 (100–200 in^2) 1300–2000 cm^2 (200–300 in^2)	0.38 mm (0.015 in) 0.50 mm (0.020 in) 0.75 mm (0.030 in) 1.0 mm (0.040 in)
3. Surfaces formed by two moving die members	**Projected area of moving die member, e.g., *D* or *E***	**Additional tolerance added to basic tolerance for each die member**
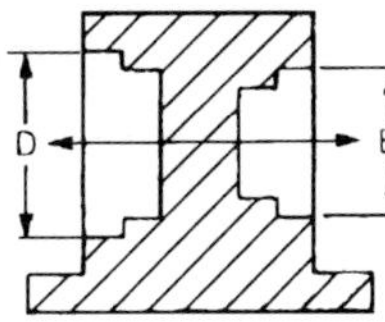	Less than 60 cm^2 (10 in^2) 60–130 cm^2 (10–20 in^2) 130–320 cm^2 (20–50 in^2) 320–650 cm^2 (50–100 in^2)	0.15 mm (0.006 in) 0.30 mm (0.012 in) 0.45 mm (0.018 in) 0.63 mm (0.025 in)

always err on the safe side if they adhere to the principles and recommendations covered above, there are also significant differences between regular-size and miniature die castings:

1. Machines for miniature die casting (virtually always hot-chamber machines) are very often special machines that are highly automatic, combining gate and flash removal in the same operation. Tooling also is often special, in that it is designed for a specific machine and not of standard configuration.
2. Machines are usually quick-cycling, having output rates up to 60 cycles/min.
3. Zinc miniature die castings can, in some instances, be cast flash-free with little or no draft.
4. Again with zinc (the normal material for miniature die castings), small cored holes down to about 0.5 mm (0.020 in) can be produced. Specialized techniques permit coring center bores with walls parallel within 0.01 mm (0.0005 in).
5. Tolerances with miniature zinc die castings can be held to closer values than those presented in Tables 5.4.7 through 5.4.10. Tolerances of ±0.025 mm (±0.001 in) are routinely feasible. Cored holes can be located within 0.025 mm (0.001 in) of true position.

FIGURE 5.4.23 Selection of flash-free miniature zinc die castings. (*Courtesy Fisher Gauge Limited.*)

6. All the preceding factors—high precision, fast cycling, and special tooling and equipment—point to high production quantities as being most advantageous. Production quantities from a few thousand to tens of millions are the province of miniature die castings. Figure 5.4.23 illustrates typical miniature die castings.

S • E • C • T • I • O • N • 6

NONMETALLIC PARTS

CHAPTER 6.1
THERMOSETTING-PLASTIC PARTS

Robert W. Bainbridge
Consultant
Occidental Chemical Corp.
Durez Resins & Molding Materials
North Tonawanda, New York

THERMOSET-MOLDING PROCESSES

Molding compounds of thermosetting materials, or thermosets, when subjected to heat and pressure within the confines of a mold, cure or set into an infusible mass. An irreversible chemical change, involving the cross-linking of molecules of the material, takes place, and a solid part is formed.

Three basic methods of molding may be selected to process thermosets: compression, transfer (pot or plunger), and injection molding. Compression molding is illustrated in Fig. 6.1.1.

The molding operation requires a press large enough to generate the required pressure and a mold mounted into the press and heated by steam, electricity, or hot oil to the recommended temperature. Auxiliary equipment, for preheating the molding compound, loading it, and unloading and finishing the molded parts, may be required.

The molding presses used for the compression or transfer methods of molding are generally vertical. Specially designed rotary presses permit automatic operation. The injection method of molding utilizes a press similar to the thermoplastic-injection design illustrated in Fig. 6.2.1 in Chap. 6.2, except that a heated mold and a special heated barrel and reciprocating screw are used. Transfer molding is a cross between compression and injection molding. A charge of molding material is loaded into a chamber from which the material is forced, by plunger pressure, into the mold cavities.

Molding pressures may range from a low of 350 kPa (50 lbf/in^2) to 80,000 kPa (12,000 lbf/in^2), depending on the type of material or the method of molding. Molding temperatures range from 140°C (280°F) to over 200°C (400°F).

Thermoset-molding compounds may be supplied to the processor in one of the following forms: granular, nodular, pelletized, or flaked. With polyester compounds, gunk, rope, logs, or sheet of various widths, lengths, or thicknesses may be used.

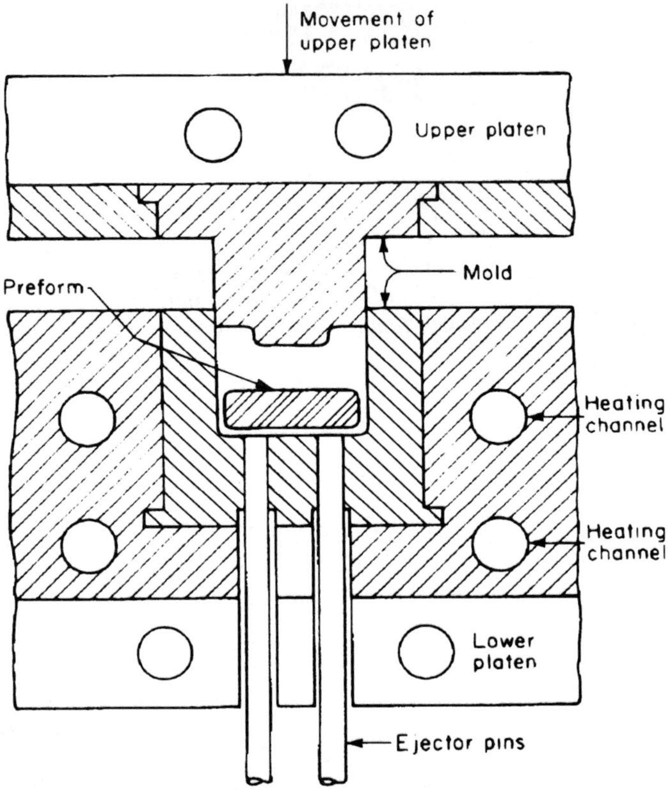

FIGURE 6.1.1 Schematic illustration of compression molding. A preform of thermosetting material is placed in a mold cavity. Heat softens the preform, and the ram pressure forces the material into all portions of the mold cavity, where it takes the shape of the cavity. The material then sets into a solid part that can be removed from the mold.

TYPICAL CHARACTERISTICS OF THERMOSET-MOLDED PARTS

Thermoset compounds, when molded by the preceding processes, produce parts having a wide range of characteristics. As compared with thermoplastic materials, they are generally used in applications that require resistance to higher temperatures and little or no creep. While sometimes flexible or resilient (rubber is a thermosetting material), thermoset parts are more often rigid with a fairly hard surface.

Parts may be molded with side draws or side undercuts, although these present problems in mold design and increase mold costs. Very intricate part designs are feasible and may incorporate molded-in inserts. Miniature electrical connectors are a good example.

Thermoset-molding compounds are formulated to provide specific end results. Phenolic, the workhorse of the thermosets, may be supplied for close-tolerance applications and used in place of die castings or metal stampings on such applications as

FIGURE 6.1.2 Automotive automatic-transmission reactor and thrust washers molded of phenolic resin replace a far larger number of metal parts. (*Courtesy Durez Division, Hooker Chemicals and Plastics Corp.*)

automotive power brakes and transmission parts such as reactors. The latter molded part replaced an assembly of a number of metal parts (see Fig. 6.1.2).

Polyester resins, when combined with various percentages of glass fibers, produce parts of varying degrees of high physical strength (see Chap. 6.6). Bulk-molding-compound (BMC) molded parts range from a simple, small potting case to structural automotive parts, appliance housings, sanitary trays, stall showers, etc. They combine good electrical and physical characteristics with superior surface finish. Polyester and other glass-filled thermoset materials, having very low shrinkage, can be molded within very close tolerances and require little or no finish machining. Because of this and their good strength characteristics, molded parts are competitive with and often cost less than zinc, aluminum, or ferrous castings.

Part sizes of thermoset-molded parts vary from miniature electronic insulators to large structural parts. Wall sections vary from 1.5 mm (0.062 in) to 50 mm (2 in). The thinner the section, the shorter is the cure and the faster is the overall cycle. Depending on mold design, many extremely small openings may be molded in a part.

Thermoset parts are sometimes excluded from decorative applications because they are not normally supplied in a wide range of colors, the exceptions being urea, melamine, alkyd, and polyester compounds. Phenolics can be molded only in black or brown but may be painted, used with melamine overlays, or plated.

Thermoset-molded parts duplicate the mold finish and are relatively free from flow lines, sink marks, and other surface defects. Depending on design selection, the part may have a high-luster finish, a satin finish, or a sand-blasted effect. Raised lettering may be molded in the top or bottom surface of the part. See Figs. 6.1.3 and 6.1.4.

ECONOMIC PRODUCTION QUANTITIES

The cost of the necessary mold to produce the quantity involved is the prime deciding factor for the economics of parts molded from thermosets as compared with other materials. Another factor is the choice of molding process.

On some occasions, when mold cost is very low, compression-molding processes may be adapted to extremely short runs. Normally, however, thermoset-molding methods are most suitable for high-production applications and can be economical for the production of millions of identical parts, especially when automatic transfer or injection molding is used.

The most economical molded part is one produced with minimal finishing costs. A high-quality mold is required to produce identical parts, dimensionally precise, having minimum flash, and free of molded defects. A well-designed and -constructed mold is the key to profitable production, and producers must be prepared to expect costly

FIGURE 6.1.3 Phenolic automotive-carburetor spacer. (*Courtesy Durez Division, Hooker Chemicals and Plastics Corp.*)

FIGURE 6.1.4 Molded hamburger-fryer housings withstand the heat of cooking. (*Courtesy Durez Division, Hooker Chemicals and Plastics Corp.*)

molds. These may range from hundreds of dollars for a simple single-cavity compression mold to many thousands of dollars for an intricate multicavity injection mold.

With captive operations or those which have molding facilities in house, producers also must consider the availability of as well as the capital expenditure required for molding equipment. Depending on this consideration, a short-run job may not be economical. A custom molder, however, generally has press capacity covering small to large tonnage requirements.

Production costs are generally based on a set press-time value for labor and overhead. The larger the number of parts produced per hour or day, the lower is the cost per part. Single-cavity production rates are very costly. Depending on the method of molding and the cross-sectional area of the part, overall cycle time may vary from a low of 15 to 20 seconds to several minutes.

SUITABLE THERMOSETTING MATERIALS

Thermoset-molding compounds consist of a basic resin, one or more of a wide range of fillers, mold-release agents, dye, a catalyst, and perhaps an accelerator. The basic resins used are phenolic, urea, melamines (melamine and phenolic), diallyl phthalate (iso and ortho), polyester (granular, bulk, or sheet molding compounds), alkyds, epoxies, silicones, polyimides, polyurethane, bismaleimides, and some specialty products. These basic resins have been developed for specific market areas, and in a minor number of applications, available materials overlap.

Phenolic compounds

These are low in cost and have been developed to cover a multitude of general-purpose applications. They have very good moldability, excellent heat resistance, good electrical properties, and satisfactory mechanical properties (albeit with limited impact strength). There are a wide range of fillers, each intended to fulfill particular end-product service requirements. Typical among them are wood flour (for general-purpose applications), cotton flock (for slightly better impact strength in general-purpose applications), mica, glass, and other minerals (for better electrical properties, heat resistance, and dimensional stability). General-purpose materials process more easily than mineral-filled varieties. Long-fiber-filled materials may present problems in loading molds, and finished parts may be difficult to deflash. Compounds are formulated with a one-stage or a two-stage phenolic resin. In general, one-stage resins are slightly more critical to process. Although phenolics have properties somewhat inferior to those of some more expensive thermosets, they are usually more easily molded. Phenolic compounds are molded by all three molding methods. For injection molding, however, special formulations may be used.

Urea Compounds

These may be produced in a wide range of pastel shades. They have fair arc resistance and are used on electrical applications, including house-wiring devices. They are also used for closures.

Urea compounds are normally processed by compression molding and may present problems if transfer- or injection-molded. They have a relatively short shelf life and will stiffen on storage.

Molded parts do incur postmold shrinkage, and this should be considered when designing parts.

Melamine Compounds

These also are produced in a wide range of colors and can be formulated to produce parts with a hard surface. Their main markets are dishware and electrical applications. They have superior arc track resistance and, when filled with glass fibers, are used in heavy switchgear.

In general, melamines are molded by using the compression method, but they also may be transfer-molded. Injection molding may present problems. Highly loaded glass-filled materials are difficult to load into molds. Short shelf life and a large post-mold shrinkage are disadvantages of these compounds.

Diallyl Phthalate Molding Compounds

These were developed primarily for superior electrical-insulation values in high-moisture and elevated-temperature environments. Their main market area is electronic apparatus.

These compounds are formulated by using glass or mineral fillers and normally are supplied in a granular form. They process easily by all three methods of molding, but with high-volume miniature electronic components, the transfer and injection methods are used most extensively.

Polyester Molding Compounds

These compounds are invariably supplied with fiberglass reinforcement and are covered in Chap. 6.6.

Alkyds

These constitute a family of materials formulated from polyester-type resins and other materials. They produce hard, stiff parts. Alkyd parts have excellent electric-arc resistance and are used in apparatus such as switchgear, motor brush holders, TV tuners, and automotive ignition parts.

Alkyds can be used at temperatures up to 220°C (425°F). They are supplied in granular form and are processed by compression, transfer, and parting-line injection molding. Molding is easy, and cure rates are rapid.

Epoxy Molding Compounds

These are generally supplied in a granular form, which presents few problems in processing. They are used primarily for the encapsulation of electrical coils and various electronic components. The compounds are processed by transfer molding at extremely low pressures (in the range of 350 to 700 kPa, or about 50 to 100 lbf/in^2).

Silicone Molding Compounds

These have superior long-term heat resistance, excellent electrical properties, and good resistance to moisture and chemicals. They are used in high-temperature electrical-switch or electronic-apparatus applications. They can be cast as well as compression- or transfer-molded. Flexible varieties are used in sealants and rubber-like parts.

Bismaleimides

These have similar applications as epoxies but are particularly suited to high-temperature applications. They are also used in the manufacture of printed circuit boards.

Polyimides

These also have excellent high-temperature capability and are used for insulation, printed-circuit boards, and chip carriers for integrated circuits.

DESIGN RECOMMENDATIONS

Effects of Shrinkage

Shrinkage on curing and cooling of the thermosetting material is a factor to be considered by the design engineer. It is a function of the type of resin, the fillers used in the formulation, the plasticity of the compound, the degree of preheat of the molding compound, the mold temperature, and the molding pressure. Wall thickness and variations within a part and the direction of flow of material in the cavity may cause variations within the mold cavity, which, if severe, may cause warpage and shrinkage. Internal shrinkage caused by poor design may cause internal stresses that result in warpage or weak parts. If extremely close tolerances are required, glass- or mineral-filled materials should be specified.

Table 6.1.1 indicates normal minimum and maximum shrinkage rates during molding for various thermosets.

TABLE 6.1.1 Minimum and Maximum Shrinkage Rates during Molding for Various Thermosetting-Plastic Materials

Material	Percent shrinkage during molding
Phenolic	0.1–0.9
Urea	0.6–1.4
Melamine	0.8–1.2*
Diallyl phthalate	0.3–0.7
Alkyd	0.5–1.0
Polyester	0–0.7
Epoxy	0.1–1.0
Silicones	0–0.5

*Except with special glass fillers.

Wall Thickness

Uniform wall sections will help to produce warp-free and strain-free molded parts. Dimensional variations caused by induced stresses from material shrinkage are also accentuated by uneven thicknesses or abrupt wall-thickness changes. The coring of thick areas to maintain uniform wall thickness, as illustrated in Fig. 6.1.5, is advisable.

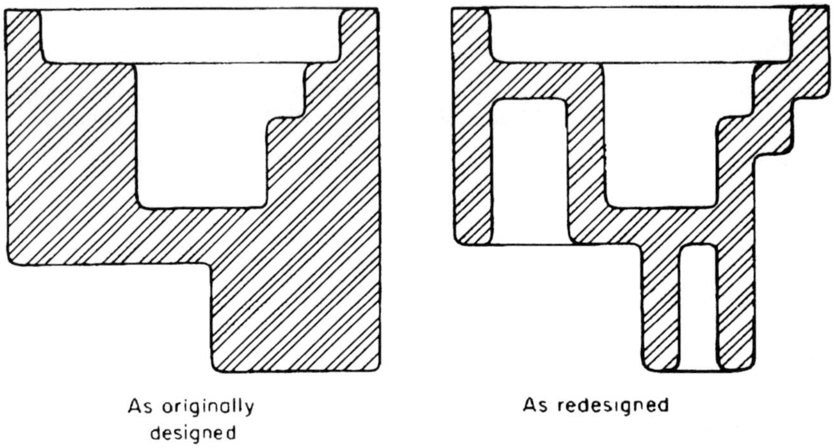

FIGURE 6.1.5 Core out thick sections to maintain uniform wall thickness in thermoset parts.

Sections that are too thin are more apt to break in handling, may restrict the flow of material, and may trap air or volatiles, causing a defective part and staining the mold. Too heavy a wall, on the other hand, will slow the curing cycle and add to materials cost. Recommended wall thicknesses for various thermosetting materials are shown in Table 6.1.2.

TABLE 6.1.2 Suggested Wall Thickness for Parts Molded from Various Thermosetting-Plastic Materials

Material	Small parts, mm (in)	Average-size parts, mm (in)	Large parts, mm (in)
Phenolic	1.5–3.0 (0.062–0.125)	2.4–4.8 (0.093–0.187)	4.8–25 (0.187–1.000)
Urea	1.5–3.0 (0.062–0.125)	2.4–4.8 (0.093–0.187)	4.8–9.5 (0.187–0.375)
Melamines	1.5–3.0 (0.062–0.125)	2.4–4.8 (0.093–0.187)	4.8–9.5 (0.187–0.375)
Diallyl phthalate	1.1–2.4 (0.045–0.093)	2.0–4.0 (0.078–0.156)	3.0–9.5 (0.125–0.375)
Alkyd	2.0–3.0 (0.078–0.125)	2.5–4.8 (0.100–0.187)	4.8–12.7 (0.187–0.500)
Polyester			
Granular	2.0–3.0 (0.078–0.125)	2.5–4.8 (0.100–0.187)	4.8–12.7 (0.187–0.500)
Bulk	1.1–2.4 (0.045–0.093)	2.0–4.0 (0.078–0.156)	3.0–9.5 (0.125–0.375)
Sheet	1.5–2.4 (0.062–0.093)	2.4–4.8 (0.093–0.187)	4.8–9.5 (0.187–0.375)
Epoxy	1.0–2.0 (0.040–0.078)	2.0–4.8 (0.078–0.187)	
Silicones	1.0–2.0 (0.040–0.078)	2.0–3.0 (0.078–0.125)	

Undercuts

Thermoset-molded parts must be designed so that they may easily be removed from the mold. If external undercuts are essential, straight draw is not possible, and side-activated split molds or removable mold sections are required. This results in increased mold and piece-part cost. Also inevitable is additional parting or flash line, which may require additional finishing operations for removal. Internal undercuts in a part are almost impossible to mold and should be avoided. (See Fig. 6.1.6.)

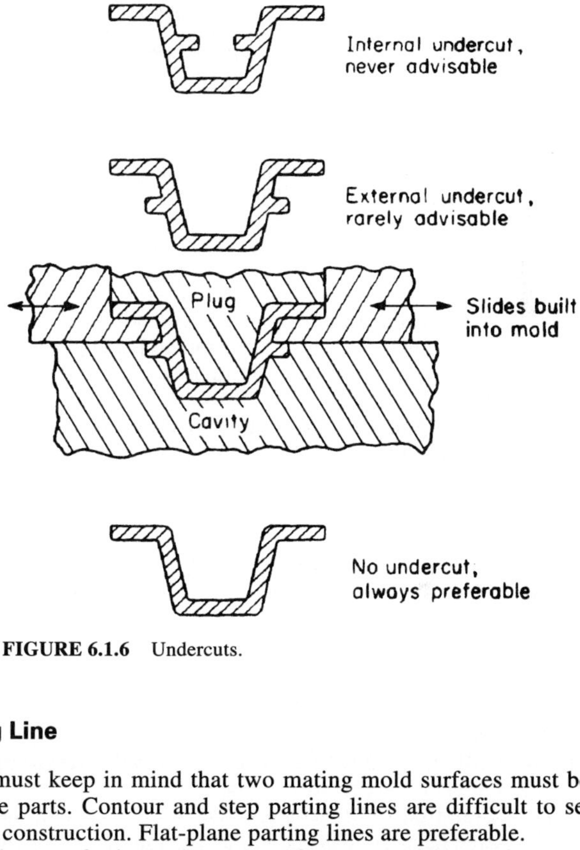

FIGURE 6.1.6 Undercuts.

Mold Parting Line

The designer must keep in mind that two mating mold surfaces must be sealed off to mold flash-free parts. Contour and step parting lines are difficult to seal off and are costly in mold construction. Flat-plane parting lines are preferable.

If there is danger of mismatch between force and cavity, design a bead or specific mismatch (see Figs. 6.2.27 and 6.2.28).* When a part is molded by the transfer or injection process, the parting-line problem is greatly simplified because the two halves of the mold are in a closed or locked position before material enters it. Sharp edges at parting lines simplify mold construction in many cases by avoiding the need for matching the position of mating cavities. However, brittle thermosets may chip when edges are sharp. The design shown in Fig. 6.1.7, though more costly to tool, may be more satisfactory in the long run.

*Some design principles, noted in this chapter, that are common to both thermosets and thermoplastics are illustrated in Chap. 6.2, "Injection-Molded Thermoplastic Parts."

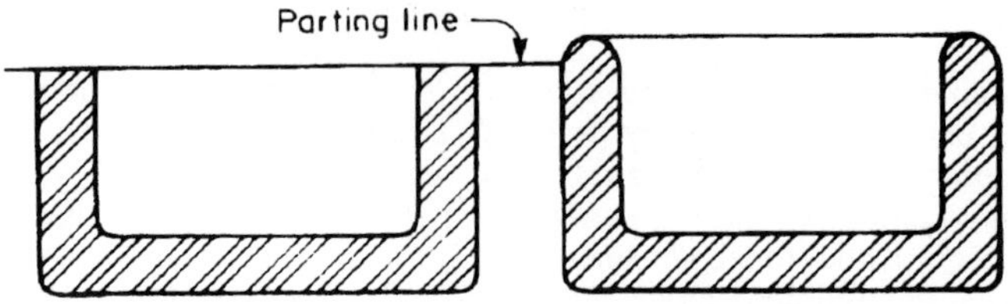

FIGURE 6.1.7 The part on the left, if molded from brittle material, could chip at sharp edges. The part on the right avoids this problem, albeit at a higher mold cost.

Vertical decoration or fluted design on the side of a part should stop short of the parting line to facilitate flash removal. (See Fig. 6.2.25.)

Sharp Corners

All corners should have a radius or fillet except at set-in sections of the mold or at the parting line. Radii and fillets incorporated in the molded part reduce chipping, simplify mold construction, add strength to both mold members and the molded part, and in general, improve appearance.

Certain thermosets may be improved for flow of material in the mold if rounded corners are incorporated. The flow will be unretarded and will result in uniform density and less molded-in stresses.

Specify fillets and corner radii of 0.8 mm to 1.1 mm (0.032 to 0.045 in).

Holes or Openings

Steel pins or sections are needed in the mold to incorporate holes or irregularly shaped openings into a thermoset part. Depending on the method of molding, these mold sections are subjected to varying pressures that may cause them to break or bend. As the plastic material flows around these mold sections, a "weld line" or "knit line" may be formed adjacent to the mold section. (See Fig. 6.2.8.)

The spacing between holes and next to sidewalls should be as large as possible. Recommended minimum values are shown in Fig. 6.1.8.

Through holes are preferred to blind holes, but in compression molding they are difficult to mold flash-free because material enters the mold cavity before the pins are seated. Flash may occur inside or outside the hole, depending on the mold pin design.

Blind-hole depths, especially when the compression-molding process is used, should not be more than $2\frac{1}{2}$ times the diameter. For holes of 1.5 mm ($\frac{1}{16}$ in) or less, a depth limit of one diameter is suggested. (See Fig. 6.1.9.)

To aid in adding strength to the mold section, permit pins to have a generous fillet or conical surface at the base. This will permit the molded hole to have a radius or countersink at the entrance, which is often desirable.

Side holes should be avoided since they require activated side cores. See Fig. 5.2.4 for a possible design change to avoid this problem.

When end-butting pins are used to produce through holes, it is advisable to allow the portion of the hole produced by one of the pins to be slightly oversize. This per-

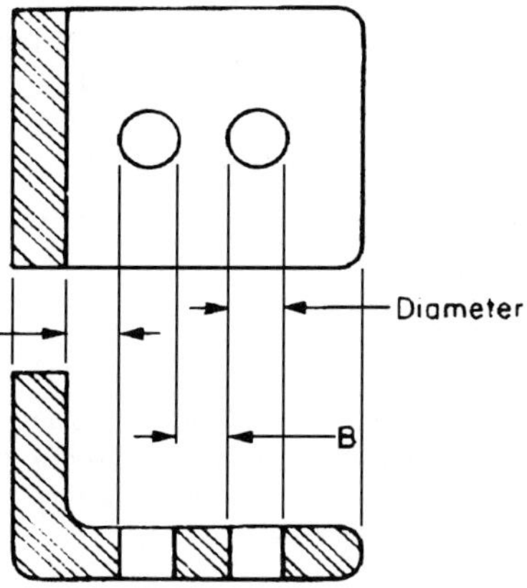

FIGURE 6.1.8 Minimum recommended hole spacing in thermosets.

Diameter of hole, mm (in)	Minimum distance *A* to sidewall	Minimum distance *B* between holes
1.5 (1/16)	1.5 (1/16)	1.5 (1/16)
3.0 (1/8)	2.4 (3/32)	2.4 (3/32)
4.8 (3/16)	3.0 (1/8)	3.0 (1/8)
6.3 (1/4)	3.0 (1/8)	4.0 (5/32)
9.5 (3/8)	4.0 (5/32)	4.8 (3/16)
12.7 (1/2)	4.8 (3/16)	5.6 (7/32)

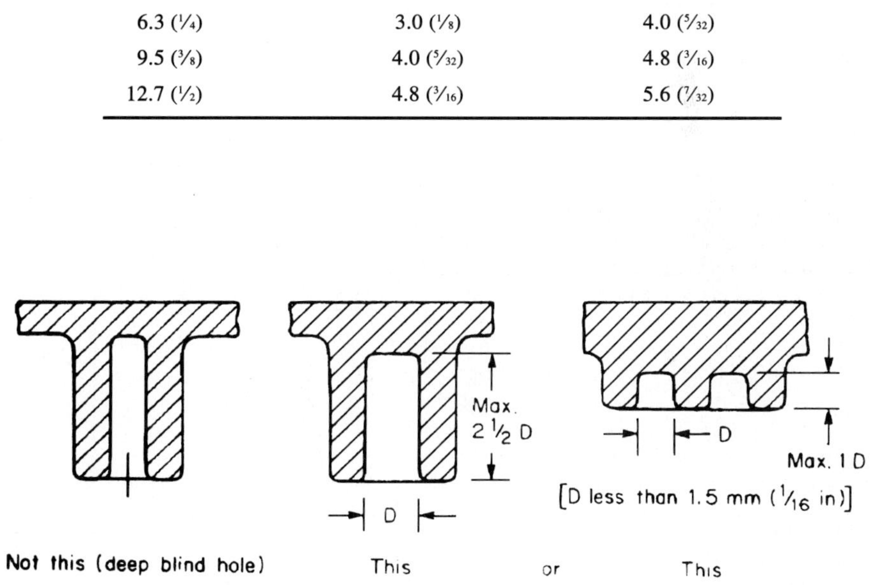

FIGURE 6.1.9 Depth limits for compression-molded blind holes.

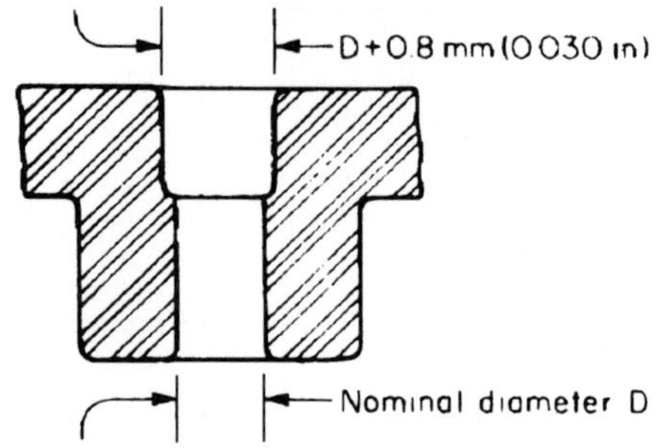

FIGURE 6.1.10 Suggested design for through holes to be produced in a compression-molded part by end-butting mold pins.

mits the minimum through-hole dimension (as for fastener clearance) to be maintained even if there is a slight mismatch or shifting of mold halves. A suggested design is shown in Fig. 6.1.10.

Ribs

Molded ribs may be incorporated to increase strength or decrease warpage of thermoset parts. The width of the base of the rib should be less than the thickness of the wall to which it is attached. Use of two or more ribs is better than increasing the height of a single rib. In this case, center-to-center spacing should be twice the base width of the ribs. In all cases, provide general draft to ease removal from the mold. (See Fig. 6.1.11.)

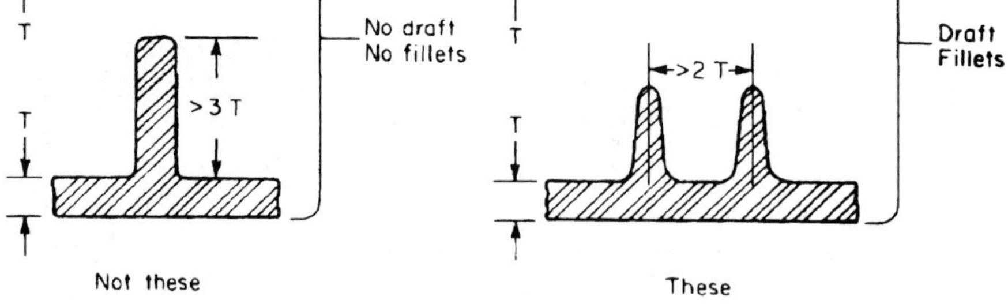

FIGURE 6.1.11 Design rules for ribs in thermoset parts.

Mounting Bosses

Protruding bosses or pads are used for mounting and support around holes. The boss height should not be more than twice the diameter. Provide a generous radius at the base of the boss for strength and ample draft for easy part removal from the mold. If increased height is absolutely necessary, add ribs to the outside diameter of the boss. Incorporate radii at the junction of the boss to the wall section to aid in the flow of material and supply additional strength. (See Fig. 6.1.12.)

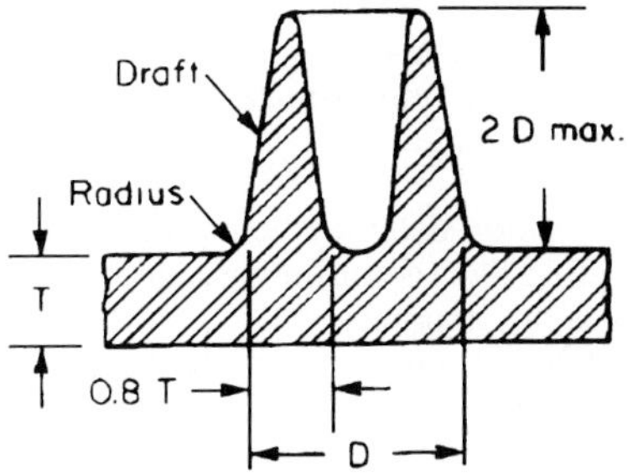

FIGURE 6.1.12 Design rules for bosses in thermoset parts.

If the molded part is to fit on a flat surface, it may be advisable to provide pads or bosses at the corners, which may be machined more easily to a flat surface. (See Fig. 6.1.13.)

Draft

Taper, or draft, should be provided both inside and outside the thermoset part. Inside surfaces have greater need for draft because a molded part tends to shrink toward the mold surface rather than away from it. When using the compression-molding process, especially in deep-drawn shapes, converging tapers assist by creating a wedging or compression action while the material flows up the sidewall as the mold is closed. This action increases the density of the part in the parting-line area. At least $\frac{1}{2}°$ and preferably 1° of taper is suggested on sidewall sections, provided the mold is polished to a high-luster finish free from all tool marks. With certain low-shrinkage materials, greater drafts may be required. Sidewalls of deep parts usually do not require as large a draft angle as those of shallow parts.

Screw Threads

Both internal and external threads may be molded in most thermosets, but increased part and mold costs may be expected. Articles with molded male or external threads

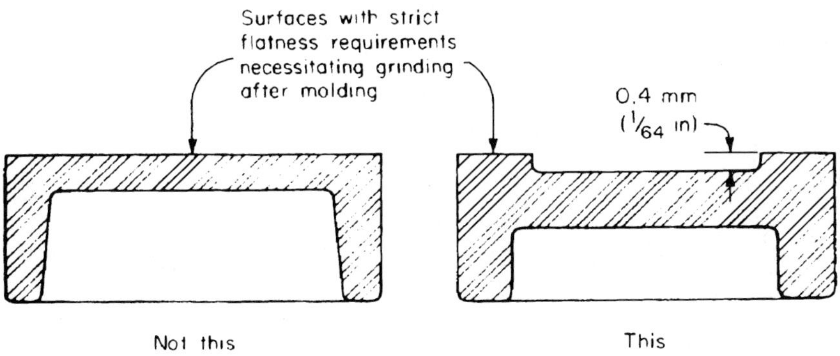

FIGURE 6.1.13 Surfaces to be sanded or ground flat should be kept small in area.

may be removed from the mold by unscrewing the part from the mold, or the threaded section may be in a removable unscrewable section of the mold.

Some thermoset materials may be formulated to permit stripping the part from the threaded mold member. The threads must be well rounded on the resisting side and shallow in depth. (See Fig. 6.2.18.)

Threads also may be molded in thermosets (as they can be with thermoplastics) by locating the axis of the threaded section on the mold-parting plane. Figure 6.2.17 illustrates this approach.

Molded screw threads in thermosets should be as coarse as possible with rounded-tooth-form edges. The finest moldable thread pitch is 32 pitch, but far coarser threads are preferable. Fits should be specified to be no closer than Class 1 or Class 2. In many cases, it is more practical to machine the threads after molding than to incorporate them in the molding operation.

If screws are to be removed or replaced several times in a part's lifetime, or if high holding strength is needed, it is suggested that metal inserts be used.

Inserts

Metal reinforcing members can be incorporated into parts molded from thermosetting plastics. However, molded-in inserts can cause problems. External metal inserts may become loose owing to shrinkage of the plastics material. Internal inserts may present problems with material cracking around them because of postmold shrinkage. Table 6.1.3 indicates the minimum wall thickness to be used with plain round inserts to avoid this problem.

Molded-in inserts slow production rates because of the time required to position them into the mold during each cycle. In the injection-molding process, a vertical press or a specially designed shuttle press may be required for ease of insert loading. Investigate the possibility of pressing an insert into the part immediately after molding, and let material shrink around the outside diameter of the insert. Epoxy cement may ensure better anchorage.

Design standards for inserts are found in Chap. 6.2.

TABLE 6.1.3 Suggested Minimum Wall Thickness around Metal Inserts Molded in Thermosetting Plastics

	Wall thickness, mm (in)					
Insert diameter	3.0 (⅛)	6.3 (¼)	9.5 (⅜)	12.7 (½)	19 (¾)	25 (1)
General-purpose phenolic	2.4 (0.093)	4.0 (0.156)	4.8 (0.187)	5.5 (0.218)	7.9 (0.312)	8.7 (0.343)
Medium-impact phenolic	2.0 (0.078)	3.6 (0.140)	4.0 (0.156)	5.2 (0.203)	7.1 (0.281)	7.9 (0.312)
High-impact phenolic	1.6 (0.062)	3.2 (0.125)	3.6 (0.140)	4.8 (0.187)	6.3 (0.250)	7.1 (0.281)
Urea	2.4 (0.093)	4.0 (0.156)	4.8 (0.187)	7.1 (0.281)	7.9 (0.312)	8.7 (0.343)
Melamine	3.2 (0.125)	4.8 (0.187)	5.5 (0.218)	7.9 (0.312)	8.7 (0.343)	9.5 (0.375)
Alkyd	3.2 (0.125)	4.8 (0.187)	4.8 (0.187)	7.9 (0.312)	8.7 (0.343)	9.5 (0.375)
Epoxy	0.5 (0.020)	0.8 (0.030)	1.0 (0.040)	1.3 (0.050)	1.5 (0.060)	1.8 (0.070)
Diallyl phthalate	3.2 (0.125)	4.8 (0.187)	6.3 (0.250)	7.9 (0.312)	8.7 (0.343)	9.5 (0.375)

Source: From *Plastic Product Design,* © Litton Educational Publishing, Inc., R. D. Beck, Van Nostrand Reinhold, New York, 1974.

Surface Flatness

Perfectly flat molded surfaces are not feasible with thermosets because of warpage due to shrinkage and trapped air or gas. For improved surface finish and strength of parts molded from rigid thermosets, a slight dome of 0.3 percent should be incorporated in the surface.

Part Drawings

The final piece-part drawing should clearly indicate the area where no parting line may appear, the area where gate marks are not permissible, those areas where cavity numbers and mold marks may appear (these are really for the convenience of the molder and provide a positive means of identification in troubleshooting or in checking defective parts), and those molded surfaces where imperfections are not permissible. Knockout pins, vent pins, etc., are necessary to mold and remove the part from the mold properly, but all leave surface marks. The mold finish should be specified, and if available, the Society of the Plastics Industry/Society of Plastics Engineers (SPI/SPE) mold-finish comparison kit should be used as a guide.

DIMENSIONAL FACTORS

Thermoset compounds can be molded by any of the three processes to what are considered practical close tolerances. However, in no way can they be molded to tolerances as close as those held in machining metal parts. The reasons for this are as follows:

1. Shrinkage of material as it sets
2. Postmold shrinkage (particularly high with urea and melamine)
3. Time, temperature, and pressure variations during molding
4. Mold variations and mold wear
5. Variations in the shot or preform size
6. Gating and material-flow variations
7. Variations in material from batch to batch

Close dimensional tolerances may greatly increase the cost of the mold and the molded part. Increased part cost may be the result of the necessity for shrink fixtures, extra finishing operations, the necessity for postmolding baking operations, and perhaps, higher scrap rates. It is suggested that the designer consult the molder or processor for tolerance values on critical dimensions.

TOLERANCES

Table 6.1.4 presents data taken from SPI tolerance standards for common thermosetting materials. (See Fig. 6.1.14.) Note that the trade customs outlined in these standards represent the historic and customary practices prevailing in the plastics-molding industry. Contract forms or other agreements of individual molders may vary.

TABLE 6.1.4 Recommended Dimensional Tolerances for Thermosetting Plastic Parts

Drawing code (see Fig. 6.1.14)	Dimensions, in	Alkyd		Diallyl phthalate		Epoxy		Melamine urea		General-purpose phenolic		Fiber-filled phenolic	
		Commercial	Fine	Commercial	Fine	Commercial	Fine	Commercial	Fine	Commercial	Fine	Commercial	Fine
A =	To 1.000	0.004	0.003	0.0035	0.0015	0.005	0.0035	0.006	0.004	0.006	0.003	0.006	0.0035
diameter*	1.000–2.000	0.005	0.003	0.004	0.002	0.006	0.004	0.008	0.005	0.008	0.004	0.008	0.004
B =	2.000–3.000	0.006	0.004	0.0045	0.0025	0.006	0.005	0.010	0.006	0.009	0.005	0.009	0.005
depth†	3.000–4.000	0.007	0.005	0.005	0.003	0.007	0.0055	0.013	0.008	0.010	0.006	0.011	0.006
C =	4.000–5.000	0.009	0.006	0.0055	0.0035	0.008	0.006	0.015	0.009	0.011	0.007	0.012	0.007
height†	5.000–6.000	0.010	0.007	0.006	0.004	0.009	0.007	0.017	0.010	0.013	0.008	0.014	0.008
	6.000–12.000, for each additional inch add	0.0015	0.001	0.0015	0.001	0.0015	0.001	0.003	0.002	0.002	0.001	0.003	0.002
D = bottom wall†		0.002	0.001	0.005	0.003	0.002	0.001	0.005	0.003	0.008	0.005	0.006	0.004
E = side-wall		0.002	0.001	0.003	0.002	0.002	0.001	0.004	0.002	0.005	0.003	0.004	0.003
F = hole	0.000–0.125	0.002	0.001	0.002	0.001	0.002	0.001	0.003	0.002	0.002	0.001	0.002	0.001
diameter	0.125–0.250	0.002	0.002	0.002	0.001	0.002	0.002	0.003	0.002	0.002	0.001	0.003	0.002
	0.250–0.500	0.002	0.002	0.002	0.001	0.002	0.002	0.004	0.003	0.003	0.002	0.004	0.003
	0.500 and over	0.004	0.003	0.0025	0.002	0.004	0.003	0.005	0.004	0.003	0.002	0.005	0.003
G = hole	0.000–0.250	0.002	0.002	0.002	0.001	0.002	0.002	0.003	0.002	0.004	0.002	0.004	0.002
depth	0.250–0.500	0.002	0.002	0.003	0.002	0.002	0.002	0.004	0.002	0.004	0.002	0.005	0.003
	0.500–01.000	0.002	0.002	0.005	0.003	0.002	0.002	0.005	0.002	0.005	0.003	0.007	0.004
Flatness	0.000–3.000	0.010	0.010	0.010	0.005	0.010	0.010	0.012	0.008	0.010	0.005	0.014	0.008
	3.000–6.000	0.012	0.015	0.012	0.008	0.010	0.015	0.018	0.013	0.012	0.010	0.021	0.014
Concentricity	TIR	0.005	0.005	0.005	0.003	0.005	0.005	0.007	0.005	0.005	0.003	0.007	0.004

*These tolerances do not include an allowance for aging characteristics of the material.

†Parting line must be taken into consideration.

Source: Data taken from *Standards and Practices of Plastics Custom Molders,* The Society of the Plastics Industry, Inc., New York, 10017.
All values are in plus or minus inches unless otherwise specified.

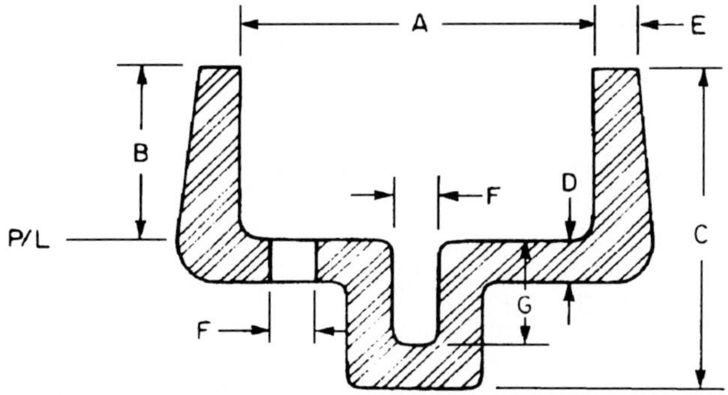

FIGURE 6.1.14 Cross-sectional view of a typical plastics part for use in identifying dimensions tabulated in Tables 6.1.4, 6.2.4, and 6.2.5.

CHAPTER 6.2
INJECTION-MOLDED THERMOPLASTIC PARTS

Thermoplastic materials are synthetic organic chemical compounds of high molecular weight that soften or liquefy when they are heated and solidify when they are cooled. When cooled, they are relatively tough and durable and suitable for a wide variety of product applications.

THE PROCESS

These materials are formed to specific shapes by injecting them when heated into a mold from which they take their final shape as they cool and solidify. The plastic material normally is received by the molder in granular form. It is placed in the hopper of an injection-molding machine, from which it is fed to a heated cylinder. As the granules heat in the cylinder, they melt, or plasticize. A typical melting temperature is about 180°C (350°F), although this varies with different materials and molding conditions. The mold, usually of steel, is clamped in the machine and is water-cooled. A plunger forces plasticized material from the cylinder into the mold. There it cools and solidifies. The mold is opened, and the molded part with its attached runners is removed. The process, with the occasional exception of part removal, is automatic. It requires about 45 s/cycle, more or less, with most of that time being devoted to cooling of the material in the mold. Very high pressures, on the order of 70,000 kPa (10,000 lbf/in^2) or more, are required during injection. Figure 6.2.1 illustrates the injection-molding process.

TYPICAL CHARACTERISTICS OF INJECTION-MOLDED PARTS

Injection molding is particularly advantageous when intricate parts must be produced in large quantities. Although there are limitations, as discussed below, generally the more irregular and intricate the part, the more likely it is that injection molding will be economical. In fact, one major advantage of the injection-molding method is that one molded part can replace what would otherwise be an assembly of components. (See Fig. 6.2.2.) In addition, color and surface finish often can be molded directly onto the part, so that secondary finishing operations are not necessary.

Injection-molded parts are generally thin-walled. Heavy sections and variable wall thicknesses are possible, though they are normally not recommended.

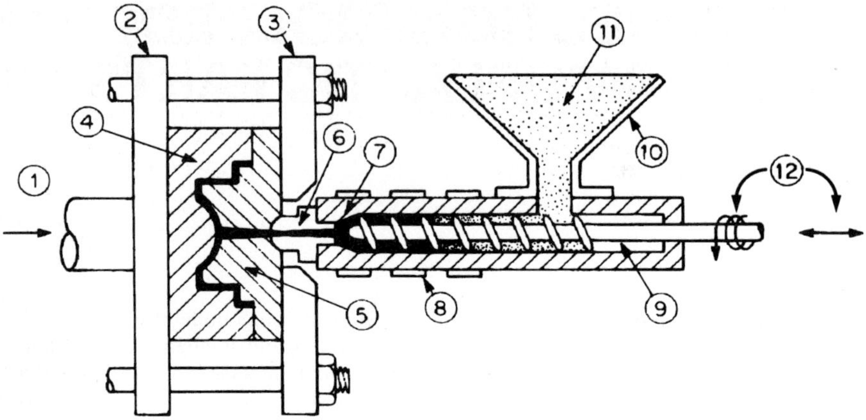

FIGURE 6.2.1 Injection molding. (1) Mold-clamping force. (2) Movable mold platen. (3) Fixed platen. (4) Cavity half of mold. (5) Force half of mold. (6) Nozzle. (7) Cylinder. (8) Electric band heaters. (9) Reciprocating screw. (10) Hopper. (11) Granulated-plastic material. (12) Rotary and reciprocating motion of screw.

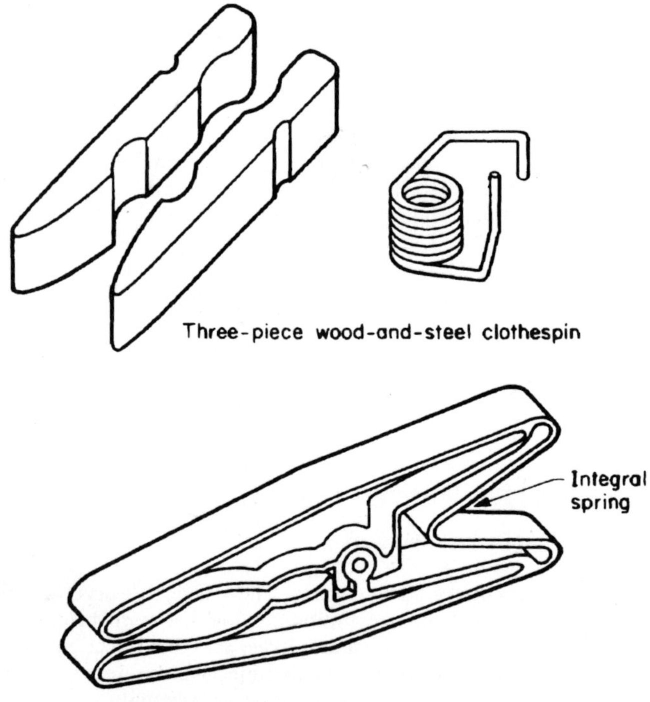

FIGURE 6.2.2 One-piece injection-molded clothespin. Two sections, connected by an integral spring, are assembled together at the pivot point after molding. (*Courtesy Celanese Plastics & Specialties Co.*)

Because thermoplastics are generally less strong than metals, they are more apt to be found in less highly stressed applications. Housings and covers are common uses rather than, for example, frames and connecting rods. However, thermoplastic materials are gradually being developed with better and better strength characteristics and are increasingly finding themselves used for moving parts and in more structural applications. The "engineering plastics," nylon, polycarbonate, acetal, phenylene oxide, polysulfone, thermoplastic polyesters, and others, particularly when reinforced with glass or other fibers, are functionally competitive with zinc, aluminum, and even steel.

Effects of Shrinkage

All thermoplastics exhibit shrinkage on cooling and solidification. Table 6.2.1 summarizes the extent of this shrinkage for common materials. In addition to effecting a reduction in most dimensions, shrinkage of plastic material causes various irregularities and warpage in the molded part. The most common such defect is the sink mark, or surface depression, opposite heavy sections. (See Fig. 6.2.3.)

TABLE 6.2.1 Shrinkage Rates of Common Thermoplastics upon Solidification in Mold

Thermoplastic	Percent
Acetal	2.0–2.5
Acrylic	0.3–0.8
Acrylonitrile butadiene styrene	0.3–0.8
Nylon	0.3–1.5
Polycarbonate	0.5–0.7
Polyethylene	1.5–5.0
Polypropylene	1.0–2.5
Polystyrene	0.2–0.6
Polyvinyl chloride, rigid	0.1–0.5
Polyvinyl chloride, flexible	1.0–5.0

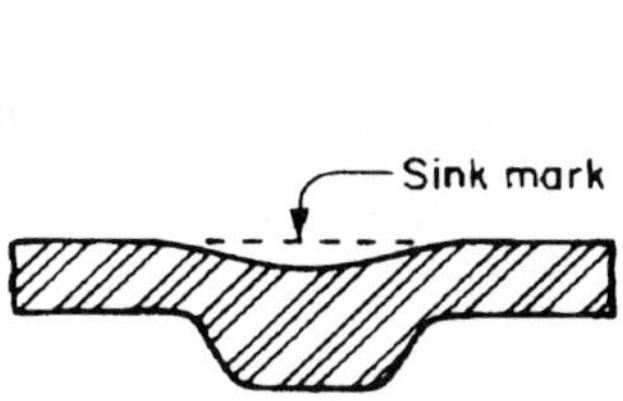

FIGURE 6.2.3 Typical sink mark opposite a heavy section.

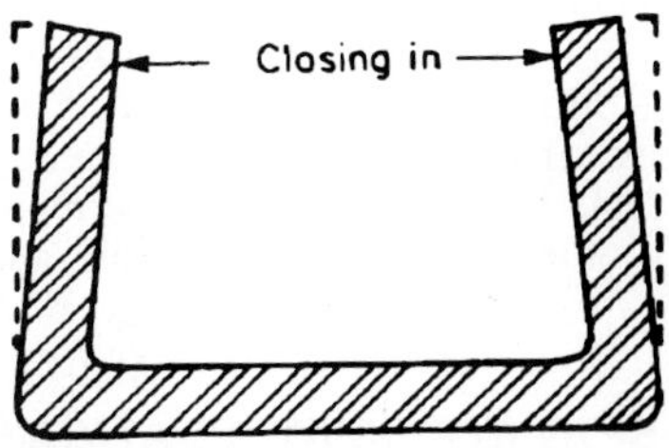

FIGURE 6.2.4 Shrinkage of plastic material on cooling causes the closing in of U-shaped sections.

Another common effect is the closing in of a U-shaped cross section, particularly if there are reinforcing ribs. (See Fig. 6.2.4.) A third common effect of shrinkage is the

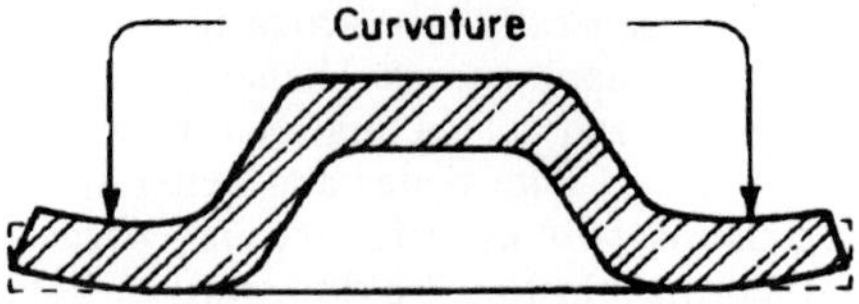

FIGURE 6.2.5 Curving of flat surfaces caused by shrinkage of material.

occurrence of curvature on a flat surface in the direction of a boss, protuberance, or added material. Figure 6.2.5 illustrates this problem.

ECONOMIC PRODUCTION QUANTITIES

Injection molding is a mass-production process. It is generally not applicable unless 10,000 or more identical parts are to be produced. The reason for this limitation is the necessity for constructing a unique mold for each part. Production must be large enough so that the mold cost can be amortized over the quantity manufactured. Even for smaller parts, molds can be costly, on the order of several thousands of dollars. For larger, intricate parts, they can cost tens or hundreds of thousands of dollars.

SUITABLE MATERIALS

A large number of suitable thermoplastics are available to the injection molder. Some of the more commonly used are polyethylene, polypropylene, polystyrene, polyvinyl chloride (vinyl or PVC), nylon, acrylonitrile butadiene styrene (ABS), and acrylic.

Because of the importance of injection molding to the commercial sale of thermoplastic materials, producers of these materials engineer them to be processible by injection molding. Physical properties and cost rather than processibility are normally the determining factors in the selection of materials for injection-molded parts.

Generally, the high-property engineering plastics are not as easy to mold as the commodity plastics such as polyethylene, polypropylene, and polystyrene. In addition, polyvinyl chloride, though low in cost and having very good physical properties, is more difficult to injection-mold than many other materials. Its prime drawback is a narrow temperature range between its melting and degradation points.

Table 6.2.2 lists common thermoplastics used for injection molding and indicates some of their properties, their cost, and their typical applications.

DESIGN RECOMMENDATIONS

Gate and Ejector-Pin Locations

The designer should consider the location of these elements because they can impair surface finish. Ejector pins usually can be located on the underside of a part if it has an outside and an underside. Gates can be located in a number of locations, as illus-

TABLE 6.2.2 Common Thermoplastics Materials Used for Injection Molding

Material	Tensile strength, MPa (lbf/in^2)	Maximum service temperature, °C (°F)	Specific gravity	Approximate cost index*	Applications and remarks
Polystyrene	48 (7000)	82 (180)	1.04	1.0	Toys, containers
Polypropylene	34 (5000)	110 (230)	0.91	1.1	Housewares, appliances
High-density polyethylene	28 (4000)	125 (260)	0.96	1.1	Refrigerator parts, housewares
Low-density polyethylene	14 (2000)	88 (190)	0.92	1.3	Kitchen utility wear, toys
Acrylonitrile butadiene styrene (ABS)	48 (7000)	93 (200)	1.06	2.1	Housings, housewares; can be chrome-plated; handle grips; electrical plugs
Polyvinyl chloride, flexible	21 (3000)	82 (180)	1.40	1.0	Seals, electrical plugs, footwear components
Nylon	76 (11,000)	120 (250)	1.14	3.9	Bearings, gears, rollers
Styrene acrylonitrile	69 (10,000)	82 (180)	1.05	2.7	Kitchenware
Cellulosics	41 (6000)	70 (160)	1.22	4.0	Optical parts, tool handles
Polyacetal	69 (10,000)	90 (195)	1.43	3.3	Appliance parts, gears, etc.
Polycarbonate	62 (9000)	110 (230)	1.22	4.1	Portable-tool housings, automotive parts
Acrylic	69 (10,000)	78 (170)	1.18	2.0	Lenses, automotive trim

*1997 prices for large-quantity purchases; polystyrene = 1.0.

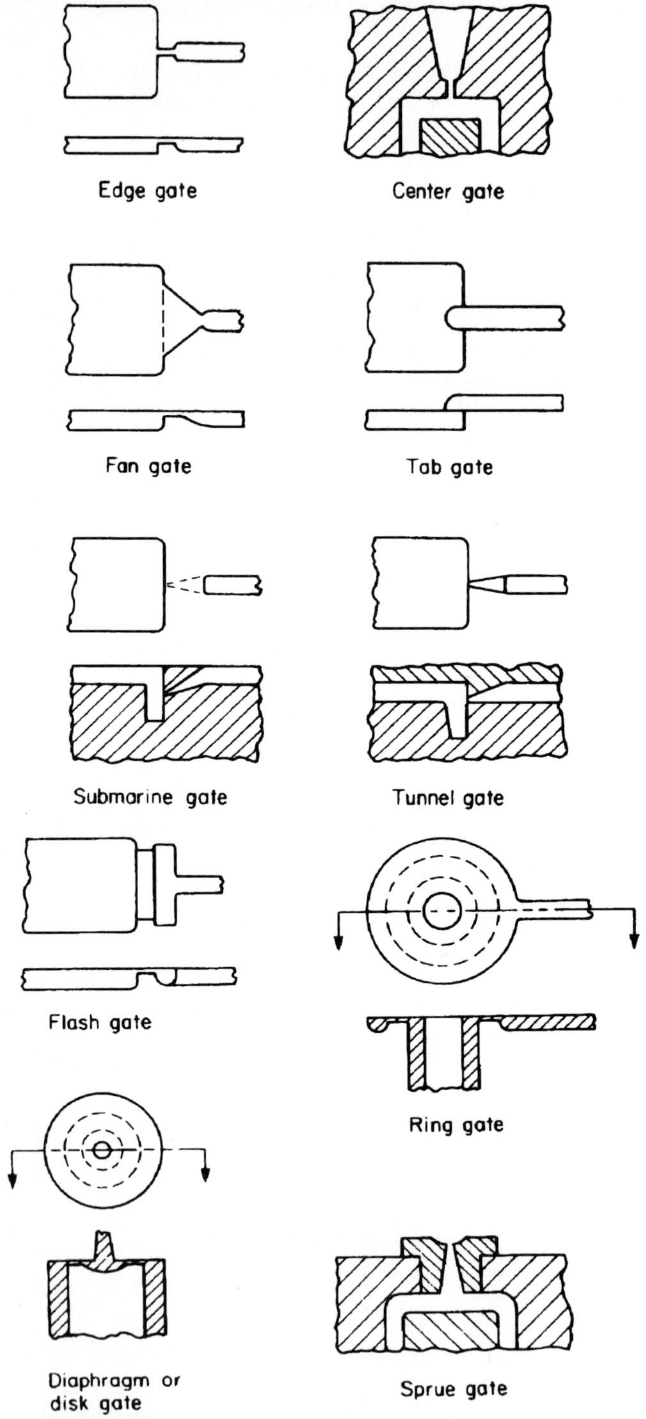

FIGURE 6.2.6 Various gating systems.

trated in Fig. 6.2.6. Center gating of round and cylindrical parts and near-center gating of other large-area parts are desirable for trouble-free mold filling.

Suggested Wall Thickness

Table 6.2.3 provides recommended normal and minimum wall thicknesses for common thermoplastics when injection-molded. Generally, thinner walls are more feasible with small parts rather than with large ones. The limiting factor in wall thinness is the tendency for the plastic material in thin walls to cool and solidify before the mold is filled. The shorter the material flow, the thinner the wall can be. Walls also should be as uniform in thickness as possible to avoid warpage from uneven shrinkage. When changes in wall thickness are unavoidable, the transition should be gradual, not abrupt. (See Fig. 6.2.7.)

TABLE 6.2.3 Suggested Wall Thicknesses for Various Thermoplastic Materials

Material	Short sections	Small sections	Average sections	Large sections
Acetal	0.6 (0.025)	0.9 (0.035)	1.9 (0.075)	3.2–4.7 (0.125–0.185)
Acrylic	0.6 (0.025)	0.9 (0.035)	2.3 (0.090)	3.2–6.3 (0.125–0.250)
Acrylonitrile butadiene styrene	0.9 (0.035)	1.3 (0.050)	1.9 (0.075)	3.2–4.7 (0.125–0.185)
Cellulose acetate butyrate	0.6 (0.025)	1.3 (0.050)	1.9 (0.075)	3.2–4.7 (0.125–0.185)
Nylon	0.3 (0.012)	0.6 (0.025)	1.5 (0.060)	2.4–3.2 (0.093–0.125)
Polycarbonate	0.4 (0.015)	0.8 (0.030)	1.8 (0.070)	2.4–3.2 (0.093–0.125)
Polyethylene				
Low-density	0.9 (0.035)	1.3 (0.050)	1.6 (0.062)	2.4–3.2 (0.093–0.125)
High-density	0.9 (0.035)	1.3 (0.050)	1.9 (0.075)	3.2–4.7 (0.125–0.185)
Polypropylene	0.6 (0.025)	0.9 (0.035)	1.9 (0.075)	3.2–4.7 (0.125–0.185)
Polystyrene	0.8 (0.030)	1.3 (0.050)	1.6 (0.062)	3.2–6.3 (0.125–0.250)
Polyvinyl chloride				
Flexible	0.6 (0.025)	1.3 (0.050)	1.9 (0.075)	3.2–4.7 (0.125–0.185)
Rigid	0.9 (0.035)	1.6 (0.062)	2.4 (0.093)	3.2–4.7 (0.125–0.185)

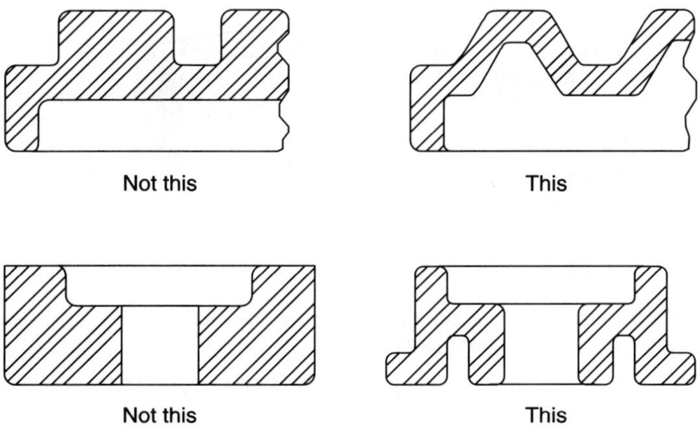

FIGURE 6.2.7 Maintain uniform wall thickness insofar as possible, and if changes in wall thickness are unavoidable, make them gradual rather than abrupt.

Holes

1. Holes are feasible in injection-molded parts but are a complicating factor in mold construction and part quality. "Knit" or "weld" lines adjacent to the hole often develop, and flashing also may occur at the edge of the hole. Figure 6.2.8 illustrates how weld lines are formed.
2. The minimum spacing between two holes or between a hole and sidewall should be one diameter. (See Fig. 6.2.9.)
3. Holes should be located three diameters or more from the edge of the part to avoid excessive stresses. (See Fig. 6.2.10.)
4. A through hole is preferred to a blind hole because the core pin that produces the hole can then be supported at both ends, resulting in better dimensional location of the hole and avoiding a bent or broken pin.
5. Holes in the bottom of the part are preferable to those in the side because the latter require retractable core pins.
6. Blind holes should not be more than two diameters deep. If the diameter is 1.5 mm ($^1\!/_{16}$ in) or less, one diameter is the maximum practical depth. (See Fig. 6.2.11.)
7. To increase the depth of a deep blind hole, use steps. This enables a stronger core pin to be employed. (See Fig. 6.2.12.)
8. Similarly, for through holes, cutout sections in the part can shorten the length of a small-diameter pin. (See Fig. 6.2.13.)
9. Use overlapping and offset mold-cavity projections instead of core pins to produce holes parallel to the die-parting line (perpendicular to the mold-movement direction). Figures 5.4.9 and 5.4.14 illustrate this approach, which is applicable to injection moldings as well as to die castings.

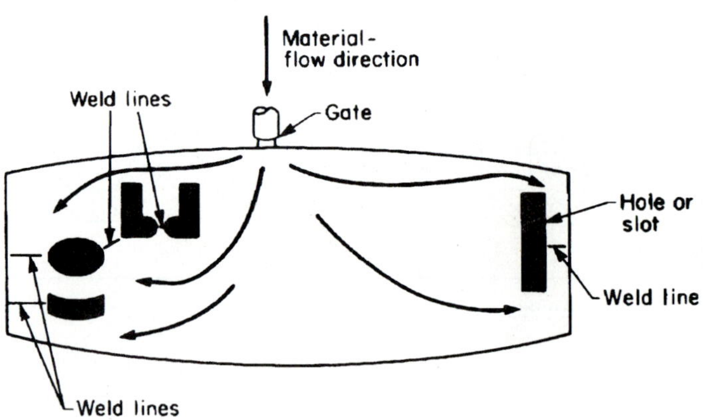

FIGURE 6.2.8 Weld or knit lines are caused when material flowing around hole core pins does not fuse together.

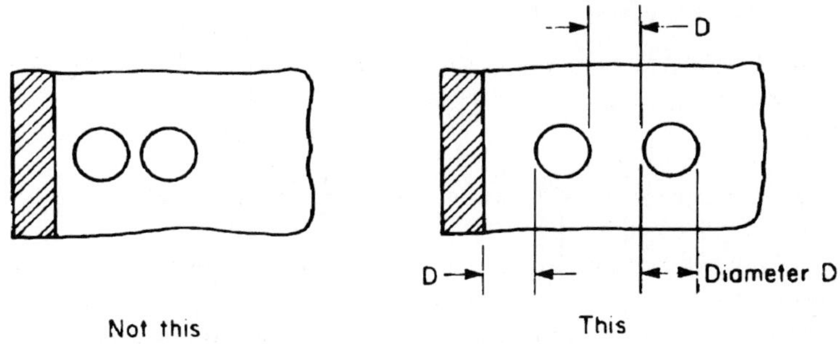

FIGURE 6.2.9 Minimum spacing for holes and sidewalls.

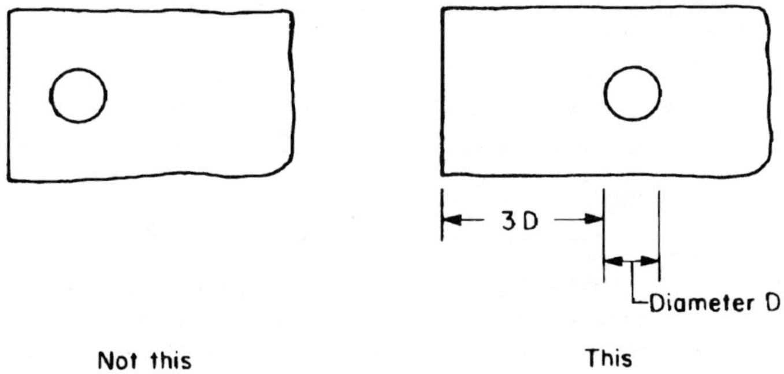

FIGURE 6.2.10 Minimum distance between a hole and the edge of the part.

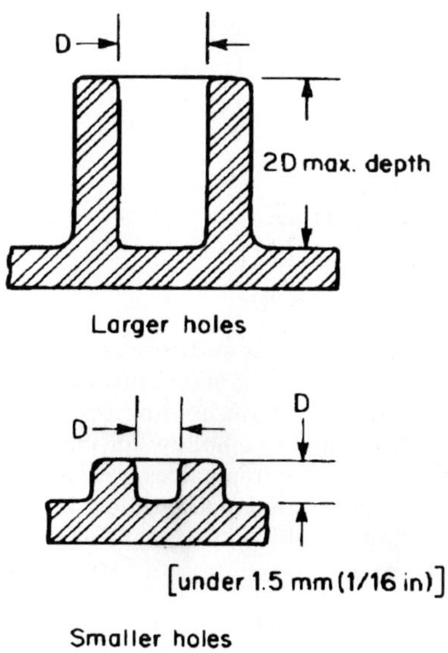

FIGURE 6.2.11 Recommended depth limits for blind holes.

Ribs

1. Reinforcing ribs should be thinner than the wall they are reinforcing to prevent sink marks in the opposite side of the wall. A good rule of thumb is to keep rib thickness to between 40% and 60% of the wall thickness.
2. Ribs should not be higher than 2.5 to 3.0 times the wall thickness.
3. Two ribs may be used, if necessary, to provide the extra reinforcement that would otherwise be provided by a high rib. The ribs should be two or more wall thicknesses apart.
4. Ribs should be perpendicular to the parting line to permit removal of the part from the mold.
5. Sink marks caused by ribs can be disguised or hidden by grooves or surface texture opposite the rib. (See Fig. 6.2.14.)

6. Ribs should have a generous draft of 0.5° to 1.5° per side.
7. There should be a radius at the base of 25% to 40% of the wall thickness.

See Fig. 6.2.15 for an illustration of most of these rules.

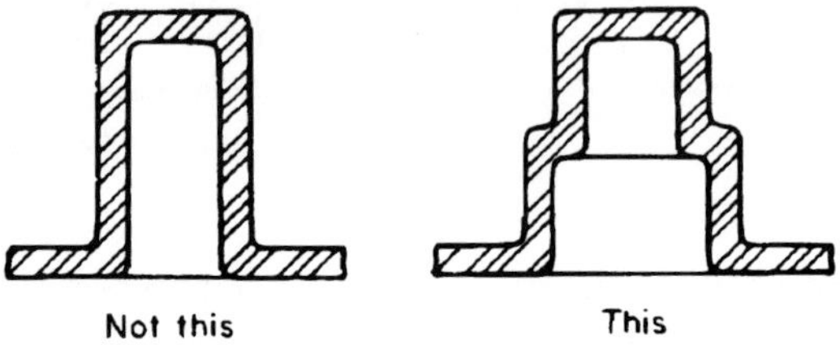

FIGURE 6.2.12 If a blind hole must be deep, use a stepped diameter.

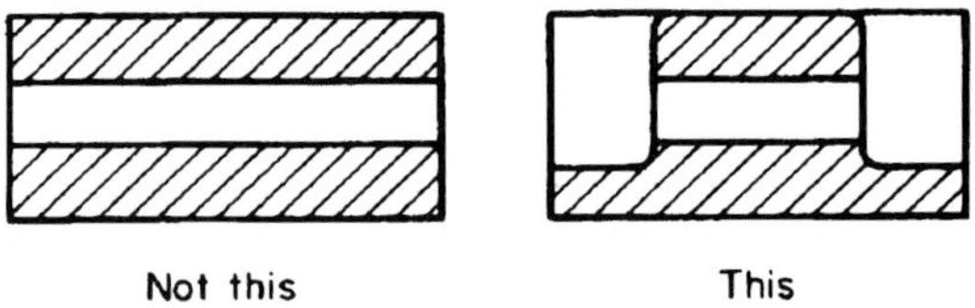

FIGURE 6.2.13 The improved design on the right provides better rigidity of the mold core pin.

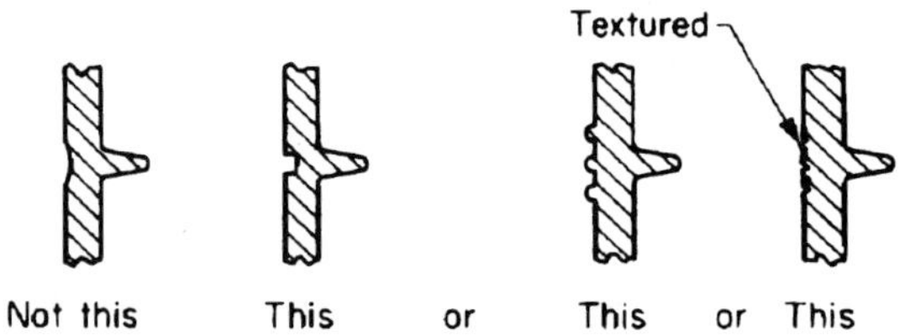

FIGURE 6.2.14 Methods for disguising sink marks.

Bosses

Bosses are protruding pads that are used to provide mounting surfaces or reinforcements around holes. The rules above for maximum height, draft, and radius of ribs are generally applicable to solid or hollow bosses. However, those holding inserts may require heavier walls. If large bosses are needed, they should be hollow for uniformity of wall thickness. Bosses in the upper portion of the mold can trap gasses and should be avoided, if possible. Locate bosses in corners, if possible, to aid material flow in filling the mold. If a detached boss is necessary, a connecting rib will aid material flow.

Undercuts

Undercuts are possible with injection-molded thermoplastic parts, but they may require sliding cores or split molds. (See Fig. 6.1.6.) External undercuts can be placed at the parting line or extended to the line to obviate the need for core pulls.

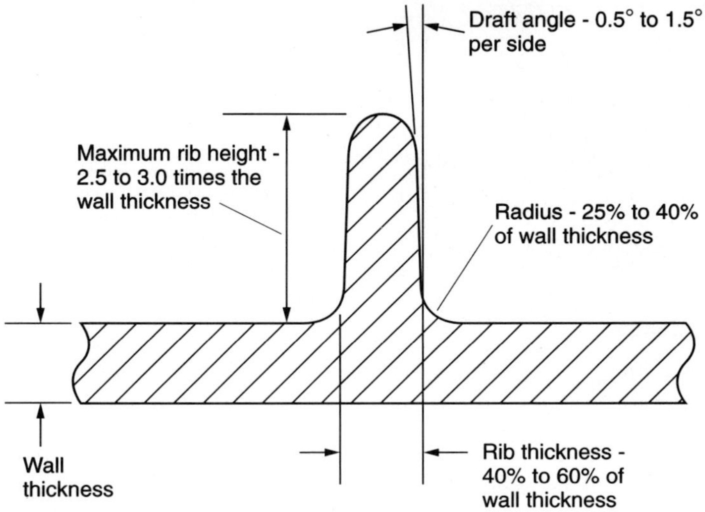

FIGURE 6.2.15 Design rules for reinforcing ribs and solid bosses.

Shallow undercuts often may be strippable from the mold without the need for core pulls. If the undercut is strippable, the other half of the mold must be removed first. Then the mold ejector pins can act to strip the part. Figure 6.2.16 shows the average maximum strippable undercut for common thermoplastics.

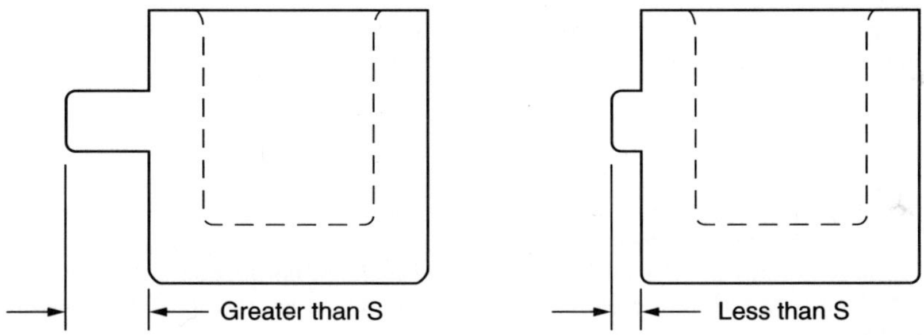

FIGURE 6.2.16 Allowable undercut for common materials.

Material	Average maximum strippable undercut *S*, mm (in)
Acrylic	1.5 (0.060)
Acrylonitrile butadiene styrene	1.8 (0.070)
Nylon	1.5 (0.060)
Polycarbonate	1.0 (0.040)
Polyethylene	2.0 (0.080)
Polypropylene	1.5 (0.060)
Polystyrene	1.0 (0.040)
Polysulfone	1.0 (0.040)
Vinyl, flexible	2.5 (0.100)

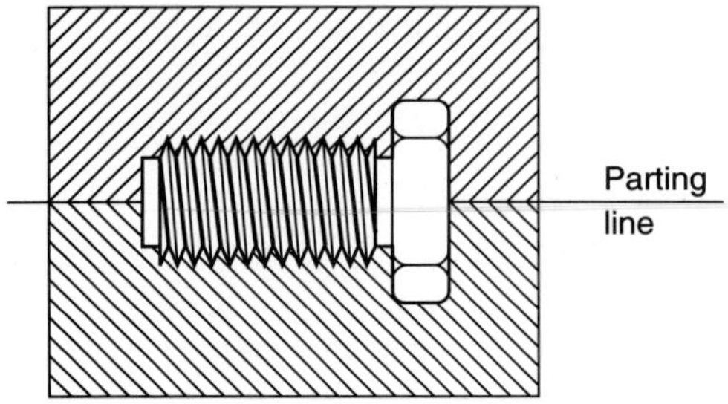

FIGURE 6.2.17 External screw threads can be molded without the need for a core pull if the threaded element is placed on the mold-parting plane. However, removal of flash from the threads may be required unless threads are excluded from the parting-line area.

Screw Threads

It is feasible, though a complicating factor, to mold screw threads in thermoplastic parts. Three basic methods can be used:

1. Use a core that is rotated after the molding cycle has been completed. This unscrews the part and enables it to be removed from the mold.
2. Put the axis of the screw at the parting line of the mold. This avoids the need for a rotating core but necessitates a very good fit between mold halves to avoid flash across the threads. This method is applicable only to external threads and generally leads to higher-cost parts unless the threads can be omitted in the area of the parting line. Often this is feasible. If this approach is not used, an extra operation probably will be required to remove parting-line flash from the threads. (See Fig. 6.2.17.)
3. Make the threads few, shallow, and of rounded form so that the part can be stripped from the mold without unscrewing. (See Fig. 6.2.18.) A coarse thread with a somewhat rounded form is preferred for all screw threads because of ease of filling and avoidance of featheredges even if it is removed by unscrewing.

Continuation of threads to the ends of threaded sections should be avoided because they create featheredges, make mold fit more critical, and promote flashing. (See Fig. 6.2.19.) If threads with strong holding power are needed, use metal inserts.

Internal threads can be tapped in almost all thermoplastics, and if the thread diameter is small [5 mm ($^3/_{16}$ in) or less], tapping is usually more economical than molding. Tapped or molded threads finer than 32 threads per inch are not practical with thermoplastics. Self-tapping screws are preferable to tapped or molded threads and a conventional screw.

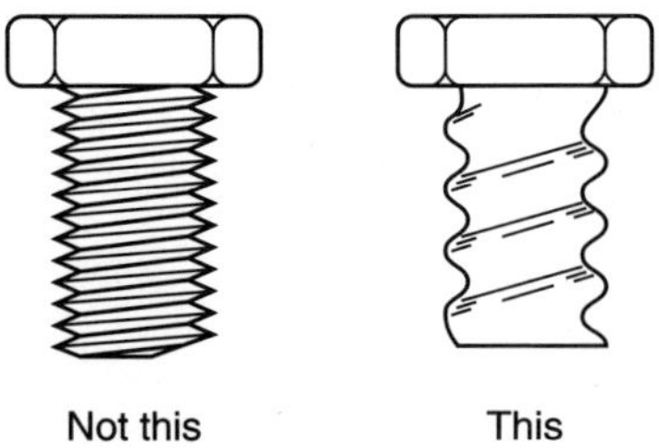

FIGURE 6.2.18 Shallow screw threads can sometimes be stripped directly from the mold.

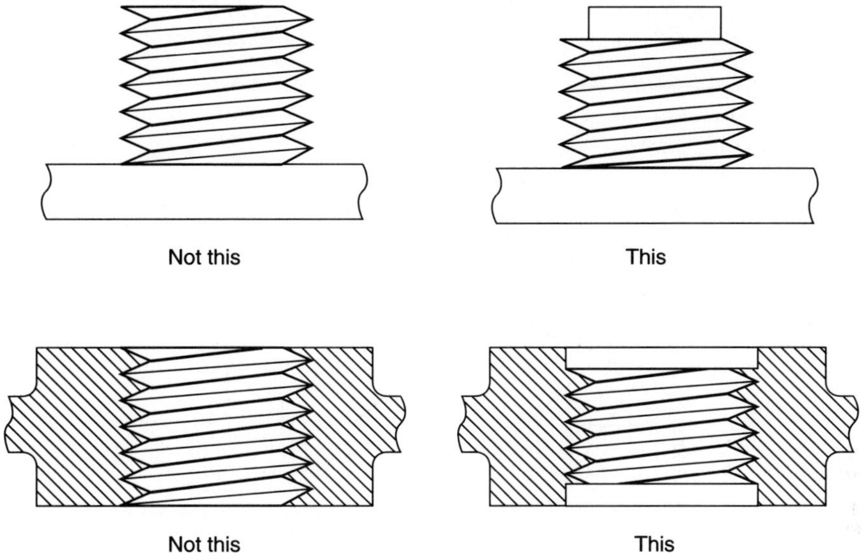

FIGURE 6.2.19 Screw threads should not extend to the ends of the threaded element.

Inserts

Inserts are useful and practical to provide reinforcement where stresses exceed the strength of the plastic material. Although they are economical, they are not without cost and should be used only when necessary for reinforcement, anchoring, or support.

Sharp corners should be avoided on the portion of the insert that is immersed in the thermoplastic. Figure 6.2.20 shows acceptable designs for the inserted ends of hooks and anchoring rods. Figure 6.2.21 shows unacceptable and acceptable designs and locations for inserted ends of screw-machine parts. Knurls on machined inserts should be relatively coarse to permit the material to flow into the recesses. There should be a smooth surface where the insert exits from the plastic.

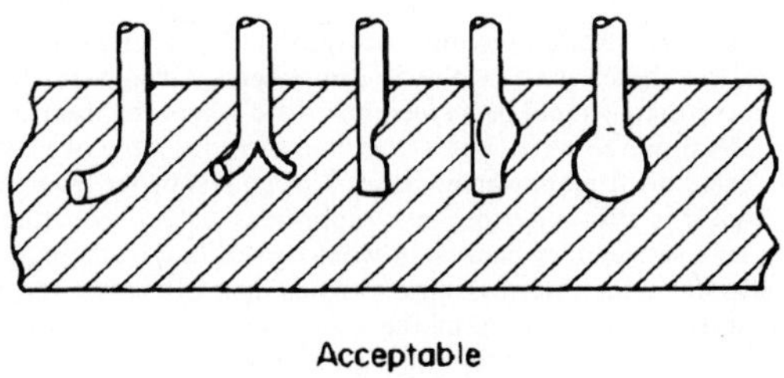

FIGURE 6.2.20 Recommended designs for the ends of hooks or rods to be inserted in plastic parts.

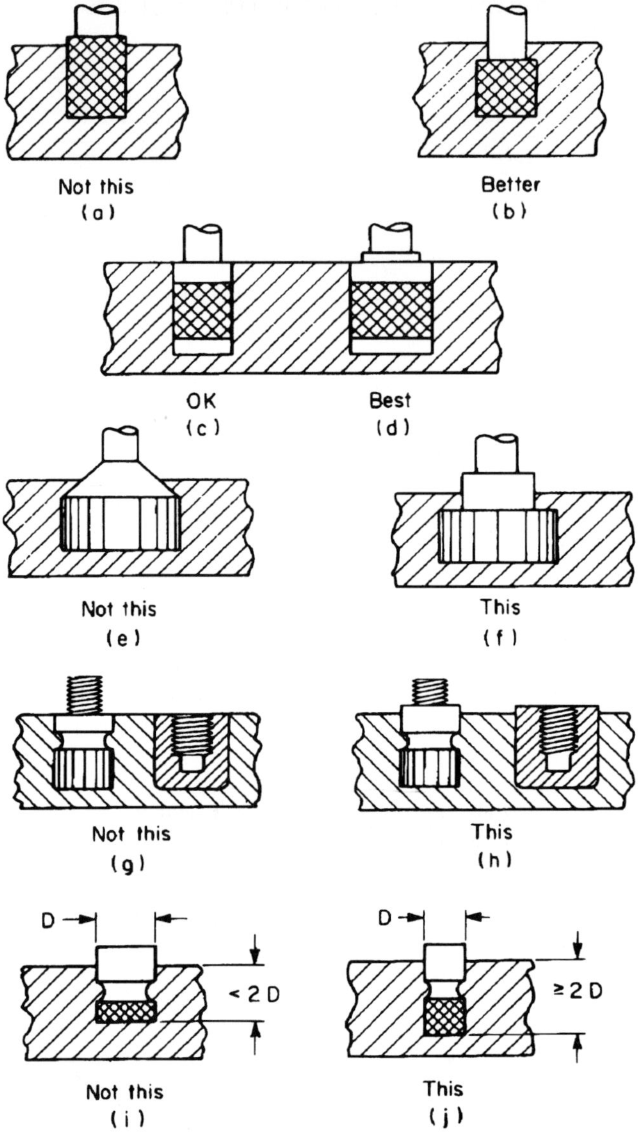

FIGURE 6.2.21 Recommended designs and placements for screw-machine parts to be inserted in plastic components. Design *a* is not a good design because there should be a smooth surface where the insert exits from the plastic. Design *b* is better, but it does not allow the flow of plastic material to be sealed off from unwanted areas. Design *c* is better from this standpoint. Design *d* is still better because it provides a double-sealing surface. Design *e* is not satisfactory because it would result in a featheredge of plastic material around the insert. Design *f* avoids this. In view *g,* the two parts are apt to suffer contamination of the screw threads with plastic material. Design *h* avoids this problem by raising the threaded portion of the inserts above the surface of the part. In view *i,* the part is not sufficiently embedded in the plastic material of the part. As shown in *j,* the depth of insertion should be at least 2 times the insert diameter.

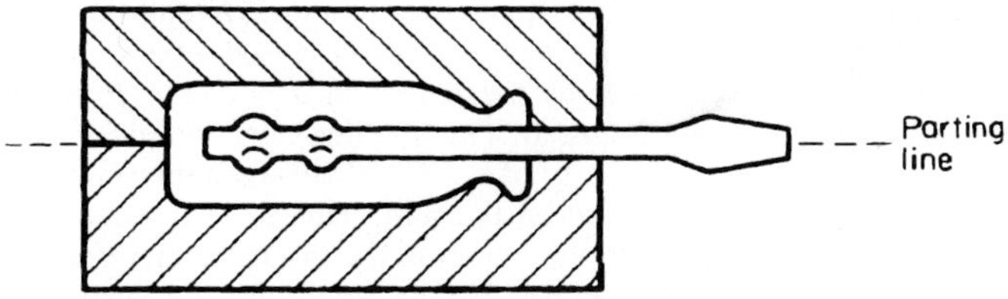

FIGURE 6.2.22 Irregularly shaped inserts are placed on the parting line of the mold.

Screw-machine inserts should be placed perpendicularly to the parting line of the mold to facilitate placement and avoid complications in mold construction. Irregular inserted parts such as screwdriver blades with noninserted widths or diameters larger than the inserted diameter are placed with their axes on the parting line of the mold. Otherwise, side cores must be provided to permit removal of the insert from the mold. (See Fig. 6.2.22.)

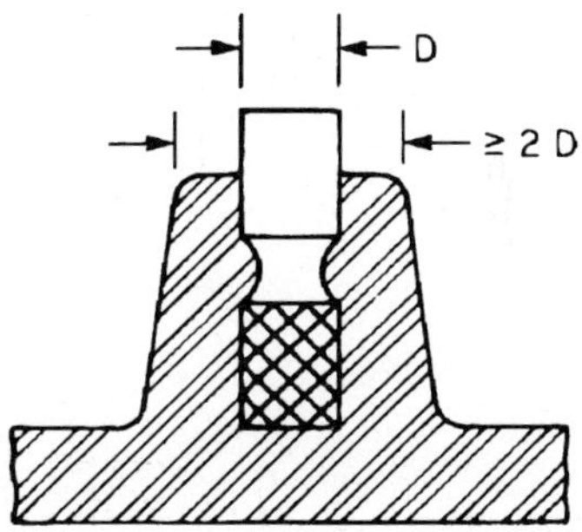

FIGURE 6.2.23 Ample supporting material must be provided around an insert. If the diameter of the insert is ¼ in or less, the boss diameter should be at least twice that of the insert.

Inserts very often are incorporated in a boss that provides supporting material for the insert. If the outside diameter of the insert is less than 6 mm (¼ in), the outside diameter of the boss should be twice that of the insert. (See Fig. 6.2.23.) If the outside diameter of the insert is larger than 6 mm (¼ in), wall thickness should be 50 to 100 percent of the insert diameter. Whenever the configuration permits, it is desirable to design the insert so that the flow of plastic is sealed off from threaded areas and other areas where plastic material is not intended to be. A good rule of thumb is to make the embedded length of an insert twice its diameter.

It is often advisable to press in the insert after molding. This avoids problems of contamination of the exposed surface of the insert with plastic material. It also avoids the possibility of damaging an injection mold with a misplaced insert. Ultrasonic techniques are particularly reliable for inserting metal inserts into plastics parts after molding. (See Chap. 6.9.)

Threaded portions of inserts should be raised from the surface of the molded part to avoid contamination of the threads with material. Also, featheredges of plastic material around inserts should be avoided.

Lettering and Surface Decorations

Lettering and other raised or depressed surface decorations and textures are easily incorporated into plastic parts. Once the lettering has been incorporated in the mold, each part will automatically show the lettering with few or no extra steps.

In most cases, mold cavities are machined rather than hubbed (pressed) or cast. With machined molds, the lettering in the *part* should be *raised,* i.e., formed by depressed, engraved letters in the mold. It is easier to engrave lettering in a mold cavity than it is to machine away the background and leave raised letters.

Hubbed (or hobbed) dies are made by pressing a hardened-steel form into annealed die steel. Very high pressures are required. This method is economical for producing a number of identical cavities, particularly of a difficult-to-machine shape. With hubbed dies it is best to have *depressed* letters in the part. This is true because the letters in the hub are depressed (engraved) and the letters in the mold are therefore raised.

With either raised or depressed letters, the letters should be perpendicular to the parting line of the mold. Otherwise, the part will have an undercut.

Sometimes it is desired to have depressed letters on the part and to fill them with paint that contrasts with the color of the plastic material. Cavities for filled lettering should be sharp-edged and 0.13 to 0.8 mm (0.005 to 0.030) in) wide. They should be one-half as deep as wide and should have a rounded bottom. (See Fig. 6.2.24.)

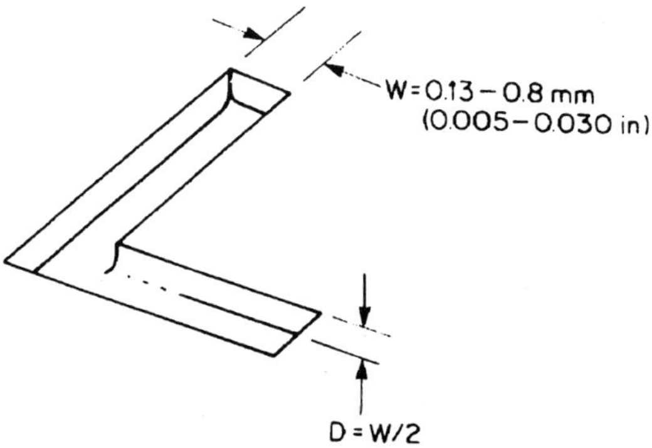

FIGURE 6.2.24 Dimensional rules for depressed lettering which is to be filled with paint.

Draft

It is highly desirable to incorporate some draft, or taper, in the sidewalls of injection-molded parts to facilitate removal of the part from the mold. Draft may not be necessary if ejector pins can be placed properly, but it still is wise, if the design permits, to make an easily removable part with draft and generous radii.

Drafts as low as $\frac{1}{4}°$ are often adequate. Usually, deep parts require less draft angle than shallow parts. For shallow parts draft should average $\frac{1}{2}°$ or more; for deep parts $\frac{1}{8}°$ can often be satisfactory. Textured surfaces require greater draft.

The following are recommended minimum drafts for some common materials:

Polyethylene	$\frac{1}{4}°$
Polystyrene	$\frac{1}{2}°$
Nylon	$0–\frac{1}{8}°$
Acetal	$0–\frac{1}{4}°$
Acrylic	$\frac{1}{4}°$

Corners: Radii and Fillets

Sharp corners should be avoided except at the parting line. They interfere with the smooth flow of material and create possibilities for turbulence with attendant surface defects. Sharp corners also cause stress concentrations in the part that are undesirable from a functional standpoint.

Fillets and radii should be as generous as possible. A desirable minimum under any circumstance is 0.5 mm (0.020 in), while 1.0 mm (0.040 in) is a preferable minimum if part requirements permit.

Surface Finish

One significant advantage of the injection-molding process is the fact that surface polish or textures can be molded into the part. No secondary surface-finishing operations (except, of course, plating, hot stamping, or painting, if desired) are necessary.

High-gloss finishes are feasible if the mold is highly polished and if molding conditions are correct. However, dull, matte, or textured finishes are preferred to glossy finishes, which tend to accentuate sink marks and other surface imperfections.

Painting of most thermoplastics is feasible but is not recommended if the color can be molded into the part. The latter approach obviously is more economical and gives superior results. If contrasting colors are required, masks can be fabricated and a portion of the part left unpainted. See Chaps. 6.12 and 8.5 for further information.

Plating of plastic materials is feasible for some plastics but is a specialized operation. See Chap. 8.2 for more information.

Surface decorations such as flutes, reeds, and textures should stop short of the parting line so that any parting-line flash is easy to remove. (See Fig. 6.2.25.)

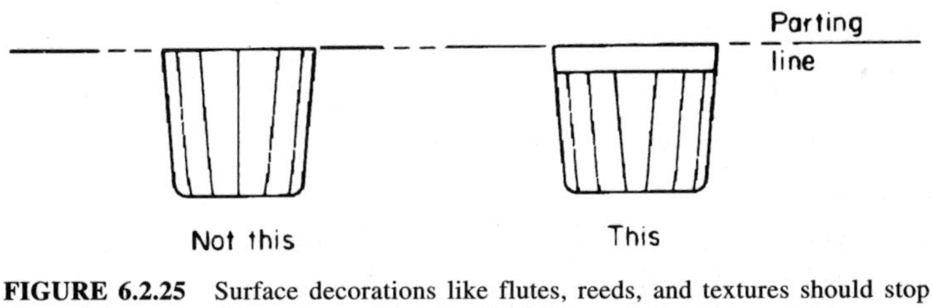

FIGURE 6.2.25 Surface decorations like flutes, reeds, and textures should stop short of the parting line so that parting-line flash is easy to remove.

Flat Surfaces

Flat surfaces, although feasible, are somewhat more prone to show irregularities than gently curved surfaces. Since the latter also produce more rigid parts, they are preferable.

Mold Parting Line

Every injection-molded part shows the effect of the mold parting line, the junction of the two halves of the mold. The part and the mold should be designed so that the part-

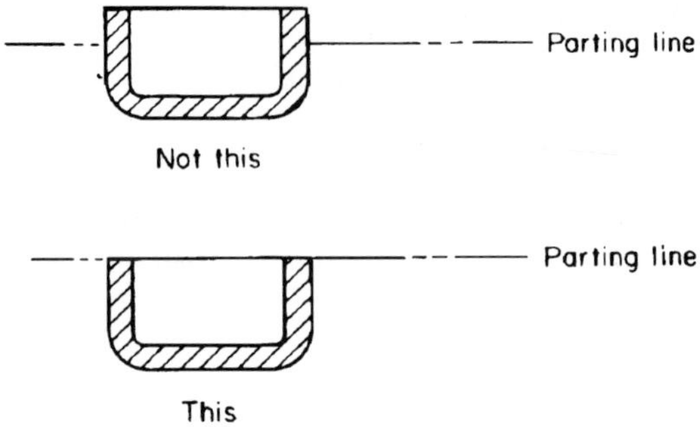

FIGURE 6.2.26 If possible, put the mold-parting line at the edge of the part.

ing occurs in an area where it does not adversely affect the appearance or function of the part. One easy way to do this is to put the parting line at the edge of the part where there is already a sharp corner. (See Fig. 6.2.26.) However, removal of parting-line flashing may destroy the sharpness of the corner. The part drawing specification should permit this.

Parting lines should be straight; i.e., the two mold halves should meet in one plane only. This obviously provides more economical mold construction, but it may not be possible if the part design is irregular.

If it is not possible to put the parting line at the edge of the part, cleaning parting-line flash is facilitated by having a bead or other raised surface at the parting line as shown in Fig. 6.2.27. Deliberately offset cavities are helpful in avoiding appearance defects, which may occur if the two mold halves do not line up properly. (See Fig. 6.2.28.)

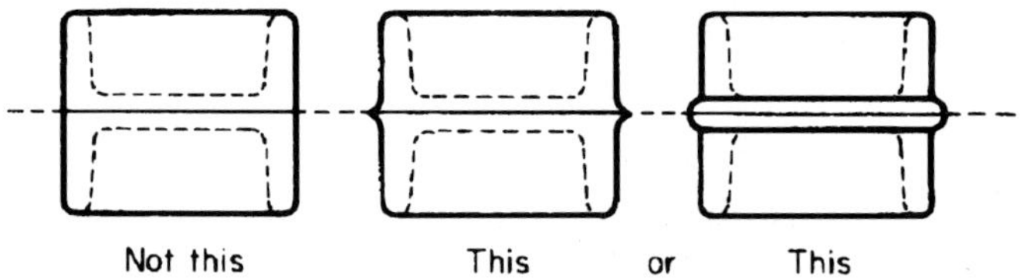

FIGURE 6.2.27 A bead at the parting line facilitates removal of mold flash.

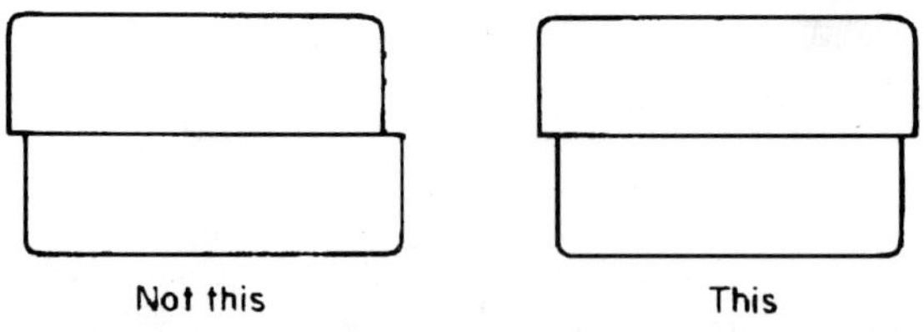

FIGURE 6.2.28 Deliberately offset sidewalls help avoid appearance defects if mold halves do not line up properly.

Computer Injection-Molding Simulations

These can be invaluable to both the product designer and tool designer. They aid in optimizing the material flow in the mold and, in the process, help avoid or minimize molding problems such as shrinkage, warpage, weld lines, and sink marks. Changes in both the mold and the part may result from such a simulation. Problems can be identified before the mold is actually machined. The need for prototypes is reduced. Software for such simulations is available from a number of sources.

Support Design for Assembly

Designers should endeavor to incorporate several functions into plastic parts so as to minimize the total number of parts to be assembled. By incorporating hinge, spring, fastener, and other functions in the injected-mold part, its cost may be only slightly increased, but separate parts for these functions will not be needed. The costs of designing, manufacturing, and stocking separate parts will be eliminated, and the assembly operation will be greatly simplified. Figure 6.2.29 illustrates one common step of this kind, the consolidation of a snap-fit element, which provides for easy assembly and eliminates the need for separate fasteners.

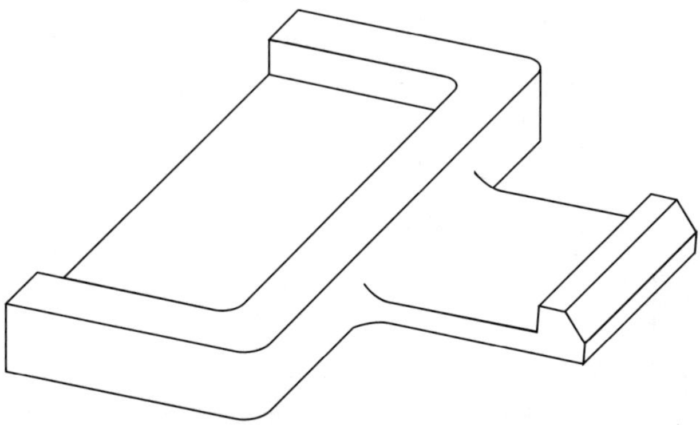

FIGURE 6.2.29 Snap-fit cantilever beams are often used in injection-molded parts to facilitate assembly.

DIMENSIONAL FACTORS AND TOLERANCE RECOMMENDATIONS

Though surprisingly tight tolerances can be held when molding thermoplastics parts, dimensions cannot be held with the precision obtainable with close-tolerance machined metal parts. There are several reasons for this:

1. There is materials shrinkage, as discussed above, including variation and unpredictability in the shrinkage.
2. Plastics exhibit a high thermal coefficient of expansion. As a result, if the toler-

ances are extreme, designers should specify the temperature at which the measurements should be taken.

3. Despite automatic-control apparatus for pressure, temperature, and time settings, there is some variation in these factors from cycle to cycle. These variations result in slight dimensional variations in molded parts.
4. Mold runners, cooling channels, and gates cannot always be located in the optimal position, leading to differences in how uniformly the material is "packed" in the mold and how uniformly it cools. Some distortion or built-in stresses are unavoidable.
5. Plastic parts are usually more flexible than metals.

A corollary of the flexibility factor is a lessened need for very close tolerances. Plastic parts, when assembled, often can be deformed slightly if this is necessary to ensure a good fit. Knowledgeable designers take advantage of this fact by designing lips and locating bosses on plastic parts to ensure alignment with mating-part surfaces when necessary. (See Fig. 6.2.30.)

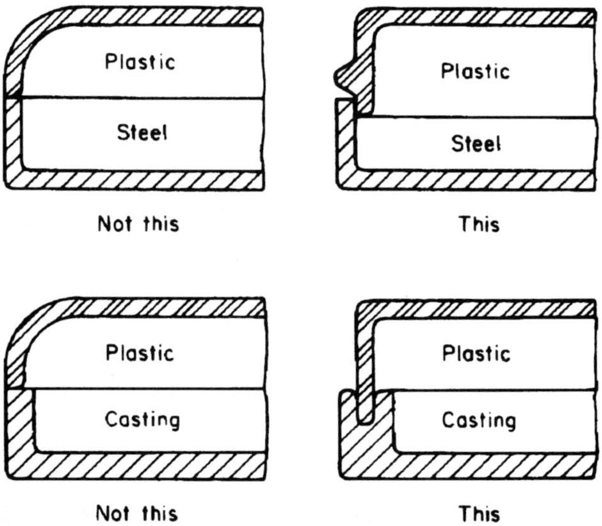

FIGURE 6.2.30 Lips and locating bosses aid in alignment when plastic parts are assembled.

As with other processes, close dimensional tolerances can greatly increase the cost of injection-molded parts. Fine-tolerance molds are more costly than looser-tolerance molds. There are processing-cost increases as well when extra tight dimensional control is needed. For example, closer process controls are needed for pressure, temperature, and cycle time; cycle time may be increased; shrink fixtures may be required after the part has been removed from the mold; and scrap rates will be higher.

Different plastics materials have different tolerance capabilities. Low-shrinkage materials can invariably be molded with closer tolerances. Glass-or mineral-filled materials can be molded more accurately than unfilled materials.

The use of a greater number of mold cavities tends to reduce the closeness of dimensional control over the molded parts. As a rule of thumb, for each cavity after the first, allowable dimensional tolerances should be increased by 5 percent. For example, a single-cavity mold with an allowable tolerance of $\pm$ 0.1 mm (0.004 in) on a particular dimension should have $\pm$ 0.15 mm (0.006 in) if the number of cavities is 10 (10$\times$5 percent = 50 percent increase in tolerance).

Tables 6.2.4 and 6.2.5 show suggested values for dimensional tolerances for various plastic materials. These tables, developed from data supplied by the Society of the Plastics Industry, represent historic and customary practices prevailing in the plastics-molding industry. Contract forms or other agreements of individual molders may vary.

TABLE 6.2.4 Recommended Dimensional Tolerances for Injection-Molded Thermoplastic Parts

Drawing code (see Fig. 6.1.14)	Dimensions, in	ABS		Acetal		Acrylic		Nylon		Polycarbonate	
		Commercial	Fine	Commercial	Fine	Commercial	Fine	Commercial	Fine	Commercial	Fine
A = diameter*	To 1.000	0.005	0.003	0.006	0.004	0.005	0.003	0.004	0.0 02	0.004	0.0025
B = depth†											
C = height†	1.000–2.000	0.006	0.004	0.008	0. 005	0.006	0.004	0.006	0.003	0.005	0.003
	2.000–3.000	0.008	0.005	0.009	0.006	0.007	0.005	0.007	0.005	0.006	0.004
	3.000–4.000	0.009	0.006	0.011	0.007	0.008	0.006	0.009	0.006	0.007	0.005
	4.000–5.000	0.011	0.007	0.013	0.008	0.009	0.007	0.010	0.007	0.008	0.005
	5.000–6.000	0.012	0.008	0.014	0.009	0.011	0.008	0.012	0.008	0.009	0.006
	6.000–12.000, for each additional inch	0.003	0.002	0.004	0.002	0.003	0.002	0.003	0.002	0.003	0.015
D = bottom wall		0.004	0.002	0.004	0.002	0.003	0.003	0.004	0.003	0.003	0.002
E = sidewall		0.003	0.002	0.004	0.002	0.005	0.003	0.005	0.003	0.003	0.002
F = hole diameter*	0.000–0.125	0.002	0.001	0.002	0.001	0.003	0.001	0.002	0.001	0.002	0.001
	0.125–0.250	0.002	0.001	0.003	0.002	0.003	0.002	0.003	0.002	0.002	0.015
	0.250–0.500	0.003	0.002	0.004	0.002	0.004	0.002	0.003	0.002	0.003	0.002
	0.500 and over	0.004	0.002	0.006	0.003	0.005	0.003	0.005	0.00 3	0.003	0.002
G = hole depth	0.000–0.250	0.003	0.002	0.004	0.002	0.004	0.002	0.004	0.002	0.002	0.002
	0.250–0.500	0.004	0.002	0.005	0.003	0.004	0.002	0.004	0.003	0.003	0.002
	0.500–1.000	0.005	0.003	0.006	0.004	0.006	0.003	0.005	0.004	0.004	0.003
Flatness	0.000–3.000	0.015	0.010	0.011	0.006	0.010	0.007	0.010	0.004	0.005	0.003
	3.000–6.000	0.030	0.020	0.020	0.010	0.015	0.010	0.015	0.007	0.007	0.004
Concentricity	TIR	0.009	0.005	0.010	0.006	0.010	0.006	0.010	0.006	0.005	0.003

*These tolerances do not include an allowance for aging characteristics of the material.

†Parting line must be taken into consideration.

Source: Data taken from *Standards and Practices of Plastics Custom Molders,* The Society of the Plastics Industry, Inc., New York 10017. All values are in plus or minus inches unless otherwise specified.

TABLE 6.2.5 Recommended Dimensional Tolerances for Injection-Molded Thermoplastic Parts

Drawing code (see Fig. 6.1.14)	Dimensions, in	Polyethylene, high-density		Polyethylene, low-density		Polypropylene		Polystyrene		Vinyl, flexible		Vinyl, rigid	
		Commercial	Fine	Commercial	Fine	Commercial	Fine	Commercial	Fine	Commercial	Fine	Commercial	Fine
A = diameter*	To 1.000	0.008	0.006	0.007	0.004	0.007	0.004	0.004	0.0 025	0.011	0.007	0.008	0.0045
B = depth†													
C = height†	1.000–2.000	0.010	0.008	0.010	0. 006	0.009	0.005	0.005	0.003	0.012	0.008	0.009	0.005
	2.000–3.000	0.013	0.011	0.012	0.008	0.011	0.007	0.007	0.004	0.014	0.009	0.010	0.006
	3.000–4.000	0.015	0.013	0.015	0.010	0.013	0.008	0.008	0.005	0.015	0.011	0.012	0.007
	4.000–5.000	0.018	0.016	0.017	0.011	0.015	0.009	0.010	0.006	0.017	0.012	0.013	0.008
	5.000–6.000	0.020	0.018	0.020	0.013	0.018	0.011	0.011	0.007	0.018	0.013	0.014	0.009
	6.000–12.000, for each additional inch	0.006	0.003	0.005	0.004	0.005	0.003	0.004	0.002	0.005	0.003	0.005	0.003
D = bottom wall		0.006	0.004	0.005	0.004	0.006	0.003	0.0055	0.003	0.007	0.003	0.007	0.003
E = sidewall		0.006	0.004	0.005	0.004	0.006	0.003	0.007	0.0035	0.007	0.003	0.007	0.003
F = hole diameter*	0.000–0.125	0.003	0.002	0.003	0.002	0.003	0.002	0.002	0.001	0.004	0.003	0.004	0.003
	0.125–0.250	0.005	0.003	0.004	0.003	0.004	0.003	0.002	0.001	0.005	0.004	0.004	0.003
	0.250–0.500	0.006	0.004	0.005	0.004	0.005	0.004	0.002	0.0015	0.006	0.005	0.005	0.004
	0.500 and over	0.008	0.005	0.006	0.005	0.008	0.006	0.0035	0.002	0.008	0.006	0.006	0.005
G = hole depth	0.000–0.250	0.005	0.003	0.003	0.003	0.005	0.003	0.0035	0.002	0.004	0.003	0.004	0.003
	0.250–0.500	0.007	0.004	0.004	0.004	0.006	0.004	0.004	0.002	0.005	0.004	0.005	0.004
	0.500–1.000	0.009	0.006	0.006	0.005	0.009	0.006	0.005	0.003	0.006	0.005	0.006	0.005
Flatness	0.000–3.000	0.023	0.015	0.020	0.015	0.021	0.014	0.007	0.004	0.010	0.007	0.015	0.010
	3.000–6.000	0.037	0.022	0.030	0.020	0.035	0.021	0.013	0.005	0.020	0.015	0.020	0.015
Concentricity	TIR	0.027	0.010	0.010	0.008	0.016	0.013	0.010	0.008	0.015	0.010	0.010	0.005

*These tolerances do not include an allowance for aging characteristics of the material.

†Parting line must be taken into consideration.

Source: Data taken from *Standards and Practices of Plastics Custom Molders,* The Society of the Plastics Industry, Inc., New York 10017. All values are in plus or minus inches unless otherwise specified.

CHAPTER 6.3
STRUCTURAL-FOAM-MOLDED PARTS

DEFINITION

Structural plastic foam-molded parts possess an integral skin encapsulating a cellular core. (See Fig. 6.3.1.) Both skin and core usually are of the same material and formed in the same molding operation. Structural-foam parts possess a high strength-to-weight ratio, other attractive properties (see below), and economical use of material.

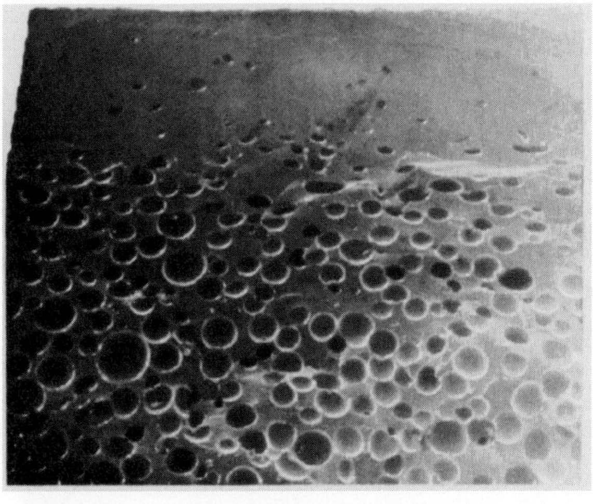

FIGURE 6.3.1 Cross section from the wall of a structural-foam-molded part showing the solid skin and cellular core, both molded in the same operation from one material. (*Courtesy Mobay Chemical Corp.*)

STRUCTURAL-FOAM-MOLDING PROCESSES

A number of processes are in commercial use for the production of structural-foam plastic parts. They are

1. Low-pressure injection molding
2. High-pressure injection molding
3. Reaction injection molding
4. Gas counterpressure molding
5. Coinjection or sandwich molding

Additionally, gas-assisted injection molding produces parts with similar advantages to those of structural foam. Profile extrusions of foam plastics are also produced and are discussed in Chap. 6.7.

Both low- and high-pressure injection molding use pelletized thermoplastic material plus a blowing agent. A chemical blowing agent that emits gas at the molding temperature can be added to the material before the molding operation. With low-pressure injection molding, another procedure is to mix a physical blowing agent (normally nitrogen) with the molten thermoplastic. For reaction injection molding and most casting processes, the material used is a two-component thermosetting plastic (most commonly urethane) in liquid form.

Low-Pressure Injection Molding

In this process, the amount of material injected into the mold is carefully controlled and is less than that required to fill the mold. The blowing agent in the plastic material causes it to expand. The mass of plastic and blowing agent expands and fills the mold cavity. The portion of the material that contacts the mold cavity forms a dense skin, while the interior portion becomes cellular.

Since the mold cavity is not filled during injection, high injection pressure is not transmitted to the mold. The mold experiences only the pressure developed by the foaming action of the polymer–blowing-agent mix. This pressure is almost always quite low, rarely exceeding 3400 kPa (500 lbf/in^2). When a chemical blowing agent is used, standard injection-molding equipment can be employed. The blowing agent is blended with the resin before it is introduced to the machine hopper. The heat of the machine's barrel activates the blowing agent, but barrel pressure prevents premature expansion from taking place.

When nitrogen gas is used as a blowing agent, the required equipment is specialized. The gas is introduced to the polymer melt in an extruder, which mixes the two and meters an amount of the mixture into an accumulator, where it is stored under sufficient pressure to prevent expansion of the nitrogen gas. When sufficient material is contained in the accumulator, it is forced into the mold cavity, partially filling it. Expansion of the nitrogen then provides the foaming action and causes the mold cavity to be filled.

High-Pressure Injection Molding

In this process, normal injection-molding action and pressure fill a mold completely with a mixture of polymer and chemical blowing agent. The size of the mold cavity is then increased by opening the mold platens or by retracting a core element. The material in the mold then foams and fills the expanded cavity. The high-pressure process uses more complex tooling but provides a smoother part surface than that obtained with low-pressure injection molding.

Reaction Injection Molding (RIM)

This process involves injection into a mold of highly reactive liquid components of thermosetting material. The components are fed to a mixing chamber just prior to injection. They polymerize and foam in the mold, forming a part with a smooth skin and a cellular core. Pressures are high in the mixing chamber but low (340 kPa, or 50 lbf/in^2) in the mold to allow foaming to take place.

Gas Counterpressure Molding

The principle of this process is to delay the foaming until after a solid skin has formed on the plastic entering the mold. This is accomplished by pressurizing the mold with nitrogen just prior to injection. The pressure is slightly above the expansion pressure of the blowing agent in the injected plastic. At the proper instant, before the mold is completely filled, the pressure is released, and the foaming action commences. However, it does not break through the surface of the injected material. Therefore, no swirling pattern occurs on the part's surface.

Coinjection or Sandwich Molding

Two materials are injected simultaneously into the mold through a special nozzle. One material forms the skin of the part and the other forms the core. The core material contains a blowing agent to provide foaming action. Two different plastics can be used. The core material can consist of a lower-cost material or even recycled plastic. Surface details can be produced accurately, and there is no swirl pattern on the surface.

Gas-Assisted Injection Molding

Parts made with this process do not have a foam core but do have a hollow interior due to the injection of nitrogen into the plastic charge as it enters the mold. The gas does not break through the surface of the plastic but does form channels in the hotter, less viscous material in the interior of the charge. The result, when the mold is filled, is a part with a smooth surface that matches the shape and finish of the mold but which is not solid. Less material is used, and the part has similar stiffness advantages, per unit of weight, as a structural foam part.

CHARACTERISTICS AND APPLICATIONS

Structural-foam molding is most advantageous for larger parts. Parts of 110 g ($\frac{1}{4}$ lb) and up to 50 kg (110 lb) are feasible. Whereas with conventional injection molding large parts require heavy, rigid molds to withstand the high injection pressures, both low-pressure injection molding and reaction injection molding permit lighter and much less costly molds to be used. This makes the molding of large structural-foam parts much less costly than solid parts of the same size.

The densities of structural-foam parts molded by high-pressure injection molding range from 50 to 100 percent of the weight of the solid object. With low-pressure injection molding, densities range from about 60 to 80 percent. With reaction injection molding of urethanes, densities range from about 40 to 60 percent. Skin thickness varies from 0.5 to 1.5 mm (0.020 to 0.060 in).

Surfaces of structural-foam parts molded with the low-pressure method tend to have a swirl or splay pattern caused by the foaming action of the polymer. This effect is virtually eliminated or greatly reduced with the other structural-foam processes mentioned above. When the condition exists, extra finishing operations involving three to five coats of paint are required if a smooth, glossy surface is desired.

The prime benefits of structural-foam components are their high stiffness-to-weight ratios and their freedom from sink marks in thick sections, a problem with solid molded parts. Hence stiffening ribs, bosses, and handles can be included in the part without so much concern that they will cause molding problems. They also have improved sound-deadening properties and high impact strengths but poor tensile strengths. Molded-in stresses are low.

Reaction injection-molded parts typically have wall thicknesses of about 6 mm (0.25 in) with a skin thickness of 1.5 mm (0.0625 in). Parts can be rigid or flexible, depending on the material formulation used.

Low-pressure injection-molded structural-foam parts typically have $1\frac{1}{2}$ to 3 times the rigidity of solid injection-molded parts of the same weight. High-pressure injection-molded parts can have excellent surface definition and very good uniformity. Portions of the parts can be solid.

Parts commonly molded of structural foam include business-machine housings, pallets, battery cases, simulated-wood furniture components, frames for upholstered chairs, drawers and drawer fronts, TV cabinets and backs, containers, boat ladders, oars, paddles, musical instrument parts, planters, gun stocks, dishwasher tubs, automotive fan shrouds, and fender liners. Reaction injection molding is used to make automobile bumpers and, in some cases, door panels. Figures 6.3.2 through 6.3.4 illustrate typical structural-foam parts.

ECONOMIC PRODUCTION QUANTITIES

All the structural-foam processes noted above except high-pressure injection molding use lower injection pressures and lighter and lower-cost molds than conventional solid-piece injection molding. Reaction injection molding and low-pressure molding have the greatest advantages in this regard. Molds of lower-strength, softer material can be employed. Aluminum is a common mold material. Kirksite (zinc alloy), beryllium copper, and sheet metal also are used. These molds cost only about one-third as much as conventional injection molds. This permits these processes to be used at much lower levels of production without cost penalties. When production levels are high, however, hardened-steel long-life tooling can be used. Equipment for reaction and low-pressure injection molding also can be lighter in construction and lower in cost than normal injection-molding equipment, further adding to their suitability for low and moderate levels of production. With RIM, it is also possible to cast liquid urethanes and other materials and utilize molds of silicone rubber, epoxy, and other materials that can be quickly and inexpensively fabricated. This low tooling cost makes it practicable to use casting for prototype and low-quantity production. In such situations, the higher labor content of the casting process is not a serious disadvantage.

FIGURE 6.3.2 Structural-foam industrial pallet, molded in two pieces that are hot-plate-welded together. (*Courtesy Union Carbide Corp.*)

High-pressure injection molding of structural foam has short cycle times but requires expensive molds and additional machine elements, necessitating high production levels to amortize their cost. Gas counterpressure injection molding uses lower injection pressures than needed for solid-piece molding, but the trickiness of the process requires more process development time and cost and a higher production quantity is needed to amortize these costs.

Cycle times for reaction injection molding of large parts range from about 1 to 7 min, which are considerably longer than those of the other injection-molding methods. Reaction injection molding is economical if the workpiece is large and intricate, combining elements that would be separate components with other processes.

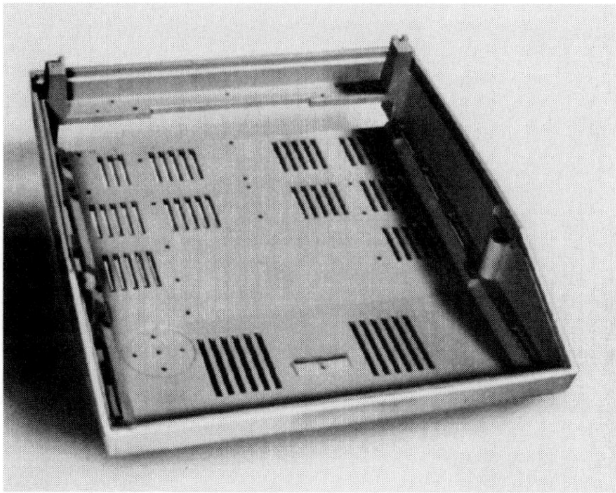

FIGURE 6.3.3 Business-machine base molded of structural-foam material. Note the integral lugs for mounting the machine. (*Courtesy Mobay Chemical Corp.*)

FIGURE 6.3.4 Metal-reinforced window frame shows the variable wall thickness feasible with structural foam. (*Courtesy Mobay Chemical Corp.*)

SUITABLE MATERIALS

Virtually any plastic material that can be injection-molded also can be used for structural-foam components. Not all materials can be used with all processes, however, and high-melting-temperature thermoplastics are less suitable.

Polyurethane dominates casting and reaction injection-molding applications, accounting for the overwhelming bulk of materials consumed. Phenolics, epoxies, and polyesters are also used.

Although polyurethane is more expensive than the commodity thermoplastics (polystyrene, polyethylene, and polypropylene), it is moldable in lower densities and also has the advantage of being available in both flexible and rigid formulations. Phenolic liquid materials for foam molding are highly viscous, contain catalysts corrosive to most mold metals, and hence are less attractive than urethanes from a processing standpoint.

Neither urethane nor phenolic can be integrally colored except with dark colors. If bright colors are desired, painting is required. However, since painting is required when smooth surfaces are needed, this is not usually a serious disadvantage.

High-density polyethylene, high-impact polystyrene, and polypropylene are the most common thermoplastics for structural-foam-molded parts. Also used are acrylonitrile butadiene styrene, cellulose acetate, acrylic, and polyvinyl chloride. Engineering resins are finding increasing use also. Polycarbonate, thermoplastic polyester, phenylene oxide–based material, and acetal, polysulfone, nylon, and ionomers are all used.

Glass reinforcement is feasible with most structural-foam materials, providing improved structural properties but not to the extent of the gain made with solid materials. Moldability is also diminished.

DESIGN RECOMMENDATIONS

Design principles for structural-foam components are similar in many respects to those for solid injection-molded parts. Material shrinkage on cooling, though a lesser factor with structural foam, still must be allowed for. Smooth flow of material into the mold is important in both cases. Specific rules for the design of structural-foam parts follow.

Wall Thickness

For most structural-foam-molding methods, 6 mm ($\frac{1}{4}$ in) is an optimum wall thickness. Thinner walls have greater density and an increased chance for sink marks, warpage, and unfilled molds. Thicker walls require longer cycle times. If the flow length is long, however, thicker walls may be required to permit good material flow. With high-pressure injection molding, walls can be thinner, if desired, down to 2.3 mm (0.090 in).

Wall-Thickness Transition

As in conventional injection molding and as illustrated in Fig. 6.2.7, abrupt changes in wall thickness should be avoided. Smooth transitions facilitate material flow and avoid molding problems and part distortion.

Fillets and Radii

Sharp corners interfere with smooth material flow and, if internal, create points of stress concentration that can lead to premature part failure. Fillets and external radii are best made equal or larger than one-half the wall thickness.

Draft

Draft angles of ½ to 3° are normally recommended for structural-foam parts. The actual amount depends on the depth of the draw, the material being used, and the closeness of mold ejection pins. If the sidewall surfaces of the part are textured, add 1° draft per 0.025-mm (0.001-in) depth of texture.

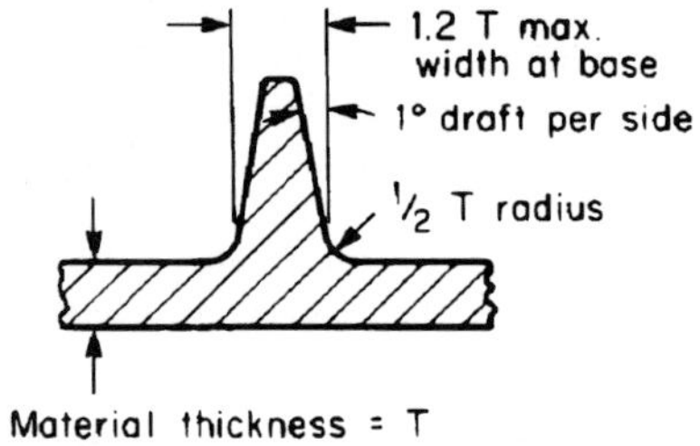

FIGURE 6.3.5 Recommended dimensional limits for reinforcing ribs for structural foam parts.

Ribs

The rigidity of a structural-foam part can be increased more economically by adding ribs than by increasing the wall thickness. Figure 6.3.5 illustrates the desirable dimensions for ribs.

Bosses

These are easily included in structural-foam parts. Bosses and other thick sections should be cored. It is good practice to attach the boss to the sidewall, as illustrated in Fig. 6.3.6. This aids material flow and provides additional load distribution for the part. Connecting ribs should be of the same thickness as the part walls. Figure 6.3.7 illustrates desirable dimensions for bosses in terms of basic wall thickness.

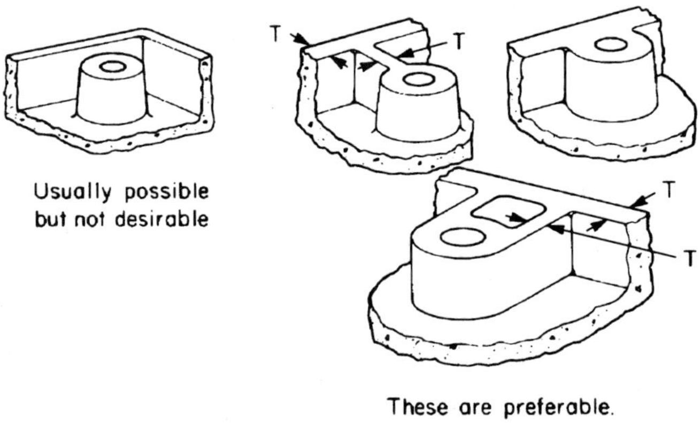

FIGURE 6.3.6 For better rigidity and material flow, connect bosses to the nearest sidewall as shown.

Holes

Both blind and through holes are more easily molded when their walls have a normal draft. However, zero-draft holes up to 50 mm (2 in) in depth can be produced if core pins are retractable or ejector pins are located close to the holes.

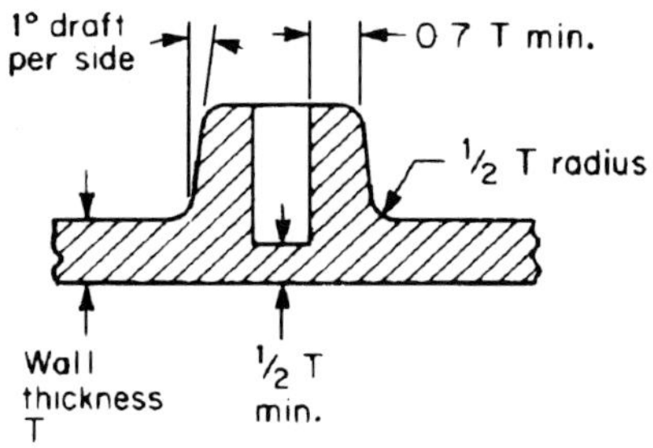

FIGURE 6.3.7 Recommended dimensional limits for bosses.

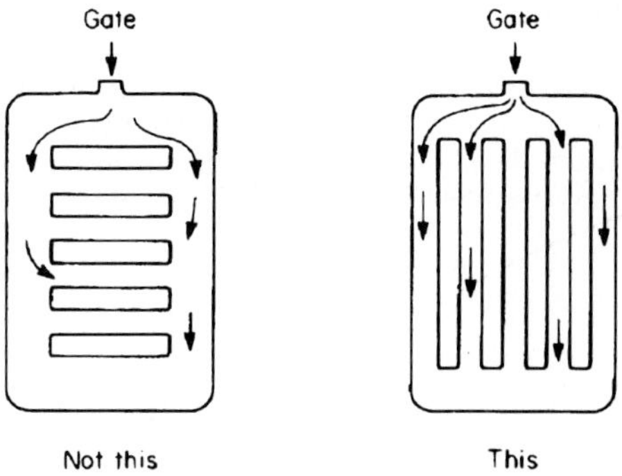

FIGURE 6.3.8 Grillwork openings should be in the direction of material flow.

Grillwork and Slots

These are feasible in structural molded parts but should be located so that they are parallel to the direction of flow of material in the mold. Otherwise, molding difficulties may arise. (See Fig. 6.3.8.)

Fasteners

Ultrasonically placed or self-tapping inserts are usually more satisfactory than molded-in inserts for most structural-foam parts. An exception is glass-reinforced material. Self-tapping screws should be of the thread-forming rather than the thread-cutting type. Screw holes should be molded, not drilled, for maximum strength and should have a diameter equal to the pitch diameter of the screw.

Snap Fits

When structural-foam parts are assembled to other parts, the use of snap-fit elements is the preferred fastening method. Snap-fit elements can be incorporated into and function well with structural-foam parts.

Hinges

Integral hinges can be utilized in structural-foam parts as they can be with solid parts.

RECOMMENDED TOLERANCES

Both favorable and unfavorable factors affect the accuracy of structural-foam components compared with the regular injection molding of thermoplastics. One favorable factor is the lower molded-in stress of structural-foam parts. An unfavorable factor is the lessened mold accuracy that usually (but not necessarily) accompanies the less rugged alternative materials used for low-pressure foam processes. The reduced injection pressure of most foam processes is another negative factor. However, the tolerances shown in Chap. 6.2 for normal injection-molded parts with various materials can be applied to structural-foam parts of the same materials.

CHAPTER 6.4
ROTATIONALLY MOLDED PLASTIC PARTS*

J. Gilbert Mohr
J. G. Mohr Co., Inc.
Maumee, Ohio

THE PROCESS

The basic elements of the rotational-molding (sometimes called *rotational-casting*) process comprise revolving on two axes a heated metal mold to which a liquid or powdered thermoplastic resin has been charged. The resin falls by gravity to the lowest portion of the rotating mold, coats the mold wall, and fuses as the mold rotates uniformly in two planes.

The equipment required to perform the rotomolding function may be of several types, but the three-spindle machine is probably the most widely used. (See Fig. 6.4.1.) The larger the part, the larger is the mold-supporting equipment required.

In carrying out the process, (1) plastic material is charged to the mold and held in a mount (spider) at a loading and unloading station, (2) the mold is revolved in a heated chamber, where the plastic fuses and builds up on the inside mold surface, (3) the mold is then subjected to a cooling force (cold air, water fog, or water spray, or a combination) and the plastic solidifies, and (4) the mold is opened and the part is removed.

TYPICAL CHARACTERISTICS

Rotationally molded parts are essentially hollow. With present improved technology, long and narrow-width parts are possible, but mold surfaces must be far enough apart to avoid bridging. Finished rotationally molded (rotocast) parts find application in tanks, food and shipping containers, automotive fender liners, machinery and instrument housings, statuary, portable lavatories, and many other products. Reinforcements

*The author gratefully acknowledges the assistance and counsel of Mr. David S. Unkefer, President, Sherwood Plastics, Inc., Fostoria, Ohio.

FIGURE 6.4.1 Widely used intermediate-sized rotational-molding machine showing loading station (center foreground), heating oven (left), and cooling area (right). (*Courtesy Ferry Industries.*)

such as fiberglass may be added to almost all the resins used in rotocasting. Several parts representative of the adaptability of the process are pictured in Fig. 6.4.2. Parts tend to be fairly large, varying in size from toy balls 75 mm (3 in) in diameter to tanks or food containers 2 m (6 ft) high by 1 m (3 ft) in diameter. The largest known production part is 6.7 m (22 ft) long, molded on equipment especially designed and built for larger parts.

Many powdered-plastic materials can be rotocast. Parts are strain-free as molded. Different types of resins may be laminated by successive charges to the mold. An insulating or honeycomb layer may be produced by incorporating a blowing agent in the powdered resin for one of the layers.

The thickness of rotocast parts is determined by the effectiveness of the cooling system in the equipment. Parts thicker than approximately 10 or 12 mm ($\frac{3}{8}$ or $\frac{1}{2}$ in) usually require some supplementary means such as inert-gas injection for internal cooling in the mold. Thickness can be varied by selective programming of the rotational cycle so that the areas to be made thicker are exposed to the low-gravity points for a larger percentage of the time. No abrupt changes in wall thickness are possible because no molding force is available. Items such as cylinders with embossed surfaces (not possible with extrusion) may be molded by rotating only in one plane or orientation. The minimum thickness moldable is between 1.3 and 1.5 mm (0.050 and 0.060 in).

External surface finish is usually not smooth and polished even though the mold surface has a high gloss. The grains or particles of plastic are not completely fused at the time when they become attached to the mold wall. Also, it is difficult to produce parts that are 100 percent free of voids throughout the cross section.

Raised or depressed portions may be incorporated into part surfaces for decoration, identification, etc. Color may be incorporated into the plastic up to a level at which it does not interfere with the melt index or flow properties. By incorporating a thermally

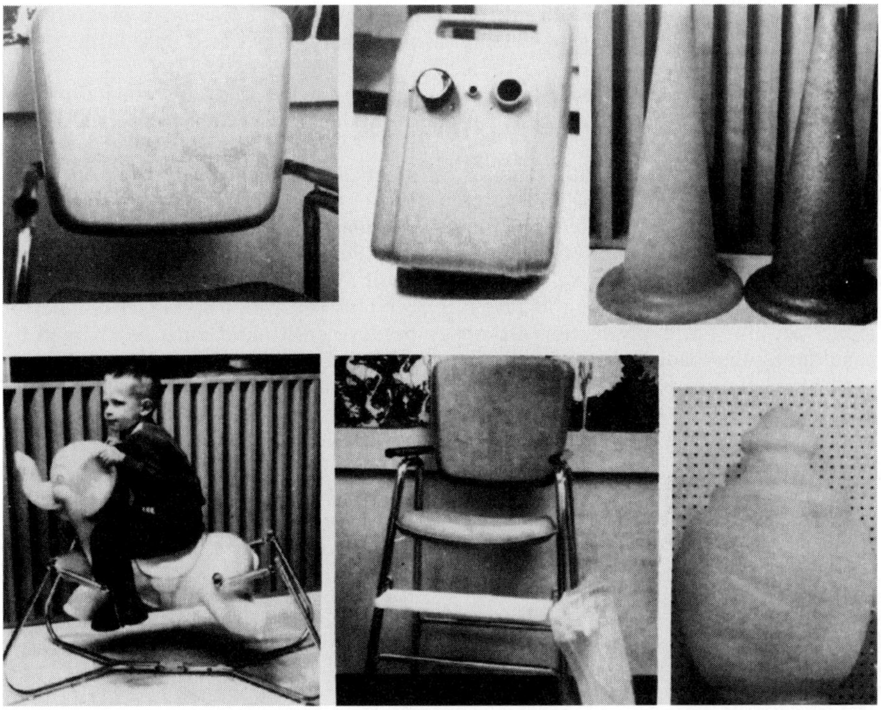

FIGURE 6.4.2 Group of typical rotationally molded end products. Parts are moldable up to lengths of 22 ft or sizes as large as 5600-gal chemical-storage tanks weighing 1650 lb. (*Courtesy McNeil Femco McNeil Corp.*)

nonconducting material such as Teflon into the mold at a particular surface or area, holes, slots, and other openings can be made in the part. Also, two rotocast parts, even containing holes, may be molded together and cut after demolding to provide two shell-like parts.

ECONOMIC PRODUCTION QUANTITIES

As compared with injection molding of thermoplastics, rotational molding is less expensive to set up and operate. Equipment costs are lower, but more working area is required. Molds are lighter, simpler, and cheaper to construct and maintain, and tool delivery time is one-half to one-third that required for injection molds. Primarily, molds are made from cast aluminum, fabricated from sheet steel, stainless steel, or sheet aluminum, or electroformed by using chromium-plated nickel or nickel copper. Cast and machined beryllium copper is also employed in mold construction. The most cost-effective molds for rotocasting are equivalent in price to molds for vacuum forming. However, the multiplicity of molds required for higher production rates of small parts may result in higher tooling costs than for a large part.

There is really no practical limit on mold life, since no excessive pressures or stresses are developed during processing. Molds must be reworked only in the event of accidental damage.

Setup is economical because single molds have quick and ready connect-disconnect fixtures. Often molds may be taken out of production and others inserted during the regular production cycle.

This is not true when multiple molds for identical parts are mounted on one individual spider. However, since production cycles for rotocasting are slower than those for injection molding (1 cycle/8 to 20 min for rotocasting), multiple molds are valuable in making the production rate economical. Parts less than 300 mm (12 in) in diameter and weighing 0.45 kg (1 lb) or less will average 12 molds per spindle arm but may be produced by using as many as 36 molds per arm. The higher number of molds per arm limits production efficiency because each mold must be charged by hand during the loading cycle. Larger parts may have two to three molds per spindle arm. These figures expand to a practical minimum of 70,000 parts per year, with a maximum of 500,000 per year. Smaller parts have been molded at the rate of 350 per hour on a six-spindle machine, yet the process is feasible for as few as 5000 parts per year.

Hence rotomolding can be satisfactory for short, intermediate, or long runs, benefited by low equipment and operating costs. However, the process is not as efficient for high production as injection molding and blow molding, which easily provide faster molding cycles and higher production rates.

SUITABLE MATERIALS

Both liquid and solid materials are used for rotomolding. Solid material is in powder form, and the preferred resin particle size is 35 mesh, held within narrow limits. However, some gainful use may be made of resins varying in particle size between 20 and 80 mesh.

Glass-fiber reinforcing can add to molded physical properties. Lengths up to 6 mm ($\frac{1}{4}$ in) may be used, but higher loadings up to 10 and 15 percent are favored by shorter lengths of 0.8 to 3 mm ($\frac{1}{32}$ to $\frac{1}{8}$ in). Glass content is limited by protrusion of bare fibers from inner-wall surfaces and/or bridging across surface lines or angles. Clay and asbestos are also in limited use as fillers and resin extenders.

Polyethylene is the most commonly used material for rotational molding. Low-, medium-, and high-density varieties all are used, depending on the application. Polyethylene is easily processed by rotational molding. Cross-linked high-density polyethylene has improved resistance to high temperature and other adverse environments.

Vinyl plastisol is the second most commonly rotomolded material. Since it is liquid in the unfused state, it can be metered automatically into unfilled molds. It is easily processed and can be molded with intricate surface detail. It is used for parts for which some flexibility is desired.

Other thermoplastics used for rotomolding are nylon, polycarbonate, cellulose acetate butyrate, ethylene vinyl acetate, acetal, acrylic, polypropylene, polystyrene, acrylonitrile butadiene styrene (ABS), and fluorocarbons. Some of these materials require thorough care to effect satisfactory molding. For example, ABS and impact polystyrene are less satisfactory than other materials in filling deep, narrow sections.

Polyurethane in two-part liquid form is the most common thermosetting plastic for rotomolding. Epoxy and polyester also are used. Efforts to rotomold powdered phenolic material have not been successful.

DESIGN RECOMMENDATIONS

Holes

Holes may be formed by molding a dome and cutting (Fig. 6.4.3), providing inwardly extending projections in the mold wall that yield scribe lines for routing in the molded part (Fig. 6.4.4), securely mounting on the inside mold wall a Teflon (TFE) sheet "plug" that prevents plastic material from adhering to the mold at that point (entire ends of a part may be left open by using TFE plugs so that there is no limit on moldable hole size or shape), and inserting machined-brass plugs, pins, or tubing through the mold wall. Though moldable, holes are a complicating factor in the mold design and molding process and may require extra postmolding operations. Therefore, the most economical designs minimize the number of holes.

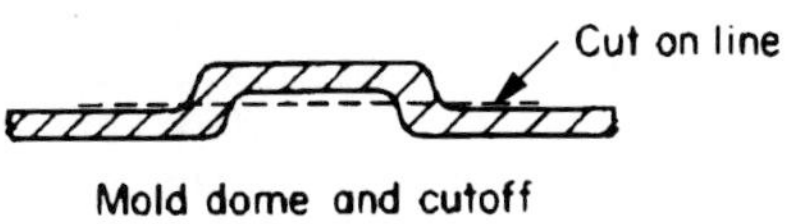

FIGURE 6.4.3 One method of making holes in rotomolded parts: mold dome and cutoff.

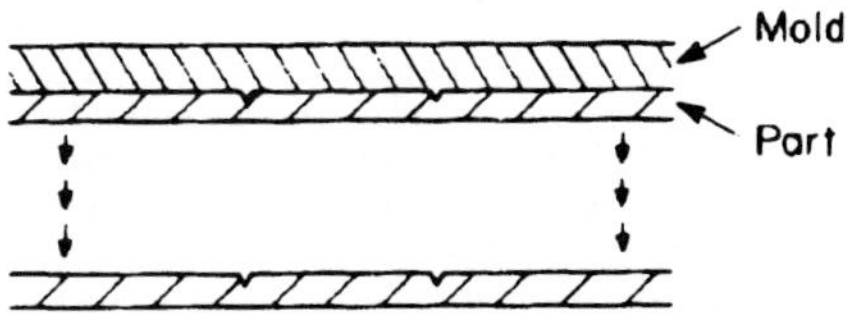

FIGURE 6.4.4 Projection of a circular pattern in the mold surface forms a scribe line in the molded part (to be routed out after demolding).

Ribs

Stiffening ribs are possible and easily moldable if the requirement of maintaining uniform wall thickness is followed. Several "don'ts" are shown in Fig. 6.4.5: A narrow rib doesn't fill and leaves inside stringers (Fig. 6.4.5*a*). A rib that is too deep presents even greater difficulties, preventing the molding powder from reaching the bottom prior to fusion (Fig. 6.4.5*b*). A small, shallow, and narrow rib fills completely but has limited strengthening effect (Fig. 6.4.5*c*). The correct rib design, illustrated in Fig. 6.4.5*d*, shows a wide gap used to form the rib, with generous draft so that the molding powder is allowed to fill uniformly without bridging. Materials requiring special rib dimensions to combat bridging are polyethylene, 13-mm ($\frac{1}{2}$-in) minimum width inside (part outside at top), and polycarbonate, 25-mm (1-in) minimum width inside.

Bosses

It is virtually impossible to produce internal or external bosses and T sections in rotationally molded parts. As explained below, a fairly uniform wall thickness must be

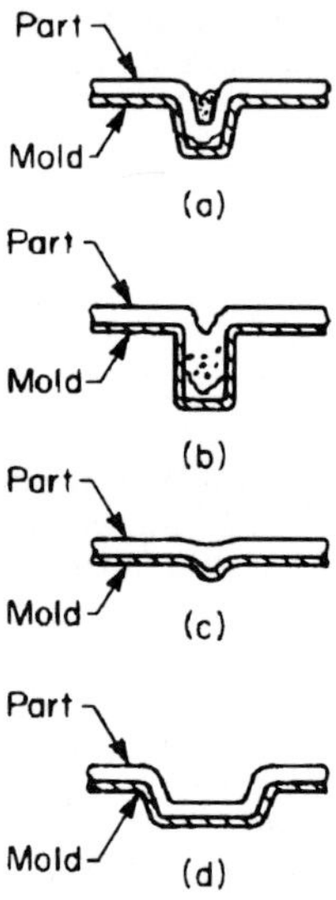

FIGURE 6.4.5 Use of stiffening ribs. (*a*) Not this—narrow rib does not fill and leaves stringers. (*b*) Not this—deep rib also prevents molding powder from reaching bottom. (*c*) Not this—small shallow and narrow rib fills completely. (*d*) Use this form—generous draft with wide gap allows molding powder to fill and provide a satisfactory stiffening rib.

maintained. It is possible to produce interior extensions by placing a metallic screen in contact with the inner mold wall. The screen heats up, attracts the plastic molding powder, and becomes covered, remaining in place after molding. (See Fig. 6.4.6.) In fact, by using this method, a hollow part may be molded so as to have two or more separate chambers, the screen material being extended entirely across the mold inside.

Wall Thickness

Production is slowed and made more difficult when differential rotational cycles must be employed to provide varying wall thicknesses in the same part. Other methods for locally changing wall thickness are also troublesome. Basically, they involve changing the amount of heat transferred by the mold to the plastic material in the designated area. Changing the mold wall thickness or the color of the mold's outer surface are two methods. Higher heat in one area causes the wall thickness to be greater in that area.

Wall-thickness variations as a result of such steps are always gradual; abrupt wall-thickness changes are virtually impossible. The maximum recommended thickness

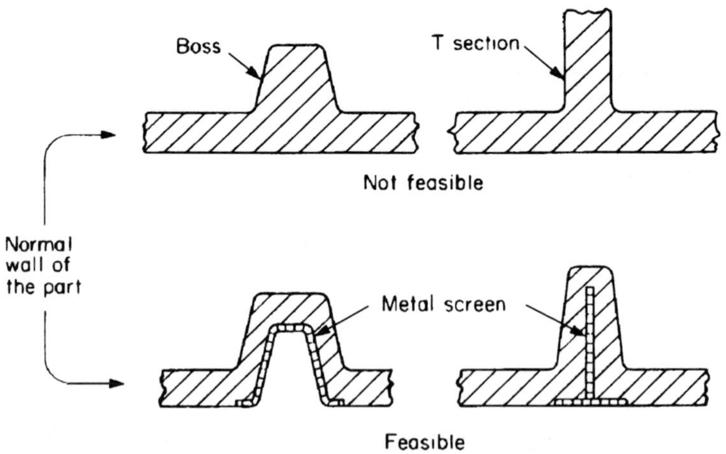

FIGURE 6.4.6 Incorporating bosses and T sections in rotomolded parts with the aid of a metal screen.

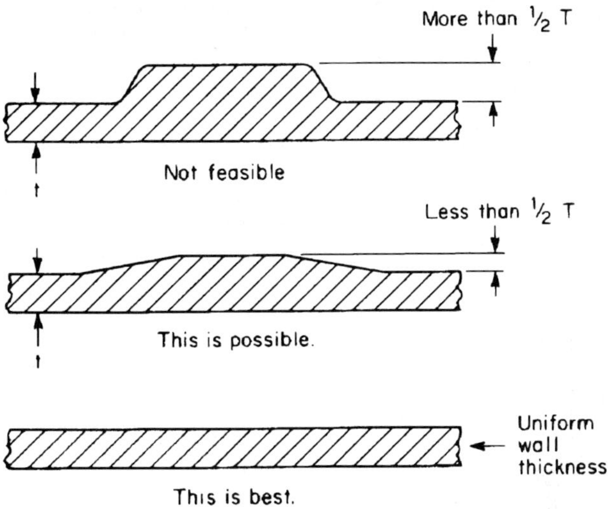

FIGURE 6.4.7 Wall thickness of rotomolded parts should be as uniform as possible.

buildup, when necessary, is 50 percent of the normal wall thickness for the part. (See Fig. 6.4.7.)

The following are recommendations for the designed wall thickness of rotomolded parts:

1. Preferred or desirable nominal wall thickness, 2.5 to 6 mm ($\frac{1}{10}$ to $\frac{1}{4}$ in)
2. Minimum wall thickness, 1.3 to 1.5 mm (0.050 to 0.060 in)
3. Maximum wall thickness, 13 mm ($\frac{1}{2}$ in)
4. Optimal thickness for fiberglass-reinforced parts, 10 mm ($\frac{3}{8}$ in)

Undercuts

Undercuts are possible but should be kept to a minimum. Provision for undercuts usually dictates higher mold costs because of having to use core pulls or splitting the mold to allow separation parallel to the undercut groove. (See Fig. 6.4.8.) Undercuts may require additional time for unloading molds.

Draft

Ample draft is recommended on the sidewalls of rotomolded parts to facilitate removal from the mold. A recommended minimum for most materials is 1°. Because of its lower cooling shrinkage, polycarbonate will require $1\frac{1}{2}$ to 2°. (See Fig. 6.4.9.)

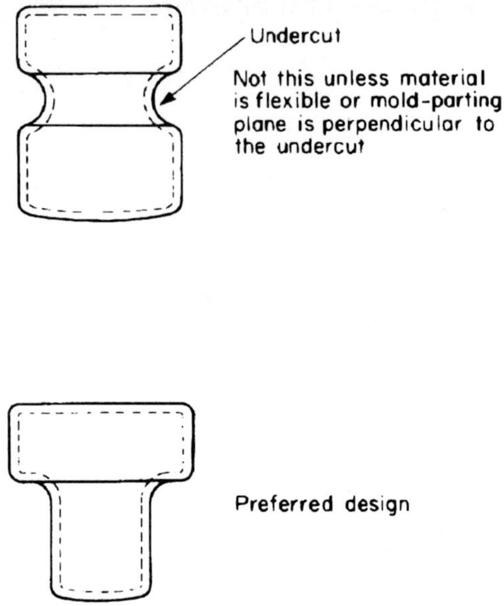

FIGURE 6.4.8 Undercuts in rotomolded parts add to manufacturing cost.

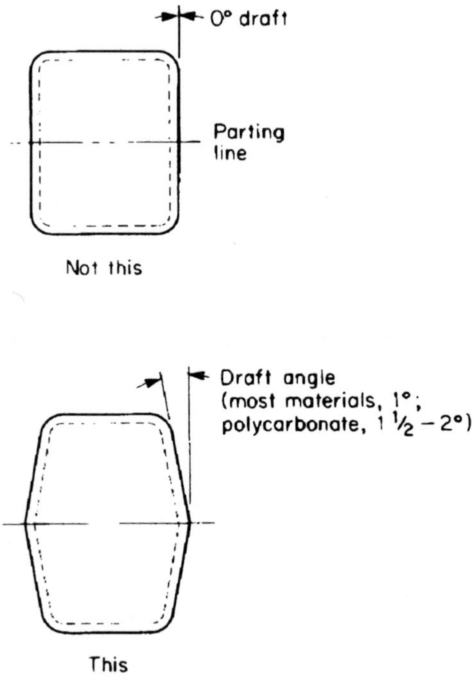

FIGURE 6.4.9 Sidewalls of rotomolded parts should be given a generous draft for easy part removal from the mold.

Screw Threads

The external contour of threads should be rounded (as for standard glass bottle designs) or a modified buttress. Sharp V threads and fine pitches do not mold well and should be avoided. Either internally or externally threaded projecting columns may be molded. Stripping the part from threaded mold members is possible if a gently rounded thread form is used. It must be done immediately after molding when the material is still soft and before full shrinkage takes place.

Threads also can be produced by threaded mold inserts that are unscrewed from the part before it is removed from the mold. Figure 6.4.10 illustrates recommendations for thread forms.

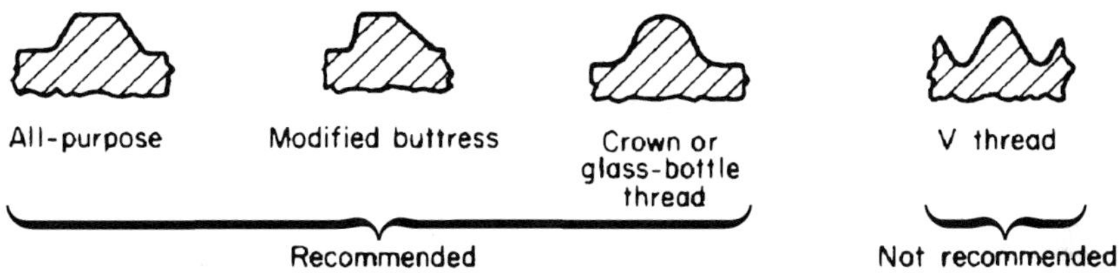

FIGURE 6.4.10 Screw threads for rotomolded parts.

For maximum resistance to localized stresses, threaded inserts should be used in a part instead of molded-in screw threads.

Inserts

Inserts may be used to fix or locate outlets, filters, valves, and the like in the walls of rotomolded parts. They may be plain or threaded, but metal units should have knurling or projections to secure them into the plastic material after it forms (one-step operation, eliminating postfinishing). Prior to molding, metal inserts may be positioned in the mold by screws, bolts, or a magnet. Plastic or plastic-metal inserts are also used, and a thin plastic flange is designed so that it melts, allowing the molding powder to adhere. (Being heavier, the body of the plastic insert does not melt during molding.) (See Fig. 6.4.11.)

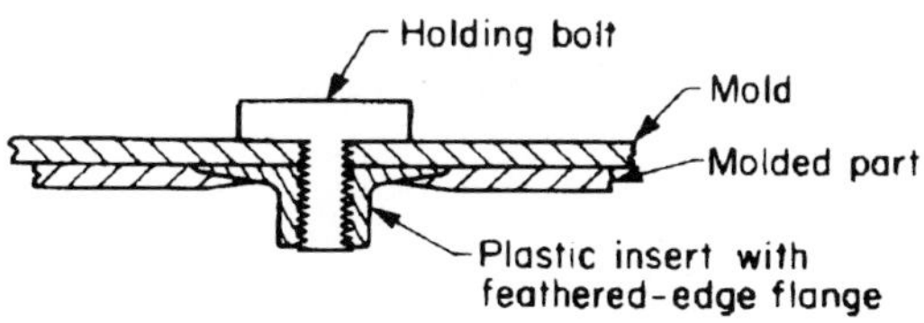

FIGURE 6.4.11 Incorporating plastic inserts into rotomolded parts.

Corner Radii and Fillets

Inside or outside corners should be well radiused and not sharp. By this means, cracking, molded-in stresses, and undesirable part thickening will be prevented. (See Fig. 6.4.12.) The smallest allowable inside radius in a part is 1.5 mm ($\frac{1}{16}$ in), with 6 mm ($\frac{1}{4}$ in) recommended for optimal filling conditions.

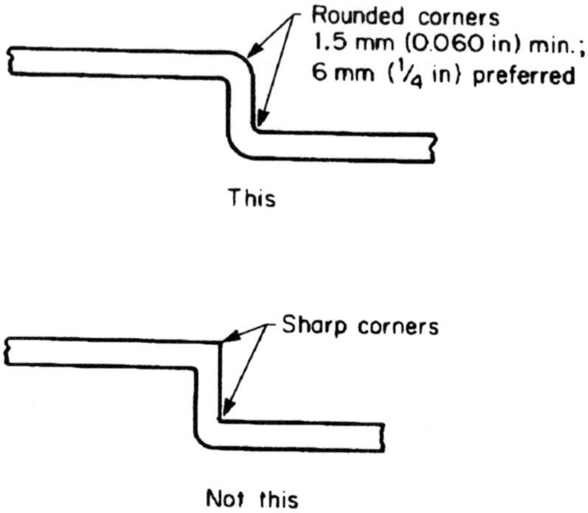

FIGURE 6.4.12 Corner radii in rotomolded parts.

Multilayered Parts

Molding of two different materials in a single part may be accomplished to combine specific properties and hence produce a better or a cheaper product. An expensive plastic may be backed with a less costly material; a skin backed with a foamed plastic may be molded in one operation. The dissimilar molding powders may possess different softening temperatures and are molded simultaneously or separately, depending on processing conditions and end-product requirements. Greater-than-normal part thicknesses usually must be designed to form multilayered parts, especially if a foam component is included.

Narrow Passages and Shape Limitations

In rotomolding, powdered material will flow in and out of a passage as small as 13 mm ($\frac{1}{2}$ in) in diameter. This is the smallest practicable internal width for rotomolded parts. (See Fig. 6.4.13.) Deep, narrow recesses and acute angles between walls of parts may present filling problems with powdered materials. (Figure 6.4.14 illustrates two shapes that should be avoided.)

Surface Finish

Owing to the inherent limitations of the surface finish of rotomolded parts, it is preferable to mold a textured surface into the parts by providing texture in the mold walls.

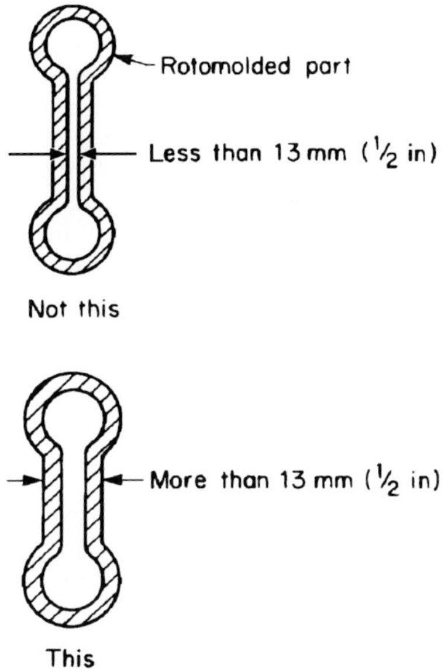

FIGURE 6.4.13 Narrow passages in rotomolded parts.

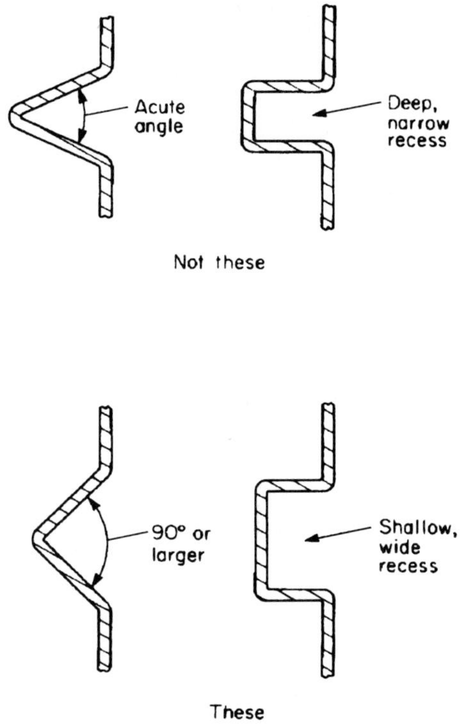

FIGURE 6.4.14 Avoid narrow projections or recesses in rotomolded parts.

Raised or depressed letters, flutes, reeds, and other decorative or utilitarian inscriptions also may be molded in by providing appropriate mold elements. Inside surfaces of molded parts are influenced by the type of molding resin and may be made smooth by selecting easily flowing polymers with a high melt index. Since such resins are sometimes chemically and mechanically inferior, inside surfaces of superior resins may be made smoother by resorting to higher molding temperatures and longer cycle times (not reaching resin-degradation conditions, of course). Incorporation of reinforcement and avoidance of its protrusion inward from inner part surfaces also may be aided by these techniques.

Parting Lines

The preferred contour for any parting line is the straightest path possible. By this means, mold construction costs will be held to a minimum, and demolding will be the easiest possible. When two parts such as a container and its lid are to be molded together, they may be separated after molding by employing a removable cutter (annular wedge) at the parting line or by molding oversize to provide a resting flange and cutting to separate (design included in Fig. 6.4.15).

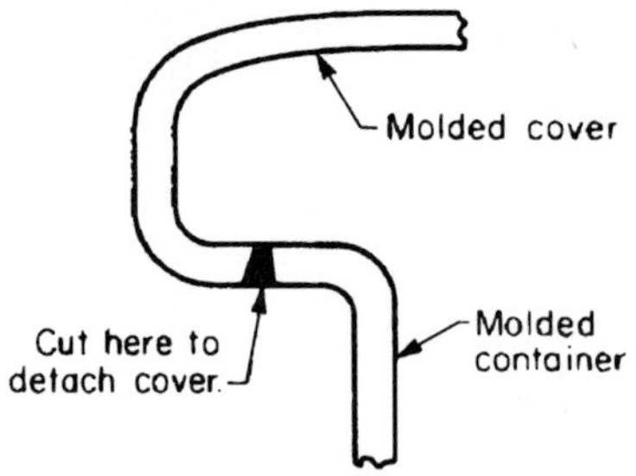

FIGURE 6.4.15 A design for molding a container and its cover in one operation.

Part Shrinkage

All rotocast parts undergo considerable shrinkage in cooling from the mold. The designer must allow for this shrinkage and its effect. Shrinkage is due in most cases to the thermal contraction of the plastic material on cooling. With thermosetting materials, molecular readjustments due to polymerization are also important. Typical shrinkages for standard rotocasting resins, under laboratory conditions, range from 1.25 to 2.5 percent. Shrinkage results not only in linear reduction of part dimensions but in distortion when wall thicknesses differ or connecting part elements pull other elements together as they shrink.

DIMENSIONAL FACTORS

Dimensional control in rotomolding is not as close as for other plastics-processing methods. Many factors influence dimensional control: polymer type, particularly its shrinkage rate, shape of the part, cooling rate (slow cooling is better), uniformity of wall thickness, mold design, and whether fillers are used (fillers reduce shrinkage and promote better dimensional control).

TOLERANCES

Recognized dimensional tolerances are listed in Table 6.4.1.

TABLE 6.4.1 Dimensional Tolerances for Rotationally Molded Plastic Parts

Dimension	Tolerance
Length and width	±0.8 mm for first 25 mm (±1/32 in for first in); ±0.4% thereafter
Wall thickness	±0.4 mm (±1/64 in)
Wall thickness buildup at corners	Usually greater by 25%

CHAPTER 6.5
BLOW-MOLDED PLASTIC PARTS

Nicholas S. Hodska
Stratford, Connecticut

THE PROCESS

Blow molding is a means of forming hollow thermoplastic objects (see Fig. 6.5.1). Air pressure applied inside a small hollow and heated plastic piece (called a *parison*) expands it like a balloon and forces it against the walls of a mold cavity, whose shape it assumes. There it cools and hardens. The mold opens, and the part is ejected.

FIGURE 6.5.1 Collection of typical blow-molded articles. (*Courtesy Air-Lock Plastics, Inc.*)

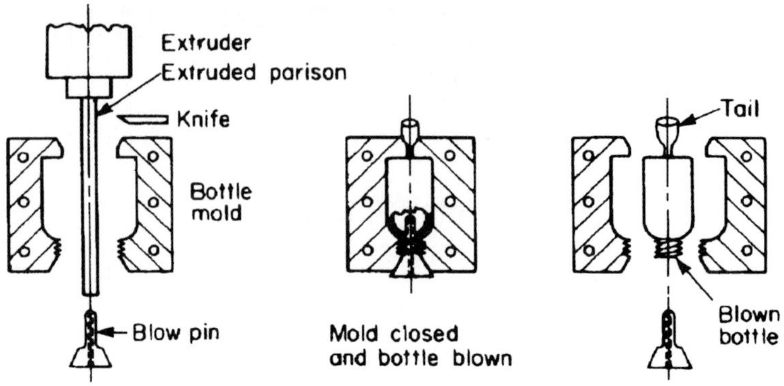

FIGURE 6.5.2 Schematic diagram of the extrusion-blow-molding process. (*From Ronald D. Beck,* Plastic Product Design, *© Litton Educational Publishing, Inc., Van Nostrand Reinhold, New York, 1970.*)

In extrusion blow molding, the parison is extruded as a tube. This is inserted in the blow-molding die with one end engaging a blow pin or needle. As the die is closed, the tube is pinched at both ends. After the pinched-off tube is expanded and the part is formed and after the die opens and the part is ejected, the surplus material adjacent to the pinched-off areas is removed. Figure 6.5.2 illustrates the extrusion-blow-molding process.

In injection blow molding, the operation is similar except that the parison is made by injection molding instead of extrusion. It is molded over a mandrel to provide the hollow shape, and this mandrel transfers the hot parison to the blow-molding die and then functions as the blow nozzle.

In stretch blow molding, a center rod stretches the parison to about two times its length. This, plus the stretching action of the inflation, produces a biaxial orientation of the molecules in the walls of the part, improving the strength and clarity of the walls.

Sometimes coextrusion or coinjection molding of more than one material is used to provide an improved barrier in blow-molded containers. Labels are also sometimes incorporated in the mold before the inflation phase to provide better adhesion and protection of the label.

TYPICAL CHARACTERISTICS AND APPLICATIONS

Blow molding produces thin-walled hollow or tubular objects. Containers for liquids and other items used in the household are the most common application. Watering cans and bottles for laundry detergent and bleach, cooking oil, shampoo, oatmeal, and various cosmetics and medicines are typical examples. The size range of blow-molded parts ranges from small pharmaceutical containers to large objects of 50 lb or more in weight.

Another large market is in blow-molded toys, ranging from simple balls and lightweight baseball bats to elaborate dolls and animal toys, large play equipment, tricycles, and rideable play vehicles. Specialized containers such as carrying cases for instruments and tools, vehicle fuel tanks, 55-gallon drums, automobile glove compartments, center consoles, bumper systems, ducts, and even modern-design desks are also made by blow molding.

Wall thicknesses for commercial household containers range from 0.4 mm (0.015 in) to about 3 mm ($^1/_8$ in). The wall thickness in the neck area of a container is normal-

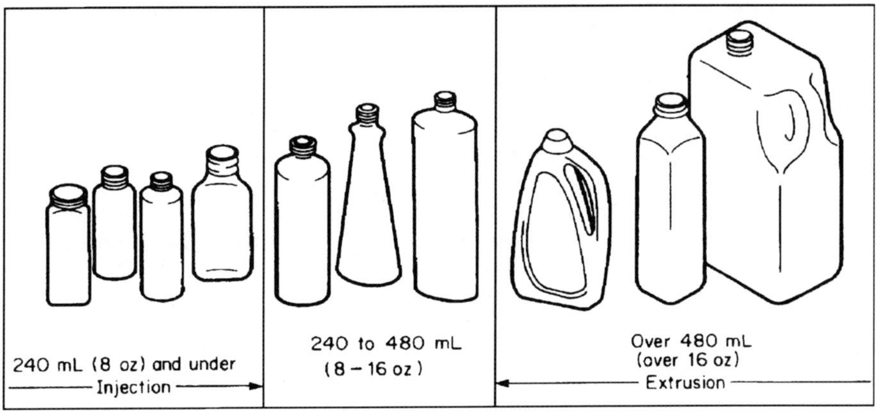

FIGURE 6.5.3 Comparison of the common size and shapes of containers made by injection blow molding and extrusion blow molding. (*From* Plastics Design & Processing.)

ly greater than in the body to provide secure sealing surfaces for caps or dispenser devices. Walls can thin considerably below nominal values in other areas where the material stretches to fill the mold.

Extrusion blow molding and injection blow molding have somewhat different capabilities when it comes to the size and shape of containers molded. Injection blow molding is more suitable for smaller containers, with fairly regular shapes, made with rigid materials. It also can produce more accurate dimensions around the neck of a container if this is necessary for the container's closure or dispenser system. Injection-blow-molded parts normally do not have any "tail" or extraneous material after molding and therefore do not require a trimming operation. Close control over parison weight, shape, and wall thickness is inherent in the process, providing a more nearly uniform and accurate blow-molded part. Extrusion blow molding is better for larger containers and those with irregular shapes. Figure 6.5.3 and Table 6.5.1 illustrate the advantageous applications for each process.

TABLE 6.5.1 Merits of Blowing Methods

Extrusion blow molding
1. More competitive for most containers over 16 oz.
2. Lower initial mold cost.
3. Can readily produce bottles with handles.
4. More effective with rigid PVC.
5. Can produce extremely irregular containers (such as ovals with maximum width-to-thickness ratios).

Injection blow molding
1. More competitive for containers under 12 oz.
2. Readily produces specialty neck finishes (safety closures).
3. Closer tolerances on critical dimensions and weight.
4. No finishing operations required.
5. Improved yield (no waste).
6. More efficient with rigid materials (polystyrene, polycarbonate).
7. More efficient with wide-mouth containers over 43 mm.

Source: From *Plastics Design & Processing.*

ECONOMIC PRODUCTION QUANTITIES

Injection blow molding is a high-volume process most commonly used for production in the millions. When less than 1 million units are to be produced, the extrusion-blow-molding process usually has the advantage, but for runs in excess of 2 million (and when part volume is less than 12 oz), injection blow molding definitely has an economic advantage. The fact that tooling for injection blow molding can be costly places a lower limit on production quantities with which the process can be competitive. Extrusion blow molds are lower in cost than injection blow molds by about one-third for the same part but still are costly. Blow-molding operations, however, can be highly mechanized, with automatic loading of the parison into the blow mold and automatic trimming if that is required. Production cycles with such equipment may be very short. For example, 175-mL (6-oz) containers can be molded in a 12-s cycle even with a multiple-cavity mold. An eight-cavity mold for such a part can produce 2400 pieces per hour.

Despite the very high production runs that are normal for blow moldings, the process can be suitable for much shorter runs in many cases. Production quantities on the order of 10,000 units are often feasible with the extrusion-blow-molding process.

SUITABLE MATERIALS

High-density polyethylene is by far the most common material used for blow-molded parts. It is used for household product containers and a wide variety of the large components listed above. Polyethylene processes well in blow molding and has good resistance to attack by solvents and corrosive materials. Polyethylene terephthalate (PET) is the second most commonly blow-molded material, with its major use being bottles for soft drinks.

Other frequently used plastics for blow-molded components are polypropylene, polystyrene, polyvinyl chloride (PVC), and cellulosics. Styrene acrylonitrile (SAN) and acrylonitrile butadiene styrene (ABS) are two additional materials, as are the following engineering plastics: polycarbonate, polyphenylene ether blends, and amorphous nylon. PVC is used for containers for edible oils, cosmetics, and household chemicals. PVC is better processed by extrusion blow molding than by injection blow molding, as are the nitrile-based materials.

DESIGN RECOMMENDATIONS

Wall Thickness

The wall thickness should be as nearly uniform as possible to ensure more rapid molding cycles, conserve material, and avoid distortion due to uneven cooling. (Thinner sections cool more quickly.)

If the neck portion of a bottle must have a thicker wall than the body to provide a rigid sealing structure, the ratio of wall thickness between the neck and body should not exceed 2:1. (See Fig. 6.5.4.)

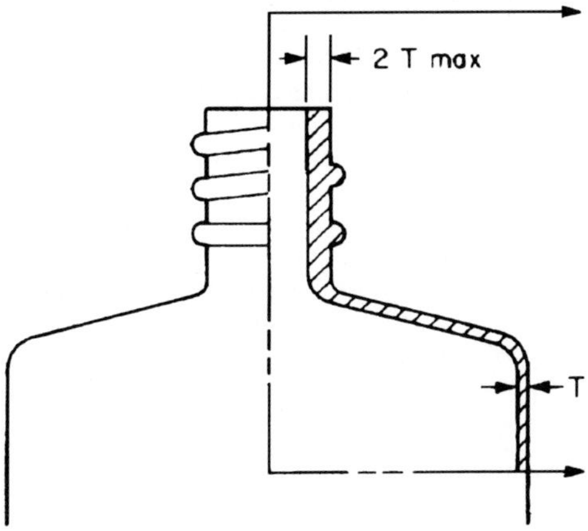

FIGURE 6.5.4 When the neck portion of a bottle must have a thicker wall to provide a rigid sealing structure, the ratio of wall thickness between the neck and the body should not exceed 2:1.

Draft

The part design should permit portions of the blow mold that are perpendicular to the parting plane to have a draft. This is preferred to a square configuration in ensuring that the molded part can be removed from the mold. (See Fig. 6.5.5.)

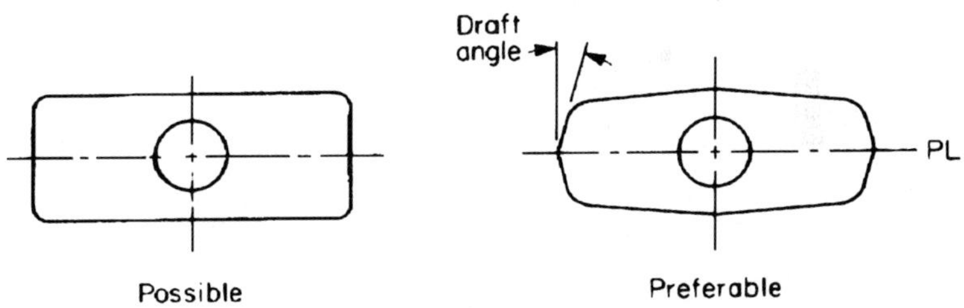

FIGURE 6.5.5 Provide a draft on surfaces perpendicular to the mold-parting line.

Corners

Generously rounded corners and large fillets are necessary in blow-molded parts to maintain more nearly uniform wall thickness and ensure easy molding and maximum strength of the molded part.

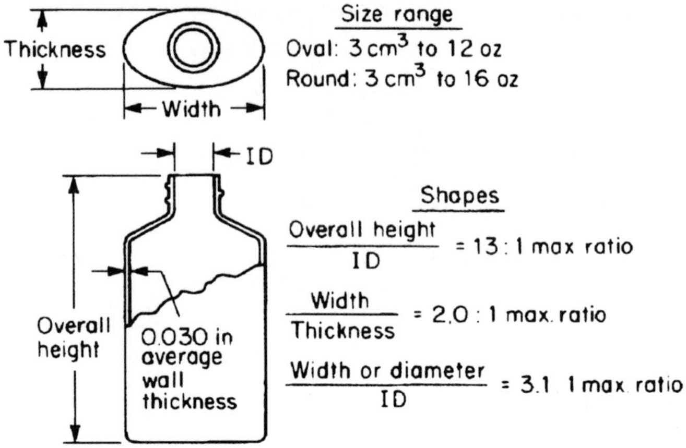

FIGURE 6.5.6 Design limitations of injection-blow-molded containers. (*From* Modern Plastics Encyclopedia, *McGraw-Hill, New York, 1970.*)

Preferred Shapes

Although somewhat intricate shapes can be blow-molded, regular, well-rounded shapes are more easily molded, are more economical, and should be specified when there is a choice. Extrusion blow molding is capable of producing containers with integral handles and other irregularities not feasible with injection blow molding. Figure 6.5.6 illustrates the practical limits of size and shape for injection-blow-molded parts.

Undercuts

Although undercuts are a complicating factor and should be avoided if not needed, they can be incorporated into blow-molded parts. Undercuts considerably deeper than those feasible with injection molding can be incorporated in blow-molded parts without difficulty.

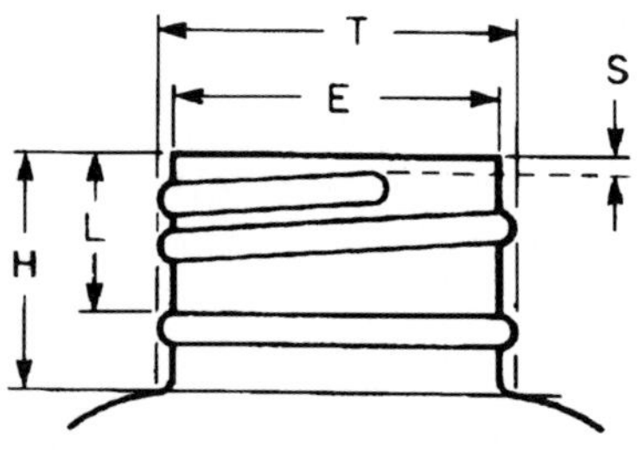

FIGURE 6.5.7 Standards for screw-threaded cap closures for blow-molded bottles ($1\frac{1}{2}$ screw threads). (*From J. Harry Dubois and F. W. John,* Plastics, *4th ed., Reinhold, New York, 1967.*)

Standard Closures

The use of standard container-closure designs helps promote manufacturing economy. The blow-molded container industry has adopted glass bottle industry standards for comparable plastic containers. Figure 6.5.7 illustrates the standards for a screw-cap closure having $1\frac{1}{2}$ threads.

Size	*T*, in	*E*, in	*H*, in	*S*, in	*L*, in (min.)	Thread turns
18	0.694 ± 0.010	0.610 ± 0.010	0.508 + 0.008 − 0.007	0.034 + 0.013 − 0.012	0.345	1½
20	0.773 ± 0.010	0.689 ± 0.010	0.539 + 0.008 − 0.007	0.034 + 0.013 − 0.012	0.345	1½
22	0.852 ± 0.010	0.768 ± 0.010	0.570 + 0.008 − 0.007	0.034 + 0.013 − 0.012	0.360	1½
24	0.930 ± 0.010	0.846 ± 0.010	0.631 + 0.008 − 0.007	0.046 + 0.016 − 0.015	0.421	2
28	1.075 ± 0.013 − 0.012	0.981 ± 0.013 − 0.012	0.693 + 0.008 − 0.007	0.046 + 0.016 − 0.015	0.405	1¼

DIMENSIONAL FACTORS

Blow-molded parts are subject to many of the same factors that cause dimensional variation in plastic parts made with other processes. Additional factors are the fact that two successive operations are involved, each with some inherent variation, the low molding pressure of the process and the stretching action that takes place. These lead to uneven wall thicknesses where the part's shape is irregular and to other dimensional variations.

For these reasons, tolerances for blow-molded parts must be quite large. The exception is the neck area of injection-blow-molded parts. Tolerances for dimensions in this area, because they are determined by the injection-molding process, are approximately equivalent to those attainable by injection molding, to values as small as ±0.1 mm (0.004 in).

Table 6.5.2 shows recommended tolerances for other areas of injection-blow-molded components and for all areas of extrusion-blow-molded components.

TABLE 6.5.2 Recommended Dimensional Tolerances for Blow-Molded Plastics Parts

1. Neck areas of injection-blow-molded containers: Use tolerances from tables in Chap. 6.2, "Injection-Molded Thermoplastic Parts."
2. Other areas of injection-blow-molded parts and all areas of extrusion-blow-molded parts:

Dimension, mm (in)	Recommended tolerance, mm (in)
0–25 (0–1.0)	±0.5 (±0.020)
25–100 (1–4)	±0.9 (±0.035)
100–200 (4–8)	±1.3 (±0.050)
200–400 (8–16)	±2.3 (±0.090)

3. Wall thickness (outside of container neck areas): ±50% of nominal wall thickness.

CHAPTER 6.6
REINFORCED-PLASTIC/COMPOSITE (RP/C) PARTS

J. Gilbert Mohr
J. G. Mohr Co., Inc.
Maumee, Ohio

PROCESSES

The molding of reinforced plastics/composites (RP/C) basically involves embedding a strong reinforcing material (fiberglass, asbestos, carbon fiber, etc.) into a liquid-that-becomes-solid polymeric matrix (polyester, epoxy, polypropylene, etc.). In the case of thermosetting plastics, the change from liquid to solid is effected by catalysts and heat. In the case of thermoplastic materials, it simply involves a cooling of the melted plastic to a temperature below its softening point. The specific shape and size of the part and the production quantity dictate which molding method will be used. A series of schematic illustrations of the various RP/C molding methods is presented in Fig. 6.6.1.

TYPICAL CHARACTERISTICS AND APPLICATIONS

Favorable characteristics of RP/C construction are a high degree of design freedom, ability to use fewer parts by combining what otherwise would be separate parts into one component, a high strength-to-weight ratio, excellent corrosion resistance, excellent appearance with a smooth surface and molded-in color, electrical- and thermal-insulation properties, and easy repairability. The physical strengths of the material depend on the glass content and method of molding and can be designed to a desired level within a wide range of possibilities. Processing advantages include low tooling costs and short tooling lead times and availability of a wide choice of molding methods.

In addition, adaptability is gained by using RP/C components together with other materials by incorporating metal inserts, by joining several parts with mechanical fasteners or adhesives, or by encapsulating as with sandwich construction.

Limitations of the material include a maximum temperature of 120 or 150°C (250

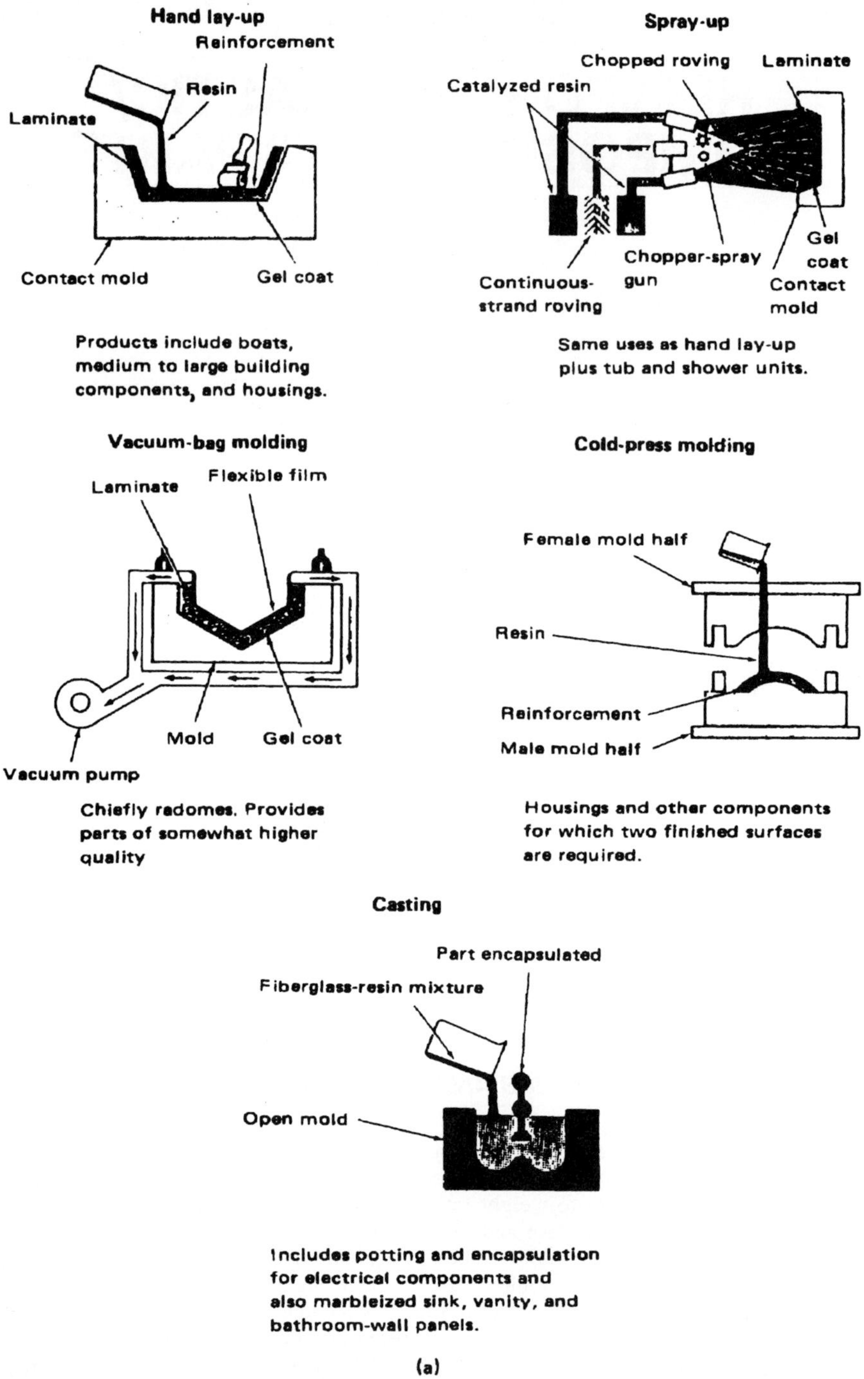

FIGURE 6.6.1*a* Room-temperature-cure RP/C molding methods. (*Sketches reproduced by permission of Owens-Corning Fiberglas Corporation.*)

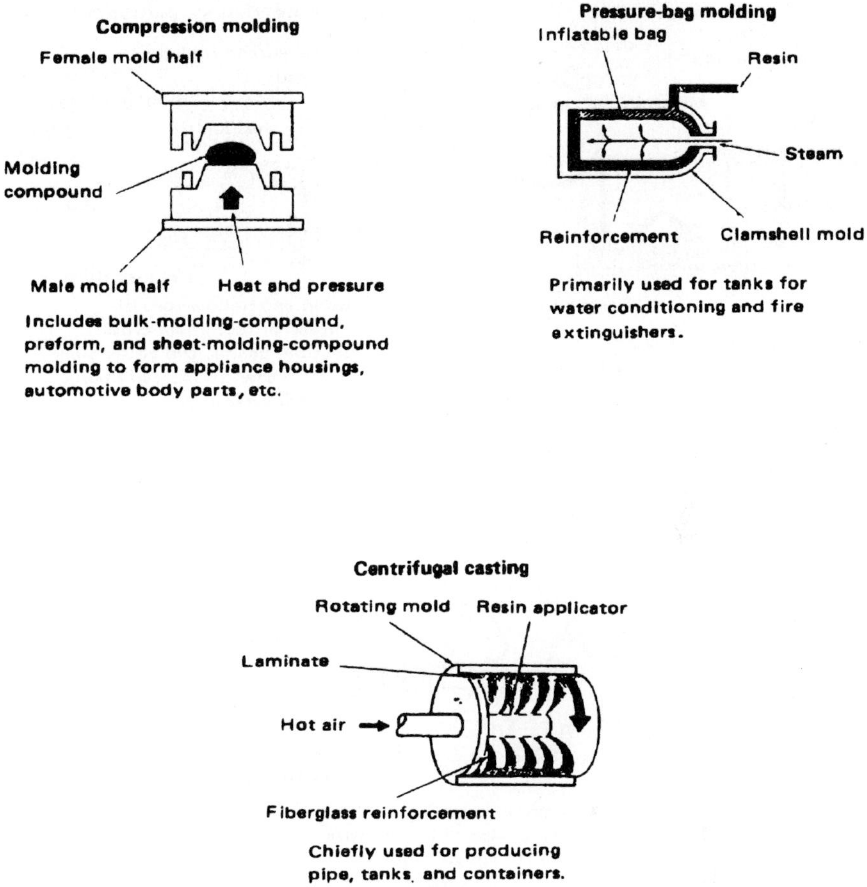

(b)

FIGURE 6.6.1*b* Thermoset processes requiring pressure. (*Sketches reproduced by permission of Owens-Corning Fiberglas Corporation.*)

or 300°F) for continuous service, creep or cold flow that precludes use for most sustained heavy-load-bearing applications, weaknesses around drilled or punched holes owing to severing of the reinforcing fibers, and poor resistance to severe abrasion or repeated heavy impact when compared with metals. However, many of these limitations may be overcome with proper design.

RP/C parts range in size from small, rapidly produced mechanical components such as gears or distributor caps, through continuously produced rods, girder stock, or architectural paneling, to extremely large tanks as much as 12.8 m (42 ft) in diameter by 12.2 m (40 ft) high.

Figure 6.6.2*a* through 6.6.2*l* illustrate representative RP/C parts or products fabricated by various molding methods.

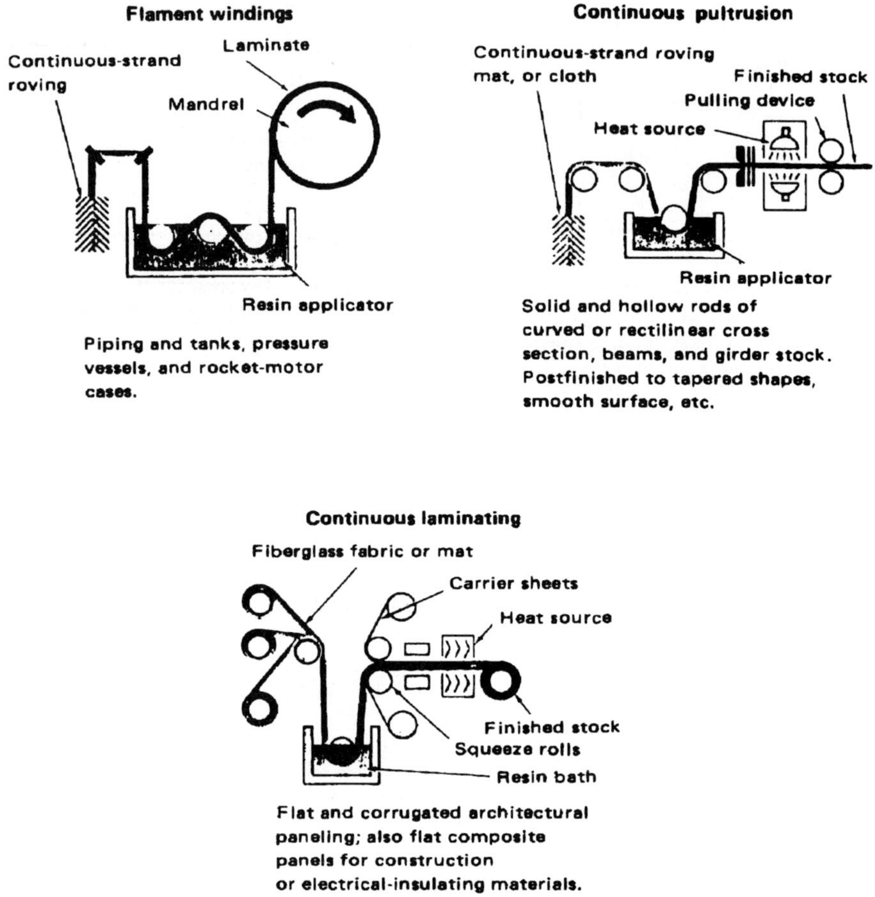

(c)

FIGURE 6.6.1*c* Continuous throughput processes. (*Sketches reproduced by permission of Owens-Corning Fiberglas Corporation.*)

Mode and Percentage of Reinforcement

Reinforcement may be random or directionally oriented and may vary in length from short milled fibers (for casting) to chopped strands from 3 mm ($\frac{1}{8}$ in) to 50 mm (2 in) or longer. It also may be sold in prepared mats or woven fabrics (for hand lay-up, etc.) or as continuous lengths (for filament winding). Mechanical strength is generally favored by increased amounts of the reinforcing agent, while good electrical properties require a dispersion of the high-dielectric-glass filaments throughout the molded part. Reinforcement quantities vary from a low of 1 or 2 percent (casting and some rotational-molding applications), through the nominal 20 to 40 percent for the majority of uses (hand lay-up, compression and injection molding, and centrifugal casting), to a high of 65 to 85 percent (for specialty compression molding, pultrusion, and fila-

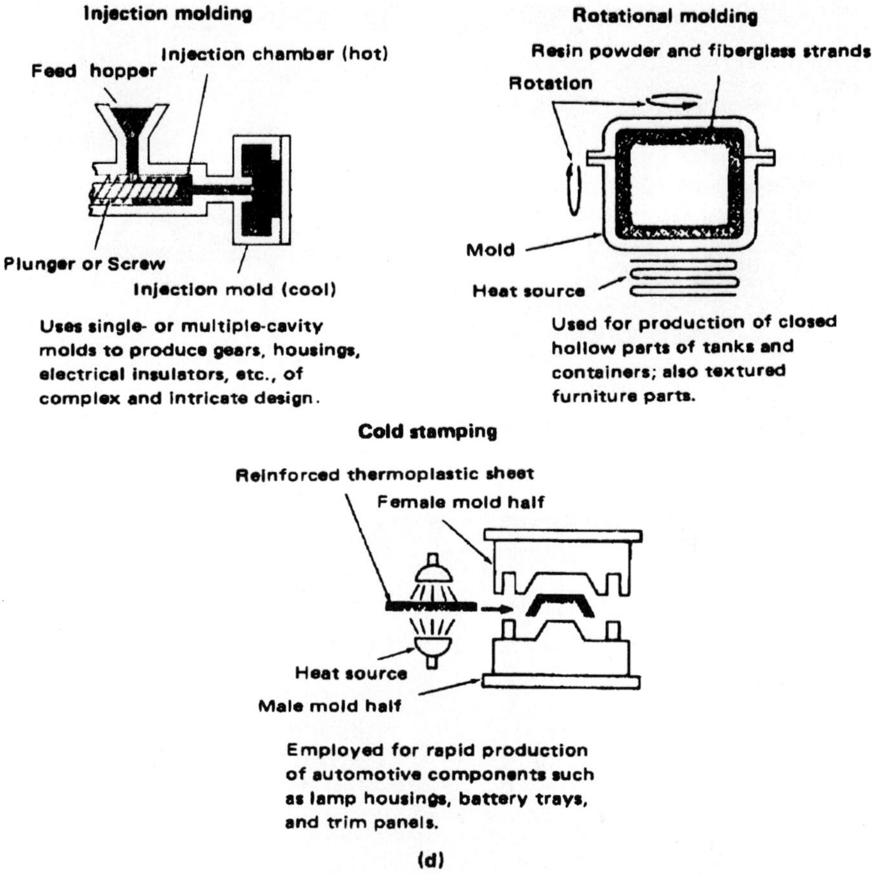

FIGURE 6.6.1*d* Processes for molding reinforced-thermoplastic resins. (*Sketches reproduced by permission of Owens-Corning Fiberglas Corporation.*)

ment winding). Naturally, highest physical strengths are provided in the direction of major glass-fiber orientation.

Surfaces

Gel coats are surface layers of material not containing reinforcing fibers. Normally 0.25 to 0.30 mm (0.010 to 0.012 in) thick, they are formed by applying and curing a highly filled resin layer. They may be designed for a range of characteristics such as improved appearance, chemical resistance, weather and craze resistance, abrasion resistance, etc.

Vacuum-formed acrylic or other thermoplastic elements form the outer skin (replacing gel coats) on many RP/C items such as sinks and vanities, with good-adher-

FIGURE 6.6.2*a* Sleighs made by the hand-lay-up process. (*Courtesy Bay View Plastic, Inc.*)

FIGURE 6.6.2*b* Tub and shower units made by the spray-up process. (*Courtesy Universal-Rundle Corp.*)

ing fiber-glass resin applied as backing for rigidizing. Mold and operating costs are greatly reduced, and some of the defects associated with gel coats are eliminated.

Incorporation of low-shrink additives in bulk molding compounds (BMC) and sheet molding compounds (SMC) has greatly enhanced acceptance and utilization of RP/C parts by avoiding the fiber pattern that formerly resulted when the fibers were revealed after molding because of resin contracting around them.

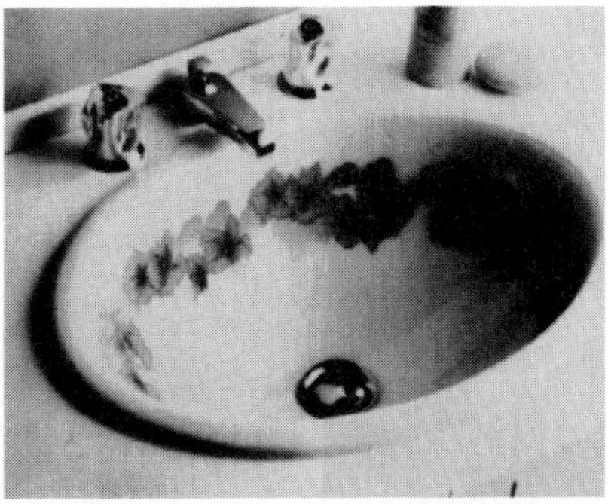

FIGURE 6.6.2c Acrylic-faced hand-bowl unit made by vacuum forming and spray-up (rigidizing). (*Courtesy J. S. Johns Co.*)

FIGURE 6.6.2d Radomes made by the vacuum-bag molding process. (*Courtesy Century Plastics, Inc.*)

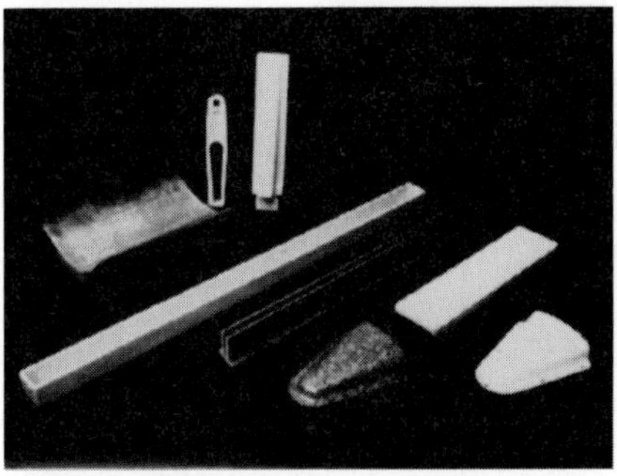

FIGURE 6.6.2e Products including motor base, channels, etc., made by the cold-molding process. (*Courtesy Bay View Plastic, Inc.*)

FIGURE 6.6.2f Sink made by the casting process. (*Courtesy Creative Marble.*)

FIGURE 6.6.2g Electrical components made by the BMC molding process. Note molded-in inserts.

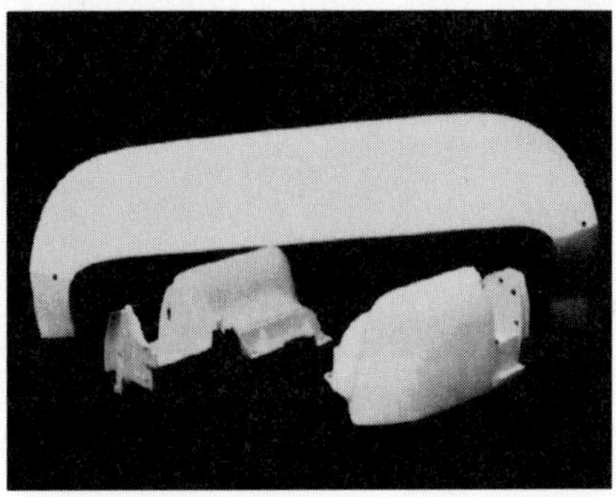

FIGURE 6.6.2h Automotive fender skirt and fender extensions made by using SMC molded-in heated matched-metal dies.

FIGURE 6.6.2*i* 5000-gal tank made by the filament-winding process. (*Courtesy Justin Enterprises and Johns-Manville Corp.*)

FIGURE 6.6.2*j* Typical cross-sectional shapes which can be produced by pultrusion. (*Courtesy Morrison Molded Fiber Glass Co.*)

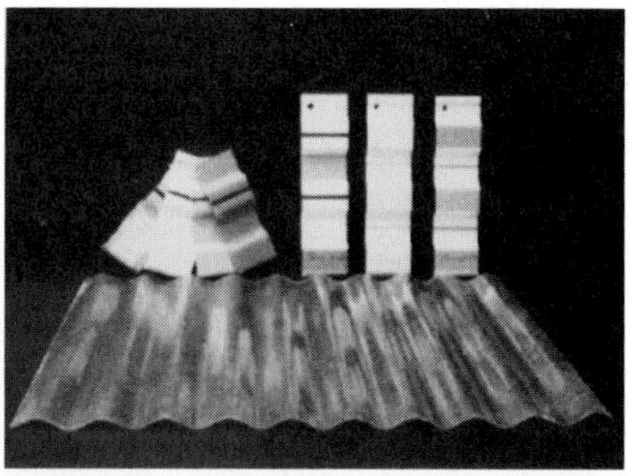

FIGURE 6.6.2*k* Corrugated architectural paneling made by the continuous-laminating process. (*Courtesy The Society of the Plastics Industry, Inc.*)

FIGURE 6.6.2*l* Automotive front-end lamp and bumper assembly (third part to rear; black) was molded by using the cold-stamping process. The other parts were molded by other processes. (*Courtesy GRTL Co.*)

Translucency

The resins used in fabricating RP/C are clear, but because they are reinforced and/or filled, they are prevented from being transparent. Translucency is readily accomplished by the use of either small amounts of filler or none and thinner sections. Flat or corrugated architectural paneling is the primary RP/C product having this property, and the panels are colored or tinted for pleasing effects. Applications include room dividers, patio covers, garage doors, insulating walls (double-glazed), decorative sandwich laminates with embedments, building lights, solar-heat collectors, greenhouse and poultry-farm panels, etc. The translucency of RP/C components is also useful in large hand-lay-up and filament-wound tanks or piping where the level of the contained liquid may be readily observed.

ECONOMIC PRODUCTION QUANTITIES

The optimal production quantity for RP/C parts (from a cost standpoint) varies tremendously depending on the process used. Hand lay-up methods are most suitable for very limited quantities, while injection molding requires mass-production levels to amortize tooling and equipment investments.

Table 6.6.1 summarizes normal economic production quantities for the common RP/C processes. Relative tooling and equipment costs and output rates are shown.

In general, larger parts will require longer production times than small parts, and increased production rates will result from mechanized and automated processes.

Table 6.6.2 shows normal mold life in terms of the number of parts producible from a single-cavity mold peculiar to each RP/C process. It also presents relative cost data for molds.

TABLE 6.6.1 Economic Production Quantities for Various RP/C Molding Methods*

Molding method	Relative investment required: Equipment	Relative investment required: Tooling	Relative production rate	Economic production quantity
Hand lay-up	VL	L	L	VL
Spray-up	L	L	L	L
Casting	M	L	L	L
Cold-press molding	M	M	L	L
Vacuum-bag molding	M	L	VL	VL
Compression-molded BMC	H	VH	H	H
SMC and preform	H	VH	H	H
Pressure-bag molding	H	H	L	L
Centrifugal casting	H	H	M	M
Filament winding	H	H	L	L
Pultrusion	H	H	H	H
Continuous lamination	H	H	H	H
Cold stamping	H	H	H	H
Rotational molding	H	H	L	M
Injection molding	VH	VH	VH	VH

*VL = very low; L-low; M-medium; H-high; VH-very high.

TABLE 6.6.2 Mold Life and Mold Cost for Various RP/C Molding Processes

RP/C process	Approximate number of parts from mold (mold life)	Relative mold cost	Type of mold
Hand lay-up	800–1000	Lowest	RP/C
Spray-up	100–200	Lowest	RP/C
Vacuum-bag molding	100–200	Low	RP/C
Cold-press molding	150–200	Medium	RP/C
Casting, electrical	3500	High	Metal
Casting, marble	300–500+	Lowest	RP/C
Compression molding: BMC	120,000	High	Metal
Matched-die molding: SMC	300,000–400,000	Highest	Metal
Pressure-bag molding	Over 1,000,000	Medium	Metal
Centrifugal casting	Over 1,000,000	Low	Metal
Filament winding	Almost unlimited	Lowest	Metal
Pultrusion	Almost unlimited	Minimal	Metal
Continuous laminating	Almost unlimited	Minimal	Metal
Injection molding:reinforced thermoplastic (RTP)	300,000–1,000,000	Highest	Metal
Rotational molding	100,000	Low	Metal
Cold stamping	1,000,000–3,000,000	Highest	Metal

GENERAL DESIGN CONSIDERATIONS

The first quest of a designer planning an RP/C part is to determine, from its size, shape, quantity to be produced, and strength requirements, which method should be used in fabrication. This is important because design principles vary considerably with the molding method. For example, there are limitations to the depth of draw obtainable with sheet molding compounds that do not apply to preforms. The component should be designed with a specific process in mind.

A second principle is to show the preliminary design to prospective molders. It is always sound advice to consult the prospective fabricator of any part, but this is especially so in the case of RP/C parts, for which a considerable variety of manufacturing processes is available.

The next major consideration, once the design shape has been finalized, is the type of resin and reinforcing materials to specify. Resins may be rigid or resilient or may be made fire-retardant, resistant to water or chemicals, light-stable for nondiscoloration, or have other special properties.

Shrinkage occurs during the cure of resins for all processes. Designers should allow for the dimensional changes and distortion that can result from shrinkage. Shrinkage may be reduced or controlled by additives in molding SMC and BMC in matched-metal dies. Shrinkage usually assists in removing parts from molds. (The exception occurs when the part shrinks onto the mold, as in filament winding.)

SUITABLE MATERIALS

Thermosetting Resins

Polyester Resins. Polyesters are lowest in cost and possess the most facile and trouble-free processibility of any thermosetting resins. They are the workhorses of the industry and are used for most general-purpose applications in room-temperature, intermediate-, or high-temperature cures. They also may be modified by composition or additives to provide improved chemical durability, electrical properties, light stability, weather resistance, high-heat resistance, or other properties. The modifications result in increased materials cost with minimum change in processing qualities. Vinyl ester resins are included in this polyester-resin category.

Epoxy Resins. These resins are medium to high in price and qualify as easy to somewhat difficult in processibility. They respond to room-temperature, intermediate-, and high-temperature cures by selection of the proper catalyst system. Epoxy resins are tougher and have lower shrinkage than polyesters (excluding additives). Prime end uses are in electrical applications and in filament winding for dimensional stability. Chemical resistance is no greater than for polyester resins and is specific. Adhesion is excellent.

Phenolic Resins. Phenolics are low in cost but difficult to process by usual RP/C standards, since most molding methods for their use call for high temperatures and pressures. A number of reinforcements and fillers are used with phenolics (see Chap. 6.1).

Alkyd Resins. Medium to high in price and difficult to process (again requiring high pressures), alkyd resins are usually sold as formulated molding compounds for use in applications requiring high electrical performance.

Dallyl Phthalate Resins. These are medium- to high-priced and difficult to process, mostly requiring high temperatures and pressures. They are purveyed mostly in molding-compound or preimpregnated-mat form and yield such molded properties as high hardness, clarity, and low shrinkage.

Furfural Alcohol Resins. Also termed *furan* resins, these are low-priced but are processed with difficulty owing to malodor and the need for a hazardous catalyst system based on H_2SO_4. Furfural alcohol resins have excellent chemical durability, even surpassing polyesters, so their processing problems are sometimes justified. Whereas they were formerly moldable only in press operations, in recent years wet or hand-lay-up types have been made available by buffering the catalyst systems or otherwise making them easier to use. Fiberglass is the main reinforcement, and fillers are used according to end requirements.

Silicone Resins. These resins are extremely high in price and possess difficult processing characteristics. Their use in RP/C is limited. Glass is the major reinforcement, and a number of fillers are used. However, the resulting molded products are comparatively low in strength. Electrical and high-temperature applications are the major end uses.

Thermoplastic Resins

At least 15 resin types are processible as reinforced thermoplastics. Their prices range from medium to high. Processibility for all is relatively easy and rapid and is carried out primarily in injection-molding equipment. Fiberglass reinforcement up to 40 percent or more by weight upgrades these materials into the engineering-plastics classification. Characteristic resin types include all molecular configurations of the nylons, acetal, polyethylenes (also the principal resins used in rotational molding and cold stamping), polypropylene, ionomer, polybutylene terephthalate (PBT), polycarbonate, polysulfone, polyphenylene oxide, polystyrene and the related styrenics (SAN and ABS), fluoroplastic polytetrafluoroethylene (PTFE; highest-priced), rigid PVC, polyurethane, and polyphenylene sulfide. The last two are the next highest priced.

Reinforcements

Fiberglass. Fiberglass, the most widely used general-purpose reinforcement, is low to medium in cost. It is also the most readily processible reinforcement, with a form for each specific type of molding of RP/C: yarns, woven tapes and fabrics, veil and random-fiber-oriented mats, rovings and woven roving, chopped strands, and milled fibers. Related glass forms are solid and hollow beads and glass flake. Reinforcement levels average 20 to 35 percent and range from 2 to 5 percent (premix and rotational molding) to 75 to 85 percent (filament winding). All fiberglass-reinforcing products supplied contain an organic sizing material (0.5 to 2.5 percent by weight) designed to promote compatibility with the resin, method of molding, and end use. This sizing should be investigated and specified in the product design.

Asbestos. This natural mineral reinforcement is low in cost and readily processible. Forms available are bulk fibers in many lengths, woven tapes and fabrics, and both parallel-fiber and random-oriented mats. Asbestos provides good uniform and hard surfaces in molding compounds such as premix for electrical applications. Because of health hazards to workers handling it, asbestos is now used much less frequently than previously.

Natural Organic Fibers: Sisal, Jute, and Cotton. Sisal prices have risen to levels above those for glass fiber in recent years, taking away one of the most desirable factors. Processibility is excellent for limited applications (BMC) because the material is a monofilament and is highly resistant to degradation in processing. Sisal and jute are available in chopped fiber lengths and also in unidirectional mats. The mats may be cross-plied for greater load-carrying capacity in hand-lay-up structures. These natural fibers possess high water absorption, which limits their use as reinforcements.

Synthetic Organic Fibers: Nylon, Polyester, Acrylonitrile, and Polyvinyl Alcohol. These fibers are medium in cost and excellent for processibility.

High-Performance Fibers: Carbon, Graphite, Metal, Boron, and Ceramic Fibers. These fibers are high-priced and medium to difficult to process, but all generate highly desirable improvements in RP/C. Carbon and graphite yield high strengths and a high modulus of elasticity plus lightness in weight. Applications are golf shafts and empennage structures for aircraft. SiO_2, Al_2O_3, other ceramic fibers, and boron also have been used.

Fillers

Clays and calcium carbonate are the fillers used in the largest volume; their level ranges from 15 to 80 percent of the weight of finished molded parts. Other fillers and their respective functions are short-fibered asbestos, which has good surface and electrical properties but imparts a characteristic greenish color; $Mg(OH)_2$ and powdered or liquid thermoplastic polymers, which respectively provide the desirable thickening and low-shrink factors in SMC; finely divided silicas, which control viscosity and prevent rundown in hand-lay-up resins; and talcs, which provide easy sanding in body putty, lightness of weight, and good electrical properties. Other filler additives reduce flammability, improve abrasion resistance, and improve corrosion and weathering resistance. Even water is used in the emulsified form as a filler-extender in casting resins.

DETAILED DESIGN RECOMMENDATIONS

Design recommendations expressed in Chap. 6.1, "Thermosetting-Plastic Parts," are generally applicable to compression, transfer, and injection molding of thermoset RP/C and need not be repeated here. Similarly, almost all design recommendations in Chap. 6.2, "Injection-Molded Thermoplastic Parts," apply equally well to injection-molded thermoplastic parts using reinforcements and fillers and are not repeated here. Design recommendations included below are applicable primarily to RP/C processes not duplicated in these chapters.

TABLE 6.6.3 Wall-Thickness Ranges and Tolerances for RP/C Parts

Molding method	Thickness range Min., mm (in)	Thickness range Max., mm (in)	Maximum practicable buildup within individual part	Normal thickness tolerance, mm (in)
Hand lay-up	1.5 (0.060)	30 (1.2)	No limit; use cores	±0.5 (0.020)
Spray-up	1.5 (0.060)	13 (0.5)	No limit; use many cores	±0.5 (0.020)
Vacuum-bag molding	1.5 (0.060)	6.3 (0.25)	No limit; over three cores possible	±0.25 (0.010)
Cold-press molding	1.5 (0.060)	13 (0.5)	3–13 mm (1/8–1/2 in)	±0.5 (0.020)
Casting, electrical	3 (1/8)	115 (4½)	3–115 mm (1/8–4½ in)	±0.4 (0.015)
Casting, marble	10 (3/8)	25 (1)	10–13 mm; 19–25 mm (3/8–1/2 in; 3/4–1 in)	±0.8 (1/32)
BMC molding	1.5 (0.060)	25 (1)	Min to max. possible	±0.13 (0.005)
Matched-die molding: SMC	1.5 (0.060)	25 (1)	Min. to max. possible	±0.13 (0.005)
Pressure-bag molding	3 (1/8)	6.3 (1/4)	2:1 variation possible	±0.25 (0.010)
Centrifugal casting	2.5 (0.100)	4½% of diameter	5% of diameter	±0.4mm for 150-mm diameter (0.015 in for 6-in diameter); ±0.8 mm for 750-mm diameter (0.030 in for 30-in diameter)
Filament winding	1.5 (0.060)	25 (1)	Pipe, none; tanks, 3:1 around ports	Pipe, ±5%; tanks, ±1.5 mm (0.060 in)
Pultrusion	1.5 (0.060)	40 (1.6)	None	1.5 mm, ±0.025 mm (1/16 in, ±0.001 in); 40 mm, ±0.5 mm (1½ in, ±0.020 in)
Continuous laminating	0.5 (0.020)	6.3 (1/4)	None	±10% by weight
Injection molding	0.9 (0.035)	13 (0.5)	Min. to max. possible	±0.13 (0.005)
Rotational molding	1.3 (0.050)	13 (0.5)	2:1 variation possible	±5%
Cold stamping	1.5 (0.060)	6.3–13 (0.25–0.50)	3:1 possible as required	± 6.5% by weight; ±6.0% for flat parts

Wall Thickness

Thickness minima and maxima for each process are given in Table 6.6.3. Thickness may be varied within parts, but prolonged cure times, slower production rates, and the possibility of warpage result. If possible, thickness should be held uniform throughout a part.

For hand-lay-up and spray-up methods, extra thicknesses can be built up by first allowing the lay-up to cool from its exotherm, but extra material should be added before 24 h have elapsed. Thicker wall sections, if needed, should be cored or sandwiched as shown in Fig. 6.6.7 to minimize weight increase and avoid distortion.

Mold Parting Line and External Edges

In all press-molding methods (BMC, SMC, preforming, cold pressing, cold stamping, etc.) a design that permits a straight mold parting line is preferred because it simplifies mold construction. Open hand lay-up and spray-up molding do not require a straight edge on the part because edge-finishing operations are carried out by hand and can easily follow steps and curves.

Draft

Draft is advisable on the sidewalls of all press-molded RP/C parts and on all interior and exterior surfaces, including all ribs and bosses. The minimum recommended draft angle is $\frac{1}{2}°$. However, angles in the range of 1 to 3° and in some cases more are preferable. (See Fig. 6.6.3.) Interior surfaces usually require more draft than exterior surfaces because shrinkage of the part prevents release rather than aiding it. Low-shrinkage materials also may necessitate greater draft on exterior surfaces. Surface textures also increase the need for draft. For each 0.025 mm (0.001 in) of depth of surface texture, 1° additional draft should be allowed.

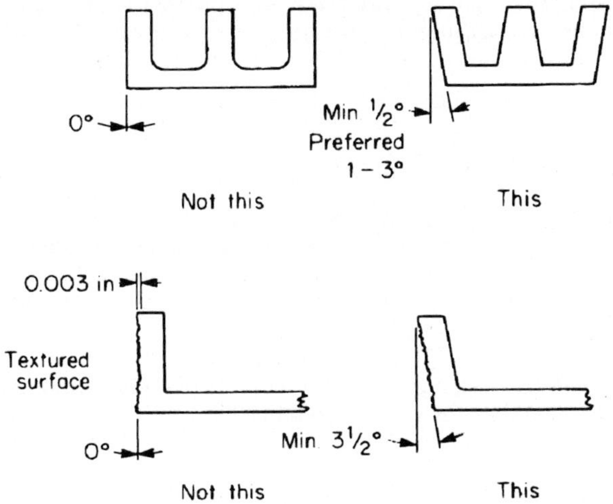

FIGURE 6.6.3 Recommendations for draft on sidewalls and other vertical members of press-molded RP/C parts. Textured surfaces require much more draft than smooth surfaces.

Open-Lay-Up (Hand-Lay-Up, Spray-Up, Vacuum-Bag) Moldings. Adequate draft is required if molds are rigidly braced. However, the flexible nature of the molds used for these processes allows zero draft or some reverse curvature to be molded in, as shown in Fig. 6.6.4.

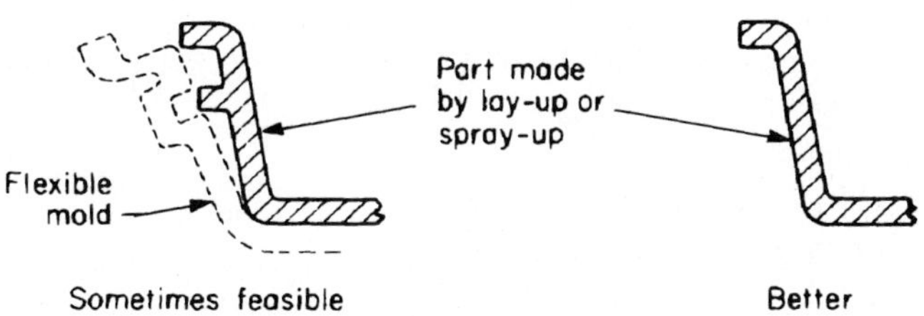

FIGURE 6.6.4 Flexible molds used for open lay-up molding allow zero draft, reverse curvature, or undercuts in the part while still permitting removal of the part from the mold.

Undercuts

Press Molding. Undercuts should be avoided. They require core pulls (sliding mold members) or extra mold sections. These increase mold costs and cause difficulty if resin flows into mold joints. BMC, SMC, and injection molding are the best methods if undercuts must be included. Internal undercuts are not feasible with any process. (See Fig. 6.1.6 in Chap. 6.1.)

Open-Lay-Up Molding. As stated, undercuts should be avoided in general but may be designed in if the mold can be flexed without damage for part removal (Fig. 6.6.4). Also, slides or pulls may be built into the molds for intricate tooling work, or a split mold may be used. These are somewhat less of a problem with open-lay-up molding than with press methods because high pressures are not involved, but they still should be avoided.

Corner Radii

Press Molding. The more generous the inside-corner radii allowed the better, with 1.5 mm ($^1/_{16}$ in) being the minimum for most methods except preform molding, for which 3 mm ($^1/_8$ in) is recommended. This will promote material flow and avoid resin-rich or resin-starved areas. Outside radii probably will be dictated by part thickness, but they should not be less than 1.5 mm ($^1/_{16}$ in).

Open-Lay-Up Molding. Again, part function probably will dictate the shape of corner radii, but they should not be less than 6 mm ($^1/_4$ in).

Edge Stiffening

A straight unsupported edge will have a tendency to warp or to be too flexible. Flanges, wall thickening, and reinforcements may all be employed to avoid this problem. Figure 6.6.5 illustrates various edge-stiffening designs that are feasible with press-molded RP/C parts.

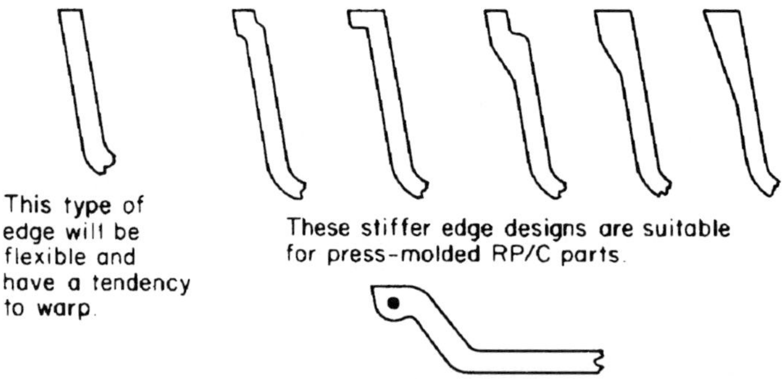

FIGURE 6.6.5 Feasible edge-stiffening designs for press-molded parts.

The large structures fabricated by using open lay-up methods require stronger bracing for the longer exposed edges even though part thicknesses are usually greater. Edges may be further braced by sandwich-type construction. Figure 6.6.6 illustrates several flange designs. Stiffness also may be imparted by impressing soft, porous wood forms of plywood into a basic laminate and laminating over them. Laminates may be substantially stiffened by incorporating layers of plywood, slab foam, balsa, etc. Figure 6.6.7 shows examples for ribs, but the same approach can be used for edge reinforcement.

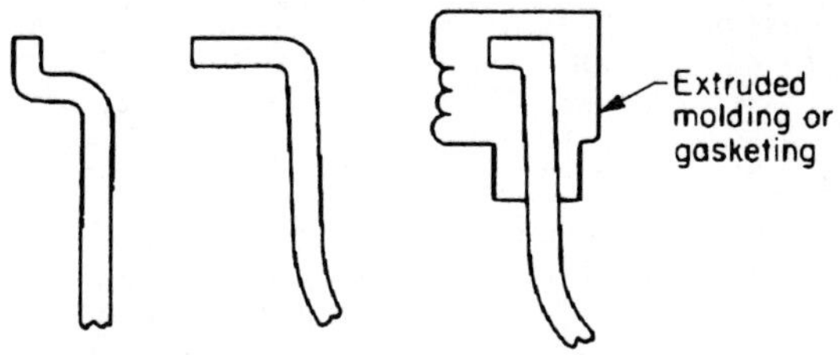

FIGURE 6.6.6 Edge-stiffening flanges for open-lay-up-molded RP/C parts.

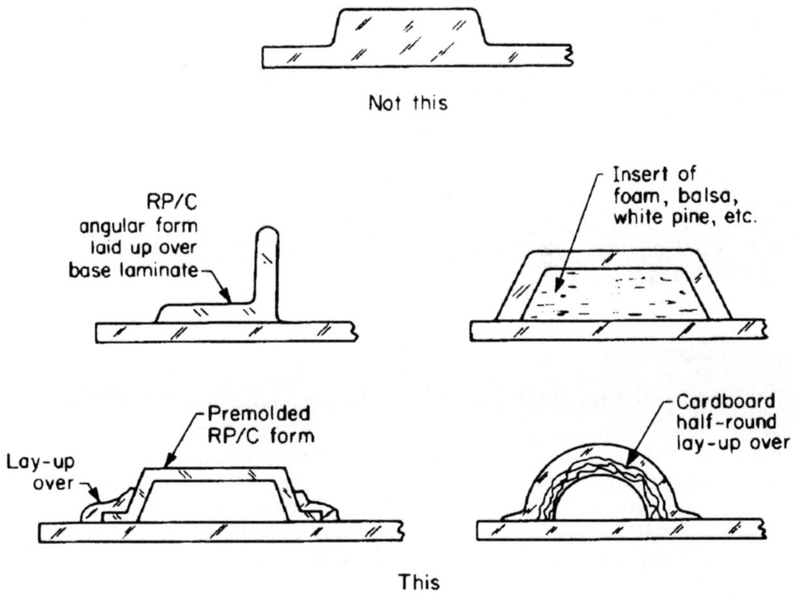

FIGURE 6.6.7 Design recommendations for ribs and other reinforcements of open-lay-up-molded parts.

Ribs

Press Molding. In press molding RP/C (SMC or BMC), a raised column or strip almost the same as nominal part thickness is preferred. Additional fillets or gussets aid flow into the rib. Since sink marks may appear in the surface opposite a rib, it is best, if appearance is important, to minimize these marks or disguise them. They can be minimized by using minimum radii at the rib base and disguised by locating the rib behind styling lines or textured surfaces, as shown in Fig. 6.2.14.

Open-Lay-Up Molding. In the larger structures possible with hand lay-up or spray-up, ribs or stiffeners are preferably added by postmolding a fin or arched stiffener around a core rather than by a thickness buildup. The need for ribs can sometimes be eliminated by monocoque construction (double layers sandwiched around a wood, foam, or other lightweight core).

Bosses

Press Molding. Bosses follow approximately the same design rules as for press-molded ribs. Blind holes should be molded in to reduce mass. Reinforcing ribs are preferable to molding a heavy mass. The width of the shoulder should be no larger than the diameter of the hole. Locate bosses in corners if possible. (See Fig. 6.6.8.)

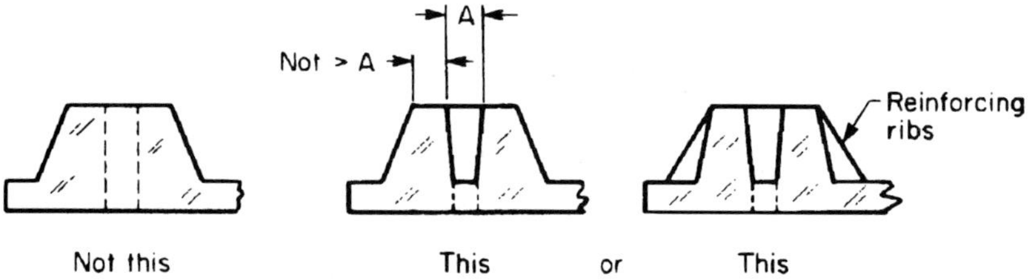

FIGURE 6.6.8 Boss design for press-molded RP/C parts.

Open-Lay-Up Molding. As in the case of ribbing, raised bosses for mounting should employ a buildup around a core material.

Holes and Openings

Holes present some difficulties in RP/C parts because they interfere with the distribution of the reinforcing fibers. Generally, the molding of holes is not recommended with preform, SMC, or cold-press molding. With these processes, holes may be drilled after demolding. Molded holes in BMC and injection-molded reinforced thermoplastics are feasible. However, holes with BMC are preferably molded as partial holes as shown in Fig. 6.6.9. In general, holes in RP/C press-molded parts should be two diameters from other holes and three diameters from part edges.

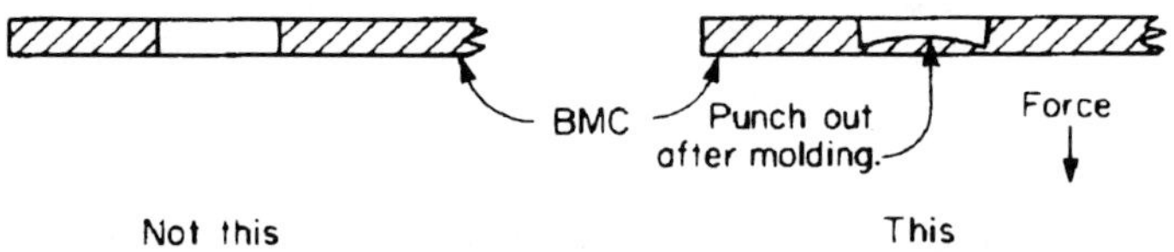

FIGURE 6.6.9 With BMC, partially molded holes for later punching out are recommended.

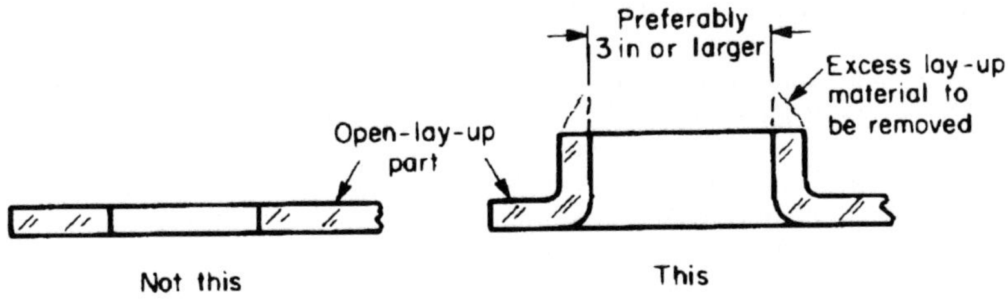

FIGURE 6.6.10 Preferred hole design in parts made by open lay-up methods.

Open-Lay-Up Molding. With open lay-up molding there is no satisfactory way to arrange the reinforcement fibers to produce a satisfactory hole in a flat wall. Instead, it is necessary to create a protrusion by draping the reinforcing mat, fabric, or roving around and over a plug in the mold. Excess material is cut off and trimmed after molding. The resulting part is stronger and more rigid than a flat section with a hole would be. (See Fig. 6.6.10.) Holes above 3-in diameter are feasible; the larger, the better.

Inserts and Threading

Press Molding. Inserts can be either molded in or set mechanically after molding, except with preform and cold-press molding, for which only after-molding assembly is recommended. A variety of interface surfaces, including diamond knurl, I-beam type, etc., may be employed for the molded-in inserts. Threads on inserts are preferable to threading molded RP/C parts.

Open-Lay-Up Molding. Inserts may be incorporated for a variety of applications such as bolting, joining, etc. Threading of open lay-up molded parts is not recommended. Insert flanges or bases should be as large as is feasible. Often inserts can be incorporated by applying additional material over the flange base of the insert, as shown in Fig. 6.6.11.

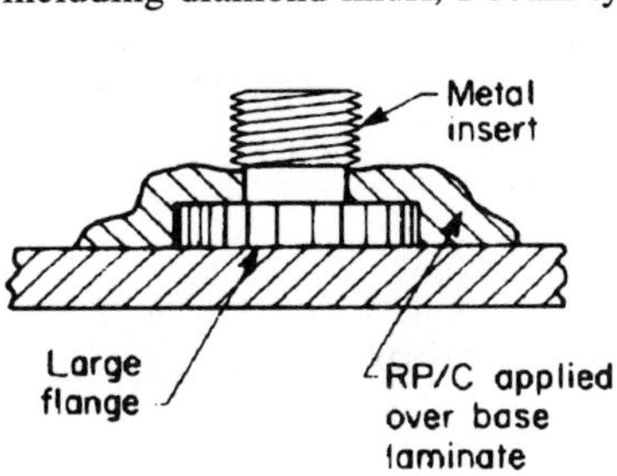

FIGURE 6.6.11 Wide-flanged metal inserts in RP/C parts can be fastened by applying material over the flange.

Surface Finish

Press Molding. The surface finish of the part is directly related to the mold finish. A variety of surface finishes can be specified for RP/C parts. The designer should bear in mind, however, that the smoother finishes normally are more costly than those in which fiber prominence is evident. However, the nonshrink additives used in SMC and BMC compounding generally ensure that the intended finish results.

Open-Lay-Up Molding. Surface finishes for hand lay-up and spray-up products again are a function of mold finish. Gel coats duplicate mold finish. Vacuum-formed thermoplastic sheets with RP/C backing provide excellent surfaces.

Joining RP/C Parts

Parts consolidation and the consequent reduction of necessary parts assembly are one of the advantages of RP/C construction. However, when the joining of separate parts is necessary, it may be handily accomplished by any of several worthwhile methods. Parts can be bonded together or held with mechanical fasteners. Figure 6.6.12 illustrates a number of recommended joint designs.

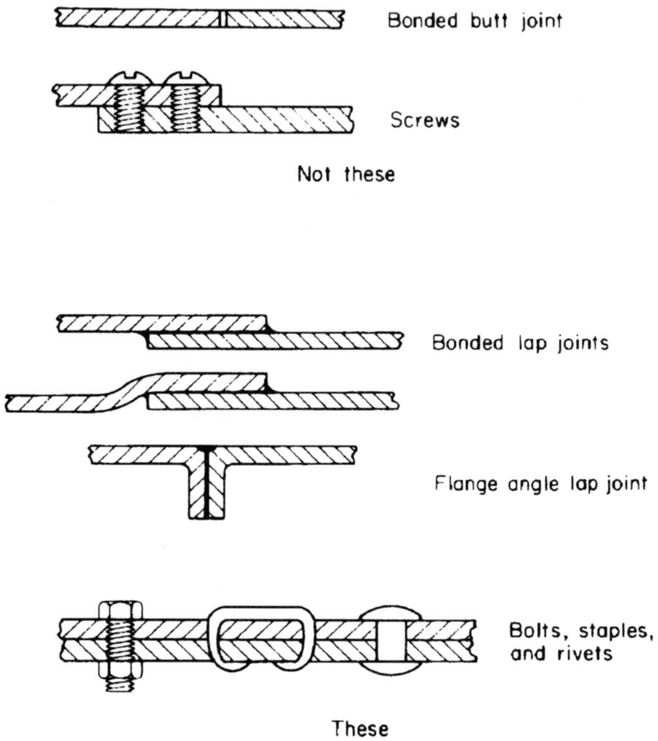

FIGURE 6.6.12 Recommendations for joint designs in assemblies of RP/C parts.

Flat Surfaces

Gently curved or rounded surfaces are preferred to flat surfaces. They are less apt to warp and flex. Normally only a small amount of curvature is necessary to improve a part's rigidity greatly, especially if flanges are also employed. If flatness is necessary in a particular surface, use ribbing or other stiffening members.

Lettering

Molded-in lettering with raised or depressed characters is feasible with all RP/C processes except filament winding, pultrusion, and continuous laminating. Raised lettering on the part is normally preferable except when the mold itself is made of RP/C materials (as in the open-mold lay-up and spray-up methods). Labels can be molded into most RP/C parts, the exceptions being pultrusions and injection moldings.

Table 6.6.4 summarizes design recommendations for RP/C parts.

TOLERANCE RECOMMENDATIONS

Allowable wall-thickness variations for the products from all RP/C molding methods are shown in Table 6.6.3. Recommended tolerances for other dimensions of parts made with various RP/C processes are included in Table 6.6.5.

TABLE 6.6.4 Design Recommendations for Reinforced-Plastic/Composite Parts

		Compression molding					
		Sheet-molding compound	Bulk-molding compound	Preform molding	Injection molding (thermoplastics)	Cold-press molding	Spray-up and hand lay-up
Minimum inside radius		1/16 in	1/16 in	1/8 in	1/16 in	1/4 in	1/4 in
Molded-in holes	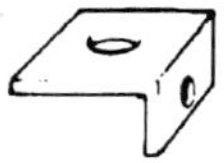	Yes*	Yes*	Yes*	Yes*	No	Large
Undercuts	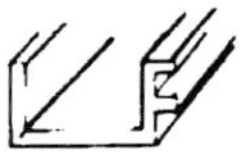	Yes †	Yes*	No	Yes †	No	Yes †
Minimum recommended draft	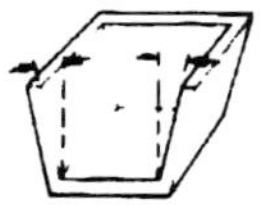	1/4- to 6-in depth: 1 to 3° 6-in depth and over: 3° or as required				2° 3°	0°
Minimum practical thickness	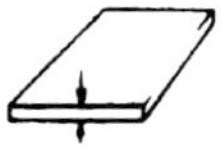	0.050 in	0.060 in	0.030 in	0.035 in	0.080 in	0.060 in
Maximum practical thickness	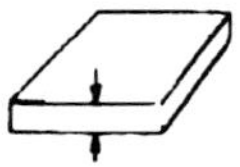	1 in	1 in	0.250 in	0.500 in	0.500 in	No limit

Maximum thickness build-up, heavy buildup, and increased cycle	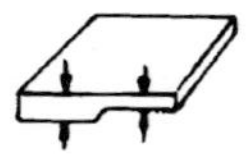	As required	As required	2:1 max.	As required	2:1 max.	As required
Corrugated sections		Yes	Yes	Yes	Yes	Yes	Yes
Metal inserts		Yes	Yes	Not recommended	Yes	No	Yes
Bosses		Yes	Yes	Yes	Yes	Not recommended	Yes
Ribs		As required	Yes	Not recommended	Yes	Not recommended	Yes
Molded-in labels		Yes	Yes	Yes	No	Yes	Yes
Raised numbers		Yes	Yes	Yes	Yes	Yes	Yes
Finished surfaces (reproduces mold surface)	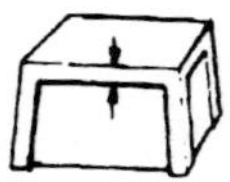	Two	Two	Two	Two	Two	One

*Parallel to ram action.

†With slides in tooling or split mold.

Source: Courtesy Owens-Corning Fiberglas Corp.

TABLE 6.6.5 Recommended Dimensional Tolerances for RP/C Parts

Dimension, mm (in)	Class A (fine tolerance), mm (in)*	Class B (normal tolerance), mm (in)†	Class C (coarse tolerance), mm (in)‡
0–25 (0–1)	±0.12 (±0.005)	±0.25 (±0.010)	±0.4 (±0.016)
25–100 (1–4)	±0.2 (±0.008)	±0.4 (±0.016)	±0.5 (±0.020)
100–200 (4–8)	±0.25 (±0.010)	±0.5 (±0.020)	±0.8 (±0.030)
200–400 (8–16)	±0.4 (±0.016)	±0.8 (±0.030)	±1.3 (±0.050)
400–800 (16–32)	±0.8 (±0.030)	±1.3 (±0.050)	±2.0 (±0.80)
800–1600 (32–64)	±1.3 (±0.050)	±2.5 (±0.100)	±3.8 (±0.150)
1.6–3.2 (64–128)	±2.5 (±0.100)	±5.0 (±0.200)	±7.0 (±0.280)

*Class A tolerances apply to parts compression-molded with precision matched-metal molds. BMC, SMC, and preform are included.

†Class B tolerances apply to parts press-molded with somewhat less precise metal molds. Cold press molding, casting, centrifugal casting, rotational molding, and cold stamping can apply to this classification when molding is done with a high degree of care. BMC, SMC, and preform compression molding can apply to this classification if extra care is not used.

‡Class C applies to hand-lay-up, spray-up, vacuum bag, and other methods using molds made of RP/C material. It applies to parts that would be covered by Class B when they are not molded with a high degree of care.

CHAPTER 6.7
PLASTIC PROFILE EXTRUSIONS

J. R. Casey Bralla

Takata Restraint Systems, Inc.
Cheraw, South Carolina

THE PROCESS

Extrusion is a process for molding thermoplastic materials into sheets, tubes, or shapes that have a constant and often complex cross section. Dry plastic material, normally in the form of pellets or powder, is placed in a hopper that feeds into a long, carefully heated chamber. In the chamber, a rotating screw mixes the plastic to produce a uniform melt and forces it through a die orifice. As the extrudate leaves the die, it is passed through a cooling medium (air or water) by a conveyor or other takeoff mechanism. It solidifies to the cross-sectional shape of the die opening. The extrudate is pulled away from the die faster than it is extruded, thus causing it to draw down to a smaller cross section. This helps to keep the extrudate straight as it cools and solidifies. It also permits slight adjustments in size to be made (by varying the drawdown rate). The ratio of the die size to the final size of the product is defined as the *drawdown ratio.*

As the molten plastic leaves the die, it must be supported as it passes through the cooling medium. One of the simplest methods to do this and provide cooling is to draw the extrudate directly into a water bath.

Tubing or other cross sections with a hollow are made from dies that have a core (or mandrel) to form the hollow. Air is introduced into the core and injected into the hollow of the extrusion as it leaves the die. The air supports the inside of the extrusion and prevents it from collapsing during cooling. Generally, tubing is also run through a sizing die or a cooling mandrel to maintain concentricity.

Secondary operations down line are often just as important in manufacturing an extruded product as the extrusion itself. The finished product is rarely the same as the original extrusion. Among typical in-line operations are cutting to length, application of films (such as simulated wood grain, foam rubber, protective tape, etc.), punching (this allows for special holes, notches, or cuts that are not possible to extrude), embossing, forming, and assembly.

CHARACTERISTICS AND APPLICATIONS

Extruded sections can be as small as a thread filament and as large as a 300-mm (12-in) pipe. Almost any shape with a constant cross section may be suitable for extrusion. The length can be as great as desired because material can be fed continually through the extruder. The finished part can be as short as a fraction of an inch if the extrusion is sliced into short pieces. Both rigid and flexible thermoplastics can be extruded, and these can be either solid or cellular.

Extrusion is often employed for door, window, and floor moldings. It finds use in wear strips for low-friction-bearing components, tubing, seals, edge guards, wire harnesses, and many other components. See Fig. 6.7.1 for examples of typical extruded shapes.

Dual extrusion (two materials being extruded side by side and joined during extrusion) is similar to simple extrusion except that two extruding machines feed the die instead of one. Each machine feeds a different color or different thermoplastic to the die. This produces a single extrusion composed of two permanently bonded plastics. One of the most common uses of dual extrusion is to bond rigid to nonrigid vinyl. The rigid portion of the extrusion is used to preserve the overall shape of the profile. (It also can provide a strong base for attachment purposes.) The less rigid portion is suitable for sealing or cushioning. Almost any number of rigid and nonrigid members may be combined in a single extrusion. The extrusion of various colors simultaneously can eliminate the need for postextrusion decorating.

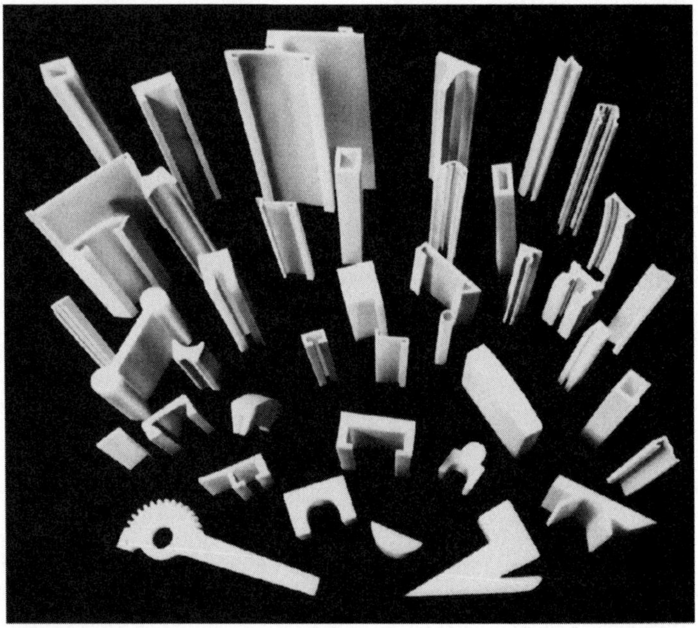

FIGURE 6.7.1 Typical cross-sectional shapes produced in plastic material by extrusion. (*Courtesy Spiratex Co., Inc.*)

Metal embedment is another common extrusion technique. It allows continuous lengths of solid wire or strips of metal to be embedded (either partially or totally) in the extruded thermoplastic profile. The advantage of metal embedment is that it provides high structural integrity while the extrusion has the warmth, color, and chemical resistance of plastics.

ECONOMICS

Plastic extrusions are most economically produced when production quantities are fairly large. Although tooling costs are very modest, a certain amount of trial-and-error development is required to achieve final dimensions with some accuracy. This process adds significantly to the cost of the extruded part if quantities are small.

Extrusion production rates are rapid. Depending on the size of extruder equipment, they range from about 5 to 200 kg/h (about 10 to 400 lb/h). Higher rates are feasible in mass-production applications.

Standard shapes of many materials are available off the shelf from local plastics distributors and from some manufacturers. Special cross sections from manufacturers require a minimum production run. Some small vendors may require only 1000 m, but for larger producers 6000 m (20,000 ft) for cross sections of 6 by 6 mm (¼ by ¼ in) is more apt to be required. For larger cross sections the minimum order length will be about one-half of this amount.

DESIGN RECOMMENDATIONS

Although virtually any profile can be extruded with thermoplastics, several design factors must be considered to produce high-quality extrusions. The most important consideration is the wall thickness of the profile. A profile with a uniform wall thickness is easiest to produce. Nonuniform wall thicknesses may cause uneven plastic flow through the die and cause different parts of the extrusion to cool at different rates. This can cause warpage toward the heavier portion of the profile. If an uneven wall thickness is unavoidable, it may be necessary to provide additional cooling for the heavier sections. This increases tooling complexity and adds to production costs. In addition, nonuniform wall thicknesses usually require twice the tolerance limits of a similar uniform product. Another disadvantage is that sink marks almost always occur in extrusions on a flat surface opposite a heavy area such as a leg or a rib. Often an unbalanced profile can be slightly redesigned, as shown in Fig. 6.7.2, to eliminate a heavy area or to undercut a leg.

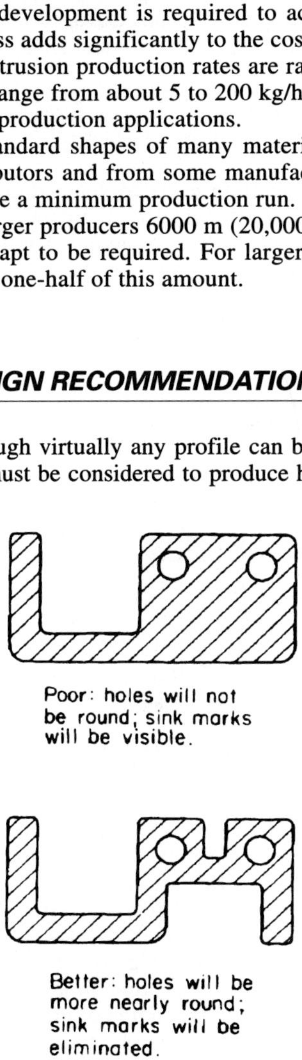

FIGURE 6.7.2 Section thicknesses should be kept as uniform as possible.

Hollows are not as difficult to extrude in thermoplastics as in metals. However,

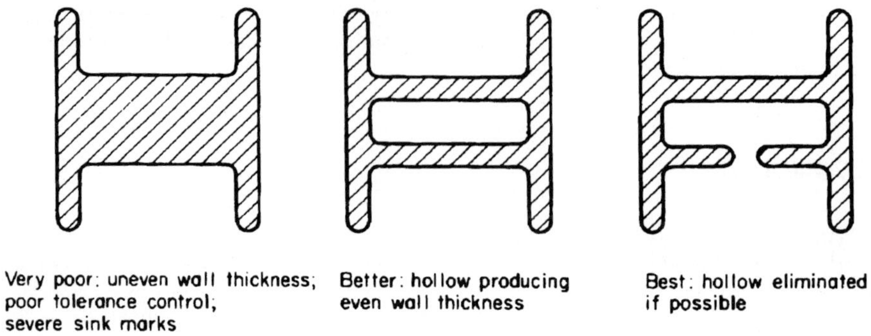

FIGURE 6.7.3 Hollow sections are preferable to heavy ones; open sections of uniform wall thickness are still better.

they increase the cost of dies and also increase operating costs by mandating compressed-air injection into the hollow. If possible, it is advisable to eliminate or minimize the hollow. If design requirements include either a hollow or an unbalanced profile, the hollow is preferable. Tolerances for hollow profiles cannot be held as closely as for profiles without them. (See Fig. 6.7.3.)

Legs or projections inside a hollow should be eliminated or minimized. They greatly increase dimensional variations because there is no way to hold their shape while they cool. If an interior leg or rib is unavoidable, its projection into the hollow should never exceed the thickness of the wall around the hollow. Figure 6.7.4 illustrates some examples.

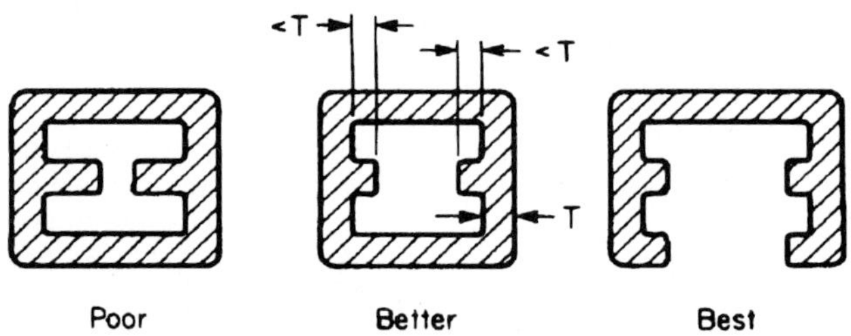

FIGURE 6.7.4 Legs or projections inside a hollow should never exceed in height the thickness of the surrounding wall.

A hollow within a hollow should be avoided at all costs because all the problems inherent in hollows are compounded by "nested" hollows. Tolerance control for nested hollows is very poor, and production is especially costly. (See Fig. 6.7.5.)

Sharp corners are very difficult to extrude because most thermoplastics bridge across sharp corners of the die and form radii. Sharp inside corners also should be avoided because they can form a notch that can be an easy breaking point for the more rigid plastics. Dies for sharp outside corners exhibit the same problem and, in addition, form a concentration point for stresses caused by heating and cooling. As a general rule, corner radii should be at least one-half the wall thickness. In addition, both inside and outside radii should have the same center so as to avoid stresses during cooling. Figure 6.7.6 illustrates an example.

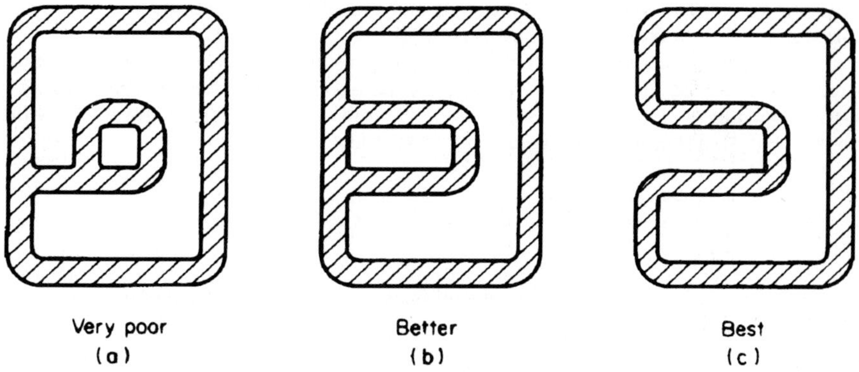

FIGURE 6.7.5 Always avoid a hollow within another hollow as in design *a*. Design *b* is better because the inner hollow is only partially enclosed by another. Design *c* is best because it eliminates the inner hollow altogether.

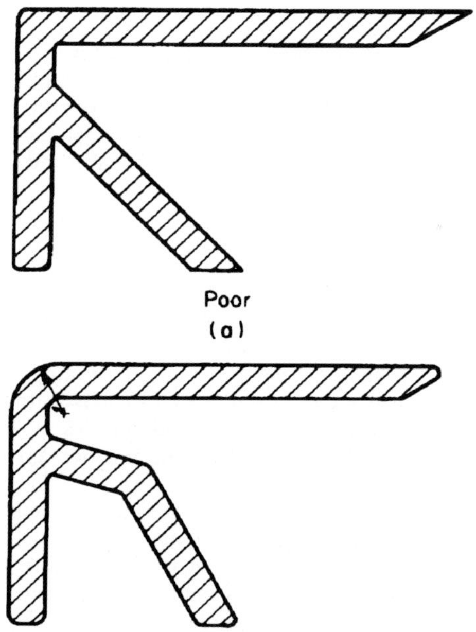

FIGURE 6.7.6 Design *a* has corners too sharp and inside and outside radii with different centers. Design *b,* with larger inside and outside radii on the same center, is preferable.

Complex hollows increase tooling cost and dimensional variation. Round, square, half-round, or rectangular shapes are preferred.

Lone unsupported die sections are more costly, requiring increased die strength and streamlining at the die entrance. To avoid this, avoid profile designs that use deep slots, especially if they are of a shape to weaken the die members. Figure 6.7.7 illustrates one design that was improved to strengthen the die members.

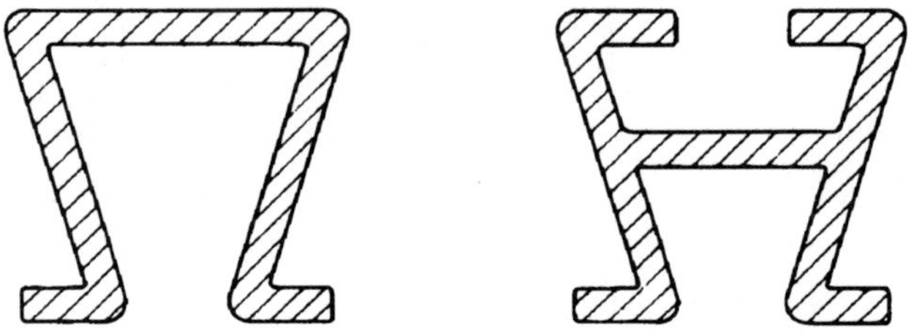

FIGURE 6.7.7 Sometimes it is necessary to modify a proposed design to ensure that die members are sufficiently strong.

Sink marks tend to occur on surfaces opposite ribs or walls. One way to avoid an appearance problem is to add serrations to the surface that is apt to be affected.

Because of the tooling development required for special profiles, it is almost always cheaper to purchase a standard extrusion from a manufacturer's stock, if at all possible, rather than use a special profile. This is especially true if production volume is small.

MATERIALS FOR EXTRUSION

The choice of thermoplastic is very important in determining the extrudability of a particular profile. High-impact styrene is the easiest plastic to extrude. Cellulosics (cellulose acetate butyrate and ethyl cellulose) and acrylics are the next easiest. The most difficult plastic to extrude is nylon. Flexible plastics are not extrudable to as tight tolerances as rigid plastics.

As a rule, rigid materials (such as polystyrene, methyl methacrylate, rigid vinyl, and cellulose acetate) are not quenched in cold water during the drawdown and cooling process. Rapid cooling in these plastics causes undesirable stresses and leaves a poor surface appearance. Crystalline plastics (such as polyethylene, nylon, vinylidene chloride, and polypropylene) are generally cooled in cold water.

Vinyls, acrylonitrile butadiene styrene, and polystyrene are easier to extrude through an unbalanced die than polyethylene and polypropylene. The latter have low melt strength and leave the die in a more fluid condition (which is harder to control) than the former.

If the two thermoplastics used for dual extrusion are not the same (e.g., not both vinyls), the bond between them probably will not be complete. Radically different materials require undercuts, dovetails, or mechanical joints to stay together. Care also should be taken when dual-extruding various plastics as some are chemically incompatible. Plasticized (flexible) vinyl, for example, is not compatible with polystyrene because of plasticizer migration to the polystyrene.

Table 6.7.1 lists commonly used thermoplastics with comments on their extrudability.

TABLE 6.7.1 Common Plastics Used for Extrusion

1. Acrylonitrile butadiene styrene (ABS). Good rigid plastic; can be extruded very easily; full range of colors; complex profiles possible; above-average tolerance control. Uses: slides, housings, handles, doorjambs.
2. Cellulose acetate. Inexpensive; available in rigid and semirigid types; full range of colors; poor weatherability; resists oils, gasoline, and many cleaning fluids; easily cleaned with soap and water; thin sections possible. Uses: edgings, clear packaging, tubes.
3. Cellulose acetate butyrate. Similar to cellulose acetate but harder, tougher, and more resistant to heat and weather; unpleasant odor; clear and full range of colors; good tolerance control. Uses: signs, letters.
4. Cellulose propionate. Very similar to cellulose acetate butyrate but without unpleasant odor.
5. Ethyl cellulose. Rigid with springiness; "chip-proof"; very easily extruded; good tolerance control; toughest and lightest of the cellulosics; stable at low temperatures. Uses: tubing, edges, decorative parts.
6. Ethylene vinyl acetate (EVA). Flexible; full range of colors; average tolerances for simple profiles. Uses: low-performance hinges, seals, gaskets, weatherstripping.
7. Nylon. Tough; light weight; very high tensile strength; only for simple profiles; requires loose tolerances; liquefies immediately from solid; high water absorption. Uses: tubing, guides, low-friction channels.
8. Polycarbonate. Best overall balance of properties; good for average profiles; good tolerance control; expensive; requires special handling because of water absorption; heat-resistant; self-extinguishing. Uses: tubing, light shades, food-related operations (can be steam-sterilized).
9. Polyethylene, high-density. Stiffer and harder than low-density polyethylene; thinner cross sections possible; difficult to extrude; poor tolerance control. Uses: belts, straps, rods.
10. Polyethylene, low-density. Somewhat flexible; nontoxic; difficult to extrude; poor tolerance control; good electrical properties; no known solvent at room temperatures. Uses: tubing, handles, straps, bumpers, edgings.
11. Polypropylene. Very light; average tolerance control; poor for complex profiles; full range of colors. Uses: high-performance hinges, slide guides, weatherstripping.
12. Polystyrene. Low cost; clear and full range of colors; limited use for extrusion because of brittle characteristics. Uses: light shields.
13. Polystyrene, high-impact. Rigid; low cost; complex profiles possible; average tolerance control; full range of colors. Uses: trim strips, sliding-door guides.
14. Polyvinyl chloride (PVC), flexible. Available in a variety of hardnesses; only average complexity of profiles possible; average tolerance control; full range of colors. Uses: gaskets, seals, trims.
15. Polyvinyl chloride, rigid. Very good for extrusion; complex profiles possible; excellent electrical properties; good tolerance control; limited color range. Uses: appliance breaker strips, electrical-conductor covers.
16. Polyurethane. Available in wide range of hardnesses; full range of colors; good abrasion resistance. Uses: conveyor belts, tubing.
17. Vinyl dichloride. Similar to rigid PVC but better high-temperature resistance; continuous-service temperature to 105°C (220°F); complex profiles possible; good tolerance control; full range of colors. Uses: hot-water pipes, high-temperature applicators.

RECOMMENDED TOLERANCES

Table 6.7.2 lists recommended dimensional tolerances for profile extrusions of various thermoplastic materials.

FOAM PROFILE EXTRUSIONS

Plastic profile extrusions with a cellular inner structure are being used increasingly as a means of saving material and increasing stiffness without a weight penalty. The process involves the addition of a blowing agent and extruding it under carefully controlled conditions. The blowing agent is normally a dry, powdered chemical that decomposes at the melting temperature of the plastic and emits gas. A sizing die or other apparatus adjacent to the extrusion die helps form a solid skin on the surface of the extrusion. Sometimes coextrusion is used, the solid surfaces being extruded from standard material, while the core material, from a separate barrel, contains the foaming agent.

Structural-foam extrusions are useful whenever the profile is large or thick enough so that the materials savings that result from the cellular interior are significant. One fairly common example is wood-like profiles that replace wood moldings. Another is foamed pipe, which is used for vent, drain, and other low-pressure applications including conduit for telephone wires. Foam sheet extruded with a solid skin is thermoformed into luggage, boats, and recreational vehicle bodies.

All thermoplastics can be extruded as foams. The most common material for structural applications is polyvinyl chloride. ABS, polystyrene, and high-density polyethylene also are in commercial use.

Foam profile extrusions present milder design restrictions than solid profiles because shrinkage of thick sections is drastically reduced by the foaming action. Designers have more latitude with foam in incorporating heavy and uneven wall thicknesses. Other design recommendations apply about equally to foam and solid profiles.

PULTRUSIONS

Pultrusions are profile shapes made most commonly from dense clusters of glass fibers saturated with polyester resin. The production process differs from extrusion in that the material is pulled rather than pushed through the die, which is heated. The die compresses the fibers and resin together to form a continuous shape, and the heat cures the resin. When the thermosetting resin hardens, a component with favorable longitudinal strength and a high strength-to-weight ratio is produced. Other fibers such as those made from boron, paper, and graphite or metal wire may be used, and epoxy or silicone resin may be employed instead of polyester, but the glass-polyester combination is by far the most common.

Pultrusions have the advantages of molded-fiberglass components. These are primarily high strength, light weight, corrosion resistance, and electrical insulation. Pultrusions are used for structural members such as angles, I beams, rods, tubes, and channels. They are good substitutes for wood or metal members when longer life and

TABLE 6.7.2 Recommended Dimensional Tolerances for Plastic Profile Extrusions

Dimension	Rigid vinyl (PVC)	Polystyrene	ABS	Polypropylene	Flexible vinyl (PVC)	Polyethylene
Wall thickness	±8%	±8%	±8%	±8%	±10%	±10%
Angles	±2°	±2°	±3°	±3°	±5°	±5°
Profile dimensions, ±mm (in)						
0–3 (0–⅛)	0.18 mm (0.007 in)	0.18 mm (0.007 in)	0.25 mm (0.010 in)	0.25 mm (0.010 in)	0.25 mm (0.010 in)	0.30 mm (0.012 in)
3–13 (⅛–½)	0.25 mm (0.010 in)	0.30 mm (0.012 in)	0.50 mm (0.020 in)	0.38 mm (0.015 in)	0.38 mm (0.015 in)	0.63 mm (0.025 in)
13–25 (½–1)	0.38 mm (0.015 in)	0.43 mm (0.017 in)	0.63 mm (0.025 in)	0.50 mm (0.020 in)	0.50 mm (0.020 in)	0.75 mm (0.030 in)
25–38 (1–1½)	0.50 mm (0.020 in)	0.63 mm (0.025 in)	0.68 mm (0.027 in)	0.68 mm (0.027 in)	0.75 mm (0.030 in)	0.90 mm (0.035 in)
38–50 (1½–2)	0.63 mm (0.025 in)	0.75 mm (0.030 in)	0.90 mm (0.035 in)	0.90 mm (0.035 in)	0.90 mm (0.035 in)	1.0 mm (0.040 in)
50–75 (2–3)	0.75 mm (0.030 in)	0.90 mm (0.035 in)	0.94 mm (0.037 in)	0.94 mm (0.037 in)	1.0 mm (0.040 in)	1.1 mm (0.045 in)
75–100 (3–4)	1.1 mm (0.045 in)	1.3 mm (0.050 in)	1.3 mm (0.050 in)	1.3 mm (0.050 in)	1.7 mm (0.065 in)	1.7 mm (0.065 in)
100–125 (4–5)	1.5 mm (0.060 in)	1.7 mm (0.065 in)	1.7 mm (0.065 in)	1.7 mm (0.065 in)	2.4 mm (0.093 in)	2.4 mm (0.093 in)
125–180 (5–7)	1.9 mm (0.075 in)	2.4 mm (0.093 in)	2.4 mm (0.093 in)	2.4 mm (0.093 in)	3.0 mm (0.125 in)	3.0 mm (0.125 in)
180–250 (7–10)	2.4 mm (0.093 in)	3.0 mm (0.125 in)	3.0 mm (0.125 in)	3.0 mm (0.125 in)	3.8 mm (0.150 in)	3.8 mm (0.150 in)

lighter weight are important, particularly when the environment is unfavorable. Handrails, ski poles, tent and sail batons, tool handles, arrow shafts, ladder side rails, cable-tray channels, and reinforcing members are typical applications.

Figure 6.6.2*j* in Chap. 6.6, "Reinforced-Plastic/Composite (RP/C) Parts," illustrates typical cross sections produced by pultrusion. Special cross sections require minimum quantities of about 8000 to 30,000 m (25,000 to 100,000 ft) to amortize tooling and development costs. Even in large quantities, however, costs of pultrusions are higher than those of similar-sized components of other materials.

Although pultrusions permit wide latitude to the designer, adherence to some guidelines is advisable. Sharp corners should be avoided in favor of generous fillets and radii. Drastic changes in section thickness are also undesirable; uniform wall thickness is preferred. Sections should not be more than 13 mm ($\frac{1}{2}$ in) thick except for solid rods and bars. The minimum practical section thickness is about 1.5 mm ($\frac{1}{16}$ in).

The surface finish on pultruded parts is usually rough. If the die contains a parting line, which may be necessary for nonround parts, the part should be designed so the parting line can be hidden at a corner or on a secondary surface.

CHAPTER 6.8
THERMOFORMED PLASTIC PARTS

THE PROCESS

Thermoforming, as the term is normally used, involves the shaping of a thermoplastic sheet by heating it and then bringing it into contact with a mold whose shape it takes. The plastic sheet or film is heated to the softening point, clamped over the mold or between mold halves, and drawn or forced into the mold by one or a combination of methods including vacuum, air pressure, gravity, and mechanical force. It cools in the mold, taking the mold's shape. It is then removed from the mold and trimmed as necessary.

The process is often called *vacuum forming* because a vacuum between the sheet and the mold is generally the means used to cause the sheet to be forced against the mold. Figure 6.8.1 shows schematically how this works.

In *straight vacuum forming,* a female mold is generally used. In *drape vacuum forming,* a male mold is used, and the plastic sheet is pulled down and "draped" over the mold before a vacuum is applied.

Plug-assist forming uses a male plug that roughly conforms to the mold cavity to prestretch the material before it is drawn into the mold. This minimizes thinning of material at the end of the formed portion.

Vacuum snap-back forming uses a vacuum to pull the sheet into a concave shape, after which a vacuum on the opposite side of the sheet pulls it tightly against the male mold. This procedure improves the uniformity of wall thickness, can reduce starting sheet size, and minimizes chill marks.

Slip-ring forming involves the use of loose clamps that allow a certain controlled amount of slippage during forming so that the material is drawn into the final shape instead of being simply stretched. This process is comparable to sheet-metal deep drawing and provides a more uniform wall thickness.

Matched-mold forming uses mechanical force and two closely fitting die halves to press the sheet to the desired form. It is the most accurate thermoforming method and can produce fine detail and special surface finishes. Molds, normally of metal, are more expensive than for other thermoforming processes.

Additional thermoforming processes involve the use of positive air pressure, often with reverse-direction preforming to provide deep forms with more uniform wall thickness.

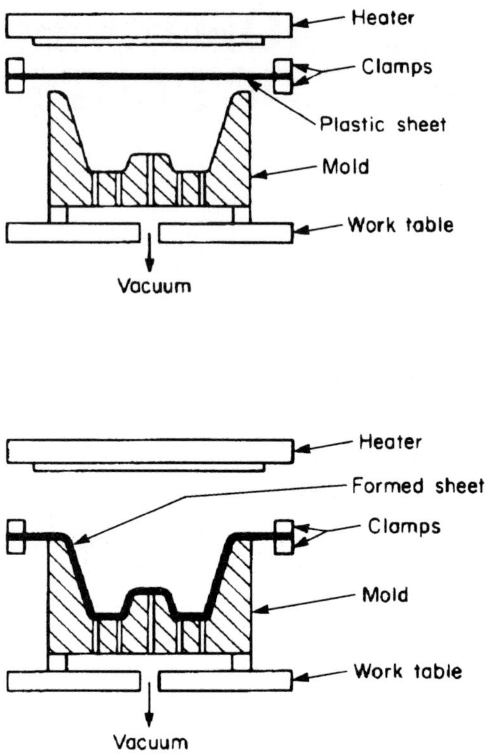

FIGURE 6.8.1 Straight vacuum forming.

APPLICATIONS AND CHARACTERISTICS

Thermoforming is most suitable for shallow-shaped articles of essentially uniform wall thickness. Large components can be made by the process, and production quantities can be very modest. Thin-walled parts are particularly suited to the process, especially if their shapes are shallow.

The most common applications are in consumer-products packaging. However, nonpackaging uses for thermoformed parts are extensive, varied, and commercially very significant. Some examples are bus and aircraft interior panels, refrigerator liners, advertising signs, truck fender liners, automobile interior liners, boats, camper tops, sports car body parts, business-machine housings, lighting panels and fixtures, tote boxes, contour maps, snowmobile hoods, TV set rear panels, bathtubs, trays, portable outdoor toilets, tabletops, door and furniture panels, and retail product display racks. Packaging applications include all kinds of containers for foods, cosmetics, hardware, and similar items. Meat trays, foam egg cartons, disposable cups, small tub containers for dairy products, candy trays, takeout food containers, and lids are examples. Blister and skin packages are important applications. See Fig. 6.8.2 for an illustration of some typical thermoformed parts.

Components as large as 3 by 9 m (10 by 30 ft) have been thermoformed. Sheet

FIGURE 6.8.2 Business-machine housings and other parts produced by thermoforming. (*Courtesy Mercury Plastics Corp.*)

thicknesses range from 0.025 mm (0.001 in) in skin packages and from 0.25 mm (0.010 in) to 13 mm (0.50 in) or more for other components. The minimum size of thermoformed components is more a matter of economics than process limitations. Postage-stamp-sized parts or blister packages are fully feasible, but injection molding becomes relatively more advantageous if parts are in that size range.

Other limitations of thermoformed parts are the following:

1. Wall thinning takes place in or adjacent to drawn sections.
2. Components cannot be as intricate as those made by molding processes.
3. Dimensional accuracy is less than that for molded plastic parts.

ECONOMIC PRODUCTION QUANTITIES

A major advantage of thermoforming is the fact that parts can be produced with a very low tooling investment. Only one mold surface (one mold half) is required. Because thermoforming is a low-pressure process, molds can be constructed of plaster of paris, wood, polyester, epoxy, and other nonmetallic materials.

The cost of thermoforming molds from these materials is a small fraction of the cost of molds for injection-molded plastics and is also cheaper than tooling for other plastics processes. This enables thermoforming to be used for prototype and small-quantity production. Lead time for tooling is short, and less skilled toolmakers are required.

When quantities are greater, cast aluminum or steel molds are used. In any case, the normal conditions in which thermoformed components are most advantageous are those of low and moderate production.

In packaging, however, thermoforming is competitive at the very highest quantity levels. Drinking cups and similar food containers can be thermoformed in multiple-cavity molds at 200 and 300 pieces per minute. Even though the cost of equipment for thermoforming, trimming, decorating, and filling may amount to several hundred thousand dollars, the high production levels of some consumer products make such an investment justifiable.

High-volume nonpackaging thermoformed parts normally do not have so many mold cavities, use heavier sheet, and have a longer production cycle. However, such parts can be produced quite rapidly when quantities necessitate it: about 20 to 45 s per forming cycle. Large parts like bathtubs or boats may require 5 or 10 min for forming, but this is still much more rapid than lay-up methods for fiberglass-reinforced polyester.

SUITABLE MATERIALS

Given proper techniques, all thermoplastics can be thermoformed. Polystyrene and other materials of high-melt viscosity are particularly desirable. High-impact polystyrene also can be stretched to over 100 times its original length. This is another property of obvious advantage in thermoforming. Polystyrene is also low in cost and available in transparent grades useful in packaging applications. Another styrenic material, ABS, is also relatively easy to form and enjoys wide use. Additional commonly thermoformed materials are acrylics, vinyls, especially rigid PVC, high-density polyethylene and other polyolefins, polystyrene foam, cellulose acetate, cellulose butyrate, polycarbonate, polypropylene, and nylon.

Polypropylene has materials-cost and final-property advantages but requires higher heat (than polystyrene) and closer temperature control and is not used in thermoforming to the extent that it is in other plastics processes. Polystyrene foam is more difficult to form than full-density sheet because the insulating property of the foam interferes with through heating and because of its limited tensile strength. Nylon 6 possesses excellent thermoformability. The cellulosics are easily thermoformed and are suitable when transparency is required. Cellulose triacetate, however, is not recommended for thermoforming. Polycarbonate has superior properties but is costly and not easily formed, requiring plug-assist molding and sometimes high pressures.

Table 6.8.1 lists commonly thermoformed materials and their typical applications.

DESIGN RECOMMENDATIONS

Designers should bear in mind the inherent nature of thermoformed components: They are essentially flat panels as contrasted with the solid, enclosed, boxlike, cylindrical, rod-like, or structural shapes of other processes. Thermoformed parts are nearly uniform in wall thickness and exhibit few undercuts or reentrant shapes. In keeping with, and in addition to, these basic limitations, the following are specific design recommendations for thermoformed parts:

TABLE 6.8.1 Commonly Thermoformed Plastics and Their Typical Applications

Material	Applications
Polystyrene	Various packaging applications including transparent meat trays, trays for cookie and candy boxes, blister packages
Polystyrene foam	Meat trays, egg cartons, takeout food containers
Acrylic	Signs and other outdoor applications like motorcycle windshields, snow-mobile hoods, and recreational-vehicle bubble tops
Rigid vinyls	Lighting panels, signs, relief maps, bus interior panels, dishes and trays for chemicals, blister packages, automotive dashboards
Acrylonitrile butadiene styren	Recreational-vehicle components, luggage, refrigerator liners, business-machine housings
Cellulose acetate	Blister packages, rigid containers, machine guards
Cellulose propionate	Machine covers, safety goggles, signs, shipping trays, displays
Cellulose acetate butyrate	Skylights, outdoor signs, pleasure-boat tops, toys
High-density poly-ethylene	Camper tops, canoes, sleds
Nylon	Reusable trays, outdoor signs, surgical equipment, meat trays
Polycarbonate	Outdoor lighting, face shields, machine guards, aircraft panels and ducts, signs
Polypropylene	Truck fender liners, drinking cups, juice and dairy product containers and lids, test-tube racks

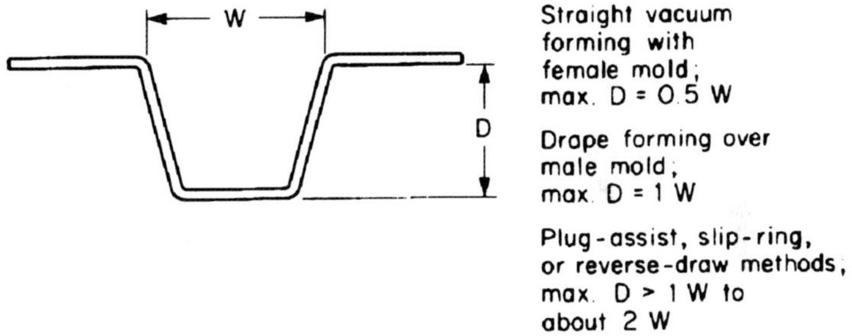

FIGURE 6.8.3 Depth-of-draw limitations with thermoformed parts.

1. Designers should be aware of and observe the depth-of-draw limitations of the thermoforming process to be used. For straight vacuum forming into a female mold, the depth-to-width ratio should not exceed 0.5:1. For drape forming over a male mold, the depth-to-width ratio should not exceed 1:1. For components with plug-assist, slip-ring, or one of the reverse-draw methods, the ratio can exceed 1:1 and perhaps reach 2:1 under normal circumstances. However, in general, shallow drafts are more easily produced than deep ones and result in more uniform wall thickness of the finished component. (See Fig. 6.8.3.)

2. Undercuts (reentrant shapes) are possible in many cases but generally should be avoided because they complicate tooling. They require movable or collapsible tool members. Such members can be incorporated into high-production applications for

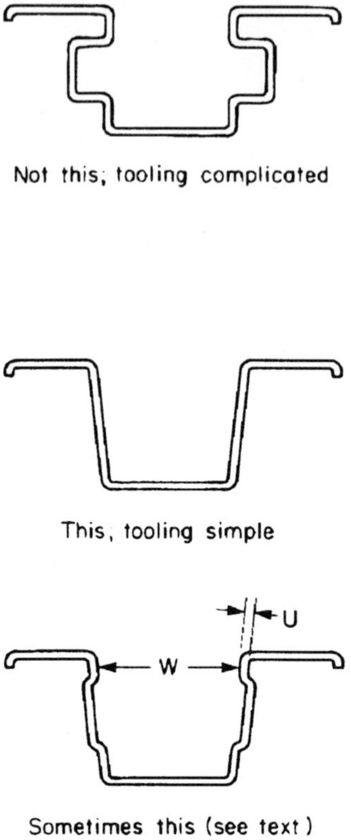

FIGURE 6.8.4 Deep undercuts should be avoided; shallow undercuts sometimes can be sprung from the mold when the workpiece is still warm.

which the cost of special tooling and machine elements can be amortized, but they are not advisable in most thermoforming applications which rely on simple, inexpensive tooling. (See Fig. 6.8.4.)

There is an exception to this rule: Small undercuts often can be sprung from a female mold when the workpiece is still warm. This works best when the workpiece material, e.g., polyethylene or nonrigid PVC, has some flexibility or is very thin. Rule-of-thumb values for the maximum amount of undercut which can be stripped from a mold are as follows:

Acrylic, polycarnonate, and other rigid materials	1 mm (0.040 in)
Polyethylene, ABS, and nylon	$1\frac{1}{2}$ mm (0.060 in)
Flexible PVC	$2\frac{1}{2}$ mm (0.100 in)

When female tooling is split to permit removal of workpieces with undercuts, a further disadvantage is that the parting line of the split halves becomes visible on the formed workpiece. If this is objectionable, the designer sometimes can incorporate the parting line in the decoration of the workpiece or at some natural line on the workpiece or can locate it in an inconspicuous place.

3. Sharp corners should never be specified because they hamper the flow of material into the mold, result in excessive thinning of materials, and cause concentrations of stress. A minimum radius of 2 times thickness is recommended. It is also more desirable from several standpoints to have large, flowing curves in the thermoformed part than to have squared corners and rectangular shapes. The best thermoformed parts have smooth natural curves and drawn sections that are spherical or near spherical in shape. Their walls will be more uniform, they will be more rigid, their surfaces will be less apt to show tool marks, and their tooling will be lower in cost.

4. Draft is required in sidewalls to facilitate easy removal of the workpiece from the mold. Female molds require less draft because the workpiece, as it shrinks during cooling, tends to pull away from the mold walls. With female tooling, for most materials, the draft on each sidewall should be at least $\frac{1}{4}°$. For male tooling, it should be 1°. (See Fig. 6.8.5.)

5. Metal inserts are not feasible with thermoformed parts because the thin walls of such parts are not sufficiently strong to hold an insert, particularly if thermal expansion and contraction take place. The best way to hold metal fittings to thermoformed parts is by bolting between washers as illustrated in Fig. 6.8.6.

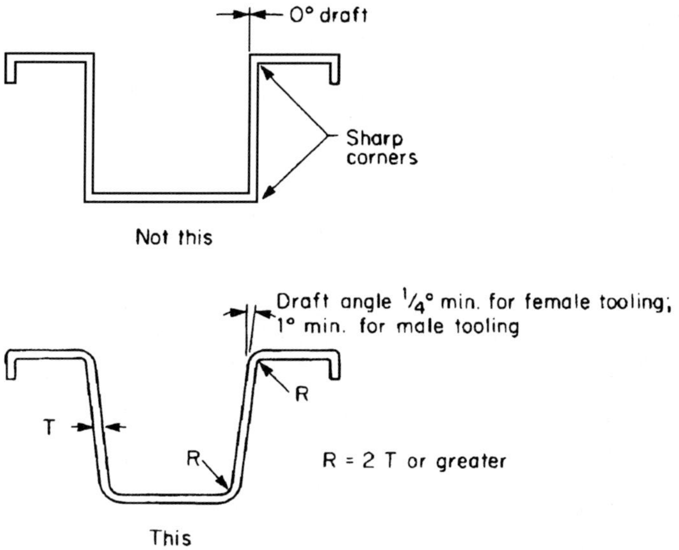

FIGURE 6.8.5 Draft is required in sidewalls to facilitate easy removal of the workpiece from the mold.

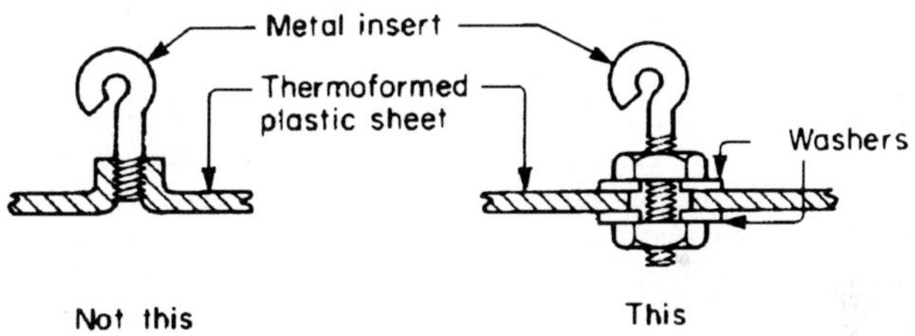

FIGURE 6.8.6 Recommended method for holding metal fittings in thermoformed parts.

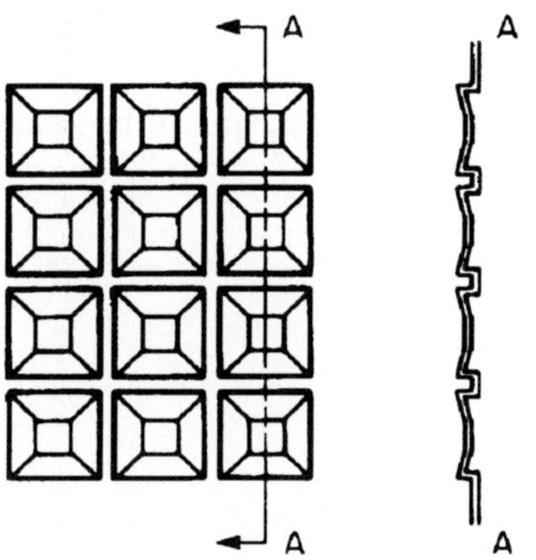

FIGURE 6.8.7 Typical stiffening pattern for flat thermoformed panels.

6. It may be desirable to increase the stiffness of the thermoformed part. Since many thermoformed parts are panel-shaped and made of thin material, they may lack rigidity. Corrugations, preferably in two directions, or an embossed pattern can add remarkably to their rigidity. Clear-polystyrene meat trays utilize this approach successfully. Figure 6.8.7 illustrates one design for such corrugations. Sometimes with short-run production it may be more economical simply to use thicker sheet material to gain additional rigidity than to go to the expense of adding corrugating elements to the tooling. If the function of the part permits, gently curved, dished, or domed surfaces can be used to add stiffness to a flat section.

DIMENSIONAL FACTORS

Thermoformed parts lack the dimensional accuracy of injection- and compression-molded parts. A number of factors contribute to this condition:

1. The low-pressure nature of thermoforming reduces the degree to which the sheet being formed is forced to conform to the molds.
2. Sheet-material variations, chiefly in the thickness of the sheet, affect the final accuracy of the part. This is particularly true because thermoforming tooling is normally one-sided. Thickness variations, including those caused by thinning during the operation, affect final workpiece dimensions.
3. Thermoformed parts are affected dimensionally by the difference between the forming temperature and the product-use temperature. Thermoforming materials have a high coefficient of thermal expansion-contraction.
4. Because thermoforming lends itself to low-quantity production, tooling is inexpensive and often not of high precision.
5. Pressure, time, and temperature variations affect final part dimensions. Of these factors, evenness of heating the sheet before forming is perhaps the most important.

It should be noted that with really precise tooling (and especially with matched-mold forming and with very careful control of temperature time and force) accuracy comparable with that of injection molding is possible. However, if this much trouble is taken and this much tooling cost is incurred, it is very often desirable to take the next step and make the part by injection molding. This gives design advantages not possible with thermoformed parts.

TOLERANCES

Table 6.8.2 presents recommended dimensional tolerances for parts to be made by thermoforming.

TABLE 6.8.2 Recommended Dimensional Tolerances for Thermoformed Plastic Parts as a Percentage of Nominal Dimension

	Normal tolerance	Close tolerance
Polystyrene, acrylonitrile butadiene styrene, cellulosics, acrylic, polycarbonate, rigid polyvinyl chloride (PVC), and nylon		
Dimensions determined by female molds	±0.6%	±0.35%
Dimensions determined by male tooling		
Dimensions under 1 m	±0.5%	±0.3%
Dimensions over 1 m	±0.8%	±0.4%
Wall thickness	±30%	±10%
Polyethylene, polypropylene, and flexible PVC	Same as above	Use "normal" values only; above "close" values not practicable.

CHAPTER 6.9
WELDED PLASTIC ASSEMBLIES

William R. Tyrrell

Director, Plastics Processing
Branson Sonic Power Division
Branson Ultrasonics Corp.
Danbury, Connecticut

It is desirable in many cases to assemble molded plastic parts further into more elaborate, permanent assemblies. This can be done to create enclosed volumes or other shapes not feasible with primary molding processes, to join components of different colors or different materials, to produce a part that is larger than can be economically produced by the primary molding process, or to enclose or contain other components.

PLASTICS-WELDING PROCESSES AND THEIR APPLICATIONS

Welding processes in this chapter are applicable to thermoplastic materials.

Ultrasonic Welding

Ultrasonic welding of thermoplastic parts is achieved by applying vibrations at an ultrasonic frequency to the assembly. This causes a temperature rise resulting from surface and intermolecular friction at the joint. When the rise is sufficient to melt the resin, material will flow and a uniform weld will result. Most systems operate at a frequency of 20,000 Hz, a frequency above the range of human hearing, and are so designated *ultrasonic*. An ultrasonic plastics-welding machine is shown in Fig. 6.9.1. Figure 6.9.2 shows a variety of ultrasonic welded parts. The components to be joined are generally injection-molded or thermoformed.

Typical ultrasonically welded assemblies include sealed taillight assemblies with clear lenses and opaque bodies, double-layer insulated drinking cups, decorative multicolor nameplate panels, photographic film cartridges, electronic calculators, small electric batteries, clock frames, electrical connectors, and toys. Many large parts such

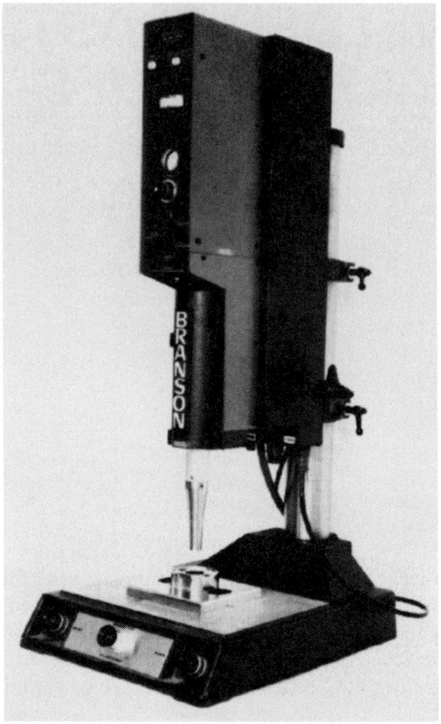

FIGURE 6.9.1 Ultrasonic welding machine. (*Courtesy Branson Ultrasonics Corp.*)

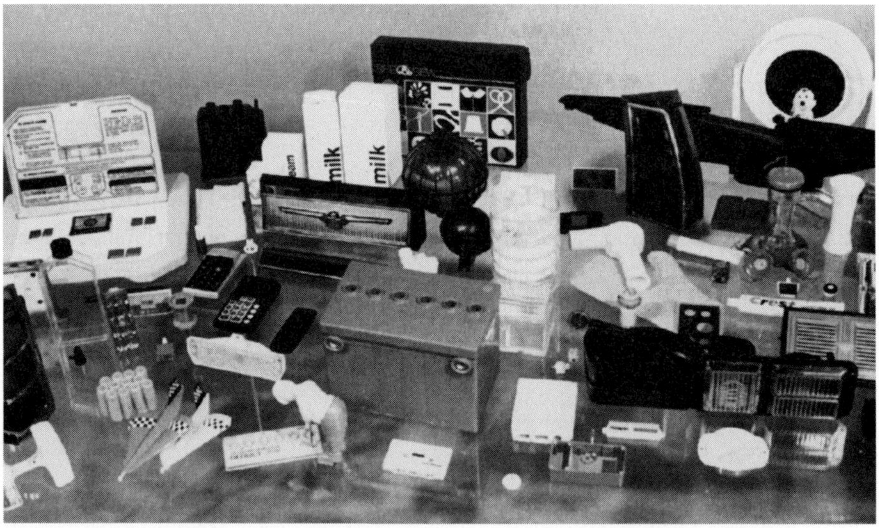

FIGURE 6.9.2 Typical ultrasonically welded parts. (*Courtesy Branson Ultrasonics Corp.*)

as automobile heater ducts, signs, or recreational-vehicle bodies are joined by spot or stud welds or seam welds of limited length.

Marking on the surface of the plastic part can occur at the horn-part contact area if the horn and fixture are not right, although this is normally an infrequent occurrence. Additionally, some materials have lower energy-transmission properties and may not be suitable for ultrasonic welding unless the horn can be located to contact the part close to the joint line.

Vibration Welding

Sometimes called *friction welding,* vibration welding is, in principle, very similar to ultrasonic welding. It differs from ultrasonic welding in that the frequency of vibration is much lower, the amplitude is much larger, and the direction is parallel to the plane of the joint rather than perpendicular. The vibration can be linear or angular as shown in Fig. 6.9.3.

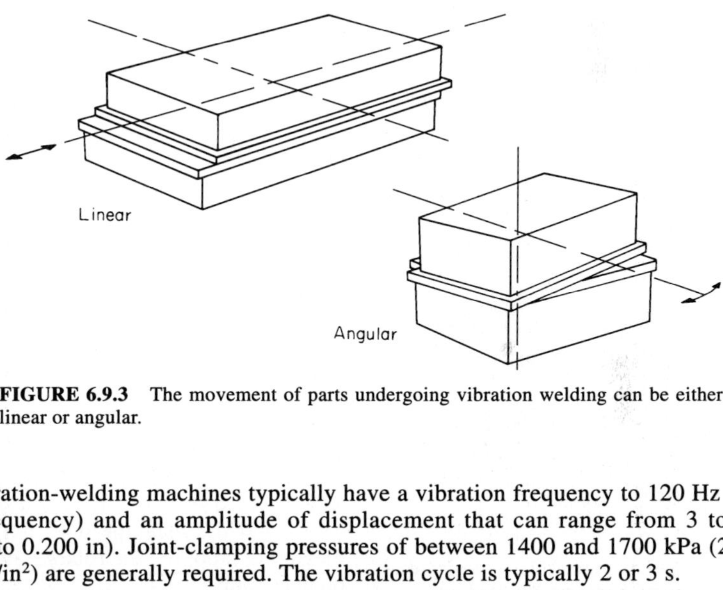

FIGURE 6.9.3 The movement of parts undergoing vibration welding can be either linear or angular.

Vibration-welding machines typically have a vibration frequency to 120 Hz (60-Hz line frequency) and an amplitude of displacement that can range from 3 to 5 mm (0.120 to 0.200 in). Joint-clamping pressures of between 1400 and 1700 kPa (200 and 250 lbf/in^2) are generally required. The vibration cycle is typically 2 or 3 s.

Vibration welding can produce strong, pressure-tight joints in all injection-molded, extruded, or thermoformed thermoplastic materials but is particularly effective for joining large parts made from crystalline resins, which can be difficult to weld with ultrasonics.

Parts for vibration welding must have a joint where the mating surfaces are free to move relative to one another in the plane of the weld. Vibration welding is generally applied to parts larger than can be assembled with a single ultrasonic-welding machine or parts that require a gastight seal but whose design prohibits the effective use of ultrasonics or other methods. Assemblies measuring 20 by 16 or 18 in in diameter can be welded on a standard machine.

Vibration welding has been used for the assembly of such products as thermoformed insulated food trays, polypropylene waterpump housings, strip filler caps for automobile batteries, automobile battery cases, automobile emission-control canisters, fuel pumps, expansion tanks, heater valves, air-intake filters, automobile taillight assemblies, vacuum reservoirs, and dishwasher spray arms.

Spin Welding

Spin welding is a form of friction welding used for the assembly of thermoplastic parts having a circular joint line. To spin-weld, one part of the assembly is rotated at high speed and pressed against the other stationary part. Frictional heat is generated, and the mating surfaces are melted. When the rotary motion is stopped, pressure is retained until the molten material solidifies. Joint strength approaching that of the parent material can be achieved.

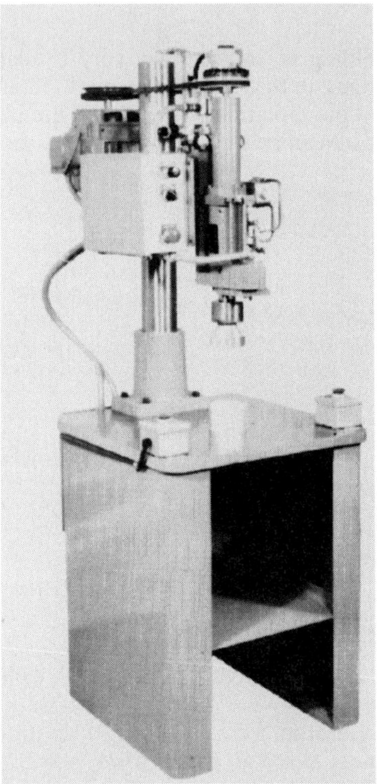

FIGURE 6.9.4 Commercial spin-welding machine. (*Courtesy Hypneumatic, Inc.*)

To reach melt temperatures, rotational velocities of from 3 to 12 m/s (10 to 40 ft/s) at the surfaces and a pressure of from 2000 to 4800 kPa (300 to 700 lbf/in^2), depending on material, are required.

Figure 6.9.4 shows a commercially available single-spindle spin-welding machine requiring hand loading and unloading. Multiple-spindle machines combined with automatic part-handling equipment can satisfy high-production requirements.

Because spin welding utilizes rotary motion, it is limited to parts with a circular joint line. The part design should enable one-half of the assembly to rotate freely relative to the other; protrusions or other eccentric components that restrict such motion prohibit use of the process. Parts requiring a specific orientation after welding also may be unsuitable for spin welding if the machine does not have a method of stopping the rotary motion at a precise point.

Typical spin-welding applications include the assembly of blow-molded polyethylene bottles, cosmetic containers, butane cigarette lighters, gasoline filters, and thermal cups and pitchers. A selection of such products is shown in Fig. 6.9.5.

Hot-Plate Welding

In the hot-plate welding of plastics, the surfaces to be joined are heated to their melting temperature by a hot blade or platen. The accurate control of heat, pressure, and

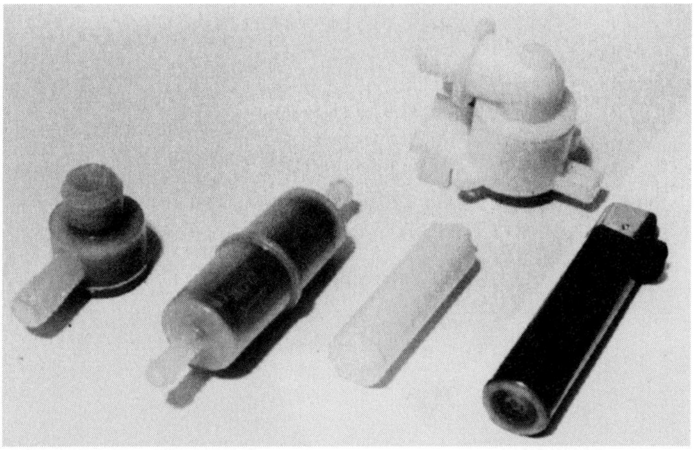

FIGURE 6.9.5 Typical spin-welded parts.

time is critical for producing quality joints with a strength from 80 to 100 percent of that of the parent material.

Most thermoplastic parts can be welded by the hot-plate method, but the process is most effective for joining large parts made from polyethylene, polypropylene, or highly plasticized PVC. Hot-plate welding can produce strong gastight or watertight seals and may be preferred for jobs for which hermetic seals are essential. It is less suitable for jobs with less stringent joint requirements. Hot-plate welding of rigid components requires a wall thickness of at least 0.75 mm (0.030 in), which in practice limits its use for assembling small or thin-walled components. One of the most common heat-welding applications is the joining of plasticized PVC gaskets to refrigerator and dishwasher doors. All applications utilize the butt-weld principle.

Among other typical hot-plate-welded parts are butane cigarette lighters, thermoplastic cups, floats for many uses, automobile batteries, taillight assemblies, and window frames (Fig. 6.9.6), waterpumps for domestic appliances, refrigerator doors, and large shipping pallets (Fig. 6.3.2). Hot-plate welding is used for the manufacture of polyethylene pipe fittings and for the joining of pipe sections in the field. Pipe diameters from 100 to 1200 mm (4 to 48 in) can be welded.

Induction Welding

Induction, or electromagnetic, welding requires the placing of additional material between the surfaces that are to be joined. This special electromagnetic layer is compatible with the thermoplastic being joined but has dispersed in it very fine micrometer-sized metal particles. The assembly to be welded is placed within or in proximity to an induction coil, through which a high-frequency alternating electric current is passed. The alternating magnetic field resulting from the electric current in the coil generates heat in the metal particles from eddy-current and hysteresis effects. The hot metal particles melt their thermoplastic binder and, in turn, the surfaces of the parts to be joined. The schematic diagram of Fig. 6.9.7 illustrates a typical installation for welding the top to a cosmetics cartridge.

FIGURE 6.9.6 Typical hot-plate-welded assemblies: window frames welded from extruded PVC sections. (*Courtesy Werner & Pfleiderer Corp.*)

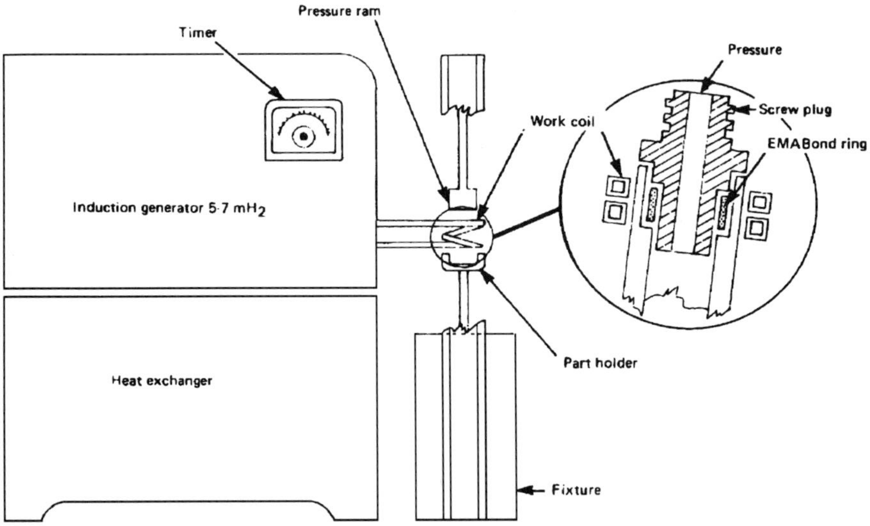

FIGURE 6.9.7 Schematic view of induction welding. (*Courtesy of EMABond, Inc.*)

Induction welding is advantageous when large open areas require bonding. Any thermoplastic part can be electromagnetically welded if its shape permits a proper joint design and the positioning of the work coil so that a uniform magnetic field can be generated in the joint area. Figure 6.9.8 illustrates components that have been assembled by using the induction-heating method.

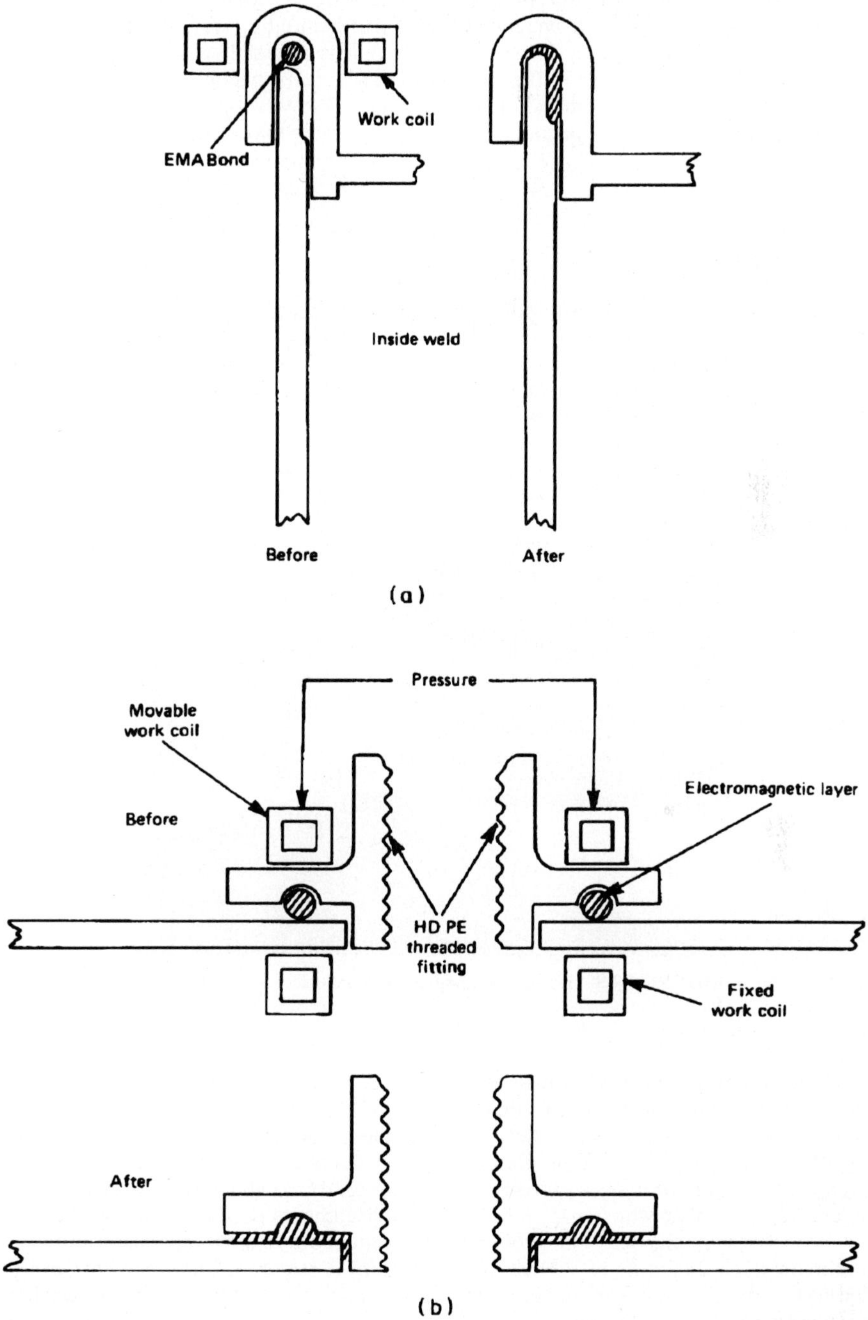

FIGURE 6.9.8 Typical induction welding. (*a*) Attaching a cover to a molded container. (*b*) Welding a hose fitting to a polyethylene drum. (*Courtesy EMABond, Inc.*)

Among the advantages of the process are its ability to produce a strong, hermetic pressure-tight joint. The electromagnetic material fills molding irregularities or cellular voids at the joint interface. The joint can be reactivated and separated if required. Applications with metal parts in the assembly may require special consideration. Any metal in the vicinity of the joint and in the magnetic field will also get hot. When such conditions are unavoidable, specially designed work coils may still permit the method to be used.

Hot-Gas Welding

Hot-gas welding is a method of joining thermoplastics that is similar in many ways to the gas welding of metals. Heat from a hand-held hot-air or hot-gas torch is applied directly to the joint area. A welding rod of the same thermoplastic material is also heated and pressed into the softened joint to form the bond. The surfaces are not heated to a liquid melt as in metal welding but softened sufficiently to allow adhesion under pressure of the welding rod. A variety of welding tips are available and can be selected to suit the particular application. Figure 6.9.9 illustrates the process.

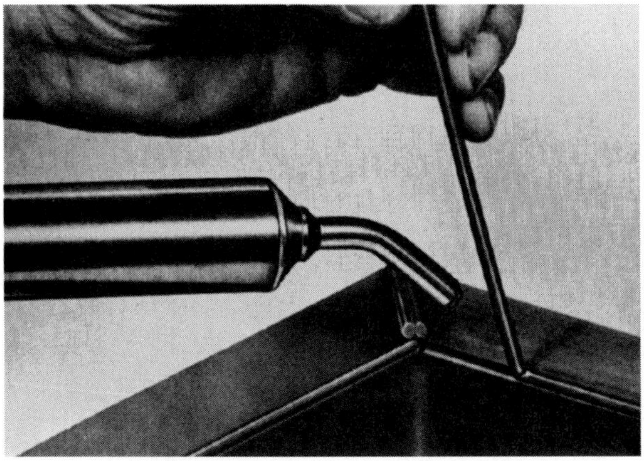

FIGURE 6.9.9 Hot-gas welding of thermoplastic. (*Courtesy Seelye, Inc.*)

Hot-gas welding is used almost exclusively for the assembly and repair of large thermoplastic parts and structures that, because of size or the necessity of assembling on site, prohibit the use of any other joining method. Some typical applications are the installation and repair of thermoplastic pipe and ductwork, the manufacture of large chemical-resistant linings, and assembly of storage bins, and many other pieces of industrial equipment whose use in corrosive environments requires that they be constructed of thermoplastics. The process is not applicable to very thin material, and a minimum stock thickness of at least $2\frac{1}{2}$ mm ($\frac{1}{10}$ in) is required. The chemical, sewage and waste disposal, nuclear, and electric power distribution fields are areas where hot-gas welding is most widely used.

ECONOMIC PRODUCTION QUANTITIES

Table 6.9.1 summarizes the economic factors relating to the five basic plastics-welding processes. Relative equipment and tooling costs, output rates, and most economical production quantities are shown.

SUITABLE MATERIALS

Most thermoplastics are weldable. Particularly if both parts to be welded are of the same material, the designer has a wide latitude of materials selection with all plastics-welding processes. When two different materials are to be welded together, the choice is more limited. Dissimilar materials must have similar melting temperatures if they are to be welded together. They also must possess a like molecular structure or have common radicals that will combine in the molten state. Table 6.9.2 shows the compatibility of various thermoplastics.

Ultrasonic Welding

Most thermoplastic materials can be welded ultrasonically. Table 6.9.3 lists the ultrasonic-welding characteristics of common thermoplastics.

Generally, the more rigid the plastic, the easier it is to weld. Low-modulus materials such as polyethylene and polypropylene can be welded if the horn is positioned close to the joint area. Polyester, in its thermoplastic form, can be welded in thin sheets and/or fabrics. Teflon, with a low coefficient of friction and a high melting temperature, is almost impossible to weld. Among the best welding materials are ABS, polystyrene, acrylic, polycarbonate, acrylonitrile styrene acrylic (ASA), and styrene acrylonitrile (SAN).

The welding characteristics of a plastic can be altered by the inclusion of additives such as plasticizers and fillers. Plasticizers impart flexibility and softness to a resin and also provide lubrication at the joint, both effects retarding the ultrasonic-welding characteristics of the plastic.

Resins with filler (e.g., glass, talc, and mica) content in the 20 percent range can be welded in the normal manner; in fact, the welding characteristics of some poor-welding resins can be improved by the addition of fillers. Abrasive-filler particles on the surface of the plastic can cause horn wear, and carbide-tipped horns may be required. Filler content in excess of 30 percent may adversely affect weld-strength because there will be more unweldable filler and less thermoplastic at the weld surface.

Vibration and Spin Welding

All thermoplastics can be vibration- and spin-welded regardless of the original molding or forming process. Polystyrene, ABS, acrylic, polycarbonate, and other amorphous plastics can be welded as well as with other joining techniques. Crystalline resins such as nylon, acetal, polyethylene, polypropylene, and thermoplastic polyester,

TABLE 6.9.1 Economic Data of Plastics-Welding Processes

Process	Equipment cost	Tooling cost	Typical output rates	Normal economic production quantities	Remarks
Ultrasonic welding	Moderately low to high	Moderate to high	1000 pieces per hour, manually loaded	High	Automatic operation possible
Vibration welding	Moderate	Moderate	240 pieces per hour from single cavity, manually loaded	Medium and high	Setup time 10 min; multiple cavities and mechanized loading possible
Spin welding	Moderate	Moderate	640 pieces per hour, manually loaded	High	Setup time ½ h; mechanization possible
Hot-plate welding	Moderately low to high	Moderate to high	120 pieces per hour per fixture cavity	Medium and high	Setup time 1 h or less
Induction welding	Low to moderate	Low	900 pieces per hour, manually loaded	High	Setup time 1 h or less
Hot-gas welding	Very low	Low (holding fixture only)	0.3–1.5 m (12–60 in) of weld seam per minute	Very low	Manual operation

which are difficult to weld by ultrasonics, are easily vibration-welded. Other suitable materials include foamed resins, injection-moldable fluoropolymers, and thermoplastic elastomers. Fillers that provide resilience or impact strength to a resin have no adverse effect on its vibration-welding properties. However, as with other joining methods, a high percentage of filler can result in a reduction in weld strength because of the high concentration of filler present at the joint.

During the vibration-welding process, a smearing and blending action occurs at the molten surfaces, thus enhancing the ability to join dissimilar plastics.

Other resins that have been bonded successfully by vibration and spin welding are PVC, thermoplastic elastomers, thermoplastic rubber (TPR), polyurethane (nesting material), polysulfone, ionomer, polycarbonate to acrylic, and acrylic to ABS.

Hot-Plate Welding

Table 6.9.4 lists some of the thermoplastic materials that can be joined by the hot-plate-welding process and the percentage of weld strength achievable. Most thermoplastics, including some dissimilar materials such as polyethylene to polypropylene, acrylic to ABS, and rigid to plasticized PVC, can be welded by this process. As with most other methods of joining thermoplastics, the percentage, content, and type of filler used in a material can affect weld strength.

Injection-moldable grades of nylon generally obtain high weld strengths. Other grades of nylon and those with flame retardants weld poorly.

Induction Welding

All thermoplastics can be joined by this process if the polymer incorporated in the electromagnetic matrix is the same or compatible with that from which the parts to be assembled are made. Some dissimilar materials also can be welded together if they are compatible in the molten state and if blends of the two dissimilar polymers are included in the electromagnetic matrix. Among thermoplastics successfully welded are polypropylene, polyethylene, polystyrene, nylon, SAN, PVC, ABS, acetals, polycarbonate, and acrylic.

Fillers have no material effect on welding; a 40 percent glass-filled polypropylene is readily welded. Cross-linked low-density and high-density polyethylenes also have been welded by using a standard low-density or high-density matrix. Table 6.9.5 shows the induction-welding compatibility of some thermoplastics.

Hot-Gas Welding

The materials most commonly used are PVC, polyethylene, polypropylene, acrylic, and some blends of ABS. With the exception of polyethylene and polypropylene, they can be welded with a hot-air heat source. Polyethylene and polypropylene require the use of nitrogen as the welding gas to minimize oxidation of the surfaces. Oxidation occurs on the surface of these materials after extrusion and should be removed by abrasion immediately before welding or welded soon after the surfaces have been cut. Table 6.9.6 shows materials and data related to their welding characteristics.

TABLE 6.9.2 Compatibility of Thermoplastics*

	ABS	ABS-polycarbonate alloy (Cycoloy 800)	ABS-PVC alloy (Cycovin)	Acetal	Acrylics	Acrylic multipolymer (XT polymer)	Acrylic-PVC alloy (Kydex)	ASA	Butyrates	Cellulosics
ABS	■	■	○		■	○	○	○	○	
ABS-polycarbonate alloy (Cycoloy 800)	■	■	○		○	○	○	○		
ABS-PVC alloy (Cycovin)	○	○	■		○	○	○	○		
Acetal				■						
Acrylics	■	○	○		■	○	○	○		
Acrylic multipolymer (XT polymer)	○	○	○		○	■	○	○		
Acrylic-PVC alloy (Kydex)	○	○	○		○	○	■	○		
ASA	○	○	○		○	○	○	■		
Butyrates	○								■	
Cellulosics										■
Modified phenylene oxide (Noryl)										
Nylon										
Polycarbonate		■			○					
Polyethylene										
Polyimide										
Polypropylene										
Polystyrene	○					○		○		
Polysulfone										
PPO										
PVC			○				○			
SAN-NAS	○	○			○	○		○		

■ denotes compatibility; ○ denotes some but not all grades and compositions compatible.

*Courtesy Branson Ultrasonics Corp. Tables should be used as a guide only since variations in resins may produce slightly different results.

TABLE 6.9.2 Compatibility of Thermoplastics* (*Continued*)

Modified phenylene oxide (Noryl)	Nylon	Polycarbonate	Polyethylene	Polyimide	Polypropylene	Polystyrene	Polysulfone	PPO	PVC	SAN-NAS
						○				○
		■								○
									○	
		○								○
						○				○
									○	
						○				○
■						■		■		○
	■									
		■								
			■							
				■						
					■					
■						■				○
							■			
■								■		
									■	
○						○				■

TABLE 6.9.3 Ultrasonic-Welding Characteristics of Thermoplastics[a]

Material	Percent weld strength	Spot welding	Staking and inserting	Swaging	Welding: Near-field[b]	Welding: Far-field[b]
General-purpose plastics						
ABS	95–100+	E	E	G	E	G
Polystyrene unfilled	95–100+	E	E	F	E	E
Structural foam (styrene)	90–100[c]	E	E	F	G	P
Rubber-modified polystyrene	95–100	E	E	G	E	G–P
Glass-filled (up to 30%) polystyrene	95–100+	E	E	F	E	E
SAN	95–100+	E	E	F	E	E
Engineering plastics						
ABS	95–100+	E	E	G	E	G
ABS-polycarbonate alloy (Cycoloy 800)	95–100+[d]	E	E	G	E	G
ABS-PVC alloy (Cycovin)	95–100+	E	E	G	G	F
Acetal	65–70[e]	G	E	P	G	G
Acrylics	95–100+[f]	G	E	P	E	G
Acrylic multipolymer (XT polymer)	95–100	E	E	G	E	G
Acrylic-PVC alloy (Kydex)	95–100+	E	E	G	G	F
ASA	95–100+	E	E	G	E	G
Methylpentene	90–100+	E	E	G	G	F
Modified phenylene oxide (Noryl)	95–100+	E	E	F–P	G	E–G
Nylon	90–100+[d]	E	E	F–P	G	F
Polyesters (thermoplastic)	90–100+	G	G	F	G	F
Phenoxy	90–100	G	E	G	G	G–F
Polyarylsulfone	95–100+	G	E	G	E	G

Polycarbonate	95–100+[d]	E	E	G–F	E	E
Polyimide	80–90	F	G	P	G	F
Polyphenylene oxide	95–100+	E	G	F–P	G	G–F
Polysulfone	95–100+[d]	E	E	F	G	G–F
High-volume, low-cost applications						
Butyrates	90–100	G	G–F	G	P	P
Cellulosics	90–100	G	G–F	G	P	P
Polyethylene	90–100	E	E	G	G–P	F–P
Polypropylene	90–100	E	E	G	G–P	F–P
Structural foam (polyolefin)	85–100	E	E	F	G	F–P
Vinyls	40–100	G	G–F	G	F–P	F–P

[a]Courtesy Branson Ultrasonics Corp. E = excellent; G = good; F = fair; P = poor.

[b]Near-field welding refers to joint ¼ in or less from area of horn contact; far-field welding to joint more than ¼ in from contact area.

[c]High-density foams weld best.

[d]Moisture will inhibit welds.

[e]Requires high-energy and long ultrasonic exposure because of low coefficient of friction.

[f]Cast grades are more difficult to weld owing to high molecular weight.

TABLE 6.9.4 Materials for Hot-Plate Welding

Material	Percent weld strength 100	90	80	70	60
Acrylonitrile butadiene styrene (ABS)			x		
Acetal		x			
Acrylic			x		
Acetate			x		
Butyrate			x		
Propionate			x		
Ethylene vinyl acetate			x		
Nylons		x			
Polycarbonate				x	
Low-density polyethylene	x				
Ultrahigh-molecular-weight polyethylene		x			
High-density polyethylene	x				
Polypropylene	x				
Polysulfone					x
General-purpose styrene			x		
High-impact styrene		x			
Flexible vinyl		x			
Rigid vinyl		x			
Thermoplastic rubber		x			

Source: Courtesy Werner & Pfleiderer Corp.

TABLE 6.9.5 Compatibility of Various Plastics to Induction Bonding

• = Compatible combinations

	ABS	Acetal	Acrylic	Nylon	PC	PE	PP	PS	PVC	PU
ABS	•		•					•	•	
Acetal		•								
Acrylic	•		•					•	•	
Nylon				•						
Polycarbonate					•					
Polyethylene						•	•			
Polypropylene						•	•	•		
Polystyrene	•		•				•	•	•	
Polyvinyl chloride	•		•					•	•	•
Polyurethane									•	•
Cellulose acetate butyrate										•
SAN	•		•				•			
Polysulfone					•					
Cross-linked PE						•				
UHMPE						•				
Paper	•	•	•	•	•	•	•	•	•	•
Polyphenylene oxide										
TFE Teflon										
FEP										
Ionomer										

Source: Courtesy EMABond, Inc.

DESIGN RECOMMENDATIONS

Ultrasonic Welding

Surfaces to be joined should be as free of distortion and warpage as possible. Therefore, designers should be careful of the location of ribs, bosses, and other elements so that they do not cause sinks or other distortions of the mating surfaces. Warped parts may require excessive pressure to bring the surfaces together for welding, possibly causing distortion of the finished assembly.

Melting will occur more rapidly when frictional heat is developed over the smallest possible surface area. Large mating surfaces will take longer to melt than when a bead or narrow raised section is molded onto one of the surfaces. Such a bead or high spot is called an *energy director.* (See Fig. 6.9.10.)

It is likely in such a basic joint design that some molten material will be expelled from the joint area and build up flash on the walls of the part. Figure 6.9.11 illustrates a step joint used in applications for which flash would be objectionable. This form of joint will produce a stronger bond than the simple joint shown in Fig. 6.9.10 because molten material will flow into the clearance provided initially for a slip fit, establishing a seal with good strength in shear as well as in tension. Even greater strength can

TABLE 6.9.5 Compatibility of Various Plastics to Induction Bonding (*Continued*)

	CAB	SAN	PSUL	X-linked PE	UHMPE	Paper	PPO	TFE	PEP	Ionomer
		●				●				
						●				
		●				●				
						●				
			●			●				
				●	●	●				
		●				●				
						●				
						●				
	●					●				
	●					●				
		●				●				
			●			●				
				●	●	●				
				●	●	●				
	●	●	●	●	●	●				
							●			
								●	●	
								●	●	
										●

TABLE 6.9.6 Material Usage for Hot-Gas Welding*

		Type I PVC	Type II PVC	Modified high-impact PVC	Type I ABS	Type II ABS	Branched PE	Medium-density PE	Linear PE	Polypropylene	Flexible tank lining	Acrylic
Service potential												
Impact strength	F–G	E	G	E	E	E	E	E	G	NA	F	
Chemical resistance		E	G–E	G–E	G	G	G	G–E	G–E	E	G	F
Hardness	E	G	G	G	E	U–F	F	G	G	NA	G	
Heat service temperature		F–G	F	F–G	G	E	U	U–F	G	E	F–G	G
Working strength	E	G	E	E	E	F	G	G–E	G–E	F	G	
Uses and availability												
Cement and fabricate		E	E	E	E	E	U	U	U	U	E	E
Demineralized water		G	G	G	G	G	G	E	E	E	F–G	G
Ductwork	E	E	E	F	G	NA	U	G	E	NA	G	
Etch stations	E	G	F	F–U	F–U	NA	U	G	E	NA	NA	
Fittings E	E	NA	E	E	NA	NA	G	E	NA	NA		
Formed parts, thermo-		F–G	G–E	G	E	E	E	E	E	E	F	E
Formed parts, injection-		G	G	NA	E	E	E	E	E	E	G	E
Gaskets F	F	NA	NA	NA	G	G	G	G	E	NA		
Hoods, exhaust	E	E	G	NA	NA	NA	U	G	E	NA	F–G	
Laboratory hoods	E	E	G	NA	NA	NA	U	G	E	NA	NA	
Laboratory utensils		E	G	G	NA	NA	G	G	E	E	NA	F

Machined parts	E	G	NA	G	G	NA	G	G	G	NA	G
Nuts and bolts	E	G	NA	NA	NA	NA	NA	F	G	NA	NA
Pipe and tubing	E	E	NA	E	E	F	E	E	E	F–G	F–G
Scrubbers	E	E	G	F	F	NA	U	F	E	NA	NA
Shapes, extruded	E	E	NA	G	G	G	G	G	G	E	G
Sheet	E	E	E	E	E	E	E	E	E	E	E
Sinks	E	E	G	G	G	NA	G	G	E	NA	G
Stress: crack	E	G	G	G	G	U–F	F	F–G	E	NA	F–G
Tanks, lined	F–G	G	F	F	F	F	U	F–G	E	E	NA
Tanks, self-supporting	E	E	F	U	U	NA	F–U	G	G–E	NA	G
Tabletop	G	E	E	E	E	F	F	F	G	NA	NA
Tumbling basket	F	F	U	U	U	NA	U	F–G	G–E	NA	F
Valves	E	F	NA	F	F	NA	NA	F–G	G–E	NA	NA
Weld and fabricate	E	E	G	F	F	G	E	E	E	E	NA

*E = excellent; G = good; F = fair; U = unsatisfactory; NA = not applicable. A tabulation of usage is always subject to interpretation and special requirements. Ratings are based on performance, availability, or cost. When lower ratings were assigned in some cases, outstanding performance capability of several other materials was considered, although the lower-rated materials would be satisfactory. An attempt was made to differentiate between nonapplicable and unsatisfactory. In the former case, the materials are never or almost never considered for such services owing to obvious limitations. In the second case, the rating designates misapplication of the material in the opinion of the writers, and use should be avoided.

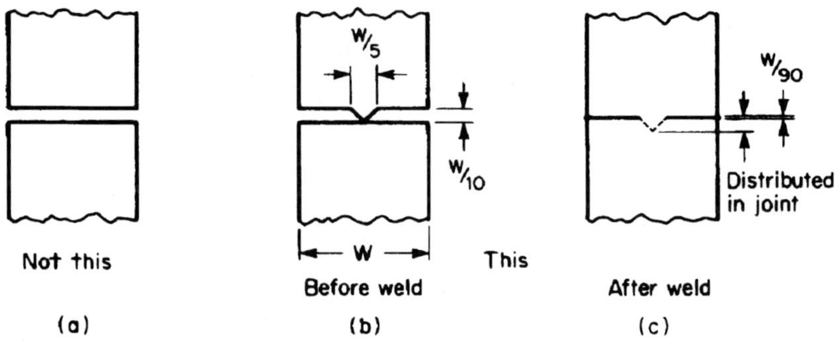

FIGURE 6.9.10 A raised bead called an *energy director* greatly facilitates ultrasonic welding. View *b* shows preferred energy-director dimensions, and *c* shows the thickness of the dissipated energy-director material after welding.

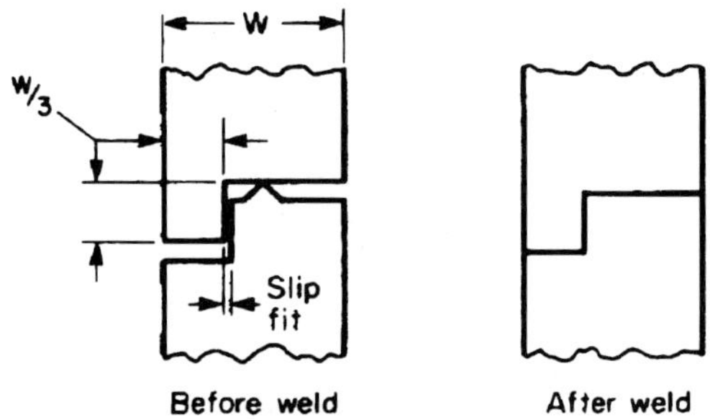

FIGURE 6.9.11 A step joint used when a weld bead on the side would be objectionable. This joint is usually much stronger than a butt joint because material flows into the clearance necessary for a slip fit, establishing a seal that provides strength in shear as well as in tension.

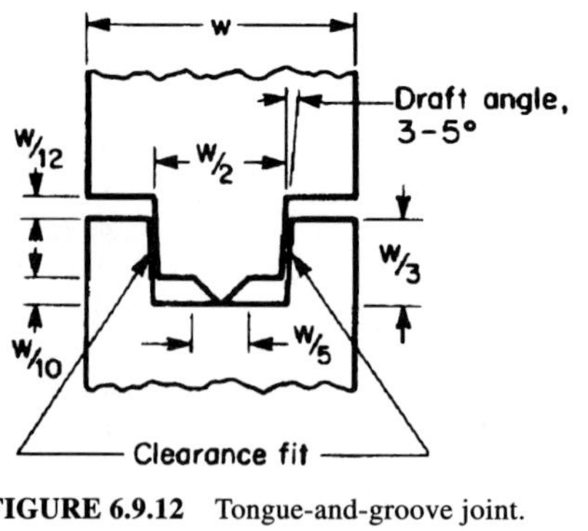

FIGURE 6.9.12 Tongue-and-groove joint.

be achieved with a tongue-and-groove joint shown in Fig. 6.9.12. However, the need to maintain clearance on both sides of the tongue may make this more difficult to mold. It is important in both step and tongue-and-groove joints to ensure clearance between the mating surfaces where shown. Tight binding fits are likely to restrict movement of one surface relative to the other and to retard or prevent heat buildup. Figure 6.9.13 shows other joint design variations. A typical mistake is beveling one joint face at a 45° angle. The problems with this practice are shown in Fig. 6.9.14.

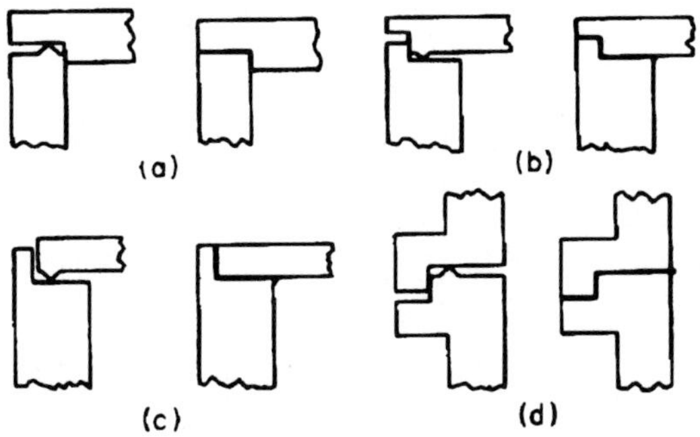

FIGURE 6.9.13 Typical ultrasonic joint designs.

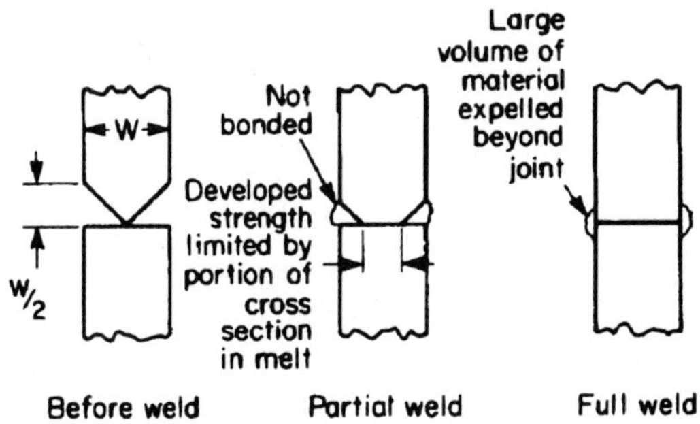

FIGURE 6.9.14 A common mistake.

The melt characteristics of crystalline thermoplastics (nylon, acetal, polyethylene, polypropylene, and thermoplastic polyester), in which transition from the solid to the molten state occurs rapidly, require special joint consideration. Figure 6.9.15 illustrates a shear joint as might be used to weld a cap to a container. This joint permits interaction between the two surfaces during the entire melt cycle by exposing progressively more area as the two surfaces shear against each other under pressure. The diagram shows the part in a fixture that supports the walls, preventing them from flexing during the shearing action.

Figure 6.9.16 shows the basic stud-weld joint before, during, and after welding. Welding occurs along the circumference of the stud. Weld strength is a function of stud diameter and depth of weld. Maximum strength in tension is achieved when the depth of weld equals one-half of the diameter. Radial interference *A* must be uniform. The hole should be a sufficient distance from the edge to prevent breakout, and a minimum of 3 mm (0.120 in) is recommended.

To help align the stud with the hole, the tip of the stud can be chamfered or the entrance of the hole countersunk. To reduce stress concentration during welding and in use, an ample fillet radius should be incorporated at the base of the stud. Recessing the fillet below the surface serves as a flash trap which allows flush contact of the parts.

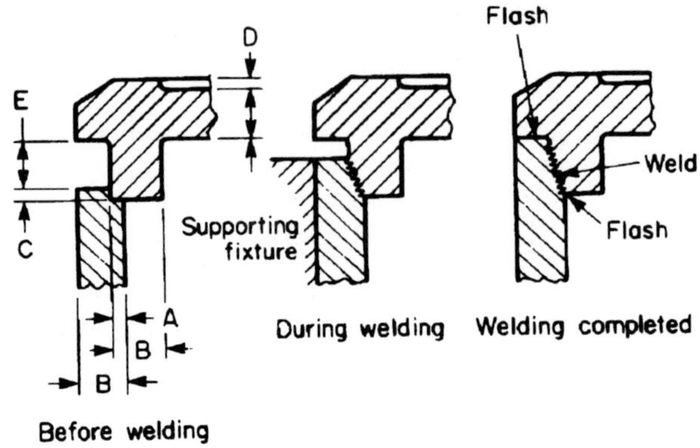

FIGURE 6.9.15 Shear joint for ultrasonically welding crystalline thermoplastics.

Dimension	
A	0.4 mm (0.016 in); suggested for most applications.
B	Wall thickness.
C	0.4–0.6 mm (0.016–0.024 in). This recess is to ensure precise location of the lid.
D	This recess is optional and is generally recommended for ensuring good contact with the welding horn.
E	Equal to or greater than dimension *B*.

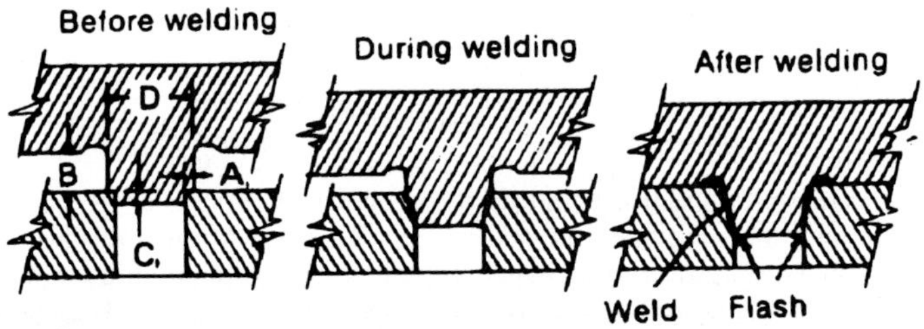

FIGURE 6.9.16 Ultrasonic stud-welding joint.

Dimension	
A	0.25–0.4 mm (0.010–0.016 in) for dimension *D* up to 13 mm (0.5 in).
B	Depth of weld. $B = 0.5D$ for maximum strength (stud to break before joint failure).
C	0.4-mm (0.016-in) minimum lead-in. The step can be at the end of the pin or the top of the hole.
D	Stud diameter.

Vibration and Spin Welding

The basic design consideration for vibration- and spin-welded assemblies is that one part of the assembly must be free to move relative to the other in the plane of the weld. This motion generates the frictional heat necessary to melt the plastic. In vibration welding, the width of the surfaces to be welded is important. In spin welding, the design must provide good part alignment. Wall thickness also is important. Thin walls will flex during welding and effectively reduce the relative motion or pressure.

A part can be made more compatible for vibration welding by adding a flange to the joining surfaces, thus providing additional weld area and a surface to which pressure can be directly applied. (See Fig. 6.9.17.) For spin welding, a tongue-and-groove design as shown in Fig. 6.9.21 provides good part alignment and greater contact surface area than could be achieved by a simple butt joint and is commonly used.

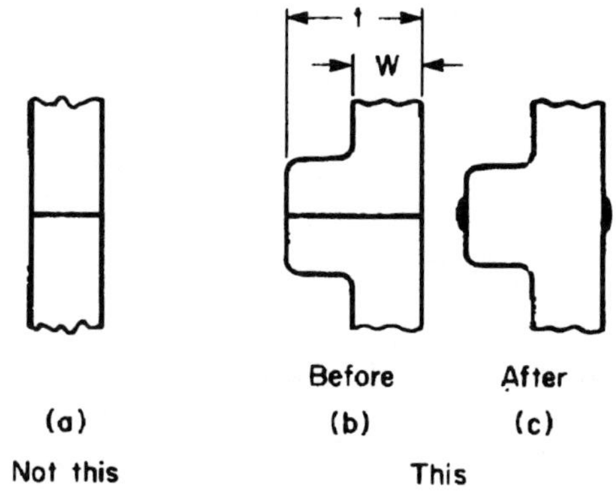

FIGURE 6.9.17 (*a*) Flange at the weld surfaces aids in vibration welding. (*b*) and (*c*) show recommended joint design. w = wall thickness; t = weld surface, 2 to 2.5 $\times$ w (for maximum strength); t minimum = 0.6 $\times$ M, where M = total relative motion.

Provided the part geometry is suitable, it must be determined how the part can be held firmly in the machine so that adequate pressure can be applied to the joint area while one part is moved against the other. A rubber friction-drive tool may suffice, but some keying points may need to be molded in the part to mate with the driving tool. The keying point, whether protrusion or recess, should be designed so that it can fit snugly in the driving tool or fixture.

For maximum strength of a vibration-welded joint, the width of the weld surface should be approximately 2 to 2.5 times the wall thickness, and when both surfaces being joined are the same, as in Fig. 6.9.17, the width of the joint should be at least 60 percent of the relative peak-to-peak vibration motion.

Parts with long, thin, unsupported walls may require special provisions such as a reverse flange or a channel-like section along the length of the joint surface to afford even greater rigidity than a single-plane flange can provide.

Special joint designs become necessary when it is required to contain the flash or surplus molten material that is squeezed to the outside of the part during the welding process. If flash is unacceptable for aesthetic reasons, traps must be included in the

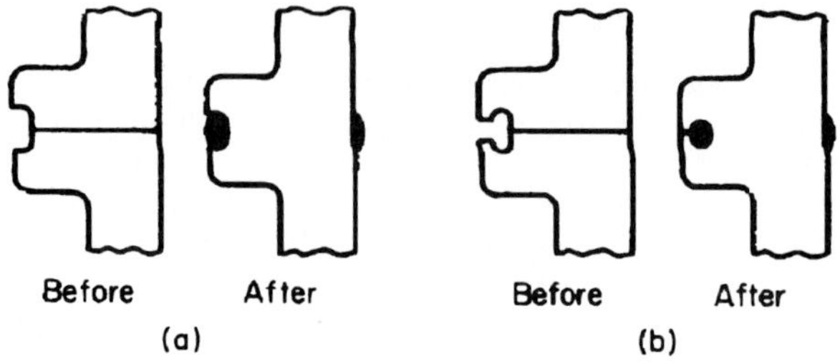

FIGURE 6.9.18 (*a*) Functional flash trap; external. (*b*) Cosmetic flash trap; external.

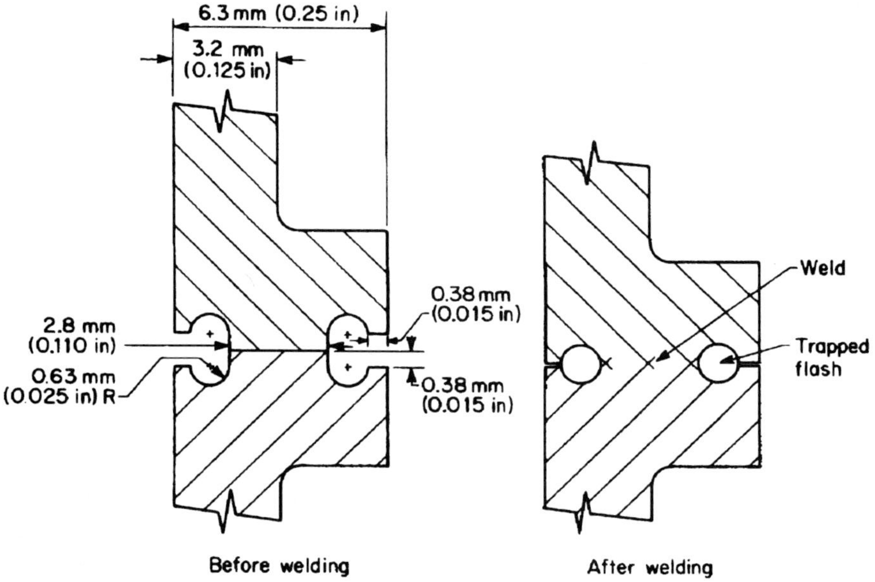

FIGURE 6.9.19 Typical flash-trap dimensions.

joint design to contain it. Two basic flash-trap designs are shown in Fig. 6.9.18. A design with a flash trap on both sides of the joint is shown and dimensioned in Fig. 6.9.19.

The displaced material that forms flash will affect the final dimension of the assembled part, and it should be understood that the height of the final assembly may be reduced by 0.4 to 0.8 mm (0.016 to 0.032 in) depending on the material and flatness of the part. If this dimensional change is critical in the final assembly, it must be allowed for.

For easy assembly alignment or fixture simplification in vibration weldments, a breakaway stud and socket may be incorporated into the part halves. Figure 6.9.20 shows examples of internal and external breakaway studs. They will normally be from 1 to 1.5 mm (0.040 to 0.060 in) in diameter or thickness, depending upon the material.

Figure 6.9.21 shows examples of typical joint designs for spin weldments, some providing more than one melting surface.

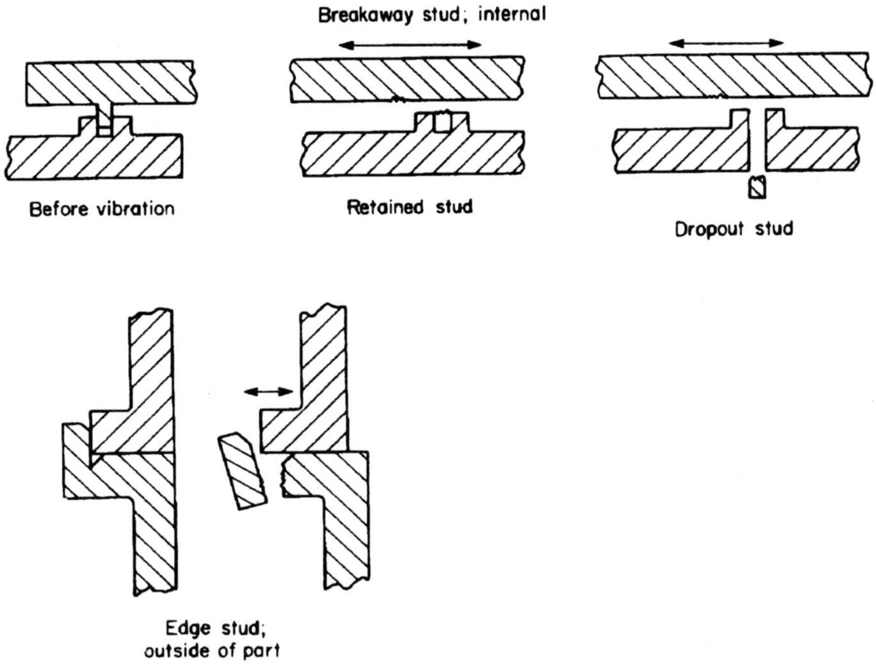

FIGURE 6.9.20 Typical breakaway studs to facilitate preassembly of parts before vibration welding.

Hot-Plate Welding

The hot-plate welding process requires the melting and expulsion under pressure of some molten material from the joint area. This loss of material at the joint should be allowed for in the design of the part. A loss of approximately 2 mm (0.080 in) can be anticipated with most injection-molded parts, the precise amount being accurately controlled by the pressure, feed rate, and mechanical stops on the machine.

A butt joint is required for hot-plate welding. Thin-walled parts may require the addition of a flange to the joining surfaces to provide increased surface area and greater part rigidity. Parts that are warped or distorted or have sinks at the joint line may benefit from having material added to thicken this section, enabling the melting process to remove high spots on the mating surfaces and allowing good contact and welding over the entire joint area.

The expulsion of molten material will, unless considered in the joint design, appear as flash on the outside of the part. Joint designs to control the flow of the displaced material similar to those shown in Figs. 6.9.18 and 6.9.19 for spin and vibration welding are also advisable for hot-plate weldments.

Induction Welding

The proper geometry and size of the joint are important considerations in designing components for induction welding. The bond line should be uniformly located in rela-

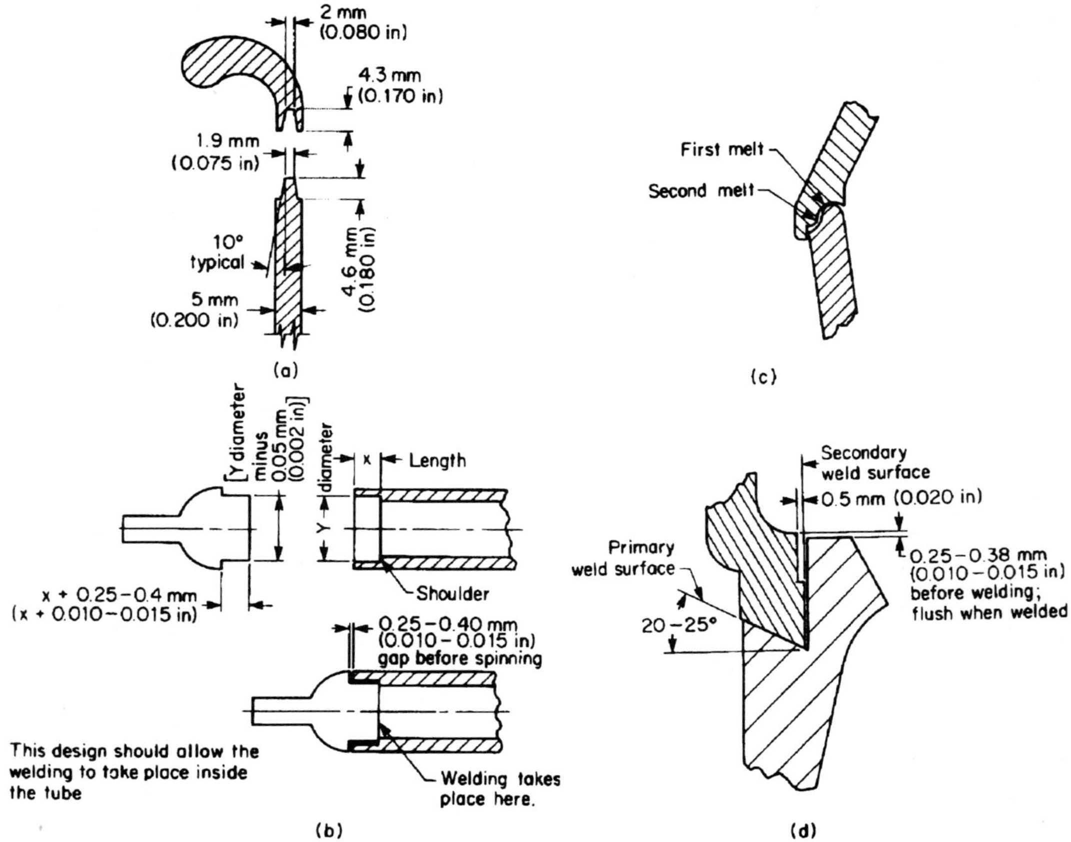

FIGURE 6.9.21 Typical joint designs for spin weldments.

tion to the outside wall. This is necessary so that the coupling distance, i.e., the space between the work coil and the bond line, is constant. The joint itself should be as regular as possible, preferably ringlike or cylindrical. Irregularly shaped joints require more complex and expensive work coils and are difficult to heat uniformly.

The joint line should be as close to the coil as possible. External lugs, protrusions, or other irregularities that prevent the work coil from being located close to the joint line should be avoided. Special reflection coils that make coupling distances up to 25 mm (1 in) feasible can be used, but these require development, are more costly, and may not work as effectively as close coils.

Figure 6.9.22 illustrates the preceding principles.

The flow of the molten electromagnetic material should be considered. It will flow in the path of least resistance and fill void areas as designed or fill shrinkage and sink

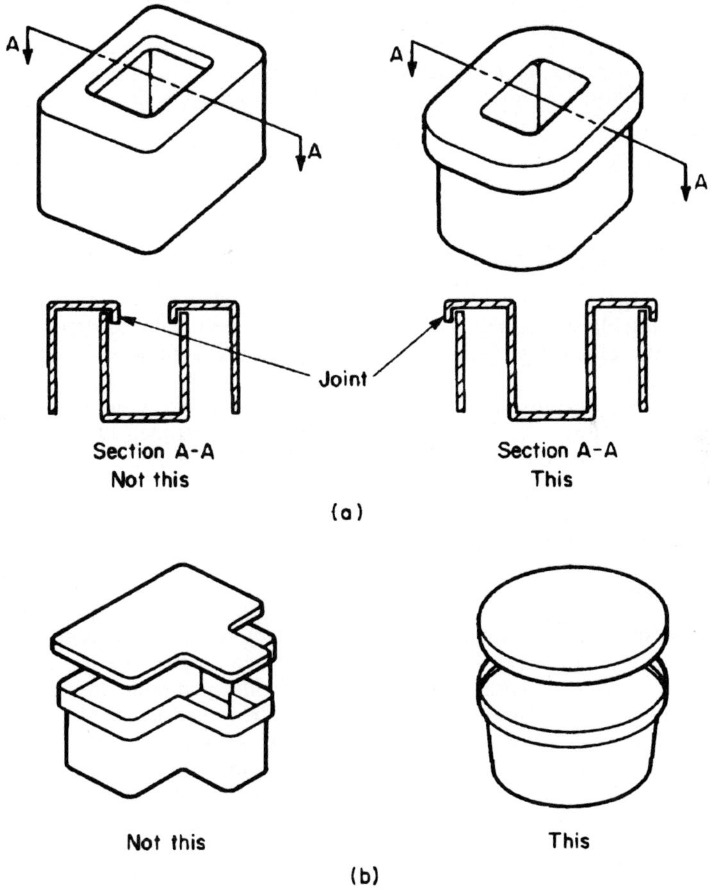

FIGURE 6.9.22 Design rules for joints to be induction-welded. (*a*) The original joint is inaccessible. The improved design is accessible and of more regular shape. (*b*) The irregular shape of the joint in the original design makes coil construction and operation difficult.

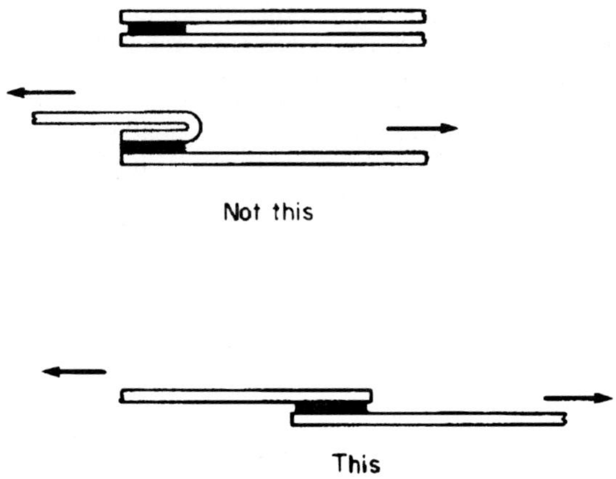

FIGURE 6.9.23 Design induction-welded joints in film or sheet metals so that shear rather than peeling stresses result.

surfaces. Joints should be designed in shear rather than in peel or butt whenever possible. (See Fig. 6.9.23.)

An increase in thickness can result from the addition of the electromagnetic material when welding sheet material. The increase will be nominal and dependent on the thickness of the electromagnetic material and pressure applied to the joint during welding.

Different joint designs require different forms of electromagnetic material. A tongue-and-groove joint will have a molded gasket or an extruded strand placed in the groove, whereas two flat surfaces require a ribbon or an extruded strand placed between the surfaces.

Hot-Gas Welding

Figure 6.9.24 shows a selection of the most common joint configurations, all of which incorporate an approximately 60° included-angle space for the welding bead. Thicker materials may require a double-V joint, enabling welding on both sides of the material. The beveling of the edge of the two surfaces to be joined is usually accomplished with a saw or a sander. For welding, the two parts of the assembly must be firmly clamped, leaving a space of 0.4 to 0.8 mm ($\frac{1}{64}$ to $\frac{1}{32}$ in) between the edges. This *root gap* allows the softened welding-rod material to be forced completely through the joint. Some applications require tack welding first to align the parts correctly prior to welding.

Joint designs for hot-gas welding are simple, but the quality of the joint will depend as much on the operator as on any other factor.

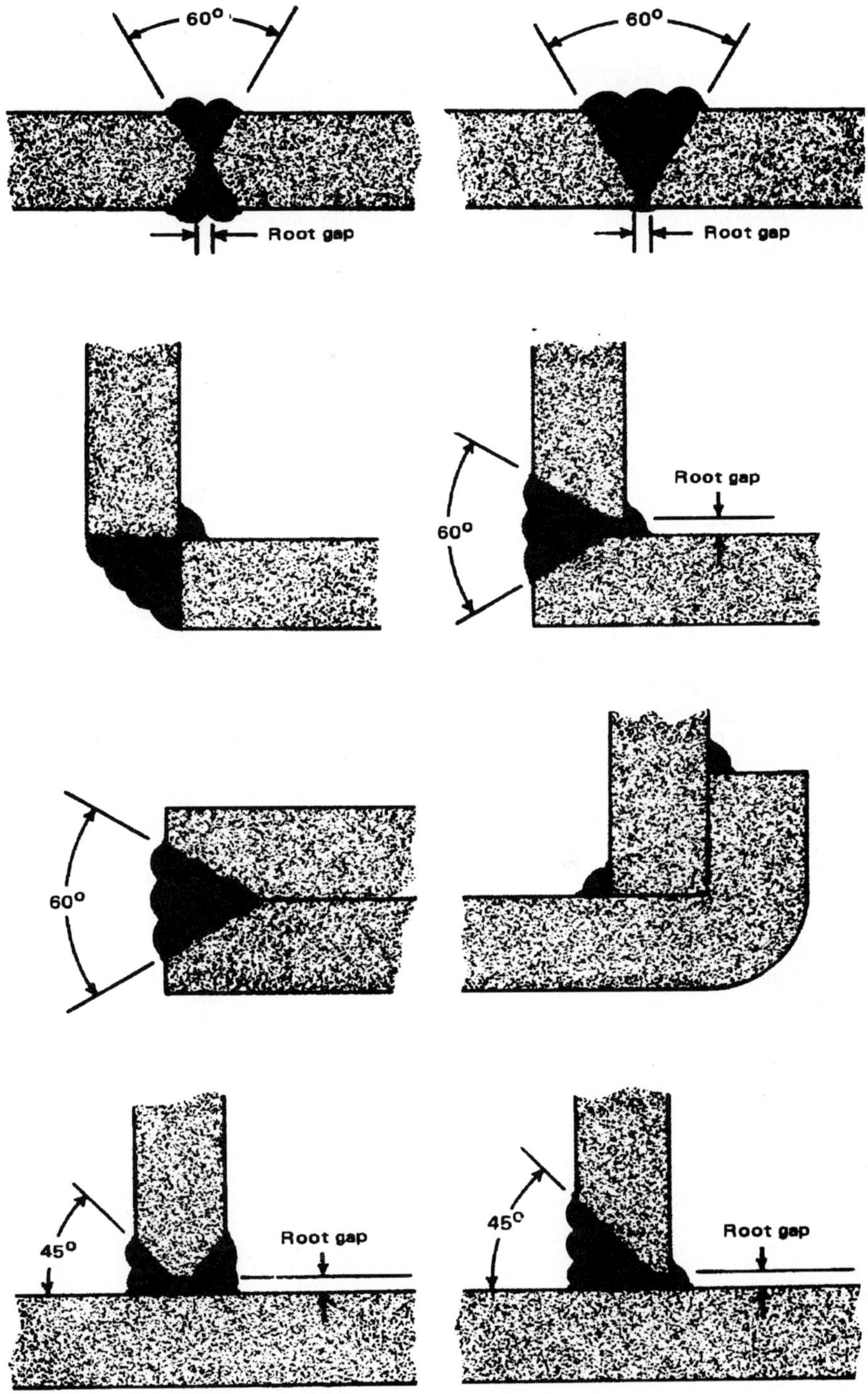

FIGURE 6.9.24 Preferred joint designs for hot-gas welding of plastics.

DIMENSIONAL FACTORS

Some factors that affect the accuracy of finished dimensions of welded plastic assemblies are (1) variations in the amount of melting at the joint, (2) the amount of pressure used to hold parts during welding, (3) dimensional variations of the parts, (4) shrinkage of plastic in the joint on cooling, (5) fixture support of parts and the assembly, and (6) the joint design.

Most molded assemblies do not involve critical dimensions across the welded joint. Often the accuracy of the molded parts that make up the assembly is close enough that any dimensional variations introduced in welding are inconsequential. Occasionally, however, welded assemblies such as film cartridges and recording-tape cassettes do have fairly close dimensional tolerances. In these cases, stops and limit switches are built into the welding fixtures to control final dimensions.

TOLERANCE RECOMMENDATIONS

Dimensions across the joints of small ultrasonically welded assemblies have been held as close as ±0.06 mm (0.0025 in). However, for consistent results under production conditions ±0.13 mm (0.005 in) should be the minimum tolerance specified, and, if possible, ±0.25 mm (0.010 in) should be specified for even carefully fixtured assembly dimensions. For across-the-weld dimensions not limited by fixture stops, use ±0.5 mm (0.020 in).

Tolerance for spin, vibration, and hot-plate-weldment dimensions across the weld should be ±0.25 mm (0.010 in) if fixture-controlled and ±0.8 m ($\frac{1}{32}$ in) if not. The horizontal alignment of vibration-welded parts to one another can normally be held to within ±0.25 mm (0.010 in) if parts are gripped tightly in the fixture and there is no loss of relative motion.

Induction-welded joints in circular and similar parts can be designed to be controlled by lips or other stops on the mating part, and across-the-weld dimensional tolerances can be based purely on the tolerances of the component parts. As a rule of thumb, hot-gas-welded assemblies can use tolerances comparable to those for arc-welded-metal assemblies of the same size.

CHAPTER 6.10
RUBBER PARTS

John G. Sommer
Research Division
GenCorp
Akron, Ohio

PROCESSES

Rubber parts are formed to the desired shape mainly by molding and extruding, with cutting and grinding being used less frequently. Most rubber products are vulcanized with sulfur at high temperatures during molding or after extruding. Thermoplastic rubber, in contrast to vulcanized rubber, has useful rubber properties only after sufficient cooling below molding or extruding temperatures.

As received by the fabricator, most rubber is a very-high-viscosity, or doughlike, material in bale form (millable rubber). Eight or more different additives, including fillers, plasticizers, vulcanizing agents, etc., are typically incorporated into millable rubber to impart the desired end-use properties. These additives are generally incorporated into the rubber by using an internal mixer, followed by forming the compound into sheets on a two-roll mill. Considerable energy is required in shearing the rubber in these steps. Smaller portions of the sheet are often used for compression and transfer molding, while long strips or pellets cut from the sheet are used to feed extruders and injection-molding machines. Injection molding is done at high temperatures, typically 180°C (360°F), using high injection pressures, typically 138,000 kPa (20,000 lbf/in^2), and in shorter cycles (typically 3 min for a single-station machine) compared with compression and transfer molding. With compression molding, a rubber blank is placed directly in an open-mold cavity. With transfer and injection molding, rubber is forced through transfer ports or channels into a closed mold, providing greater control of closure dimensions. (See Fig. 6.1.1 in Chap. 6.1 for an illustration of compression molding.)

Rubber parts are also prepared by techniques such as dipping forms in rubber latex or calendering rubber onto fabric or cord. Other techniques include the die cutting of vulcanized sheets that were formed by molding, calendering, or extruding and the forming of circular parts (washers) by cutting rubber tubes on a lathe.

Products manufactured from thermoplastic rubber are typically extruded or injection-molded from granules as received, but other materials may be added before extruding or molding.

Polyurethane parts are prepared from raw materials that are often low-viscosity liquids at room temperature. These materials are mixed at low pressures by using mechanical stirring or at high pressures by using reaction-injection molding (RIM). With the RIM method, streams of reactive components are impinged on one another by using pressures of about 13,800 kPa (2000 lbf/in^2), after which the mixed material is molded at low pressures, generally less than 690 kPa (100 lbf/in^2).

TYPICAL CHARACTERISTICS AND APPLICATIONS

An essential element of rubber behavior is that it retracts forcibly and quickly from large deformations. This property makes it extremely useful in applications that require sealing or shock absorption.

Extrusion is favored for producing long rubber parts of uniform cross section, and the cross section may be simple, like tubing, or highly complex, like door seals for automobiles (see Fig. 6.10.1). Lengths of extruded weatherstrip are often joined by splicing to produce an endless unit of uniform cross section. Lengths also can be joined in a separate molding operation, and this allows a wide range of cross sections at the junction or a different cross section at the ends of the extrusions. These procedures permit fabricating seals with long perimeters without requiring molds having a large surface area.

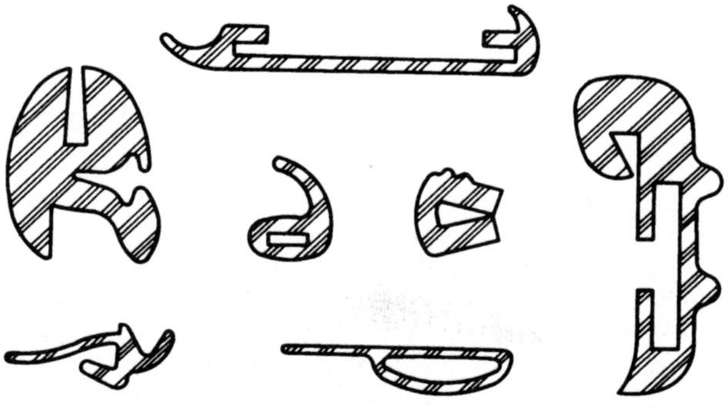

FIGURE 6.10.1 Typical cross sections of extruded rubber. (*General Motors Drafting Standards. Courtesy General Motors Corp.*)

Rubber-metal composites are often used. These include hollow rubber cylinders with an inner and an outer metal sleeve used in automobile suspension systems and other composites, as shown in Fig. 6.10.2. Bridge bearings (see Fig. 6.10.3) consist of alternate layers of rubber and metal to provide extremely high stiffness in compression but low stiffness in shear. Attaining this unique stiffness ratio justifies the extra care necessary to locate the metal plates properly in the assembly.

Rubber is often combined with cord or fabric to form composites that possess both flexibility and high strength. Typical examples are V belts and hose (see Fig. 6.10.4) and tires.

The wall thickness of rubber parts varies considerably. That of an injection-molded automotive brake diaphragm (see Fig. 6.10.5) is about 1.3 mm (0.050 in), while the

FIGURE 6.10.2 Suspension system parts and other rubber-metal composites. (*Courtesy GenCorp.*)

FIGURE 6.10.3 Section cut from a bridge bearing showing metal plates bonded to rubber. (*Courtesy GenCorp.*)

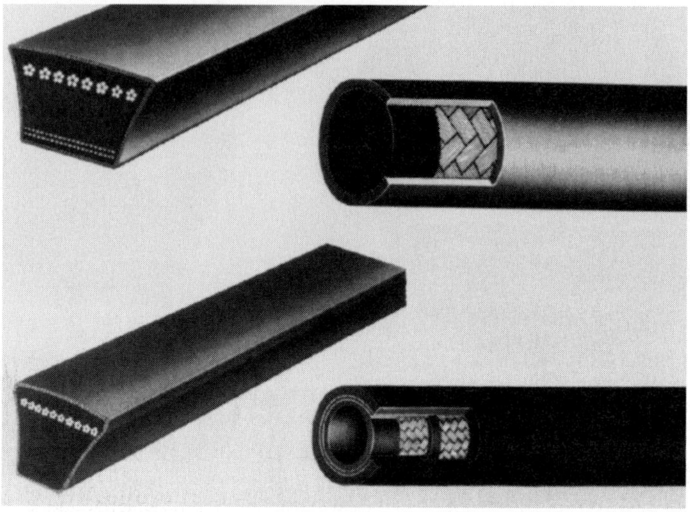

FIGURE 6.10.4 V belts (*left*) and reinforced-rubber hose (*right*), showing cross sections. (*Courtesy Gates Rubber Company.*)

FIGURE 6.10.5 Injection-molded automotive brake diaphragm. (*Courtesy GenCorp.*)

thickness of rubber bridge bearings (see Fig. 6.10.3 (may be 180 mm (7 in). As thickness increases, longer time is generally needed to cross-link rubber or to cool thermoplastic rubber before removal from the mold.

Rubber products prepared by latex dipping are typically hollow and have extremely thin walls [about 0.25 mm (0.010 in)]. Examples of latex-dipped products are shown in Fig. 6.10.6.

Gaskets are frequently prepared by die-cutting rubber sheets, while rubber tubes may be cut on a mandrel in a lathe to form washers.

O rings (see Fig. 6.10.7) are molded from a variety of rubber compounds to provide resistance to swelling in different fluids. Because they are used in critical sealing applications, tolerances are much narrower for O rings and other precision parts than for most rubber products.

Polyurethane parts manufactured by RIM typically have a dense skin and a cellular core and are usually much stiffer than conventional rubber parts. Flexible front ends for automobiles are made from RIM polyurethane.

Most rubber parts are black because of the carbon-black filler that is incorporated to impart the desired physical properties. However, colored products can be made from many types of rubber. Because of inventory considerations, it sometimes may be desirable to color a part by painting. Painted-rubber surfaces having excellent appear-

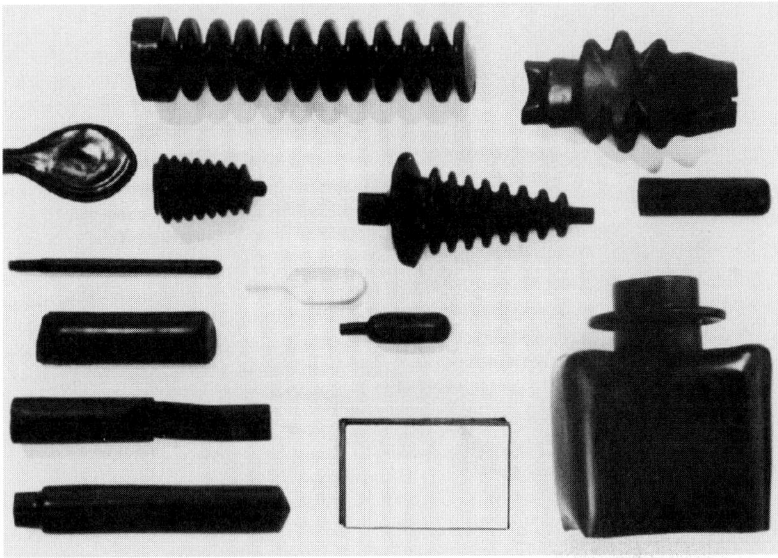

FIGURE 6.10.6 Products prepared by latex dipping. (*Courtesy Oak Rubber Company.*)

FIGURE 6.10.7 O rings and other precision parts. (*Courtesy Precision Rubber Products Corporation.*)

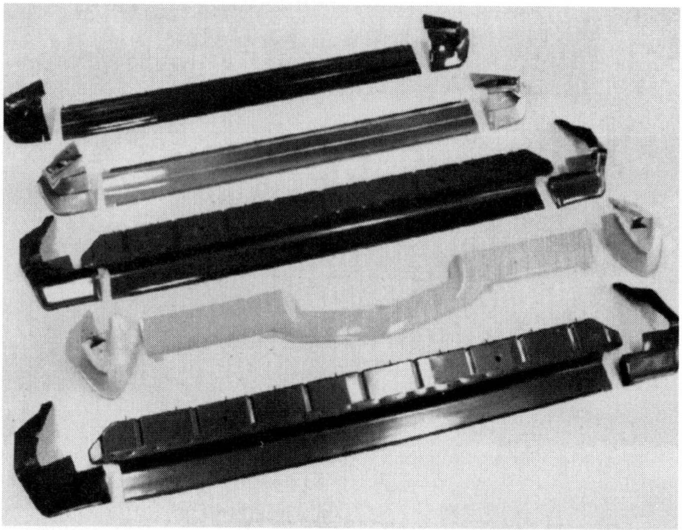

FIGURE 6.10.8 Exterior automotive parts. (*Courtesy GenCorp.*)

ance are shown in Fig. 6.10.8. The proper rubber composition, highly polished molds, and judicious selection of mold release are necessary for decorative rubber.

The surface of cellular rubber is affected by several factors. Sponge sheets or strips are often molded against fabric to provide venting during sponge expansion, and this causes a fabric impression on the sponge. Open cells will occur on parts cut from this sheet or strip. A solid-rubber skin is sometimes applied to sponge to give a water-resistant surface.

ECONOMIC PRODUCTION QUANTITIES

Fabrication costs for an extrusion die for a simple profile such as a dense round section are low, and shorter runs are therefore economical relative to complex profiles (as in Fig. 6.10.1). For complex profiles to be economical, longer runs are required because the extrusion die must be modified frequently to obtain specified dimensions. Although extrusion rates vary considerably, about 15 m/min (50 ft/min) is typical.

Compression molding is generally used for short runs and injection molding for long runs (typically 50,000 identical parts for a single-mold machine) so that the injection-mold cost can be reasonably amortized over the quantity manufactured.

GENERAL DESIGN CONSIDERATIONS

Although rubber can be molded and extruded into complex shapes, it is best to keep parts as simple as possible. Avoid projections, overhangs, undercuts, and complex shapes as much as possible because they increase mold and trimming die costs, are difficult and costly to mold, and raise the reject rate. Designs should be reviewed early with the rubber manufacturer, who may be able to suggest alternatives that reduce costs or improve product performance.

The parting line of the mold should be located to minimize air entrapment, minimize mold-fabricating costs, and provide for easy part removal. Ideally, the parting line should be in a plane.

Because of the flexible nature of rubber, draft is frequently unnecessary, and undercuts generally can be used without the need for sliding cores or split molds. However, draft is desirable for some soft thermoplastic rubber, is recommended for harder compositions, and is needed for hard rubber. Increasing draft and decreasing the depth of undercuts make removal of parts from the mold easier.

Flat and angular surfaces are more economical than curved surfaces because of easier mold fabrication. However, for rubber parts that will be strained in service, use generous radii for both internal and external corners of rubber. The corners of metal or other rigid materials also should be provided with a radius where they contact or are encapsulated with rubber that will be strained in service. Uniform wall thickness minimizes sink marks and usually provides for both shorter molding cycles and more uniform cross-linking throughout the part.

The appearance of a rubber part is affected both by the smoothness of the mold surface and by the nature of the rubber composition. Ingredients in the composition, wax, for example, may later migrate and dull the surface of an originally glossy rubber part. However, by using special rubber compositions and highly polished molds, long-lasting, smooth rubber surfaces can be made. To minimize costs, the smoothness specified should be no more critical than needed for the intended application.

Shrinkage occurs in both thermoplastic and cross-linked rubber after molding because the thermal coefficient of expansion is higher for rubber than for the metal molds that are typically used. Because shrinkage varies with different types of rubber, it is an important factor in mold design.

SUITABLE MATERIALS

Most rubber products are prepared from special compositions based on millable rubber and are designed to meet well-defined specifications. The American Society for Testing and Materials (ASTM) and the Society of Automotive Engineers, Inc. (SAE) provide suitable specifications, as described in ASTM D2000 and SAE J200. The manufacturer of rubber products often has compositions on hand meeting these specifications, obviating the need for additional material evaluation, process development, and increased costs.

Most rubber products are manufactured from very-high-viscosity rubber, and it is generally cross-linked. Cross-linking ties the long rubber molecules together, thus improving properties. Products classified as *hard rubber* do not show conventional rubber behavior because they are so highly cross-linked that they are virtually inextensible. Design of hard-rubber parts is more like that for thermosetting plastics (see Chap. 6.1).

Thermoplastic rubber is generally extruded or molded as received, but it may be modified with fillers and other additives before being fabricated into products.

Some advantages and disadvantages of different types of rubber used in relatively large volume are shown in Table 6.10.1. Building tack, the ability of un-cross-linked

TABLE 6.10.1 Advantages and Disadvantages of Some Types of Rubber

Rubber	Advantages	Disadvantages	Typical applications
Natural rubber (NR)	Building tack, resilience, and flex resistance	Reversion at high molding temperature	Tires, engine mounts
Styrene butadiene rubber (SBR)	Abrasion resistance	Poor ozone resistance	Tires, general molded goods
Ethylene propylene diene monomer (EPDM)	Good ozone resistance	Poor hot-tear resistance	Door and window seals, wire insulation
Nitrile butadiene rubber (NBR) or nitrile	Good solvent resistance	Poor building tack	O rings and hose
Thermoplastic rubber	Short injection-molding cycle (thin parts)	Poor creep characteristics	Shoe soles, wire insulation
Polyurethane	Short molding cycle (RIM) and low molding pressure	Adhesion to mold	Cushioning, rolls, exterior automotive parts
Isobutylene isoprene rubber (IIR) or butyl	Low air permeation in finished parts	Voids caused by trapped air during molding	Inner tubes body mounts for automobiles
Chloroprene rubber (CR) or neoprene	Moderate solvent resistance	Sticking during processing and premature cross-linking (scorch) with some types	Hose tubes and covers, V belts

rubber to stick to itself over relatively long time periods, is very favorable for natural rubber (NR). It is a desirable property for fabricating some products such as tires. At very high molding temperatures (used to shorten curing time), NR may revert and become sticky on the surface. Styrene butadiene rubber (SBR) is a general-purpose rubber widely used for tires and molded goods. Its inherently poor ozone resistance can be significantly improved by using special additives (antiozonants). The poor tear resistance of ethylene propylene diene monomer rubber (EPDM) at high temperatures sometimes causes demolding problems, especially for moldings having a complex shape. Mold releases generally must be used to prevent adhesion of polyurethanes to molds. These releases often build up on the mold, necessitating periodic cleaning and downtime. Sticking to mills and premature cross-linking or scorch are a problem with some chloroprene rubber (CR) compositions. Isoprene rubber (IR) has properties generally similar to NR. Some types of butadiene rubber (BR) have poor milling characteristics, and they are frequently blended with other types of rubber, such as SBR, for use in tires.

Certain specialty rubbers, for example, fluorocarbon rubber, are used to prepare automotive seals because of their favorable fluid resistance and excellent high-temperature resistance. Problems with mold sticking, mold corrosion, and difficult processing behavior are associated with some of these specialty rubbers.

DETAILED DESIGN RECOMMENDATIONS

Gating

Consideration must be given, in rubber parts that will be injection- or transfer-molded, to the location, size, and shape of the gate. (The *gate* is the restricted opening located between the runner and the mold cavity.) The rubber in the gate region should permit both easy stripping of parts from cavities and easy trimming. The gate should be extra large in the case of RIM.

Refer to Fig. 6.2.6 in Chap. 6.2 for gating arrangements that are used for injection-molding thermoplastic parts. These are generally applicable to molding millable rubber by transfer and injection. Gate selection is affected by factors such as the viscosity and scorch time of the compound being molded. Typically, gates having a larger cross section are needed for compounds having higher viscosity and shorter scorch times.

Holes

1. Holes in rubber are generally easiest to form and most economical to produce during molding. Drilling holes in cured rubber by conventional means is difficult because of its flexible nature. However, holes as small as 0.13 mm (0.005 in) can be drilled in rubber by using lasers.
2. Holes should be as shallow and as wide as possible, consistent with functional requirements. (See Fig. 6.10.9.)
3. Through holes are preferable to blind holes because the pins forming them can be anchored at both ends and are therefore less susceptible to deflection during molding.
4. Through holes can be molded as small as 0.8 mm ($\frac{1}{32}$ in) in diameter while at the same time being 16 mm ($\frac{5}{8}$ in) deep.
5. Avoid deep blind holes having a narrow diameter because the mold pin forming them tends to bend during molding. Instead, form blind holes from tapered or stepped pins. (See Fig. 6.10.10.)
6. Holes having their axes in the mold-closing direction are preferred to avoid the need for retractable core pins.

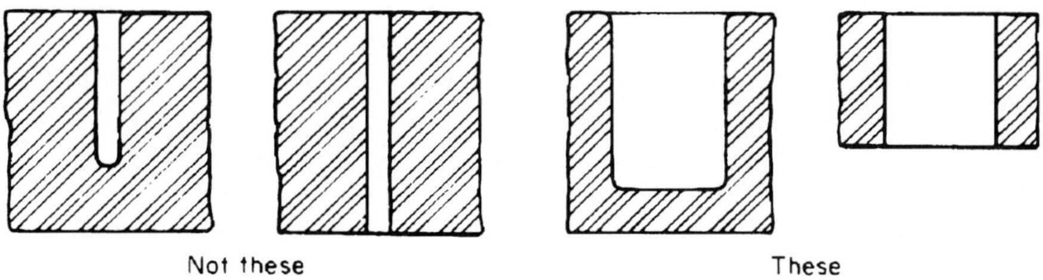

FIGURE 6.10.9 Shallow, wide holes are preferred.

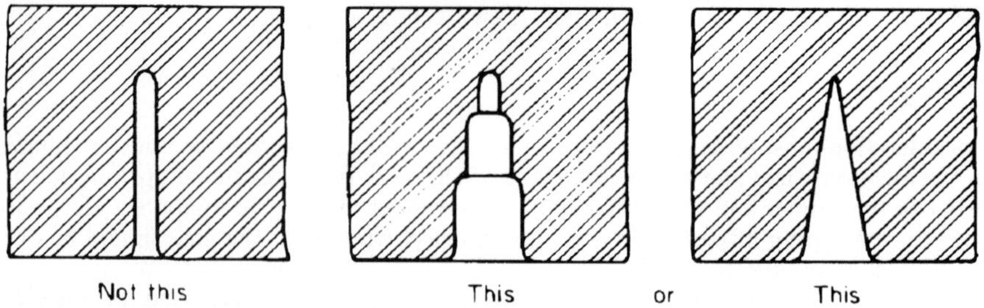

FIGURE 6.10.10 Use of stepped or tapered pins to form deep blind holes.

7. Incorporate sufficient wall thickness around holes used for mounting to minimize the danger of the rubber part tearing from the hole.
8. Keep hole-to-hole and hold-to-edge spacing adequate to prevent tearing, or a minimum of one hole diameter. (See Fig. 6.10.11.)

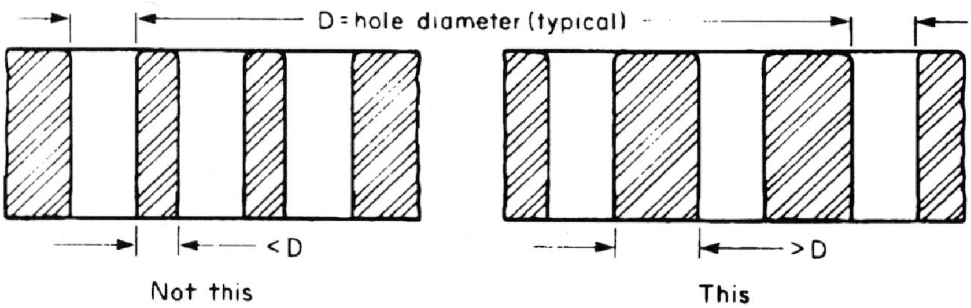

FIGURE 6.10.11 Spacing between holes, and between holes and edges, should be at least one hole diameter to minimize tearing of rubber.

Wall Thickness

Keep the wall thickness of parts as constant as is feasible for the intended application. Constant wall thickness in cross-linked rubber results in more uniform cross-linking and therefore in more uniform properties throughout the product. It also minimizes distortion.

Sink marks (see Fig. 6.2.3), caused by heavy wall thicknesses, may be more pronounced with some millable and thermoplastic compositions than with RIM polyurethane.

Undercuts

Undercuts should be avoided if possible. However, they can be incorporated in low- and medium-hardness rubber without using split molds or split cores owing to the flexible nature of rubber.

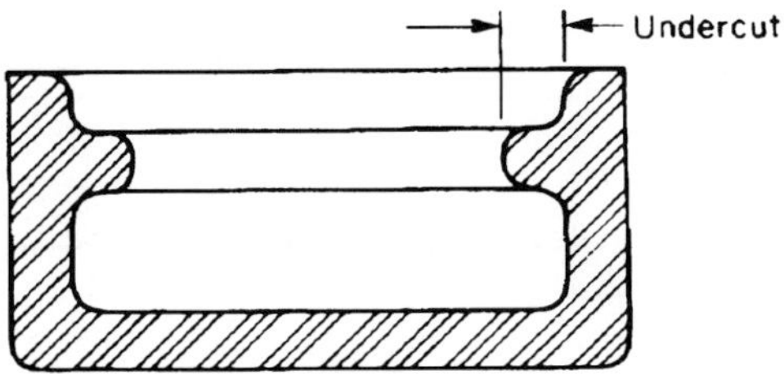

FIGURE 6.10.12 Avoid undercuts that make molding more difficult.

1. If undercuts are required in hard rubber, they are best machined in the part.
2. An internal undercut like the one shown in Fig. 6.10.12 should be avoided, if possible, because it causes difficulty during demolding and increases production costs.
3. External undercuts are satisfactory if they are not so deep that they cause demolding problems. (See Fig. 6.10.13.) An undercut with a radius aids both demolding and removal of trapped air. Softer rubbers can tolerate larger undercuts.

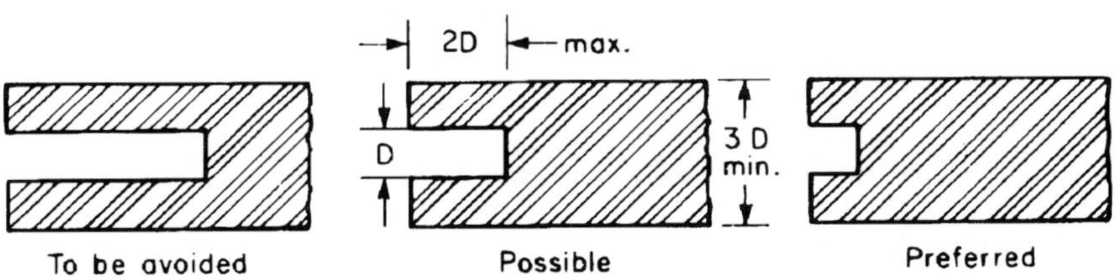

FIGURE 6.10.13 Undercuts with round corners and limited depth are preferable. (*Based on Roger W. Bolz,* Production Processes: The Productivity Handbook, *Conquest Publications, Winston-Salem, N.C., 1977.*)

Screw Threads

Screw threads are rarely molded directly on rubber. An exception is hard rubber, which has greater holding strength. Recommendations for female and male threads in hard rubber are generally the same as those for other thermosetting plastics and are given in Chap. 6.1.

For softer rubber, threaded-metal inserts are often used for attachment purposes. If these inserts are located improperly, rubber flows into the threads during molding, and this requires trimming. Locate threaded inserts away from the molded rubber to avoid this, as shown in Chap. 6.2.

Inserts

1. Stagger the location of nonflush threaded inserts to keep more uniform rubber thickness and therefore lower stress concentration, as shown in Fig. 6.10.14. Use inserts with flush heads to reduce stress concentrations further.

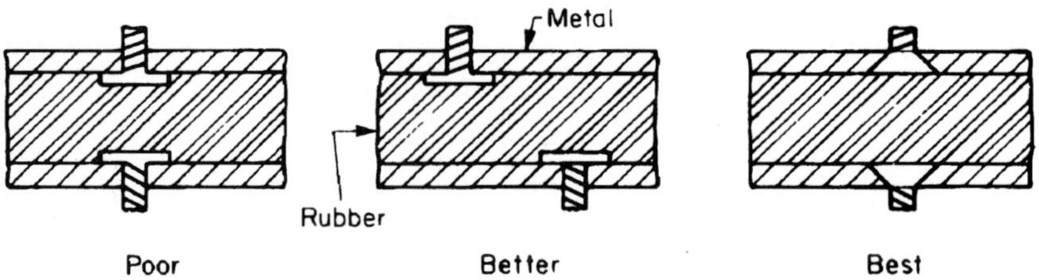

FIGURE 6.10.14 If it is not feasible to separate fasteners from the molded rubber, they should be placed so as to keep the rubber thickness as uniform as possible to avoid stress concentrations. (*From A. R. Payne and J. R. Scott,* Engineering Design with Rubber, *Interscience, New York, 1960.*)

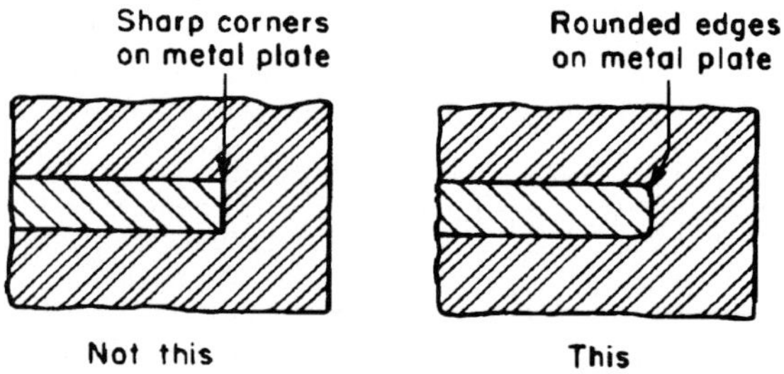

FIGURE 6.10.15 Radius metal edges that contact rubber.

2. The design should permit gate locations that avoid flowing rubber tangential to adhesive-coated metal during injection molding. This prevents scouring adhesive from the metal, causing loss of adhesion.
3. Avoid inserts with sharp edges, as shown in Fig. 6.10.15, to avoid cutting the rubber.
4. The embedded inserts shown in Figs. 6.10.3 and 6.10.16 can be used when they are needed to impart specific properties. The inserts shown in Fig. 6.10.16 are easier to register in the rubber than those in Fig. 6.10.3 because rubber in the latter part completely surrounds the metal insert. Figure 6.10.16*a* shows a somewhat less sat-

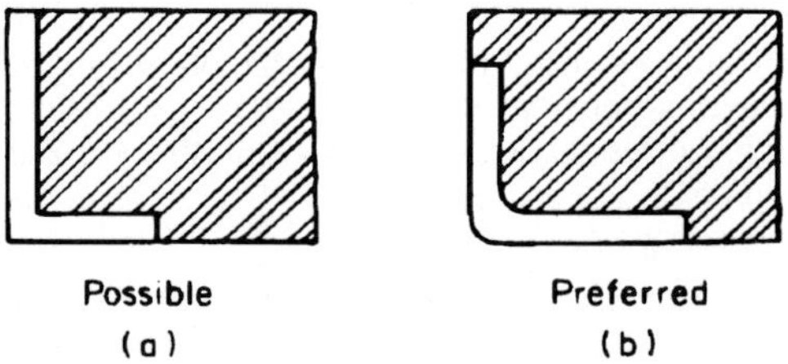

FIGURE 6.10.16 Angle inserts molded in rubber. (*From Roger W. Bolz,* Production Processes: The Productivity Handbook, *Conquest Publications, Winston-Salem, N.C., 1977.*)

isfactory approach than Fig. 6.10.16*b* because trimming at the upper-left corner where the metal edge and the molded rubber come together is more difficult. Loss of adhesion between metal and rubber is also more apt to occur with the design in Fig. 6.10.16*a* because of the sharp corner.

Draft

The need for draft in molded-rubber parts varies with both the part design and the nature of the rubber. For many parts having hardness below about 90 Shore A hardness, no draft is needed.

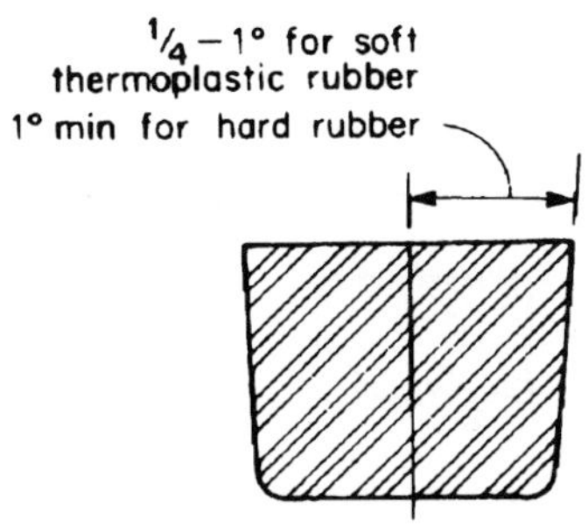

FIGURE 6.10.17 Provide draft to aid demolding.

1. An exception is some soft thermoplastic rubber, for which a draft angle of 1/4 to 1° perpendicular to the parting line is recommended. (See Fig. 6.10.17.)
2. For hard rubber, provide at least 1° draft. (See Fig. 6.10.17.)
3. Provide draft in metal attachments to facilitate demolding.

Corners: Radii and Fillets

Radii and fillets are generally desired on the corners of rubber parts to facilitate flow of the rubber during molding and to minimize stress concentrations in the part and the mold.

1. Provide a fillet of at least 0.8-mm (1/32-in) radius in rubber where it is bonded to metal. (See Fig. 6.10.18*a*.) If a fillet is not possible, extend the rubber beyond the edge of the metal. (See Fig. 6.10.18*b*.) Both designs minimize stress concentrations at the rubber-metal interface.
2. If the parting line is not involved, rounded corners are preferred because they minimize air trapping and facilitate flow of rubber. (See Fig. 6.10.19.)
3. Lower-cost parts often can be made by cutting an extruded sleeve to length instead of molding the parts. The square corner caused by cutting off is satisfactory for some applications. (See Fig. 6.10.20.)
4. Only one mold half need be machined if sharp corners rather than chamfers or radii are used at the parting line. The part shown in Fig. 6.10.21*b* is preferred over Fig. 6.10.21*a* because both mold design and flash removal are simpler for *b*.
5. The minimum radius for forming hose or tubing on a mandrel or core depends on the wall thickness and outside diameter. The minimum radius should be at least 1.5 times the outside diameter, as shown in Fig. 6.10.22.
6. Provide generous radii in the walls of rubber bellows to minimize stress concentrations and to make demolding easier.

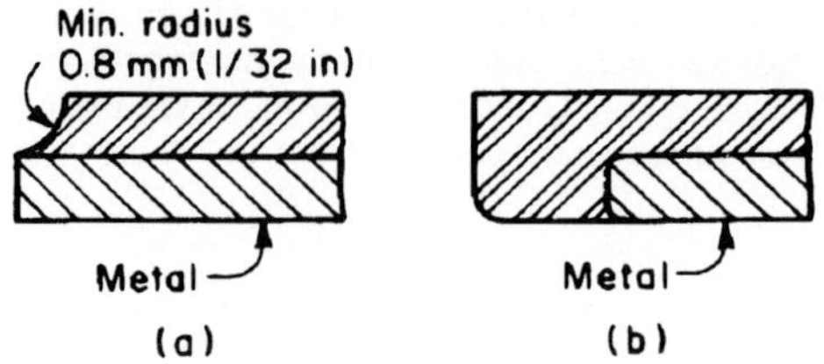

FIGURE 6.10.18 Provide radius in rubber (*from Roger W. Bolz,* Production Processes: The Productivity Handbook, *Conquest Publications, Winston-Salem, N.C., 1977.*) or metal (*from General Motors Drafting Standards. Courtesy General Motors Corp.*) to minimize stress concentrations. (*a*) Radius in rubber. (*b*) Rounded metal edges.

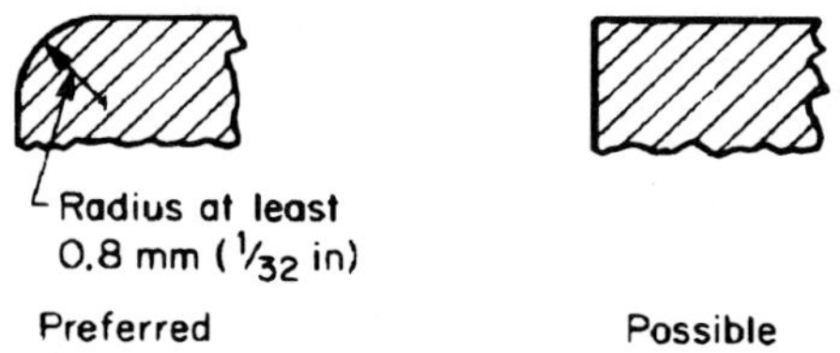

FIGURE 6.10.19 Rounded corners facilitate rubber flow and minimize trapped air. (*From Roger W. Bolz,* Production Processes: The Productivity Handbook, *Conquest Publications, Winston-Salem, N.C., 1977.*)

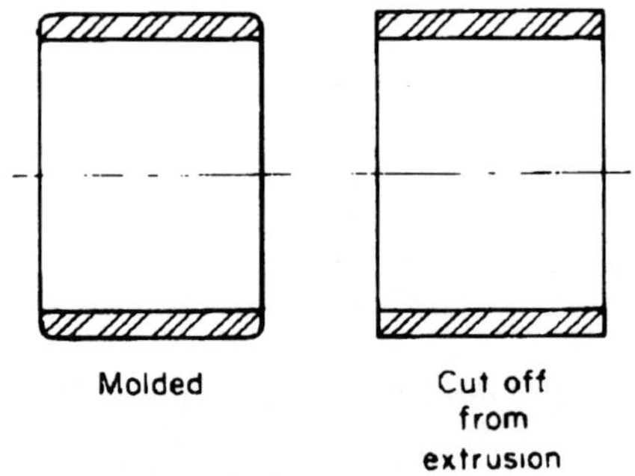

FIGURE 6.10.20 Square corners of cut extruded pieces are satisfactory for most applications.

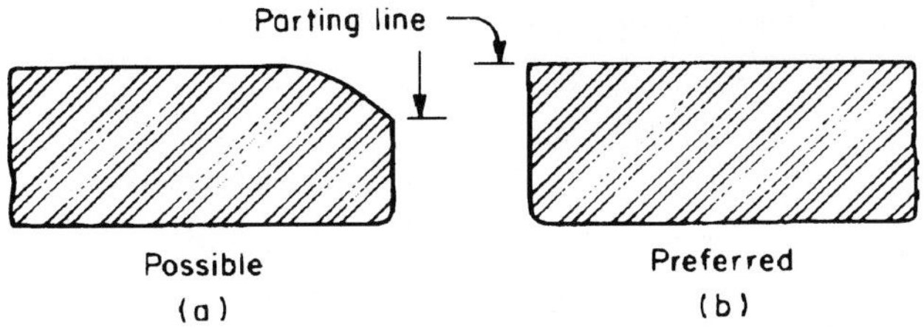

FIGURE 6.10.21 Only one mold plate is machined when square corners are used.

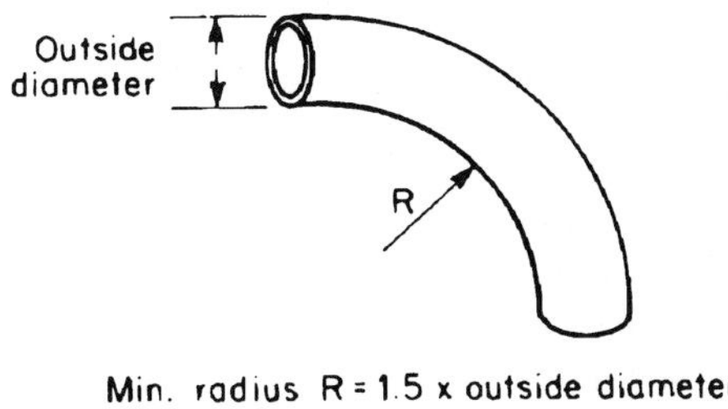

FIGURE 6.10.22 Minimum recommended radius for hose or tubing formed on a mandrel or core. (*From* RMA Handbook, *3d ed., Rubber Manufacturers Association, New York, 1970.*)

7. In forming dipped-latex products such as balloons, avoid sharp corners, as these cause weak points.

8. Polyurethane tends to build up in sharp corners of molds used for RIM and adversely affects part appearance. To minimize this tendency, provide radii of about 1.3 to 2.5 mm (0.050 to 0.100 in) at corners.

Flash

Flash tends to be formed at the edge of a molded-rubber part at the mold parting line. In compression molding, excess rubber flows into the flash groove. The amount of flash permissible depends on the function of the rubber part. Flash extension is the extension of flash beyond the part edge and should be designated on the part drawing as described in Table 6.10.2. For example, drawing designation T.000 mm means that no flash is permitted, while T.40 mm means that 0.40-mm (0.016-in) flash extension is permitted.

Typically, a mold is designed to produce thin flash or, for some precision molds, no flash. However, some molds are designed to produce flash that is thick enough to provide for removal of parts as a sheet (see Fig. 6.10.23), and the parts are later cut with a die from the sheet.

TABLE 6.10.2 Drawing Designations for Flash Extension for Molded-Rubber Parts

Drawing designation	Maximum flash permitted	Remarks
T.00 mm	None	No flash permitted on area designated. (Standard notation regarding other surfaces must accompany this notation.)
T.08 mm	0.08 mm (0.003 in)	This tolerance will normally require buffing, facing, grinding, or a similar operation.
T.40 mm	0.40 mm (0.016 in)	This tolerance will normally require precision die trimming, buffing, or extremely accurate trimming.
T.80 mm	0.80 mm (0.032 in)	This tolerance will normally necessitate die trimming, machine trimming, tumbling, or hand trimming.
T1.60 mm	1.60 mm (0.063 in)	This would be the normal tear-trim tolerance.
T2.35 mm	2.35 mm (0.093 in)	On large parts, this tolerance will normally require die trim, tear trim, or hand trim of some type.
T∞	No limit	No flash limitation.

Source: From *RMA Handbook,* 5th ed., Rubber Manufacturers Association, Washington, DC, 1992.

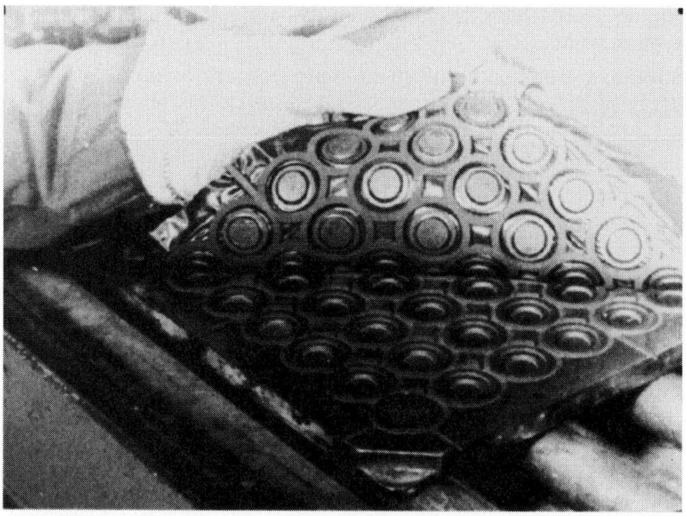

FIGURE 6.10.23 Removal of rubber parts as a sheet. (*Courtesy GenCorp.*)

Parting Line

A parting line is formed at the junction of the two halves of a curing mold. When possible, design parts so that the mold-parting line is in a plane.

Design edges so they have about 0.8-mm ($^1/_{32}$-in) minimum thickness at parting lines, as shown in Fig. 6.10.24. Molds providing featheredges on the rubber are more subject to damage, and trim cost for parts is higher.

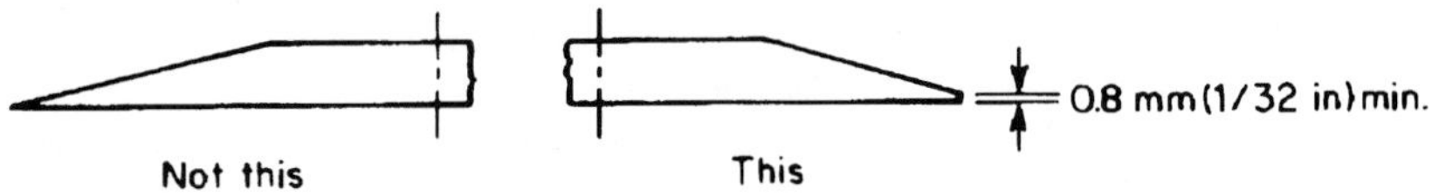

FIGURE 6.10.24 Avoid featheredges. (*From H. L. Boyce, "How to Avoid Common Pitfalls in Designing Rubber Parts,"* Materials Engineering, *October 1976, p. 86.*)

Venting

Venting at the final fill point avoids defects by releasing trapped air, especially for parts formed by high-rate injection molding, in which the air has only a short time to escape. In severe cases, the rubber can discolor from oxidation.

1. Typically, parting-line vents for injection molding might be about 0.002 in deep by 0.125 in wide for relatively high-viscosity millable compositions.
2. The depth of the vent should be decreased and the width increased for lower-viscosity millable compositions and some thermoplastic rubber.
3. Marks left on the molded part by the vent are generally acceptable because they are usually quite small.
4. Provide vents in molds for very-low-viscosity (castable) compositions to avoid trapped air at the top of the cavity.

Surface Finish

Although smooth, glossy surfaces on molded-rubber parts are feasible, they are more costly than more conventional commercial-grade finishes. The recommended commercial finish and other standard finishes of the RMA are described in Table 6.10.3. Mold-release agents may cause an oily surface on the molded part.

Lettering

Identifying letters, numbers, and surface decorations on molded-rubber parts are generally raised because it is preferable to form them from depressions in the mold. Identification generally needs to be raised no more than 0.5 mm (0.020 in) for legibility. As an alternative method, disposable aluminum tags or strips stamped with identification are sometimes used in the mold to form depressed lettering in rubber parts.

TABLE 6.10.3 Rubber Manufacturers Association Standard Drawing Designations

Drawing designation	Remarks
F-1	A smooth, polished, and even finish, completely free of tool marks, dents, nicks, and scratches, as produced from a highly polished steel mold. In areas where F-1 is specified, the mold will be polished to a surface finish of 0.25 μm (10 μin) or better.
F-2	An even finish as produced from a polished steel mold. In areas where F-2 is specified, the mold will be polished to a surface finish of 0.80 μm (32 μin) or better but with very small tool marks not polished out.
F-3	Surfaces of the mold will conform to good machine-shop practice, and no microinch finish will be specified. This is "commercial finish."
F-4	Satin finish as produced from water honing.

Source: From *RMA Handbook,* 5th ed., Rubber Manufacturers Association, Washington, DC, 1992.

Shrinkage

Shrinkage, the difference between the dimensions of a mold cavity and the rubber product at room temperature, is an important factor to be kept in mind by the design engineer. For typical cross-linked rubbers (millable) containing normal filler content, the shrinkage is about 1.5 percent, but it may be as low as about 0.6 percent for some highly filled compounds. Shrinkage is as high as 4 percent for fluorocarbon rubber that is postcured in an oven at high temperature. Shrinkage varies for different thermoplastic rubbers, but it is as low as 1 percent or less for some types.

Shrinkage for RIM polyurethane is typically about 1 to 2 percent, but it can vary outside this range depending on compositional and processing factors.

For rubber compounds containing oriented fiber, such as glass, shrinkage is lower along the fiber axis. Higher amounts of filler or fiber increase hardness of a rubber compound and reduce the shrinkage. Shrinkage also may be nonuniform for rubber in rubber-metal composites. This effect is seen in products like a bridge bearing (see Fig. 6.10.3), in which shrinkage will be considerably lower in the plane of the metal plates than perpendicular to the plates.

Flatness and Distortion

As a result of shrinkage and other factors, exposed rubber surfaces are seldom truly flat, particularly if metal reinforcing members are used. Metals have a lower coefficient of thermal expansion than rubber, and this causes distortion in the composite part. Figures 6.10.25 and 6.10.26 illustrate two examples.

If a part's surface is to be ground flat, provide sufficient rubber height for grinding. A soft or unfilled rubber composition will generally require more height for grinding than a harder composition because of the higher shrinkage of the soft rubber.

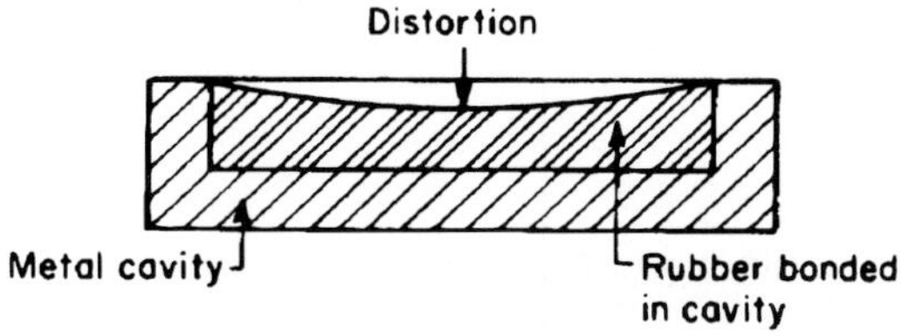

FIGURE 6.10.25 Distortion caused by bonding rubber in a metal cavity. (*From* RMA Handbook, *3d ed., Rubber Manufacturers Association, New York, 1970.*)

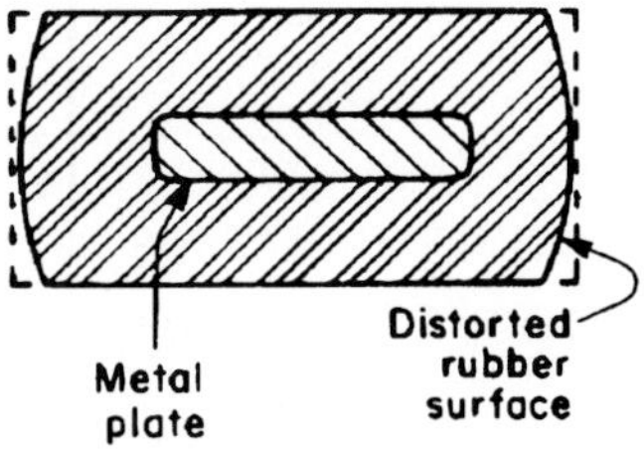

FIGURE 6.10.26 Distortion caused by a metal insert.

DIMENSIONAL FACTORS

Rubber parts with very close tolerances can be produced, but their costs will be significantly higher because of higher tool costs, more stringent process controls, and higher scrap rates. Increasing the number of cavities per mold reduces both the cost per part and the ability to hold tolerances.

The tolerances that can be held on extruded and lathe-cut parts are affected by the hardness of the compound and other factors. Some rubber parts may not be dimensionally stable in an unsupported condition. Support for them may be necessary during manufacture and subsequent packaging. The dimensions of rubber parts may change after molding, but this change is generally minimal for rubber parts stored unstressed for 1 day or longer. Exceptions are some moisture-absorbing rubbers like polyurethane because dimensions may change with varying relative humidity. Dimensions of cellular rubber may not be as important as with dense rubber because cellular rubber is typically much more compliant.

TOLERANCES

Tolerances for fixed and closure dimensions of molded solid-rubber products are shown in Table 6.10.4. The tolerances of parts designated "A1, high precision" are so critical that errors in measurement may be large in relation to tolerances. Products requiring postcuring should not be included in this designation.

Tolerances for cross sections of extruded rubber are shown in Table 6.10.5, while those for cut length for normal extrusions are shown in Table 6.10.6. Normal extrusions are those which generally require only extruding, vulcanizing, and cutting to the desired length. The tolerances for the length of lathe-cut products are listed in Table 6.10.7.

The tolerances on thickness dimensions for die-cut, open-cell, or closed-cell molded cellular rubber are shown in Table 6.10.8. In die-cutting closed-cell parts thicker than ½ in, a dished effect occurs on the edges that may further affect tolerances.

Other dimensional factors of specific importance to certain rubber products are concentricity, parallelism, and squareness. Using a 75-mm- (3-in-) diameter wheel as

TABLE 6.10.4 Standard Tolerances for Molded Solid-Rubber Products

All tolerances are plus or minus.)

Drawing designation: Size	A1, high precision, mm (in)		A2, precision, mm (in)		A3, commercial, mm (in)		A4, noncritical, mm (in)	
	Fixed	Closure	Fixed	Closure	Fixed	Closure	Fixed	Closure
0 through 10 mm (0 through 0.40 in)	0.10 (0.004)	0.13 (0.005)	0.16 (0.006)	0.20 (0.008)	0.20 (0.008)	0.32 (0.013)	0.32 (0.013)	0.80 (0.032)
10 through 16 mm (0.40 through 0.63 in)	0.13 (0.005)	0.16 (0.006)	0.20 (0.008)	0.25 (0.010)	0.25 (0.010)	0.40 (0.016)	0.40 (0.016)	0.90 (0.036)
16 through 25 mm (0.63 through 1.00 in)	0.16 (0.006)	0.20 (0.008)	0.25 (0.010)	0.32 (0.013)	0.32 (0.013)	0.50 (0.020)	0.50 (0.020)	1.00 (0.040)
25 through 40 mm (1.00 through 1.60 in)	0.20 (0.008)	0.25 (0.010)	0.32 (0.013)	0.40 (0.016)	0.40 (0.016)	0.63 (0.025)	0.63 (0.025)	1.12 (0.045)
40 through 63 mm (1.60 through 2.50 in)	0.25 (0.010)	0.32 (0.013)	0.40 (0.016)	0.50 (0.020)	0.50 (0.020)	0.80 (0.032)	0.80 (0.032)	1.25 (0.050)
63 through 100 mm (2.50 through 4.00 in)	0.32 (0.013)	0.40 (0.016)	0.50 (0.020)	0.63 (0.025)	0.63 (0.025)	1.00 (0.040)	1.00 (0.040)	1.40 (0.056)
100 through 160 mm (4.00 through 6.30 in)	0.40 (0.016)	0.50 (0.020)	0.63 (0.025)	0.80 (0.032)	0.80 (0.032)	1.25 (0.050)	1.25 (0.050)	1.60 (0.063)
160 mm and over (6.30 in and over)*			±0.4%	±0.5%	±0.5%	±0.8%	±0.8%	±1.0%

*To find fixed tolerances for part sizes greater than 160 mm or 6.30 in, multiply size by the indicated factor.

Source: *RMA Handbook,* 5th ed., Rubber Manufacturers Association, Washington, DC, 1992.

TABLE 6.10.5 Tolerances for Cross Sections of Extruded Rubber

(All tolerances are plus or minus.)

	RMA class		
	1	2	3
	High precision	Precision	Commercial
	Drawing designation		
Dimension, mm (in)	E1	E2	E3
0 up to 1.5 mm (0 up to 0.060 in)	0.15 (0.006)	0.25 (0.010)	0.40 (0.015)
1.5 up to 2.5 mm (0.060 up to 0.100 in)	0.20 (0.008)	0.35 (0.014)	0.50 (0.020)
2.5 up to 4.0 mm (0.100 up to 0.160 in)	0.25 (0.010)	0.40 (0.016)	0.70 (0.027)
4.0 up to 6.3 mm (0.160 up to 0.250 in)	0.35 (0.014)	0.50 (0.020)	0.80 (0.031)
6.3 up to 10 mm (0.250 up to 0.400 in)	0.40 (0.016)	0.70 (0.027)	1.00 (0.039)
10 up to 16 mm (0.400 up to 0.630 in)	0.50 (0.020)	0.80 (0.031)	1.30 (0.050)
16 up to 25 mm (0.630 up to 1.000 in)	0.70 (0.027)	1.00 (0.039)	1.60 (0.060)
25 up to 40 mm (1.000 up to 1.600 in)	0.80 (0.031)	1.30 (0.050)	2.00 (0.080)
40 up to 63 mm (1.600 up to 2.500 in)	1.00 (0.039)	1.60 (0.060)	2.50 (0.100)
63 up to 100 mm (2.500 up to 4.000 in)	1.30 (0.050)	2.00 (0.080)	3.20 (0.130)

Note: Tolerances for dimensions greater than 100 mm should be negotiated between designer and supplier. In general, softer materials and those requiring a postcure need greater tolerance.

Source: Data taken from *RMA Handbook,* 5th ed., Rubber Manufacturers Association, Washington, DC, 1992, and rounded off in some cases.

TABLE 6.10.6 Standard Tolerances for Cut Length for Normal Extrusions

(All tolerances are plus or minus.)

	RMA class		
	1	2	3
	Precision	Commercial	Noncritical
	Drawing designation		
Length, mm (in)	L1, mm (in)	L2, mm (in)	L3, mm (in)
0 up to 40 mm (0 up to 1.6 in)	0.7 (0.030)	1.0 (0.040)	1.6 (0.060)
40 up to 63 mm (1.6 up to 2.5 in)	0.8 (0.030)	1.3 (0.050)	2.0 (0.080)
63 up to 100 mm (2.5 up to 4 in)	1.0 (0.040)	1.6 (0.060)	2.5 (0.100)
100 up to 160 mm (4 up to 6.3 in)	1.3 (0.050)	2.0 (0.080)	3.2 (0.130)
160 up to 250 mm (6.3 up to 10 in)	1.6 (0.060)	2.5 (0.100)	4.0 (0.160)
250 up to 400 mm (10 up to 16 in)	2.0 (0.080)	3.2 (0.130)	5.0 (0.200)
400 up to 630 mm (16 up to 25 in)	2.5 (0.100)	4.0 (0.160)	6.3 (0.250)
630 up to 1000 mm (25 up to 40 in)	3.2 (0.130)	5.0 (0.200)	10.0 (0.400)
1000 up to 1600 mm (40 up to 63 in)	4.0 (0.160)	6.3 (0.250)	12.5 (0.500)
1600 up to 2500 mm (63 up to 100 in)	5.0 (0.200)	10.0 (0.400)	16.0 (0.630)
2500 up to 4000 mm (100 up to 160 in)	6.3 (0.250)	12.5 (0.500)	20.0 (0.800)
4000 (160 in)	0.16%	0.32%	0.50%

Note: Special consideration on tolerances should be given to both extremely soft and high tensile stocks.

Source: From *RMA Handboook,* 5th ed., Rubber Manufacturers Association, Washington, DC, 1992.

TABLE 6.10.7 Tolerances for Thickness (Length) of Lathe-Cut Products

(All tolerances are plus or minus.)

	Drawing designation		
	C1	C2	C3
	Precision	Commercial	Noncritical
	Tolerance		
Thickness, mm (in)	mm (in)	mm (in)	mm (in)
0 through 5 mm (0.00 through 0.200 in)	0.10 (0.004)	0.13 (0.005)	
5 through 10 mm (0.200 through 0.400 in)	0.15 (0.006)	0.25 (0.010)	0.38 (0.015)
10 through 13 mm (0.400 through 0.500 in)	0.18 (0.007)	0.38 (0.015)	0.76 (0.030)
13 through 15 mm (0.500 through 0.600 in)	0.20 (0.008)	0.51 (0.020)	1.1 (0.045)
15 through 18 mm (0.600 through 0.700 in)	0.23 (0.009)	0.64 (0.025)	1.5 (0.060)
18 through 25 mm (0.700 through 1.000 in)	0.25 (0.010)	0.76 (0.030)	1.9 (0.075)

Source: Data taken from *RMA Handbook,* 5th ed., Rubber Manufacturers Association, Washington, DC, 1992, and rounded off in some cases.

TABLE 6.10.8 Tolerances on Thickness Dimensions of Open-Cell Sponge, Die-Cut Sheet or Strip, and Open- or Closed-Cell Molded Cellular Rubber

(All tolerances are plus or minus.)

	RMA class			
	1	2	3	4
	RMA drawing designation			
	ATH 1	ATH 2	ATH 3	ATH 4
	Tolerance			
Thickness, mm (in)	mm (in)	mm (in)	mm (in)	mm (in)
0 through 3.2 mm (0 through 0.125 in)	0.32 (0.013)	0.40 (0.016)	0.50 (0.020)	0.63 (0.025)
3.2 through 6.3 mm (0.125 through 0.25 in)	0.40 (0.016)	0.50 (0.020)	0.63 (0.025)	0.80 (0.032)
6.3 through 13 mm (0.25 through 0.50 in)	0.50 (0.020)	0.63 (0.025)	0.80 (0.032)	1.00 (0.040)
13 through 25 mm (0.50 through 1.00 in)	0.63 (0.025)	0.80 (0.032)	1.00 (0.040)	1.25 (0.050)
25 through 50 mm (1.00 through 2.00 in)	0.80 (0.032)	1.00 (0.040)	1.25 (0.050)	1.50 (0.055)
50 mm and over (2.00 in and over)	2%*	2.5%*	3%*	3.5%*

*To find tolerances for thicknesses greater than 50 mm or 2.00 in, multiply by the indicated factor.

Source: Data from *RMA Handbook,* 5th ed., Rubber Manufacturers Association, Washington, DC, 1992, and rounded off in some cases.

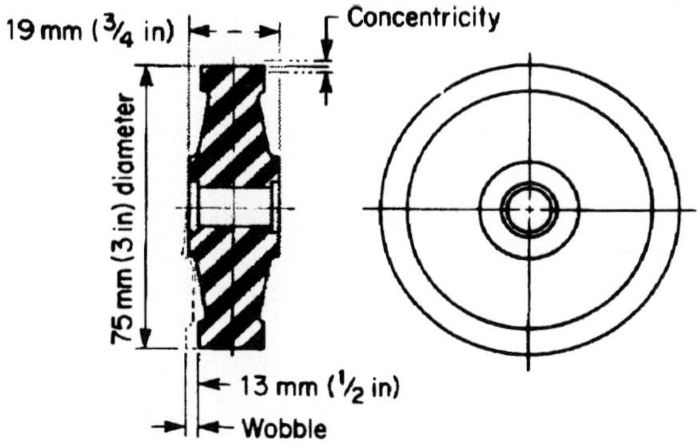

FIGURE 6.10.27 Concentricity and wobble in a rubber wheel. (*From* RMA Handbook, *3d ed., Rubber Manufacturers Association, New York, 1970.*)

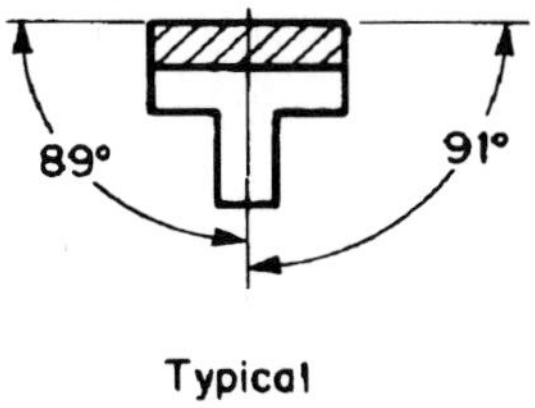

FIGURE 6.10.28 Squareness of a bonded-rubber product. (*From* RMA Handbook, *3d ed., Rubber Manufacturers Association, New York, 1970.*)

an example, both concentricity and wobble should typically be within 0.75 mm (0.030 in). (See Fig. 6.10.27.) Squareness for a bonded-rubber product typically can be held to a tolerance of 1°. (See Fig. 6.10.28.) Thicker rubber and subsequent grinding can be used to provide increased squareness.

When a rubber part is trimmed by hand, only the trimmed region is affected. In contrast, tumbling rubber parts to remove flash will dull exposed surfaces.

CHAPTER 6.11
CERAMIC AND GLASS PARTS

J. Gilbert Mohr
J. G. Mohr Co., Inc.
Maumee, Ohio

DEFINITIONS

Ceramics are defined as inorganic, nonmetallic materials and are described further in Chap. 2.4. They can be classified into the following groups:

1. *Whitewares.* These include, in addition to mechanical and electrical components, earthenware, china, tiles, and porcelain.
2. *Glass.* Glass is a mutual solution of fused, inorganic oxides cooled to a rigid condition without crystallization. It is made into a variety of hard, transparent objects.
3. *Refractories.* These include heat-resistant and insulating blocks, bricks, mortar, and fireclay.
4. *Structural-clay products.* They consist of bricks, tiles, and piping made from natural clays.
5. *Porcelain enamels.* These are ceramic coatings on cast-iron, steel, and other metal products.

MANUFACTURING PROCESSES

Ceramic Parts

To produce ceramic parts, refined powders of the basic raw materials are first thoroughly mixed with some water and small quantities of selected additives, normally metallic oxides that act as fluxing agents and inhibitors. Then the basic fabrication operation, such as pressing, extrusion, or casting, takes place. Depending on the shape and dimensions, machining or grinding of the formed part also may be involved.

The "green" ceramic part is then dried and fired at a high temperature for a specified period of time. This fuses the powders into a hard, dense, strong, and homogeneous material. A typical temperature range within which alumina and other common materials are fused is 1400 to 1800°C (2550 to 3250°F).

Pressing is the most common basic fabrication operation prior to firing. It is similar to compression molding or powder-metal pressing in that the material is compressed at high pressure into a mold cavity of the shape of the workpiece. The pressed part is then trimmed as necessary and dried.

In wet pressing, the mixture is quite moist and flows somewhat as it is pressed, similarly to the behavior of plastics being compression-molded. In dry pressing there is a minimum amount of moisture, and the ceramic powders behave very similarly to metal powders in the powder-metallurgy processes.

Many ceramic parts can be formed directly to the final shape, with allowances being made for shrinkage during firing. Often, however, turning, drilling, boring, threading, tapping, and other machining operations take place to meet some special requirement. Because of the highly abrasive nature of ceramic material, carbide tools are used. Grinding also may be employed. After firing, if dimensional tolerances are particularly close, further grinding and lapping can be performed with diamond abrasives.

Glaze may be added to the part to provide a smooth, glossy surface. It is applied sometimes before firing and sometimes afterward, followed by a second lower-temperature firing operation.

A more liquid mixture is used for casting and extrusion than for pressing. Casting is performed with plaster-of-paris molds that absorb water from the mixture, gradually building up a leathery cast that may be handled, refinished, and fired with or without a glaze.

Jiggering, a process often used for dish- or bowl-shaped parts, involves the use of a rotating form, usually of plaster, against which a putty-like clay mix is pressed with a clay knife. Separate pieces may be joined together before drying and firing.

Glass Parts

Glass components are produced from a hot, viscous, homogenized melt. They may be processed or formed by pressing, blowing, drawing, or rolling, after which the glass is cooled at a controlled rate to anneal it (remove residual stresses) prior to finishing. Figure 6.11.1 illustrates the pressing operation.

TYPICAL CHARACTERISTICS AND APPLICATIONS

Ceramic Parts

Ceramic parts are hard, extremely strong in compression, highly chemical- and corrosion-resistant, nonflammable, and suitable for use at extremely high operating temperatures. Ceramic whitewares generally have good thermal shock resistance and low thermal expansion. High modulus of elasticity and high radiation resistance are two additional properties of importance in some applications. Most ceramics are dielectrics and, except for ferrites, lack magnetic properties.

Excellent abrasion-resistant surfaces are possible. These surfaces also offer a pleasing gloss or patina and can be vitreous and nonporous. In addition to their resistance to chemical substances and corrosive materials, ceramics are relatively immune to fire, heat, and weathering.

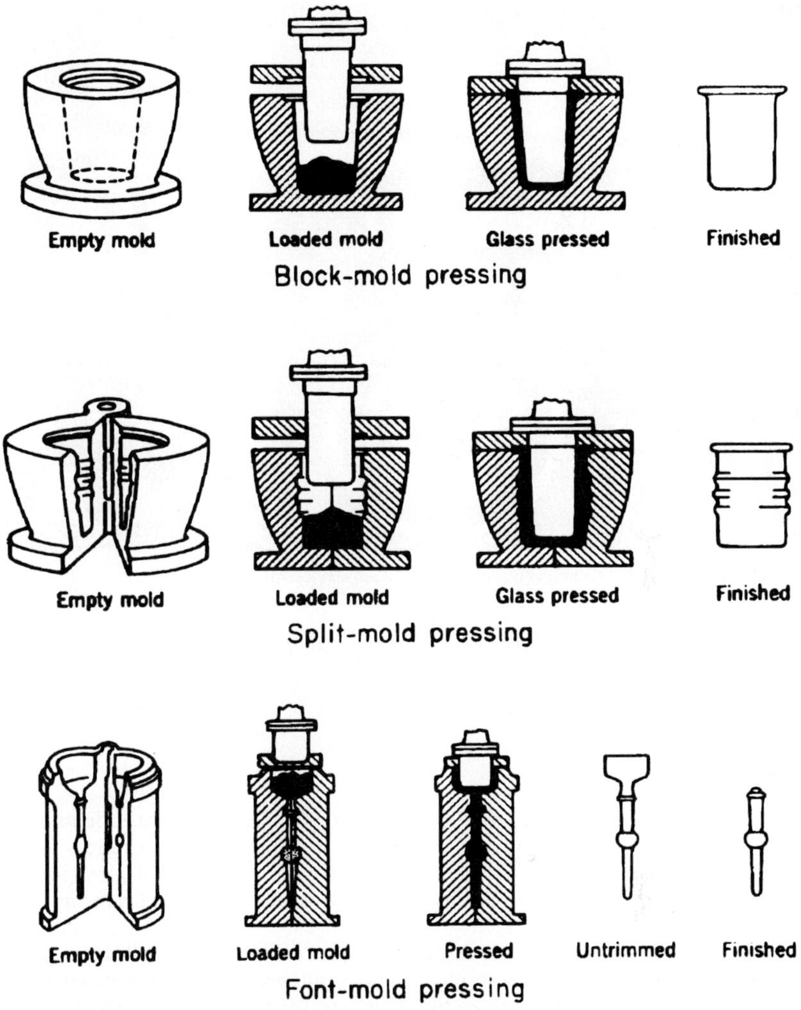

FIGURE 6.11.1 Pressing methods and mold types for glass. (*From Errol B. Shand,* Glass Engineering Handbook, *McGraw-Hill, New York.*)

Generally, all ceramic materials are brittle. Tensile strengths are somewhat limited. There also are some limitations in freedom of design because of processing complexities and inherent mechanical properties. Because of high firing temperatures, metal inserts cannot be molded in.

The size of commercial ceramic components ranges from the very small electronic components to large nose cones and radomes. Typical ceramic parts for mechanical applications are bearings, turbine blades, cams, cutting tools, extrusion dies, thread and wire guides, nozzles for abrasive materials, wear plates, seals, valve seats, filters, pump parts, crucibles, and trays. Typical parts for electrical and electronic applications include coil forms, tubes, insulators, lamp housings, printed-circuit boards,

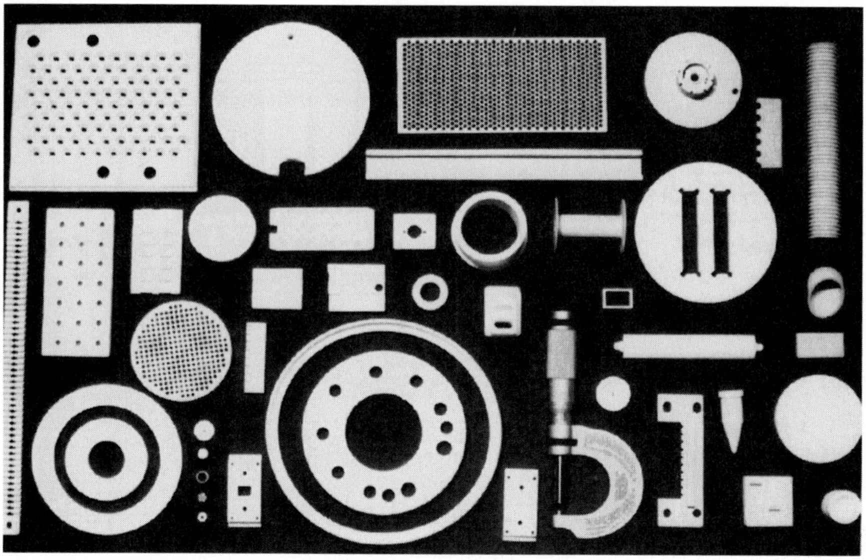

FIGURE 6.11.2 Typical technical ceramic parts. (*Courtesy Duramic Products Inc.*)

radomes, resistor bases, vacuum-tube-element supports, and terminals. Figure 6.11.2 illustrates some typical ceramic components.

Glass Parts

Transparency is the most important property of glass and accounts for most of its applications. Other properties are similar to those of whiteware but with less favorable strength and high-temperature characteristics. The poor resistance of glass to thermal shock can be improved by tempering, which also provides increased mechanical strength.

Glass products range in size from microspheres of fractional-millimeter diameter used as fillers for plastics to large plate-glass windows. Normally, pressed parts are about 9 kg (20 lb) or less in weight, while blown ware can range up to 16 kg (35 lb).

Typical pressed-glass components are electrical insulators, baking dishes, food blenders, stoppers and stopcocks for laboratory vessels, eyeglasses, and ornamental pieces. Typical blown-glass components are bottles and other containers, incandescent lamps, electron tubes, laboratory glassware, and television picture tubes.

Tubing and piping of glass, made by drawing, are used for laboratory, chemical industry, and high-temperature applications and thermometers. Flat glass for glazing, mirrors, tabletops, and other purposes is made either by drawing or by rolling, which, in the case of plate glass, is followed by grinding and polishing or by floating onto molten tin and drawing horizontally. Glass powders are sintered to make filters and other porous objects. Glass fibers are a major reinforcing medium for many products (see Chap. 6.6), for insulation and for fiber optics.

Figures 6.11.3 and 6.11.4 illustrate typical pressed- and blown-glass parts.

Cellular glass is almost invariably black or dark-colored. Pore size can be varied, depending on the method of introducing porosity. Thermal expansion is the same as that of the base glass.

FIGURE 6.11.3 Examples of pressed glassware. (*a*) Block-mold glassware. (*b*) Split-mold glassware. (*c*) Font-mold glassware. (*Courtesy Corning Glass Works.*)

Color can be incorporated into most glass, whiteware, porcelain, and other ceramic materials by introducing the proper pigmentation medium to the material before firing.

Refractories

Refractory products, being resistant to very high temperature and, generally, to thermal shock, are used in such applications as furnace linings and similar insulation. For the most part, they are molded in the shape of bricks of relatively small dimensions. They also may be fusion-cast in large shapes (e.g., 1 by 2 by 4 ft) and then cut into the required size and configuration.

FIGURE 6.11.4 Examples of blown glassware. (*a*) Paste-mold glassware. (*b*) Hot-iron-mold glassware. (*c*) Press-and-blow glassware. (*Courtesy Corning Glass Works.*)

ECONOMIC PRODUCTION QUANTITIES

Excluding the art forms—glass, pottery, porcelain enameling, etc. using ceramic media—much of the true industrial portion of the ceramics field is long-established, well stabilized, and geared for efficient large-scale production. Factors such as adaptability to mass production, costs, setup times, output rates, and equipment life are summarized for various branches of the industry in Table 6.11.1.

TABLE 6.11.1 Economic Production Quantities

Material and method	Technical ceramics, mostly machined	Technical ceramics, mostly pressed	Pressed glass	Blown glass	Flat glass	Whiteware: vitreous and semivitreous dinnerware	Refractories
Normal economic production quantities	Short to medium run	Medium to long run	Long run	Long run	Long run except for thickness change	Medium to long run	Medium to long run
Investment required							
Equipment	Moderate	High	Medium to high	Very high	Very high	High	High
Tooling	Low	High	Medium to high	Very high	Very high	High	High
Lead time to tool up new product	1 month	3–6 months	3 months	3 months		3 months	3–6 months
Typical output rate	Varies greatly; typical rate: 100 pieces per shift	To about 15,000 pieces per shift if order quantities justify automatic methods	From 100 pieces per day (hand methods) to 40,000 pieces per day (automatic machines)	To 150,000 containers per day; to 1,000,000 light bulbs per day	200 tons per day	6–10 pieces per day per mold; many molds used for long runs	40,000 bricks per day for continuous extrusion; much lower rate for fuse casting
Normal life of tooling	Cutter life very short compared with metal machining	Moderately long (some wear due to abrasive properties of material)	Long	Long	1–2 months	Plaster molds limited to 500 to 1000 parts; not reclaimable	Moderately long

SUITABLE MATERIALS

Ceramics

Technical ceramics are normally dense bodies that contain steatite aluminum oxide (alumina), beryllium oxide (beryllia), or related oxides such as mullite ($3Al_2O_3 \cdot 2SiO_2$), forsterite [$(Mg \div Fe)_2 \div SiO_4$], and cordierite ($2MgO \cdot 2Al_2O_3 \cdot 5SiO_2$). Silicon carbide, silicon nitride, and boron nitride are other materials of commercial use.

Glass

The major portion of the glass industry uses as its raw materials oxides and carbonates of silicon, calcium, and sodium, mainly as sand, limestone, and soda ash. Numerous other oxides are added to obtain special properties such as radiation resistance, hardness, controlled expansion, etc. The principal types of glass are as follows:

Silica Glass. Silica (silicon oxide) or silica quartz (sand), when fused, forms a glass with very-high-temperature resistance, high strength, chemical resistance, and resistance to thermal shock. Unfortunately, it is extremely difficult to form into useful shapes, and articles made from it are therefore expensive.

96 Percent Silica Glass. This type has somewhat easier formability and slightly reduced properties compared with silica glass because of the presence of small amounts of boric oxide and other ingredients.

Borosilicate Glass. This type contains silica as the chief ingredient but has from 13 to 28 percent of boric oxide for low thermal expansion and other oxides that provide further improvements in workability. Mechanical, electrical, and chemical resistance properties are still good, and borosilicate glass has wide usage for electrical insulators, laboratory glassware, cookware, and sight and gauge glasses.

Lead Glass. This type contains a portion of lead oxide in addition to silica and other oxides. Normally, the lead oxide content is less than 50 percent, but it can be as much as 90 percent for glass used for radiation shielding. In portions below 50 percent, lead oxide enhances the workability of glass, and lead glass is normally called for when intricate forming is required. Optical and electrical properties are also excellent, although mechanical properties (strength and abrasion resistance) are low. Lead glass is used for thermometer tubing, neon and fluorescent lights, television tubes, art glassware, and jewelry.

Soda-Lime Glass. This type contains appreciable quantities of soda, Na_2O, and lime, CaO, in addition to the chief ingredient, silicon oxide. Soda and lime lower the melting point of the glass, reduce its viscosity when melted, and thereby improve its workability. Soda-lime glass is a good general-purpose glass and is used for window and plate glass, containers, and electric-lamp bulbs. It is economical to melt and to fabricate.

Table 6.11.2 summarizes the prime characteristics of these common glasses on a comparative basis.

TABLE 6.11.2 Properties of Principal Types of Glass

	Lime glass	Lead glass	Borosilicate glass	96% silica glass	Silica glass
Weight	Heavy	Heaviest	Medium	Light	Lightest
Strength	Weak	Weak	Moderately strong	Strong	Strongest
Relative cost	Lowest	Low	Medium	High	Highest
Resistance to thermal shock	Low	Low	Good	Better	Best
Electrical resistivity	Moderate	Best	Good	Good	Good
Hot workability	Good	Best	Fair	Poor	Poorest
Heat treatability	Good	Good	Poor	None	None
Chemical resistance	Poor	Fair	Good	Better	Best
Impact-abrasion resistance	Fair	Poor	Good	Good	Best
Ultraviolet-light transmission	Poor	Poor	Fair	Good	Good

Other Ceramics

Porcelain enamels, or frits, are low-melting, lead oxide–based glasses. Ground-coat enamels, which cross-bond a metal substrate to a topcoat porcelain enamel, always contain cobalt oxide.

Whiteware used for dinnerware and other nontechnical applications is normally similar in composition to that used for mechanical and electrical parts. Combinations of clay, feldspar, and flint are used with minor variations to impart desired characteristics.

Refractory materials produced from aluminum and chromium oxides are used for large tank furnaces. Silica is used when acidic atmospheres are involved, and fireclay is employed for general nonnoxious high-temperature environments.

DESIGN RECOMMENDATIONS

Ceramic Parts

Although technical ceramics can be fabricated into complex shapes, it is always desirable to keep shapes as simple as possible for economic reasons. Tolerances also should be as liberal as the function of the component permits. It is important, from a structural standpoint, to avoid problems that result from the low tensile strength and lack of ductility of ceramics.

Specific design recommendations for technical ceramics are as follows:

1. Edges and corners should have chamfers or generous radii to minimize chipping and stress concentration and aid forming. When parts are machined, outside radii should be 1.5 mm ($\frac{1}{16}$ in) or more and inside radii at least 2.4 mm ($\frac{3}{32}$ in). For dry-pressed parts, outside edges should be beveled in a manner similar to that employed with powder-metal parts; 0.8 mm by 45° is a desirable minimum. Inside

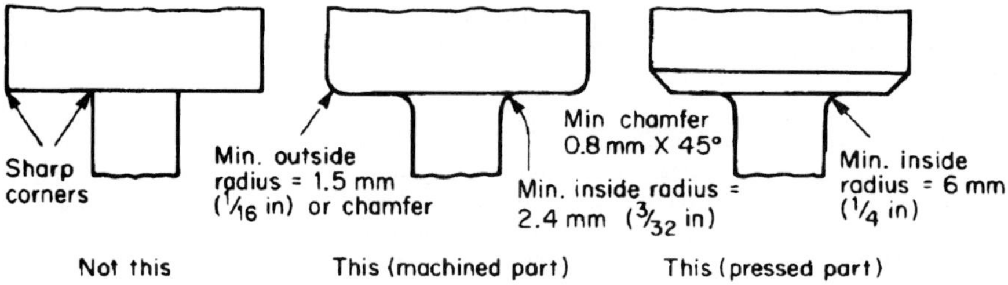

FIGURE 6.11.5 Design rules for corners of ceramic parts.

radii should be as large as possible: 6 mm (¼ in) unless the height or width of the smaller surface is less than 6 mm. (See Fig. 6.11.5 for an illustration of these rules.)

2. Since parts may sag or be distorted if not properly supported during firing, it is preferable to avoid large overhanging or unsupported sections. Otherwise, supporting-fixture costs may be excessive.
3. Pressed parts should be designed with as uniform a wall thickness as possible. Differential shrinkage of sections of nonuniform thickness during drying and firing causes stress, distortion, and cracking. Sections should not exceed 25 mm (1 in) in thickness. (See Table 6.11.3 for wall-thickness information.)

TABLE 6.11.3 Thicknesses of Ceramic Products

	Thickness range, mm (in)		Maximum practical thickness buildup within individual part, ratio
	Minimum	Maximum	
Technical ceramics			
Standard types	0.5 (0.020)	25 (1.0) or more	4:1
Glass			
Glass containers	Blown: 1.5 (1/16)*	9.5 (3/8)	4:1
	Pressed: 2.4 (3/32)	9.5 (3/8)	
Flat glass	Picture glass: 1.1 (0.043)	Preferably none	
	Doors: 22–25 (7/8–1)		
Technical glass	1.5 (1/16)or as required	4:1	
Cellular glass	As desired (cast material)		As desired; machinable
Whiteware			
Vitreous sanitary ware	6.3 (1/4)	50 (2)	2:1
Vitreous dinnerware	1 (0.040)†	3 (1/8)	3:1
Semivitreous dinnerware	1.7 (0.065)	9.5 (3/8)	3:1
Floor and wall tiles	6.3 (1/4)	13 (1/2)	Raised ridges: 1.2:1 to 1
Porcelain enameling			
Cast-iron plumbing	3 (1/8)	4.8 (3/16)	Preferably none
Steel plumbing	1.5 (1/16)	3 (1/8)	Preferably none
Appliances	1.5 (1/16)	2.4 (3/32)	Preferably none
Refractories			
Standard types	As required; bricks and heavy cast shapes		Preferably none

*Throwaway bottles; returnable bottles are slightly thicker. Light bulbs = 0.020 in thick.
†Or less, as in Japanese rice or Irish Belleek ware.

4. Other factors being equal, simple symmetrical shapes without deep recesses, holes, and projections are preferable. Gently curved surfaces without abrupt break lines or angularity are normally preferred with most ceramic-forming processes.
5. When hollow pieces are cast against a male mold (e.g., cup-shaped parts), a draft angle of at least 5° must be provided to facilitate removal of the green body. If the part is left in the mold too long, drying shrinkage will draw the material against the mold, resulting in cracking. Dry-pressed parts do not require draft on either outside surfaces or the walls of through holes. Wet-pressed parts should have at least 1° on exterior surfaces and 2° on interior surfaces. (See Fig. 6.11.6.)
6. Undercuts should be avoided in ceramic components if possible. Although some undercuts can be incorporated through the use of mold cores, machining is the normal method for producing them. With dry pressing, machining is essential if undercuts are required. In all cases, costs are added.
7. Dry-pressed ceramics are subject to other design rules of powder-metal parts (Chap. 3.12) also but cannot match their close dimensional tolerances.
8. Cavities, grooves, and blind holes in pressed parts should not be deeper than one-half the part thickness and preferably only one-third the thickness. (See Fig. 6.11.7.)

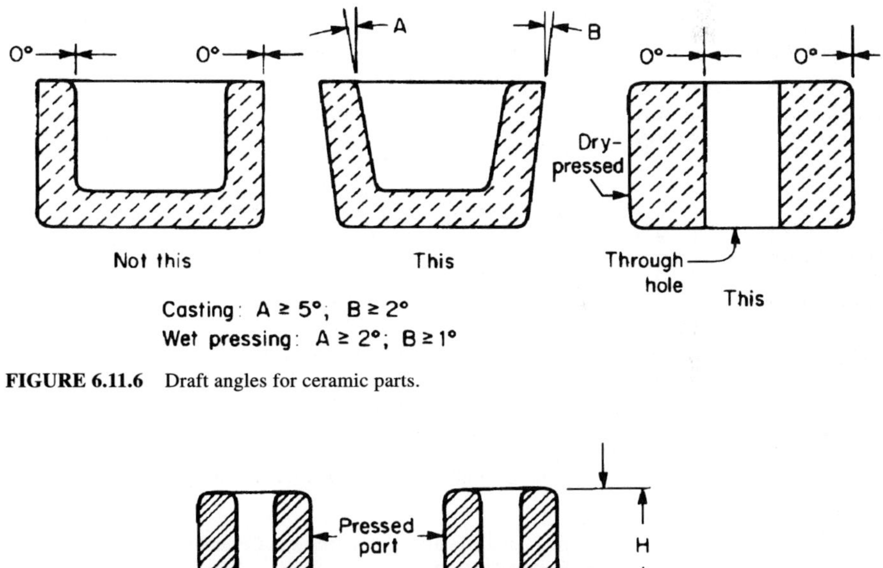

FIGURE 6.11.6 Draft angles for ceramic parts.

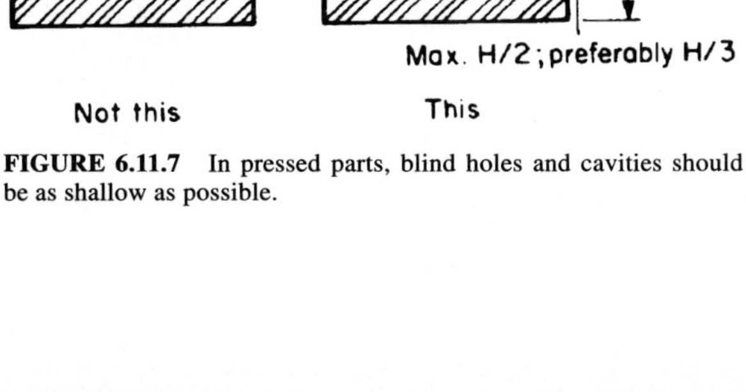

FIGURE 6.11.7 In pressed parts, blind holes and cavities should be as shallow as possible.

9. Extruded parts should be symmetrical, if possible, with uniform wall thickness. The minimum wall thickness for extrusions should be 0.4 mm ($\frac{1}{64}$ in) or, for round sections, 10 percent of the extrusion diameter. For long extrusions, 150 mm (6 in) in length or more, the wall should be thicker, at least 20 percent of the extrusion's outside diameter. (See Fig. 6.11.8.)

10. Holes in pressed parts should be large and as widely spaced as possible. Thin walls between holes, depressions, or outside edges should be avoided. These walls should be at least as thick as the basic walls of the part, especially if the part is small and thin-walled. In any case, the minimum in internal areas should be 0.8 mm (0.030 in) and, in the case of outside edges, 3 mm ($\frac{1}{8}$ in). Machined holes should be at least 1.5 mm ($\frac{1}{16}$ in) in diameter if possible, although smaller holes can be produced.

11. It must be remembered that distortions from shrinkage can cause fitting problems when holes are used for fasteners and when holes in ceramic parts are to be aligned with holes in mating parts. Holes in ceramic parts may become slightly out of round after firing. Table 6.11.4 provides recommended minimum clearances

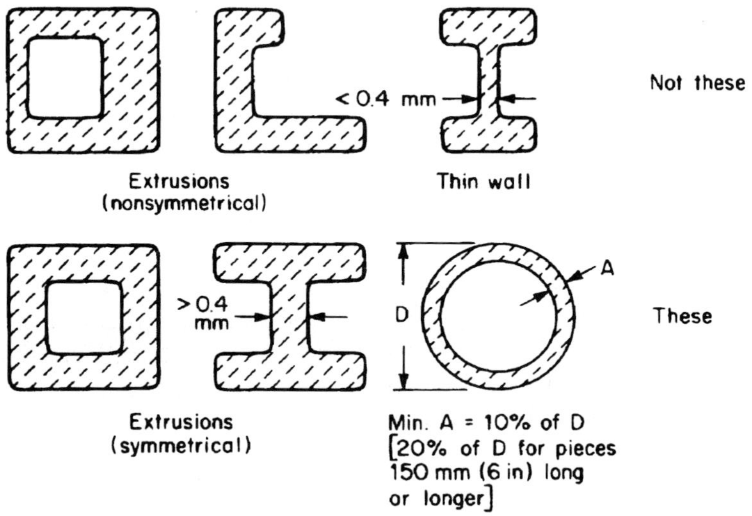

FIGURE 6.11.8 Design rules for the wall thickness of extruded ceramic parts.

TABLE 6.11.4 Recommended Clearances for Holes in Ceramic Parts When Fasteners Are to Be Used

Fastener outside diameter, mm (in)	Minimum additional hole diameter (in addition to hole tolerance), mm (in)
To 4 ($\frac{5}{32}$)	0.15 (0.006)
Over 4 to 10 ($\frac{5}{32}$ to $\frac{3}{8}$)	0.20 (0.008)
Over 10 to 16 ($\frac{3}{8}$ to $\frac{5}{8}$)	0.25 (0.010)
Over 16 to 25 ($\frac{5}{8}$ to 1)	0.30 (0.012)
Over 25 (1)	0.40 (0.016)

for holes that are designed to accept fasteners. To compensate for variations in hole spacing, multiple holes that are to be aligned with corresponding holes in other parts must be further enlarged (or elongated in the direction of the other holes). The amount of the enlargement or elongation depends on the allowable hole-to-hole tolerance of the two parts.

12. Molding of screw threads in ceramic parts is not feasible. Screw threads can be machined in green ceramic workpieces, but they constitute a potential problem, and it is better to design parts without screw threads if possible. If incorporated, threads should be coarse and not smaller than 6-32. Internal threads should be considered acceptable if they accept a Class 1A mating metal screw; external threads should be considered acceptable if they accept a Class 1 nut. Holes should not be tapped to a depth greater than six threads because dimensional variations in the thread pitch from firing shrinkage may cause fit problems if too long a thread is used. All tapped holes should be countersunk. (See Fig. 6.11.9.) External threads also should be as coarse as possible and have a well-rounded thread form to reduce edge chipping and stress cracking. Coarse-pitch threads with a truncated form also can be used to increase the strength of the threaded ceramic part. As with internal threads, it is recommended that the number of threads in engagement be limited to six.

13. Ribs and fins should be well rounded, wide, and well spaced and have normal draft. Figure 6.11.10 illustrates design rules for ribs.

14. Grinding after firing can produce ceramic parts of high accuracy, but stock-removal rates are slow, and the operation is expensive. When the operation is nec-

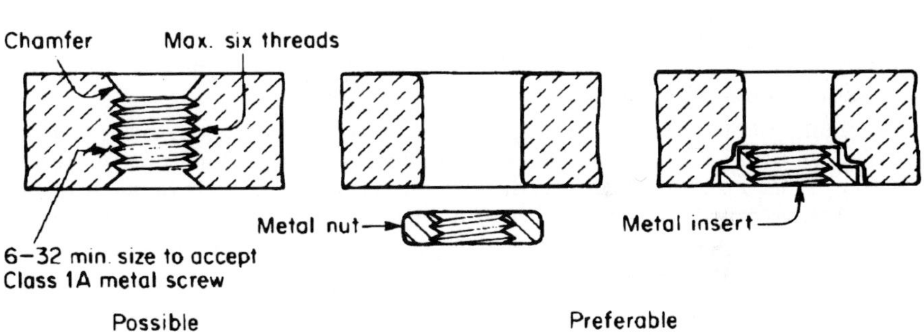

FIGURE 6.11.9 Internal screw threads in ceramic parts.

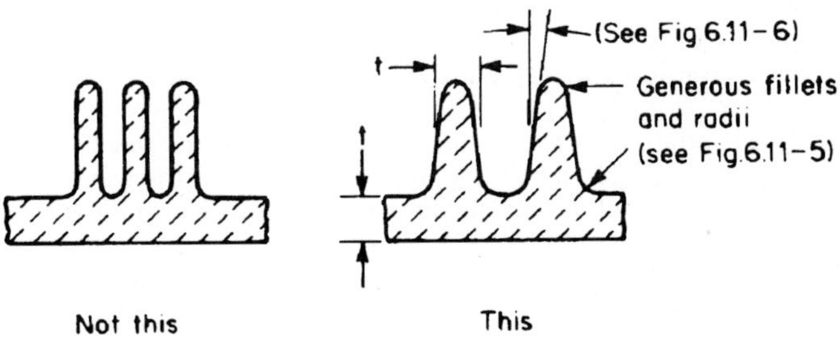

FIGURE 6.11.10 Design rules for ribs in ceramic parts.

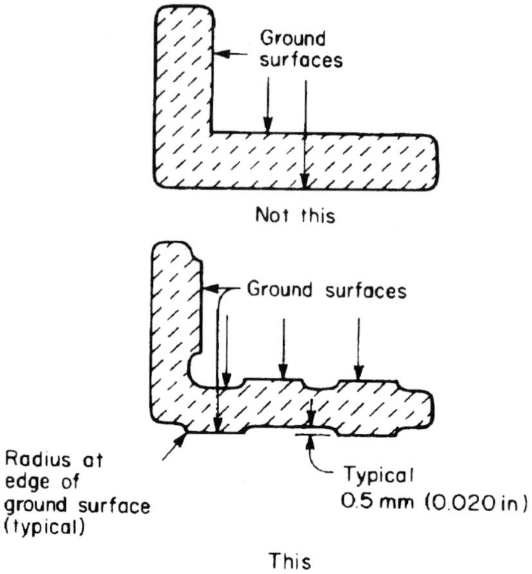

FIGURE 6.11.11 Minimize the area to be finish-ground after firing.

essary, it is advisable to reduce the area of the surface to be ground as much as possible and to provide clearance for the grinding wheel at the ends of the surface. (See Fig. 6.11.11.)

15. Ceramic parts can be permanently joined to metal components by adhesive bonding, soldering, brazing, and shrink fitting. Shrink fitting is highly satisfactory as long as the metal is on the outside (in tension) and the ceramic on the inside (in compression). Brazing is stronger than bonding or soldering and more temperature-resistant but requires a metallized layer as a base for the brazing alloy.

Glass Parts

Guidelines for the design of pressed- and blown-glass components are shown in Tables 6.11.5 and 6.11.6. Note that tolerances and minimum desirable production quantities also are shown.

Other points to bear in mind when designing glass parts are the following:

1. Holes, cavities, and deep slots can cause molding problems and should be included in a part only if absolutely necessary. Holes are normally not punched through in the pressing operation but are machined from a thin web or hollow boss, as shown in Fig. 6.11.12.
2. As in the case of whiteware parts, best results are obtained when walls are uniform in thickness, when the part is designed for compressive rather than tensile strength, and when gently curved rather than sharp-angled shapes are employed.

TABLE 6.11.5 Manufacturing Tolerance and Design Recommendations for Pressed Glassware

Method of manufacture		Size group	Smallest desirable production run, pieces	d, diameter or width, usual range, in	h, depth or height, usual range, in	h_1/d, depth versus diameter, max. ratio	d/d, length versus width, max. ratio	w, wall thickness, usual range, in	Weight
Block or open-and-shut mold	Hand-produced	Small	1,200	1–3	1–5	1.5:1	2:1	5/32–1/4	8 oz
		Medium	800	3–6	1–8	2:1	2:1	3/16–3/8	5 lb
		Large	500	6–24*	1½–12*	1.5:1	2:1	1/4–1/2	20 lb
	Machine-produced	Small	50,000	¾–1½	½–1½	1:1	1:1	3/32–5/32	2 oz
		Medium	35,000	2–6	¾–6	1.75:1	1.5:1	5/32–5/16	3 lb
		Large	25,000	6–18*	1–12*	1.5:1	1.5:1	1/4–5/8	10 lb
Front-molded	Hand-produced	Small	1,200	¼–⅝	½–3				2 oz
		Medium	800	⅝–1	1–7				8 oz
		Large	500	1–2½	2–12				5 lb
	Machine-produced	Small	50,000	¼–⅝	½–3				2 oz
		Medium	35,000	⅝–1	1–7				8 oz
		Large	25,000	1–2½	2–10				4 lb

TABLE 6.11.5 Manufacturing Tolerance and Design Recommendations for Pressed Glassware (*Continued*)

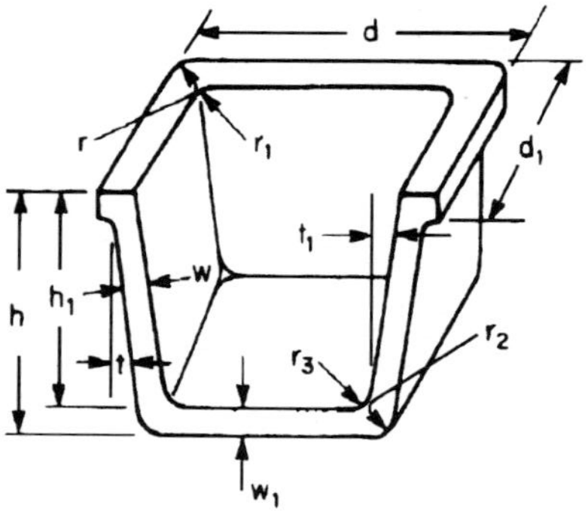

Sidewall thickness			Bottom thickness				
						Tolerance on w_1, %	
h, in	*Min.* w, in	Tolerance on w, in	h, in	Min. w_1, in	Ratio h/d	Hand-produced	Machine-produced
1–4	⅛	±0.025	1–4	3/16	Up to 1	±8 to ±12	±4 to ±8
4–8	3/16	±1/32	4–8	5/16			
8–12	¼	±3/64	8–12	⅜	Over 1	±25 to ±50	±15 to ±30

Normal tolerances on outside dimensions		Minimum desirable fillets and radii			Desirable side taper or drafts					
					Exterior (block molds)			Interior		
d or h, in	Tolerance, in	d or h, in	r, r_2, in	r_1, r_3, in	h, in	Normal t, °	Min. t, °	h_1, in	Normal t_1, °	Min. t_1, °
1–4	±1/32	1–4	⅜	¼	1–4	2	1	1–4	3	2
4–8	±3/64	4–8	½	⅜	4–8	3	1½	4–8	4	2
8–12	±1/16	8–12	1	¾	Over 8	4	1½	Over 8	5	2

*Maximum dimension.

Source: From Errol B. Shand, *Glass Engineering Handbook,* McGraw-Hill, New York.

TABLE 6.11.6 Manufacturing Tolerance and Design Recommendations for Blown Glassware

Method of manufacture		Size group	Smallest desirable production run, pieces	d, diameter or width, usual range, in	h, height, usual range, in	*h/d*, height versus diameter, max. ratio	d/d_1, length versus width, max. ratio	*w*, normal median wall thickness: Light-wall ware, in	*w*, normal median wall thickness: Heavy-wall ware, in	Wall thickness, variation piece to piece, %	*d/n*, max. ratio, body diameter to neck	*r*, min. corner radius	Weight*
Paste mold	Hand	Small	1,000	1–2	3–8	10:1		0.060	0.100	±33	3:1	$\frac{d}{5}$	8 oz
		Medium	350	2–7	8–18	6:1		0.075	0.140	±33	4:1	$\frac{d}{7}$	10 lb
		Large	50	7–20	12–30	3:1		0.090	0.250	±50	5:1	$\frac{d}{10}$	35 lb
	Machine	Small	100,000	1–3	3–5	3:1		0.060	0.100	±33	3:1	$\frac{d}{5}$	4 oz
		Medium	75,000	3–5	4–6	2:1		0.070	0.100	±33	3:1	$\frac{d}{5}$	8 oz
		Large	75,000	5–6½	5–10	1½:1		0.070	0.090	±33	2¼:1	$\frac{d}{5}$	1 lb
Hot iron	Hand	Small	2,000	1–3	3–8	10:1	2½:1	0.060	0.100	±33	3:1	$\frac{d}{5}$	8 oz
		Medium	700	3–8	8–18	6:1	1½:1	0.075	0.140	±33	4:1	$\frac{d}{7}$	10 lb
		Large	50	8–20	12–30	3:1	1:1	0.090	0.250	±50	5:1	$\frac{d}{10}$	35 lb
Press and blow	Machine	Small	150,000	¾–2	1–3	3:1	2:1	0.060	0.100	±25	3:1	⅛ in	3 oz
		Medium	100,000	2–4	3–6	3½:1	2:1	0.075	0.120	±25	4½:1	¼ in	1 lb
		Large	50,000	4–6½	6–12	3:1	2:1	0.120	0.180	±25	6:1	⅜ in	3½ lb

TABLE 6.11.6 Manufacturing Tolerance and Design Recommendations for Blown Glassware
(*Continued*)

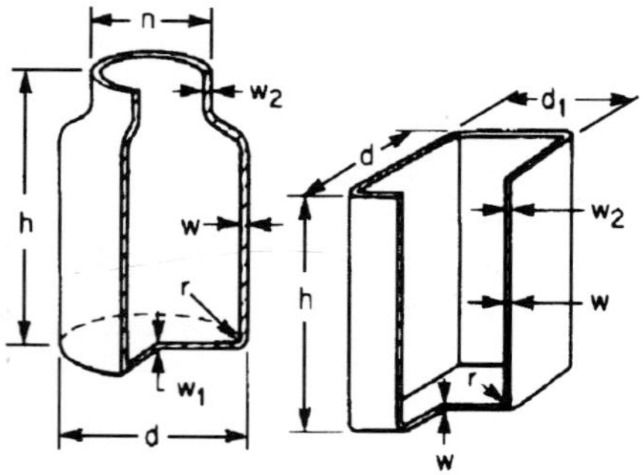

Normal tolerances on outside dimensions					
Height tolerance			Diameter or width tolerance, in†		Max. out-of-round major axis less minor axis
h, in	Tolerance, in	d or d_1, in	Paste-mold ware, in	Iron-mold ware, in	
1–3	$\pm 1/16$	1–3	$\pm 1/32$	$\pm 1/32$	$0.02 \times d$
3–6	$\pm 1/16$	3–6	$\pm 3/64$	$\pm 3/64$	$0.02 \times d$
6–9	$\pm 1/16$	6–9	$\pm 1/16$	$\pm 1/16$	$0.0175 \times d$
9–12	$\pm 1/16$	9–12	$\pm 3/32$	$\pm 1/16$	$0.0175 \times d$
12–20	$\pm 1/4$	12–20	$\pm 1/8$	$\pm 1/16$	$0.0175 \times d$

*Weight based on density of commercial glasses.

†Tolerance on diameter for circular pieces is on mean diameter and does not include out-of-round.

Notes: Wall thickness cannot be positively controlled. It depends on glass distribution in the blank and on the shape into which it is blown. Note in table that wall thickness of blown ware can be much lighter than for pressed ware. A pear-shaped piece is ideally suited to blowing. An inverted cone is undesirable. Long, thin necks make it difficult to handle the blank during the blowing operation. In hot-iron ware, a circular section at the cutoff point permits a flame burn-off which is considerably cheaper than other methods.

Source: From Errol B. Shand, *Glass Engineering Handbook,* McGraw-Hill, New York.

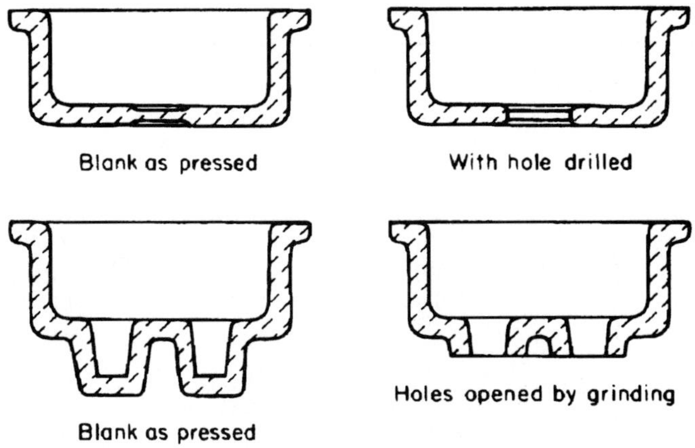

FIGURE 6.11.12 Two designs for holes in pressed glassware. (*From Errol B. Shand,* Glass Engineering Handbook, *McGraw-Hill, New York, 1958.*)

3. Lettering or other irregular surface features may be incorporated as long as they are aligned in the direction of, and not perpendicular to, the mold opening.
4. Ribs and flanges can be incorporated in pressed-glass components, but they are not practicable in blown parts.
5. While bosses may be incorporated in some items like electrical insulators, they are normally not practicable for general-purpose design and manufacture.
6. Threads for bottle caps or similar connecting devices may be incorporated in blown-glass parts as they are with blow-molded plastics, and the same screw-thread designs (see Chap. 6.5) are recommended.

DIMENSIONAL FACTORS AND TOLERANCES

Ceramic Parts

These parts are affected dimensionally primarily by drying shrinkage and firing shrinkage, which can total as much as 25 percent for high-clay ceramics and about 14 percent for porcelains. Other factors affecting the accuracy of ceramic parts are mold accuracy and mold wear. Processing variables, such as the amount of material pressed, pressing time, and pressure, affect the dimensions of pressed parts. Machining variations affect green-state machined and finish-ground parts.

Table 6.11.7 presents recommended dimensional tolerances for technical ceramic parts.

Glass Parts

These parts are dimensionally affected by gob weight, temperature of the melt and mold, mold tolerance and wear, and shrinkage of the glass on cooling. Shrinkage rates

TABLE 6.11.7 Recommended Dimensional Tolerances for Technical Ceramic Parts*

	Porcelain; cast ceramics	Standard tolerance for technical ceramics	Tightest tolerance for precision electronic and mechanical applications
As-fired lengths and widths, unglazed	±1½%, NLT 0.38 mm (0.015 in)	±1%, NLT 0.13 mm (0.005 in)	±½%, NLT 0.08 mm (0.003 in)
As-fired lengths and widths, glazed	±3%, NLT 0.75 mm (0.030 in)	±2%, NLT 0.30 mm (0.012 in)	±1%, NLT 0.13 mm (0.005 in)
Angles	±2°	±2°	±1°
As-fired thickness	±10%	±10%	±5%
Ground thickness	±0.10 mm (0.004 in)	±0.025 mm (0.001 in)	±0.025 mm (0.001 in)
Other ground dimensions	±0.10 mm (0.004 in)	±0.025 mm (0.001 in)	±0.013 mm (0.0005 in)
Hole diameter, unglazed			
0 to 13 mm (to ½ in)	±0.13 mm (0.005 in)	±0.08 mm (0.003 in)	±0.05 mm (0.002 in)
Over 13 mm (over ½ in)	±2%	±0.13 mm (0.005 in)	±0.10 mm (0.004 in)
Hole diameter, glazed			
0 to 13 mm (to ½ in)	±0.30 mm (0.012 in)	±0.20 mm (0.008 in)	±0.10 mm (0.004 in)
Over 13 mm (over ½ in)	±2%	±1%	±1%
Hole locations center-to-center	±2%, NLT 0.13 mm (0.005 in)	±1%, NLT 0.08 mm (0.003 in)	±½%, NLT 0.08 mm (0.003 in)

*NLT = not less than.

vary from the equivalent of that of steel [92×10^{-7} in/(in ÷ °C)] down to that for pure silica glass [7×10^{-7}in/(in · °C)].

Recommended tolerances for various dimensions of pressed- and blown-glass parts are shown in Tables 6.11.5 and 6.11.6. Tolerances for locations and diameters of holes are shown in Table 6.11.8.

TABLE 6.11.8 Recommended Location and Diametral Tolerances for Holes in Pressed-Glass Components

		Recommended tolerance, mm (in)	
Method	Hole diameter	Location within	Diameter
Drilled or pressed and ground	3–6 (⅛–¼)	0.8 ($\frac{1}{32}$)	±0.4 (±$\frac{1}{64}$)
	6–25 (¼–1)	0.8 ($\frac{1}{32}$)	±0.8 (±$\frac{1}{32}$)
Burned through and punched	3–6 (⅛–¼)	0.50 (0.020)	±0.25 (±.011)

CHAPTER 6.12

PLASTIC-PART DECORATIONS*

Ronald D. Beck
Fisher Body Division
General Motors Corp.
Warren, Michigan

Molded plastic parts are generally decorated for eye appeal. However, decorating can be functional as well as attractive. Functional improvements include resistance to wear, scratching, and marring and increased resistance to heat, light, and chemical and electrical exposure. A molded plastic part can be decorated to transmit part identification and product information as described in Chap. 8.6. Most plastic materials are decorative in themselves and can be molded with raised or depressed designs that may be accented with paint or lacquer. Decorative overlays provide contrast and novelty. All decorating depends on applying a marking or coating permanently to the plastic surface without damage to the original shape or appearance. Virtually every type of molded plastic product can be decorated in one way or another. The majority of all decorating on plastic parts is done with painting, dyeing, printing, silk screening, metallizing and plating, labeling, hot stamping, and polishing.

SURFACE FINISH OF PLASTIC PART PRIOR TO DECORATING

The decorating of any plastic part is a surface treatment, and the first prerequisite is that the surface be clean and prepared in the right manner for decorating. Some of the common causes of failure in decorating plastic parts are contamination from mold lubricants, dust, natural skin greasiness and excess surface plasticizers, surface moisture and humidity conditions, and strains locked into the molded part. Water washing with alkaline cleaners, solvent cleaning from spraying, immersion, or wiping, and vapor degreasing may be used to remove these contaminants.

Some plastic materials are more difficult to decorate than others. Thermoplastic materials are prone to solvent attack, while thermosetting plastics have excellent solvent-resistance characteristics. Each family of plastic materials must be considered separately when selecting coatings, thinners, foils, tapes, etc.

---

*This chapter is adapted and updated in Chap. 9, "Decorating Plastics," of the author's book *Plastic Product Design,* © 1970 by Litton Educational Publishing, Inc. Reprinted by permission of Van Nostrand Reinhold Co.

Because the part surface is reproduced by the paint, film, or foil overlay, the finished decorative item is no better than the initial molded part. It is desirable to mold for high gloss and a low stress level in the part. The use of fillers in the plastic material causes some difficulty. Fiberglass does not always give a good luster finish, nor do cloth-filled thermosetting materials. A clear thermoplastic material may show flow marks.

Buffing of thermosetting parts produces a highly polished surface, but a buffed surface does not wear as well as a nonbuffed surface because buffing removes a small portion of the outer resinous skin. Buffing helps, however, to disguise flow marks on thermosetting materials. Buffing of thermoplastic materials, however, does not disguise flow marks, but it helps to remove outside blemishes. Flow marks may be disguised nicely by using a mottled two- or three-color material.

If a large, flat area is required, flow marks may be hidden by graining the surface. Since graining of the part is done by the mold, the grained surface always should be made on a plane perpendicular to the draw, or adequate taper should be allowed so that the molded part can be removed from the mold. This is true because a grained surface contains undercuts.

Frequently, it is necessary to pretreat a plastic surface before it can be painted or decorated. This is true of polyolefins and other plastics with low critical surface tension. Flame, corona discharge, or chemical pretreatment may be required to ensure good bonding on the slippery surface.

Static electricity on plastic surfaces has always caused problems. Airborne impurities become attracted to and settle on the plastic surface. Wiping or rubbing with a cloth may only increase the static charge and aggravate the problem. Ionized air blowers will remove these static charges and dust that may have accumulated on the surface. The decorating of plastics should be carried out in clean, controlled room conditions.

SPRAY, FLOW, AND DIP COATING WITH PAINT

Painting is an important approach used in changing the appearance and other properties of a plastic part. Color, gloss level, and abrasion and weather resistance can be improved. A plastic substrate may be painted by several methods: spraying, dipping, flow and curtain coating, and roller coating. For further information, see Chap. 8.5.

MASK-SPRAY PAINTING

A paint spray mask is used to provide sharp lines of demarcation when different colors are designated. Masks are generally made of electroformed nickel, but sometimes paper, rubber, and plastic can be used for masks for short production runs.

Three types of spray masks are in use today. (See Figs. 6.12.1 and 6.12.2.) The *plug mask* is used to fill and keep depressed areas clean while the surrounding area is being painted. The plugs are suspended from wire bridging and must fill the depressed sections to prevent paint leakage. Plug masks are also used for protecting areas to be vacuum-plated. The *block or cutout mask* is used to confine paint to a shallow area that has no paint steps. A secondary wiping is usually required after removal of the

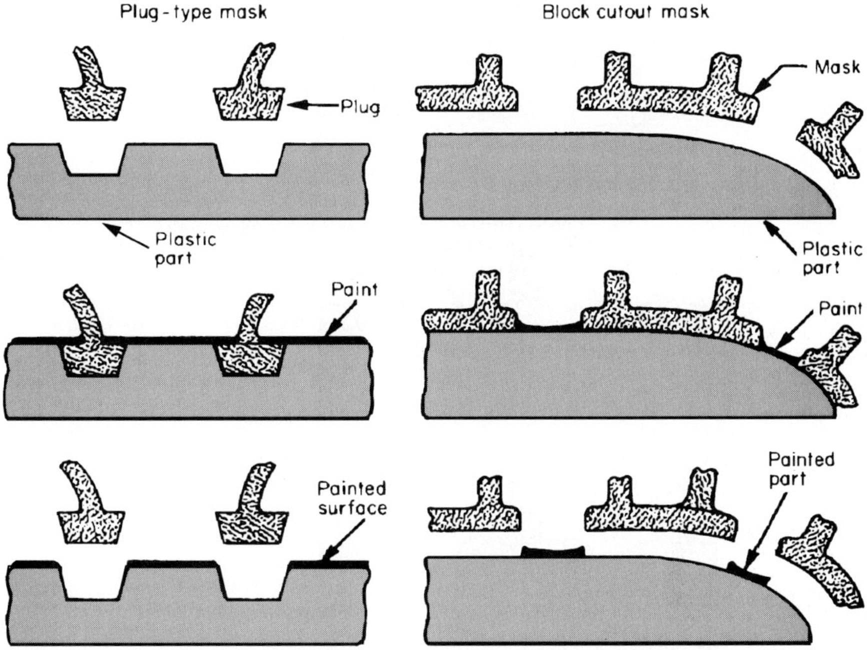

FIGURE 6.12.1 The plug mask is used to fill and keep depressed areas clean while the surrounding area is being painted. The block or cutout mask is used to confine paint to an area that has no paint step.

mask. This type of mask is used when lines and lettering are too small for practical individual masking. The *cap or lip mask* covers raised areas and vertical sidewalls to allow painting of recessed areas with a fine line of definition and without a secondary wiping step. The lip keeps the top clean and all, or as much as may be desired, of the sidewalls. The mask must have a lip of metal that extends down the sidewalls.

If spray masks are to be worthwhile, the molded part must be held to close tolerances. This calls for good mold design and molding technique. Slight variations in dimensions from part to part will allow paint to blow by the mask and cause rejects. Good, consistent molded parts are a must in spray-mask painting.

ROLLER COATING

This is a process for coating plastic surfaces with fluid paint by contacting the surface of the part with a roller on which the paint is spread (see Fig. 6.12.3). The roller surface is usually neoprene.

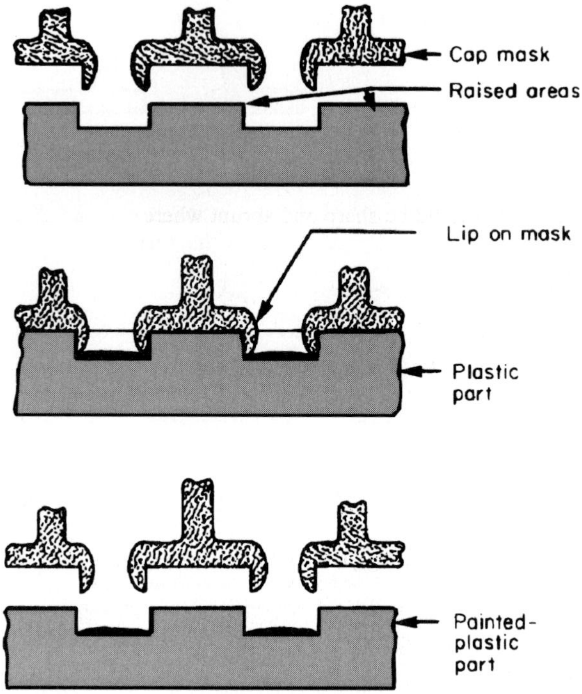

FIGURE 6.12.2 The cap or lip mask covers raised areas and/or vertical sidewalls to allow painting of recessed areas with a fine line of definition and without a secondary wiping operation.

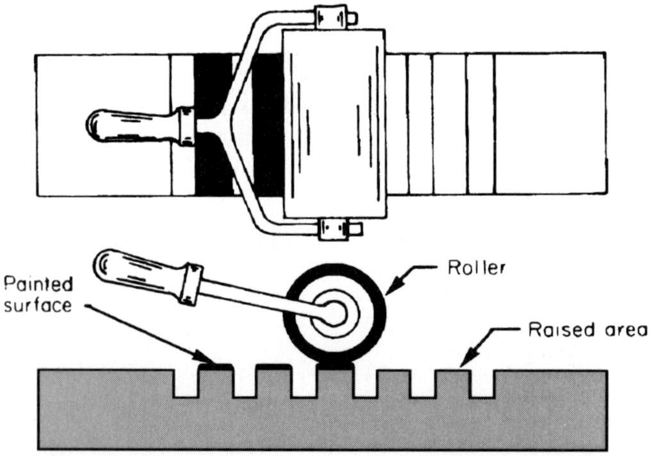

FIGURE 6.12.3 Small raised areas of lettering, figures, and other designs may be decorated by roller coating.

SPRAY-AND-WIPE DECORATING

In this process, paint is sprayed over a recessed area in the plastic part, and the excess paint is wiped away. Wiped-in letters are indented in the part, and either first or second surfaces can be sprayed and wiped.

Recess or surface depressions for wipe-in decorations should have a depth-to-width ratio of $1\frac{1}{2}$:1 (Fig. 6.12.4). Recesses are seldom over 0.8 mm ($\frac{1}{32}$ in) wide. Indentations or recesses should be sharp and abrupt where they meet the surface of the part. This prevents streaking of the paint across the surface as it is wiped off (Fig. 6.12.5). Wiped-in letters or decorations will not chip or rub away as easily as stamped letters. This is a more expensive method, but it does improve visibility of letters and decorations.

An injection-molded part that has depressed letters that are to be filled in with paint should be gated in such a manner that weld or knit lines do not form between the letters. If weld or knit lines are present, paint may run into them when the letters are being filled, thus resulting in a scrap part.

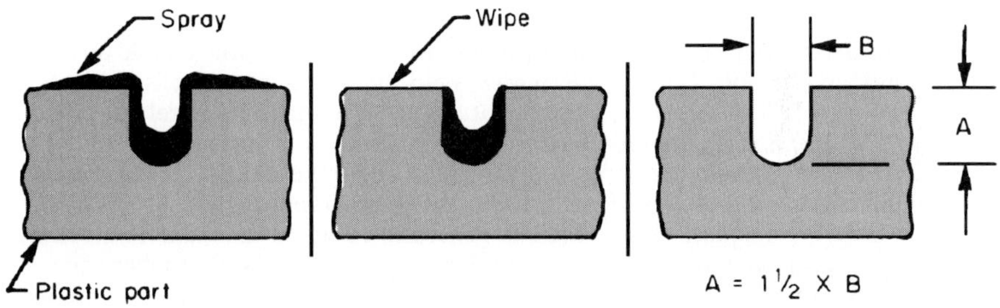

FIGURE 6.12.4 The proportions of depth to width for letters or depressions in the spray-and-wipe paint process.

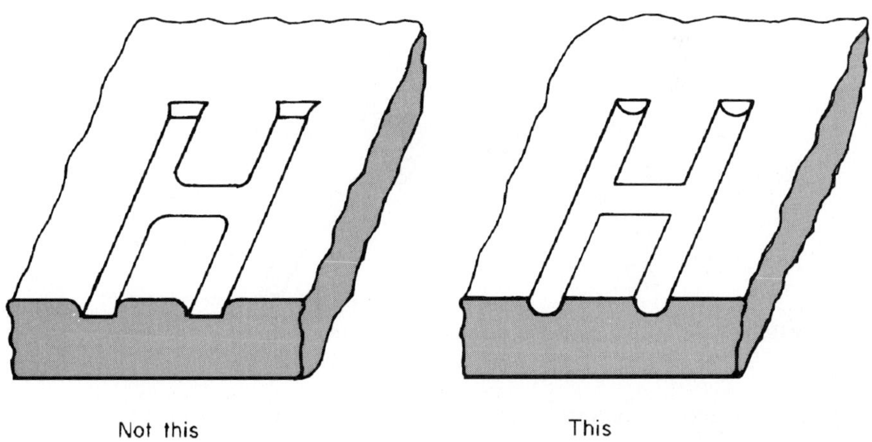

FIGURE 6.12.5 Wiped-in letters are depressed and should be sharp where they meet the surface of the part.

DYEING

The operation involves immersing the part in a dye solution that penetrates the part, providing permanent coloring. Eyeglass frames, buttons, buckles, and other clothing accessories are frequently colored by this method. Dyeable plastics include cellulose acetate, acrylates, nylon, and thermosetting polyesters.

PRINTING

Printing is a coating process in which ink or paint is applied to the surface of the part in selected small areas.

SILK-SCREEN DECORATING

Silk-screen decorating or printing is a process used to force paint or ink through a stencil fabric, commonly called a *silk screen,* onto the plastic part that is being decorated (see Fig. 6.12.6). The screen consists of a taut woven fabric securely attached to a rectangular frame and carefully masked with a stencil in a manner that allows the paint to be pressed through the screen only at areas where the stencil is open. For most applications, a screen with 9 perforations per millimeter (approximately 230 perforations per inch) is adequate. The stencil is generally of nylon or stainless steel. The screen is placed above the plastic part to be decorated, and a flexible rubber squeegee forces the ink or paint through the openings in the screen onto the surface of the plastic part.

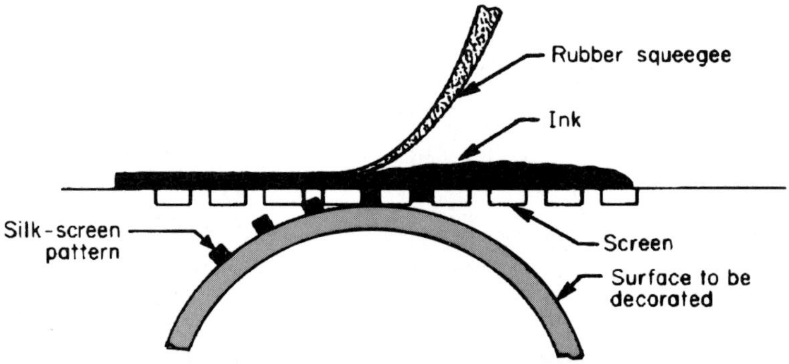

FIGURE 6.12.6 Silk screening is a process of transferring paint through an open screen onto a flat or gently rounded surface.

Multicolor decorating and intricate designs can be achieved in this process by employing a series of different screens. Silk-screen decorating is done on flat or curved surfaces and does not give the three-dimensional-depth effect. Although the inks used for this process are expensive, overall costs are generally economical.

Figure 6.12.7 shows a lampshade that has been silk-screen decorated. The injection-molded shade is made from high-density polyethylene, and the surface is flame-treated before being silk-screen-painted.

FIGURE 6.12.7 Injection-molded high-density polyethylene lampshade. The decoration on the lampshade has been silk-screen-painted.

PAD PRINTING

This process, shown in Fig. 6.12.14, is useful for printing small areas. It uses very light pressures and can be employed to print on highly sensitive, thin-walled products. It can apply multicolor designs. The operation is rapid and can be carried out without drying between color applications. The cost of tooling is normally less than that of other decorating methods. Acid-etched, chromium-plated copper plates normally are used. Eyeglass lenses and frames, sporting goods, pens, medical devices, spools, electrical devices, computer parts, and others with irregular surfaces are printed with this method.

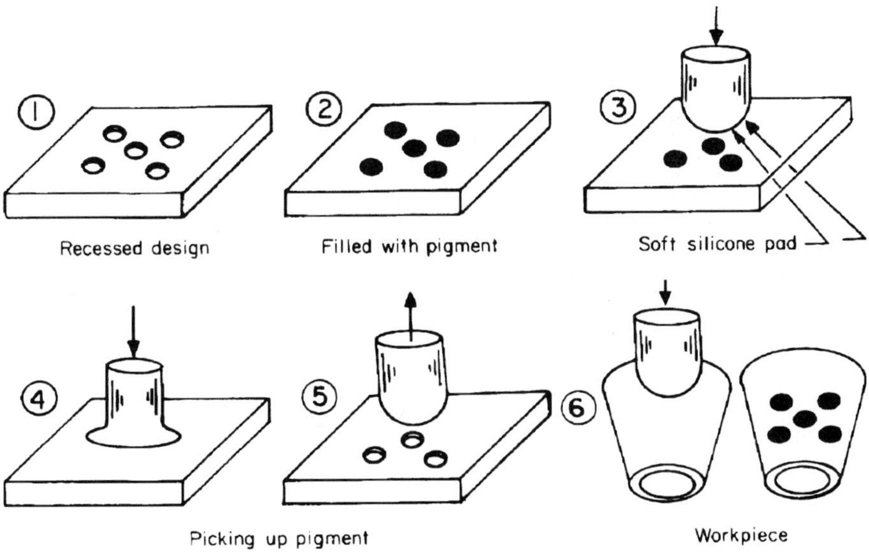

FIGURE 6.12.14 This illustrates schematically the pad-printing method of decorating.

HOT STAMPING

This is a process for marking plastics by transferring the decorative coating (paint or bright metallized plastic) from a carrier foil to the part by heat and pressure. The foil is pressed against the plastic surface by means of a heated die, thus transferring and welding selected areas of the foil to the plastic surface. (See Fig. 6.12.8.) The pressure from the die creates a recess for marking, which protects the stamped letter from abrasion, and heat causes the marking medium to adhere to the plastic.

Block lettering can be hot-stamped on either raised or flat areas. A flat metal or silicone rubber plate can be used to transmit heat and pressure to the foil for stamping raised areas. Silicone rubber dies are used to cover irregular surfaces. The flexibility of the die allows good transfer over erratic surfaces (see Fig. 6.12.9). Silicone rubber does not conduct heat as well as metal. If a silicone rubber die face must have a temperature of 135°C (275°F), the backup metal should be heated to 163°C (325°F). Surface pretreatment of the plastic part is not a prerequisite in hot stamping.

The tapes and foils that are used in hot stamping are made up specially for each

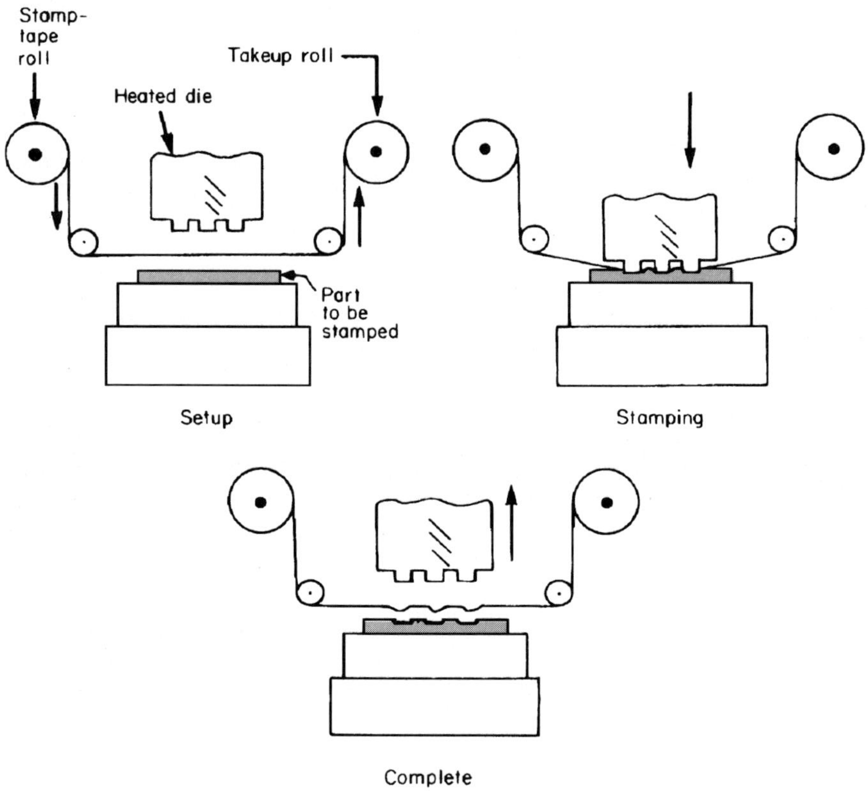

FIGURE 6.12.8 This illustrates schematically the process of hot stamping by using a heated metal die.

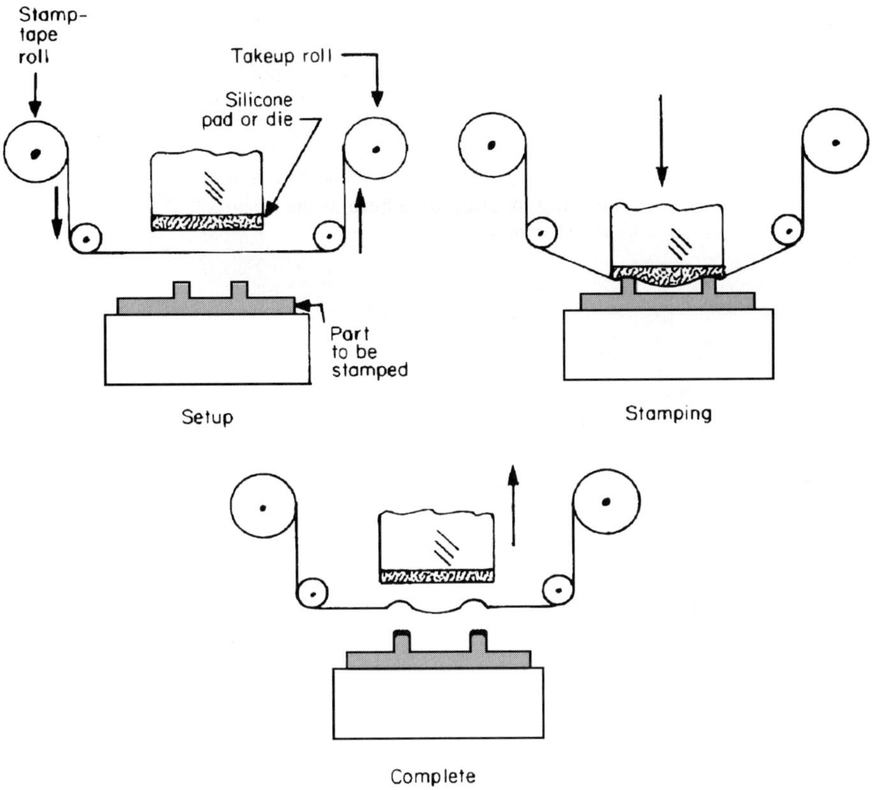

FIGURE 6.12.9 This illustrates schematically the process of hot stamping by using a hot-silicone pad or die.

type of plastic. Foils typically are made in four plies: (1) a thin film carrier, generally polyester, (2) a release coating, (3) a decorative coating, pigment or metallic, and (4) an adhesive, specific to the substrate material. During the stamping process, the heated die activates release agents in the foil and causes the coating to transfer to the plastic surface. Depending on the plastic and the tape makeup, either thermal bonding or activation of the adhesive occurs. Metallized foils are much more complicated, having as many as five layers. Figure 6.12.10 illustrates a typical engraved hot-stamping die. Note the intricate details that have been engraved into this die.

Metal hot-stamping dies are made from magnesium, brass, or steel. Inexpensive photoetched dies work well for prototype or short runs on flat areas. However, production stamping dies are normally pantographed and deep-cut from either brass or steel.

Figure 6.12.11 illustrates the following design recommendations for hot-stamping raised letters, borders, etc.:

1. The raised section to be hot-stamped should be at least 0.8 mm (0.030 in) high.
2. The width of a section to be hot-stamped should be at least 0.25 mm (0.010 in).

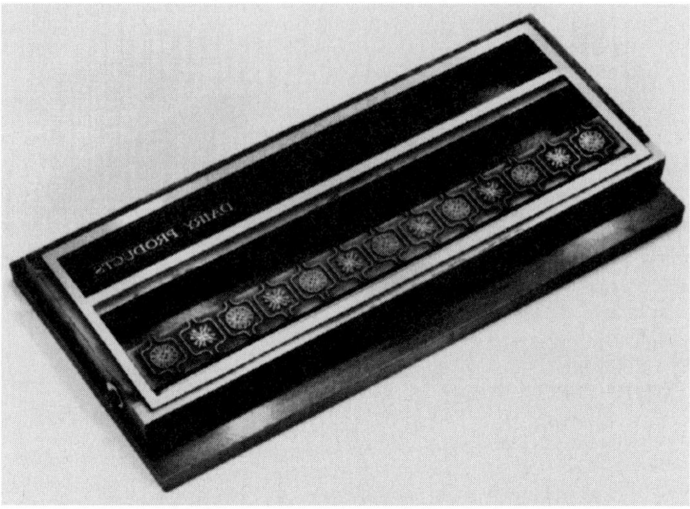

FIGURE 6.12.10 An engraved hot-stamp die. The die will be heated to the transfer temperature of the foil (about 275°F, or 135°C). When it is pressed against the part, an exact duplicate of the die pattern will transfer. Note the intricate detail that this process can transfer. (*Courtesy Service Tectonics Instrument, Detroit, Mich.*)

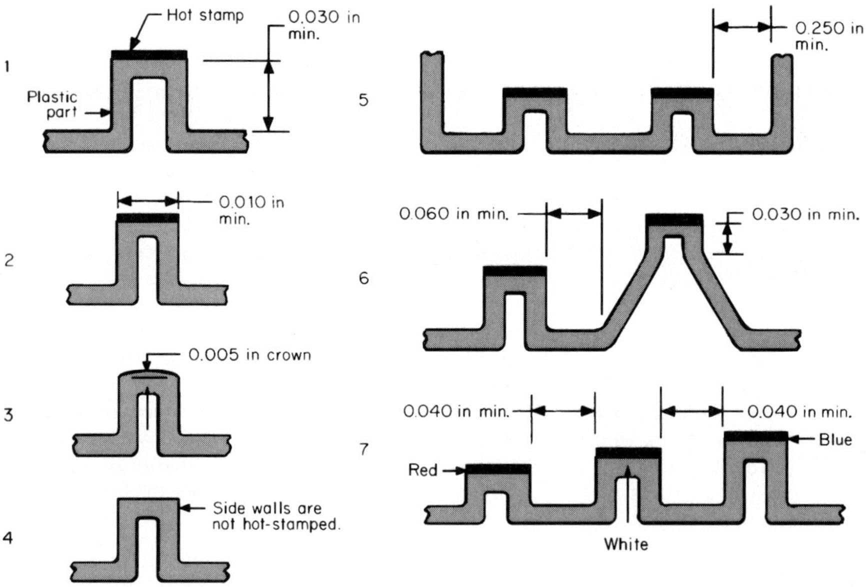

FIGURE 6.12.11 Design recommendations for hot stamping. Raised sections on plastic parts can be hot-stamped by using silicone rubber.

3. Raised sections to be hot-stamped should have a slight crown if possible.
4. Sidewalls of raised sections cannot be hot-stamped.
5. Raised sections located at the bottom of a straight sidewall of a part should be at least 6.3 mm (0.250 in) from the sidewall.
6. Raised sections located at the bottom of an angled sidewall should have a minimum clearance of 1.5 mm (0.060 in) from the sidewall to allow for roll-leaf clearance and silicone rubber squeeze-out.
7. In raised sections that are to be hot-stamped with several colors, the sections should be separated by a minimum of 1 mm (0.040 in) to avoid overlap of the hot silicone stamp dies.

See Fig. 6.12.12.

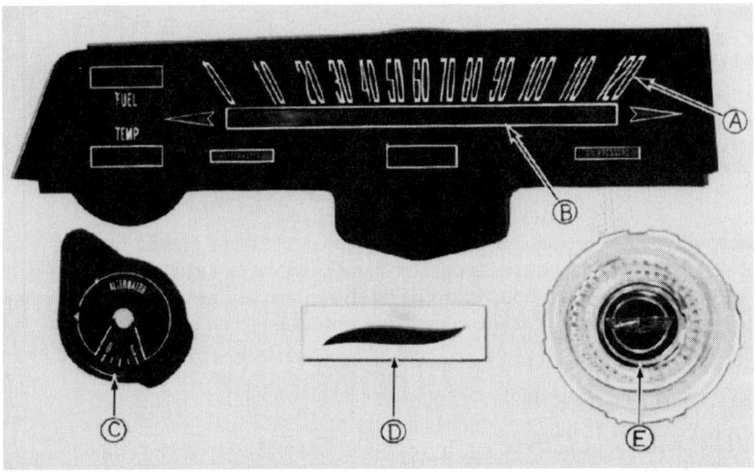

FIGURE 6.12.12 Molded plastic parts decorated by hot stamping. (*A*) Raised-dial numbers hot-stamped white with a silicone die. (*B*) Raised openings hot-stamped silver with a silicone die. (*C*) Hot-stamped white with a steel die. (*D*) Silicone rubber die. (*E*) Clear acrylic medallion hot-stamped on the second surface with a silicone rubber die.

DECALS

By printing in several colors on lacquer film bonded to a paper or plastic backing, multicolor decorative effects can be produced in one operation. The method is low in cost and produces high-quality effects. Heated rubber pads or rollers are used but at lower pressure than with the hot-stamping method described above. The transfer or decal adheres to the plastic part, and the backing sheet is removed. High-speed gravure presses are used to print the decoration on the decal film, and all colors are transferred to the plastic part. Wood-grained finishes on television and loudspeaker cabinets and many kinds of container labels are applied with this method.

Figure 6.12.13 illustrates two proprietary variations of this method: Electrocal and Di-Na-Cal. The Electrocal process is most effective on plastic parts that are rigid rather than those made from thin-walled, softer plastic materials. The Di-Na-Cal process can print two different multicolored designs on both sides of a plastic part

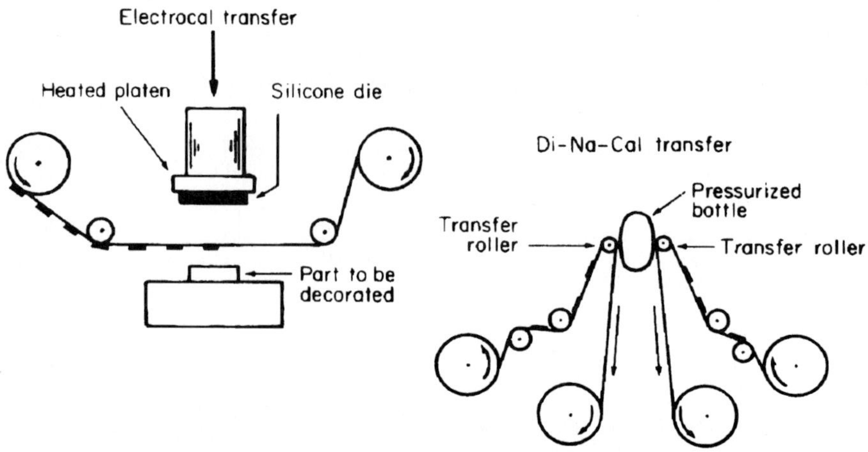

FIGURE 6.12.13 Two methods of heat-transfer decorating. The Electrocal transfer has a printed design on a roll-fed release-coated paper and is transferred to the part by heat and pressure. The Di-Na-Cal transfer method has parallel feed and takeup lines with printed transfer designs on roll-fed release-coated paper. Both sides of the article to be decorated are done at one time.

simultaneously. This process is used mostly for decorating plastic bottles. The maximum area on cylindrical bottles is approximately 135° per side, or about 75 percent of the total area. Thin-walled bottles are generally internally pressurized with air to provide the support necessary for the decorating process.

VACUUM METALLIZING

This process for depositing a thin layer of metal onto a plastic surface is covered in Chap. 8.3. Also see Fig. 6.12.15.

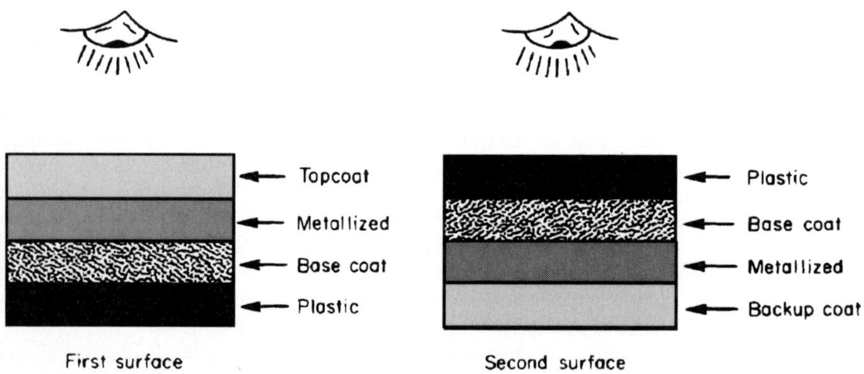

FIGURE 6.12.15 The difference between first- and second-surface metallizing on plastic parts. Second-surface metallizing is often preferred for clear plastic parts.

LETTERING PLASTIC PARTS

Preferred methods for molding lettering into plastic parts are discussed in Chaps. 6.1 and 6.2. A raised letter is visible if it is only 0.08 mm (0.003 in) high. Letters normally 0.4 mm ($^1/_{64}$ in) high are easily read because they catch sharp highlights. Letters over 0.8 mm (0.030 in) high should be tapered and have fillets at the base. Raised letters can be decorated or painted by roll coating.

ELECTROPLATING PLASTIC PARTS

Many common plastics can be electroplated with methods similar to those used to plate metal. A plated surface gives plastic parts greater surface hardness and solvent resistance in addition to the metallic appearance. Plated plastic parts are used in faucets, kitchenware, marine hardware, furniture, automotive components, hospital accessories, and toys. Processes and design rules are discussed in Chap. 8.2.

TWO-COLOR MOLDING

This is sometimes called *double-shot molding.* It is an injection-molding process for making two-color molded parts by means of successive molding operations. This is accomplished by first molding the basic case or shell and then, using this as an insert, molding numerals, letters, and designs in and around the insert. There are two variations of two-color injection molding, as shown in Fig. 6.12.16.

Figure 6.12.17 shows the two-color injection molding of adding-machine numerals.

In this type of molding, mold design is critical, and gating is specialized. Typical parts currently being molded include caps for typewriter and business-machine keyboards, automotive-taillight lenses, push buttons, and telephone-dial components.

DECORATING MOLDED-MELAMINE PARTS

Melamine dinnerware (Fig. 6.12.18) is made by molding a color-printed foil pattern on the top surfaces of the part. In making a decorative melamine part, the mold is first charged with melamine powder and then closed until the material has cured sufficiently so that it is blister-free when the mold is opened. The part also must be rigid enough to stay in the cavity as the mold is opened. The printed foil, saturated with resin, is inserted on the partially cured melamine molded part, and the mold is again closed. The printed overlay is molded into the melamine part and becomes a part of it. No fillers or pigments are used in the foil paper because the overlay must disappear into the molding, leaving only the design on the molded product.

Applications include dinnerware, wall plates for light switches, utensil handles, clock faces, instrument-dial faces, and color-coded control knobs.

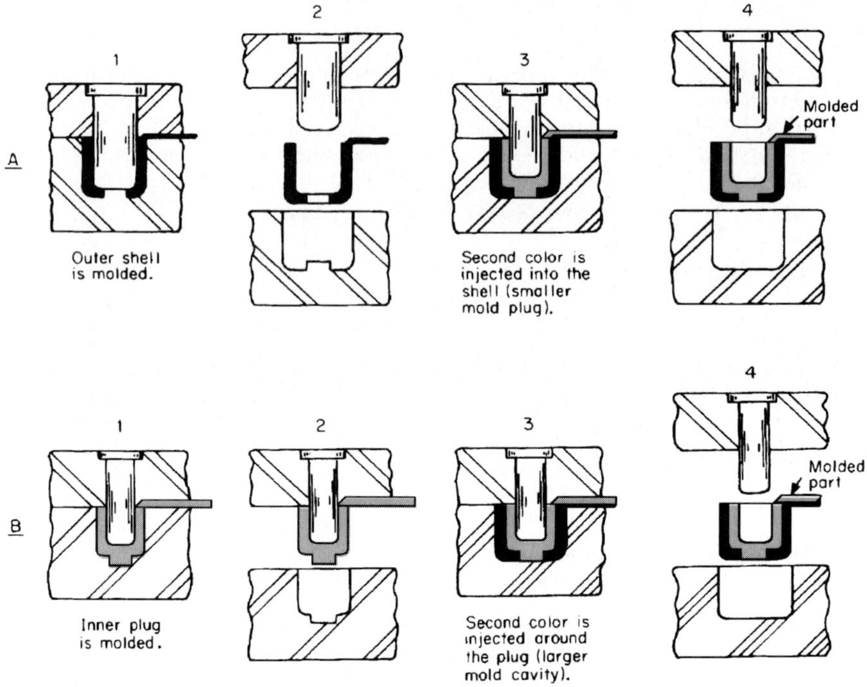

FIGURE 6.12.16 Two methods of making two-color molded parts.

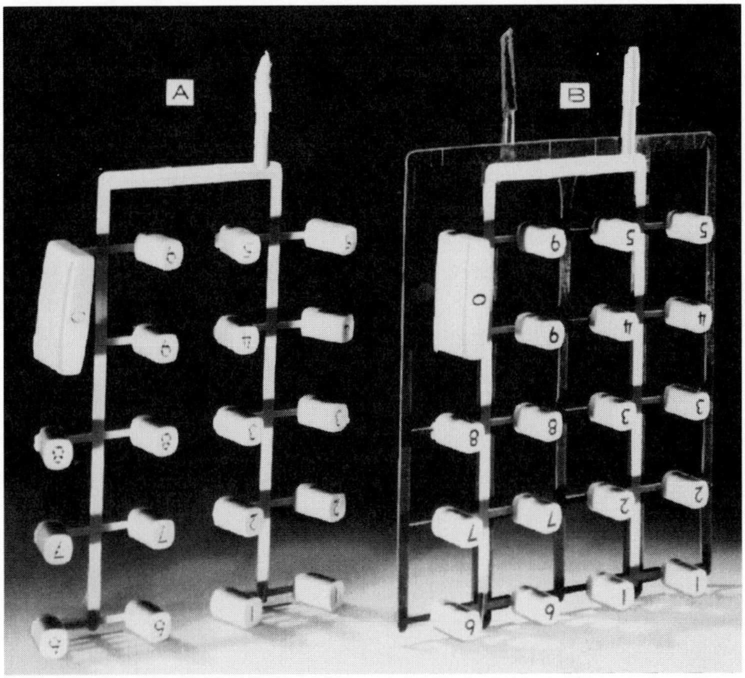

FIGURE 6.12.17 Two-color injection molding of adding machine keys. (*A*) An injection-molded shot of the shells with impressions for the numerals. (*B*) An injection-molded shot of finished key tops with the second contrasting color injected into the shells.

FIGURE 6.12.18 Materials that make melamine dinnerware. (*A*) Melamine powder in a preform. (*B*) Molded-melamine plate without the overlay sheet. (*C*) The printed overlay sheet. (*D*) Molded decorated melamine plate.

S • E • C • T • I • O • N • 7

ASSEMBLIES

CHAPTER 7.1
DESIGN FOR ASSEMBLY (DFA)

Very often, the most significant benefits with design for manufacturability (DFM) come from *designing for assembly* (DFA), simplifying the product so that it has fewer parts and its assembly is easier and faster. One reason for this is the fact that final assembly labor and overhead together may comprise the largest single item of cost in the manufacture of a product, and if this item can be reduced, the benefits are substantial. However, designing for assembly has ramifications far beyond the cost of the assembly operation, itself. A product designed for ease of assembly can provide benefits throughout the company. Service and recycling are facilitated when a product is simplified; one that is easy to assemble in the factory is normally easier to disassemble when maintenance, repair, or disassembly for recycling takes place. Simpler assemblies often can be brought to market sooner because of fewer parts to design, procure, inspect, and stock with less probability that a delay will occur. Products with fewer parts have an opportunity for better quality and reliability, although these attributes are not ensured.

Additional benefits, when the number of parts is reduced, are fewer engineering and production-control documents, lower inventory levels, reduced need for inspection and quality-control documents, fewer setups, less materials handling, and probably, reduced purchasing workload.

From this, it may be seen that the most advantageous sequence of analysis of a product or an assembly is to attack the overall assembly first. Not only does this provide the most benefits but when this is done, it will be known which parts will remain and what their function and general configuration will be. These parts can then be simplified, but we need to know which parts are retained and their general shape before they are analyzed for DFM/DFX improvements.

THE ASSEMBLY PROCESS

The assembly process involves the placement and fastening of one or more parts in or on another. Often, the operation is manual, although increasingly it is being performed by automatic equipment, particularly when production volumes are large. Very often, fixtures are used to hold one or more parts conveniently during the operation or to locate parts in precise relationship to one another.

Most assembly work is performed at individual workbenches, sometimes with the assistance of a conveyor to move parts or assemblies in process from one workstation to another. In the mass-production industries, a conveyor moves an assembly past a number of workstations, at each of which certain parts are added. This is the common assemblyline.

CHARACTERISTICS AND APPLICATIONS

Mechanical assemblies involve the use of multiple parts held together by a variety of means: fasteners such as screws, rivets, bolts, pins, wire staples, spring clips, and other parts that have the function of holding other pieces together. The pieces also may be held together by the tightness of fit or by the interlocking of the assembled parts themselves.

Mechanical assemblies may consist of only two parts (e.g., a kitchen salt shaker) or thousands of parts (e.g., an automobile). They can be as small as a woman's wristwatch and as large as a wide-bodied passenger airplane. They can have components of metal, wood, rubber, paper, plastics, ceramics, or a combination of these materials. Almost all household and commercial products including electronic products such as televisions, computers, etc., are to some degree mechanical assemblies. If a product's package is considered as well, it is only an extremely rare item that does not have at least one assembly operation in its manufacturing sequence.

ECONOMIC PRODUCTION QUANTITIES

There are no economic limits to the lot size for an assembly operation. Quantities can range from one to millions. However, the greater the quantity and the more often the assembly operations are repeated, the easier it is for product designers and manufacturing engineers to develop design and methods changes that simplify assembly operations even to the point of eliminating them altogether with a one-piece design. (For example, the coat hanger, which probably started as a four- or five-piece wood-and-wire assembly, is now commonly made from a single wire or plastic part.)

Usually, production quantities on the order of millions per year are required before a fully automatic assembly operation with dedicated equipment can be justified. The most common examples are packing operations for low-cost variety store items. These are almost invariably automated to some degree. However, computer-controlled robotic apparatus can assemble automatically when quantities are much lower. From a product design standpoint, if the designer assumes automatic assembly, it does not make a great deal of difference if the assembly is actually automatic or manual. By designing with automatic assembly in mind, the product designer will provide an assembly that is also more suitable for manual methods.

DESIGN CONSIDERATIONS

Product designers, when designing parts for assembly, should visualize how the parts are put together. This in itself will help ensure that they consider design alternatives that facilitate assembly. The designer should understand the assembly method to be used and know what tools, fixtures, and gauges will be used during assembly.

Designers should look on the assembly or subassembly as a means to reduce the cost of production. An assembly should be used when the desired results and cost (including tooling investment) can be achieved better with a grouping of parts than with a more complex individual part. Each component of assembly should be designed to reduce the number of manufacturing and assembly operations to a minimum.

The best assembly is usually the one that has the fewest parts and the least costly type of fastening (consistent, of course, with the functional requirements of the product). In the long run, the lowest-cost assembly is the one that minimizes the total costs for parts, assembly labor, finishing, tool amortization, and product service and warranties.

Most of the design suggestions that follow are applicable to assemblies fastened by the purely mechanical methods mentioned above and to those welded, soldered, brazed, and bonded as well.

Minimizing the Number of Parts

Reducing the number of parts is the prime approach in the improvement of an assembly, far overshadowing in impact any other changes in design that improve manufacturability and further other important design objectives.

There are a number of design approaches that lead to a reduction in the number of component parts in an assembly. The important principles are as follows:

1. *Combine parts.* Integrate the functions of several parts into one. For example:

- ***a.*** *Incorporate hinges.* Hinges can be incorporated in many plastics parts if the plastic material is flexible and the wall section is thin. This eliminates the need for a two- or three-part hinge and the fasteners required to attach it to two other parts. Many storage containers for consumer products are made with integral hinges. Both injection molding and thermoforming permit this design approach. Figure 7.1.1 illustrates a typical example, a container for a spare key, whose lid is held with a "living hinge" of the same material as the container and lid.
- ***b.*** *Use integral springs.* Springs can be incorporated in metal and plastic parts and sometimes those made of other materials such as fiber. The result is a simpler, faster assembly. Separate springs are often troublesome from a handling standpoint because they can tangle easily and because their flexibility makes them difficult to handle and insert in the assembly. The integral spring, therefore, provides very significant assembly advantages. Figure 7.1.2 illustrates the principle.
- ***c.*** *Use snap fits.* Replace screw-type and other separate fasteners with integral snap-fit elements, tabs, or catches. An example is shown in Fig. 7.1.3. It illustrates the kind of element which, when incorporated in a plastic or sheet metal part, snaps into place in an undercut in the mating part to hold the two together. Figure 7.1.39 shows a number of snap-fit assemblies of metal parts.
- ***d.*** *Incorporate guides, bearings, covers, etc.* With some manufacturing processes, these elements can be incorporated in the basic part with a tremendous reduction in the number of components. Many plastic materials have natural lubricity that makes them quite suitable for applications involving bearing surfaces, particularly if the velocity and pressure involved are low. For more demanding applications where bearings, slides, cranks, etc., are needed, powder-metal parts can be made with sufficient precision for these functions. They also can be made with porosity so that lubricating oil is retained in the part itself.
- ***e.*** *Put electrical and electronic components in one location,* and consolidate components as much as possible. For example, one combination printed circuit board is preferable to several in separate locations; a light switch and ventilation switch on the same mounting plate is preferable to locating them separately, each with its own mounting hardware.

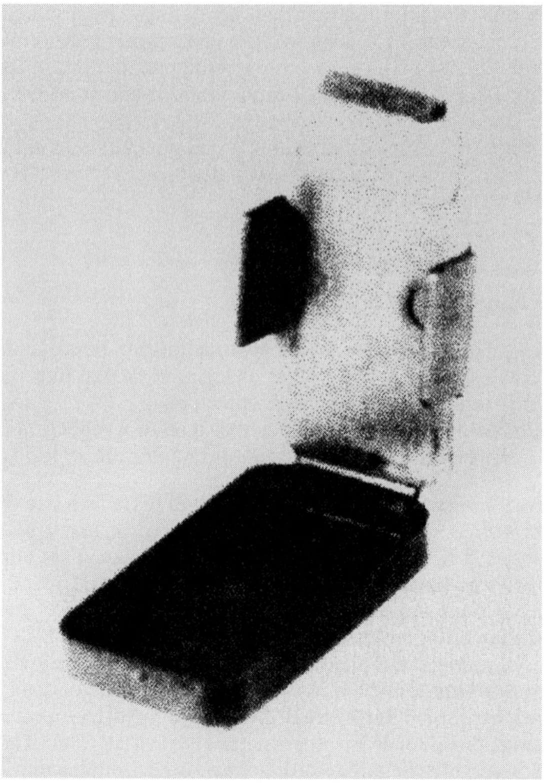

FIGURE 7.1.1 An example of an integral "living" hinge. The box, cover, and latch of this container for a spare automobile ignition key are all elements of one injection-molded plastic part. The flexibility of polyethylene provides sufficient hinge action for the thin-walled section that connects the box and the lid. (*From James G. Bralla,* Design for Excellence, *McGraw-Hill, New York, 1996.*)

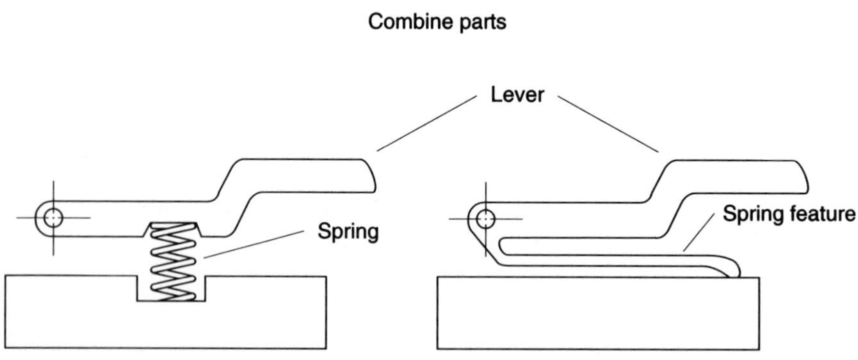

FIGURE 7.1.2 By incorporating the spring function in the lever, the need for a separate coiled spring is eliminated. (*From James G. Bralla,* Design for Excellence, *McGraw-Hill, New York, 1996.*)

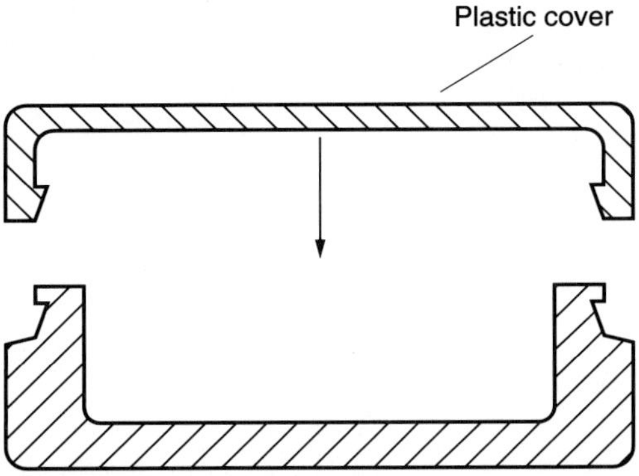

FIGURE 7.1.3 This product illustrates the snap-fit principal to attach the cover, eliminating the need for screw fasteners. (*From James G. Bralla,* Design for Excellence, *McGraw-Hill, New York, 1996.*)

Some processes allow very complex parts that result when separate parts are combined into one. Injection molding, die casting, and investment casting are examples. With these processes, a complex part, as would result from the combination of several simpler parts, primarily requires only a more complex mold or die. The extra complexity adds only marginally to the cost of the part made by the tooling.

Some other specific guidelines for combining parts are the following:

1. Use bent tabs or crimped sections instead of separate fasteners to hold several parts together. (See Fig. 7.1.4.)
2. Use combined fasteners, i.e., those with integral washers.
3. Instead of using nuts or threaded holes in mating parts, use self-tapping screws. This eliminates costly machining operations otherwise needed to provide a precision hole with internal screw threads.
4. Use cast or molded-in identification instead of attached labels. Such identification is more reliable because it is more permanent. It completely eliminates the costs involved in purchasing, stocking, and affixing a separate label.
5. Use integral locators, hooks, or lips to replace some of the fasteners holding one part to another. (See Fig. 7.1.5 for an example.)
6. Press fits, integral tabs, or rivets sometimes can be used to replace threaded or other fasteners that may be more complex. Press fits with flexible or grooved components are normally less expensive and as effective as precision-machined parts.

Use the following guideline questions to evaluate the feasibility of combining parts:

1. *When the product is being used, does the part move with respect to mating parts?*
2. *Must adjacent parts be made of different materials?*
3. *If parts were combined, would assembly of other parts or field service be made more difficult or unfeasible?*

If the answer to any of the preceding questions is "Yes," then it is probably not feasible to eliminate the part, but if the answer to all is "No," the part is a good candidate to be combined with others.

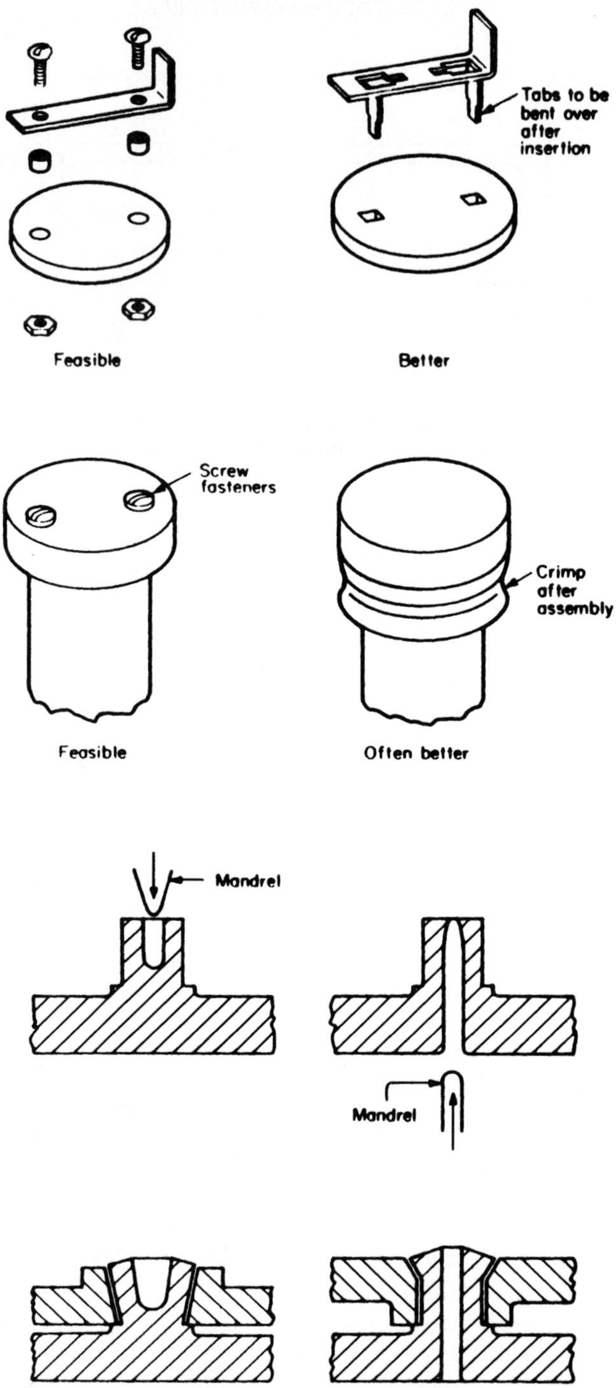

FIGURE 7.1.4 Bent tabs, rivet-like extensions, and crimped sheet-metal members are often less costly than separate fasteners. (*Items at the bottom courtesy American Die Casting Institute.*)

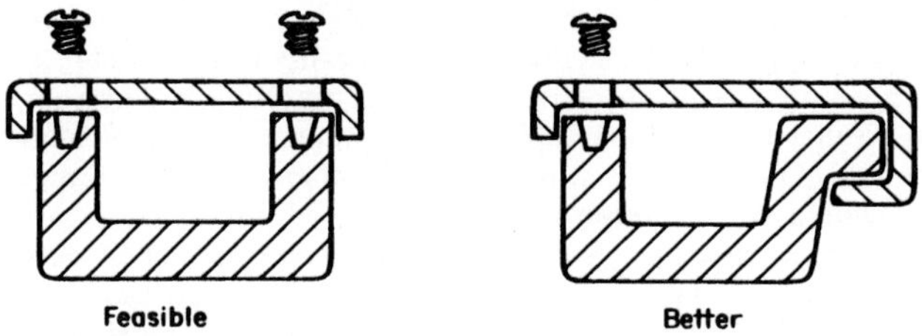

FIGURE 7.1.5 Minimize the number of fasteners by incorporating lips or other hooking elements in the basic parts.

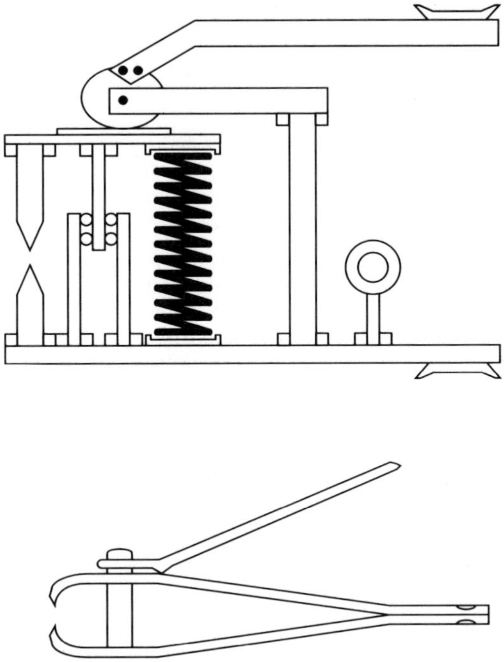

FIGURE 7.1.6 Two designs for a fingernail clipper. (*Top*) With single-function elements. (*Bottom*) With function-sharing elements. (*Courtesy of Prof. Karl Ulrich of The Warton School, University of Pennsylvania.*)

Figures 7.1.1 through 7.1.4 illustrate some simple examples of how two or more parts can be combined into one. Figure 7.1.6 illustrates the large number of functions that have been consolidated into a few parts in a common everyday product, a fingernail clipper.

2. *Make an outright reduction in the number of parts.* Sometimes a designer, not tuned to manufacturability, will incorporate more fasteners or other elements than the product really needs. (This is sometimes evident when a value analysis is made of an existing product and, with the benefit of hindsight, it is seen that there are more fasteners or other elements than needed to meet all reliability and other performance objectives.) Sometimes, the number of elements can be reduced by increasing the size of those which remain; e.g., replace a series of small machine screws with a

smaller number of heavier ones. In other cases, a planned or existing redundancy can be eliminated by upgrading the performance capability of the primary component.

3. *Make a major or full-product redesign.* This occurs when an assembly is redesigned so that the function supplied by a separate component is achieved by another method. One example would be the replacement of a flanged and bolted pipe system with a threaded pipe system.
4. *Use a different technology.* Sometimes great benefits can be achieved when a drastic design change enables a product function to be performed in a completely different manner. This occurs, for example, when a mechanical device is replaced by an electronic microcircuit.

OTHER MAJOR GUIDELINES FOR ASSEMBLY IMPROVEMENT

General Recommendations

1. Standardize designs. Use standard fasteners and other parts. Use as few sizes and styles as possible. Reduce the total number. Create a preferred parts list, and minimize the number of varieties on the list. Subassemblies usable on several products also should be standardized as much as possible.
2. Use subassemblies, particularly modular subassemblies, especially if the subassembly process is different from the final assembly process (e.g., a PC board in an electromechanical product). Modular assembly provides quality and reliability advantages (when the module is used previously and pretested) as well as serviceability advantages. Final assembly is also greatly simplified if it involves only the placement and attachment of major modules. Product service also is usually simpler with this approach. In many cases, a particular module can be applicable to a number of different final assemblies and thereby gain the benefit of economy of scale of production.
3. At the same time, avoid too many levels of subassembly, since extra subassemblies add overhead in the form of manufacturing specifications, floor space, and inventory and tend to increase manufacturing throughput time.
4. Design parts so that they cannot be inserted incorrectly. This aids in manufacturability but is particularly important for quality reasons. Figure 7.1.7 provides an example.
5. Avoid the use of flexible parts, if possible, because it is more time-consuming to handle and place them in position. Such parts are also susceptible to tangling. One common example is the replacement of connecting cables for electronic circuit boards with a design that provides a plug-in connection from a board to another element.
6. When the design permits and the part can be contained, open-end slots are preferable to holes or closed slots because they allow shafts or other mating parts to be assembled from the top (or side) instead of endways. (See Fig. 7.1.8.)
7. Use layered, top-down assembly. In other words, design the product so that each successive part can be added to the assembly from above rather than from the side or bottom. This approach is virtually essential for robotic assembly and has been found to be very beneficial for manual assembly as well. In addition to the benefit of having robotic motions standardized, there is a benefit from gravity assist to the assembly.

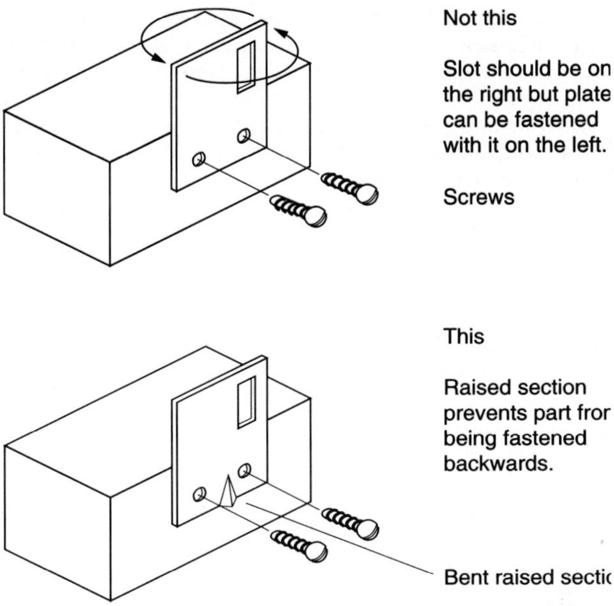

FIGURE 7.1.7 Design parts so that they cannot be assembled incorrectly. The lower plate has a raised section that prevents it from seating securely to the base if it is turned incorrectly. (*From James G. Bralla,* Design for Excellence, *McGraw-Hill, New York, 1996.*)

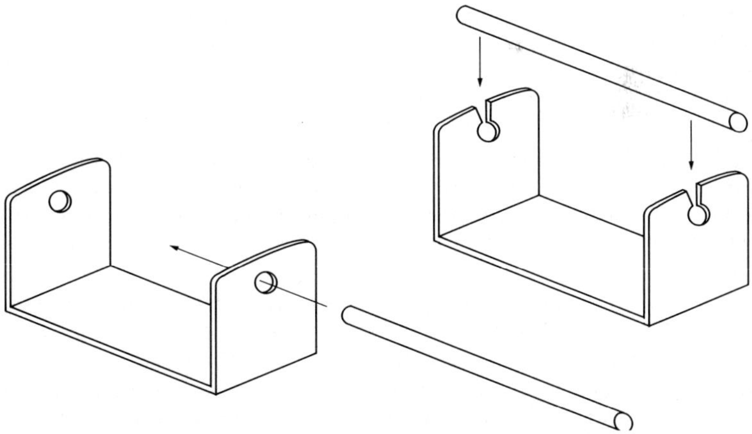

FIGURE 7.1.8 The sketch on the left shows undesirable from-the-side assembly. The other sketch shows the desirable top-down assembly and a snap fit for the shaft. (*From James G. Bralla,* Design for Excellence, *McGraw-Hill, New York, 1996.*)

If parts cannot be assembled from the top down, an effort should be made to ensure that simple straight-line motions can be used during the insertion of the part. Parts that must be inserted by "snaking" them around other parts make the assembly operation more costly.

Parts also should be designed so that each one acts as a nest for the part that follows. Avoid designs that require reorientation of assembled parts or some subassembly in order for an additional part to be added.

8. Avoid designs that require the simultaneous addition of several parts that must be kept in alignment as they are added to the main assembly, particularly when space is limited.

9. Design parts to be self-aligning. (See Fig. 7.1.9.)

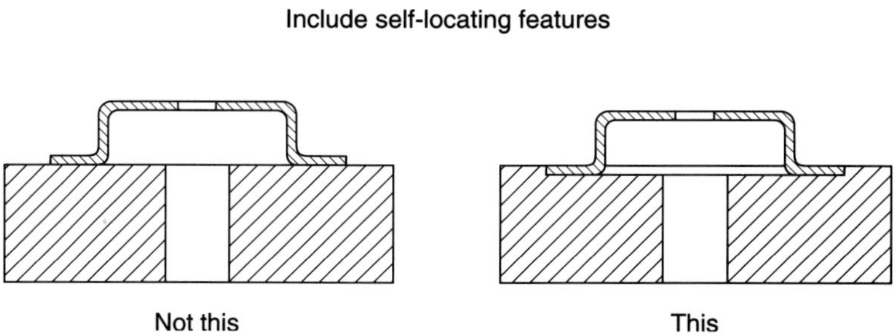

FIGURE 7.1.9 Design parts to be self-aligning, as in the assembly on the right. (*From James G. Bralla,* Design for Excellence, *McGraw-Hill, New York, 1996.*)

10. Eliminate adjustments as much as possible. One way to do this is to use resilient parts to take up the slack when the natural fit between components is not exact and an adjustment of position would otherwise be necessary. If parts dimensions are controlled well enough, it is better to rely on surface-to-surface positioning of components to establish assembly dimensions than to use an adjustment during assembly. Tabs, shoulders, notches, or other locators on the parts should be used whenever possible. However, if requirements for accuracy of the dimensions of the finished assembly are greater than those which can be provided by the parts, it may be preferable to control the assembly dimension by means of a fixture.

11. Use funnel-shaped openings of holes and slots whenever possible (without adding manufacturing operations) because this simplifies insertion of mating parts. Similarly, if possible without added operations, put a taper or bullet nose on parts that are to be inserted into other parts. (See Fig. 7.1.10.)

12. Use fasteners that lend themselves to hopper, strip, or other automatic feeding methods, especially if assembly quantities are large. This reduces handling labor and provides greater assurance or correct placement.

13. Design parts so that they are easy to handle. This may involve adding a grasping element or projection to very small or highly irregular parts. For parts that are automatically fed, it may involve having them fastened to a feeding strip.

14. When mating parts have through holes for shafts, fasteners, etc., use slots or oversized holes on one of the parts to allow for possible misalignment. (See Fig. 7.1.11.)

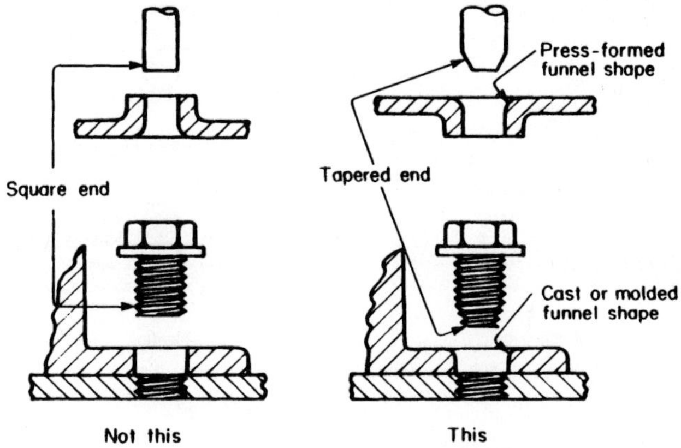

FIGURE 7.1.10 Use funnel-shaped openings and tapered ends to facilitate insertion of parts.

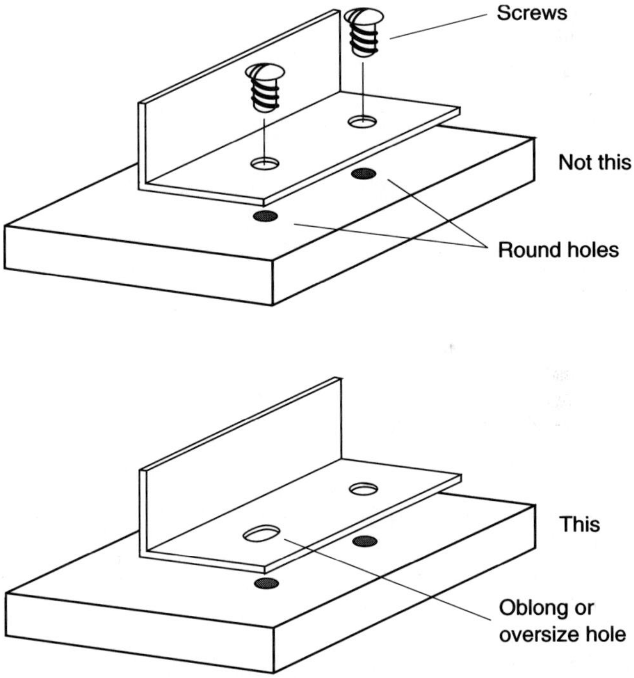

FIGURE 7.1.11 Use oblong or oversized fastener holes to reduce the need for accurate alignment when two parts are fastened together. This reduces tolerance requirements of the mating parts and simplifies assembly. (*From James G. Bralla,* Design for Excellence, *McGraw-Hill, New York, 1996.*)

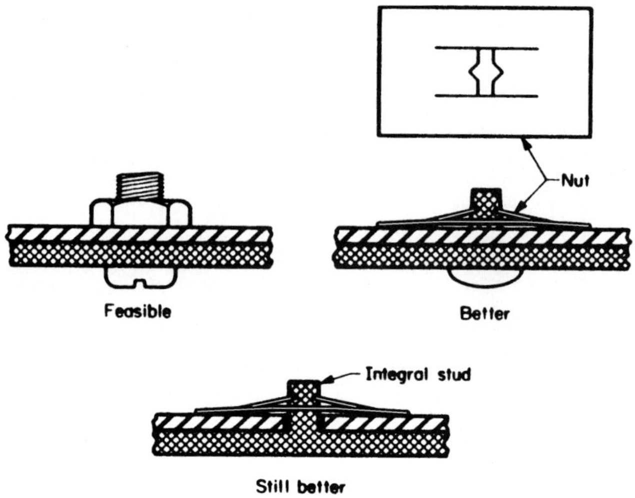

FIGURE 7.1.12 Push-on fasteners often can be used to speed assembly. Costs are reduced further if integral studs are employed.

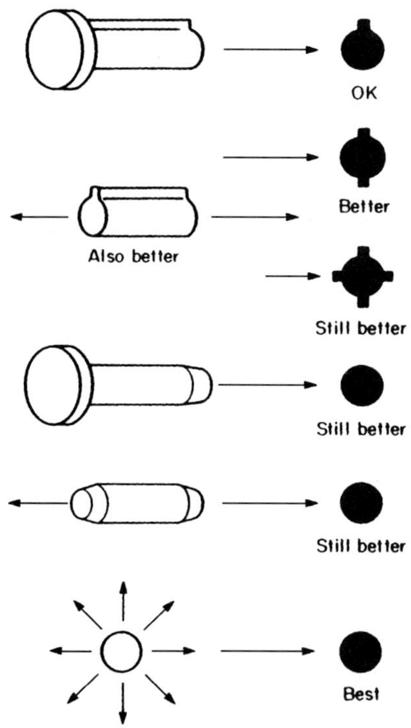

FIGURE 7.1.13 Design parts so that they can be inserted in as many ways as possible. Reduce the amount of axial and end-to-end turning required.

15. As much as possible, avoid assembly designs that require parts to be manually held in place until other parts are inserted because this kind of situation has some risk of quality problems as well as additional assembly time and cost.
16. Use the loosest fits possible between mating parts, consistent with product function, unless the purpose of the tight fit is to hold the parts together.
17. Keep internal mechanisms accessible, or use a design that permits a housing cover to be installed after all other assembly and adjustment operations are complete.
18. Use push-on fasteners, if snap fits are not feasible, instead of threaded fasteners to reduce assembly time. (See Fig. 7.1.12.)
19. Design small parts so that they can be inserted in as many ways as possible, from both ends, if possible, with the least amount of angular orientation. (See Fig. 7.1.13.)
20. Consider the use of plastic molding or zinc die casting to hold metal parts together. (See Fig. 7.1.14.)

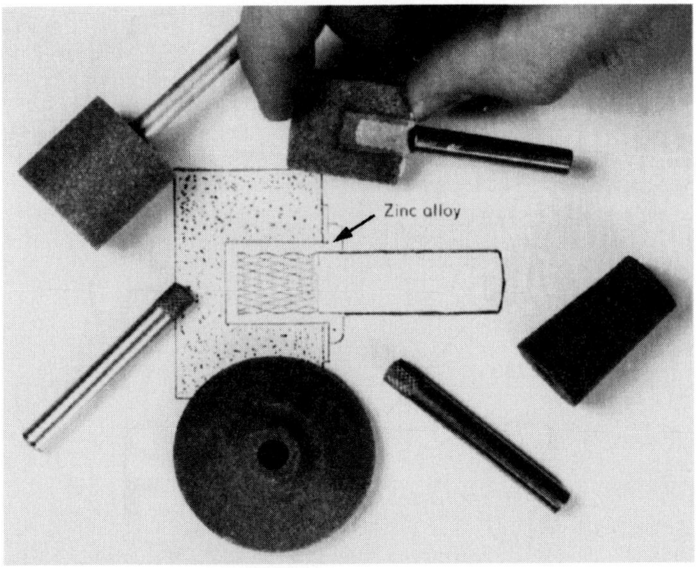

FIGURE 7.1.14 Injected zinc or lead provides a permanent means of fastening parts together. A combination casing die-holding fixture is required. (*Courtesy Fisher Gauge Limited.*)

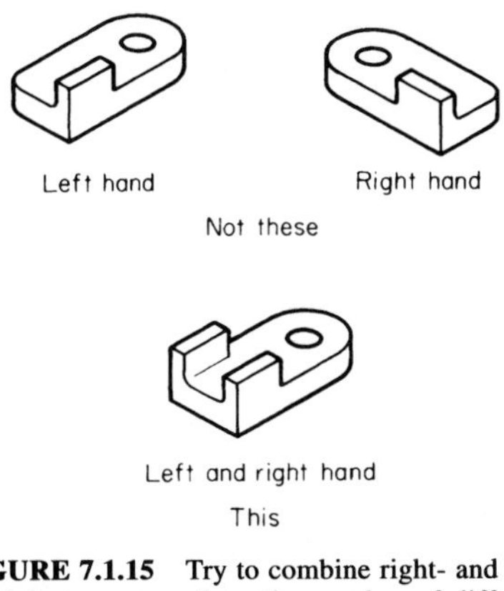

FIGURE 7.1.15 Try to combine right- and left-hand features to reduce the number of different parts.

21. Check right- and left-hand parts and subassemblies to see if they can be made identical, thus avoiding the need for extra part designs with separate inventories, etc. (See Fig. 7.1.15.)

22. As much as possible, avoid the use of components that can tangle when in a mass prior to assembly. This means that hooklike projections should be avoided, surfaces should be smooth, and holes and slots should be avoided. (See Fig. 7.1.16.)

23. It is usually better, from a manufacturing-cost standpoint, to use fewer large fasteners instead of a larger number of small fasteners.

24. If possible, avoid subassemblies of loosely held and flexible parts that can be damaged or entangled in handling. When such assemblies are unavoidable, fixtures should be provided to facilitate assembly and holding boards used to facilitate handling.

25. Snap rings often provide an inexpensive way to fasten parts, particularly when freedom of movement is desired, as in the case of rotating shafts. A separate retaining ring is often more economical than the use of a headed pin because considerable machining can be eliminated. (See Fig. 7.1.17.)

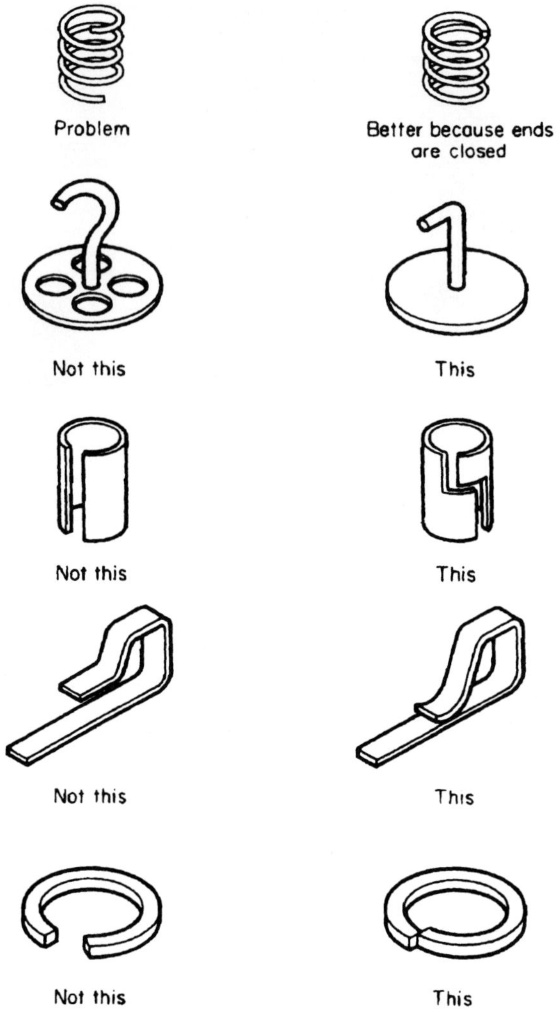

FIGURE 7.1.16 The use of parts that easily become entangled should be avoided.

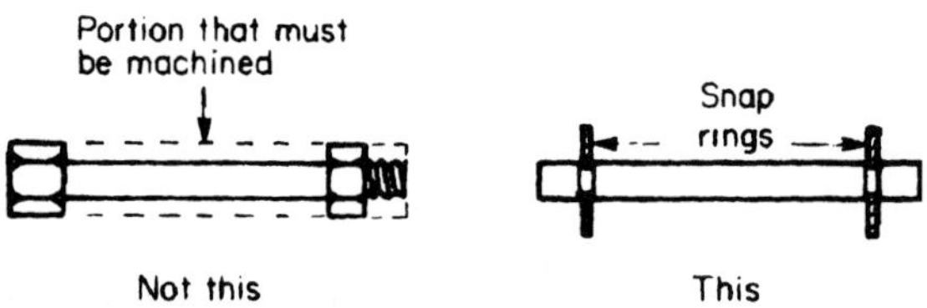

FIGURE 7.1.17 Snap rings can function as shoulders or heads on pins and shafts as well as fasteners.

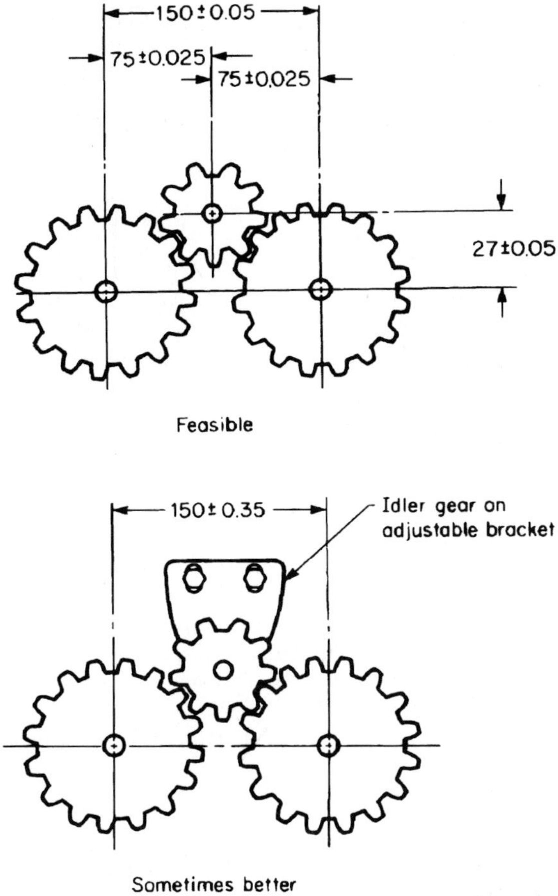

FIGURE 7.1.18 Sometimes added parts in an assembly like this adjustable idler-gear bracket can reduce costs by reducing the machining precision that would otherwise be required.

26. Occasionally, it pays to *add* parts to an assembly if doing so allows more liberal tolerances in the component parts. Figure 7.1.18 shows a gear train with an idler gear whose position is adjustable, thus obviating the need for extreme tolerances on the location of the gear-shaft holes.

Rivets

The major virtue of rivets is the strength and permanence of the joint they produce. When hopper-fed and clinched automatically, however, they also constitute a low-cost means for fastening parts together.

The following design rules will aid in the economical use of rivets:

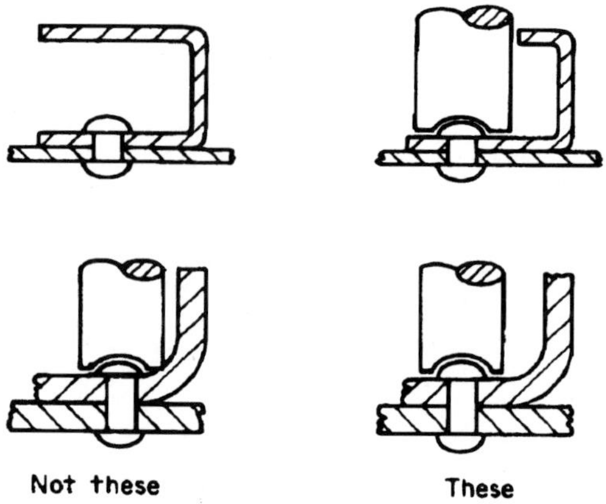

FIGURE 7.1.19 Provide sufficient room in the assembly for rivet-clinching tools.

1. Provide sufficient clearance around rivet locations to allow room for a standard riveting gun and avoid marring the workpiece. (See Fig. 7.1.19.) The inclusion of an access hole sometimes permits a rivet to be placed in an otherwise unclinchable location.
2. Use eyelets and tubular rivets whenever they provide sufficient holding power for the application. Tubular and semitubular rivets and eyelets require much lower clinching forces and can be hopper-fed, inserted, and set automatically on inexpensive equipment. (See Fig. 7.1.20.)

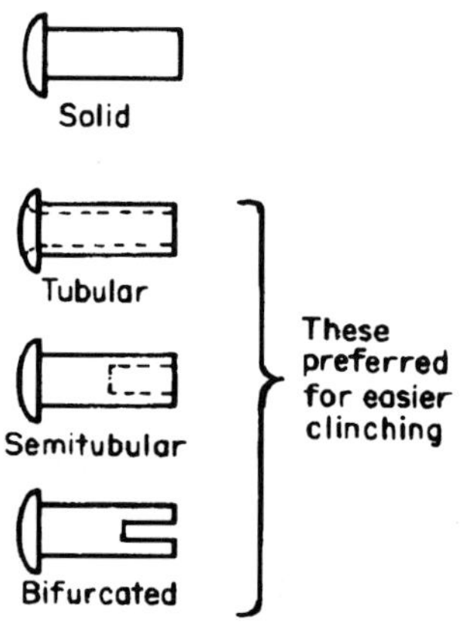

FIGURE 7.1.20 Tubular and semitubular rivets and eyelets are preferred.

3. Hole diameters must be correct for best results when rivets join two workpieces. The recommended diametral clearance is 5 to 7 percent. If the rivet hole is too large, the rivet may buckle during clinching and the joint will be loose and weak.
4. Blind rivets are valuable when the back side of a riveted assembly is inaccessible, but such rivets are more expensive than the conventional type and should not be substituted for conventional rivets unless these cannot be clinched.
5. When riveting thick materials, buckling of rivets can be avoided by counterboring the rivet holes. (See Fig. 7.1.21.) Counterbores must be wide enough to permit access of the riveting tool.
6. Rivet holes should not be too close or too far from the edges of the parts being joined. Holes should be from $1\frac{1}{2}$ to 8 stock thicknesses from the edge. This provides good support for the riveting tool and ensures that edges are held together. (See Fig. 7.1.22.)

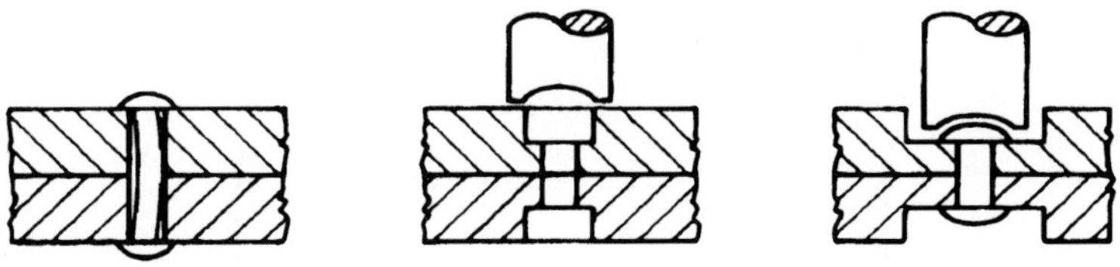

FIGURE 7.1.21 Use wide counterbores when riveting thick components to avoid buckling the rivets.

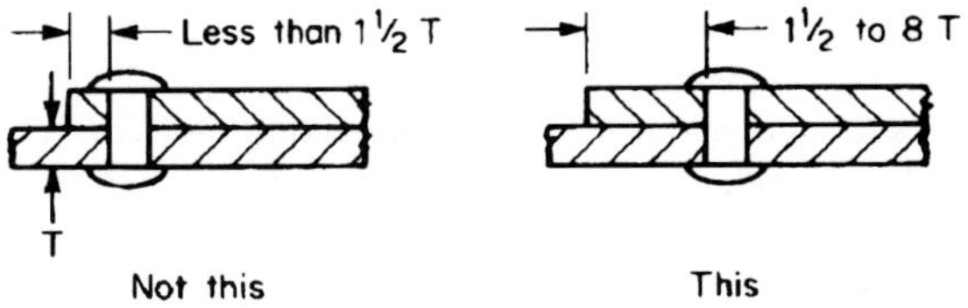

FIGURE 7.1.22 Recommended rivet-to-edge dimensions.

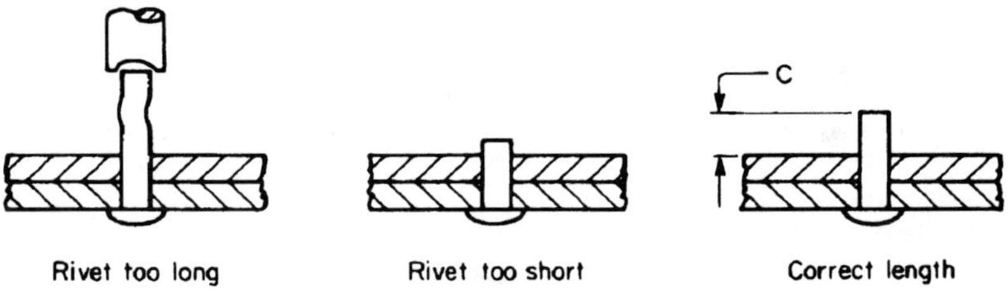

FIGURE 7.1.23 Proper rivet length is important.

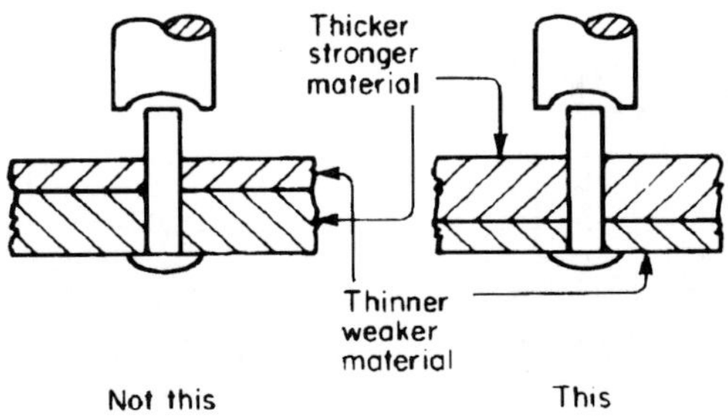

FIGURE 7.1.24 Upset rivets against the thicker, stronger material.

7. Rivets should be the proper length to avoid incorrect clinching. (See Fig. 7.1.23.) Rule-of-thumb clinching allowances *C* are as follows: solid rivets, 200 percent of shank diameter; semitubular rivets, 50 to 70 percent of shank diameter; and full tubular or bifulcated rivets, 100 percent of shank diameter.

8. When joining pieces of different thickness, it is preferable to upset the rivet against the thicker, stronger material. (See Fig. 7.1.24.)

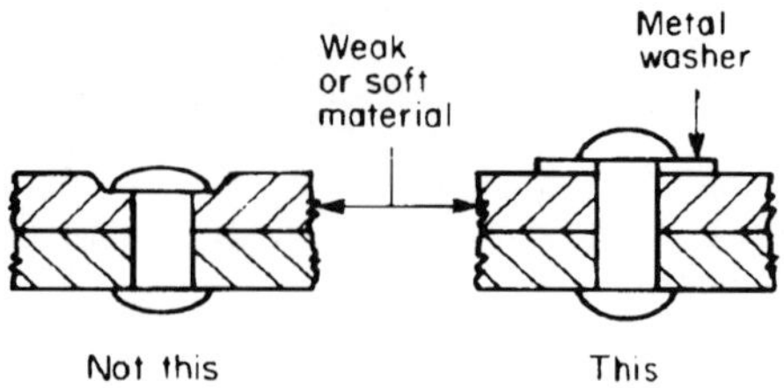

FIGURE 7.1.25 Metal washers distribute the force of upsetting.

9. When joining soft or fragile materials with rivets, it is desirable to use metal washers to distribute the force of upsetting and prevent damage to the weak part. (See Fig. 7.1.25.)
10. When joining a weaker material (such as leather, plastic, or wood) to a stronger material (such as sheet steel or aluminum) with a blind rivet, it is best to use a large head that bears against the weaker material and to clinch against the stronger material.
11. Blind rivets also require adequate tool clearance for the clinching tool on the side from which the rivet is placed. If space is limited, it may be advisable to rivet from the other side.
12. Blind rivets, though they do not have the strong axial forces of conventional rivets, still should not be set against insufficiently supported surfaces. (See Fig. 7.1.26.)
13. Sometimes it is feasible to have a rivet-type fastening by incorporating integral lugs on one of the assembled parts. This eliminates the need for rivets as separate parts. Figure 7.1.4 shows examples.

FIGURE 7.1.26 The surface against which blind rivets are set must be well supported.

Screw Fasteners

Threaded fasteners—screws, bolts, and machine screws—are common devices used to secure parts together. The following recommendations should assist in achieving lower assembly cost when threaded fasteners are required:

FIGURE 7.1.27 Drive screws for metal (*left*) and for wood (*right*).

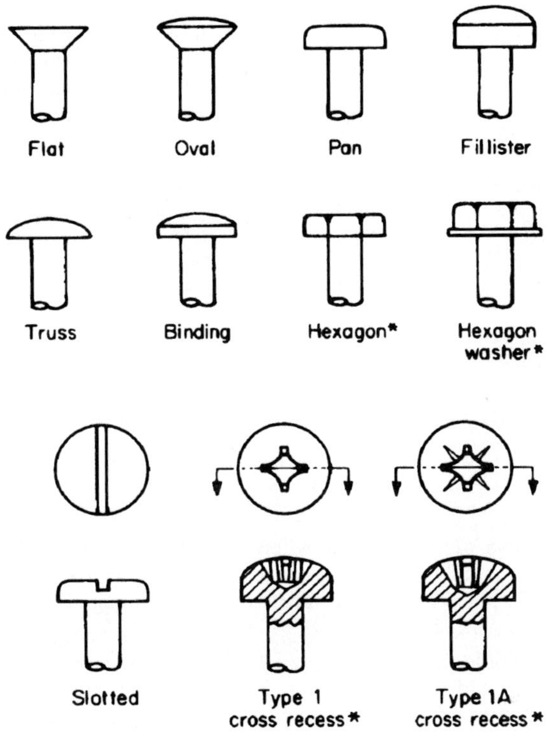

FIGURE 7.1.28 Various screw-head styles and driving provisions. *Preferred types for easier driving. (*From General Motors Drafting Standards.*)

1. When strong holding forces are not required, use drivescrews as shown in Fig. 7.1.27 to reduce assembly and hole-machining costs.
2. Figure 7.1.28 illustrates various types of screw heads commercially available. The preferred types, for ease of driving, are the hexagonal head and the cross-recess (Phillips) head. These are less susceptible to driver slippage and marred surfaces. They are generally lower in cost than types such as the hexagonal socket (Allen head).
3. For applications in which mating parts are subject to misalignment, it may be advisable to use machine screws with points that provide a piloting action and avoid cross threading. Dog and cone points as illustrated in Fig. 7.1.29 are suitable for this purpose.

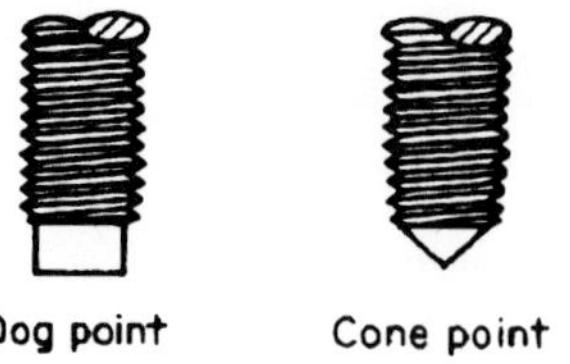

FIGURE 7.1.29 Dog- and cone-pointed screws can accommodate some misalignment of holes.

4. Screw and washer assemblies as shown in Fig. 7.1.30 are desirable as a means of reducing assembly labor, since only one part need be handled. Procurement and stock handling also are simplified.
5. Self-tapping screws are preferable to conventional screws because they eliminate the need for tapping operations on the parts to be joined. Figure 7.1.31 shows various types of thread-cutting and thread-forming screws.

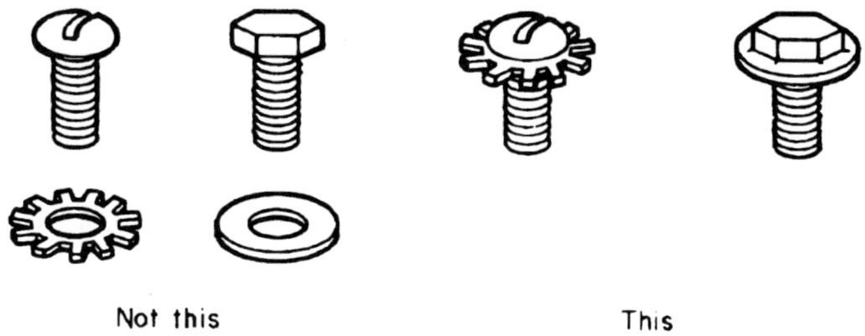

FIGURE 7.1.30 Screw and washer assemblies reduce assembly time.

6. Consider the use of spring nuts whenever torque requirements are not high, because this type of nut is inexpensive, is easier to assemble, and in certain designs stays in place even when the screw is not engaged or not tight. (See Fig. 7.1.32.)
7. The designer should make allowance for access to the screw fastener by the most efficient driving and tightening tools. Powered screwdrivers should have access whenever possible. If not, the design should permit the use of hand-powered socket wrenches. Open-end and box-end wrenches should be used only for holding a bolt head while the nut is being tightened.
8. When box-end, open-end, or hand ratchet-lever socket wrenches are employed, the design should permit at least 60° of lever swing so that sufficient tightening per stroke can take place.
9. Whenever possible, avoid the use of slotted nuts and cotter pins. Instead, when locknuts are required, use the single-component types as illustrated in Fig. 7.1.33. These lock satisfactorily in all but the most extremely demanding applications.
10. For smaller production quantities it usually is more economical to employ separate nuts to hold fastening screws than to tap screw threads into the base part. For high-production applications, on the other hand, the tapping operation often will prove to be more economical.

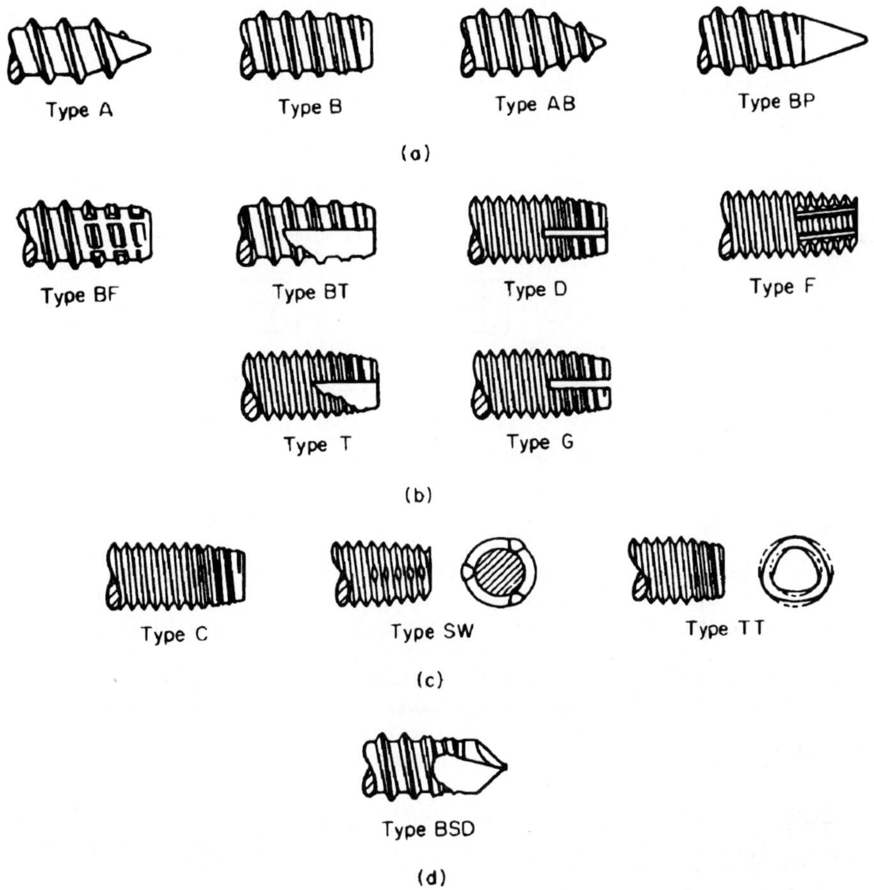

FIGURE 7.1.31 Self-tapping screws. (*a*) Thread-forming types. (*b*) Thread-cutting types. (*c*) Thread-forming types for unified threads. (*d*) Hole-drilling type. (*From General Motors Drafting Standards.*)

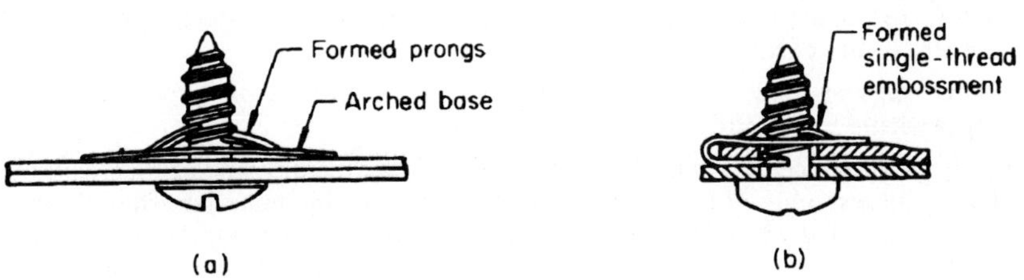

FIGURE 7.1.32 Spring nuts. (*a*) Flat type; (*b*) J type.

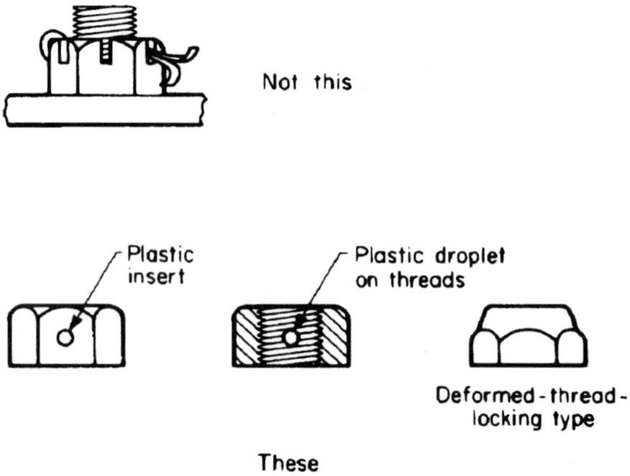

FIGURE 7.1.33 Preferred locknuts.

Metal Stitching

This is often an economical method for fastening various nonmetallic and even some metallic materials, usually in sheet form. It should be considered whenever any of the following conditions are present: (1) Thin materials are involved. (2) It is difficult or costly to drill or punch holes in the materials to be joined. (3) The materials to be joined are somewhat soft. (4) Wire or tubular parts are joined to sheet materials.

Table 7.1.1 shows the recommended maximum thickness of various materials that can be fastened successfully by metal stitching. Figure 7.1.34 illustrates typical stitched assemblies and some rules of thumb for the placement of stitches.

Gaskets and Seals

When a seal is required in an assembly of metal or other rigid parts, it is invariably cheaper to provide it by a gasket rather than through the use of extremely accurately fitted mating surfaces. The following are design suggestions intended to make gaskets as economical as possible to incorporate in an assembly:

1. The gasket shape should be kept as simple as the product design permits. This reduces tooling costs and simplifies assembly.
2. Use O rings or other standard commercial shapes instead of a specially blanked gasket whenever possible.
3. Sometimes the need for a gasket can be avoided by making one of the basic members of the assembly of a plastic, rubber, or other flexible material. This is most apt to be practical if the member in question is naturally in compression during the function of the assembly. A narrow sealing element integral with one part also can provide adequate sealing, especially if the material has some resiliency.
4. Liberal thickness and width tolerances should be specified to minimize the cost of gaskets. Tight width and length dimensions increase tooling and unit costs. Table 7.1.2 provides normal commercial thickness and recommended blank size tolerances for common gasket materials.

TABLE 7.1.1 Recommended Maximum Thickness for Assembly of Two Pieces by Metal Stitching

Metals	Loop-clinched metal stitches, in				Flat-clinched metal stitches, in				
	Metal-to-metal; any combination of these		Metal-to-nonmetal; any combination of these		Metal-to-metal; any combination of these		Metal-to-nonmetal; any combination of these		
					Metal	Nonmetal	Metal	Nonmetal	Nonmetals
Aluminum (soft) 3003; 5052; 6061; Alclad 2024-O, -RT; 2014-T; 7075-T	0.093	0.093	0.125	1/2	0.093	0.093	0.125	1/8	Sheet cork
Aluminum (half-hard) 3003, 5052	0.064	0.064	0.080	3/8	0.064	0.064	0.080	3/16	Leather
Aluminum (hard) 6061-T; Alclad 2024-T, -RT; 2014-T; 7075-T	0.040	0.040	0.064	1/4	0.040	0.040	0.064	3/16	Sheet asbestos
Aluminum extrusion	0.062	0.062	0.093	1/2	0.062	0.062	0.093	1/4	Fiberboard
1010 cold-rolled steel	0.0475 (18 gauge)	0.0475* (18 gauge)	0.0800 (14 gauge)	1/2	0.0348 (20 gauge)	0.0348 (20 gauge)	0.0348 (20 gauge)	3/8	Sponge rubber
Hot-rolled steel	0.0475 (18 gauge)	0.0348* (20 gauge)	0.0625 (16 gauge)	1/4	0.0348 (20 gauge)	0.0348 (20 gauge)	0.0348 (20 gauge)	1/4	Solid rubber
Galvanized sheet	0.0348 (20 gauge)	0.0348* (20 gauge)	0.0475 (18 gauge)	1/8	0.0312 (21 gauge)	0.0312 (21 gauge)	0.0312 (21 gauge)	1/8	Phenolics†
Stainless (Type 302), full-hard	0.010	0.010	0.020	3/16	0.010	§	0.010	3/16	Plastics†
Stainless, half-hard	0.012	0.012	0.025	3/8	0.012	§	0.012	3/16	Standard Masonite
Stainless, quarter-hard	0.015	0.015	0.030	1/4	0.015	0.015	0.015	1/4	Tempered Masonite
Stainless, annealed	0.020	0.020	0.040	3/8‡	0.020	0.020	0.020	3/16	Solid wood
Sheet brass, soft	0.030	0.030	0.050	3/8‡	0.030	0.030	0.040	1/4	Plywood
Sheet copper	0.035	0.035	0.064		0.035	0.035	0.045		

Note: UNS designations for the above alloys are as follows: 3003, A93003; 5052, A95052; 6061, A96061; 2024, A92024; 2014, A92014; 7075, A97075; 1010 steel, G10100; 302 stainless, S30200.

*Rockwell B 50 or softer.

†Must be soft enough to penetrate without cracking.

‡Grain structure may cause leg wander in thickness over 3/8 in.

§Stitching full-hard or half-hard stainless to itself is not recommended.

Source: From D. B. Dallas (ed.), *Tool and Manufacturing Engineers Handbook,* 3d ed., McGraw-Hill, New York. Compiled by the Stitching Wire Division, Acme Steel Company, Chicago.

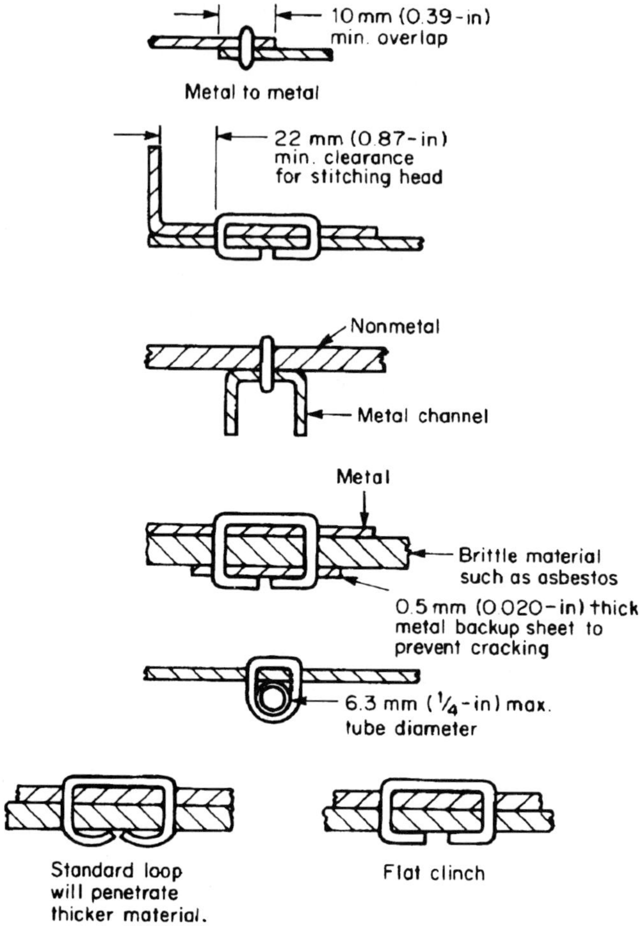

FIGURE 7.1.34 Typical metal-stitched assemblies and some design suggestions for metal stitch placement.

TABLE 7.1.2 Commercial Thickness Tolerances and Recommended Blank Size Tolerances for Common Sheet Gasket Materials*

Material	Commercial thickness tolerance, mm (in)	Recommended blank dimension tolerance, mm (in)
Cork composition	±0.25 (0.010)	±0.40 (0.015)
Cork and rubber	±0.40 (0.015)	±0.40 (0.015)
Vegetable fiber	±0.15 (0.006)	±0.40 (0.015)
Molded sheet rubber	±0.40 (0.015)	±0.80 (0.030)
Calendered stock rubber-sheet plastics, etc.	±0.13 (0.005)	±0.40 (0.015)

*The table is based on 1.5-mm- (1/16-in-) thick material. Thicker sheets will require a correspondingly greater thickness tolerance.

5. When using an O ring as a seal for a shaft or tube, it is better to machine an external groove for the O ring in the shaft or tube rather than an internal recess in the female part.

6. Gasket widths of at least 3 mm ($\frac{1}{8}$ in) should be specified to prevent damage from handling or assembly. Notches for bolt holes are sometimes satisfactory and are less costly than projections with holes. (See Fig. 7.1.35.)

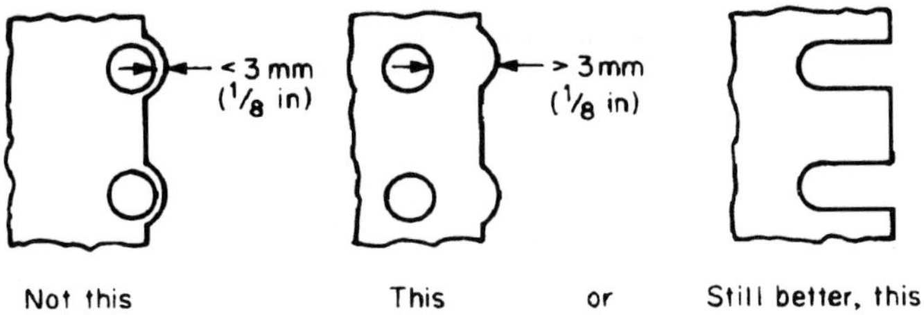

FIGURE 7.1.35 Recommended gasket width around holes.

7. In threaded joints, at least one of the flange faces should have a smooth surface to avoid cutting or tearing the gasket. (See Fig. 7.1.36.)

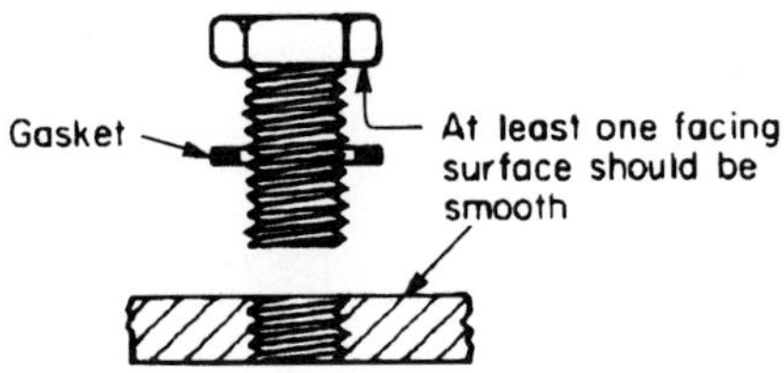

FIGURE 7.1.36 In flange joints sealed by a gasket, at least one of the mating surfaces should be smooth to avoid damage to the gasket.

8. When using molded-rubber seals, it is desirable to have them held in place by a snap-in action rather than with fasteners.

9. For maintenance and low-quantity production, formed-in-place gaskets made from silicone rubber or anaerobic-plastic adhesives can be very useful since they eliminate the need to cut and fit sheet gaskets.

Press-Fits

Press-fits provide a low-cost method for fastening parts together securely and permanently. The disadvantage of the method is that it requires precise dimensional control of mating parts. Table 7.1.3 lists standard dimensional tolerances applicable to pins and holes for various classes of press-fit.

Even though tolerances are close, part machining costs may not be excessive, especially if the pin is suitable for through-feed centerless grinding and the hole is reamed to size. Metal-to-metal press-fits are found most commonly in heavier apparatus involving castings and plates, etc.

The most economical use of press-fits occurs when spring force or the natural

TABLE 7.1.3 Dimensional Tolerances for Press Fits

Limits are in 0.001 in. Limits for hole and shaft are applied algebraically to the basic size to obtain limits of size for the parts.

Nominal size range, in		Class 1*			Class 2†			Class 3‡			Class 4§			Class 5§		
		Limits of inter-ference	Standard limits		Limits of inter-ference	Standard limits		Limits of inter-ference	Standard limits		Limits of inter-ference	Standard limits		Limits of inter-ference	Standard limits	
Over	To		Hole	Shaft		Hole	Shaft		Hole	Shaft		Hole	Shaft		Hole	Shaft
0.04	0.12	0.05	+0.25	+0.5	0.2	+0.4	+0.85				0.3	+0.4	+0.95	0.5	+0.4	+1.3
		0.5	−0	+0.3	0.85	−0	+0.6				0.95	−0	+0.7	1.3	−0	+0.9
0.12	0.24	0.1	+0.3	+0.6	0.2	+0.5	+1.0				0.4	+0.5	+1.2	0.7	+0.5	+1.7
		0.6	−0	+0.4	1.0	−0	+0.7				1.2	−0	+0.9	1.7	−0	+1.2
0.24	0.40	0.1	+0.4	+0.75	0.4	+0.6	+1.4				0.6	+0.6	+1.6	0.8	+0.6	+2.0
		0.75	−0	+0.5	1.4	−0	+1.0				1.6	−0	+1.2	2.0	−0	+1.4
0.40	0.56	0.1	+0.4	+0.8	0.5	+0.7	+1.6				0.7	+0.7	+1.8	0.9	+0.7	+2.3
		0.8	−0	+0.5	1.6	−0	+1.2				1.8	−0	+1.4	2.3	−0	+1.6
0.56	0.71	0.2	+0.4	+0.9	0.5	+0.7	+1.6				0.7	+0.7	+1.8	1.1	+0.7	+2.5
		0.9	−0	+0.6	1.6	−0	+1.2				1.8	−0	+1.4	2.5	−0	+1.8
0.71	0.95	0.2	+0.5	+1.1	0.6	+0.8	+1.9				0.8	+0.8	+2.1	1.4	+0.8	+3.0
		1.1	−0	+0.7	1.9	−0	+1.4				2.1	−0	+1.6	3.0	−0	+2.2
0.95	1.19	0.3	+0.5	+1.2	0.6	+0.8	+1.9	0.8	+0.8	+2.1	1.0	+0.8	+2.3	1.7	+0.8	+3.3
		1.2	−0	+0.8	1.9	−0	+1.4	2.1	−0	+1.6	2.3	−0	+1.8	3.3	−0	+2.5
1.19	1.58	0.3	+0.6	+1.3	0.8	+1.0	+2.4	0.8	+1.0	+2.4	1.5	+1.0	+3.1	2.0	+1.0	+4.0
		1.3	−0	+ 0.9	2.4	−0	+1.8	2.4	−0	+1.8	3.1	−0	+2.5	4.0	−0	+3.0
1.58	1.97	0.4	+0.6	+1.4	0.8	+1.0	+2.4	1.2	+1.0	+2.8	1.8	+1.0	+3.4	3.0	+1.0	+5.0
		1.4	−0	+1.0	2.4	−0	+1.8	2.8	−0	+2.2	3.4	−0	+2.8	5.0	−0	+4.0
1.97	2.56	0.6	+0.7	+1.8	0.8	+1.2	+2.7	1.3	+1.2	+3.2	2.3	+1.2	+4.2	3.8	+1.2	+6.2
		1.8	−0	+1.3	2.7	−0	+2.0	3.2	−0	+2.5	4.2	−0	+3.5	6.2	−0	+5.0
2.56	3.15	0.7	+0.7	+1.9	1.0	+1.2	+2.9	1.8	+1.2	+3.7	2.8	+1.2	+4.7	4.8	+1.2	+7.2
		1.9	−0	+1.4	2.9	−0	+2.2	3.7	−0	+3.0	4.7	−0	+4.0	7.2	−0	+6.0
3.15	3.94	0.9	+0.9	+2.4	1.4	+1.4	+3.7	2.1	+1.4	+4.4	3.6	+1.4	+5.9	5.6	+1.4	+8.4
		2.4	−0	+1.8	3.7	−0	+2.8	4.4	−0	+3.5	5.9	−0	+5.0	8.4	−0	+7.0
3.94	4.73	1.1	+0.9	+2.6	1.6	+1.4	+3.9	2.6	+1.4	+4.9	4.6	+1.4	+6.9	6.6	+1.4	+9.4
		2.6	−0	+2.0	3.9	−0	+3.0	4.9	−0	+4.0	6.9	−0	+6.0	9.4	−0	+8.0
4.73	5.52	1.2	+1.0	+2.9	1.9	+1.6	+4.5	3.4	+1.6	+6.0	5.4	+1.6	+8.0	8.4	+1.6	+11.6
		2.9	−0	+2.2	4.5	−0	+3.5	6.0	−0	+5.0	8.0	−0	+7.0	11.6	−0	+10.0
5.52	6.30	1.5	+1.0	+3.2	2.4	+1.6	+5.0	3.4	+1.6	+6.0	5.4	+1.6	+8.0	10.4	+1.6	+13.6
		3.2	−0	+2.5	5.0	−0	+4.0	6.0	−0	+5.0	8.0	−0	+7.0	13.6	−0	+12.0

*Light-drive fits are those requiring light assembly pressures and produce more or less permanent assemblies. They are suitable for thin sections or long fits or in cast-iron external members.

†Medium-drive fits are suitable for ordinary steel parts or for shrink fits on light sections. They are about the tightest fits that can be used with high-grade cast-iron external members.

‡Heavy-drive fits are suitable for heavier steel parts or for fits in medium sections.

§Force fits are suitable for parts which can be highly stressed or for shrink fits when the heavy pressing forces required are impractical.

Source: Reprinted from *American Machinist's Handbook*, McGraw-Hill, New York, 19XX.

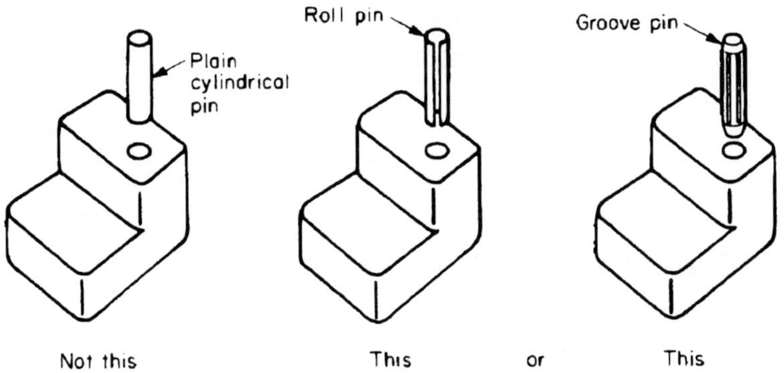

FIGURE 7.1.37 Use commercial roll and groove pins instead of plain cylindrical pins to simplify machining and reduce pin costs.

resiliency of one of the parts instead of precise machining is relied on to provide the holding force. Roll pins and groove pins provide an inexpensive means for holding parts together or in a fixed position to one another. They are inexpensive in themselves and do not require an accurate hole diameter. (Drilling accuracy usually is satisfactory without reaming or boring.) Figure 7.1.37 illustrates these pins.

Snap-Fits

Spring force can be used in other ways to fasten parts together. Figure 7.1.38 illustrates the type of fastening effect that can be achieved with spring-steel holding devices. The assembly operation requires only that the spring be snapped into place. Reaming and tapping operations are eliminated. This approach requires either the use of a standard commercially available spring clip or sufficient production volume to justify tooling a four-slide machine or punch press to make the clip.

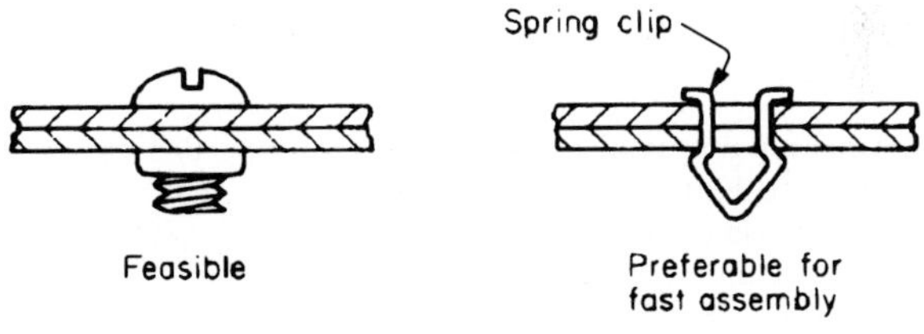

FIGURE 7.1.38 Typical spring-clip application.

A still better approach is to use the flexibility of the parts themselves rather than that of a separate fastener to hold the parts together. Figure 7.1.39 illustrates typical snap-in-place assemblies with metal parts. Figure 7.1.3 shows the same approach with a plastic cover. Covers and lids are particularly suitable for this approach. Springed elements of plastic or sheet metal (or lips, hooked surfaces, or undercuts) provide an easy holding method without the use of fasteners.

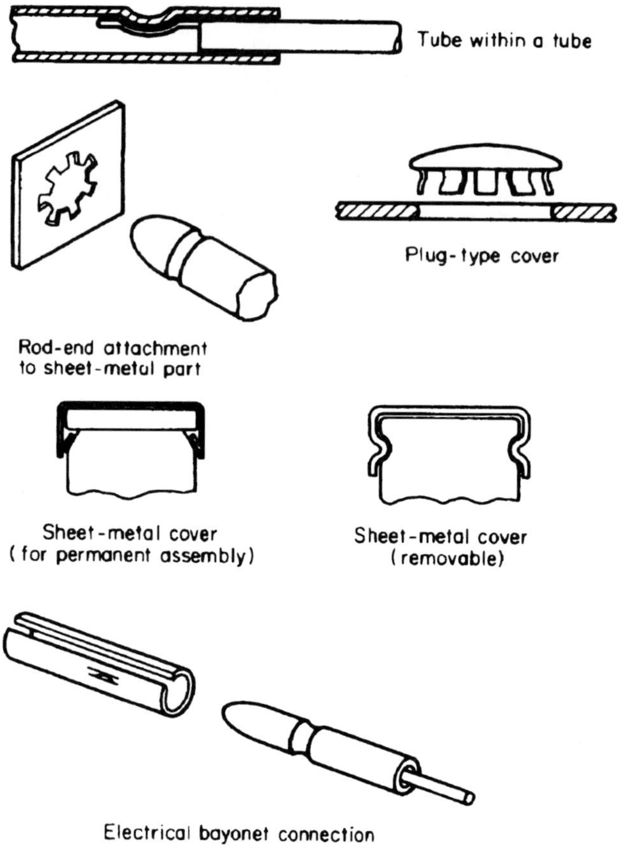

FIGURE 7.1.39 Typical snap-in-place assemblies.

Automatic Assembly

Assembly operations have proved to be among the most difficult to mechanize. The human operator is capable of many subtle manipulations, adjustments, and compensations for component variations. These can be extremely difficult to incorporate into mechanical assembly equipment. There has, however, been very significant progress in developing machines that automatically put together mechanical components. Notable among them have been computer-controlled robot mechanisms. One other factor in this progress has been improvement in the level of knowledge of designing components to facilitate assembly.

Invariably, design improvements made to facilitate automatic assembly also facilitate manual assembly. Conversely, design modifications (like bullet-nosed parts, funnel-shaped openings, nontangling components, etc.) made to ease manual assembly also aid automatic assembly. The difference between the two is that automatic assembly has more demanding part-design requirements, although with computer-controlled equipment the difference is decreasing.

The following design recommendations are particularly adapted to automatic assembly:

1. Components should be as symmetrical as possible (around as many axes as possible) to aid in positioning them in feeding apparatus. They should be stocky and basically simple in shape for best results. (See Fig. 7.1.40.)
2. When symmetry is not feasible, the nonsymmetrical attribute (weight, length, width, etc.) should be emphasized or exaggerated to aid the mechanism that orients parts during automatic feeding. (See Fig. 7.1.41.)
3. If the part is to be marked in some way, don't rely only on visual characteristics to determine its orientation; include some physical difference so that the feeding

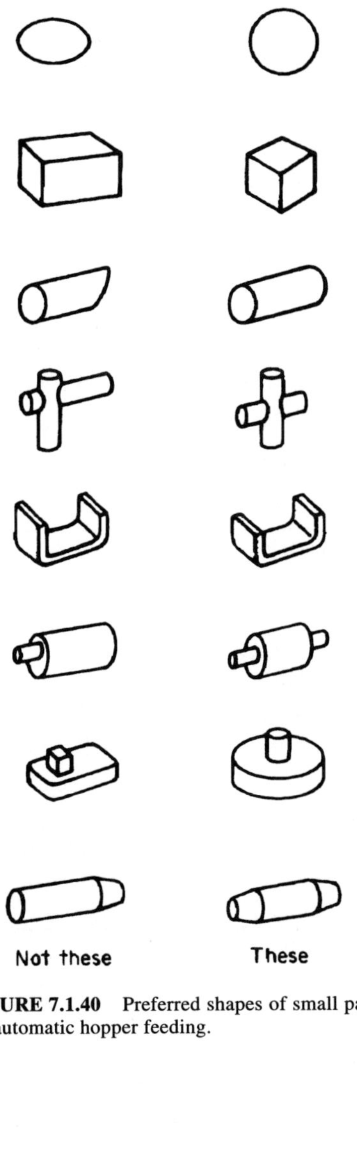

FIGURE 7.1.40 Preferred shapes of small parts for automatic hopper feeding.

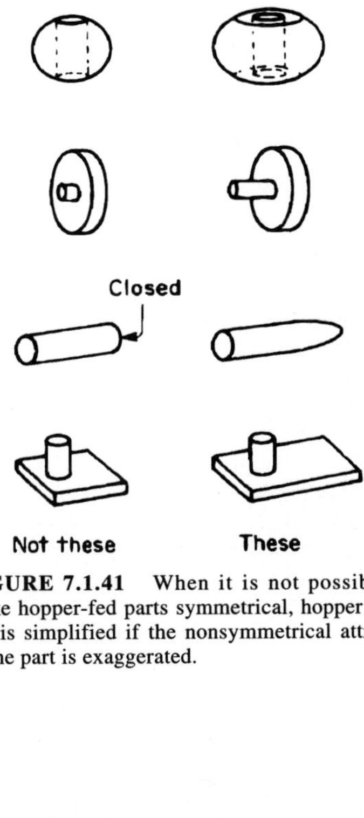

FIGURE 7.1.41 When it is not possible to make hopper-fed parts symmetrical, hopper feeding is simplified if the nonsymmetrical attribute of the part is exaggerated.

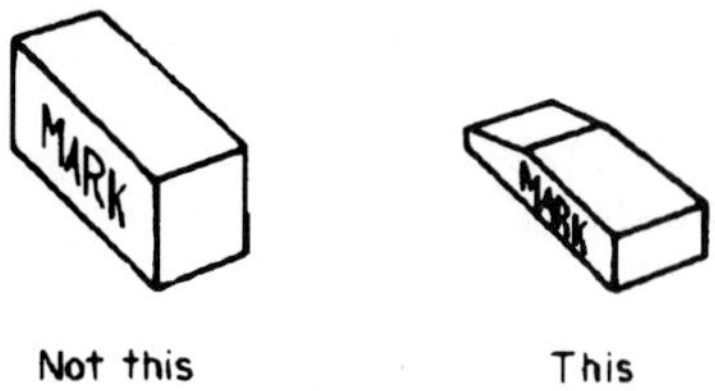

FIGURE 7.1.42 Since hopper-feeding apparatus cannot normally locate the marked surface, it is necessary to provide some other attribute that can serve to orient the part coming from the hopper so that the marking is in the desired location.

mechanism can be designed to position the part with the marking in the specified direction. (See Fig. 7.1.42.)

4. Dimensions used to locate the part during automatic feeding and placement must be held to consistent values. The designer should visualize both the feeding and orienting mechanisms to be used for the part and the manufacturing method to be followed in fabricating it to be sure that the critical dimensions will be controlled.
5. Flexible parts such as springs and other parts made from wires, thin strip metal, or rubber are the most difficult to feed, orient, and place automatically and usually are best excluded from automatic assemblies.
6. Press fits are a particularly suitable fastening method when automatic assembly is involved. This and other attachment methods which do not require separate fasteners should be used whenever possible. Other methods include spot welding, twisted or bent lugs, and snap-in-place parts shown in Figs. 7.1.3 and 7.1.39.

TOLERANCE RECOMMENDATIONS

The values in Table 7.1.4 can be considered a guide to designers in specifying dimensional tolerances of finished assemblies.

TABLE 7.1.4 Recommended Dimensional Tolerances for Mechanical Assemblies

Dimensions	Recommended tolerances, mm (in)	
	Normal	Tightest
Dimensions which result from face-to-face fit of mating parts	+0.13 (0.005), −0.00 (0.000) plus sum of tolerances of the parts	Sum of tolerances of the parts
Dimensions dependent on visual alignment of assembled parts	±0.8 (0.030)	±0.4 (0.015)
Dimensions controlled by fixture stops	±0.25 (0.010)	±0.13 (0.005)
Dimensions controlled by dial-indicator gauges on assembly fixtures	±0.05 (0.002)	±0.025 (0.001)

REFERENCES

1. G. Boothroyd, C. Poli, and L. Murch, *Automatic Assembly,* Marcel Dekker, New York, 1983.
2. G. Boothroyd and P. Dewhurst, *Design for Assembly,* University of Rhode Island Press, Providence, 1983.
3. David M. Anderson, *Design for Manufacture,* CIF Press, Lafayette, CA, 1991.
4. Gordon Lewis and Helen Connelly, *Product Design for Assembly: The Methodology Applied.* Privately published, 1990.
5. James L. Nevins and Daniel E. Whitney, *Concurrent Design of Products and Processes,* McGraw-Hill, New York, 1989.

CHAPTER 7.2

ARC WELDMENTS AND OTHER WELDMENTS

Jay C. Willcox

Toro Company
Minneapolis, Minnesota

ARC WELDING AND SIMILAR PROCESSES

Arc welding is a method of permanently joining two or more metal parts. A homogeneous joint is produced through the melting and fusing together of adjacent portions of the originally separate pieces. The final welded joint has unit strength approximately equal to that of the base material.

Intense heat is applied to the joint by means of an electric arc that passes between a welding rod and the work. The arc temperature is approximately 4400°C (8000°F). With most processes, the welding rod is consumable. As it melts, it provides filler material for the weld. To prevent oxidation of the molten or heated metal, one of two basic alternative methods is used. The first method involves use of a flux material that decomposes under the heat of welding and releases a gas that shields the arc and the hot metal. The flux also melts to provide a protective coating of slag. The second basic method employs an inert or nearly inert gas to form a protective envelope around the arc and the weld. Helium, argon, and carbon dioxide are the most commonly used gases.

The following are the specific welding processes commonly in use in industry.

Shielded-Metal Arc (SMAW) or Stick Welding

This is a manual process. The electrode is in rod form, covered with flux. Flux and rod are consumed at approximately the same rate. Figure 7.2.1 illustrates the process.

Submerged Arc Welding (SAW)

This process uses solid flux in loose granular form. The flux is applied as a layer that literally submerges the end of the electrode, the arc, and the weld puddle with a pile of

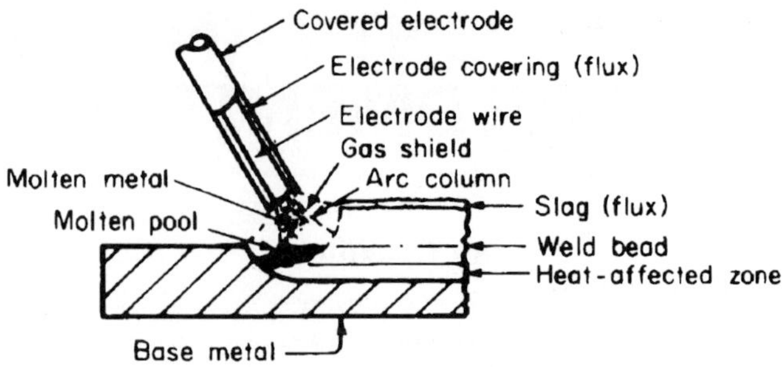

FIGURE 7.2.1 The shielded-metal arc or stick welding process. (*From R. L. Little,* Welding and Welding Technology, *McGraw-Hill, New York, 1972.*)

flux. The electrode is in wire form and is continuously fed from a reel. The movement of weld gun and the dispensing of the flux are usually automatic. Pickup of surplus flux granules behind the gun is by vacuum and is also automatic.

Flux-Cored Arc Welding (FCAW)

This process is similar to normal shielded-arc stick welding except that the flux, instead of being on the outside of the welding rod, is inside. The electrode is tubular, and the flux is contained inside. Coiled, continuously fed electrodes are used, and the cost of changing welding rods is saved. Sometimes, externally supplied gas is used to assist in shielding the arc.

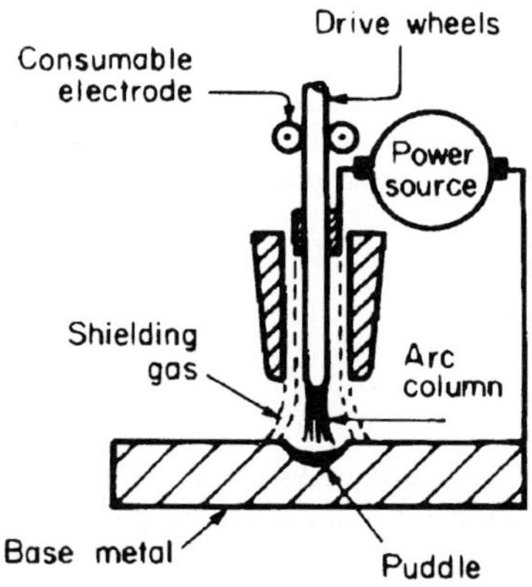

FIGURE 7.2.2 The gas-metal arc (GMAW) welding process. (*From R. L. Little,* Welding and Welding Technology, *McGraw-Hill, New York, 1972.*)

Gas-Metal Arc Welding (GMAW)

Formerly called *metal–inert gas (MIG) welding,* this process uses a shield of gas—argon, helium, carbon dioxide, or a mixture of them—to prevent atmospheric contamination of the weld. The bare, solid-wire electrode is coiled or reeled and fed continuously to the weld at a fixed rate. The wire is consumed during the process and thereby provides filler metal. The shielding gas flows through the weld gun. The GMAW process is rapid and requires less labor than stick welding because welding-rod changes are not required and there is no slag to be chipped away. Figure 7.2.2 illustrates this common process.

Gas-Tungsten Arc Welding (GTAW)

Also known as *tungsten–inert gas (TIG) welding,* this is quite similar to the GMAW process. The difference is that the electrode in the GTAW process is not consumed and does not provide filler metal. A gas shield (usually inert gas) is used as in the GMAW process. If filler metal is required, an auxiliary rod is used.

Plasma Arc Welding (PAW)

This is similar to GTAW welding. In both processes, a nonconsumable electrode is used. In both there can be an arc between the electrode and the work. In both cases there is a plasma, a zone of ionized gas around the arc. Plasma arc welding differs from GTAW welding in that the amount of ionized gas is greatly increased, and it is this that provides the heat of welding. The plasma consists of free electrons, positive ions, and neutral particles. It is of a higher temperature than a normal arc and is formed by constricting the orifice of the shielding gas at the location of the arc. The arc's path is not necessarily from the electrode to the work; it may exist entirely within the welding gun. Figure 7.2.3 illustrates the process

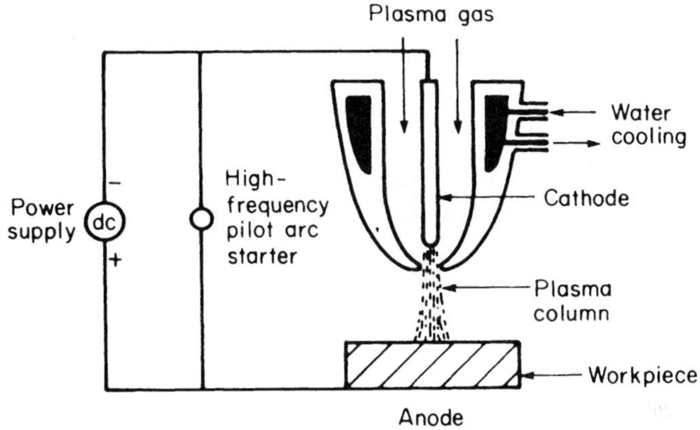

FIGURE 7.2.3 The plasma-arc-welding process. (*From R. L. Little,* Welding and Welding Technology, *McGraw-Hill, New York, 1972.*)

Oxyfuel Gas Welding (OFW)

This process is also known as *oxyacetylene welding.* Heat is supplied by the combustion of acetylene in a stream of oxygen. Both gases are supplied to the torch through flexible hoses. Heat from this torch is lower and far less concentrated than that from an electric arc.

Electron-Beam Welding (EBW)

This process is accomplished when a high-velocity, narrow stream of electrons is directed against the joint, causing localized heating and melting of the weld metal.

The electrons are emitted by a heated cathode and are accelerated and focused by electrostatic and magnetic elements. When the rapidly moving electrons strike the work, their kinetic energy is transformed into heat. The heating is intense and localized.

The electron-beam generation takes place in a vacuum, and the process works best when the entire operation and the workpiece are also in a high vacuum of 10^{-4} torr or lower. However, systems in which the workpiece is in a soft-vacuum (10^{-1} to 10^{-3} torr) or nonvacuum environment are also used. The welding operator must be protected from the x-ray, infrared, and ultraviolet radiation that the process generates.

Laser-Beam Welding (LBW)

This process uses the energy in an extremely concentrated beam of coherent, monochromatic light to melt the weld metal. The word *laser* is an acronym for *l*ight *a*mplification by *s*timulated *e*mission of *r*adiation. Figure 7.2.4 schematically illustrates the operation of laser welding equipment. When triggered with high-intensity light, the gas or solid-state laser emits pulses of coherent (in-phase), monochromatic (single-wavelength) light, the laser beam. This beam is focused to concentrate it within a very small area, which then has a tremendous concentration of energy. The pulse duration is only about 0.002 s and normally occurs about 10 times per second.

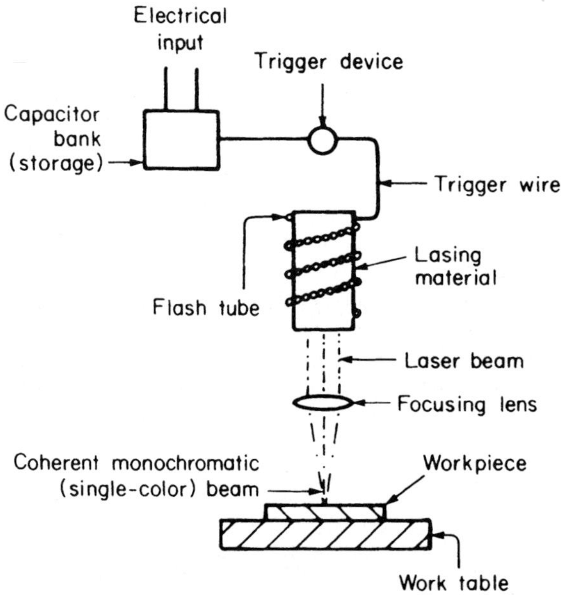

FIGURE 7.2.4 Laser-beam welding. (*From R. L. Little,* Welding and Welding Technology, *McGraw-Hill, New York, 1972.*)

Friction Welding (FRW)

This process relies on the frictional heat developed when one part is rotated and pressed against another that is stationary. When the temperature at the interface of the two parts is sufficiently high, the rotation is stopped, and increased axial force is applied. This fuses the two parts together. Either a strong motor or a flywheel is used to provide the rotational force. In the latter case the process may be called *inertia welding*. Rotational speed, axial force, and time of contact must be developed for each assembly.

Other Welding Processes

Other processes used industrially are the following:

1. *Diffusion bonding (DB)*. Parts are pressed together at an elevated temperature below the melting point for a period of time.
2. *Explosion welding (EXW)*. The parts to be welded are driven together at an angle by means of an explosive charge and fuse together from the friction of the impact.
3. *Ultrasonic welding (USW) for metals*. This process utilizes transverse oscillation of one part against the other to develop sufficient frictional heat for fusion to occur. It is very similar to the process used for plastics and described in Chap. 6.9.
4. *Electroslag (ESW) and electrogas (EGW) processes*. These use a molten pool of weld metal contained by copper "shoes" to make vertical butt welds in heavy plate.

CHARACTERISTICS AND APPLICATIONS OF ARC WELDMENTS

Almost any imaginable size or shape of assembly can be arc-welded from the appropriate parts. Parts as simple as a pipe with a welded seam and as complex as the structure and hull of a seagoing vessel are within the capabilities of the process. However, complex and large assemblies are more typical because the advantages of arc welding in comparison with other processes are enhanced as the finished product becomes more complex and unwieldy.

A wide variety of ferrous and nonferrous metals, as detailed below, is employed in arc weldments. Devices commonly produced by arc welding also include tube fittings, storage tanks, pressure vessels, machine frames, structures for industrial equipment, railroad cars, agricultural machinery, aircraft components, mining equipment, chemical-process equipment, trucks, metal furniture, stairways, platforms, electric-motor lamination stacks, buildings, and bridges.

Welded assemblies can utilize the following types of fabricated parts: (1) sheet metal and stampings—sheared, slit, blanked, formed, drawn, etc.; (2) structural shapes—angles, channels, I-beams, etc.; (3) standard warehouse or mill sizes—flats, hexagons, rounds, squares, etc.; (4) machined parts—turned, milled, drilled, tapped, etc.; (5) pipe and tubing—round, square, and rectangular; (6) castings and forgings; and (7) extrusions, commercial fasteners, and other off-the-shelf hardware.

Production-welded components, properly designed, can exhibit excellent strength characteristics, light weight, economical cost, and pleasing appearance. However, it may be somewhat difficult to attain the long, sweeping curves, rounded contours, and relatively smooth surfaces sometimes exhibited by castings and forgings.

A welded joint can be made as strong as or even stronger than the base components. However, for ordinary butt joints, when sophisticated x-ray or ultrasonic inspection devices are not used, it is common in design calculations to consider that the weld will have 80 percent of the strength of the base material.

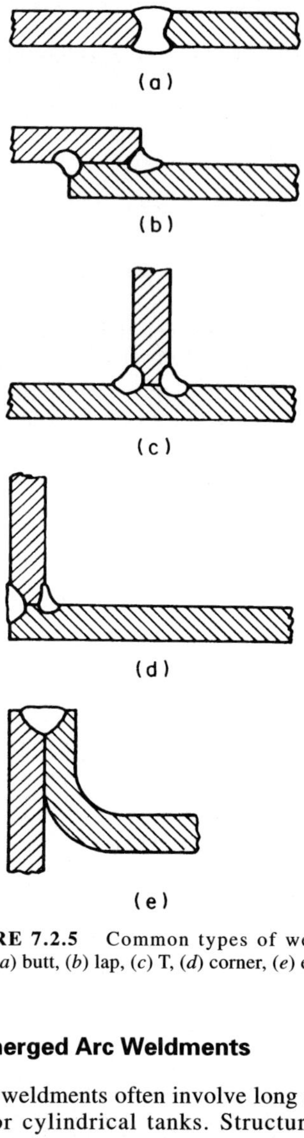

FIGURE 7.2.5 Common types of welded joints: (*a*) butt, (*b*) lap, (*c*) T, (*d*) corner, (*e*) edge.

There are five basic classes of weld joints in common use: butt joints, lap joints, T joints, corner joints, and edge joints, which are shown in Fig. 7.2.5. Butt, lap, T, and corner joints are shown with two fillets, although it is possible and not uncommon to use only one with each type of weld.

The six arc-welding processes described above and gas welding each have particular areas of advantageous application. These are as follows:

Shielded-Metal Arc or Stick Weldments

Often, large outdoor structural assemblies are welded by this process, which works well even if there is some wind (as contrasted with processes which use a shielding gas) and can be applied in vertical, angular, or upside-down locations. Inaccessible welds can be made most easily with this process, and since equipment is simple and portable and can be located at some distance from the point of weld, remote welds are possible.

The minimum normal thickness of sheet metal for stick welds is 1.5 mm (0.060 in), although material as thin as 1.0 mm (0.040 in) is sometimes welded. There is no maximum material thickness as long as repeated weld passes can be made.

Submerged Arc Weldments

These weldments often involve long seams such as those required in the fabrication of pipe or cylindrical tanks. Structures outdoors can be welded with this process, although it is not nearly as portable as stick welding. Remote locations of welds are not feasible, nor are vertical or underside welds. Usually, thick plates are involved because the process gives deep penetration. However, stock as thin as 4.5 mm ($^3/_{16}$ in) can be incorporated in submerged arc weldments.

Flux-Cored Weldments

These weldments are often similar to those made with the submerged arc process. Both processes have similar applicability. Sections over 12 mm (0.5 in) and up to 400 mm (16 in) thick or more are practical and common applications. The minimum practical sheet thickness is about 1.5 mm (0.060 in). Vault doors, pump housings, special welded beams, bulldozer blades, and pipe are typical applications.

Gas-Metal Arc Weldments

This process has a number of advantages compared with stick welding: higher speeds and deeper penetration are possible, and less operator skill and less postweld cleaning are required. GMAW is used in shipbuilding and railroad car fabrication and is quite suitable for thinner sheet stock. With careful current control, sheet as thin as 0.5 mm (0.020 in) and even thinner can be welded.

The GMAW process is widely used in production but is not so well suited for outdoor work, where there is a possibility of wind, nor is it suitable for remote or inaccessible joint locations. Vertical and underside welds are possible. High-quality welds are possible because of the absence of slag. However, freedom from slag causes more rapid cooling of the weld, and if air-hardening steel is involved, GMAW welding may not be as satisfactory as one of the solid-flux methods.

Gas-Tungsten Arc Weldments

These weldments may employ continuous-, skip-, or spot-welded joints. Assemblies using thin sheet metal (e.g., metal bellows) are typically welded with the GTAW process. The process is suited to stock less than 3 mm (0.120 in) thick down to as little as 0.12 mm (0.005 in) thick. For sections over 6 mm ($\frac{1}{4}$ in) thick, other processes are preferable. The process is advantageous when there are no bevels at the adjoining edges of a butt joint or in applications in which the joint does not require filler metal. Joints in all positions can be GTAW-welded.

Plasma Arc Welding

This process is used widely in welding stainless steels. It is well suited for very thin sheets and when high-quality welds are essential, particularly if only a narrow heat-affected zone is desired. Metals as thin as 0.025 mm (0.001 in) can be welded by this process, although 0.075 mm (0.003 in) is a more practical minimum. It is also used for a first pass in the welding of heavy-wall stainless steel pipes and tanks because up to 12.5 mm (0.5 in) can be welded in a single pass.

Oxyfuel Gas Welding

This is another process used for sheet-metal components. With thin sheets, the slower heating rate of gas welding is not a real disadvantage. The fact that the weld-puddle position can be controlled by flame pressure is an advantage. A typical application is the welding of air-conditioner drip pans. If there is a lack of good fit-up of the edges to be joined, the welder can compensate easily and still produce a leakproof joint.

Electron-Beam Weldments

These welds can have the following characteristics:

1. Deep, narrow welds with almost parallel sides. Depth-to-width ratios of 10:1 are common in high-vacuum welding, and ratios of 25 or 30:1 are possible.
2. Minimum workpiece distortion and a narrow heat-affected zone owing to low total-heat input.
3. Weld joints close to heat-sensitive components.
4. In vacuum welding, freedom from atmospheric contamination and possible use of reactive and refractory metals.
5. More feasible welding of hardened metals.
6. Less need for machining or straightening after welding because there is reduced shrinkage and distortion.
7. Welds in otherwise inaccessible locations such as at the bottom of a hole.

Electron-beam weldments can involve components as thin as foil and as thick as 150 mm (6 in). Maximum penetration depths for high-vacuum welds are steel, 90 mm ($3\frac{1}{2}$ in); aluminum, 150 mm (6 in); and copper, 40 mm ($1\frac{1}{2}$ in). For medium-vacuum welds, maximum penetration with steel is 60 mm ($2\frac{1}{2}$ in); with nonvacuum welds, it is 12 mm ($\frac{1}{2}$ in).

The welds from nonvacuum electron beams are considerably wider as well as shallower than vacuum welds. However, penetration is still better than from conventional welding processes.

Joints for electron-beam weldments must be close-fitting because of the narrower width of the melted zone. Butt joints are more suitable than lap joints. There is also a limit on size of the welded assembly, if vacuum electron-beam welding is used, since the assembly must fit into the vacuum chamber. The maximum chamber size in equipment currently available is 3.7 by 1.4 by 1.7 m (144 by 56 by 66 in).

Laser-Beam Weldments

These welds are used when precise control of energy input is important. The joining of microelectronic components and the welding of thin-gauge parts like bellows and watch springs are typical applications. However, laser-beam welding is also used for heavy plate. High-powered lasers can penetrate 50 mm (2 in) in 304 stainless steel. Because of the small size of the heat-affected zone, weldments with heat-sensitive components are particularly suitable. Inaccessible joints and flimsy assemblies are also especially applicable.

Friction Weldments

These welds normally involve the ends of circular or nearly circular parts. Pipe, tubing, or solid-bar ends are particularly applicable to the processes. Machines currently available can handle parts up to about 100 mm (4 in) in diameter.

ECONOMIC PRODUCTION QUANTITIES

Arc and gas weldments are found in products produced in a broad range of quantities, from one of a kind to the highest levels of mass production. However, the more normal and common uses of these processes are in the lower portion of this range. As manufacturing quantities increase, castings, brazed assemblies, and resistance weldments become more cost-competitive even though they may involve a greater investment for tooling or equipment.

There is a real advantage of welding for repair or other low-quantity work because the operation can commence with inexpensive equipment and only the barest minimum of tooling. Tooling also can be adapted easily to design changes. With higher production, the low tooling cost for weldments becomes less of an advantage and the high labor content becomes more of a disadvantage.

The following is a summary of pertinent factors relating to economic production quantities for common welding processes:

Oxyfuel Gas Welding

For this process, equipment and any tooling required are low in cost. However, welding time per length of weld seam is high because of the inherently slower heating rate. This tends to make the process costly except for repair and low-quantity work.

Stick Welding

This process also requires only low-cost equipment and tooling. Welding speed, though much faster than that of gas welding, is nevertheless slower than with other arc-welding processes (because of electrode changes and slag removal as well as welding time per se). Stick welding is best suited to low-quantity production.

Submerged Arc Welding

This process involves much more elaborate equipment and fixturing than are needed for simple stick welds. Equipment and tooling costs are high compared with those of other welding processes. The metal deposition rate, however, is quite rapid. This process therefore is justifiable for large-quantity work, particularly when seams are long.

Flux-Cored Welding

This process requires relatively expensive equipment, particularly if shielding gas also is used. Although slag-removal labor is required, welding rates are high, and with automatic equipment, the process fits medium- and higher-volume situations.

Gas-Metal Arc Welding

This process can be suitable for higher production levels. It is low in labor cost because it is free from slag removal and because the electrode is fed continuously. However, equipment is more costly than for stick welding, particularly if it is made automatic. Unless welding time is only a small portion of total-cycle time, GMAW is apt to be the most economical process for low-quantity weldments also if it is suitable for metallurgical and other reasons.

Gas-Tungsten Arc Welding

This process is slower than other processes but does not require as much investment for equipment as do the GMAW and submerged arc processes. While this would seem to indicate applicability to low-production work, the choice of GTAW is much more a question of technical suitability than of economical lot size.

Plasma Arc Welding

This process involves much more costly equipment (2 to 5 times the cost of GTAW equipment, for example). However, it is very rapid, producing welds at four or more times the rate of other arc processes. The benefit of this high welding speed is offset to some degree by the high costs of shielding gas and nozzle maintenance. These cost patterns would indicate suitability for larger production volumes. Actually, technical considerations such as the capability to make deep-penetration welds are more of a factor in the choice of plasma arc welding than production quantities.

Electron-Beam Welding

Electron-beam equipment is very expensive. Fixturing is also more costly than for other welding processes because it must locate the joints more precisely. Cycle times for full-vacuum electron-beam welding (EBW) are long even though the welding itself is rapid because an extended time is required to evacuate the welding chamber to high-vacuum levels. Offsetting the longer cycle is the probability that neither machining nor straightening will be required after welding.

While the use of high-vacuum electron-beam welding is more apt to be dictated by metallurgical and weld-strength factors, the nonvacuum approach has advantages on purely economic grounds. When production volumes are very large, as in the automotive industry, the high investment of EBW equipment can be amortized and the high welding speed, single-pass welding, and freedom from the need for secondary operations can mean minimum unit welding costs.

Laser-Beam Welding

This process also requires costly equipment, though not quite so costly as that required for electron-beam welding. Therefore, high volume or critical welding conditions are required to justify the necessary investment. Welding speeds for heavier section thickness are rather slow, but the low heat distortion of the process usually obviates the need for stress-relieving, straightening, or machining operations after welding.

Friction Welding

This process normally requires special equipment and some development of process conditions. Medium to high levels of production are most suitable from the economic standpoint.

SUITABLE MATERIALS

Almost all metals are to some degree arc-weldable. It is not practical, however, to weld some of them because of strength, brittleness, or processing problems. In general, metals which are less highly alloyed and metals of moderate to high melting temperatures are the most suitable for arc welding. Welding dissimilar metals is invariably difficult.

The following is a review of the weldability of various common commercial metals.

Carbon Steels

Low-carbon steels are easily welded and constitute by far the most widely used materials for arc-welded assemblies. They are easily welded by all the commercial arc-welding processes. In general, arc weldability decreases as carbon content increases. Increasing alloy content also tends to decrease weldability, but carbon content is probably the most important factor. Except for some high-strength–low-carbon structural steels that have a narrow heat zone, steels with a carbon content of 0.30 percent or less present no welding problems.

Medium-carbon steels are usually reasonably easy to process and give good results but may require precautions or extra process steps to avoid a tendency toward stress cracking. Normally, plain carbon steels up to about 0.80 percent carbon present no problems unless ingredients to promote hardenability (manganese, chromium, molybdenum) are present.

High-carbon steels (over 0.80 percent carbon) are more apt to exhibit a tendency toward stress cracking at the weld and cannot be rated well suited for welding. The shielded-metal arc (stick), flux-cored arc, and submerged arc processes are not as good for welding high-carbon steel as are the gas-metal arc welding (GMAW) and gas-tungsten arc welding (GTAW) processes.

Alloy Steels

These vary greatly in their ability to be welded satisfactorily and economically. Steels that form a martensitic structure on cooling from the transformation temperature are more apt to exhibit high hardness and hence brittleness and cracking of the weld metal. The most critical alloying elements are phosphorus and sulfur (the total of which should be below 0.30 percent).

GMAW and GTAW welding are the best processes for higher-alloy steels, providing better visual control, usually better shielding, and more localized heating.

Stainless Steels

Stainless steels of the austenitic type (300 Series) are the most weldable (except for formulations that promote free machining). The ferritic and martensitic types of stainless (400 Series) are less weldable. Types 318, 321, 347, and 348 are less susceptible to carbon precipitation and are recommended grades for arc welding. Types 304L and 316L are quite suitable for welding but are not recommended for applications involving service temperatures over 425°C (800°F).

GTAW, GMAW, and shielded-metal arc (stick) welding are most common for stainless materials, but plasma, submerged arc, and flux-cored arc processes are also used.

Aluminum Alloys

These alloys are welded by the GMAW or GTAW process; other processes are rarely used. Most aluminum alloys are suitable for welding. Those which are not exhibit brittleness or loss of ductility in the weld area. Table 7.2.1 lists common aluminum alloys in four categories of weldability. Both wrought and cast alloys are included.

TABLE 7.2.1 Weldability of Aluminum Alloys by the GMAW and GTAW Processes

Readily weldable	Wrought alloys: pure aluminum, 1060 (A91060), 1100 (A91100), 1350 (A91350), 2219 (A92219), 3003 (A93003), 3004 (A93004), 5005 (A95005), 5050 (A95050), 5052 (A95052), 5083 (A95083), 5086 (A95086), 5154 (A95154), 5254 (A95254), 5454 (A95454), 5456 (A95456), 5652 (A95456), 6061 (A96061), 6063 (A96063), 6101 (A96101), 6151 (A96151), 7005 (A97005), 7039 (A97039); casting alloy: 443.0 (AO4430)
Weldable in most applications*	Wrought alloys: 2014 (A92014), 4032 (A94032); casting alloys: 413 (A04130), 319 (A03190), 333 (A03330), 355 (A03550), C355 (A33550).
Limited weldability†	Wrought alloy: 2024 (A92024)
Welding not recommended	Wrought alloys: 7075 (A97075), 7178 (A97178); casting alloys: 222.0 (A02220), 242.0 (A02420), 520.0 (A05200)

*May require special techniques for some applications.

†Requires special techniques.

Source: Updated from *Metals Handbook* vol. 6, 8th ed., American Society for Metals, Metals Park, OH, 1971.

Copper and Copper Alloys

These alloys are most commonly welded with the GTAW process. If the material is 12 mm (0.5 in) or more in thickness, GMAW welding is employed. Occasionally conventional stick welding is also used.

A number of copper alloys are not suited to welding. These are alloys with high percentages of zinc or with lead, tellurium, sulfur, or oxygen. Zinc gives off toxic vapors and reduces weldability. However, low-zinc brasses (15 percent zinc or less) are normally weldable. The elements added to brass to promote free machining—lead, sulfur, and tellurium—tend to cause stress cracking at the weld area. Oxygen reduces weld strength, while beryllium, aluminum, and nickel all necessitate additional cleaning of the weld area prior to welding. Table 7.2.2 lists copper alloys suitable for welding.

TABLE 7.2.2 Weldability of Copper and Selected Copper Alloys

Alloy	Name	Nominal composition, %	Weldability (GMAW or GTAW)*
C12000	Phosphorus deoxidized copper (low P)	99.9 Cu, 0.008 P	A
C12200	Phosphorus deoxidized copper (high P)	99.9 Cu, 0.02 P	A
C65100	Low-silicon bronze B	98.5 Cu, 1.5 Si	A
C65500	High-silicon bronze A	97 Cu, 3 Si	A
C70600	Copper nickel, 10%	88.6 Cu, 10 Ni, 1.4 Fe, 1.0 Mn	A
C71500	Copper nickel, 30%	70 Cu, 30 Ni	A
C61300	Aluminum bronze D (Sn stabilized)	89 Cu, 7 Al, 3.5 Fe, 0.35 Sn	A*
C61400	Aluminum bronze D	91 Cu, 7 Al, 2.5 Fe, 1 max. Mn	A*
C10200	Oxygen-free copper	99.95 Cu	B
C17000	High-strength beryllium copper, 1.7%	98.3 Cu, 1.7 Be	B
C17200	High-strength beryllium copper, 1.9%	98.1 Cu, 1.9 Be	B
C21000	Gilding, 95%	95 Cu, 5 Zn	B
C22000	Commercial bronze, 90%	90 Cu, 10 Zn	B
C23000	Red brass, 85%	85 Cu, 15 Zn	B
C24000	Low brass, 80%	80 Cu, 20 Zn	B
C50500	Phosphor bronze, 1.25% E	98.7 Cu, 1.3 Sn, 0.2 P	B
C51000	Phosphor bronze, 5% A	95 Cu, 5 Sn, 0.2 P	B
C52100	Phosphor bronze, 8% C	92 Cu, 8 Sn, 0.2 P	B
C52400	Phosphor bronze, 10% D	90 Cu, 10 Sn, 0.2 P	B

Note: In most cases, shielded-metal arc (stick) welding is not recommended for the above metals.

A = excellent weldability; B = good weldability. A for GTAW, B for GMAW.

Source: Updated from *Metals Handbook,* vol. 6, 8th ed., American Society for Metals, Metals Park, OH, 1971.

TABLE 7.2.3 Weldability of Magnesium Alloys

Alloy designation		
Common	UNS	Rating
	Casting alloys	
EZ33A	M12330	A
KIA	M18010	A
AM100A	M10100	B+
AZ81A	M11810	B+
AZ91C	M11914	B+
HK31A	M13310	B+
AZ92A	M11920	B
QE22A	M18220	B
AZ63A	M11630	C
HZ32A	M13320	C
ZE41A	M16410	C
ZH62A	M16620	C−
ZK51A	M16510	D
ZK61A	M16610	D
	Wrought alloys	
AZ31B	M11311	A
HK31A	M13310	A
HM21A	M13210	A
HM31A	M13312	A
ZE10A	M16100	A
AZ61A	M11610	B
AZ80A	M11800	B
ZK21A	M16210	B

Magnesium Alloys

For the most part, these alloys are weldable. The GTAW and GMAW processes are used. Table 7.2.3 summarizes the weldability of common magnesium alloys. When ratings are lower than A (excellent), the reason usually is susceptibility to stress cracking. A second factor is a lessening of strength at the welded joint.

Cast Iron

This metal normally is not welded as a production operation. The repair of cast-iron parts, especially machinery or automotive components, is, however, quite common. Cast iron tends to be embrittled in the heat-affected zone adjacent to the weld. Preheating, postheating, and peening the weld area, as well as the use of high-nickel welding rods, can minimize the tendency sufficiently so that satisfactory repairs can be made.

Low-Melting-Temperature Metals

Metals such as lead, zinc, tin, and others are not suitable for arc welding. The temperature difference between the arc and the metal's melting point is great, and in some cases the metal will vaporize in the arc. Steel coated with such metals also is not recommended for arc welding.

Other Metals

Cast-nickel alloys, particularly if they are high in silicon, are not easily welded. Some wrought nickel and nickel alloys such as nickels 200 (N02200) and 201 (N02201) and Monels 400 (N04400), R405 (N04405), and K500 (N05500) can be arc-welded. Methods similar to those used for stainless steel are applied, with the GTAW process predominating.

Titanium is not easy to weld because careful shielding, including shielding the side of the joint opposite the arc, is required. However, it is regularly welded in production. Unalloyed titanium is suitable for arc welding, as are the alpha alloys and the alpha-beta alloy Ti-6Al-4V (R56401).

Heat-resistant alloys such as the nickel-base (Hastelloy, Inconel) iron-chromium-nickel and iron-nickel-chromium, cobalt-base (Discaloy, Incoloy, Thermalloy), and refractory (HS, S-816, UMCo50, W152) alloys are not easily welded because of stress cracking and because the service requirements that they face are often severe. Nevertheless, arc welding of these metals, especially by the GTAW process, is fairly common.

Electron- and Laser-Beam Welding

These processes have much greater ability to produce satisfactory results with difficult-to-weld materials because their heat input is very highly concentrated. Alloy and high-carbon steels that may become brittle in the weld area with other processes can be electron- and laser-beam-welded. Hardened steels can be welded with a smaller adversely affected area than with more conventional processes. With EBW, the use of a vacuum chamber avoids the contamination that can occur in some metals even with shielding gas.

Heat-resistant alloys Inconel 625 (N26625), Hastelloy X (N06002), and Hastelloy N (N10003) are easily electron-beam-welded. René 41 (N07041), alloy 718 (N07718), and Inconel X-750 (N07750) are rated good in weldability by EBW.

Refractory metals can be vacuum-electron-beam-welded, with their weldability ranked in the following order: tantalum, niobium, molybdenum, and tungsten.

Copper alloys are somewhat more weldable with EBW than with other processes. Aluminum and magnesium alloys are not easily weldable with EBW but can be processed.

Laser welding is also suitable for refractory metals and for glass-to-metal welds and minimizes the problems that can occur when heat-sensitive materials are welded. Table 7.2.4 summarizes the weldability of various dissimilar metals by laser beam. Materials to be laser-welded should not have reflective, polished surfaces since these do not absorb the light energy of the laser beam.

TABLE 7.2.4 Laser-Beam Weldability of Binary Metal Combinations

	W	*Ta*	*Mo*	*Cr*	*Co*	*Ti*	*Be*	*Fe*	*Pt*	*Ni*	*Pd*	*Cu*	*Au*	*Ag*	*Mg*	*Al*	*Zn*	*Cd*	*Pb*
W	E																		
Ta	E																		
Mo	E	E																	
Cr	E	P	E																
Co	F	P	F	G															
Ti	F	E	E	G	F														
Be	P	P	P	P	F	P													
Fe	F	F	G	F	E	F	F												
Pt	G	F	G	G	E	F	P	G											
Ni	F	G	F	G	E	F	F	G	E										
Pd	F	G	G	G	E	F	F	G	E	E									
Cu	P	P	P	P	F	F	F	F	E	E	E								
Au			P	F	P	F	F	F	E	E	E	E							
Ag	P	P	P	P	P	F	P	P	F	P	E	F	E						
Mg	P		P	P	P	P	P	P	P	P	P	F	F						
Al	P	P	P	P	F	F	P	F	P	F	P	F	F	F	F				
Zn	P		P	P	F	P	P	F	P	F	F	G	F	G	P	F			
Cd				P	P	P		P	F	F	F	P	F	G	E	P	P		
Pb	P		P	P	P	P		P	P	P	P	P	P	P	P	P	P	P	
Sn	P	P	P	P	P	P	P	P	F	P	F	P	F	P	P	P	P	P	F

Note: E = excellent; G = good; F = fair; P = poor.

Source: From *Design News.*

DESIGN RECOMMENDATIONS

Cost Reduction

Costs can be reduced by the following design recommendations:

1. Welded assemblies should be made up of as few parts as possible. Bending and forming operations are almost always less costly than welding, and designers should endeavor to develop at least part of the configuration of their assemblies by these methods instead of welding. Sometimes machining also can provide a lower-cost method of producing a particular shape than the welding together of several components.
2. Weld joints should be placed so that there is room for easy access of the welding nozzle. This is particularly important for the GMAW, GTAW, plasma, and other welding methods that use a wire feed and a shielding gas. The wire feed requires more room than stick welding, and when shielding gas is used, it is important that the nozzle be close to the welding point so that the molten metal is well shielded.
3. The design that requires the least weld metal and the least arc time usually provides the least costly welded assembly. Designers should specify the minimum amount of weld filler, with respect to both fillet size and length, that meets functional requirements of the assembly. Tack welds and intermittent welds should be specified if the application does not involve high stresses or a leakproof construction. However, the time required to deposit the largest fillet that can be deposited in one pass by the equipment is only slightly greater than that required for a small fillet. Greater economy is obtained through the use of intermittent large fillets than with continuous smaller fillets. Designers have the responsibility for making whatever calculations, analyses, or tests necessary to specify the sizes and types of welded joints rather than simply to specify "Weld parts together."
4. Whenever possible, the assembly should be designed so that the welded joint is horizontal, with the stick or electrode holder pointing downward during welding. This position is the most rapid and convenient with all welding methods.
5. Good fit-up of parts at the weld joint is essential not only for welding speed but also for minimizing distortion of the finished weldment. Especially with butt joints, edges of workpieces should be straight and uniform. Often, the extra operation required to provide a straight edge will be less costly than the extra welding labor required when the fit is not correct. Figure 7.2.6 illustrates examples of good and poor fit-up.
6. The buildup of weld fillets should be kept to a minimum. Additional material in the convex portion of the fillet's cross section does not add significantly to the strength of the joint. (See Fig. 7.2.7.)
7. When forgings or castings are part of a welded assembly, care should be taken to ensure good fit-up of the parts to be welded. Untrimmed parting-line areas should not be included in the welded joint. The casting should also be designed so that the wall thickness of both parts to be joined is equal at the joint. This ensures more rapid and less distortion-prone welding. (See Fig. 7.2.8.)
8. It is preferable to locate welds out of sight rather than in locations where special finishing operations are required for the sake of appearance.
9. The joint should be designed so that it requires minimal edge preparation. It is often advisable to use slip or lap joints in welded assemblies to avoid the cost of

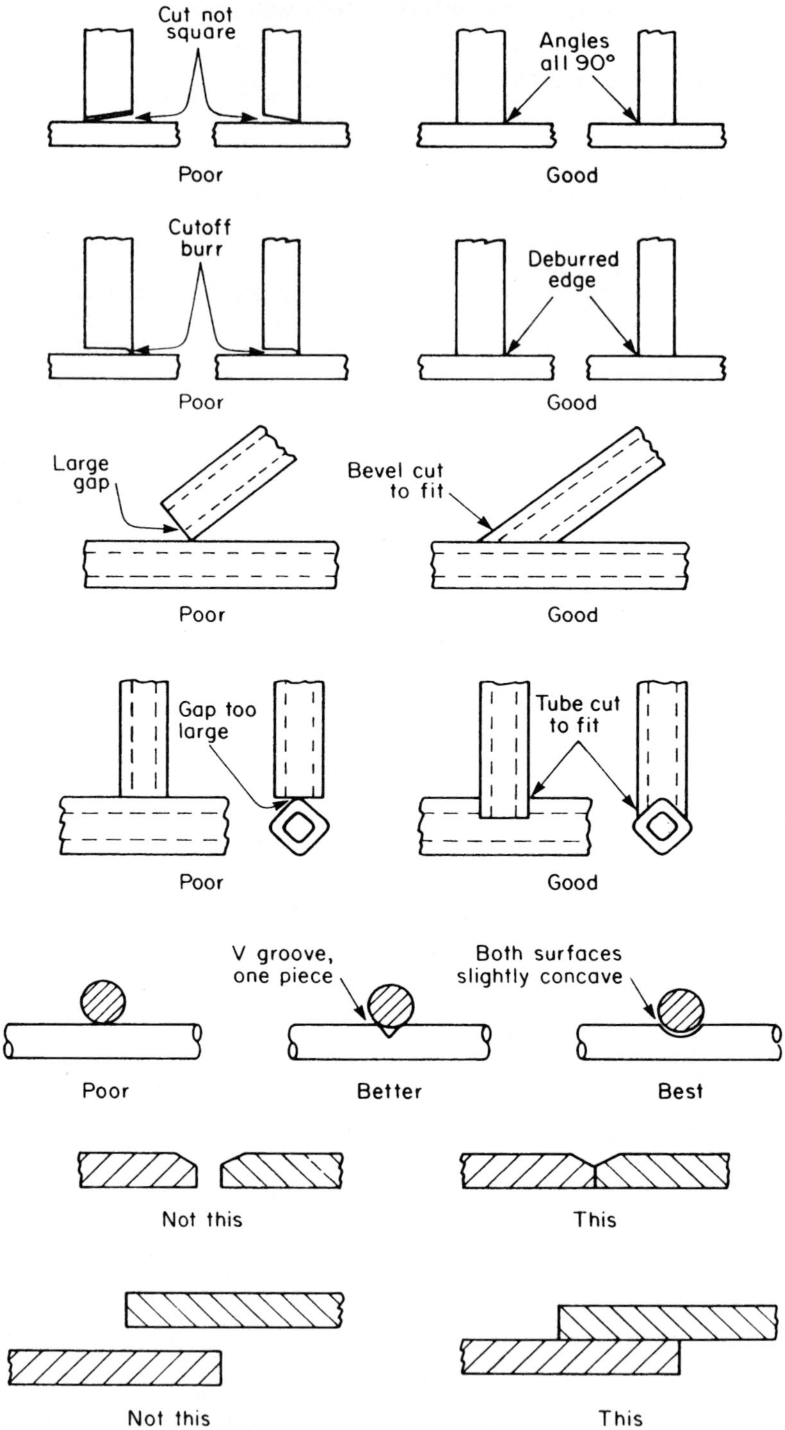

FIGURE 7.2.6 Poor and good fit-up of weld joints.

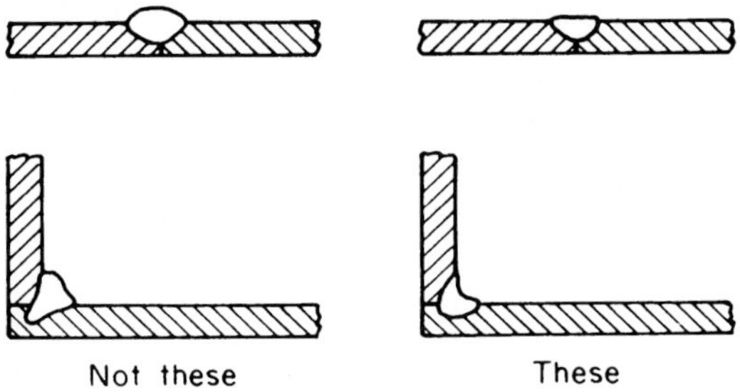

FIGURE 7.2.7 A buildup of filler material in weld joints does not add materially to joint strength.

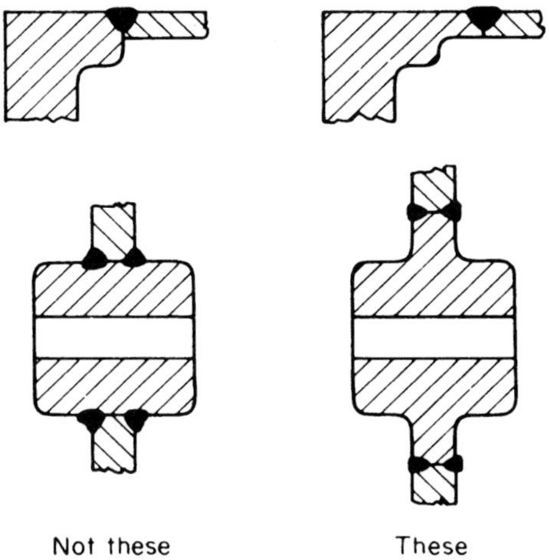

FIGURE 7.2.8 The wall thickness of parts to be joined should be equal at the joint.

close edge preparation and to simplify fit-up problems. However, lap joints are more difficult to clean, finish, and repair and frequently have root defects. They are therefore not necessarily less costly than butt joints. It also is necessary to avoid gaps between the plates of a lap joint by clamping them together or ensuring flatness beforehand. Figure 7.2.9 illustrates the use of lap or slip joints.

10. In some cases, it is possible to use the curved edges or sides of parts comprising the assembly to provide the equivalent of a grooved edge for the welded joint. Little or no edge preparation is therefore needed, and total operation time is reduced. Figure 7.2.10 shows some examples.

11. If machining after welding is required, welds should be placed away from the material to be machined if possible. This will avoid machining problems which can occur in the heat-affected zone and the extra cost of making a larger fillet only to have part of it removed. Figure 7.2.11 illustrates some alternatives.

12. It is often advisable to utilize a number of welded subassemblies in the fabrication of a large, complex final assembly. Subassemblies can be handled more easily;

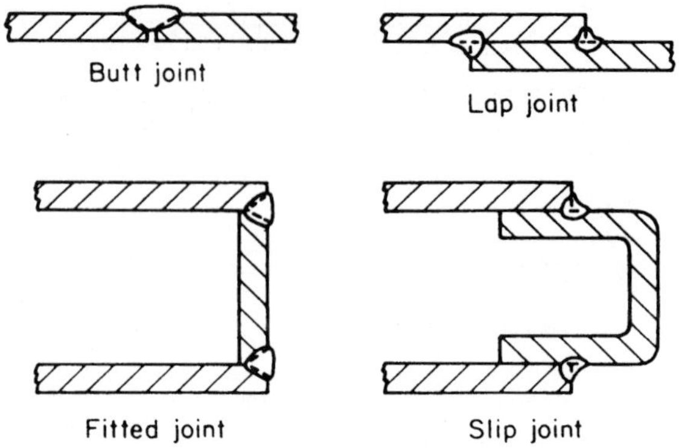

FIGURE 7.2.9 The joints on the right require less edge preparation.

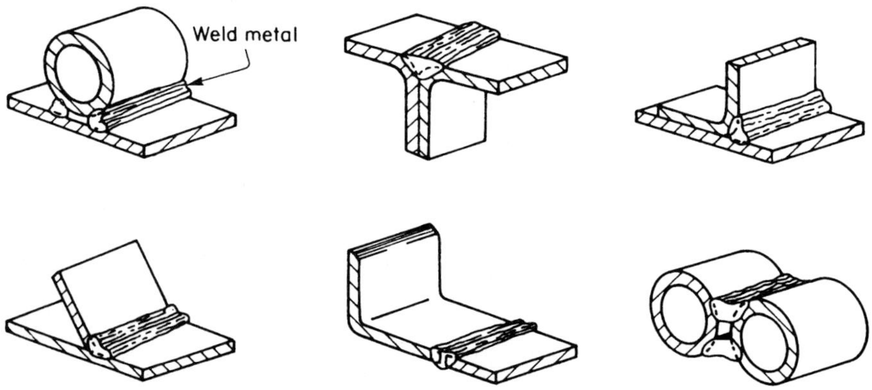

FIGURE 7.2.10 Joints that have natural grooves and thus need little or no edge preparation. (*From* Metals Handbook, *vol. 6, 8th ed., American Society for Metals, Metals Park, Ohio, 1971.*)

usually they can be positioned for easy access of the electrode, and the joint can be kept horizontal during welding.

13. When machining a groove on the end of a cylindrical component to be welded by submerged or shielded arc, it sometimes is advantageous to include a backup strip as an integral part of the component to be welded. (See Fig. 7.2.12.)

Minimizing Distortion

Distortion can be minimized by the following design recommendations:

1. Good fit of parts is important not only for minimum welding time but also for control of distortion. As illustrated in Fig. 7.2.6, maximum contact of all mating surfaces is desirable. The more gap to fill, the greater the possible weldment distortion.

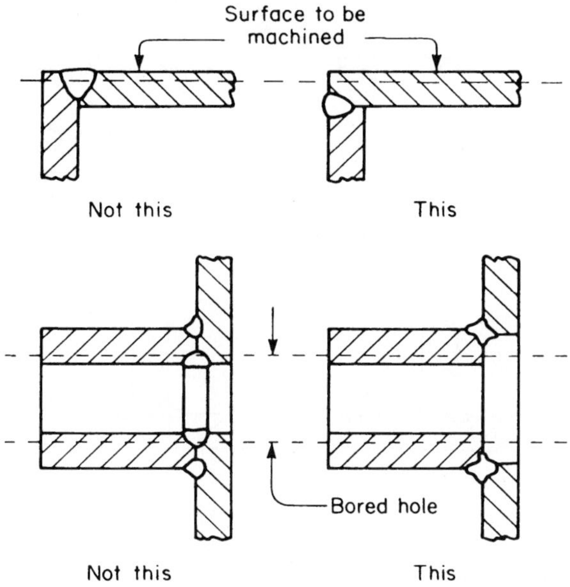

FIGURE 7.2.11 If machining after welding is required, keep the weld metal outside the portion of the weldment which will be machined.

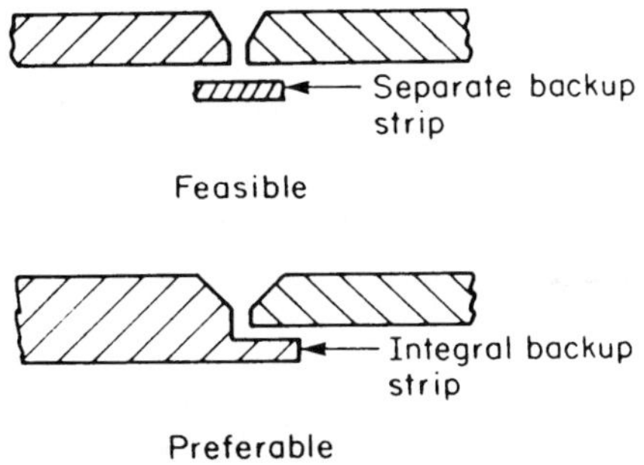

FIGURE 7.2.12 Sometimes it is advantageous to include a weld backup strip as an integral part of one of the components to be welded.

2. Generally, heavier sections are less prone to distortion from welding. If distortion prevention is important to the application, designers should consider the use of thicker, more rigid components.
3. Long sections of thinner material, when welded together, are apt to be distorted and to buckle unless there is a good rigid support for the joint. One method of supplying this for butt joints is to use a short-flanged butt joint as illustrated in Fig. 7.2.13. This joint is well suited to GTAW, gas, and plasma welding because it requires no filler metal.

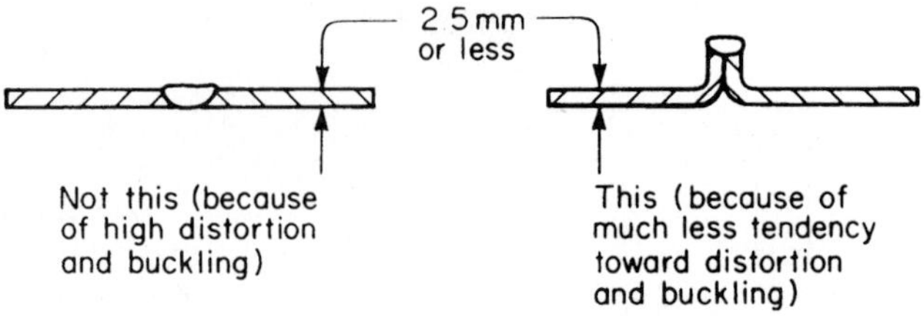

FIGURE 7.2.13 A short-flanged butt joint is often preferable for joining thin material.

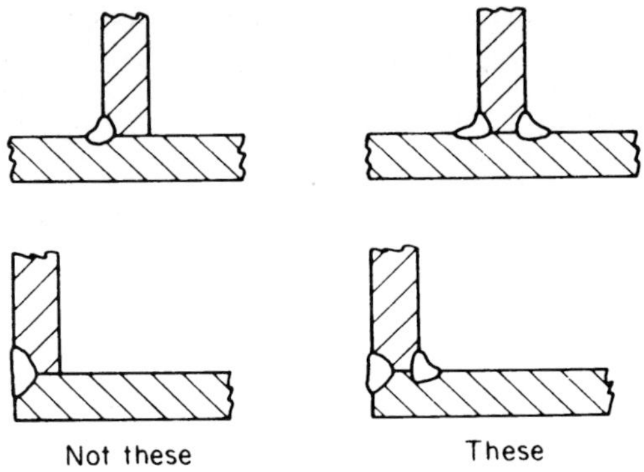

FIGURE 7.2.14 Use opposing welds to reduce angular distortion.

4. Whenever possible, place welds opposite one another to reduce distortion. In this way the shrinkage forces in the weld fillets are balanced and tend to offset one another. Figure 7.2.14 illustrates some examples. If extra length of weld fillets is caused by this procedure, use intermittent opposite welds if possible.
5. If sections of unequal thickness must be welded together, distortion can be reduced by machining a groove in the thicker piece adjacent to the weld joint. (See Fig. 7.2.15.)
6. When dimensioning welded assemblies, it is essential that consideration be given to the shrinkage inherent in each weld. The following are specific examples intended to give guidelines to designers.

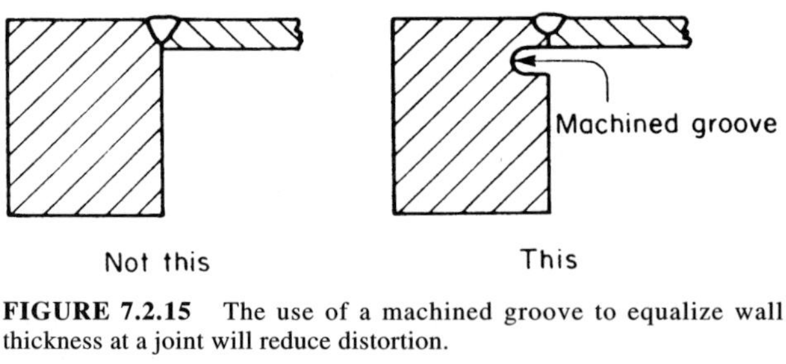

FIGURE 7.2.15 The use of a machined groove to equalize wall thickness at a joint will reduce distortion.

Example 7.2.1 illustrates requirements for stock allowances for machining after welding.

Example 7.2.1 Machining after Welding. A common oversight in design is to forget to allow enough stock on component parts of a weldment that must be machined subsequently to enable the finished assembly to fall within specified tolerances. In Fig. 7.2.16, three parts are to be welded together to form an assembly that has an overall length of 324 mm ±0.5 mm (12.58 in ±0.020 in). This close tolerance necessitates machining after welding. If the material thickness of the two side pieces is 12 mm (0.47 in) or more, there may be enough material to machine, but if, as often happens, the material is less than the nominal 12-mm thickness, and if the center piece is of nominal length or shorter, then no stock will be available to machine the sides to produce the 324-mm final dimension within the tolerance allowed. The solution here is to dimension part *X* so that there will always be machining stock available on the weldment regardless of stock and process tolerance variations on all parts.

Example 7.2.2 deals further with concentricity and alignment when close tolerances are required.

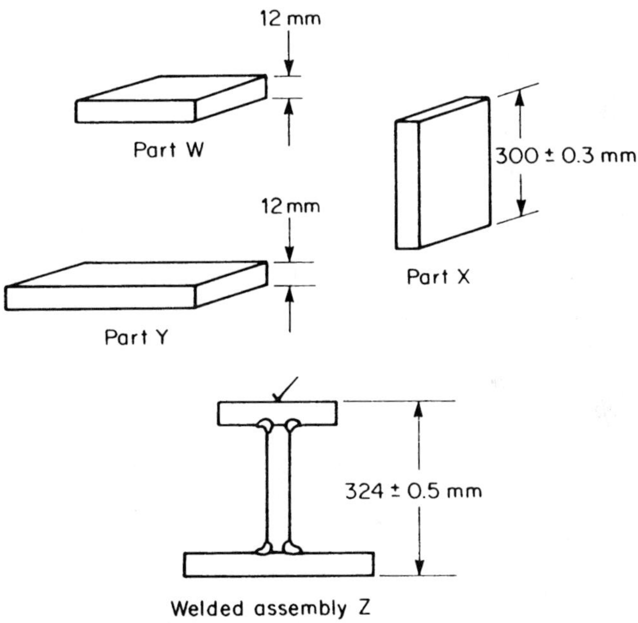

FIGURE 7.2.16 See Example 7.2.1.

	Maximum, mm	Mean, mm	Minimum, mm
W	12.3	12.0	11.7
X	300.3	300.0	299.7
Y	12.3	12.0	11.7
Z	324.9	324.0	323.1

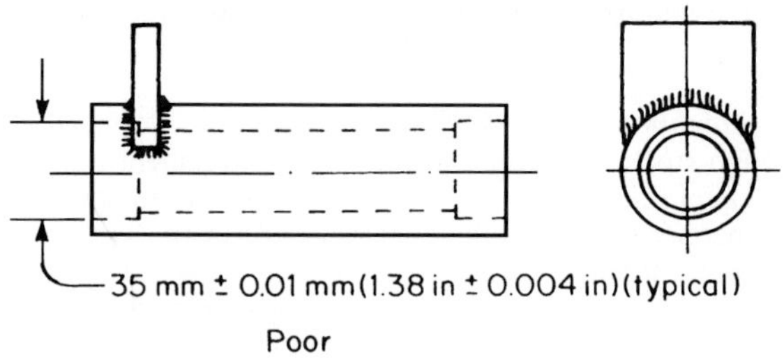

FIGURE 7.2.17 See Example 7.2.2.

Example 7.2.2 Out-of-Round Conditions. Figure 7.2.17 shows a lug welded to a piece of tubing that has enlarged holes at the ends bored to very close tolerances. The designer's intent was to machine the tube first, then weld. However, the close-tolerance bores will become egg-shaped (elliptical) as a result of welding warpage.

Figure 7.2.18 shows a somewhat better approach. The lug extends all around the tube, and the weld stresses are balanced. If the bore tolerance is fairly large (±0.06 mm or more), the designer has a good chance of maintaining hole roundness within tolerances. Very close tolerances such as ±0.01 mm (±0.0004 in) will require machining after welding or further design changes to eliminate egg-shaped distortion.

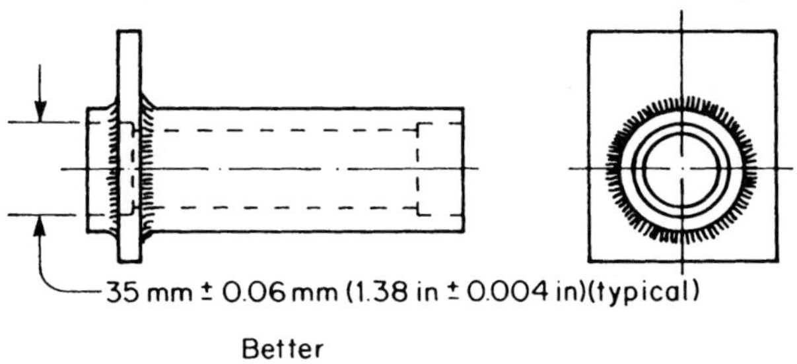

FIGURE 7.2.18 An improved design compared with Fig. 7.2.17. See Example 7.2.2.

Two such design improvements to reduce the likelihood of the need to machine after welding are shown in Figs. 7.2.19 and 7.2.20. In Fig. 7.2.19, heavy-wall tubing is

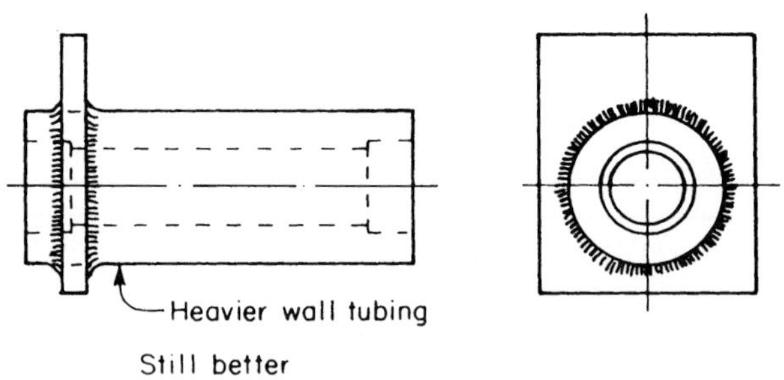

FIGURE 7.2.19 A further improvement compared with the design shown in Fig. 7.2.18. See Example 7.2.2.

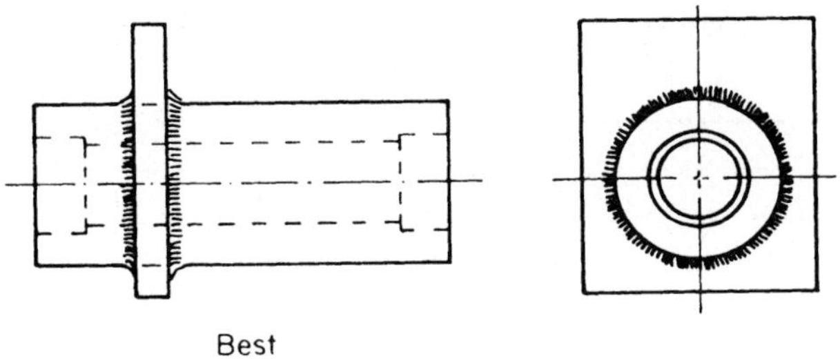

FIGURE 7.2.20 A still better design. See Example 7.2.2.

used, providing a thicker and stronger wall at the bore and reducing internal-diameter shrinkage and eccentricity. In Fig. 7.2.20, a further improvement is made by placing the lug to one side of the bored area, farther from the critical surface and in a location where the wall is still heavier.

Weld Strength

Welds can be strengthened by the following design recommendations:

1. The butt joint (Figs. 7.2.5 and 7.2.9) is the most efficient type. If deep-penetration welding is used or the stock thickness is not great, the square-edged butt joint can be employed and edge-preparation time therefore saved. Thicker stock or less penetrating methods may require grooved edges.
2. For efficient and economical welding, minimize the stress that the joint must carry. This can be achieved by locating weld joints away from areas of stress or designing the assembly so that the parts themselves rather than the weld joints bear the load. See Fig. 7.2.21 for examples.
3. Groove welds should be designed to be in either compression or tension. Fillet welds should be in shear only. (See Figs. 7.2.22 and 7.2.23.)
4. When intermittent welds are used in place of continuous welds for cost reduction, the length of each fillet should be at least 4 times the fillet thickness and not less than 40 mm ($1\frac{1}{2}$ in). If the joint is in compression, the spacing of the welds should not exceed 16 times thickness. If the joint is in tension, the spacing may be as much as 32 times thickness but not over 300 mm (12 in). (See Fig. 7.2.24.)
5. When using grooves for welds, follow the standard American Welding Society weld-groove dimensions, as shown in Fig. 7.2.25.

Electron- and Laser-Beam Weldments

The narrow width and deep penetration inherent in these welding processes make butt joints preferable to lap joints. Beveled edges are not needed and, in fact, should be avoided. However, good fit-up of the mating pieces is essential because of the narrow beam.

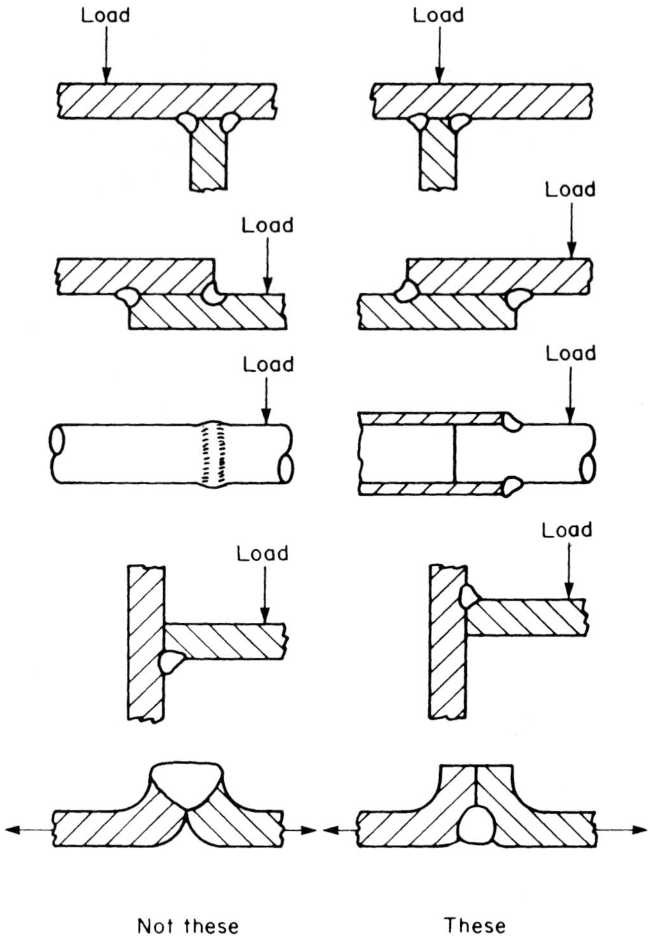

FIGURE 7.2.21 Weldments should be designed so that welds are placed to minimize stress concentration in the weld fillet.

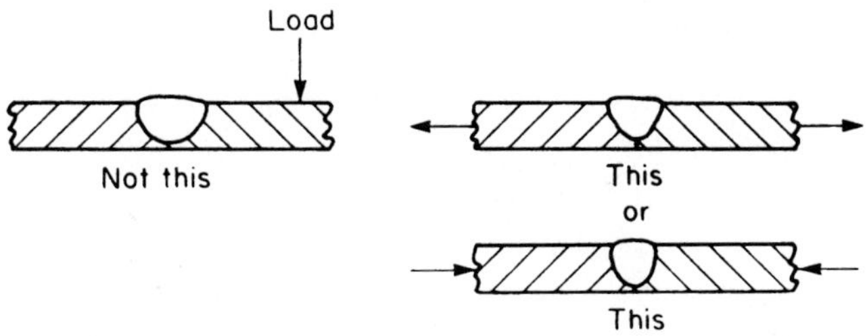

FIGURE 7.2.22 Groove welds should be designed to be in either tension or compression only.

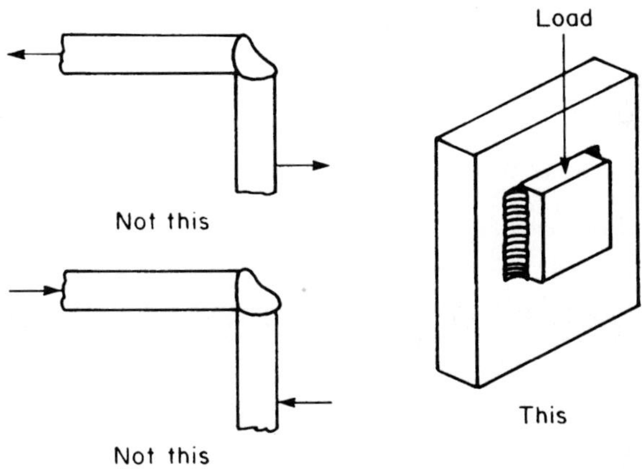

FIGURE 7.2.23 Fillet welds should be designed to be in shear only.

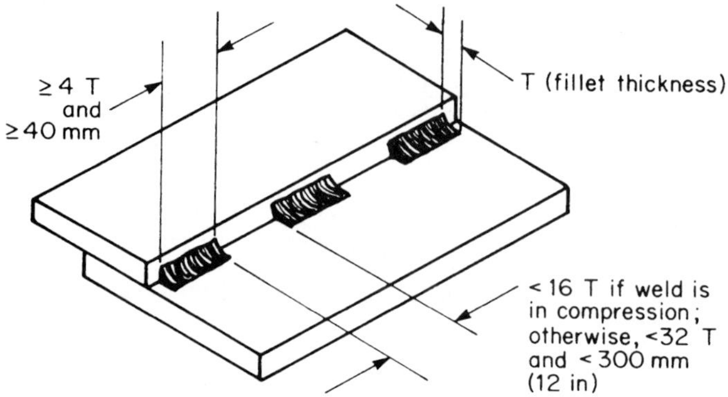

FIGURE 7.2.24 Recommended length and spacing of intermittent welds.

Electron-beam weldments to be processed in vacuum chambers should be self-fixturing to permit batches of assemblies to be placed in the welding chamber with minimum space occupancy and to avoid costs of multiple fixtures.

Weldments and Heat Treatment

Designers should remember a few basic rules concerning the use of heat treatment and weldments:

1. Welding of carburized or hardened steels requires controlled conditions and proper equipment and supplies. Designers should not specify such welding unless it is unavoidable.
2. Welding will reduce or remove completely the hardness of carburized or nitrided mild (low-carbon) steels in the area of the weld.
3. Carbon in welded areas will affect the physical and chemical characteristics of the weld bead, resulting in possible cracking or weld failure in or adjacent to the weld.
4. Weldments may be heat-treated after the welding has been completed without undesirable effects other than possible distortion from stress relief and heat treatment. Designers should remember that any straightening operation on carburized and hardened parts may result in some surface cracking in the welded area.

DIMENSIONAL FACTORS AND RECOMMENDED TOLERANCES

By far the most significant factor that affects the dimensions of welded assemblies is the shrinkage of the weld fillet on cooling. A steel weld bead cools from about 1510°C (2750°F) when it solidifies after being deposited to a usual ambient tempera-

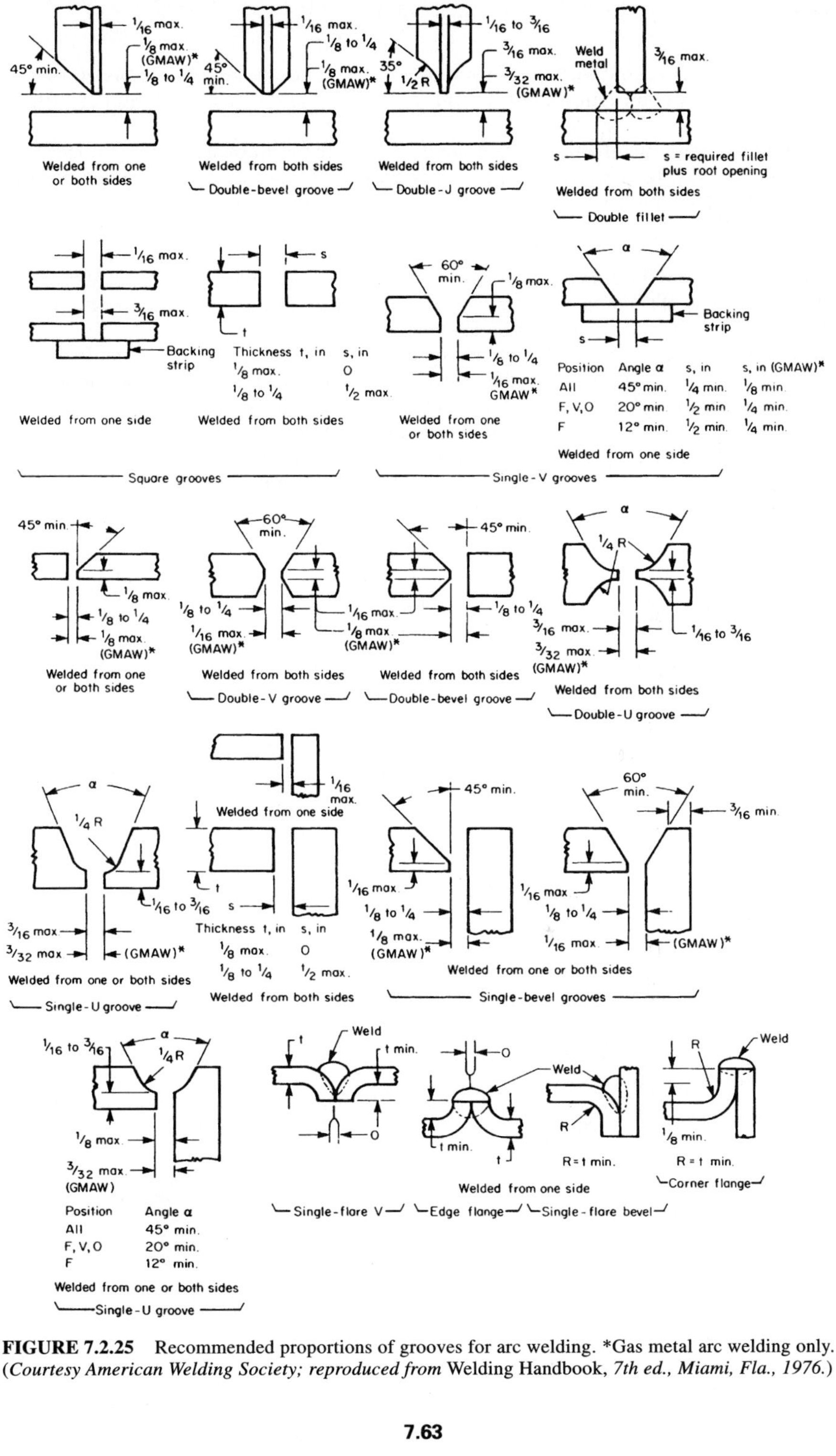

FIGURE 7.2.25 Recommended proportions of grooves for arc welding. *Gas metal arc welding only. (*Courtesy American Welding Society; reproduced from* Welding Handbook, *7th ed., Miami, Fla., 1976.*)

ture of about 20°C (68°F). Since steel has a thermal coefficient of expansion of 15.1 $\times 10^{-6}$/°C, the contraction between these two temperatures is 0.22 mm/cm (0.022 in/in).

Only the weld metal and adjacent areas and not the bulk of welded components go through this vast temperature change. Only these areas exhibit the thermal shrinkage. This different amount of shrinkage within the assembly causes internal stresses and distortion.

Thermal shrinkage cannot be eliminated by fixturing, but it can be reduced somewhat when heavy fixtures are used. However, cooling and the inherent shrinkage still take place. The shrinkage forces are very strong and can even cause the weldment to be "frozen" in the fixture if it is not designed properly.

The dimensional changes take place both across the weld head and along its length. They are, of course, usually less than the amount indicated by thermal-contraction factors alone because of the restraining effect of the attached metal parts.

Table 7.2.5 presents recommended dimensional tolerances for arc weldments produced under average production conditions. Closer limits will involve additional manufacturing cost, usually in the form of extra straightening and/or machining after welding.

TABLE 7.2.5 Recommended Dimensional Tolerances for Arc Weldments

	Assembly with					
	Little welding		Moderate welding		Heavy welding	
Basic dimension, m (in)	Close tolerance, mm (in)	Normal tolerance, mm (in)	Close tolerance, mm (in)	Normal tolerance, mm (in)	Close tolerance, mm (in)	Normal tolerance, mm (in)
0–0.3 (0–12)	±0.4 (±0.015)	±0.8 (±0.030)	±0.8 (±0.030)	±1.5 (±0.060)	±1.5 (±0.060)	±3 (±0.125)
0.3–1 (12–36)	±0.8 (±0.030)	±1.5 (±0.060)	±1.5 (±0.060)	±3 (±0.125)	±3 (±0.125)	±6 (±0.25)
1–2 (36–80)	±1.5 (±0.060)	±3 (±0.125)	±3 (±0.125)	±6 (±0.25)	±6 (±0.25)	±12 (±0.5)

CHAPTER 7.3
RESISTANCE WELDMENTS

Nicholas S. Hodska
Stratford, Connecticut

THE PROCESS

Resistance welding is a process in which localized coalescence of metal is accomplished by the heat induced from the metal's resistance to heavy electric-current flow. The electrodes that supply the electric current also apply pressure to the area to be joined, but there is no external heat source. No fluxes or filler material is required.

The process has four major subdivisions: (1) resistance spot welding (RSW), (2) seam welding (RSEW), (3) projection welding (PW), and (4) flash (FW) or upset (UW) welding.

Resistance-spot welding utilizes opposed electrodes to apply pressure to the parts to be joined, provide electric contact with these parts, and conduct the electric current to them. High current at low potential, from a transformer secondary, is localized by the blunt points of the electrodes and under the applied pressure creates the weld spot. (See Fig. 7.3.1).

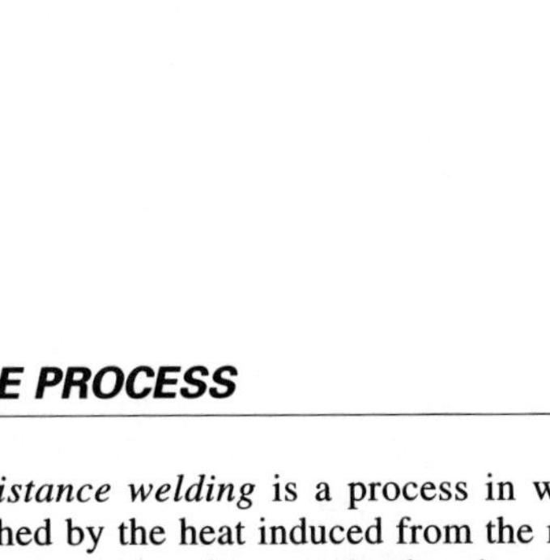

FIGURE 7.3.1 Spot welding. (*From D. B. Dallas (ed.),* Tool and Manufacturing Engineers Handbook, *3d ed., McGraw-Hill, New York, 1976.*)

Seam welding utilizes circular wheel-like electrodes. These press the parts to be welded together and, while rolling, conduct a series of high-current, low-voltage pulses to the work. These produce overlapping spot welds that become a continuous seam. The process is illustrated by Fig. 7.3.2.

In projection welding, embossments or projections of the parts themselves, rather than separate narrow-shaped electrodes, localize the current flow from one workpiece to the other. The projected metal is heated sufficiently to soften and fuse it to the mating part.

In flash welding, parts are brought together with only a slight pressure against one

FIGURE 7.3.2 Seam-welding machine tooled for welding automotive gasoline tanks. (*Courtesy Acro Welder Manufacturing Co.*)

another or are kept slightly separated. Electrodes are clamped to each part, and heavy current is made to flow through them. Arcing takes place where the parts come together, and the whole area of the ends is raised to a high temperature. Heavy pressure is then applied to the two parts, forcing them together where they fuse when the current is stopped. A thin fin around the joint is formed, and this normally contains any metal oxides, leaving only sound metal in the weld itself.

Upset welding is similar to flash welding. Parts are butted together in firm contact, and electric current flows between them. However, no flashing occurs; joint material remains in a solid state but softens from the heat generated by the electric current. The surfaces in contact fuse together.

CHARACTERISTICS AND APPLICATIONS

Resistance welding is most advantageously and commonly used in assemblies made from sheet metal. Heavier sections require more power and usually much more pressure and are best fabricated by arc or gas welding processes.

The following products utilize spot and projection weldments:

1. *Office furniture.* File cabinets, desks, in-and-out trays, storage cabinets, bookcases, typewriter tables, and business machine housings.

2. *Appliance housings.* Refrigerator cabinets and doors, stoves (covers, doors, etc.), toasters, freezers, and washing machine and dryer cabinets.
3. *Automobile components.* Chassis, body panels, doors, fan blades, and floor and dashboard assembles for cars and trucks.
4. *Aircraft assemblies.* Structural components, cowlings, doors, fuselage outer skin, decks, ribs, and seat frames.
5. *Farm machinery.* Frames, panels, seat buildups, housings, and other components of tractors and other machines.

Seam welding is used when fluid-carrying ability or pressure tightness is required. Notable examples are fuel tanks for vehicles, refrigerator evaporator coils, and ductwork.

Spot welding has several advantages over rivet or screw fastening. It is performed more rapidly and requires no drilling, punching, or separate fasteners. It produces assemblies that can be light in weight.

Spot welding can be used for foil materials as thin as 0.025 mm (0.001 in). At the other extreme, plate as thick as 25 mm (1 in) can be spot-welded with the proper technique and equipment. Normal stock thickness, however, ranges between 0.25 mm (0.010 in) and 8 mm (0.315 in). Although spot welding is normally performed on sheet metals, it is not necessary that both pieces to be joined be of sheet material. As long as a localized contact can be made between the two pieces, spot welding can take place. Thus spot welds are used to fasten sheet material to machined parts, castings, or plate parts of considerable thickness.

Spot welds characteristically produce a depression in the surface of the workpiece where the electrode pressure is applied. This can be avoided on one surface by using indirect welding, in which the two electrodes are positioned on the same side of the workpiece with a smooth backup plate on the opposite side. (See Fig. 7.3.3.)

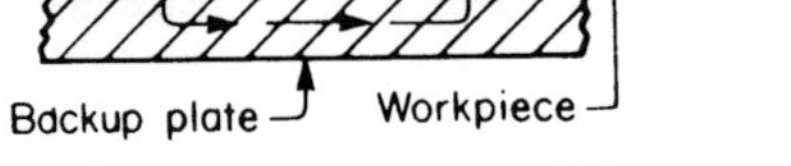

FIGURE 7.3.3 Indirect welding provides a smooth surface with no depression on one side of the spot-welded joint.

Some seam-type welds, when pressure tightness is not required, are intermittent; i.e., individual welds do not overlap. These welds are called *stitch welds* or *roll spot welds*. They have the advantage of producing less warpage, which sometimes occurs in the work after seam welding. Normal seam welds involve sheet stock within a thickness range of 0.25 to 3 mm (0.010 to 0.120 in).

Projection welds are not limited to sheet material. Machined pieces, forgings, castings, heavy stampings, powder-metal parts, and wire components can be welded by this method. The only requirement is a projection or small area of contact to concentrate the welding current. Projection welding is ideal for long-run parts like stampings in which the projection-making elements can be incorporated in the part tooling.

Projection welds can be more closely spaced than spot welds. With the latter, the electric current may shunt to a nearby existing weld, causing poor fusion in the second weld if the welds are too close together.

Flash and upset welds usually involve the end-to-end joining of wire, rods, bars, tubes, and other parts of circular cross section. Sheets, plates, flats, strips, angles, and sometimes components of other shapes also can be joined. There are two basic requirements:

1. Cross sections of both parts at the joint should match in size and shape.
2. The parts must be clampable so that they can be pressed together, and the electric current must be large enough to fuse the metal at the joint.

Flash and upset welds, properly made, are strong and uniform with relatively little tendency toward warpage because stresses are balanced. Joints are normally free of oxides and other impurities. However, there is a thickening at the joint area because of the upsetting action, and the excess metal usually must be trimmed or machined away.

ECONOMIC PRODUCTION QUANTITIES

Resistance welding is a productive operation. Output rates are high, surface-preparation requirements are minimal, and no filler metal is required. Although resistance welding often can be used profitably at all levels of production, it is usually most suitable for moderate and high levels.

Spot welding, however, is often applied to job-shop and other low-quantity work because tooling costs are minimal or nonexistent and setup is generally fast. Jobbing sheet-metal shops customarily include spot welders. If standard materials and gauges are used in such shops or if the operator is skilled in setup, one-piece lots are practicable. General-purpose spot-welding machines can be inexpensive.

Contributing to the attractiveness of resistance welding, especially for high production, is its simplicity and the high speed of the operation; speeds of press-type spot welders average from 12 to 180 spot welds per minute, the higher speeds being accomplished on lighter gauges. Output rates are even higher on special-purpose multiple spot welders, which may have as many as 50 pairs of electrodes operating simultaneously. Such equipment can be justified if production quantities are very high.

Seam welding is also a high-speed operation; linear welding speeds customarily range from 50 to 400 cm/min (20 to 160 in/min) and sometimes higher. Both general-purpose and special-purpose seam-welding machines are available. Tooling and setup are simple enough to make moderate-sized production runs practicable.

Projection welding is more applicable to higher production levels. It is considerably faster than spot welding because multiple welds can easily be made simultaneously. Tooling is required to provide projections, and equipment costs are higher than for spot welding because much heavier machines are required to provide sufficient pressure and current for multiple welds. These factors necessitate high quantities so that fixed costs can be amortized. Special-purpose equipment is employed if the weldment is unusual in shape or if quantities are large enough to make such equipment justifiable.

Flash and upset welding require moderate or higher production levels to amortize the cost of current-carrying clamping dies and the development of processing pressures, currents, and time cycles. Automatic power-operated machines are used for quantity production, although manual machines are available for smaller quantities of smaller-size weldments.

TABLE 7.3.1 Resistance Weldability Ratings of Various Metals

Metal or alloy			
Common designation	UNS designation	Rating*	Typical welding current, A
Low carbon steel SAE 1010	G10100	A	100
Stainless steel 302, 303, 304	S30200, S30300, S30400	B	75
Aluminum 1100	A91100	C	250
Aluminum 5052	A05052	C	300
Magnesium MIC	M15102	B	290
Nickel		B	130
Nickel alloys (Monel, etc.)	N044XX	B	65
Nickel silver (10% Ni)	C74500	B	100
Copper		F	350
Silver		F	350
Low brass (80% Cu)	C24000	E	150
Yellow brass (65% Cu)	C27000	C	150
Phosphor bronze (95% Cu)	C51000	C	120
Silicon bronze (up to 3% Si)	C65500 C65100 C65400	B	80
Aluminum bronze (8% Al)	C61000	C	80

*A = excellent; B = good; C = fair, D = poor; E = extremely poor; F = impractical.

Source: Adapted from Roger W. Bolz, *Production Processes: The Productivity Handbook,* 5th ed., Industrial Press, New York, 1981.

SUITABLE MATERIALS

Table 7.3.1 presents ratings on the weldability of various metals.

Carbon Steel

The most commonly resistance-welded material is low-carbon steel. It is the most weldable commercial material, and its weldability is the basis of comparison for other metals and alloys. Its favorable resistance-welding characteristics result from three factors: (1) It has a broad plastic temperature range during which fusion can take place, (2) it is free from embrittlement and other metallurgical problems in the weld area, and (3) it has sufficient electrical resistance so that adequate heating occurs.

Best results with carbon steel are achieved when carbon content is between 0.05 and 0.15 percent. As carbon content is increased above these levels, embrittlement of the weld gradually starts to occur. With carbon content up to about 0.60 percent, it is advisable to provide a postweld heat treatment, which is possible with most spot-, projection-, flash-, or butt-welding machines. This involves a lower-temperature reheating cycle shortly after welding, which reduces or eliminates the brittle condition.

Steel with higher carbon content than 0.60 percent requires annealing after resistance welding unless the function of the welded assembly permits hard and brittle welds.

Material to be welded should be free of scale, rust, paint, and other coatings. Otherwise, hot-rolled material is comparable with cold-finished material.

Wrought Iron

This material has a low yield point and is easy to resistance-weld.

Alloy Steel

Alloying materials decrease the breadth of the temperature range within which steel is plastic. They also increase the tendency of the weld areas to harden. However, if carbon content is low, spot and seam welding are only slightly more critical in technique than with plain low-carbon steel. If the finished assembly requires greater strength than that obtainable with low-carbon steel, the better approach, from a weldability standpoint, is to use a low-alloy steel rather than a medium-carbon steel of the same strength.

Low-alloy steels are also easily weldable by projection, flash, and upset methods. However, welding or upset pressures must be increased somewhat compared with those used for low-carbon steels.

Medium- and high-alloy steels, when resistance-welded according to normal methods, produce a very hard and brittle weld nugget. Postweld heat treatment is essential if such materials are used, and sometimes preweld heating is also used. Once treated, however, these steels provide stronger welded joints.

Alloy steels are more difficult to flash-weld in that the upset pressure applied is quite critical.

Stainless Steel

Most stainless steels, when heating to welding temperatures, have a tendency to precipitate carbon at grain boundaries. This leads to intergranular corrosion and reduced strength. Other than this problem, which is controllable by the addition of niobium or titanium, stainless steel is easily resistance-welded, although more exact control, higher pressures, and more rapid cooling are necessary. Stainless steels also can be upset-welded, but grades with over 0.08 percent carbon are not recommended for other resistance-welding methods.

Aluminum

Aluminum alloys are resistance-weldable but with greater difficulty than steel. Aluminum has a narrow plastic temperature range and high electrical conductivity. The first of these characteristics necessitates precise control of welding conditions; the second requires large weld currents, $2\frac{1}{2}$ to 3 times those of low-carbon steel. Another problem is the oxide coating that forms on aluminum. This film is electrically insulative and must be removed by chemical or abrasive means prior to welding.

Projection weldability of aluminum is quite limited. Small, simple joints are most easily weldable. Aluminum alloys 6061 (A96061), 6053 (A96053), 3003 (A93003), and 1100 (A91100) normally can be flash-welded if the stock thickness is greater than 1.3 mm (0.050 in). Alloy 2024 (A92024) is among the aluminum alloys that can be upset-welded.

Copper and Its Alloys

The high conductivity of copper and its alloys presents problems when these materials are spot-, seam-, projection-, or upset-welded. The most highly conductive alloys are the most difficult to weld. Nickel silver and silicon bronze, having lower conductivity, can be spot- or seam-welded. With other copper alloys, spot and seam welds are feasible if stock thickness is 1 mm (0.040 in) or less.

Copper and copper alloys are feasible for flash weldments provided flashing time is short. Exacting control of welding conditions and strongly designed projections are necessary if projection welding is used.

Spot welding of copper can be facilitated if the joint area is tin-, nickel-, or manganese-plated.

Magnesium Alloys

These alloys have a narrow plastic range, tend to adhere to the welding electrodes, and often are received with a chromium-nickel corrosion-resistant coating that must be removed prior to welding. All these factors hamper the resistance weldability of magnesium.

Magnesium alloys, however, can be spot-welded with high-current equipment. Seam welding requires very precise control of welding conditions. Projection and upset welding of these alloys is not recommended. Flash welding is possible if special methods are employed.

Nickel and Nickel Alloys

Nickel and nickel alloys can be resistance-welded according to processes applicable to their particular characteristics. Welding methods vary with different alloys. Monel is satisfactory for projection welding.

Coated Steel

Coatings of aluminum, cadmium, zinc, chromium, copper, nickel, terne, and tin on steel necessitate different resistance-welding techniques. However, the weldability of steel with such coatings is still generally satisfactory.

Dissimilar Materials

Dissimilar metals may be resistance-welded if they can combine to form an alloy and if their melting points are not too far apart. Extreme differences in electrical conductivity can present problems. Table 7.3.2 indicates the spot weldability of various metal combinations.

Upset welding can be used to weld dissimilar metals and those which are difficult to weld, including those commonly considered unweldable.

DESIGN RECOMMENDATIONS

Spot Weldments

Spot-welded assemblies should be designed so that simple, inexpensive electrodes can be used. Joint areas should be easy to reach. When special electrode shapes are required, every effort should be made to avoid extremes of shape and provide as accessible a joint as possible. Deep, narrow joint areas should be avoided. Figure 7.3.4 provides rules of thumb for accessibility of joints.

Other design recommendations primarily for spot weldments are as follows:

1. Multiple thicknesses of sheet up to about eight layers and 25-mm (1-in) stack height are possible for spot-welding steel in one operation. With aluminum, the maximum recommended stack height is 8 mm ($\frac{5}{16}$ in) and with magnesium 6.3 mm ($\frac{1}{4}$ in). With these materials, a maximum of three layers is recommended. Dissimilar sheet thicknesses usually reduce the number of steel sheets which are spot-weldable.
2. When multiple layers are employed, the use of a folded double sheet as shown in Fig. 7.3.5 provides a shunt for the welding current and can prevent good-quality welding.
3. Even when welds of multiple stock thicknesses are feasible, they are more troublesome than simple welds of two sheet thicknesses. Therefore, it is sometimes advisable to provide scalloped edges as shown in Fig. 7.3.6 to permit alternate welding of spots in two thicknesses only.
4. Resistance welding permits the easy addition of lightweight reinforcing members when desired for increased strength and rigidity. Reinforcements should be simple in shape for economy of fabrication and be welded to existing flanges or other flat areas for best results and ease of welding. (See Fig. 7.3.7.)
5. The economy of spot welding often makes it advantageous to fabricate special sections or members rather than using a rolled, drawn, forged, or machined component. Added stiffening of a resistance weldment also can be obtained by bending up the edge of a sheet before welding it to another sheet.
6. In dealing with curved parts, avoid putting the weld joint in the area where the parts are curved. Otherwise, special electrode ends are required. Also provide a means by which the parts can be self-locating if possible.
7. Resistance welds, like other welds, are subject to shrinkage on cooling. To avoid distortion, welds should be placed as much as possible so that shrinkage forces are balanced. Figure 7.3.8 gives one example.
8. There is a minimum practicable spacing for spot welds. If welds are attempted too close together, welding current will shunt through the nearest previous weld and fusion will not be adequate. Table 7.3.3 presents recommendations for minimum spacing for welds in low-carbon steel. For aluminum and magnesium, minimum spot spacing should be 8 times stock thickness, with 16 times being still better if the strength requirements of the joint permits.

TABLE 7.3.2 Spot-Welding Metal Combinations and Weldabilities*

Metals	Aluminum	Stainless steel	Brass	Copper	Galvanized iron	Steel	Lead	Monel	Nickel	Nichrome	Tinplate	Zinc	Phosphor bronze	Nickel silver	Terneplate
Aluminum	B	E	D	E	C	D	E	D	D	D	C	C	C	F	C
Stainless steel	F	A	E	E	B	A	F	C	C	C	B	F	D	D	B
Brass	D	E	C	D	D	D	F	C	C	C	D	E	C	C	D
Copper	E	E	D	F	E	E	E	D	D	D	E	E	C	C	E
Galvanized iron	C	B	D	E	B	B	D	C	C	C	B	C	D	E	B
Steel	D	A	D	E	B	A	E	C	C	C	B	F	C	D	A
Lead	E	F	F	E	D	E	C	E	E	E		C	E	E	D
Monel	D	C	C	D	C	C	E	A	B	B	C	F	C	B	C
Nickel	D	C	C	D	C	C	E	B	A	B	C	F	C	B	C
Nichrome	D	C	C	D	C	C	E	B	B	A	C	F	D	B	C
Tinplate	C	B	D	E	B	B		C	C	C	C	C	D	D	C
Zinc	C	E	E	E	C	F	C	F	F	F	C	C	D	F	C
Phosphor bronze	C	D	C	C	D	C	E	C	C	D	D	D	B	B	D
Nickel silver	F	D	C	C	E	D	E	B	B	B	D	F	B	A	D
Terneplate	C	B	D	E	B	A	D	C	C	C	C	C	D	D	B

*A = excellent; B = good; C = fair; D = poor; E-very poor; F-impractical.

Source: Reprinted with permission from Roger W. Bolz, *Production Processes: The Productivity Handbook,* 5th ed., © 1981, Industrial Press, New York.

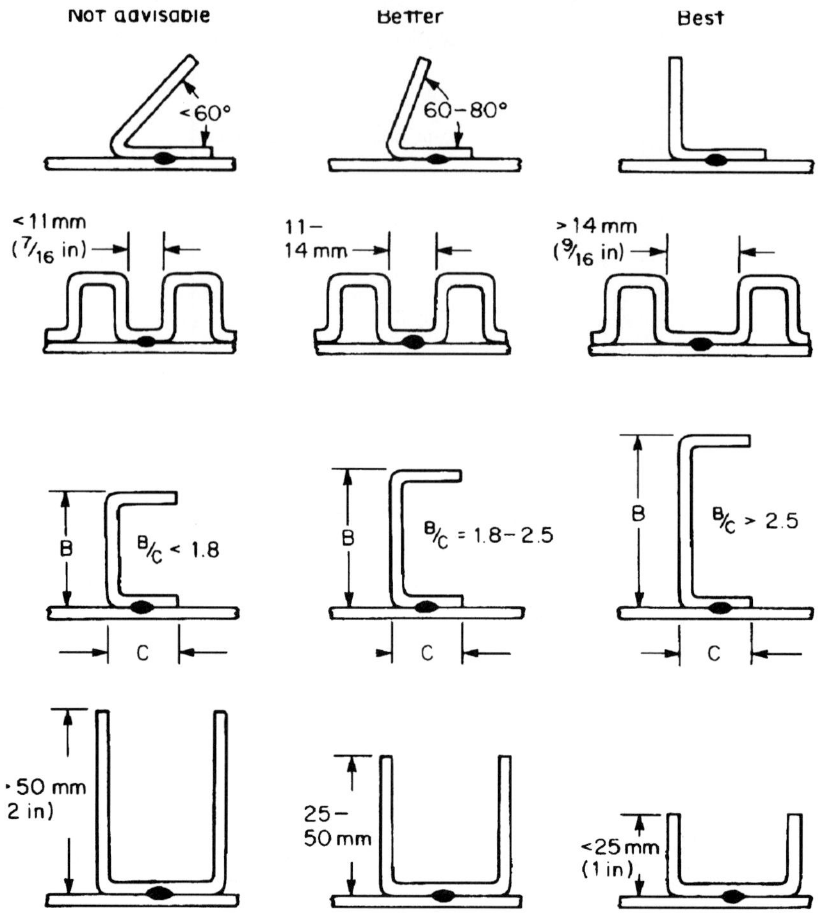

FIGURE 7.3.4 Rules of thumb for accessibility of joints.

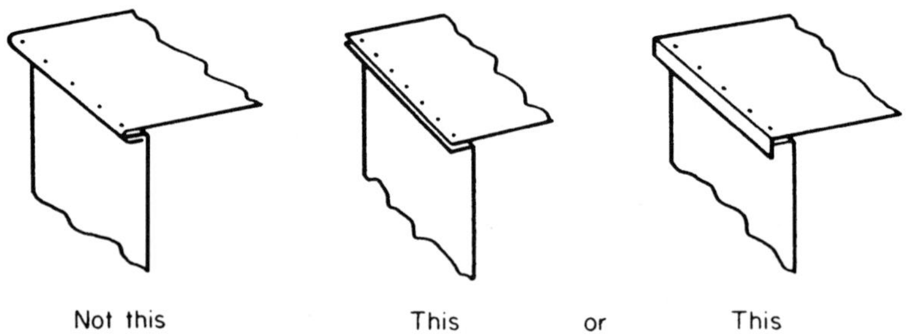

FIGURE 7.3.5 When a spot weld is used to join two sections of a folded-over sheet, some welding current passes through the folded sheet and bypasses the joint, resulting in inadequate fusion.

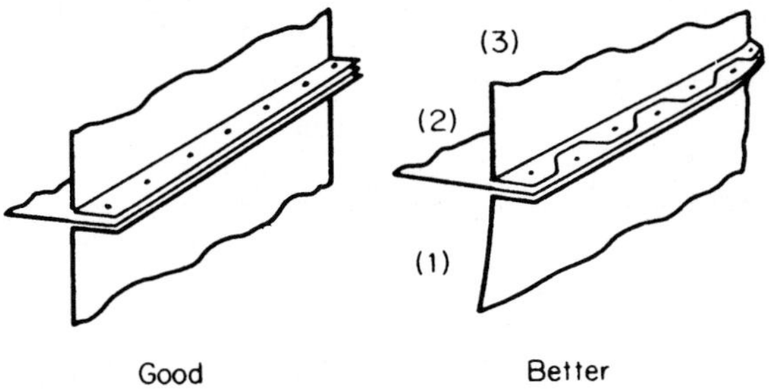

FIGURE 7.3.6 Scalloped edges permit alternate welding of spots in two thicknesses only.

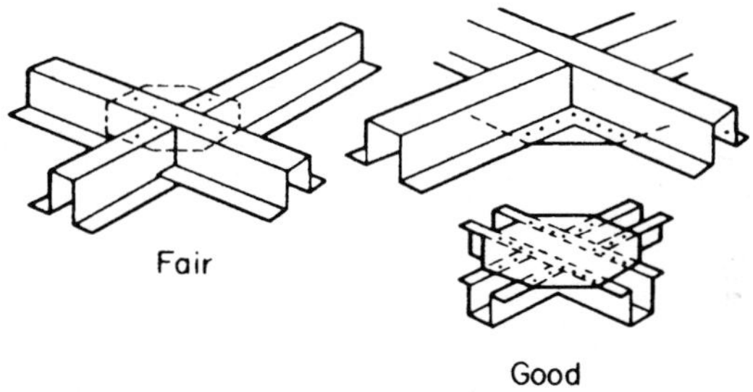

FIGURE 7.3.7 Spot-weld reinforcements to flat accessible areas.

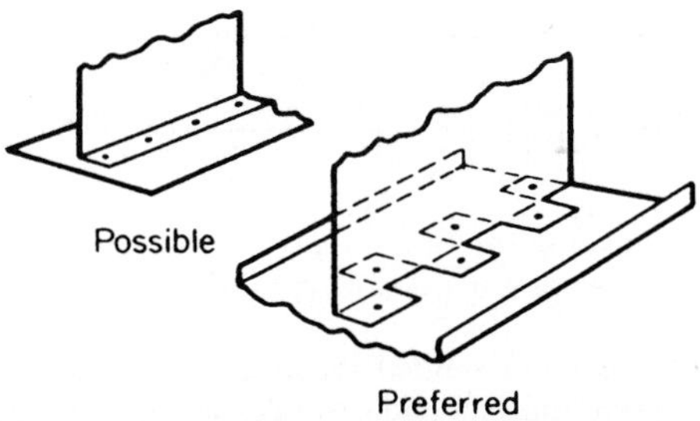

FIGURE 7.3.8 Place welds so that shrinkage forces are balanced and distortion is minimized.

9. Sufficient overlap of individual sheets being spot-welded is also necessary for adequate joint strength and for use of standard electrodes. Recommended minimum overlap and minimum spot-to-edge dimensions for low-carbon steel are shown in Table 7.3.3. For aluminum and magnesium, the minimum recommended distance from spot to the edge is 4 times stock thickness, with 6 times being still more preferable.

TABLE 7.3.3 Minimum Dimensions for Spot Welds (Single-Impulse Welds) in Low-Carbon Steel

Stock thickness, mm (in)	Minimum weld spacing, S, mm (in)	Minimum overlap of material, L, mm (in)
0.25 (0.010)	6.3 (0.250)	9.5 (0.375)
0.53 (0.021)	9.5 (0.375)	11 (0.433)
0.80 (0.031)	13 (0.500)	11 (0.433)
1.0 (0.040)	19 (0.750)	13 (0.500)
1.3 (0.050)	22 (0.875)	14 (0.551)
1.6 (0.063)	27 (1.062)	16 (0.625)
2.0 (0.078)	35 (1.375)	17 (0.669)
2.4 (0.094)	41 (1.625)	19 (0.750)
2.8 (0.109)	46 (1.813)	21 (0.827)
3.2 (0.125)	51 (2.000)	22 (0.875)

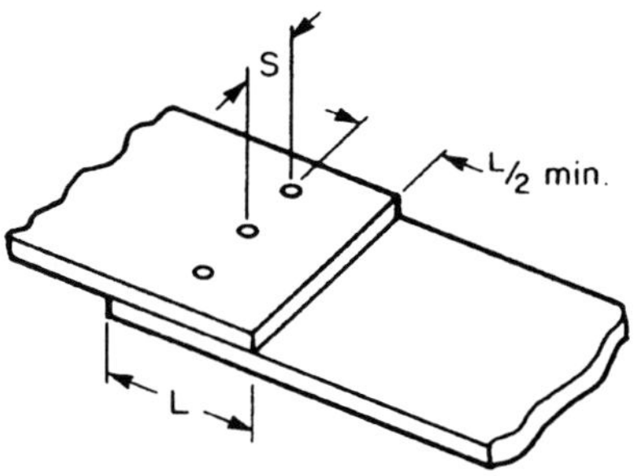

Seam Weldments

Table 7.3.4 provides data on the minimum advisable overlap of low-carbon sheets to be seam-welded. The following other design recommendations apply to seam welds:

1. When flanged parts are seam-welded together and the seam is not straight, there are minimum practicable internal and external radii of curvature of the weld seam. Minimum recommended values vary with material and stock thickness and are shown in Fig. 7.3.9.
2. Clearance must be provided for the electrode wheels, the minimum diameter for which is about 50 mm (2 in). If necessary, some seam welders permit canting electrode wheels by up to 10° to provide clearance of obstructing members.
3. Circumferential welds, in tubular parts, as illustrated in Fig. 7.3.10, require a minimum tubing inside diameter of 100 mm (4 in) to provide room for the electrode wheel. If the tubular part is not circular, the minimum internal radius at any point should be 50 mm (2 in) but preferably considerably larger. With circumferential seam welds it is also important that the two parts to be joined fit closely together

TABLE 7.3.4 Minimum Recommended Overlap of Low-Carbon-Steel Sheets Being Seam-Welded

Stock thickness, mm (in)	Normal minimum overlap, mm (in)	Extreme minimum overlap mm (in)
0.25 (0.010)	9.5 (0.375)	6.3 (0.250)
0.53 (0.021)	11 (0.438)	8.0 (0.315)
0.80 (0.031)	13 (0.500)	8.0 (0.315)
1.0 (0.040)	13 (0.500)	9.5 (0.375)
1.3 (0.050)	14 (0.563)	11 (0.433)
1.6 (0.063)	16 (0.625)	11 (0.433)
2.0 (0.078)	17 (0.688)	13 (0.500)
2.4 (0.094)	19 (0.750)	13 (0.500)
2.8 (0.109)	21 (0.813)	14 (0.551)
3.2 (0.125)	22 (0.875)	16 (0.625)

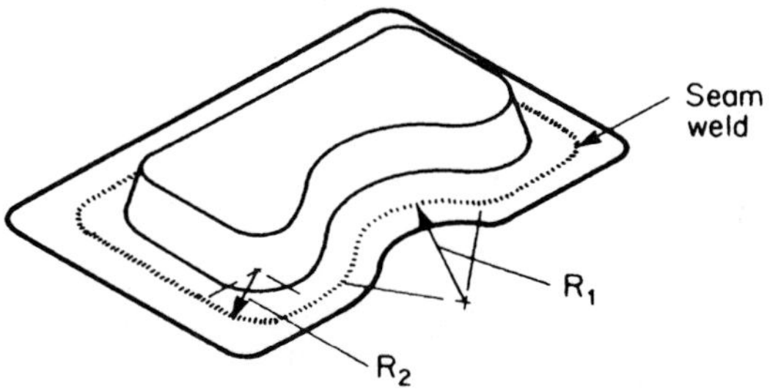

FIGURE 7.3.9 Minimum practicable radii of curvature of weld seams.

Minimum recommended radii R	
Inside radius R_1	
Aluminum	250 mm (10 in)
Steel	150 mm (6 in)
Outside radius, R_2	
Aluminum	75 mm (3 in)
Steel, 0.8–1.3 mm thick	50 mm (2 in)
Steel, to 1.6 mm thick	57 mm ($2\frac{1}{4}$ in)
Steel, to 2.0 mm thick	64 mm ($2\frac{1}{2}$ in)
Steel, to 3.0 mm thick	75 mm (3 in)

without the necessity for strong electrode-wheel force to bring them together. Therefore, such parts should be designed and fabricated to have a press or close-sliding fit.

4. Seam welds can be used to join metal sheets of unequal thickness, although it is preferable that the thickness be the same. With carbon or stainless steels, maintain a maximum ratio between the two gauges of 3:1. With aluminum, the ratio should not exceed 2:1.

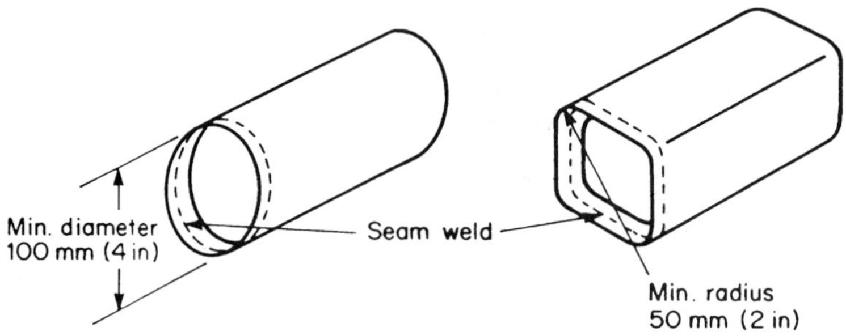

FIGURE 7.3.10 Rules for circumferential seam welds.

5. It is possible to produce a seam weldment when the seam ends at a shoulder, lip, or other protuberance. However, such designs should be avoided because they necessitate partial or notched electrode wheels that are more expensive and less usable. The wheels also must be geared together if the obstruction extends to both sides of the assembly.

Projection Weldments

Figures 7.3.11 and 7.3.12 illustrate recommended configurations of projections in sheet metal of varying thickness. The rounded shape shown in Fig. 7.3.12 allows initial point contact. This provides high initial resistance, which is desirable. The joint area grows to the full size of the projection as the projection softens and fusion takes place. Recommended overlap dimensions for projection-welded lap joints and minimum spacing between projections are shown in Table 7.3.5.

When the parts to be joined are curved or located at an angle to the stroke of the welding press, elongated and tapered projections, as shown in Fig. 7.3.13, are recommended because these permit the area of contact to grow progressively as the projection softens.

Projection welding works best when the parts are designed so that the proper fusing temperature is created at precisely the same instant in both elements being welded. This usually means that when parts of unequal thickness are being projection-welded together, it is better to make the projections in the heavier part.

Projections in heavier, nonsheet parts can be made economically by forging or cold heading. When projections are required on a screw-machine part, annular-ring types are recommended because they are relatively easy to machine. Figure 7.3.14 illustrates some common examples. Ring-type projections are also used when a pressure-tight joint is required.

Despite the advisability, from a welding standpoint, of having projections in the heavier part, it is often more economical, when a sheet-metal piece is projection-welded to a heavier piece, to put the projections in the sheet-metal piece. This is true because forming members for embossed projections in a sheet-metal part can be incorporated in blanking and forming dies. Forming of projections also can be performed very quickly as a separate press operation. In such cases, projections should be heavy or elongated. Ring-type projections as shown in Fig. 7.3.15 also may be used.

Projections must be strong enough so that they do not collapse from the welding-machine pressure. For this reason, the projection welding of sheets less than 0.4 mm

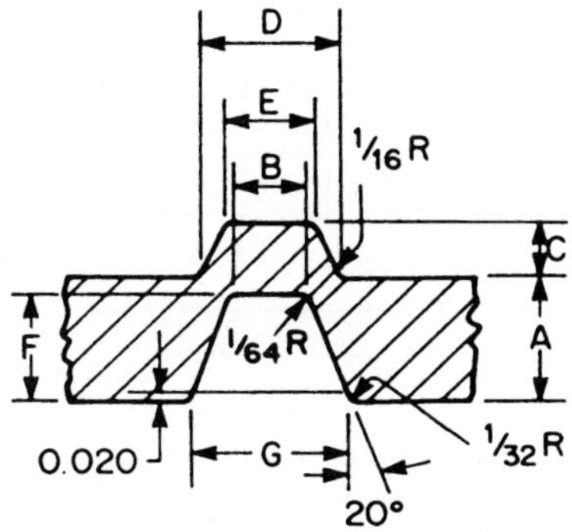

FIGURE 7.3.11 Recommended configurations of weld projections (inches) in 5- to 12-gauge sheet metal. (*Reprinted with permission from Roger W. Bolz,* Production Processes: The Productivity Handbook, *5th ed., © 1981, Industrial Press, New York.*)

Gauge	*A*	*B*	*C*	*D*	*E*	*F*	*G*
5	0.218	0.080	0.075	0.210	0.100	0.200	0.220
6	0.2031	0.080	0.070	0.203	0.094	0.182	0.206
7	0.1875	0.080	0.070	0.203	0.094	0.166	0.185
8	0.1718	0.080	0.060	0.190	0.080	0.138	0.166
9	0.1562	0.080	0.060	0.172	0.080	0.122	0.155
10	0.1406	0.080	0.060	0.172	0.080	0.110	0.145
11	0.125	0.080	0.055	0.172	0.080	0.100	0.138
12	0.1093	0.080	0.055	0.172	0.080	0.090	0.131

(0.016 in) thick is not generally recommended, nor is the use of semipiercings as weld projections. (See Fig. 7.3.16.)

Sometimes the intersection of a sharp corner with a beveled surface can be used to function in the same way as a weld projection. Figure 7.3.17 illustrates two examples.

Sawtooth edges can furnish an economical method for providing weld projections in heavy stampings. Weld strength is high, but pushed-out weld metal will be visible at the edges of the joint. (See Fig. 7.3.18.)

Lips, flanges, or other elements often can aid in the correct positioning of parts that are to be welded together. In many cases, such elements can be produced in the same punch-press operation that produces the weld projections on a part. When the concentricity, alignment, or position of the mating parts is critical, designers should consider such an approach. Figure 7.3.19 shows some examples.

Standard fastening devices made with projections for ease of welding to flat surfaces are available in a wide variety for the designer and are economically priced because they are produced in large quantities. Some examples are shown in Fig. 7.3.20.

Flash and Upset Weldments

Parts to be flash- or upset-welded together should have mating joint surfaces as nearly identical as possible. The shape and cross-sectional areas at the joint should be the same. Figure 7.3.21 illustrates examples of both good and poor designs.

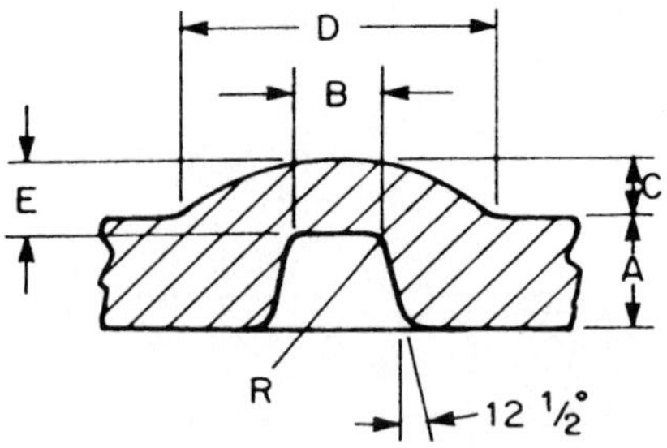

FIGURE 7.3.12 Recommended configurations of weld projections (inches) in 13- to 24-gauge sheet stock. (*Reprinted with permission from Roger W. Bolz,* Production Processes: The Productivity Handbook, *5th ed., © 1981, Industrial Press, New York.*)

Gauge	*A*	*B*	*C*	*D*	*E*	*R*
13	0.0937	0.075	0.050	0.180	0.065	1/64
14	0.0781	0.075	0.050	0.180	0.065	1/64
15	0.0703	0.075	0.045	0.172	0.055	1/64
16	0.0625	0.060	0.045	0.172	0.050	1/64
17	0.0562	0.055	0.040	0.156	0.045	1/64
18	0.050	0.050	0.040	0.156	0.040	1/64
19	0.0437	0.050	0.040	0.125	0.035	1/64
20	0.0375	0.050	0.035	0.125	0.035	1/64
21	0.0344	0.050	0.030	0.125	0.030	1/64
22	0.0312	0.050	0.030	0.125	0.030	1/64
23	0.0281	0.050	0.025	0.109	0.025	1/64
24	0.025	0.050	0.025	0.109	0.025	1/64

Gripping pressures of electrodes on parts in upset and flash welding are very high. It is important that the design for each part allow both gripping space and internal strength to resist the gripping forces. [Clamping pressures can range from 80 MPa (12,000 lbf/in^2) to 550 MPa (80,000 lbf/in^2).]

The shape of the parts should be such that the butting force can be squarely applied. (See Fig. 7.3.22.) Rings or cylinders are made circular after the ends or edges have been welded together.

It is necessary in designing upset- and flash-welded assemblies to allow for the reduction in length of the weldment compared with the sum of the length of the individual parts. This reduction is due to displacement of material at the joint. Figure 7.3.23 shows recommended allowances for the amount of length lost in steel flash weldments.

Parts to be upset- or flash-welded together must be designed so that they can be held in accurate alignment during welding.

Ends to be flash-welded, unless the diameter or wall thickness is less than 6 mm ($^1/_4$ in), should be beveled as shown in Fig. 7.3.24. This helps start flashing and ensures that any slag that is formed during welding will be expelled outward and not get trapped in the joint, causing it to be weak. The same approach is used in tubing with a wall thickness greater than about 5 mm ($^3/_{16}$ in).

The wall thickness of parts to be flash-welded should be sufficient to withstand the considerable forces involved. Tubing with a large ratio of circumference to wall thick-

TABLE 7.3.5 Recommended Projection-Weld Spacing and Overlap

Stock thickness		Minimum overlap, in (mm)	Minimum spacing, in (mm)
USS gauge	in (approx. mm)		
31	0.011 (0.28)	0.25 (6)	0.30 (8)
26	0.019 (0.48)	0.25 (6)	0.30 (8)
25	0.022 (0.56)	0.25 (6)	0.38 (10)
23	0.028 (0.71)	0.25 (6)	0.38 (10)
21	0.034 (0.86)	0.38 (10)	0.50 (13)
19	0.043 (1.1)	0.38 (10)	0.50 (13)
18	0.049 (1.2)	0.38 (10)	0.75 (19)
16	0.061 (1.6)	0.50 (13)	0.75 (19)
14	0.077 (2.0)	0.50 (13)	0.85 (22)
13	0.092 (2.3)	0.62 (16)	1.1 (28)
12	0.107 (2.7)	0.75 (19)	1.3 (33)
11	0.123 (3.1)	0.75 (19)	1.5 (38)
10	0.135 (3.4)	0.85 (22)	1.6 (41)
9	0.153 (3.9)	0.90 (23)	1.7 (43)
8	0.164 (4.2)	0.95 (24)	1.8 (46)
7	0.179 (4.6)	1.00 (25)	1.9 (48)
6	0.195 (5.0)	1.05 (27)	2.0 (51)
5	0.210 (5.3)	1.15 (29)	2.1 (53)
4	0.225 (5.7)	1.20 (30)	2.3 (58)
3	0.245 (6.2)	1.30 (33)	2.5 (64)

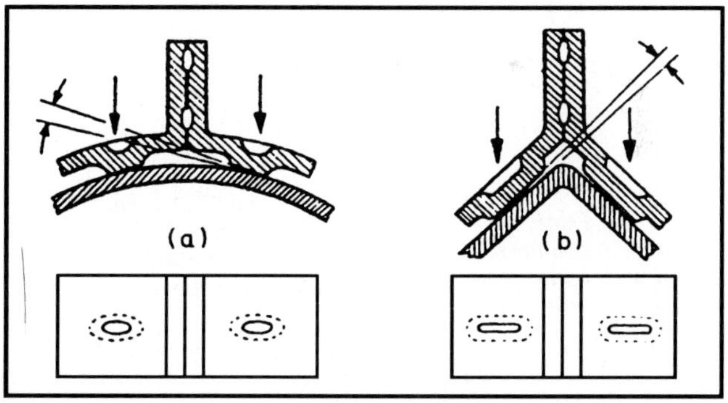

FIGURE 7.3.13 Use elongated and tapered projections for welds on angular or curved surfaces.

ness should be avoided, as should long joints in flat plate. Tables 7.3.6 and 7.3.7 chart the recommended limits.

Angle welds should be avoided if possible because the upsetting force is not applied so squarely to the joint. For tubing and other sections, the joint angle should

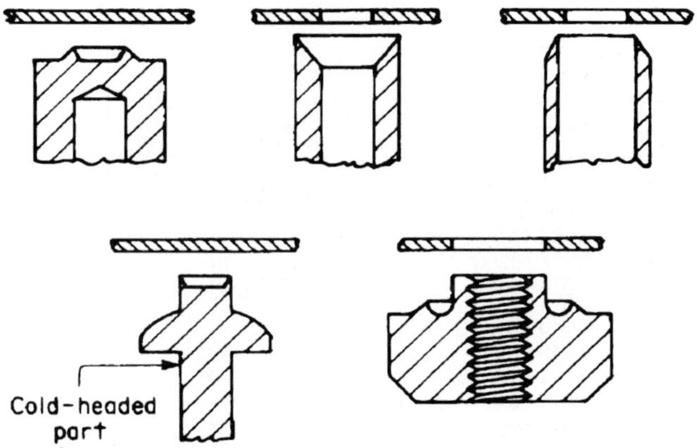

FIGURE 7.3.14 Commonly used weld projections on screw-machine and cold-headed parts.

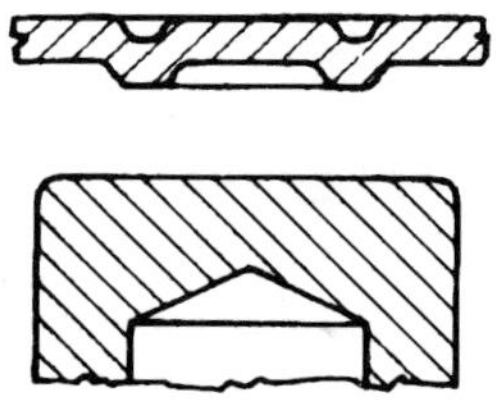

FIGURE 7.3.15 Annular-ring-weld projection on sheet-metal parts.

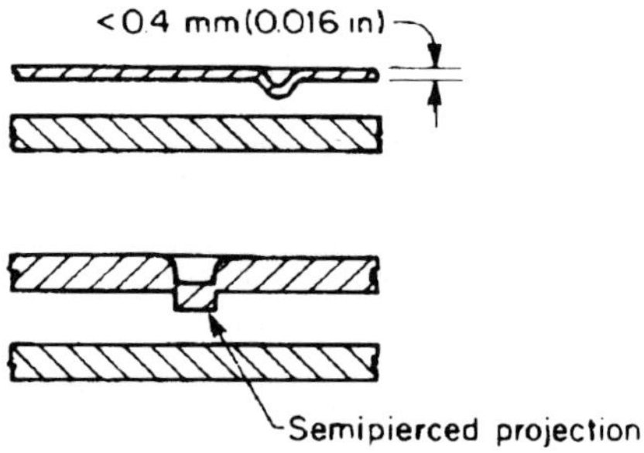

FIGURE 7.3.16 These designs are not recommended because the projections are too weak.

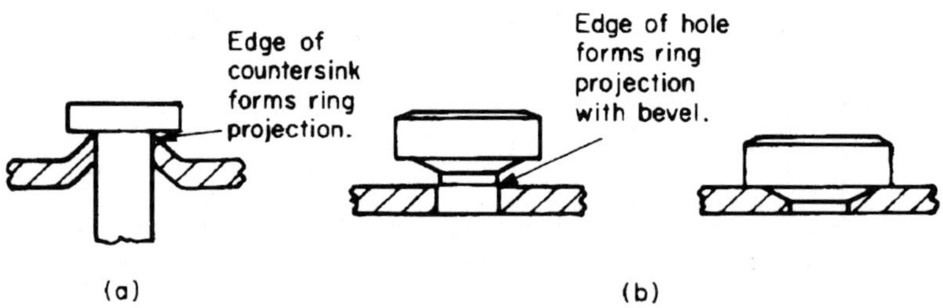

FIGURE 7.3.17 Two examples of sharp-edged intersections which provide the same function as weld projections. (*From V. H. Laughner and A. D. Hargan,* Handbook of Fastening and Joining of Metal Parts, *McGraw-Hill, New York, 1956.*)

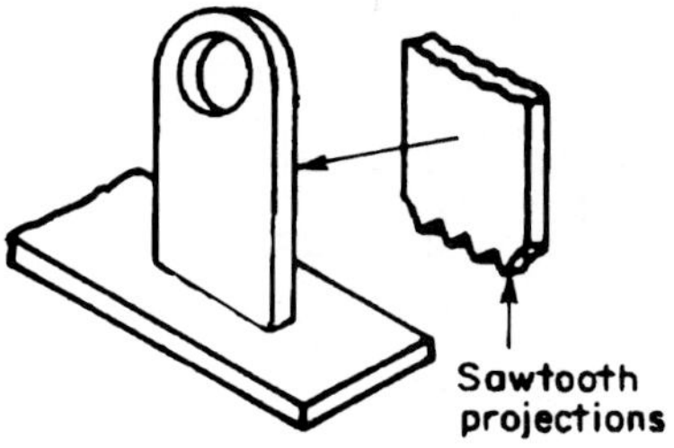

FIGURE 7.3.18 Sawtooth edges can serve as weld projections. (*From V. H. Laughner and A. D. Hargan,* Handbook of Fastening and Joining of Metal Parts, *McGraw-Hill, New York, 1956.*)

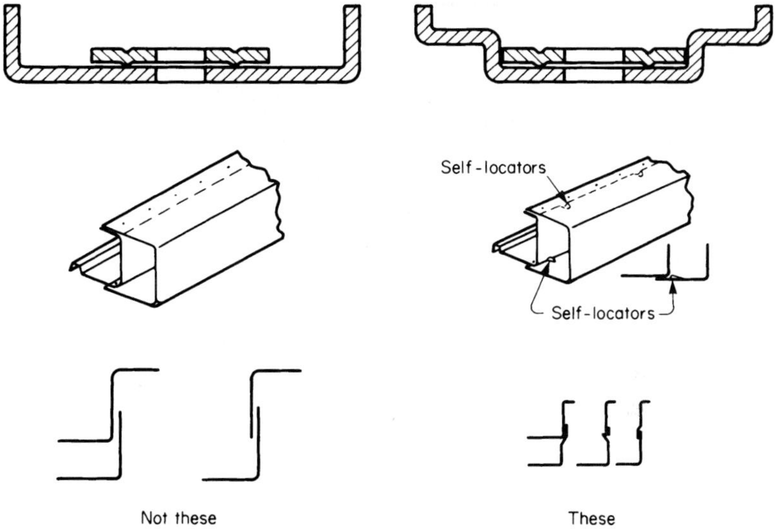

FIGURE 7.3.19 Self-locating elements often can be incorporated easily in stamped parts which are designed for spot, seam, or projection welding.

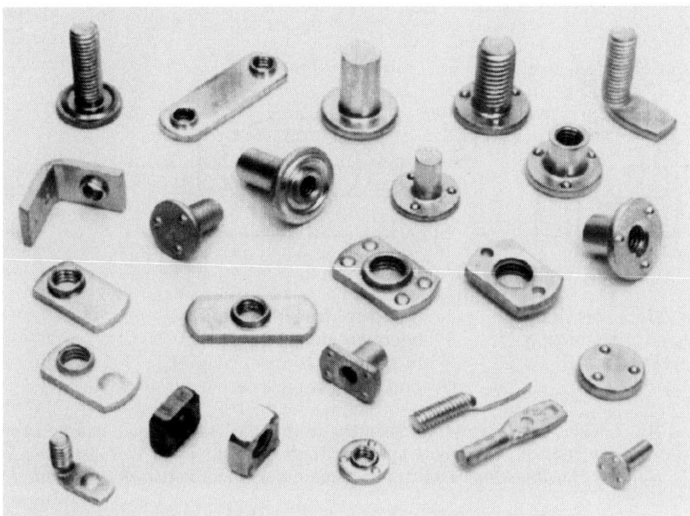

FIGURE 7.3.20 A variety of nuts and bolts for spot and projection welding. (*Courtesy Ohio Nut & Bolt Co.*)

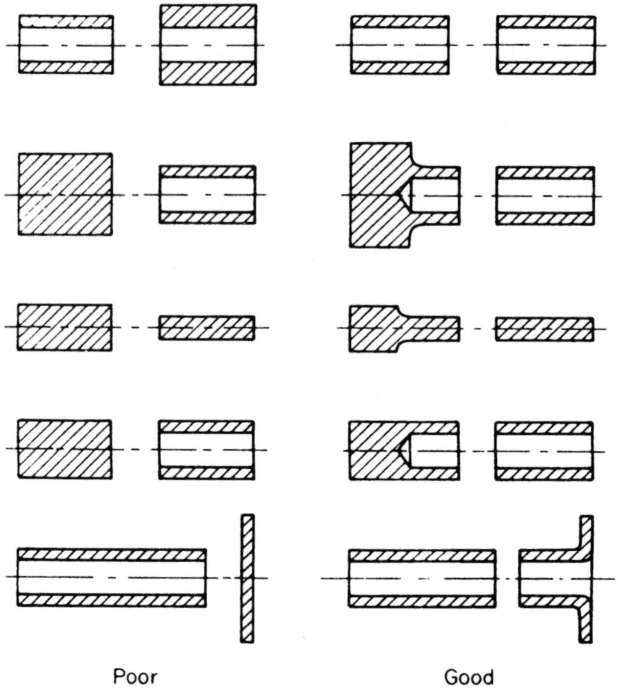

FIGURE 7.3.21 Good and bad design practice for mating surfaces of upset- and flash-welded assemblies. (*Reprinted with permission from Roger W. Bolz,* Production Processes: The Productivity Handbook, *5th ed., © 1981, Industrial Press, New York.*)

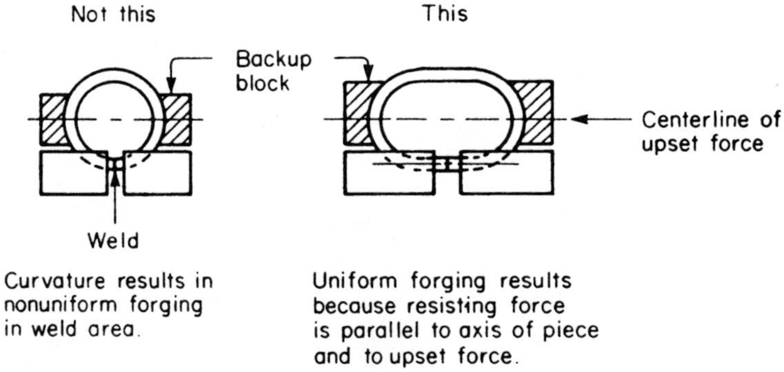

FIGURE 7.3.22 Parts should be designed so that the butting force can be applied squarely to the ends or edges being upset- or flash-welded. (*From V. H. Laughner and A. D. Hargan,* Handbook of Fastening and Joining of Metal Parts, *McGraw-Hill, New York, 1956.*)

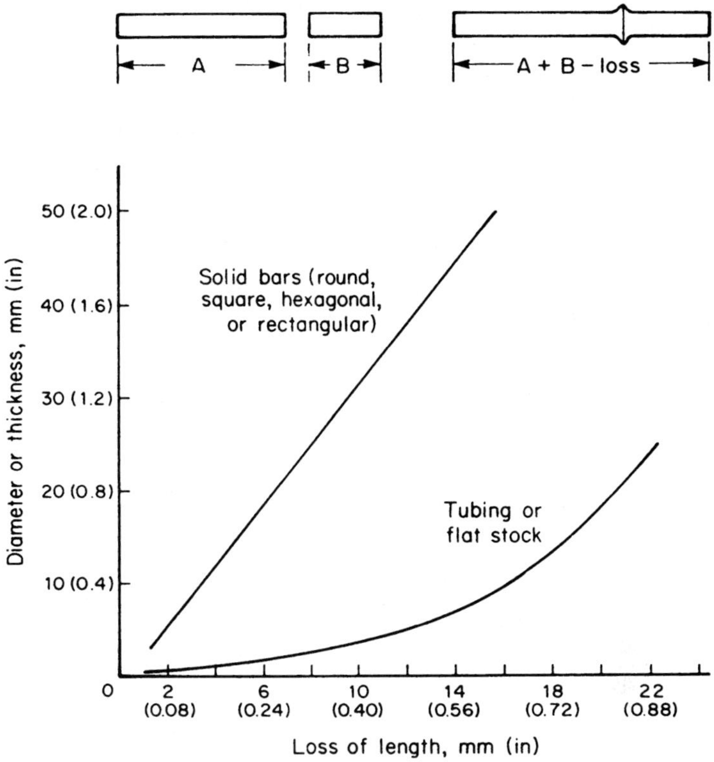

FIGURE 7.3.23 Loss of length of flash-welded assemblies due to displacement of metal at the joint (steel materials).

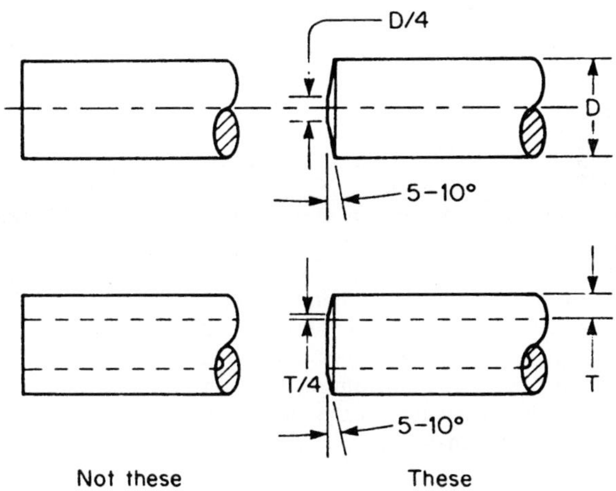

FIGURE 7.3.24 Bars over 6 mm in diameter and tubing with a wall thickness over 5 mm should be beveled as shown before flash welding.

TABLE 7.3.6 Recommended Maximum-Diameter Tubing to Be Upset- or Flash-Welded

Wall thickness, mm (in)	Maximum diameter, mm (in)
0.8 (0.031)	19 (0.75)
1.1 (0.043)	25 (1.0)
1.5 (0.059)	38 (1.5)
2.0 (0.079)	58 (2.3)
3.0 (0.120)	100 (3.9)
4.0 (0.157)	135 (5.3)
6.3 (0.250)	235 (9.3)
9.0 (0.354)	335 (13.2)
12.7 (0.500)	500 (19.7)

TABLE 7.3.7 Recommended Maximum Length of Flat Plate Upset- or Flash-Welded Joints

Sheet thickness, mm (in)	Maximum length of welded joints, mm (in)
0.5 (0.02)	125 (5)
1 (0.04)	375 (15)
1.5 (0.06)	625 (25)
2 (0.08)	875 (35)
3 (0.12)	1375 (55)
4 (0.16)	1875 (75)

not normally be less than 150°. For plate, it can be as little as 90° with satisfactory results. Figure 7.3.25 illustrates these limits.

DIMENSIONAL FACTORS

Resistance welds are subject to the same shrinkage effects as arc welds. (See Chap. 7.2.) Other dimensional factors are also the same. However, the effect with spot, projection, and seam welds is apt to be less than with arc welds because of the smaller size of the resistance-weld nugget compared with the arc-weld bead and its slightly lower temperature at the time of fusion. However, designers of resistance weldments must be aware of these shrinkage forces and the distortions which they can cause. Dimensional accuracy and freedom from residual stresses of components before being resistance-welded are prime factors in the closeness with which the welded assembly can be held to specified dimensions.

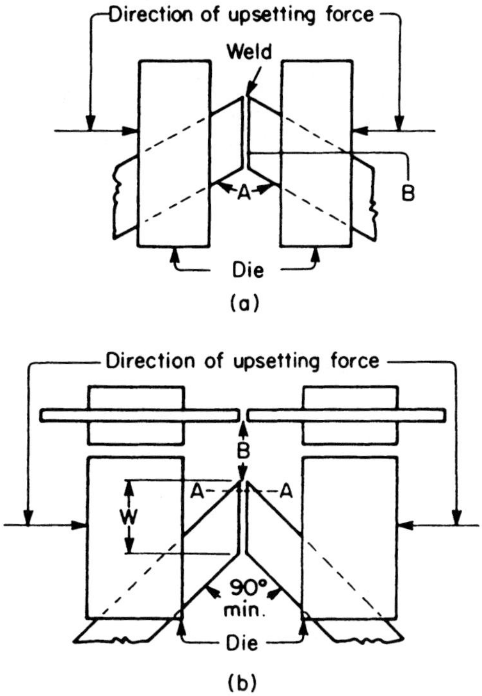

FIGURE 7.3.25 Limits for angular flash welds. For solid compact sections (*a*), the angle at *A* must be greater than 150° for good joint design. Poor fusion may occur at *B*. For thin rectangular sections (*b*), *W* should be at least 20 times wall thickness. Metal should be removed at end *B* to line *A-A* to prevent weld defects. (*From V. H. Laughner and A. D. Hargan,* Handbook of Fastening and Joining of Metal Parts, *McGraw-Hill, New York.*)

CHAPTER 7.4
SOLDERED AND BRAZED ASSEMBLIES*

John Casteras
Alpha Metals, Inc.
Jersey City, New Jersey

THE PROCESS

Soldering and brazing are closely related processes in which metal components are joined together by means of a filler metal. The filler metal, which has a melting point lower than that of the base metal, is introduced to the heated joint. It melts, wets the surface to be joined, and is distributed in the joint by capillary attraction.

In the case of soldering, the filler metal, called *solder,* has a melting point (liquidus) below 425°C (800°F). Common solders consist chiefly of tin and lead. In the case of brazing, the filler metal has a melting point (liquidus) above 425°C (800°F). The most common brazing metals are alloys of either silver or copper.

Heat can be applied to the joint in a number of ways. The more significant heat sources in use in industry are (1) heated iron (solder only), (2) gas torch, (3) furnace, (4) induction coils, (5) dipping in (a) molten filler metal or (b) a molten salt bath, (6) electrical resistance (of the workpiece itself), and (7) infrared lamps.

Wetting action is crucial to successful soldering and brazing. It requires that the joint surfaces be chemically clean. A first step in these processes is invariably a cleaning operation. Cleanliness is maintained during heating by means of flux or a protective atmosphere.

TYPICAL CHARACTERISTICS

Brazed and soldered assemblies represent configurations that are impractical or uneconomical to make from a single piece. This may occur when:

*This chapter deals primarily with mechanical applications. Electronics assemblies are covered in Chap. 9.4.

1. Dissimilar metals are involved, e.g., a carbide tool bit and a steel-alloy shank.
2. Light weight is important, but the shape is intricate, e.g., for an assembly of bent tubing and fittings.
3. The part is too intricate to machine from one piece, especially because of thin sections, and when high strength and accuracy are important.
4. Hollow shapes such as tanks, floats, and evaporators are involved. (Leaktight joints often dictate the use of soldering or brazing.)

Brazed and soldered assemblies most often are composed of one or more sheet-metal components. Additional components may be machined parts, forgings, and in some cases, castings.

Most brazed assemblies are not large because heating, cleaning, and handling costs are compounded with larger size. Resistance-heated joints are normally less than 1 in^2 in area. Furnace-brazed assemblies are seldom larger than 2.5 kg (5.5 lb). Figures 7.4.1 and 7.4.2 show typical production-brazed assemblies.

Soldered assemblies are used when strength requirements are not as great, when a capability of later disassembly is important, or when electrical conductivity or fluid sealing is the prime requirement. Brazing is employed rather than soldering when strength requirements are greater. Strength capabilities of brazed joints are significant. Tensile strengths of 830 MPa (120,000 lbf/in^2) are feasible with silver brazing alloys,

FIGURE 7.4.1 Typical brazed assemblies. (*Courtesy Handy & Harmon.*)

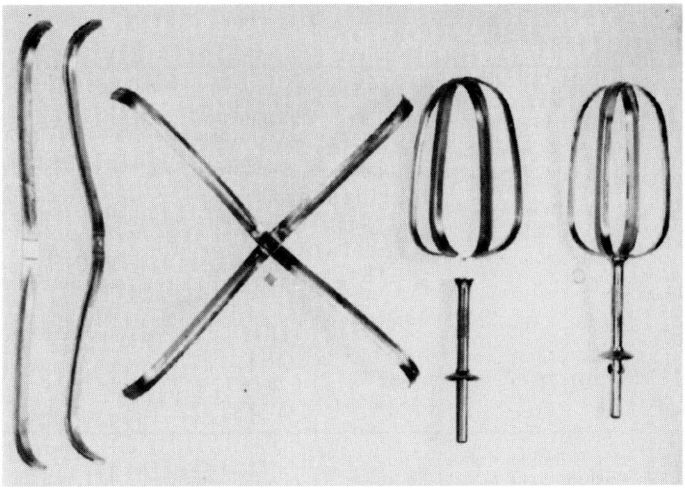

FIGURE 7.4.2 This light and strong brazed beater assembly would be difficult to fasten with other methods. (*Courtesy Handy & Harmon.*)

for example. Brazing may be used when postjoining heat treating is required. Brazed assemblies also can offer good corrosion resistance and a neat appearance.

ECONOMIC PRODUCTION QUANTITIES

Brazing and soldering are suitable for a broad range of production quantities. Hand-torch brazing requires a certain amount of operator skill but very little investment for equipment or tooling. It is therefore a suitable technique for the maintenance shop and few-of-a-kind production.

Other heating methods require some investment for equipment and tooling and therefore require some production quantity for amortization. In general, however, these investments are moderate, and brazing and soldering have proved to be good processes at medium as well as higher production levels, with a variety of heat-application methods. The exception is conveyorized-furnace brazing, which requires substantial production volume to be justifiable because of the cost of furnaces, atmosphere generators, and sometimes fixtures.

Table 7.4.1 summarizes, in general terms, the investment cost and production-level applicability of various heating methods.

SUITABLE MATERIALS

Soldering

Solders are available in bar, wire (solid or flux-cored), and paste forms. The bar forms may be used in tinning pots or wave soldering machines for electronic circuit boards.

TABLE 7.4.1 Economic Production Quantities and Relative Investment Costs for Various Soldering and Brazing Methods*

Heating method	Relative investment required: Equipment	Tooling	Economic production quantities	Remarks
Hand torch	VL	L†	VL–L	
Automatic torch	M–H	M–H	M–H	Fixtures required
Induction	M	M	M	Special coil and fixture required for each assembly
Metal dipping	L–M	M	L–M	Fixture required
Resistance	M–H	M	M–H	
Furnace	H	H‡	M–VH	
Infrared	M	M	M–H	

*VL = very low; L = low; M = medium; H = high; VH = very high.
†If a fixture is required; otherwise, very low.

Wire forms are used for hand or automatic soldering for assembly and repair. Pastes are comprised of solder particles blended in a rosin-based or organic acid flux vehicle.

Recent concerns about leaching of the lead component of solders in copper water supply systems and scrapped electronic and other products have fostered legislative efforts to encourage the use of lead-free solders. Several commercially available lead-free solders are listed in Table 7.4.2, which lists commercially available solders, their compositions, melting temperatures, and common applications.

Base metals vary in their ability to accept solder. The following metals are solderable and are listed in order of decreasing solderability:

1. Tin
2. Cadmium
3. Silver
4. Copper
5. Brass
6. Bronze
7. Lead
8. Nickel
9. Monel
10. Zinc
11. Steel
12. Inconel
13. Stainless steel
14. Chromium
15. Nichrome
16. Silicon bronze
17. Alnico
18. Aluminum

Metals numbered from 1 to 10 can be soldered with noncorrosive flux. Difficult-to-solder base metals can be electroplated or dip-coated with tin, copper, silver, lead-tin, nickel, zinc, or cadmium to provide a base for solder.

Brazing

The brazing process is applicable to a wide variety of base metals. Many common materials used in production—carbon and alloy steel, copper, brass, aluminum, cast

TABLE 7.4.2 Common Commercial Solder Alloys

ASTM classification	Major composition, %	Melting range, °F	Tensile strength, lbf/in^2	Typical applications
Sb	5 Sb–94 Pb (min)	450–464		For coating and joining metals subject to high operating temperatures
Sn 5A	5 Sn–95 Pb	586–594	4,190	General purpose (higher temperature)
Sn 20A	20 Sn–80 Pb	361–535	5,140	Automobile body repair
Sn 20B	20 Sn–1 Sb–79 Pb	363–517	5,300	Auto body repair, stronger, but avoid galvanized
Sn 30A	30 Sn–70 Pb	361–491	6,140	Auto body repair, dip solder wire stripping
Sn 30B	30 Sn–1.6 Sb–68.4 Pb	364–482	6,460	Auto body repair
Sn 40A	40 Sn–60 Pb	361–455	6,320	Cable sheaths, dip soldering
Sn 40B	40 Sn–2 Sb–58 Pb	365–448	7,100	General purpose, not for galvanized
Sn 50	50 Sn–50 Pb	361–421	6,450	General purpose, plumbing, noncritical electrical
Sn 60	60 Sn–40 Pb	361–374	6,400	Electrical and electronic
Sn 62	62 Sn–36 Pb–2 Ag	354–372		Electronics, for silver-coated surfaces
Sn 63	63 Sn–37 Pb	361–361		Electronics, printed circuit boards
Ag 5.5	94.5 Pb–1.5 Ag	579–689	4,000	For copper, brass at higher temperature (to 350°F)
Ag 1.5	97.5 Pb–1.5 Ag–1 Sn	588–588	3,600	
Lead-free alloys:				
Sn 95	95 Sn–5 Ag	430–473	10,100	Medical, potable water, jewelry
	95 Sn–5 Sb	455–464	5,900	Plumbing joints
Sn 96	96.5 Sn–3.5 Ag	430–430	14,000	Food, nontoxic applications, higher joint strength
	80 Sn–20 Zn	390–518		For soldering aluminum
	40 Cd–60 Zn	509–635		For soldering aluminum
	50 Sn–50 In	243–260	1,200	For soldering glass to metal or itself

iron, nickel, and nickel alloys—all can be brazed successfully. Some base metals require special processing steps. The following is a discussion of the brazability of common base metals.

Low-Carbon Steel. This metal is very suitable for brazing. When quantities are sufficient to justify furnace brazing, copper brazing alloy is the recommended filler metal because of its low cost.

High-Carbon and Alloy Steels. These steels are fully brazable. Special steps or added operations may be necessary, however, to preserve the metallurgical condition of the steel. Grain coarsening can occur as a result of copper furnace brazing of such steels. Cooling after brazing can cause hardening, and subsequent annealing may be required. Fortunately, copper-brazed assemblies can be subjected to the temperatures required to anneal the base metal without difficulty. Other heat-treating operations also can be performed on copper-brazed assemblies.

Stainless Steel. This metal is frequently brazed. Generally, the weldable grades of stainless steel are also the most brazable. Stainless steel tends to be more difficult to clean fully of surface oxides before brazing and therefore may produce incompletely filled joints. It is more difficult to braze than low-alloy or carbon steels.

Cast Iron. Cast iron is difficult to braze because of its graphite content, which necessitates special cleaning procedures. The brazing heat also can adversely affect the strength and hardness of the iron. Malleable iron is the most brazable cast iron, ductile iron is next, and conventional gray iron is least brazable, especially the high-silicon grades.

Copper and Copper Alloys. These metals, including brass, are relatively easily brazed. Possible exceptions are leaded brasses and silicon bronze.

Aluminum. Aluminum is difficult to braze because of the closeness of the melting temperature of the base and filler metals and because of the rapidity of formation and adherence of aluminum oxide. The following classes of alloys are more easily brazed: 1XXX, 3XXX, and 6XXX. The 2XXX and 7XXX grades are more difficult.

Titanium. Titanium requires special brazing alloys and processing steps.

Table 7.4.3 lists typical brazing alloys, their melting temperatures, and their recommended applications.

DETAILED DESIGN RECOMMENDATIONS

Joint Design

The design of a joint, of course, depends on a number of factors such as strength and service requirements, whether pressure-tightness and electrical conductivity are involved, composition of the metals to be joined, and the brazing alloy. There are some constant principles that can be applied, however. For example, the prime requirement is to provide opportunity for filler metal to flow into the joint by capillary attraction. This requires a close gap between surfaces of the joint. The strength of the

TABLE 7.4.3 Common Commercial Brazing Alloys

Classification	Composition, %	UNS designation	Melting range, °F	Optimal gap, in	Typical applications
B Ag-1 (AWS)	45 Ag–15 Cu–15.5 Zn–24 Cd	P07450	1125–1145	0.0015–0.002	General purpose; OK for cast iron
B Ag-1a (AWS)	50 Ag–15.5 Cu–16.5 Zn–18 Cd	P07500	1160–1175	0.0015–0.002	General purpose; higher strength
B Ag-2 (AWS)	35 Ag–26 Cu–21 Zn–18 Cd	P07350	1125–1295	0.003–0.006	For building up, bridging, and filleting
B Ag-3 (AWS)	50 Ag–15 Cu–15.5 Zn–16 Cd–3 Ni	P07501	1170–1270	0.003–0.006	For building up, bridging, and filleting; for carbide tool bits
B Ag-4 (AWS)	40 Ag–30 Cu–28 Zn–2 Ni	P07400	1240–1435	0.003–0.006	For tungsten, stellite, refractory alloys
B Cu-1 (AWS)	99 + Cu	C14180	1980–1980	0.000	For furnace-brazing ferrous metals
B Cu Zn-1 (AWS)	60 Cu–40 Zn	C28000	1650–1660	0.002–0.006	General-purpose torch and induction
B Cu P-2 (AWS)	93 Cu–7 P	C55181	1305–1485	0.001–0.003	For copper, brass, and bronze; not steels
B Cu P-5 (AWS)	15 Ag–85 Cu–5 P	C55284	1185–1500	0.003–0.005	
B Al Si-3 (AWS)	86 Al–10 Si–4 Cu	A94145	970–1085	0.006–0.025	General purpose for dip and furnace brazing of aluminum
B Al Si-4 (AWS)	88 Al–12 Si	A94147	1070–1080	0.002–0.010	
B Mg-1 (AWS)	89 Mg–9 AI–2 Zn	M19001	770–1110	0.004–0.010	General purpose for magnesium
B Ni-1 (AWS)	Ni–14 Cr–4 Si–3.5 B–4.5 Fe–0.75 C		1790–1900	0.001–0.003	For stainless and heat-resistant alloys
B Ni-2 (AWS)	Ni–7 Cr–4.5 Si–3.1 B–3.0 Fe		1780–1830	0.001–0.003	For stainless and heat-resistant alloys
B Ni-3 (AWS)	Ni–4.5 Si–3.1 B–0.5 Fe		1800–1900	0.001–0.003	For stainless and heat-resistant alloys

joint lies more in the film of filler metal between the parts than it does in fillets at the edges of the joint.

Lap joints should be used whenever possible because they provide an easy means for controlling the joint area and gap and usually do not present assembly or fixturing problems. The joint area (overlap) should be sufficiently large so that the joint is as strong as the weaker member of the assembly. Allowance must be made for the usual lower unit strength of the filler metal compared with the base metal and the possibility of voids within the joint. One rule of thumb is to allow an overlap of at least three times the thickness of the thinnest member. (See Fig. 7.4.3.)

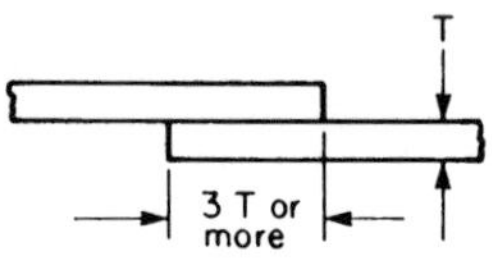

FIGURE 7.4.3 Lap joint (recommended).

FIGURE 7.4.4 Butt joint (not recommended).

The clearance between joint surfaces can be critical. Its amount depends on the fluidity of the filler metal. Normally from 0.025 to 0.20 mm (0.001 to 0.008 in) is specified, with an average value of perhaps 0.10 mm (0.004 in).

Butt joints (see Fig. 7.4.4) are not recommended unless strength requirements are very low and there is no need for a pressure seal at the joint. The joint area in a butt joint is small, and there is a greater possibility of variations from squareness and flatness within the joint. These variations interfere with the filling action of the solder or braze material. Butt joints may be satisfactory for electrical applications, but they do add to the resistance of the circuit.

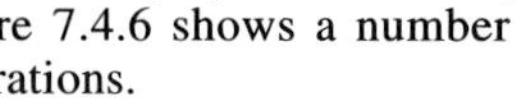

FIGURE 7.4.5 Scarf joint (not recommended).

Scarf joints (Fig. 7.4.5) are occasionally used in place of butt joints because they can be produced with a greater joint area. If the joint area is three or more times the stock thickness, they are satisfactory from a strength standpoint. From a manufacturability standpoint, however, they are not recommended because of the cost of the machining necessary for the diagonal surfaces and the difficulty in holding the parts in the proper assembled position during the brazing or soldering operation.

Figure 7.4.6 shows a number of both recommended and not-recommended joint configurations.

Assembly

Parts that comprise brazed assemblies should be designed so that they are easily assembled to one another. Chapter 7.1 covers this subject and gives design suggestions to facilitate assembly.

The most economical soldered and brazed assemblies are those which are self-jigging, i.e., those in which the assembled parts and the filler metal hold together during the heating cycle without any external fixture. This principle applies regardless of the heating method for the following reasons:

1. Self-jigging avoids the cost of building and maintaining fixtures.

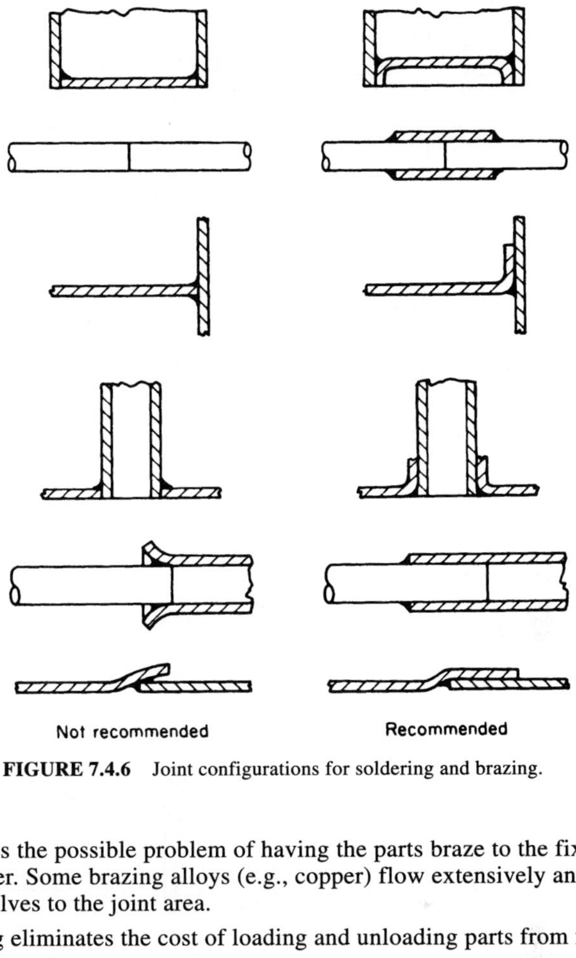

FIGURE 7.4.6 Joint configurations for soldering and brazing.

2. It minimizes the possible problem of having the parts braze to the fixture as well as to each other. Some brazing alloys (e.g., copper) flow extensively and may not confine themselves to the joint area.
3. Self-jigging eliminates the cost of loading and unloading parts from fixtures.
4. Self-jigging reduces the heating load and therefore the heating cost.
5. It gives better dimensional control than is usually obtained with fixtures, which must allow for both part-to-part dimensional variations and thermal expansion. It is often possible, when self-jigging is used, to accommodate such variations with no loss of accuracy.

There are a number of self-jigging methods. They involve the following approaches for holding parts together: gravity, spot and tack welding, friction and press fits, staking and peening, swaging, crimping and forming, threading, and riveting. See Fig. 7.4.7 for illustrations of these methods.

Placement of Braze Metal

Solder and braze metal can be in a number of forms that permit preplacement before heating. These are wire (commonly formed into rings for circular joints), slugs, shims,

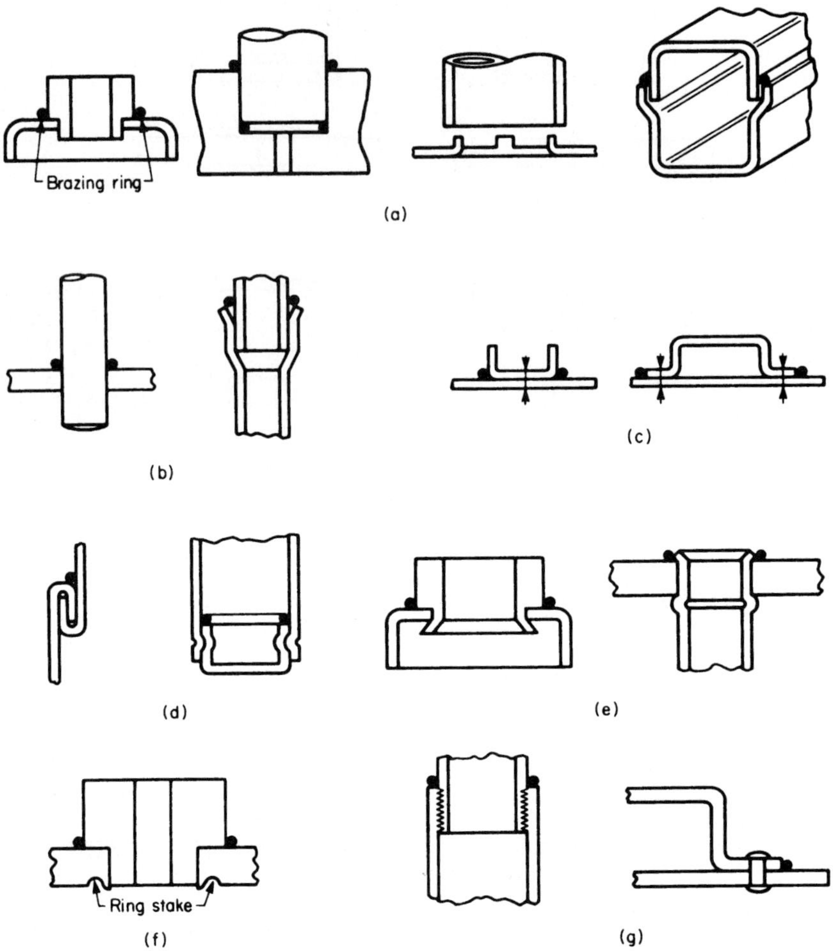

FIGURE 7.4.7 Various self-jigging methods for brazed assemblies. (*a*) Gravity is the most convenient force to use for self-jigging parts to be brazed. (*b*) Press-fit joint is used when the parts are not self-locating in all planes. (*c*) Spot and tack welding may be used when part configurations preclude the use of press fits. (*d*) Crimping or forming. (*e*) Expanding and swaging. (*f*) Staking and peening. (*g*) Threading and riveting form excellent self-locking and self-jigging joints for brazing and soldering operations. (*Illustrations reprinted from* Assembly Engineering, *by permission of the publisher, © 1972, Hitchcock Publishing Company. All rights reserved.*)

paste, and sprayed or plated coatings. The last named are especially applicable to copper brazing.

Because of copper's strong capillary action, it can be placed some distance from the joint and still produce satisfactory results. Nevertheless, for copper and especially for more viscous filler metals, the joint should be designed so that the filler metal stays put during the operation.

When filler metal is used in shim form, the assembly should be such that the parts

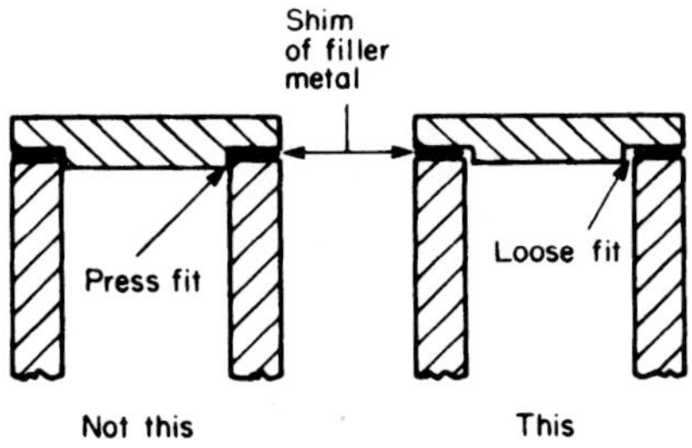

FIGURE 7.4.8 When filler metal in shim form is used, allow looseness between parts so that they can settle together when the filler metal flows, providing a stronger joint.

are free to move when the filler metal melts. This allows a stronger, narrow-gap joint. (See Fig. 7.4.8.)

Joints should be designed to avoid gas entrapment. The best way to ensure this is to provide a vent hole in the joint.

Other Considerations

The high temperature of brazing can cause distortion of the parts. Large, unsupported flat areas should be replaced by curved areas if possible, because the latter are more self-supporting. (See Fig. 7.4.9.)

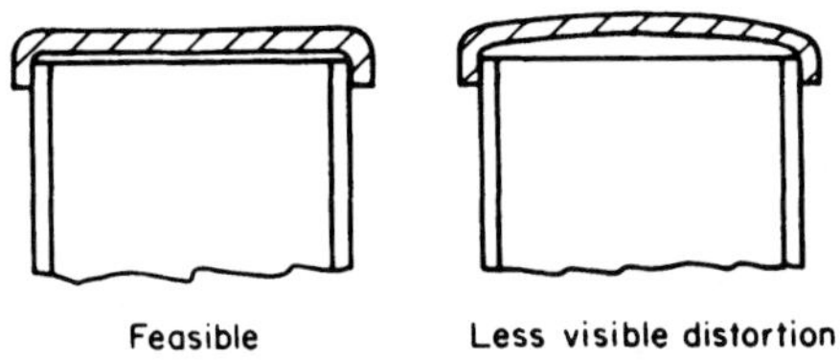

FIGURE 7.4.9 Use curved areas if possible.

When brazing tanks or other enclosed assemblies, a vent must be provided to allow the escape of gas generated when brazing flux is heated. It also must be kept in mind that the heat of brazing causes the parts involved to expand, and if different metals are involved, different expansion rates may cause movement or slippage of the parts. When this is apt to occur, provide steps, stakes, or some holding means to keep the parts in the correct position. (See Fig. 7.4.7.)

If induction heating is used, the joint must be designed to allow space for proper location of the induction coil. (See Fig. 7.4.10.) Dip-brazed assemblies must be designed so that flux from the bath is not trapped in the joint. (See Fig. 7.4.11.)

Clearances for brazed metal or solder will cause a variation in the location of the parts to be joined. If critical, this variation can be controlled by knurling the parts,

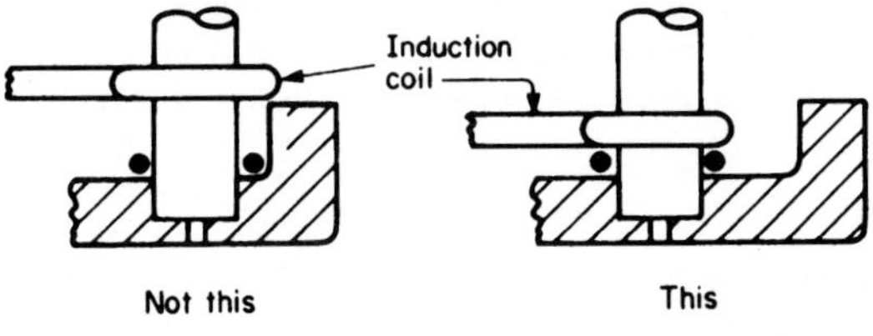

FIGURE 7.4.10 If induction heating is to be used, make sure that the design allows room for the coil.

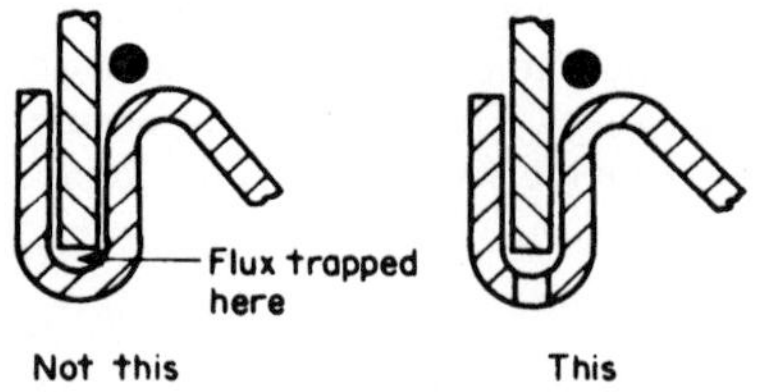

FIGURE 7.4.11 If dip brazing is used, avoid designs which entrap flux.

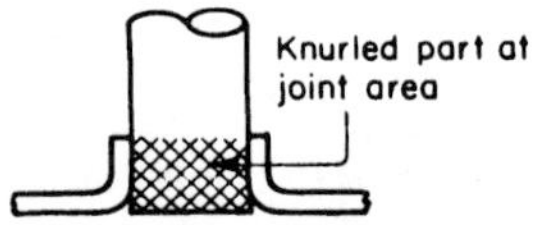

FIGURE 7.4.12 The tolerance to be applied to a dimension across a brazed joint depends on the tolerances required by the parts to be joined as well as on the variations in thickness of the film of filler metal. Knurling permits concentricity of the assembly to be maintained while still allowing room for filler metal to flow by capillary action.

providing other means of minimizing the clearance before assembly, or swaging or crimping after assembly but before brazing. (See Fig. 7.4.12.)

DIMENSIONAL FACTORS

Tolerances of brazed and soldered assemblies are very largely determined by the tolerances of the component parts of the assemblies. However, it is often possible to perform finish-machining operations after brazing if some dimension must be held to closer limits than are possible in the component parts or the joined assembly. Thermal effects are paramount in causing the dimensional variations that do result from brazing (or soldering) operations. There is, of course, some across-the-joint dimensional variation that shows up as variations in filler-metal thickness in the assembly, but this is usually a result of part variation or fixturing factors rather than variations inherent in the solder or braze metal itself. Across-the-joint dimensions sometimes can be controlled by fixturing.

RECOMMENDED TOLERANCES

Across-the-Joint Dimensions

If no heat distortion is involved (with solder and low-temperature brazing alloys), the assembly tolerance should approximately equal the stack-up of tolerances of the individual parts plus an allowance for gap variations. A good average figure for this is +0.05 mm (0.002 in). If heat distortion is involved [likely if temperatures exceed 700°C (1300°F)], add the expected distortion variation to this total tolerance.

Concentricity across the Joint

Allow the variation permitted by the tolerances of the individual parts except when, with preapplied *paste* filler metal, it is possible to rotate one of the parts to be joined and thereby to distribute the paste uniformly and reduce eccentricity. In this case, apply one-half the tolerance that would result from the parts' spacing. The same reduction in tolerance can be made if torch brazing is used and one part can be rotated manually while the filler metal is molten.

CHAPTER 7.5
ADHESIVELY BONDED ASSEMBLIES

Originally prepared by
Dr. Gerald L. Schneberger

Training Resources, Inc.
Grand Blanc, Michigan

Updated by
John Cocco, P.E.

Loctite Corporation
Rocky Hill, Connecticut

Adhesives are compounds capable of holding objects together in a useful fashion by surface attraction. Adhesive joints are often less costly, more easily produced, or better able to resist fatigue and corrosion than mechanical fasteners or welds. In some cases, adhesives are the only practical means of assembly.

THE PROCESS

Bonds are made by positioning a film of liquid or semiliquid adhesive between the parts and immobilizing the assembly until the adhesive solidifies. The adhesive may be applied in the solid or molten state or as a pure liquid or solution, as summarized in Table 7.5.1. Conversion to the final solid film (curing) may involve heating, cooling, evaporation, or a combination of these. The cure may require seconds, hours, or days and is usually accelerated by increasing temperature. Ease of assembly is closely related to the technique of adhesive application and cure, as shown in Table 7.5.2.

TYPICAL CHARACTERISTICS

Adhesive bonding is apt to be employed in preference to other joining methods when one or more of the following characteristics are important: (1) when there are limitations to the weight of the finished assembly, (2) when porous, fragile, or heat-sensitive materials must be joined, (3) when the appearance of other fastening methods would

TABLE 7.5.1 Typical Application Methods for Common Adhesives

Application method	Solvent cements	Epoxy	Phenolic	Silicone	Polyesters	Urethanes	Vinyls	Anaerobics	Polyamides	Cyano-acrylate	Acrylics	Ultraviolet or radiation curing
Brush or spray	x	x	x		x	x	x	x			x	x
Dry film		x	x						x			
Hot melt									x			
Pressure-sensitive				x		x						
Roller	x						x				x	x
Hand pump		x		x	x	x	x	x		x	x	x
Powered or pressurized dispensing*		x		x	x	x		x		x	x	x

*Positive-displacement pump or pressure-time dispensing valve.

TABLE 7.5.2 Manufacturing Aspects of Various Adhesive-Application Methods*

	Hot melt	Pressure-sensitive	Brush	Spray	Roller	Dry film	Hand pump	Powered or pressurized dispensing†
Skill required	N	N	N	Y	N	N	N	N
Speed of assembly	F	F	S	S	M	F	M	F
Messy operation	Y	N	Y	Y	M	N	M	N
Clamping required	N	N	V	V	V	Y	Y	Y
Heating after assembly	N	N	V	V	V	Y	V	V
Cure time	F	F	V	V	V	S	S	V

*N = no; Y = yes; F = fast; S = slow; M = medium; V = variable (depends on the adhesive).

†Positive-displacement pump or pressure-time dispensing valve.

not be satisfactory, (4) when it is necessary to provide sound deadening or vibration attenuation in the finished assembly, (5) when the parts to be joined must be electrically insulated from each other to avoid galvanic corrosion or for other reasons, and (6) when materials of dissimilar composition, thickness, or modulus must be joined together.

Although adhesively bonded joints can be engineered for high strength, adhesive bonding may not be suitable if strength requirements or temperature variations are extreme. Other fastening methods also may be indicated if provision must be made for disassembly and reassembly of the component.

Adhesives are used routinely to bond parts of extremely diverse size, shape, and composition to one another. Parts generally must tolerate a surface-cleaning operation prior to bonding, and they must be rather stable chemically in the intended service environment. The majority of adhesively bonded parts are not load-bearing because other assembly techniques historically have been more economical for these applications. This situation is changing, however, and there is now increasing use of load-bearing adhesives in the automotive, aircraft, and construction industries. Adhesive use is unrelated to the size of the assembled objects. Adhesives perform well in applications ranging from minute electronic assemblies to large building panels. Part geometry is usually not a problem because special applicators are available for hard-to-reach surfaces.

Typical adhesively bonded assemblies include brake-band and brake-disk assemblies, helicopter blades, plywood and wood furniture, aircraft honeycomb structures, paper bags and other paper products, and pulley-shaft and gear-shaft assemblies. Adhesives are used to fasten bushings, nameplates, decorative appliance panels, insulating and sound-deadening parts, floor tiles, and automobile rear-view mirrors to glass windshields and to lock threaded fasteners permanently. Figure 7.5.1 illustrates typical adhesively bonded assemblies.

ECONOMIC PRODUCTION QUANTITIES

Adhesive assembly is practical at virtually any production level. However, the number of joints to be made usually determines the type of application and curing equipment used. Thus prototype quantities or very short production runs may involve hand brushing or troweling, whereas high-volume production of the same items might involve automatic spray or roller coating equipment to apply the same adhesive.

Since adhesives are used as thin films, the cost of the adhesive material itself is usually minimal. It is common to pay more for part handling, surface preparation, fixturing, and curing than for the adhesive.

SUITABLE MATERIALS

Nearly any solid material can be bonded to any other. Teflon (registered trade name for Du Pont's polytetrafluoroethylene), polyethylene, and polypropylene are inherently difficult to bond because they have little tendency to stick to anything, including adhesives. Such surfaces usually must be modified by chemical treatment prior to bonding. If the bonded materials have greatly different coefficients of thermal expansion, an adhesive primer or a double-face adhesive-tape system may be advisable.

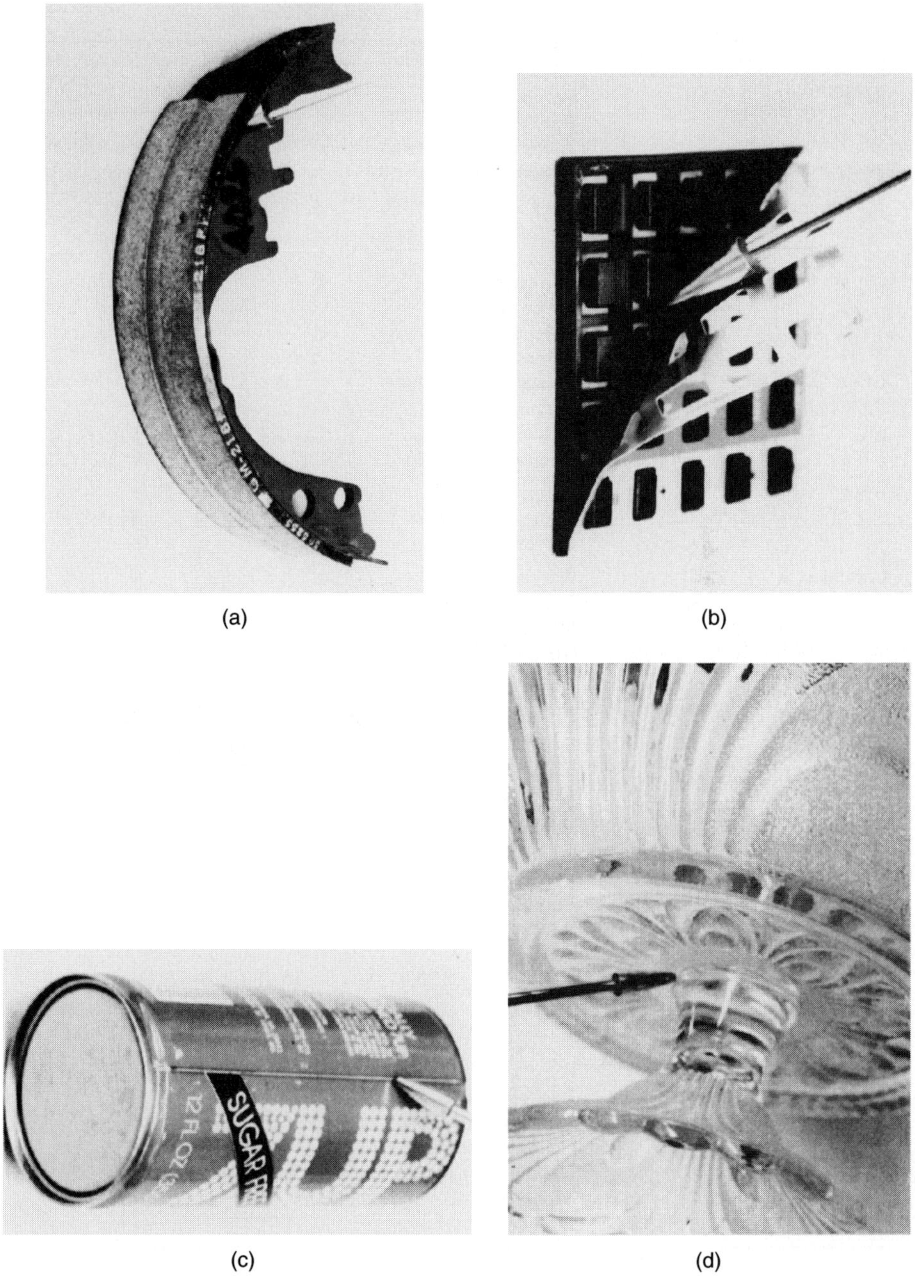

FIGURE 7.5.1 Some typical adhesively bonded assemblies. (*a*) Automotive brake lining. (*b*) Pocket calculator faceplate. (*c*) Beverage can with bonded side seam. (*d*) Glass fruit bowl with adhesively bonded stem. (*Courtesy GMI Technical Institute.*)

TABLE 7.5.3 Adhesives Commonly Used for Joining Various Materials

Materials to be joined	Adhesive	Table 7.5.4 reference
ABS	Polyesters	*a*
	Epoxy	*e*
	Alpha-cyanoacrylate	*c*
	Nitrile phenolic	*b*
Aluminum and its alloys	Acrylics	*w*
	UV/radiation curing adhesives	*x*
	Epoxy	*e*
	Epoxy-phenolic	*d*
	Nylon-epoxies	*f*
	Polyurethane rubber	*g*
	Polyesters	*a*
	Alpha-cyanoacrylate	*c*
	Polyamides	*h*
	Polyvinyl-phenolic	*i*
	Neoprene-phenolic	*b*
Brick	Epoxy	*e*
	Epoxy-phenolic	*d*
	Polyesters	*a*
Ceramics	Acrylics	*w*
	Epoxy	*e*
	Cellulose esters	*j*
	Vinyl chloride–vinyl acetate	*k*
	Polyvinyl butyral	*l*
Chromium	Epoxy	*e*
Concrete	Polyesters	*a*
	Epoxy	*e*
Copper and its alloys	Polyesters	*a*
	Epoxy	*e*
	Alpha-cyanoacrylate	*c*
	Polyamide	*h*
	Polyvinyl-phenolic	*i*
	Polyhydroxyether	*m*
Fluorocarbons	Epoxy	*e*
	Nitrile-phenolic	*b*
	Silicone	*t*
Glass	Epoxy	*e*
	Acrylics	*w*
	Epoxy-phenolic	*d*
	Alpha-cyanoacrylate	*c*
	Cellulose esters	*j*
	UV/radiation curing adhesives	*x*
	Vinyl chloride–vinyl acetate	*k*
	Polyvinyl butyral	*l*

Table 7.5.3 lists a number of materials that are commonly bonded with adhesives and indicates the adhesives commonly used. Table 7.5.4 describes the characteristics of common adhesives.

The bond strength of adhesives that use a catalyst or curing agent is often inversely

TABLE 7.5.3 Adhesives Commonly Used for Joining Various Materials (*Continued*)

Materials to be joined	Adhesive	Table 7.5.4 reference
Lead	Epoxy	*a*
	Vinyl chloride–vinyl acetate	*k*
	Polyesters	*a*
Leather	Vinyl chloride–vinyl acetate	*k*
	Polyvinyl butyral	*l*
	Polyhydroxyether	*m*
	Polyvinyl acetate	*n*
	Flexible adhesives	*g*
	Alpha cyanoacrylate	*c*
Magnesium	Polyesters	*a*
	Epoxy	*e*
	Polyamide	*h*
	Polyvinyl-phenolic	*i*
	Neoprene-phenolic	*h*
	Nylon-epoxy	*f*
Nickel	Epoxy	*e*
	Neoprene	*g*
	Polyhydroxyether	*m*
Paper	Animal glue	*o*
	Starch glue	*p*
	Urea, melamine, resorcinol, and phenol formaldehyde	*q*
	Epoxy	*e*
	Polyesters	*a*
	Cellulose esters	*j*
	Vinyl chloride–vinyl acetate	*k*
	Polyvinyl butyral	*l*
	Polyvinyl acetate	*n*
	Polyamide	*h*
	Flexible adhesives	*g*
Phenolic and melamine	Epoxy	*e*
	Alpha-cyanoacrylate	*c*
	Flexible adhesives	*g*
Polyamide	Epoxy	*e*
	Flexible adhesives	*g*
	Phenol and resorcinol formaldehyde	*q*
	Polyesters	*a*
Polycarbonate	Polyesters	*a*
	Acrylics	*w*
	UV/radiation curing adhesive	*x*
	Epoxy	*e*
	Alpha-cyanoacrylate	*c*
	Polyurethane rubber	*g*

related to the cure time. Fast-curing adhesives (including hot melts) may cause production problems by setting up too quickly. Cleanup also may be a problem with fast-curing adhesives.

TABLE 7.5.3 Adhesives Commonly Used for Joining Various Materials (*Continued*)

Materials to be joined	Adhesive	Table 7.5.4 reference
Polyester, glass-reinforced	Polyesters	*a*
	Epoxy	*e*
	Acrylics	*w*
	Polyacrylates	*t*
	Nitrile-phenolic	*b*
Polyethylene	Polyester, isocyanate modified	*a*
	UV/radiation curing adhesives	*x*
	Butadiene-acrylonitrile	*g*
	Nitrile-phenolic	*b*
	Alpha-cyanoacrylate	*c*
Polyformaldehyde	Polyester, isocyanate modified	*a*
	Butadiene-acrylonitrile	*g*
	Nitrile-phenolic	*b*
Polymethyl methacrylate	Epoxy	*e*
	Alpha-cyanoacrylate	*c*
	UV/radiation curing adhesive	*x*
	Polyester	*a*
	Nitrile-phenolic	*b*
Polypropylene	Polyester, isocyanate modified	*a*
	Nitrile-phenolic	*b*
	Butadiene acrylonitrile	*g*
	Alpha-cyanoacrylate	*c*
Polystyrene	Vinyl chloride–vinyl acetate	*k*
	UV/radiation curing adhesives	*x*
	Polyesters	*a*
	Alpha-cyanoacrylate	*c*
Polyvinyl chloride, rigid	UV/radiation curing adhesives	*x*
	Polyesters	*a*
	Epoxy	*e*
	Polyurethane	*g*
	Acrylics	*w*
	Alpha-cyanoacrylates	*c*
Rubber, butadiene styrene	Butadiene-acrylonitrile	*g*
	Epoxy	*e*
	Urethane rubber	*g*
	Alpha-cyanoacrylates	*c*
Rubber, natural	Epoxy	*e*
	Flexible adhesives	*h*
	Alpha-cyanoacrylates	*c*
Rubber, neoprene	Epoxy	*e*
	Flexible adhesives	*h*
	Alpha-cyanoacrylates	*c*

TABLE 7.5.3 Adhesives Commonly Used for Joining Various Materials (*Continued*)

Materials to be joined	Adhesive	Table 7.5.4 reference
Rubber, silicone	Silicone	*t*
Rubber, urethane	Flexible adhesives	*h*
	Alpha-cyanoacrylate	*c*
	Silicone	*t*
Silver	Epoxy	*e*
	Neoprene	*g*
	Polyhydroxyether	*m*
Steel	Epoxy	*e*
	Acrylics	*w*
	Polyesters	*a*
	Polyvinyl butal	*l*
	Alpha-cyanoacrylate	*c*
	Polyamides	*h*
	Polyvinyl-phenolic	*i*
	Nitrile-phenolic	*b*
	Neoprene-phenolic	*b*
	Nylon epoxy	*d*
Stone	See "Brick"	
Tin	Epoxy	*e*
Wood	Animal glue	*o*
	Polyvinyl acetate	*n*
	Ethylene–vinyl acetate	*u*
	Urea, melamine, resorcinol, and phenol formaldehyde	*q*

Source: Adapted from Robert O. Parmley, *Standard Handbook of Fastening and Joining,* McGraw-Hill, New York, 1989.

DESIGN RECOMMENDATIONS

1. Design for shear, tension, and compression, not cleavage or peel. Adhesive bonds resist shear, tensile, and compressive forces better than cleavage or peel. Thus the designs shown in Fig. 7.5.2*b* are preferable to those of Fig. 7.5.2*a*.
2. The width of the joint overlap is more important than its length. Bond strength is not proportional to bond area except in cases of pure tension and compression. In a lap joint loaded in shear, the stresses are concentrated at the bond ends. Joint strength therefore is increased more by widening the joint than by lengthening it.
3. Match expansion coefficients. Large shear stresses are generated when materials with different thermal expansion coefficients are thermally cycled after bonding. Ideally, an adhesive should have an expansion coefficient midway between those of the adherends. Plastics-to-metal bonds may be a problem because of the large differences in thermal expansion of these materials. Fillers are often added to an adhesive to control (usually reduce) its coefficient of expansion. Faster heat-up rates may be possible when the expansion behavior of the adhesive is close to that of the parts.
4. Thin bond lines are preferred. A thin layer of adhesive is usually advantageous; 25 μm (0.001 in) is typical; thick glue lines consume excess adhesive, have a statistically greater chance for cracks and voids, and do not respond as quickly to tem-

TABLE 7.5.4 Adhesive Characteristics

Adhesive type	Table 7.5.3 reference	Comments	Typical cure conditions
Polyesters and their variations	*a*	Used primarily for repairing fiberglass-reinforced polyester resins, ABS, and concrete. Generally unsaturated esters are polymerized with a catalyst such as methyl ethyl ketone (MEK) peroxide and an accelerator such as cobalt naphthenate. A coreactant solvent such as styrene may be present. Bonds are strong. Sometimes combined with polyisocyanates to control shrinkage stresses and reduce brittleness. Unreacted monomer, if present, keeps viscosity low for application, provides good wetting, and enhances cross-linking. Occasionally used on metals.	Minutes to hours at room temperatures
Nitrile-phenolic, neoprene-phenolic	*b*	These adhesives are a blend of flexible nitrile or neoprene rubber with phenotic novolac resin. They combine the impact resistance of the rubber with the strength of the cross-linked phenolic. They are inexpensive and produce strong, durable bonds that resist water, salt spray, and other corrosive media as well. They are the workhorses of the adhesive-tape industry, although they do require high-pressure and relatively long high-temperature cures. They are used for metals and some plastics, including ABS, polyethylene, and polypropylene. Airframe components and automotive brakes are typical examples.	Up to 12 h at 250-300°F (120–150°C)
Alpha-cyanoacrylate	*c*	These single-component gels or liquids polymerize or cure rapidly when confined between most substances. Thus most surfaces can be bonded. The bonds are fairly strong but somewhat brittle. Used widely for the assembly of many diverse components ranging from jewelry to microelectronics.	0.1–5 min at room temperature
Epoxy-phenolic	*d*	A combination of epoxy resin with a resol phenolic. Noted for strength retention at 300–500°F (150–250°C), strong bonds, and good moisture resistance. Normally stored refrigerated. Used for some metals, glass, and phenolic resins.	1 h at 350°F (175°C)

Epoxy, amine-, amide-, and anhydride-cured	*e*	As a class epoxies are noted for high tensile and low peel strengths. They are cross-linked and in general have good high-temperature strength, resistance to moisture, and little tendency to react with acids, bases, salts, or solvents. There are important exceptions to these generalizations, however, that are often the result of the curing agent used. Primary amines give faster-setting adhesives that are less flexible and less moisture-resistant than is the case when polyamide curing agents are used. Anhydride-cured epoxies generally have high-temperature strength but are subject to hydrolysis, especially in the presence of acids or bases. Other important features of epoxies are their low shrinkage on cure, their compatibility with a variety of fillers, their long life when properly applied, and their easy modification with other resins. Cross-link density is easily varied with epoxies; thus some control over brittleness, vapor permeation, and heat deflection is possible. These resins are widely used to bond metals, ceramics, and rigid plastics (not polyolefins).	
Nylon-epoxy	*f*	Tensile shear strengths above 6000 lbf/in^2 (41.4 MPa) and peel strength above 100 lb/in (18 kg/cm) are possible when epoxy resins are modified with special low-melting nylons. These gains, however, are accompanied by loss of strength on exposure to moist air, a tendency to creep under load, and poor low-temperature impact behavior. A phenolic primer may increase bond life and moisture resistance. Used primarily for aluminum, magnesium, and steel.	1 h at 300–350°F (150–175°C)
Flexible adhesives: natural rubber, butadiene acrylonitrile, neoprene, polyurethane, polyacrylate, silicone	*g*	These adhesives are flexible. Thus their load-bearing ability is limited. They have excellent impact and moisture resistance. They are easily tackified and are used as pressure-sensitive tapes or as contact cements. Urethane and silicone adhesives are lightly cross-linked, which gives them reasonable hot strength. They are also compatible with many surfaces but are somewhat costly and must be protected against moisture before use. They have good low-temperature tensile, shear, and impact strength. The urethanes are two-part products that require mixing before use. Silicones cure in the presence of atmospheric moisture (see entries *r* and *t*).	Pressure-sensitive tape or solvent cements; low-temperature bake for urethane; ambient cure for silicones

TABLE 7.5.4 Adhesive Characteristics (*Continued*)

Adhesive type	Table 7.5.3 reference	Comments	Typical cure conditions
Polyamides	*h*	These adhesives, which are chemically similar to nylon resins, have good strength at ambient temperatures and are fairly tough. They are available in a variety of molecular weights, softening ranges, and melt viscosities. Often applied as hot melts, they have good adhesion to a variety of surfaces The higher-molecular-weight varieties often have the best tensile properties. Lower-molecular-weight polyamides may be applied in solution.	Hot-melt; cures by cooling
Polyvinyl phenolic	*i*	These resins, which combine a phenolic with polyvinyl formaldehyde or polyvinyl butyral, were the first important synthetic structural adhesives. A considerable range of compositions is available, with hot strength and tensile properties increasing at the expense of impact and peel strength as the phenolic content rises. The durability of vinyl phenolics is generally excellent. They are often selected for low-cost applications in which heat and pressure curing can be used.	1 h at 300°F (150°C)
Cellulose esters	*j*	Cellulose ester adhesives are usually high-viscosity, inexpensive rigid materials. They do not have high strength and are sensitive to heat and many solvents. Normally used for holding small parts or repairing wood, cardboard, or plastic items. Model airplane cement is a common example.	Air dry
Vinylchloride–vinyl acetate	*k*	This is a combination of two resins that are sometimes used alone.They may be used as hot melts or as solution adhesives. Since thin films of vinyl chloride–vinyl acetate are somewhat flexible, they are often used for bonding metal foil, paper, and leather. A range of compositions is available with a corresponding variety of properties.	Cooling (hot-melt) or solvent loss
Polyvinyl butyral	*l*	A tough, transparent resin that is used as a hot-melt or heat-cured solution adhesive. It has good adhesion to glass, wood, metal, and textiles. It is flexible and can be modified with other resins or additives to give a range of properties. Not generally used as astructural adhesive, although structural phenolics sometimes incorporate polyvinyl butyral to give better impact resistance.	Cooling (hot-melt), heating under pressure

Polyhydroxyether	*m*	These are resins based on hydroxylated polyethylene oxide polymers. Generally used as hot melts, they have only moderate strength but are flexible and have fairly good adhesion.	Hot-melt; cures by cooling
Polyvinyl acetate	*n*	This adhesive is generally supplied as a water emulsion (white glue) or used as a hot-melt. It dries quickly and forms a strong bond. It is flexible and has low resistance to heat and moisture. Porous substrates are required when the resin is used as an emulsion	Hot-melt; cures by cooling; emulsion by air drying
Animal glue	*o*	Chemically, animal glues are proteins; they are polar water-soluble polymers with high affinity for paper, wood, and leather surfaces. They easily form strong bonds but have poor resistance to moisture. They are being replaced in many areas by synthetic resin adhesives, but their low cost is often an important advantage. They are usually applied as highly viscous liquids.	Air dry under pressure
Starch glue	*p*	These products, based on cornstarch, have high affinity for paper but are used for little else. They are moisture-sensitive and are applied as water dispersions.	Low-temperature dry
Urea formaldehyde, melamine formaldehyde, resorcinol formaldehyde, phenol formaldehyde	*q*	These thermosetting resins are widely used for wood bonding. Urea formaldehyde is inexpensive but has low moisture resistance. It can be cured at room temperature if a catalyst is used. Melamine formaldehyde resins have better moisture resistance but must be heat-cured. Phenol formaldehyde adhesives form strong, waterproof wood-to-wood bonds. The resorcinol formaldehyde resin will cure at room temperature, while phenol formaldehyde requires heating. These resins are often combined, resulting in an adhesive with intermediate processing or performance characteristics	Up to 300°F (149°C) and 200 lbf/in^2 (1.38 MPa)

TABLE 7.5.4 Adhesive Characteristics (*Continued*)

Adhesive type	Table 7.5.3 reference	Comments	Typical cure conditions
Polyacrylate esters	*r*	These resins are *N*-alkyl esters of acrylic acid. They have good flexibility and find frequent use for high-quality pressure-sensitive tapes and foams. They are not suitable for structural applications because of their poor heat resistance and their cold-flow behavior. Frequently used on flexible substrates.	Pressure-sensitive
Polysulfides	*s*	These resins have good moisture resistance and can range from thermoplastic to thermosetting, depending on the degree of cross-linking that is developed during cure. They are two- or three-part systems, the third part being a catalyst. Ventilation is generally required. They make excellent adhesive sealants for wood, metal, concrete, and glass. Polysulfide resins may be combined with epoxies to flexibilize the latter.	Low pressures, moderate temperatures
Silicones (see also "Flexible adhesives")	*t*	These adhesives have high peel strength and excellent property retention at high and low temperatures. They resist all except the most corrosive environments and will adhere to a wide range of materials. They are usually formulated to react with atmospheric moisture and form lightly cross-linked films.	Low pressure, room temperature
Ethylene–vinyl acetate	*u*	This copolymer is widely used as a hot-melt adhesive because it is inexpensive, adheres to most surfaces, and is available in a range of melting points. It is widely used for bookbinding and packaging.	Hot-melt; cures by cooling

Urethanes, rigid	*v*	Rigid urethanes are highly cross-linked. While somewhat expensive, they adhere well to most materials, especially plastics, and have good impact strength. Structural urethanes are two-part systems and have good low-temperature strength retention.	Low pressures, up to 300°F (149°C)
Acrylics (also rubber-toughened acrylics)	*w*	These formulations have evolved in performance with the addition of rubber. Toughening agents yield excellent property retention at high and low temperatures and high peel strength. They exist as two-component no-mix adhesives and as two-part mixed formulations depending on the process requirements of the specific application.	Room-temperature cure in 20 s to 20 min, depending on formulation
Ultraviolet radiation-curing adhesives	*x*	A wide variety of chemistries can cross-link when exposed to electromagnetic radiation of variable wavelengths. Recent developments have resulted in visible-light-curing formulations that are suitable for plastics. Formulations are single-component and 100% reactive.	Room temperature with light exposures of less than 1 min

Source: Adapted from Robert O. Parmley, *Standard Handbook of Fastening and Joining,* McGraw-Hill, New York, 1989.

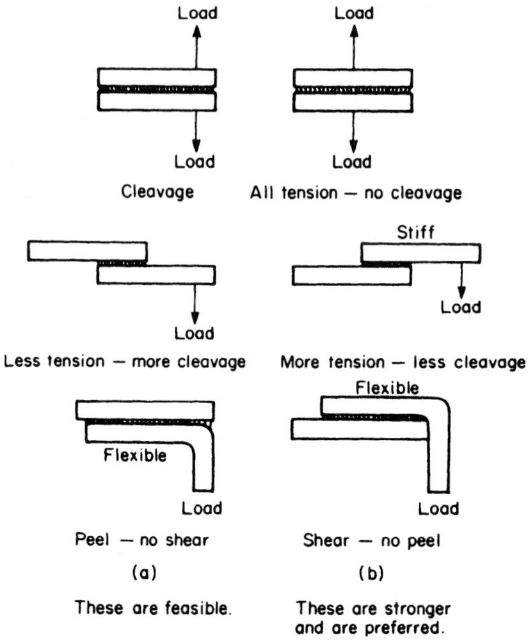

FIGURE 7.5.2 The designs shown in *b* utilize shear, tensile, and compressive stresses instead of cleavage and peel stresses and are therefore preferred to those of *a*.

perature changes. An exception would occur when high-impact strength is desired. Then a thicker, more flexible adhesive might well be preferred.

5. Design for easy cleaning. Dirty surfaces are a major cause of poor joint performance. Vapor degreasing is a preferred method of preparing surfaces. Solvent wiping may be sufficient if the wipe rags are not allowed to become dirty. Dip cleaning is risky because of the possibility of gradual or sudden contamination of the dip tanks. (See Chap. 8.1 for information on designing parts for easy cleaning.)

6. Smooth surfaces are preferred. Smooth surfaces are more easily wet by a spreading liquid adhesive. A greater percentage of the area of part surfaces can contact the adhesive when each surface is smooth. Surfaces are often roughened by abrasive treatment before bonding to remove loosely held surface material, even though surface contact with the adhesive is thereby reduced. Loosely held surface layers are the greater evil.

7. Simple butt joints should be used only when fairly large bond surfaces are involved and when cleavage stresses are not anticipated. Figure 7.5.3 illustrates some of the ways in which a butt joint can be modified to increase resistance to cleavage failure.

8. When simple lap joints are stressed in tension, they tend to deform as shown in Fig. 7.5.4. This deformation introduces cleavage stresses (for rigid adherends) or peel stresses (for flexible adherends) at the joint ends. Commonly used designs to minimize these stresses are shown in Fig. 7.5.5.

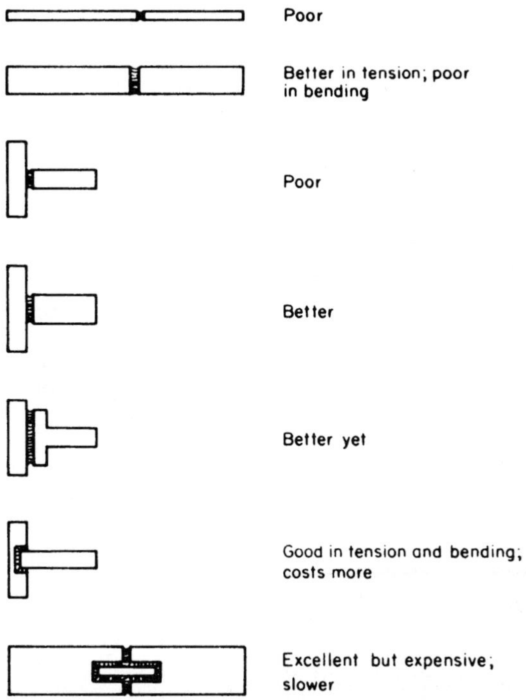

FIGURE 7.5.3 Butt joints. Simple butt joints leave little resistance to cleavage stresses. Some common modifications are shown here.

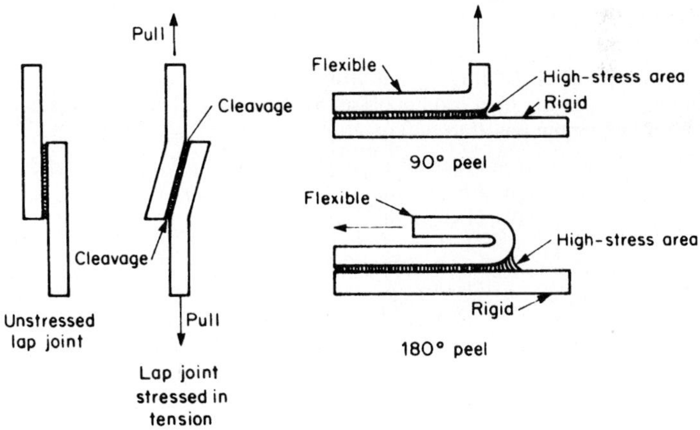

FIGURE 7.5.4 Generation of cleavage or peel stresses in simple lap joints in tension.

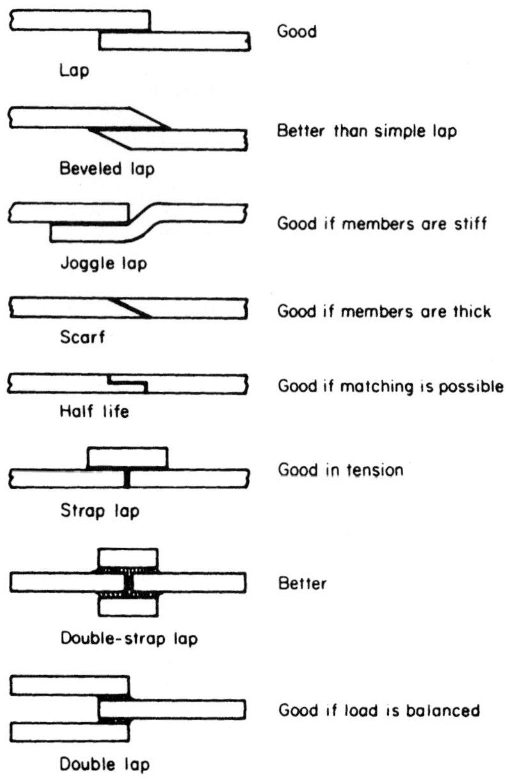

FIGURE 7.5.5 Joint modifications that increase lap shear strengths.

9. Corner joints involving members of various thicknesses are shown in Fig. 7.5.6. Many variations are possible. The preferred design is usually the one that involves the least preparation (including handling) cost. Sometimes the cost of an extra machining or forming operation is offset by the easier cleaning or assembly methods that may then be possible.

10. Figure 7.5.7 depicts a number of common techniques for joining rods and tubes adhesively. That design is best which requires the least machining and assembly time. Corners are usually best handled with elbows.

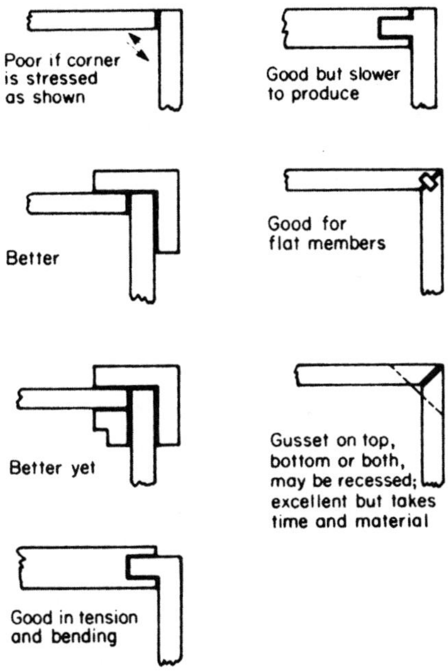

FIGURE 7.5.6 Common corner-joint designs.

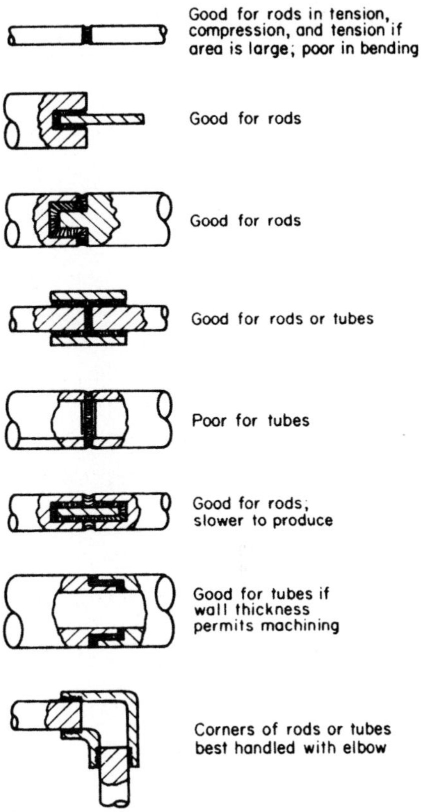

FIGURE 7.5.7 Designs for joining rods and tubes.

S • E • C • T • I • O • N • 8

FINISHES

CHAPTER 8.1
DESIGNING FOR CLEANING

INTRODUCTION

Cleaning consists of the removal of contaminating or otherwise unwanted liquid, solid, or semisolid matter from the surfaces of a component. Though too often not considered a part of the manufacturing process, cleaning operations are very important in industrial production. They are also very common, perhaps the most frequently occurring of all manufacturing operations.

Cleaning operations are specified to put a component into proper condition for subsequent events. Most cleaning is for in-process use: to prepare the part for the next manufacturing operation, e.g., further machining, painting, plating, or assembly. Some cleaning is performed prior to storage to remove corrosives. Some is performed as a final operation to put the product in proper condition for sale and for use.

The choice of cleaning process to be used depends on three major factors: (1) the soil to be removed, (2) the degree of cleanliness required, and (3) the cost of cleaning. Part design is another factor, although a lesser one. It may influence the method of application of the cleaning agent.

Specifying the degree of cleanliness required is not easy. Normally, painting, plating, and other finishing operations require the highest degree of cleanliness. However, the degree of cleanliness required for a finishing operation is seldom specified on a parts drawing. Instead, it is left to the discretion of the finishing engineers, who provide whatever cleaning process is necessary for a finish of adequate quality.

SOILS TO BE REMOVED

Six basic types of soil are removed industrially: (1) rust and scale, (2) chips and metal-cutting fluids, (3) oil and grease (unpigmented), (4) pigmented drawing compounds, (5) polishing and buffing compounds, and (6) miscellaneous soils such as lapping residue or residue from magnetic-particle inspection.

Rust and scale result from surface oxidation of hot-worked mill products, forgings, castings, and weld fillets. They interfere with heat transfer during heat-treating operations and increase the wear rate of cutting tools and press-working dies. Rust and scale provide an unsatisfactory surface for painting, plating, galvanizing, and other metal finishes.

Chips and metal-cutting fluids are residues from machining operations and are usually easy to remove. Cutting fluids of the plain or sulfurized-oil type, the soluble-oil type, or the water-based type all can be removed with a variety of processes that also flush away the chips.

Unpigmented oil and grease consist of rust-preventive oils, quenching oils, unpigmented drawing lubricants, and general lubricating greases and oils. They are easily removable with a variety of economical cleaning methods.

Pigmented drawing compounds such as are used in cold-finishing steel bars or other shapes are difficult to remove. White lead, molybdenum disulfide, graphite, and soaps are the most tenacious pigments, particularly if high pressure and high temperature are used. Other somewhat less troublesome pigments are whiting, bentonite, flour, zinc oxide, and mica. The drying of a drawing compound because of a delay between drawing and cleaning increases the difficulty of cleaning. Because of the problems involved in cleaning these compounds, it sometimes pays to perform tests to see if their use can be eliminated or reduced.

Polishing and buffing compounds leave a residue consisting of abrasive particles, metal particles, and waxes and greases. They are difficult to remove because of chemical changes which result from the high local heat generated during polishing or buffing.

Miscellaneous soils such as lapping and magnetic-particle-inspection residues can be difficult to remove, particularly if the particles have dried to the surface of the workpiece. Another complicating factor is the fact that parts subject to such operations are for high-precision applications that also demand high cleanliness levels.

IN-PROCESS CLEANING OPERATIONS

The following are the common in-process industrial cleaning operations:

1. *Cleaning prior to machining, forming, assembly, or other production operations or prior to gauging.* The prime purpose of such cleaning is to remove grease or oil that could interfere with handling and to remove chips or dust that could prevent accurate fixturing of the component for the subsequent operation or in other ways interfere with it. Cleanliness requirements for this application are not severe.
2. *Cleaning prior to phosphating or other surface-treatment operations.* Phosphate coating requires a sufficiently clean workpiece surface so that the phosphating solution can wet it enough for the chemical reaction to take place. Some contaminants restrict contact of the solution with the part surface; others act as a complete barrier. Contaminants such as drawing compounds, rust preventives, cutting oils, and coolants can react with the base metal and form a surface layer that adversely affects the characteristics of the phosphate coating.
3. *Cleaning prior to painting operations.* Paint films will not adhere permanently if the workpiece surface is not thoroughly cleaned. The appearance of the paint finish also can be affected adversely if there are soils such as oil, dust, grease, rust, scale, water, or salts on the part.
4. *Cleaning prior to bonding* presents requirements similar to those for painting. Any foreign matter or film that prevents molecular contact between the adhesive and the substrate materials compromises the strength of the joint. Surfaces should be clean and especially free of oils and greases. Structural adhesive joints are more demanding of surface cleanliness than are painted coatings because of the heavier load imposed on bonding surfaces.

5. *Cleaning prior to brazing.* The cleanliness of the base-metal joint surfaces is a major factor in the success of a brazed joint. Surfaces must be free of oil, grease, dirt, and oxides. The degree of cleanliness depends on the metals being joined. With stainless steels and aircraft-quality joints, cleanliness requirements are extremely high.
6. *Cleaning prior to electroplating* involves higher levels of cleanliness than for other applications. Multistage cleaning is usually employed. Solvent, alkaline, electrolytic, and acid cleaning comprise a typical four-step sequence.

CLEANING PROCESSES AND THEIR APPLICATIONS

Mechanical Processes

Brushing. This process is performed with wire or fiber brushes that are power-driven in industrial applications. Different brush materials, bristle thicknesses, and lengths provide a wide variety of abrasive action. Milder action is produced by nonmetallic bristles, moderate action with fine wire of nonferrous metals, and the strongest abrasive action with stiff, heavy wire.

Brushing is used to remove solid material such as rust, loose paint, and caked dirt. Normally it is only a first operation in a sequence that includes one or more steps with liquid cleaners. It is most applicable to flat parts, those made from tubing, or those with regular contours. Castings or forgings or other parts with surface irregularities and recesses are less suitable. This limitation is particularly true of automatic brushing equipment, which is costly to develop for irregularly surfaced components.

Wire brushing is used most often at low and moderate production levels.

Abrasive Blasting. This consists of bombarding the workpiece surface with abrasive particles at high velocity. The abrasive can be a hard material such as aluminum oxide, sand, silicon carbide, or steel shot or a soft material like rice hulls, corncobs, nut shells, or plastic beads. Blasting is particularly suited to the removal of rust, scale, and other corrosion products, dry surface dirt, and paint. It is not particularly satisfactory for the removal of grease. However, it is used for deburring, surface roughening, and surface improvement. (See also Chaps. 8.7, "Shot-Peened Surfaces," and 4.23, "Designing Parts for Economical Deburring.") When the abrasive material is mixed with water, a finer abrasive can be used and certain processing advantages result. Dry-abrasive blasting is less satisfactory when surfaces are irregular or when tight dimensional tolerances and fine surface finishes are required. It is, however, a process that lends itself well to high-production applications.

Steam-Jet Cleaning. This is an effective method for removing oil, grease, and dirt from equipment and other bulky objects too large to be immersed in cleaning tanks. Flame jets are used to remove scale or old paint from large iron or steel components.

Tumbling. This is an effective and low-cost method for removing rust and scale and performing similar cleaning of small parts as well as for surface finishing as described in Chap. 8.2. It is used for in-process cleaning and as part of a prepainting, preplating sequence. It is well suited to moderate and high production levels.

Chemical Processes

Solvent Cleaning. This process involves the use of liquid hydrocarbons. The workpiece can be cleaned with one or a combination of three application methods: immersion or soak-tank cleaning, spray degreasing, or vapor degreasing.

The solvents customarily used are petroleum solvents (Stoddard solvent, mineral spirits, or kerosene) and chlorinated hydrocarbons (trichlorethylene or perchlorethylene). Petroleum solvents are most commonly used for immersion cleaning; chlorinated hydrocarbons are standard for vapor degreasers. The chlorinated hydrocarbons have the advantage of nonflammability, although they are more toxic than petroleum solvents.

Immersion Cleaning. This process removes oil, grease, and oil-borne dirt, but because of rapid solvent contamination, it does not clean to high levels of cleanliness. It is most effective for in-process cleaning.

Vapor Degreasing. This process is very effective in removing oil and grease. However, a plain vapor-degreasing cycle lacks enough flushing action to remove much solid soil material. Therefore, immersion and spraying are commonly incorporated in the degreasing cycle prior to the vapor step. In the vapor degreaser, a small amount of solvent vaporizes, but the vapor is contained in the tank by the tank walls and a cooling coil. When a workpiece is lowered into the vapor, the vapor condenses on its surface and drips back to the reservoir below. As the workpiece is heated by the warm vapor, condensation stops and the workpiece dries. It is removed from the vapor warm and completely free of oil and grease.

Solvent and solvent-vapor cleaning tend to be expensive because of the high cost of solvents. These processes are also normally limited to lower and medium production levels because for real mass production, for which unit costs must be tightly controlled, alkaline cleaning and other processes are more economical. Solvent disposal, toxicity, and fire prevention are also problem areas for solvent cleaning.

Ultrasonic Cleaning. This process, which uses ultrasonic vibrations to agitate the cleaning solution, is usually employed with chlorinated solvents. However, alkaline solutions also can be used. The process is effective for the thorough cleaning of small, intricate parts that may be difficult to clean with other methods.

Emulsion Cleaning. This is a good means for removing caked materials such as heavy buffing compounds or grease loaded with solid particles. The liquid cleaner is a mixture of water, a hydrocarbon solvent, and emulsifying agents. The hydrocarbon solvent is generally petroleum-based and is dispersed in the water as fine globules in a ratio of 1 to 10 parts solvent to 100 parts water. The emulsifying agents normally used are soaps, glycerols, polyethers, and polyalcohols. The emulsion cleaner is applied by spray or immersion, normally at a temperature of 54 to 60°C (130 to 140°F) and sometimes up to 82°C (180°F). If immersion is used, agitation of the liquid is advisable. The cleaning cycle is followed by a hot-water rinse.

Emulsion cleaning is economical because of the water extension of the solvent and is safe for the same reason. It is most suitable for in-process cleaning. The cleanliness level of the workpiece after emulsion cleaning is only moderate. A thin film of oil is normally left on the cleaned parts. This is desirable for rust prevention in the shop, but if a high level of cleanliness is required, other cleaning operations should follow the emulsion operation. Emulsion cleaning is safer and cheaper than solvent cleaning but more expensive than alkaline cleaning.

Alkaline Cleaning. Probably the most widely used cleaning method, alkaline cleaning is also the least expensive for high-production purposes. The cleaner consists of a solution of certain alkaline salts and detergents in water. The salts are usually mixed and consist of caustic soda, trisodium phosphate, silicates, borates, or carbonates. Cleaning is done in soak tanks, often with agitation and heat of 60 to 93°C (140 to 200°F), and by pressure spray. Alkaline cleaning is effective for oil, grease, shop dirt, and compounds from polishing, buffing, and drawing operations. Light scale, rust, and carbon smut also can be removed by this method but at a slower rate and at a higher cost than by acid pickling. Water rinsing usually follows the alkaline-cleaning operation.

Acid Cleaning. This process uses methods and solutions similar to those of alkaline cleaning except that the cleaner is a solution of acids or acid salts instead of alkaline salts. Mineral or organic acids may be used. A wetting agent and a detergent are also part of the solution. Soak-tank immersion, power spray, and wiping are the normal methods of application of the heated or room-temperature solution. Acid cleaning is effective for removal of light rust, tarnish, scale, and similar deposits. Drawing compounds, oil, and grease also can be removed. For heavy coatings of oil and grease, however, acid cleaning is not the most effective method.

Some etching of the metal surface takes place with acid cleaning. This is desirable for good adhesion of paint.

Aluminum is commonly cleaned by the acid method. The method is also suitable for ferrous metals and copper but is seldom used on nickel, magnesium, lead, or tin.

Pickling. A stronger form of acid cleaning, pickling is used for the removal of surface oxides, scale, and dirt from metals. The cleaning solution consists of an aqueous solution of acid with a wetting agent. Choice of acid depends on the metal to be cleaned. For ferrous metals, sulfuric, hydrochloric, and sometimes phosphoric acids are used.

The method, which involves immersion of the part in the pickling bath, is the most effective method for scale removal. It works because the pickling solution attacks the surface of the metal, destroying the adhesion of scale. The pickling operation is followed by an alkaline rinse to neutralize the pickling acid.

Although mill products, forgings, and castings of steel and iron are the most commonly pickled materials, aluminum, copper, stainless steel, magnesium, and nickel alloys are also cleaned by this method. The cost of pickling equipment can be high, restricting the use of the process to larger shops or higher-production applications.

Overpickling can pit or roughen the workpiece surface objectionably. The disposal of used pickling solution requires extra care, normally neutralization of the pickling acid.

Salt-Bath Cleaning. This is another method for removing scale, oxides, sand, carbon, and graphite from a variety of metals. Sometimes known as *sodium hydride descaling,* the process involves immersion of the workpiece in a bath of the molten salt. The temperature of the salt is 440 to 524°C (825 to 975°F). The cleaning process is as follows: The salts in the bath reduce the scale and loosen it somewhat from the workpiece surface. When the workpiece is placed in a water rinsing bath after the salt bath, the water turns to steam because of the heat of the part and blasts the scale from its surface.

The process is effective for recesses and other surface irregularities. A neutralizing-acid dip and a final rinse complete the process, which leaves parts unaffected dimensionally and with a bright surface. Suitable metals are plain carbon steel, alloy steel, stainless steel, and alloys of chromium, copper, nickel, tungsten, and cobalt. Titanium, however, can pick up hydrogen, which results in embrittlement. Low-melting alloys such as magnesium, zinc, tin, aluminum, and lead are not suitable.

The heat of the salt bath can distort thin materials. It also can sometimes affect the hardness of the workpiece.

Both the high cost of the salt-bath equipment and the high cost of starting up again when a bath is allowed to cool down demand that the process be confined to high-production applications.

Electrochemical Processes

Alkaline cleaning, acid cleaning, pickling, and salt-bath cleaning can all be performed with an electrolytic (electrochemical) assist. In electrolytic-alkaline cleaning, the workpiece becomes one electrode and the tank or a steel plate the other. The alkaline solution is the electrolyte. When current is applied, hydrogen gas is released at the cathode and oxygen at the anode. The gas bubbles provide a scrubbing action at the workpiece surface, causing the soil to break up rapidly. The formation of like electrical charges on the work and the dirt also causes them to separate. Because dirt contamination of the electrolyte quickly reduces the effectiveness of the operation, it is normally performed as a final cleaning operation after regular alkaline cleaning.

Electrolytic pickling is another common electrolytic-cleaning operation. The workpiece is the cathode, and the acid pickling solution is the electrolyte. Gas liberated on the workpiece loosens the scale and speeds its removal. Care must be taken in closely controlling the bath temperature and concentration and in avoiding a dimensional change of the workpiece or pitting of its surface.

TABLE 8.1.1 Cleaning Methods Suitable for In-Process-Inspection Operations

Soil	Cleaning method
Pigmented drawing compounds	Low production Hot emulsion, hand slush, spray emulsion Vapor-slush degrease High production Automatic spray emulsion
Unpigmented oil and grease	Low production Emulsion dip or spray Vapor degrease Cold-solvent dip Alkaline dip, rinse, dry High production Automatic vapor degrease Emulsion, tumble, spray, rinse, dry
Chips and cutting fluid	Low production Alkaline dip and surfactant Solvent Steam High production Alkaline dip or spray and emulsion surfactant

Source: From AMCP 706-100, *Engineering Design Handbook,* 1971.

TABLE 8.1.2 Cleaning Methods in Preparation for Plating

Soil	Cleaning method
Pigmented drawing compounds	Low production Alkaline soak, hot rinse, hand wipe High production Hot emulsion or alkaline soak, hot rinse, electrolytic alkaline, hot rinse
Unpigmented oil and grease	Low production Emulsion soak, barrel rinse, electrolytic alkaline, rinse, hydrochloric acid dip, rinse High production Automatic vapor degrease, electrolytic alkaline, rinse, hydrocholoric acid dip, rinse
Chips and cutting fluid	Low production Alkaline dip, rinse, electrolytic alkaline, rinse, acid dip, rinse High production Same as low production except for soak rather than dip
Polishing and buffing compounds	Low production Surfactant, rinse, electroclean Emulsion spray or soak, rinse, alkaline spray or soak, rinse, electroclean Solvent presoak, alkaline soak or spray, electroclean High production Surfactant, alkaline soak, spray rinse, electrolytic alkaline Emulsion spray or soak, rinse, alkaline spray or soak, rinse, electroclean Solvent presoak, alkaline soak or spray, electroclean

Source: From AMCP 706-100, *Engineering Design Handbook,* 1971.

Although electrolytic-cleaning processes are rapid, the equipment is considerably more costly than that required for nonelectrolytic methods. Hence higher levels of production are generally necessary to justify using the process.

Tables 8.1.1 through 8.1.4 summarize normal cleaning processes for the major subsequent operations at both low and high levels of production.

SUITABLE MATERIALS

Table 8.1.5 lists both suitable and unsuitable materials for the common cleaning processes.

TABLE 8.1.3 Cleaning Methods in Preparation for Phosphating

Soil	Cleaning method
Pigmented drawing compounds	Low production Hot-emulsion hand slush, spray emulsion, hot rinse, wipe High production Alkaline soak, hot rinse, alkaline spray, hot rinse
Unpigmented oil and grease	Low production Emulsion dip or spray, rinse Vapor degrease High production Emulsion power spray, rinse Vapor degrease Acid-clean
Chips and cutting fluid	Low production Alkaline dip, emulsion surfactant Solvent or vapor rinse High production Alkaline dip or spray and emulsion surfactant
Polishing compounds	Low production Surfactant, rinse Emulsion soak, rinse High production Surfactant, alkaline spray, spray rinse Emulsion spray, rinse

Source: From AMCP 706-100, *Engineering Design Handbook,* 1971.

DESIGN RECOMMENDATIONS

1. The part should provide easy access to cleaning media and implements. Undercuts, narrow recesses, and deep blind holes are problems when mechanical or electrolytic cleaning methods are used and are also less satisfactory for chemical cleaning methods. (See Fig. 8.1.1.)
2. When liquid cleaners are to be used, drain holes must be provided whenever the part configuration or method of hanging would cause the cleaning solution to be trapped. (See Figs. 8.5.3 and 8.5.4.)
3. Parts subject to liquid cleaning operations should have some hole or protuberance that can be used for a hook or other rack member that supports the workpiece during cleaning. (See Fig. 8.1.2.)
4. In some cases it may be advisable to avoid incorporating a large workpiece in the product but instead to specify two or more smaller parts. The smaller parts may be easier to handle and clean, especially if the cleaning process involves immersion tanks or tumbling barrels of limited size. Assembly of the smaller components together after cleaning and finishing may be more economical than cleaning and finishing a large single component.
5. Parts that nest may cause difficulties in some cleaning operations, e.g., those which involve immersion in a bath. Designers must weigh the handling advantages of nestable parts against this potential disadvantage.

TABLE 8.1.4 Cleaning Methods in Preparation for Painting and Bonding

Soil	Cleaning method

Pigmented drawing compounds	Low production
	  Hot alkaline, blow off, wipe
	  Vapor-slush degrease, wipe
	  Acid-clean
	High production
	  Alkaline soak, rinse, alkaline spray rinse

Unpigmented oil and grease	Low production
	  Vapor degrease
	  Phosphoric acid-clean
	High production
	  Automatic vapor degrease

Chips and cutting fluid	Low production
	  Alkaline dip and emulsion surfactant
	  Solvent or vapor
	High production
	  Alkaline dip or spray and emulsion surfactant

Polishing compounds	Low production
	  Agitated soak and rinse
	  Emulsion soak, rinse
	High production
	  Surfactant, alkaline spray and rinse

Source: From AMCP 706-100, *Engineering Design Handbook,* 1971

CLEANLINESS SPECIFICATIONS

There is no common, universal system for specifying cleanliness of industrial components. There are some cases, however, in which cleanliness specifications by the product designer are called for. One common example is in the aerospace industry, in which cleanliness of certain equipment is critical to its reliable functioning.

The following methods can be used to specify cleanliness requirements:

1. Specifications of the maximum allowable weight of soils on a part
2. Specifications of the maximum particle size of allowable soils on a part
3. Various specifications based on some kind of performance test, e.g.,
  a. Water-break test
  b. Acid-copper test (electroless-copper-plating coverage)
  c. Atomizer test (blue-dye coverage)
  d. Wiping test

Laboratory methods are required to measure performance against most of these specified requirements. The water-break test consists simply of immersing the cleaned part in water and observing if the retained water on the part forms a continuous, unbroken film. This is a fairly satisfactory means of verifying moderate levels of cleanliness. The wiping test is simply a verification that visual soils are absent from workpiece surfaces. The workpiece is wiped with a clean white cloth or paper, which is then examined for discoloration.

TABLE 8.1.5 Suitable Materials for Common Cleaning Processes

Process	Suitable materials	Marginal, difficult, or unsuitable materials
Brushing	Steel, cast iron, and most other metals, ceramics, and glass	Plastics and rubber require soft brushes. Stainless steel and aluminum require stainless brushes. Magnesium requires dust control.
Abrasive blasting	Steel and all harder metals	Ductile metals require special abrasives. Plastics and other soft materials may entrap abrasive. Stainless steel, copper, brass, zinc, aluminum, lead, and tin require nonmetallic abrasive. Magnesium can be hazardous.
Steam cleaning	Steel, cast iron, and other metals	Thermoplastics may suffer from heat effects. Glass and ceramics can crack if heating is uneven.
Solvent cleaning: petroleum solvents	All metals, glass, and ceramics; many plastics*	Plastics and elastomers.*
Solvent cleaning: chlorinated solvents	Virtually any metal, glass, and ceramics	Plastics and elastomers,* organic dyes, sometimes aluminum and magnesium.
Emulsion cleaning	Steel, cast iron, aluminum, brass, glass, and ceramics	Powder-metal parts, sand-core brass castings, zinc and aluminum if pH is not controlled.
Alkaline cleaning	Steel, stainless steel, aluminum, brass, copper, magnesium, zinc, epoxy, fluorocarbons, glass, and ceramics	Zinc, aluminum, lead-tin solders, brass, and other copper alloys require inhibited alkaline cleaners. Painted surfaces may be contaminated. Powder-metal parts. Plastics.*
Acid cleaning	Steel, iron, copper, fluorocarbons, glass, and ceramics	Nickel, manganese, lead and tin solders, magnesium, most plastics.* Aluminum may be etched by some acid cleaners. Powder-metal parts
Salt-bath cleaning	Steel, nickel, copper, iron, manganese, refractory and heat-resistant alloys, and stainless steel	Magnesium, zinc, tin, lead, aluminum, titanium, zirconium, cadmium, precipitation-hardenable stainless steel, powder-metal parts
Pickling	Aluminum including castings, copper, iron, steel, stainless steel, and magnesium	Under some conditions, steel can be subject to hydrogen embrittlement. Not applicable to nonmetallic materials.

*Since there is a wide variety of available plastic and elastomer materials, each with its own unique properties, it is recommended that a test be made to verify that the particular plastic or elastomer used is not adversely affected by the chemical cleaning agent used.

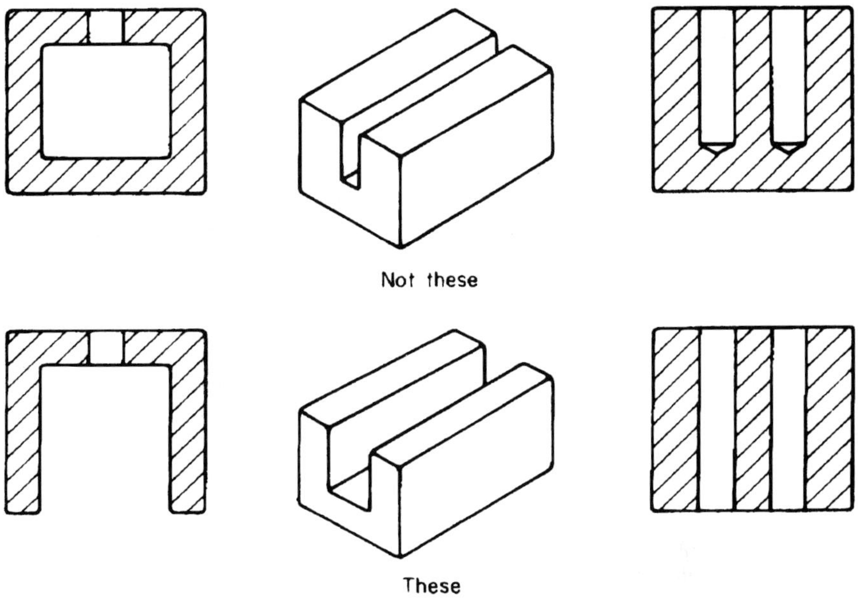

FIGURE 8.1.1 The design of parts to be cleaned should permit easy access to cleaning implements and full flow of cleaning fluids.

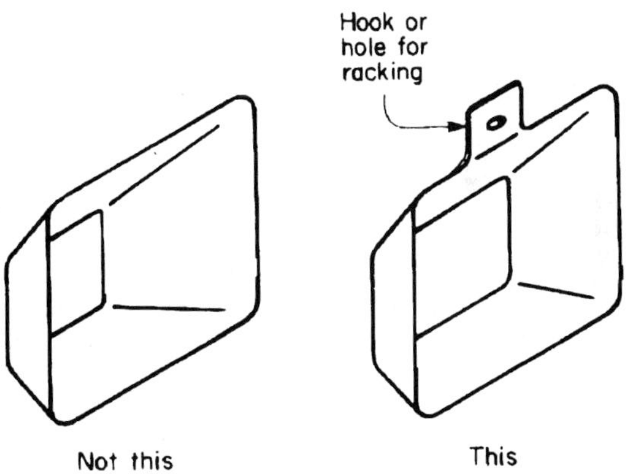

FIGURE 8.1.2 Parts that are to be subjected to cleaning operations should be designed so that they can be racked easily.

CHAPTER 8.2
POLISHED AND PLATED SURFACES

Albert J. Gonas

Fisher Body Division
General Motors Corp.
Warren, Michigan

Alan J. Musbach

MCP Industries, Inc.
Detroit, Michigan

INTRODUCTION

Polishing and plating are companion processes, usually part of a total finishing sequence. Polishing prior to plating is important in providing a lustrous plated finish. It also tends to promote uniformity of plating thickness and improves the corrosion protection of a plated finish. Conversely, for most metals plating enhances the appearance and helps ensure the permanence of a polished finish.

POLISHED SURFACES

Polishing Processes

Conventional Polishing. In this operation, abrasive particles, glued to a flexible wheel or a belt, remove surface irregularities from a workpiece. Conventional polishing is often an intermediate operation, being preceded by casting, forging, machining, stamping, or some other basic operation and followed by buffing. Sometimes referred to as *flexible grinding,* the process can remove metal either slowly or rapidly and causes some plastic deformation of the metal surface. Generally, less flexible wheels cut more rapidly, and softer wheels cut more slowly but conform better to contoured shapes. Wheels can be made from fabric, hemp, leather, wood, or felt. Polishing belts have contact wheels of various materials, the most common being rubber. Various durometers are employed, depending on the speed of cutting required and the shape of the parts normally processed. Hard wheels are better for fast cutting and soft wheels for contoured shapes and blending operations.

Lubricants—tallow, wax, and fatty acids—are used with both wheel and belt polishing to extend the life of the abrasive surface. Both wheel and belt polishing are often carried out in several steps with progressively finer abrasive. *Roughing, fining,* and *oiling* are common terms for three steps of polishing.

Buffing. A secondary operation that follows polishing, buffing is performed with a milder or finer abrasive. Very little material is removed from the workpiece surface. Buffing also differs from polishing in that the abrasive, rather than being glued to the wheel surface, is loosely held by the wheel. The abrasive is applied to the wheel from a bar of abrasive compound.

Buffing further smooths the surface after polishing and leaves a lustrous, grainless finish. Two steps of buffing may be involved: "cutting down" and "coloring," the latter being a finer-abrasive, final operation. Buffing wheels are usually of fabric but may be of kraft paper, hemp, or sheepskin.

Barrel Polishing or Tumbling. Though primarily a deburring process (see Chap. 4.23), this is also an effective means of surface polishing. With proper conditions and equipment, parts can be polished to finishes comparable with those achieved with conventional polishing and buffing.

The process involves a rotating barrel or vibrating hopper, water, a compound (cleaners or detergents plus fine abrasives), a medium (chunks of ceramic, stone, or metal), and the parts to be polished. The rotation of the barrel or vibration of the hopper causes the medium to rub against the part with the abrasive, providing a polishing action. The cleaner and water separate the abraded particles from the work.

Electropolishing. This is an electrolytic process, the reverse of electroplating. Whereas in electroplating particles of metal are added to the workpiece surface by electrochemical action, in electropolishing the workpiece is the anode and metal is removed from the surface. The workpiece is connected to the positive side of a low-voltage direct-current power source, while a cathode is connected to the negative side. Both are immersed in a conductive solution. The nature of the electrolytic action is such that the higher points including microprojections are subjected to more intense electrolytic action and more rapid metal removal. The result is a gradual smoothing and an increase in luster of the metal surface. (The process also is very useful for deburring and is covered for that purpose in Chap. 4.23.) The final result is a metal surface with a smooth, glossy appearance.

Characteristics and Applications

Polishing operations have the purpose of surface-finish improvement rather than dimensional refinement. However, parts and surfaces of irregular contour that require some refinement of shape with minimum stock removal are suitable for polishing operations. If dimensional-accuracy requirements are high or the amount of metal to be removed is large, it is better to use conventional grinding or machining operations. By far the most important function of polishing operations is the improvement of appearance. This involves the removal of surface imperfections like scratch marks, gates, stretch marks, cutting-tool marks, pits, parting lines, etc. It also includes surface smoothing prior to plating, anodizing, painting, or other surface finishes. There are, however, other reasons for polishing workpieces:

1. To improve the contour of a part for functional reasons, as, for example, to improve fluid around propeller, turbine, or fan blades
2. To provide clearance for assembly, especially when gates, runners, or flashing is removed from a part by polishing operations
3. To remove burrs
4. To clean a surface for brazing, soldering, or surface finishing
5. To improve resistance to corrosion by removing pockets in which contaminants can collect and stressed areas that could promote stress corrosion (Electropolishing is the best method for removing surface-stress concentrations because it is a completely non-stess-inducing method of metal removal. All machining operations, including grinding and mechanical polishing, induce surface stresses because of tearing, smearing, and cutting actions.)
6. To facilitate inspection of surface imperfections in forged, cast, or formed metal parts because such defects are more visible if the surface is smooth

Some common manufactured components that have polished surfaces are plumbing fittings, cutlery, door hardware, automotive bumpers and trim, firearms parts, wrenches, pliers, and other tools, bicycle handlebars, golf-club heads and shafts, fishing-reel parts, stainless-steel and aluminum panels for appliances, turbine blades, nose cones, hydraulic cylinders, square and round tubing, bar stock, fountain pens, and cast cooking utensils. Many of these parts, if they are of a high-luster finish, are both polished and buffed, as are reflectors and other mirror-like surfaces.

Conventional wheel polishing and buffing are most suitable for parts that can be hand-held during the operation at a wheel mounted in a stationary location. Very large parts, such as equipment-housing panels, therefore, are not so suitable unless production quantities are large enough to justify specialized equipment. Very small parts can be mounted in holding fixtures or tongs but often are more easily processed by barrel finishing.

Belt polishing is most practical for workpieces with flat, cylindrical, or gently contoured surfaces. Parts should be over about 12 mm ($\frac{1}{2}$ in) in diameter and sufficiently stiff so as to maintain uniform contact pressure across the surface to be polished. Parts with bosses, attachments, depressions, or even moderately sharp internal corners are not applicable. Small parts are better handled by electropolishing or with polishing wheels.

Parts most suitable for barrel polishing are those that are small enough to slide freely against the medium in the tumbling barrel. For conventional barrel polishing, the major dimension of the workpiece should be less than one-third of the diameter of the barrel. However, rodlike parts longer than this are processible provided they are placed lengthwise in the barrel and are longer than the barrel diameter; i.e., they are long enough to remain lengthwise in the barrel. Examples of barrel-polished parts are textile needles, small gears, cleats, clips, etc.

In vibratory tumbling, larger parts can be processed. As long as the part will fit into the barrel, surrounded by the medium, and the vibratory action is strong enough to create relative motion between the part and the medium, polishing can take place. Vibratory polishing also has the advantage of finishing holes, slots, and recesses better than barrel tumbling, which provides little abrasive action in such workpiece locations. Hence parts that require polishing in such recessed locations are candidates for vibratory polishing.

Parts not suitable for barrel polishing, in addition to those which are too large, are parts that are apt to become intertwined, like open-coil springs or parts with deep, nontapered, or undercut recesses that will trap media. Fragile and delicate parts may not be suitable either, particularly if conventional barrels rather than vibratory hoppers are used.

Electropolishing is used for parts as small as 1 cm^2 in area, such as miniature electrical contacts, rivets, and screws found in the electronics industry. It is used also for components as large as refrigerator doors with a polished surface area of 1 m^2 or more. The maximum-size workpiece processible depends on the size of tank available. A wide variety of household, commercial, and military products are electropolished. (The corrosion-inhibiting and surface-stress-removal characteristics of the process may be particularly valued in military or aircraft components.)

Economic Production Quantities

Mechanical Polishing and Buffing. These are normally manual operations, most suitable from a cost standpoint when production quantities or labor costs are low. Equipment and tooling costs are very modest, and any complexities in the operation are provided through the skill of the operator rather than through sophistication of tooling and equipment.

When volumes are high, buffing and polishing can be mechanized. Mechanized buffing equipment can be expensive, particularly if there are many surfaces or contours on the workpiece or if it is large.

Flat, cylindrical, or other simple surfaces are much more easily polished by automatic means than are multiple contoured or irregular surfaces. Belt polishing can cover large surfaces in each pass. Cylindrical parts can be automatically polished with an approach similar to centerless grinding but using abrasive belts instead of rigid grinding wheels. Costs for semiautomatic polishing equipment for high production can easily run to the high hundreds of thousands of dollars if conveyorized special holding fixtures and a large number of polishing heads are involved.

The major cost factors for conventional polishing and buffing are labor, equipment, wheels or belts, compound, and electric power. Direct labor is usually the largest item when hand methods are employed; equipment is the largest item for fully mechanized arrangements.

Barrel Polishing. This process requires a medium to high production quantity to amortize the cost of setting up the best processing conditions for each part. Considerable time and experimentation may be required to determine the optimum mix of parts and medium, which compounds and medium to use, the cycle time, and the need for any preliminary or subsequent operations. The cost of the engineering and setup time required to establish these processing conditions usually will be greater than one part's share of the depreciation of the basic equipment.

The cost of tumbling equipment is modest. Tooling is not required unless some portion of the workpiece must be shielded from the polishing medium or protected from impact with other parts. Typical unit costs for barrel polishing will have a distribution similar to the following:

	Percent
Depreciation of tumbling equipment	8
Electrical power	4
Polishing medium and compound	40
Direct labor	40
Other costs (maintenance, water, etc.)	8
Total	100

Typical cycle times range from $\frac{1}{4}$ to 24 h, with common times from $\frac{1}{2}$ to 2 h. Normal barrel loads may range from one piece to hundreds of thousands of pieces depending on barrel size and workpiece size. The volume of parts in the barrel is normally only 25 to 33 percent of the volume of the medium. In the case of fragile parts, it may be as small as 5 percent. Since the total volume of parts and medium normally occupies only 50 to 60 percent of the barrel's volume, it can be seen that the number of workpieces processed per barrel load can, in some cases, be rather small.

Electropolishing. This process does not normally require a tooling investment for any unique part. (Only if special racking or masking is required will there be a one-time cost for a particular part.) Hence the process is economical for all levels of production. Equipment costs are modest, being comparable with those required for electroplating. If production levels are high, the process can be mechanized fairly easily as well.

The process time for any one part ranges from about 3 to 10 min, plus the time required for cleaning, rinsing, and other preliminary and subsequent steps. Because no real setup is required, it is practical to electropolish only one piece at a time if necessary. Skill requirements for the operator are also minimal. The full manufacturing cost for electropolishing includes equipment depreciation, labor, chemicals, power, and maintenance.

Suitable Materials

Few limitations are imposed on the choice of materials by the need for polishing the workpiece. Since conventional polishing and buffing and barrel polishing are primarily abrasive cutting operations, any material that is suitable for abrasive machining is also easily polished by these processes. This, of course, covers virtually all metallic and rigid nonmetallic materials.

The best materials for polishing are those with a uniform, fine grain structure. Nonductile machinable metals are perhaps best. However, since part of the polishing and buffing process involves a burnishing of the surface with an accompanying displacement of surface material, ductile materials are also fully polishable.

The most difficult materials are those with nonuniform hardness caused by work hardening or other factors. Composite materials with a reinforcement of high hardness and a soft matrix are difficult to polish to a high surface smoothness because of different rates of cutting of the two materials. Thermoplastics are more critical to polish because of their tendency to soften or melt under the frictional heat of polishing. When barrel polishing is accomplished by means of a burnishing medium, a softer or more ductile material will give more rapid results. However, softer workpieces are more subject to knicks or dents. Electropolishing is also applicable to a diverse list of metallic materials but is not, of course, applicable to plastics, ceramics, or other nonconductive materials.

Fine-grained, homogeneous single-phase metal structures are best for electropolishing. Multiphase alloys are also suitable if the phases dissolve at approximately the same anodic potential. Unalloyed metals and alpha solid-solution alloys provide the best finishes. Aluminum die-casting alloys do not electropolish well because they contain alloying ingredients which do not dissolve in common commercial electrolytes. The same problem exists with free-machining brass alloys.

Some materials that are suitable for electropolishing are stainless steels 202 (S20200), 302 (S30200), 304 (S30400), 316 (S31600), 317 (S31700), 347 (S34700), 410 (S41000), 420 (S42000), 430 (S43000), 440 (S44000), AMS 5366 (J94211) and AMS 5648 (J31600); mild steels and low-alloy steels 4130 (G41300), 4140 (G41400), and similar formulations; aluminum alloys 3003 (A93003), 5457 (A95457), 5557 (G95557), and 6463 (A96463); brass alloys 95-5 (guilding, C21000), 70-30 (cartridge brass, C26000), 60-40 (Muntz metal, C28000), and 85-15 (red brass, C23000); zinc die-casting alloys; beryllium copper; nickel silver; molybdenum; Waspalloy (N07001); and tungsten.

Design Recommendations

For best results with conventional polishing and buffing, workpieces must be of such a shape that the polishing wheel or belt can contact all points of the surface uniformly without interference. The following are specific rules for the design of parts for wheel or belt polishing and buffing:

1. Mild contours are preferable to compound curves, inside or outside sharp corners, deep recesses, or other irregular shapes, especially if belt polishing is used. (See Fig. 8.2.1.)
2. Avoid parts with hooked edges or sharp projections, because these are likely to snag or cut the polishing wheel or belt. (See Fig. 8.2.2.)
3. Bosses, handles, and other obstructions to the free access of the wheel or belt to the surface to be polished should be moved out of the way or eliminated. (See Fig. 8.2.3.)
4. Providing a uniformly polished surface over a large area with a polishing wheel is difficult; it is preferable, if the design permits, to avoid large surfaces which require polishing.
5. Very small parts or those difficult to hold are not easily polished. Preferred designs are those with polishing on only one plane and with a means for easy holding by hand or fixture.

For barrel polishing, the following design guidelines should be observed:

1. Small holes, slots, or recesses are undesirable features of parts to be barrel-polished because they can trap pieces of the tumbling medium either directly or by "bridging." (See Fig. 8.2.4.)
2. Larger holes, recesses, or shielded areas, even if big enough to avoid trapping medium pieces, are not polished well in the barrel-polishing process because there is less opportunity for abrasive motion of the medium in such spaces. Designers should not specify a polished surface in these locations. (See Fig. 8.2.5.)

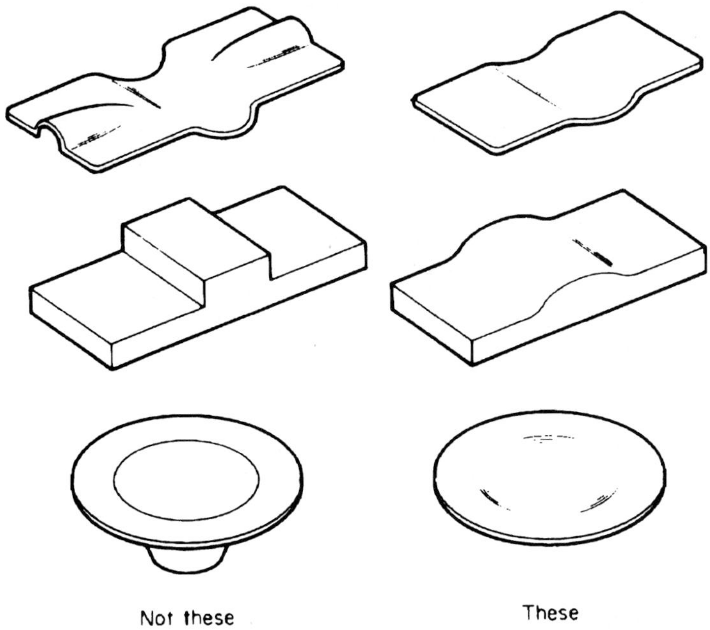

FIGURE 8.2.1 Polishing is facilitated when the design avoids deep recesses, irregular shapes, and sharp corners.

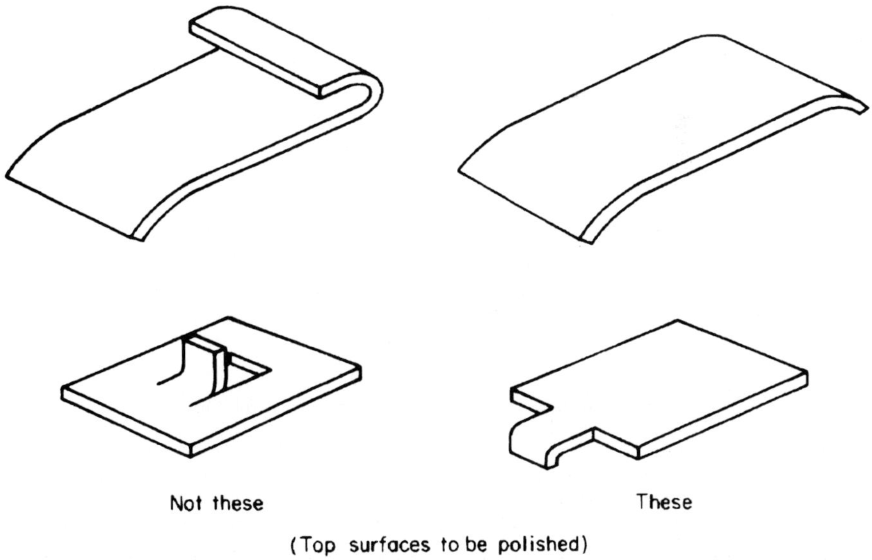

FIGURE 8.2.2 Avoid designs which incorporate hooked edges or sharp projections that are likely to snag the polishing wheel or belt.

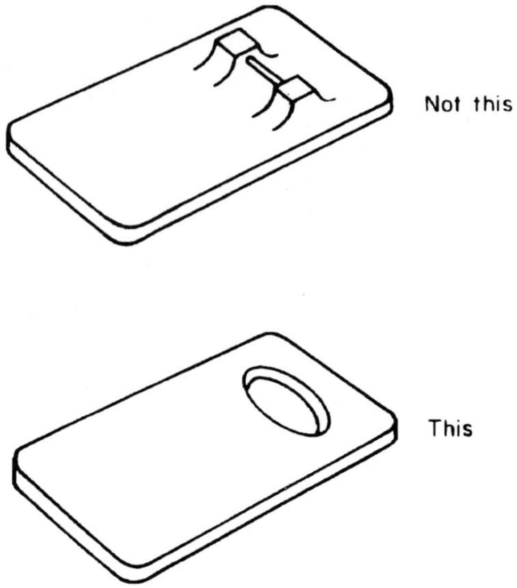

FIGURE 8.2.3 Bosses, handles, and other obstructions to the free access of the polishing wheel or belt should be avoided.

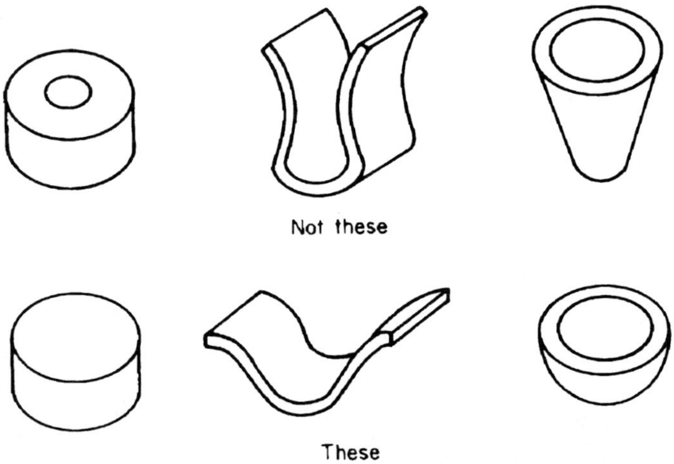

FIGURE 8.2.4 Small holes, slots, or recesses are undesirable in parts to be barrel-polished.

3. Large flat surfaces do not lend themselves to effective barrel polishing. If other methods cannot be employed, such surfaces should not be specified to be polished.

4. Be careful of designing parts that can interlock and tangle during the barrel-finishing operation. Springs and other wire or strip parts are susceptible to this.

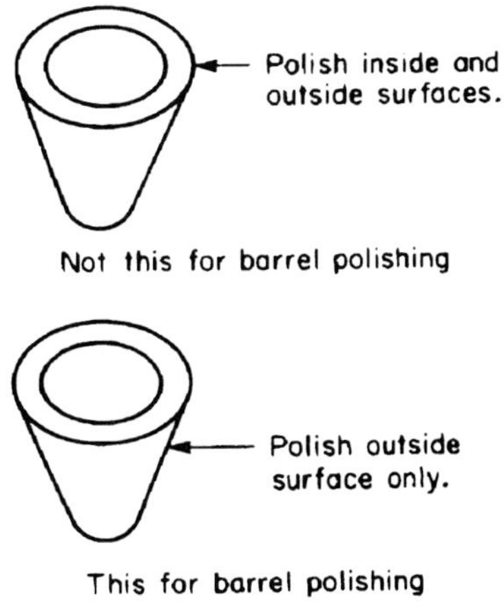

FIGURE 8.2.5 Interior surfaces normally do not get barrel-polished well. It is best not to specify that such surfaces have a polished finish.

Electropolished surfaces possess a brilliant luster and superior heat and light reflectance. The surface, however, has less mirror-like reflectance than conventionally polished and buffed surfaces and a slightly different appearance. Designers should not place electropolished and mechanically polished surfaces adjacent to one another in an assembly if uniform appearance is required.

Lesser polishing will occur in holes, recesses, and slots of workpieces being electropolished than in more prominent surfaces because the nature of the process is to attack predominantly the high spots. Designers should be aware of this circumstance and not specify as fine a polish in these recessed areas. If a fine finish on such surfaces is essential, specially shaped and placed electrodes can be used, but this, of course, is more costly.

In all polishing operations the designer and the process engineer should be aware of the stock removed by the operation. Although normally small, it can affect the accuracy of critical dimensions. The amount of stock that will be removed depends on the surface finish before and after the polishing operation. As a rule of thumb in electropolishing, the removal of 0.025 mm (0.001 in) in the 0.2- to 1.2-μm (8- to 50-μin) range reduces surface roughness by about one-half. Stock removal in barrel polishing is normally on the order of 5 μm (0.0002 in). In mechanical polishing it may be far more. While 0.05 mm (0.002 in) is a fair target figure, in heavy polishing 10 times this amount of stock may be removed. With ground surfaces receiving a final polish, the amount removed is far less.

Recommended Tolerances

All three polishing processes described in this subsection can, if necessary, produce surfaces with extremely high luster and smoothness. As with other operations, extreme values are achieved only at a much higher cost.

Table 8.2.1 tabulates the recommended surface finish to be specified with the three processes.

TABLE 8.2.1 Recommended Surface-Finish Values for Mechanically Polished, Barrel-Polished, and Electropolished Surfaces

Most economical	0.4 μm (16 μin)
Normal	0.2 μm (8 μin)
Finest	0.05 μm (2 μin)

PLATED SURFACES

Electroplating Process

Plating is a family of processes that are used to apply metallic coatings to parts. The parts are usually but not necessarily metallic. Actual procedures vary, depending on the material to be deposited and the substrate to be coated, but, in general, can be described as the transfer of metal from the anode (source of metal) to the cathode (part) in a solution medium using electrical energy as the moving force.

Thorough surface cleaning as described in Chap. 8.1 always precedes the plating operation. Thorough rinsing after cleaning and after plating is a standard procedure.

Processing can be accomplished by rack methods, as shown in Fig. 8.2.6, or by bulk methods (barrel plating), as shown in Fig. 8.2.7. Each method has advantages in certain situations, and selection is generally a matter of economics.

FIGURE 8.2.6 Automatic rack-plating equipment. (*Courtesy Progalvano s.n.c., Milan, Italy.*)

Sometimes, more than one metal may be applied in successive steps. Chemical conversion or other types of coatings may be applied over the electroplated metal.

FIGURE 8.2.7 Automatic barrel-plating equipment. (*Courtesy Progalvano s.n.c., Milan, Italy.*)

Typical Characteristics

Plating is used to impart desirable properties to parts or their surfaces: improved corrosion resistance, permitting use of base materials which would not be satisfactory without such protection, improved appearance, greater wear resistance, higher electrical conductivity and better electrical contact with other parts, greater surface smoothness, better light reflectance, and a more suitable base for bonding other materials, to change a part's dimensions or for a combination of these reasons.

Electroplating is used to coat parts with the simplest of shapes as well as those which are most complex. An example of the former is aluminum wire plated with copper to protect it from the environment and promote better electrical connections. At the other extreme, the B-70, an experimental supersonic aircraft, was completely coated with nickel to a thickness of 0.025 to 3 mm (0.001 to 0.125 in) to protect the exterior shell from the effects of the high speed.

While the size and shape of parts that can be produced are unlimited, each plating shop has equipment that limits the type of work it is able to perform satisfactorily. Excluding special setups, the parts must physically fit into processing tanks. At the other end of the spectrum, when bulk procedures are used, the parts must be large enough not to fall through the openings in the processing cylinders. Common parts processed range from automotive bumpers to needles and pins.

By specifying appropriate plating over selected substrates, almost any physical characteristics may be achieved. For example, automotive radiator grilles may be formed from steel, cast from zinc, or molded in polypropylene plastic. All these materials may be plated with conventional copper-nickel-chromium plating to achieve nearly identical appearance. The selection of the basis material is thus a factor of the forming requirements of part manufacture independent of appearance factors.

Printed-circuit boards, used in the electronics industry, are typically made by electroplating copper to a phenolic board in the pattern required by the specific circuitry. This copper is often overplated with tin as an aid to consistent soldering.

Typical electroplate thicknesses range from a few molecules to a normal maximum of 0.05 mm (0.002 in) for common work, with thinner coatings being far more usual. The actual thickness of the coating is a controllable variable and may be increased or decreased according to product requirements.

While the statement "Anything can be electroplated" is literally true, many families of parts require special processing procedures. If a particular part is unwieldy or has unusual plating requirements, the cost may be quite high unless a shop equipped to handle these unusual situations can be found.

Economic Production Quantities

While it is true that electroplating can be used for quantities ranging from one to millions, the method of processing often will be affected by the total quantity to be processed and the rate at which the job must be completed. For very small quantities, the parts would most typically be racked or wired onto fixtures ("racks") for processing. Movement of racks between cleaning, rinsing, and plating tanks would be manual. As production requirements increase, racks specially designed for the part become economically feasible. Such racks increase the ability of the racker to attach the parts without reducing plating capability, and they actually improve it in some cases. With increased production, automatic movement of racks between tanks, as shown in Fig. 8.2.6, would be used.

Bulk work involves small parts, typically less than 90 g (0.2 lb) per part. This kind of work normally does require a minimum quantity for processing. In almost any shop of this type, 100 lb is adequate for economic production. Larger lots, however, certainly improve the economic aspects of the procedure. Automatic handling of barrels between tanks, as shown in Fig. 8.2.7, is common in shops equipped for high-quantity production.

Suitable Materials

While any material can be electroplated, production applications are normally limited to metals and some specific plastics. Since plating is normally accomplished by transfer of electrical energy, the part to be plated must be conductive or made conductive in order to be electroplated. From this consideration, it becomes obvious that metallic parts are simplest to process. Plastics require preplating to generate a conductive surface prior to plating. Only certain plastics may be plated without resorting to conductive paints or other stopgap methods. The most commonly plated plastics are acrylonitrile butadiene styrene (ABS), chlorinated polyethylene (PEC), and ABS/

polycarbonate. Polyphenylene oxide, polysulfone, polypropylene, nylon, polyester, polyacetal, polystyrene, and polycarbonate are also plated. Other plastics such as epoxy/glass are plated for printed-circuit boards. Plastic, when used as a plating substrate, does not typically retain flexibility after plating, but it does offer lighter weight and a reasonable degree of impact resistance.

Table 8.2.2 indicates combinations of basis materials and plating metals commonly used. This table is representative of frequent processing and is not an attempt to catalog all possible applications.

Luster is a function of the smoothness of the surface to be plated. Materials such as hot-rolled carbon steel will appear rough and often dull after plating. By appropriate preplate mechanical finishing, smoothness and luster can be enhanced. Cold-rolled carbon steel, which typically has a higher-quality surface, generally will require less preparation for equivalent final results.

Aluminum is more difficult to process than either steel or zinc and is a specialized segment of the plating industry. Because plating reflects the condition of the basis material, the appearance of plated-aluminum castings is often not as nearly perfect as more easily processed materials such as zinc.

Design Recommendations

Surface Finish Prior to Plating. Electroplated coatings follow the contours of the basis material with a high degree of fidelity. Rough or discontinuous surfaces on the unplated part will be reflected in the final product. There are available plating solutions, termed *high leveling,* with enhanced ability to cover scratches, pits, or rough spots, but this ability is relative, and the best possible surface that can be achieved economically for product requirements should be the goal of the preplating finisher (i.e., the polisher or buffer).

Area to Be Plated. Many parts do not require plating coverage over their entire surface area. For platers to do their jobs adequately, designers must specify which areas are critical and which do not require coverage. Platers can then proceed accordingly, and if the partially plated part is large or unwieldy, savings may be very significant.

Flat Surfaces. Surfaces that are totally flat will appear hollow after electroplating. However, by using a technique known as *crowning,* the specification of a positive radius in the shortest direction across the face of such nominally flat surfaces, this hollowness can be eliminated. The rise is generally about 0.1 mm/cm (0.010 in/in) of width. This curvature not only eliminates the hollow appearance but hides tool marks and other defects in the surface of the part as well and is generally preferred by the toolmaker.

Inside Corners. The larger the inside radius can be, the better is the plating condition. For small surfaces, such as the bottom of a groove, full radiusing is best. As the groove or step becomes larger, a radius of 0.5 to 1.0 mm (0.020 to 0.040 in) is appropriate. These radii will permit adequate electroplate deposit in the corner to prevent subsequent corrosion or the appearance of misplating.

Outside Corners. Electroplating will build up on sharp corners to a point where nodules form. (This is called *treeing* and is shown in Fig. 8.2.8.) The use of generous radii on such corners will minimize this effect.

TABLE 8.2.2 Commonly Used Combinations of Basis Materials and Plating Metals*

	Basis material										
Plate coating	Carbon steel	Zinc	Stainless steel	Aluminum	Copper	Brass	Nylon	Polyphenylene oxide	Polypropylene	Bronze	Alloy steels
Cadmium	1,2										1,2
Zinc	1,2										1,2
Brass	2	2			2						2
Nickel	1,2		2		1,2	1,2				2	1,2
Copper	6										6
Tin	1,2,5,6				6	6					1,2,5,6
Chromium	4	4									4
Silver					2	2				2	
Gold					2	2				2	
Bronze	1,2	2			1,2						1,2
Copper-nickel	1,2	1,2					2	2	2		
Copper-nickel-chromium	1,2,3	1,2,3	2,3	1,2,3	1,2,3	1,2,3	2,3,4	2,3	2,3,4	2,3	1,2,3

*1 = corrosion protection; 2 = appearance; 3 = tarnish resistance; 4 = physical properties; 5 = in-process protection; 6 = simplication of subsequent processing.

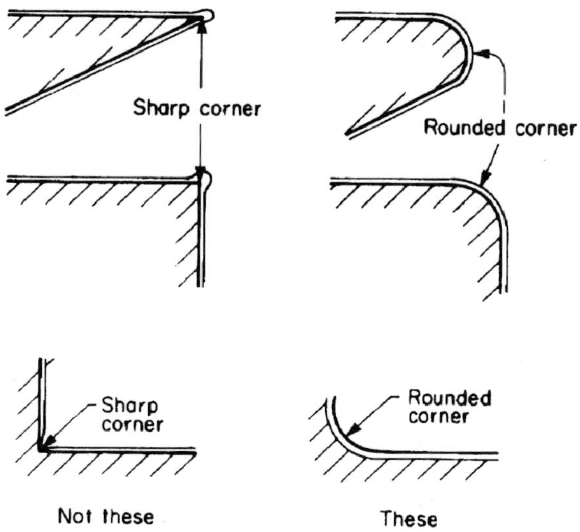

FIGURE 8.2.8 External sharp corners receive excess plating deposit, whereas sharp internal corners receive insufficient deposit; large radii promote a more uniform plating thickness.

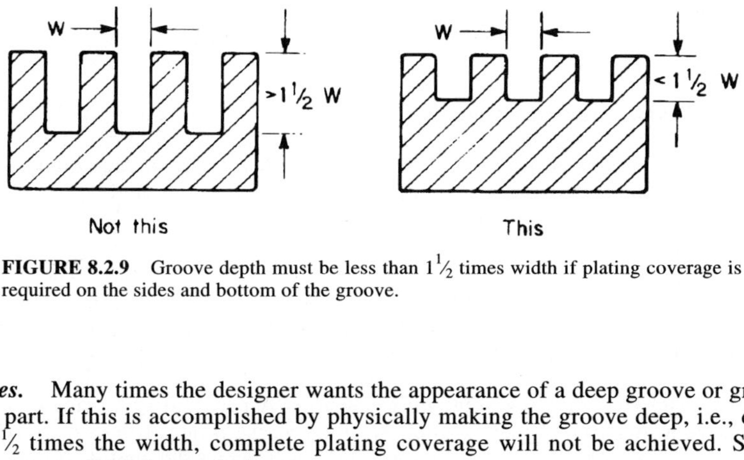

FIGURE 8.2.9 Groove depth must be less than 1½ times width if plating coverage is required on the sides and bottom of the groove.

Grooves. Many times the designer wants the appearance of a deep groove or grooves on the part. If this is accomplished by physically making the groove deep, i.e., deeper than 1½ times the width, complete plating coverage will not be achieved. Several options are available to the designer:

1. Reduce the depth of the groove, as shown in Fig. 8.2.9.
2. After plating, paint the groove with dull black paint. After painting, it is almost impossible to determine visually the depth of the groove, and this permits depth reduction to 1½ mm ($^1/_{16}$ in) or so with the appearance of a very deep groove.
3. Use a through slot instead of a groove. If in the final product there is another surface behind the part, this often will permit a significant improvement in platability.

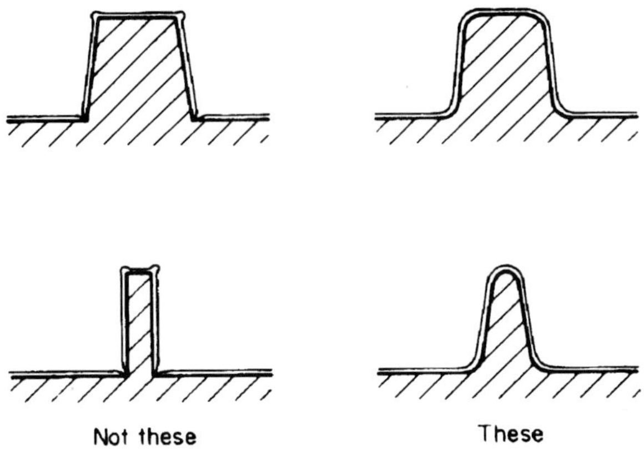

FIGURE 8.2.10 Bosses, ribs, and ridges should be well rounded to avoid excessive plating buildup.

Bosses and Ridges. To avoid excessive plating buildup on outside corners, radii must be used. For decorative designs, such radii are more pleasing to the eye than the sterile sharpness of "razor edges" and square corners. They are also less subject to physical damage during fabrication and use. (See Fig. 8.2.10.)

Pockets. Pockets can be thought of as stopped grooves. The 1:1½ depth-to-width ratio restriction should be followed, with the maximum radii permitted by the design. If greater depths are required, subsequent painting or a multiple-piece assembly can be used to reduce effectively the actual depth of the pocket in the plating process.

Holes and Tubes. The best design for holes for screws uses countersink screws (to avoid a sharp edge at the hole) and a minimum depth of hole. Often, the minimum depth must be disregarded because of the physical fit of the part. A common solution to this problem in die castings is to form a tube for the screw to guide it to the attachment point. After plating, the interior of this tube will not be coated with electroplated metal. This would not matter except that a chemical attack of the basis metal occurs during processing. The gas formed during this attack causes lines or misplate on the face of the part if it is permitted to exhaust onto those surfaces. By slotting the tube, the production of such gas is reduced, and the likelihood of the formation of the misplate lines is reduced. (See Fig. 8.2.11.)

Cupping. The electroplating solution is valuable. A part that is designed so that it will remove a quantity of that solution with each insertion not only increases cost unnecessarily but contaminates the solution and thus reduces the quality of the plating. A drain hole permits the solution (or air if the part is in the solution with its open top down) to drain from the part. Such holes also improve overall plating in other ways.

Other Considerations. To be bulk-plated, parts must be free-rolling, i.e., must not lock together. For example, if nuts and bolts are plated together, some of them will "self-assemble" in the processing and often be locked or plated together so that disassembly is difficult or impossible. For parts for which this specialized processing is not appropriate, racking or fixturing is commonly used to keep parts separated during the

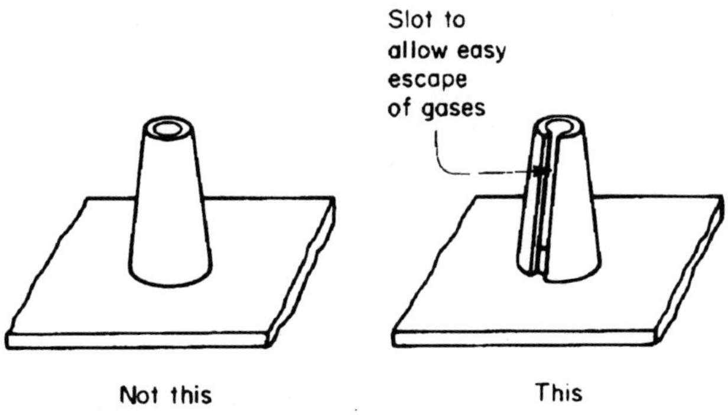

FIGURE 8.2.11 Screw-holding bosses in die castings should have a slot in the sidewall to allow gases formed from the chemical action of the plating solution to escape easily. Otherwise, misplate lines may appear on adjacent surfaces.

process steps. Then means must be provided on the parts so that they can be hung on the plating rack. Holes, attachment studs, spring contacts, or tabs that are required for assembly of parts often can be used for these connections. Parts will not be plated in the contact zone, however, and this must be kept in mind when planning the rack-point location. (See Fig. 8.1.2.)

Alternatives. For some components, cost and quality can be optimized by using preplated stock. This is material that is plated prior to fabrication of the part. Galvanized steel is a common example of this approach, but chromium and other plated materials are available. Obviously, the edges of parts made from such materials will not be plated, but by appropriate design the edges often can be hidden and made noncritical.

Dimensional Considerations

Electroplating is usually not of sufficient thickness to cause assembly problems because of changes in size. Exceptions include parts that are constructed to extremely close tolerance or for which, because of sharp edges or corners, thicker-than-normal coatings are produced. The most common difficulty encountered in this area is with threaded fasteners. Because plating covers the two opposite sides of a threaded part and because the flank angle of the threads magnifies the diametral effect of surface coatings, the dimensional effect of plating on pitch diameter is about 4 times the thickness of the plating on one surface. Therefore, if a normal plating thickness of 0.005 mm (0.0002 in) were applied to a screw thread, the pitch diameter would change by 0.02 mm (0.0008 in), often enough to cause a precision threaded part to fail to fit the "go" gauge.

Plating thickness can vary markedly over the area of a workpiece, particularly if sharp corners and recesses are involved. For this reason, it is preferable to specify plating thickness in terms of minimum values necessary to achieve the desired results rather than average values.

Table 8.2.3 presents recommended minimum electroplating thickness for nickel or chromium finishes on steel, iron, and zinc products.

TABLE 8.2.3 Recommended Minimum Thicknesses for Nickel or Chromium Finishes on Steel, Iron, and Zinc Products

Service conditions	Minimum thickness: Nickel	Minimum thickness: Chromium	Typical applications
Exposure indoors in normally warm, dry atmospheres with coating subject to minimum wear or abrasion	10 μm (0.0004 in)	0.13–0.8 μm (0.000005–0.00003 in)	Toaster bodies, rotisseries, waffle bakers, oven doors and liners, interior automobile hardware, trim for major appliances, hair dryers, fans, inexpensive utensils, coat and luggage racks, standing ashtrays, interior trash containers, light fixtures
Moderate: exposure indoors in places where condensation of moisture may occur, for example, in kitchens and bathrooms	15–20 μm (0.0006–0.0008 in)	0.25–8 μm (0.00001–0.00032 in)	Steel and iron: stove tops, oven liners, home, office, and school furniture, bar stools, golf-club shafts Zinc alloys: bathroom accessories, cabinet hardware

Source: From *Quality Metal Finishing Guide,* Metal Finishing Suppliers' Association.

Electroless Plating

Electroless and immersion plating are processes that do not use external electric current. The most significant commercial application of these processes is electroless plating of nickel. Copper also is commercially plated by this method.

Electroless plating utilizes a chemical-reducing-agent substitute for electric current in reducing a metallic salt. The metal is then deposited on a catalytic surface of the workpiece. In the case of electroless nickel plating, the salt is nickel chloride, and the reducing agent is sodium hypophosphate. The nickel is in an aqueous solution heated above 70°C. The workpiece surface is catalytic, as is the nickel plating. Hence the process can continue indefinitely. The plated deposit is an alloy with from 4 to 12 percent phosphorus.

Electroless plating is used when it is either not possible or impractical to use normal electroplating because of physical factors or special properties needed. It provides a uniform thickness of coating even on the inside of tubes, recesses, and cavities not easily electroplated.

Electroless nickel is uniformly hard to a diamond-pattern hardness (DPH) of 425 to 575. It can be heat-treated to 1000 DPH. Both values are significantly higher than those of pure nickel as electrolytically deposited. Electroless nickel and copper plating can be applied to plastics, ceramics, and other nonconductive base materials if these are properly pretreated.

Parts are plated electrolytically for four purposes:

1. *For corrosion resistance.* Electroless nickel plating has markedly better corrosion resistance than standard electrolytic nickel plating. It also provides uniform coverage in recessed areas that normally would receive little plated material.
2. *For wear resistance.* Electroless nickel is desirable for this application because of its high hardness and its superior adherence.

3. *For surface buildup.* Plating thickness can be as much as 0.15 mm (0.006 in) or more.
4. *For electrical conductivity.* Electroless copper is used for this purpose on circuit boards, plastics, and ceramics parts. Coating thickness is as little as 0.5 μm (20 μin) and is followed by normal copper electroplating.

Another metal that can be plated nonelectrolytically is cobalt. Palladium and gold also can be plated by electroless deposition, but this is not common.

Table 8.2.4 lists base metals in order of decreasing suitability for electroless nickel plating.

TABLE 8.2.4 Suitability of Base Metals for Electroless Nickel Plating

Best	Nickel, cobalt, palladium
Good	Aluminum, low-alloy steel
Fair	Requiring pretreatment first: copper, glass, ceramics, plastics (ABS, polypropylene, and polyphenylene oxide best)
Not suitable	Requiring copper plating first: lead, tin, zinc, cadmium, antimony, arsenic, molybdenum

Electroless plating is considerably more costly than electroplating: about 50 percent more on the average for nickel. Deposition rates are much slower, and chemical costs are higher. Equipment costs, however, are lower for electroless plating. Despite these cost differences, the choice between the two processes is a result of physical rather than cost factors or production quantities.

Electroless copper is significantly more expensive than electrolytically deposited copper owing to high chemical costs and very slow deposition rates. Designers should not specify electroless plating unless the application demands the special properties obtainable with the process because the cost is higher than that of conventional plating. If welding is required after plating or if brittleness of the plating would be disadvantageous, it is better not to specify electroless nickel but rather to stay with conventional electrolytic nickel deposition.

Plating thickness varies less with electroless plating than with the electrolytic process, in that the coating is much more uniform throughout the surface of the part. If dimensional accuracy of the plated part is critical, electroless plating should be specified.

Table 8.2.5 gives suggested thickness limits for various types of electroless plating.

TABLE 8.2.5 Suggested Plating-Thickness Limits for Electroless Plating

Application	Thickness limits
Preplating of copper on circuit boards, plastics, or ceramics	0.5–1.0 μm (20–40 μin)
Nickel applied for corrosion resistance (relatively mild environments)	25–50 μm (0.001–0.002 in)
Nickel applied for corrosion resistance (relatively severe environments)	50–100 μm (0.002–0.004 in)
Nickel applied for wear resistance	5–50 μm (0.0002–0.002 in) as required
Nickel applied to rebuild worn areas	10–150 μm (0.0004–0.006 in) as required

CHAPTER 8.3

OTHER METALLIC COATINGS

PART 1

HOT-DIP METALLIC COATINGS

Daryl E. Tonini

Manager of Technical Services
American Hot Dip Galvanizers Association, Inc.
Washington, D.C.

THE PROCESS

Hot-dip metallic coatings are applied by immersing the workpiece in a bath of molten metal. Typical of all these processes and most prevalent is that for applying molten zinc to fabricated steel.

The first step involves cleaning (See Chap. 8.1). Rust, scale, oil, paint, and other surface contaminants must be removed by a suitable preliminary treatment. Caustic-bath degreasing followed by chemical cleaning in an acid bath is common. Ferrous castings are often shot-blasted and pickled.

Before entering the molten-zinc bath, the material is given a fluxing treatment by immersing it in an aqueous-flux solution, by maintaining a flux blanket on the bath, or by a combination of both techniques.

Coating involves immersion in the molten metal for periods ranging from a fraction of a minute for small objects such as fasteners to several minutes for structural shapes. The bath temperature for galvanizing (zinc coating) is about 450°C (842°F), and for aluminum approximately 700°C (1290°F). When an object that exceeds the size of the molten-metal bath is to be coated, a technique known as *double dipping* may be employed if the item cannot be redesigned to fit into the bath in a single dipping operation.

Postcoating treatment varies according to the needs of the coated component. Slow cooling, quenching, conversion coating, and painting (preceded by prepainting treatments) may be employed.

Figures 8.3.1, 8.3.2, and 8.3.3 illustrate hot-dip-coating operations for steel structural members.

CHARACTERISTICS

When properly selected, hot-dip metallic coatings provide excellent long-term corrosion protection for the coated object. They provide this in two distinct ways: (1) The coating functions as a barrier that shields the base material. (2) The coating may provide sacrificial galvanic protection. Coatings of metals higher in the electromotive series than the basis metal will corrode in preference to the basis metal. Protection will be provided even if there is a break in the continuity of the coating. Figure 8.3.4 lists common metals in accordance with their places in the electromotive series.

For atmospheric corrosion, zinc exhibits a linear relationship between weight of coating and both expected service life and severity of the environment. (See Fig. 8.3.5.)

In general, the size of objects that can be hot-dip-coated is limited by the size of melting pot that is available. Hot-dip-tinning operations are normally confined to

FIGURE 8.3.1 The galvanizing process. Caustic cleaning to remove dirt and organic contaminants is a first step. (*Courtesy St. Joe Lead Company.*)

FIGURE 8.3.2 In galvanizing, the steel is immersed in molten zinc covered by a flux layer that ensures a chemically clean surface. (*Courtesy St. Joe Lead Company.*)

FIGURE 8.3.3 The galvanized steel is withdrawn from the clean molten-zinc bath. The zinc layer is now metallurgically bonded to the steel. (*Courtesy St. Joe Lead Company.*)

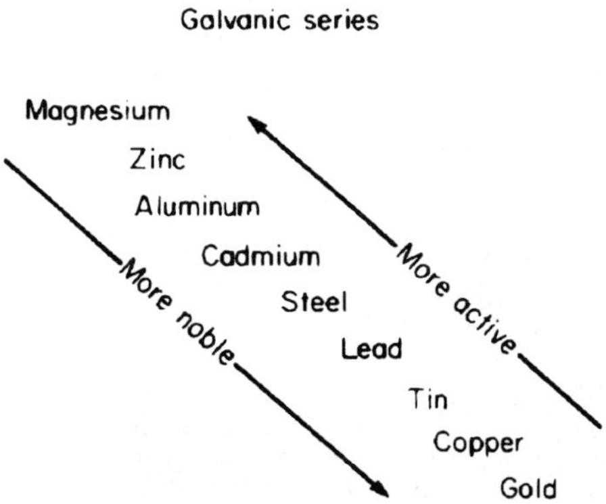

FIGURE 8.3.4 An abbreviated galvanic series ranking metals according to their tendency (activity) to enter solution.

small pieces. At the other end of the scale, galvanizing kettles approaching 18 m (60 ft) in length are in service.

Coating thickness varies by process as well as by coating type. In the case of hot-dip-galvanized material coating, thickness is customarily specified in grams per square meter or ounces per square foot as either a minimum or an average coating weight. Normal values range from about 150 to 920 g/m^2 (0.5 to 3 oz/ft^2). This is equivalent to a thickness of 0.01 to 0.13 mm (0.0004 to 0.005 in).

Figure 8.3.6 illustrates typical components protected by hot-dip coatings.

ECONOMIC PRODUCTION QUANTITIES

Depending on the degree of material-handling sophistication in the hot-dip-coating plant, hot-dip-applied coatings can be applied on either a short-run or a mass-produced basis. In judging whether the hot-dip process is economical, engineers should

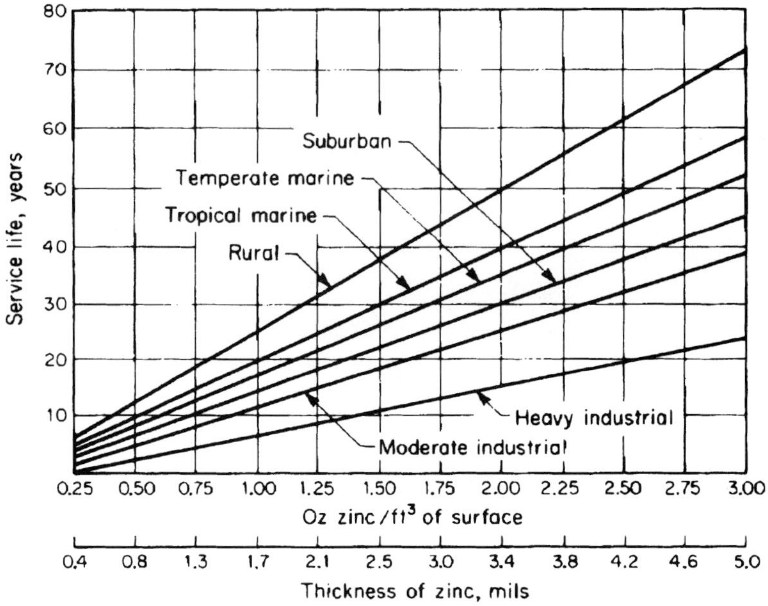

FIGURE 8.3.5 Anticipated life of zinc coatings in the atmosphere. (*Courtesy American Hot Dip Galvanizers Association.*)

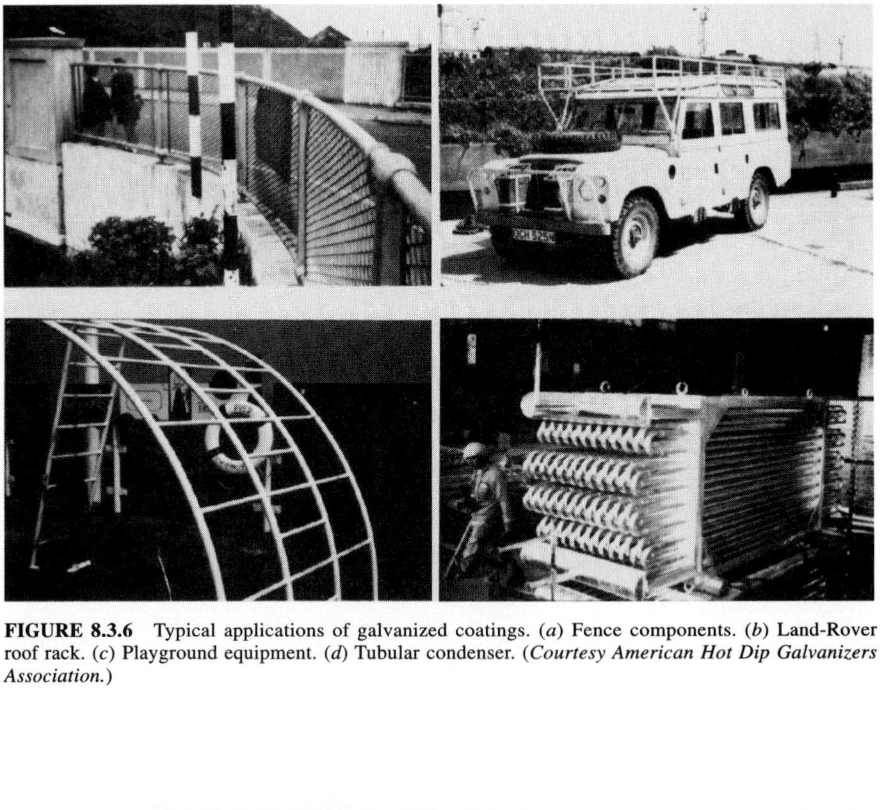

FIGURE 8.3.6 Typical applications of galvanized coatings. (*a*) Fence components. (*b*) Land-Rover roof rack. (*c*) Playground equipment. (*d*) Tubular condenser. (*Courtesy American Hot Dip Galvanizers Association.*)

keep in mind that coaters normally use a throughput of tons of material per hour in their kettles as their measure of equipment utilization. A smaller kettle operation may operate on a completely different economic basis than a larger kettle operation.

Thus, for short runs of small objects, a rule of thumb would be to investigate costs with a hot-dip-coating applicator having an efficient operation and a kettle suited to coating the object in question. However, it must be noted that economical production quantities will vary from coater to coater, as will costs to handle small lots.

SUITABLE MATERIALS

Substrates

By far the most common substrate materials for hot-dip-applied metallic coating are cast iron and steel. In some specialized situations, materials such as high-strength low-alloy steel also may be used.

In selecting a steel substrate material, it is most beneficial if the designer keeps the following considerations in mind:

1. Steels containing carbon below 0.25 percent, phosphorus below 0.05 percent, and manganese below 1.35 percent, either individually or in combination, are normally most suitable and will usually develop a "typical" galvanized coating when conventional galvanizing techniques are used, provided the silicon is 0.05 percent or less or between 0.15 and 0.30 percent.
2. Combinations of different steel materials and different manufacturing processes (e.g., welded, cast, or hot-rolled) may produce coatings of different thicknesses and with a variable surface appearance. While this may not be detrimental to the corrosion performance of the coating for many applications, the designer and fabricator should be aware of this circumstance in advance.
3. The effect of the heat of the molten bath on the substrate material should be considered. If the steel is susceptible to strain-age hardening owing to being severely cold-worked, it may be brittle after it has been coated because of accelerated aging. Very often, strain-age embrittlement is an indication that the material was incorrectly chosen or that it was subjected to extreme fabrication conditions. In this regard, it is best to select steels having a low transition temperature.

Coatings

Zinc, aluminum, tin, lead, and terne (lead alloyed with 10 to 20 percent tin) are commonly used for hot-dip coatings. Generally, zinc, aluminum, and terne perform well in providing corrosion protection under most atmospheric conditions. For other service conditions such as are encountered in soils or aqueous media, the choice can involve complex technical considerations.

Aluminum is considered to be resistant to distilled and ordinary fresh waters, hard as well as soft, under most conditions. However, although hot-dip aluminum coatings can be applied provided proper care is taken, they are generally not suited for fabricated objects and job-shop operations. This is due to the formation of a tenacious film of aluminum oxide which interferes with the iron-aluminum interface during the dipping operation.

Lead and tin do not provide galvanic protection, but lead coatings may be very resistant to soils which are highly aggressive to zinc coatings. Lead has good resistance to sulfuric and hydrochloric acids, brines, etc. However, owing to its tendency to coat unevenly and to pit, it is not used extensively as a coating for steel in the atmosphere or in most soils.

Alloying lead with other elements such as tin, antimony, cadmium, mercury, and arsenic produces coatings that have been used with success. Terne coatings are applied in much the same manner as are hot-dip tin coatings.

DESIGN RECOMMENDATIONS

Access and Drainage of Molten Metal

To achieve complete corrosion protection, the molten coating metal must be able to flow freely to all surfaces of the part. Coating hollow sections greatly reduces the possibility of hidden corrosion at some later date. Some general principles to ensure full, uniform coverage are the following:

1. Holes for venting and draining should be as large as possible. Full cutouts are preferred for pipe assemblies and base plates.
2. Holes for venting and draining should be diagonally opposite at the high and low points of the fabrication as it is suspended for the coating process.
3. Internal and external stiffeners, baffles, diaphragms, and gussets should have their corners cropped to aid the free flow of molten metal.

Figures 8.3.7 through 8.3.9 illustrate these guidelines.

Minimizing Distortion

Because of the temperatures involved in the coating process, there can be stress relief, which can result in distortion. To minimize distortion, design engineers should observe the following recommendations:

1. When possible, use symmetrical rolled sections in preference to angle or channel frames.
2. Specify the minimum welds possible in order to reduce thermal stresses.
3. When bending component members, use the largest acceptable radii, not less than 3 times material thickness.
4. When possible, build assemblies and subassemblies in units suitable for single dipping.
5. The extent to which an assembly will distort is largely dependent on the design and can be appreciably reduced by avoiding combinations of extremes in weight and cross section of structural members such as shown in Fig. 8.3.10.

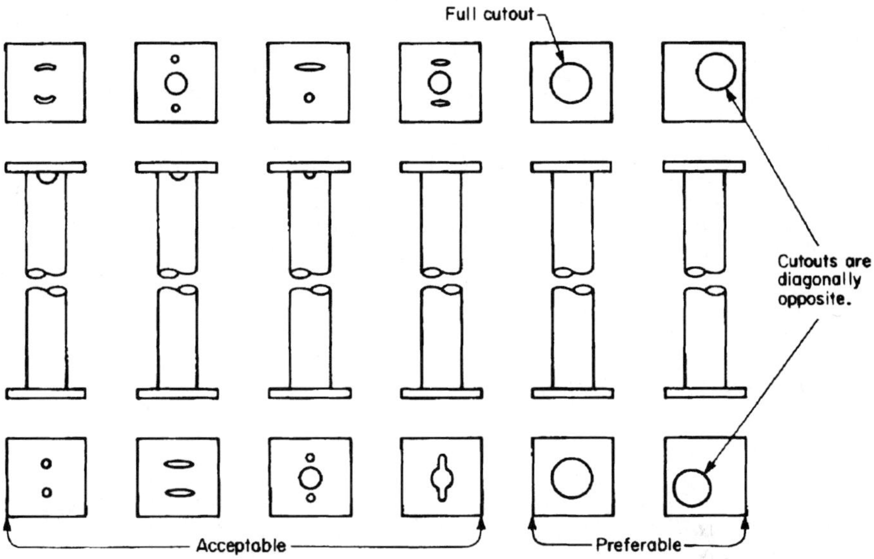

FIGURE 8.3.7 Pipe-column venting. Any of the possible venting methods shown is acceptable. Vent holes should be off center, as close to the inside wall of the pipe as possible, as large as possible, and diagonally opposite the ends of the pipe plates.

Screw Threads

When the product to be hot-dip-coated includes threaded members, the pitch diameter must be adjusted to allow for the thickness of the coating. Recommended oversize tappings for female threads to be used with galvanized bolts are given in Table 8.3.1. On threads greater than 40 mm (1½ in) in diameter, it is often more practical, design and load requirements permitting, to cut the male thread 0.8 mm (0.032 in) undersize so that a standard tap can be used on the nut. Threads on long rods and faces of flanges may be brushed while hot. This reduces the coating thickness but allows a better mechanical fit. Small threaded parts can be centrifuged to remove excess zinc and produce smoother coatings.

Postcoating Treatments

If designers anticipate that galvanized components will be subjected to conditions conducive to wet-surface stain (e.g., prolonged stacking in a moist or wet environment), they should specify a protective postcoating treatment. One of the most effective of such treatments involves immersion of the workpiece in a solution of sodium dichromate. Galvanized material also can be painted if the proper pretreatment steps are carried out.

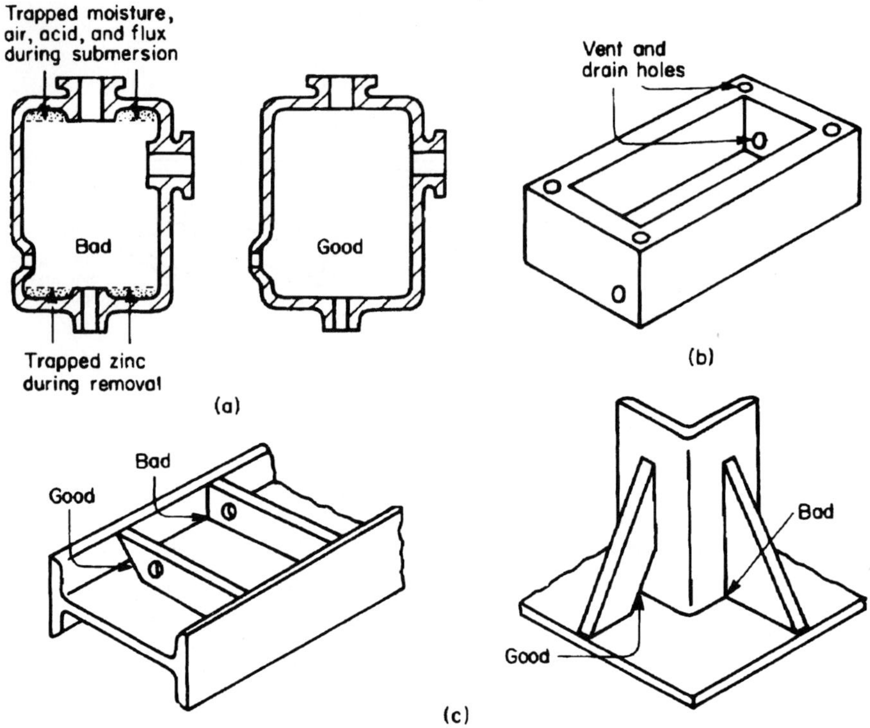

FIGURE 8.3.8 Venting and draining of enclosed and semienclosed products. (*a*) Tank design should always provide for flush interiors to prevent trapped moisture, air, acid, and flux during submersion and trapped zinc during removal. (*b*) When a housing design requires return flanges, provide drain holes in all corners and through holes for venting in at least two opposite surfaces. (*c*) When gussets and stiffeners are used, generous clipping provides for free drainage.

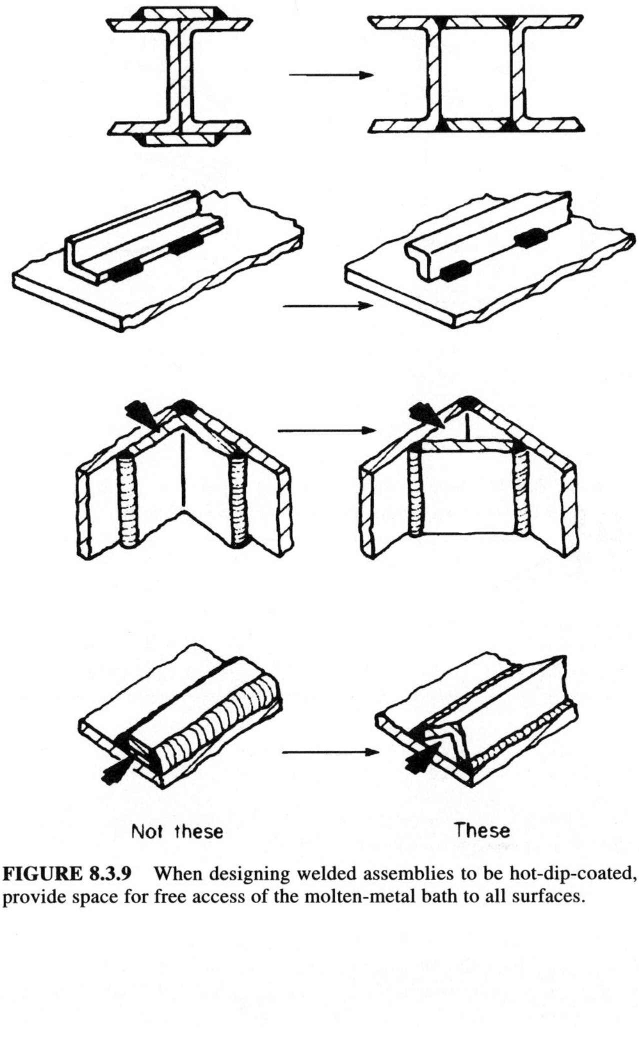

FIGURE 8.3.9 When designing welded assemblies to be hot-dip-coated, provide space for free access of the molten-metal bath to all surfaces.

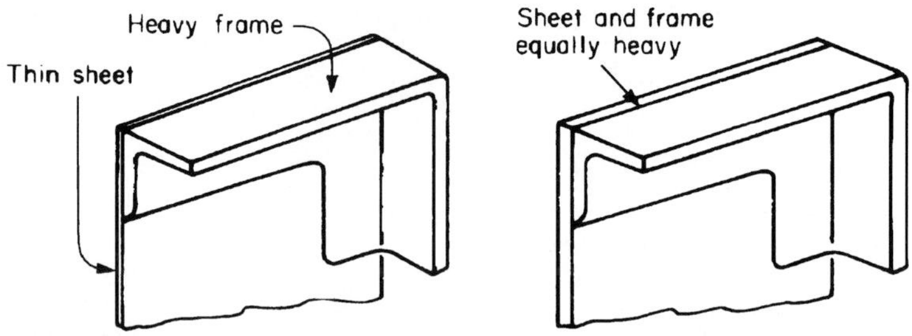

FIGURE 8.3.10 To avoid distortion, avoid a combination of extremes in the weight and cross section of structural members.

TABLE 8.3.1 Recommended Oversize Tapping for Threaded Holes to Be Used with Galvanized Bolts (ASTM A563)

Bolt or stud diameter, mm (in)	Minimum oversize tapping required, mm (in)
11 (7/16) and smaller	+0.4 (+0.015)
Over 11 (7/16) to 25 (1)	+0.5 (+0.020)
Over 25 (1)	+0.8 (+0.032)

Coating-Thickness Recommendations

Table 8.3.2 presents ASTM specifications for zinc-coating-weight thickness in job coating applications. Normal minimal coating thicknesses for precoated sheet run from 0.05 to 0.08 mm (0.002 to 0.003 in) for aluminum, 0.006 to 0.026 mm (0.0002 to 0.001 in) for terne, and 0.007 to 0.043 mm (0.0001 to 0.0017 in) for zinc.

TABLE 8.3.2 Recommended Coating Weight for Hot-Dip-Galvanized Products (ASTM Specifications)

Class of material	ASTM specification	Minimum weight of zinc coating of any individual specimen, g/m² (oz/ft²)
Products fabricated from rolled, pressed, and forged steel shapes, plates, bars, and strip	A123	
3–5 mm (1/8–3/16 in) thick		550 (1.80)
6 mm (1/4 in) and heavier		610 (2.00)
Hardware	A153*	
Class A: castings—gray iron, malleable iron, steel		550 (1.80)
Class B: rolled, pressed, and forged articles (except those that would be included under Classes C and D)		
Class B-1: 5 mm (3/16 in) and over in thickness and over 200 mm (8 in) in length		550 (1.80)
Class B-2: under 5 mm (3/16 in) in thickness and over 200 mm (8 in) in length		380 (1.25)
Class B-3: 200 mm (8 in) and under in length and any thickness		340 (1.10)
Class C: bolts and drivescrews over 9.5 mm (3/8 in) in diameter and similar articles; washers 5 and 6 mm (3/16 and 1/4 in) thick		305 (1.00)
Class D: screws, stove bolts, and bolts 9.5 mm (3/8 in) and under in diameter, rivets, nails, and similar articles; washers under 5 mm (3/16 in) thick		260 (0.85)
Assemblies	A386	
Class A: castings—gray iron, malleable iron, steel		550 (1.80)
Class B: rolled, pressed, and forged steel		
Class B-1: 5 mm (3/16 in) and over in thickness		550 (1.80)
Class B-2: under 5 mm (3/16 in) in thickness		380 (1.25)
Class C: bolts and drivescrews over 9.5 mm (3/8 in) in diameter and similar articles		305 (1.00)
Class D: bolts and drivescrews 9.5 mm 3/8 in) and under in diameter and similar articles		260 (0.85)
High-strength bolts	A153, Class C	305 (1.00)
Tower bolts	A394	305 (1.00)

*Length of the piece, stated in Classes B-1, B-2, and B-3, refers to the overall dimension and not to its developed length; 1 oz zinc/ft² surface corresponds to coating thickness of 0.043 mm (0.0017 in).

PART 2

THERMAL-SPRAYED COATINGS

F. N. Longo

F. N. Longo Associates
East Northport, New York

THERMAL-SPRAYING PROCESS

In the thermal-spray-coating process (sometimes called *flame-spray-coating process*) a gun is used which, with either a combustion flame, a plasma arc, or an electric arc, melts and propels metal, ceramic, or other material in finely divided form toward the article to be coated. The particles strike the object and flow into thin, flat droplets that interlock and overlap as they solidify. Multiple passes are usually necessary to build the coating to the desired thickness.

Wire metallizing is the oldest and best-established thermal-spraying method. Wire is fed axially through the gun and through a ring of flame (oxygen and acetylene, propane, or other fuel gas), which melts the wire tip. An outer sheath of compressed air "atomizes" the molten metal to droplets roughly 10 to 100 μm in diameter and propels these to velocities ranging from 91 m/s (300 ft/s) to 240 m/s (800 ft/s). The droplets strike the substrate at about 10 cm (4 in) to 25 cm (10 in) from the gun and flatten out while freezing, thus building up a coating.

Powder is sprayed similarly with a combustion flame-spray gun. It is carried from a hopper and fed axially through the ring of flame. The flame accelerates the molten particles to about 90 m/s (300 ft/s).

In the plasma arc-spray gun, a direct-current arc similar to a welding arc heats a gas (argon or nitrogen in combination with hydrogen or helium) above 4400°C (8000°F), at which point it is dissociated and partially ionized to create the plasma state, which issues from the nozzle. Power is usually fed radially into the plasma near the exit point.

The electric-arc process also has an electric arc as the heat source but uses wire instead of powdered metal. The wire serves as the electrode.

Detonation-gun spraying, another process variation, employs a mixture of nitrogen, oxygen, acetylene, and metal powder in a chamber open at one end. When an electric spark detonates the mixture, the gas and powder particles leave the chamber at an explosive velocity. The particles, in a plastic state from the force of the detonation, strike and adhere to the base metal. The process is automatic, repeating 4 to 8 times per second. It takes place within a concrete structure because of the extreme noise that accompanies the detonations.

HVOF (high-velocity oxygen fuel) has become an important metal-spraying process. HVOF devices have an internal combustion chamber in which a mixture of fuel (propane, acetylene, hydrogen, or kerosene) is combusted with oxygen. The resulting products are expanded through a nozzle, where they attain supersonic velocity. Powder to be processed is axially injected into the hot, expanding gas, where it is heated and accelerated toward the substrate. Particle velocities can be in excess of 300 m/s (1000 ft/s) with gas temperatures of approximately 2100°C (3800°F). Because of this high velocity, the particles form a nearly fully dense coating on the substrate. The high temperature of the combustion process limits the materials that can be sprayed. Figure 8.3.11 illustrates an HVOF spray gun in operation.

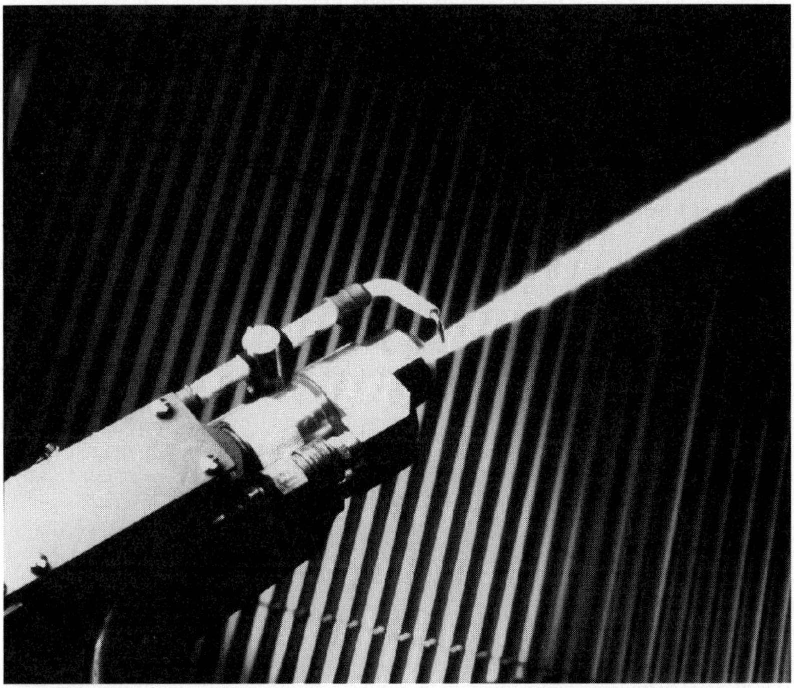

FIGURE 8.3.11 An HVOF (high-velocity oxygen fuel) gun in operation. (*Courtesy Sulzer Metco.*)

To accomplish adequate and reliable bonding with all these methods, the substrate must be conditioned properly. The first step is to clean the surface of dust, oil, and other foreign matter. Once cleaned, the surface is further prepared by one or more of several methods: (1) roughening by grooving, threading, undercut machining, knurling, or otherwise rough-machining with metal-cutting tools, (2) coarse-grit blasting, (3) application of an initial bond-coat material (grit blasting may precede bond coating if especially strong bonds are required), (4) placement of a number of studs in the area to be coated, and (5) sanding or grinding (used for hard-base materials). Preheating of the workpiece is also sometimes employed to improve adhesion of the coating.

If the materials used are reactive, these processes may be carried out in an inert-atmosphere chamber, sometimes at reduced pressure.

Masking is required to prevent the coating from adhering to areas where none is wanted. It is performed before surface roughening by grit blasting. Mask materials include tape, sprayed compounds, and metals.

Postcoating operations may be performed, depending on the nature and function of the coating and the part. Fusion of the coating reduces porosity and improves uniformity and adhesion. It involves heating after coating to a temperature below the melting point of the coating and is a necessary step if combustion flame spraying of powdered material is used. Sealing with organic coatings is used to offset the porosity of most coatings, primarily when it is necessary to provide corrosion protection. Machining, grinding, and lapping provide better dimensional accuracy and surface finish than those of the unfinished coating. Sometimes nondimensional polishing is also carried out to provide a smoother coating.

TYPICAL CHARACTERISTICS AND APPLICATIONS

The sprayed deposit is usually between 0.05 mm (0.002 in) and 2.5 mm (0.100 in) thick, but it may be as much as 6 mm ($\frac{1}{4}$ in) and occasionally as much as 25 mm (1 in) thick.

Sprayed coatings are somewhat porous, the degree of porosity depending on the spraying method. Porosity is desirable in bearing applications because the pores in the coating act as reservoirs for lubricants. Porosity in combustion-flame coatings ranges from 6 to 13 percent, in plasma-arc coatings from 1 to 12 percent, and in detonation-gun coatings from $\frac{1}{4}$ to 1 percent.

Thermal-sprayed coatings tend to be considerably harder than the parent coating material and have superior wear properties. They generally possess low ductility and reduced tensile strength. Tensile stresses are present. Bonding is mechanical rather than metallurgical or chemical except with the detonation-gun process, which gives high bond strengths. Typical bond strengths of thermal-sprayed coatings range from 34 MPa (5000 psi) to over 69 MPa (10,000 psi).

Part size is seldom a limitation to the applicability of thermal-sprayed coatings. Large and unwieldy parts can be coated with portable equipment. Some part shapes, however, may not be suitable if the surface to be sprayed is inaccessible. For parts requiring coatings on inside surfaces, there are gun extensions that can be used if the bore is 50 mm (2 in) or greater.

Common uses for thermal-sprayed coatings are to provide corrosion protection, to provide wear-resistant surfaces, to salvage worn or undersized parts, to provide electrical contact and electrically conductive surfaces, and to provide heat-oxidation protection.

Wear-resistant coatings are an important application. Wear-resistant properties are of several forms:

1. *Resistance to adhesive wear.* This is the wear that occurs when two materials come into intimate contact. Many soft bearing materials that resist this kind of wear can be applied by thermal spraying. Typical applications are bronze and babbitt bearings, piston guides, thrust-bearing shoes, hydraulic-press sleeves, and crosshead slippers. In general, compared with poured babbitt, thermal-sprayed babbitt has a higher load capacity, greater integrity, and better bond to base. It also allows thinner coatings.

There are also hard bearing surfaces that provide resistance to adhesive wear. Hard bearing materials are used when self-alignment is not important and when lubrication may be marginal. Typically, a molybdenum or molybdenum-containing coating is used to increase wear resistance and resist scoring. A typical hard bearing application is the piston-ring facing shown in Fig. 8.3.12. Other hard bearing applications are roll journals, antigalling sleeves, fuel-pump rotors, rudder bearings, and armature shafts.

2. *Resistance to abrasive wear.* Flame coatings are used to resist the abrasive effect of hard foreign particles such as metal debris, metallic oxides, and dust. Typical appli-

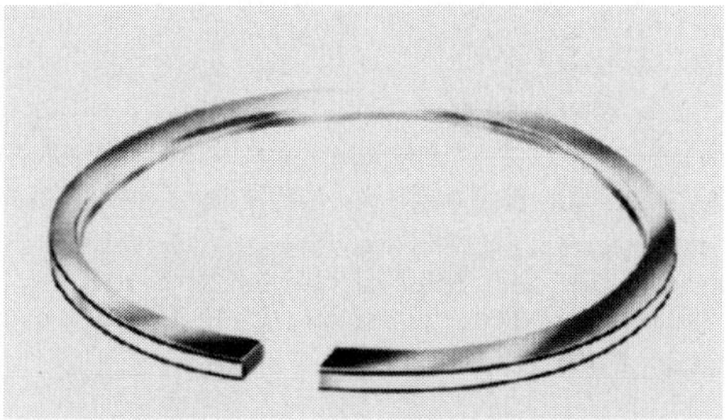

FIGURE 8.3.12 Piston ring with flame-spray-coated hard bearing material. (*Courtesy Metco, Inc.*)

cations of this type of coating are grinding hammers, pigtails, slush-pump piston rods, buffing and polishing fixtures, reciprocating-pump plungers, and pump drive shafts.

3. *Resistance to surface-fatigue wear and cavitation.* Flame-sprayed coatings that resist the surface cracking and chipping that can result from repeated cyclic stresses are available. These coatings also resist the mechanical shock caused by cavitation in liquid flow. Typical applications of these coatings are valve faces, jet engine compressor-blade midspan supports, rocker arms, expansion joints, water turbine nozzles, pumps, and impeller blades.

4. *Resistance to erosion.* This is closely analogous to resistance to abrasion, except that the abrasive particles are carried by a gas or a liquid. Coatings for these conditions are available for temperatures from ambient to 840°C (1550°F). Typical applications are cyclone dust collectors, hydroelectric valves, exhaust-valve seats, and exhaust fans.

Table 8.3.3 lists typical flame-spraying applications with the coating material and process used and the coating thickness.

ECONOMICS OF PRODUCTION

Initially, thermal spraying (metallizing) was used almost entirely as a repair process for replacing metal on worn parts. However, recognition gradually was given to its applicability to production conditions as well. Production applications are becoming more common as original parts are designed for sprayed coatings.

Equipment costs for flame spraying are modest. Tooling expenses are low also because only simple holding fixtures and masks, if required, are involved. Therefore, high production levels are not required to amortize investments for production apparatus. However, when high production levels are required, the operation can be mechanized to minimize labor costs.

Metal deposition rates normally range from 1.6 to 4.5 kg/h (3.6 to 9 lb/h). This is equivalent to a range of 1.3 to 6.6 m^2/h per 0.1 mm of coating thickness (57 to 280 ft^2/h per 0.001 in of thickness). Note that these figure do not include an allowance for overspray. With electric-arc deposition of nickel alloys, however, rates as much as 10 times these have been achieved.

TABLE 8.3.3 Typical Thermal-Spray-Process Applications

Part name	Process	Coating material	Coating thickness, mm (in)	Surface finish, μm rms (μin rms)	Operating conditions
Feed rolls	Plasma	Tungsten carbide	0.05–0.10 (0.002–0.004)	3.8 (150)	Low load, dry rubbing
Compressor rods	Detonation gun	Tungsten carbide	0.25–0.30 (0.010–0.012)	0.13–0.25 (5–10)	Sliding wear
Vanes	Plasma	Chromium carbide	0.08–0.13 (0.003–0.005)	3.8 (150)	High temperature, fretting wear
Seal spacers	Detonation gun	Chromium carbide	0.20–0.25 (0.008–0.010)	0.025–0.08 (1–3)	High temperature, sliding wear
Pump sleeves	Oxyacetylene	Aluminum oxide	0.38–0.46 (0.015–0.018)	0.50 (20)	Ambient temperature, water and sand abrasion
Carbon graphite seals	Plasma	Aluminum oxide	0.20–0.25 (0.008–0.010)	0.25 (10)	High load, fretting wear
Snick plates	Detonation gun	Aluminum oxide	0.05–0.08 (0.002–0.003)	Various	Abrasive wear by synthetic fibers
Shaft sleeves	Oxyacetylene	Chromium oxide	0.38–0.46 (0.015–0.018)	0.63 (25)	Ambient temperature, various chemical solutions
Wire-drawing sheaves	Plasma	Chromium oxide	0.20–0.33 (0.008–0.013)	0.63–0.88 (25–35)	High-speed sliding wear
Abradable sheals	Oxyacetylene	Nickel-graphite	2.5 (0.100)	10 (400)	Temperatures up to 800°F
Various	Plasma	Nickel	Various		Used for buildup of salvageable parts

Thermal spraying, including preparatory and finishing operations, has relatively high unit costs. It is most applicable when the workpiece itself is costly or critical to the function of the final product.

Metallizing (combustion thermal spraying) is the least costly thermal-spraying method. However, the bond strength of the coating is lower than with the plasma-arc and detonation-gun methods. The electric-arc method is in common use for zinc corrosion-resistant coatings and other low-cost applications. Detonation-gun spraying is most costly but provides the best coating properties. It requires a large-scale operation to be cost-effective because of the greater investment required, largely for soundproofed protected facilities, and the need for remote-controlled automatic operation.

SUITABLE MATERIALS

Coatings

Almost any material that can be melted without decomposing can be thermal-sprayed. Standard sprayable materials can be classified as follows:

1. *Pure metals.* Aluminum, zinc, nickel, refractory metals, etc.
2. *Alloys.* Steel, cobalt- and nickel-based superalloys, hard-surfacing and self-fluxing alloys, etc.
3. *Compounds.* Aluminum oxides and other metal oxides, carbides, cermets, and nitrides, etc.
4. *Composites and blends.* Admixtures and combinations of two or more of the above in a single powder. Cobalt-bonded tungsten carbide and nickel-clad graphite are two examples.

Any material that melts into droplets (glass forms strands rather than droplets) and which can be put into wire form can be sprayed by the combustion-flame-spraying (wire-metallizing) method. Metals commonly sprayed by this method include zinc and aluminum (the most common), brass, bronze, copper, tin, lead, magnesium, molybdenum, nickel, babbitt, cadmium, carbon steels, alloy steels, and stainless steels.

Flame coating from powder is applicable to hard, corrosion-resistant alloys, notably those cobalt- and nickel-based alloys that contain chromium, silicon, and boron. Powder spraying is also suitable for low-melting-temperature ceramic materials and plastics. Low-melting temperature metals such as zinc, tin, and aluminum are sprayed from powder, but wire-metallizing of these materials is faster and more efficient.

Plasma coating is applicable to practically an unlimited series of materials. Refractory metals such as tungsten, tantalum, and molybdenum are routinely handled by the plasma methods. Other materials that can be coated by plasma-arc spraying include carbides, borides, oxides, nitrides, beryllides, silicides, and sulfides, pure metals and alloys, plastics, glass, and various blends. Plasma spraying is particularly useful when high-melting-temperature materials such as carbides and ceramics are to be sprayed or when it is necessary to avoid oxide contamination of metal coating particles.

Detonation-gun spraying is best for hard surfacing with tungsten carbide, aluminum oxide, chromium carbide, and similar materials.

Substrates

These are usually metal but can be ceramic, glass, concrete, plaster, carbon, or even wood, plastic, rubber, or cloth. The cooling effect of the airstream prevents heat damage to temperature-sensitive materials. However, with the detonation-gun process nonmetallic substrates are not suitable because of erosion from high-velocity particles.

Both nonferrous and ferrous metals are thermal-spray-coatable. Hard surfacing of nonferrous substrates is a common application. Resulfurized and leaded steels are not recommended because the additives embrittle the metallic coating. Hardened steel can be coated but is more difficult to surface-roughen prior to coating than unhardened steel. Also, low-carbon steels are generally preferable to high-carbon steels as substrate materials.

Bonding Coatings

These materials adhere to a clean, smooth surface that does not have to be mechanically roughened or grit-blasted, although grit blasting may be performed. The top coating is then sprayed over the bonding coating. This approach has a major advantage besides the reduction of precoating surface preparation, which can be costly and can adversely affect the workpiece. Added insurance against bond failure is provided. This is important in many applications, for example, when it is necessary to machine or grind the coating to a featheredge. Molybdenum is the major bond-coating material. Nickel-aluminum composites (Metco M405-10, T or P404-10, and T or P450-10, 11) are also used quite extensively.

These bonding coatings adhere well to the following materials: all common steels including hardened alloy steels and 300 and 400 Series stainless steel, nickel and cobalt and their alloys, cast iron and cast steel, titanium and aluminum and their alloys, brass and Monel. They do not reliably self-bond to copper and some copper alloys, molybdenum, tantalum, niobium, and tungsten.

DESIGN RECOMMENDATIONS

The following comments are offered to guide the design engineer in specifying sound coatings. A sprayed coating will not impart structural strength to the base material. Proper design should anticipate this and the part made to withstand anticipated forces without the coating. Generally, the designer should use a bond coat, but this is not always practical; therefore, a few guidelines are listed. However, do not consider these to be hard and fast. There may be specific instances in which they cannot be followed.

1. When the primary coating *is not self-bonding,* the coating is applied with the plasma gun, and the design requires a coating thickness less than 0.25 mm (0.010 in), use a blasted surface only. If the application requires a coating thicker than 0.25 mm, use a bond coat. Then apply the topcoat.
2. For machine-design work, when the primary coating is not self-bonding and is applied with a wire or powder combustion gun, always design with provision for a bond coat. If less than a 0.25-mm (0.010-in) coating thickness is needed, use an undercut greater than 0.25 mm to allow for the thickness of a bond coat.

3. When very thick coatings are required, it is sometimes advisable to use a low-cost intermediate coat between the bonding coat and the functional topcoat. The bond coat should be 0.075 to 0.12 mm (0.003 to 0.005 in) thick. Coating materials should be selected to avoid galvanic corrosion between layers.
4. Some coatings that possess both the favorable adhesion of bonding coatings and the functional properties of topcoatings are now available. With these materials, no separate bonding coating is required. Some of these materials are listed in Table 8.3.4. Their use is recommended.

Because of the brittleness of flame-sprayed coatings, they should not be used when the function requires point contact or when the coating will be exposed to sharp blows, continued pounding, or severe strain at an edge or in a small area. Examples of this kind of condition are gear teeth, journals running against needle bearings, impact faces of hammers, and rolling-mill ways. Journals, plain bearings, hydraulic rams, crankshafts, piston sleeves, and machine-tool ways are examples of components which have the load on the coating evenly distributed.

Sprayed coatings are also susceptible to chipping at the edges because of sudden blows either in service or during handling. For this reason, it is best to provide an undercut at the end of the coating. Thus, when the coating has been machined flush, no raw edges are exposed. (See Fig. 8.3.13.)

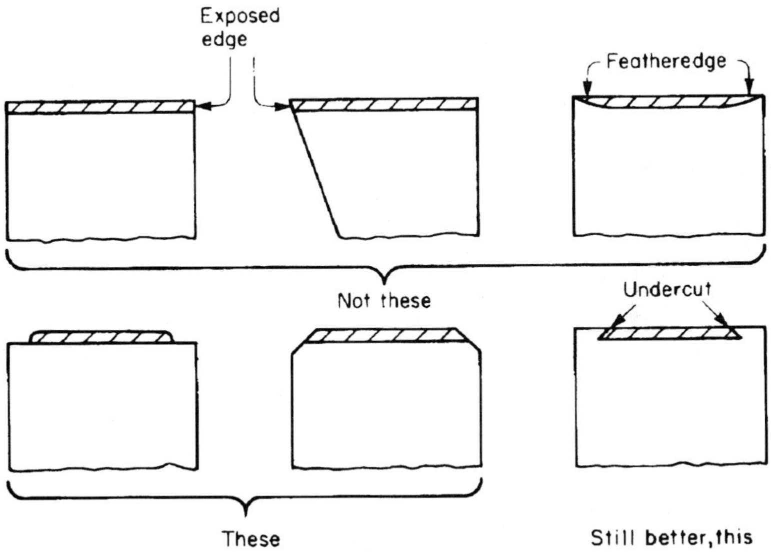

FIGURE 8.3.13 Avoid exposed edges of flame-sprayed coatings and also featheredges. Surfaces held by undercuts at the edges are best.

When coatings are needed on internal surfaces, there are limitations to the depth that can be sprayed. If the opening is large enough for a spray-gun extension, there is

TABLE 8.3.4 Coatings for Hard Bearing Applications*

Coating process†	Description and chemistry	Hardness	Finish, μin AA†	Maximum service temperature, °F	Comments
APS PC	Nickel, 55 iron, 5 Mo, 5 Al, composite powder	R_b 90 to R_c 25	AS, 400–700 G, 15–30 L, 4–10 M, 25–45	700	Excellent bond and machinability, APS coating powder, harder than PC
PC	Nickel 7% Mo, 6% Al, composite powder	R_b 90	AS, 300–400 M, 40–60	900	Best bonding, poor machinability, good corrosion resistance
APS PC	Nickel, 20% chromium, 5% molybdenum, 6% Al, composite powder	R_b 90	AS, 400–500 M, 20–30 G, 15–20 L, 4–7	1200	High-temperature service, good bonding
HVOF	Tungsten carbide, 12% Co, composite 41-μm particles	R_c 60	AS, 175–200 L, 1–3	1000	Best wear resistance, high surface finish, needs diamond grinding
APS	Tungsten carbide, 12% Co, blended with 50% nickel-base fluxing alloy	R_c 50–55	AS, 200–300 L, 2–4	1000	Can be HVOF sprayed to get improved properties
WC	Molybdenum wire, 99.9% Mo	R_c 40	G, 50–100	650	One coat of self-bonding high-scuff-resistant bond coating

PC NC	Iron, chromium, nickel, stainless steel, powder or wire	R_b 90	AS, 300–400 M, 40–50 G, 10–20 L, 5–10	1400	Machinable 315 stainless steel coating
WC EA	High-chromium stainless steel wire, 0.35% C, 0.02% P, 0.02% S, 2% Mn, 8% Ni, 18% Cr, 0.75% Si, balance Fe	R_c 33	G, 25–50	1000	Hard grindable 420 stainless steel coating, bond coat or other surface preparation required

*Including one-coat self-bonding coatings.

†AS = as sprayed; G = ground; M = machined; L = lapped; APS = air plasma spray; HVOF = high-velocity oxygen fuel spray; EA = electric-arc spray; WC = wire combustion spray; PC = powder combustion spray (low velocity).

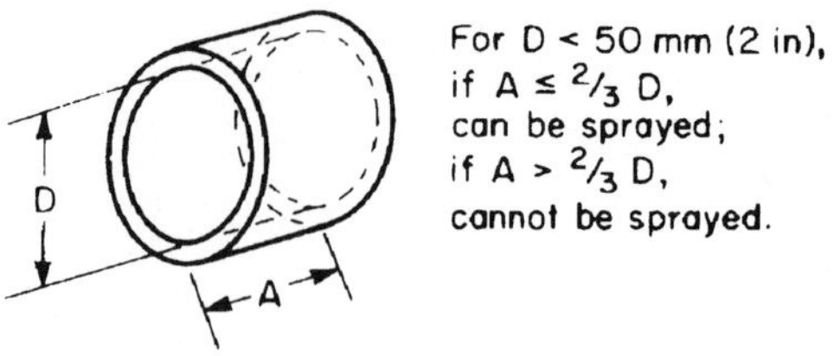

FIGURE 8.3.14 Limitations on the depth of internal surfaces that can be flame-sprayed.

no length limitation; otherwise, the maximum coating depth is two-thirds of the diameter of the opening, as shown in Fig. 8.3.14.

For best results in production, the workpiece should be designed to facilitate handling. For example, a cylindrical part should, when possible, include in its design a means by which it can be rotated: centers, flanges, etc. A sheet-metal stamping could be only partially sheared through to allow handling a continuous strip rather than individual parts at the spraying operation. Irregular parts should include a locating hole or other means for easy orientation.

To reduce the costs of oversprayed material and increase the rate of production, it is common practice to design parts for nesting as illustrated in Fig. 8.3.15.

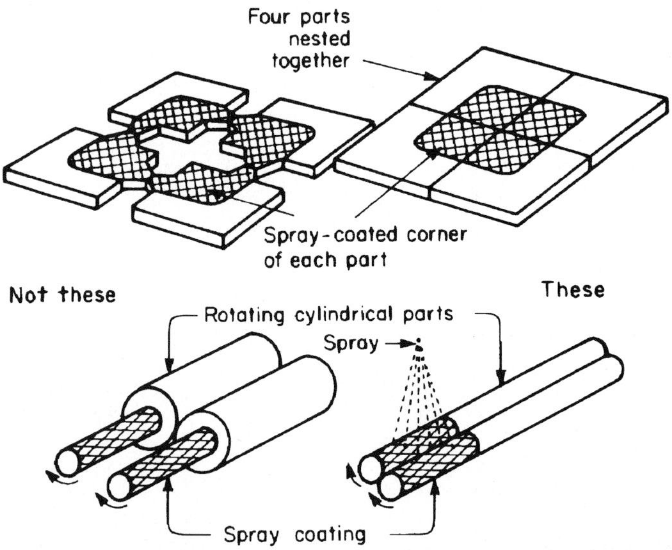

FIGURE 8.3.15 Design parts so that they can be nested together during flame spraying to reduce overspray losses.

TABLE 8.3.5 Recommended Minimum Thickness of Thermal-Sprayed Coatings on Shafts

Shaft diameter, mm (in)	Minimum coating thickness per side, mm (in)
25 (1) or less	0.25 (0.010)
25–50 (1–2)	0.38 (0.015)
50–75 (2–3)	0.50 (0.020)
75–100 (3–4)	0.63 (0.025)
100–125 (4–5)	0.76 (0.030)
125–150 (5–6)	0.89 (0.035)
150 (6) or more	1.0 (0.040)

Source: Data taken from ASM Committee on Metal Spraying, "Metal Spraying," *Metals Handbook,* vol. 2, 8th ed., Lyman Taylor (ed.), American Society for Metals, Metals Park, Ohio, 1984, p. 509. Values are based on average conditions of wear and include an allowance for finish machining after flame spraying.

For coated parts to be press-fitted, the coating should be at least 0.13 mm (0.005 in) thick to ensure adequate strength. For rotating shafts, recommended minimum coating thicknesses for average conditions of wear and finish machining are given in Table 8.3.5.

Deposits on internal surfaces should be kept to minimum thickness. Because of the shrinkage forces inherent in flame-sprayed coatings, thick coatings on internal surfaces are apt to separate from the workpiece. However, extremely thin coatings should not be specified for internal or external surfaces. Coatings under about 0.08 mm (0.003 in) are not recommended because thickness uniformity cannot be controlled sufficiently.

As indicated above, a roughened or grooved surface is advisable to ensure adequate adhesion of the flame-sprayed coating. Figure 8.3.16 illustrates the recommended groove design for a shaft that is to be coated.

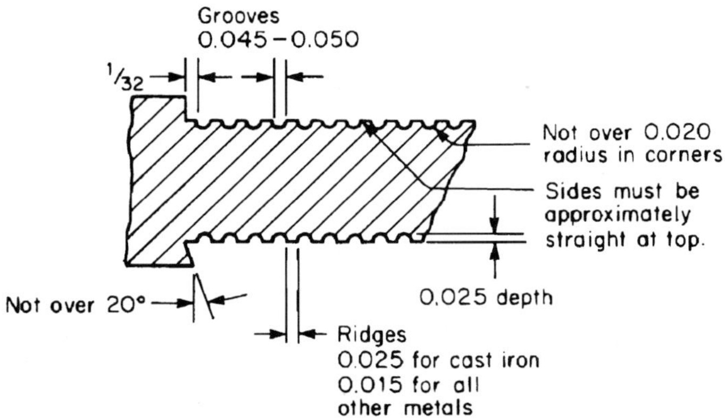

FIGURE 8.3.16 A recommended design for grooving a shaft to ensure strong adhesion of flame-sprayed coatings. Dimensions shown are in inches. Grooves can be either annular or helical. Knurling at the tops of the ridges produces a dovetail undercut that further ensures a strong bond.

DIMENSIONAL FACTORS

The surface finish of flame-sprayed coatings ranges from 1.0 to 10.0 μm (40 to 400 μin) rms. Careful machining, grinding, lapping, etc., can reduce the roughness after machining to normal values achievable with materials of lower machinability.

The surface finish and the accuracy of coating thickness depend on a number of factors. One major factor is the uniformity of speed of spray-gun movement. Hand-held guns produce much more variation than machine-mounted guns. The nature of surface preparation is an important determinant, with blasted surfaces providing more uniformity than deeply grooved surfaces. Shrinkage of coating material also must be allowed for. This varies with different materials but ranges from about 0.15 percent to about 1.2 percent.

The normal variation of coating thickness under controlled production conditions ranges from about 0.05 mm (0.002 in) to about 0.13 mm (0.005 in) in total variation.

PART 3

VACUUM-METALLIZED SURFACES

THE PROCESS

Vacuum metallizing is a process in which thin coatings of metals or metal compounds are deposited on workpiece surfaces in a vacuum chamber. The coating material is heated to the point of evaporation. The vapors condense on the cooler workpiece surface and solidify. The entire process takes place within a closed chamber having a vacuum of 10^{-3} to 10^{-5} mbar. Workpieces are located on racks surrounding the source of the vapor, which is centrally located. The vapor travels only in a straight line, but the workpiece can be rotated if coverage is required on more than one side. Proper positioning of the workpiece is very important.

The material to be deposited is heated by electrical resistance, induction, or electron beam. The resistance method, however, is most common.

Because of the high vacuum levels required, the process must operate on a batch basis except for strip materials, which can be passed through vacuum seals.

Proper adhesion of the coated material demands that the workpiece surface be thoroughly cleaned. Vapor degreasing is the most common method. Many parts are precoated prior to metallizing. The precoat provides a smoother substitute for the metal coating and, for some base materials, seals the escape of volatile substances which would interfere with the coating process. After the metal-coating operation, a protective lacquer or other nonmetallic coating is often applied for corrosion and abrasion protection. Plastic parts receiving decorative coatings do not usually require extensive precleaning but are given a base coat and a clear protective coating.

Vacuum chambers range from one to several hundred cubic feet in size.

TYPICAL CHARACTERISTICS AND APPLICATIONS

There are five basic applications for vacuum coatings:

1. *For decoration.* A bright, lustrous metallic coating is applied to plastic components, film, or paper.
2. *For optical applications.* A thin film is applied, normally to optical glass to provide reflectance or, with closely controlled film thickness, light attenuation (e.g., sunglasses), filtering, or beam splitting.
3. *For corrosion protection.* Although less used for this application and usually inferior for corrosion protection to electroplated coatings, vacuum coatings are used for some corrosion-protection applications.
4. *For electronics and electrical applications.* Capacitors, integrated circuits, contacts, and photoconductors utilize vacuum coatings.
5. *For encapsulation of powdered materials.*

The coating finish can be brilliant and requires no buffing if the substrate surface is smooth. Coatings are normally on the first surface (top side) of a component, but in the case of clear materials they can be applied to the second surface (rear side). This provides full abrasion protection to the coating and a very rich, jewel-like look to the decorated product.

Typical components receiving decorative vacuum coatings are costume jewelry, toys, home appliances, hardware, automotive trim (normally interior), decorative nameplates, and various sheet materials. Optical applications include mirrors, automobile headlights, flashlights and other reflectors, telescopes, microscope filters, sunglasses, optical filters and other instruments, beam splitters, and other objects requiring some degree of light reflection. Also included in this classification are reflective coatings for infrared and longer-wave thermal energy.

The most notable anticorrosion application of vacuum coating is in food containers, for which tin can be applied by vacuum methods more economically than by electroplating. Aluminum-coated steel is used for packaging applications. An advantage of vacuum coating over electroplating for highly stressed parts is the absence of hydrogen embrittlement. Cadmium and aluminum vacuum-applied anticorrosion coatings average from 5 to 10 μm (0.2 to 0.4 mil) in thickness. Vacuum-applied aluminum improves the corrosion resistance of galvanized steel.

Electrical and electronic applications of vacuum coating are quite varied and are growing rapidly. Capacitors are perhaps the largest-volume application, using tantalum, titanium, silicon monoxide, and magnesium fluoride coatings on paper, plastic film, and glass. Integrated circuits lean heavily on vacuum-coating techniques to deposit conductive, semiconductive, and resistive materials. Magnetic tapes are another major application. Magnetic iron-nickel and other alloys are vacuum-deposited on plastic film. For conductors and contacts in various electrical and electronic devices, conductive metals like gold, silver, copper, tin, platinum, and iridium are vacuum-deposited. An additional application is the deposition of copper on nonconductive surfaces to provide a base coat for electroplating or electroforming.

Vacuum-deposited coatings are generally of high purity and density. While thin coatings are lustrous, thicker coatings of 2.5 μm (0.1 mil) or more tend to be gray or wavy in appearance. Normal coatings are from 0.05 to 10 μm (0.002 to 0.4 mil), with the thin coatings for decorative applications and the thick coatings for corrosion-protection applications. Metal coatings 0.08 to 0.1 μm (0.003 to 0.004 mil) thick are opaque.

The maximum size of component that can be vacuum-coated depends on the size of vacuum chamber available. See Tables 8.3.6 and 8.3.7 for summaries of applications of vacuum coatings.

ECONOMIC PRODUCTION QUANTITIES

For maximum economy of production of vacuum-coated parts, quantities should be large. The process is not limited to mass production, and for many applications, primarily in the optical and electronics industries, vacuum coating is the only process that can provide coatings of the correct controlled thickness. However, for decorative applications electroplating is a competing process that may be more economical in some situations, especially when production quantities are small. For large production volumes, if the product configuration lends itself to vacuum coating, the vacuum process is invariably lower in cost.

TABLE 8.3.6 Typical Applications of Vacuum-Deposited Coatings

Coating material	Function	Coating thickness, mil	Substrate material	Pretreatment	Post-treatment
Aluminum	Decorative	0.001–0.005	Metal[a]	Lacquered	Lacquered
Aluminum	Reflective	0.001–0.005	Glass	None	[b]
Aluminum	Protective	0.5	Steel	None	Anodized[c]
Aluminum	Decorative	0.001	Plastic sheet	None	Laminated
Aluminum	Decorative	0.5	Aluminum[d]	None	Anodized
Aluminum	Electrodes[e]	0.0005–0.001	Plastic shot	Baked	None
Cadmium	Protective	0.5	Steel	None	Painted[e]
Cadmium	Electrical resistance	0.0001–0.001	Glass or plastic	None	Multiple layers
Chromium	Electrical resistance	0.0001–0.001	Glass or plastic	Cleaned	Air-baked
Gold	Electrodes for piezoelectric crystals	0.001–0.005	Organic or m inorganic crystals	Cleaned	None
Magnesium fluoride	Nonreflective	0.004	Glass	None	None
Silicon monoxide	Abrasion resistance	0.004	Glass	[b]	Oxidized
Titanium dioxide	Decorative; optical	0.004	Glass	None	Oxidized
Zinc	Electrical conductivity	0.001–0.005	Paper	Lacquer	None

[a]Automotive trim. [b]Aluminum coating for reflectance followed by silicon monoxide coating for abrasion resistance for mirrors. [c]If required. [d]Die casting. [e]For capacitors.

Source: ASM Committee on Vapor Deposition Coating, "Vacuum Coating," *Metals Handbook,* vol. 2, 8th ed., Lyman Taylor (ed.), American Society for Metals, Metals Park, Ohio, p. 517.

Unit costs for decorative coatings comprise primarily the handling labor to load and unload the chamber and to apply the precoat and topcoat. Lacquer coating materials are far more costly than the aluminum usually employed for decorative coating. However, tungsten filaments used to provide the heat for vaporizing the coating material also often comprise a significant portion of the total unit cost.

Processing times for decorative coatings run about 15 min per batch in large (2-m-diameter) chambers, most of which is pump-down time for the vacuum. The actual coating cycle on the chamber (e.g., for aluminum) is about 60 s. Heavy coatings for corrosion protection require more time, with a total cycle of approximately 50 to 90 min.

SUITABLE MATERIALS

Aluminum is by far the most common vacuum-coated material. It accounts for over 90 percent of decorative coatings and may be used for optical, electronic, or corrosion-protection purposes also. Aluminum has a high vapor pressure; i.e., it vaporizes easily at lesser vacuum levels. It also has an attractive appearance and high reflectivity when used as a coating. After aluminum, in order of decreasing vapor pressure (the most important factor in vacuum coatability), are selenium, cadmium, silver, copper, and

TABLE 8.3.7 Electronic Applications of Vacuum-Deposited Metals and Metal Compounds

Coating material	Application	Coating thickness, μm
	Metals	
Aluminum	Conductor[a]	0.01–0.2
Bismuth	Conductor[b]	0.05–0.5
Cadmium	Conductor	0.05–1
Chromium[c]	Resistor	0.002–0.1
Columbium (niobium)[d]	Superconductor	0.05–0.1
Copper	Conductor	0.01–0.2
Germanium[d]	Semiconductor	0.5–1.0
Gold	Conductor	0.01–0.2
Indium	Conductor	0.05–0.2
Lead	Conductor	0.05–0.2
Molybdenum	Conductor[e]	0.05–0.2
Nickel	Conductor	0.05–0.2
Platinum[f]	Conductor	0.01–0.2
Selenium	Semiconductor	0.5–100
Silicon[d]	Semiconductor	0.5–10
Silver	Conductor	0.01–0.2
Tantalum[f]	Resistor	0.01–0.2
Tin	Superconductor	0.05–0.2
	Metal compounds	
Aluminum oxide[f]	Capacitor	0.1–2
Cadmium sulfide	Semiconductor; photoconductor	0.1–2
Cerium oxide	Capacitor	0.1–2
Silicon oxide	Capacitor; insulator	0.1–2
Tantalum oxide[g]	Capacitor	0.01–0.2
Titanium oxide[h]	Capacitor	0.03–0.2
Zinc oxide	Semiconductor	0.1–2

[a]Good counterelectrode for capacitors. [b]Good counterelectrode for rectifiers. [c]Adheres exceptionally well to glass. [d]Difficult to obtain purity desired. [e]Good conductor for high-temperature applications. [f]Difficult to vaporize except by electron-bombardment heating. [g]Produced by anodizing tantalum films. [h] From thermal oxidation of titanium metal.

Source: ASM Committee on Vapor Deposition Coating, "Vacuum Coating," *Metals Handbook,* vol. 2, 8th ed., Lyman Taylor (ed.), American Society for Metals, Metals Park, Ohio, p. 517.

silicon monoxide. A second group, again in order of decreasing vapor pressure, are gold, chromium, palladium, nickel-chromium alloy, magnesium fluoride, and titanium. Other materials, considerably less volatile, that have been vapor-deposited are platinum and niobium.

Actually, almost all metals, many alloys, and many semimetallic elements can be vacuum-deposited. However, materials that do not vaporize easily present difficulties in processing because very high vacuum levels are necessary and processing costs multiply.

Certain compounds of metals and other elements are useful as coatings and can be deposited by vapor-deposition methods. Silicon monoxide is the most commonly used of such compounds. It provides an abrasive-resistant finish over aluminum and is often applied immediately after the deposit of aluminum without interrupting the vacuum. Magnesium fluoride and aluminum oxide are used in the manufacture of electronic capacitors.

Alloys and compounds sometimes present difficulties in uniformity of coating composition. These coating materials can fractionate during heating; more volatile constituents tend to vaporize first and condense on the workpiece separately before less volatile materials.

A variety of substrate materials can be vacuum-coated. The fact that electrical conductivity is not required permits many nonmetallic materials to be coated. The one essential requirement of a base material is that it be relatively stable in a vacuum. It must not give off gases or vapor. Gases may come from contaminants on the surface of some parts, may be trapped in the pores of the base materials, or may result from the evaporation of an ingredient in the base material.

Metals usually are satisfactory as substrates. However, zinc-based alloys and brass (because of the zinc content) are not suitable for heavy protective deposits, although they can be coated with thin decorative coatings. Many nonmetallic base materials that would not be satisfactory normally can be made suitable for the application of a base coating. Common substrate materials are glass, ceramics, silicon and germanium (semiconductors), most plastics (polystyrene, acrylic, polycarbonate, acrylonitrile butadiene styrene, styrene acrylonitrile, urea, melamine, phenolic, polyester, and epoxy), steel, copper, and other metals, and paper. Normally, vinyl and cellulosics are not recommended because of gassing, and polyolefins (polyethylene and polypropylene) require pretreatment of the surface to improve the adhesion of the coating.

DESIGN RECOMMENDATIONS

Design recommendations for vacuum-coated components stem from two factors:

1. The need for an enclosed vacuum chamber with its inherent size limitations
2. The line-of-sight travel path of coating vapor from source to workpiece

The following are specific design recommendations arising from these factors:

1. Parts should be as compact as possible to enable a large number to be placed in the vacuum chamber. Arms and other extensions that may interfere with the placement of other workpieces or with holding fixtures should be avoided. It may be desirable to separate a large component into two parts for later assembly if only one portion of the large component is to be vacuum-coated.
2. Deep recesses and holes should be avoided in surfaces to be coated. Holes up to two diameters deep generally coat satisfactorily. Holes and recesses up to five diameters (or widths) deep often can be coated also but with a thin layer only. (See Fig. 8.3.17.)
3. When a decorative coating is applied to the second surface of a component, ribs, bosses, or other appendages that affect the thickness of the part as measured from the first surface will affect its appearance. Thicker sections will show up with a darker shade. For best appearance, the thickness of the part between the first and second surfaces should be uniform. (See Fig. 8.3.18.)
4. Coating the interior surfaces of hollow parts is difficult unless the interior of the workpiece is large enough to contain the filament or other source of evaporated metal. This is normally a more expensive procedure because more vapor sources are required (except perhaps in the case of large parts), and exterior surface coating is therefore usually preferable.

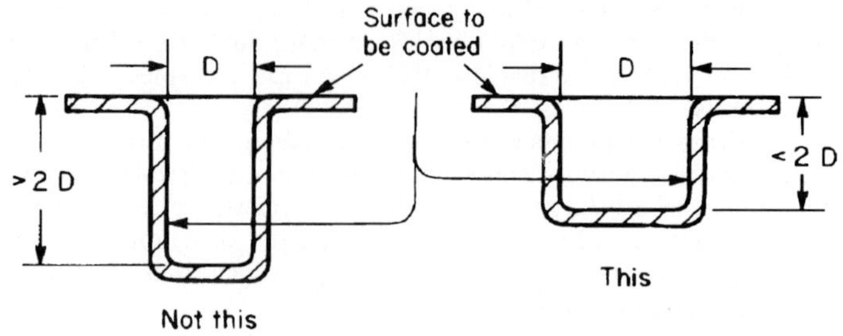

FIGURE 8.3.17 Deep recesses in the surface to be coated should be avoided.

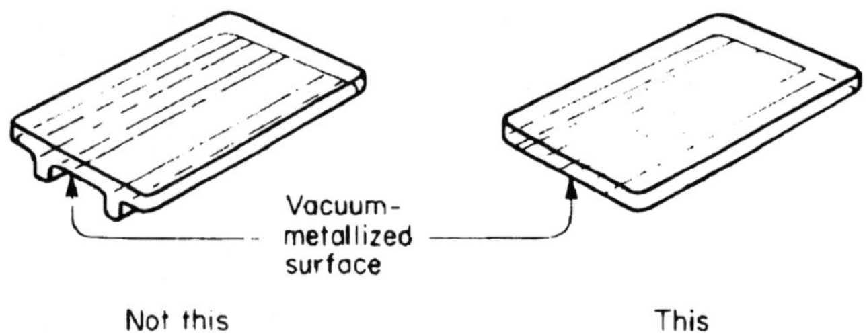

FIGURE 8.3.18 When the second (rear) surface is to be coated, the thickness of the part between both surfaces should be uniform.

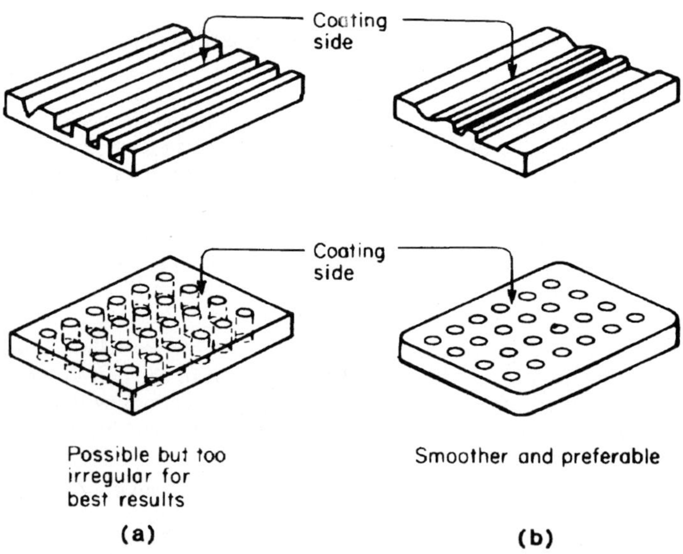

FIGURE 8.3.19 Regular surfaces are preferred for vacuum coating and should be specified whenever possible.

5. Irregular surfaces such as are illustrated in Fig. 8.3.19*a* require multiple sources of metal evaporation to ensure coverage of all areas. In addition, they are more difficult to spray with precoat lacquer. More regular surfaces as illustrated in Fig. 8.3.19*b* are preferable and should be specified whenever possible.
6. Fixtures that rotate the workpiece so that it is coated on more than one side are more costly than those that simply hold the workpiece for coating on one side. One-side coatings should be specified whenever possible.
7. Angles of incidence between the line-of-sight path of the vapor and the workpiece surfaces should not be shallower than 45°. Recesses should have as gently sloping walls as possible. (See Fig. 8.3.20.)
8. Optical parts and others with a surface smooth enough so that the vacuum coating can be deposited directly on the substrate without necessity for a precoat should be designed so that precleaning solvents can drain easily from the surface.
9. While highly irregular surfaces are not desirable for vacuum coating, large flat surfaces may cause problems too if the material is such that a lacquer precoat is required. It is sometimes difficult to spray a smooth coating over a large surface and maintain a uniform appearance. Either extra care must be taken in the lacquer application, or the surface should be made smaller or lightly textured in certain areas.

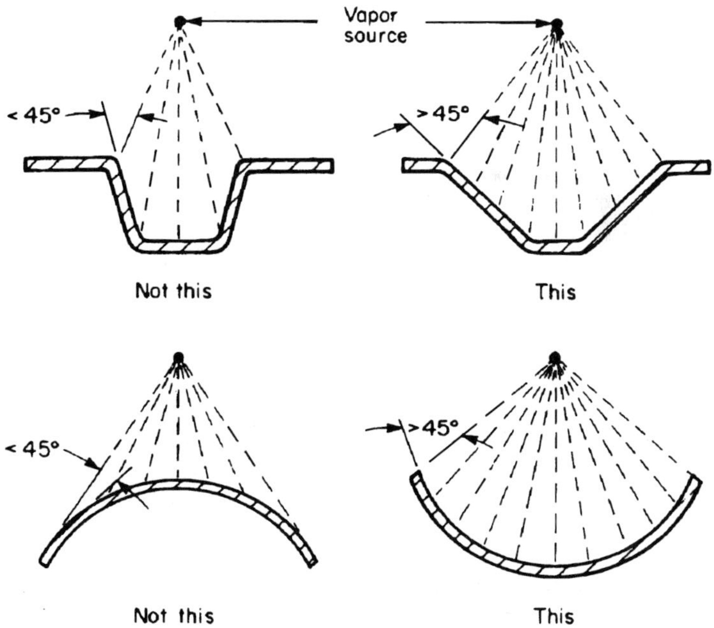

FIGURE 8.3.20 The angle of the part's surface and the line-of-sight path of the vapor should not be less than 45°.

DIMENSIONAL FACTORS AND TOLERANCE RECOMMENDATIONS

While coating thickness can be controlled quite closely, there are normally variations in part-to-part coating thickness and in the thickness on one workpiece. These result from a number of process variables. The major variables are an inherent nonuniformity in the dispersement pattern of vapor from the source, irregularities in shape of the workpiece surface, and the shadow effect which results when the supporting frame comes between the workpiece and the vapor source. Rotation of the workpiece averages out the latter variations. Thickness is usually controlled by controlling the process parameters: filament size and current, coating-metal clip size, degree of vacuum, and vaporization time.

Decorative and simple reflective coatings usually have a wide allowable thickness latitude, as much as ±50 percent. Electronic and optical filter applications, on the other hand, require very close thickness control, often ± 5 percent and, in some cases, to ±1 or ±2 percent. Such close control is more expensive and involves extra process steps to develop and maintain process parameters within narrow limits. For normal product applications, the following nominal dimensional ranges are recommended: decorative and simple reflective coatings, 0.05 to 0.13 μm (0.002 to 0.005 mil); corrosion-protection coatings, 5 to 10 μm (0.2 to 0.4 mil).

CHAPTER 8.4

DESIGNING FOR HEAT TREATING

Donald A. Adams
Warren, New Jersey

Heat-treating operations, though usually secondary in sequence to other parts-making operations and having less obvious effects on the workpiece, nevertheless have a primary effect on its eventual success or failure in use. Although the number of heat-treating processes is large and the details of each process almost infinite, designers should be aware of the general types of heat treatment available and know enough of their details so that they can make reasonable decisions as to how their parts should be heat-treated. While it is not good practice for designers to spell out the method of heat treatment, it is highly desirable that they recognize the process involved and any limitations that influence its use on the parts they are designing.

HEAT-TREATING PROCESSES IN GENERAL

Heat treating is a general term. It refers to the process of heating a material to change its physical properties. This heating (and eventual cooling to room temperature) either (1) softens the material or (2) hardens it. The results of the process depends, first, on the specific material, second, on the temperature to which the material is heated, and, third, on the method and rate of cooling. The atmosphere in which the heating takes place and the length of the heating cycle are also significant factors which influence the final workpiece properties.

Specific heat-treating processes are as follows.

HEAT-TREATING PROCESSES FOR STEEL

Softening Processes

Stress Relieving. This process is used to reduce the internal stresses that are the result of mechanical cold working on the metal during machining, swaging, drawing, bending, etc. This reduction is accomplished by raising the temperature to some point below the critical temperature (the temperature at which the material structure changes) and then allowing the material to cool slowly to room temperature.

Annealing. Annealing is performed to soften the material so that it can be mechanically processed more easily by cold drawing, bending, machining, etc. It is accomplished by heating the workpiece to a temperature above the austenizing temperature (the temperature at which the structure of the material starts to change), holding at this temperature for sufficient time to allow the structure to transform, and cooling at a slow rate.

Normalizing. This process is carried out to return the structure to a "normal" condition and is generally performed after hot-working processes such as forging by heating the workpiece to a temperature above that at which the structure changes, holding it at this temperature to complete transformation, and cooling it in air. The resulting structure is not as soft as if fully annealed, but it is uniform.

Hardening Processes

To harden steel, it is necessary to heat the material to a temperature above the austenizing temperature and then to cool it rapidly. The length of time at the elevated temperature and the cooling rate depend on the type of steel. Subsequent heating and cooling at lower temperatures are used to modify the as-hardened structure.

Case Hardening. To case-harden steel, carbon or nitrogen must be diffused into the surface at an elevated temperature. The workpiece is then cooled at a rate sufficient to harden the surface. Four techniques used are carburizing, cyaniding, nitriding, and carbonitriding, the choice depending chiefly on the properties desired in the case-hardened part.

Surface Hardening. Whereas with case-hardening heat treatment the entire part is heated through yet only a shallow surface becomes hard owing to its higher carbon content, surface hardening accomplishes a similar effect by heating only the surface. Surface heating is usually accomplished with torches (flame hardening) or by high-frequency electrical induction. In either instance, the surface is rapidly heated and then quenched in water or oil. In many applications, long symmetrical parts such as shafts are automatically fed through a series of torches or an induction coil and immediately after emerging from the heating area are spray-quenched with water.

Through Hardening. This involves heating the part to a point above the critical temperature, holding it at this point until the part is uniformly heated, and then quenching it in a suitable medium. Figure 8.4.1 shows a typical transformation curve to illustrate the cooling rate required for conventional hardening.

The quenching medium depends on the size and shape of the part as well as on the grade of steel. Water, oil, or air, with varying degrees of agitation, is most commonly used to provide the cooling rate necessary to harden the part.

Through-hardened parts are normally tempered after hardening to reduce the hardening stresses and increase the toughness. This is done by heating to a temperature below the critical temperature and then air cooling.

Martempering. This is a special technique which, by quenching to a temperature above the point at which the steel hardens and then by slow cooling, provides a through-hardened structure with less distortion.

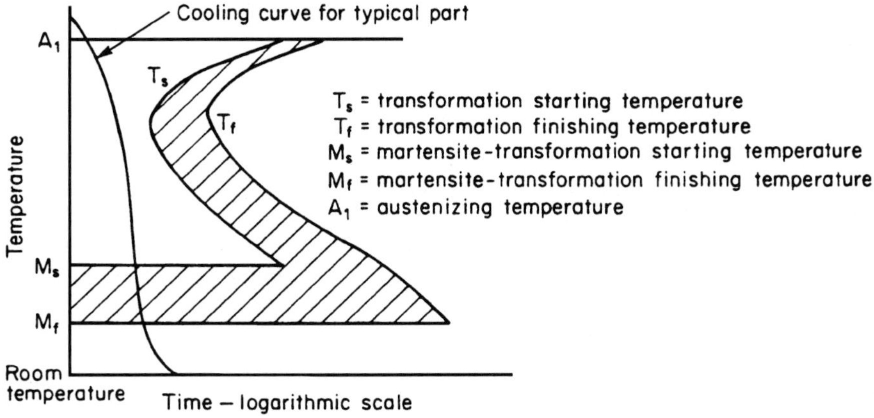

FIGURE 8.4.1 Typical *S* transformation curve for one grade of steel. A cooling curve for conventional heat treatment is superimposed. Line T_s indicates the starting point of transformation of the steel to pearlite, bainite, or ferrite, while T_f indicates the point where the transformation is finished. The portion of the curve that indicates the transformation of the material to hard-constituent martensite is designated as M_s (start) and M_f (finish). The cooling curve shows quenching, i.e., cooling from the austenizing temperature A_1 at a rapid enough rate so that the "nose" of the curve is not intersected. When M_s is reached, the transformation of the workpiece material to martensite commences.

Austempering. This process also uses an interrupted quench to produce higher ductility at approximately the same hardnesses obtained with conventional hardening.

HEAT-TREATING PROCESSES FOR CAST IRON

Similar to steel, cast iron can be annealed to soften it and improve its machinability. The soaking temperature and the cooling rate determine the annealed properties. Temperatures between 700 and 815°C (1300 and 1500°F) decompose the iron carbide to ferrite, temperatures from 790 to 900°C (1450 to 1650°F) give a full anneal to high-alloy irons, and temperatures from 900 to 950°C (1650 to 1750°F) transform massive carbides (white iron) to a machinable structure after slow cooling (normally less than 110°C/h). Normalizing involves the cooling of the casting in still air from the higher above-temperature range. This restores the as-cast properties and improves hardness and tensile strength.

Gray iron is hardened by heating it above the transformation temperature and quenching it in oil. Austempering and martempering treatments also can be given to gray iron to provide heat-treated properties without the high stresses associated with a full quench. Hardness depends on the carbon and alloy content of the austenite at the austenizing temperature.

Flame hardening is the most commonly applied surface heat treatment for gray iron. It is normally followed by stress relieving at 150 to 200°C (300 to 400°F) to reduce subsequent cracking of the hard surface.

HEAT-TREATING PROCESSES FOR STAINLESS STEELS

Martensitic stainless steels are hardened with processes similar to those used for plain carbon or low-alloy steels. Owing to the high alloy content of these steels, maximum hardness can be produced by air cooling.

Austenitic and ferritic grades are not hardenable by heat treating but are annealed after work hardening. Austenitic grades are annealed by heating thoroughly to about 1100°C (2000°F) and then cooling rapidly in water or an air blast. Ferritic grades are heated to the 700 to 800°C (1300 to 1450°F) range and then water-quenched or air-cooled.

Semiaustenitic stainless steels are hardened by first heating to 1040°C (1900°F) and then quenching in air to put the material in the solution-treated condition. Precipitation hardening is then performed by heating the workpiece to a subcritical-temperature range, i.e., 480 to 620°C (900 to 1150°F).

HEAT-TREATING PROCESSES FOR NONFERROUS METALS

These materials can be stress-relieved, annealed, solution-treated, and precipitation-hardened. Solution treating and precipitation hardening (aging) commonly are performed in sequence. Solution treatment involves heating the workpiece material to a temperature just below the eutectic melting temperature. At this temperature, alloying constituents go into a supersaturated solid solution. Rapid quenching then preserves the solution. Precipitation hardening then involves heating the workpiece to a somewhat lower temperature for a period of time. Alloying constituents precipitate throughout the workpiece, providing it with increased strength. With some alloys, this precipitation occurs over a period of days at room temperature.

Copper and Copper Alloys

Stress relieving to remove manufacturing stresses involves heating the material to 80 to 100°C (150 to 200°F) above the temperature used to produce mill products of the same alloy. Age hardening of fabricated parts requires a time-temperature cycle that depends on the specific copper alloy and its work-hardened condition. Temperatures ranging from 320 to 480°C (600 to 900°F) and times from ¼ to 3 h are common for beryllium copper flat-rolled material commonly used for flat springs and clips. The age hardening can be carried out in various types of furnaces, but the parts must be protected from oxidation. Normally an atmosphere of exothermic (reducing) gas or disassociated ammonia is used.

Magnesium Alloys

Stress relieving involves heating to temperatures that range from 205 to 260°C (400 to 500°F) for periods from 15 to 60 min, depending on the alloy. Solution heat treatments are performed at temperatures between 370 and 540°C (700 to 1000°F), depending on the alloy, and require up to 24-h soaking time before air cooling. Precipitation hardening takes place with low temperatures in the 150 to 260°C (300 to 500°F) range for periods ranging up to 24 h.

The strength of cast-magnesium alloys decreases at elevated temperatures to such an extent that it is often necessary to prevent castings from sagging from their own weight during solution treatment. Flat castings also will warp during treatment owing to the release of casting stresses. Special fixtures are used to prevent excessive warpage; however, straightening also may be required. Straightening is most readily done after solution treating and prior to aging.

Nickel and Nickel Alloys

Annealing usually requires temperatures between 700 and 1200°C (1300 to 2200°F), depending on alloy composition and degree of work hardening. Stress relieving is used to reduce stresses in work-hardened, non-age-hardenable alloys and is done at temperatures from 430 to 870°C (800 to 1600°F), depending on alloy composition and degree of work hardening. Stress equalizing is a low-temperature heat treatment used to balance stresses in cold-worked material without significantly reducing the mechanical strength produced by cold working. Normally, temperatures from 260 to 480°C (500 to 900°F) and heating periods of 1 to 3 h are required.

Solution treatment is performed at a high temperature to put age-hardenable constituents and carbides into solid solution. Solution-treatment temperatures range from 820 to 1260°C (1500 to 2300°F) with soaking times from ½ to 4 h, depending on the alloy and thickness of the part. Age hardening takes place at temperatures of 430 to 870°C (800 to 1600°F) with time cycles ranging up to 18 h.

Aluminum Alloys

Solution-treating temperatures vary with each specific alloy but generally involve 480 to 510°C (900 to 950°F). The parts are held at temperature for up to 60 min, depending on thickness, and then quenched in water so that cooling below the precipitation range of 400 to 260°C (750 to 500°F) takes place very rapidly. The precipitation-hardening treatment depends on the alloy and can range from no treatment (when precipitation takes place at room temperature over a few days) to a long-time process at low temperatures in the 150°C (300°F) range.

Special treatments for relieving quenching stresses are used, and they range from repeated cycling from −70°C to +100°C (−100°F to +212°F) to quenching at −200°C (−320°F) and then *uphill quenching* in a blast of live steam. Reduction of stresses by 25 percent with cycling to 80 percent with uphill quenching may be obtained.

Titanium and Titanium Alloys

Stress relieving without affecting strength or ductility is carried out by heating to from 480 to 700°C (900 to 1300°F) for periods from ¼ to 50 h, depending on the alloy and the temperature. Higher temperatures permit shorter time cycles.

Solution-treating temperatures range from 760 to 980°C (1400 to 1800°F) with soaking times in the ½- to 1-h range. Heating is followed by a water quench. The aging treatment involves temperature ranges of 480 to 600°C (900 to 1100°F) and heating periods of 2 to 8 h for most alloys. Special fixturing is required during the aging process to prevent distortion.

CHARACTERISTICS AND APPLICATIONS OF HEAT-TREATED PARTS

Case-Hardened Steel Parts

These incorporate a thin surface layer of high hardness with an inner core of unhardened material, although hardening of the core is incorporated if extra strength is required. Common case depths are 0.025 to 1.5 mm (0.001 to 0.060 in), and case hardness ranges up to R_c 65. (See Fig. 8.4.2.) Case hardening provides wear resistance without loss of toughness and enables low-cost low-carbon steel to be used for parts that otherwise would require more costly heat-treatable steels.

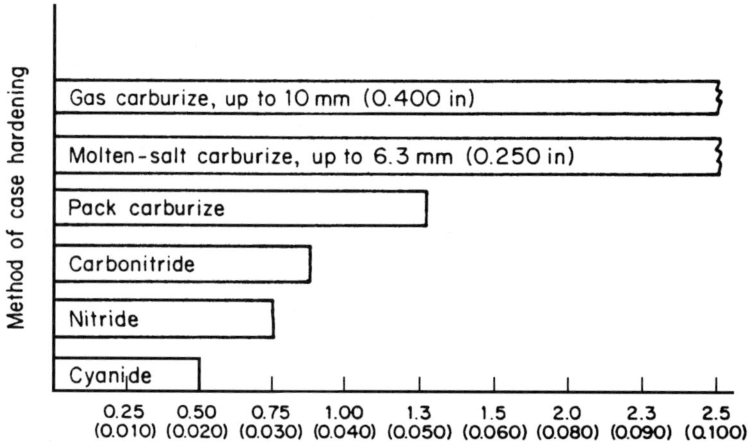

FIGURE 8.4.2 Various case-hardening methods and the range of case depths normally produced by each.

Surface-Hardened Steel Parts

These parts generally have a somewhat thicker layer of hardened surface material than when case hardening is involved. Typical thicknesses range to 6.3 mm (0.250 in). Surface-hardness values are typically between R_c 40 and R_c 60. Parts requiring higher tensile strength as well as a hardened surface can be through-heat-treated for high strength and then surface-hardened for wear resistance.

Flame and induction surface hardening are very fast and can be applied to sections of a part. Since only the surface is heated above the critical temperatures, overall distortion is low. The process has many applications when only local hardening is required. Gear teeth are one example, as are long, thin actuating levers or cam followers which would distort severely if hardened all over.

Through-Hardened Steel Parts

The maximum physical properties of steel are developed by through hardening. Specifically, steel's yield strength is greatly increased, enabling it to be used in high-stress, high-strain applications in which it would otherwise deform and not function. Examples of through-hardened parts include springs, cutting tools of various kinds for wood and metal, dies and molds, and various machine parts subject to high loads.

Martempering provides high hardness with less distortion and fewer harmful residual stresses than does conventional quenching. Austempering provides reasonably high hardness (to R_c 50) with retained ductility and toughness and decreased likelihood of cracking and distortion.

Hardened Cast-Iron Parts

Gray iron is hardened to improve its mechanical properties such as strength and wear resistance. Cast iron can be compared with steel in its reaction to hardening, although it contains graphite and a higher silicon content, which require higher austenizing temperatures along with longer carbon-absorption times at temperature. The hardness depends on the carbon and alloy content of the austenite at the austenizing temperature. Wear resistance of up to 5 times that of soft iron can be obtained, and, as with steel, the iron can be tempered at a lower temperature to increase its strength and toughness.

Flame hardening is the most commonly applied heat treatment for gray iron. It produces a hard, wear-resistant outer layer of martensite and a core of softer gray iron.

Stainless Steel Parts

Stainless steel is heat-treated to produce changes in physical condition, mechanical properties, and residual stress level or to restore maximum corrosion resistance after fabrication. Precipitation-hardened semiaustenitic steels provide tensile strengths up to 1310 MPa (190,000 lbf/in^2). Ferritic and austenitic stainless steels are annealed to restore ductility and softness after work hardening and to ensure maximum corrosion resistance.

Copper-Alloy Parts

Normally, wrought material is supplied by the mill in the solution-treated or age-hardened condition. Parts fabricated from age-hardened material need no further treatment other than stress relief to remove manufacturing stresses. Parts fabricated from solution-treated material must be age-hardened.

Beryllium copper distorts during age hardening, and the material moves in the direction in which it was plastically formed or elastically deflected. Close control of finished dimensions requires fixturing during age hardening.

Magnesium-Alloy Parts

Solution heat treatment improves strength, toughness, and shock resistance. Artificial aging (precipitation heat treatment) after solution treatment produces maximum hardness and yield strength but sacrifices toughness somewhat. Annealed magnesium possesses increased ductility with reduced tensile strength, making the material more suitable for some types of further fabrication.

Aluminum-Alloy Parts

Aluminum parts are heat-treated to increase physical properties. Solution- and precipitation-hardened aluminum has increased yield and tensile strength and hardness and decreased ductility. Annealing softens aluminum and increases its ductility, providing for easier fabrication with forming methods.

Titanium-Alloy Parts

Titanium and titanium-alloy parts can be stress-relieved to reduce residual stresses, annealed to produce optimum ductility, machinability, or dimensional stability, and age-hardened to increase strength. Solution treating, quenching, and aging of certain alloys produce high strength and useful ductility.

Nickel-Alloy Parts

These parts are age-hardened to develop maximums strength. Annealing provides softening necessary for cold-working operations.

SELECTION OF MATERIAL

Generally, the most common grade of material that will produce the physical properties specified should be selected because it will have favorable availability and cost. There are always special grades and special heat treatments that can be used to ensure that the final product is acceptable, but the cost of the material and special heat treating can be so great as to preclude their use under normal manufacturing conditions. The knowledge of what is required for heat treatment and the results that may be expected on the specific part under consideration should provide a limited number of grades from which to make a final determination.

Low-Carbon Steel

When heat-treated, low-carbon steel (up to 0.25 percent carbon) is normally case-hardened to provide a hard, wear-resistant surface. If higher strength is required, the use of alloy grades or higher-carbon steels is necessary.

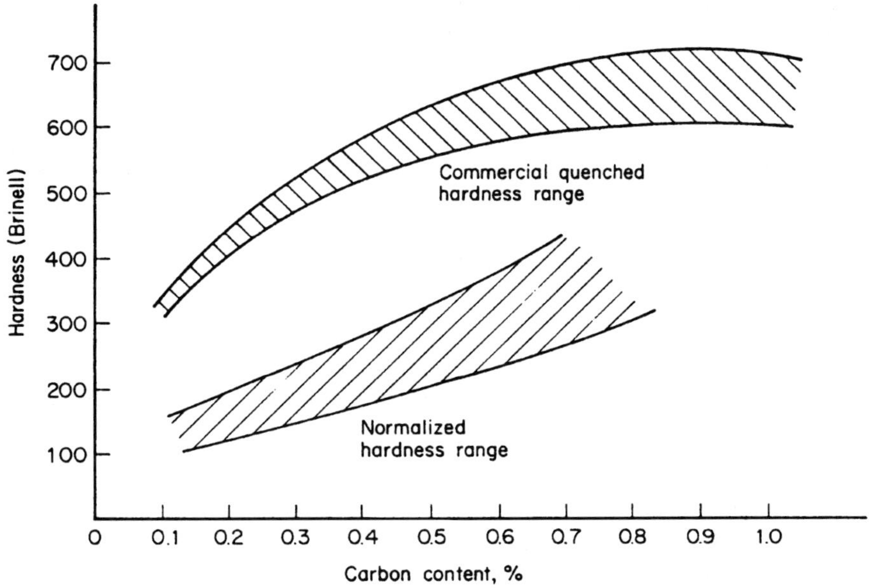

FIGURE 8.4.3 Relation between hardness and carbon content for steel.

Medium-Carbon Steel

The selection of a medium-carbon steel (0.25 to 0.55 percent carbon) presumes that low-carbon grades will not provide the strength necessary for the part and that a steel that will respond to heat treatment for a higher strength is required. Medium-carbon steels are normally through-hardened and tempered to tensile strengths ranging from 690 to 1380 MPa (100,000 to 200,000 lbf/in^2) and are widely applied to parts requiring medium strength and high toughness. Normally, they are quenched in water solutions to develop maximum hardness. If higher strength or hardness is required, alloy-grades or grades with higher carbon content would be selected. (See Fig. 8.4.3.) Medium-carbon steels are available in hot- or cold-finished forms; both are heat-treatable.

High-Carbon Steel

Applications of high-carbon steel (0.55 to 1.00 percent carbon) are more restricted because of decreased ease of fabrication and higher cost. These grades of steel are specified when an application requires high strength and hardness, high resistance to fatigue (e.g., springs), or high abrasion resistance (e.g., knives).

These steels are normally hardened by quenching in water or oil (thin sections) and tempering to the required hardness range for the application. Special heat-treating techniques such as austempering or martempering can be used to obtain the desired physical properties without the distortion associated with the rapid quench necessary to harden the high-carbon-steel grades. These steels are by nature shallow-hardening, and section sizes to which they can be applied are limited. The problem of quench cracking is critical, and designs involving changes in section thickness must be avoided.

Parts requiring different physical properties in the same piece (e.g., tools) can be made by using high-carbon steel since an entire part can be hardened and tempered to provide the toughness required and then rehardened locally to produce an area of higher hardness. Parts with sections too thick for hardening or with changes in section thickness which could cause breakage or with shapes that would cause excessive distortion require the use of alloy steel instead of high-carbon steel.

Alloy Steels

Alloying elements are added to plain carbon steels to provide one or more of the following: (1) greater strength in larger sections, (2) less distortion during hardening, (3) greater resistance to abrasion at the same hardness, (4) higher toughness at the same hardness in small sections, or (5) greater hardness and strength at elevated temperatures. These characteristics are obtained by (1) changing the hardening characteristic, (2) altering the nature and amount of the carbide phase of the steel, or (3) changing the tempering characteristics.

All the alloying elements added to steel (except cobalt) tend to give greater hardenability. Hardenability is the measure of the depth of hardening produced by quenching. A steel with high hardenability will harden more deeply than a steel with low hardenability.

Alloying elements retard the transformation on cooling, thereby allowing a slower (less severe) quench to obtain the same results as with a plain carbon steel. The maximum properties that can be obtained from the alloy-steel grades are the same as those obtained from plain carbon steel of the same carbon content.

SAE Standards list five groups of hardenable alloy steels based on carbon content.

1. *Alloy steels with carbon content of 0.30 to 0.37 percent.* For parts requiring moderate strength and great toughness; frequently used for water-quenched parts of moderate section size or oil-quenched parts of small size. Low hardenability: SAE grades 1330 (G13300), 1335 (G13350), 4037 (G40370), 4042 (G40420), 4130 (G41300), 5130 (G51300), 5132 (G51320), and 8630 (G86300). Medium hardenability: SAE grades 2330, 3130, 3135, 4137 (G41370), 5135 (G51350), 8632, 8635, 8637, (G86370), and 8735 (G87350).
2. *Alloy steels with carbon content of 0.40 to 0.42 percent.* For heat-treated parts requiring higher strength and good toughness; used for medium- and large-sized parts; generally oil-quenched. Most of these steels can be machined in the heat-treated condition under 300 Brinell hardness. Low hardenability: SAE grades 1340 (G13400), 4047 (G40470), 5140 (G51400) (best machinability), and 9440. Medium hardenability: SAE grades 2340, 3140 (G31400), 3141, 4053, 4063, (G40630), 4140 (G41400), 4640 (best machinability), 8640 (G86400), 8642, 8740 (G87400), and 8742 (G87420). High hardenability: SAE grades 4340 (G43400) (best machinability) and 9840 (G98400).
3. *Alloy steels with carbon content of 0.45 to 0.50 percent.* For heat-treated parts requiring fairly high hardness and strength along with moderate toughness. Low hardenability: SAE grades 5045, 5046 (G50460), and 5145 (G51450). Medium hardenability: SAE grades 2345, 3145, 3150, 4145 (G41450), 5147 (G51470), 5150 (G51500), 8645 (G86450), 8647, 8650, (G86500), 8745, 8750, and 9445. High hardenability: SAE grade 4150 (G41500).

4. *Alloy steels with carbon content of 0.50 to 0.62 percent.* Used primarily for springs and tools. Medium hardenability: SAE grades 4068, 5150 (G51500), 6150 (G61500), 8650 (G86500), 9254 (G92540), 9255 (G92550), 9260 (G92600), and 9261. High hardenability: SAE grades 8653, 8655 (G86550), 8660 (G86600), and 9262 (G92620).
5. *Alloy steels with carbon content of 1.00 percent.* Used for parts requiring high hardness and wear resistance such as ball bearings or rollers. Medium hardenability: SAE grade 52100 (G52986).

Cast Iron

The major determining factor in flame hardenability is the combined carbon content, which should range between 0.50 and 0.70 percent. Alloy gray iron can be flame-hardened with greater ease than unalloyed irons. This is due to the wider temperature range and the increased martensitic depth owing to the increased hardenability.

Stainless Steels

Austenitic and ferritic grades are not hardenable, so heat treatments are confined to annealing processes. Martensitic steel can be hardened similarly to plain carbon or low-alloy steels. Maximum strength and hardness depend basically on carbon content.

The high cost of material and of processing stainless steels precludes their use unless corrosion resistance is required. The hardening procedures used therefore are normally restricted to those which do not interfere with corrosion resistance.

Copper and Copper Alloys

Copper and copper alloys are susceptible to the following heat-treating processes: stress relieving, annealing, solution treating, and precipitation hardening. Precipitation hardening can be carried out on the following alloys: aluminum bronze, beryllium copper, copper-nickel-silicon, and copper-nickel phosphorous.

DISTORTION

Heat-treating operations, while producing the necessary properties required in a part, also produce undesirable changes in size and shape. The amount of dimensional change that takes place during heat treatment is the result of the stresses introduced by the manufacturing operations, the thermal stresses from heating and cooling, the stresses set up by the transformation of the hardened structure, and distortion that may result from handling during heating, particularly at the austenizing temperature. Typical dimensional changes due to heating and cooling are shown in Fig. 8.4.4.

Rapid heating and cooling intensify these stresses when there are thermal expansion and contraction at the same time in different portions of the part. (See Fig. 8.4.5.) A steel part cooled in water can develop stresses high enough to crack it. The same part cooled in air will show minimal distortion but will not be hard.

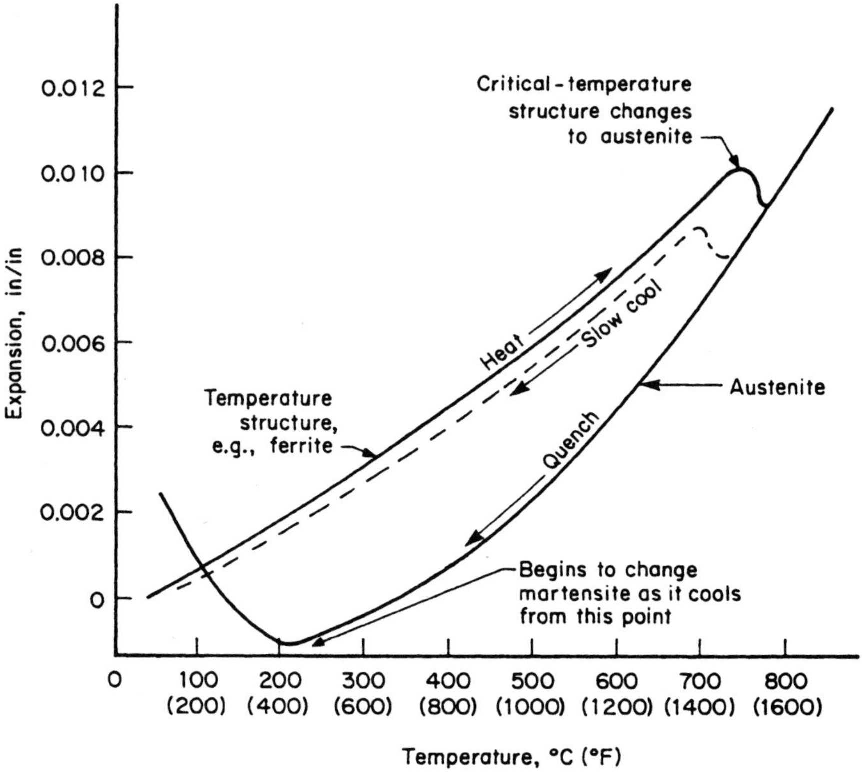

FIGURE 8.4.4 Expansion of 1 percent carbon steel on heating and quenching (solid line) or slow cooling (dotted line). Note that there is a reduction in volume when the steel reaches the critical temperature (about 1400°F) and changes to austenite. There is an expansion after the steel cools to about 400°F and transforms to martensite.

Case-hardened parts are subject to dimensional changes and distortion depending (1) on the configuration of the part, (2) on the specific process employed, (3) on the previous stresses in the part, and (4) on the grade of steel used. Good design will reduce the amount of distortion no matter which method is used, but if the case-hardening method used causes too great a distortion, the heat treater must select a less severe method to produce the specified results. This could involve a simple change in the way in which the part is handled or a completely different case-hardening method, e.g., cyaniding instead of carbonitriding.

As with steel, the heat treating of aluminum and other nonferrous metals causes distortion because of the thermal stresses set up during heating and rapid cooling. The magnitude of the stresses increases with section size and severity of quench. The stress pattern consists of compression in the outer layers and tension in the center portions. The compressive stresses in the surface are desirable to reduce failure by fatigue or stress corrosion. Finishing operations that remove metal tend to expose metal that has tensile stresses, thereby reducing strength in fatigue and corrosion resistance. If the metal is not removed symmetrically, the part will distort.

Warpage of heat-treated parts varies considerably even within the same charge of

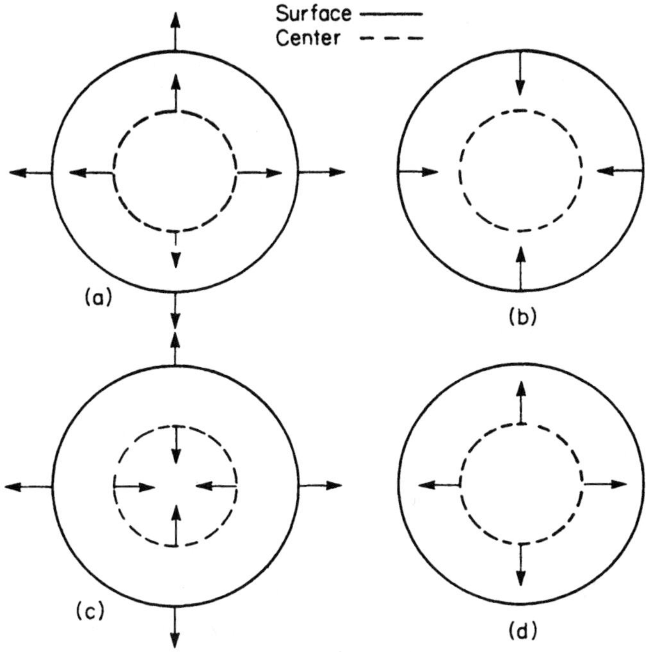

FIGURE 8.4.5 The distribution of stresses in a typical part during heating and quenching. Arrows show direction of stress. (*a*) Part heated uniformly throughout. Center area and surface both are expanded. (*b*) Start of quench; surface contracted more than center. (*c*) Midway during quench; surface transformed, center contracting. (*d*) Near end of quench: surface cold and rigid, center transforming and expanding.

identical parts. In many cases, this necessitates costly hand straightening. For parts that require machining and are subject to objectionable warpage, it is sometimes necessary to rough-machine, heat-treat, and then finish-machine. Special racking to provide uniform quenching under fixed conditions is also employed to minimize and control warpage.

For cast parts, the same heat-treating principles apply, but the stresses developed during heat treatment are compounded by casting stresses and irregularities normally associated with castings. Exact heat-treating cycles are generally tailored to the specific casting and ultimate properties required. Longer soaking times are required for solution of larger alloy constituents, and milder quenches such as boiling water are required to reduce distortion.

In addition to the distortion caused by the severe thermal stresses during cooling, a metallurgical transformation takes place in hardened steel that causes the hardened part to increase in volume. (See Fig. 8.4.6.) Symmetrical parts of uniform thickness increase in volume when hardened without much distortion, but if adjacent thin and heavy sections are hardened, the difference in temperature within the two sections might cause the thin section to be expanding (transformation at hardening point) while the heavy section is still contracting (thermal contraction due to cooling). The resulting stresses could be large enough to exceed the strength of the material, resulting in a cracked part.

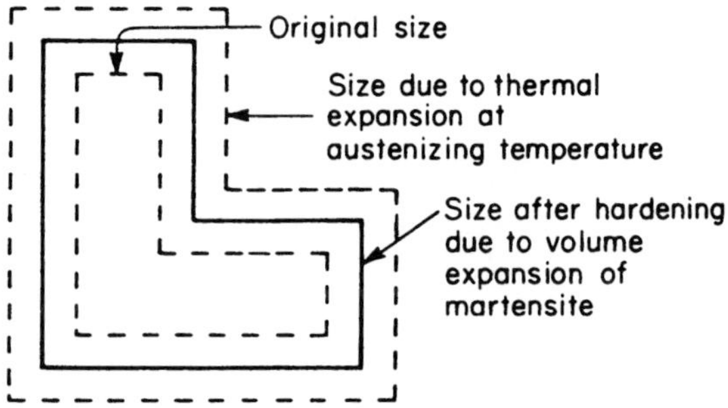

FIGURE 8.4.6 Size changes during and after hardening for a typical part. The size changes are shown greatly exaggerated.

DESIGN RECOMMENDATIONS

The principles of good design for heat treatment are (1) that the properties required in the heat-treated part be obtained without the part's being distorted beyond acceptable limits and (2) that the heat-treated part withstand external stresses during service without failure.

In designing a part for heat treatment, the overall rule is to make the part as simple as possible. Whereas the causes of distortion and failure are many and very complicated, the methods used to overcome these forces are very few and very simple. First and foremost, the part being designed must fulfill its basic function. Once the overall size and shape of the part have been decided, the design for heat treatment can be made. The fundamental rule is to make it as simple as possible, keep it symmetrical, have uniform cross sections, and balance the weight. The ideal design would be a shape that when heated (or cooled) would have the same temperature at every point within the part.

Abrupt changes in section sizes must be avoided by making the change as gradual as possible. (See Fig. 8.4.7.) Long, thin parts (such as connecting rods) should have symmetrical, rounded cross sections to reduce stresses. (See Fig. 8.4.8.)

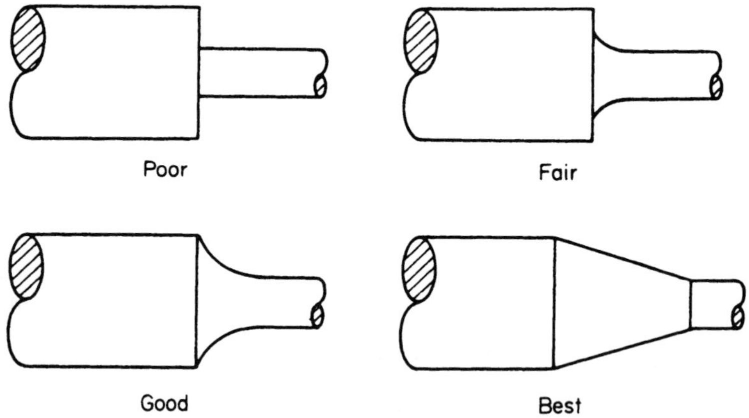

FIGURE 8.4.7 Avoid abrupt changes in sections of parts to be heat-treated.

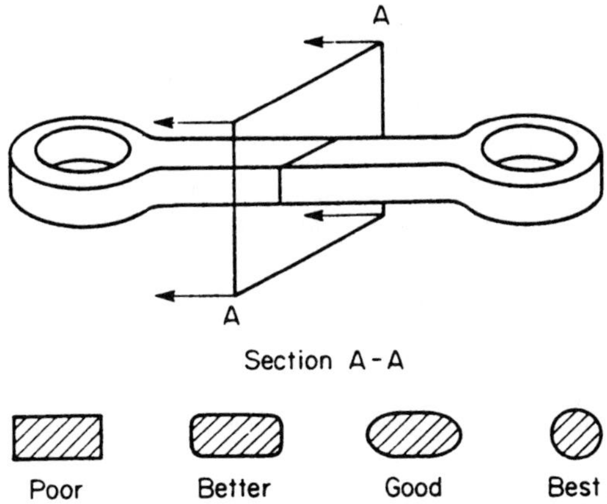

FIGURE 8.4.8 Rounded, symmetrical cross sections reduce heat-treating stresses.

The same principle of symmetrical cross sections applies to parts with holes. Holes should be centered so that the mass of metal surrounding them is equally balanced. (See Fig. 8.4.9.)

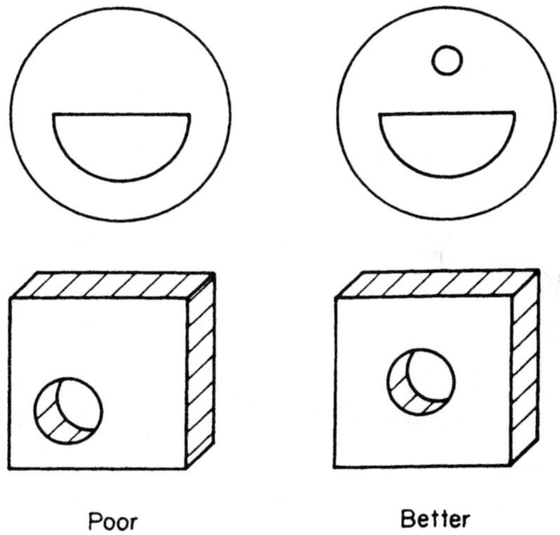

FIGURE 8.4.9 Maintain uniform section thickness around holes.

The location of holes or cutouts is important. They should not be located closer than $1\frac{1}{2}$ diameters from the edge in order to avoid unequal mass distribution. (See Fig. 8.4.10.) All internal corners should be rounded, and noncutting holes should have radii at the top and bottom surfaces. (See Figs. 8.4.11 and 8.4.12.)

Press-metal dies are normally heavy, flat sections designed to withstand high stresses during operation. They require extreme accuracy after hardening. The config-

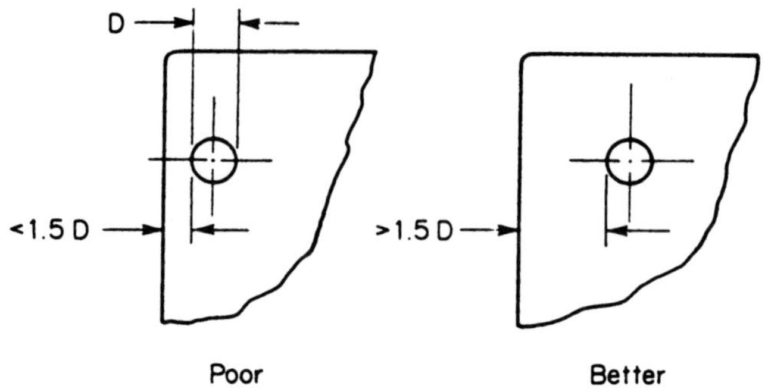

FIGURE 8.4.10 Do not locate holes closer than $1\frac{1}{2}$ diameters from an edge.

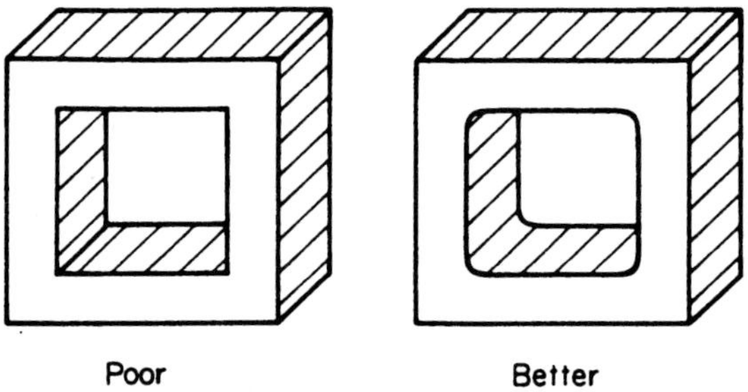

FIGURE 8.4.11 Avoid sharp internal corners, which concentrate heat-treating stresses.

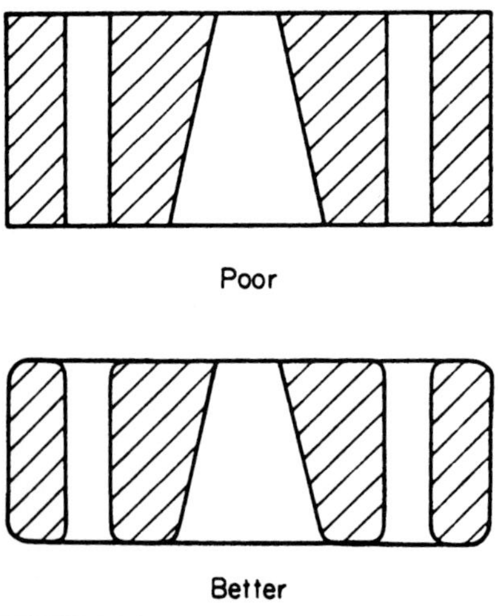

FIGURE 8.4.12 Entry and exit edges of holes should be radiused whenever possible.

uration of the working shape of the die cannot be altered, but the remainder of the die should be designed with balanced mass (see Fig. 8.4.13), ample radii, proper hole locations, and a peripheral width equal to twice the thickness of the die block (see Fig. 8.4.14). If the die section is irregular in shape or must have unbalanced sections, it should be considered for two-piece construction (see Fig. 8.4.15).

Teeth, whether gear, cutter, or in the form of splines, stand out from the heavier mass of the part and are subject to thermal stresses disproportionate to those of the body. Individual teeth should have ample radii at the root and edges (see Fig. 8.4.16) and be located so as to avoid thin sections (see Fig. 8.4.17). Sharp edges, as on a cutter, should be ground after hardening.

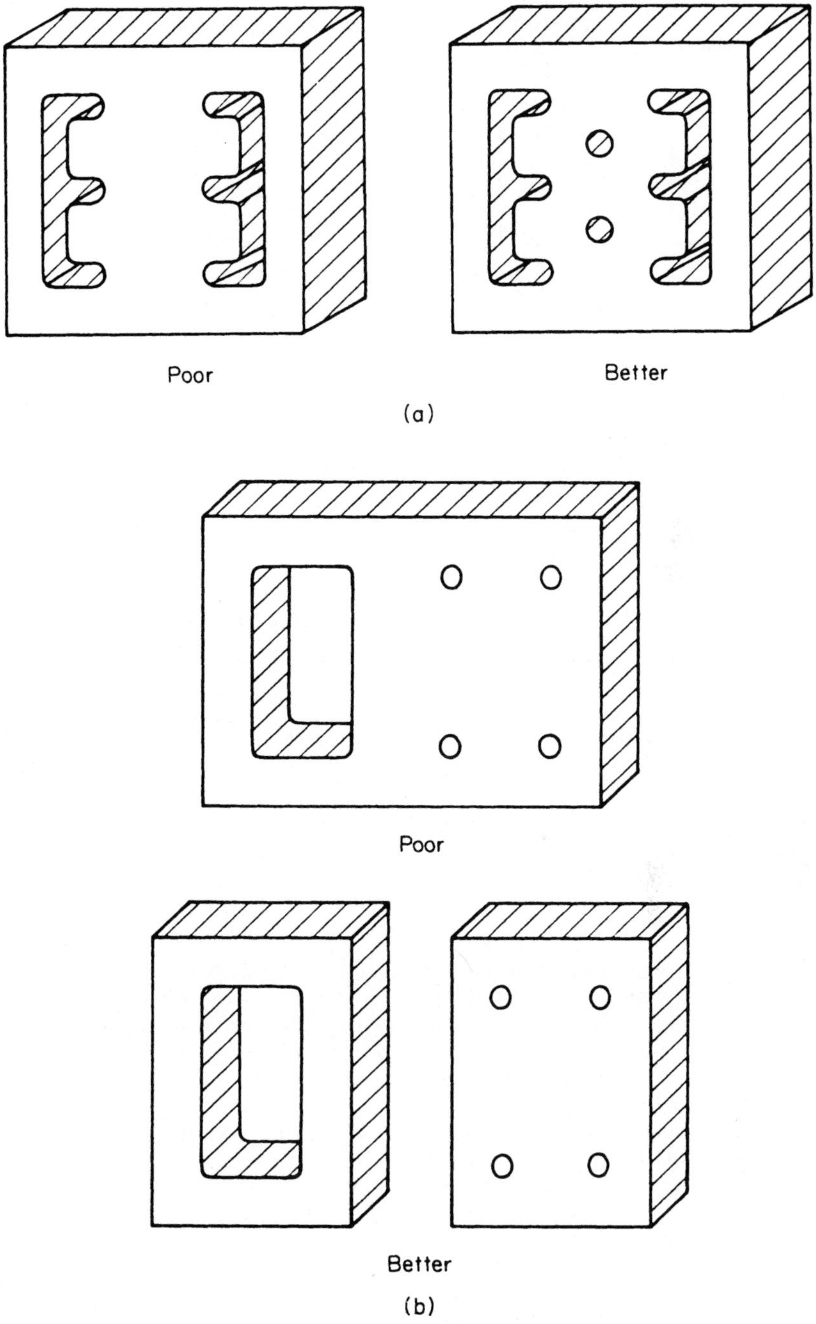

FIGURE 8.4.13 Die sections should have balanced mass to avoid heat-treating problems.

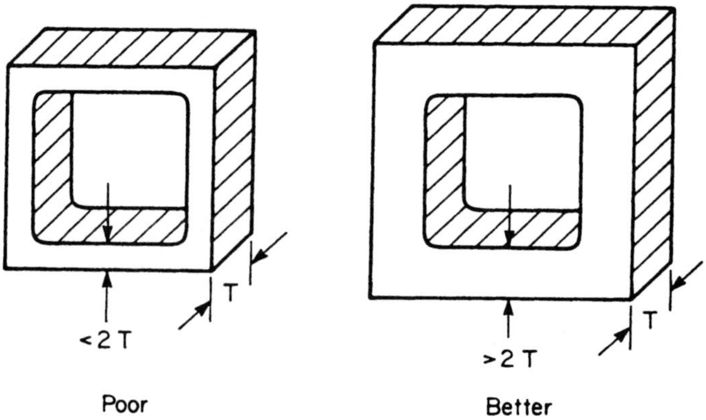

FIGURE 8.4.14 Section width around large openings should be at least two times stock thickness.

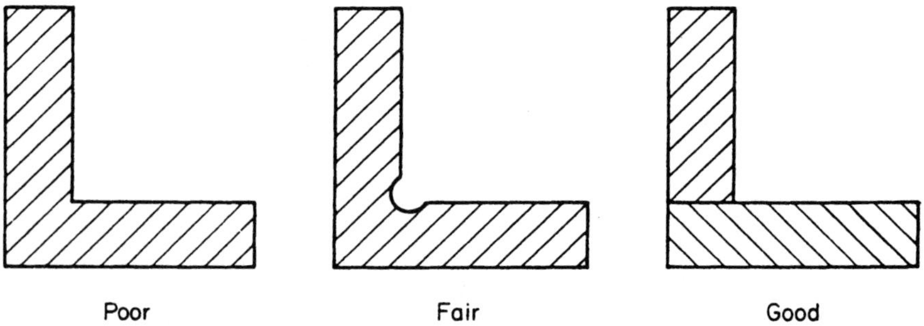

FIGURE 8.4.15 Two-piece construction may be advisable to eliminate stress concentrations in internal corners.

Keyways, whether external or internal, should have generous radii and be located symmetrically. Keyways used with gear or cutter teeth should be positioned in line with the base of the tooth so as to maintain uniform cross-sectional areas.

The hub of a gear is also subject to distortion if the mass is unbalanced, and if function requires unequal mass distribution, a sectional design should be considered (see Fig. 8.4.18).

Parts with through holes concentric with the axis do not present a serious distortion problem, but blind holes that are deep cause distortion owing to unequal mass and additional difficulty if the internal surface must be hard (see Fig. 8.4.19).

When specifying heat treating on a drawing, it is preferable to denote the end condition desired rather than the process. In this way, both the heat treater and the part user can check to the same standard. Spelling out the complete process will not guarantee the required end condition.

When stipulating heat treatment, be specific as to the condition required rather than simply specifying "anneal" or "heat-treat and temper." Refer to specific hardness tests

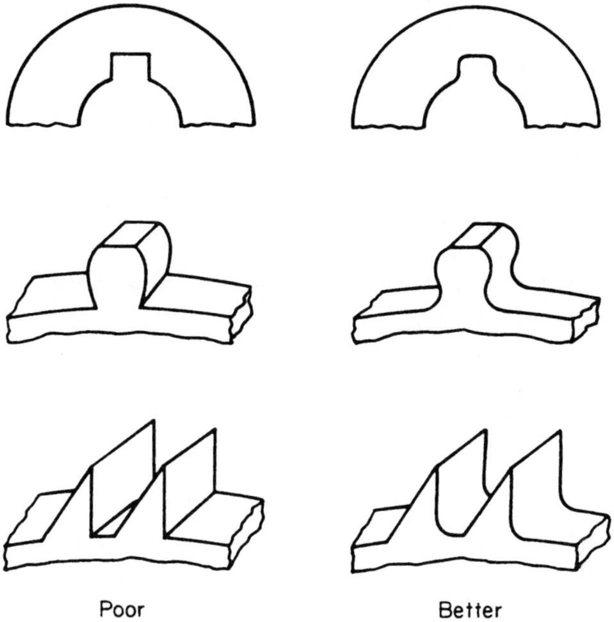

FIGURE 8.4.16 Rounded corners at the base of gear and ratchet teeth and at the bottoms of keyways reduce heat-treatment stress concentrations.

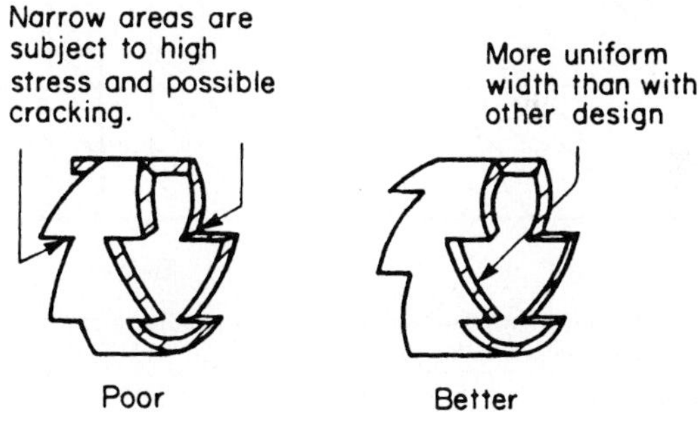

FIGURE 8.4.17 Teeth should be located so as to avoid thin sections.

(e.g., Brinell, Rockwell, Rockwell superficial, Vickers, etc.) and specific hardness scales and hardness values.

When specifying case hardening, state either the total case depth or the effective case depth. If the part is to be ground, state whether the depth applies before or after grinding. Allow as liberal a tolerance as possible on both case depth and hardness.

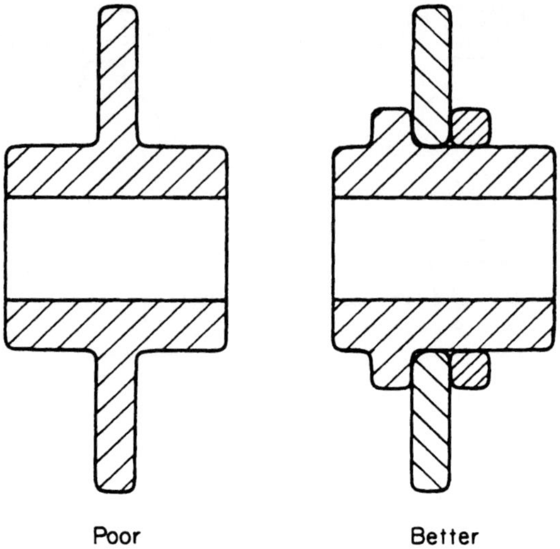

FIGURE 8.4.18 A sectional sprocket or gear design avoids stresses due to unbalanced sections.

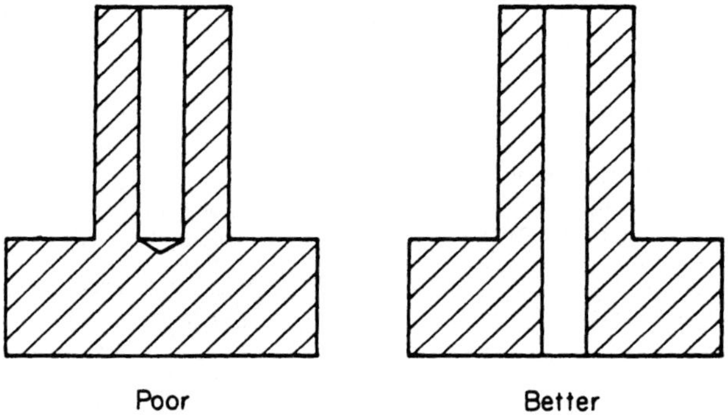

FIGURE 8.4.19 Through holes provide more uniform section thickness and reduce heat-treating problems.

CHAPTER 8.5
ORGANIC FINISHES

Dr. Gerald Schneberger

G. L. Schneberger and Associates
Grand Blanc, Michigan

THE PROCESS

Organic coating is the application of a film-forming composition to the surface of a part or product usually for the purpose of protection and/or decoration. The coating is applied as a liquid or, sometimes, as a dry powder. Conversion to the final thin film is accomplished by evaporation or chemical reaction, often with the aid of heat. There are a number of specific coating processes:

1. *Brushing.* Liquid paint is applied with a hand-held brush.
2. *Roller coating.* This is similar to brushing except that a cylindrical roller is used to spread the paint over the part's surface.
3. *Curtain coating.* Application takes place by passing the part beneath a trough that has a bottom slit of controllable width through which the coating material flows. The amount of paint applied is controlled by varying the slit width or the part's speed.
4. *Dipping.* The part is submerged in a container of paint, with the excess allowed to drain away after the part has been removed from the container.
5. *Flow coating.* A stream of paint is directed against the part by one or more nozzles. The paint is not atomized, and the excess is allowed to flow from the part and drip back into a storage tank by gravity.
6. *Air spray.* Compressed air is used to atomize liquid paint and to propel it toward the part.
7. *Airless spray.* Pressure is applied to liquid paint, forcing it at great velocity through a nozzle. The pressure drop experienced by the paint on leaving the nozzle is sufficiently great to break it up into fine droplets that are carried forward away from the nozzle.
8. *HVLP* (*high volume, low pressure*). Similar to air spray except that a high volume of low-pressure air is used to atomize the paint instead of a lower volume of high-pressure air. HVLP is often slower but more transfer-efficient than air spray.

9. *Electrostatic spray.* Electrical energy is used to charge air or airless paint-spray droplets and thereby cause an attraction between the charged droplets and the grounded workpiece. The part must be somewhat conductive.
10. *Powder coating.* Powdered-paint particles are directed toward the part and adhere to its surfaces by electrostatic attraction and compaction. The coated part is heated to fuse the powder particles and create an integral thin film. The powder may be electrostatically sprayed or suspended as a fluidized cloud into which the heated part is dipped.
11. *Electrodeposition.* Charged paint, in an aqueous medium, is electrically plated out on the surface of the submerged part.

Table 8.5.1 summarizes those aspects of paint-application processes which are of importance to part designers.

TYPICAL CHARACTERISTICS

Organic coatings are used on parts of nearly every description. These parts may range in size from bobby pins to airplanes and in complexity of shape from plates to bedsprings. Objects to be coated can be soft or hard, tough or brittle. They can be formed of a single piece or be an assembly of hundreds. The entire surface or only a portion may require coating. A single coat or a combination of three or more different coats may be required.

Most paints are applied in thicknesses ranging from 0.025 to 0.100 mm (0.001 to 0.004 in). The purpose of the paint coating can be any of the following: (1) to provide corrosion protection, (2) to hide surface defects, (3) to change the color or surface gloss of a part, (4) to provide increased surface hardness or strength for soft or fragile base materials, (5) to make a surface either more or less slippery, e.g., on ship bottoms, steps, or blackboards, or (6) to provide identification or other marking.

TABLE 8.5.1 Paint-Process Characteristics Which Influence Part Design

	Relative speed	Coats recesses	Requires conductive parts	Requires good drainage	Requires flat surfaces
Brushing	S	Y?	N	N	N
Roller coating	F	N	N	N	Y
Curtain coating	F	N	N	Y	Y
Dipping	F	N?	N	Y	N
Flow coating	F	Y?	N	Y	N
Air spray	S	Y	N	N	N
Airless spray	S	Y	N	N	N
HVLP	S	Y	N	N	N
Electrostatic spray	S	Y?	Y	N	N
Powder-fluidized bed	F?	N?	N	N	N
Powder-electrostatic spray	S	Y?	Y	N	N
Electrodeposition	F	Y	Y	Y	N

Note: S = slow, F = fast, N = no, Y = yes, and ? = depends on the particular design.

Improved appearance and improved corrosion protection are the most frequent reasons for applying organic coatings. A properly applied organic coating can provide a high level of corrosion protection.

Organic finishes can be clear or colored. The surface can be glossy or dull or have various degrees of texture. Textured finishes have the advantage of hiding surface irregularities better than glossy finishes do.

Primer coats are used to provide better adhesion to the base material and better corrosion protection, filling or hiding power, or sandability than the topcoat material does. Topcoat materials, conversely, usually have better appearance, toughness, or hardness than primer materials.

ECONOMIC PRODUCTION QUANTITIES

There really is no limit, either upper or lower, to the quantity of parts that may be coated successfully. Custom-built, one-of-a-kind items are coated routinely by individuals or job shops. The number of items to be coated, however, may have a very great effect on the process chosen to apply the paint. Therefore, the design of the part is frequently affected by the number of parts to be produced, as indicated in Table 8.5.1.

SUITABLE MATERIALS

Substrates

Virtually any solid material can be finished with an organic coating, but some materials may require more extensive precleaning and special pretreatment before the application of the paint. Materials that are inherently difficult to paint include brightly plated metals and some plastics, e.g., polyolefins (polyethylene and polypropylene), polytetrafluoroethylene (Teflon*), and acetal, and some silicones. With these materials, the fundamental difficulty is one of paint adhesion and has little or nothing to do with the part's configuration. Table 8.5.2 summarizes the paintability of common thermoplastics. Most thermosetting materials are paintable.

Coating Materials

Selecting a paint for production use is a complex task usually handled as a joint effort with the paint supplier. Table 8.5.3 lists important characteristics of common coating materials.

Traditionally, industrial finishes have been organic solvent–based lacquers or enamels, the distinction being that lacquers remain soluble while enamels become insoluble after curing. In recent years, water-borne coatings and powder finishes have become more common because of their reduced organic solvent content and because they are sometimes more economical. These systems are compared with conventional finishes in Table 8.5.4. The performance of a given paint is determined more by its

*Trademark of E. I. DuPont de Nemours & Co.

TABLE 8.5.2 Paintability of Common Plastics

ABS	Paintable with a broad range of paint systems.
Acetals	Exhibit poor adhesion to most paints; an acid-etched surface is recommended for good paint adhesion, followed by a primer coat.
Acrylics	Paintable; lacquer-type coating materials are most common.
Cellulosics	Migration of internal plasticizers can cause tackiness in the film after a period of time; generally coated with air-dry lacquer.
Ionomer	Can be painted; an epoxy prime coat is recommended.
Nylons	Can be painted with many coating systems.
Polycarbonate	Parts can be painted with either air-drying or baked coatings; however, some solvents craze polycarbonate, so any planned coating should be tested on samples.
Thermoplastic polyester	Can be painted with lacquer or enamel; requires baked-alkyd primer.
Polyethylene	Requires flame treatment or chemical-oxidation treatment prior to painting.
Polypropylene	Can be painted without pretreatment if one of several available paint systems is used.
Polyphenylene oxide	Does not require a primer and can be painted with enamel or lacquer and baked at 120°C (250°F).
Polystyrene	Can be painted with alkyd-, acrylic-, and urea-based coatings; others may contain solvents that attack the polystyrene.
Vinyls	Can be painted, but plasticized vinyls may exhibit tackiness or loss of adhesion due to plasticizer migration.

Source: Material for this table supplied by Ronald D. Beck.

chemical structure, thickness, and cure than by whether it is applied from a solvent, from water, or as a powder.

Every paint requires good surface preparation. Ferrous materials should be phosphated for maximum adhesion and corrosion resistance. Plastics must be thoroughly freed of mold-release agents.

The relative cost of various paint materials is influenced by far more than simply the cost per gallon. Factors such as transfer efficiency (percentage of paint reaching the part), thickness, surface preparation required, energy and time for cure, solvent and solids disposal, floor space, and ease of repair also must be considered.

DESIGN RECOMMENDATIONS

General Recommendations

1. Always design parts for easy cleaning because cleaning is a vital prerequisite to successful painting. (See Chap. 8.1.)
2. Avoid holes, slots, gaps, or other recesses with depth-to-width ratios greater than 1, as shown in Fig. 8.5.1, for these are difficult to coat on the inside except by dipping.

TABLE 8.5.3 General Characteristics of Common Finishing Materials

Chemical type	Abrasion resistance	Brittle-ness	Weather-ability	Ease of repair	Cure energy	Solvent toxicity	Fire hazard	Moisture resistance	Solvent resistance
Alkyd enamel	M	M	M	M	H	H	H	M	H
Acrylic enamel	M	M–H	H	L	H	H	H	M	H
Acrylic lacquer	L	H	H	M	L	H	H	M	V
Polyester enamel	M	M	M–H	V	M	H	H	L	H
Polyurethane enamel	H	L	M	H	M	H	H	V	H
Vinyl lacquer	H	L	M–L	M	M	H	H	H	V
Epoxy enamel	M	M	L	L	M	H	H	H	H
Thermoset powder	M–H	M–H	M–H	L–M	M	N	L	V	M–H

Note: H = high; M = medium; L = low; V = variable; N = none.

TABLE 8.5.4 Comparative Properties of Solvent-Based, Water-Borne, and Powder Coatings

Type	Fire hazard	Nonferrous piping	Gloss limitations	Overspray recovered	Humidity temperature control critical	Easy color change	Heat cure
Organic solvent	Y	N	N	Sometimes	Sometimes	Y	Usually
Water-borne	N	*	N	N	Often	Y	Usually
Powder	†	N	Y	Y	Y	‡	Always

Note: Y = yes; N = no.

*Water-borne coatings usually require plastic or stainless-steel piping.

†Although airborne powder can explode if sparked, the ignition energy required is higher than for solvent-based paint vapor.

‡Changing powder color sometimes can be a problem.

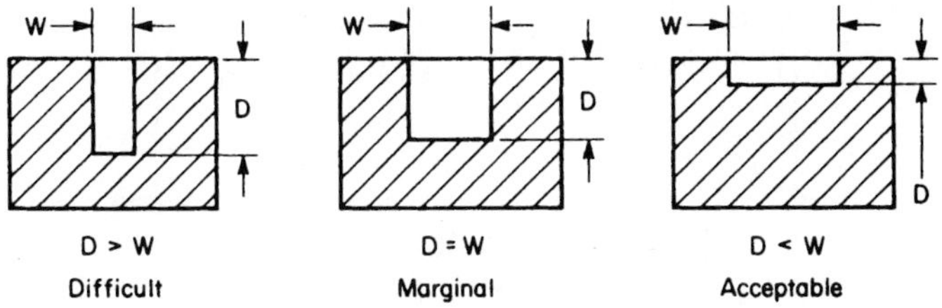

FIGURE 8.5.1 Deep holes, slots, or other recesses are difficult to coat with most painting processes.

3. Avoid small holes of any depth because they may have a tendency to clog unless the electrodeposition process is used.
4. Avoid part designs that call for lines of demarcation between paint colors in the middle of a panel, as shown in Fig. 8.5.2. It is much better to join different colors at a naturally occurring seam in the product.
5. Avoid designs that require painting of both plastic and metal sections at the same time. Often the proper technique for metal surfaces is inappropriate for plastics, and vice versa.
6. Avoid sharp edges and points, whenever possible, because paint tends to creep away from these areas and leave too thin a film.
7. When the product includes a casting, forging, weldment, or other component that is apt to have a surface that requires polishing prior to painting, consider the use of a textured, hammertone, or wrinkled finish. These finishes have the property of hiding surface defects and can greatly reduce the cost of prepainting.
8. Although hardness and corrosion resistance may be lowered, use of air-drying rather than oven-curing paint coatings will usually reduce costs and should be considered when the application permits.
9. The designer should avoid specifying more paint coverage than required. Fewer or thinner coats usually will reduce costs. Often nonorganic coatings (e.g., anodizing) are lower in cost than painting, and many materials may require no organic coating at all.

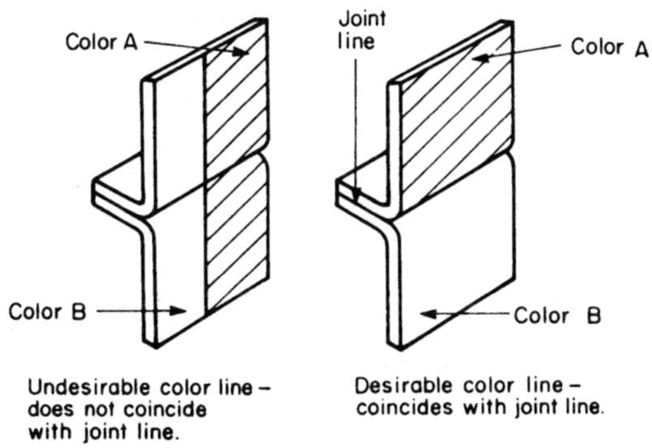

FIGURE 8.5.2 Avoid part designs that require paint masking.

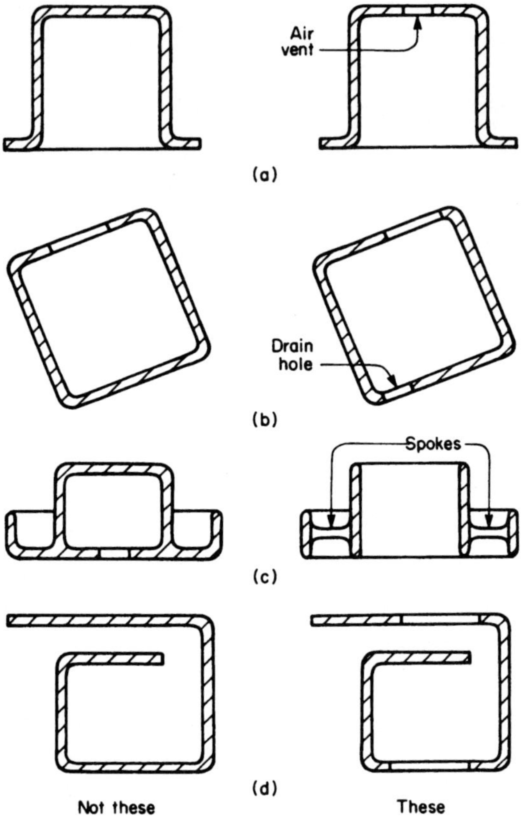

FIGURE 8.5.3 In view *a,* the air vent allows paint to reach the upper inner surface if the part is dip-painted in the position shown. In view *b,* the small drain hole prevents the entrapment of flow-coated or dip-coated paint. The more open design of the right-hand part in view *c* provides for unimpeded flow of coating material. In view *d,* the design on the right is better for flow coating because the openings make inner surfaces accessible.

Dip and Flow Coating

Parts to be coated by these methods should allow for adequate drainage and provide surfaces accessible to the paint. Dip-coated parts should not have air pockets. Recessed areas that cannot be reached by a flow-coating stream should be avoided. Figure 8.5.3 illustrates examples of good and bad design. Many times the designer needs only to be certain that the part can be racked with the proper orientation, as shown in Fig. 8.5.4.

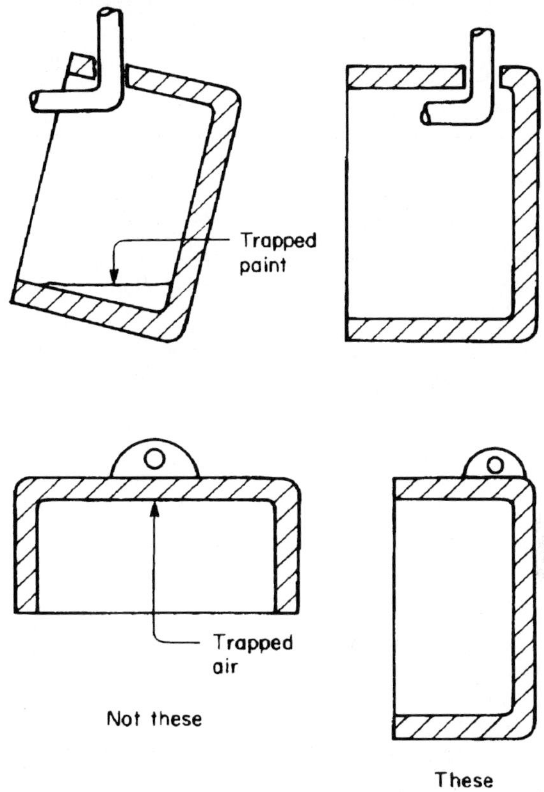

FIGURE 8.5.4 Parts to be coated by flow coating or dipping in liquid paint or a fluidized bed should be designed so that they can be racked in a position that allows air venting and easy drainage.

Air Spray

For parts to be sprayed, it is important to avoid dead air recesses, blind spots, and sharp edges. Recesses, particularly if they are deep, may refuse to accept atomized paint if the air carrying that paint develops a backpressure in the recess. Also avoid blind areas, i.e., areas in which paint cannot approach the surface on a direct line of flight from the spray gun, as shown in Fig. 8.5.5. Air spray is best for large, solid objects for which the amount of overspray will be minimal.

Airless Spray, HVLP

Because airless-spray clouds are less turbulent and involve moving smaller volumes of air than conventional air spray, it is somewhat easier to coat recesses with the airless

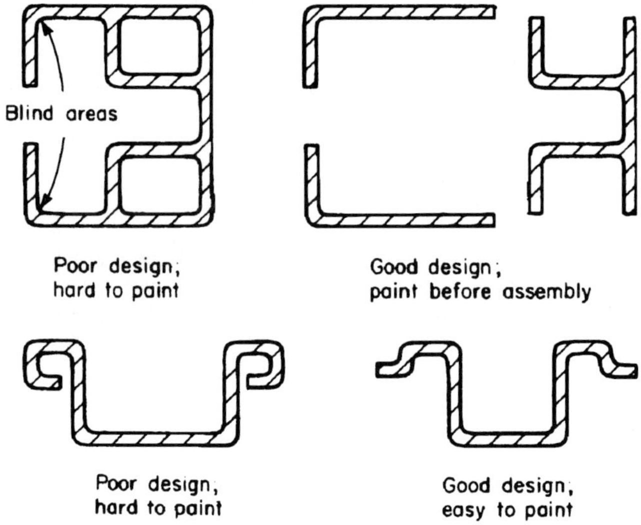

FIGURE 8.5.5 Avoid blind areas, i.e., surfaces that are behind corners or protrusions. If blind areas are unavoidable, paint before assembly if possible.

process. Deep recesses, however, should be avoided, particularly if they are narrow. Intricate part shapes also should be avoided because these shapes are most efficiently coated if the rate of paint application can be controlled. Airless-spray equipment requires that paint delivery be either totally on or totally off, so it is impossible to cut back on the delivery rate and spray only a little (at the same pressure).

Electrostatic Coating

Electrical charges tend to concentrate at sharp edges and points and repel paint; thus these features should be avoided if parts are to be electrostatically coated. Since electrical fields do not penetrate recesses well, it is also desirable to keep corner angles as large as possible, as shown in Fig. 8.5.6. Frequently, the angle is of small importance

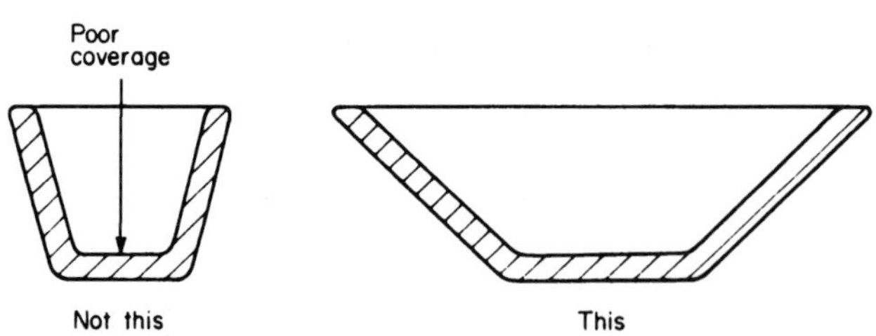

FIGURE 8.5.6 Electrostatic painting requires recesses to be shallow and wide and corner angles to be large as shown to ensure adequate coverage.

to the function of the part but can make considerable difference in the ease with which electrostatically sprayed paint can be deposited in recesses.

If metallic parts of an assembly are electronically insulated from the grounded portion, the electrostatic field may charge them to the point at which they arc to the nearest ground. These insulated subassemblies must be grounded to avoid fire hazards.

Roller Coating

For effective roller coating, the surfaces of the part should be flat, with a minimum of embossing, in order to avoid puddling. In addition, curved part edges should be absent or at least not critical to the finish because the roller-coating process is essentially effective only for flat surfaces.

Curtain Coating

Parts to be curtain-coated should be flat and without depressions. Sharp edges and vertical sides also should be avoided, as shown in Fig. 8.5.7.

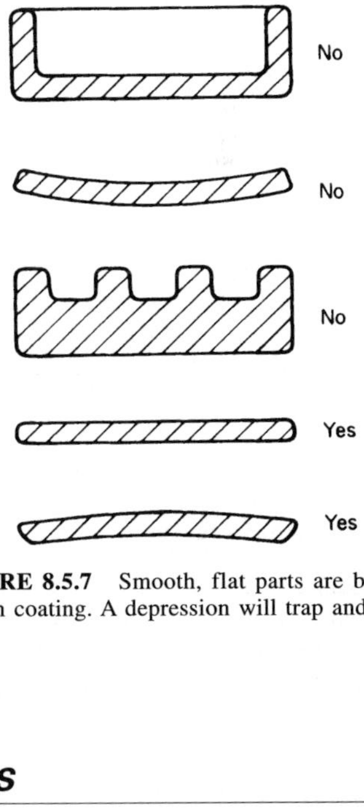

FIGURE 8.5.7 Smooth, flat parts are best for curtain coating. A depression will trap and waste paint.

Powder Coating

If parts are to be coated by the fluidized-bed process, they should be designed without air pockets and so that there are no dead spaces, as shown in Figs. 8.5.3 and 8.5.4, to trap unnecessary powder. If an electrostatic powder-spray process is used, sharp edges and recesses should be avoided. Inside angles of much less than 90° are also undesirable because of the inability of electrostatic fields to penetrate recesses.

DIMENSIONAL AND QUALITY FACTORS

Paint-film thickness is important because it influences appearance, durability, function, and cost. Many paint specifications call for only a minimum thickness, but too thick a coating can cause problems also. Films that are too thick often crack when subjected to sudden temperature changes, and they also result in high paint consumption. The cold-cracking problem is especially serious if the paint has been overbaked, leaving a brittle film. Extra paint thickness sometimes adversely affects assembly or leads to seizing or undue friction in service. Thin films, however, often have inadequate hiding power and poor moisture resistance. In addition, they simply wear away faster in service.

Minimum desirable film thickness depends on a number of factors such as the environment the product will experience and the hiding, filling, and decorative properties required. One U.S. manufacturer of consumer-durable products used indoors specifies the following minimum thicknesses: primer coat, 0.013 mm (0.0005 in); exterior coat, 0.025 mm (0.001 in).

The maximum-thickness specification should be generous, 150 to 200 percent of the minimum, unless adverse functional factors such as those mentioned above are present. Most coating processes and product configurations have led to wide inherent variations in coating thickness (25 to 100 percent).

Inexpensive instruments are available for measuring paint-film thickness. Their frequent use should be specified as an integral part of the controlled manufacturing operation. Frequently, equally important as thickness is adhesion and hardness (scratch resistance) of coatings. Other important characteristics are gloss; flexibility; color; abrasion, water, oil, and impact resistance; and resistance to other environmental factors. Tests are available to measure these properties.

CHAPTER 8.6
DESIGNING FOR MARKING

Ralph A. Pannier
The Pannier Corporation
Pittsburgh, Pennsylvania

REASONS FOR MARKING

The widespread practice of marking products, parts, and materials has many purposes. They include the following:

1. *For production identification.* Product and brand names and model names and numbers are normally included on finished products.
2. *For advertising.* Trademarks, the manufacturer's name, and model identification all have advertising value if attractively presented.
3. *To indicate grade, size, or class of a material or component.*
4. *To indicate that an inspection or other manufacturing operation has been carried out.*
5. *To indicate the part number in case product repair is later required.*
6. *For patent protection.*
7. *To indicate the date of manufacture or the product's serial number.*
8. *For decoration.*
9. *To indicate dimensions of components before selective assembly.*
10. *To provide graduations or other functional marks.*
11. *To provide instructions on the operation of the product.*
12. *To display product specifications such as horsepower, voltage, etc.*
13. *To provide automatic data collection through the use of bar codes or other readable markings.*

MARKING METHODS

The designer has a wide range of marking methods to choose from. They include the following.

Freehand Marking

This is a fundamental method of applying visible markings to many kinds of surfaces, using brush, crayon, chalk, pencil, pen, or a similar hand-impelled marking device. The marking action always involves the transfer of part of the marking medium (e.g., ink, stain, paint, pencil graphite, etc.) to the marked surface.

Printing

Printing, either directly to a workpiece or to a nameplate, label, or tag, is a common marking method. In industrial marking, it most frequently involves transferring an image from an inked printing die, usually rubber, onto the surface being marked. This is called *flexographic* or *contact printing.* Noncontact printing uses paint or ink applied by means of a spray. This is commonly referred to as *ink-jet marking.*

Stenciling

When ink, paint, dye, stain, or other fluid coloring is applied to any surface through the openings of a marking device so that the shape of the aperture controls the form of the mark, the method of application is known as *stenciling.* The fluid coloring may be applied by brush, spray, roller, or a similar implement, and the stencil may be prepared from paper, steel, copper, rubber, plastic, etc. Also, abrasives can be sprayed through a rubber or plastic stencil to mark materials such as marble, glass, and ceramics. Silk-screen printing (described in Chap. 6.12) is another stencil method particularly useful when decorative effects or attractive brand or model names are marked on a finished product.

Etching and Engraving

These two kinds of marking produce marks on the surface of metal or other material by removal of part of the material in such manner that the material remaining forms the desired marking. In the case of *engraving,* the material removed is cut away by sharp or abrasive instruments; in *etching,* the material is removed by chemical or electrolytic action. The resulting mark may be positive or negative depending on whether the material removed or the material remaining forms the image. Generally speaking, an engraved mark is much deeper than an etched one.

Laser Marking

This can involve several different approaches: (1) direct material removal (laser engraving), (2) removal of only the surface coating when the coating has a high contrast to the base material, and (3) inducing a color change in the surface coating from the energy of the laser. (Laser material removal is discussed in Chap. 4.21.) Computer control provides flexibility and automatic operation.

Casting and Molding

The mold cavity from which the component takes its form includes the required mark (often accomplished by inserting an engraved die or an embossed tag), and the final component thus includes the mark as an integral part of its surface. Molded or cast marks may be either cameo (raised) or intaglio (depressed) in form.

Branding

The application of a symbol by the use of a heated marking device that burns the mark into the receiving surface is known as *branding*. This is one of the oldest kinds of marking and is still used for identifying materials such as wood, leather, fiber, plastic, composition materials, etc. Hot stamping is a form of branding in which the heat and dwell time transfer the image onto or into the material. This process, described in Chap. 6.12, uses a plastic film between the workpiece and the stamping die to provide the marking medium.

Embossing and Coining

These two methods are covered in Chap. 3.2. In embossing, raising in relief (cameo) from one surface causes the opposed surface to be equally displaced in debossed (intaglio) form because the material is squeezed between mating male and female dies. In coining, both cameo and intaglio effects are produced on one surface without necessarily affecting the opposite surface, for in this process the workpiece to be marked is struck or pressed against the dies with such force that the metal or other material flows into conformity with the die's shape.

Embossing is commonly used for marking metal tags. Individual tags or tags in continuous coils may be embossed by manually activated or automatic embossing machines.

Stamping

In this common method of marking, the lettering or other symbols are impressed into the surface of the workpiece. A hardened die is forced into the workpiece surface either by a steady press force or by impact. The latter may be applied manually by a hammer blow, one letter or symbol at a time. Pneumatic, hydraulic, mechanical, or solenoid presses can stamp combinations of characters in one stroke. Roll-marking devices are also economical for many applications.

Programmable stampers that generate characters or graphics by indenting multiple dots or impressions to form a recognizable shape are widely used. Flexible software enables the stamper to imprint numbers, diagrams, or logos on a wide variety of substrate materials.

Nameplates, Labels, and Tags

These are often used to provide the same kind of information that would otherwise be marked directly on the product. All the marking methods noted above are also applicable to nameplates, labels, and tags, provided the base material is suitable. A variety of fastening methods are available for nameplates, but threaded fasteners, adhesives, and rivets are used most commonly.

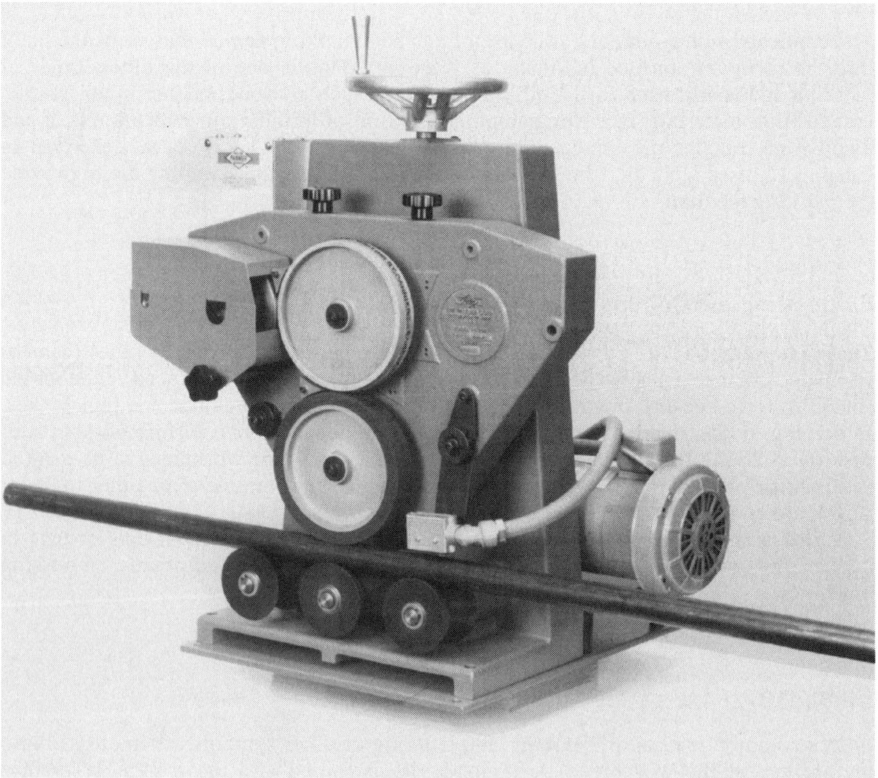

(a)

FIGURE 8.6.1 Common methods for product marking. (*a*) Offset print marking metal bar stock with rubber *type*. (*Courtesy Pannier Corporation.*)

Decals

The transfer of printed material from plastic film or paper to the product's surface provides a means for displaying brand names, trademarks, and decorative material in an attractive manner.

Machine-Readable Identification

This includes markings of several types that can be read by computer-connected optical scanners and high-resolution video devices. Bar codes or characters designed specifically to be machine-readable can be used.

Radio-frequency identification (RFID) encompasses a wide range of applications, notably railroad cars and highway toll plazas where information can be read at high vehicular speeds, despite fog, mist, dust, grease, oil, snow, ice, darkness, walls, and other obstacles. RFID requires separate "tags" that are applied to products by various methods.

Figure 8.6.1 illustrates a number of these marking methods.

CHARACTERISTICS AND APPLICATIONS

Table 8.6.1 lists some dimensional characteristics and applications of the common kinds of markings.

ECONOMICS

Table 8.6.2 summarizes the economic factors attendant on the common marking methods.

SUITABLE MATERIALS

Table 8.6.3 summarizes commonly used or advantageous materials for each listed marking method.

DESIGN RECOMMENDATIONS

Perhaps the most important question facing the design engineer with respect to marking is its location on the product. Several considerations are applicable:

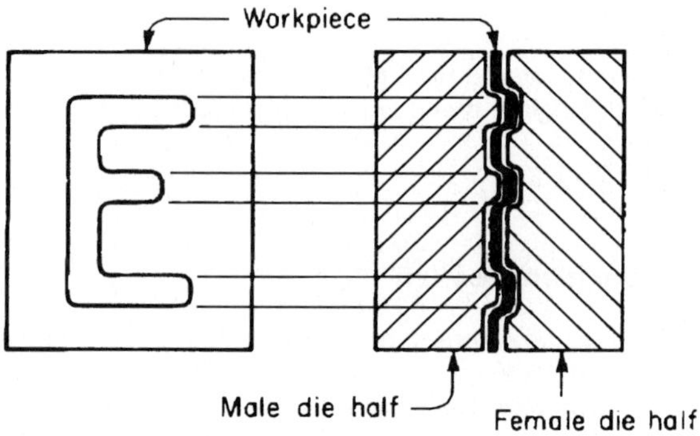

(b)

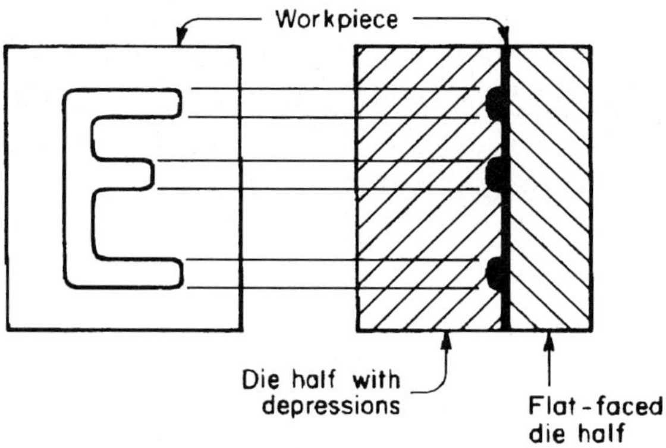

(c)

FIGURE 8.6.1 (*Continued*) Common methods for product marking. (*b*) Embossing; material remains essentially uniform in thickness. (*c*) Coining; material flows into die depressions.

1. The marking location should be accessible to the marking device. Overhangs and deep and narrow recesses should be avoided.
2. The marking should be visible. Brand identification and trademarks must be prominent; nameplates and other identifying material should not be located in an obscure position.
3. When the final marked component or product is bulky, it is usually more convenient and economical to mark an individual part before assembly.

Other considerations are the following:

1. The surface to be marked should be flat. Cylindrical surfaces are also satisfactory for stamp indenting and some other methods, but spherical or irregular surfaces always present a problem. Hand-finished surfaces may be difficult because of poor flatness.

(d)

FIGURE 8.6.1 (*Continued*) Common methods for product marking. (*d*) Apparatus for noncontact ink-jet marking of components and cartons. (*Courtesy Pannier Corporation.*)

2. When it is desirable to mark without displacing the metal of the workpiece, the following methods can be employed: acid etching, grit blasting with stencils, silk screening, and stencil and other printing methods.
3. Surfaces to be used for freehand or stenciled lettering should not be too small.
4. Decorative designs—etched, silk-screened, printed, embossed, or otherwise applied to an open surface—can serve to hide surface imperfections such as nicks, scratches, etc.
5. When two or more areas of different color are incorporated in a silk-screened, lithographed, or stenciled marking, the designer should allow for possible mismatch of the colors. If one color is not completely opaque when overprinting another, a dividing line of 1.2 mm ($\frac{3}{64}$ in) should be allowed between the two areas if silk screening is used and 0.8 mm ($\frac{1}{32}$ in) if lithography is used. (See Fig. 8.6.2.)

The following design recommendations apply to embossed markings:

1. To avoid an oilcan effect and to simplify tooling, limit the depth of embossed lettering to one material thickness. The stroke (letter leg width) should be at least three times the material thickness. Figure 8.6.3 illustrates these rules.
2. The angle of bevel of embossed lettering, as also illustrated in Fig. 8.6.3, should normally be between 35 and 40°. For sharpness and clarity of lettering, the angle

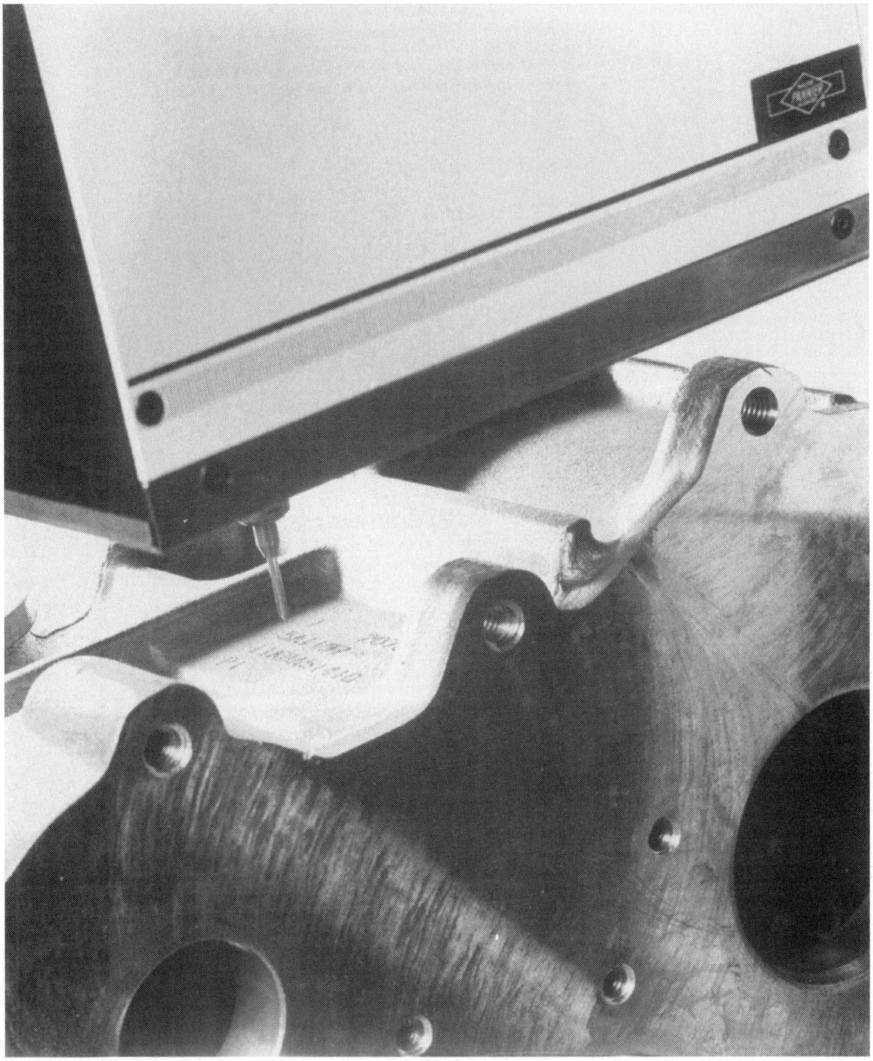

(e)

FIGURE 8.6.1 (*Continued*) Common methods for product marking. (*e*) Marking a casting with a programmable stamper that generates a series of depressed dots. (*Courtesy Pannier Corporation.*)

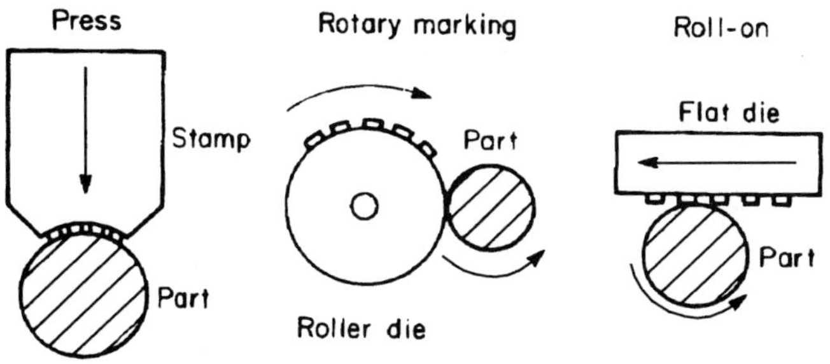

(f)

FIGURE 8.6.1 (*Continued*) Common methods for product marking. (*f*) Methods of stamp-indenting cylincrical parts.

can be held within 25 to 30°, but the wider angle provides easier metal flow and longer die life.

The following design recommendations apply to stamp-indented markings:

1. A number of character styles and faces are available in standard commercial stamping dies. Figure 8.6.4 illustrates popular lettering styles. Sharp-faced gothic characters are recommended for best clarity of impression and easiest operation.
2. Round-faced and interrupted-dot marking dies are advisable if there is a need to minimize induced stresses in the workpiece. (See Fig. 8.6.5.)
3. Generally speaking, the smallest lettering commensurate with good legibility should be specified even when ample marking surface is available. Small characters require less force and more easily produce a clear, well-defined impression.
4. A series of high-legibility numbering dies has been developed by the Steel Marking Tool Institute for hand stamping in situations in which part of any character may not be distinct. This type of numbering is recommended for miscellaneous hand-stamping applications. (See Fig. 8.6.6.)

RECOMMENDED TOLERANCES

Table 8.6.4 presents recommended tolerances for various marking elements including nameplates. The values shown are for normal, noncritical applications. When necessary, closer tolerances than those shown can be held for product appearance or other reasons. If so, however, tooling and processing costs usually will be increased.

TABLE 8.6.1 Characteristics and Applications of Markings Made by Various Methods

Marking method	Minimum recommended letter height, mm (in)	Permanent or easily removable	Remarks	Typical applications
Freehand	2.5 (0.100)	R	Appearance and legibility poor; limited information	Raw-material grade (color code), various identification tags
Contact printing	1.5 (0.060)	R	Multicolors feasible	Raw-material grade, identification labels, tags, nameplates
Noncontact printing	1.5 (0.060)	R	High print quality	Raw material, shipping containers
Stencil	6 (1/4)	R	Good for larger-sized lettering	Shipping cartons; with abrasive blasting, glass, and ceramics components
Silk screening	1.5 (0.060)	R	Good appearance; can be detailed and decorative	Signs, instrument panels, clock faces
Etching	1.5 (0.060)	P	Normal depth of etch, 0.07–0.3 mm (0.003–0.012 in); much finer work possible with photographic techniques	Ruler and instrument graduations
Engraving	2.5 (3/32)	P		Nameplates, plaques
Cast-in, molded-in	Sand mold, 4 (0.160) Plastic or die-cast, 1.5 (0.060)*	P	May be raised or depressed	Automotive engine block, many plastic and die-cast products
Branding	1.5 (0.060)	P		Wooden tool handles
Hot stamping	1.5 (0.60)	P	Decorative with metallic and other colors	Small appliances, cosmetic containers, golf balls
Embossing	4.8 (3/16)	P	Good appearance and legibility	Metal identification tags
Programmable stamper	1.0 (0.04)	P	Good resolution	Part marking of work in process
Laser marking	1.0 (0.04)	P	Excellent legibility on small parts	Electronic components

*1.5 mm recommended for legibility; smaller lettering possible.

TABLE 8.6.2 Economic Factors for Various Marking Methods

Marking method	Relative tooling cost	Relative equipment cost	Relative direct unit cost	Remarks; economic production quantity
Freehand	None	None	High	Suitable for low-quantity applications or with limited information
Contact printing	Moderate	High	Low	For medium to high production levels
Noncontact printing	Moderate	High	Low	For medium to high production levels
Rubber stamp	Low	None	Moderately low	For low-quantity production
Stencil	Low	None	Moderate	For low-quantity production
Silk screen	Moderately low	None	Moderate	Can be mechanized for high production
Etching	Moderately low	Low	Moderate	For low and medium quantities
Engraving	Low	Moderately high	High	For low quantities except when appearance of engraving is essential
Cast-in, molded-in	Moderately high	None additional	Normally no additional direct cost	Best permanent marking, especially for medium- and high-quantity production
Branding	Moderate	Low	Moderate	For medium production and high production if mechanized
Hot stamping	Moderate	Moderately high	Moderate	For medium and high production
Embossing	High	Moderately high	Low	For medium and high production
Coining	High	High	Low	For high quantities only
Stamp indenting	Moderate	Low to high	*	Hand stamping for low production; power stamping for moderate or high production
Nameplates	Moderate	†	High	Used at all levels of production for larger products

TABLE 8.6.2 Economic Factors for Various Marking Methods (*Continued*)

Marking method	Relative tooling cost	Relative equipment cost	Relative direct unit cost	Remarks; economic production quantity
Labels	Moderate	†	Low to moderate	Best for mass production
Tags	Moderate	†	High	Often used for temporary markings
Bar-code tags	Moderate	High	Moderate‡	Machine readable
Decals	Moderate	High	High	Very attractive; best for high production
Programmable stamper	Low to moderate	Moderately high	Moderate	For low to medium production
Laser	Low	Very high	Low	For high production or difficult-to-mark products

*No additional direct cost if impression tooling is incorporated in a normal punch-press or screw machine operation; otherwise moderate.

†Equipment cost may be high but can be amortized over large quantities of units for a number of different products.

‡Tags can be purchased preprinted.

TABLE 8.6.3 Suitable Materials for Various Marking Methods

Marking method	Coating materials	Base materials
Freehand	Ink, paint, dye, pencil, chalk, crayon	Any clean surface capable of being marked with some marking medium, but nonslippery nonpolished surfaces best
Contact printing and noncontact printing	Inks of various types	Paper, cardboard, wood, fiberboard, and other resilient materials best with metal-type process; all clean, smooth surfaces suitable with rubber type
Rubber stamp	Ink	Paper, metals, wood, fiberboard, fabric, plastics
Stencil	Ink, paint, dye	Paper, wood, metal, painted surfaces, fabric, plastics
Silk screen	Ink, paint	Painted surfaces, metal, wood, glass, plastics, paper, fabric
Etching	None	Glass, ceramics, all metals
Engraving	None	All machinable materials
Cast-in	None	All castable materials
Molded-in	None	All plastics and rubber
Branding	None	Wood, leather, fiber, plastics, fiberboard
Hot stamping	Special leaf-pigmented coatings carried to the work by a plastic film	Plastics, paper, fiberboard, leather, wood, hard rubber
Embossing	None	Sheet steel, aluminum, brass, other formable metals, coated-vinyl sheet
Coining	None	Low-carbon steel, aluminum, brass, other formable metals
Stamp indenting	None	Steel, brass, aluminum, cast iron (machined surfaces), stainless steel
Decals	Special paints, inks, and dies on a plastic film or paper carrier	Any smooth-surfaced material
Tags	Various materials, depending on which process is used	Paper, cardboard, plastic, sheet metals
Labels	Same as for tags	Paper, plastic sheet
Nameplates	Same as for tags	Various sheet metals, plastics
Programmable stamper	None	All machinable metals
Laser	Coating not required but may be used to obtain desired appearance	Plastics, metals, ceramics

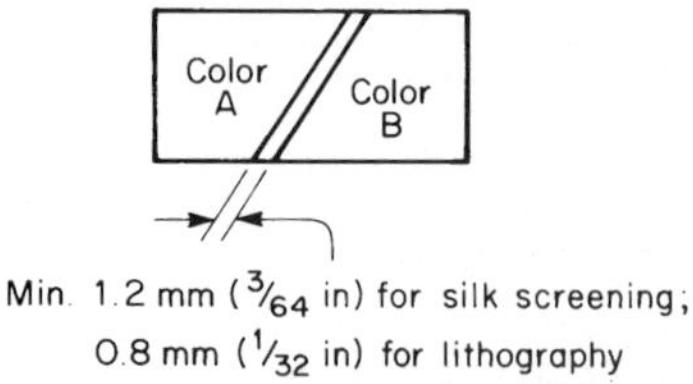

FIGURE 8.6.2 A dividing line of the width shown is advisable whenever two adjacent areas are marked with contrasting colors.

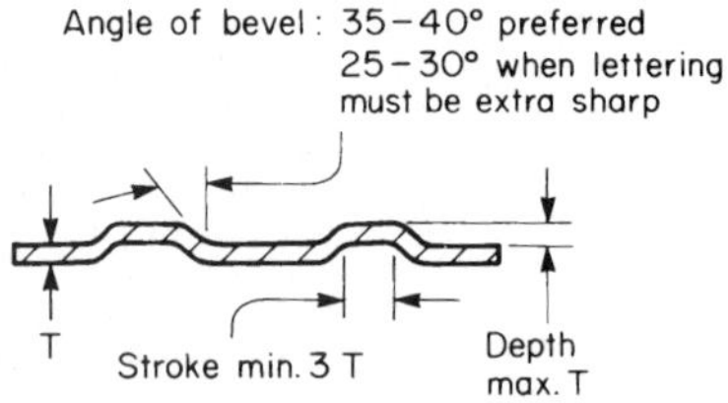

FIGURE 8.6.3 Recommended depth and stroke width and sidewall bevel angle for embossed letters.

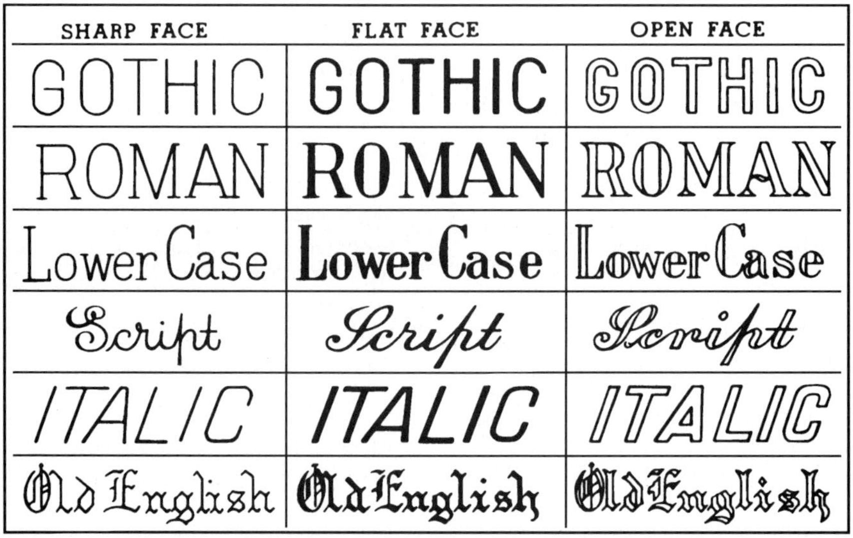

FIGURE 8.6.4 Sharp-faced gothic is the recommended typeface for stamp-indented markings. (*Courtesy Pannier Corporation.*)

FIGURE 8.6.5 Low stress-inducing type for stamp-indented markings. (*Courtesy Pannier Corporation.*)

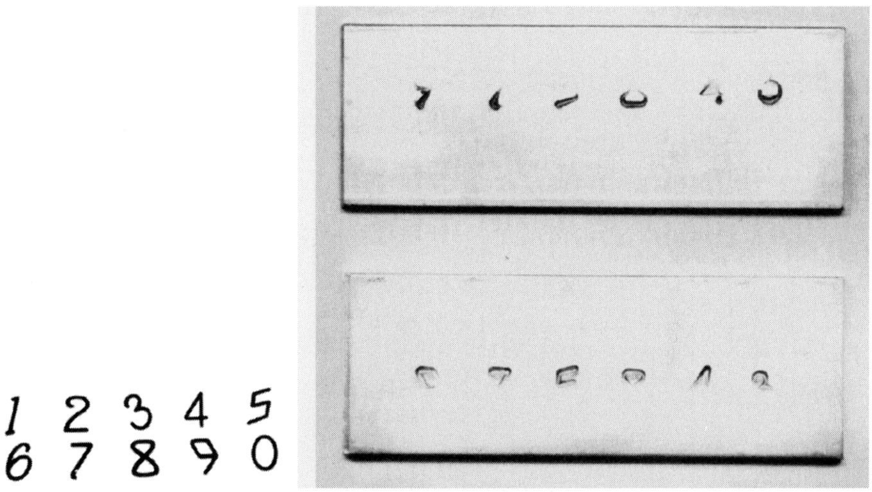

FIGURE 8.6.6 High-legibility typeface enables each digit to be identified even if there is only a partial impression. In the upper piece, the top portion of each number is indistinct, and in the lower piece, the bottom portion of each number is indistinct. In both cases, however, all numbers are readable. (*Courtesy Pannier Corporation.*)

TABLE 8.6.4 Recommended Tolerances for Product Markings

Lettering height	±10%
Lettering stroke	±10%
Nameplate size (length and width)	Normal tolerance, ±0.4 mm (±0.015 in) Closest tolerance, ±0.13 mm (±0.005 in)
Marking with respect to nameplate border	±0.4 mm (±0.015 in)
Nameplate holes, center to center	Normal tolerance, ±0.4 mm (±0.015 in) Closest tolerance, ±0.13 mm (±0.005 in)
Marking location on component if fixturing is used (excluding freehand methods)	Normal tolerance, ±0.4 mm (±0.015 in) Closest tolerance, ±0.13 mm (±0.005 in)

CHAPTER 8.7
SHOT-PEENED SURFACES

Henry O. Fuchs
Stanford University
Stanford, California
and
Consultant and Former President
Metal Improvement Co.
Paramus, New Jersey

THE PROCESS

Shot peening produces a skin of compressively stressed metal by the impact of many small balls on the surface of the workpiece. The balls are impelled by air through a nozzle or by a wheel. For some special purposes, captive balls, resembling small ball-peen hammers, are used. Proper control of the process is all-important to obtain the desired results.

Wheels provide a broad stream of medium and are used to peen large areas or large quantities of simple parts. They cannot reach the interior of deep holes or similar recesses. Nozzles are more flexible. They can peen even the inside of a small pressure vessel or different surfaces of a part to different specifications.

Peening is normally the last operation before assembly, painting, or plating. Heating to too high a temperature or plastic deformation from forming or straightening operations usually destroys the effect of peening. Fine finishing after peening is permissible, provided it does not go below the bottom of the peening dimples and does not create excessive heat, as grinding often does.

Shot peening is most often performed with cast-steel shot in screened sizes from 0.25-mm (0.010-in) to 1.5-mm (0.060-in) diameter. For larger shot, forged-steel balls from 1.5 mm (0.060 in) to 5 mm (0.200 in) are sometimes used. Shot made from cut wire is sometimes used, as are glass beads when very small shot is required. Special hard balls are used for hardened steel for greater effect and easier inspection. Peening balls break in time and must then be removed.

Masking is sometimes employed when peening of selective areas is required. Plastic masks or resilient tape are two common methods.

APPLICATIONS AND CHARACTERISTICS

Designers specify peening most often to increase the fatigue strength of parts. For example, the improvement that is obtained on notched parts of hard steel of 2000 MPa (290-ksi) tensile strength can be 300 percent and for smooth parts of the same material 40 percent. Similar improvements have been observed on other high-strength materials. On smooth specimens or softer material, the improvement is less. Static strength is not improved.

The fatigue-strength improvement results from compressive stresses that are generated in a shallow surface layer of the workpiece. The compressive stress is greatest near the surface. It decreases to zero at a depth that depends on peening intensity. The compressive stresses prevent the growth of fatigue cracks as long as the applied tensile stress is less than the compressive self-stress. Shot peening thus compensates for the damaging effect of notches, surface weakness, roughness, chromium plating, or small cracks. Stress corrosion and intergranular corrosion are retarded or prevented by shot peening as long as the applied tensile stress is less than the compressive self-stress.

Surface hardening can be accomplished by shot peening on certain materials that become harder by plastic deformation, e.g., austenitic stainless steel.

The self-stresses produce a small amount of elongation and can produce a change of shape. If a part is peened on one side only, that side will expand and the part will bend. Forming of simple curves in slender or thin parts (e.g., aircraft wings as shown in Fig. 8.7.1) is an application of this effect. Salvaging of distorted parts is another.

FIGURE 8.7.1 Forming an aircraft-wing panel by peening. (*Courtesy Metal Improvement Co.*)

Double curvatures and curving crosswise to ribs also can be achieved but not as easily as simpler bends. Surface modification by peening is used to improve the wear qualities of cams and seal seats, to change the quality of optical reflectors, and to close pores in cast vessels.

The size of parts shot-peened varies from small valve springs to large bull gears. Portable peening equipment can be used for peening parts too large to be moved into the shop.

Shot peening is not used to make surfaces smooth. It leaves instead a pattern of overlapping dimples. On tappets and under oil seals, the peened surface improves performance.

Typical shot-peened parts include gears, coil and leaf springs, valve plates, shafts, crankpins, torsion bars, and other highly stressed components.

ECONOMIC PRODUCTION QUANTITIES

The quantity of parts to be processed is relatively unimportant in peening. Single special gears and millions of automobile transmission gears have been peened economically. When quantities are large, mechanized handling of parts and automatic operation of nozzles produce uniform repetitive results at low unit costs.

DESIGN RECOMMENDATIONS

Fillet Radii

Shot must reach the surface to do its job. A fillet radius equal to the shot radius would barely permit shot to touch the bottom of the fillet; the close fit would diminish or prevent the desired plastic deformation. A good rule is to keep fillet radii larger than twice the shot diameter, as shown in Fig. 8.7.2. Note that small shot size limits the intensity which can be obtained.

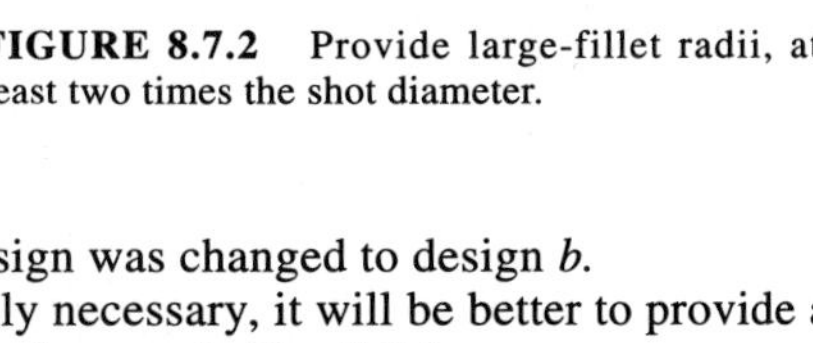

FIGURE 8.7.2 Provide large-fillet radii, at least two times the shot diameter.

Figure 8.7.3 shows drawings of splines that gave trouble by fatigue failures as originally designed according to design *a*. They were satisfactory when the design was changed to design *b*.

If a shoulder with very small radius is really necessary, it will be better to provide a separate part, mounted after peening the workpiece, as in Fig. 8.7.4.

Grooves

A narrow, deep groove will prevent some of the shot from hitting the bottom. Width equal to depth is a good proportion for peening. Smaller widths or deeper depths can be handled by special setups, which may increase cost. (See Fig. 8.7.5.) Groove width should be more than four times the shot diameter.

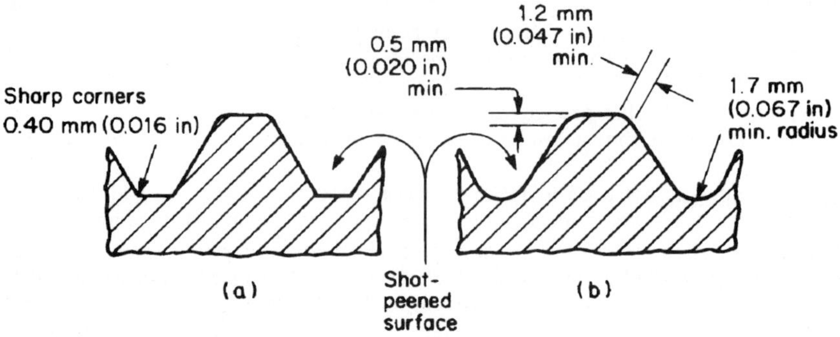

FIGURE 8.7.3 Splines. Design *a* failed after short service; design *b* was satisfactory.

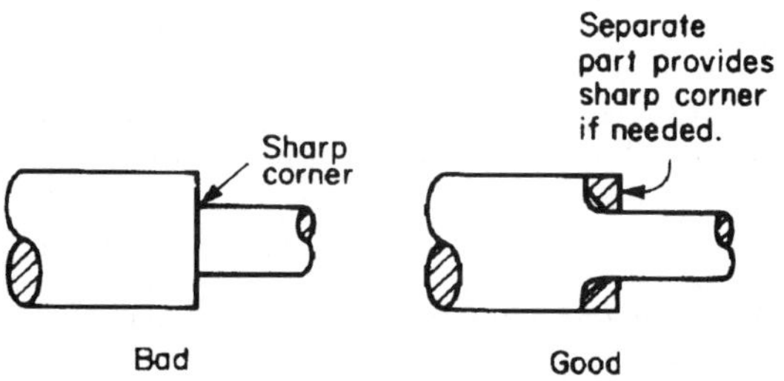

FIGURE 8.7.4 Avoid sharp internal corners.

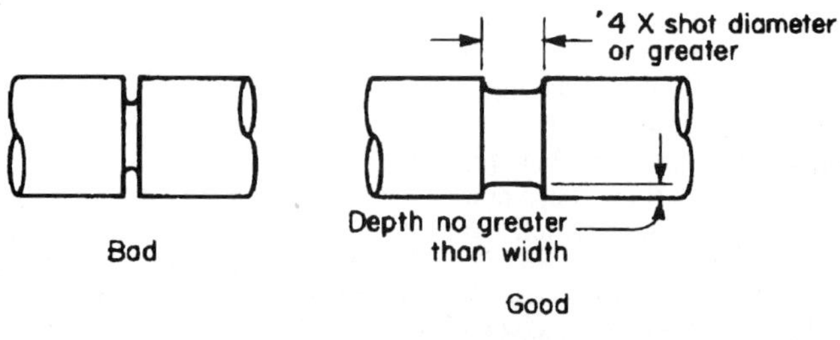

FIGURE 8.7.5 Design rules for grooves that are to be shot-peened.

Holes

Deep holes to be peened require the use of a "lance" nozzle, especially if the holes are blind. Shallow through holes with a depth no greater than the diameter can be peened without using lances.

Sharp Edges

A peened surface tries to expand. When two such surfaces meet at a sharp edge, they will push material out to form a fin. Sharp edges should therefore be broken before peening.

Selective Peening

When only certain areas of a part require peening, the least expensive job will result if areas that must be peened are specified and shot impressions on other areas are permitted but not required. The cost of masking is thus avoided. Careful definition of the transition zone between peened and unpeened areas is seldom required. There is no evidence of adverse product performance from such transition zones or from scattered shot impressions.

Peening Intensity

Saturation of peening should be specified for most general applications unless cost or performance dictate otherwise. Saturation is indicated by near-maximum deformation of a thin strip-test specimen subjected to the same processing conditions as the workpiece. With uniform impingement by uniform shot, one can use coverage of the surface as a sign of saturation. It is checked by magnified viewing or by coating parts before peening with adherent fluorescent dye, which is removed by peening impacts, and by checking under ultraviolet black light.

Sometimes it is necessary to limit the intensity of peening to avoid deformation. Crankshafts are an example.

Specification of the process conditions and effect on Almen* test strips are the common methods for specifying peening intensity because inspection of finished parts for depth of compressive stress is destructive and expensive.

TOLERANCE RECOMMENDATIONS

A tolerance of ± 15 percent Almen intensity* is reasonable on most parts. A typical roughness value for steel of hardness 400 BHN peened with hard-steel shot of 1-mm (0.039-in) diameter would be 2.5 μm (100 μin) rms. For equal intensity, a harder workpiece or a larger shot size would produce less roughness. Softer shot would produce less roughness, but it would be difficult to check coverage visually. Peening smooth surfaces to low intensities, as with glass shot, produces a satin finish of low roughness, typically on the order of 1 μm (40 μin) rms.

Close-tolerance specifications on the location of masking will increase costs. A tolerance of ±1 mm (0.039 in) is quite close, while ±2 mm (0.079) is easy.

*Almen intensity is the standard method of specifying shot peening. See military specification S-13165 or SAE Standard J442.

S • E • C • T • I • O • N • 9

ADDITIONAL DEVELOPMENTS

CHAPTER 9.1
DFM FOR LOW-QUANTITY PRODUCTION*

INTRODUCTION

Many products, notably equipment, machines, tooling, and devices for industrial use, products for niche markets, and pilot lots, are inherently limited in production volume. The question arises, "Is it worthwhile, even possible, to apply DFM techniques successfully to products that are manufactured in these limited quantities?" The answer is "Yes," but the approach and the improvement guidelines in such cases will vary from that used in high-production applications.

The cost of tooling is perhaps the most important factor in choosing a design suitable for lower-quantity production, but it is not the only factor. Others are the cost and lead time for development of the manufacturing process, the selection of production equipment, and the selection of material. These factors may be compounded if the design is innovative. Tooling cost is significant because it must be amortized over the production life of the product. High tooling costs mean high manufacturing costs unless the production run is long enough so that the unit amortized cost is no longer a major factor. The same is true of any development costs for production equipment, materials, or other design characteristics. Development of new production equipment and of innovative materials and designs requires time and cost. If these innovations apply only to one product, that product must bear the cost. Lower production volumes limit the degree to which they can be amortized.

When expected production quantities are not large, the designer cannot afford to use components or configurations that require high-cost tooling or extensive process or product development. He or she must concentrate on simpler components or those which are already in production and available elsewhere—from other products already in production in his or her firm or from commercially stocked components available from vendors.

Another interesting point is that design for speed-to-market and design for low-quantity production have much in common. Often, it is desirable to use low-quantity methods for initial production of a new product. This minimizes the lead times that would be required for procurement of high-production tooling and equipment, enabling the product to be brought to market sooner. Later, if the product is successful, higher-production tooling can be used, with worthwhile savings in manufacturing cost.

*Adapted from Chapter 22 of *Design for Excellence,* by James G. Bralla, McGraw-Hill, New York, 1996.

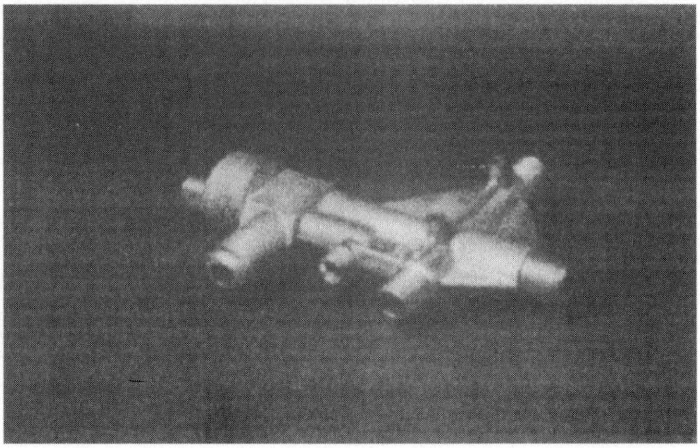

(a)

(b)

FIGURE 9.1.1 (*a*) This zinc die-cast appliance fluid manifold provides a low-cost, quality design for mass-production conditions. Note that the die-casting process allows screw threads to be cast into the part, avoiding expensive machining operations. This design is much less costly than the approaches shown in parts *b* and *c* but requires a complex die-casting die, the cost of which cannot be easily amortized unless production quantities are high. (*Photo courtesy Chicago White Metal Casting, Inc., Bensenville, IL.*) (*b*) This machined and mechanically assembled appliance fluid manifold provides a workable component with low tooling cost. It is a suitable predecessor of the die-cast manifold in part *a* during prototype or pilot production conditions. (*Photo courtesy Chicago White Metal Casting, Inc., Bensenville, IL.*)

Figure 9.1.1 depicts an interesting case of an appliance fluid manifold. For mass production, the zinc die casting of Fig. 9.1.1*a* provided a low-cost, quality connection for the various hoses in the equipment. For earlier pilot production, the machined manifold block shown in Fig. 9.1.1*b* was used. It had a higher unit cost than the die casting but avoided the need to invest in a rather expensive die-casting die before the

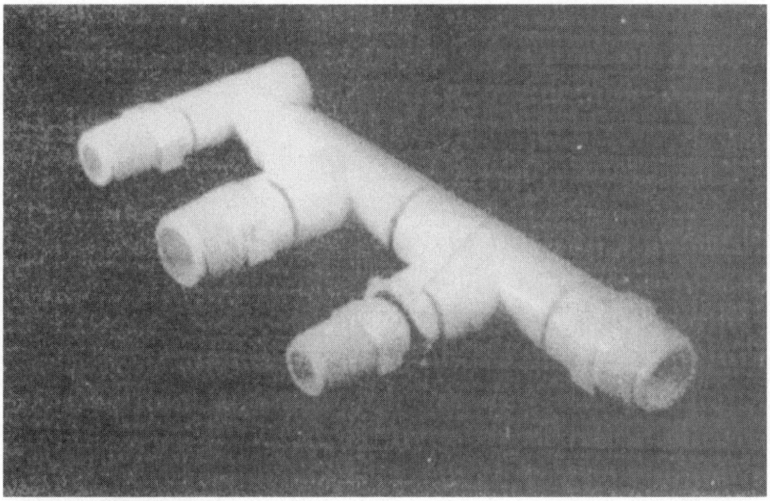

(c)

FIGURE 9.1.1 (*Continued*) (*c*) Assembled and bonded manifold made from inexpensive, commercially available PVC plastic pipe fittings. This is another possible approach to the design of the manifold of part *a* for prototype production conditions.

product design was fully finalized for mass production. Another approach that would be suitable for prototypes and possibly pilot production is shown in Fig. 9.1.1*c,* where inexpensive catalog-available PVC plastic fittings were bonded together to provide the same function. Figure 9.1.1*b* and *c* then, represents designs suitable for low production. Figure 9.1.1*a* shows a mass-production design.

GUIDELINES APPLICABLE TO LOW-QUANTITY PRODUCTION

The recommendations that follow are intended to provide more detailed guidance to the designer in complying with the principles discussed above.

1. An important rule for design and manufacturing managers is, "Standardize, standardize, standardize!" Standardization is even more important for limited-quantity production than it is for mass production because, in the former, there is not the opportunity to amortize the tooling and developmental cost of special items.

Standardization should encompass all facets of the company's manufacturing operations—product drawings, design features, fasteners, all other kinds of parts, materials, organization, procedures, processes, etc. When they follow repetitive, predictable patterns, they provide much more powerful benefits at lower production levels than they do when mass production is involved. They eliminate all the one-time costs that accrue when something new is started. In the case of component parts, these one-time costs include the tooling investment required. For other items such as engineering drawings, procedures, etc., they include the training, setup, defect prevention, and other expenses attendant on starting something new.

FIGURE 9.1.2 The legs for this movable fixture were made from commercially available railing pipe and fittings. This compares with a previous design that used welded tubing that required several secondary operations. The new design is preferable, especially considering the limited production run involved. (*Courtesy PowerHouse Exhibits.*)

2. "Never design a part that you can take out of a catalog." (David M. Anderson, *Design for Manufacturability,* CIM Press, Lafayette, CA, 1990.) Catalog-standard parts often have advantages beyond that of company-standard parts in that they are made by the supplier in large quantities with efficient production methods, providing favorable costs. In addition, quality and reliability are already established and field service, if ever needed, is facilitated.

Figure 9.1.2 shows the frame of an industrial fixture assembled from pipe components of the type used to make fencing and railing. These are off-the-shelf, available commercial components. This design replaced a previous one that required the welding of square steel tubing. With the off-the-shelf pipe and fittings, relatively low-skilled assembly labor replaced that of a higher-skilled welder. Secondary operations that often occur with welding (slag removal, filling, straightening) were eliminated, and appearance was improved.

3. Design for manufacturing processes that are suitable for low-production levels, i.e., those with low tooling costs. Examples are assemblies that are assembled mechanically, soldered, welded, brazed, or adhesively bonded. For smaller quantities, this is often more economical than using castings or molded parts or performing intricate machining. Thermoforming is a desirable approach for plastic parts that would otherwise be injection-molded in mass production. Thermoforming tooling is almost always inexpensive and has short lead times. Sometimes, wooden forms are suitable. Figure 9.1.3 shows a thermoformed housing that, in mass production, probably would be injection-molded. Injection molding provides opportunity for more design complexity that, when production volume permits, can be advantageous in reducing parts count and

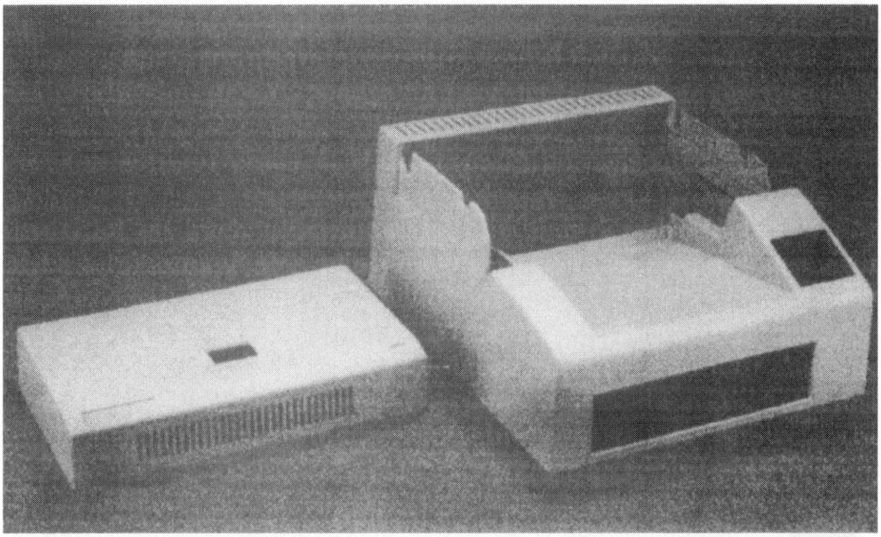

FIGURE 9.1.3 This thermoformed plastic housing provides attractive appearance and utility to the product, a color printer, but does not require a sizable tooling investment nor the lead time that would be required if it were injection-molded. It is therefore a favorable design alternative for a product made in low or moderate quantities. It was fabricated for pilot production. When the product was successful, an injection-molded enclosure took its place. (*Courtesy Profile Plastics Corporation.*)

providing useful features. However, for short-run production or for speeding a product to market, the thermoformed approach can be quite satisfactory.

For machined parts, if the company possesses computer-controlled machine tools, it is desirable to design parts so that they can be processed on such equipment rather than on production equipment that requires special cutting tools and holding fixtures. Providing the proper configuration electronically is less costly for shorter runs than providing it with hardware. Machining centers with tool changers can perform a whole series of operations in one setup. Such equipment is particularly advantageous when it is part of a CAD/CAM system.

4. It may prove advantageous to "hog" the part solely from solid stock material rather than to produce a sand-mold casting or other casting and finish machine it, particularly when computer-controlled machining equipment is available. Sand molding and other casting processes require some tooling investment for patterns and molds. This can be avoided if a purely machining approach is used.

5. When plastic injection-molded parts would be favorable for the product but production levels are not high enough to justify the tooling cost, consider the use of short-run injection-molded parts. Walls of such parts are usually thicker than those made from high-production tooling. This simplifies molding somewhat, since lower pressures can be used and mold filling is easier. Aluminum molds are often used because of their easier machinability, but cast or cast-and-machined molds also may be used. These may require disassembly between shots to avoid the need for core pulls and ejector pins. The labor content as well as the materials content of such parts may be rather high, but the total cost, including mold amortization, is far lower than would be the case if more sophisticated injection molding were employed.

FIGURE 9.1.4 This cast-urethane fixture block replaced an earlier part that was machined from solid stock. Note the pattern and silicone rubber mold. This type of casting is a desirable process for low-quantity production levels. (*Courtesy PowerHouse Exhibits.*)

Another technique is to use cast plastic parts. Urethane material can be cast in a variety of flexibilities with or without a foaming effect. Other materials such as polyester, epoxy, acrylic, and nylon also can be cast. One advantage of casting, when silicone rubber or other rubber molds are used, is that large undercuts can be incorporated in the configuration of the part. The flexibility of the mold allows the cast part to be removed from the mold despite the existence of substantial undercuts. Most silicone molds can withstand from 100 to 250 shots, depending on the material and the complexity of the part.

Figure 9.1.4 shows a cast-urethane block in a fixture used in the processing of golf clubs. The illustration also shows the silicone rubber mold used to cast the block. The mold also was cast from a wooden pattern of the part. In this example, the previous design relied on machined plastic blocks for the same application. Machining was expensive, but the quantity involved was too low for amortization of the cost of injection molds. The cast blocks were far less costly than those made by machining. Because of the casting process—pouring liquid resin into rubber molds—undercut and sidewall draft limitations of injection molding could be bypassed.

6. If machined components are used (which is more likely at low production levels), utilize materials with good machinability ratings. When machining at high production levels, the development of optimal machining techniques—tool grinding angles, coolants, feeds, speeds, etc.—can aid in the use of lower-cost but less machinable materials. However, when runs are short, such developmental costs cannot be absorbed, and the designer is advised to select materials that can be processed easily, even if the material cost rises somewhat as a result.

7. For other processes, try to use materials that are formulated for easy processibility. For example, for stampings or other formed-metal parts, use metal alloys or grades that are specifically intended for easy formability, e.g., certain grades of brass that bend and flow easily without problems. The brass may be more expensive per unit weight or unit volume than mild steel, but when quantities are low, the added materials cost may be negligible in comparison with the overhead costs that can be saved.

8. If metal stampings are used, design parts suitable for short-run methods such as those involving turret punches, nibblers, and press brakes rather than punch presses with complex dies. This means that operations such as deep drawing and complex forming should be avoided in favor of simpler bent and joined surfaces.

9. Use stock material shapes as much as possible to avoid machining. This is a good DFM rule for all levels of production, but it may be particularly beneficial if production runs are short because in such cases setup times are a larger portion of the total manufacturing cycle. Figure 2.2.1 (in Chap. 2.2) shows some examples.

SAVINGS PRIORITY

The savings priority in your design of a product produced in low quantities should be the following:

1. Save tooling costs.
2. Save other overhead costs.
3. Save labor costs.
4. Save materials costs.

Normally, for high-production manufacture, direct labor and materials costs are more important, and it is justifiable to expend a greater investment in tooling and engineering in order to reduce them. This, however, normally will not be true at low production levels because the higher investment cannot be amortized.

CHAPTER 9.2

DFM IN ELECTRONICS*

ELECTRONIC PRODUCTS

The electronics industry involves three overlapping categories of products: consumer electronics products such as television sets, stereos, camcorders, and, more recently, home computers; industrial electronics such as business and industrial computers and other control apparatus, lasers, and instruments; and military products such as radar and navigation equipment, etc. The industry is characterized by relatively short product lives, a high rate of design changes, high material and overhead costs, and low labor costs.[10]

The growth of the industry has been spurred by the development of the integrated circuit, a small "chip" containing up to many millions of circuit elements. Integrated circuits permit products to incorporate many desirable functions at low cost. Another benefit is their high reliability, no matter how many individual transistors and other devices the circuit contains. Since individual components and their interconnections have a higher failure rate, the reliability of an electronic product is actually improved when more circuit elements can be placed in a chip instead of in separate elements on a circuit board.

PRINTED-CIRCUIT BOARDS

The printed-circuit board (sometimes called either the *printed-wiring board* or the *printed-wiring assembly* and often referred to by the initials *PCB* or *PWB*) has become the standard design approach for electronic circuits, even when production quantities are small. The board provides both a mounting surface for electronic devices, resistors, capacitors, and semiconductors and a wiring path between them and the external circuit. A PCB is a reinforced-plastic board (typically glass-reinforced epoxy) on which a thin coating of copper is electroplated and etched to provide the "wire" pathways, commonly called *traces*. The economic and quality advantages of this approach are quite powerful. Although the approach was developed originally for mass-production applications, computer-controlled methods and equipment also enable automation for small quantities.

*Adapted from Chapter 21 of *Design for Excellence,* by James G. Bralla, McGraw-Hill, New York, 1996.

Boards can be single-sided, double-sided, or multilayered,[10] e.g., they can have wiring paths on one side, both sides, or both sides and "buried" on inner layers. Electrical connections between sides and/or layers are achieved by means of *vias*—holes drilled through the board and conductively plated.

Technological advances are very rapid in this industry, with the thrust of the advances being achieved with integrated circuits. Advances with printed-circuit boards are also rapid, the result of both being an ever-increasing density of devices per unit of area of board surface.

One advance of recent years is the use of surface-mount technology (SMT). This involves the attachment of electronic devices to circuit boards by direct soldering onto circuit *pads* rather than by inserting connecting wires through holes in the board before soldering. Figure 9.2.1 illustrates the two types of board construction. Hybrid boards with some devices surface mounted and others assembled with the through-hole technique are also used. SMT allows boards to be smaller, which provides both compactness and faster circuit response due to shortened distances between devices. For further compactness, some circuit boards have devices attached to both sides of the board. This requires some design differences to permit soldering of both sides of the board while still holding all devices in place. Circuit boards using SMT with a dense layout are sometimes referred to as *fine-pitch boards. Fine pitch* typically refers to leads on component edges spaced linearly 25 mils (0.63 mm) or less center to center.

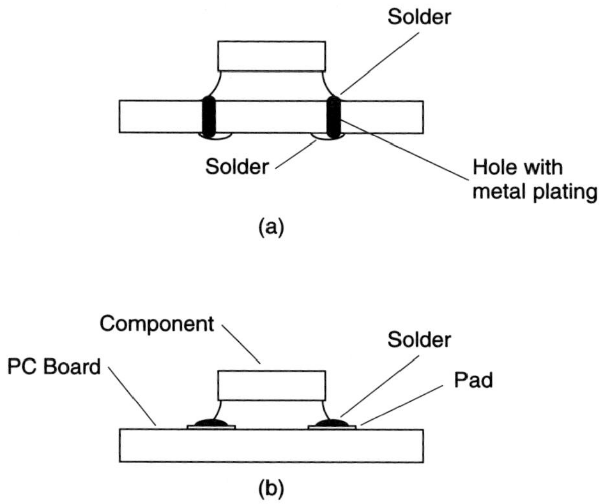

FIGURE 9.2.1 The two types of printed-circuit boards shown in cross section. (*a*) The conventional through-hole attachment of devices to a board. (*b*) The more recent surface-mounted construction. The advantages of the latter approach is a greater density of devices per unit of board area and a consequently faster operation of the electronic circuits. There are also hybrid board designs, those which include both through-hole and surface-mounted devices.

Many of the design approaches for these high-density circuit boards are quite sophisticated. The design requirements for SMT boards, particularly those with fine pitch, are much more demanding than those for the traditional board assemblies. Panel and component layout are critical elements. So is the geometry of components. Both electronic and mechanical factors are involved in the design of a typical board.

SOLDER JOINTS

By far the most common method for securing a sound electrical connection between diverse circuit devices is by soldering. One product may have hundreds or thousands of solder connections.[12] Thorough wetting with solder of the surfaces to be joined is essential. The shape of the solder fillet, the amount of material it contains, and its freedom from voids are also important. Common failure modes of solder joints are shown in Fig. 9.2.2. Pads for surface-mounted devices should be large enough so that the entire solder

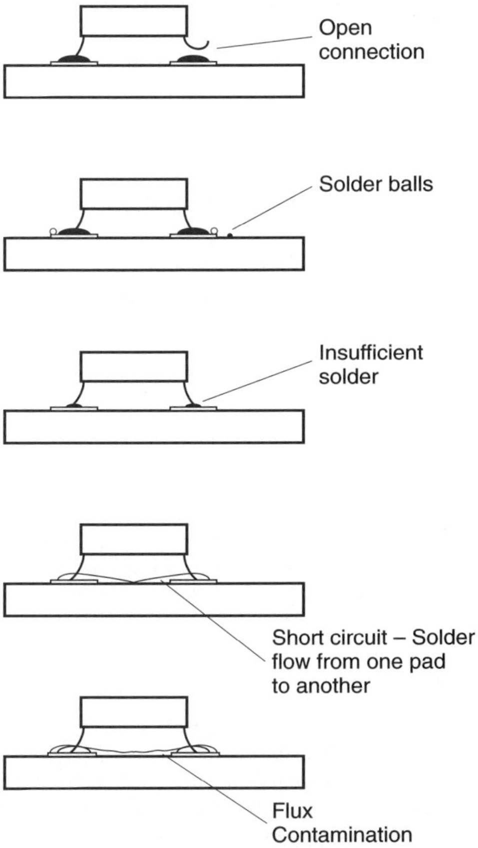

FIGURE 9.2.2 Common faults with electronic solder joints: (*a*) open connection; (*b*) solder balls; (*c*) insufficient solder; (*d*) short-circuit solder flow from one pad to another; and (*e*) flux contamination that can permit unwanted leakage current.

fillet is visible, even if the component has a shape that makes this difficult, so that inspection of the solder joint is possible. Many circuit devices are heat-sensitive; the amount of time that they are subjected to temperatures above the melting point of solder must be minimized (less than 15 to 20 seconds). This limitation makes it more difficult to achieve good soldering, so other design and process steps must be taken to enhance the solderability of the component to the board. Heat sinks incorporated in the board assembly to prevent overheating of circuit elements during the product's operation may make it more difficult to obtain high-quality solder joints because they tend to limit the temperature of adjacent soldering points when the board is heated.

Cleanability of flux must be provided to avoid performance and reliability problems from current leakage due to flux contamination. Cleanability requires access for cleaning solvents during the flux-removal step of the manufacturing process. Easy cleaning under each component device is necessary, and room must be provided for flushing action of the solvent. The exception to this is the design of board assemblies that uses a "no-clean" solder flux. Use of no-clean fluxes is a design for the environment issue because it reduces the volume of contaminated waste solvents that needs to be handled and disposed of.

Solders for printed-circuit boards and other electronic applications are included in Table 7.4.2 (Chap. 7.4).

SOLDERING PROCESSES FOR PRINTED-CIRCUIT BOARDS

The following processes are used to apply and/or "reflow" solder to PCBs to provide mechanical fastening and electrical connections.

Wave Soldering

The circuit board is momentarily put in contact with a fountain or wave of molten solder. The solder, which contacts only one side of the board, heats and wets exposed metal surfaces of the board and the devices mounted on it. It also fills plated through-holes in the board. The operation can provide very rapid soldering of the entire board. The process is illustrated schematically in Fig. 9.2.3.

Vapor-Phase Soldering

This process provides precise temperature control. It is used primarily to solder surface-mounted components to the board. The board, with solder paste applied at each joint, is passed through saturated vapor of a fluorocarbon fluid. The vapor condenses on the board, heating it rapidly and uniformly and melting the solder paste. The process can be used with different solder alloys and vapors—with different condensation temperatures—to solder parts of the board assembly sequentially. The process is adaptable to very-fine-pitch boards, with center-to-center lead spacings of 0.38 mm (0.015 in).

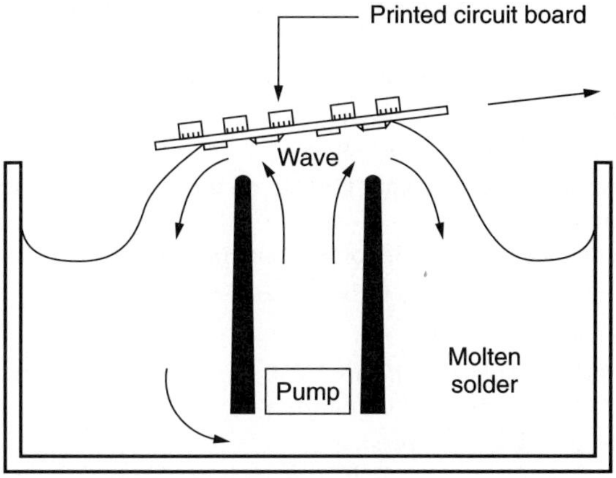

FIGURE 9.2.3 The wave soldering process shown schematically. The printed-circuit board, with components attached, is moved against a "wave" of flowing molten solder. The solder heats and wets metal pads, leads, and wire paths, coating them with solder.

Laser Soldering

A controlled laser beam is focused on each solder joint for a precise length of time to melt the preplaced solder and wet the joint. The process avoids overheating the connected devices because the heat is narrowly localized, and it provides a high-quality solder joint. However, production rates are lower than with other processes because joints are soldered one at a time and equipment costs are high.

Infrared Oven Soldering

Infrared energy is another method used to provide the heat for melting solder paste at the joints of circuit boards. It has the advantage of being suitable to high production rates, but time and temperature profiles must be set carefully to avoid overheating of components. The different heat-absorption characteristics of various components also must be considered.

THE PRINTED-CIRCUIT BOARD OPERATION SEQUENCE

The following is a typical operation sequence for assembling and soldering a printed circuit board:[6]

1. Solder paste is silk screened onto the "bare" board, one that has metal wiring paths and pads.
2. SMT components are placed on the solder paste, and the solder paste is then dried.
3. Solder paste is heated to its flow temperature by infrared or vapor-phase methods.
4. Various nonsoldered components ("dips" or other similar devices) are inserted automatically.
5. Components that are not suitable for machine insertion are placed by hand.
6. The board is masked to prepare for wave soldering.
7. The board is inspected.
8. Wave soldering takes place. (This includes flux application, preheating, the solder wave, and air-knife application to remove excess solder.)
9. Flux is removed by vapor degreasing.
10. After-solder components are installed, inspected, and their solder joints touched up.
11. As necessary, excess flux is removed.
12. Further inspection is made.
13. The circuit is tested.
14. Environmental testing takes place.
15. Boards are separated from the panel that typically contains a multiple number.
16. Inspection and alignment take place.
17. Final defluxing is carried out.
18. The boards are masked for conformal coating.
19. Conformal coating is applied.
20. Masking is removed.
21. Final inspection is made.

Current machines that place components on the board use machine vision to locate devices more accurately. Currently available equipment can test each component, bend and trim its leads, insert it in the board, and then crimp the leads at speeds as high as 25 pieces per minute.[11] Machines for assembling devices to SMT boards have much higher speeds—over 80 devices per minute. Variations in device dimensions, bent leads, and board distortions all complicate the task of locating the device accurately on the board. One important principle for printed-circuit board DFM, therefore, especially when fine pitch is involved, is to provide a design that facilitates automatic assembly of components to the board.

TESTABILITY

Testing is an important part of the manufacturing process for circuit boards to ensure quality and to avoid conditions that will adversely affect reliability. Testing does not produce quality, but it provides the opportunity to uncover faults that may exist. The extreme complexity of many electronic circuits increases the possibility of faults somewhere in the circuit. A fault in one device or subcircuit may mask a problem in another. Sometimes rework to replace a defective component or solder joint can create another quality defect. Defect isolation may be a time-consuming procedure.[12] The short product life of many electronic products is another complication. These factors also make testing for defects both difficult and essential. Design for testability is one of the responsibilities of the product designer and has gained much attention as a worthwhile endeavor in the electronics industry.[12] The designer may have to allow for circuit function testing, module testing, electromagnetic interference (EMI) testing, and environmental testing to ensure that the product is correct.[6] Radiographic (x-ray) inspection of solder joints may be used to complement circuit testing.

The designer must visualize how and where in the manufacturing process the board will be tested. Sufficient test points must be provided. The testing system and device should be designed concurrently with the circuit itself. Testability considerations should be addressed at the design concept stage and thereafter. Otherwise, proper testing may not be possible, or the product's introduction may have to be delayed while a test apparatus is constructed. Paying attention to testability can enhance product quality and reliability, aid in reducing time-to-market by reducing the time lost due to late discovery of product problems, and reduce manufacturing and service costs.

Incorporating self-testing circuitry in an electronic product is another approach to testability that is sometimes advisable, particularly as individual components become more complex. Other approaches are the partitioning of the circuit into smaller, more measurable blocks and the provision of connection points that make it easier to connect test devices to the board.

DESIGN GUIDELINES

The following is a list of some of the more important guidelines for designers of electronic products. They can provide both direct cost reductions and those which result from higher-yield manufacturing.

1. *Standardization.* It makes good sense for a company to develop a standard design manual for its products and adopt standard manufacturing processes and standard procedures. The design standards can be keyed to the requirements of the product line and can include standards for such items as printed-circuit board material, board thickness, hole diameters, component spacing, etc., as well as preferred materials, components, and component tolerances. Most of this information can be incorporated in computer-aided design (CAD) data libraries, and the manufacturing sequence then can be specified automatically. If company-applicable DFM guidelines are also incorporated in the CAD system, further advantages can be achieved. Maximum benefits are realized when designs are standardized and when production equipment and tooling are keyed to the design standard values.

a. Standardize circuit board sizes and dimensions. Make them all the same size as much as possible (24×18 in is a standard size for bare-board manufacture, but assembly machine manufacturers have individual standard sizes). This will reduce fixturing time and fixture expense. When smaller boards will suffice, design them so that multiple quantities can be cut from a standard-size board without waste. The board can then run in standard fixtures in wave soldering, cleaning, and other operations. Use interlocking board shapes if this facilitates multiple-board processing. Also standardize the size and location of tooling

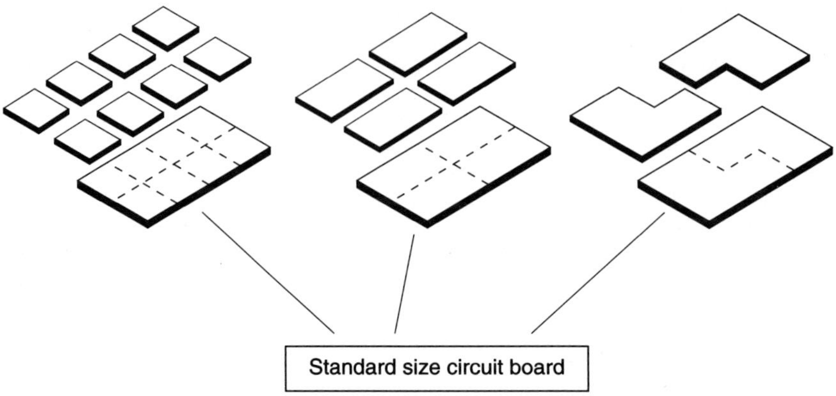

FIGURE 9.2.4 Make printed-circuit boards of standard sizes or to fit, without waste, on a standard size. Making several smaller-sized boards from a larger board provides the advantages of multiple processing.

holes and board thickness. Figure 9.2.4 illustrates some of these recommendations.

b. Utilize standard components. Minimize the number of varieties of resistors, capacitors, and semiconductor devices used in circuits by standardizing on certain sizes, values, and tolerances. The standard list of components can concentrate on those which have established quality and reliability records.[2] The use of parts of proven quality and reliability is a first step in providing these attributes in the product. Fewer varieties provide better opportunities for larger-quantity procurement at favorable cost. Equally important, when automatic component insertion is available—with its quality and cost advantages—the standard devices can be loaded in the automatic assembly equipment. Nonstandard devices may have to be subjected to the less satisfactory manual assembly. The greater the use of standard components, the greater are the benefits.

c. Standardize the orientation of devices on the board.[2] They should be oriented at 0 or 90° from the edge of the board, preferably in one direction only.[3] When components have polarity, maintain it all in the same direction for such components, and the polarity should be printed on the board. This layout also should be on a grid compatible with the CAD system in order to simplify design steps. Quality problems due to misorientation of components can be minimized if such a consistent orientation is followed. Some soldering problems also may be avoided. Individual parts should be made so that uniform orientation is facilitated.

d. The number of different hole sizes on the printed board should be minimized.[1] Different hole sizes require different drill sizes, necessitating drill changes during board fabrication. Such drill changing may not be automatic, adding to the labor content of the operation. Design mounting holes of such diameter that they cannot be confused with tooling holes. Also limit the hole sizes to standard wire drill sizes.[6]

e. The location of identification labels on components should be standardized.[1] This ensures their readability and, if bar-code labels are used, simplifies the fixturing of the bar-code reading device. Product markings on individual devices should be such that they conform to the board layout (all readable from the same direction) and facilitate placement of the parts.

2. *Product simplification.* The most economical and most reliable electronic products are normally those with the fewest number of circuit boards and the fewest number of components on the boards.

a. Using integrated circuits to perform the function of discrete devices that would otherwise be on a circuit board should be a design objective. Use integrated circuits as much as possible, and put as much circuitry in them as possible. The cost advantages of such an approach stem from the fact that manufacturing operations are much the same no matter how complex the chip may be; the reliability advantages are noted above.

b. For mechanical elements in an electronic product (such as the housing, hinges, clips, fasteners, etc.), use the same approaches as are used in mechanical products, i.e., reduce the number of parts, eliminate the need for costly machining, reduce or eliminate fasteners, and design for easy top-down assembly.

c. Minimize or eliminate adjustments. Mechanical adjustments of potentiometers and other elements are a source of potential error and are labor-intensive. Consideration should be given to eliminating the need for adjustment by utilizing electronic circuits—phase-lock loops, feedback loops, voltage regulators, nonvolatile memories, and autoconfiguration registers—or by tightening specifications of components. Positional adjustments of components are also costly and a potential source of quality problems. Footprint designs for SMT boards should provide space for a good solder joint even if the part shifts slightly or is not positioned precisely.[2]

d. It is advisable to simplify the assembly design as much as possible, utilizing only one placement method of circuit components. A fully SMT design is more easily manufactured than a hybrid through-hole and SMT approach; one-sided boards are preferable to two-sided boards. This is not to imply that the more complex configurations should not be used. The point is to avoid the complexity unless it provides demonstrable advantages that offset the additional manufacturing steps required.

e. Minimize the use of sockets. Mechanical connections have diminished reliability compared with soldered connections.[1]

f. Avoid the use of connecting cables. Substitute direct plug-in circuit boards, or combine several boards into one. Because of their flexibility and possibility of tangling, cables are costly to handle and insert.

g. "Avoid switches or jumpers; configure instead with software."[1] If switches are retained, use rubber-dome or other types partially built into the circuit board rather than separate switch subassemblies.

3. Provide sufficient spacing between leads, vias, and traces. This is to avoid solder short circuits between these elements. See Fig. 9.2.5.

4. "Make sure that the component selection and placement conforms to the design guidelines of the equipment."[1] This involves designing the printed-circuit board with the clearances and spacings needed by the automatic-insertion equipment. This improves soldering results—by more widely spacing components—and reduces the kinds of errors to which manual assembly is more prone. Specifically, the components should be compatible with the handling capability of the equipment and must be spaced regularly and liberally. (*Note:* The objective of wide spacing of components for manufacturability conflicts with the objective of closely spaced, "fine-pitch" circuit elements for faster operating digital equipment. High-speed computer or microprocessor operation needs closely spaced circuit

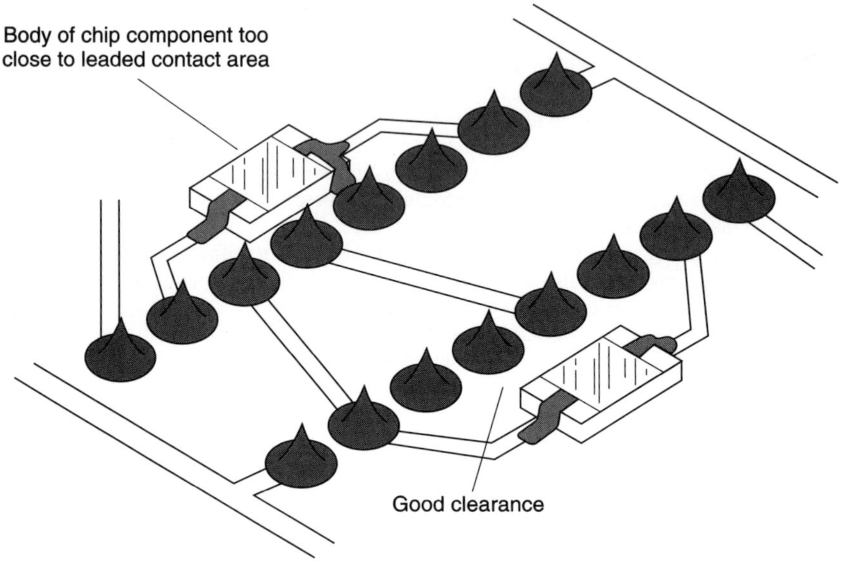

FIGURE 9.2.5 Solder bridging is common when inadequate spacing is provided between epoxy-attached surface-mount devices and the leads of devices mounted from the opposite side. (*From Vern Solberg,* Design Guidelines for Surface Mount and Fine Pitch Technology, *2d ed., McGraw-Hill,*[13] *New York, 1996.*)

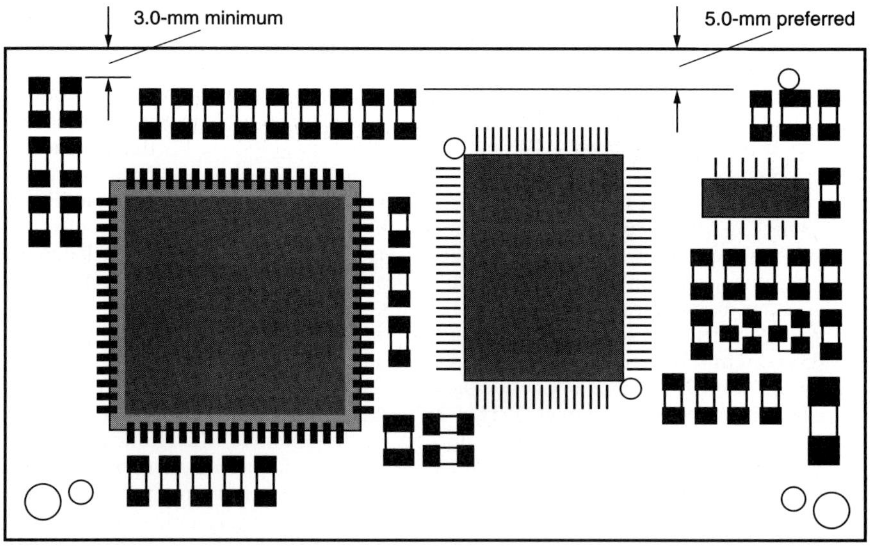

FIGURE 9.2.6 Space at the edges of printed-circuit boards should be allowed to avoid interference with or shorting to other components. (*From Vern Soldberg,* Design Guidelines for Surface Mount and Fine Pitch Technology, *2d ed., McGraw-Hill*[13]*, New York, 1996.*)

board devices. This poses a challenge to manufacturing. Design tradeoffs are involved.)

5. Design electromechanical parts for automatic assembly (handling, aligning, inserting)[5]:
   a. Specify parts in connected form for easier, more positive and accurate automatic feeding.
   b. If advantageous, leave an extra portion on each part to aid in assembly. The extra portion can be removed after the insertion, although typically it is not worth the effort required to do so.
6. There should be enough space between the edge of the circuit board and components and traces on the board so that shorting does not take place to exterior conductors. Figure 9.2.6 illustrates a recommended minimum.
7. When raised bosses or brackets are required to hold certain components on a printed-circuit board, consider molding such elements into the circuit board as one piece rather than using raised brackets. (Make the board a three-dimensional injection-molded part.)
8. Designers should be aware of the stack-up of positional tolerances—deviations due to fixture clearances and tolerances, board size variations, machine positioning accuracy, deviations in the dimensions of drilled holes, accuracy of tooling holes, and variations in lead wire diameter. Holes must be large enough for insertion of leads. However, holes too large cause soldering problems.[3]
9. Components that require a press to install should be avoided, if possible. Press fits can cause damage to circuit boards and are time-consuming in themselves.[1]
10. When flexible circuit boards are used, avoid bends that are sharp so as not to crack copper conduction paths. A bend radius of 30 to 40 times board thickness may be required for boards subjected to flexing or for multilayer boards, while in other cases, when flexing is not involved, 3 to 6 times board thickness may be adequate. Where flexible boards are bent adjacent to a stiffener, it is advisable to use a rounded-edge rubber fillet or a foam-rubber pad between the board and the stiffener to prevent the bend from being too sharp.[4] Maintain a uniform hole density over the entire board.
11. Designers should allow for some variations in the properties and dimensions of the components used. Lead lengths, widths, or diameters and positions can vary somewhat from part to part, and the board layout should allow for this as much as possible.
12. Consider the environment in which the product will be manufactured and used. Components that cannot withstand solder bath or reflow temperatures or cleaning processes require hand assembly. The temperature, humidity, vibration, electrostatic charges, dirt, dust, vapors, etc., that are in the product's environment all can have a tremendous effect on its function, its reliability, and its product life. The product must be designed to withstand any adverse conditions that it may be subjected to.
13. The following are additional guidelines from Anderson,[1] all providing potential means for improving printed-circuit board quality:
    a. Via holes should not be placed under metal components. This is to avoid having solder flow through via holes and short-circuiting metal components.
    b. For through-hole boards, most traces should be on the component side of the board and the fewest on the solder side. This is so because solder-side traces are most prone to damage and shorting.
    c. It is preferable to use an even number of circuit board layers to reduce the possibility of the circuit board warping when the heat for soldering is applied.

- *d.* To avoid shorting, solid metal layers used for power and ground planes should be on symmetrical layers.
- *e.* Use polarized cable connectors to ensure proper connections.
- *f.* Placing components under other soldered or socketed components makes it difficult to test and repair the device that is hidden.

14. If radiographic (x-ray) inspection of solder joints is used, SMT pads on the top and bottom sides of the circuit boards should not overlap, nor should radiologically opaque devices (some capacitors and transformers) be located directly opposite pads to be x-ray inspected.[7]

15. SMT boards require proper size and spacing of mounting pads. Pads should be designed to match the components being used. Suppliers of circuit devices should be consulted for the nominal dimensions and tolerances of their leads. Adequate pad-to-pad and pad-to-trace space also must be provided to prevent voltage leakage between traces and components and to allow sufficient room for coverage of mask material if the process calls for it.[8]

16. Consider the following guidelines for through-hole boards that receive wave or drag soldering:[9]

- *a.* Provide even spacing of components to minimize heat-sink effects.
- *b.* Reduce ground planes to a minimum size to minimize unwanted heat-sink effects on soldering.
- *c.* Leads in plated through-holes should have a diameter no more than 3/4 the hole diameter nor less than 1/3 to avoid soldering problems. (See Fig. 9.2.7.)

17. Finally, with electronic apparatus as with all other kinds of products, close collaboration between design and manufacturing people is the most effective approach for ensuring trouble-free and economical manufacturing.

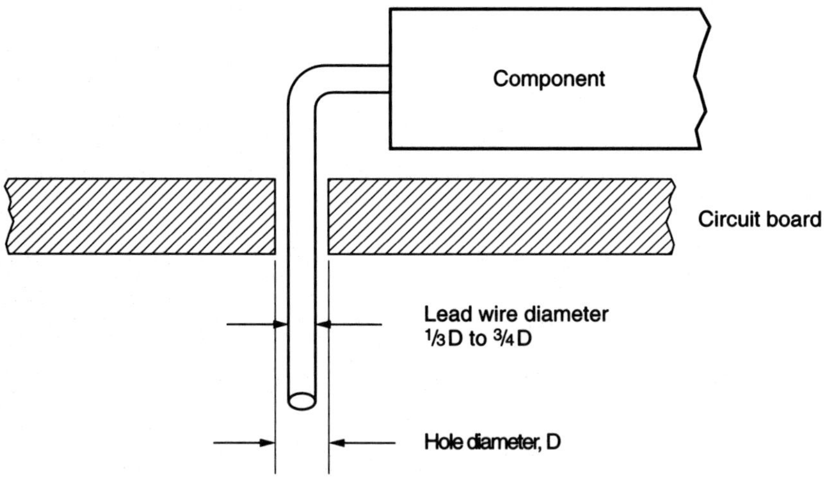

FIGURE 9.2.7 Lead wires in through-hole connections to printed-circuit boards should occupy from 1/3 to 3/4 of the hole diameter. Otherwise, soldering problems may result.

REFERENCES

1. David M. Anderson, *Design for Manufacturability,* CIM Press, Lafayette, CA, 1990.
2. *Tool and Manufacturing Engineers Handbook,* Vol. 6, *Design for Manufacturability,* Chap. 16, "Design for Electronic Assembly," Society of Manufacturing Engineers, Dearborn, MI, 1992.
3. Frank Riley (ed.), *The Electronic Assembly Handbook,* IFS Publications, United Kingdom, 1988.
4. Joseph Fjelstad, "Design Considerations for Surface Mount on Flexible Circuitry," *PC FAB Magazine,* October 1989.
5. M. Hirabayashi, K. Bito, and K. Nakanishi, "Design for Manufacturing and Its Implementing Production Facilities," SME technical paper, SME International Conference and Exposition, Detroit, MI, 1989.
6. Theodore J. Day, "Producibility: Design Considerations for Printed-Circuit Board Assemblies," SME E'Assembly Conference, Anaheim, CA, 1986.
7. Thomas W. Stroebel, "Designing PWB Assemblies for Automated Inspection," *Circuits Assembly,* July 1992.
8. Ron Daniels and Pete Waddell, "Design for Assembly," *Circuits Assembly,* July 1991.
9. Ben J. Cunningham, "The Return of Drag Soldering," *Circuits Manufacturing,* December 1988.
10. W. K. Hodson (ed.), "Electronics," Chap. 15-2 in *Maynard's Industrial Engineering Handbook,* McGraw-Hill, New York, 1992.
11. J. L. Nevins and D. E. Whitney, *Concurrent Design of Products and Processes,* McGraw-Hill, New York, 1989.
12. Johanson A. Edosomwan and Arvind Ballakur, *Productivity and Quality Improvement in Electronics Assembly,* McGraw-Hill, New York, 1989.
13. Vern Solberg, *Design Guidelines for Surface Mount and Fine Pitch Technology,* 2d ed., McGraw-Hill, New York, 1996.

CHAPTER 9.3
DFX*

DEFINITION

DFX, or *design for X,* can be defined as a knowledge-based approach that attempts to design products that maximize all desirable characteristics—such as high quality, reliability, serviceability, safety, user friendliness, environmental friendliness, and short time-to-market—in a product design while at the same time minimizing lifetime costs, including manufacturing costs.

Historically, designers have tended to underemphasize or overlook the preceding factors and have concentrated their efforts on only three factors: the *function* (performance), *features,* and *appearance* of the product that they develop. They have tended to neglect the "downstream" considerations that affect the usability and cost of the product during its lifetime.

AT&T Bell Laboratories recognized the need to satisfy these objectives and used the term *DFX* to designate designing for all desired factors.[2] DFX was described as a design procedure in which the objective broadly covers cost-effective "downstream" operations: distribution, installation, service, and customer use. Reliability, safety, conformance to environmental regulations, and liability prevention are also objectives. These are in addition to low manufacturing costs. DFX is "the process where the full life-cycle needs of the product are addressed during the product's design." AT&T made note of the value of incorporating DFX knowledge into CAE/CAD (computer-aided engineering/computer aided design). Education was stated to be essential.[3]

THE ATTRIBUTES OF A GOOD DESIGN

The following design objectives have been recommended as being most important.

Function and Performance

These are still vital. The product must perform the task for which it is designed.

*Adapted from Chap. 2 of *Design for Excellence,* by James G. Bralla, McGraw-Hill, New York, 1996.

Safety

Those involved in the manufacture, sale, and use of the product and other persons must be protected from physical injury and illness. A sound design from the safety standpoint is one whose manufacturing process does not involve hazards to workers; it is one whose operation poses the minimum risks to the user and those in the vicinity; it is one which, when the product is discarded after its useful life, does not entail hazardous waste.

Long-Term Quality

That is, *quality, reliability,* and *durability* (the customer tends to group these objectives together; the designer should also). Will the product continue to provide its desired function over a period of time? Will it retain its appearance, its accuracy, its ease of use, etc.? Quality and reliability result from care and attention at a number of stages, but perhaps the most important stage is the design stage. Quality and reliability cannot be built in if the basic design is not conducive to them.

Manufacturability

This includes testability, shipability, etc.—all the objectives of DFM.

Environmental Friendliness

This is closely related to safety but affects all living creatures and plant life. Will the product, its manufacturing process, its use, and its disposal avoid the release of pollutants and other environmental hazards? The manufacturing process should be one that generates minimal pollution. The product itself should be nonpolluting and, as noted above, nonhazardous in its operation and disposal. Even if nonhazardous, are its components configured so that they can be recycled easily? *Design for the environment* (DFE) has been used as a term to describe this approach. *Design for disassembly* is the name given to the system of product design that emphasizes recyclability of components. Primarily, this involves designs that ensure that recyclable components can be separated easily from the rest of the product.

Serviceability (Maintainability and Repairability)

This involves the ease with which the product can be returned to use after some failure has occurred, or the ease with which it can be attended to to avoid future failures. This objective is closely related to reliability. Easy serviceability may compensate for what otherwise would be a reliability problem.

User Friendliness, or Ergonomics

This involves how well the product fits its human users, how easy it is to use. (*Human factors engineering* was a previously common term for the discipline that this involves.) Is the product easy for the user to install and operate? Are all functions and controls clear? User unfriendliness can lead to safety and reliability problems as well as make the product less functional.

Appearance (or Aesthetics)

This is the attractiveness of the product, which may be a very important factor in its salability, particularly with many consumer products.

Features

The accessories, attachments, and peripheral functions, like the stereo, air-conditioning, and cruise control in an automobile, may be more important to the buyer than its basic function, i.e., in the case of an automobile, transportation.

Short Time-to-Market

This is how suitable the design is for short lead-time production. This normally means whether the design is one that requires unique long lead-time tooling for some of its components. Short time-to-market has important implications in the current era where product designs change rapidly and where commercial success often hinges on being the first supplier to market a product with particular features. The company that puts an innovation on the market first often reaps ongoing benefits in the form of increased market share for its product.

Other objectives such as *installability, testability, shipability, upgradeability, easy customizing,* etc., also may be important in many cases.

REFERENCES

1. James G. Bralla, *Design for Excellence, DFX,* McGraw-Hill, New York, 1996.
2. David A. Gatenby, "Design for `X' (DFX): Key to Efficient, Profitable Product Realization," in J. A. Edosomwan and A. Ballakur (eds.), *Productivity and Quality Improvement in Electronics Assembly,* McGraw-Hill, New York, 1989, Chap. 45.
3. R. A. Layendecker and B. Suing Kim, "From DFMA to DFX: An AT&T Example," paper presented at the 1993 DFM Conference at the National Design Engineering Conference, Chicago, March 1993.

CHAPTER 9.4

GLOSSARY OF TERMS*

The following terms describe management systems that are either part of DFM or DFX, related to them, or provide alternative means of improving product designs and manufacturing operations.

Benchmarking A "systematic process for evaluating the products, services and work processes of organizations that are recognized as representing best practices...."[3] The objective of the study is to gain information that leads to improvements in the organization making the study. Almost any organizational activity can be the subject of a benchmark study. Both broad functional areas such as quality control, service, manufacturing, etc., and narrow factors within them such as the method used for a specific operation, production yields, mean time to failure of products, cycle times, sales territory assignments, etc., can be covered. Comparison studies can be made of different branches of an organization, of competitors, and of a company unrelated to the one making the study but one that performs some function in an outstanding manner.

Concurrent design Another name for *concurrent engineering.*

Concurrent engineering The approach that brings both design and manufacturing engineers together during the design phase (often along with product managers, quality controllers, production people, and others). The product, the manufacturing process, and supporting activities are engineered at the same time. This facilitates DFM and speeds the product-realization cycle. The use of DFM tends to promote and facilitate concurrent engineering because the product designers, who are seldom expert in manufacturing constraints, need to work hand in hand with manufacturing engineers to ensure that their designs are optimally manufacturable.

Continuous improvement A philosophy that is inherent in a number of currently used management approaches such as TQM (total quality management). It emphasizes that industrial competitiveness does not come from one or a few massive improvements that simply have to be maintained after implementation. Rather, competitiveness is better served by the installation of a series of improvements that may be incremental but which are part of an improvement process that is ongoing.

Controlled-experiment method A test method applicable when there are a number of variables, all of which have some effect on the process or product being studied. This approach allows all the variables to be evaluated at one time. *Design of experiments (DOE), factorial experiments, directed experimentation, orthogonal arrays,* and *statistically designed experiments* are all terms for essentially the same approach. By mathematical analysis of test results, the person making the tests determines which setting of each variable is optimal. The number of test runs needed for full optimization is greatly reduced compared with the traditional method in which the variables are tested one at a time while other variables are held constant.

*Adapted from various chapters of *Design for Excellence,* by James G. Bralla, McGraw-Hill, New York, 1996.

DFA, design for assembly This is an important product design approach whose objective is to simplify the product and its overall assembly. A reduction in parts count is the hallmark of successful DFA. (See Chap. 7.1.)

DFM, design for manufacturability In the broadest sense, DFM includes any step, method, or system that provides a product design that eases the task of manufacturing and lowers manufacturing cost. In a somewhat more specific sense, DFM is primarily a knowledge-based technique that invokes a series of guidelines, principles, recommendations, or rules-of-thumb for designing a product so that it is easy to make. These guidelines tend to aid many common product attributes—proper function, reliability, good appearance, serviceability, etc.—but their primary objective is to improve manufacturability.

DFX (See Chap. 9.3.)

Design to cost This includes design efforts to reduce operating, maintenance, and manufacturing or acquisition costs of a product.[1] It utilizes a variety of analytical and goal-setting and control methods to achieve this, including value analysis, cost estimating, design simplification, work simplification, pareto analysis, etc.[7]

Failure mode and effects analysis (FMEA) This is a quality analysis technique useful in improving product designs and manufacturing processes by eliminating or minimizing existing and potential quality problems. The approach prioritizes problems by considering the seriousness of the problem, its frequency, and the probability of its being undiscovered. The combined effect of these three factors determines the priority of corrective action.[6]

Fractional factorial experiments These are controlled experiments in which not all levels of all variables are tested in combination with all levels of all other variables. A sample, or fraction, of all theoretical combinations is tested. This approach is taken when a full series of factorial experiments would require a prohibitively large number of test runs.[4]

Group technology This is a production arrangement wherein parts are grouped together into families with similar characteristics, and production equipment is laid out in the sequence needed for the parts to progress from operation to operation with minimum transport distance. The production unit for the parts family is then self-contained and is sometimes called a *cell.* This differs from the traditional job-shop factory layout, in which equipment of each type is grouped together in various departments and each part moves from department to department for processing. The advantage of group technology is much reduced throughput time, simpler production control, reduced materials handling, and better operator understanding of quality requirements.

Life-cycle costs These are all the costs involved not only in the manufacture and distribution of a product but also in its ownership, its operation, and its disposal at the end of its useful life. Service and repair costs, warranty costs, energy costs for operation, and medical costs of persons injured by it, if any, are included. Costs borne by persons other than the buyer or user of the product are included, as are costs to the general public for environmental damage, etc.

Manufacturability This is the ease with which a product or part can be produced.

Producibility This is another term for *manufacturability.* With some practitioners, the term refers only to the ease of manufacture of parts and components rather than to assemblies of them.

Quality function deployment (QFD) This is a systematic approach for improving product quality by concentrating on what the customer wants and will continue to buy in the product. The approach utilizes the skills within the organization on a team basis to design, manufacture, and market products that incorporate the customers' desires.[5]

Quality loss function A Taguchi concept, this is the relationship between undesired variations in some characteristic of a product and the financial loss that is borne by society as a result of it.

Society's financial loss includes the cost of service, repair, warranty, lost goodwill, etc.[2] (See Life-cycle costs above.) The greater the deviation in a quality characteristic, the greater is the financial loss.

Simultaneous engineering Another name that has been given to *concurrent engineering.*

Statistical process control (SPC) This is a form of quality control that uses statistical methods to help control dimensions and other characteristics of manufactured products and their components. Its purpose is to ensure that variations in dimensions and other characteristics are random rather than due to some correctable process flaw and that they remain within acceptable limits so that the product's quality is ensured.

Synchronized manufacturing A system of manufacturing advocated by Eliyahu Goldratt.[8] It involves small lots of production, run on as continuous a basis as possible by supplementing bottleneck operations where necessary, balancing production to customers' orders, and avoiding stocking of work-in-process. When implemented, it may greatly reduce factory throughput time, providing both customer service and manufacturing advantages.

Total quality management (TQM) A managerial approach that emphasizes product and service quality improvement. Its elements include full company organizational involvement in promoting quality, quality measurement, and a focus on customers' wants, the use of teams, thorough training, and continuous improvement.

Value analysis This involves a systematic review of cost of producing a component or product compared with the value it provides. The objective is to increase or maintain its value while reducing its manufacturing cost.

Value engineering This involves value analysis applied during the design phase of a product rather than after the product has gone into production.

REFERENCES

1. Marvin A. Moss, *Designing for Minimal Maintenance Expense,* Marcel Dekker, New York, 1985.
2. Steven Ashley, "Applying Taguchi's Quality Engineering to Technology Development," *Mechanical Engineering,* July 1992.
3. Michael J. Spendolini, *The Benchmarking Book,* Amacon Division of American Management Association, New York, 1992.
4. George Box and Soren Bisgaard, "The Scientific Context of Quality Improvement," technical paper, University of Wisconsin, Madison, WI, 1987.
5. John Hauser and Don Clausing, "The House of Quality," *Harvard Business Review,* May–June 1988.
6. S. Choka, "Failure Mode and Effects Analysis," IBM internal paper, 1992.
7. Jack V. Michaels and William P. Wood, *Design to Cost,* John Wiley & Sons, New York, 1989.
8. E. M. Goldratt and J. Cox, *The Goal,* North River Press, Croton-on-Hudson, NY, 1986.

INDEX

ABOUT THE EDITOR

James G. Bralla is a manufacturing consultant and noted authority on manufacturing and design. He has more than 40 years experience in the field as a line manager, consultant, industrial, and project engineer. He has served as Industry Professor at Polytechnic University in New York; Vice President, Operations for Alpha Metals, Inc.; and Director Manufacturing, Asia for the Singer Company. He is a graduate of Princeton University and a registered professional engineer. Mr. Bralla is also the author of McGraw-Hill's *Design for Excellence*.